MODERNIST DESIGN COMPLETE

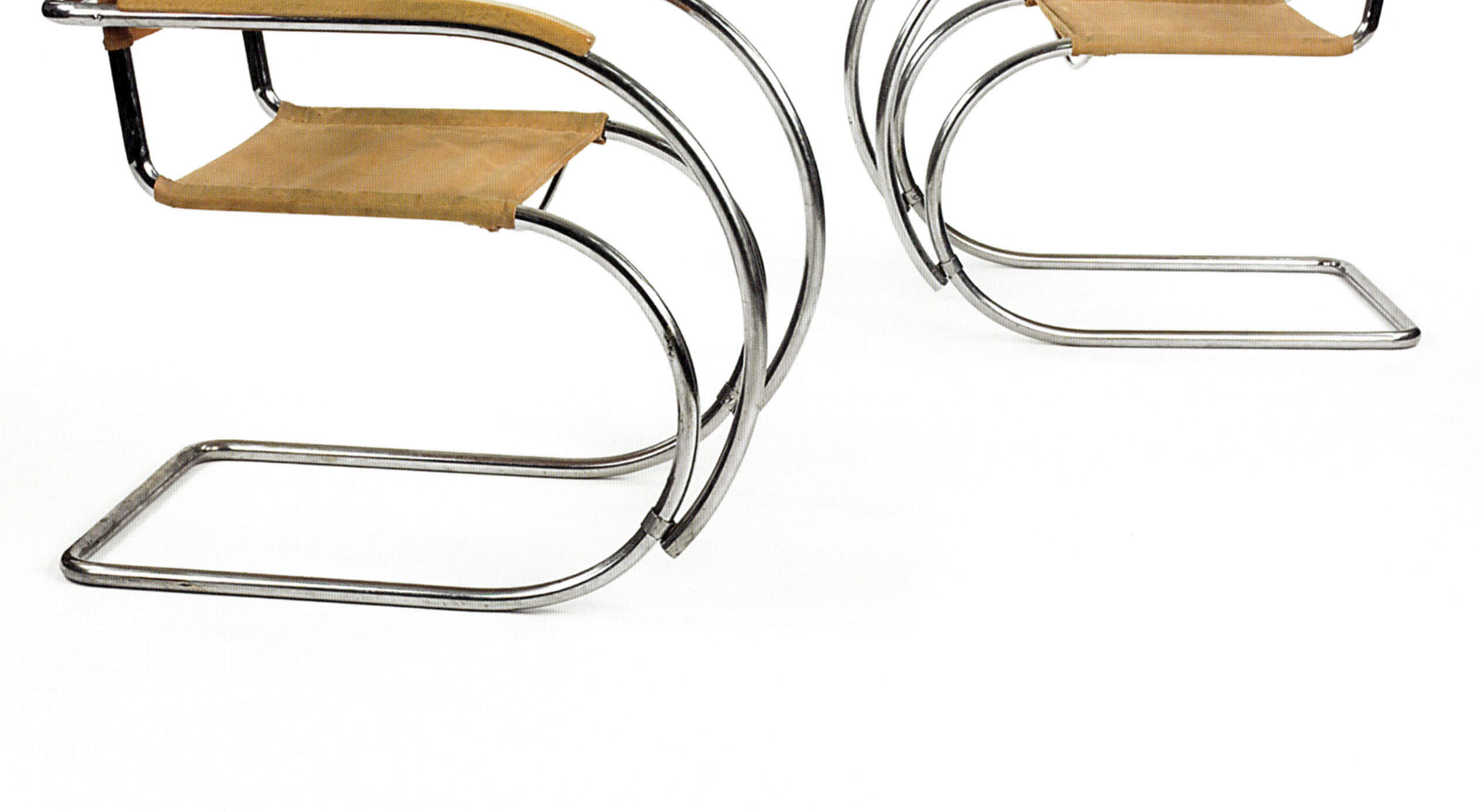

MODERNIST DESIGN COMPLETE

Dominic Bradbury

Thames & Hudson

CONTENTS

6 Introduction

MEDIA & MASTERS

FURNITURE

16 Introduction

20 **The Advent of Modernity: Modernist French Furniture and the New Materiality** by Patrick & Laurence Seguin

24 **Bauhaus By Design: The Ideas and the Influence of the Bauhaus Masters** by Jolanthe Kugler

28 Erik Gunnar Asplund

32 Marcel Breuer

38 Donald Deskey

44 Jean-Michel Frank

48 Eileen Gray

54 René Herbst

58 Kaare Klint

62 Mogens Koch

64 Le Corbusier, Pierre Jeanneret & Charlotte Perriand

72 Bruno Mathsson

78 Ludwig Mies van der Rohe & Lilly Reich

84 Jean Prouvé

92 Gerrit Rietveld

98 Gilbert Rohde

106 Kem Weber

LIGHTING

112 Introduction

116 **The Age of Enlightenment: The Evolution of Lighting from Art Deco to Modernism** by Richard Wright

120 Marianne Brandt

124 George Carwardine

128 Christian Dell

132 Poul Henningsen

140 Walter von Nessen

146 Gio Ponti & Pietro Chiesa

CERAMICS & GLASS

156 Introduction

160 **The New Aesthetic: The Evolution of Modernist Ceramics in the Inter-War Years** by Mel Buchanan

164 Bernard Leach & Michael Cardew

168 Otto Lindig & Theodor Bogler

174 Grete Marks

178 Orrefors

182 Paolo Venini & Carlo Scarpa

190 Marguerite Wildenhain

196 Russel Wright

HOUSES & INTERIORS

296 Introduction

300 **Welcome to the Machine: The Power of the Automobile and New Technology in Shaping the Modern Home** by Tim Benton

304 **The Bauhaus in America: The Modernist Émigrés and their Influence on American Architecture, Interiors and Design, 1920–1940** by Margret Kentgens-Craig

308 Alvar Aalto: Villa Mairea, Noormarkku, Finland

314 Brinkman & Van der Vlugt: Sonneveld House, Rotterdam, Netherlands

A–Z OF DESIGNERS, ARCHITECTS & MANUFACTURERS

INDUSTRIAL & PRODUCT DESIGN

204 Introduction

208 **'Flash-and-Gleam Beauty': The Streamlined Decade in America** by Anne H. Hoy

214 Peter Behrens

218 Harold van Doren

222 Henry Dreyfuss

226 Richard Buckminster Fuller

230 Norman Bel Geddes

236 Raymond Loewy

240 Ferdinand Porsche

242 Walter Dorwin Teague

246 Wilhelm Wagenfeld

GRAPHICS & POSTERS

254 Introduction

258 **The Birth of Modern Type: Typographical Revolution and the Advent of Modernist Graphic Design** by Rick Poynor

262 Josef Albers

268 Herbert Bayer

272 A. M. Cassandre

276 Theo van Doesburg

280 László Moholy-Nagy

286 Alexander Rodchenko

290 Jan Tschichold

324 Pierre Chareau: Maison de Verre, Paris, France

328 Serge Chermayeff: Bentley Wood, East Sussex, UK

336 Ernő Goldfinger: Willow Road, London, UK

340 Eileen Gray & Jean Badovici: E-1027, Roquebrune, France

348 Walter Gropius: Gropius House, Lincoln, Massachusetts, USA

358 Arne Jacobsen: Rothenborg House, Klampenborg, Denmark

366 Le Corbusier: Villa Savoye, Poissy, France

374 Robert Mallet-Stevens: Villa Noailles, Hyères, France

380 Ludwig Mies van der Rohe: Villa Tugendhat, Brno, Czech Republic

384 Richard Neutra: Lovell Health House, Los Angeles, USA

388 Auguste Perret: Studio-Residence for Chana Orloff, Paris, France

396 Gerrit Rietveld: Rietveld Schröder House, Utrecht, Netherlands

400 Eliel Saarinen: Saarinen House, Cranbrook, Michigan, USA

408 Hans Scharoun: Haus Schminke, Löbau, Germany

412 Rudolph Schindler: Schindler House, Los Angeles, USA

420 Giuseppe Terragni: Villa Bianca, Seveso, Italy

428 Frank Lloyd Wright: Fallingwater, Bear Run, Pennsylvania, USA

438 A–Z of Designers, Architects & Manufacturers

469 Contributors' Biographies

470 Notes

472 Select Bibliography

474 Picture Credits

475 Acknowledgments

476 Index

INTRODUCTION

Modernism offered its architects and designers a broad church, made of glass, concrete and steel. It provided space for a complex congregation – an assembly of movements, sub-movements and genres from many disciplines of art, architecture and design. There was room enough under the high shelter of the crisp, rectilinear roofline for regional representation, such as the Italian Futurists*, the disciples of Dutch De Stijl*, the Russian Constructivists* and more. There was a place within its generous, open volume for Functionalists*, Rationalists*, Purists* and Expressionists*, as well as the followers of the International Style*. And there was even an inviting annex, with timber-panelled walls, for the soft modernists, particularly the Scandinavians, who believed that there was still a place for a more organic aesthetic. Modernism welcomed them all, whatever their many differences, and certainly there were common elements that united everyone.

There was, above all, the ambition to make it new, to create a fresh kind of architecture and design, which reflected the many advances in materials, technology and engineering on the one hand, and the dynamics of the machine age on the other. To some – such as the radical Futurists – this meant a firm and decisive break with the past. For others, it was another step upon an evolutionary path, drawing upon the lessons offered by neoclassicism and other period styles, while moving towards a new aesthetic more representative of the 20th century.

Modernist architects and designers generally concurred on the need to strip away excessive ornamentation and decoration in favour of a purer aesthetic, rejecting the flamboyant excesses of Art Nouveau and other decorative styles. Many embraced Adolf Loos's* dictum that ornament was a crime, first made in a 1908 lecture, entitled 'Ornament & Crime', which was partly a response to Art Nouveau. The result was a purity of line and a renewed emphasis on form, which was combined with a corresponding emphasis on function.

The majority of modernists embraced the machine age and the possibilities offered by factory production, which had the potential to make good design available to a far wider spectrum of consumers than ever before. In this way, modernist design can be seen as democratic in some respects, rather than elitist, with products that were better made and more affordable. It was a key point of difference from the Art Deco* style, which shared an interest in machine-age streamlining and aesthetics, yet was primarily a movement of elitism and luxury, with a focus on hand-crafted design and fine, costly materials.

Modernists – particularly those influenced by the Bauhaus school and the views of its director, WALTER GROPIUS – believed in making their work available to the masses through the advantages of assembly-line manufacturing (although some, such as LUDWIG MIES VAN DER ROHE, continued to explore the use of fine, expressive materials within a modern context, and others, such as the soft modernists of the Nordic countries, explored a more organic approach, while advocating many modernist tenets). The super-efficient assembly-line systems pioneered by car manufacturer Henry Ford* and his industrialist followers helped reduce costs by speeding up production, drawing upon a standardized kit of components and parts. It was a system that was soon being applied in many other spheres of design.

To a greater or lesser extent, nearly all modernists saw the opportunity to make the world a better place through the medium of architecture and design. They believed in progress allied to developments in technology, materiality and engineering, which could lead to better homes and better standards of living. Such homes were sometimes seen – in Le Corbusier's famous phrase – as 'machines for living in', suggestive of the efficiency of design and layout, but also of materiality and means of construction. These were spaces, too, that were healthier, with the theme of hygiene becoming a repeated preoccupation for many architects, designers, manufacturers and advertising agencies during the inter-war period.

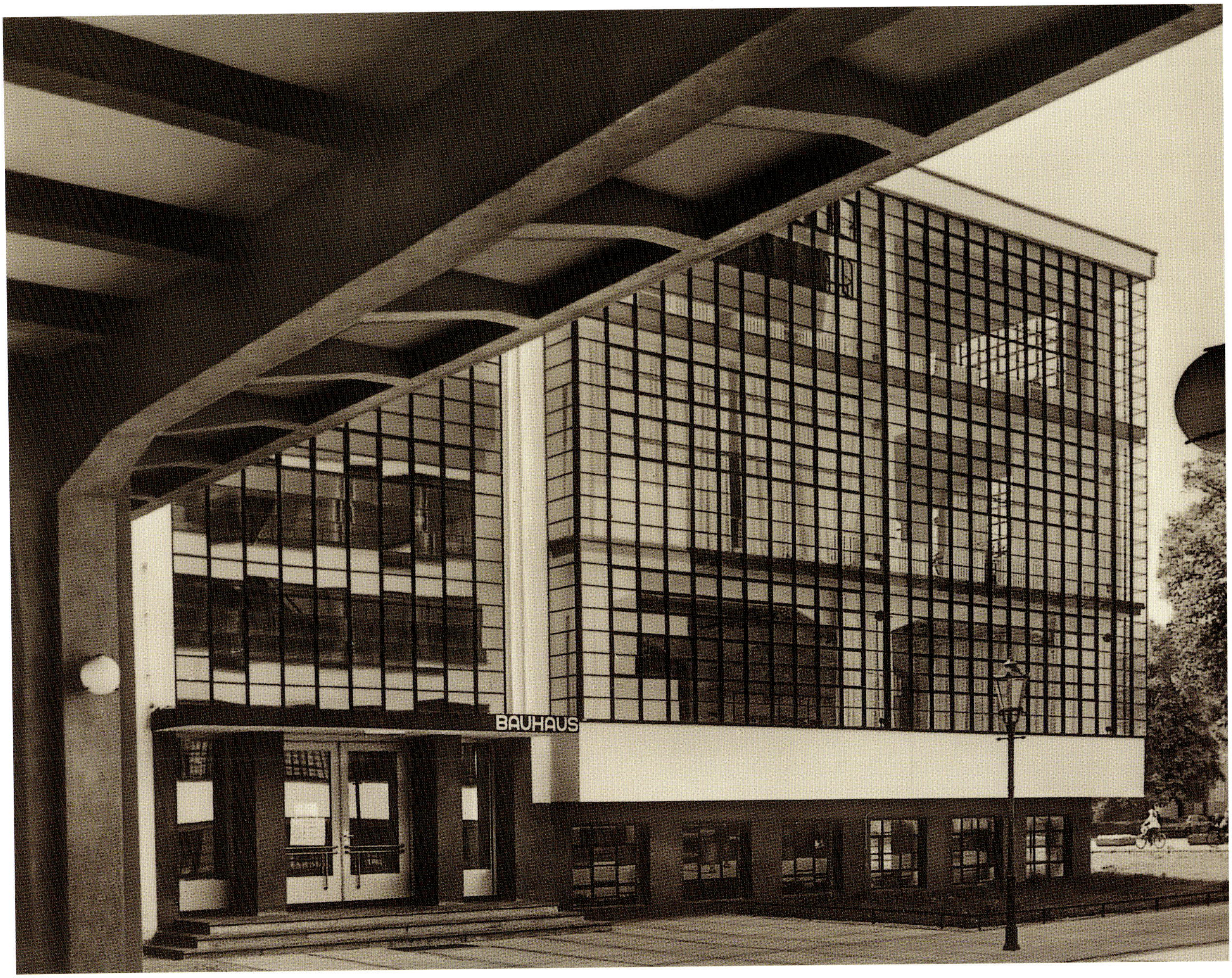

ABOVE View of the northern façade and entrance of the Bauhaus workshop wing in Dessau, completed in 1926 to a design by Walter Gropius.

Note to the reader

Names that appear in CAPITAL LETTERS indicate cross-references to main entries in the book (pp. 6–437).

Names that appear with asterisks* indicate cross-references to entries in the A–Z section (pp. 440–68).

For some, such as the Russian Constructivists, the progressive social dimension of modernism was a vital part of their philosophy. For others it was simply one of many elements within the wider ambition of creating a better world, particularly at a time when political views in certain regions were beginning to polarize and become extreme, with the possibility for renewed conflict, even while the memories of World War I – with all its destructive force and horror – were still fresh.

In Ayn Rand's* powerful novel *The Fountainhead*, first published in 1943, the tragic hero, Howard Roark, is presented as an ardent modernist battling the forces of conservatism (his fatal flaw being his all-enveloping pride and an inability to compromise). For Roark, as for many early modernists, there was a key element of struggle within the overriding ambition to 'make it new' and forge a new outlook and aesthetic for a new age. 'I inherit nothing. I stand at the end of no tradition. I may, perhaps, stand at the beginning of one,' says the young Roark.[1] Many pioneering modernists would have sympathized with his ongoing fight against the forces of tradition, which often launched vivid attacks upon modernist projects and buildings, particularly in more conservative parts of the world, such as England. In some cases, the ambition to make it new had to be tested in the courts and through legal appeal.

Modernists in Germany, in particular, found themselves persecuted by the state itself. The Bauhaus was famously closed down by the fascist authorities, and many leading Central European modernists went into exile, firstly in England and then, more commonly, in the United States, which took a far more open and welcoming attitude to the modernist dream. There was, therefore, a commonly held belief during the Twenties and Thirties that modernism was born of adversity and in opposition to the forces of tradition.

Given the context of growing political extremism on both right and left, perhaps it is not surprising that at times modernists tended to enjoy manifestos, grand statements and associations of all kinds, partly as a way of combatting resistance from conservatives. Many of these associations and their manifestos had a semi-political dimension, arguing for the most part in favour of social progress through design.

Among the most prominent of these associations stood the Deutscher Werkbund* in Germany, which actively promoted an alliance between designers, architects and industry; one of its most famous successes was the Weissenhof Estate* in Stuttgart (1927), where a roll call of leading modernists created a series of show houses and apartments, providing a full-scale paradigm of 'contemporary' – or perhaps futuristic – living.

The Congrès International d'Architecture Moderne (CIAM)*, founded in Switzerland in 1928, was another leading talking shop for modernist architecture and design, with Le Corbusier, Pierre Chareau and Gerrit Rietveld among its founding members. CIAM organized a series of conferences on key themes and challenges, with a number of meetings during the late Twenties and Thirties. In France, the Union des Artistes Modernes (UAM)*, founded in 1929, was also highly influential, while an English equivalent was known as the Modern Architectural Research Group (MARS)*, established in 1933.

A key aspect of so many of these associations, groups and conferences was that their members saw themselves as part of a wider, international cultural movement. The relationship between art, architecture and design was particularly intimate during the modernist period, with many sub-movements, such as De Stijl and Constructivism, founded by artists and sculptors as well as designers. The Bauhaus placed a strong emphasis upon providing its students with a proper grounding in art and design in the widest sense, with many artists – including Wassily Kandinsky* and Paul Klee* – among its tutors; it was a model that was later applied in other design schools. Many architects and designers were also artists, including Le Corbusier, who painted and exhibited his work, as well as applying colourful murals to a number of his projects. More broadly, there

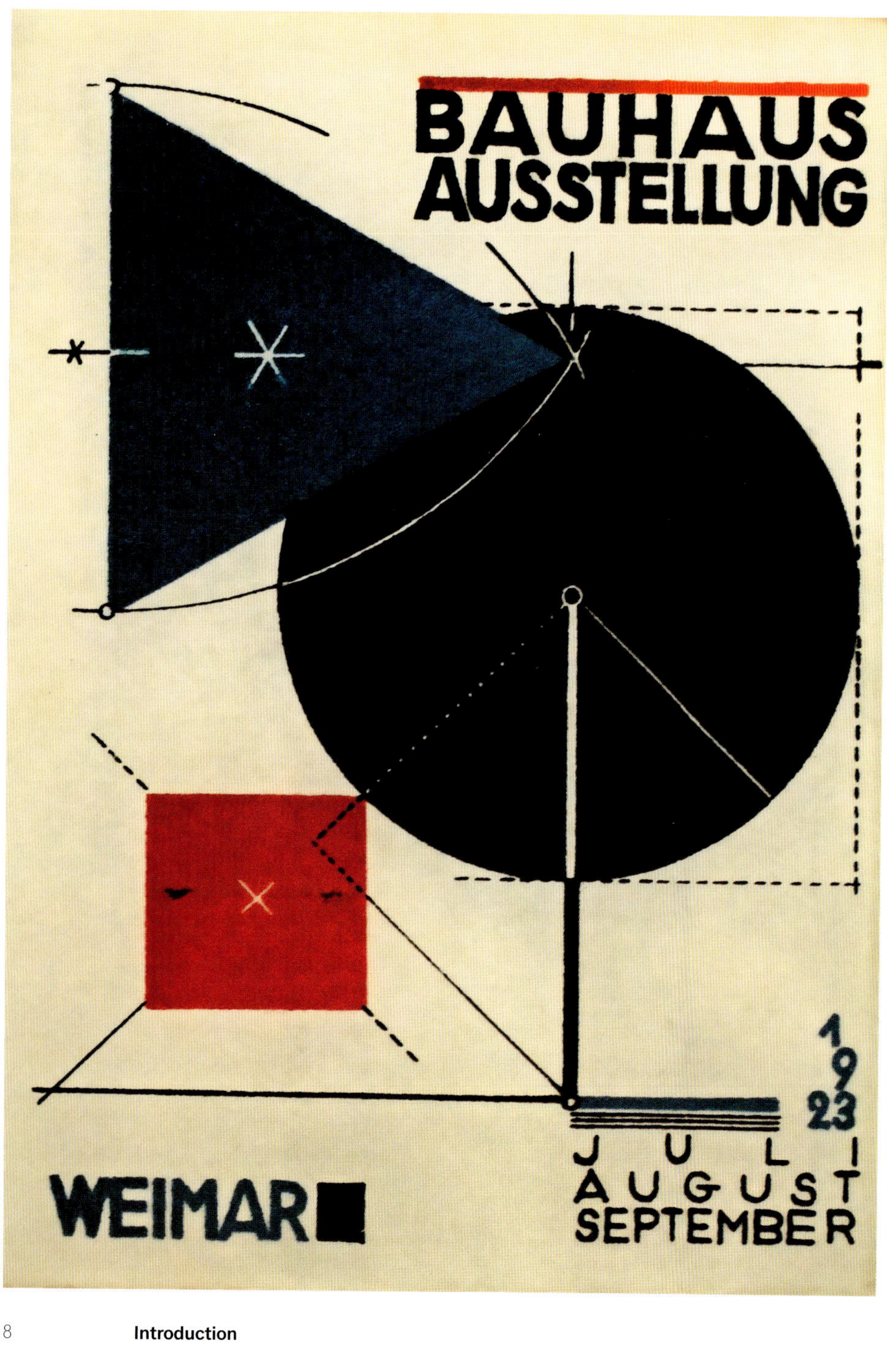

LEFT A postcard designed by Herbert Bayer to commemorate the Bauhaus exhibition of 1923.

was continued dialogue with modernist writers, poets and musicians, many of whom shared similar preoccupations and concerns.

These preoccupations were wide-ranging and often social and political, as well as artistic and cultural. At the core of so many of the conversations throughout the modernist period was the rise of the machine and the associated growth of industrialization, a preoccupation that continues to be part of modern life today, as we move into an era of high technology, automation and robotics. Then, as now, technology was seen as a force for progress, but with many caveats and concerns. World War I, in particular, had shown in graphic terms the destructive power of the machine with the evolution of machine guns, tanks, airships and fighter planes; it was, arguably, the first truly 'modern' war. During the Twenties and Thirties, the rise of the machine and factory production continued to provoke nervousness about loss of employment, in particular, and potential over-reliance on technology. Disasters relating to these new technologies became powerfully resonant and symbolic, from the sinking of the *Titanic* in 1912 through to the Hindenburg airship catastrophe of 1937.

The machine was an inter-war obsession, explored in art, literature and also feature films, such as Fritz Lang's famous movie *Metropolis* (1927) and, later, Charlie Chaplin's *Modern Times* (1936). The rapid evolution of the automobile and the aeroplane, especially, were repeated motifs explored in so many different artistic mediums. For the Italian Futurists, in particular, the machine was emblematic of speed, progress and vitality, yet these were also ideas explored by artists, sculptors and designers of all kinds.

In the world of architecture, certainly, the influence of the car, the plane and also the ocean liner was very real and often profound. The automobile played an important role within the evolution of a number of landmark buildings of the period, with Le Corbusier's Villa Savoye (1931) and Mies van der Rohe's Villa Tugendhat (1930), for instance, recognizing the needs of the car within the process of arrival as well as providing integrated garaging and a self-contained chauffeur's apartment. A number of other houses – such as Hans Scharoun's Haus Schminke (1933) and Eileen Gray's E-1027 (1929) – possess a clear maritime influence, with decks and streamlined forms reminiscent of a transatlantic liner; such features also made themselves apparent within the designs of many Twenties and Thirties apartment buildings.

The wider vogue for streamlining – which crossed the boundaries of both Art Deco and modernist design – was largely derived from the aesthetics of transportation and speed. Streamlining was soon being applied to all kinds of products, becoming emblematic of the machine age itself. This was particularly true of a new generation of labour-saving household appliances, from refrigerators and washing machines to vacuum cleaners. In a bid to make these appliances more attractive and desirable to growing ranks of consumers, manufacturers increasingly turned to the talents of an entirely new profession born of the modernist age, that of the industrial designer.

The steady push towards industrialization and factory production had profound implications for many different disciplines of design, as well as for the product or industrial designer. Increasingly, furniture designers began to explore the possibilities offered by industrially made or manipulated materials, such as tubular steel and plywood. Ceramicists were tempted by the opportunity to take their work to a mass audience by designing pieces that could be made in large numbers on a factory scale, rather than hand-crafted in the studio. Thus, a tension developed – even within the modernist movement – between advocates of craft and artisanal talent and those who believed that factory production provided the best way of bringing well-made, functional products to a wide audience at an affordable price.

The Great Depression that began in 1929 made the relationship between man and machine rather more complex and problematic. Some of those who lost their livelihoods in

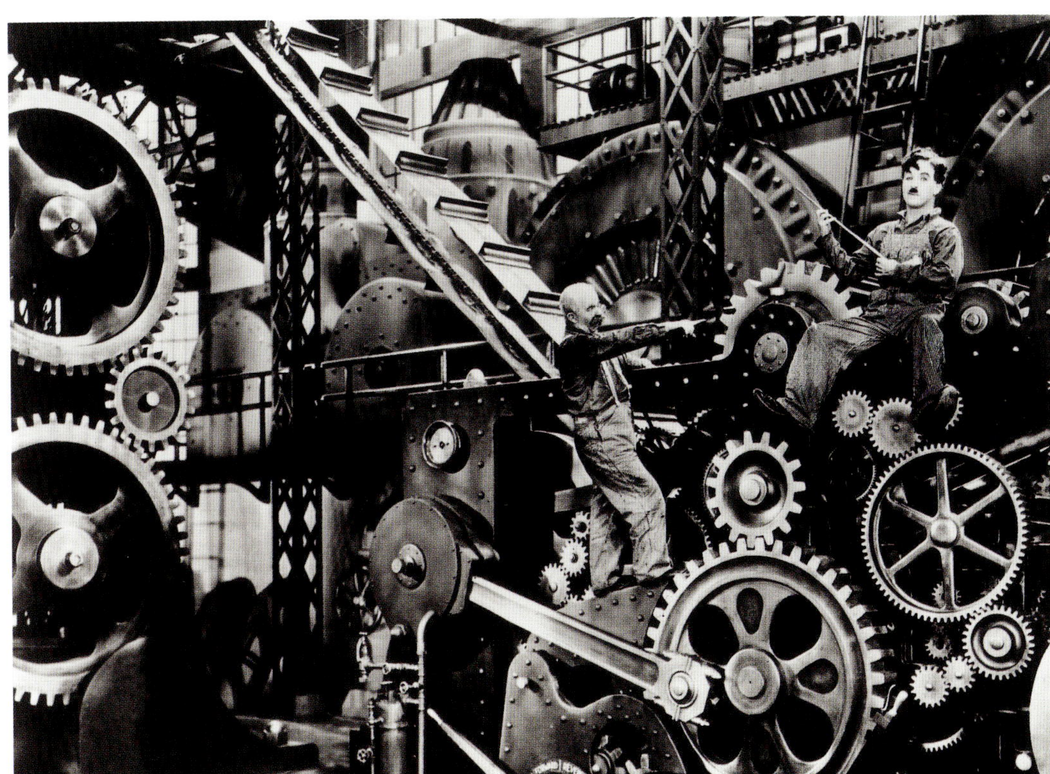

RIGHT A film still image from Charlie Chaplin's classic 1936 film, *Modern Times*, written and directed by its star.

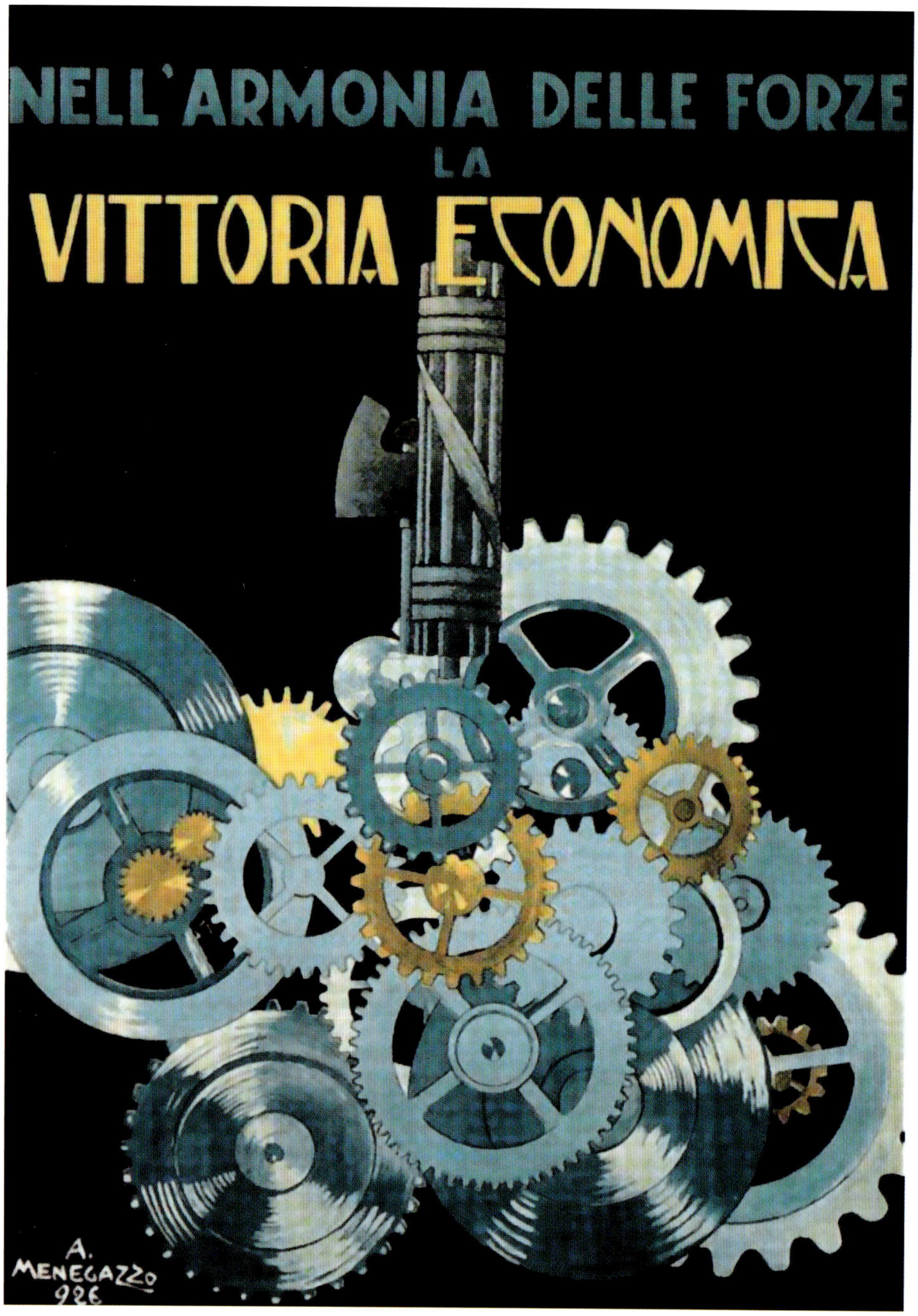

ABOVE A *c.* 1930s Italian propaganda poster designed by Antonio Menegazzo.

OPPOSITE Gerrit Rietveld's iconic Red & Blue chair, designed in 1918, produced by Cassina in stained pine and plywood.

the Depression blamed increasing mechanization – both on the farms and in the cities – for the loss of their jobs. Others saw industrial production and the growth of factory-made consumer products as a golden opportunity to create new jobs and new professions.

In Germany, Adolf Hitler's National Socialist German Workers' party, which came to power during the austerity years provoked by the Depression, partially embraced industrialization through programmes such as the state-sponsored effort to build a people's car, or Volkswagen, which ultimately led to the birth of the VW Beetle. When it came to war production and the creation of a new generation of war machinery, Hitler and his chief architect, Albert Speer*, were technocrats.

Yet at the same time Hitler and the Nazi party viewed modernists in general as suspect, avant-garde and potentially degenerate. Increasingly, the Nazis sought to undermine and suppress modernist architects and designers, forcing the closure of the Bauhaus and pushing many leading German modernist architects and designers into exile. Hitler, like Napoleon before him, favoured neoclassicism, seeing within it a way of recreating a new empire on the Roman model, while playing with ideas of architectural scale and monumentality to both impress and intimidate.

Mussolini and the Italian fascists took a more ambiguous view of modernism. Again, Mussolini enjoyed the symbolism of modernity and machine-age technocracy, seeing within it a way of underlining his commitment to the modern world and Italy's future ambitions. The Italian fascists collaborated with a number of essentially modernist architects and designers, yet were also tempted towards neoclassical monumentality. The result, typically, was a dangerous muddle.

As Europe began to polarize and drift towards extremism in the mid- to late Thirties, and eventually everyone was forced to take sides, modernism was increasingly seen as a movement of the centre left. Many artists and writers regarded as essentially modernist supported the socialist camp during the Spanish Civil War of 1936 to 1939, the terrible dress rehearsal for the global conflict to come. The bombing of the Basque town of Guernica by German Junkers bombers and the Italian air force – an extreme demonstration of fascist air power – provoked particular outrage among artists and intellectuals, including Picasso, who depicted the attack in his famous painting of 1937 and presented it at the World's Fair in Paris that same year.

During the following years many modernist architects and designers – along with their clients – were caught up in the chaos and destruction of World War II. Among the countless victims of political oppression in Germany and Central Europe were seminal modernist buildings, such as Hans Scharoun's Haus Schminke and Mies van der Rohe's Villa Tugendhat, which were confiscated.

In the United States, the Art Deco influence upon early modernism was strong and significant, particularly on the East Coast. A number of key designers, including Gilbert Rohde and Donald Deskey, transitioned from a predominantly Deco style to modernism, carrying machine-age themes and preoccupations with them. On the West Coast, the influence of European émigrés, such as Richard Neutra and Rudolph Schindler, took architecture and design in a more experimental direction from the mid-Twenties onwards and played a significant part in the evolution of 'Desert Modernism'. A later influx of former Bauhaus designers and other European-trained modernists in the mid- to late Thirties helped give shape to what eventually became known as the International Style.

On both sides of the Atlantic, in the years following the war, mid-century designers were able to build upon the foundations laid by the pioneering modernists of the Twenties and Thirties, while establishing a new aesthetic of their own. Partly in reaction to a long period of wartime austerity and hardship, mid-century architecture tended to be rather more playful, sculptural, colourful, organic and expressive. Mid-century architects and designers began to reinvent modernism in their own way, creating a distinct aesthetic rooted in the Fifties and Sixties context.

In this book, the scope is international, focusing primarily on Europe and the United States, where modernism first took hold, with a particular spotlight on key design disciplines including residential architecture and interiors, furniture, graphic design, industrial design, ceramics and glass, while a range of other themes, such as textile design, are covered in the A–Z section. At the same time as recognizing the contextual reality of male-dominated professions between the wars, the book seeks to respect the important role of innovative women designers who came to prominence during the period, including Eileen Gray, Lilly Reich, Charlotte Perriand and Marguerite Wildenhain. The groundbreaking style established by such icons of modernism represented a new beginning.

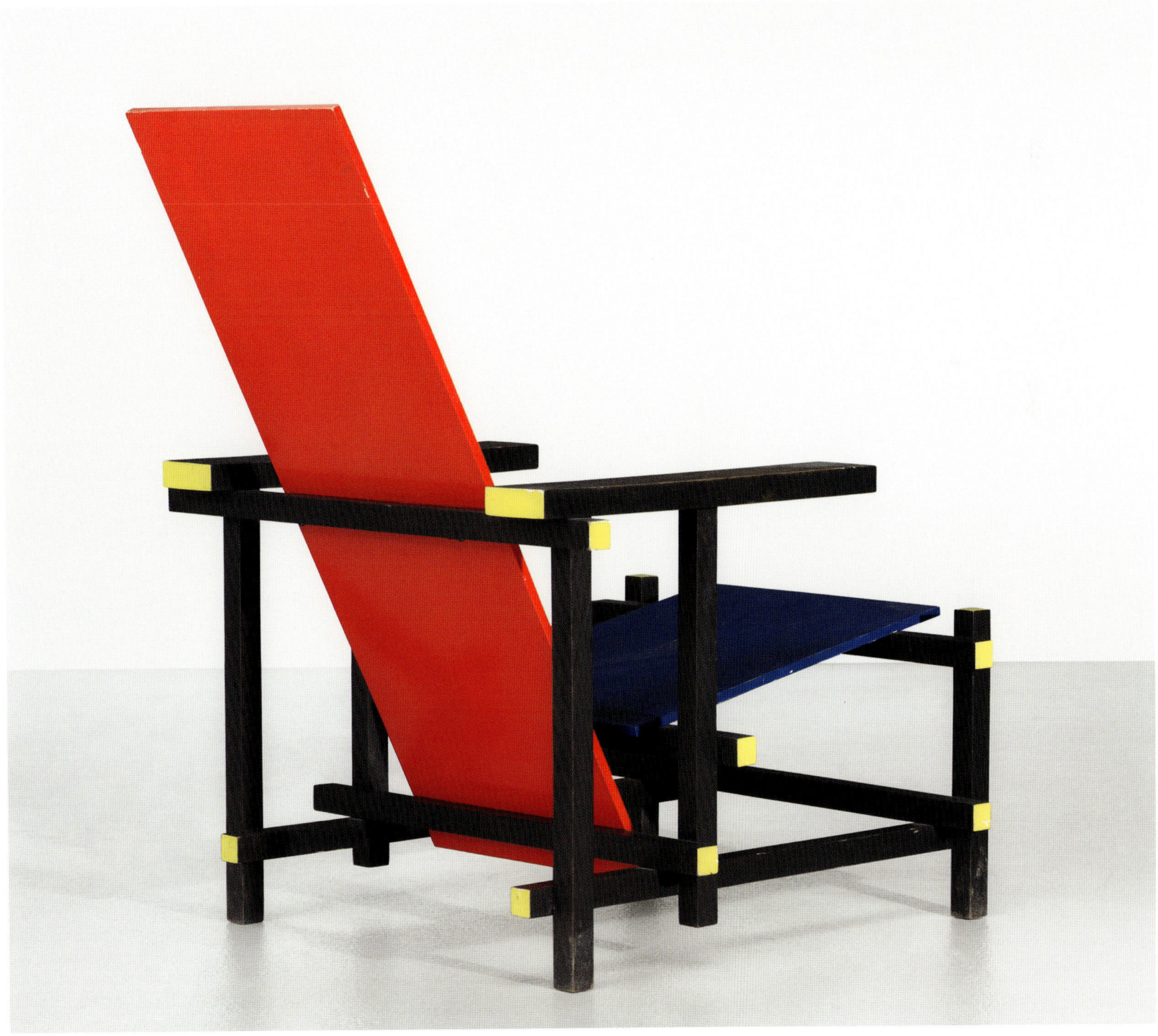

MEDIA & MASTERS

FURNITURE

LIGHTING

CERAMICS & GLASS

INDUSTRIAL &
PRODUCT DESIGN

GRAPHICS & POSTERS

FURNITURE

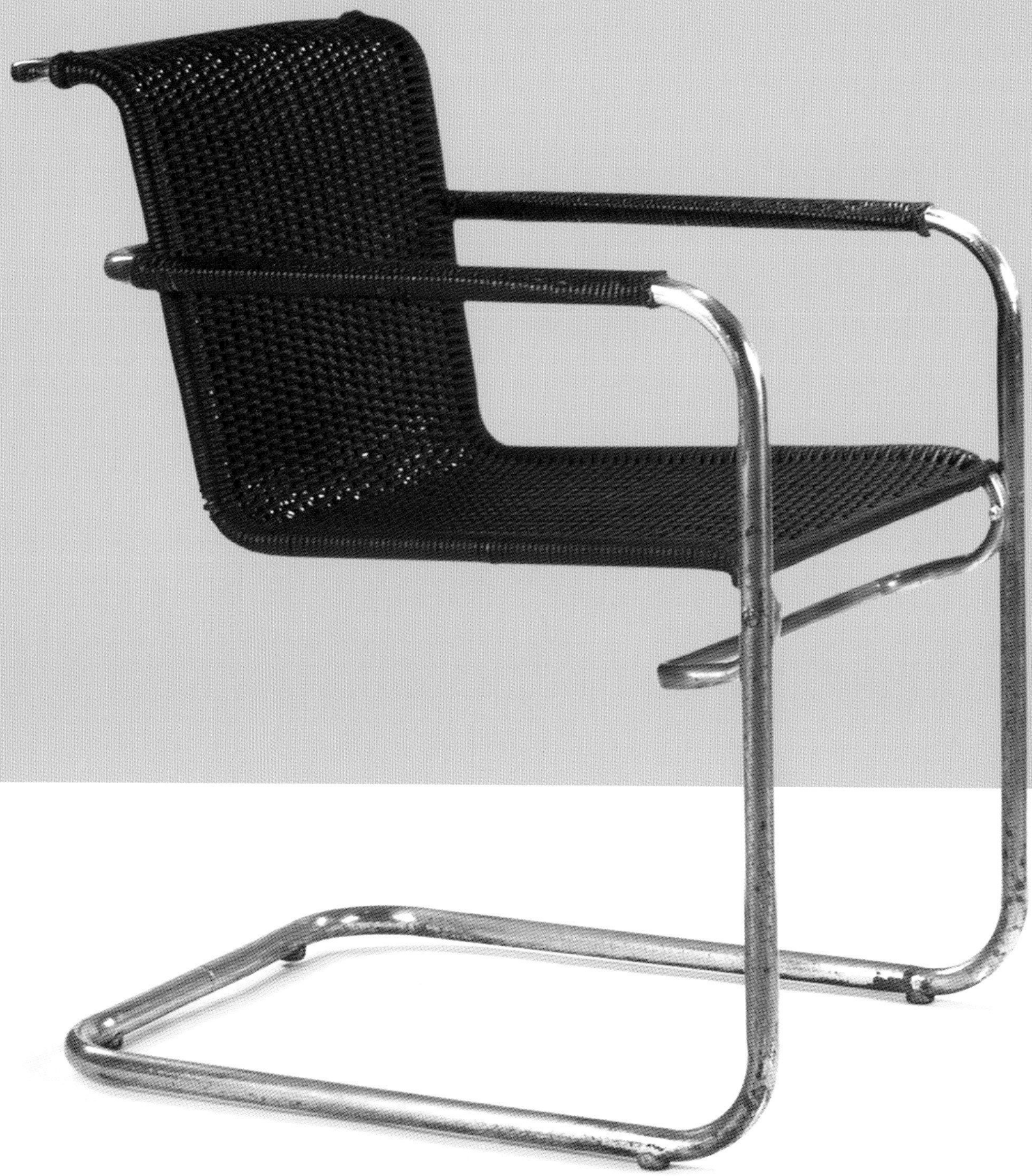

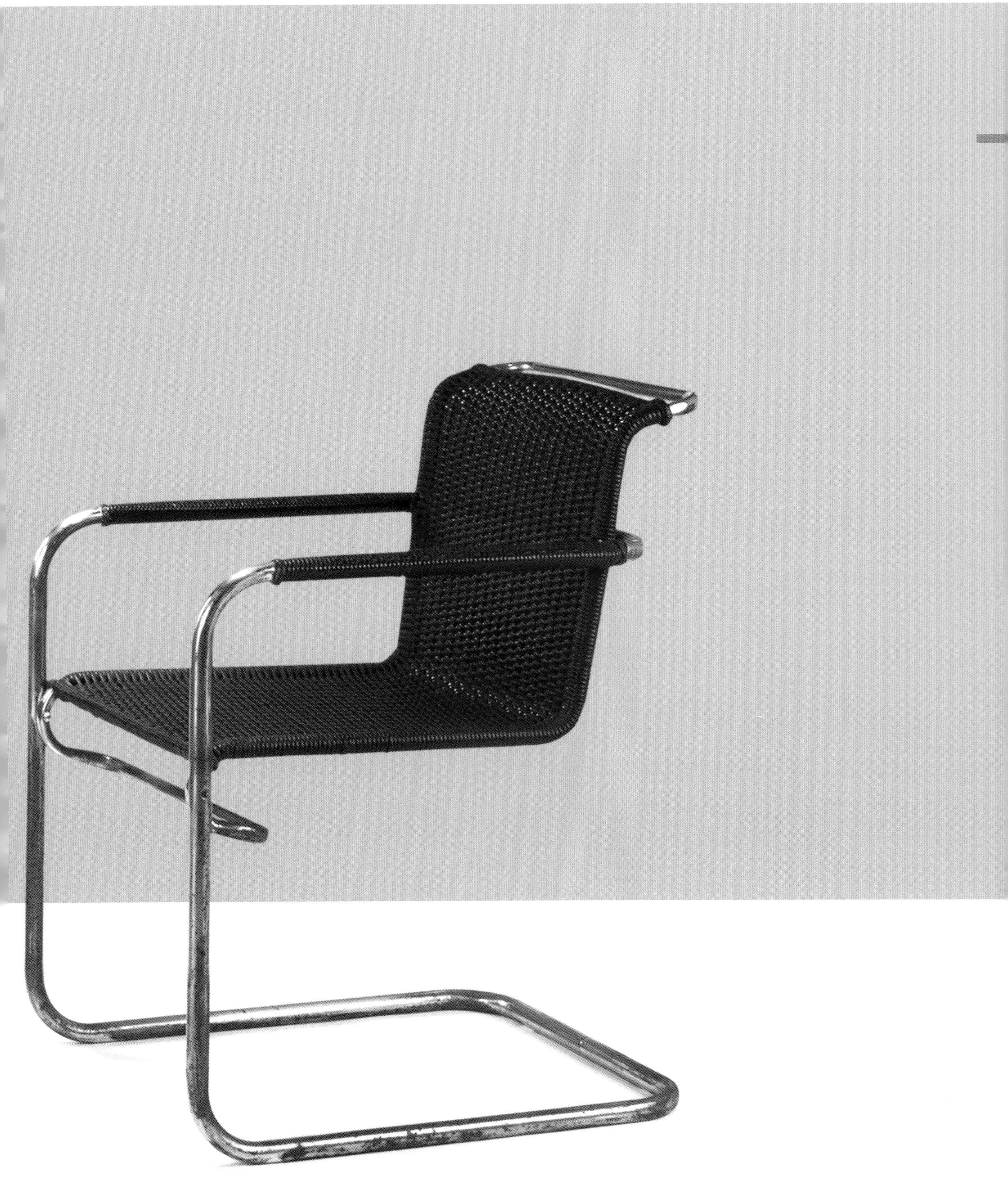

INTRODUCTION

Many iconic modernist pieces designed in the Twenties and Thirties now feel like familiar friends. They have become part of the common language of design, speaking of modernity itself and a vibrant machine-age aesthetic. For pieces conceived so many decades ago, it is remarkable that designs such as Ludwig Mies van der Rohe and Lilly Reich's Barcelona collection, Eileen Gray's E-1027 glass and chrome coffee table, Marcel Breuer's Wassily chair and the famous LC4 reclining chaise longue designed by Le Corbusier, Pierre Jeanneret and Charlotte Perriand still feel fresh and current. Their enduring success speaks of the strength of their design – with a powerful combination of style, substance and functionality – and they have become striking symbols of the 'modern', used as visual proof of design-consciousness and a progressive approach to form and function. It is fascinating to consider that such pieces, born between the wars, still manage to feel 'contemporary' in some respects, sitting easily within 21st-century spaces and buildings.

Early modernist furniture evolved from fresh thinking about materials, methods of production, ergonomics, functionality and aesthetics. There was a gradual process of questioning accepted wisdoms, ways of making and period styles, combined with radical advances in engineering and technology. Together, these factors created the foundations for a new approach to furniture design, largely focused upon pared-down, practical pieces with the potential for factory production, which would make the products more affordable and 'democratic'.

This was, essentially, the philosophy of the influential Bauhaus masters, including Walter Gropius and his successor as director of the Dessau school, Hannes Meyer*. They argued for a well-designed, functional collection of 'standard' products that could be mass-produced at an affordable price, enhancing the everyday lives of their users. In the case of a chair or other piece of furniture, such a design required a limited number of robust components that could be easily assembled in a workshop or factory, or even at home. In many respects, the Bauhaus philosophy was exemplified by the collection of tubular steel furniture (including the Wassily chair of 1925) designed by Marcel Breuer, who served as master of the furniture workshop; many of Breuer's designs were used to furnish parts of the Bauhaus campus and the masters' houses.

Famously, Breuer was inspired by the tubular steel frame of his own bicycle and began thinking about fresh applications for this industrial but highly versatile material. Other designers and architects embarked on a similar process of experimentation with industrial materials, turning away from a more craft-led approach to furniture design. They included Eileen Gray, René Herbst, Le Corbusier/Jeanneret/Perriand and, in the States, Donald Deskey and Gilbert Rohde. Jean Prouvé – fusing a talent for engineering, invention and design – pioneered his own range of innovative, factory-made, steel furniture and building components. Others experimented with a fresh palette of commercially produced materials and early plastics such as Bakelite and Formica, as well as glass products including Vitrolite and Lucite.

'A true modern piece of furniture is designed to incorporate all the advantages of modern materials, production facilities and construction, and if it is a good design it will give a beautiful result,' said West Coast designer Kem Weber, a German-born émigré. 'It will be good to look at, practical in its uses and becoming to ourselves. It will be less expensive and of better quality.'[1]

Weber also developed pieces using plywood and bentwood laminates, as did Breuer and, in Scandinavia, soft modernists such as Bruno Mathsson and architect and designer Alvar Aalto. An engineered wood, made up of many different layers – or veneers – of timber glued together, plywood was first produced in the late 19th century but not widely used for furniture or as a building material until the 1920s. Commercially produced, ply is relatively cheap but also flexible, strong and malleable enough for making chairs and other kinds of furniture, yet at the same time it benefits from an organic warmth. It was also widely used by mid-century designers such as Charles & Ray Eames*.

The ingenious Weber pioneered fresh ways of making and assembling laminated timber furniture, including his flat-pack, $25 Airline chair (*c.* 1934), which could be sent out by post and assembled at home. German designer and architect Ferdinand Kramer*, similarly, developed a range of flat-pack – or 'knock down' – plywood furniture in the Forties and early Fifties after emigrating to the States in 1938.

New materials and 'democratic' methods of production were not the only concerns, however. Ergonomics and functionality were key priorities for early modernist pioneers, with designers such as Mathsson and Herbst returning to first principles and examining the shape, contours and 'mechanics' of the human body, analysing position and posture in the hope of finding solutions that offered comfort, relaxation and repose as well as pure functionality.

With this analysis in mind, designers turned their attention to a range of different furniture staples and typologies, reassessing the function and form of each in the greatest detail, while adopting the kinds of innovative materials touched upon above. Mogens Koch and Kaare Klint invented new takes on the Safari chair, while Jean Prouvé also experimented with various kinds of folding chairs. Breuer, Aalto and Le Corbusier/Jeanneret/Perriand offered fresh interpretations of the chaise longue, while Mies van der Rohe/Lilly Reich and Eileen Gray reinvented the daybed. Even relatively simple constructs like the deck chair were revisited and rethought by Gray and Klint, among others, with the inspiration of the transatlantic liners fresh in their minds.

As with architecture, the inspiration of machine-age dynamism – seen in the form of liners, airships and automobiles – was tangible among modernist furniture and product designers. The links between the age of the machine and modernist furniture were numerous, from industrial-style kitchens and home appliances – exemplified by Austrian designer Margarete Schütte-Lihotzky's* functional Frankfurt Kitchen (1926) – through to Eileen Gray's Bibendum armchair of 1926, with the thick, comfortable

PAGES 14–15 A pair of B33 armchairs designed by Marcel Breuer, manufactured by Thonet, *c.* 1929, in chrome-plated steel with lacquered cane seats.

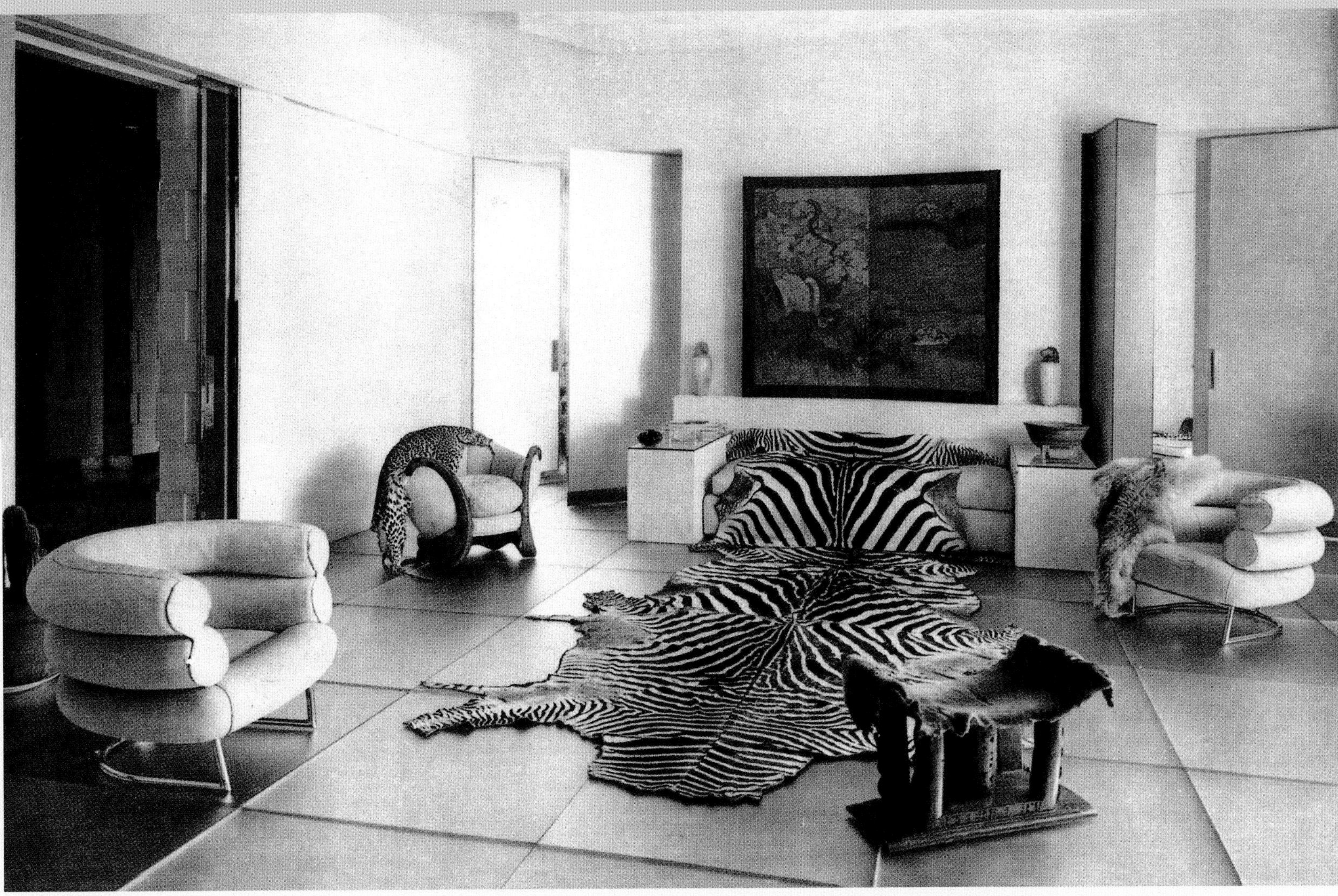

BELOW The sitting room of Madame Mathieu-Lévy's Paris apartment on the rue de Lota, *c.* 1920; the apartment features Eileen Gray's Bibendum and Dragon armchairs.

RIGHT A marketing and publicity photograph documenting the assembly of designer Kem Weber's flat-pack Airline chair, c. 1934.

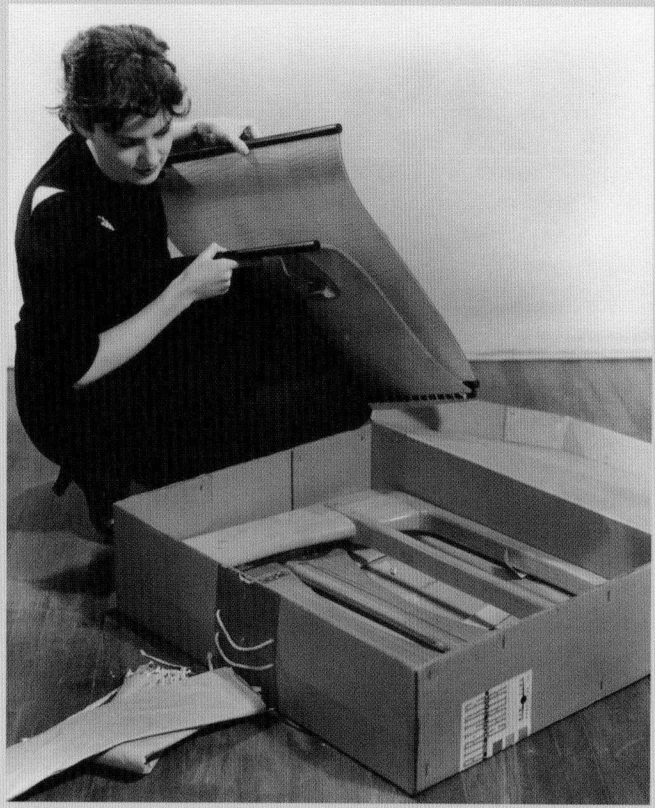

rings of its upholstered back and seat (resting on a tubular steel frame) supposedly inspired by the inflated car tyres that formed the body of Michelin's famous mascot.

'Machines, our detractors say, have killed inner life,' said René Herbst of the influence of the machine upon domestic design. 'And why? Because we want furniture doors to operate in new ways, windows to slide silently, furniture to be made in steel because it is easier to produce in large quantities and at low price? Because we want chairs to be more comfortable by the incorporation of new, more flexible springs, and lighter thanks to tubular steel frames?'[12]

Aesthetically, there was an emphasis on the modernist mantra that 'form follows function' – a phrase attributed to American architect and skyscraper pioneer Louis Sullivan* in 1896 (within a publication entitled 'The Tall Office Building Artistically Considered'). Modernist furniture typically adopted a pared-down, simplified outward appearance, whereby ornament and decoration were stripped back or dispensed with altogether, with the focus remaining on the materials and tectonics of the piece in question.

There were certain points of common interest between designers generally seen as part of the Art Deco* movement and – in America – Streamline Moderne, as seen in the careers of Donald Deskey and Gilbert Rohde, as well as French masters such as JEAN-MICHEL FRANK and Jacques-Émile Ruhlmann*. This overlap expressed itself in some shared use of materials – particular metallics, such as chrome or polished steel – and automotive-inspired streamlining (see Anne H. Hoy's essay, p. 208). Yet, whereas Art Deco was seen as largely a movement of the elite, employing luxurious materials, subtle craftsmanship and fine detailing, modernist designers were more interested in a more readily available and democratic form of production using industrial materials and assembly-line methods. At the same time, many modernists turned their backs upon some of the superfluous ornament associated with more flamboyant strands of the Deco style, advocating a far simpler aesthetic approach.

At the same time, there were other important elements within early modernist aesthetics that were to become increasingly significant and vital in the post-war, mid-century period. There was a strong connection between the worlds of modern art and design, seen at the Bauhaus and in the murals and colour theories of Le Corbusier, but also in movements such as De Stijl* in Holland. Furniture by one of the leading proponents of De Stijl, designer GERRIT RIETVELD, was intricately imbued with an artistic energy fed by painters such as Piet Mondrian. Rietveld's famous Red & Blue chair of 1918 could be seen as a painting that you could also happen to sit upon. Rietveld's work, in particular, reminds us that the world of modernist design was intricately and intimately linked with other strands of cultural modernism, including art, literature, theatre and music (see the general Introduction, p. 9). In the Fifties and Sixties, the relationship between art and furniture design became even more pronounced, with artist-designers such as Harry Bertoia and Joaquim Tenreiro working within the fields of both sculpture and furniture design.

Despite the focus on industrial production, there remained an influential group of designers such as Alvar Aalto, Bruno Mathsson and others of the Scandinavian warm modernist wing who retained a powerful interest in craft and

OPPOSITE A Pernilla Long Chair, designed by Bruno Mathsson and produced by Firma Karl Mathsson, 1944, in laminated steam-bent beech, canvas, brass and leather.

its complex relationship with new approaches to democratic design, seen most explicitly in their plywood and laminate furniture. Again, the romanticism of this soft modernist outlook became highly influential in the Fifties, seen in the work of Nordic designers Hans Wegner and Poul Kjaerholm as well as the powerful craft-based furniture of Wharton Esherick* and George Nakashima.

Another interesting thread that carried over from the pre- to the post-war period was the link between furniture design and architecture. Many key pieces of pioneering modernist furniture, such as Mies van der Rohe/Lilly Reich's Barcelona collection, and a number of Eileen Gray's designs, along with pieces by Alvar Aalto and ARNE JACOBSEN, evolved from architectural and interiors commissions, first designed to sit within a specific project and later put into production. The same was true of many key furniture designs of the Fifties, when architects such as Gio Ponti, Eero Saarinen, Carlo Mollino and others crossed the borderlands between architecture, furniture and product design with easy regularity.

The level of experimentation and innovation set by the pioneering modernists of the Twenties and Thirties offered profound inspiration for the mid-century masters who followed in their wake. By the Sixties, with the revolution in plastics, a fresh wave of exuberant artistry captured the attention of the design world. The dynamism of the early design revolutionaries remains inspirational and profoundly influential, while their work manages to remain current.

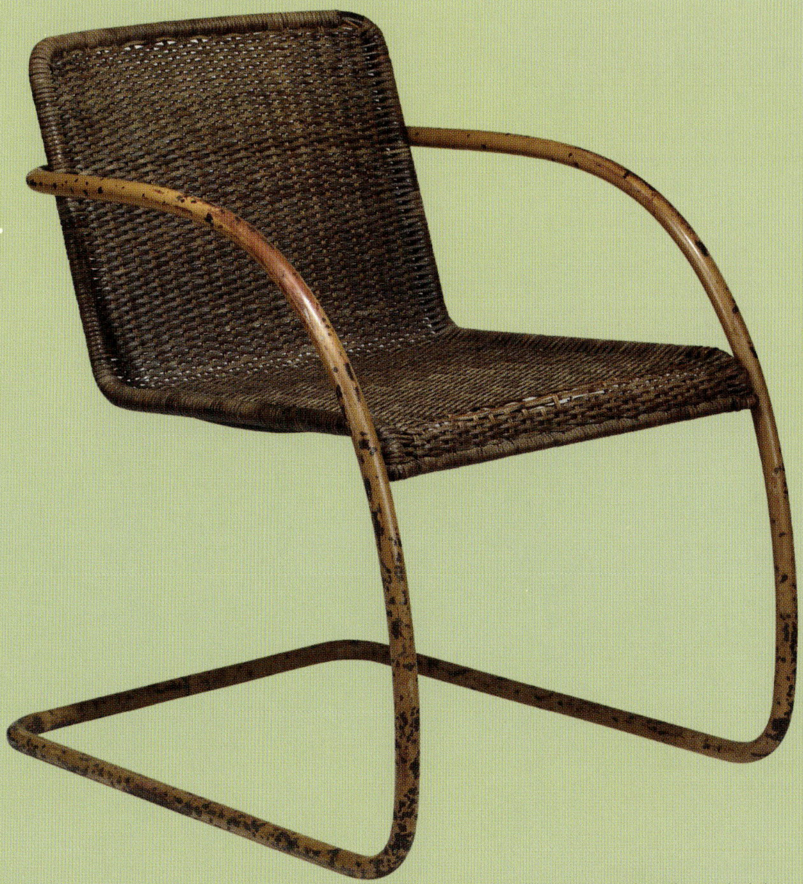

ABOVE Cantilever chair by Ludwig Mies van der Rohe and Lilly Reich, c. 1929, in enamelled tubular steel with a wicker seat, manufactured in Berlin.

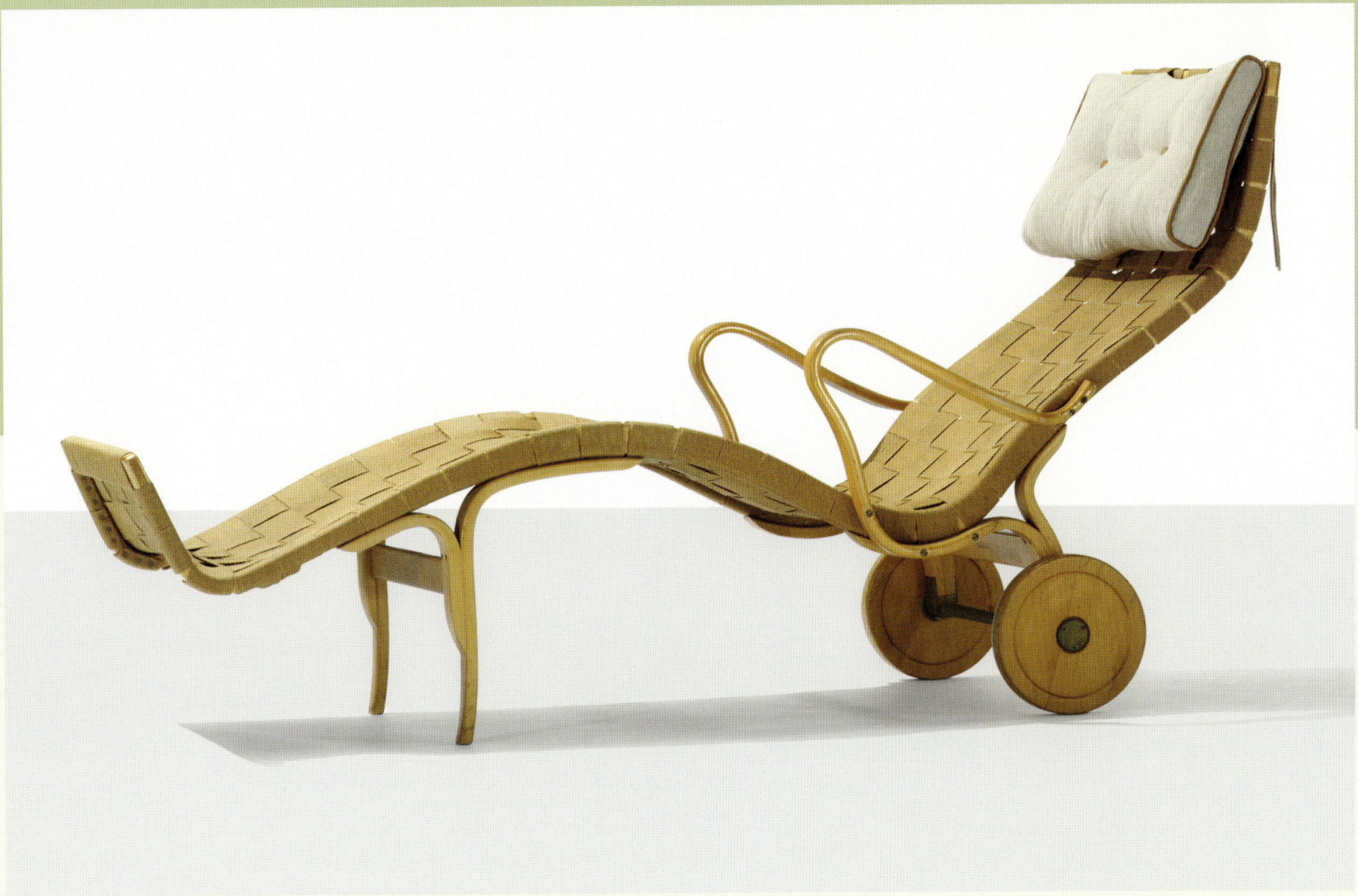

From the early years of the 20th century, European ideas about furniture focused on combining necessity with functionality and aesthetics. This fashion for 'beauty in usefulness' first emerged in England during the 1860s thanks to the efforts of John Ruskin and William Morris, founders of the Arts and Crafts movement, which championed 'total art', art that encompassed all the arts – architecture, furniture, the decorative and the plastic arts.

In France, between 1900 and 1910, these same aspirations were embraced by the École de Nancy, identified with the movement known as Art Nouveau and counting among its members Émile Gallé, Hector Guimard and Victor Prouvé (father of JEAN PROUVÉ). Art Nouveau saw the birth of a completely new aesthetic that drew its inspiration from plant and animal forms and favoured curves and flowing lines, while its champions continued fiercely to support traditional craftsmanship and the work of the ateliers (cabinet-making, glass, bronze and enamel work, and work in precious metals). The creations of the École de Nancy, in terms of furniture and the decorative arts, consequently laid the foundations for a wholly modern approach but without involving any new materials.

After 1910, the desire for radical change gained momentum, initially in the sphere of art thanks to the Cubist movement and the influence of Pablo Picasso and Georges Braque. Cubism, with its insistence on geometric forms, left its mark on the decorative arts, furniture, architecture and the general aesthetic of the age, but World War I was to interrupt the creative efforts of a great many men who left for the Front.

Paradoxically, the war also accelerated the impetus towards modernity by provoking a series of far-reaching changes on the economic, industrial and social levels. Considerable advances were being made in terms of industrial manufacturing, and with industrialization came mass production. Electricity, the automobile, air travel and long-distance communication: each technological advance was life-transforming, and these profound changes of the very fabric of civilization led to an aesthetic revolution that made itself felt in France and across the whole of Europe during the Twenties and Thirties.

The World's Fairs accelerated the phenomenon by bringing artists and designers into contact with sources of inspiration from other parts of the globe. Three of these fairs took place in Paris in 1900, 1925 and 1937 (together with the Exposition Coloniale in 1931). During the Twenties, two opposing tendencies developed in Europe: on the one hand, the production of luxury goods destined for a well-heeled clientele, and, on the other, medium-size production runs aiming to offer high quality to a major section of the consumer market. Production in France reflected the general model, even if the dividing line between the different approaches and types of production was a great deal more subtle, with certain French designers – such as Jacques-Émile Ruhlmann* and JEAN-MICHEL FRANK – succeeding in

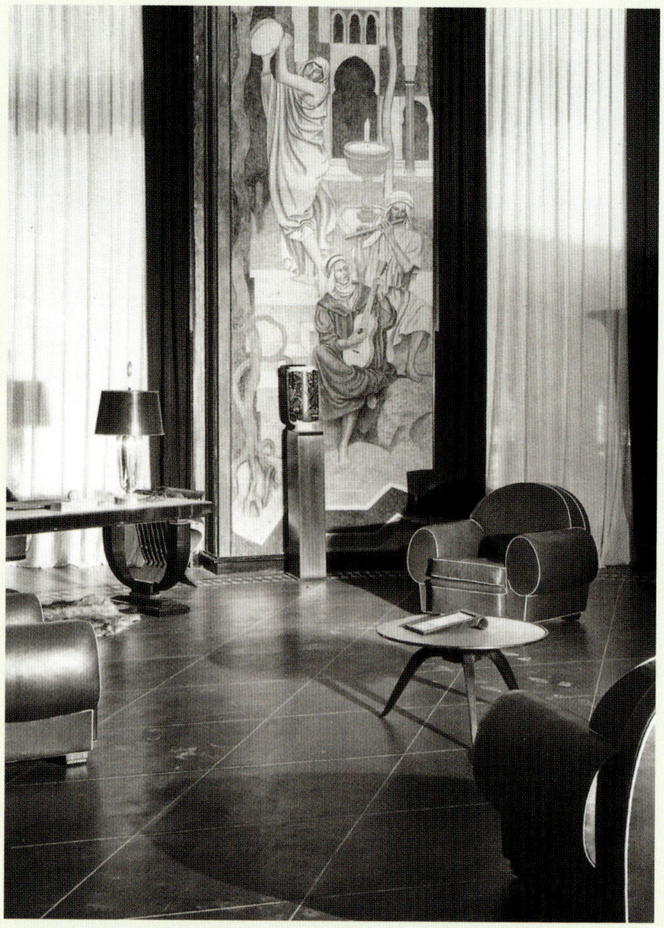

ABOVE An office displayed at the Exposition Coloniale, Paris, 1931, with furniture by Jacques-Émile Ruhlmann.

OPPOSITE Design for the interior of a first-class cabin on a liner, by René Herbst, from *Acier* magazine, July 1935.

THE ADVENT OF MODERNITY: MODERNIST FRENCH FURNITURE AND THE NEW MATERIALITY

PATRICK & LAURENCE SEGUIN

reconciling technical innovation and luxury, or simple styles and sophistication.

At the 1925 Exposition Universelle in Paris, the two tendencies coexisted, putting into opposition artists who were openly drawing on the past but inventing a contemporary signature style in the spirit of the great designers (in the manner of Ruhlmann, Jean Dunand, Edgar Brandt, Süe and Mare) with 'modern' makers who were influenced by industry and technology, and radically focused on the future (the likes of PIERRE CHAREAU, EILEEN GRAY and ROBERT MALLET-STEVENS), spearheading a movement of aesthetic reform in France during the Twenties. The rejection of ornamentation became gradually more widespread and went hand in hand with a geometric rigour that was common to architecture, furniture, and the decorative and plastic arts.

It was this same decade that saw the inception of the Bauhaus in Germany, the De Stijl* movement in the Netherlands and Constructivism* in Russia – France's answer to which was the Union des Artistes Modernes*, more commonly known as the UAM, founded in 1929 by RENÉ HERBST, Robert Mallet-Stevens, Pierre Chareau and Jean Prouvé, who were later joined by, among others, CHARLOTTE PERRIAND, LE CORBUSIER, PIERRE JEANNERET, Eileen Gray and Sonia Delaunay*. Embracing functionality and seeking to make a clean sweep of the past, the UAM marked a rift between the modernist movement and the lavishness of Art Deco*. Its protagonists (many of whom combined architecture with furniture design) were keen to replace expensive luxury materials with more functional ones, better adapted to standardization, such as glass and steel – and, in architectural terms, concrete.

The work of Jean Prouvé perfectly embodies the UAM's innovative spirit. Born in Nancy in 1901, Prouvé was a great champion of modernity. A metalworker by training, he was quick to embrace the new repertoire, with its deliberately rational geometric forms, and eager to experiment with new techniques, readily acquiring the latest materials and tools. While his early years were steeped in the spirit of the École de Nancy, and he remained dedicated to a workshop mentality, Prouvé was always ready to push the boundaries, shifting his focus towards a more 'industrial' style of production. In the mid-Twenties, for instance, he became interested in electric welding and started working with thin sheet steel, which relied on principles very different from those that underlay his work with wrought iron. In 1929, he started making furniture, including an adjustable reclining armchair, a pedestal table with a rubber top and folding chairs that also reclined, and filed his first patents for movable partitions, metal doors and sash windows. He showed examples of his work, principally furniture, at the UAM's inaugural

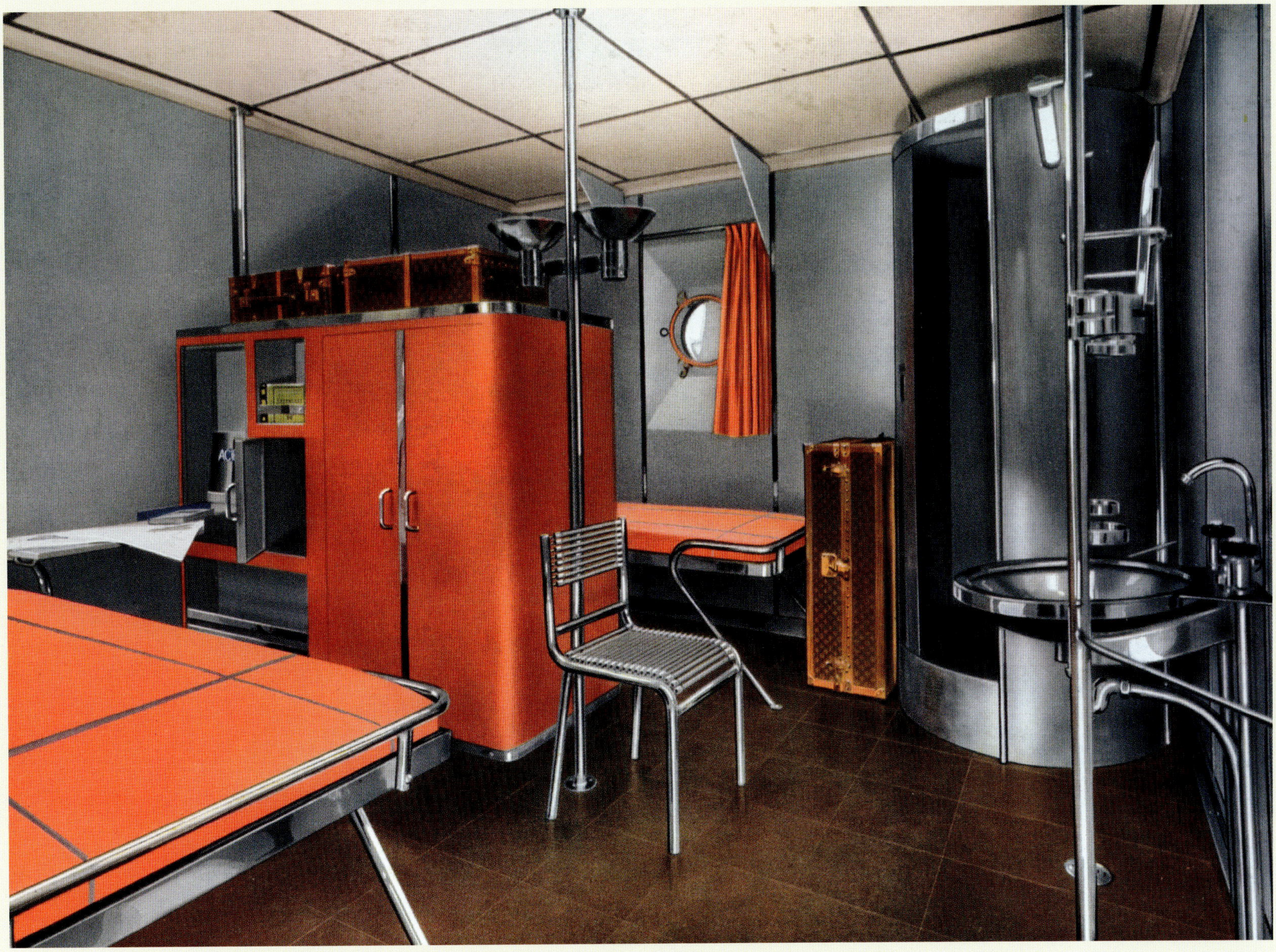

exhibition in 1930, and in January 1931 he founded the Ateliers Jean Prouvé, in Nancy's rue des Jardiniers, whose premises were large enough to accommodate the construction of industrialized building elements and medium-scale production runs of furniture.

One of the most innovative aspects of Prouvé's work was his approach to materials, with wood and metal featuring in both his furniture and his architectural designs. The way he saw it, making a piece of furniture was no different from constructing a building, and he applied the same principles to both, several of his designs demonstrating very clearly that a great many elements envisaged for assembling a piece of furniture work equally well for architecture.

Prouvé's familiarity with his materials led him to exploit their full potential, really pushing them to their limits. Experimenting with thin sheet steel and aluminium, he was able to show that bending the metal made it more rigid, and that by adding solder it was possible to produce a hollow component with essentially the same resistance as a solid one, but much lighter – a discovery that opened up all sorts of possibilities for both furniture and architecture.

BELOW Standard chair by Jean Prouvé (1951–52), in enamelled metal and birch, first designed in 1934.

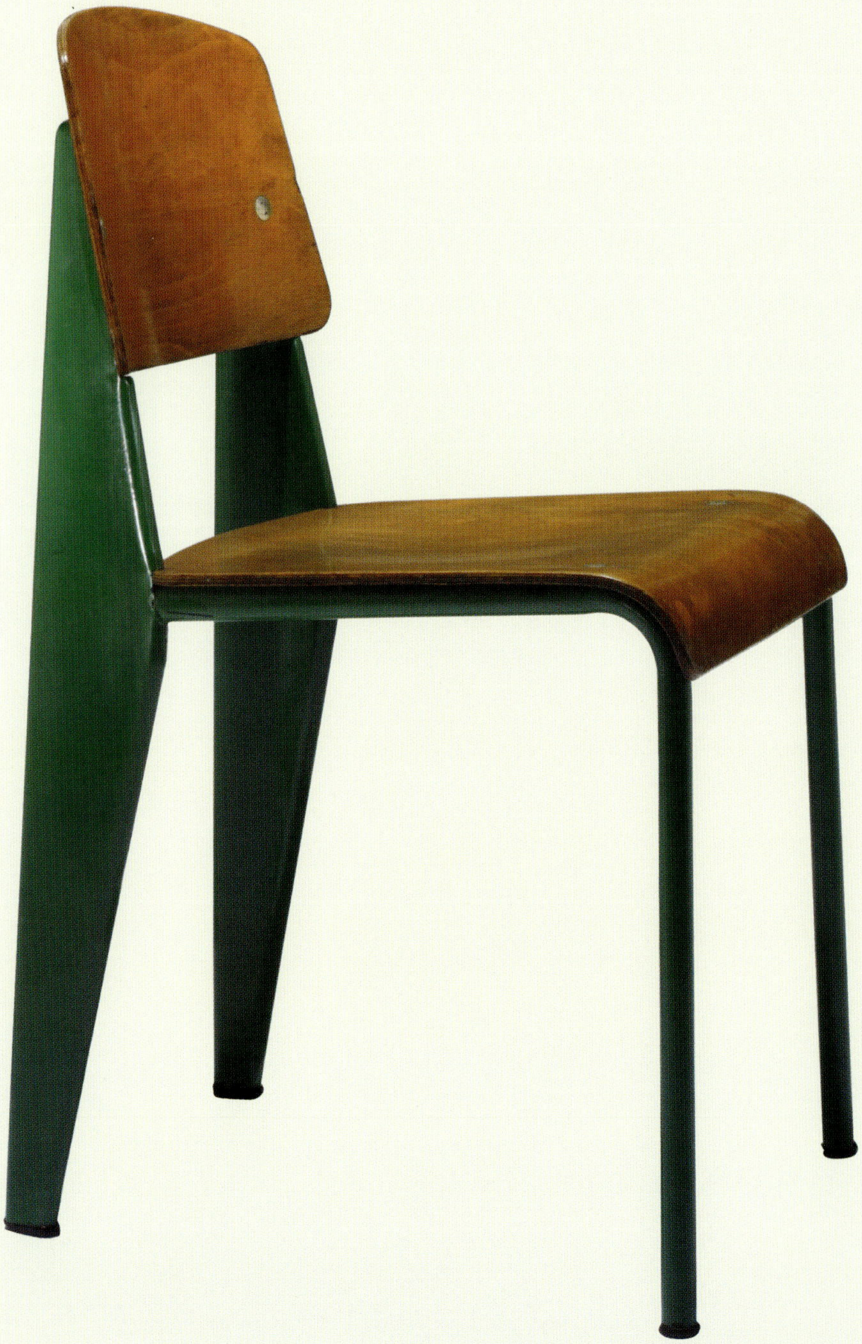

Prouvé's perfect understanding of his material enabled him to put it to the best possible use: he used more material at the point of greatest mechanical stress and less where the stresses were reduced. His Standard chair, with its tubing and bent steel base, which he first started producing in 1934, illustrates this principle perfectly: rather than existing for aesthetic effect, the relatively solid back legs (a characteristic feature of the design) are warranted by Prouvé's own habit of leaning back in his chair so that the front legs lifted off the floor – a position requiring the rear supports to be sturdier than the front.

Prouvé's originality also derives from the fact that he never made unique pieces – unlike most of his fellow associates within the UAM, including Robert Mallet-Stevens and Pierre Chareau. Everything he designed, right from his very first creations, was intended to serve the public at large and to meet the daily needs of schools, universities, hospitals and sanatoria, offices and public buildings. This explains why, when these pieces were created in the Thirties, Forties and Fifties, they were seen as purely utilitarian, and why it would be several decades before design enthusiasts were able to appreciate an aesthetic so absolutely rooted in functionality. Prouvé himself may have laid no claim to the beauty and elegance of his creations, but beautiful and elegant they are, thanks to the perfect marriage between form and function.

During the Thirties, architecture and furniture began to adapt to society's changing needs. With the advent of paid holidays in 1936 came increased mobility, and the open-air life was now something that everyone could enjoy, not just the elite few, as previously. Architects and designers began envisaging furniture that folded or could readily be taken outside. However, war – this time the 1939–45 conflict – halted the momentum of these new developments.

The question of a type of architecture and a type of design that dovetailed happily with modern living resurfaced at the end of hostilities and was all the more urgent now, given the need for reconstruction. It was at this point that Jean Prouvé started working again on the idea of 'demountable' houses – something he had experimented with prior to the war, notably with his BLPS weekend house in 1937. He created emergency living accommodation for people whose homes had been destroyed in the bombings (houses measuring 6 x 6 m and 6 x 9 m for refugees in Lorraine), designed to be erected by three people in a single day, each element light enough to be carried by one person without assistance. He also designed demountable and prefabricated houses that could be transported by lorry or boat, including the Maison des Jours Meilleurs commissioned by the clergyman Abbé Pierre, and a Maison Tropicale destined for the colonies in West and Central Africa.

By the beginning of the Fifties, mass-produced furniture was proving hugely popular, in both private homes and civic buildings. The direct relationship between form and function – something the UAM had been envisaging since the early Thirties – had become quite simply a given for the majority of people.

When, in the early Fifties, Le Corbusier and Pierre Jeanneret were tasked by Pandit Nehru with the job of constructing the new city of Chandigarh in India, they set about designing buildings and furniture that reflected the local customs and lifestyle and were ideally suited to the climate. Charlotte Perriand, already a keen advocate of metal in the early Thirties, designed a new type of mountain dwelling, demonstrating her interest in the simplicity of forms drawn from rural life on the eve of war, and at the start of the Fifties she returned to her earlier experiments with renewed authority, strengthened in her convictions by lengthy periods of study in Japan during the Forties. Perriand's furniture and fittings are both pleasing and functional; they are perfectly in tune with the aspirations of the age, combining elegance and simplicity, and the need for 'standardization' with an eye to producing something unique. Meanwhile Jean Prouvé, too, was continuing to embark on different projects and to explore the possibilities inherent in his materials, principally metal.

Jean Prouvé, Charlotte Perriand, Le Corbusier and Pierre Jeanneret, and other artists associated with the UAM, were the veritable precursors of modernity. They were visionaries, capable of projecting themselves well beyond their time and anticipating what the century was to generate in terms of new ways of living and thinking. It is to this modernity, no doubt, and to a refined and timeless aesthetic – enabled by a certain rigour – that all these creations owe their iconic status in the history of design. These highly desirable pieces have naturally – and deservedly – found their way into the finest public and private collections across the globe.

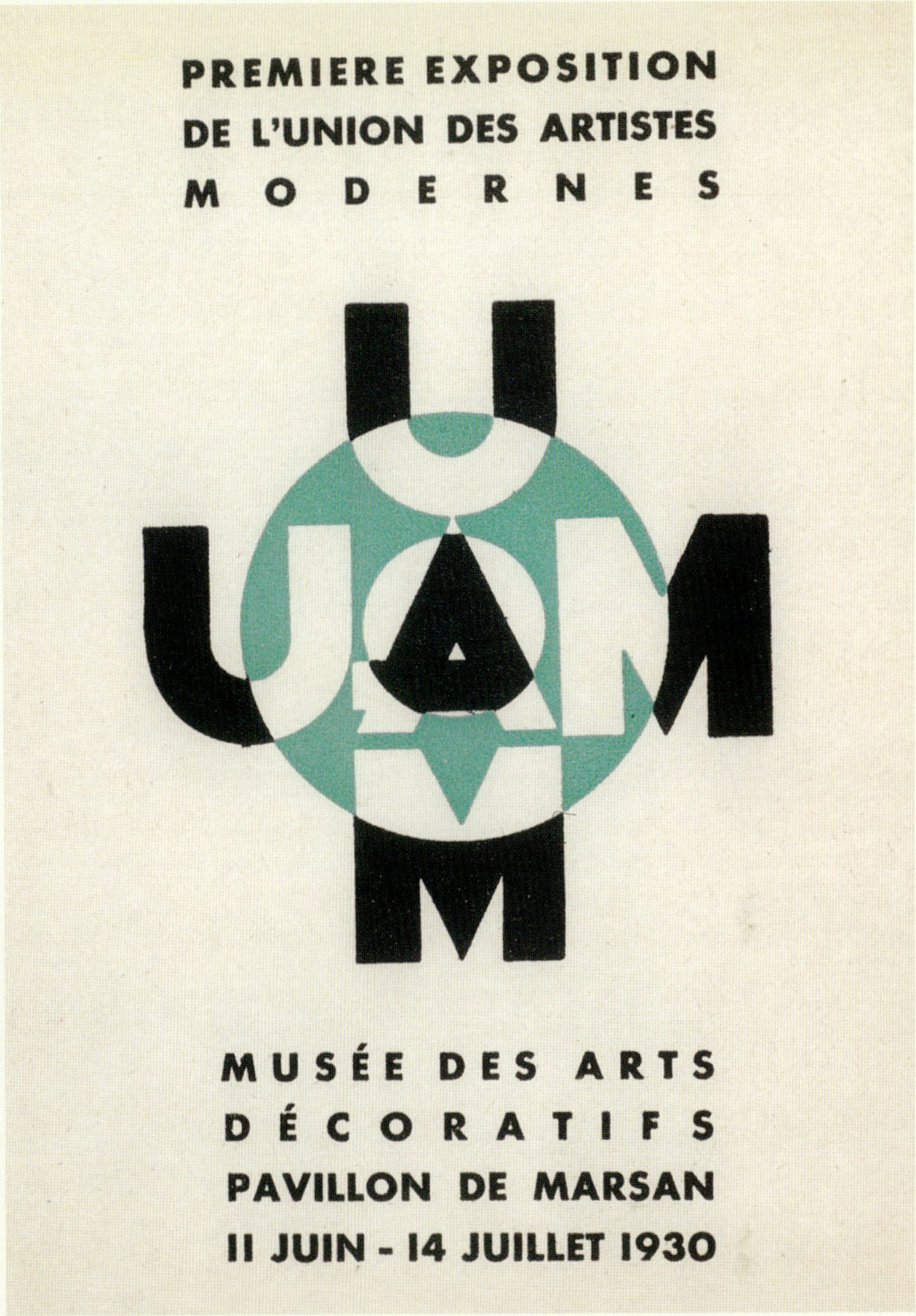

ABOVE Cover to the catalogue for the first exhibition of the Union des Artistes Modernes (UAM) at the Musée des Arts Décoratifs, Paris, 1930; the UAM logo was designed by Pierre Legrain.

The Bauhaus was one of the most influential design schools of the 20th century. During its short existence (1919–33), it succeeded in setting new standards for avant-garde art as well as playing a major role in social change. It became a kind of melting pot, bringing together the dominant trends of the European avant-garde; and, under the direction of the architect WALTER GROPIUS, many of the most important artists of the day, such as Paul Klee*, Wassily Kandinsky*, LÁSZLÓ MOHOLY-NAGY and Oskar Schlemmer*, made their way to Weimar in pursuit of an idea, a common vision – the creation of a new, fairer and more humane society through the artistic redesign of every facet of life.

This vision required a new kind of designer, who on the one hand would harmonize art and craft, and on the other would use art to bridge the wide gap between humans and technology. This modern designer would be able to create everyday objects, furniture, houses and even cities by combining an artistic approach, training in the crafts, scientific knowledge and industrial mass production as a new form of manufacture. The Bauhaus set out to train this new designer, but did not claim experimentality, nor even originality; neither did the establishment see itself as a school that was setting out to break with tradition.[3] Its aim was to take up movements that had emerged during the 19th and 20th centuries (such as Arts and Crafts, the Wiener Werkstätte, and later the Deutscher Werkbund*), to develop them further and, where necessary, to adapt them to new conditions.

The origin of an idea

The Bauhaus came into being during a period that was racked with uncertainty and burning with a desire for radical change. At the turn of the century, the prevailing view was that progressive industrialization heralded a process of humanization, and people could expect a happier, more prosperous future. The outbreak of World War I in 1914 therefore came as a rude awakening. Everything people had believed in – the dignity of the individual, fraternity, humanity – now lay buried in the trenches. By the end of the war in 1918, there was nothing but economic and political chaos. The war had undermined or even destroyed the existing order, but there was nothing new to replace it. The former powers-that-be had gone, and now the middle classes emerged as the new powerhouse. Against a background of buzzwords such as pacifism, the League of Nations, social reform and industrialization, the bourgeoisie joined forces with the artists of the avant-garde to look for new, more equitable ways of life and self-expression. These, then, were the basic principles behind the Bauhaus, together with the conviction that all areas of society could be designed. And so the 'new designer' was given a special task: he or she was to be the driving force of social transformation.

The Bauhaus grew out of a merger between two schools in Weimar: the Hochschule für Bildende Kunst (College of Fine Arts) and the Kunstgewerbeschule (School of Arts and Crafts). The former had been well known since 1860 for the 'Weimar School of Painting' – a movement of landscape art in Germany – whereas the latter had only been founded in 1908 by Henry van de Velde*, who was its director until his enforced resignation in 1915. It was suggested that his successor should be Walter Gropius, a student of PETER BEHRENS and member of the Deutscher Werkbund. Following an invitation from the Ministry for Home Affairs, in January 1916 he submitted his proposals 'for the foundation of an educational establishment that would also be an artistic advice centre for industry, trade and crafts'. It took another three years for the two institutions to merge under the name 'Staatliches Bauhaus in Weimar' (State Bauhaus in Weimar), and during this period Germany engaged in the greatest catastrophe of its short history. After the war negotiations resumed, and on 11 April 1919 Gropius signed a contract to become director. The next day, the new name of the school was approved, and this completed the foundation of the Bauhaus without any further formalities or fanfares.

Not long afterwards, the school published its programme and manifesto, issued for promotional purposes with a cover designed by Lyonel Feininger. This depicted a cathedral with three towers, the tops of which were crowned with three stars that symbolized the fine and the applied arts and architecture. The programme provided for dual training: after a preparatory course, the prospective designer would be trained in handicrafts

BAUHAUS BY DESIGN: THE IDEAS AND THE INFLUENCE OF THE BAUHAUS MASTERS

JOLANTHE KUGLER

and in art and theory. A course on harmonization theory remained compulsory until 1923 and served to liberate artistic creativity, while at the same time helping the student to find an appropriate placement in a workshop.

The reality of an idea: Weimar, Dessau, Berlin

What greeted Walter Gropius on his arrival in Weimar in 1919 was a situation far from conducive to the rapid realization of his ideas. Leaders in industry and crafts, many teachers from the former college of fine arts, and the conservative people of Weimar were all hostile towards the school. There was a shortage of funds and an inadequate infrastructure, and the lack of materials together with an ever-worsening hyperinflation crisis made life into what Lyonel Feininger described as 'daily torture'.[4] Nevertheless, the Bauhaus masters adapted to the 'postwar psyche' (Feininger again) and with impressive vigour set about translating the ideas of the programme into reality. In between hunting for accommodation and lunch tables for the students, procuring money to pay the salaries of the new teachers, equipping the workshops, and warding off attacks from the government and population (and some of the avant-garde), the masters strove to put together an educational concept which, despite their different artistic positions, they could all identify with.

It is therefore all the more astonishing how much they achieved in the short period between April 1919 and August 1923, when the products of their work were presented at an exhibition. The provincial government had insisted on this exhibition, and further funding was made dependent on its success. Gropius was unhappy at having to go public after so short a time, because from his standpoint the Bauhaus was still in its infancy and was a long way from achieving the high standards he himself had set. The works on display could therefore only show the best results thus far, and could not claim to provide a solution to every problem. Nevertheless, a large number of works were produced which today rank as Bauhaus icons: they included MARCEL BREUER's slatted chair, MARIANNE BRANDT's objects for everyday use, WILHELM WAGENFELD's table lamp, Anni Albers's wall hangings, and Oskar Schlemmer's stage productions. Although the exhibition was enthusiastically received by the press, and brought national and international recognition to the Bauhaus, criticisms of the school continued unabated. Eventually, in December 1924, the increasingly powerful right-wing authorities gave notice to all the Bauhaus masters, whowere finally dismissed on 31 March 1925. The school was then closed.

With hindsight, however, this proved to be a blessing in disguise, because the Bauhaus found a new home in the

BELOW Walter Gropius in front of a development plan, 1930.

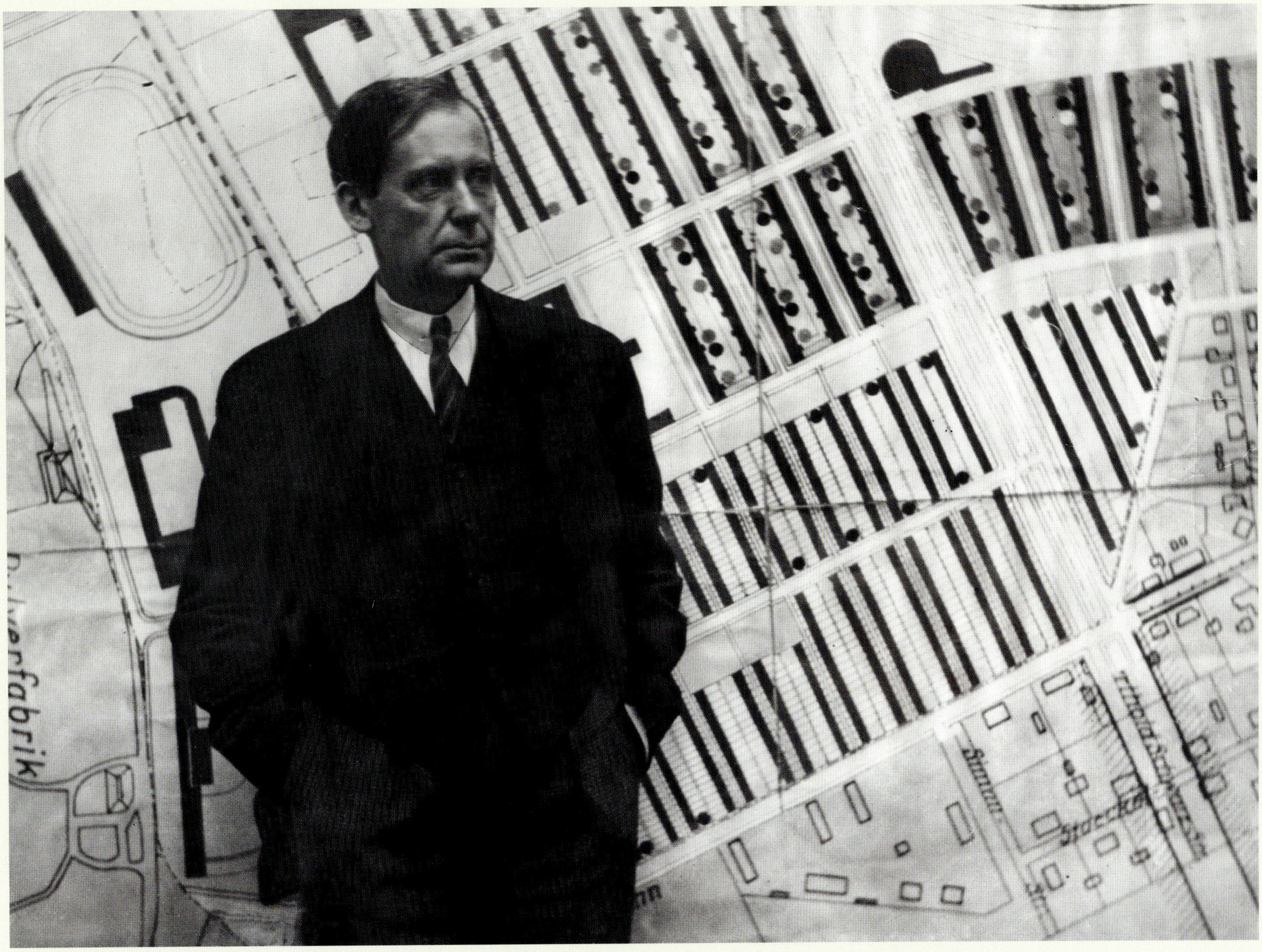

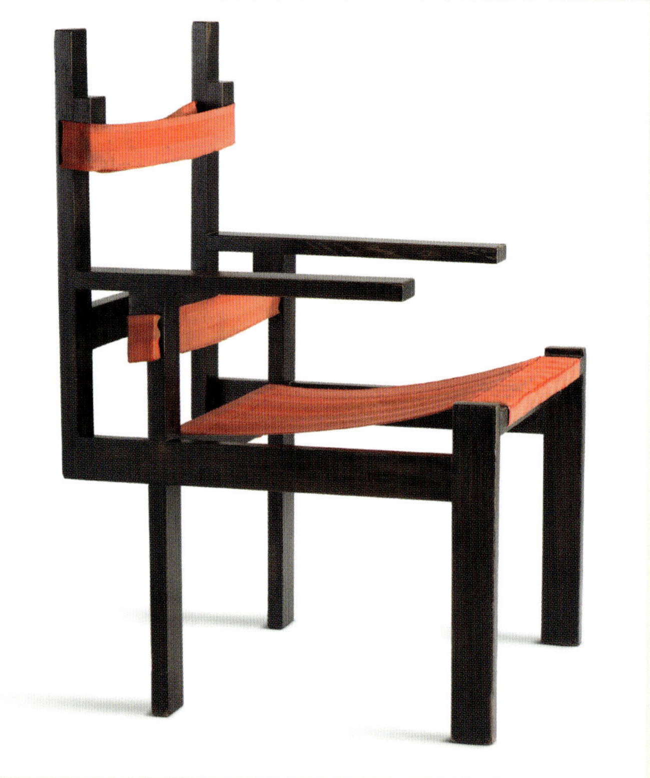

ABOVE The ti 1a, or Slatted chair, designed by Marcel Breuer, 1922.

progressive industrial town of Dessau. The town welcomed the 'Bauhäusler' with open arms, provided money for a new school building and for houses to accommodate them, and went out of its way to promote the sort of cooperation with industry that had hitherto been sadly lacking. The courses and their contents were restructured, and the Romantic or 'universalistic'[5] trends of the earlier years were largely abandoned. The training now took the form of studies culminating in a diploma. Accordingly, from 1926 onwards the Bauhaus carried the additional tag of 'Hochschule für Gestaltung' (College of Design). In 1927, it opened an architecture department, thereby fulfilling one of its original aspirations. The head of this department was the architect Hannes Meyer*, and the following year, on Walter Gropius's own invitation, Meyer took over from Gropius as director of the Bauhaus itself.

Under Meyer, the Bauhaus moved resolutely in the direction Gropius had advocated back in 1919. In close cooperation with industry, household products and forms of architecture were developed that met the needs of ordinary people but at the same time were high quality and affordable.[6] The Bauhaus even succeeded in selling its designs to industry: for instance 'Bauhaus wallpaper', which was very successfully marketed by the firm of Rasch in 1930. However, after just two years the town council summarily dismissed Meyer on 1 August 1930 because of his 'communist machinations'. In the same year, the architect LUDWIG MIES VAN DER ROHE was appointed director. He had become famous as artistic director of the Weissenhof Estate* that was designed for the Stuttgart Werkbund exhibition, and as designer of the German Pavilion for the 1929 International Exposition in Barcelona. During the years that followed, he was careful to keep the Bauhaus out of politics, and under his directorship it became a school of architecture without any socio-political ambitions.

Despite these measures, the powers-that-be in Dessau also began to turn against the Bauhaus: the town council became increasingly dominated by the right wing, who suspected the masters of sympathizing with the left and rejected their internationalism. In 1932, the school was closed. Initially, Mies tried to keep it going as a private institution in Berlin, but political opposition to a free-thinking, internationally orientated school had now grown too great, and on 20 July 1933, under pressure from the National Socialists and with the consent of the masters' council, he closed the school once and for all. It was the premature end of an institution whose utopian, socialist ideas had been shattered by reality. Those ideas, however, were taken all over the world by the migrating masters and students, and they became the model for countless international centres of education.

Not a style but a way of thinking

In the fourteen years of its existence, the Bauhaus never stopped developing and redefining itself. Because of its comprehensive approach, it was always essential to match the work with changing realities, and so to incorporate new discoveries and advances in the structure of the teaching and in the goals the school set itself. As a result, the teaching was never dogmatic, and from 'productive disunity' (Josef Albers) it derived astonishing energy.

The manner in which the teaching principles and goals developed can be vividly illustrated by the work of Marcel Breuer. In many respects, Breuer was an exemplary 'Bauhäusler'. He was an outstandingly gifted artist, but he also had a sense for practicality and for the crafts. Having first come to the Bauhaus as a student, he then became a master in Dessau.

Breuer began his studies during the early Expressionist phase of the school, which was strongly influenced by Johannes Itten*, who gave the introductory course. Breuer's first two designs, Chair with Woven Seat and African Chair (both 1921), regard the chair less as an everyday object than as a study in form and materials. Functionality is secondary to originality and the pedagogical purpose. The two designs clearly reflect the counter-currents of Romantic ebb and technical flow that Walter Gropius tried to reconcile, though without dogmatically prescribing any one direction.

Breuer's Slatted chair ti 1a (1922) already paved the way for the metal furniture that is now regarded as typical Bauhaus aesthetics. Clearly this was part of the quest to find a type of chair that would provide an economic solution to the problem of comfortable seating. While the shape and proportions still correspond to those of a traditional chair, the structure is reduced to a mechanical minimum and the upholstery is replaced by stretched material, which adapts itself to the human body. This principle reaches its apogee in Breuer's Wassily chair (B3), which he regarded as the most extreme of his designs.[7] According to Breuer, it would not be possible to reduce the essentials of sitting any further – unless one day someone could invent 'columns of air'.[8]

Breuer's view was that furniture should be 'part of a modern room', without any direct connection to the architecture of the house and geared only to the human body. In this manner it would remain 'styleless', because apart from its purpose and the construction needed to fulfil that purpose, no other factors should influence the design.[9] Such furniture had the requirements of the person as its central consideration, and it sought to be both functional and aesthetically pleasing. The Bauhaus always took account of the fact that form and beauty fulfilled psychological functions, and so in addition to physical needs, one must always bear in mind the importance of human emotions, intuition, speculative irrationalism and dreamlike fantasy.

The future of an idea

The fascination and the influence of the Bauhaus have never waned, and even a hundred years after it was founded, it continues to serve as a model for schools and designers all over the world. This is due to the complexity and diversity of ideas, the international orientation, the need for interdisciplinary cooperation and, perhaps above all, faith in the power of design to exert a positive influence on society. It is not merely a matter of particular designs or of any supposed Bauhaus style. This has never existed, and the 'Bauhäusler' themselves vehemently resisted any attempt to reduce Bauhaus thinking to any one style.[10] The aim was to establish a method whereby objects and buildings suited to their time and their society could be developed from an artistic approach. The starting point for all considerations had to be humans and their living conditions.

Mies was also convinced that the lasting influence of the Bauhaus was due to the fact that it was not a style and it was not just a school. Its power lay, rather, in the fact that it was an idea and a concept of life. It required a new way of thinking that would focus on society with all its needs and aspirations. The central point was therefore 'life design' and not 'object design', just as the formation of the 'new human' was far more important than that of the pedagogical programme. This method – or, to be more precise, this approach – was promoted from the very beginning and reaches far beyond what modern times have interpreted much too narrowly as mere functionalism. It is this approach that remains as relevant today as it was a hundred years ago.

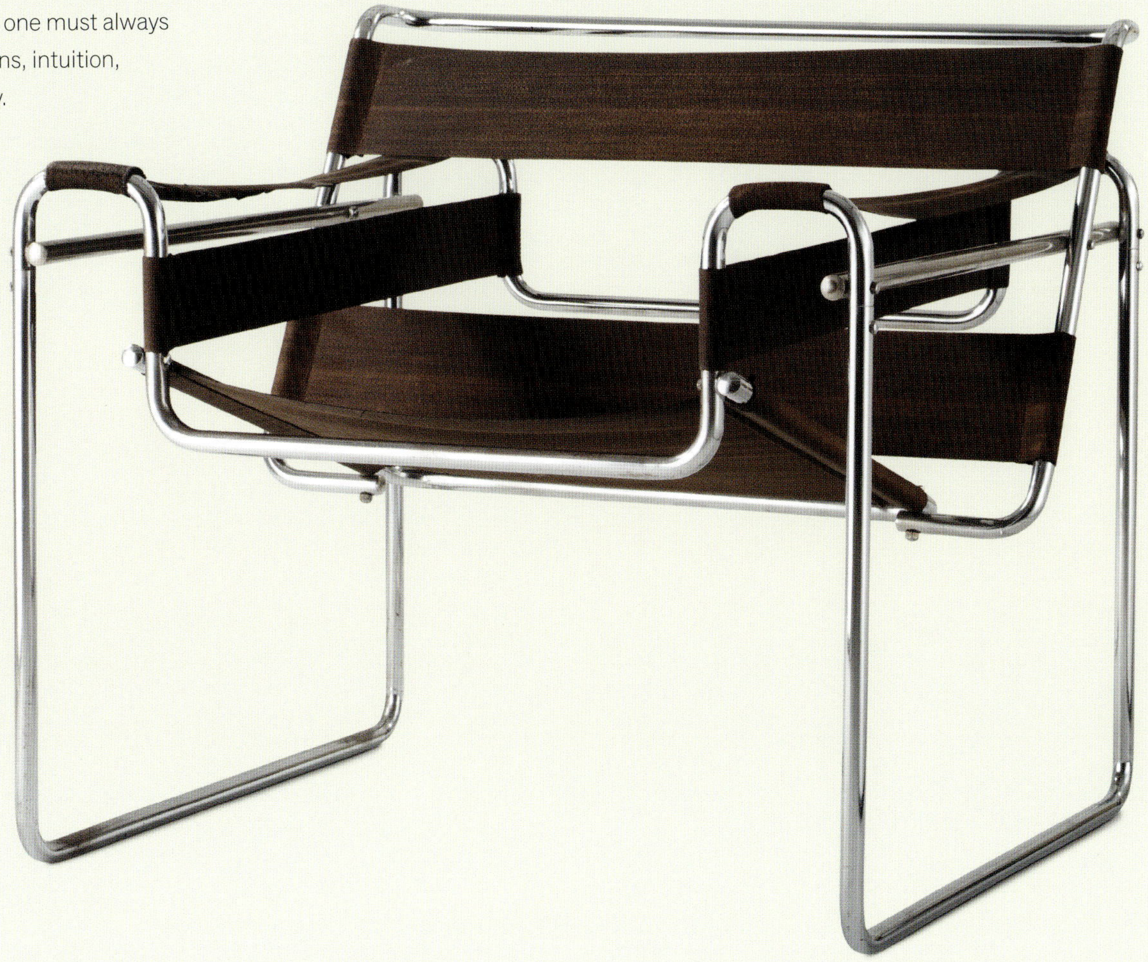

RIGHT The Wassily chair (B3) by Marcel Breuer, 1925.

ERIK GUNNAR ASPLUND
(1885–1940)

So many designers and architects of the Twenties and Thirties grew up in a world before modernism. They were immersed, during their formative years, in a different world – most often neoclassicism, or perhaps Arts and Crafts. There was, therefore, a process of conversion and immersion based upon a decision about whether to embrace the future and a new approach to design or to remain devoted to the familiar precepts of the past. For some, this was a difficult path to navigate, while others sought some kind of accommodation between past and future.

The Swedish architect and designer Erik Gunnar Asplund built his career upon a neoclassical foundation, yet started to question its prevalence as he began to look outwards, noting the design revolution instigated by the Bauhaus pioneers and early modernist prophets. Asplund embraced functionalism and the modernist style, tentatively at first but then decisively. One of his greatest projects, Stockholm City Library (1928), has been described as an inspired and epic act of schizophrenia, conceived in the neoclassical idiom but completed at a time when Asplund was shifting over to modernism, with fittings, furniture and lighting exhibiting a strong thread of Scandinavian functionalism. Indeed, Asplund's furniture shows him at his most experimental, embracing fresh materials, forms and aesthetics in a push towards the new.

Asplund was born in Stockholm and was the son of a tax official. He showed an early talent for drawing and painting, but his father encouraged him to switch his attention to architecture and so he went on to study the subject at the Royal Institute of Technology. After graduating in 1910, Asplund moved on to the Royal College of Art, but soon found that he was dissatisfied with the prosaic curriculum and methods of its tutors. He and a number of fellow students broke away to set up a new college, the Klara School, and sought out progressive local architects to come and teach them. While still completing his studies, Asplund began entering architectural competitions, winning a commission for a school in 1913 and securing a project to extend Gothenburg Law Courts a year later.

In the years that followed, Asplund won a number of other major competitions, which were balanced with smaller projects; he also served as editor of the architectural journal *Arkitektur* for three years, up to 1920. In 1918, he first became involved in the ten-year project that would result in his most famous building, Stockholm City Library, with its dramatic circular tower, or drum, floating upon a square service base, often compared to an enigmatic fortress hovering in the middle of the city.

Even before the completion of the library, Asplund was reassessing furniture and lighting influenced by the spirit of experimentation

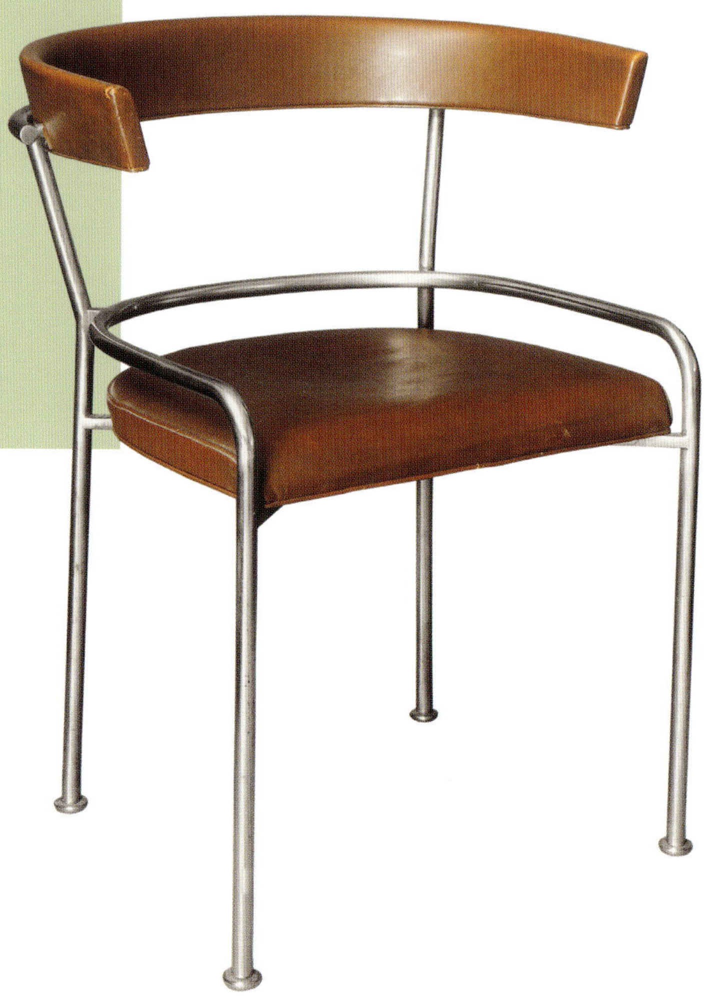

LEFT The Karmstol chair, 1931, produced by Nordiska Kompaniet.

OPPOSITE The Senna armchair, designed in 1925; this piece was manufactured by Cassina in leather, fruitwood, plastic and brass.

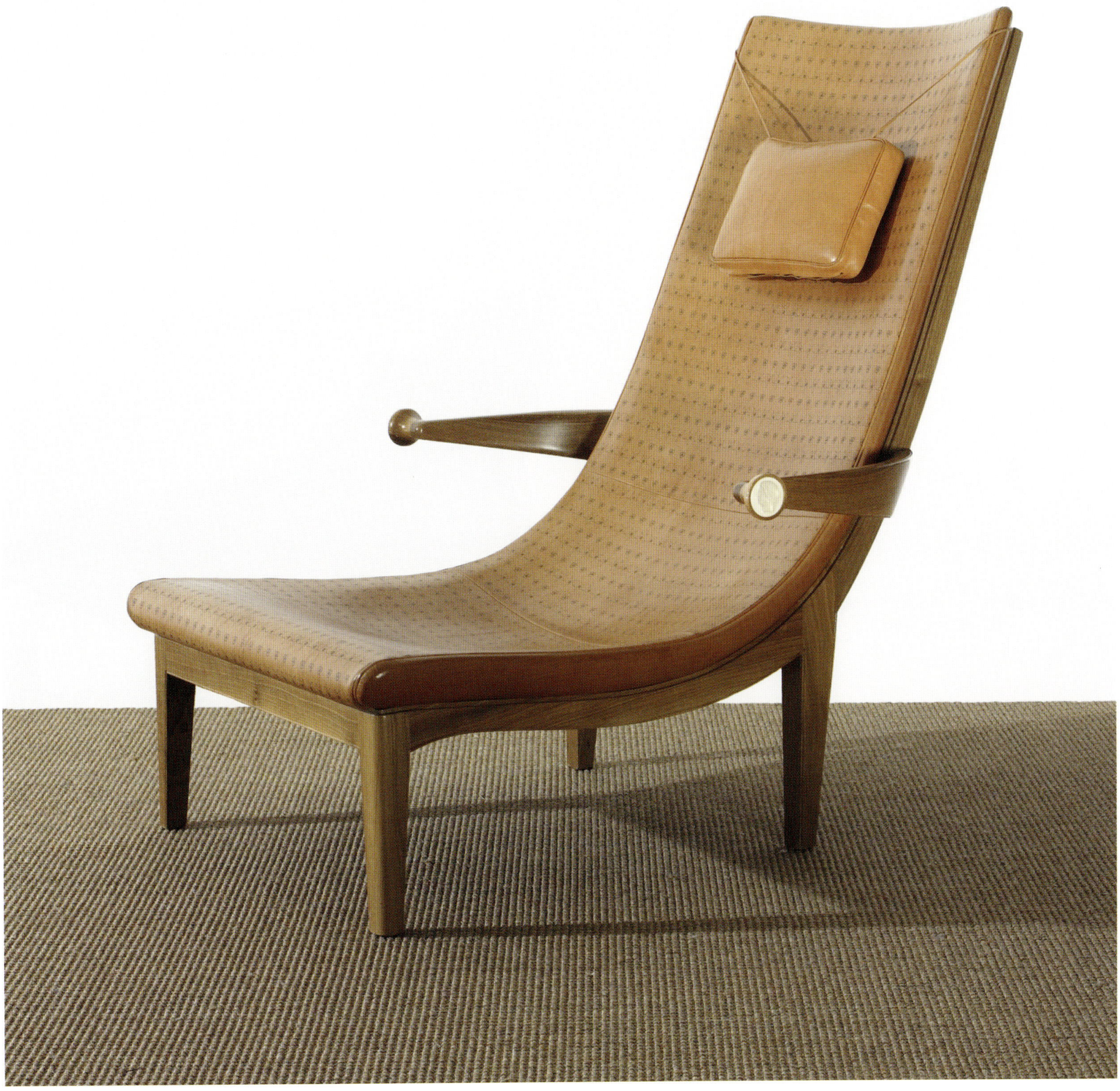

pioneered by the early modernists in Germany and France. His Senna armchair was produced for the Paris World Expo of 1925 – a dynamic, sculptural piece with a high curving back, short protruding brass armrests, and a slim continuous leather cushion pad with an additional headrest; Cassina reissued the design, briefly, in the Eighties.

In 1930, Asplund broke decisively with the past and with neoclassicism. He was appointed Chief Architect of the Stockholm Exhibition, and during the design process visited the modernist-inspired Weissenhof Estate* exhibition in Stuttgart and also Le Corbusier's studio in Paris. Asplund's design for the temporary Paradise Restaurant at the Stockholm Exhibition saw him take his commitment to the modernist style to a new level, with its steel-framed structure, expanses of glass, balconies and a streamlined aesthetic.

Just a year later the Swedish department store Nordiska Kompaniet issued a small collection of Asplund's modernist furniture, including his Karmstol chair – a dining chair in exposed tubular steel with a leather upholstered seat and a slim semi-circular backrest. The GA-2 armchair also used tubular steel, with a thick padded leather seat and cushion floating upon light, steel legs with a distinctive, elongated twist to the rear sections of these supports. The combination of the chunky cushion and light frame in the GA-2 is somewhat reminiscent of Le Corbusier, Pierre Jeanneret and Charlotte Perriand's Grand Confort armchair of 1928. The same year, 1931, Asplund co-authored a design manifesto called *Acceptera*, a 'call to arms' for Scandinavian modernists.

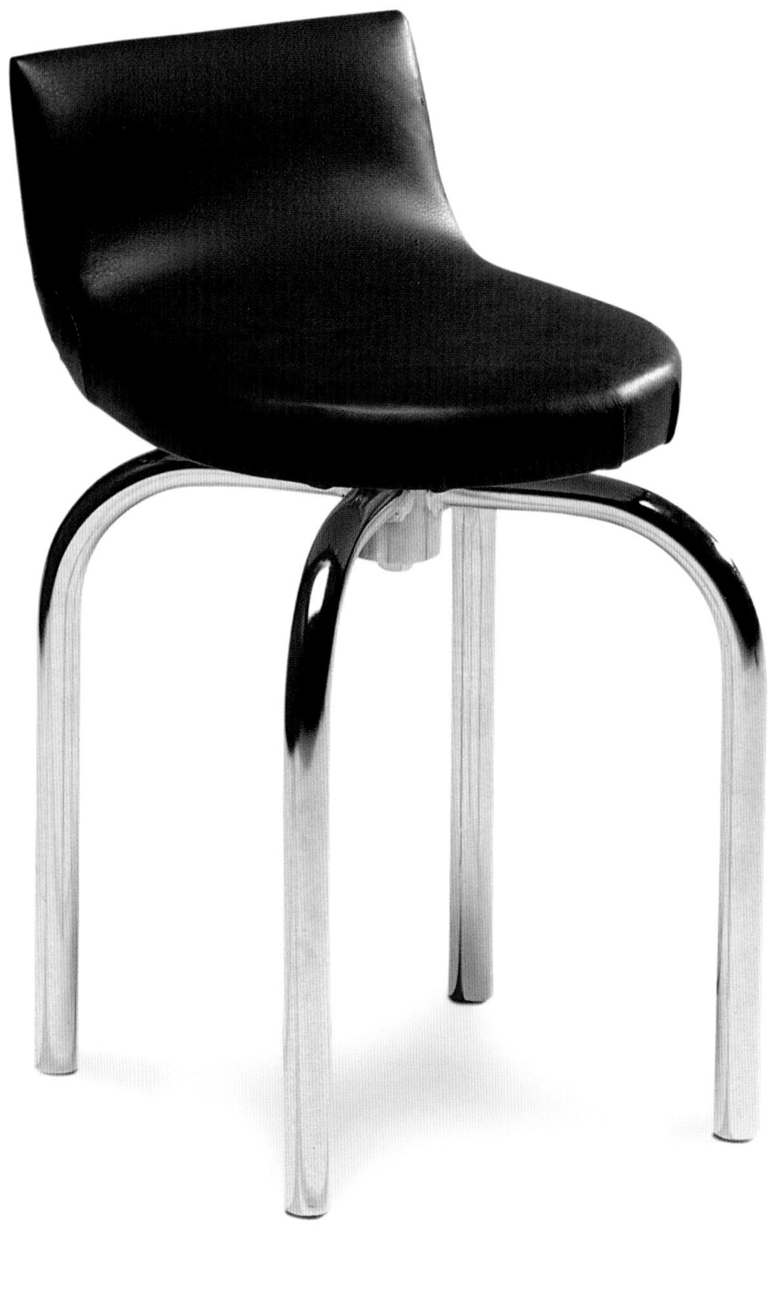

LEFT A low-backed stool with chrome-plated tubular steel legs; reissued by Källemo.

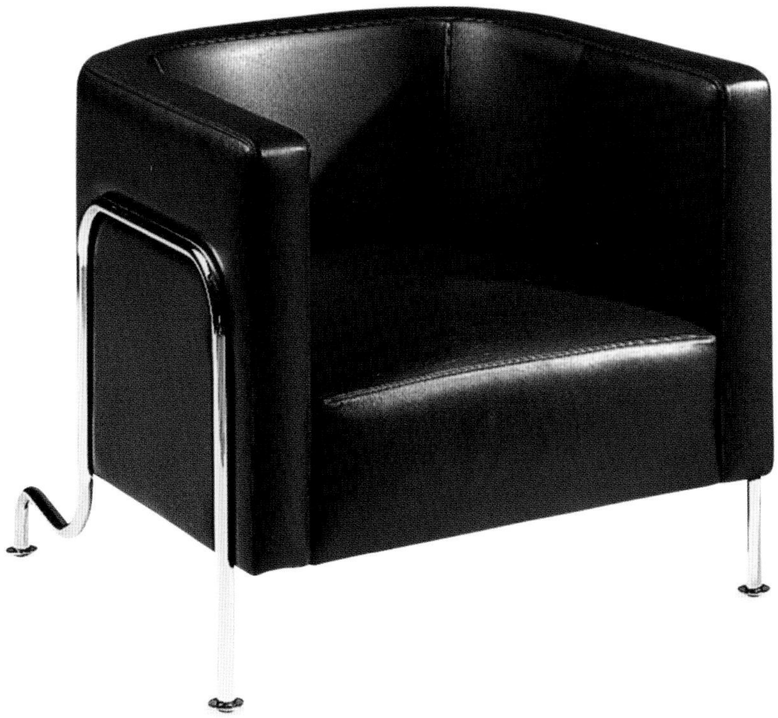

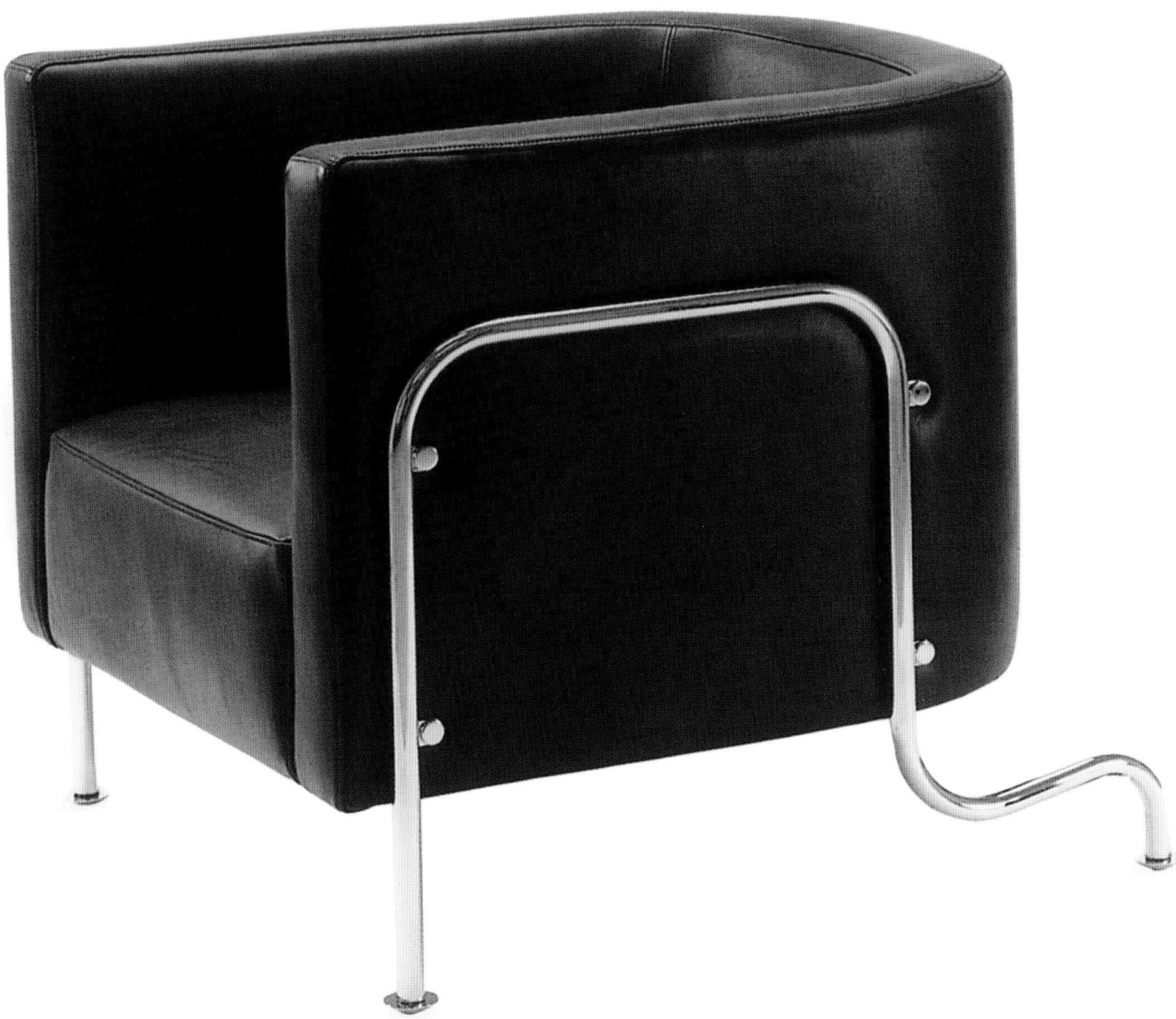

ABOVE & RIGHT The GA-2 armchair, designed in 1931, and originally produced by Nordiska Kompaniet; reissued by Källemo.

MARCEL BREUER
(1902–1981)

As both a designer and architect, Marcel Breuer proved himself to be one of the greatest innovators of the early modernist era. Never afraid to experiment, he pioneered new approaches in his buildings and furniture, while also dedicating himself to fresh materials – such as tubular steel, aluminium and plywood – and giving them new relevance. Within his work there was a strong degree of ingenuity, helped by an appreciation of engineering, which allowed him to explore new structural solutions in a fearless way. Educated at the Bauhaus and immersed in its approach and methodology, Breuer helped to encapsulate its ethos with a range of furniture that also became emblematic of modernism itself. Key pieces – such as his iconic Wassily/B3 chair – remain in production and still feel fresh and relevant in a contemporary context.

Marcel Breuer was born in the Hungarian town of Pécs, the son of a dental technician with an interest in the arts. During his childhood, Breuer sketched and painted, and in 1920 won a scholarship to the Academy of Fine Arts in Vienna. However, he walked away from the academy after just one day, disillusioned by the tutors and teaching. Following a few months in Vienna, he decided to enrol at the Bauhaus, founded in Weimar by Walter Gropius one year earlier. After taking the Preliminary Course – a foundation curriculum in art and design that lasted six months – Breuer decided to enter the carpentry workshop headed by Johannes Itten*. Recognizing Breuer's talents, Gropius befriended his student, becoming his mentor and greatest supporter, and helping him navigate a number of dramatic shifts and changes in career and circumstance over the years.

The first fruits of Breuer's experimental approach in the furniture department were made largely in timber. The most striking of these designs is the Lattenstuhl – or ti 1a – slat armchair (1922), infused with a spirit of deconstructed abstraction influenced by the work of Gerrit Rietveld and the Dutch De Stijl* movement (see p. 26). The chair was made of oak with a leather seat, complemented by two leather bands to support the back, yet the piece felt like a work of modern art, as though two separate chairs had been fused together to create a single sculpture.

After graduating from the Bauhaus, Breuer decided to pursue his architectural interests and travelled to Paris, where he served an apprenticeship with Pierre Chareau. After the Bauhaus move from Weimar to Dessau in 1925, Gropius invited Breuer to return as Master of the furniture workshop. During this second period at the Bauhaus, Breuer embraced the shift in emphasis from craft towards design solutions suited to mass production, and began to experiment with more industrial materials. One famous source of inspiration was Breuer's own bicycle: on his trips around the city he noted the strength and versatility of the bike's frame and contacted the makers of the steel tubes. The result was a collection of tubular steel pieces developed from 1925 to around 1928 that was to prove hugely influential. Most famously, there was the B3 chair of 1925, which was later named the Wassily, in honour of Breuer's friend and Bauhaus colleague, the painter Wassily Kandinsky*. This was a startlingly sophisticated design, made of interlocking steel tubes forming a lightweight cube that contained a floating seat in canvas or leather connected to a slim backrest (see p. 27).

The tubular collection also included tables and nesting tables, desks, stools and armchairs, which were used to furnish parts of the Bauhaus complex and the masters' houses designed by

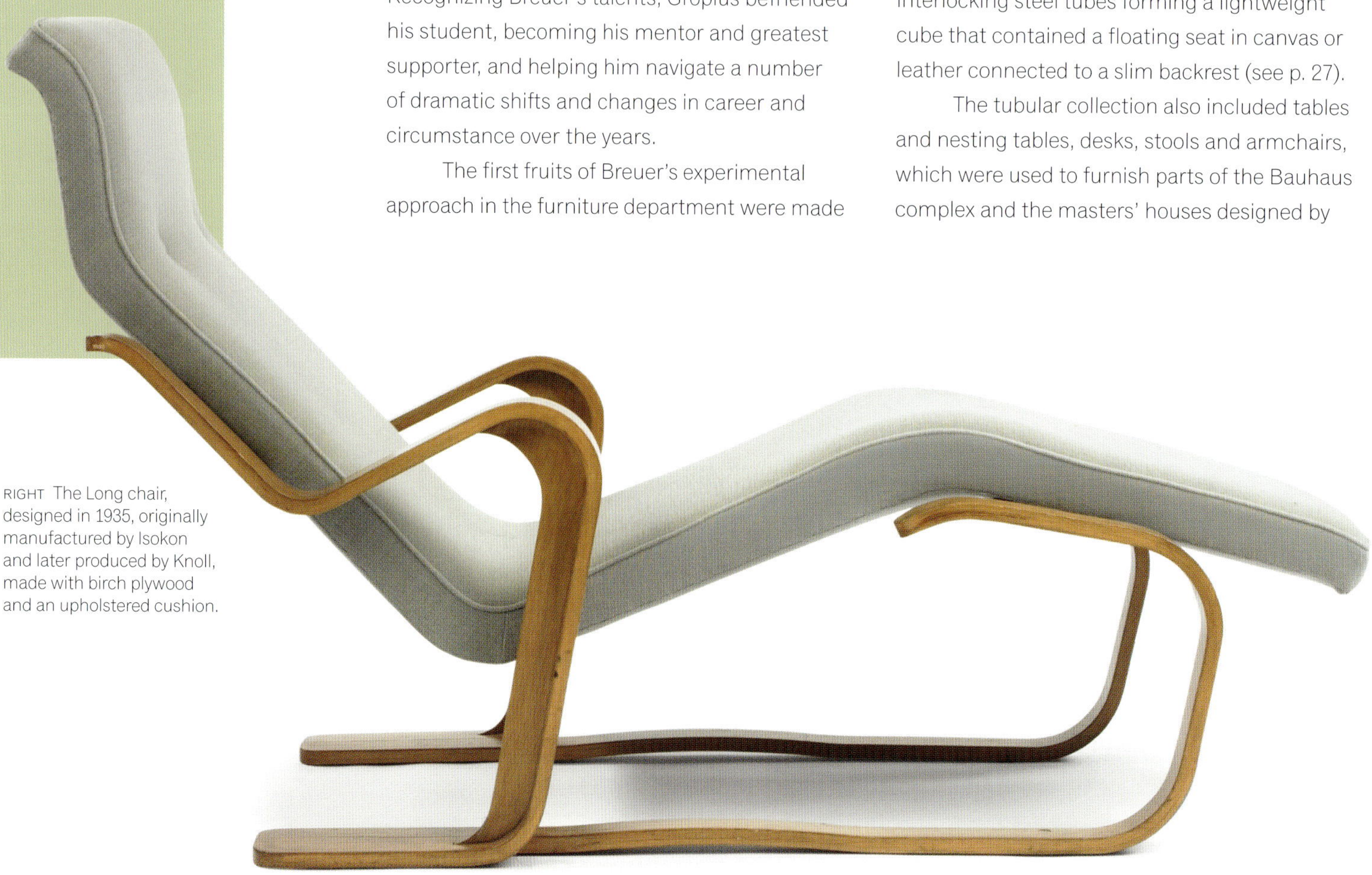

RIGHT The Long chair, designed in 1935, originally manufactured by Isokon and later produced by Knoll, made with birch plywood and an upholstered cushion.

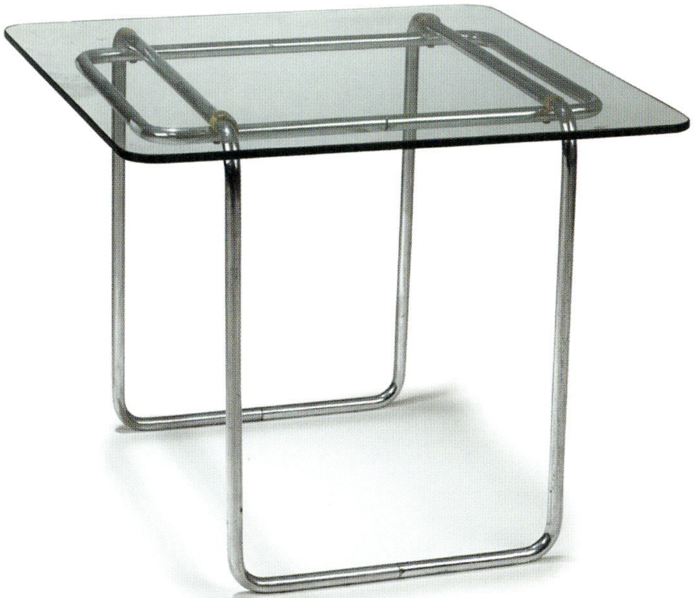

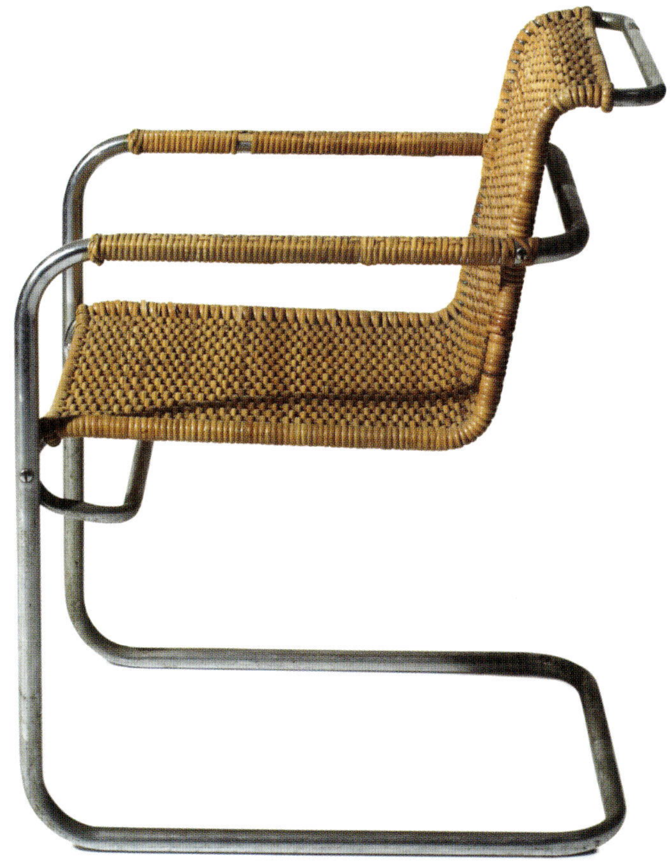

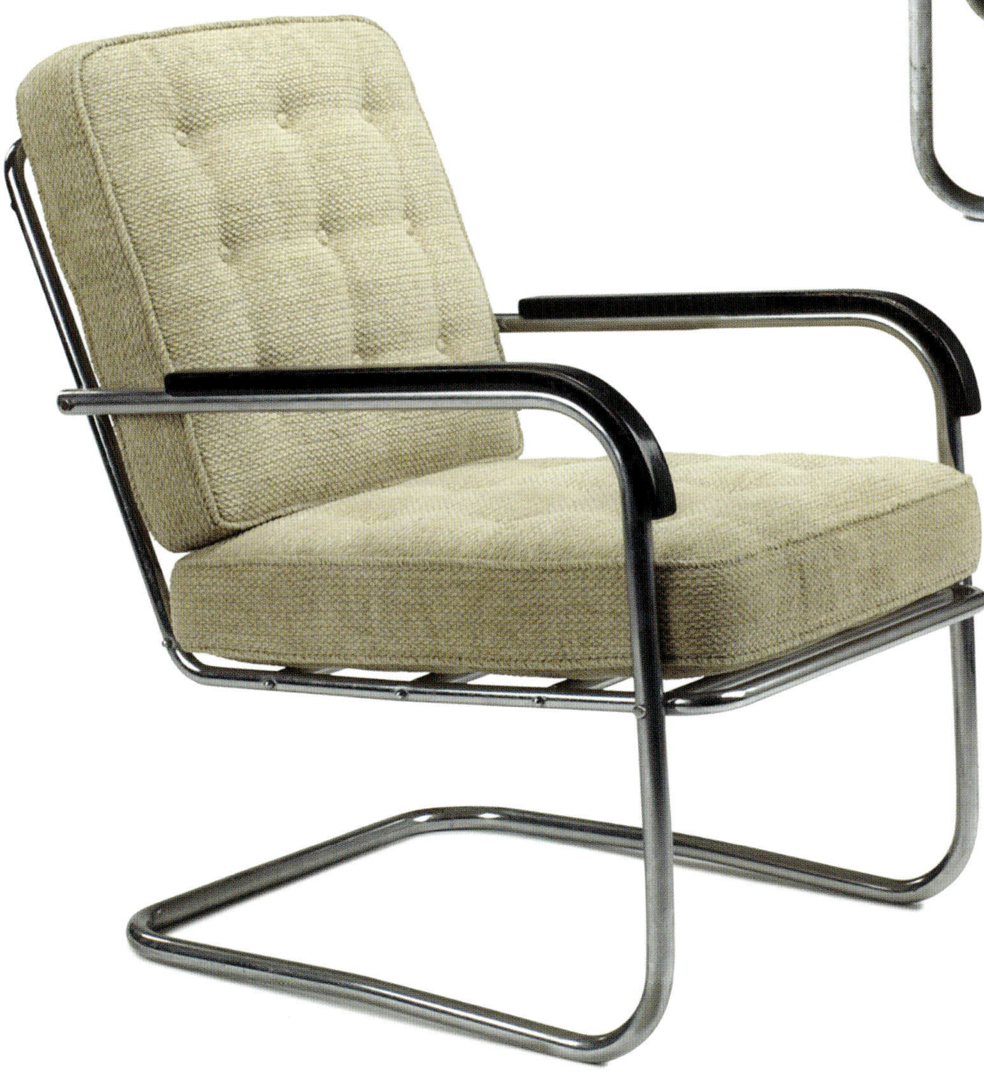

TOP LEFT Breuer's B18 occasional table, manufactured by Thonet, *c.* 1924, in chrome-plated steel and glass.

ABOVE The B34 armchair, produced by Thonet, *c.* 1928, made with chrome-plated steel and a cane seat and back.

LEFT A model B36 lounge chair manufactured by Thonet, *c.* 1928, with a chrome-plated tubular steel frame and upholstered cushions.

Gropius. It was a highly accomplished and original collection, created with the lightest of touches, only marred by a long-running controversy over the authorship of the B30, B33 (see pp. 14–15) and B34 cantilevered chairs, with the copyright in the designs contested by Dutch designer Mart Stam*, who was working on similar ideas around the same time. In late 1926/early 1927, Breuer co-founded a company, Standard-Möbel, to make his tubular collection, which was later manufactured by Thonet* and, later still, by Knoll*; when Knoll began making the B32 and B64 cantilevered chairs, they were renamed 'Cesca'.

In the late Twenties, with the future of the Bauhaus under threat, Gropius resigned and Breuer left Weimar, founding an architectural practice in Berlin. There were a small number of commissions and also experiments with a new range of furniture, made with slatted aluminium frames, which were manufactured by Embru-Werke in Zurich. The best-known design from this period is the Chaise Longue 313, a streamlined design well suited to outdoor use and contoured to support the reclining shape of the body.

A few years later, as the political situation in Germany worsened, Gropius encouraged his friend to follow him to London, where Breuer began a partnership with architect F. R. S. Yorke*, which resulted in a few British commissions. He also formed a working relationship with entrepreneur Jack Pritchard, who developed the Lawn Road Isokon* apartment building in North London (see p. 444), where both Gropius and Breuer lived for a time. Pritchard's Isokon furniture company, which specialized in plywood production, produced a number of key pieces designed by Breuer, who embraced the new material. Breuer's Isokon collection included the famous Long chair of 1935, a reinterpretation of the earlier aluminium recliner. There was also a stacking chair, nesting tables and a dining table, all in plywood.

In 1937, Gropius encouraged another move, this time inviting Breuer to join him at Harvard University, where they both taught. Here, there was a second flowering of Breuer's career, focused on architecture. There were a number of influential mid-century houses designed with Gropius and alone, combining contextuality and local materials with structural and spatial innovations. It was, in a way, a new beginning that saw Breuer author a significant catalogue of buildings in the post-war period, including the Whitney Museum of American Art in New York (1966). His tubular steel furniture, in particular, developed while he was still a young designer, has continued to be sought out by new audiences and is still in production today.

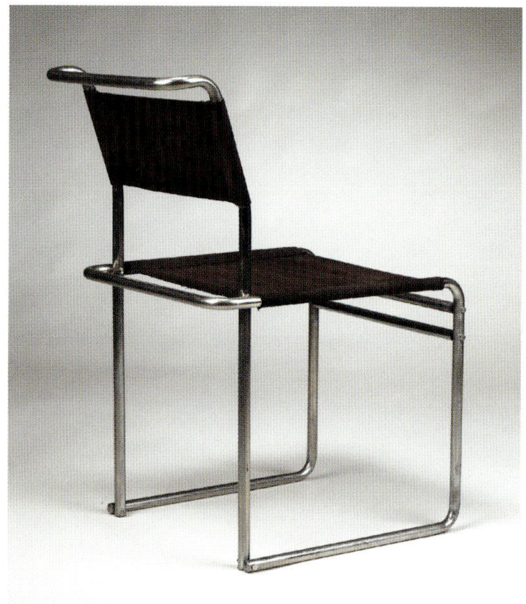

ABOVE A B5 side chair, produced by Thonet, 1926, in chrome-plated steel and Eisengarn (waxed cotton thread).

BELOW A pair of B35 lounge chairs, in chrome-plated steel and leather, designed in 1928 and manufactured *c.* 1965 by Knoll.

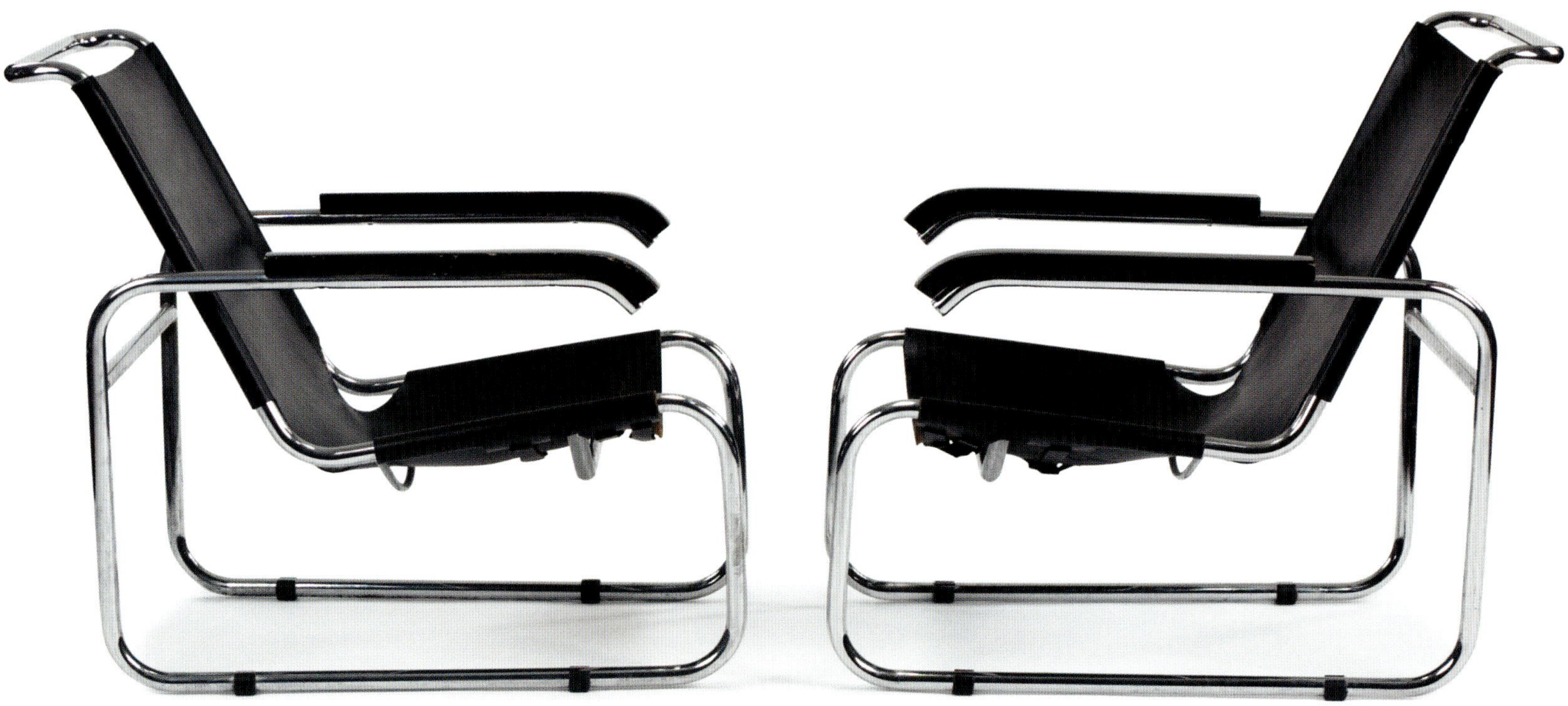

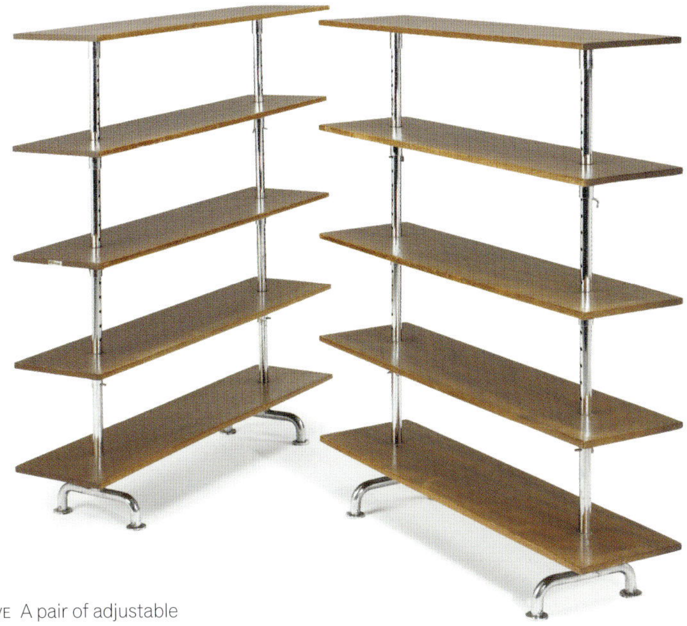

ABOVE A pair of adjustable bookshelves, made by Embru, *c.* 1931, in mahogany and chrome-plated steel.

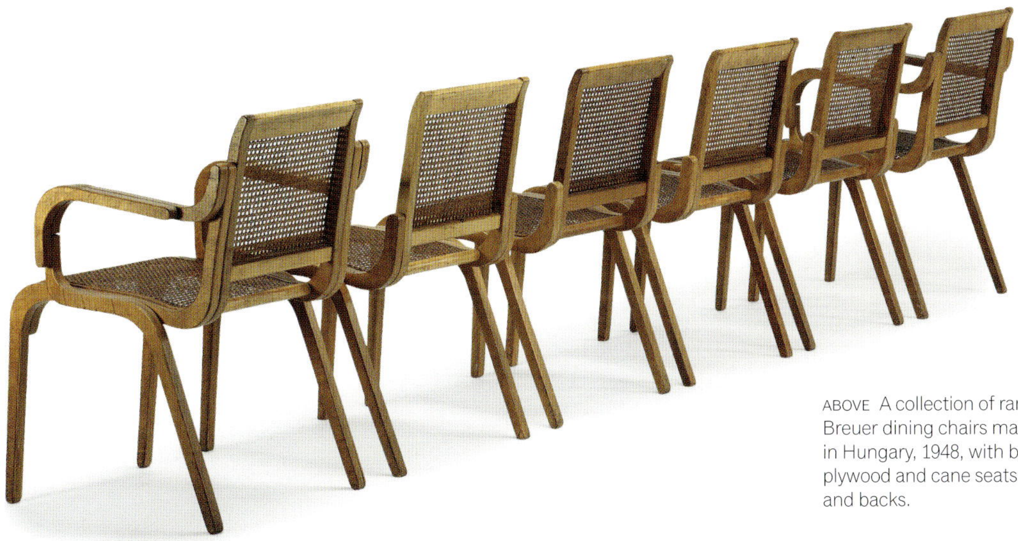

ABOVE A collection of rare Breuer dining chairs made in Hungary, 1948, with birch plywood and cane seats and backs.

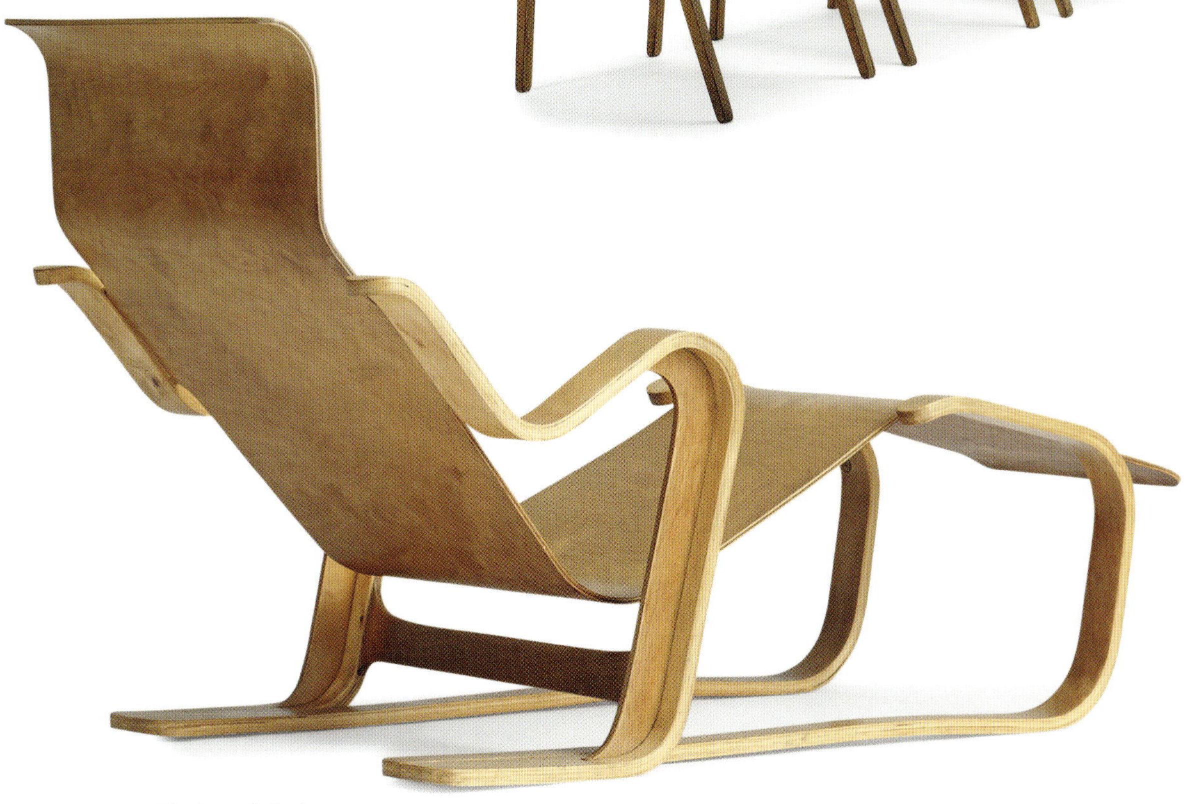

ABOVE The Long chair of 1935, made in Hungary for the English furniture company Isokon, using birch plywood.

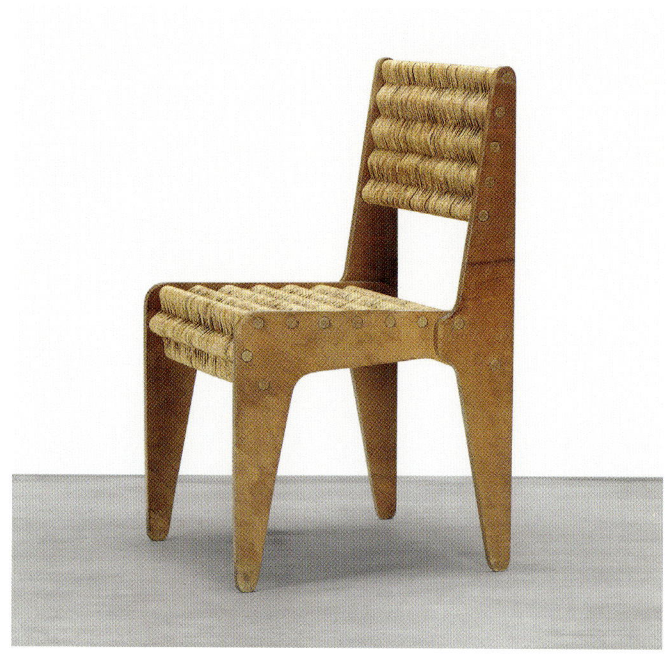

ABOVE Breuer's 1938 prototype chair, in birch plywood and jute, for Bryn Mawr College in Pennsylvania.

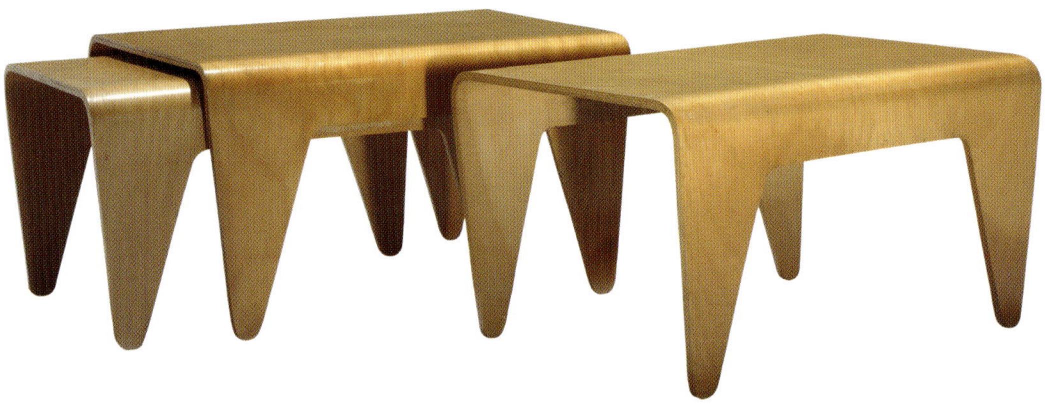

LEFT Three nesting tables, produced by Isokon, in beech plywood, 1936.

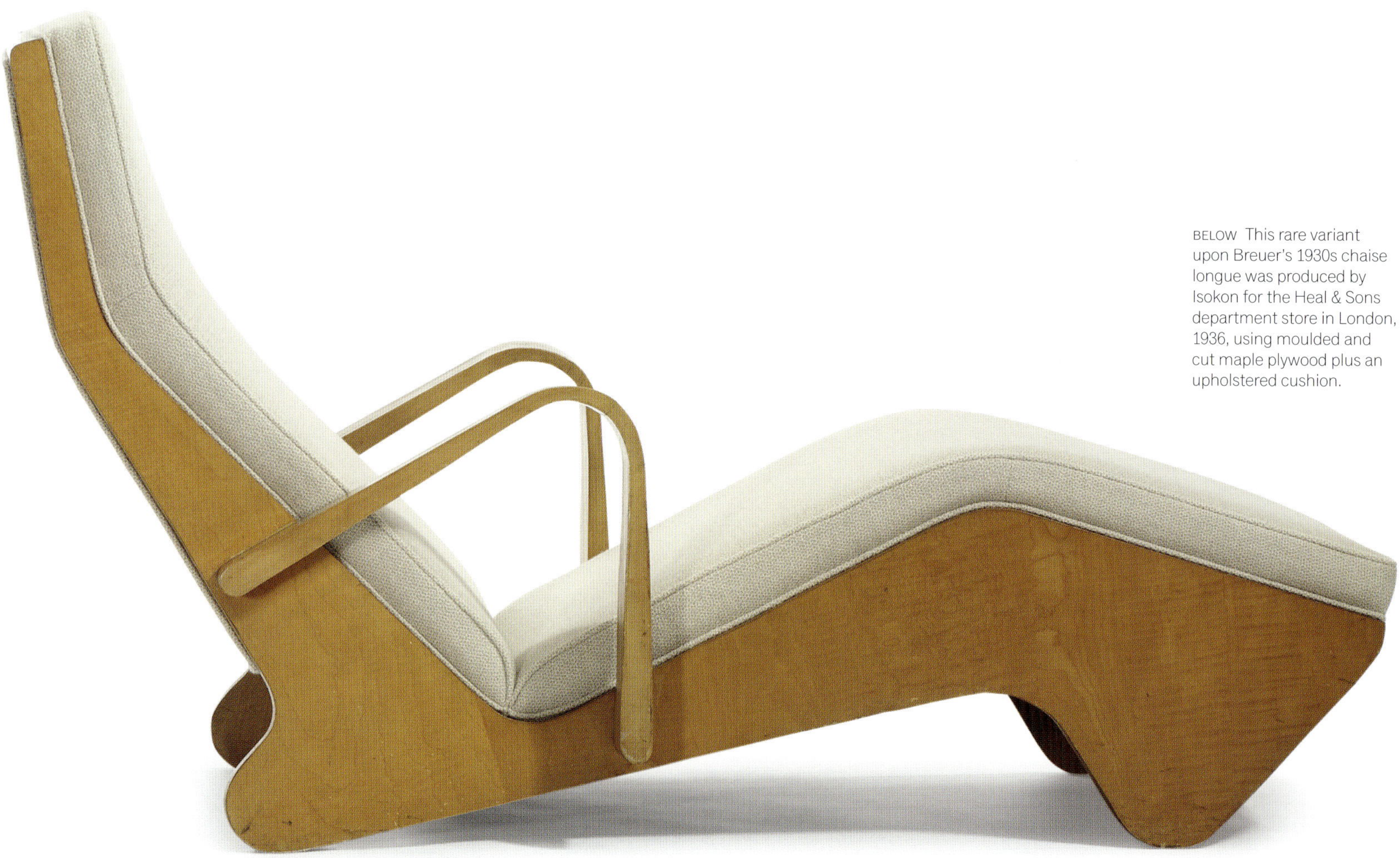

BELOW This rare variant upon Breuer's 1930s chaise longue was produced by Isokon for the Heal & Sons department store in London, 1936, using moulded and cut maple plywood plus an upholstered cushion.

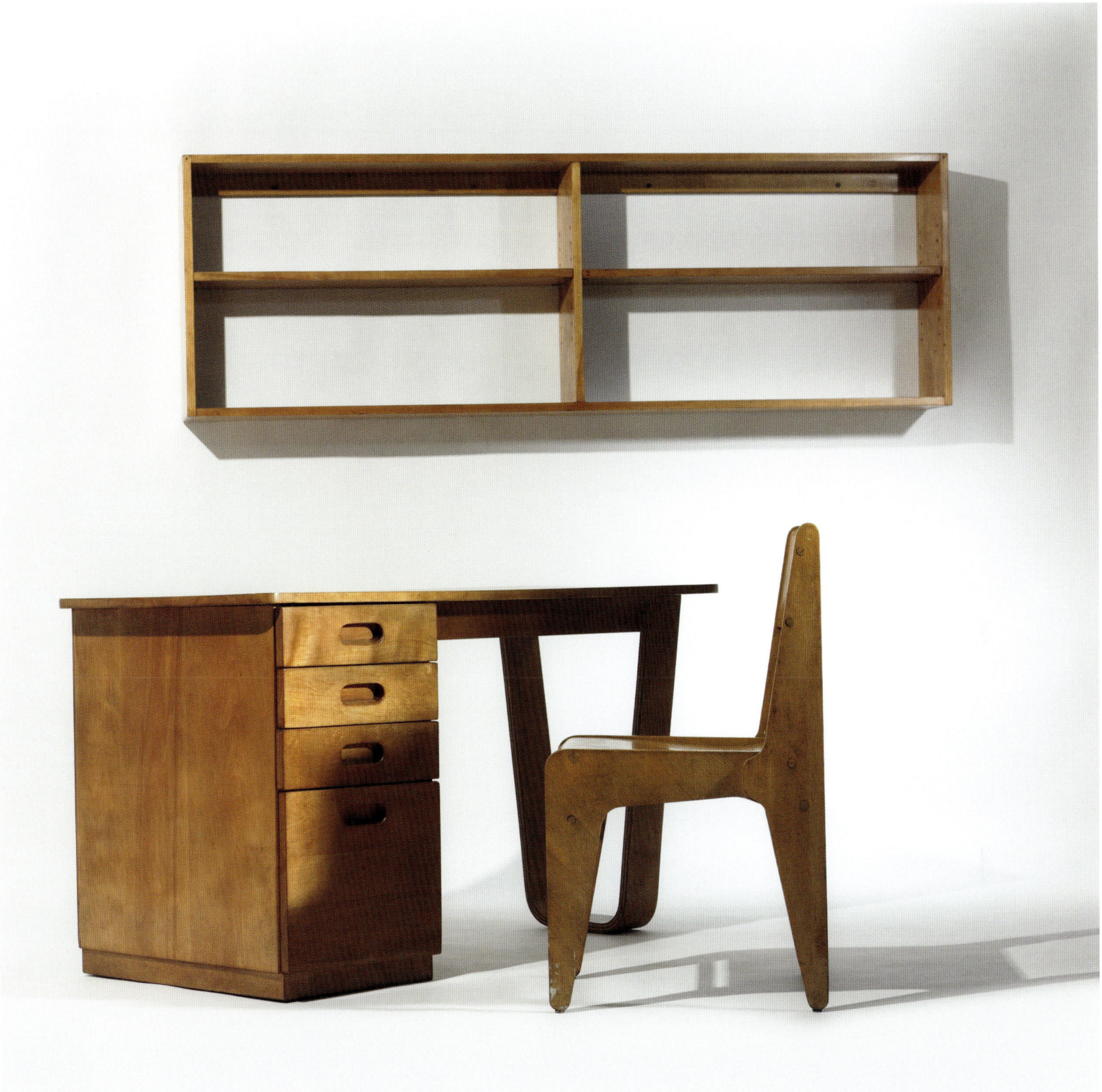

ABOVE A matching wall-mounted shelf and desk set, in birch, 1938, designed for Bryn Mawr College, Pennsylvania.

DONALD DESKEY
(1894–1989)

Over the course of a long and prolific career Donald Deskey reinvented himself many times over. His early work of the pre-war period was largely focused on furniture and interior design, but he also expanded into industrial design and in the post-war years moved into branding, marketing and graphic design, becoming one of the pioneers of this growing field. At the same time, he took a wide-ranging view of the evolution of design itself, drawing inspiration from both the European practitioners of Art Deco* and the modernist pioneers. His early work was infused with the spirit and dynamism of Deco – although he rejected the label in his later years – and gradually he moved towards a style often described as 'Streamline Moderne', experimenting with new materials and a pared-down aesthetic.

Deskey was born in Minnesota and studied architecture at the University of California in Berkeley, followed by studies in fine arts and painting in New York, Chicago and Paris. He began working in graphic design initially, but moved into furniture and interior design after a visit to the 1925 Exposition Internationale des Arts Décoratifs et Industriels Modernes in Paris. Based in New York from 1926 onwards, Deskey designed window displays for the Saks Fifth Avenue department store, screens and lighting for designer Paul Theodore Frankl* and interiors for prestigious private clients who included Helena Rubinstein and the Rockefeller family. Deskey's most famous commission was for the interiors of Radio City Music Hall, designed in a predominantly Deco style; the project also included an apartment for Radio City's developer, the theatrical entrepreneur Samuel L. 'Roxy' Rothafel.

These interior commissions usually featured Deskey's own furniture designs and from the Twenties to the Thirties he produced a vast collection totalling around 400 different designs. In 1927, he co-founded a company, Deskey-Vollmer, in association with entrepreneur Phillip Vollmer, to produce his furniture and lighting. The Twenties pieces have a Deco quality, including glass-topped tables with chrome bases made up of crescent ribbons floating upon shining metal blocks, or lacquered screens. But at the same time there is an engaging lightness, restraint and transparency to many of the chrome and glass pieces, echoed by Deskey's lighting of the period, including table lamps and floor lights.

During the Thirties Deskey became increasingly interested in the possibilities offered by new and more industrial materials, as he moved away from the luxurious finishes and aesthetics of the Deco style and stepped into the path of modernism, influenced by the furniture developed by MARCEL BREUER, EILEEN GRAY and other early European modernists. He began experimenting with tubular steel, Vitrolite, Bakelite, Formica, Transite, Fabrikoid and a variant on plywood known as Weldtex, and he also embraced aluminium wallpaper and other fresh finishes for his interiors, although at the same time he continued to work in

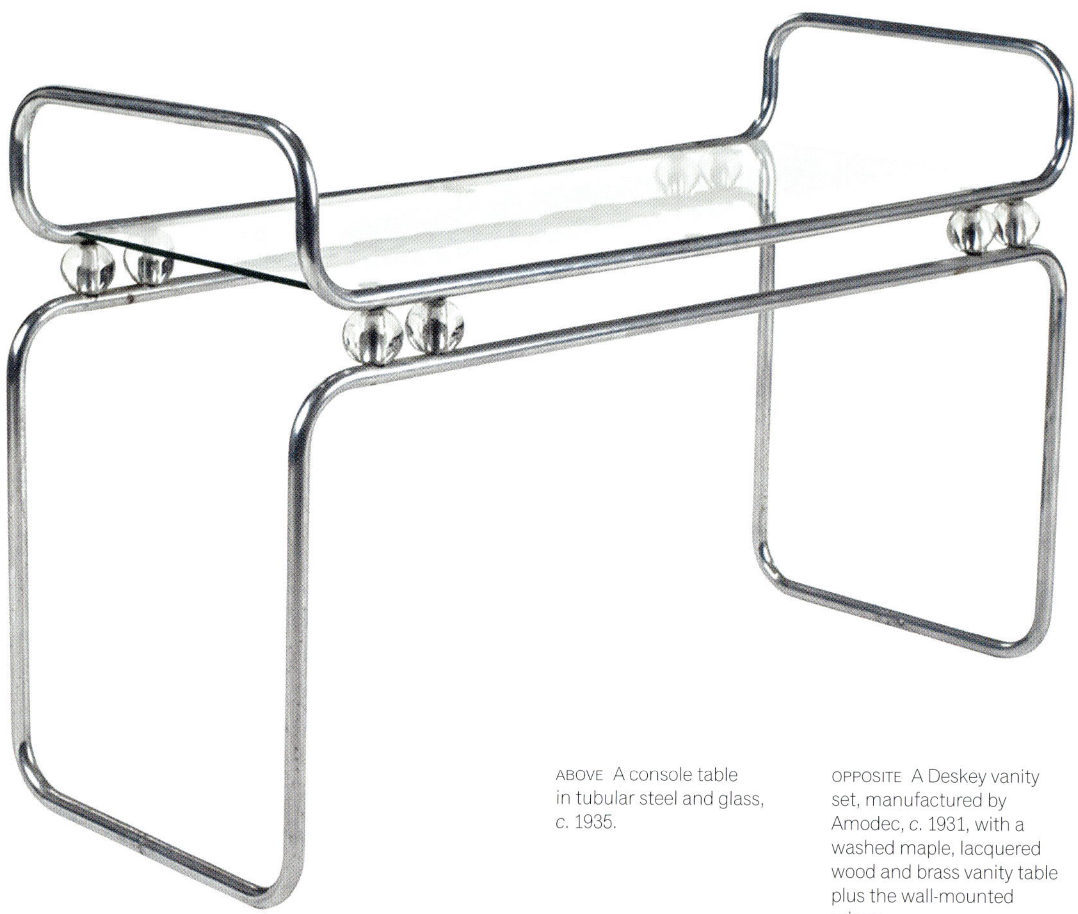

ABOVE A console table in tubular steel and glass, *c.* 1935.

OPPOSITE A Deskey vanity set, manufactured by Amodec, *c.* 1931, with a washed maple, lacquered wood and brass vanity table plus the wall-mounted mirror.

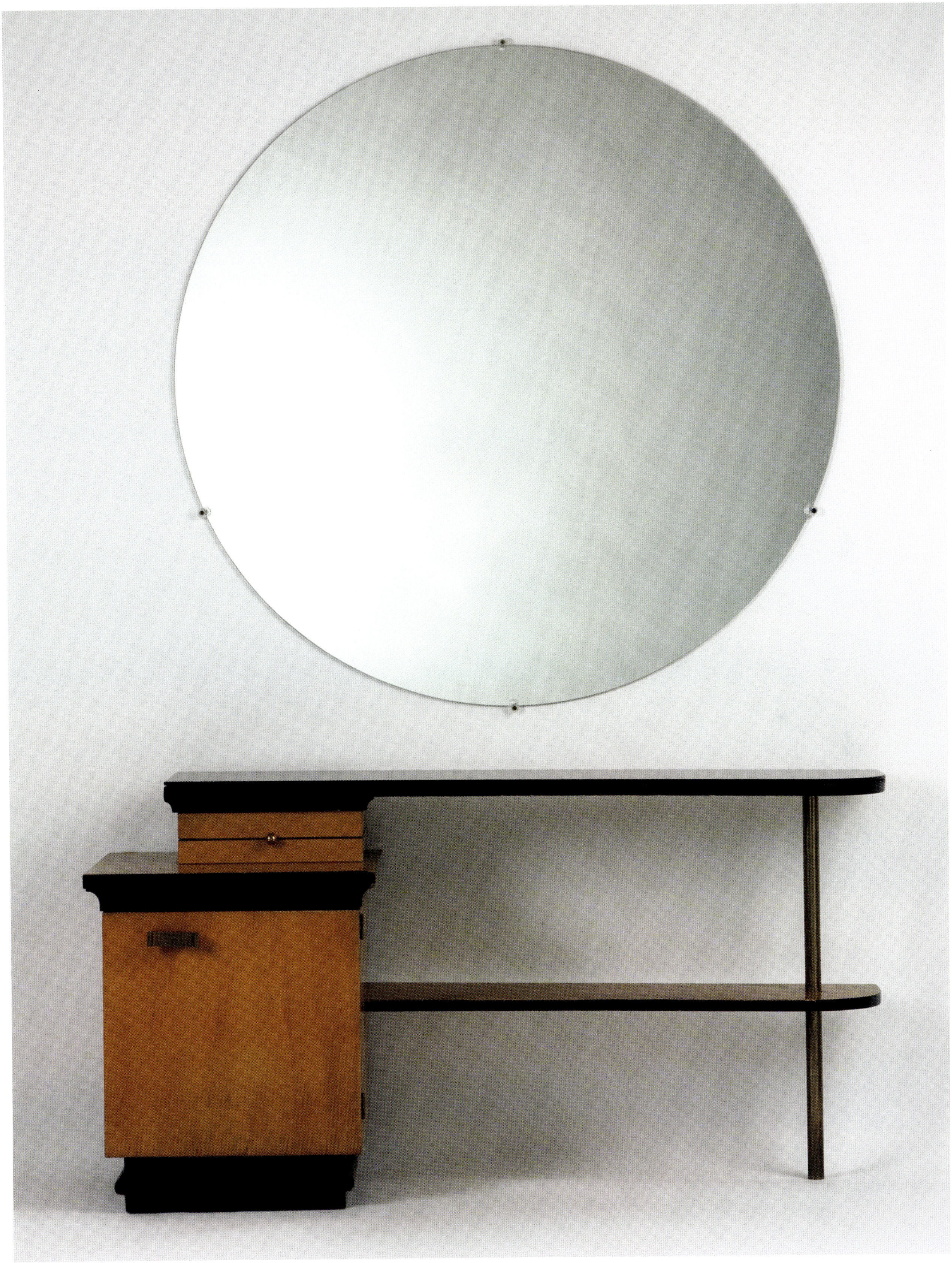

mahogany, tiger maple, macassar ebony and other, more familiar materials.

The tubular steel pieces included a striking console table, dated 1935, with a sculpted steel frame and glass top – a design with a striking lightness of structure and character in comparison with some of Deskey's more crafted timber pieces. There were nesting tables (1931) with simple aluminium bases and lacquered timber tops, echoing Breuer's Bauhaus collection, and other machine-age pieces, such as a circular table with an S-shaped aluminium base and Bakelite top (*c.* 1928). With Samuel Marx, Deskey designed an original upholstered armchair (1939) for Marx's Chicago home, with a light chrome-plated steel frame and delicate peg armrests.

There was an openness to innovation threaded throughout Deskey's work that carried him into industrial design during the Thirties, helping to feed the growing American appetite for new domestic products and appliances.

In the post-war era, Deskey changed direction once again, turning back to graphic design but now in combination with a fresh focus on brand consulting and packaging, comparable to the mid-century work of Paul Rand.

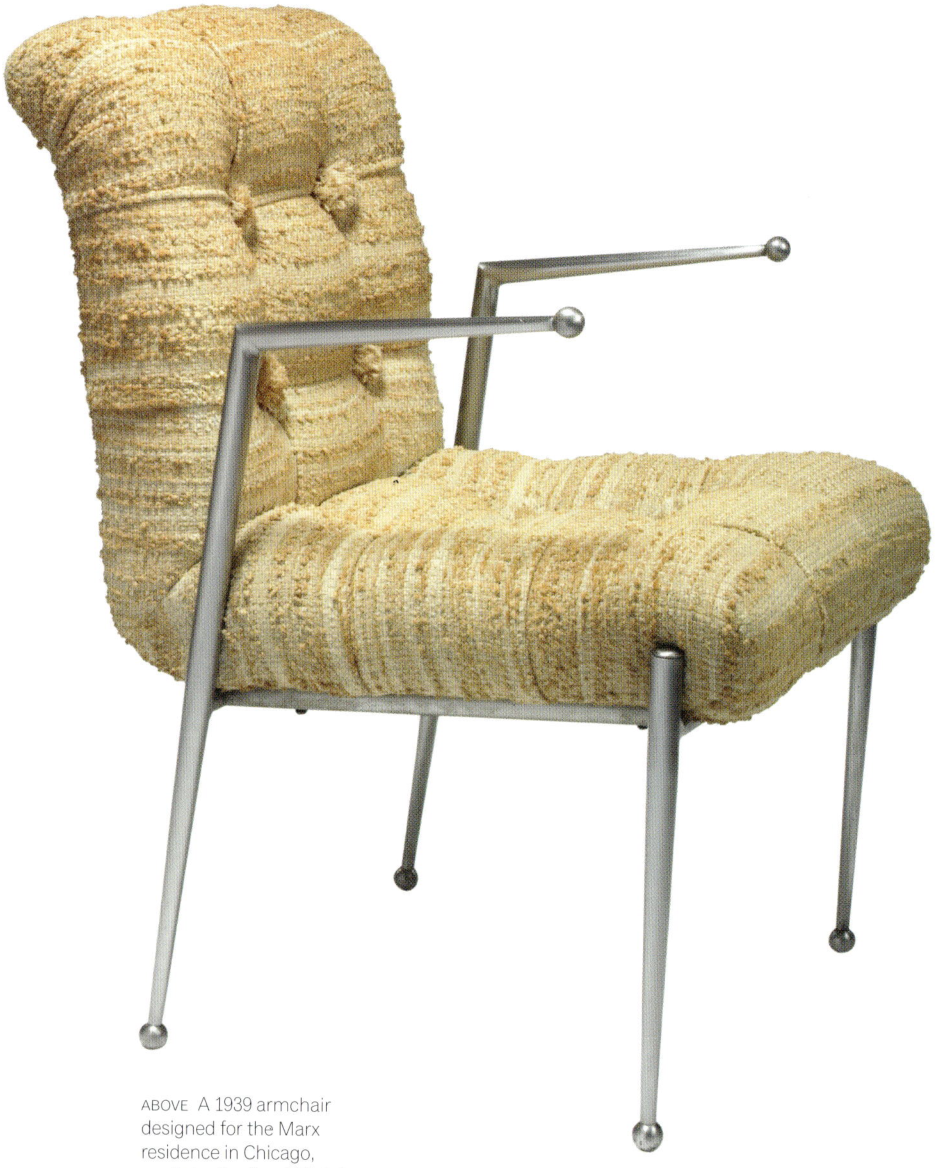

ABOVE A 1939 armchair designed for the Marx residence in Chicago, made by the Royal Metal Manufacturing Company, in chrome-plated steel and upholstery.

RIGHT A machine-age table produced by Deskey-Vollmer, *c.* 1928, in aluminium and Bakelite-coated wood.

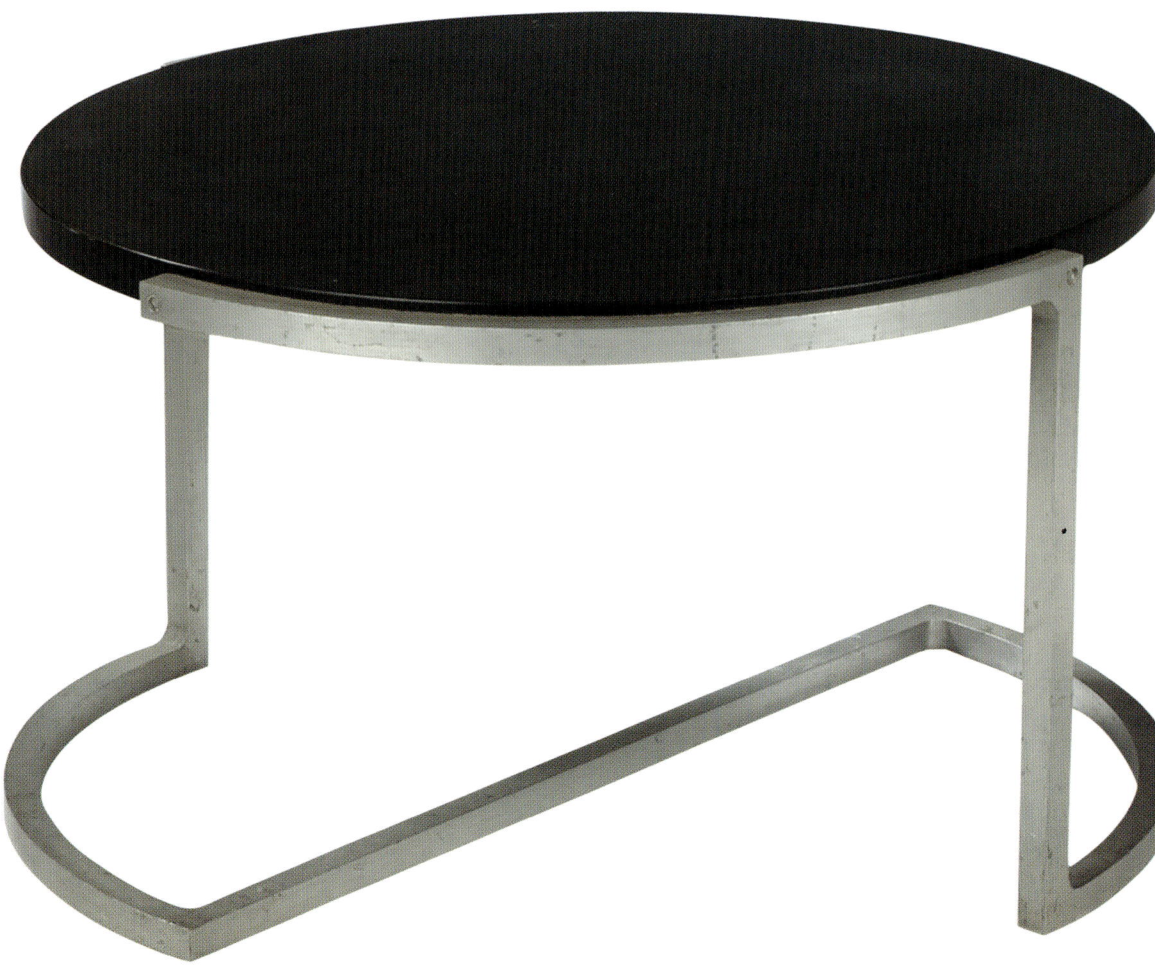

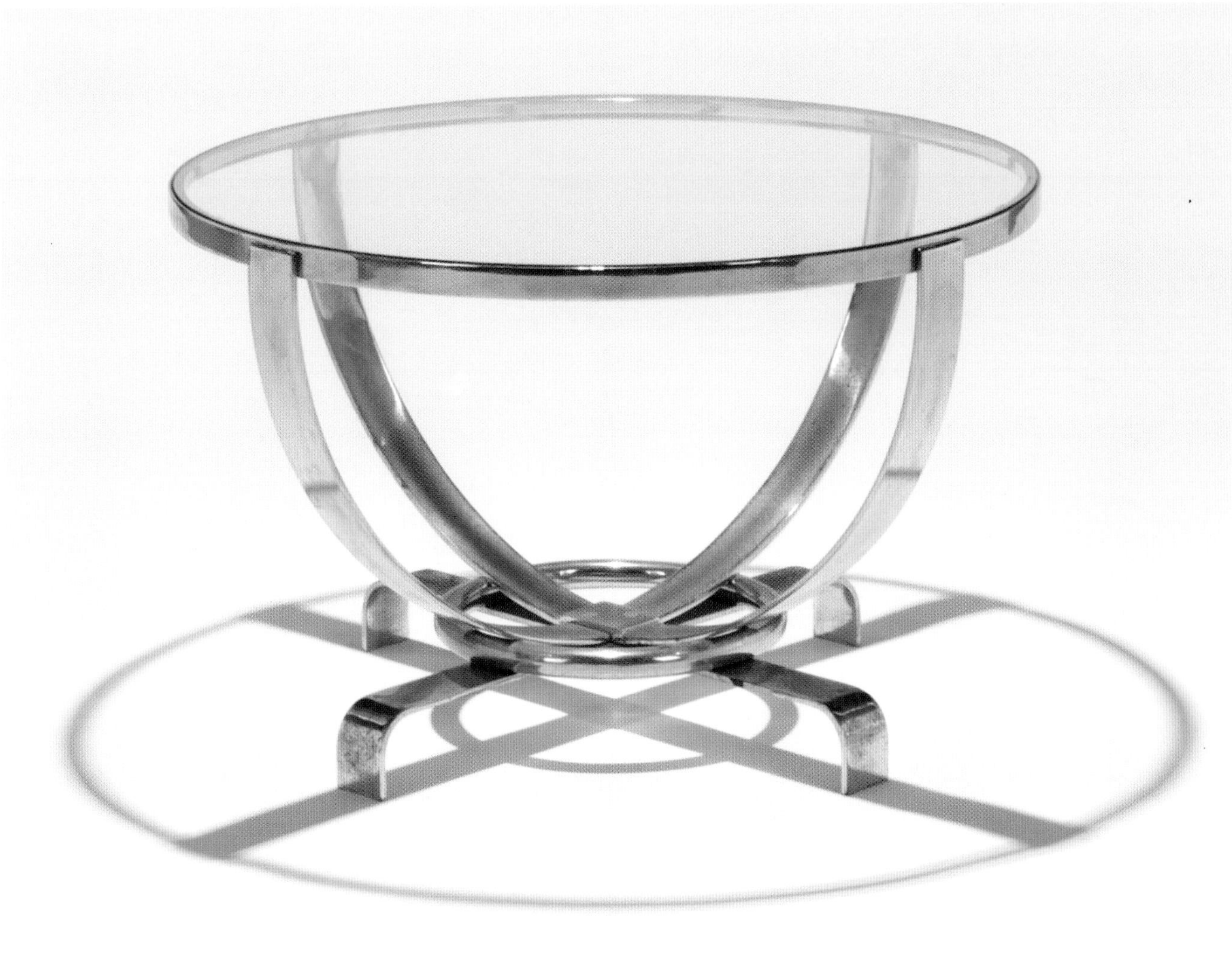

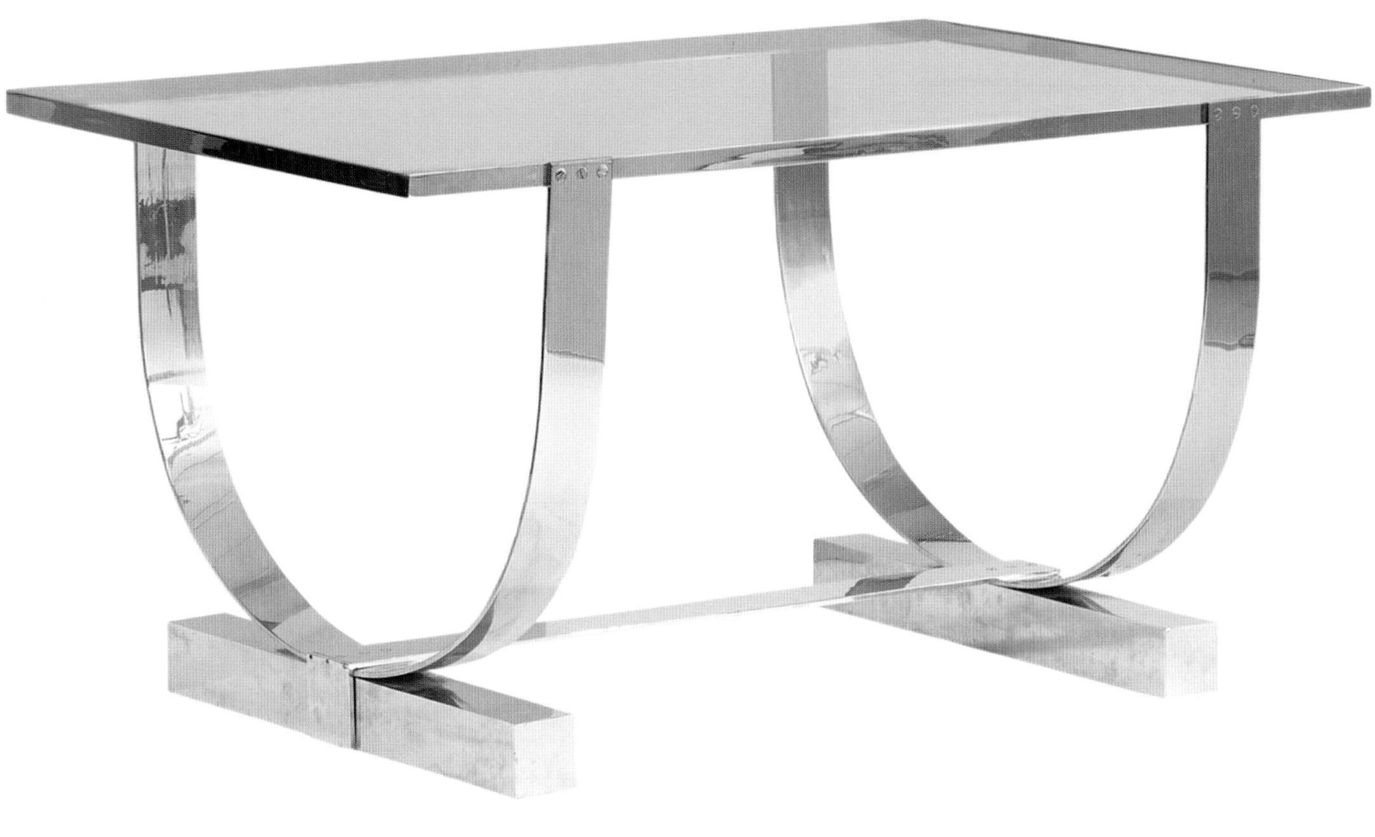

ABOVE A Donald Deskey coffee table, *c.* 1930, in chrome-plated steel and glass.

LEFT A 1927 Deskey-Vollmer table in chrome-plated brass and glass.

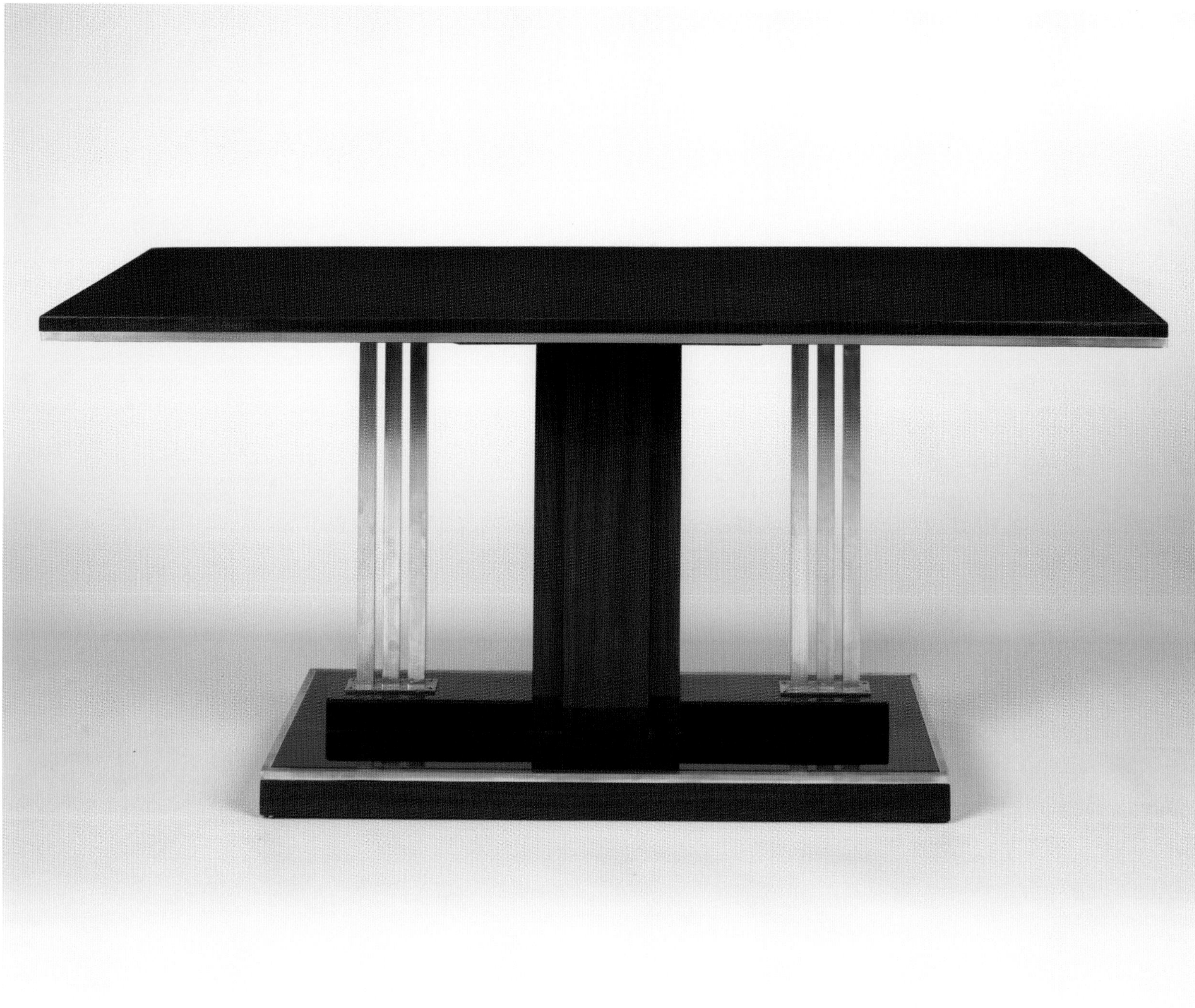

ABOVE A console table, manufactured by Deskey-Vollmer, *c*. 1929, in maple burl, steel, lacquered wood and Bakelite.

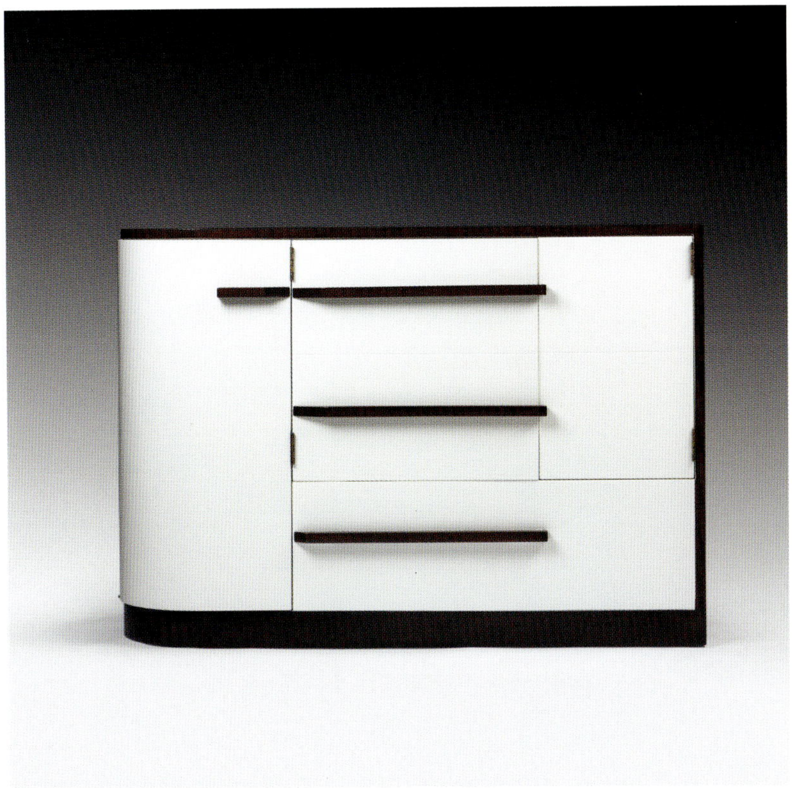

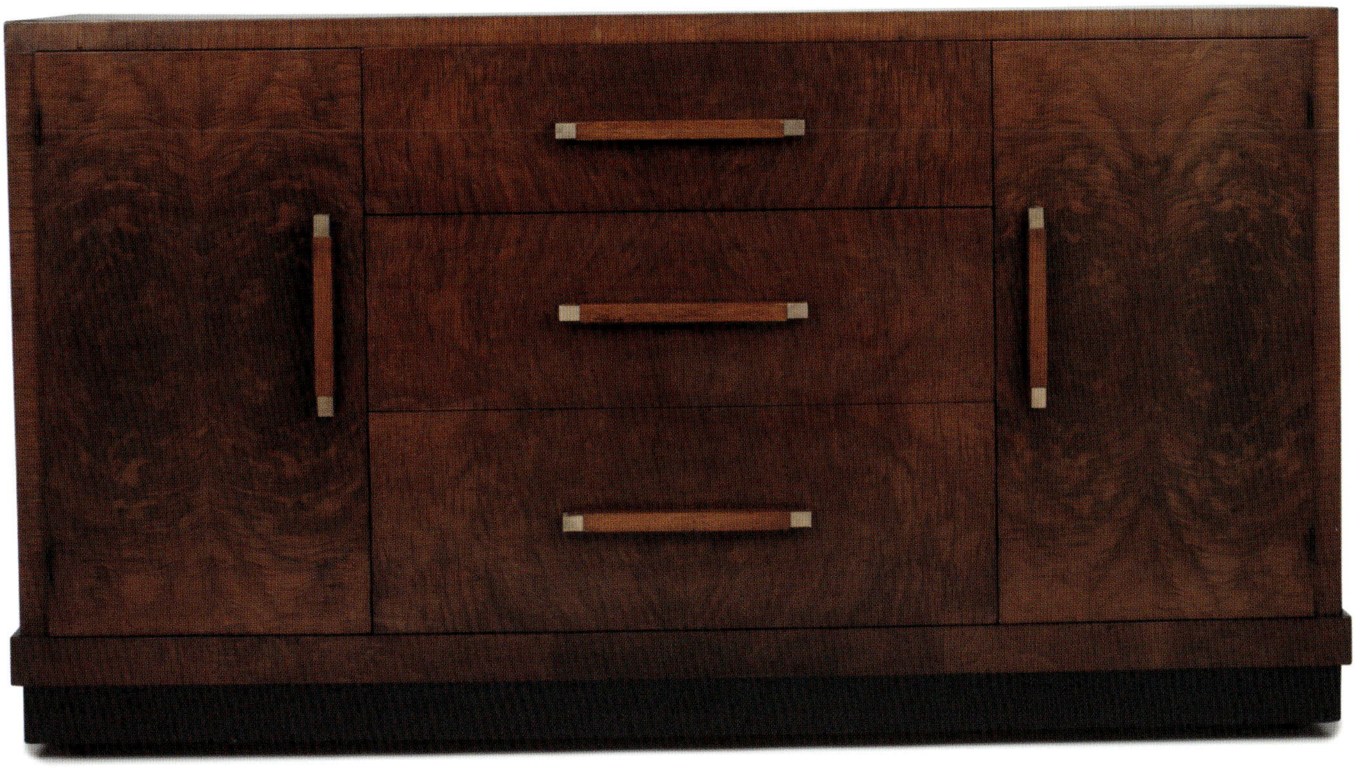

ABOVE The white cabinet is a 1936 design in mahogany and lacquered wood for the Brown Palace Hotel in Denver, Colorado.

LEFT The mahogany and brass cabinet was produced by American Modern Decoration, c. 1935.

JEAN-MICHEL FRANK
(1895–1941)

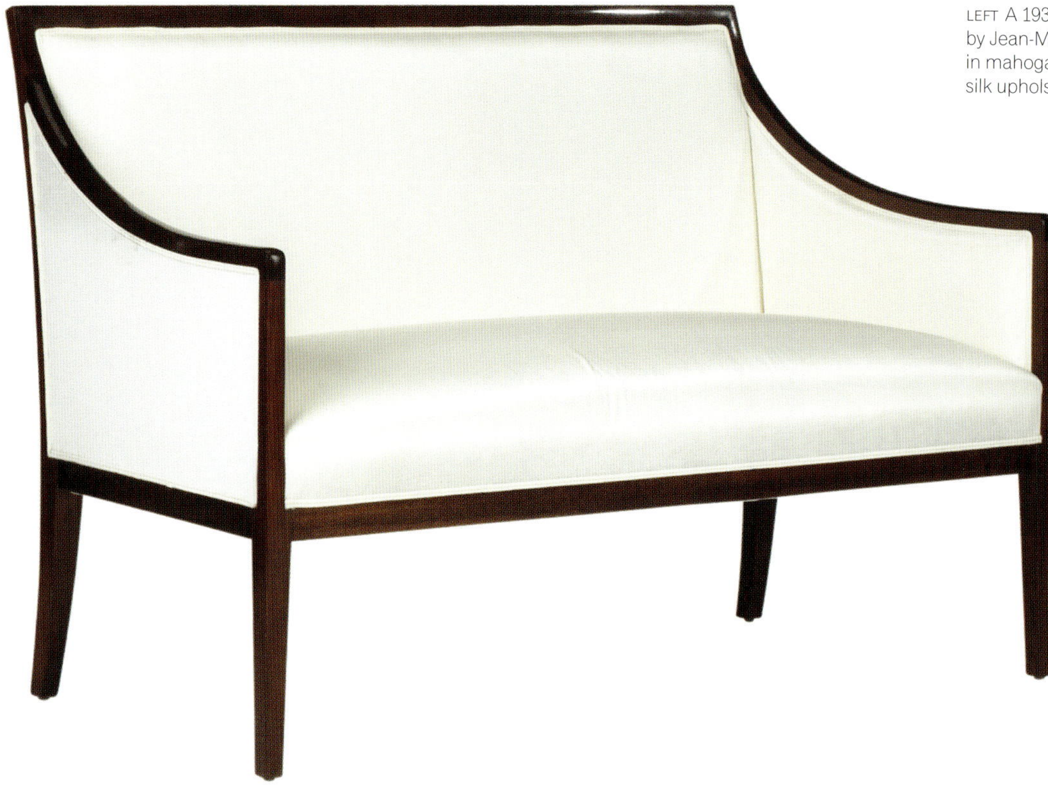

LEFT A 1930s settee designed by Jean-Michel Frank in mahogany with silk upholstery.

It is almost impossible to categorize the work of Parisian designer Jean-Michel Frank. For many, Frank is a leading light of the Art Deco* movement, yet that is to neglect the almost minimalist purity of his work and his predilection for clean lines and restraint, both in his furniture and interiors. Others call him a modernist, yet there is a focus on luxurious materials and high craftsmanship that does not sit easily with mainstream modernism's focus on mass-production techniques and machine-age aesthetics. It is a unique and highly influential combination that is sometimes called 'luxe pauvre', while Frank also drew upon many historical and global influences, from neoclassicism to African tribal furniture and Asian crafts. In the end, it becomes easier to place him in a category of his own.

Frank was born in Paris to a family of Germanic Jewish descent and his father Léon was involved in the banking industry, along with other members of the extended family. The young Frank proved a capable student at the Lycée Janson-de-Sailly and went on to study law. With the arrival of World War I the family – like so many others – was thrown into turmoil. Frank's two older brothers were killed fighting for the French side, leading – in 1915 – to their father's suicide; his mother also plunged into depression and spent periods institutionalized until her death in 1928.

This turbulent family history left Frank with a legacy of vulnerability, with the designer prone to his own phases of depression and a susceptibility to substance abuse. Yet he was also granted a degree of financial freedom by his inheritance and began to indulge his passion for art and design. He started moving in circles that were affluent, progressive and artistic, through which he met many of his early clients.

In 1921 Frank embarked upon his first interior design commissions, with projects for the publisher and printer Charles Peignot and the writer Pierre Drieu La Rochelle. In the years that followed, Frank won other commissions from leading society figures such as the expat English writer and heiress Nancy Cunard, fashion designer Elsa Schiaparelli and composer Cole Porter. Among his most important clients were the art patrons Charles and Marie-Laure de Noailles, for whom Frank designed the interiors of their Parisian mansion on the Place des États-Unis.

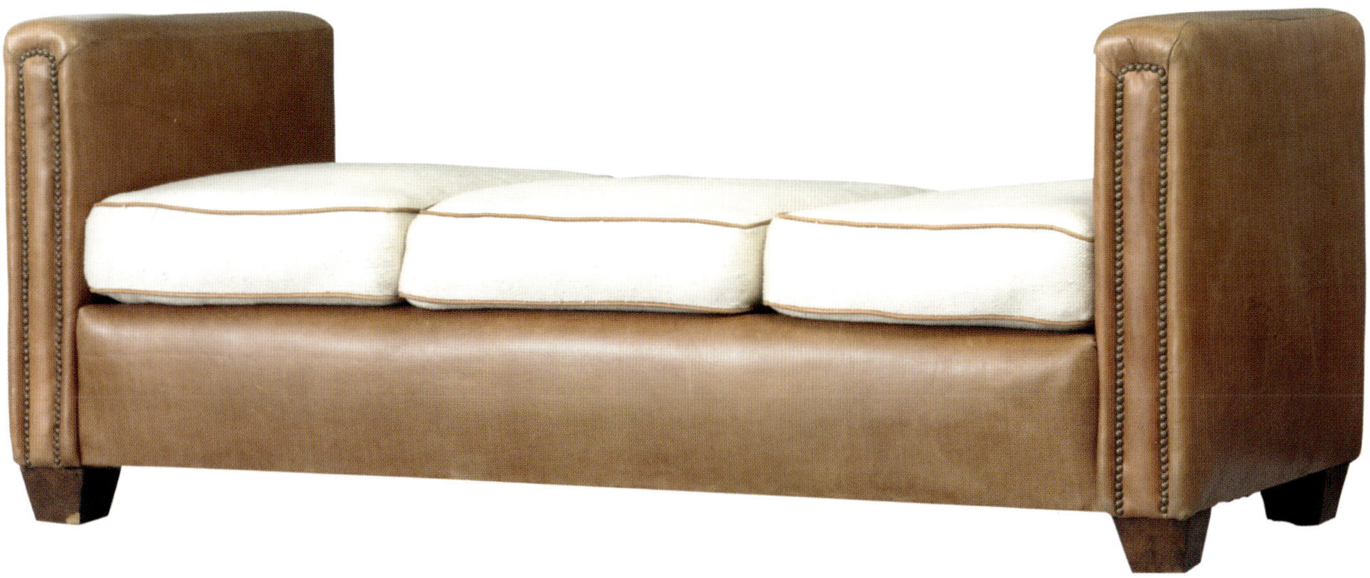

ABOVE A Frank daybed, produced by Chanaux & Co., *c.* 1930, in leather, stained oak and brass, with upholstered cushions.

LEFT An occasional table, made by Chanaux & Co., *c.* 1930, in wrought iron and vellum over oak; Frank served as artistic director of Chanaux & Co.

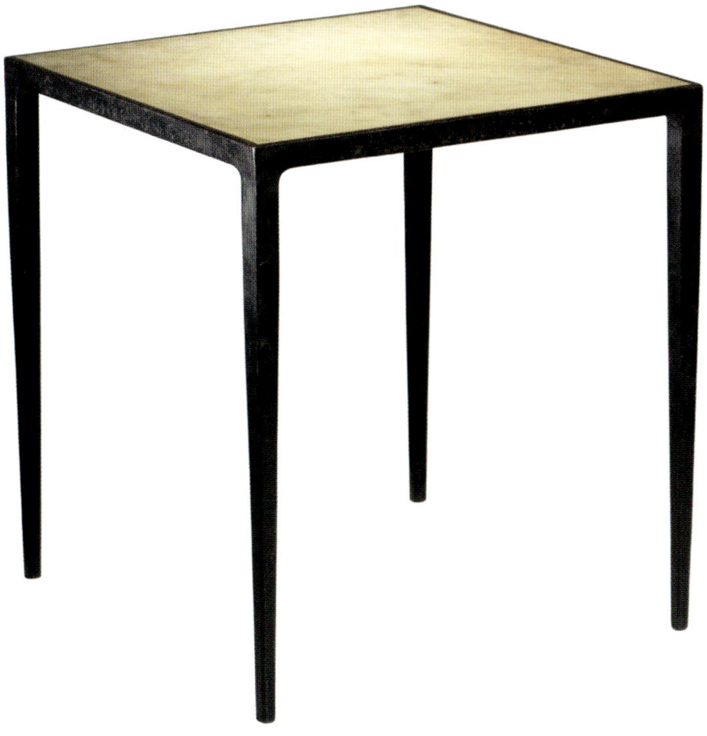

The design of these avant-garde 'luxe pauvre' interiors naturally involved the creation of many individual pieces of furniture and lighting design. Frank began working with a loosely affiliated team of artists and artisans, who included the artist and sculptor Alberto Giacometti* – who created a series of lamp bases for Frank – and artist Christian Bérard, as well as designer-architect Emilio Terry. The materials chosen by Frank were fine, luxurious and expensive, including shagreen, vellum, silk, bronze, mahogany and other crafted timbers, often with inlaid detailing and marquetry.

Frank formed an important alliance with the cabinetmaker Adolphe Chanaux and in 1930 became the artistic director of Chanaux & Company, with a team of around thirty artisans largely dedicated to producing Frank's designs. At the same time, he began experimenting with a broader spectrum of influences and working with a wider colour palette, although his work is still associated with his 'trademark' off-whites and beige tones.

The look of Frank's furniture – like his interiors overall – is pared down, with a minimum of applied ornament or pattern. The emphasis is upon the fine quality and texture of the materials themselves and the line and form of the pieces, with much of the work adopting a strict, linear character rather than a streamlined dynamism. The club chairs designed for philanthropist Charles Templeton Crocker's penthouse home in San Francisco (1929) have a typical Frank quality – finely made and detailed, but also ordered and linear.

In 1935, Frank opened a Parisian boutique selling his work; that same year he also designed an apartment for perfumier Jean-Pierre Guerlain and a showroom for fashion designer Lucien Lelong. A year later he began an association with the Argentinian furniture producers Comté, and then in 1938–39 he created the interiors for business magnate Nelson Rockefeller's apartment at the Rockefeller Center in New York. After the outbreak of World War II and the closure of his Parisian business, Frank headed to Argentina and then New York, where the family tendency towards fatalistic depression finally caught up with him and he threw himself to his death from a Manhattan apartment building.

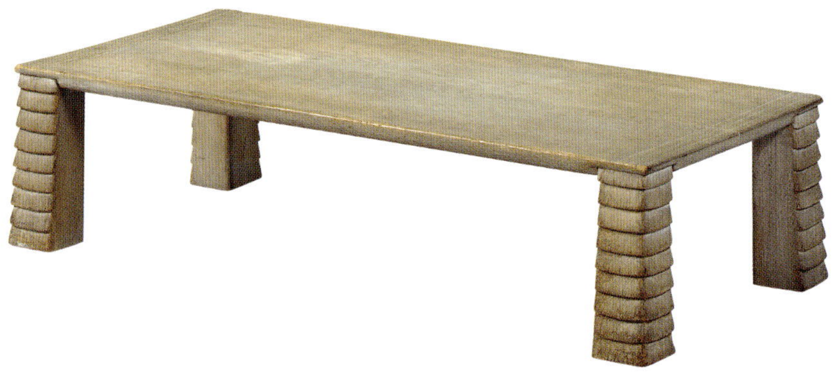

ABOVE Frank's Pineapple table, *c.* 1938–40, in sanded and limed oak.

BELOW A pair of Frank lounge chairs, produced by Chanaux & Co., *c.* 1930, in leather, oak and brass.

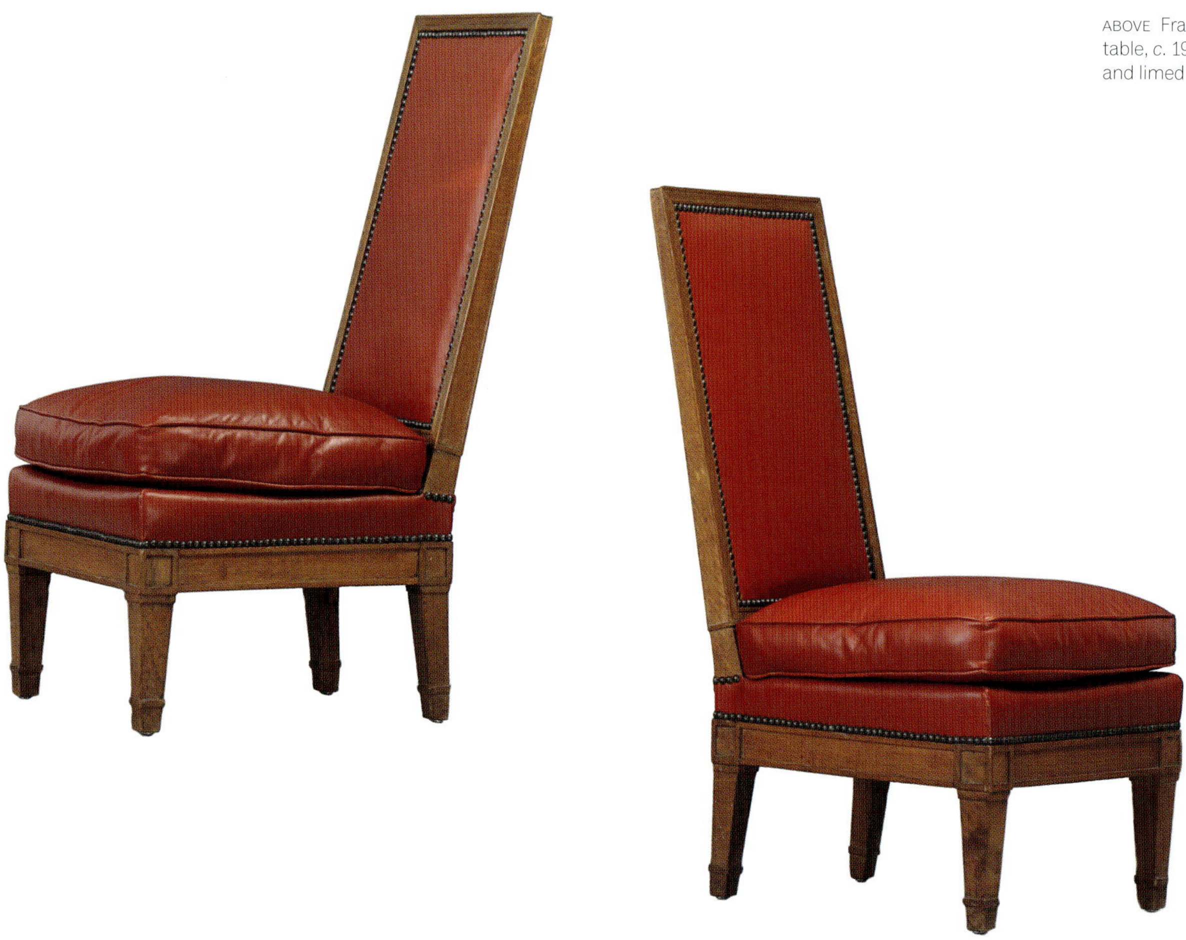

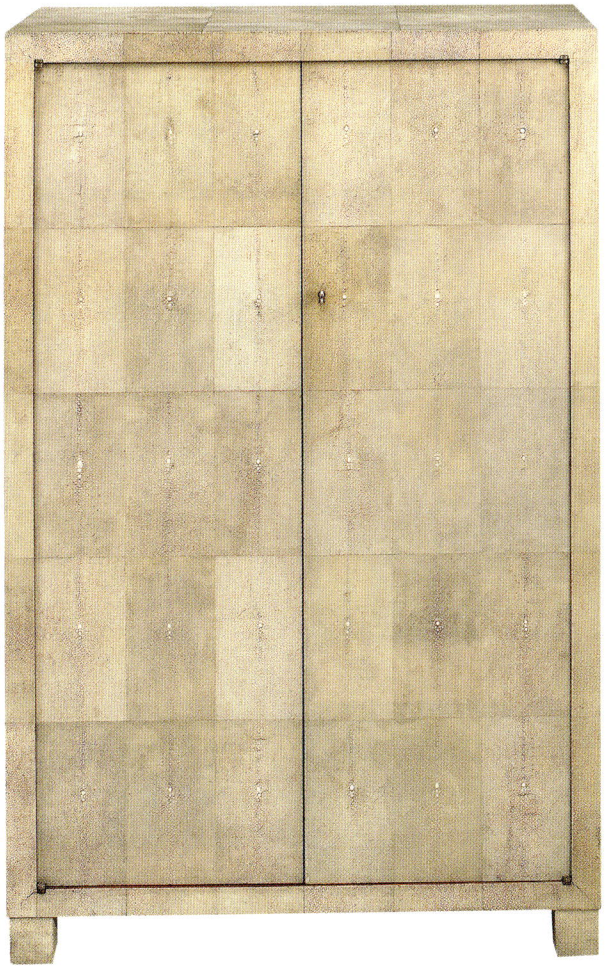

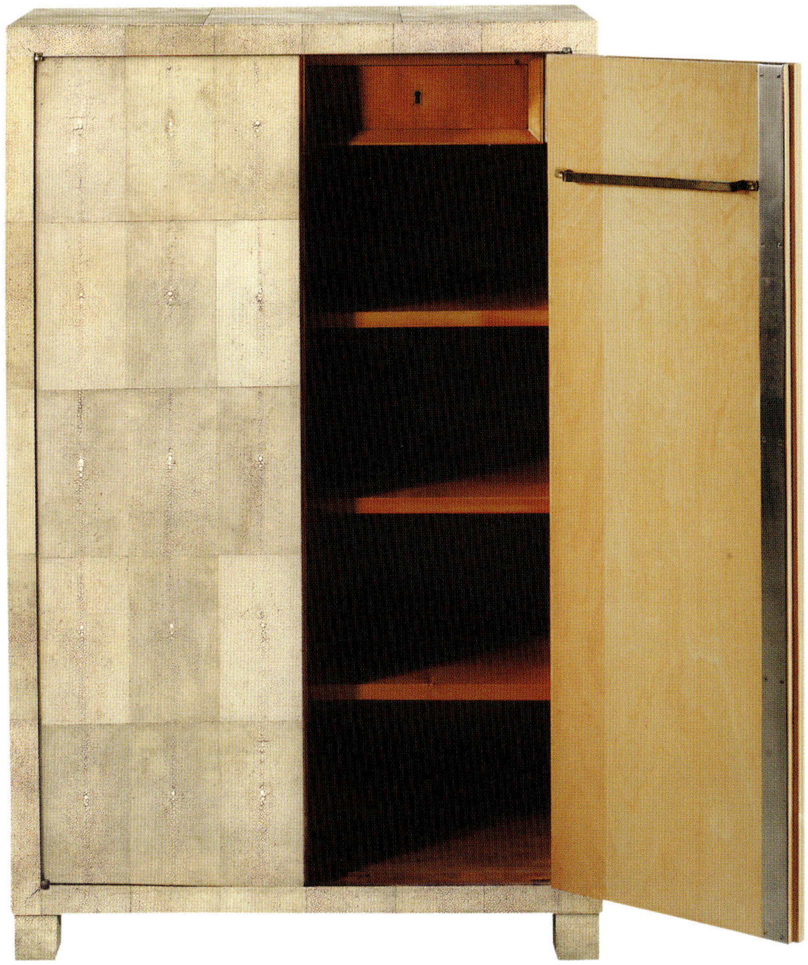

ABOVE Two views of a late 1920s Frank cabinet, manufactured by Chanaux & Co., in wood overlaid with shagreen.

BELOW An upholstered club chair and ottoman, *c.* 1938; this is one of a number of variants on the club chair designed by Frank.

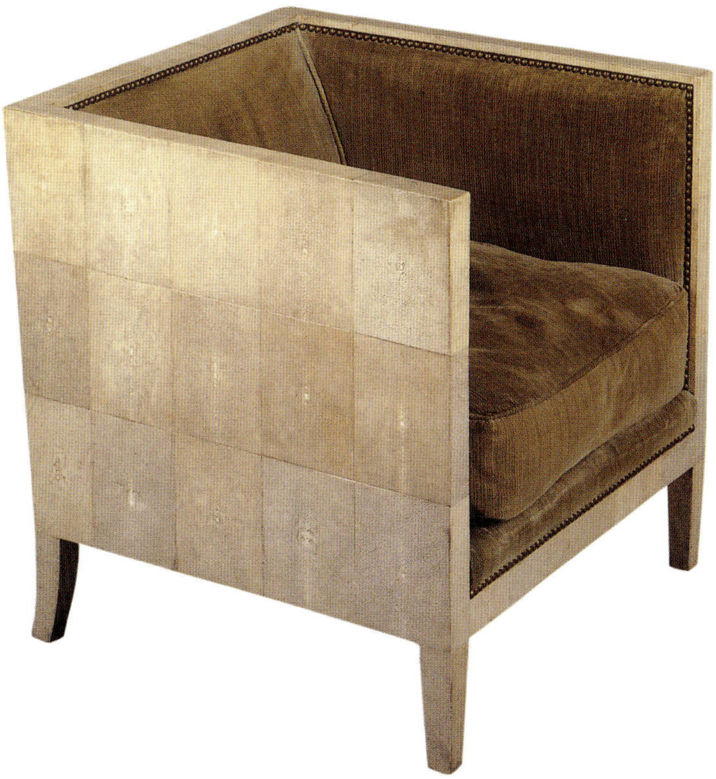

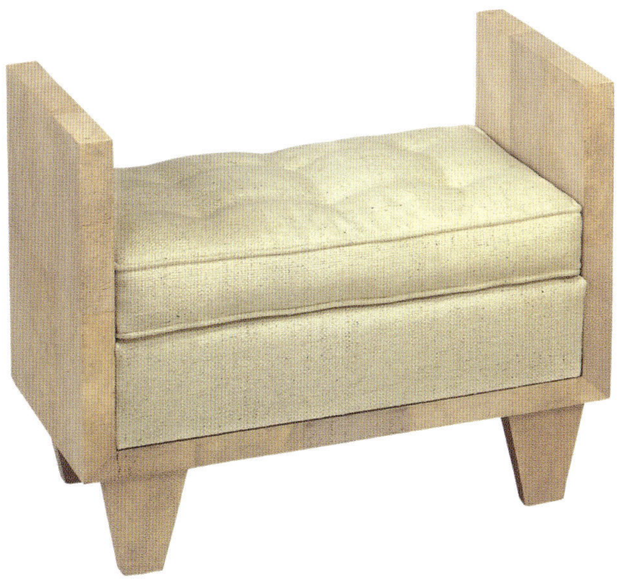

EILEEN GRAY
(1878–1976)

As a designer, Eileen Gray was respected but seldom fêted during her lifetime. In the post-war years she almost faded from view, with just a handful of projects, many of them unrealized. She divided her time between her apartment in Paris and her country retreat near Saint-Tropez, known as Lou Pérou, which is listed as her last completed architectural design project. Free-spirited but modest, she was surprised when interest in her work revived towards the end of her life. Today, her work has been reassessed and has found a new and appreciative audience, with her furniture still much in demand and her early pieces highly collectable.

Gray was born in Ireland, in County Wexford, and was one of five children from a wealthy family with homes both in Ireland and London. Her father was a painter, who travelled widely and spent a good deal of time in Italy and Switzerland, but died when Gray was in her early twenties. Gray studied at the Slade School of Fine Art and then moved to Paris, where she continued her studies in drawing and painting at the Académie Colarossi art school and then the Académie Julian.

She became interested in lacquer and its craft, enlisting the help of a Japanese lacquer master, Seizo Sugawara, in Paris. She acquired an apartment on the rue Bonaparte, which she kept until the end of her life and where she designed the interiors and furnishings. She also established both a lacquer studio with Sugawara and a textile workshop with Evelyn Wyld, designing carpets.

Gray's early furniture designs explored lacquer finishes and techniques, splicing Art Deco* and early modernist influences. Clients included Charles de Noailles, the Duchess of Clermont-Tonnerre, and art collector and patron Jacques Doucet, who first commissioned her Lotus table (1915) and Le Destin screen (1914). During World War I, she served as an ambulance driver before retreating to London, taking Sugawara with her.

After the war there were interior design projects, including an apartment for Madame Mathieu-Lévy – the owner of a milliners – on the rue de Lota in Paris (1924; see p. 17). The project included a pair of lacquered screens made up of a series of cubist panels (Brick Screen, completed 1925) and Gray's Lota sofa (1924), featuring a rectangular lacquered timber frame and thick cushions spilling over the back and sides.

In 1922, Gray opened her own furniture shop at 217 rue du Faubourg Saint-Honoré, known as Jean Désert. The name may have referenced JEAN BADOVICI, whom she met in the early Twenties. An architect and magazine editor, Badovici also became her lover, and encouraged Gray to expand her work, including architectural commissions. They collaborated on a number of house projects, including E-1027, their landmark home by the sea in Roquebrune-Cap-Martin (see p. 340).

By this time, Gray's work was decidedly modernist in character, influenced by the work of De Stijl* designers such as GERRIT RIETVELD and

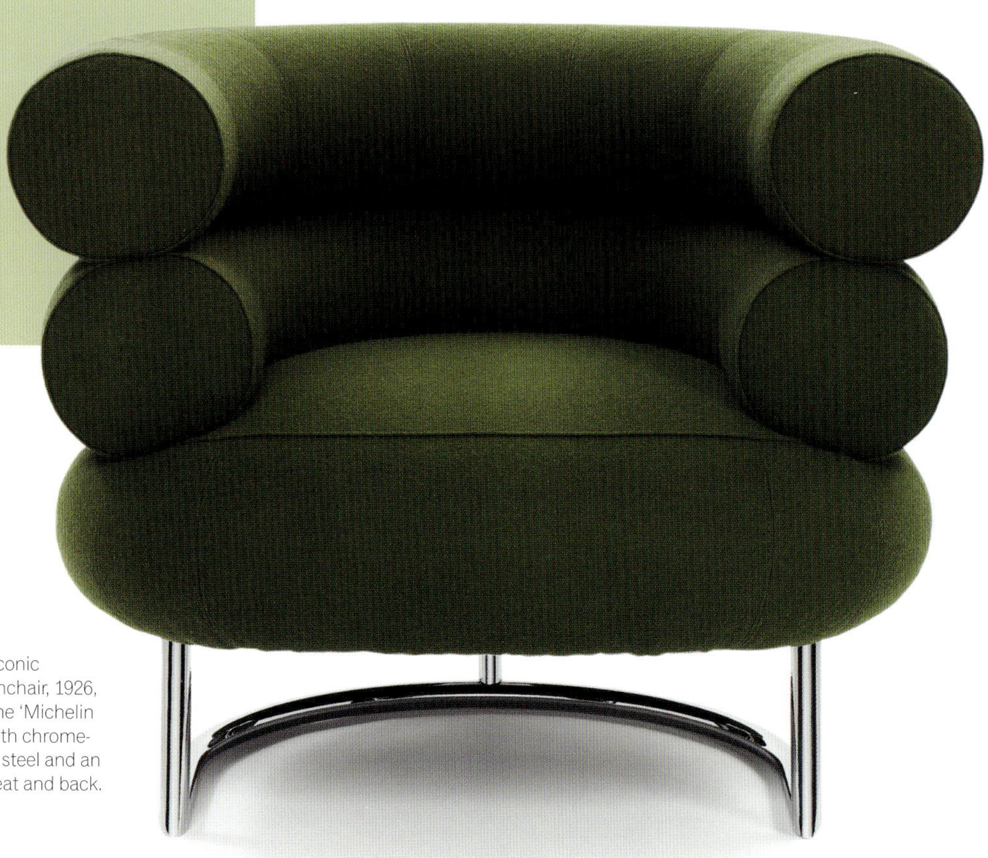

RIGHT Gray's iconic Bibendum armchair, 1926, named after the 'Michelin Man', made with chrome-plated tubular steel and an upholstered seat and back.

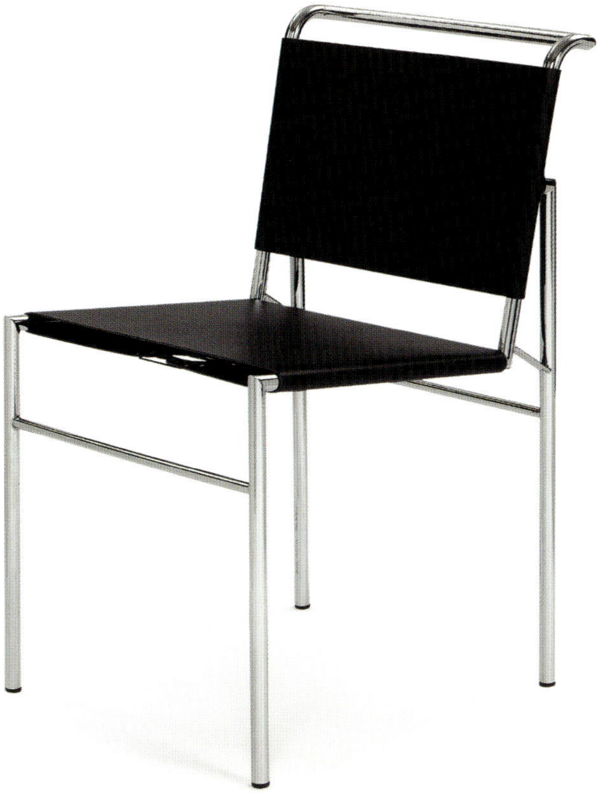

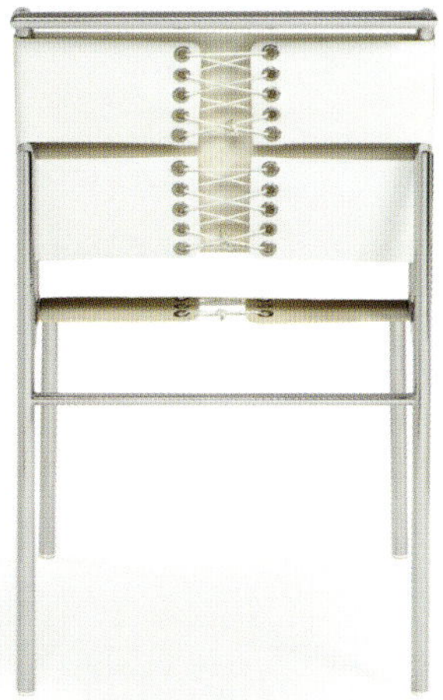

LEFT Front and rear views of the Roquebrune chair, *c.* 1927, in chrome-plated tubular steel with a laced coach hide seat and back.

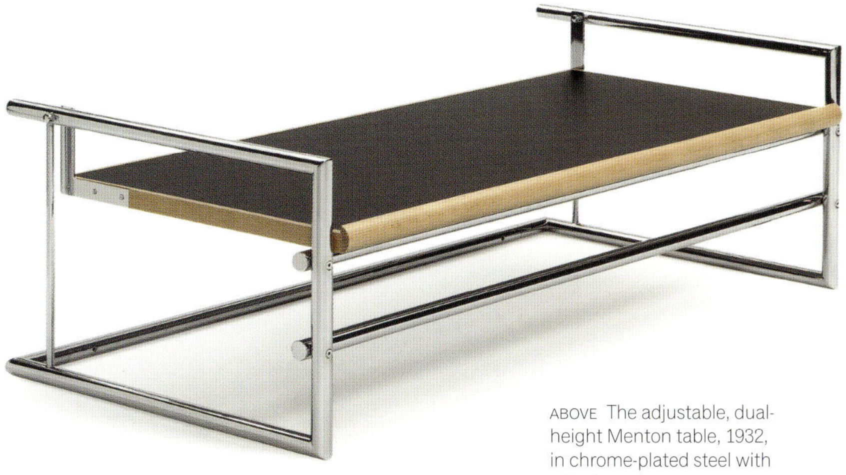

ABOVE The adjustable, dual-height Menton table, 1932, in chrome-plated steel with a linoleum top.

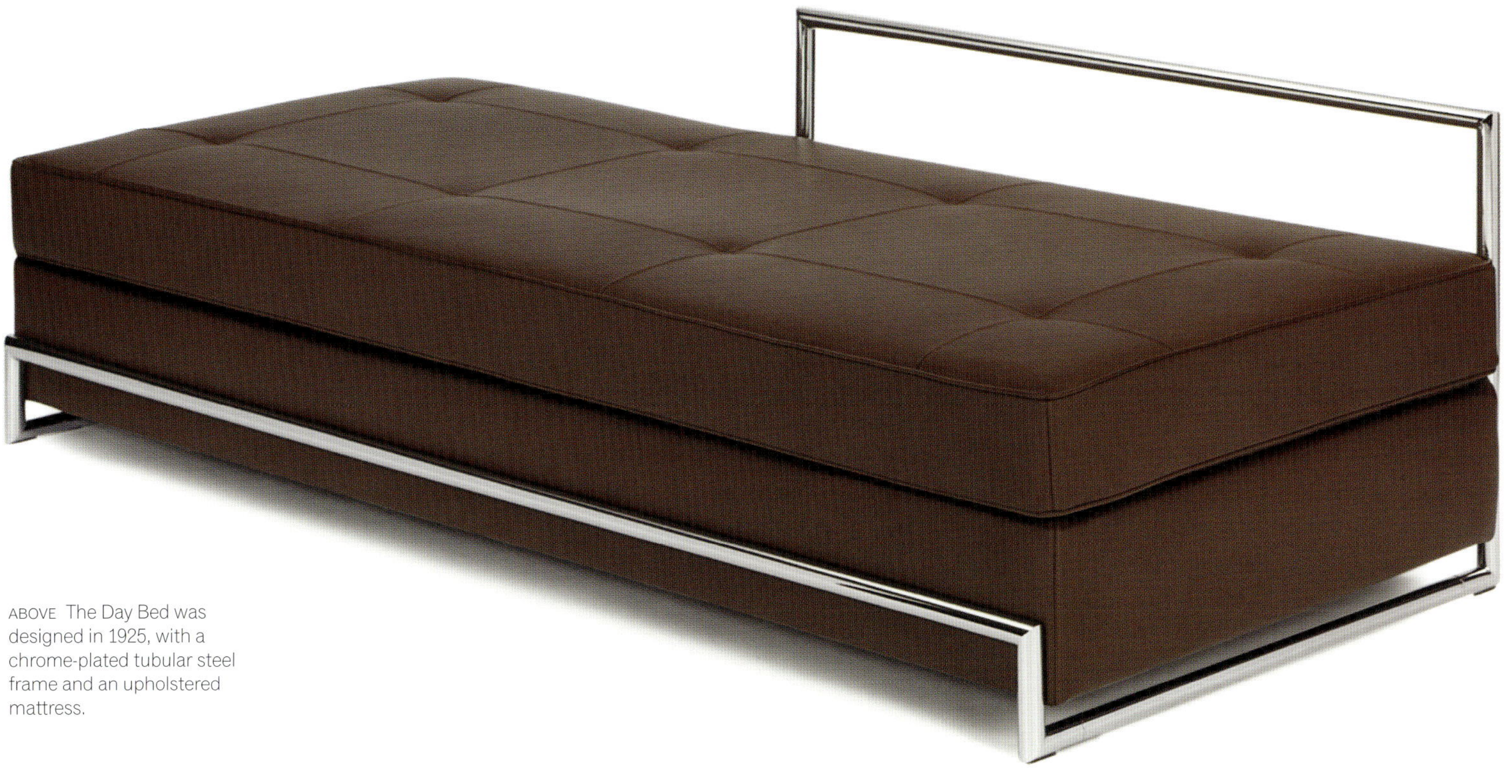

ABOVE The Day Bed was designed in 1925, with a chrome-plated tubular steel frame and an upholstered mattress.

the work of Le Corbusier, who became a friend of Badovici's. A number of key furniture designs from the Twenties were initially conceived for E-1027 but also sold in small editions at Gray's store, including her Transat chair (*c.* 1927) with a lacquered wooden frame and leather sling seat, reminiscent of a deck chair or 'transatlantique'. There was also her iconic E-1027 circular table in steel and glass (1927), which featured an adjustable top that could be raised or lowered to suit and has become one of the instantly recognizable staples of the modernist design canon.

A number of Gray's chair designs contrasted lightweight steel or tubular steel frames with comfortable leather upholstery, including her Non Conformist chair of 1926, the S chair of 1932, and the famous Bibendum armchair (1926), which featured a series of bulbous upholstered bands forming the back of the seat, echoing the Michelin Man's famous tyre-like figure. The Roquebrune dining chair (*c.* 1927) was, by contrast, a more reserved and rational piece, with an ordered steel frame and a tight leather seat and back. Gray's Day Bed (1925) and Bonaparte chair (1935) put comfort to the fore, as well as modernist forms and styling, with thick mattresses or chunky upholstered leather seats and backs.

A number of Gray's reissued classics – which also include rugs and lighting designs – are pieces originally designed for Gray's own homes, including E-1027, Tempe à Pailla and Lou Pérou, all in Provence. They include her Menton table (1932), the Rivoli tea table (1928) and her flexible, folding table, Jean (1929). They are pieces of both grace and ingenuity, which still feel fresh and full of delight within the context of the 21st century.

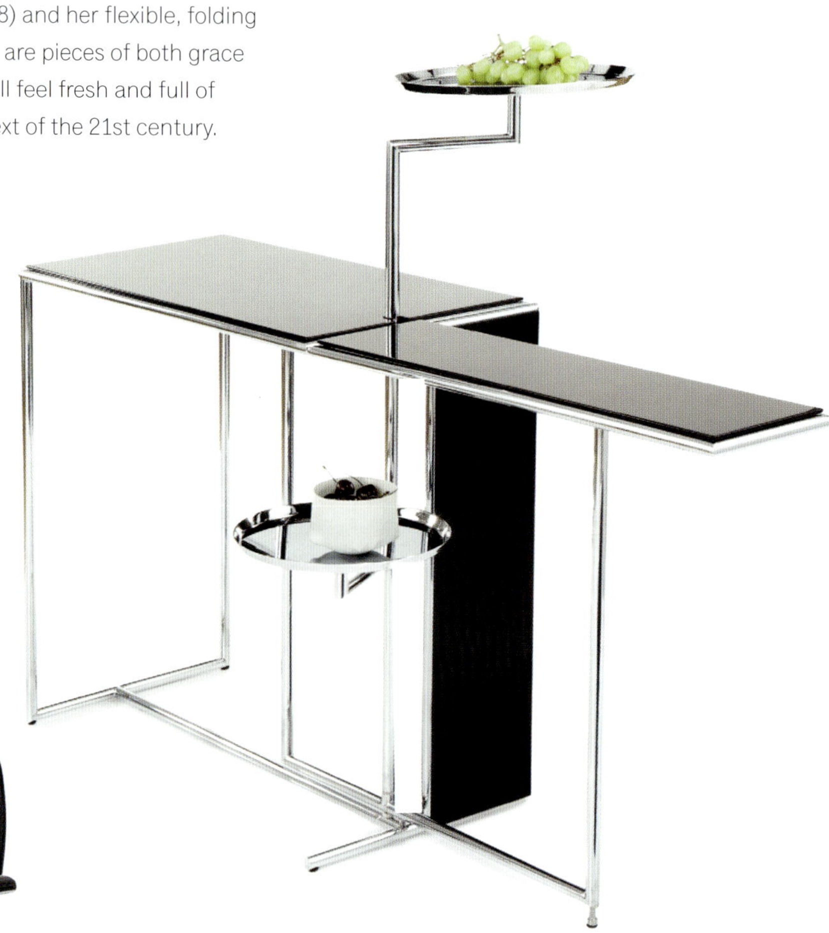

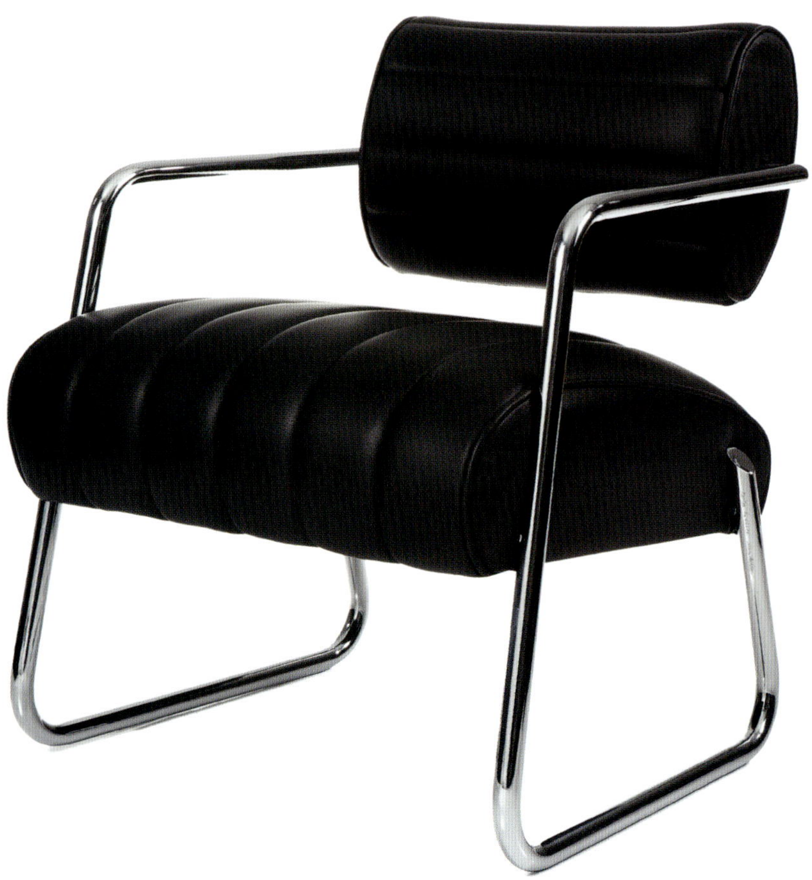

ABOVE The Rivoli table, 1928, in chrome-plated tubular steel with integrated lacquer table top plus integrated trays; the piece was initially designed as a serving table for Eileen Gray and Jean Badovici's Roquebrune home, E-1027.

LEFT Gray's Bonaparte chair, 1935, in chrome-plated tubular steel with an upholstered, leather-coated seat and backrest.

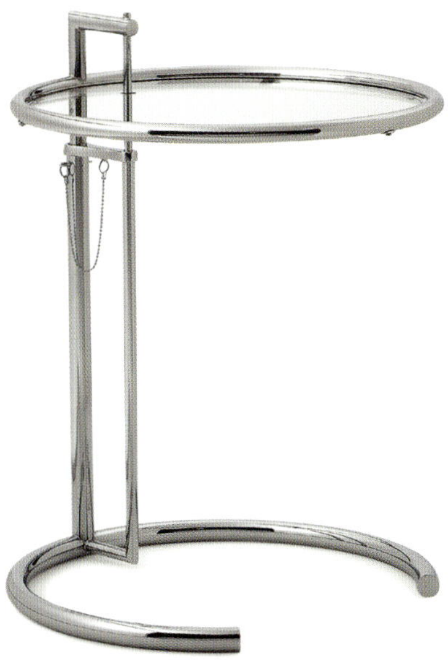

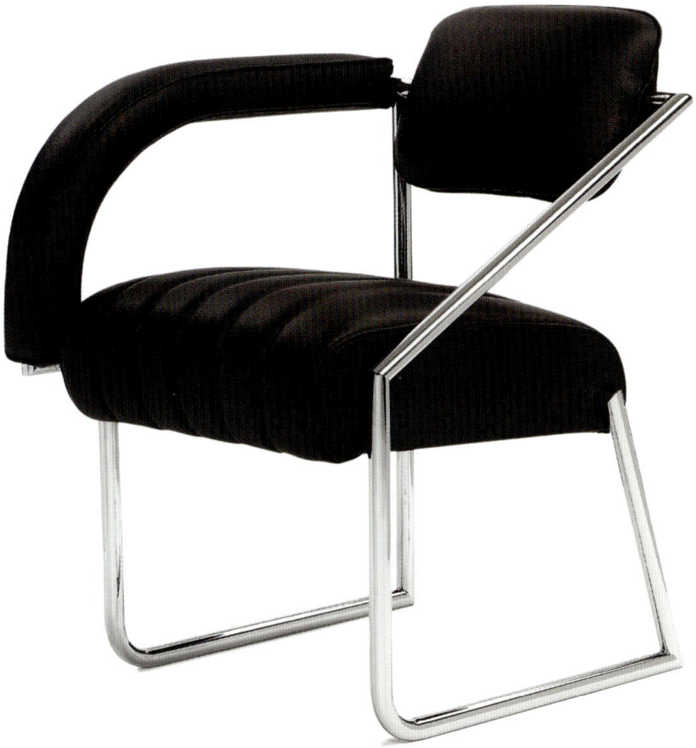

ABOVE LEFT The E-1027 occasional or bedside table, 1927, also designed for the house at Roquebrune; the height of the table, made of chrome-plated tubular steel and glass, is adjustable.

ABOVE RIGHT The Non Conformist chair, 1926, in chrome-plated tubular steel with an upholstered, seat, backrest and singular arm.

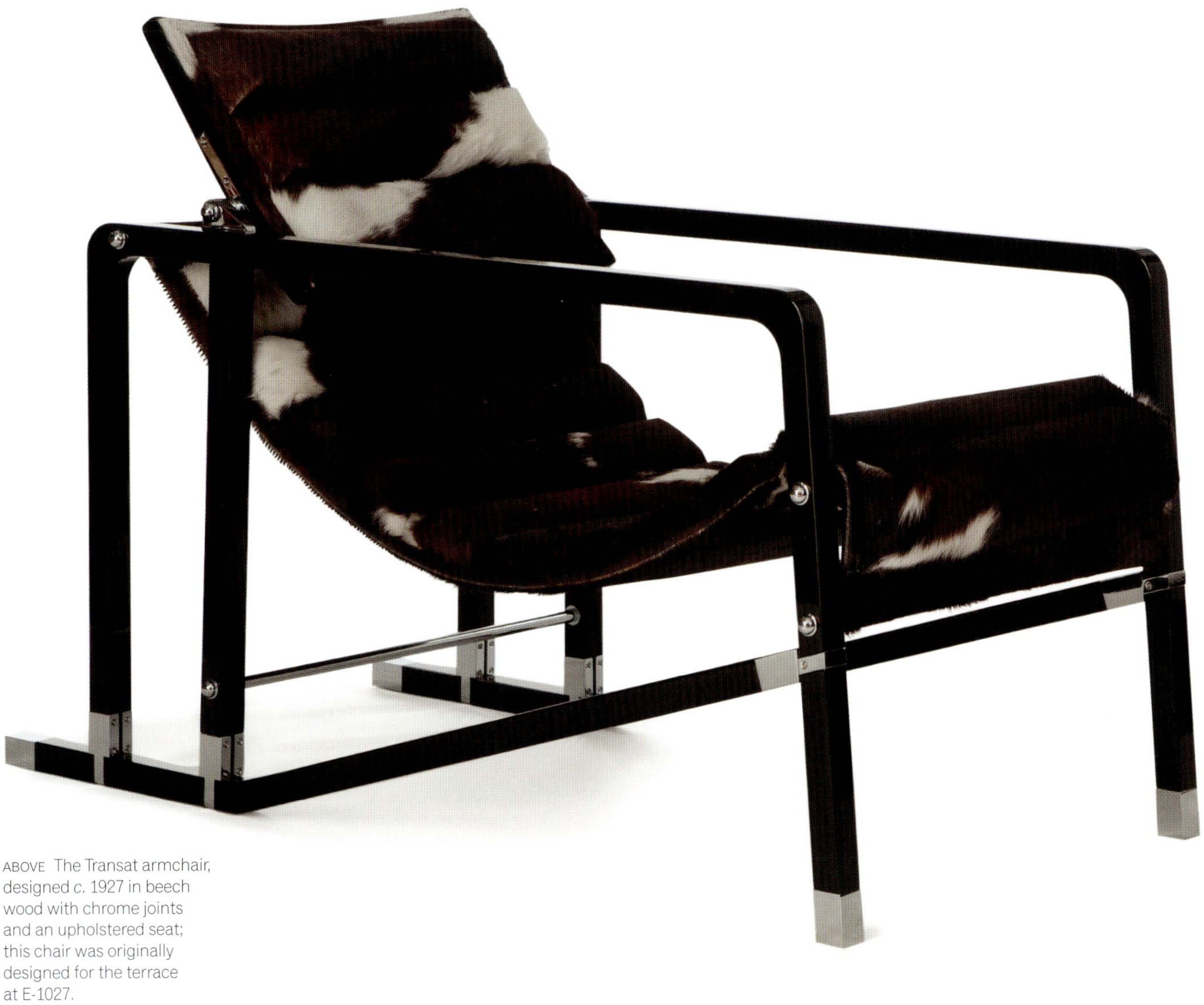

ABOVE The Transat armchair, designed *c.* 1927 in beech wood with chrome joints and an upholstered seat; this chair was originally designed for the terrace at E-1027.

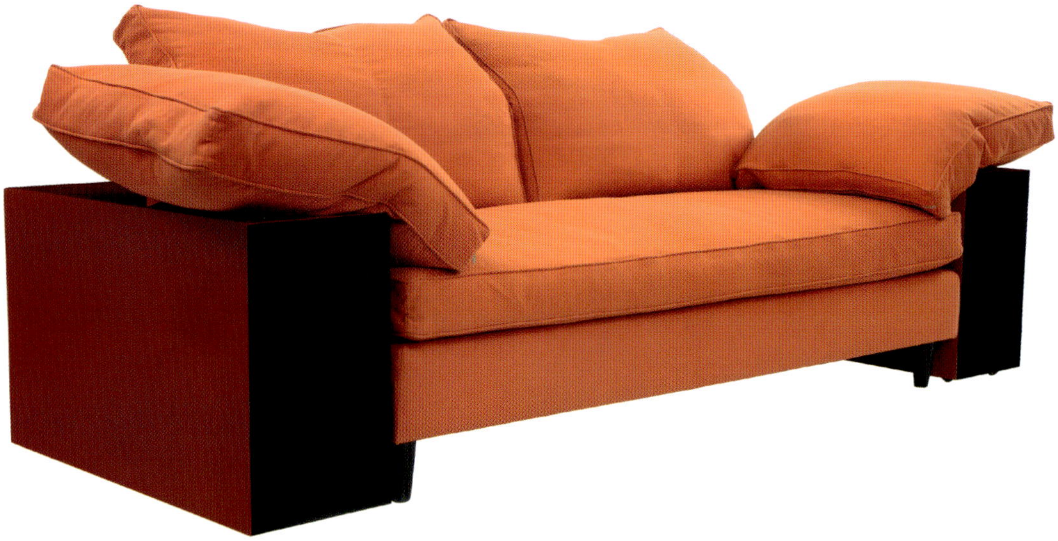

ABOVE The Lota sofa of 1924, with two lacquered and detachable units at either end of the upholstered settee; the piece was originally designed as part of Gray's scheme for the interiors of an apartment at the rue de Lota in Paris.

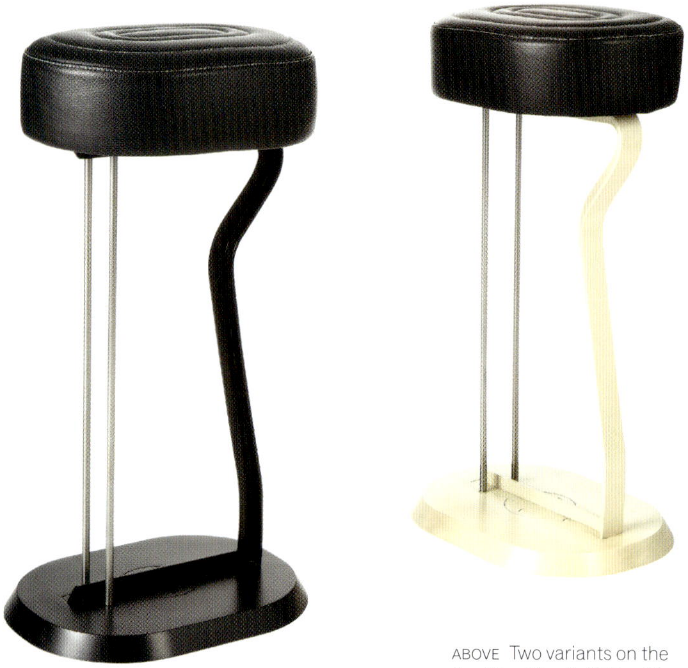

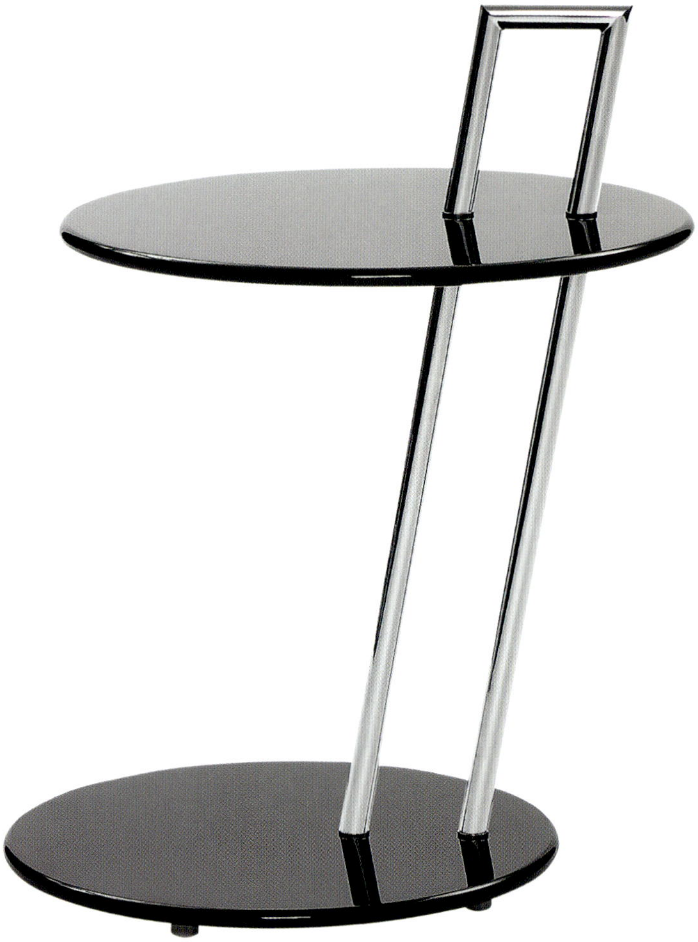

ABOVE Two variants on the Bar Stool No. 2, 1928, with a black or cream lacquered base and upholstered leather seat.

LEFT Occasional table, 1927, in chrome-plated tubular steel with a lacquered top and base.

RIGHT Gray's Brick Screen, originally designed *c.* 1922, with a steel rod frame and lacquered wooden panels, both fixed and flexible.

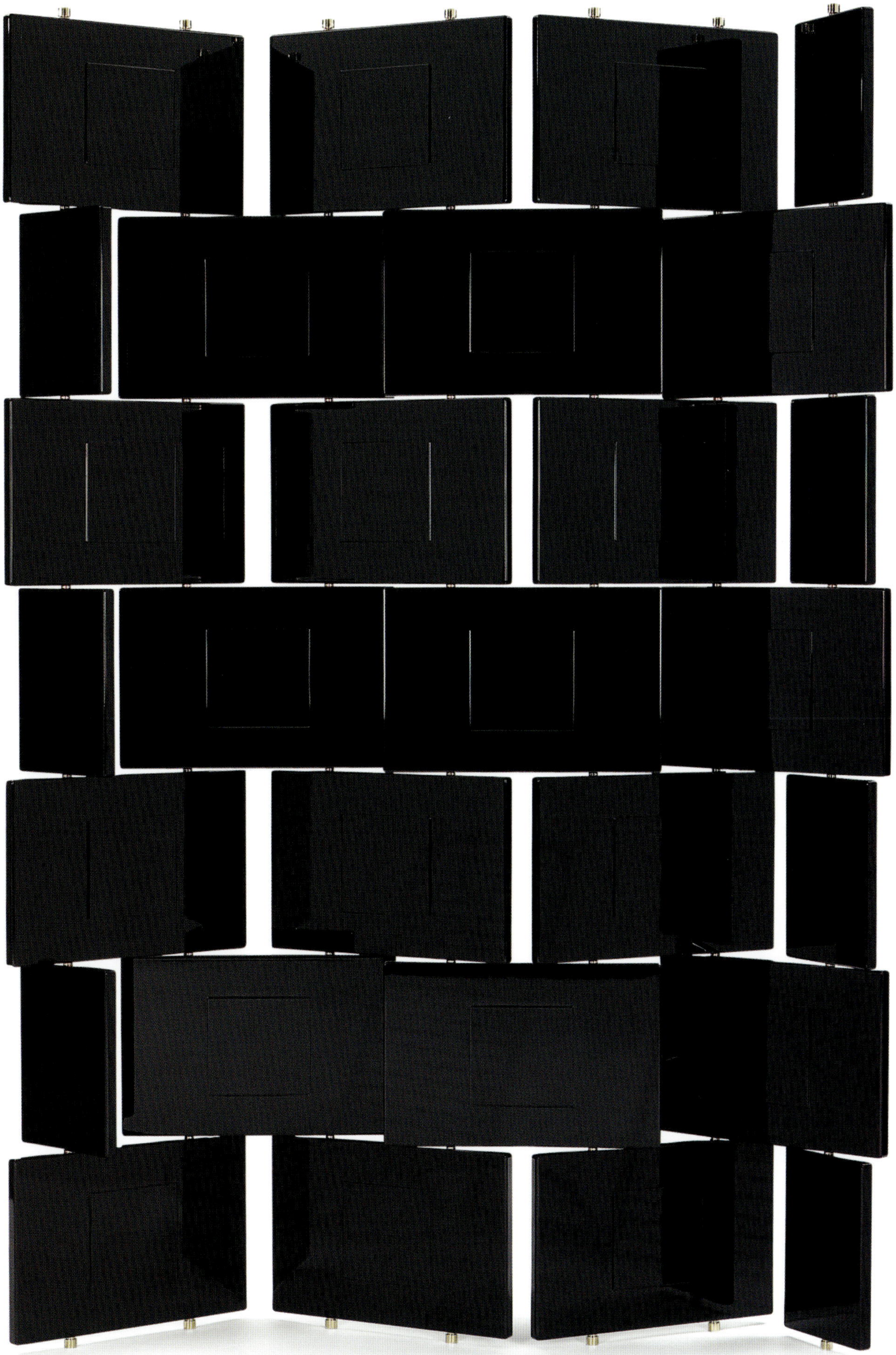

RENÉ HERBST
(1891–1982)

The designers of the machine age were well aware of comfort. These early modernists were not simply didactic aesthetes determined to 'make it new' at any cost, including the health of our bones and backs. They were interested in new forms and materials, but they were also interested in ergonomics and the ambition of creating furniture and interiors that would enhance quality of life and everyday pleasures. This approach was true of the French designer René Herbst.

Herbst noted the stone seats of ancient amphitheatres and how the stone had been worn away by thousands of spectators over countless years until the seat echoed the contours of the body itself. Working with new materials such as tubular steel, he aimed to produce seating suited to mass, industrial production that would be a delight within the home, adapting to the weight and profile of the individual sitter.

'In seating design, originality for originality's sake is always a mistake,' said Herbst. 'The chair provides an envelope for the body and, ideally, should be conceived to impart an agreeable sense of repose. It was for this reason that ... I came upon the idea of devising an elastic support which would mould exactly to the form of the body. I attained the ideal solution using rubber [straps] placed almost side by side.'[11]

The most famous example of Herbst's series of flexible chairs was the Chaise Sandows of 1929, which used a simple, rounded, tubular steel frame and rubber stretcher belts, or expanders, with the tension belts hooked directly into the frame. The chair – named after the turn-of-the-century bodybuilder Eugen Sandow and the chest expanders that he invented – is light and elegant, modest and appealing, and at the same time it readily submits to the body and will of each and every guest. Among the owners of Chaise Sandows was, reputedly, the celebrated modernist artist Tamara de Lempicka.

Herbst was born in Paris, where he studied architecture. In 1908, he served architectural internships in both London and Frankfurt, and later served in the French air force. In the early Twenties, he concentrated primarily upon store designs, window displays and exhibition stands, while also presenting his work at showcases such as the Salon d'Automne. Herbst began designing his own furniture and lighting for these interiors and later commissions including homes for prestigious clients such as the Aga Khan and his Parisian residence on rue Scheffer (1930–33). These commissions provided Herbst with fresh opportunities to develop designs for both fitted and free-standing furniture, along with lighting.

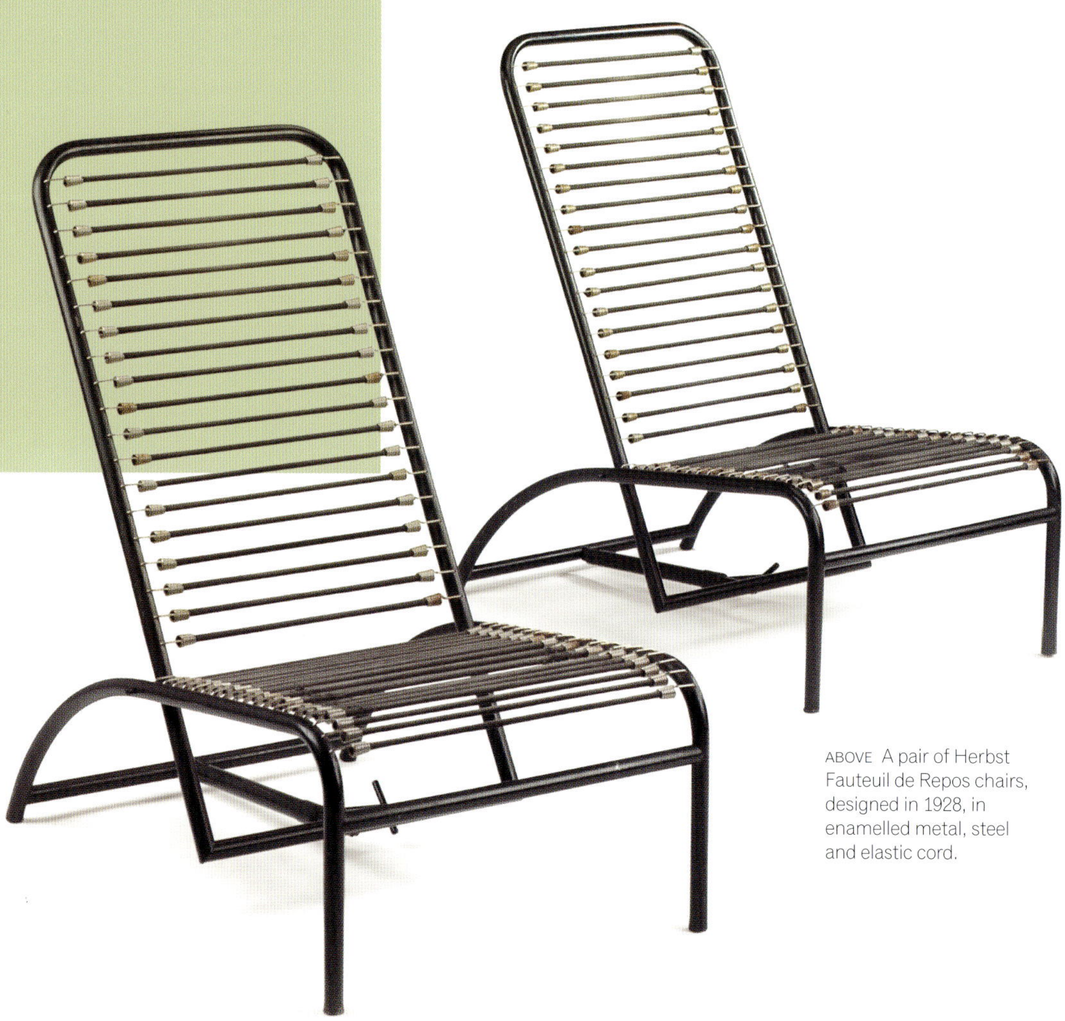

ABOVE A pair of Herbst Fauteuil de Repos chairs, designed in 1928, in enamelled metal, steel and elastic cord.

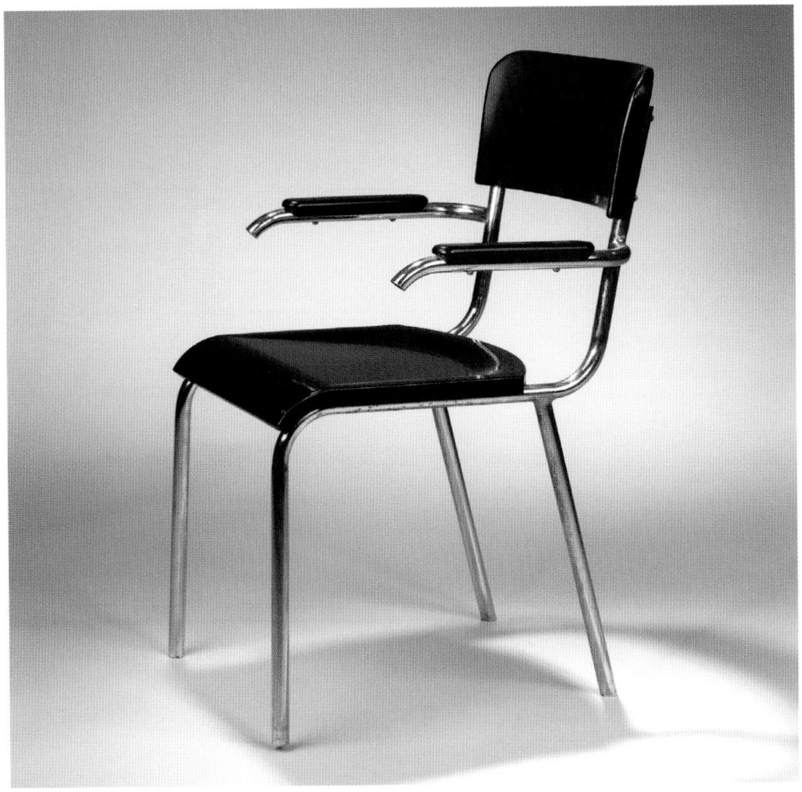

LEFT A rare prototype armchair designed by Herbst c. 1930, with a chrome-plated tubular steel frame and a Bakelite seat, backrest and armrests.

BELOW A set of six Herbst dining chairs, 1943, in chrome-plated steel and Bakelite.

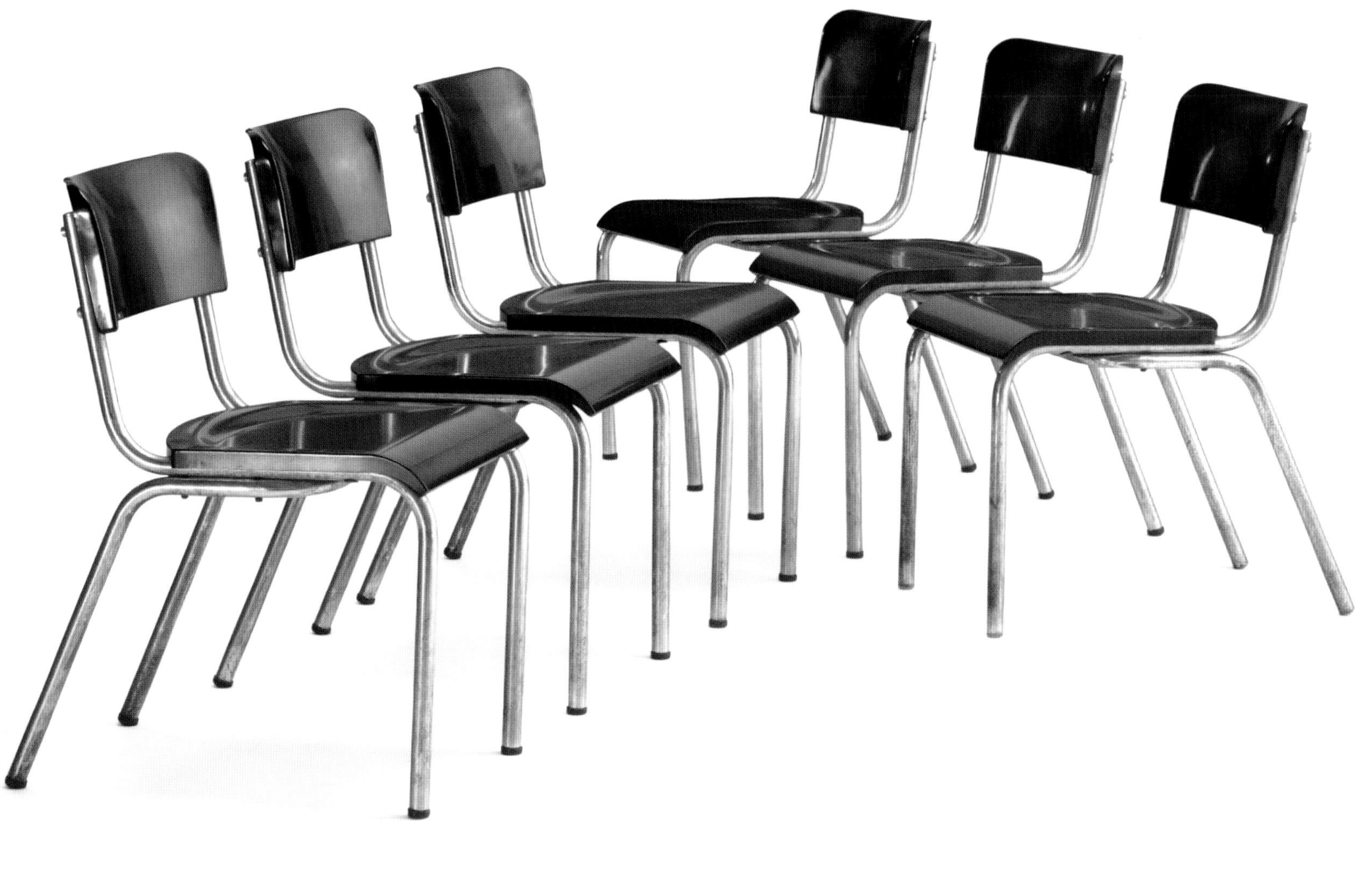

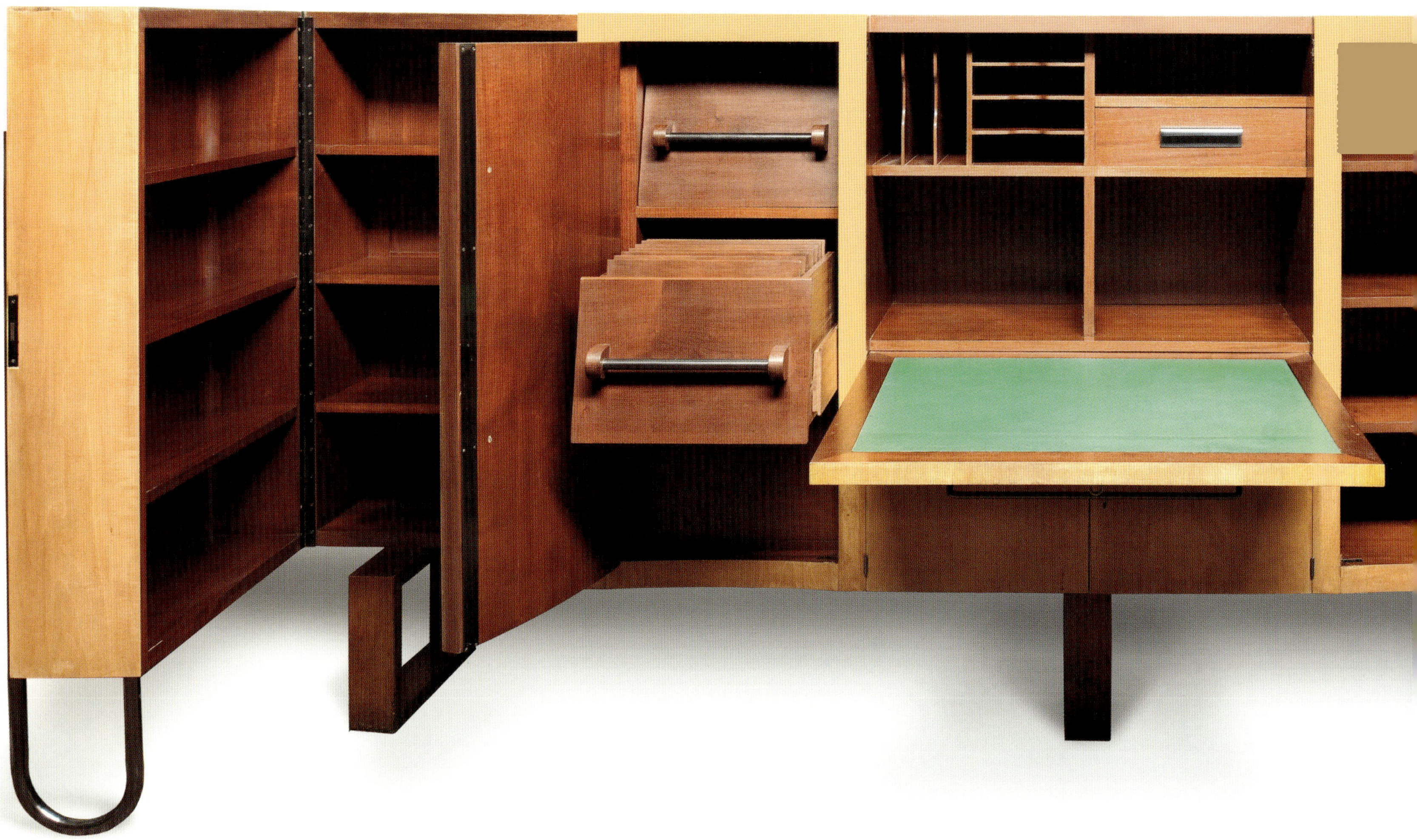

Furniture, for Herbst, was primarily functional rather than decorative, and many of his designs evolved from specific needs yet also spoke of his fascination with the possibilities offered by new materials such as tubular steel, rubber and Bakelite. Herbst noted the strong elasticity of the chest expanders and exercise systems used in gyms and began experimenting with these rubber straps as seating and back supports for his chairs.

In addition to the Chaise Sandows, Herbst developed a number of other designs using the elasticated strap seating attached to a steel frame, which together create a semi-transparent piece allowing the light to filter through the gaps between the horizontal supports. There was a high-backed Fauteuil de Repos of 1928 and a streamlined chaise longue from the same period, which could be coated in a seat pad made of leather or velvet.

Other Herbst pieces include a light armchair with an elegant combination of tubular steel and Bakelite, which was used to create the seat, back support and modest armrests (*c.* 1930); a set of dining chairs from 1943 also used Bakelite seats and backrests. Herbst went on to develop engaging combinations of industrial frames and more organic materials, as seen in his upholstered leather seats for armchairs, and also in a fireside chair of 1937 with a flat steel framework and a woven wicker seat and back. Many of Herbst's designs feel ahead of their time, reflecting his commitment to modernity and experimentation, combined with a sculptural approach to form.

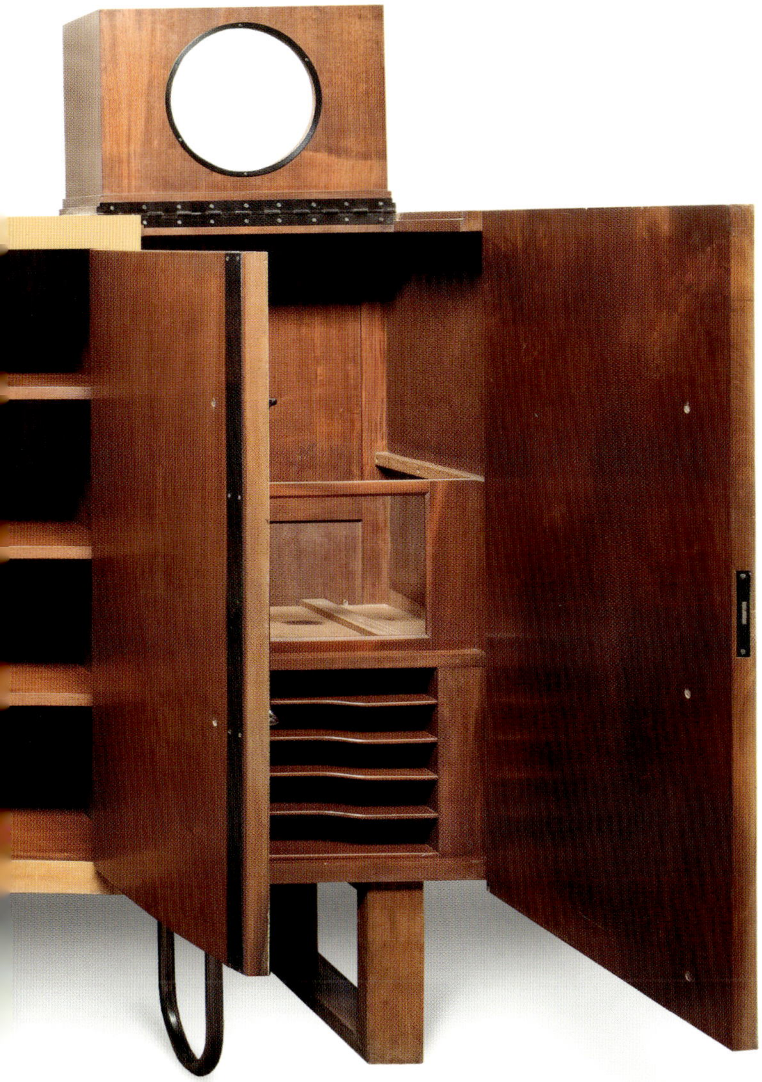

LEFT A multifunctional cabinet, 1937, in pear wood, mahogany and metal, with a leather interior.

BELOW LEFT A Herbst armchair, *c.* 1931, in chrome-plated steel with an upholstered leather seat and back.

BELOW RIGHT A vitrine, or display case, *c.* 1935, made by the Siegel Display Company, France, in nickel-plated brass, glass and maple.

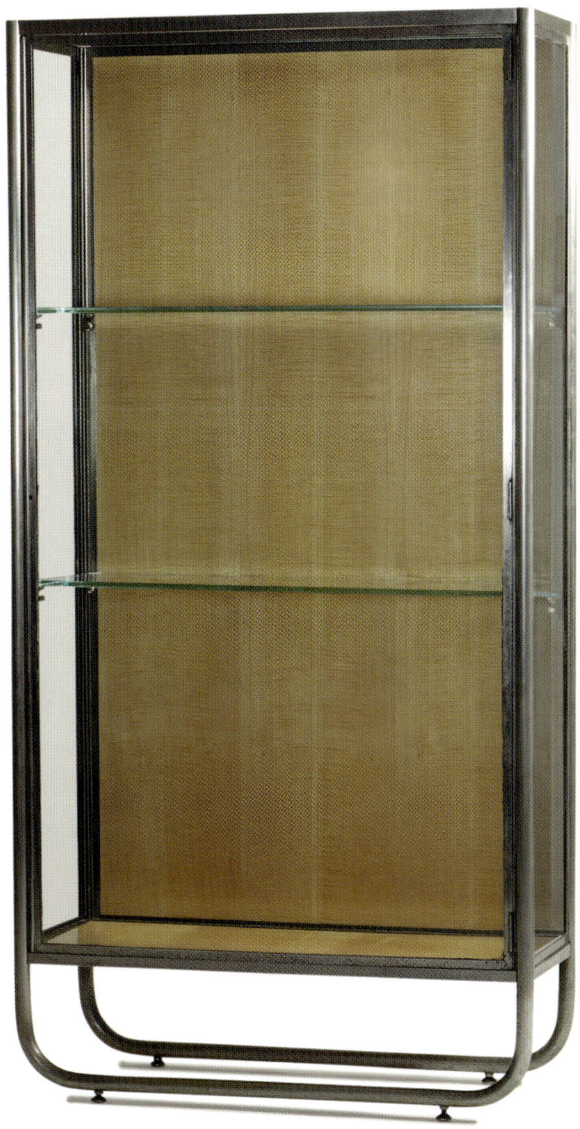

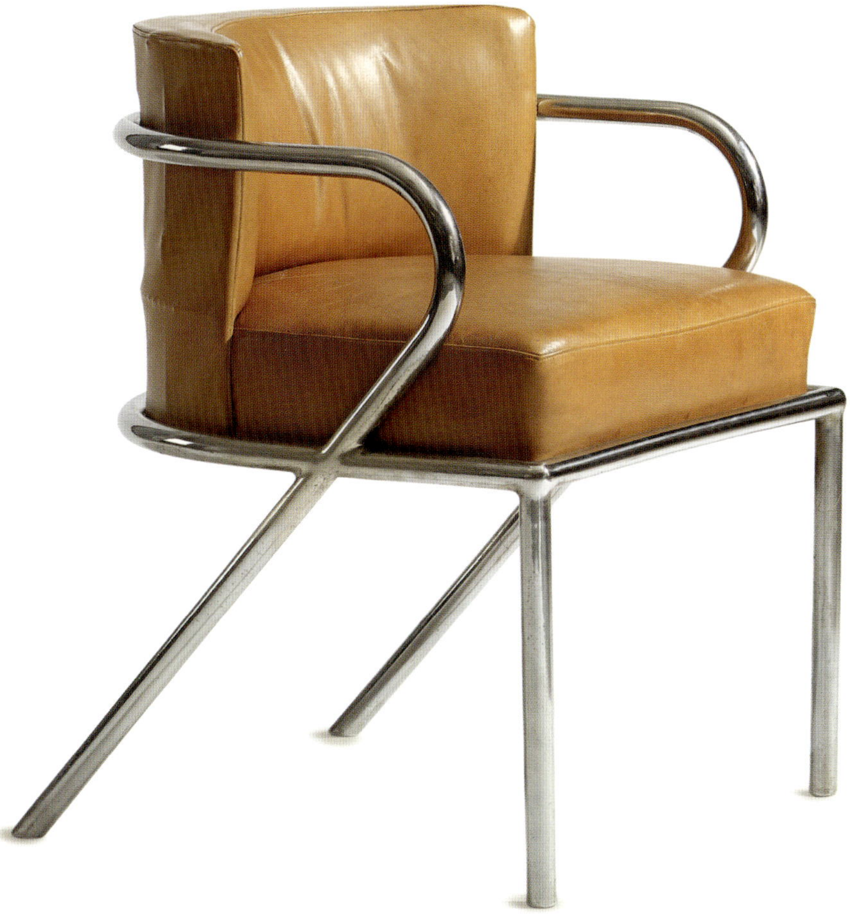

KAARE KLINT
(1888–1954)

The Danish designer and architect Kaare Klint saw modernism as part of a process of gradual evolution rather than a revolutionary break with the past. Many of his own designs were reinterpretations of familiar, generic pieces of furniture, such as the folding stool, safari chair and deck chair. He favoured natural materials rather than chrome or steel, and believed in an alliance of craft and functionality, spliced with a deep-rooted understanding of ergonomics and anthropometrics, having closely studied the relationship of the body and its dimensions to the practical necessities of furniture design. As a highly respected pioneer of soft Scandinavian modernism, Klint was lauded as the 'godfather' of a younger generation of mid-century Nordic designers, including Hans Wegner and Poul Kjærholm, along with his own students Børge Mogensen and Nanna Ditzel.

Klint grew up within a family of architects, designers and innovators. His father, Peder Vilhelm Jensen-Klint, was an architect, with whom he served an apprenticeship after initially studying painting in Frederiksberg. He went on to work with two architectural practices in Copenhagen before founding his own studio in 1920, focusing primarily on furniture design. His father began experimenting with pleated paper lampshades when Klint was a teenager and in 1943, Klint's brother, Tage Klint, established a family company making fresh versions of these pleated designs; Kaare Klint designed a number of Le Klint's most successful shades, including the LK 101 Fruit Lamp.

As an architect, Klint completed two churches in Copenhagen, largely based upon designs by his father, who died in 1930. Yet it was as a furniture designer that he secured his reputation. One of his earliest designs was the Faaborg chair of 1914, designed for Faaborg Museum on the island of Funen by architect Carl Petersen, with whom Klint worked for three years from 1914 to 1917. With its elegant timber frame and rounded back, along with a canework seat, the piece represents a transitional design in the journey from neoclassicism to functional modernism.

However, Klint's most famous designs – until recently still produced by Rud Rasmussen – date from the Thirties. His Safari chair (1933) was one of a series of modernist updates of generic classics. In the late Twenties, Klint had become fascinated by the African safari adventures of the American photographers and filmmakers Martin and Osa Johnson, noting their safari tent complete with classic English safari or camping chairs, which could be easily dismantled and transported from place to place. Klint created

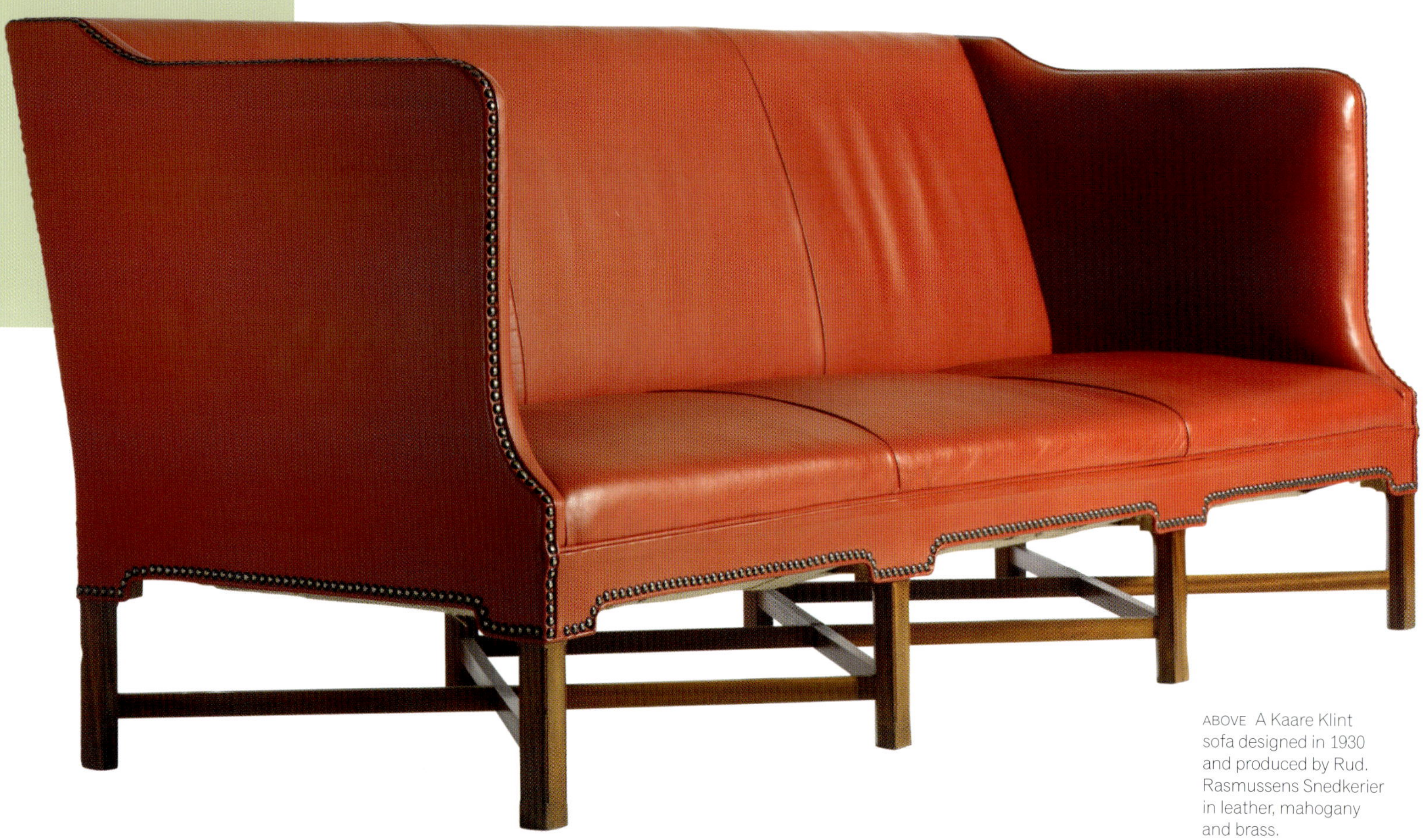

ABOVE A Kaare Klint sofa designed in 1930 and produced by Rud. Rasmussens Snedkerier in leather, mahogany and brass.

RIGHT Model 4488 lounge chairs, made by Rud. Rasmussens Snedkerier, 1932, in bleached mahogany, rosewood, cane and leather.

BELOW A pair of Model 4396 lounge chairs, manufactured by Rud. Rasmussens Snedkerier, 1930, in Cuban mahogany, leather and brass.

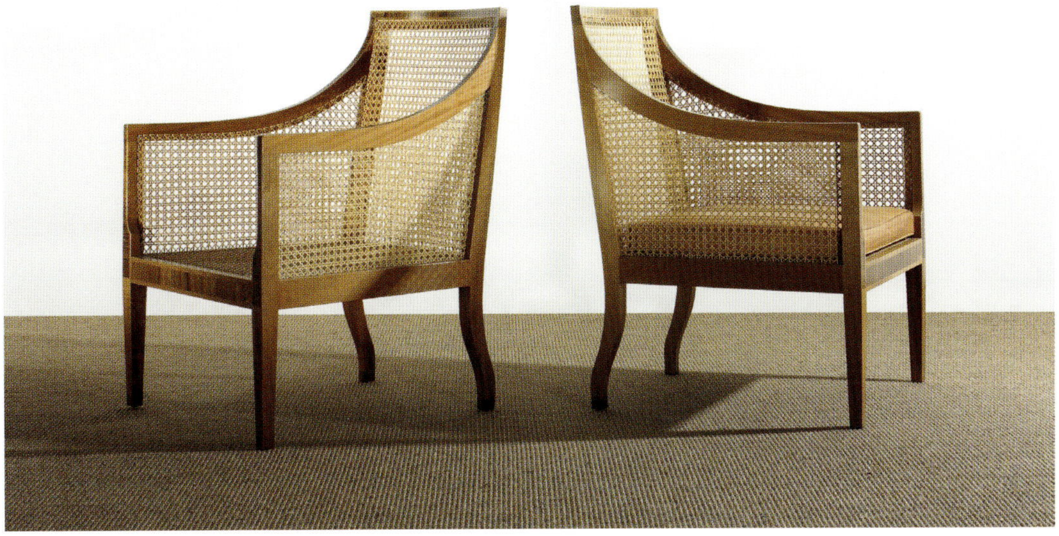

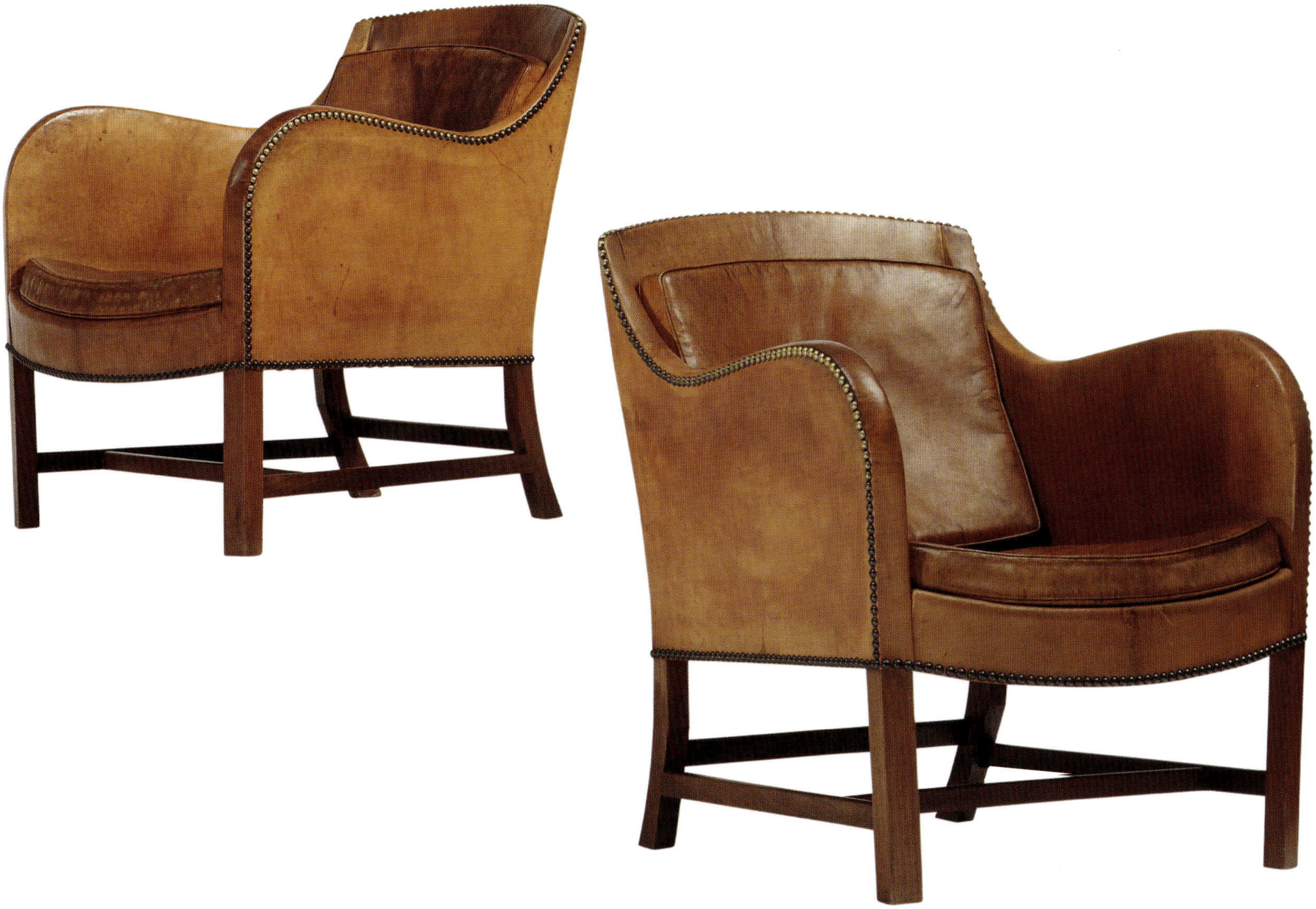

a new version of the chair using a lightweight ash frame, a leather seat and a pivoting, adjustable back; leather straps and armrests lend the chair additional strength and stability. Like the originals, Klint's chair could be used inside or outside and could easily be assembled or disassembled.

A similar thought process lay behind Klint's Propeller stool, first designed in 1930, although not put into production until the Sixties. This was a remodelling of a simple folding camping stool with a light oak and ash scissor frame supporting a canvas seat. The innovation lay in the 'propeller' legs, where the wood for each pair was cut from a piece of ash or oak with a twist that allows the legs to interlock neatly and completely when folded, creating a slim, portable and beautifully crafted solution. Klint's Model 4699 deck chair of 1933, again, gave new life, relevance and vitality to an old friend. A number of Klint's other furniture designs from the period were reworkings of familiar European staples, such as his English chair of 1931, his Wing chair of 1941 and his Greek sofa of 1941.

Klint's work – which also included cabinets, dining tables and occasional tables – was shown at international design fairs in Spain and Paris, including his Barcelona or Red chair shown at the Exposición Internacional de Barcelona in 1929. As an educator, Klint founded the furniture school at the Royal Danish Academy of Fine Arts in Copenhagen in 1924, and then in 1944 became professor of architecture.

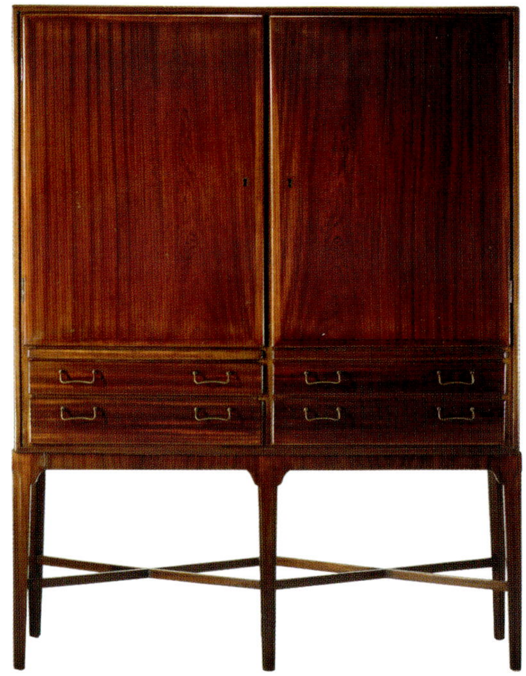

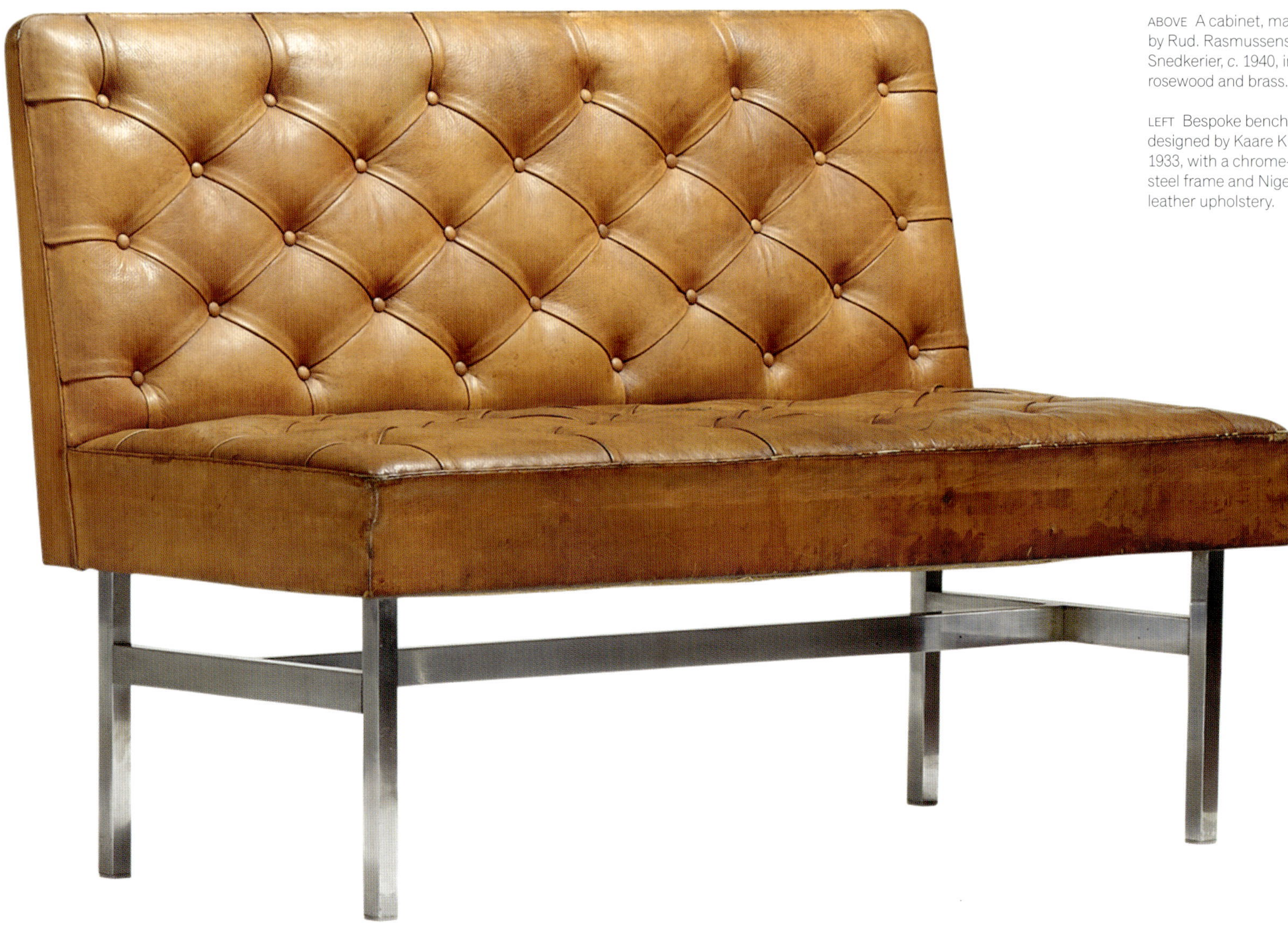

ABOVE A cabinet, made by Rud. Rasmussens Snedkerier, c. 1940, in rosewood and brass.

LEFT Bespoke bench, designed by Kaare Klint, 1933, with a chrome-plated steel frame and Niger leather upholstery.

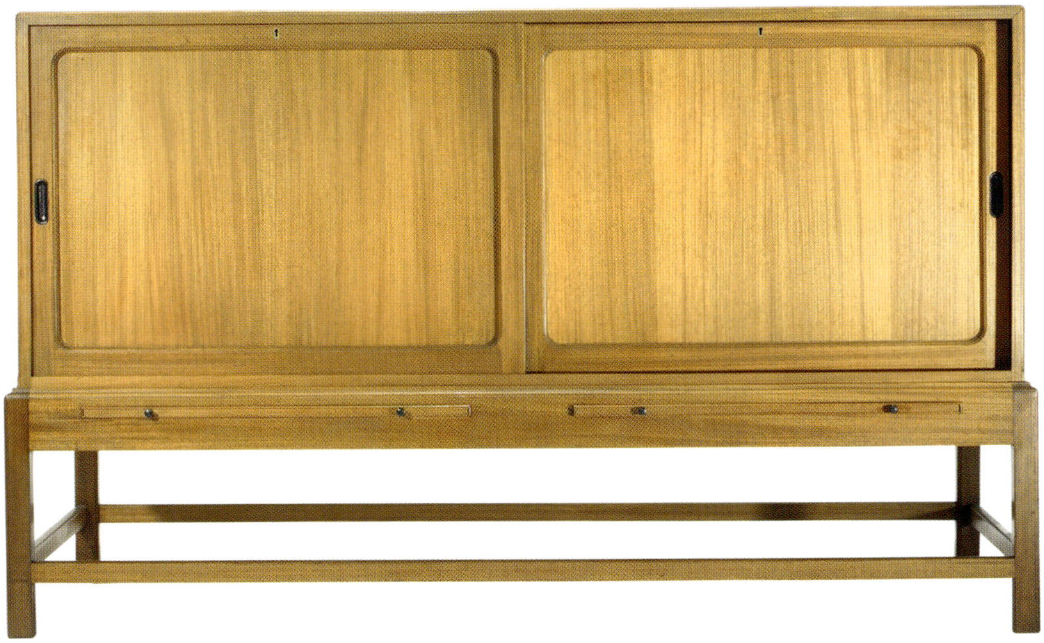

ABOVE A cabinet, made by Rud. Rasmussens Snedkerier, 1930, in ribbon mahogany, brass and ebony; the cabinet features sliding doors and drawers.

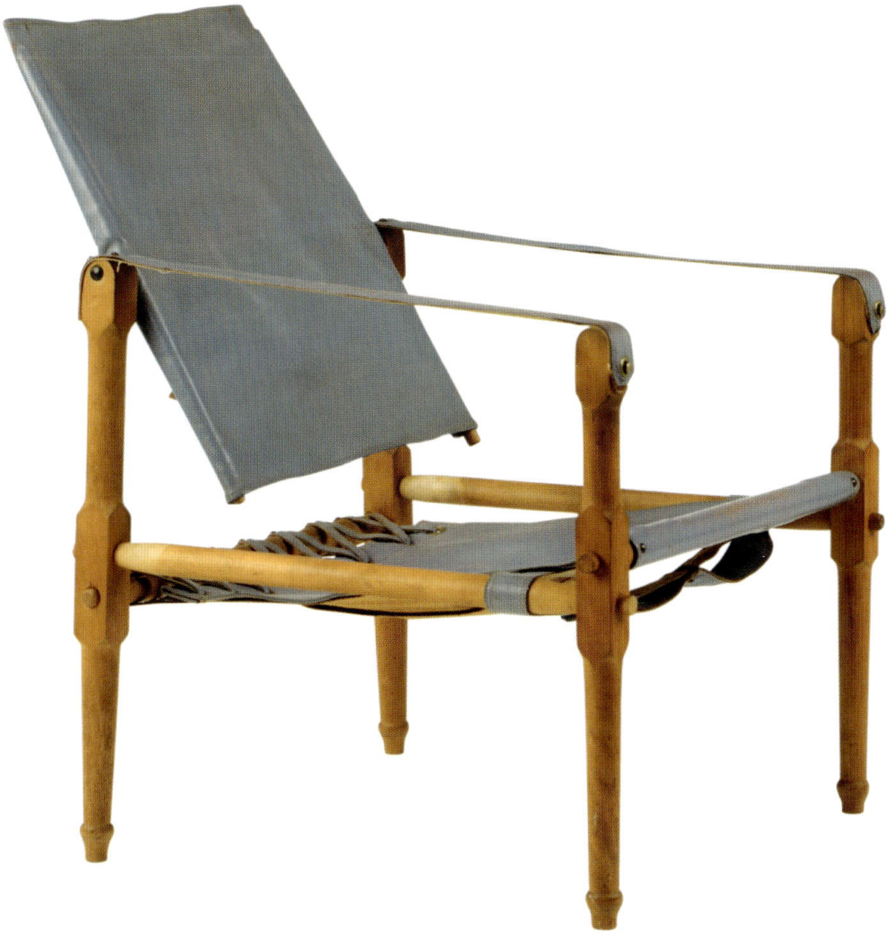

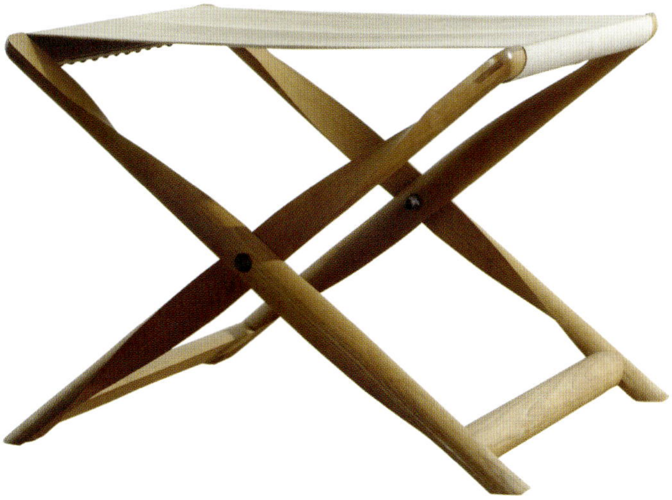

ABOVE Klint's version of the Safari chair, made by Rud. Rasmussens Snedkerier, 1933, in maple wood and leather.

RIGHT The Propeller stool, designed in 1930, manufactured by Rud. Rasmussens Snedkerier, with a folding oak frame and linen canvas seat.

MOGENS KOCH
(1898–1992)

The career of the Danish designer Mogens Koch partly mirrored that of his mentor Kaare Klint. Like Klint, Koch fused a strong foundation in historical precedent and period influences with a determination to question accepted wisdom and familiar modes of design and production. There was a love of organic, crafted materials combined with a particular talent for ingenious, original solutions and fresh ways of thinking. This embraced concepts of modularity, in particular, with Koch's bookcase system of 1932 a notable case in point.

The timber bookcases were made in units that could be easily combined and configured in a variety of different combinations, allowing the user to adapt them according to need and the dimensions of the space available. Koch played a key role in developing the notion of a modular furniture system, which was embraced by many designers in the post-war period.

Like many Scandinavian designers, Koch initially trained in architecture. Born in Copenhagen, he studied at the Royal Danish Academy of Fine Arts before travelling widely throughout Europe and the Americas. He became an assistant in Kaare Klint's studio in the late Twenties and went on to establish his own design practice in Copenhagen in 1934. Koch designed textiles and silverware, as well as light shades for Le Klint, the pleated paper lightshade business owned by Kaare Klint's family. But he is best known for his furniture designs, many of which are still produced – along with much of Klint's portfolio – by Rud Rasmussen in Denmark.

In 1932, Koch became intrigued by the notion of reinventing the campaign chair and created his folding MK99200 Safari or Folding chair. It was a very different solution from the rather more complex version invented by Klint just a year later (see p. 61) and more reminiscent of a Hollywood director's chair. Made with a beech frame with pivoting scissor legs, a simple canvas seat and back, and leather straps serving as armrests, the chair was elegantly made, light and comfortable. Importantly, the entire chair could be folded flat in an instant, making it highly portable and easy to store. Koch also designed a complementary stool and storage rack, which could accommodate a whole set of the chairs.

The Folding chair was originally designed for a competition for light and affordable church furniture, but was seen as too complex for large-scale production until 1960. Around this time Koch designed a handsome folding table to add to his flexible collection, suited to indoor or outdoor use.

The MK40880 bookcase also evolved from a specific set of circumstances, having originally been designed for the interiors of Koch's own home in Copenhagen in 1928. The designer wanted a flexible storage system so sketched a modular design, which was prototyped in 1930 and put into production by Rud Rasmussen in 1932. The modular units were made of thin cuts of traditional timbers, including pine, teak and mahogany, solidly jointed for stability. The rectangular units could be turned to accommodate different-sized books or magazines and they could be stacked in a variety of combinations; cupboard doors also feature on some versions to create a sealed cabinet. The notion of a modular system is a familiar design staple today, but at the time it was highly original, creating choice for the consumer on the one hand, and offering adaptability on the other, which made the bookcases suitable for spaces of any dimension.

Some of Koch's other designs were reinterpretations of classics, such as his MK50 Wing Back chair (1936), which featured a matching ottoman. But other pieces from the Thirties were true originals, full of thought and delight, such as his 1938 armchair in stained oak with a woven leather seat, echoing the webbed designs of chairs by Alvar Aalto, Bruno Mathsson and others.

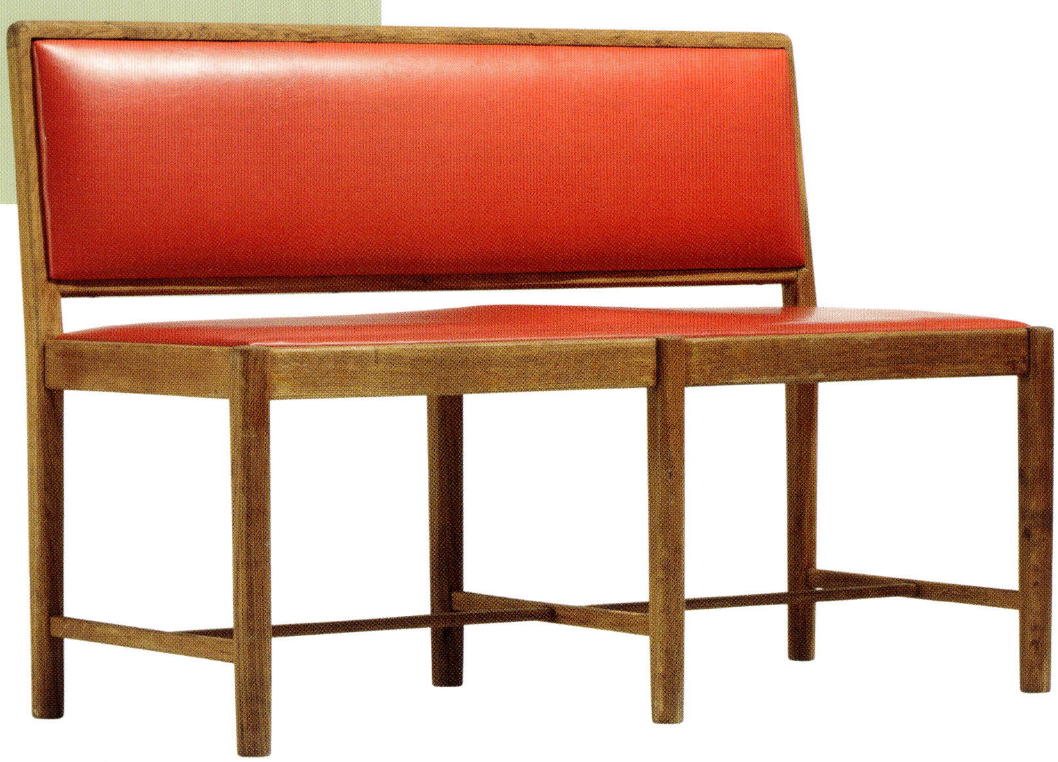

LEFT A Mogens Koch bench designed for the Sønderborg Hospital in Denmark, 1936, in stained oak with an upholstered seat and backrest.

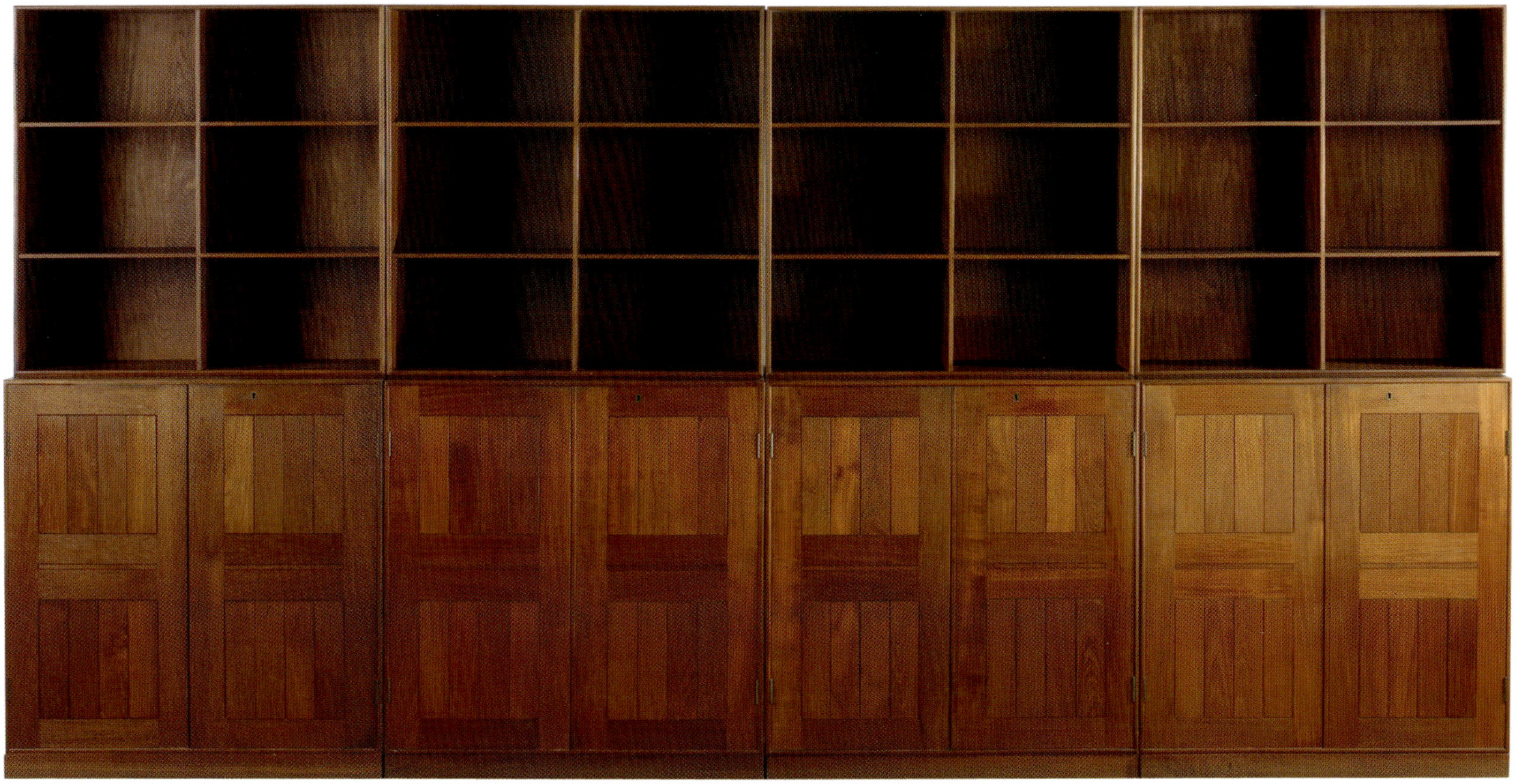

ABOVE A substantial shelving and storage system consisting of four bookcases and base cabinets, made by Rud. Rasmussens Snedkerier, Copenhagen, in teak.

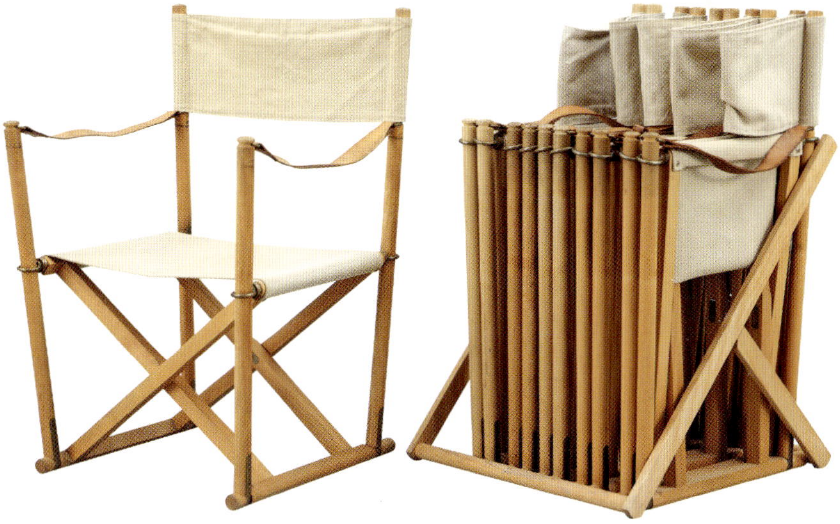

ABOVE The MK99200 Safari chair, Mogens Koch's version of a modern and lightweight folding chair, designed *c.* 1932.

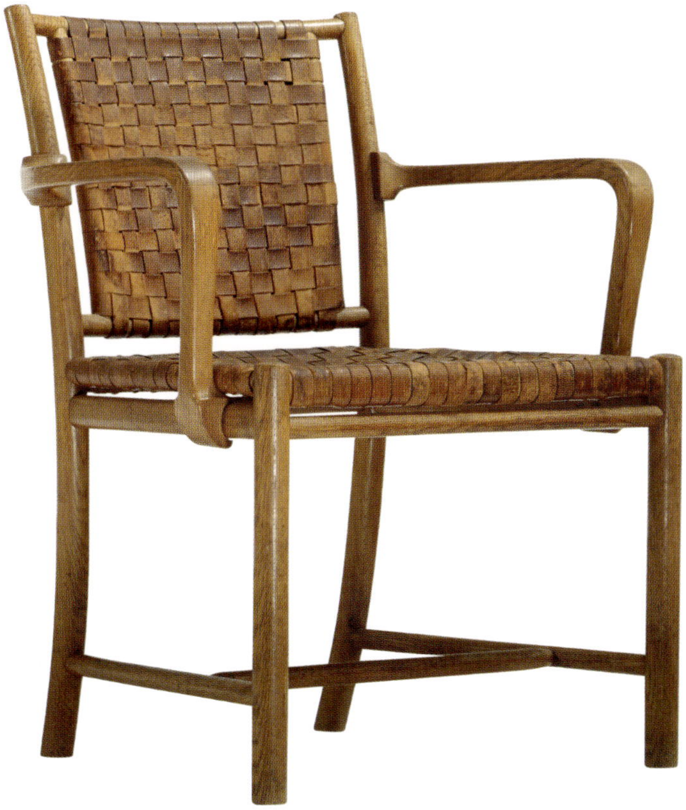

LEFT An armchair made by N. C. Jensen Kjær, *c.* 1938, in stained oak and leather.

LE CORBUSIER
(1887–1965)

PIERRE JEANNERET
(1896–1967)

CHARLOTTE PERRIAND
(1903–1999)

The master's reach was always long and his eye all-seeing. Le Corbusier (Charles-Édouard Jeanneret), the godfather of modernism, involved himself not just in architecture, but in so many different strands of art and design. There was painting, furniture, interiors and the written word, his books forming the closest thing we have to a modernist manifesto. No wonder, with a reputation that burned so brightly, his colleagues lived and worked in his shadow.

Le Corbusier's closest collaborator throughout much of his career was his cousin, Pierre Jeanneret, who formed a key part of the atelier and made a significant contribution to many of its major works, including the seminal Villa Savoye of 1931, on the outskirts of Paris (see p. 366). In an interview from the Eighties, Charlotte Perriand, who worked closely with both Le Corbusier and Jeanneret, summed up the relationship between the two men: 'It wasn't just the atelier ... nor was it only Le Corbusier – it was Le Corbusier and Pierre Jeanneret, and that is crucial. Corbu was the symbol, he had his ideology, he acted as a catalyst – but Pierre Jeanneret spent all his time at his drawing board, from morning to night. He drew out everything very precisely, in great detail – he drew like Aalto. So there were the two of them: they were complementary. Corbu was the publicist, of course, but Jeanneret was his shadow.'[12]

There is no doubt that Le Corbusier relied upon his cousin and trusted his judgment. When Perriand came to see Le Corbusier at the age of just 24, having been inspired by his writings, the master was dismissive and sent her away with the famous riposte: 'We don't embroider cushions here.' It was Jeanneret who, a few months later, took the master to see Perriand's accomplished interiors for a rooftop bar at the Salon d'Automne of 1927. Le Corbusier, for once, was humbled and apologized to Perriand, inviting her to join the atelier.

Perriand, who had studied design at the École de l'Union Centrale des Arts Décoratifs in Paris and embraced the challenge to 'make it new', was put to work straight away on the design of furniture and interiors, working alongside Le Corbusier and Jeanneret. Le Corbusier had tended to favour simple furniture by Thonet* and others within spare interiors, where the emphasis was upon architectural purity and integrated elements. He tasked Perriand with the challenge of coming up with a small furniture collection of simple, modern, functional, ergonomic pieces: a recliner, an armchair, a swivel chair, an easy chair and a table. He sketched out some ideas, chiefly concerned with posture rather than detail.

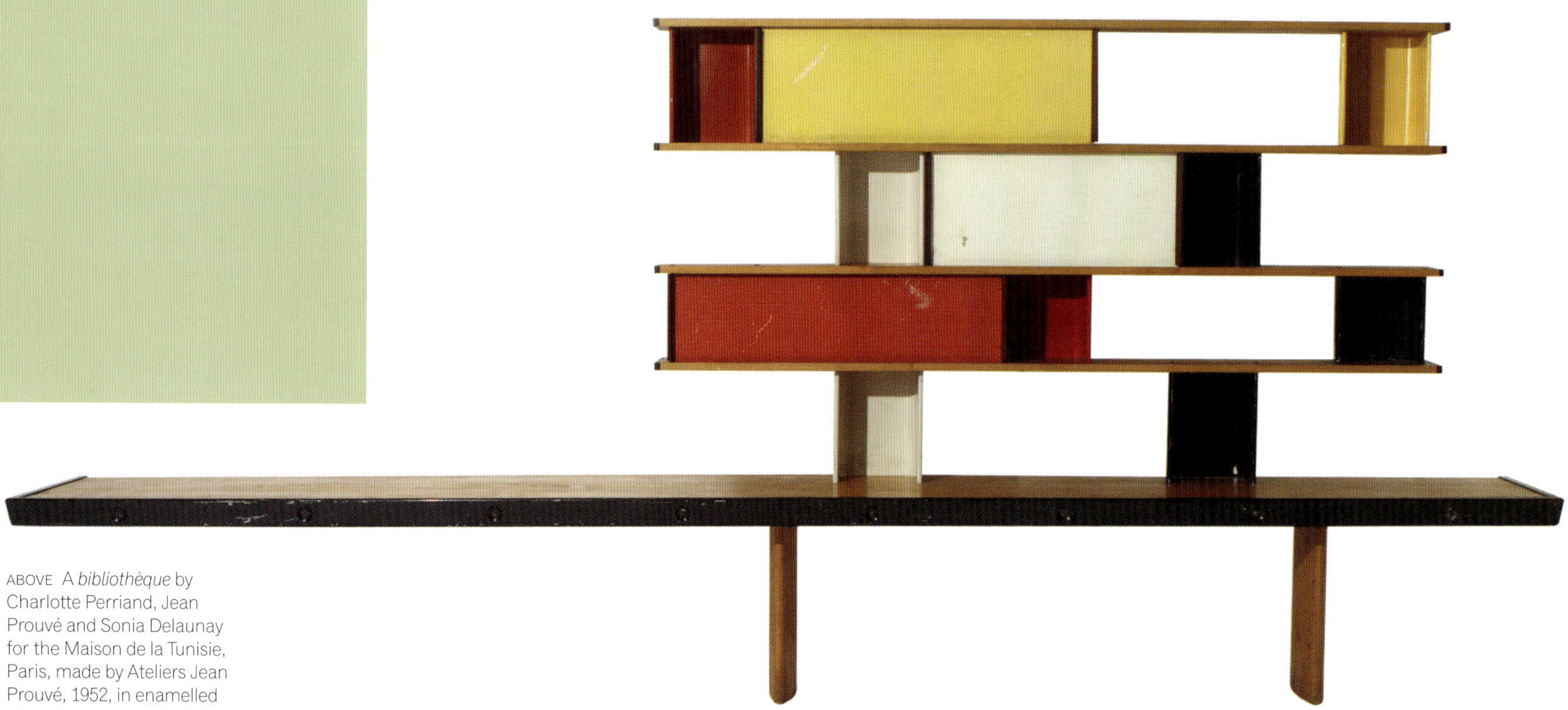

ABOVE A *bibliothèque* by Charlotte Perriand, Jean Prouvé and Sonia Delaunay for the Maison de la Tunisie, Paris, made by Ateliers Jean Prouvé, 1952, in enamelled aluminium and steel, pine and mahogany.

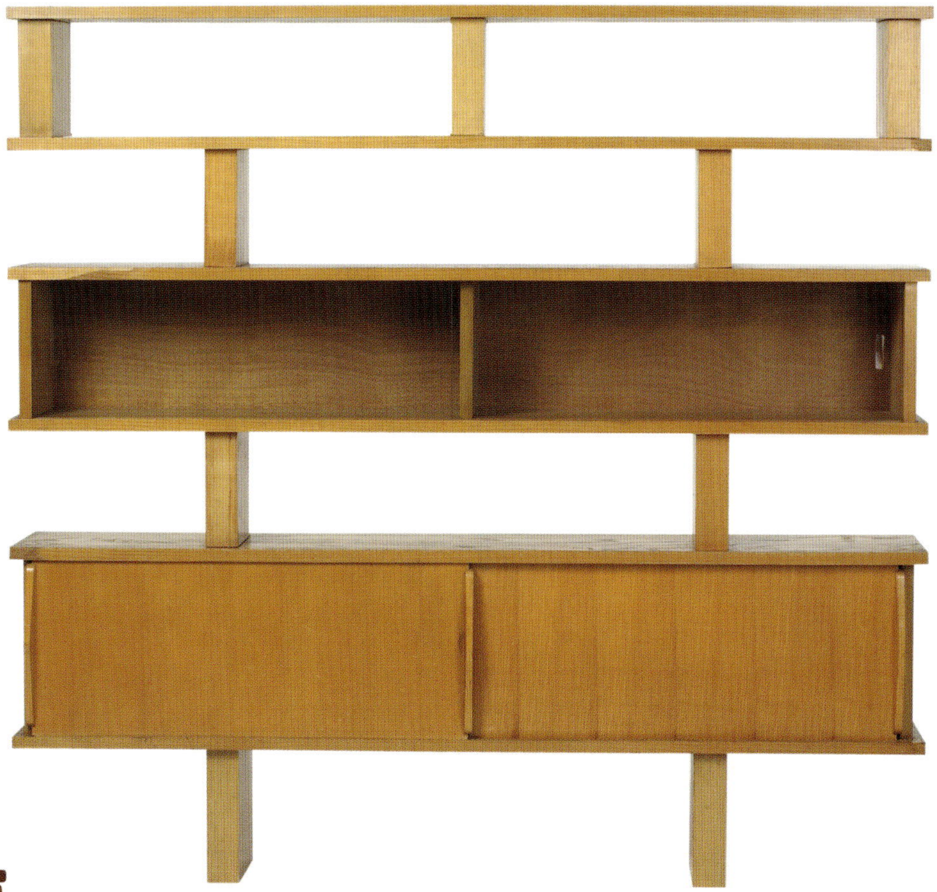

RIGHT *Bibliothèque* designed by Perriand and Jeanneret, made by L'Equipement de la Maison, *c.* 1948, in oak and glass.

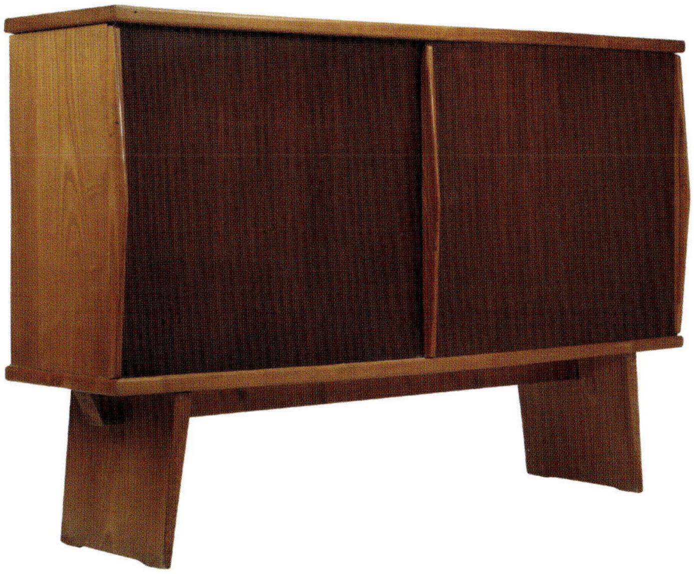

ABOVE Perriand and Jeanneret's Bahut No. Two, produced by Georges Blanchon for L'Equipement de la Maison, 1945, in oak, mahogany and pine.

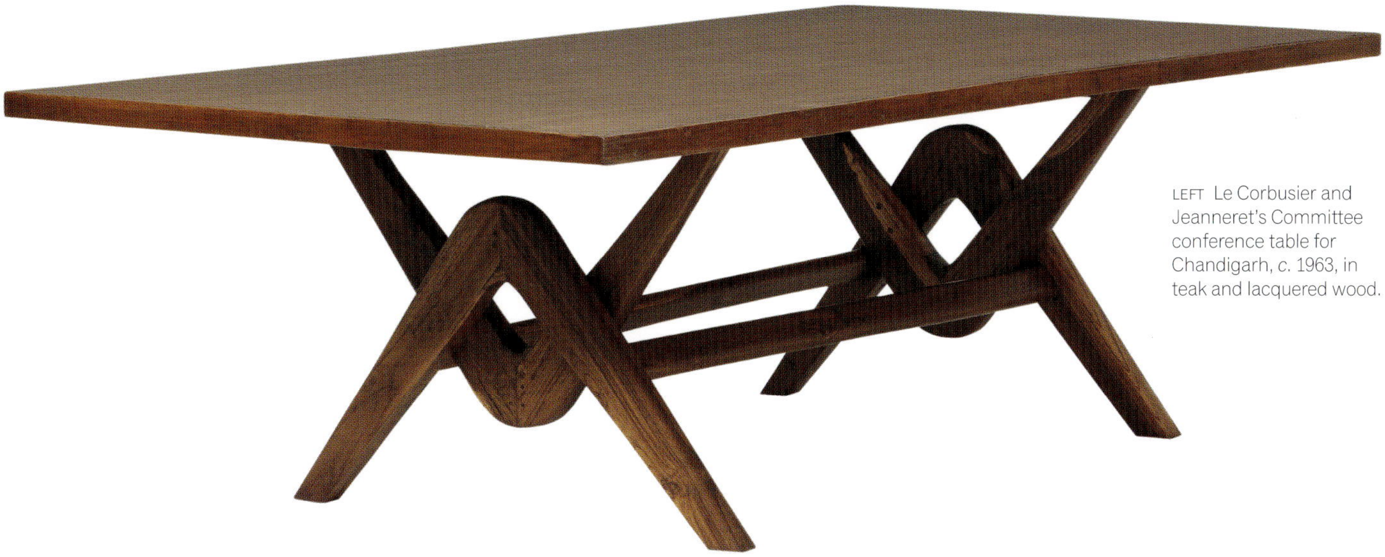

LEFT Le Corbusier and Jeanneret's Committee conference table for Chandigarh, *c.* 1963, in teak and lacquered wood.

The pieces that Perriand produced, with Jeanneret's support, have become familiar staples of 20th- and 21st-century interiors, as well as icons of modernist design. Each one represented a process of re-evaluating a standard typology, deconstructing its purpose, form and function, and then recreating it with new materials, aesthetics and ergonomic practicality. It was a purposeful, thoughtful and methodological approach with extraordinary results.

The reclining chaise longue, known as the B306 or LC4 (1928), marked the remodelling of a Twenties-Thirties favourite. It came in two parts, with a steel-framed seat – echoing the shape of the resting body – floating upon a separate metal base; the floating seat could be easily adjusted by sliding it up or down upon the H-shaped supports at either end of the base. A padded seat in leather or cow hide reinforced the feeling of comfort, with an additional support cushion for the head, while the organic quality of the seat contrasts with the modern character of the tubular framework and the steel base unit.

The B301/LC1 Basculant chair was – more or less – a reinterpretation of the generic safari chair: a lightweight piece with a slender frame and a suspended seat. Rather than a timber frame, Perriand turned again to tubular steel, with a leather/calfskin seat and back floating upon a minimal, sculptural support. The backrest pivots upon the frame, allowing the user to adjust position with ease, while the armrests are made of leather bands, which mould themselves to the elbows and arms.

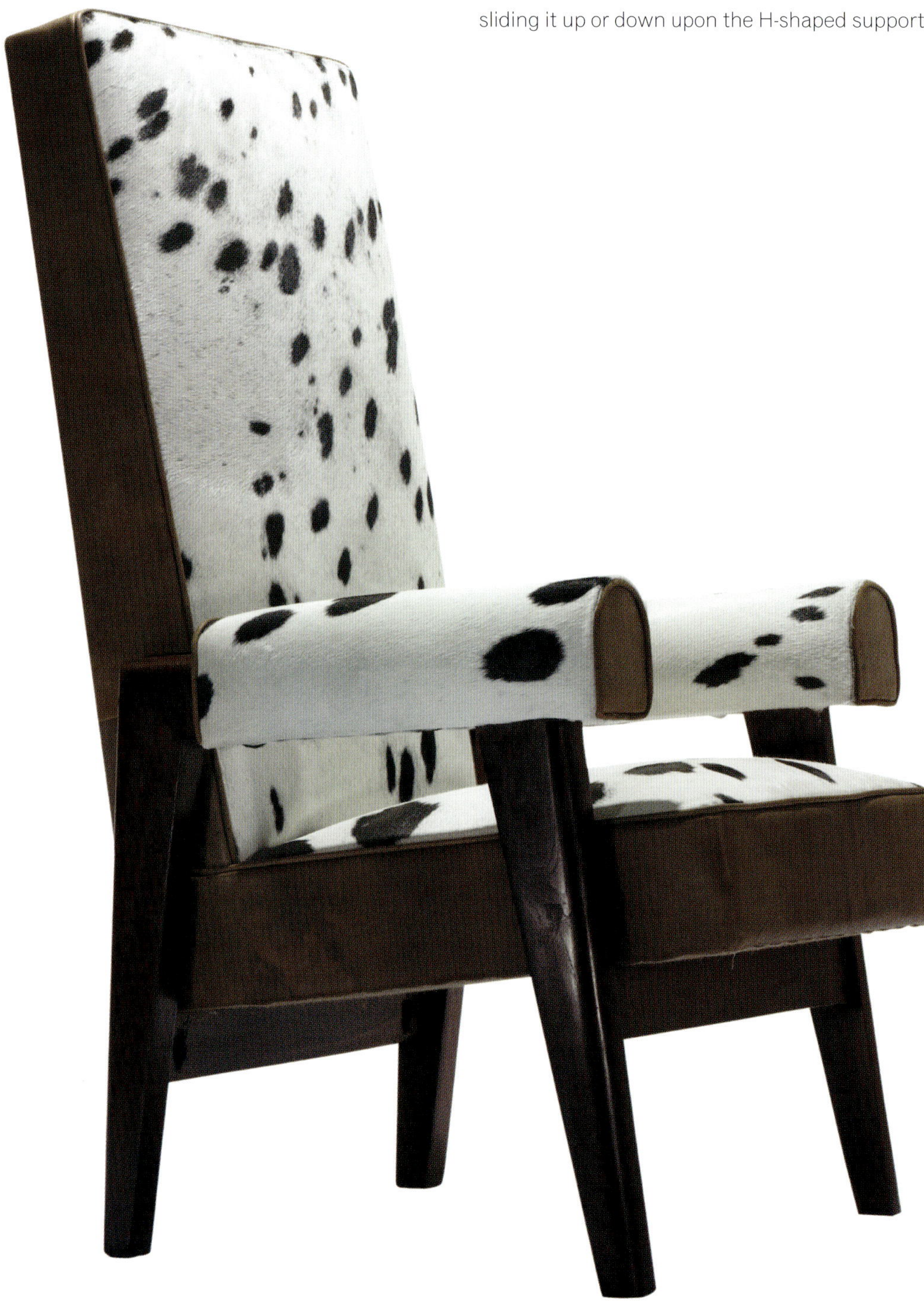

ABOVE Le Corbusier and Jeanneret's Judge's armchair from the High Court at Chandigarh, c. 1955, in teak, steel, leather and horse hide.

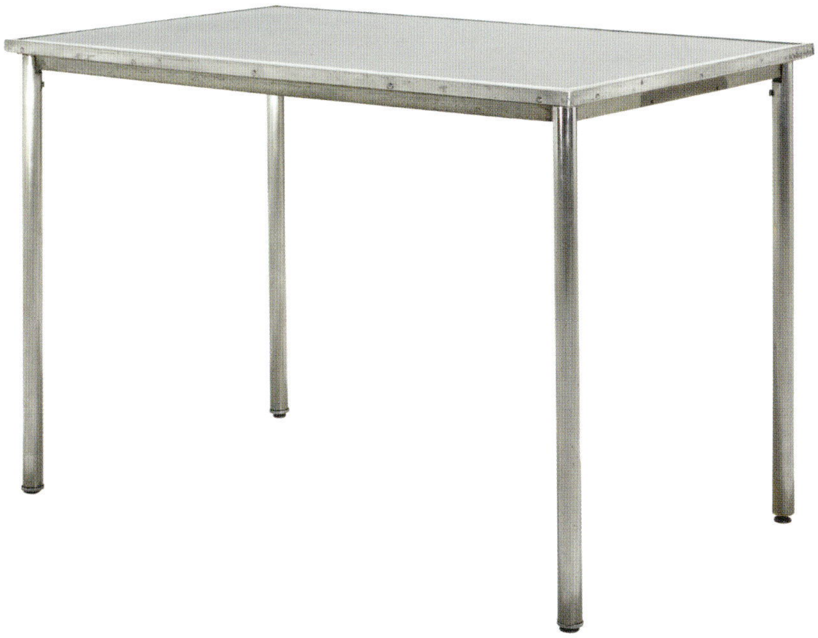

LEFT The B307 dining table by Jeanneret, Perriand and Le Corbusier, manufactured by Thonet, *c.* 1930, in aluminium, steel and reverse-painted glass.

BELOW An early example of Jeanneret, Perriand and Le Corbusier's Basculant chair, manufactured by Thonet, *c.* 1935, in chrome-plated tubular steel, leather and pony hide.

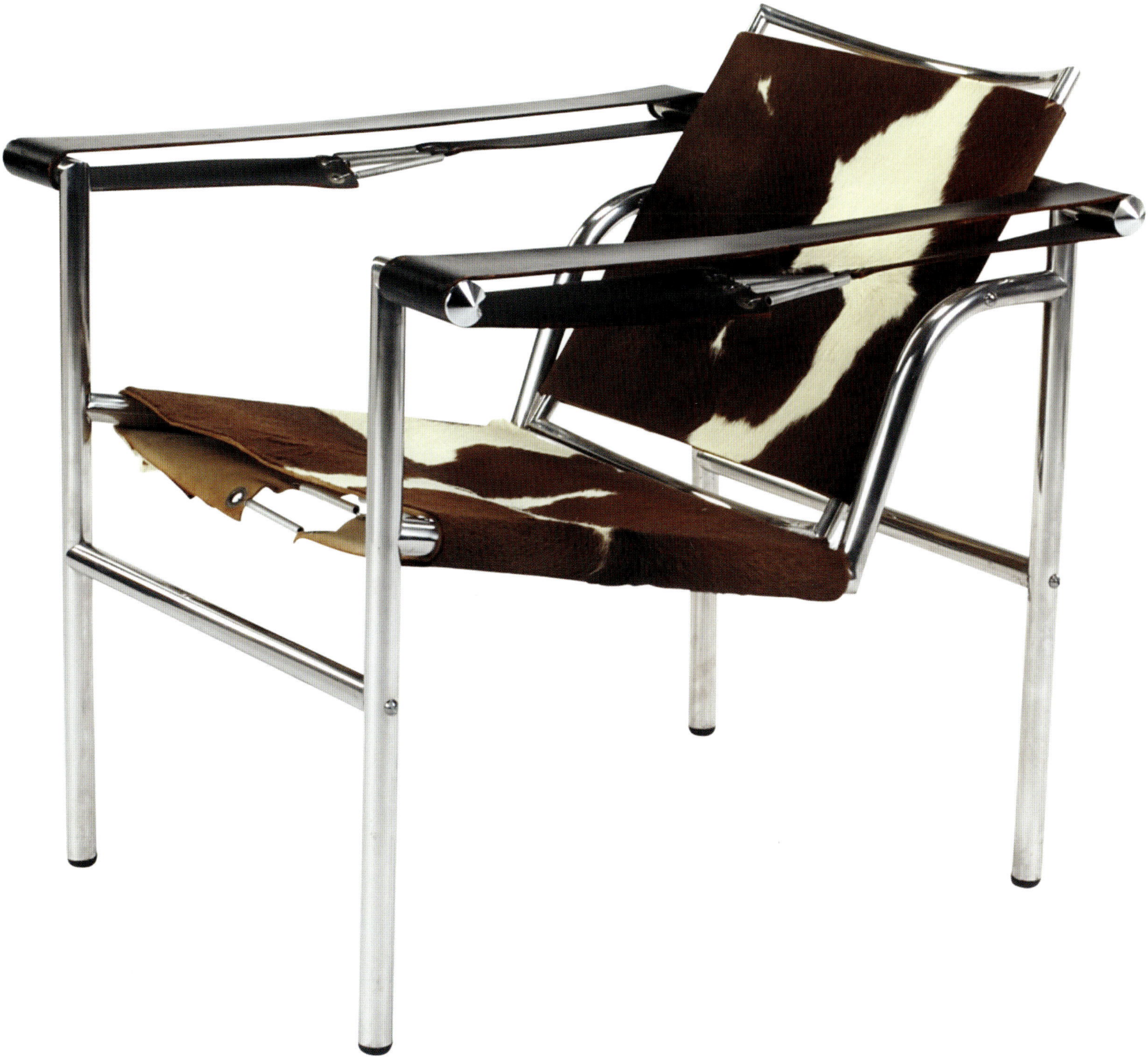

Other pieces in the collection included the B302/LC7 Revolving armchair, the LC6 table with a metal base and glass top, and the LC2 Grand Confort armchair. The LC2 was the trio's reimagining of the easy or club chair, which remains a perennial favourite, largely because it places the emphasis upon comfort as well as style and function. The framework is also in lightweight tubular steel, forming a kind of elegant cage for chunky leather cushions that form the seat, sides and back, enveloping the user; in later years the design was adapted to create a complementary sofa.

The pieces were prototyped and used within the interiors of two villas (Maison La Roche and Villa Church) before being shown at the Salon d'Automne of 1929. Thonet produced the first production runs of the series in the Twenties, and then in the Sixties Cassina reissued the collection to great success. The pieces have often been attributed to Le Corbusier alone, yet the authorship of Jeanneret and Perriand is beyond dispute, with Perriand – who became increasingly interested in more natural materials as her career progressed – establishing the intriguing contrast between the industrial and the organic within the LC series.

In 1937, Perriand left Corbusier's atelier and worked with the artist Fernand Léger* and – more significantly – began a collaboration with JEAN PROUVÉ that continued in the post-war period. Pierre Jeanneret, 'the shadow', continued working with his cousin but it was only with their momentous Chandigarh development in India – for which Jeanneret became chief architect – that he really established his own reputation.

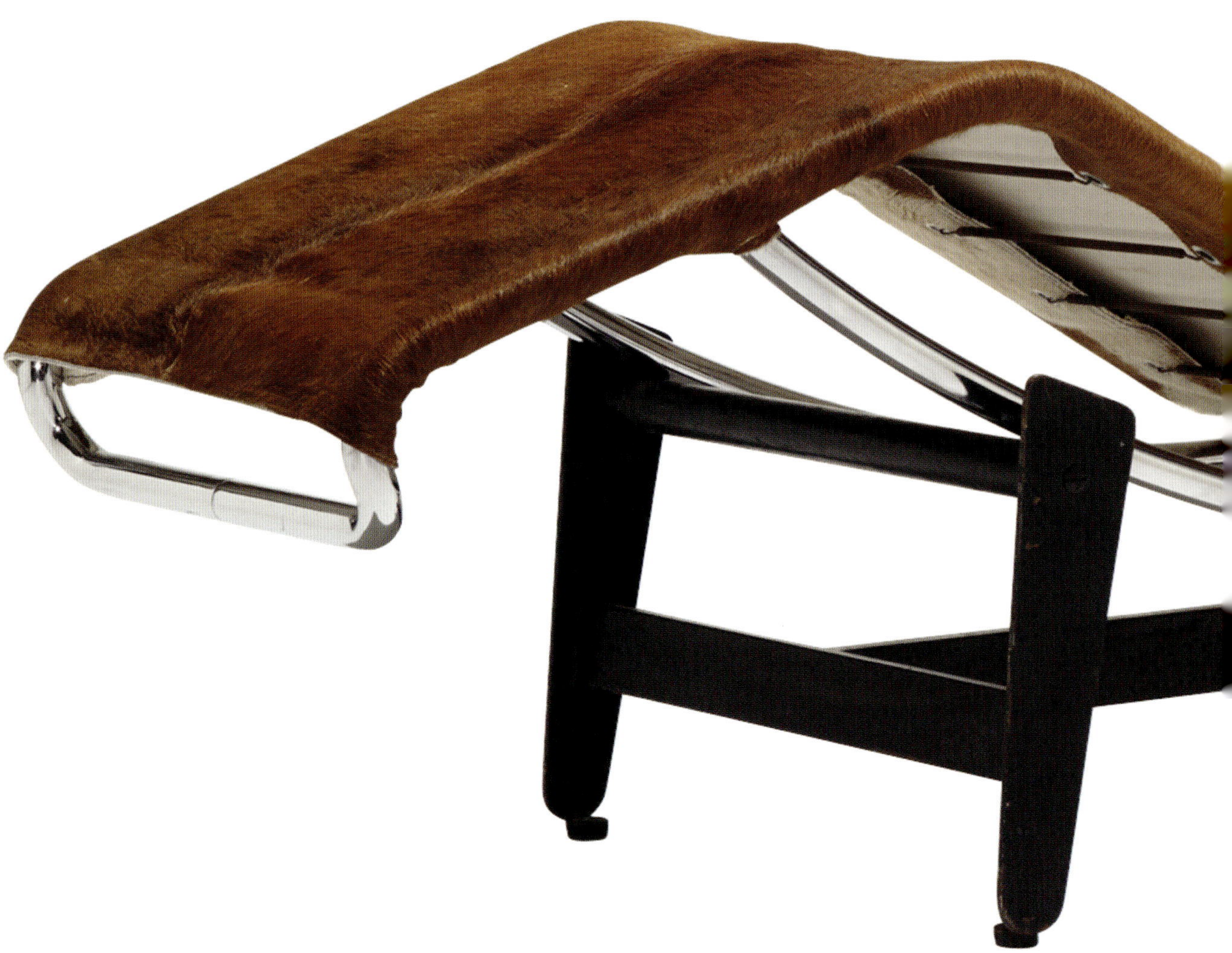

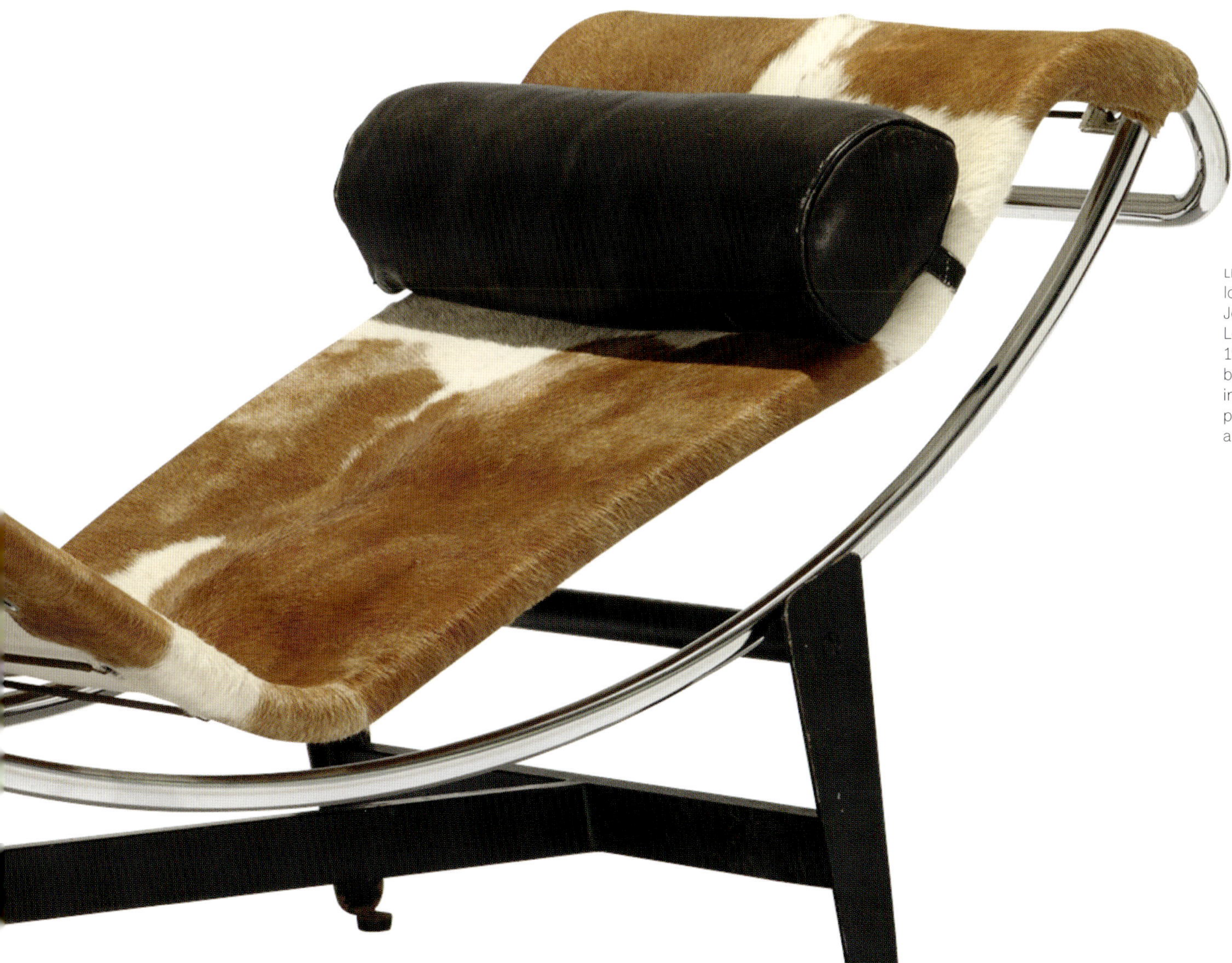

LEFT The iconic LC4 chaise longue, designed by Jeanneret, Perriand and Le Corbusier, designed in 1928; this version was made by Wohnbedarf *c.* 1955, in enamelled and chrome-plated steel, leather and hide.

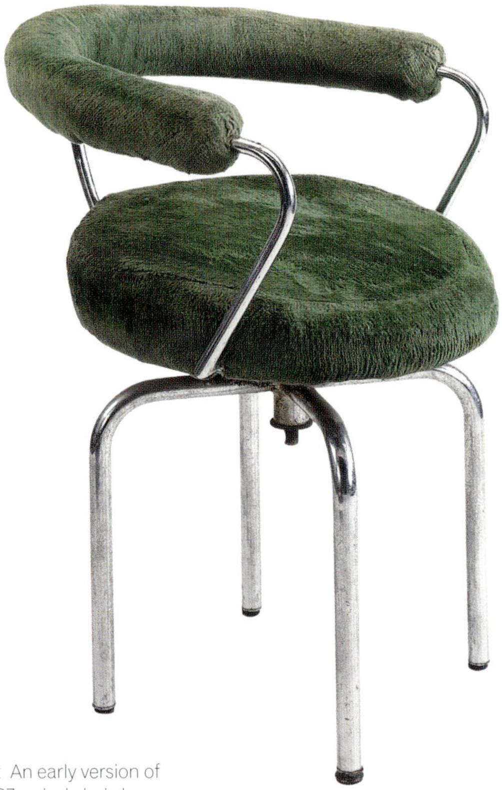

ABOVE An early version of the LC7 swivel chair, by Jeanneret, Perriand and Le Corbusier, manufactured by Thonet, 1928, and exhibited at the Salon d'Automne, Paris, in 1929.

BELOW A lounge chair designed by Perriand and Jeanneret, *c.* 1937, in oak and bamboo; the chair was one of a collection of pieces designed for the Union des Artistes Modernes exhibition of 1937.

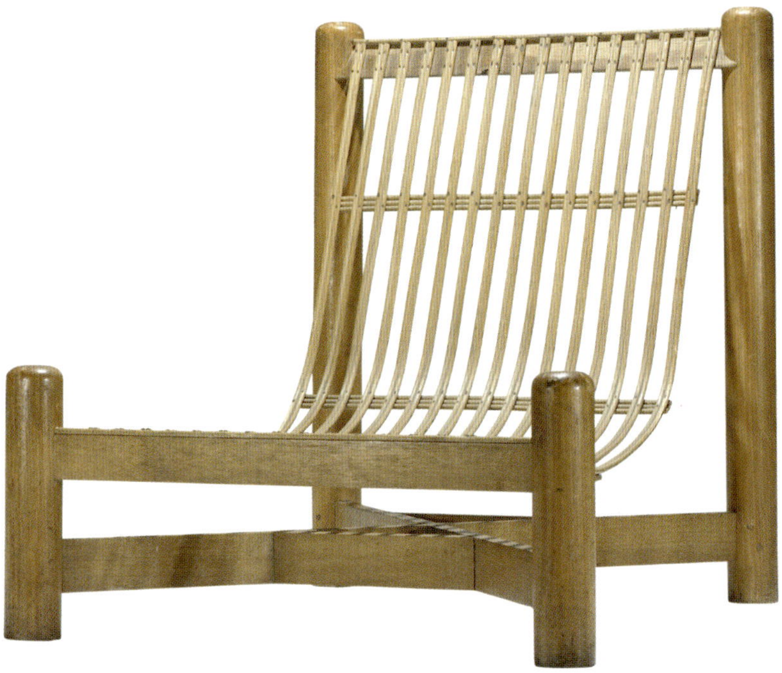

BELOW A Charlotte Perriand daybed, produced by Editions Steph Simon, 1959, with an oak frame and upholstered mattress and cushion.

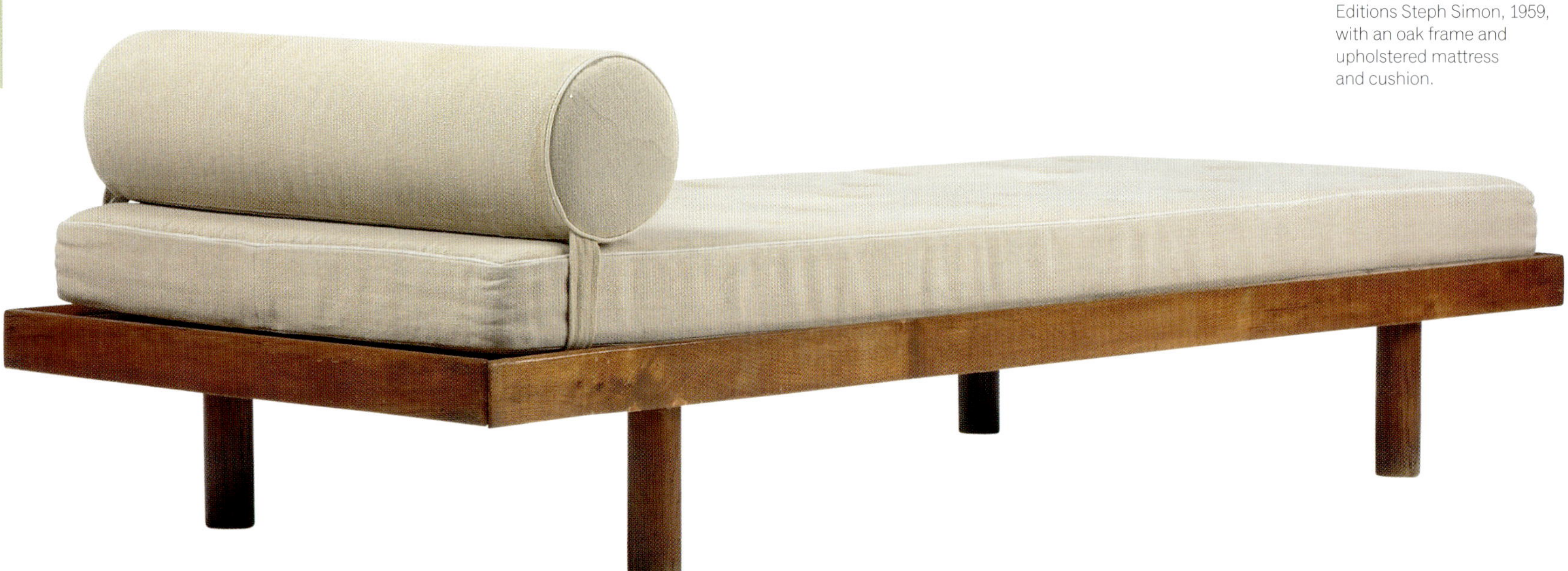

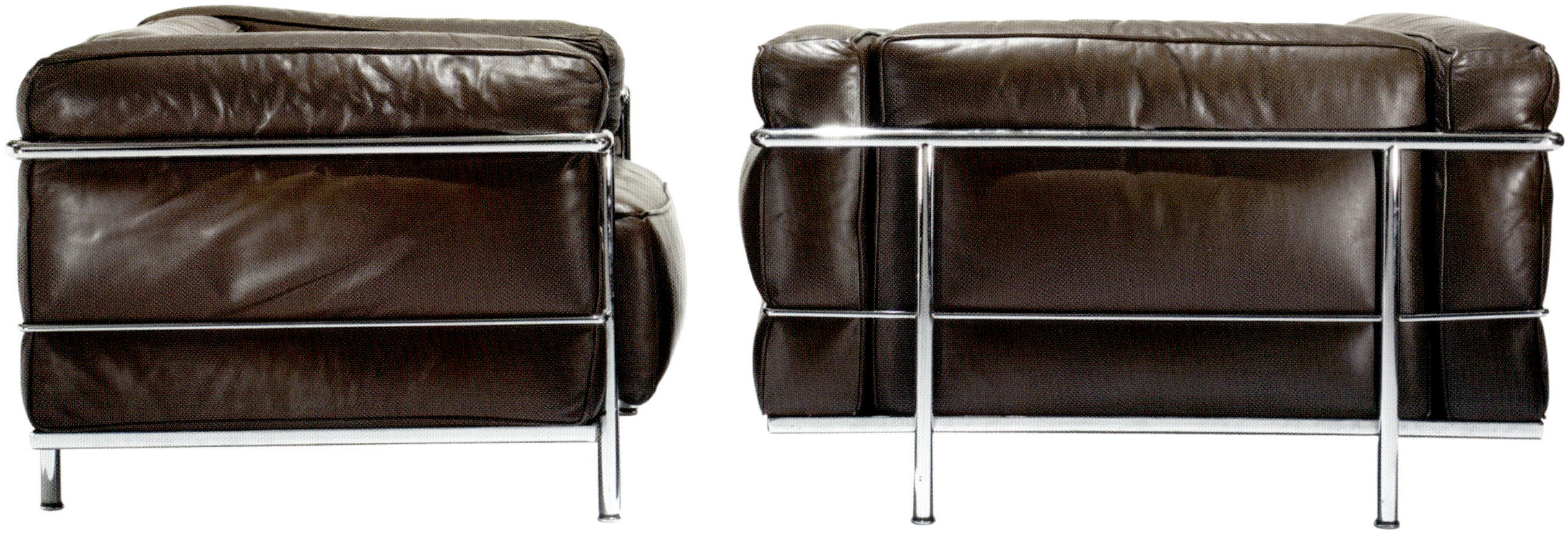

ABOVE A pair of Grand Confort LC2 armchairs by Jeanneret, Perriand and Le Corbusier, designed in 1928, made with chrome-plated tubular steel and upholstered leather cushions.

BELOW A Demountable dining table and six dining chairs by Perriand and Jeanneret, 1940, in oak with rush seats.

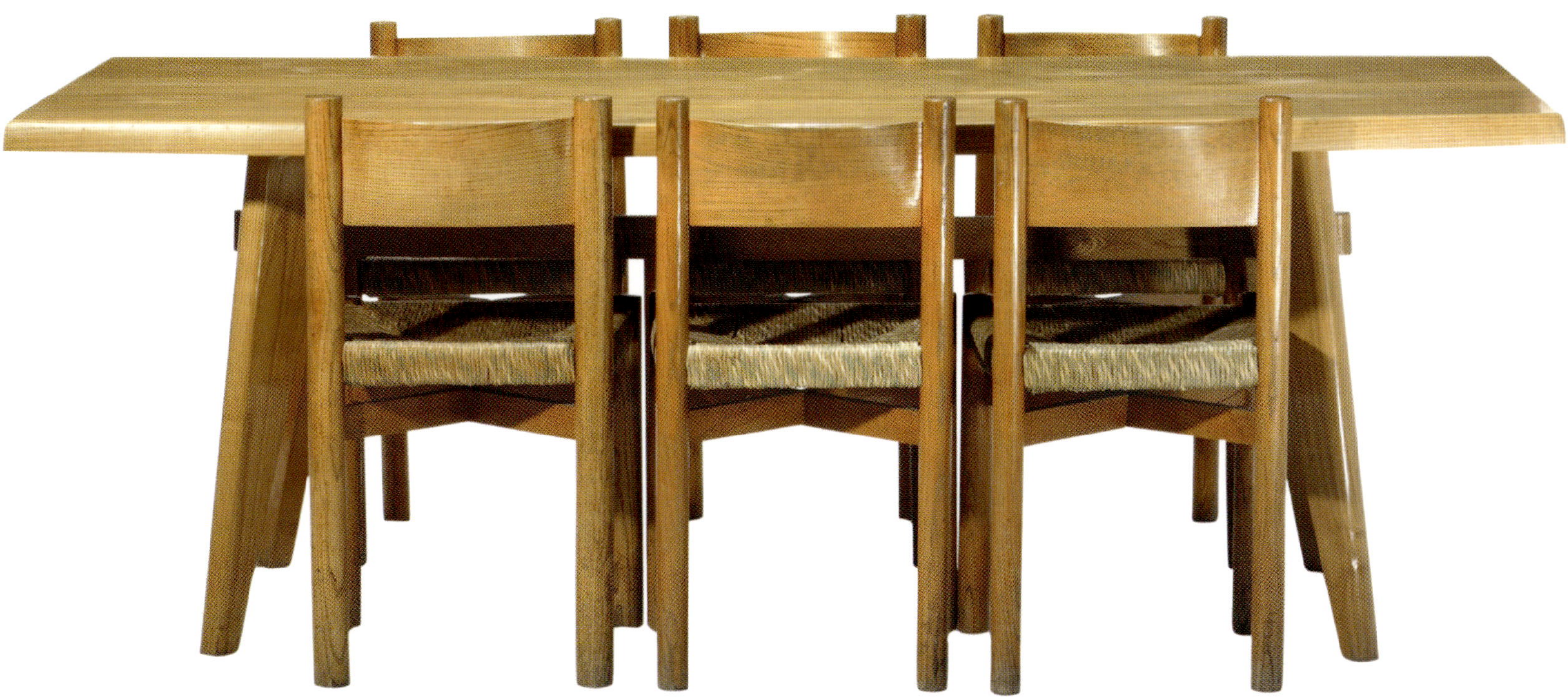

BRUNO MATHSSON
(1907–1988)

The Swedish designer Bruno Mathsson was one of the early pioneers of warm modernism. His work was characterized by an emphasis on crafted, natural materials and a deep understanding and appreciation of ergonomics and the principles of comfort. Added to this was a playful quality, with a love of biomorphic references – most famously with his Grasshopper easy chair of 1931 – and a habit of christening his pieces with endearing names such as 'Eva', 'Miranda' and 'Pernilla', which helped him to market his work and entice his audience. As such, Mathsson was not a designer in the International Style*, but a progressive modernist rooted in a sense of place and deeply influenced by the natural world and an appreciation of the relationship between the individual and his or her everyday surroundings.

Mathsson grew up immersed in a world of craft. He came from a family of artisans and was a fifth-generation cabinetmaker in the town of Värnamo, where he continued to base himself for much of his working life, even as he reached out to international markets and influences. He served an apprenticeship in the family business, learning his trade with his father, Karl Mathsson, while supplementing his studies with his own reading. He began designing furniture for the family company, initially with an Arts and Crafts influence, yet became increasingly interested in Swedish functionalism.

From around this time, Mathsson began to adopt a semi-scientific approach to ergonomics, looking at the way the body's shape and needs might be accommodated by a particular piece of furniture. During his studies in the 'mechanics of seating' he would lie down in the snow, examining the contours of his body's imprint. He combined this work with a passion for bentwood frames and laminated timbers, and also began experimenting with plaited canvas webbing for seats; the webbing moulded itself easily to the sitter, ensuring comfort and support.

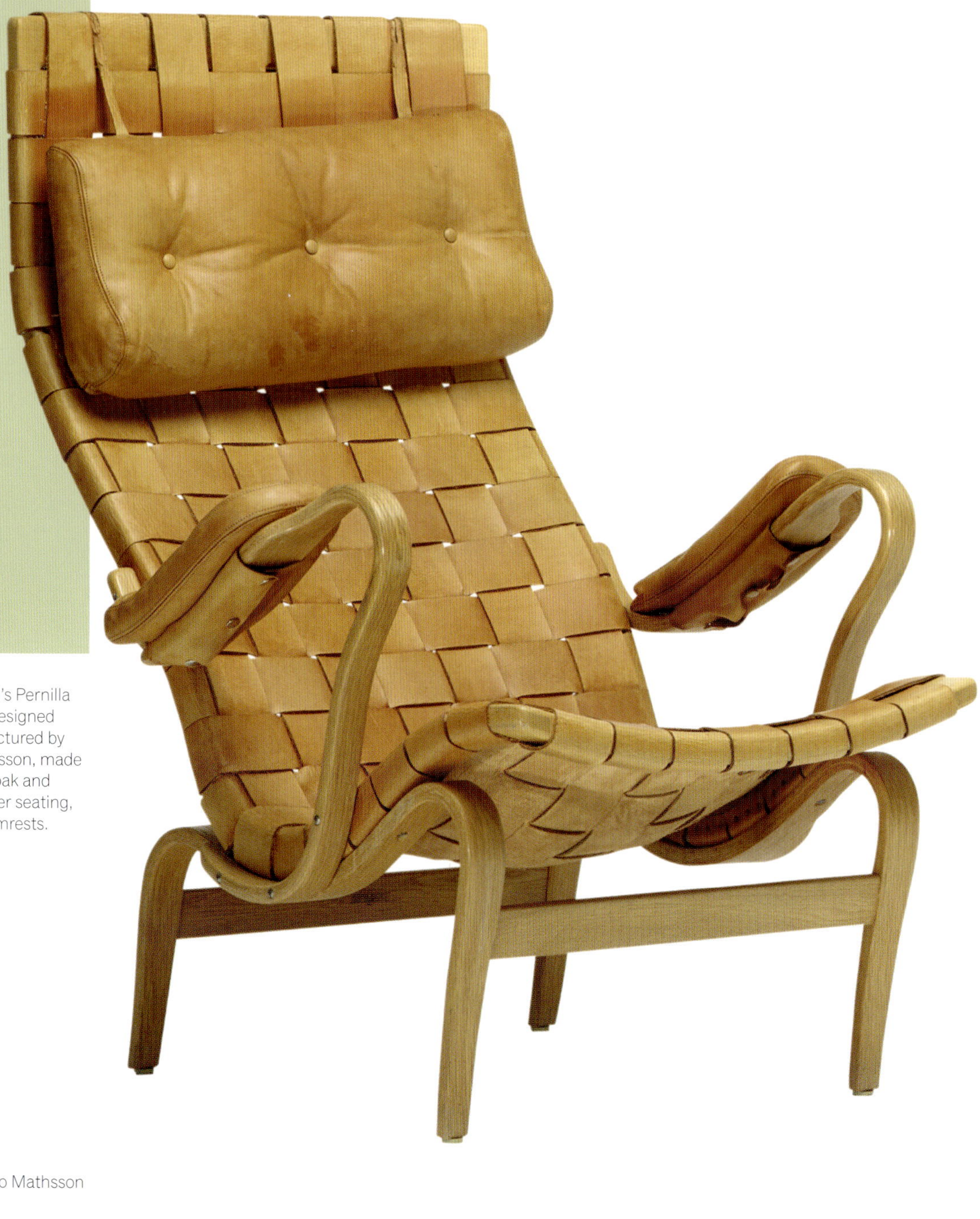

RIGHT Mathsson's Pernilla 2 lounge chair, designed in 1941, manufactured by Firma Karl Mathsson, made with laminated oak and birch, with leather seating, headrest and armrests.

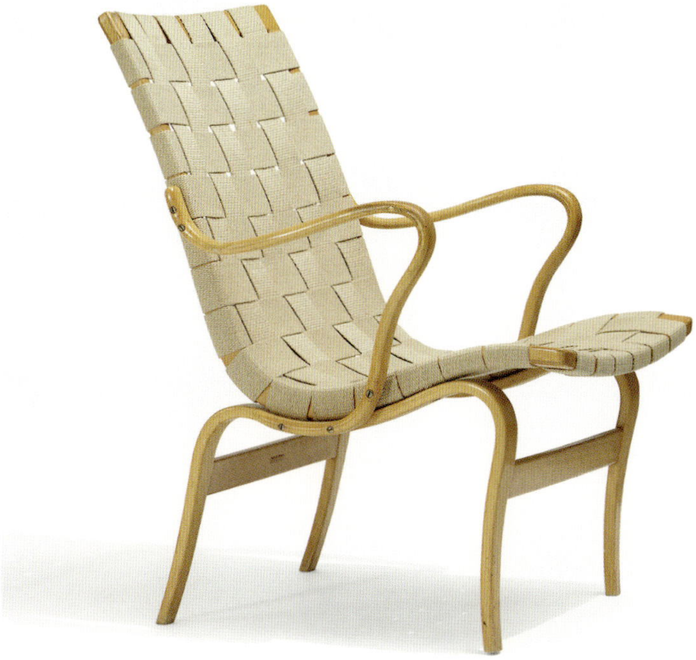

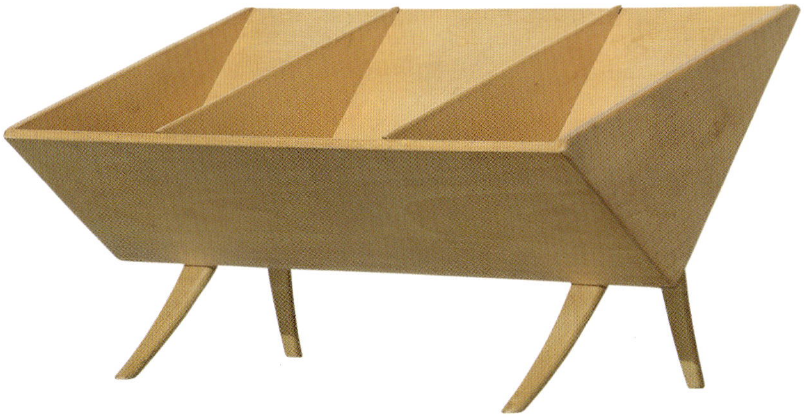

ABOVE The Eva chair, this version designed in 1935, produced by Firma Karl Mathsson, in laminated steam-bent beech with webbed canvas seating.

RIGHT A book crib, produced by Firma Karl Mathsson, designed in 1941, made in ash and birch plywood.

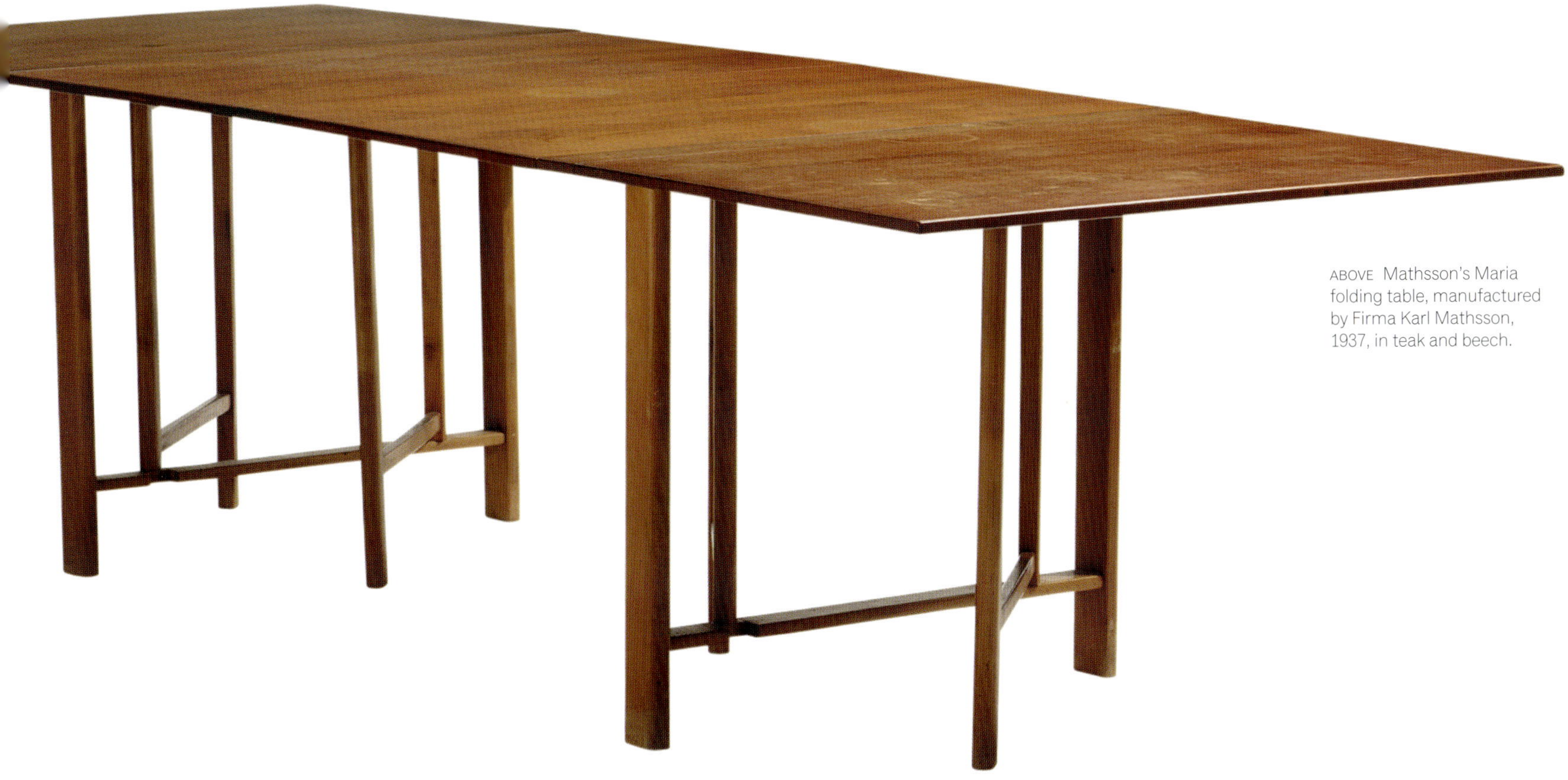

ABOVE Mathsson's Maria folding table, manufactured by Firma Karl Mathsson, 1937, in teak and beech.

During the early Thirties Mathsson developed a number of designs that combined bentwood laminated frames (chiefly in beech or birch) and webbed seating. One of the first designs produced by the family workshop in Värnamo was the Grasshopper chair, initially designed as an easy chair for the local hospital (although allegedly tucked away by hospital staff, who had given it its nickname of 'Gräshoppan', until well after Mathsson had become a famous figure). This was followed by the Arbetsstol or Working chair (1934) – eventually renamed Eva – which became one of his most famous designs. The chair was sinuous and accommodating, with the seat and high back cradling the body but also wholly organic, with its sculptural, rounded armrests inviting to the touch.

Other Mathsson pieces of the Thirties included a simple but elegant daybed (1937) with a plywood frame and webbed seat. The Pernilla chair of 1934 was a variant upon Eva, but with a higher and longer back, more reminiscent of a chaise longue; the design was adapted a number of times to include an ottoman and a 'Pernilla Long Chair' – a veritable chaise longue complete with rear wheels for ease of movement to a terrace or deck, as well as around the home itself.

These pieces, all produced and controlled by the family firm, established Mathsson's reputation both as a craftsman and a design innovator. He participated in the Paris Expo of 1937, and in 1939 the Museum of Modern Art in New York acquired a number of his chairs for a new extension; in the same year Mathsson's work was shown at the World's Fair in New York.

In the post-war years, Mathsson stepped up his involvement in architecture, designing a number of pioneering glass houses and experimenting with green energy solutions, including the notion of taking warmth from the ground (as in the contemporary equivalent, the ground source heat pump). One of his key furniture designs of the post-war period was the Super-Elliptical table (1968) designed with the Danish inventor, mathematician and poet Piet Hein; the table was later produced by Fritz Hansen.

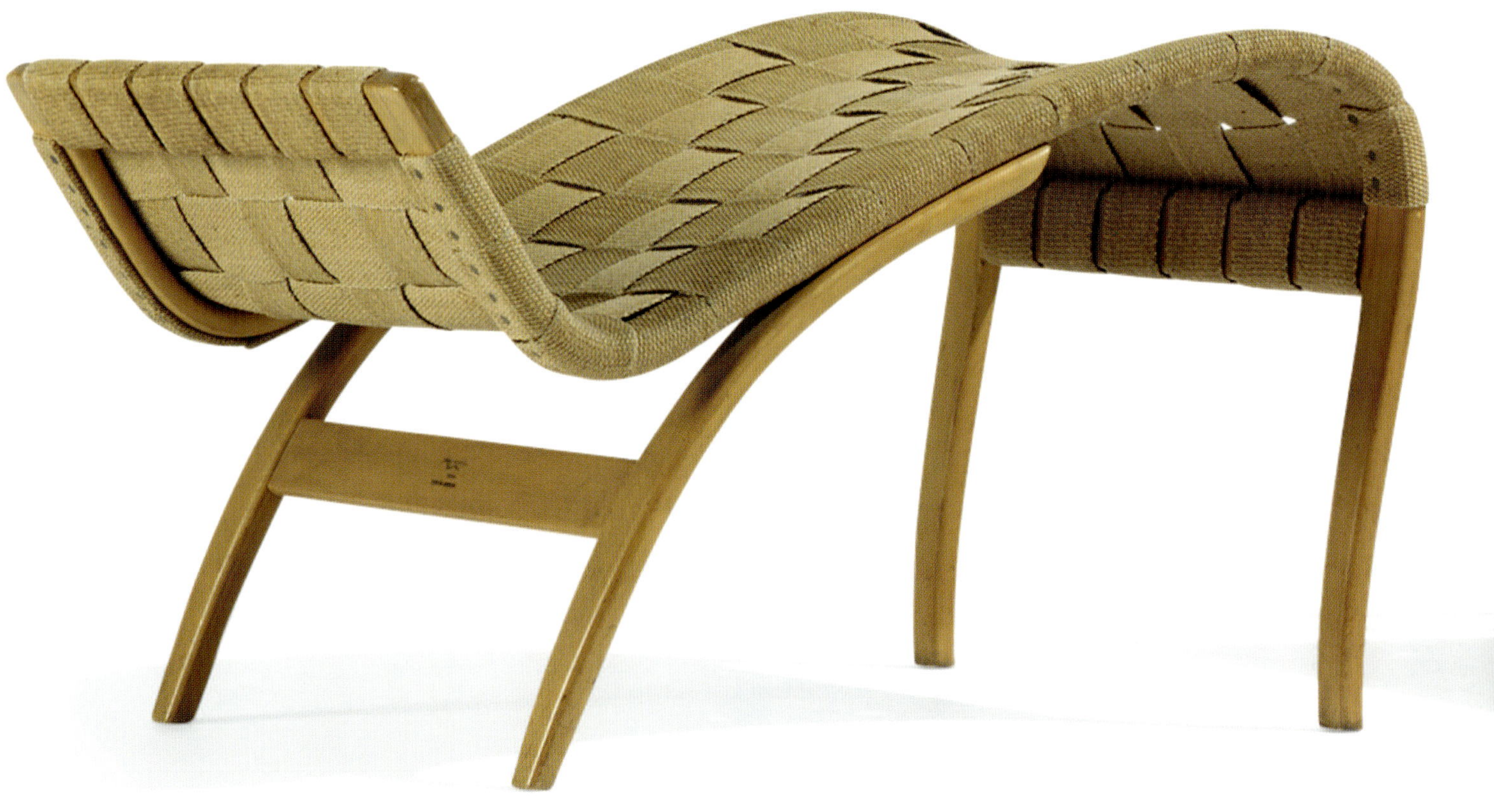

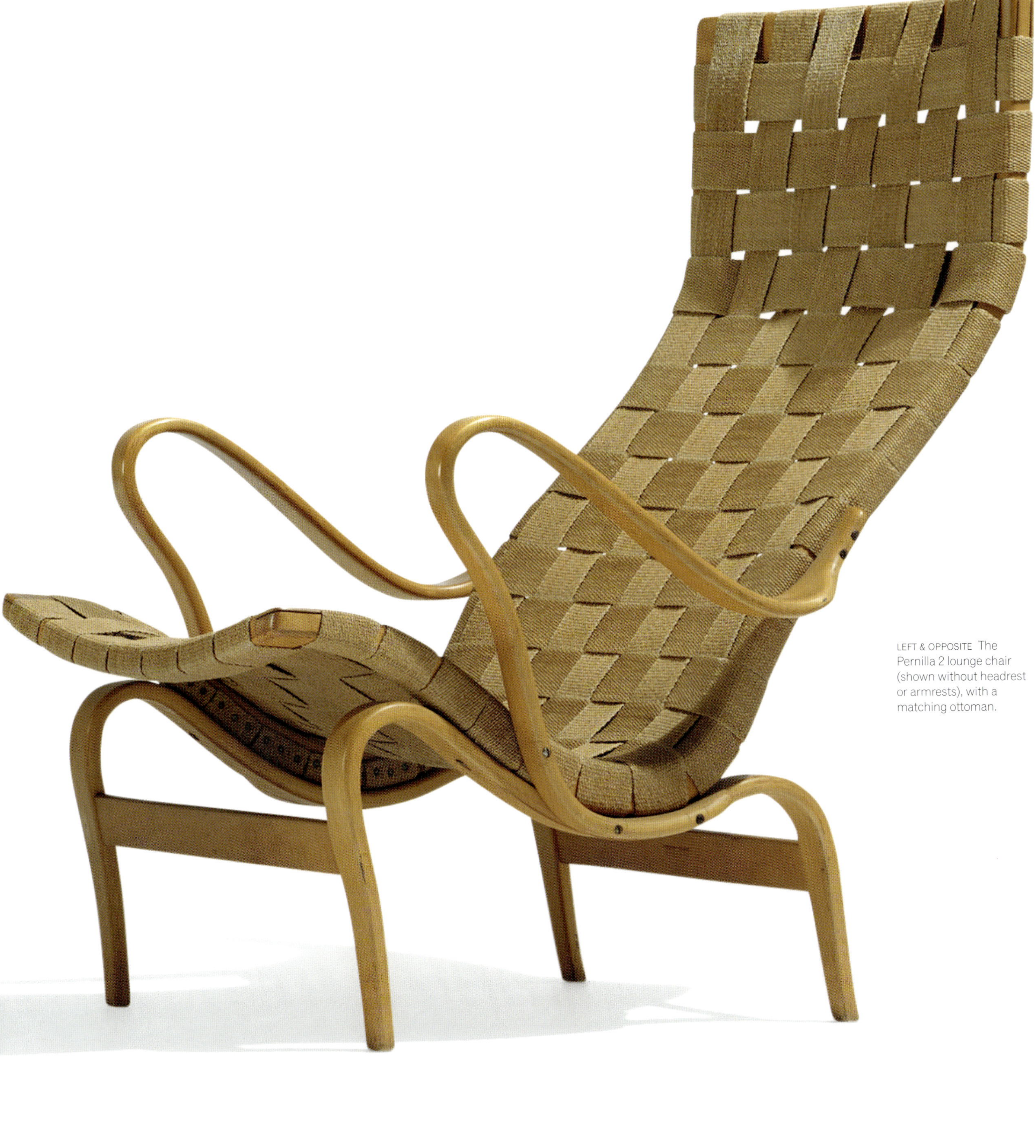

LEFT & OPPOSITE The Pernilla 2 lounge chair (shown without headrest or armrests), with a matching ottoman.

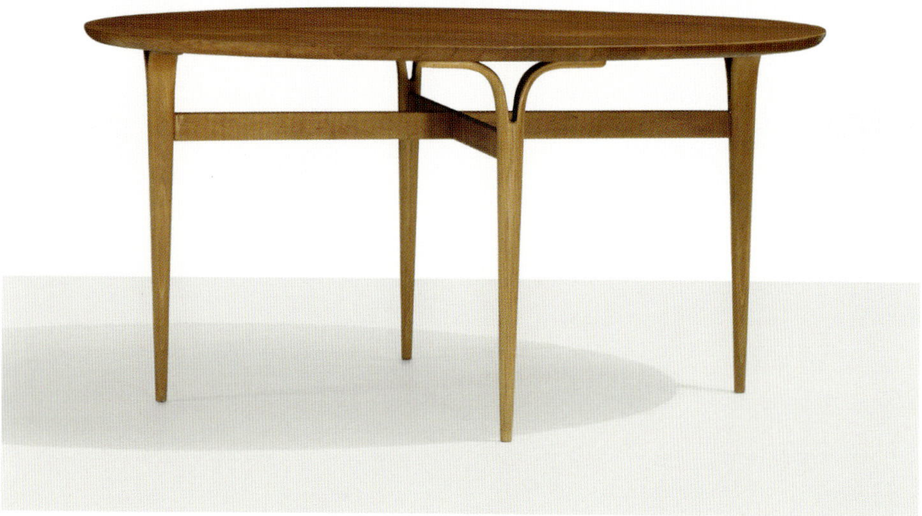

LEFT An occasional table, made by Firma Karl Mathsson, *c.* 1945, in teak, beech, and laminated and steam-bent ash.

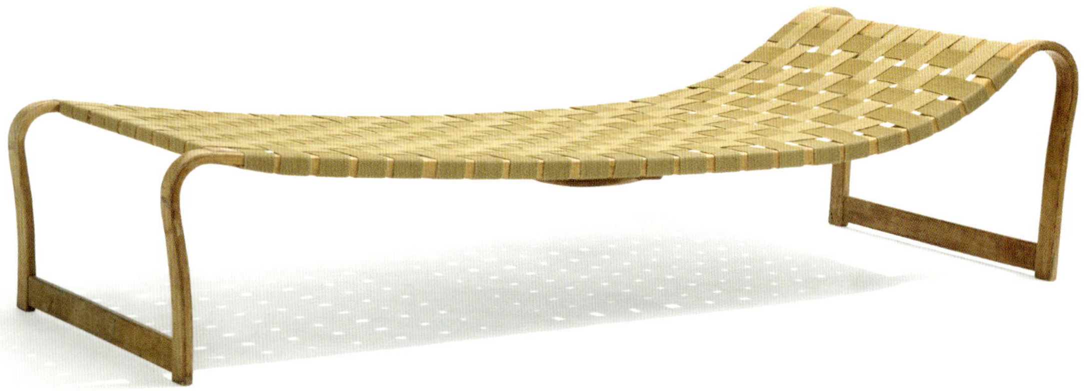

BELOW A Mathsson daybed, produced by Firma Karl Mathsson, 1937, in birch plywood and webbed canvas.

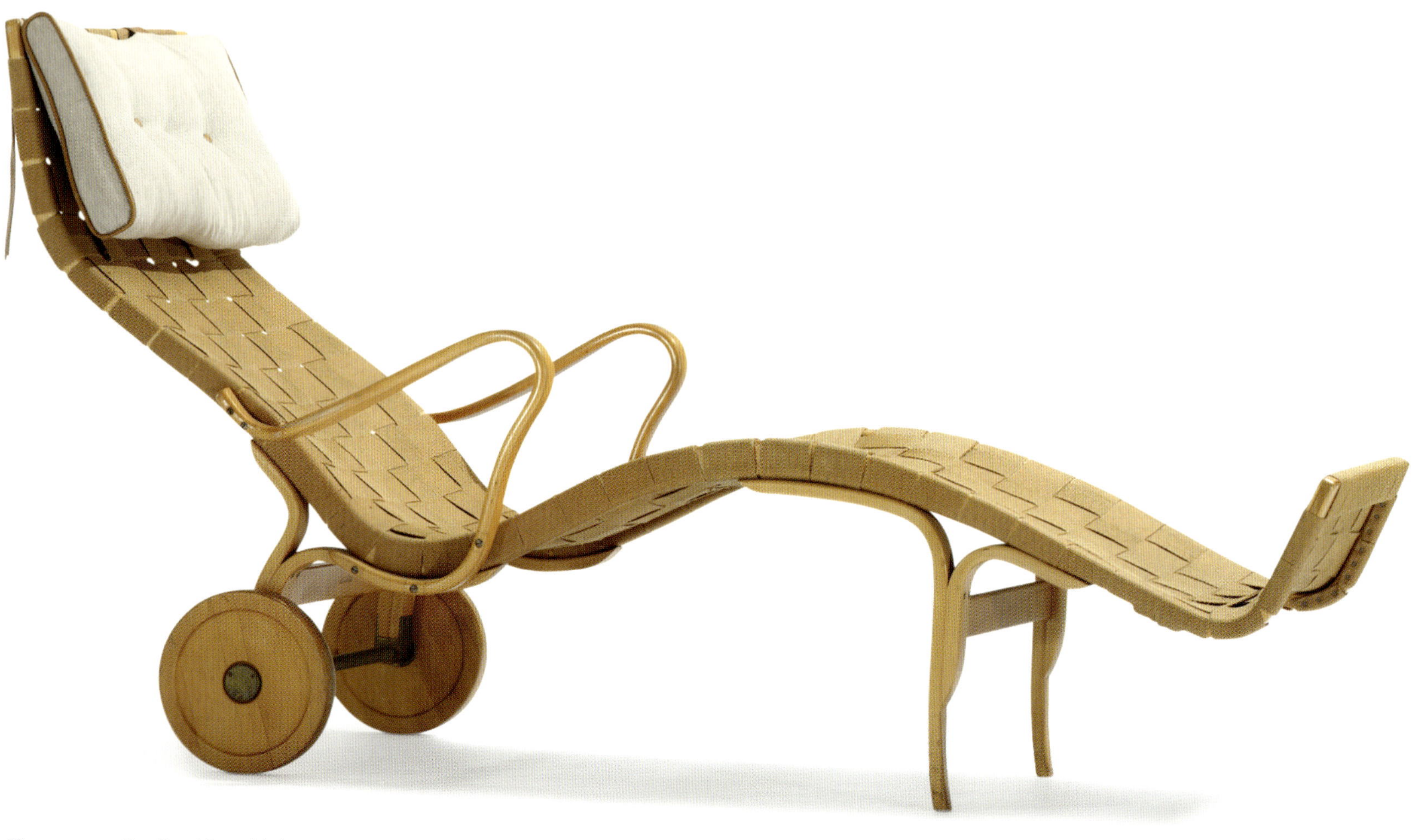

BOTTOM A rare example of the wheeled Pernilla Long Chair, made by Firma Karl Mathsson, in laminated steam-bent beech, canvas, brass and leather.

OPPOSITE A lounge chair with reading stand, made by Karl Mathsson, 1936, in laminated and steam-bent beech, birch, ash, brass and canvas.

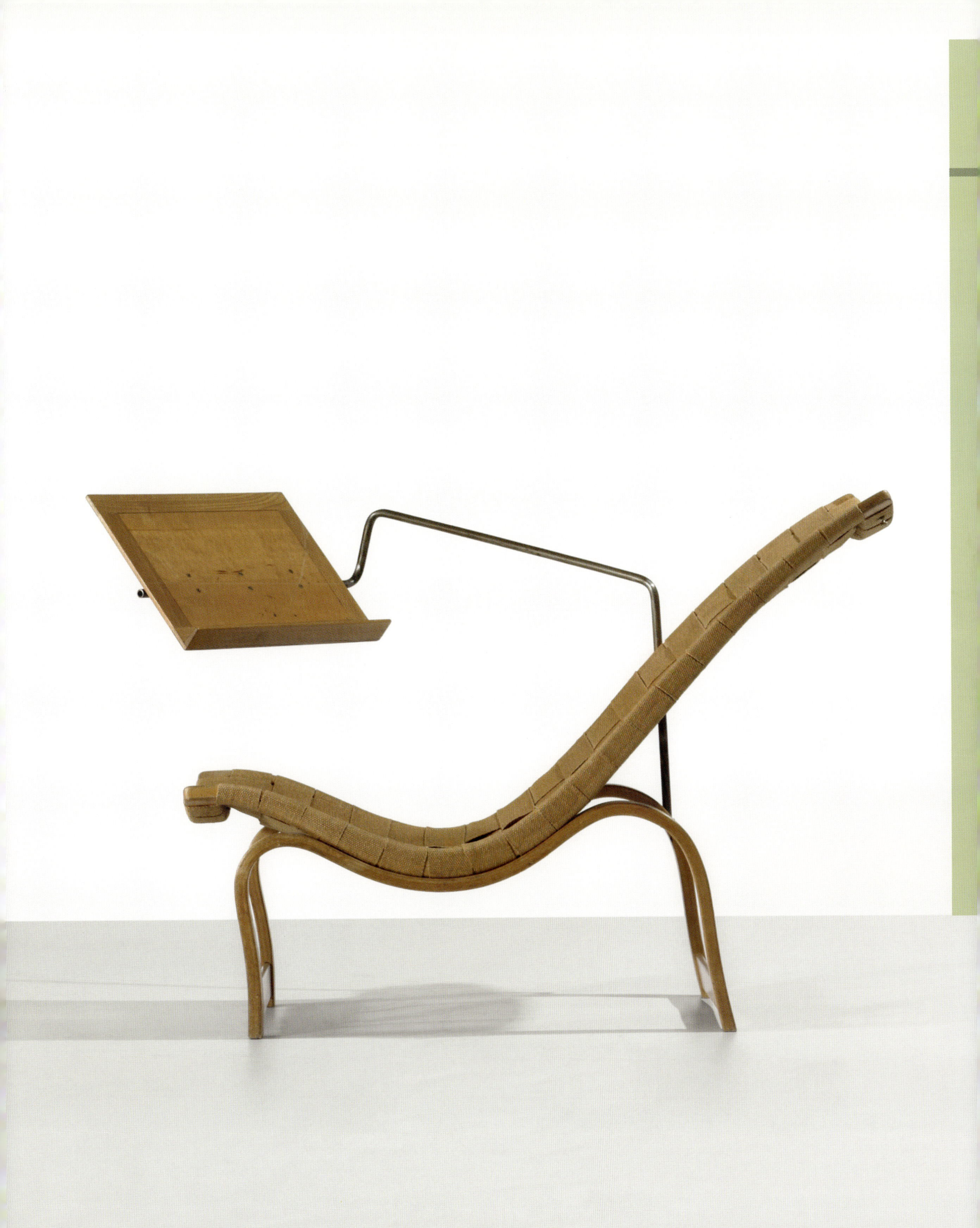

LUDWIG MIES VAN DER ROHE
(1886–1969)

LILLY REICH
(1885–1947)

Ludwig Mies van der Rohe managed to fit many careers into one extraordinary lifetime. He was one of the most influential of the early German modernists, creating a successful architectural practice of his own and leading the Bauhaus for three years, from 1930 until its dissolution in 1933. His early buildings spliced innovative engineering and a pioneering spatial approach, using load-bearing structural frames and curtain walls to create fluid spaces bathed with natural light flooding in through glass façades. Yet at the same time there was a particular love of fine and expressive materials, which brought texture and warmth to the interiors.

Later, Mies began a second career in the United States, becoming an American citizen in 1944. Here he continued to push at the boundaries of experimental modernism, creating fresh masterpieces such as the Farnsworth House (1951) and inventing new-generation skyscrapers such as the Lake Shore Drive Apartments (1951) in Chicago and the Seagram Building (1958) in New York, which took the concepts of the universal space and steel-framed structures with glazed coats to a new level.

There was also another strand to Mies's work that was to have a profound influence. This was his work in the field of furniture design, concentrated in the highly creative period of the late Twenties and early Thirties, when he collaborated on the interior realm and furniture design with Lilly Reich. Key pieces of this period include Mies's iconic Barcelona chair of 1929, made with a lightweight, sinuous skeleton of chrome-plated flat steel and upholstered leather cushions. This chair, which was manufactured by the American firm Knoll* from 1948 onwards, also manages to combine – on a small scale – innovative engineering and fine, warm materials fusing the industrial and the organic.

Mies's understanding of materials, craft and tectonics stretches back into his own childhood. He was the son of a stonemason and worked as a young apprentice on building sites in his home town of Aachen, as well as developing his skills as a draughtsman. He moved to Berlin to continue his studies and later began working in the office of PETER BEHRENS. Some four years later, in 1913, he established his own practice and married Ada Bruhn. After military service in World War I he separated from his wife, and in the mid-Twenties Lilly Reich became both his companion and work colleague.

Reich was an accomplished designer in her own right. She trained as an embroiderer and textile designer, and then worked in Josef Hoffman's* atelier in Vienna before returning to her home city of Berlin, where she designed furniture, shop windows, interiors and clothes. She became an influential figure in the Deutscher Werkbund*, a federation looking to promote the work of German designers, architects and manufacturers, and helped plan and organize a number of Werkbund exhibitions. Reich collaborated with Mies from 1926 onwards, although she always retained her own studio.

The Mies and Reich collaborations included a key Deutscher Werkbund exhibition in Stuttgart in 1927, known as the Weissenhof*, which featured a series of show houses and buildings designed

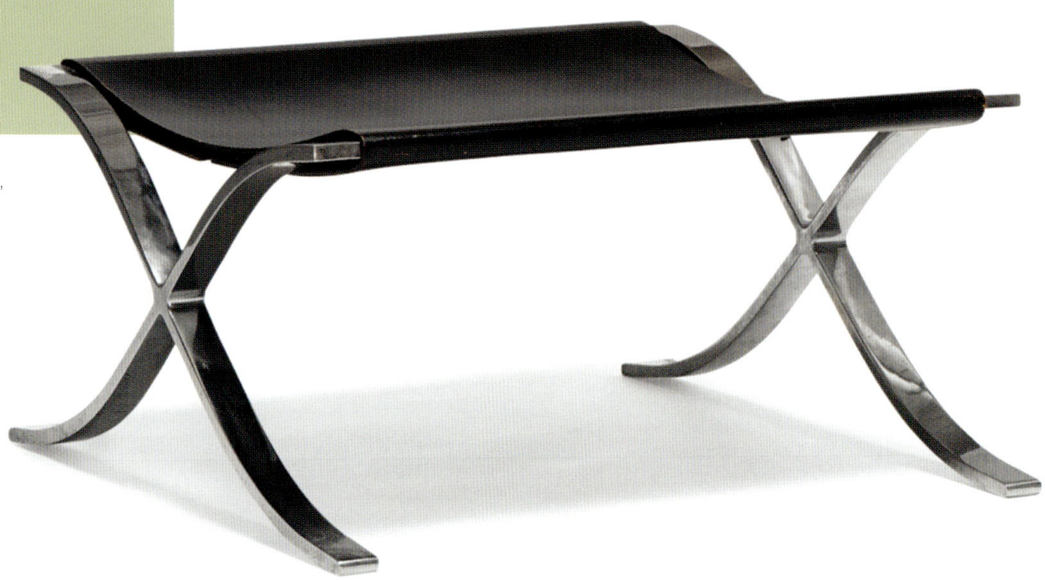

RIGHT A Barcelona stool, designed in 1925, later produced by Knoll, in chrome-plated brass and leather.

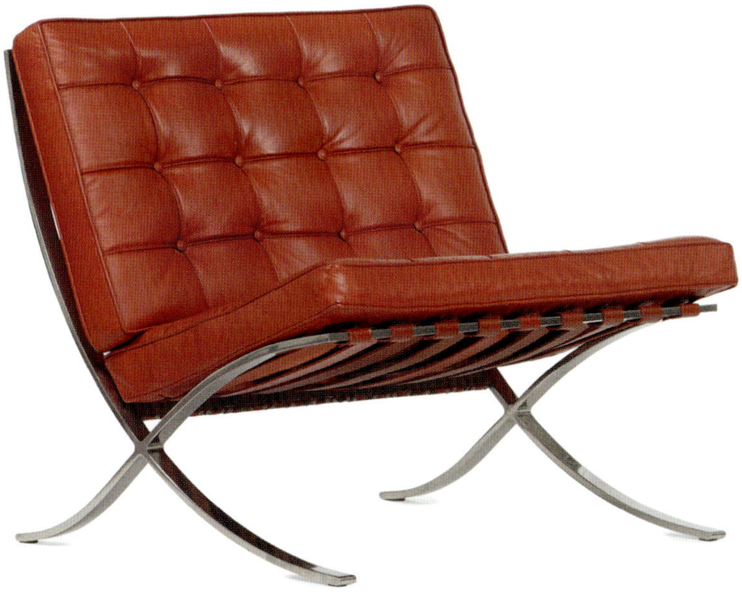

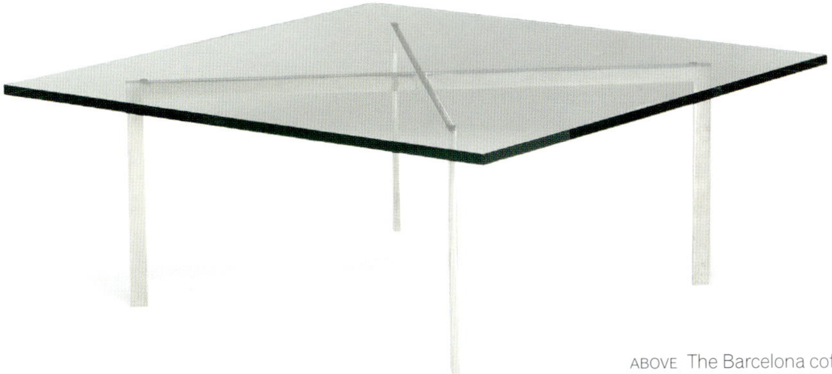

ABOVE The Barcelona coffee table, designed in 1928, in stainless steel and glass.

ABOVE & BELOW A pair of Barcelona chairs in red, designed in 1928 and later produced by Knoll, in stainless steel and spinneyback leather.

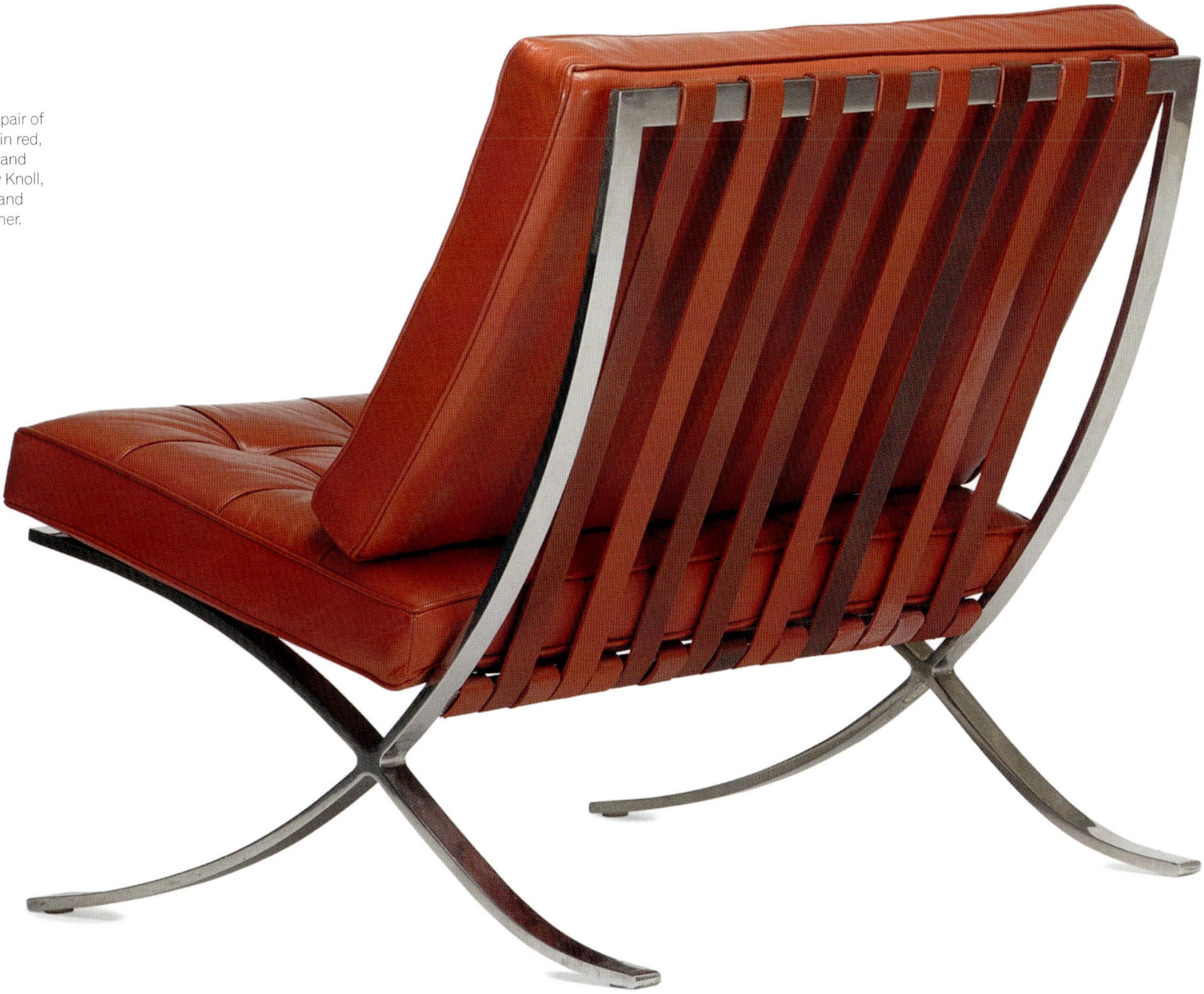

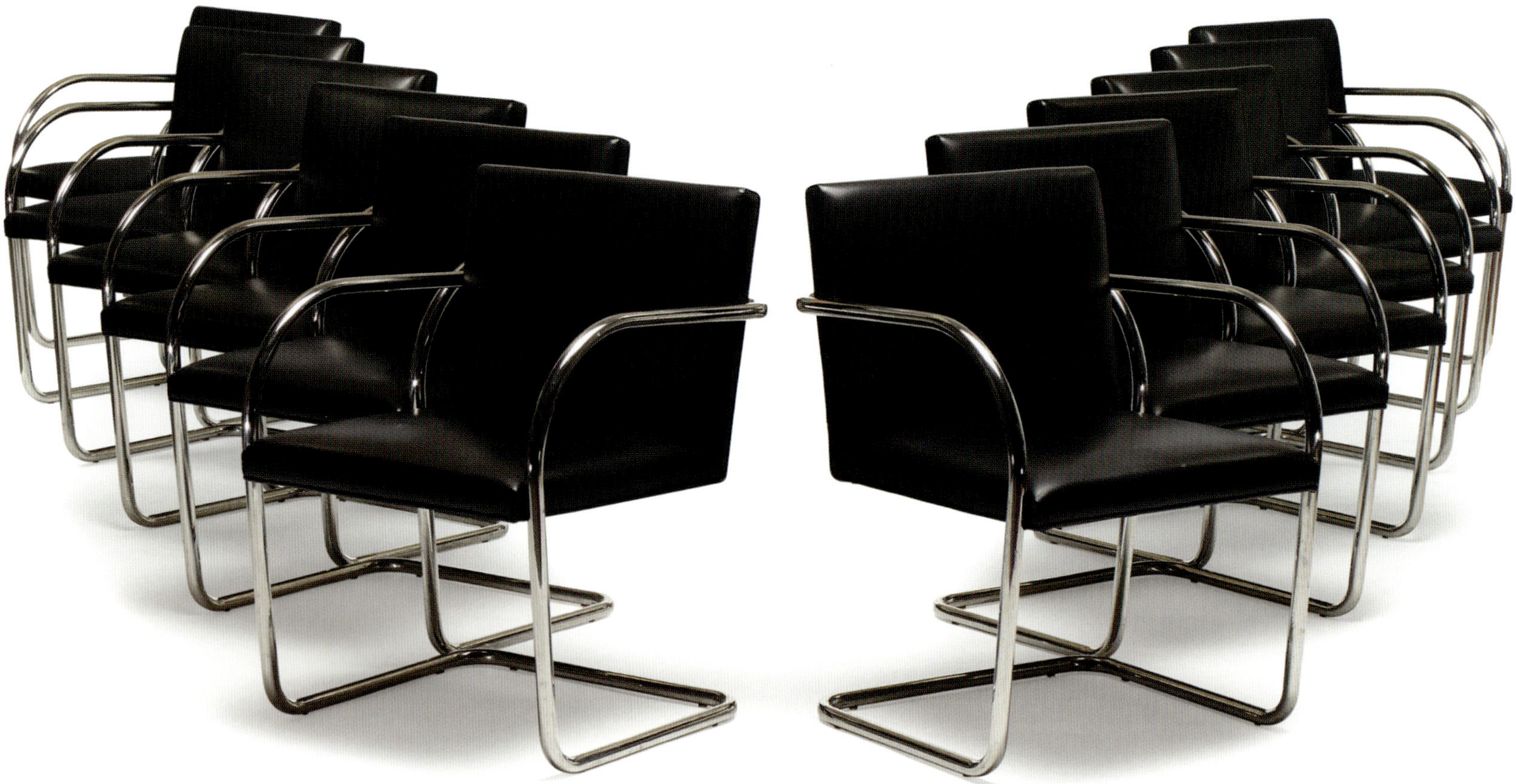

BELOW A set of twelve Brno chairs, designed in 1929, and later manufactured by Knoll; these particular chairs are from the Arts Club of Chicago, for which Mies van der Rohe designed the interiors in 1951.

by a range of modernist architects; Mies was the artistic director of the project. Soon after, there followed two key projects: the Barcelona Pavilion of 1929 and the Villa Tugendhat (see p. 380) in the Czechoslovakian city of Brno, completed in 1930. The projects ran in tandem and shared some common architectural elements, including the use of steel frames, curtain walls and universal space or free plans. The Pavilion was intended as an introduction and reception centre for the German segment of the International Exhibition staged in Barcelona, while the Villa Tugendhat was a private home for a wealthy family who had made their fortune in the textile industry. Yet both combined architectural prowess with crafted, imaginative interiors and a range of bespoke furniture, with Mies and Reich working closely on the development of the interiors and furnishings.

For the Barcelona Pavilion there was the famous Barcelona chair and a matching ottoman stool; there is also a Barcelona daybed or couch, and a Barcelona table with a criss-cross steel base and a simple glass top. For the Tugendhat project, Mies and Reich developed a small number of pieces including the Brno chair, which featured an innovative cantilevered steel framework in bent flat steel or tubular steel, along with a comfortable leather seat and back (sometimes seen in untreated cow hide). There was also a side table – known as the MR table – with a tubular steel base and a circular glass top. The pieces sat perfectly with the architecture and had elements in common, using light metal frames and transparency.

Despite her input into the collection, Lilly Reich is seldom credited as a co-designer. After Mies emigrated to the United States, Reich stayed in Germany and looked after his affairs there while trying to continue her own atelier, which was bombed during the war. She was forced to join the wartime architectural and engineering service known as Organisation Todt, led by Albert Speer*. She reopened her studio after the war, but died in 1947.

Mies himself did not develop any further furniture designs. The Barcelona and Brno pieces were put into production by Hans and Florence Knoll* and graced many of Mies's US buildings.

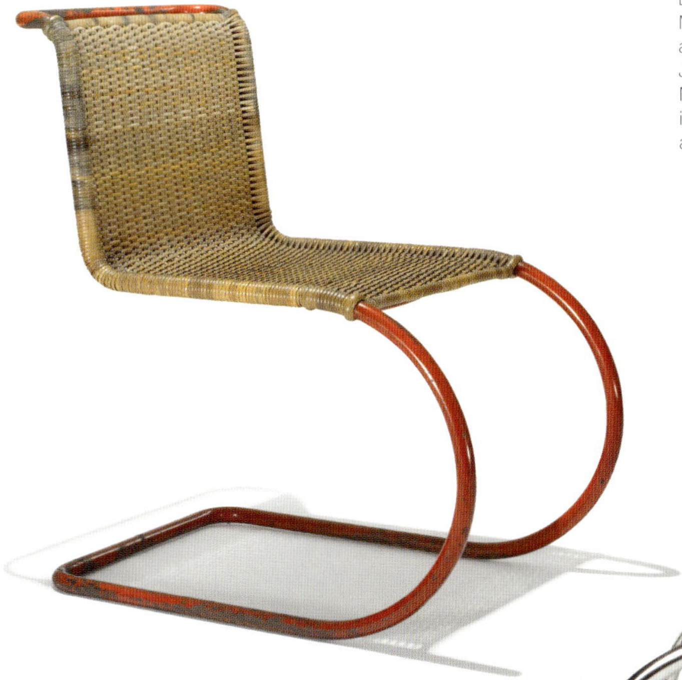

LEFT An MR10 chair by Mies van der Rohe and Reich, made by Joseph Müller Berliner Metallgewerbe, 1927, in lacquered steel with a cane seat.

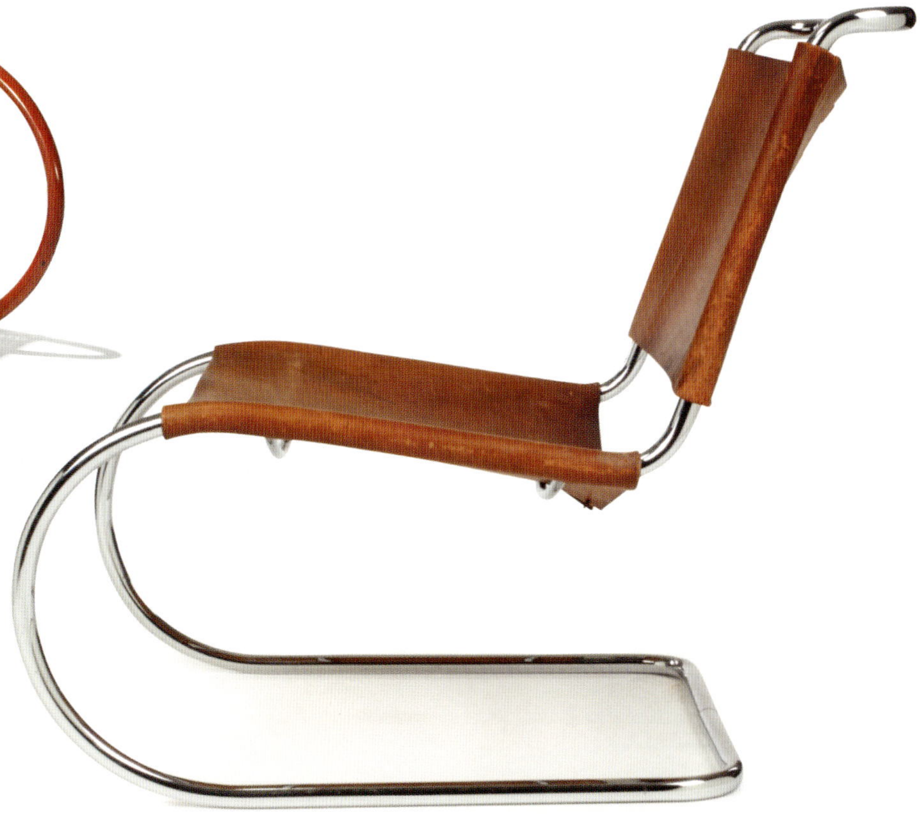

ABOVE An MR30/5 chair produced by Bamberg Metallwerkstatten, *c.* 1930, in chrome and nickel-plated tubular steel with a leather seat and backrest.

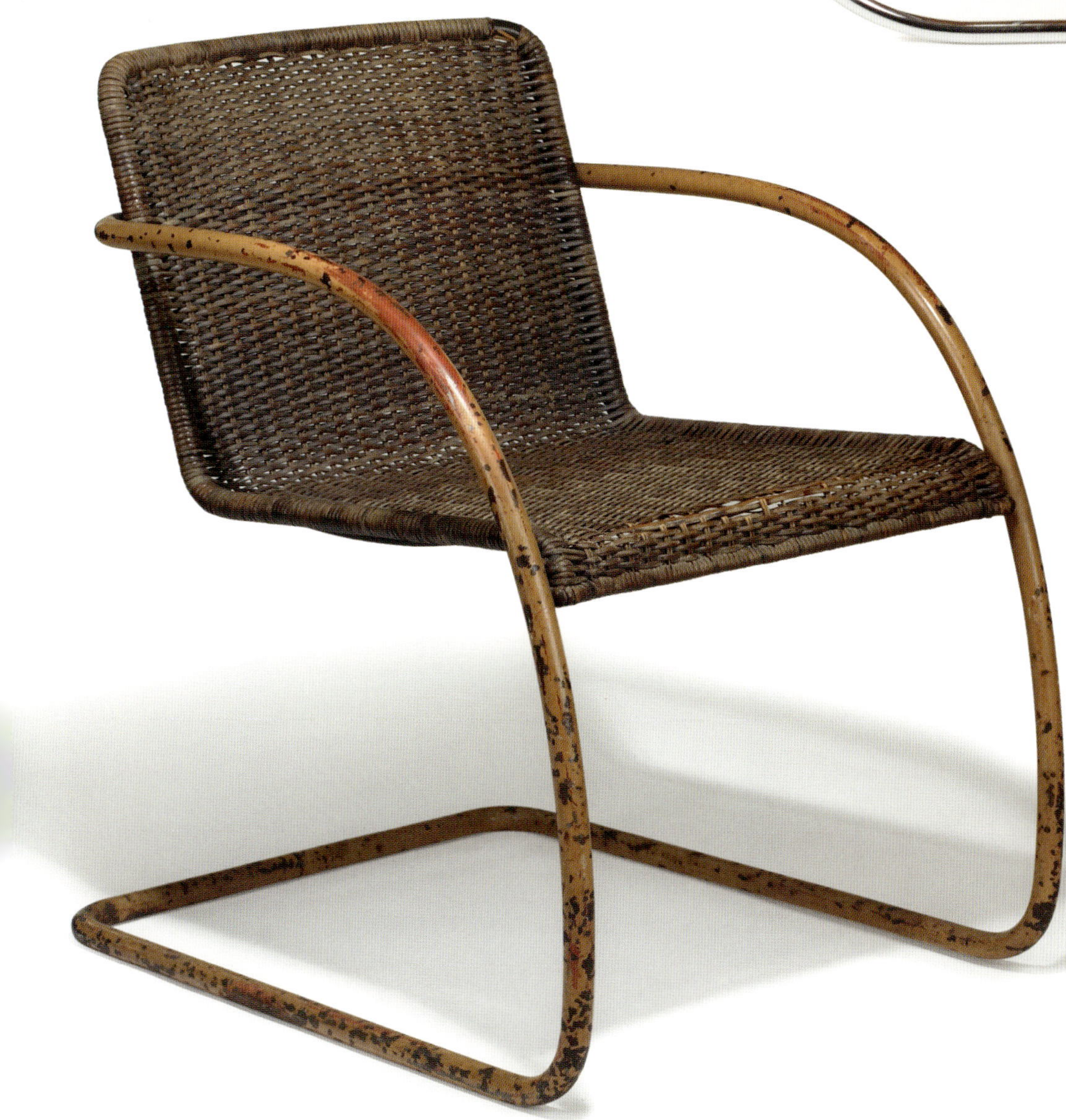

LEFT An early cantilever chair made by Joseph Muller or Bamberg, *c.* 1929, in enamelled tubular steel with a wicker seat.

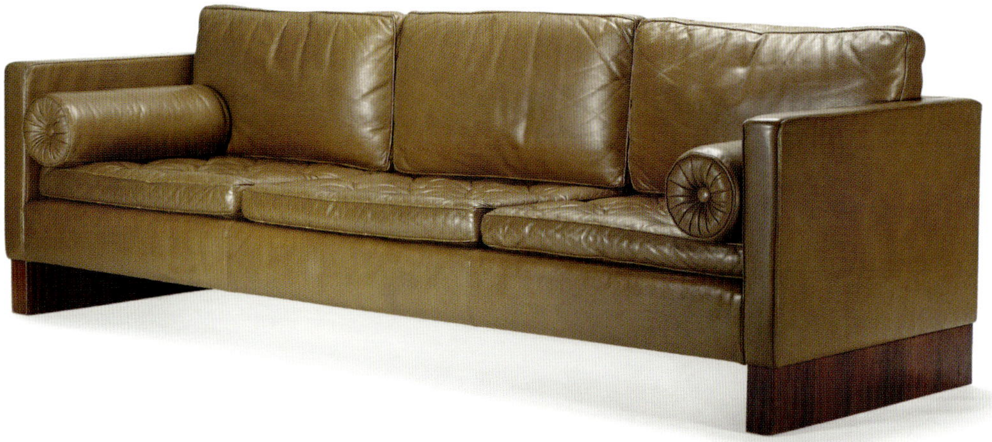

RIGHT This sofa was designed in 1930 and later produced by Knoll, in rosewood and upholstered leather.

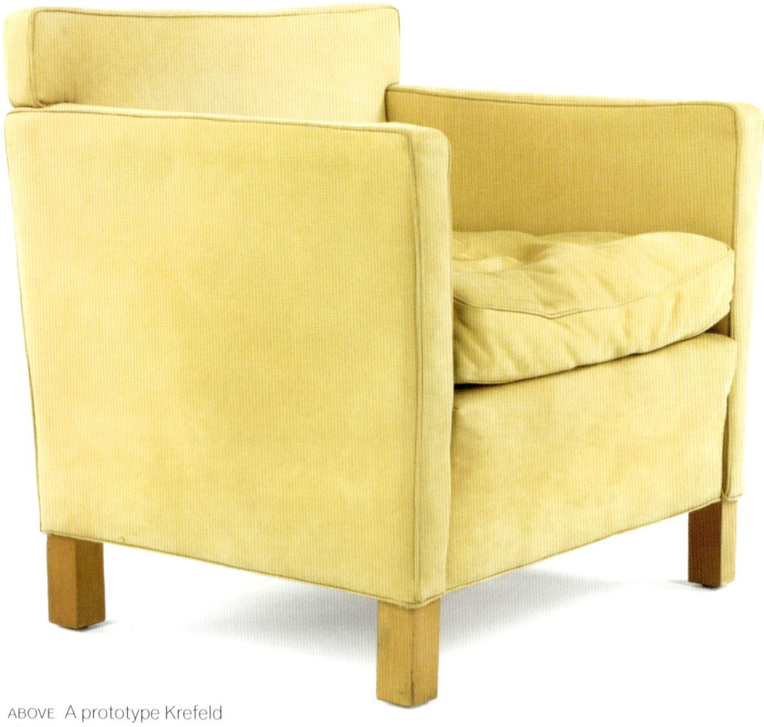

ABOVE A prototype Krefeld chair originally designed in 1929 for the Esters House in Krefeld, Germany; this piece was made by Knoll.

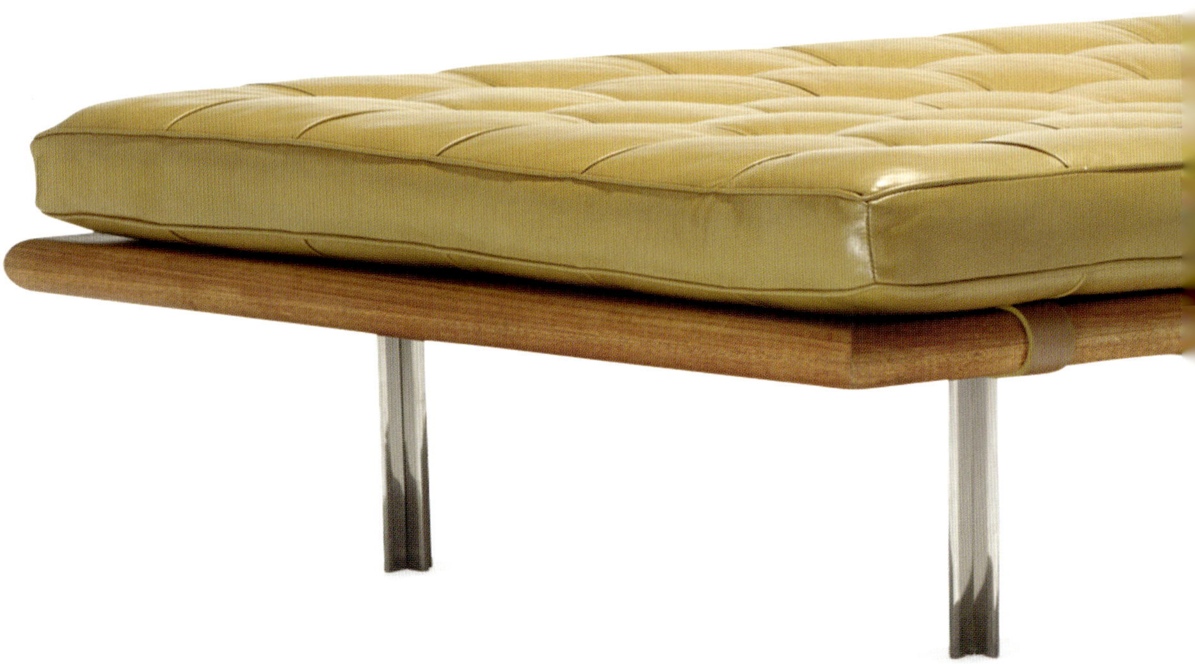

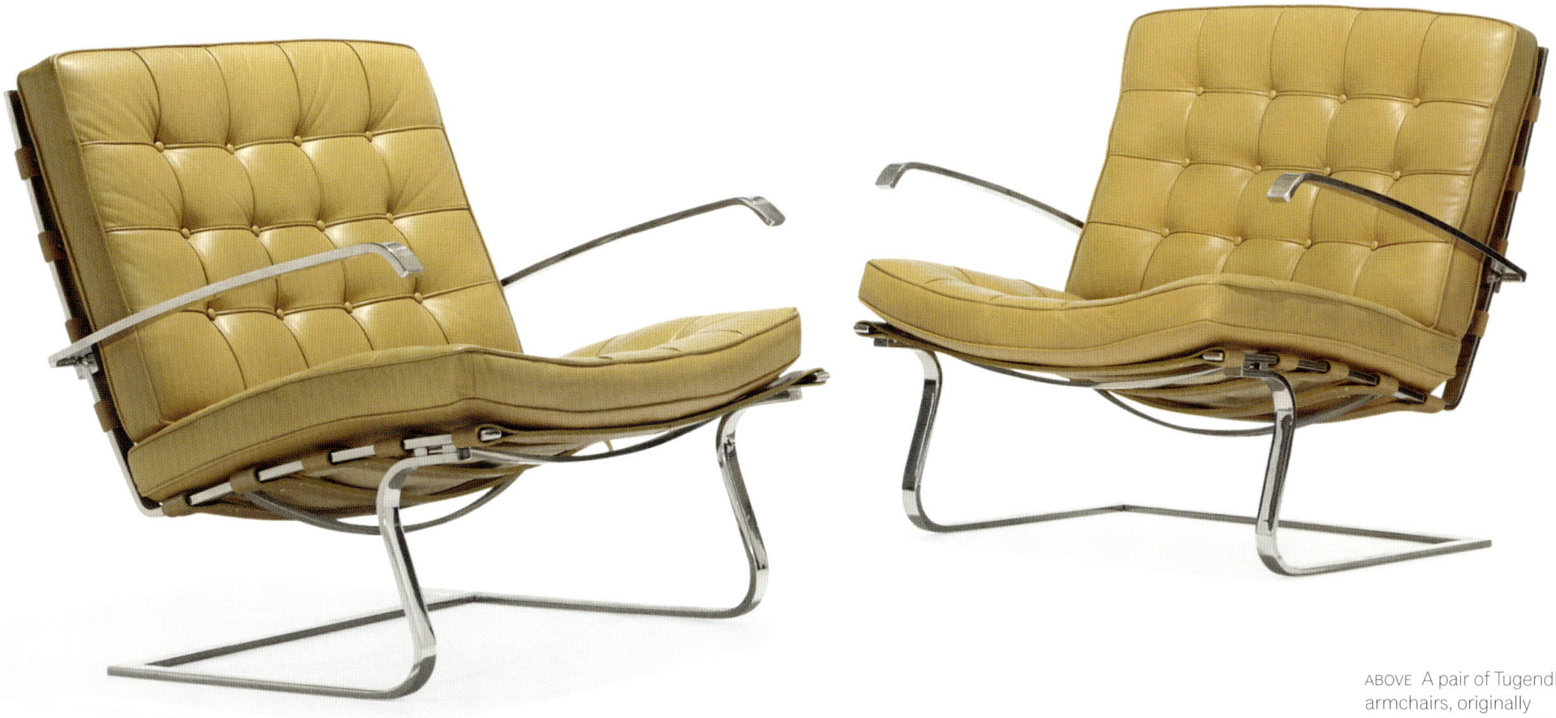

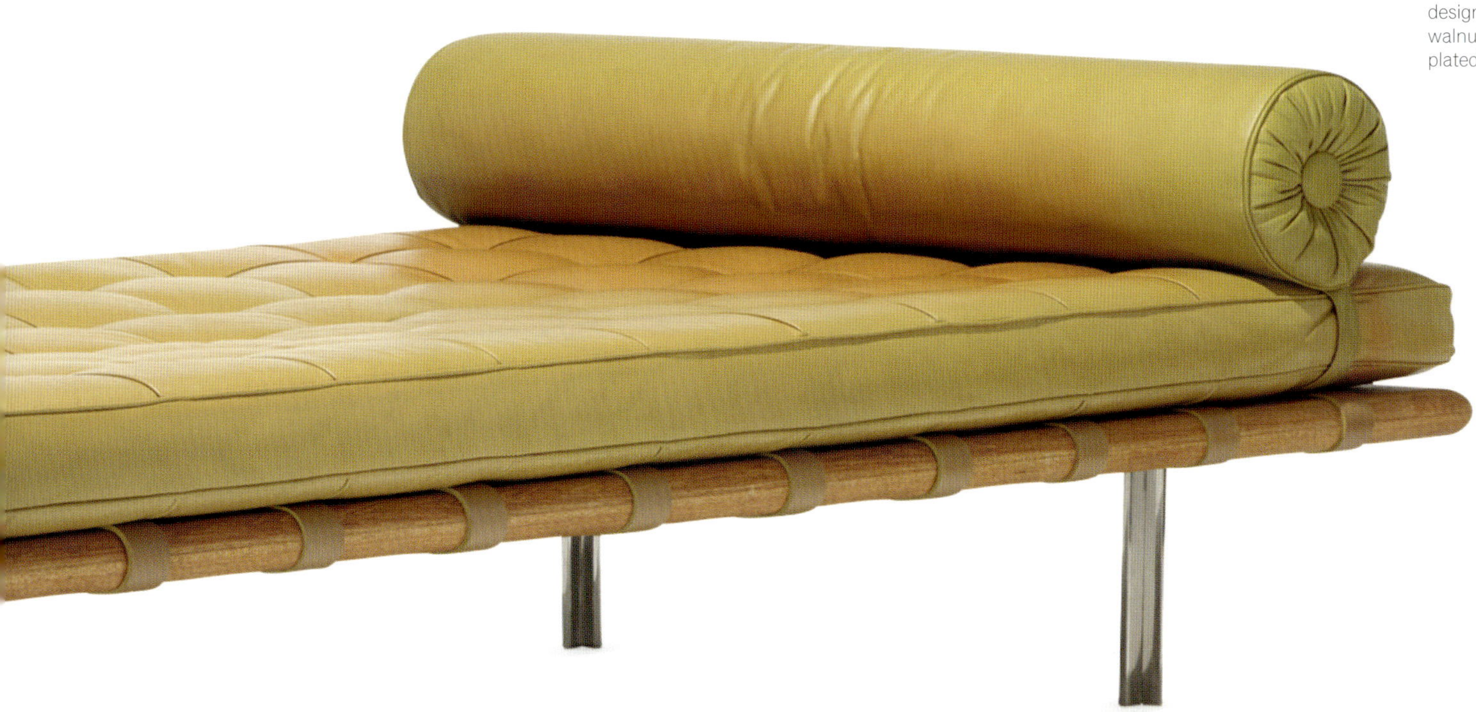

ABOVE A pair of Tugendhat armchairs, originally designed in 1928 for the Villa Tugendhat in Brno, made of chrome-plated brass and with upholstered leather cushions.

LEFT The Barcelona daybed, designed in 1928, made with walnut, leather and chrome-plated brass.

JEAN PROUVÉ
(1901–1984)

The architectural designer and engineer Jean Prouvé described himself as a '*constructeur*'. For Prouvé, the design of a building or a piece of furniture went hand in hand with the process of working out how they would be made, with prototypes and working models created in his own workshops in Nancy and Maxéville. He saw no point in designing anything that could not be made, taking a practical approach combined with a powerful appetite for innovation and fresh thinking.

Although he was not formally trained as an architect, Prouvé was to have a profound influence upon the world of architecture. He was a highly imaginative designer, seeking out methods for building lightweight and prefabricated structures with components that could be factory-made and then shipped and assembled on site. Over the years Prouvé developed a number of prefabricated buildings, including the Maisons à Portiques (1939–47) intended for use as emergency accommodation by wartime refugees, the low-cost housing known as the Standard Houses (1949–52) and the Maisons Tropicales (1949), which were shipped to Niamey in Niger and Brazzaville in the Congo.

Yet Prouvé is perhaps more widely known as a furniture designer – or 'maker' – and proved himself just as innovative and free-thinking in the world of modernist furniture design. Here, too, he adopted an engineering-led approach, re-examining the mechanics, materials and methodology of each and every piece, aiming largely for designs that could be potentially mass-produced and at a relatively low cost.

Prouvé was born in the city of Nancy in northeastern France, not far from Metz, Luxembourg and Strasbourg. His father, Victor, was a painter and artisan and an influential local figure who co-founded the École de Nancy – a loose grouping of Art Nouveau artists and producers. Prouvé hoped to study engineering but the family finances were too stretched, so he took an apprenticeship in a blacksmith's workshop near Paris, followed by a position with a metalworking atelier in Paris run by Adalbert Szabo. After military service Prouvé opened his own small studio in Nancy, initially producing ironmongery with an Art Nouveau influence. Within just a few years his interests shifted to modernist architecture and design, and he began to experiment with stainless steel and welding techniques.

One of his earliest commissions, in 1927, was for a bespoke steel entrance portal for a house in Paris – the Villa Reifenberg – designed by the leading modernist architect ROBERT MALLET-STEVENS. The commission opened many

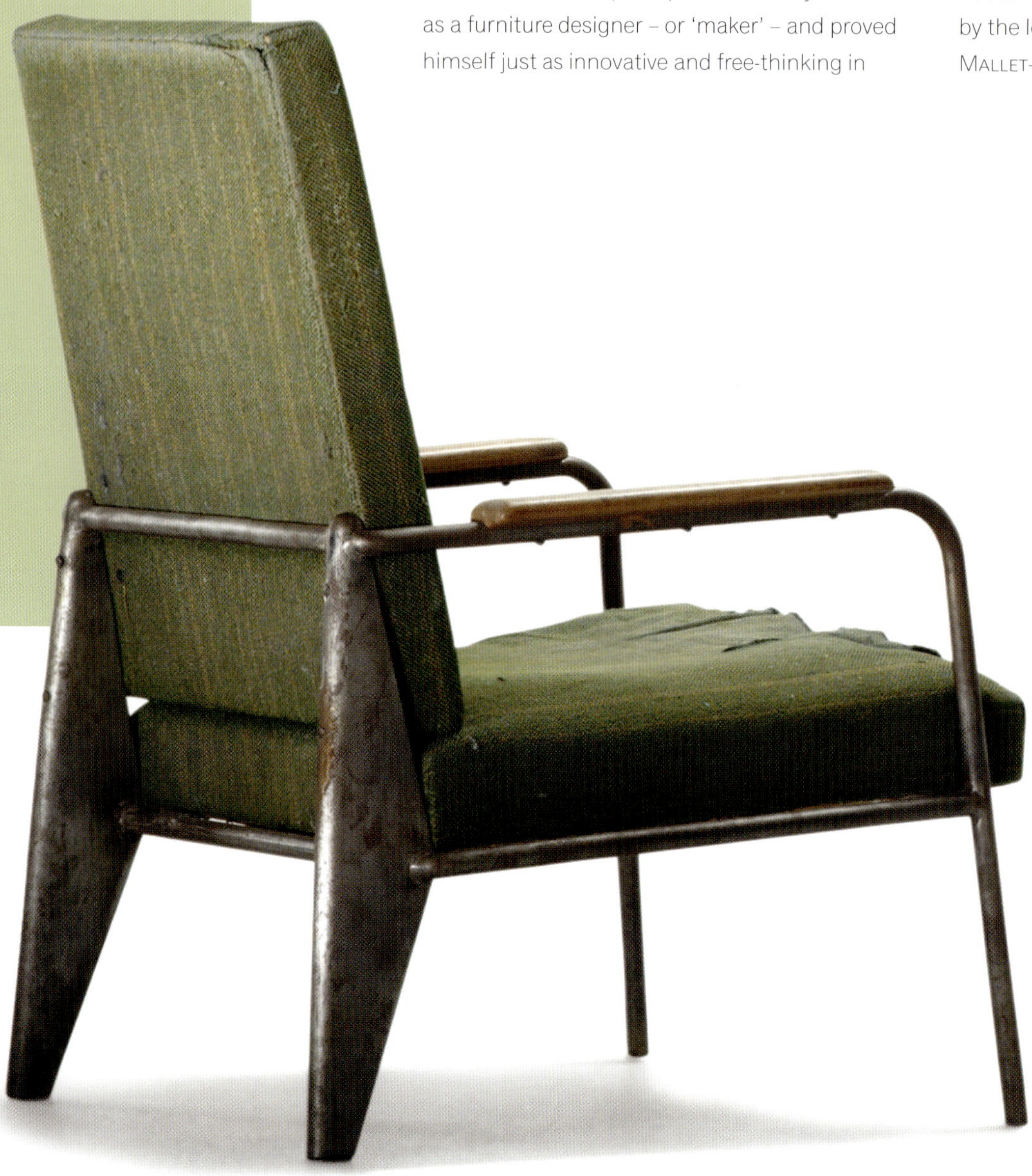

LEFT A lounge chair produced by Ateliers Jean Prouvé, 1939, in steel, oak and upholstery filled with horsehair.

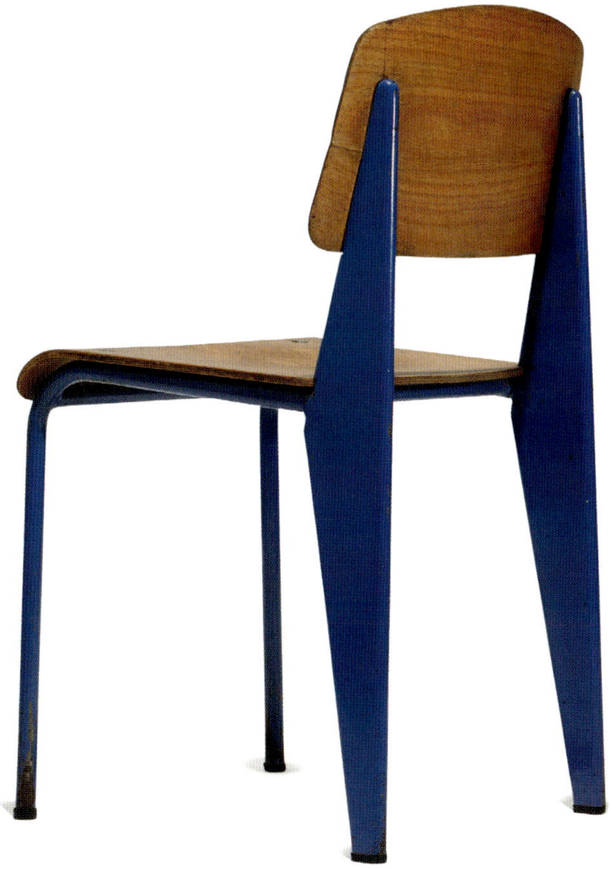

ABOVE Prouvé's Standard chair 305 in blue, made by Ateliers Jean Prouvé, 1950, in enamelled steel and beech plywood.

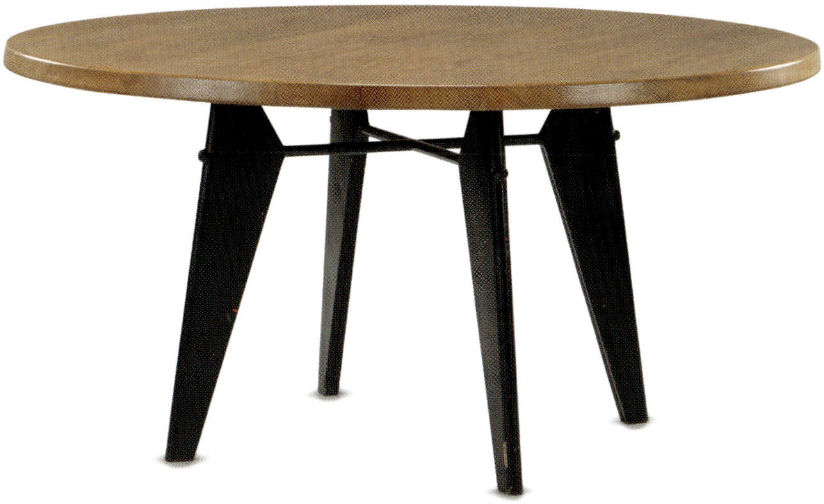

ABOVE A round dining table by Jean Prouvé and Charlotte Perriand, *c.* 1950, in enamelled steel and oak.

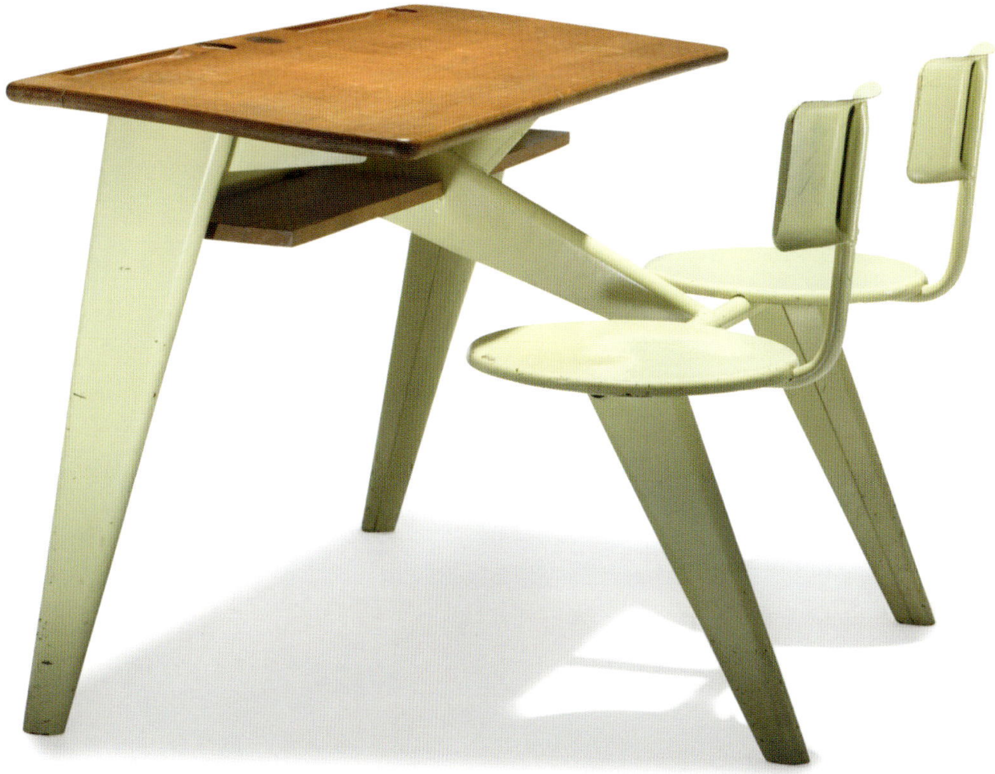

RIGHT A school desk manufactured by Ateliers Jean Prouvé, 1946, in bent and enamelled steel with an oak desk top.

BELOW Bed No. 102 for the Lycée Fabert in Metz, made by Ateliers Jean Prouvé, 1936, in enamelled steel, stained oak and leather.

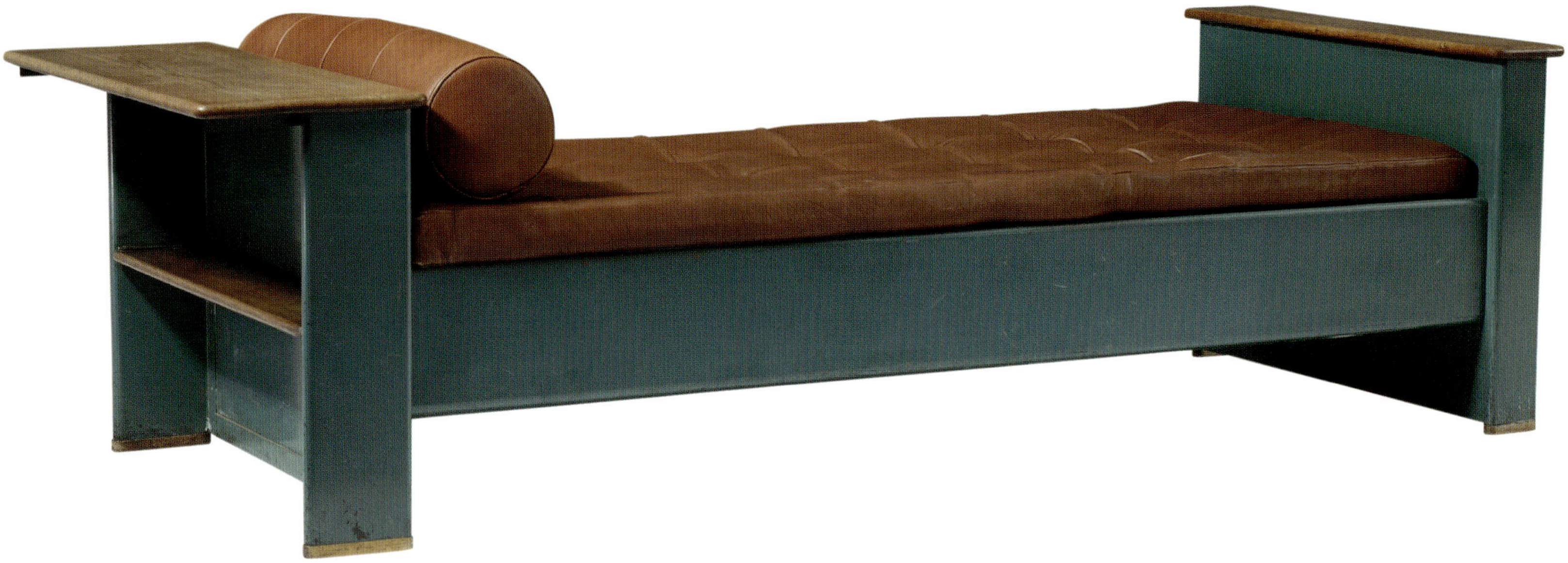

doors, introducing Prouvé to other designers and architects, including LE CORBUSIER and CHARLOTTE PERRIAND, and his atelier expanded rapidly in the late Twenties and early Thirties. He also became a co-founder of the Union des Artistes Modernes* (UAM), which included a number of the key Parisian modernists of the period.

During the Thirties there were commissions for buildings such as the Aéro Club Roland Garros (1936) and the Maison du Peuple in Clichy (1939). But it was also a period when Prouvé began developing some of his early furniture designs, which were made in his own atelier in Nancy. Among the key pieces there was the Grand Repos armchair (1930), a sculptural, dynamic and streamlined design in varnished steel with canvas cushioning for the seat and armrests. A variant of a chaise longue, the chair – which featured in a UAM exhibition in 1930 – features a sliding seat with a spring system that allows the user to recline by simply leaning backwards. The Folding chair of 1930 was an early attempt to create a piece for mass production – a light and easily movable chair in flat and tubular steel with a canvas or upholstered seat and back; Prouvé explored the idea of folding seats in a number of other pre- and post-war designs.

Two of Prouvé's most recognizable designs, having both been reissued, also date from this period. The Cité armchair (1931) was developed as part of a competition-winning entry for new furniture for student accommodation at the Cité Universitaire in Nancy. As with the Grand Repos, the shape was dynamic and sculptural, using a light and simple steel 'sledge' frame subtly attached to an ergonomic high-backed seat made of sheet metal coated in canvas; the armrests were made of leather straps. The armchair, with its floating seat, was one of a series of pieces made for the university.

In 1934, Prouvé produced his famous Standard chair, which brought together a number of ideas and themes that he had experimented with in earlier designs and prototypes. Here was a chair that was both simple and refined. The back legs were in flat steel, rather like metallic fins or propellers, while the seat support and front legs were lighter and subservient, made of tubular steel, supporting a rounded plywood seat and backrest (see also p. 22). Later, Prouvé created a variant on the design that could be flat-packed and self-assembled in a matter of minutes. The Standard chair was a highly versatile and hardwearing piece, suited to many different contexts, and is now produced by Vitra.

Prouvé never stopped innovating. His career re-started after World War II, with new furniture designs and construction projects, tempered by the loss of control of his own factory in the Fifties after a dispute with shareholders. There were many other significant furniture designs, including the Guéridon table of 1949, the Compas table and desk (1953), and the Antony chair of 1954. Today, interest in this remarkable '*constructeur*' is more acute than ever.

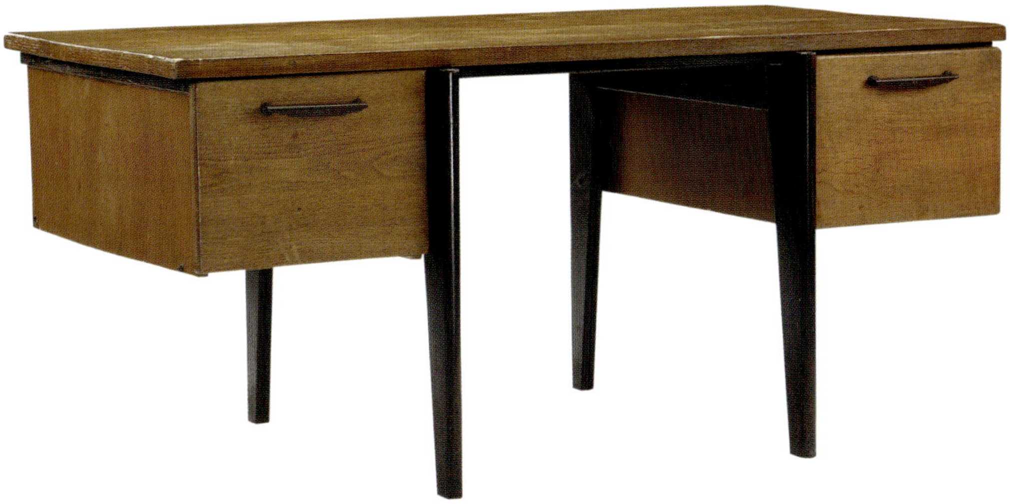

ABOVE Prouvé's Standard desk, designed in 1941, made by Ateliers Jean Prouvé, in enamelled steel and oak.

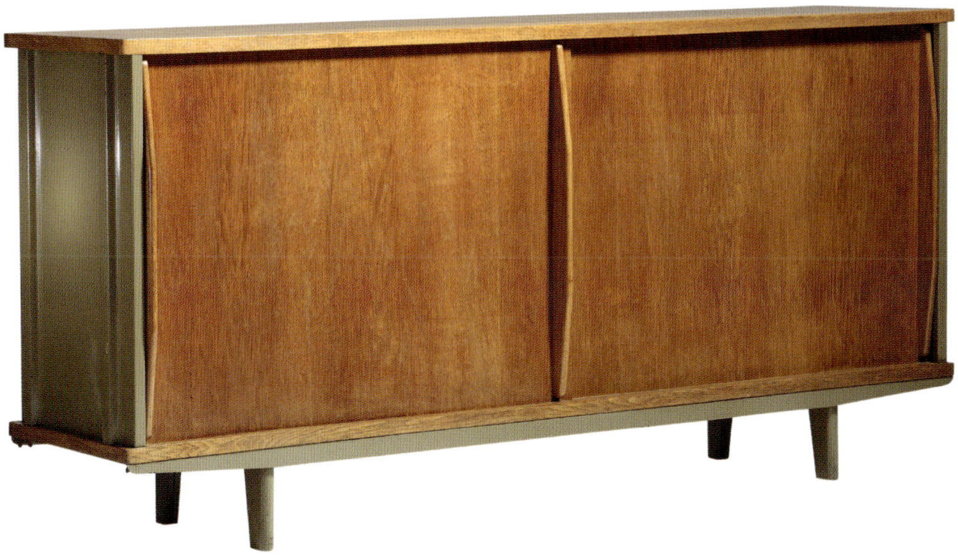

ABOVE Cabinet No. 150, produced by Ateliers Jean Prouvé, 1950, in oak and enamelled steel.

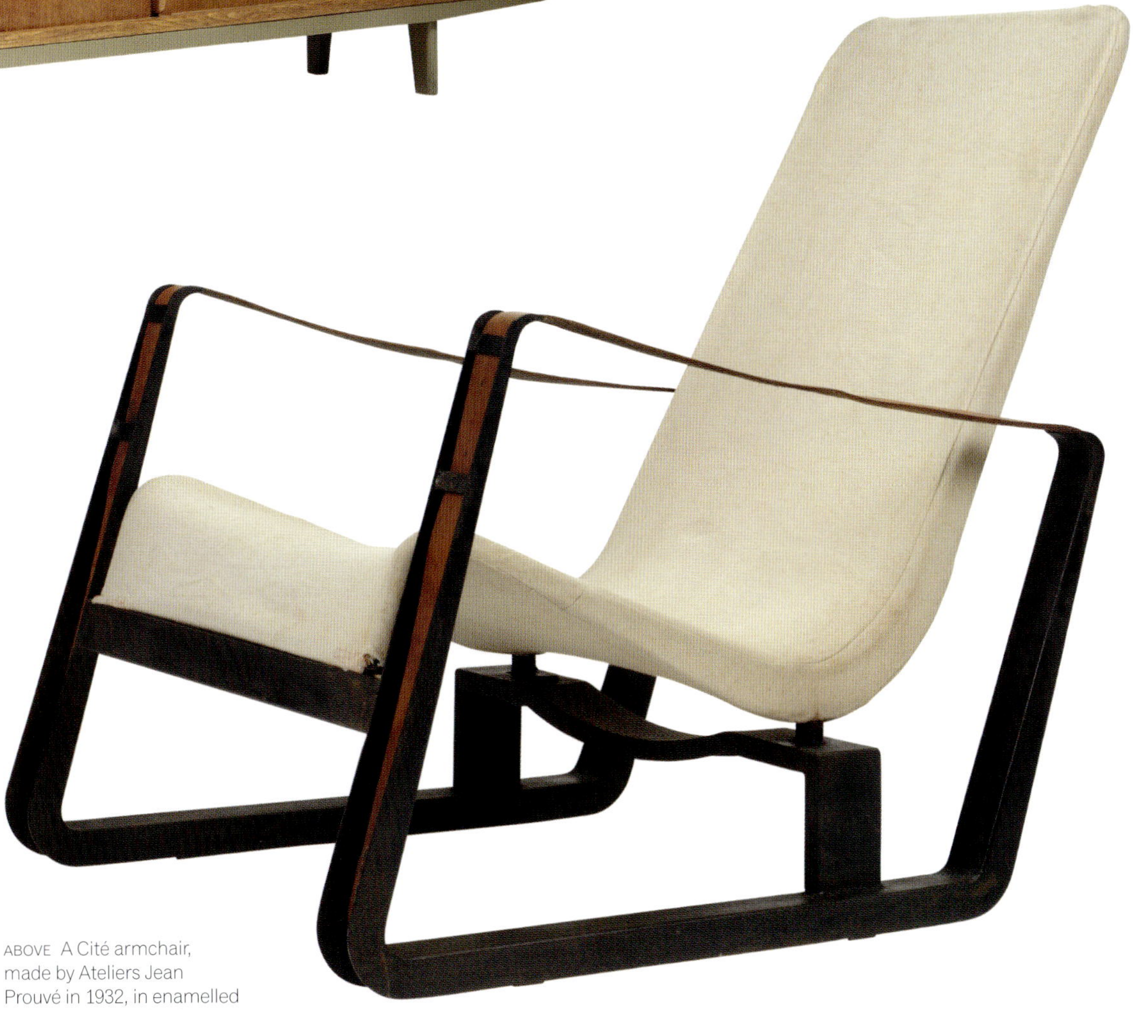

ABOVE A Cité armchair, made by Ateliers Jean Prouvé in 1932, in enamelled steel, canvas and leather.

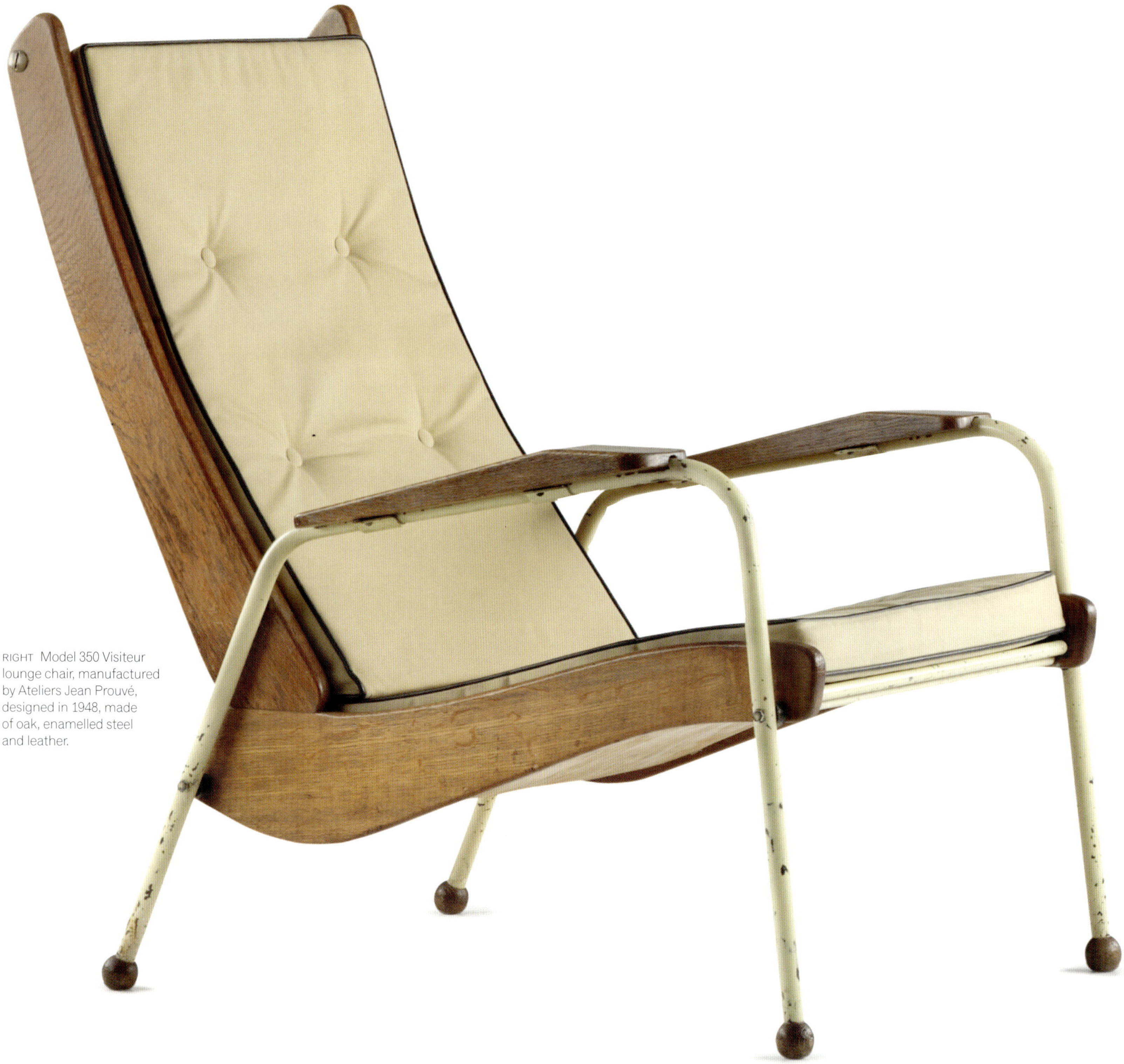

RIGHT Model 350 Visiteur lounge chair, manufactured by Ateliers Jean Prouvé, designed in 1948, made of oak, enamelled steel and leather.

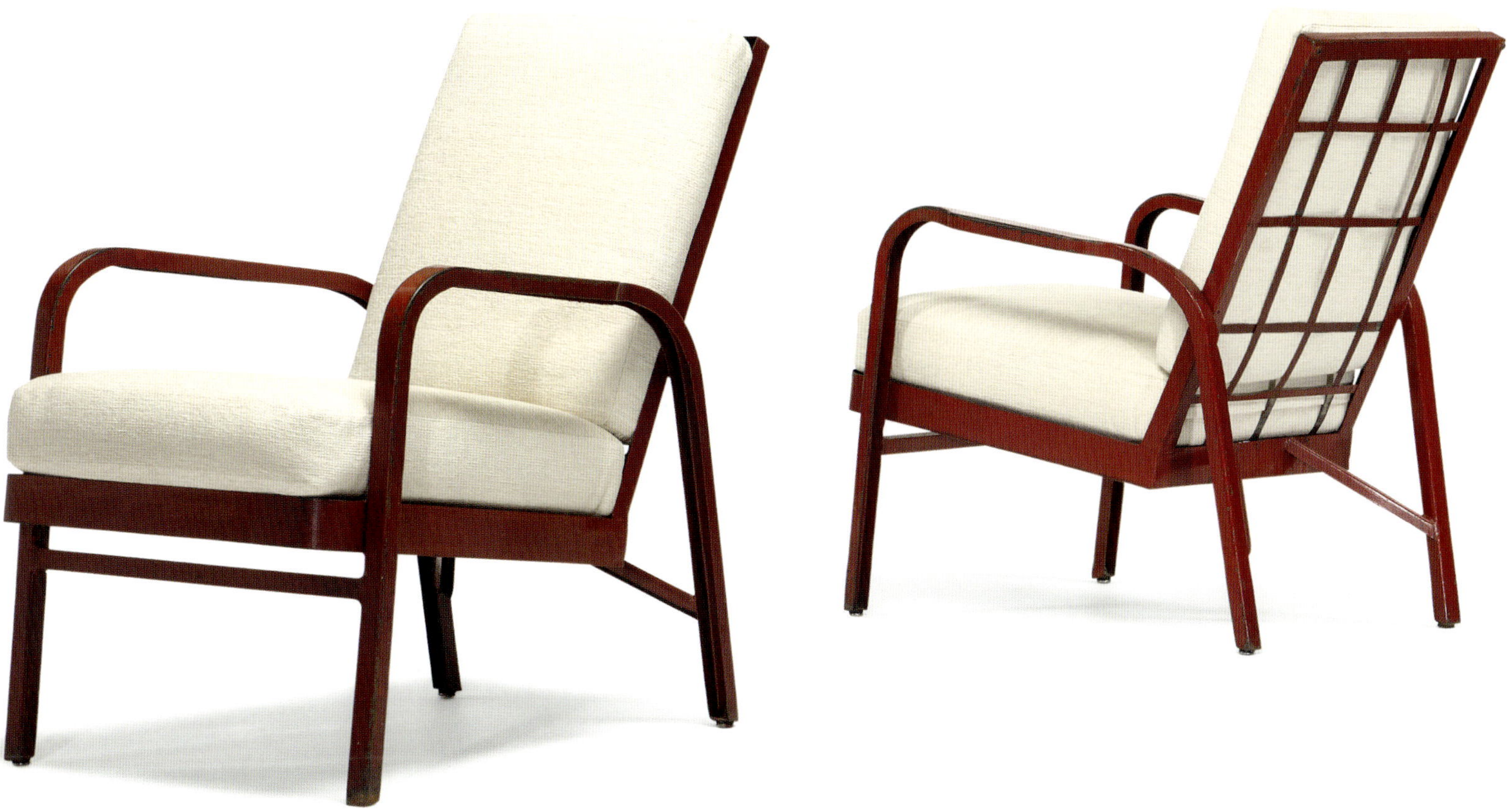

ABOVE A pair of lounge chairs by Jean Prouvé and Jules Leleu, 1936, in enamelled steel with upholstered cushions.

RIGHT A pair of occasional tables, produced by Ateliers Jean Prouvé, *c.* 1935, in enamelled steel and oak.

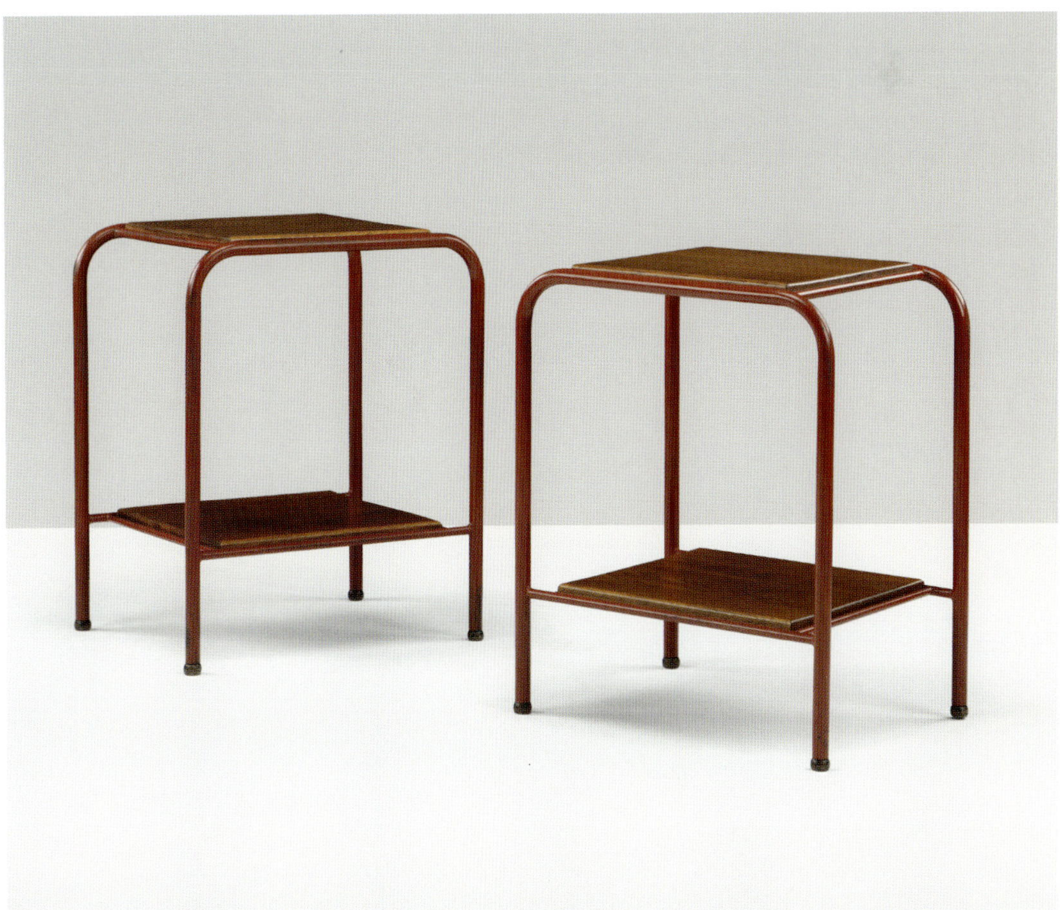

BELOW Prouvé's Dactylo desk, Model BD41, made by Ateliers Jean Prouvé, 1946, in enamelled steel and stained oak.

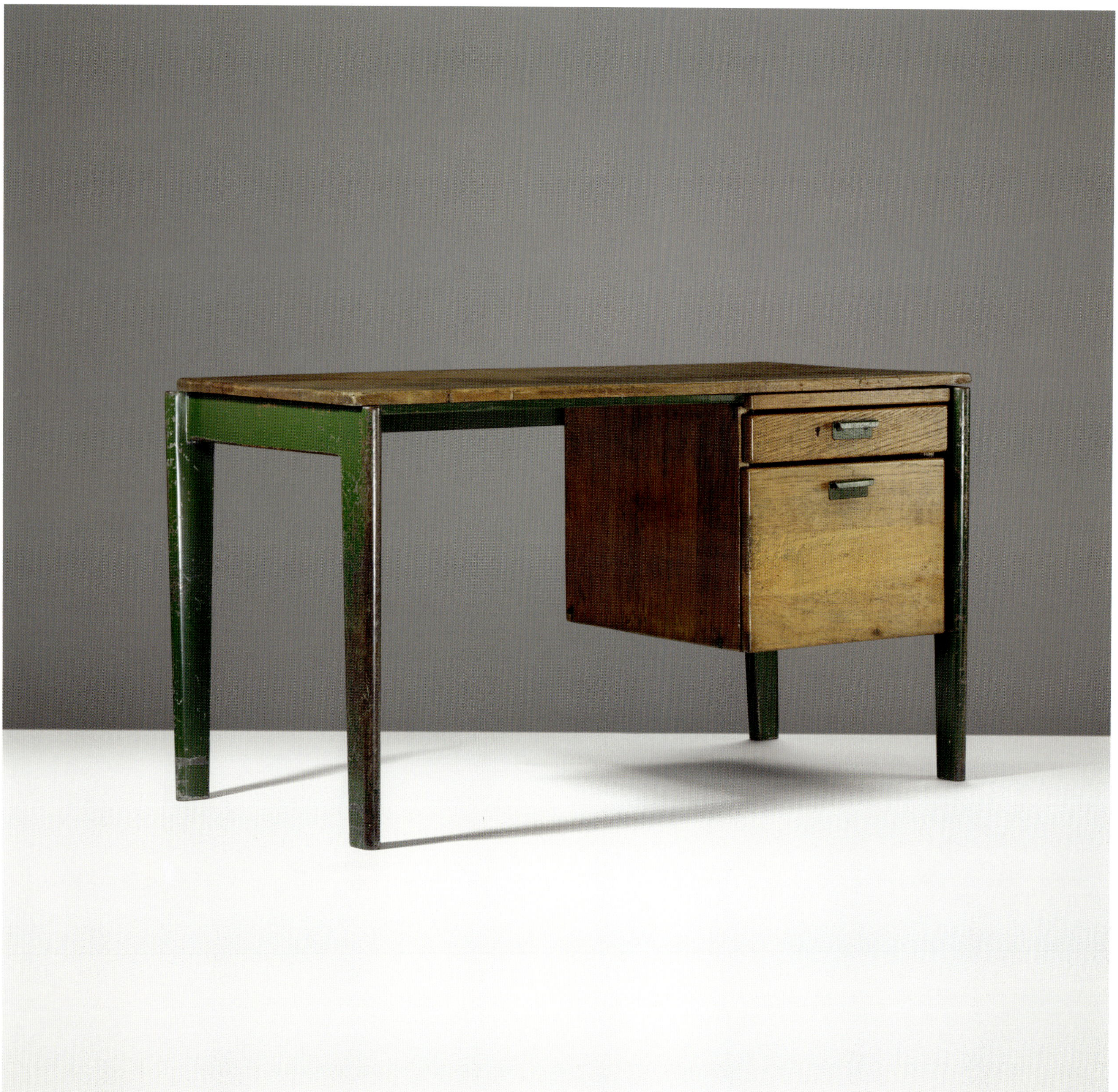

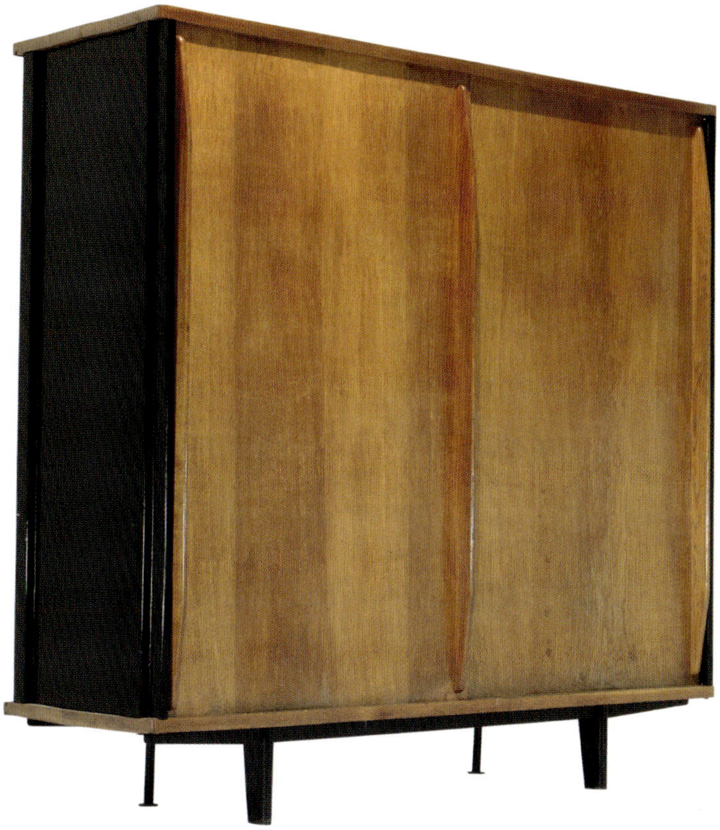

LEFT An armoire with two doors and three adjustable interior shelves, produced by Ateliers Jean Prouvé, 1945, in oak, enamelled steel and aluminium.

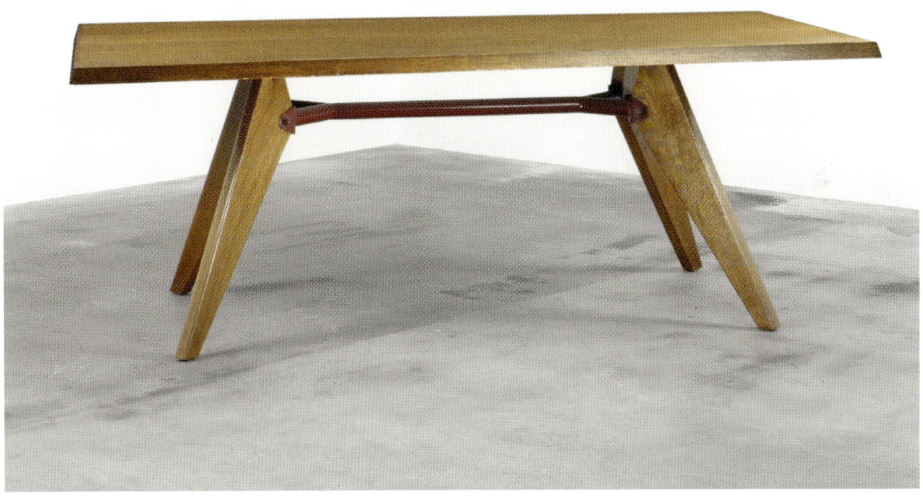

ABOVE A dining table made by Ateliers Jean Prouvé, 1942, in enamelled steel and white oak.

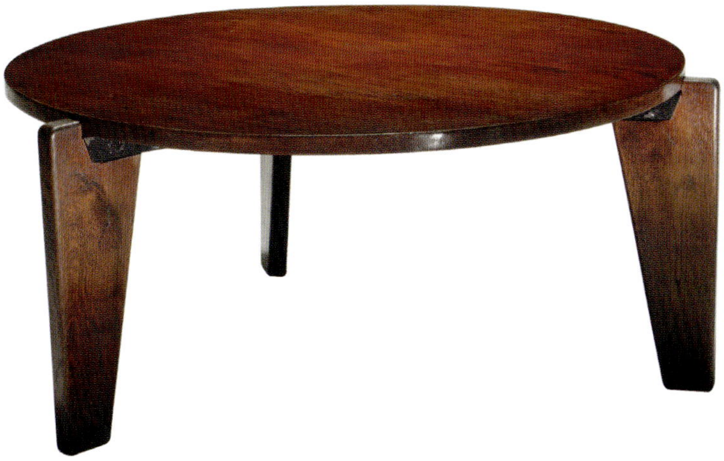

LEFT A Guéridon Bas coffee table, produced by Ateliers Jean Prouvé, designed c. 1942, made of oak and enamelled steel.

GERRIT RIETVELD
(1888–1964)

With the design of both a chair and a house Gerrit Rietveld was able to encapsulate an entire design movement, known as De Stijl*. His Red & Blue armchair of 1918 has often been described as the manifesto of De Stijl expressed in three-dimensional form, with its startling geometrical abstraction, intersecting planes and vivid use of primary colours. The Schröder House of 1924, in Utrecht (see p. 396), applied the same principles and themes on a larger and broader scale, creating a highly individual and influential home, full of ideas, bespoke elements and fitted furniture. The chair and the house are undoubted icons of De Stijl and modernism itself, of which De Stijl formed one of many sub-divisions.

Rietveld, like so many designers of the modernist period, grew up immersed in a world of tradition but broke away from its conditioning in dramatic fashion. He was born in Utrecht, where he lived and worked for much of his life. His father – Johannes Rietveld – was a respected cabinetmaker who produced work in traditional styles, according to fashion and need. The young Rietveld served an apprenticeship in his father's atelier, supplemented by evening classes in the industrial arts.

By 1915 Rietveld was producing furniture himself and gathering a small and progressive clientele; by 1917 he had established his own workshop in Utrecht. Just a year later he produced the first unpainted prototypes of what became the Red & Blue chair – a piece of furniture reduced down to a distinctive linear wooden frame painted black, with yellow detailing for the ends of each piece of timber, while a blue plank forms the seat and a red one provides a high back. These planks appear to float upon the frame in a disconnected way, while all the joints and tectonic elements of the chair are hidden. Here, magical abstraction, artistic form and bold colours are everything, like a painting turned into a piece of furniture.

'With this chair an attempt has been made to allow each component simply to be what it is, and that in the most elementary form, according to function and materials, in the form that is the most responsive, through proportionality, to attaining harmony with the whole,' said Rietveld. 'The construction helps to interconnect the parts without mutilating them, to ensure that no part dominates or is subordinate to the others, in order that above all the whole stands free and clear in space and the form wins from the material.'[13]

The Red & Blue chair was taken up by the pioneers of De Stijl and published in its journal, helping to promote Rietveld's work and bring in new clients; in 1924 the chair was put into production by Gerard van de Groenekan and, much later, in the 1970s, by Cassina. It was one of a series of De Stijl-inspired pieces that also included the Hogestoel of 1919 and the Berlin chair of 1923. Just a few years later Rietveld's passion for experimentation led him to tubular steel, taking inspiration from Marcel Breuer's pioneering Bauhaus pieces. There was a tubular steel dining chair with a plywood seat designed for the Schröder House (1924) and the Beugelstoel 2 armchair (1927), also in tubular steel and plywood, which was put into production by Metz & Co.

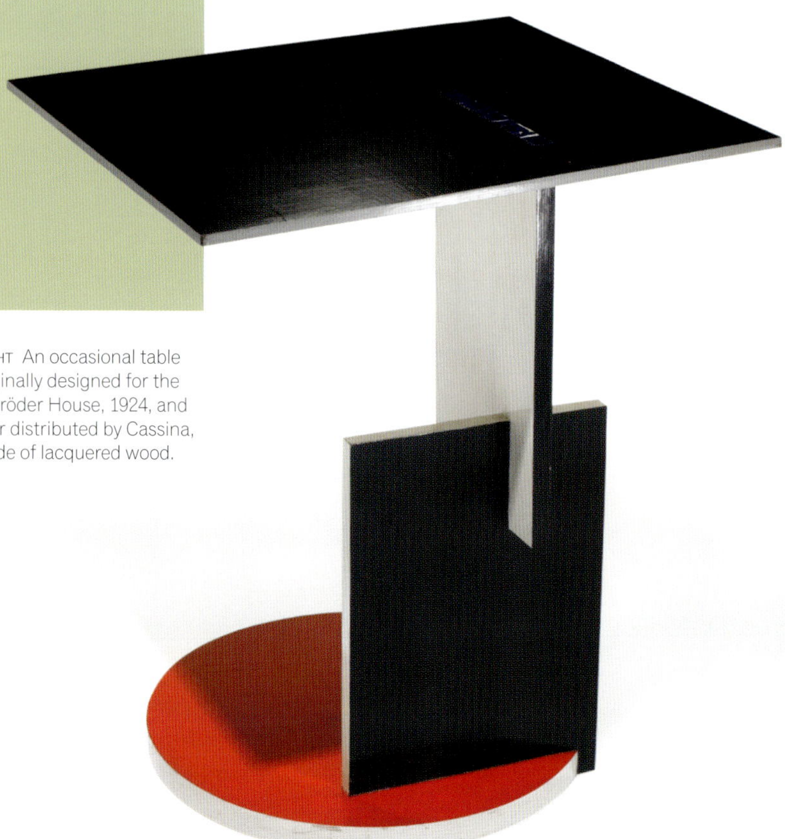

RIGHT An occasional table originally designed for the Schröder House, 1924, and later distributed by Cassina, made of lacquered wood.

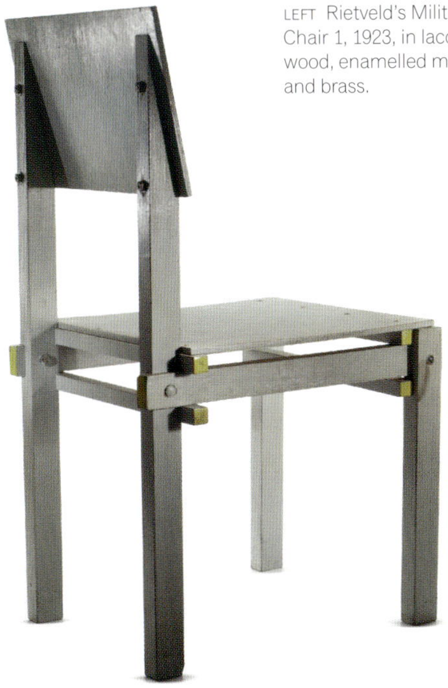

LEFT Rietveld's Military Chair 1, 1923, in lacquered wood, enamelled metal and brass.

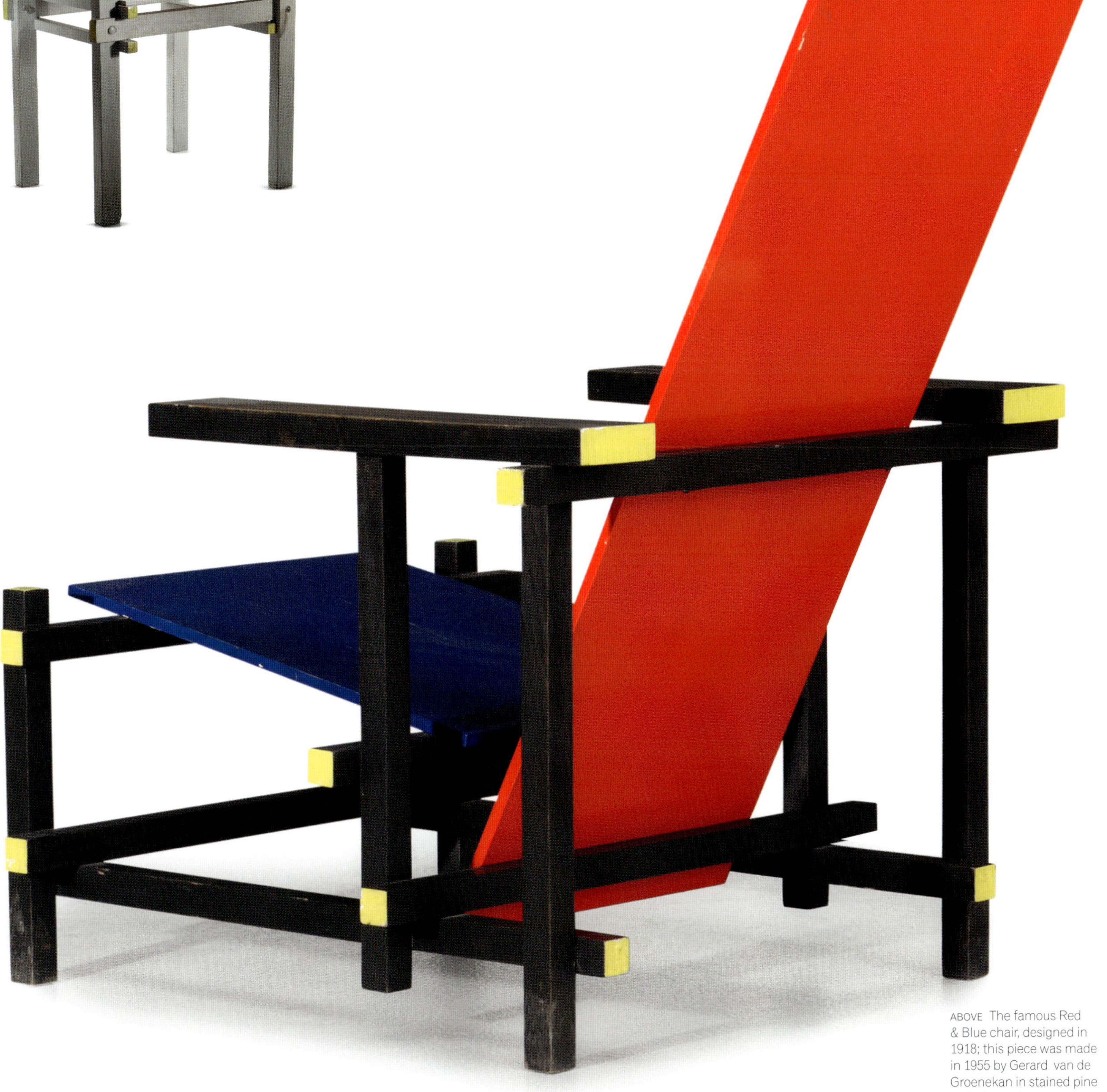

ABOVE The famous Red & Blue chair, designed in 1918; this piece was made in 1955 by Gerard van de Groenekan in stained pine and plywood.

In addition, Rietveld experimented with lighting, creating his famous Hanging lamp in 1922, which featured in the Schröder House. This was a combination of three or four lighting tubes arranged in a vertical and horizontal formation and held in place by wooden fittings and rods, which were attached to a ceiling plate. Again, it represented a process of reduction combined with artistic abstraction.

Increasingly, from the Thirties onwards, Rietveld began to focus on architecture, with the shape of his career echoing Breuer's, although none of his later buildings had the radical impact of the Schröder House. But there were also other significant pieces of furniture, both in the pre- and post-war periods. Chief among these is the Zig Zag chair (1934), which represented another process of paring down to the essential elements of form, structure and materiality. Rather like Breuer's cantilevered steel chairs, Rietveld's timber zig-zag has an air of impossibility, as though it should collapse the moment you sit down upon it. Only closer inspection reveals the wedges and brass fittings that keep the four intersecting planes of oak that form the zig-zag from collapsing.

Rietveld's Crate chair of 1934 was made of the kind of rough red spruce and raw timber used for packaging, with a purposefully unrefined finish to the wood and exposed fixings. It has served as a highly inspirational piece for designers interested in the 'poverty' of materials and employing recycled or salvaged timbers. There was also the upholstered Utrecht armchair of 1935, initially made by Metz & Co. and most famously coated in felt; the chair was later reissued by Cassina. In the early Forties, Rietveld turned back to metallics, producing his perforated Aluminium chair of 1942, with his love of experimentation and making it new continuing into the post-war period.

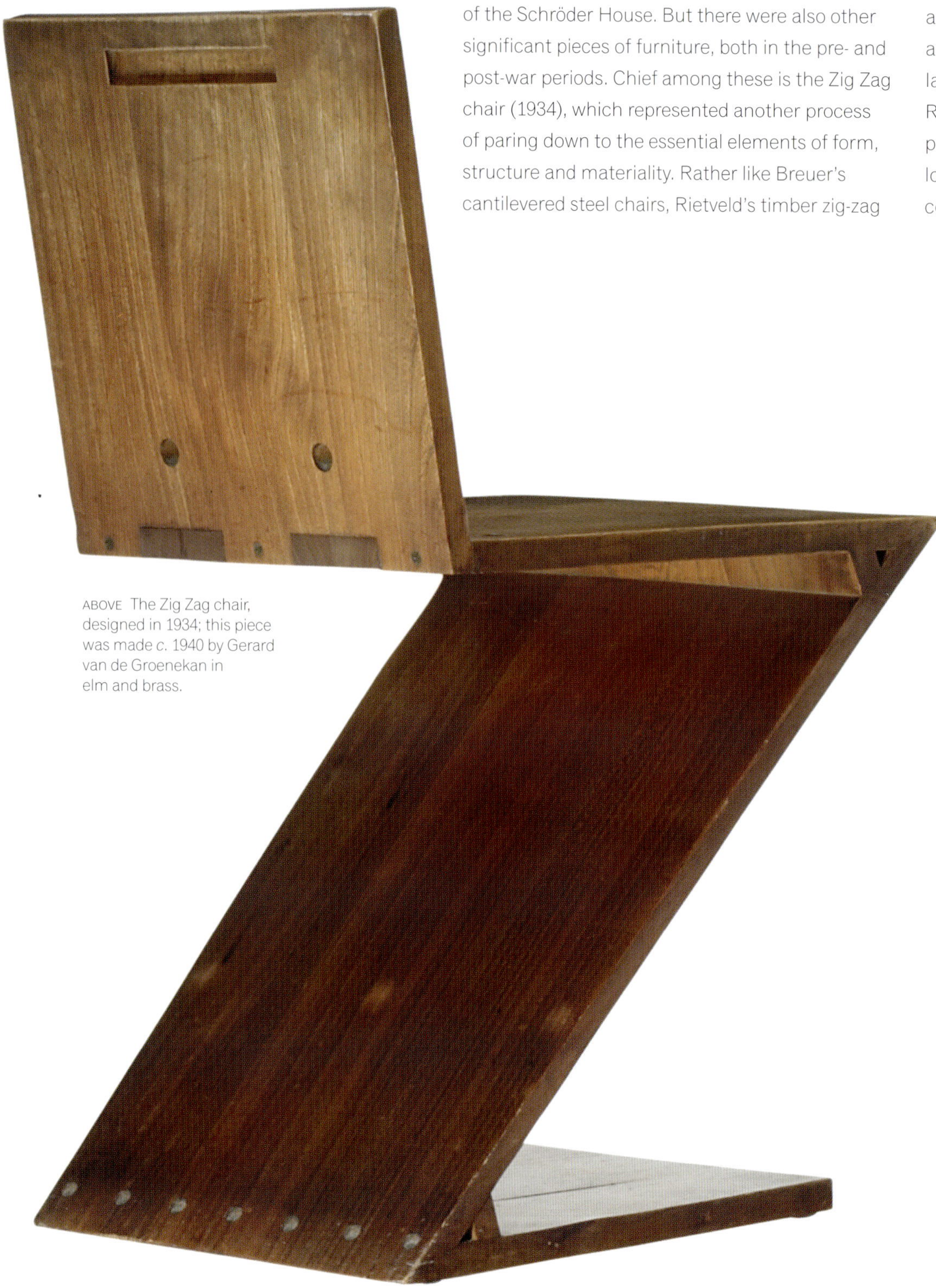

ABOVE The Zig Zag chair, designed in 1934; this piece was made *c.* 1940 by Gerard van de Groenekan in elm and brass.

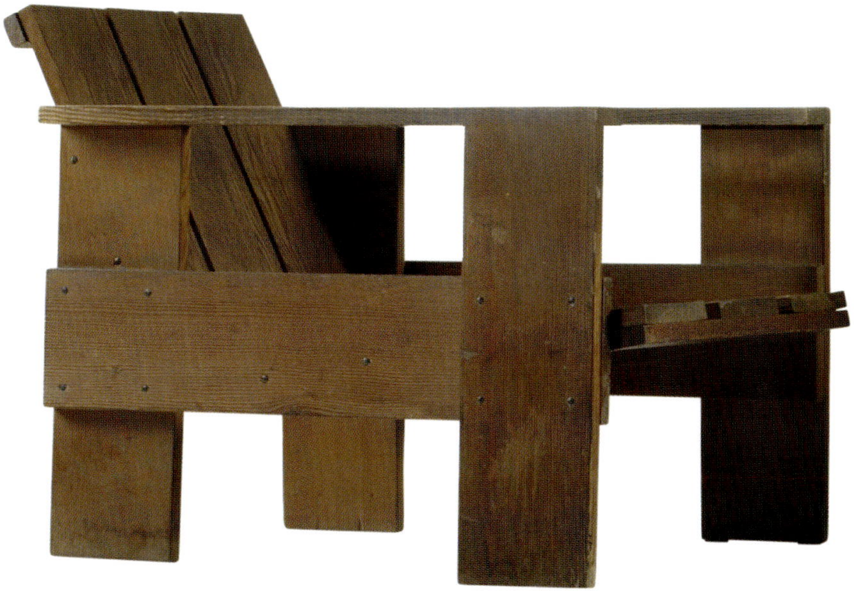

ABOVE A pair of Rietveld Crate chairs, designed in 1934, made with stained pine.

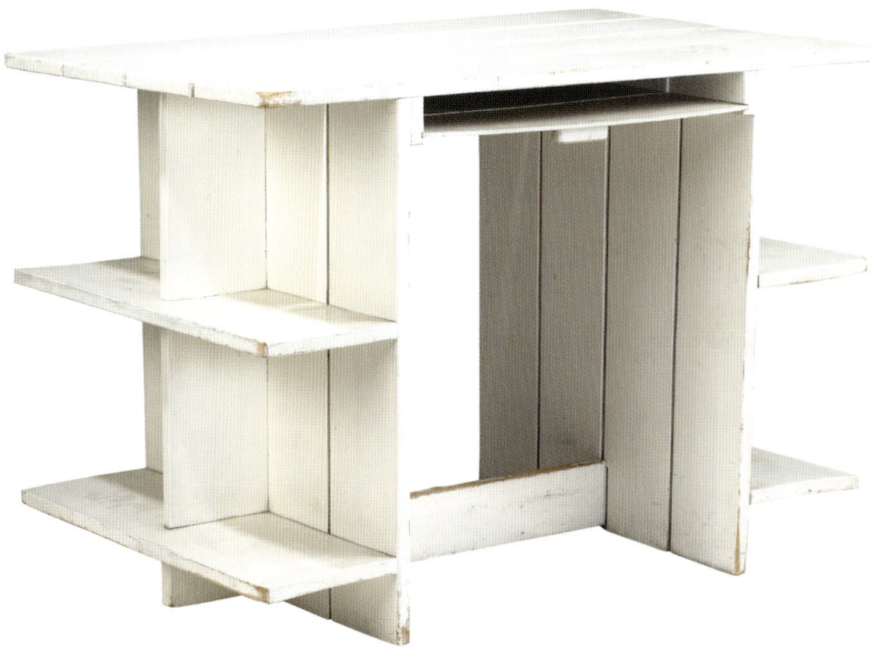

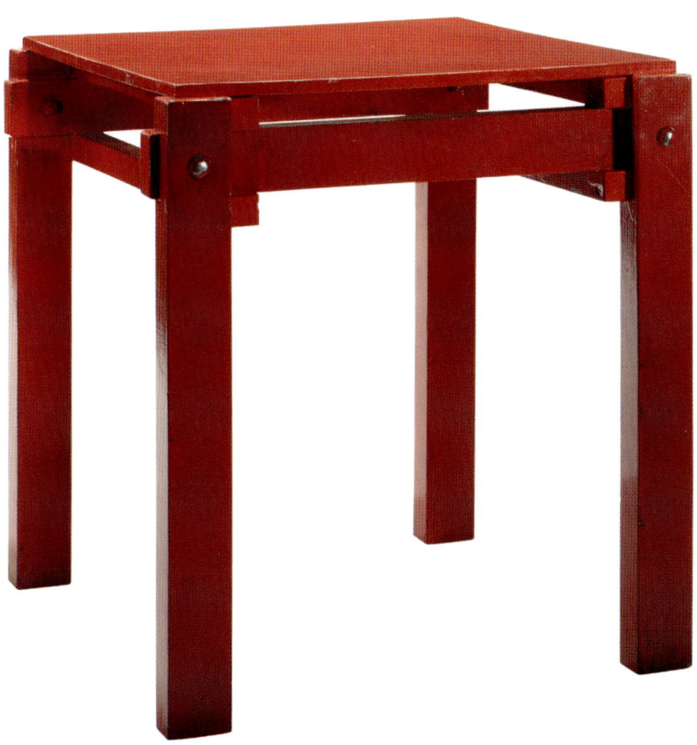

ABOVE A Crate desk, made by Metz & Co., c. 1934, in painted pine.

LEFT A Military stool, 1924, made of lacquered wood and enamelled metal.

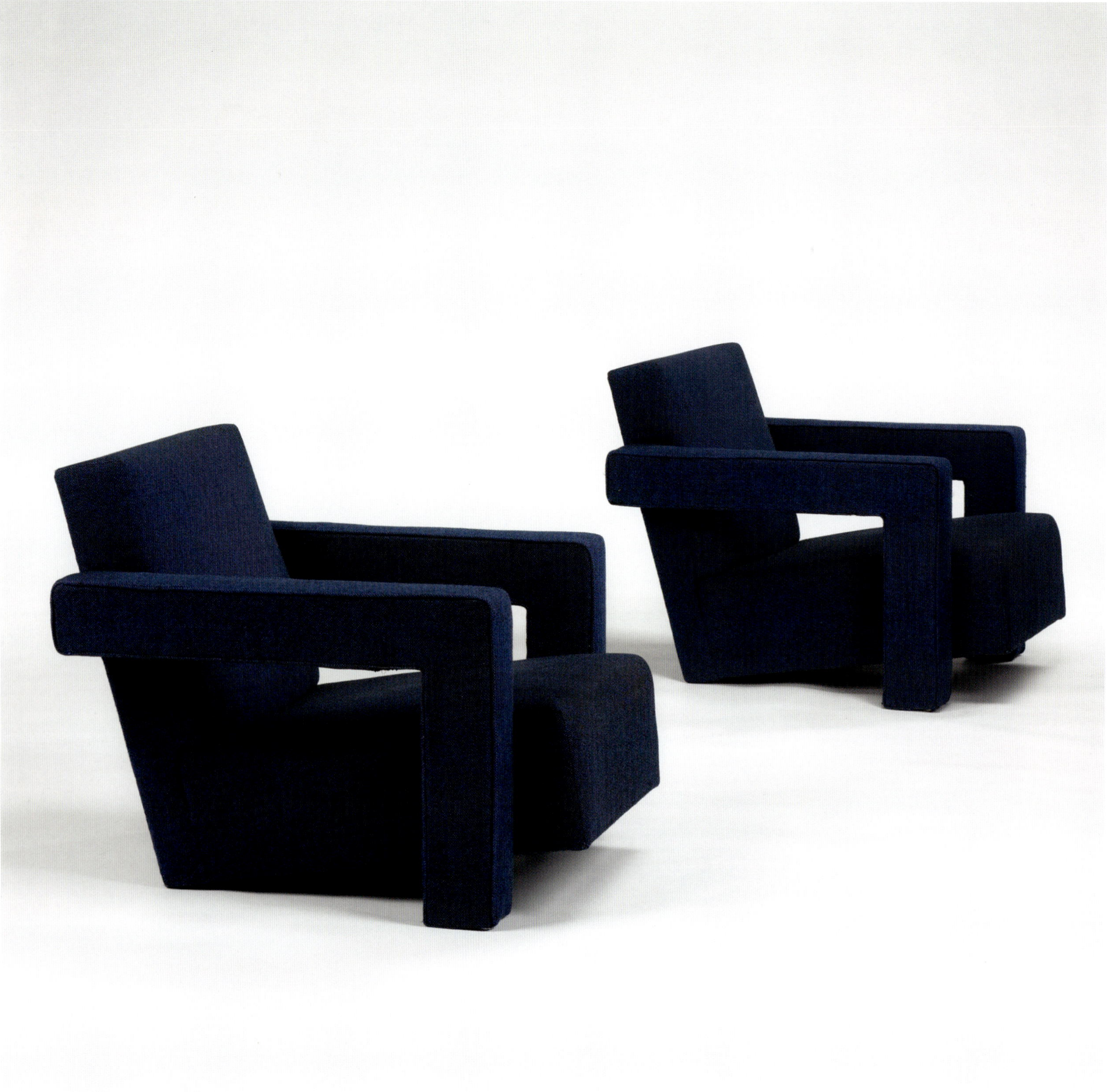

ABOVE A pair of Utrecht armchairs, designed in 1935, made by Metz & Co. with felt-coated upholstery over wood.

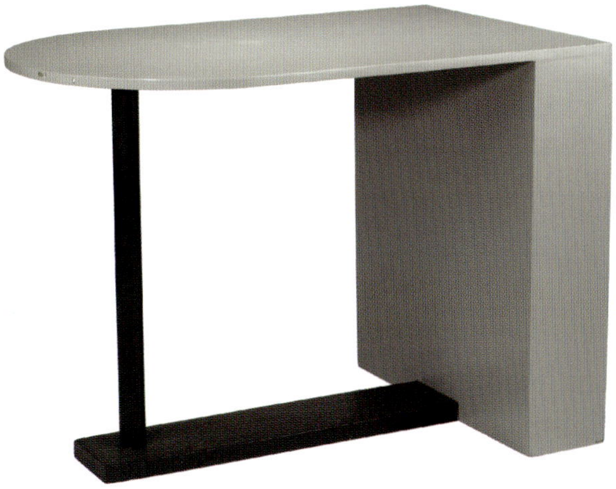

ABOVE A writing desk, made by Metz & Co., *c.* 1940, in lacquered wood.

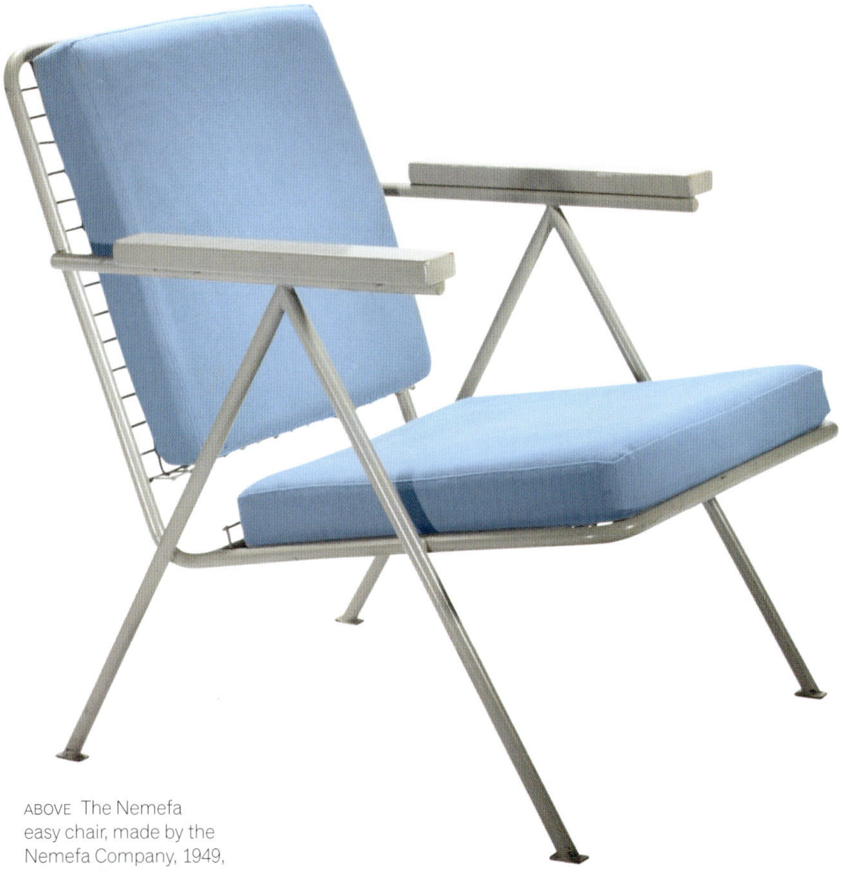

ABOVE The Nemefa easy chair, made by the Nemefa Company, 1949, in enamelled steel and lacquered wood with upholstered cushions.

RIGHT A Beugelstoel chair, produced by Gerard van de Groenekan, *c.* 1927, in steel, aluminium and lacquered wood.

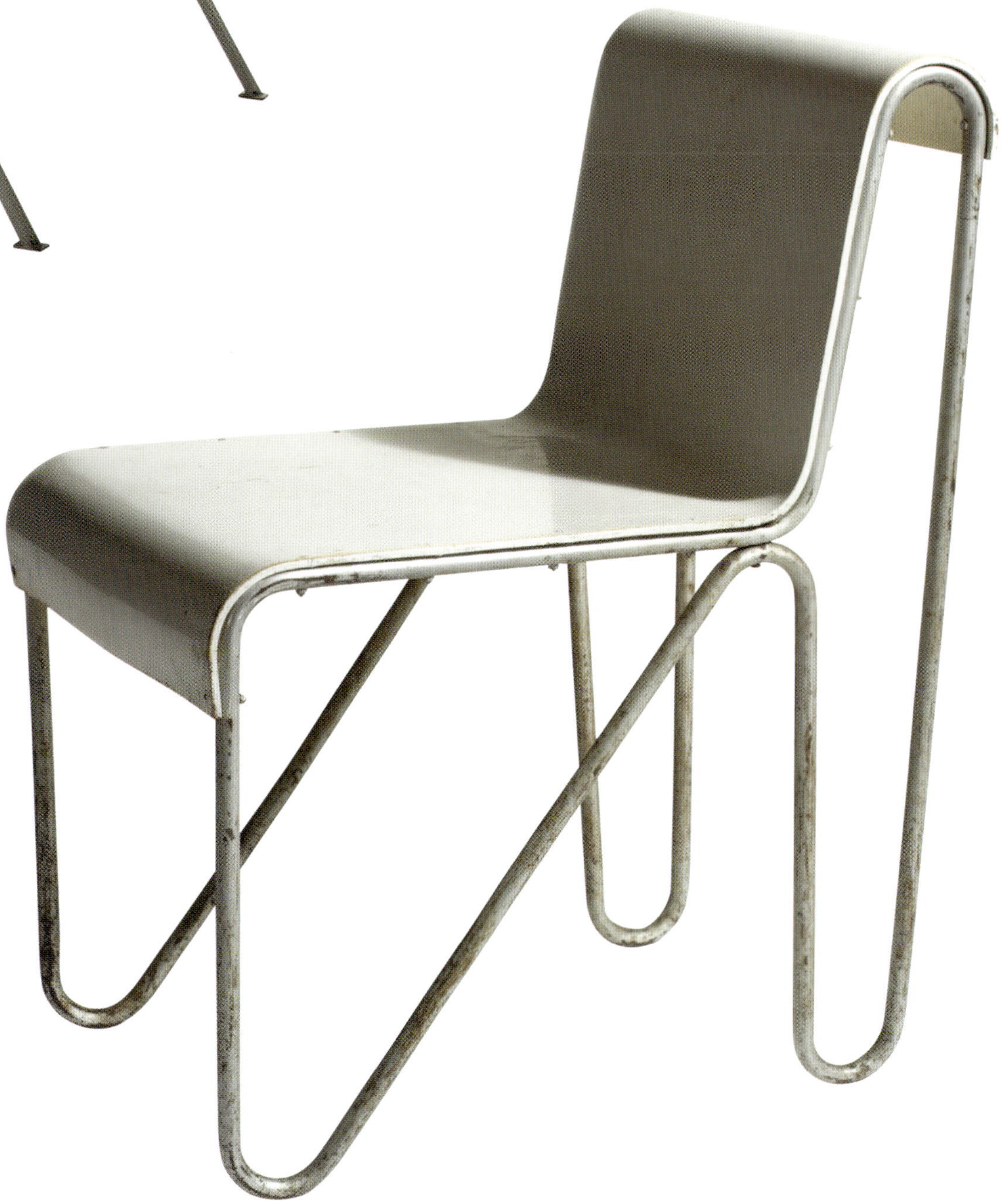

GILBERT ROHDE
(1894–1944)

New York designer Gilbert Rohde played a key part in the evolution of modernist furniture in the pre-war era, before an untimely death cut a prolific career short. He was at the forefront of a new approach to furniture design that on the one hand involved exploring new materials and mass-production techniques but, on the other, sought to create modular systems of furniture and carefully edited groups of complementary pieces for home and office use. In doing so, he was able to offer his audience the opportunity to gradually build a range of pieces with a common synergy and develop a rounded, cohesive, modern look for the home. Rohde helped to revive the Herman Miller* furniture company with his focus on flexible modernity and paved the way for the growth of modular systems, sectional sofas and the notion of an annual or seasonal 'collection' of furniture.

Gustav Rohde was born in the Bronx district of New York to a cabinetmaker father from East Prussia and a German mother; his parents had arrived separately in America in the 1880s. His father died before Gustav – who eventually changed his name to Gilbert – was ten years old and he was largely brought up by his aunt. Rohde attended Stuyvesant High School, which offered vocational classes in fields such as woodworking, metalwork and draughtsmanship. After leaving school he worked briefly as an illustrator for a local newspaper and then moved into illustration work for department stores and advertising agencies, while also attending evening classes.

Following the failure of a brief first marriage, he met his second wife – Gladys Vorsanger – through his illustration work. It was Vorsanger who suggested that Rohde travel to Europe in 1927 to see, at first hand, the work being produced by the influential early modernists in Germany and France. He spent four months overseas, including an extended stay in Dessau, where he spent time at the Bauhaus and was much impressed by the work of Marcel Breuer and the furniture department, particular Breuer's experiments with tubular steel and industrialized production techniques.

Back in New York, Rohde started his new career as a furniture designer with a fresh impetus, embracing the Bauhaus example. He set up a home studio and began producing his first pieces for the Lord & Taylor department store, starting with a series of side tables with chrome bases and Bakelite or Fabrikoid tops.

During the early Thirties Rohde's career as a designer began to take off. There were a number of tubular steel pieces designed for the Troy Sunshade Company, such as a sofa with a tubular steel frame (*c.* 1930) and a cantilevered S chair

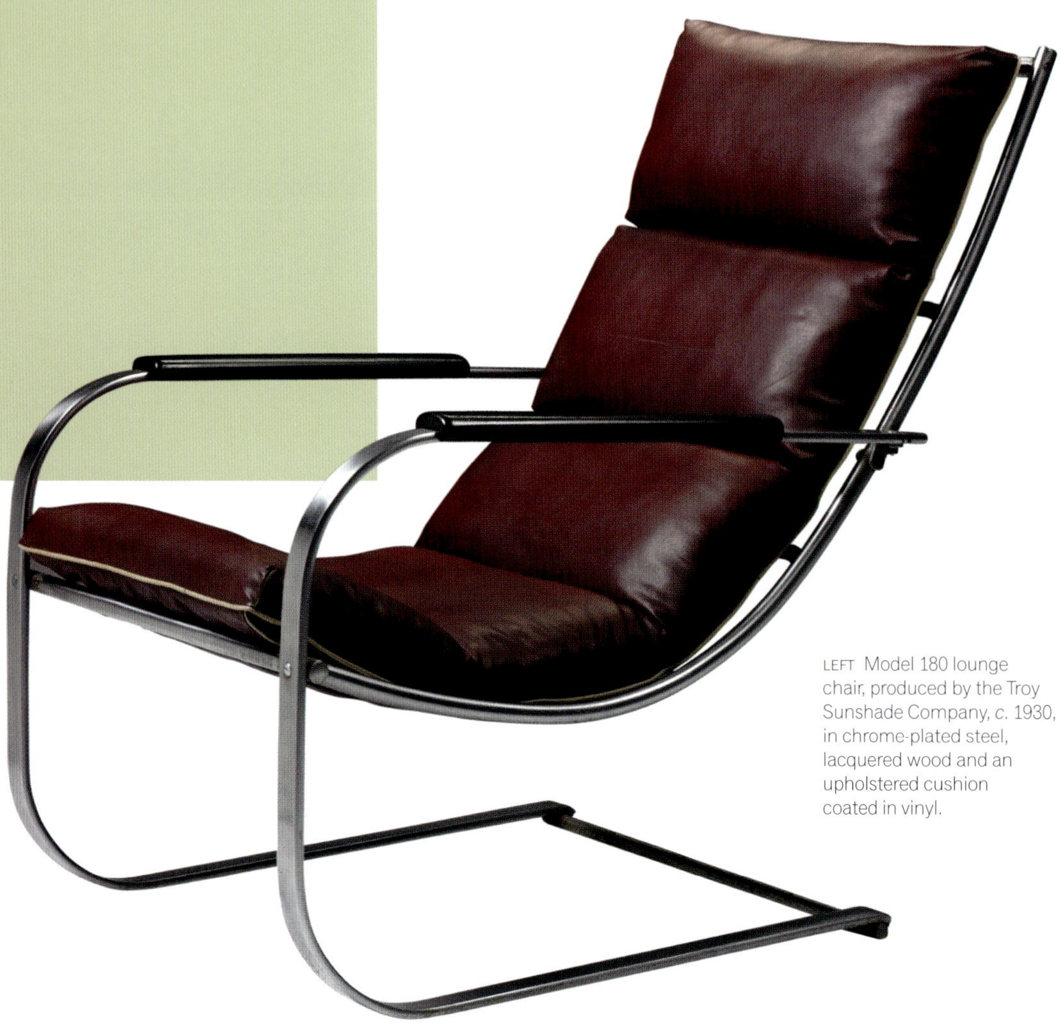

LEFT Model 180 lounge chair, produced by the Troy Sunshade Company, *c.* 1930, in chrome-plated steel, lacquered wood and an upholstered cushion coated in vinyl.

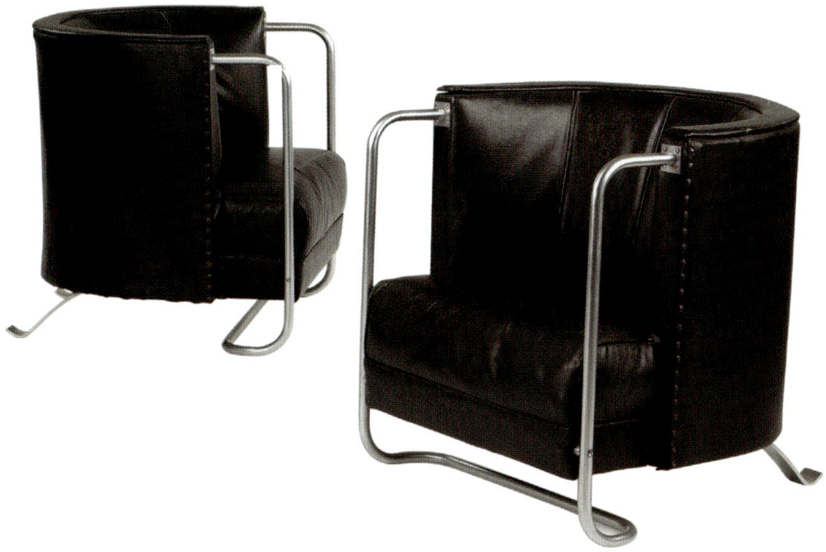

ABOVE A pair of lounge chairs, manufactured by the Troy Sunshade Company, *c.* 1930, in chrome-plated tubular and flat steel and upholstery.

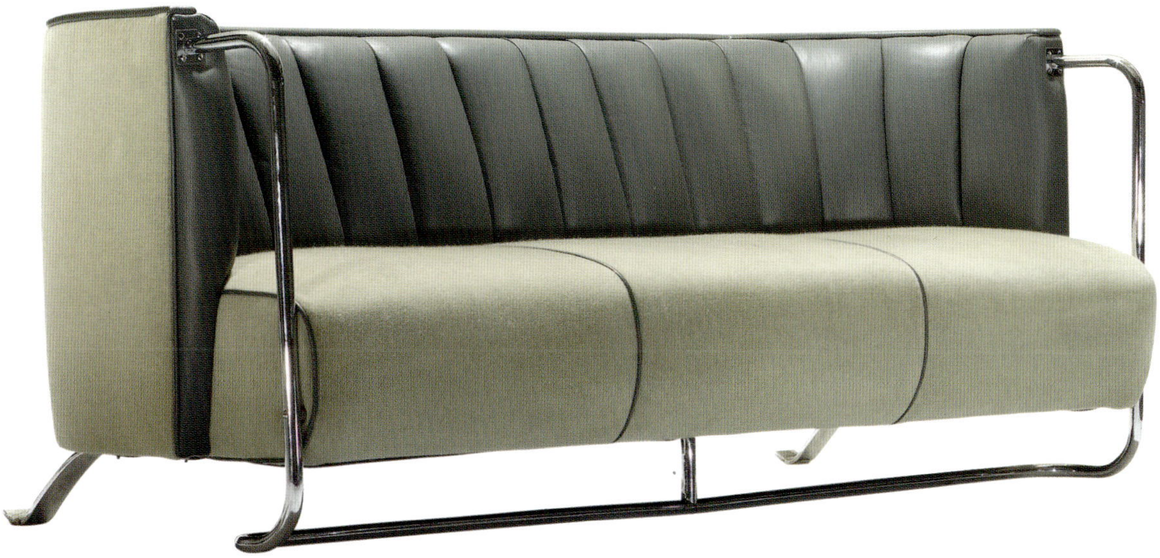

ABOVE A Gilbert Rohde sofa, manufactured by the Troy Sunshade Company, *c.* 1930, in chrome-plated steel, leather and mohair.

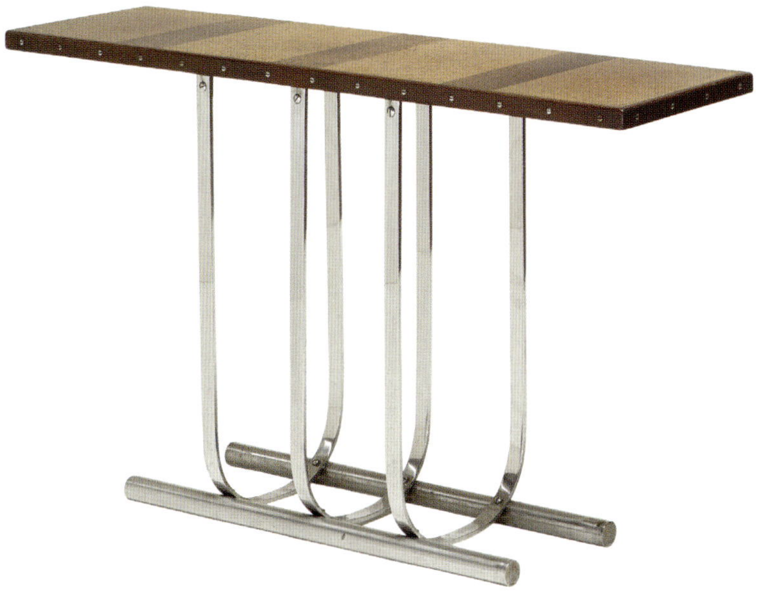

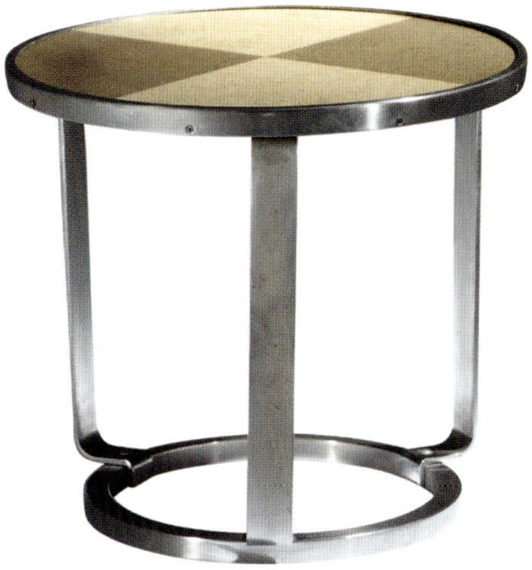

LEFT An occasional table, made by the Troy Sunshade Company, *c.* 1935, in chrome-plated steel, cork and glass.

ABOVE Model 56-B console table, produced by the Troy Sunshade Company, *c.* 1935, in chrome-plated steel, lacquered cork and stained oak.

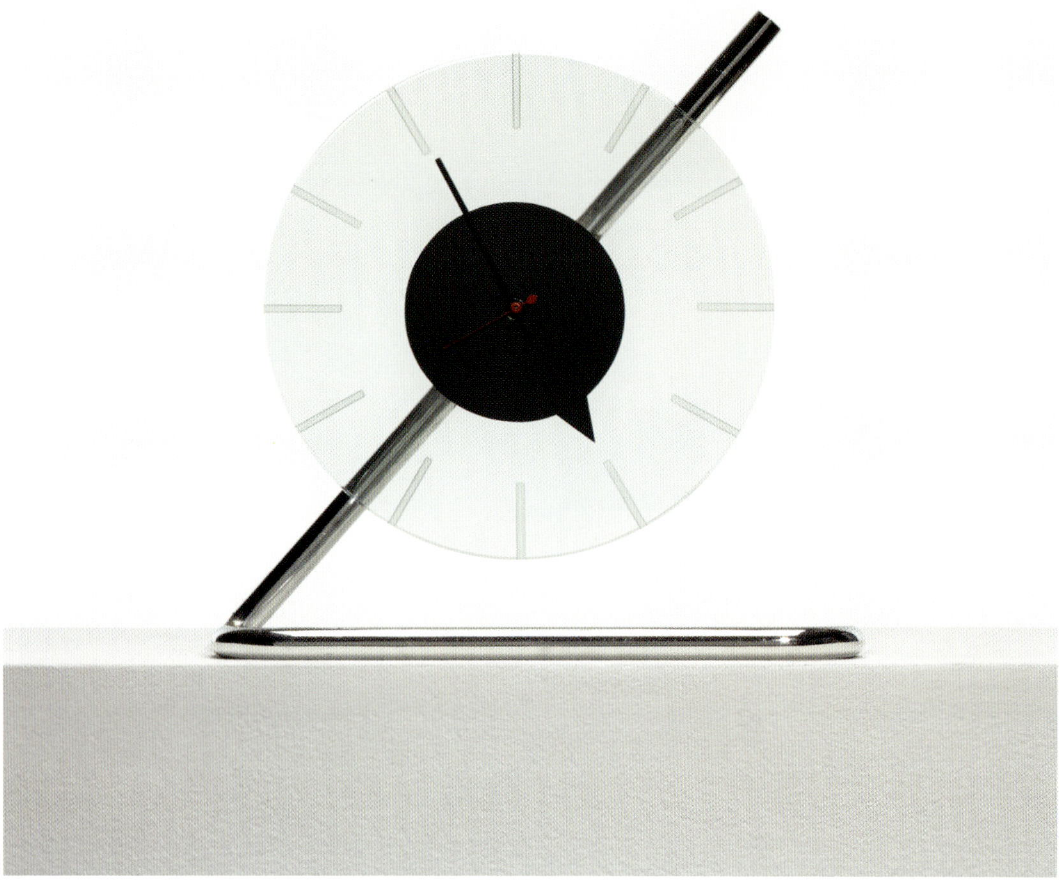

LEFT Rohde's Z-clock, Model 4090, made by Herman Miller, *c.* 1933, in chrome-plated steel, etched glass and aluminium.

(1934). Other pieces were more Art Deco* in style, influenced by Jacques-Émile Ruhlmann* and his contemporaries, while some designs have been categorized as Streamline Moderne.

What soon became apparent was the breadth of Rohde's frame of reference, combined with a commitment to modernity and innovation. He worked with more crafted materials such as burled white ash, mahogany and paldao timber, but he also used chrome-plated steel, cork, glass, Bakelite, Formica, Lucite, Plexiglas and other more industrial materials.

In 1932, Rohde began working with the Herman Miller furniture company, a relationship that would define Rohde's design and revive a business that had become stuck in its ways. The company had been founded in 1923 by Miller and his son-in-law D. J. De Pree, and in the early Thirties was still manufacturing traditional furniture with a period influence while struggling to cope with the impact of the Great Depression. De Pree, the company president, suggested that the business may have been a year away from bankruptcy when Rohde persuaded him to go modern.

Over the following ten years, Rohde transformed Herman Miller's entire approach to design with a series of innovations. The designs themselves were modern in character, but, more than this, Rohde introduced modular concepts little by little. There were multi-functional pieces – such as combined bookshelves and cabinets, or vanity units with integrated mirrors – and also suites of furniture for different parts of the home, such as the bedroom or kitchen. Encouraged by their successes, Rohde and De Pree introduced entire groups of complementary pieces of furniture designed with similar materials and aesthetics, pieces that could be mixed and matched with ease by the client.

By 1936, Rohde had designed around 350 modern pieces for Herman Miller, set out in detail within a company style book and presented in room sets and showrooms that gave a complete impression of how the furniture could be combined in the home. 'You are not making furniture anymore,' Rohde told De Pree. 'You are making a way of living – a lifestyle.'[14] Rohde expanded the concept of modern modularity into office furniture, introducing the Executive Office Group of designs. By the time of his death Herman Miller were offering 300 Rohde designs in their catalogue. A sister company, the Herman Miller Clock Company, produced Rohde's extraordinary collection of timepieces, including the Z-clock (1933) in chrome-plated brass, glass and enamelled metal, which still feels fresh and original today.

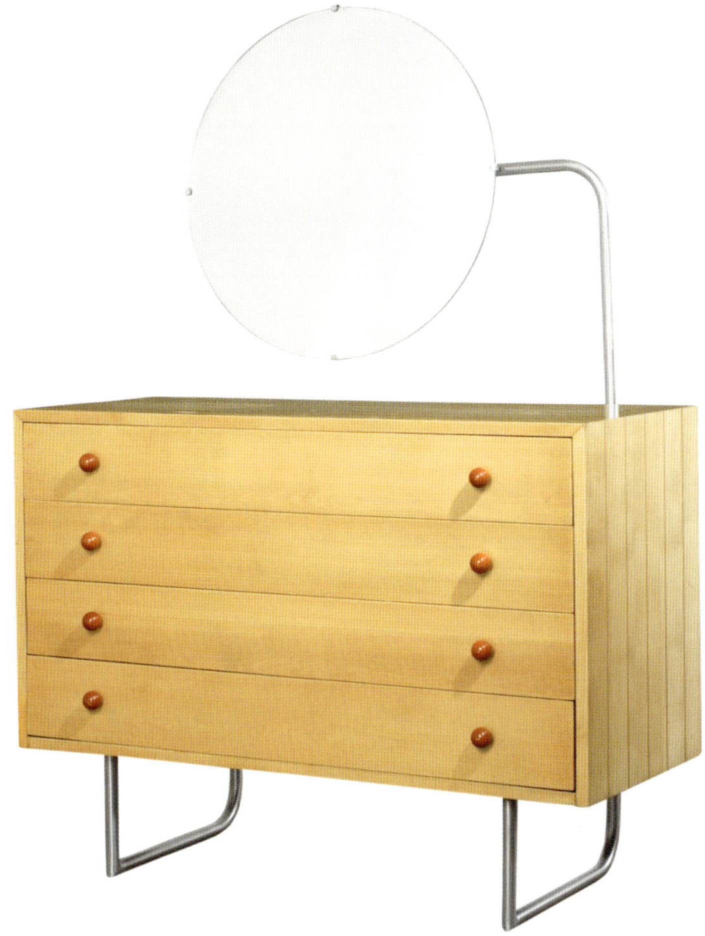

ABOVE Model 3323 dresser, produced by Herman Miller, 1933, in beech, mirrored glass, chrome-plated steel, stained wood and Bakelite.

BELOW A Model 3321 cabinet, made by Herman Miller, 1933, in maidou burl, mahogany, chrome-plated steel and brass.

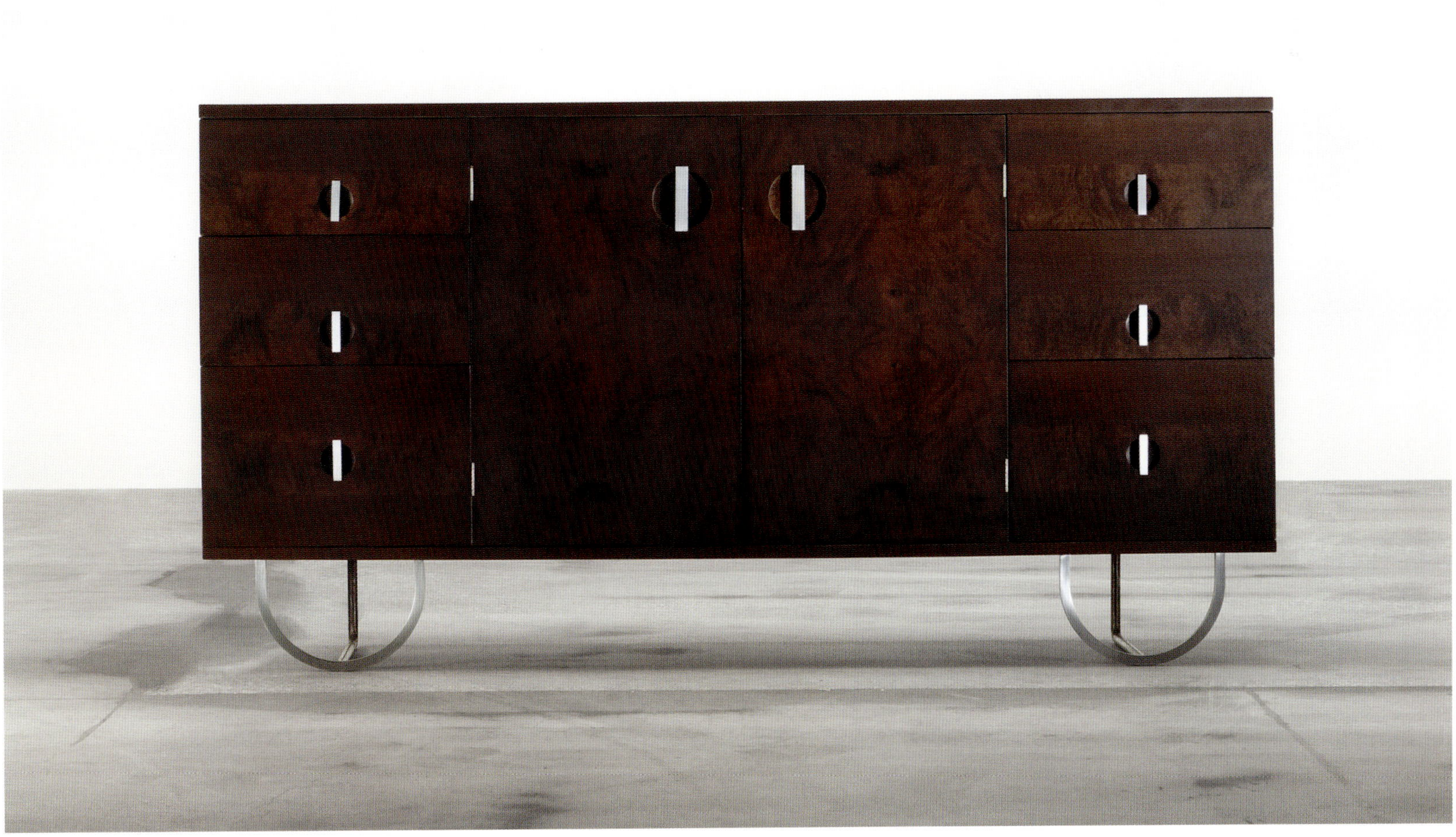

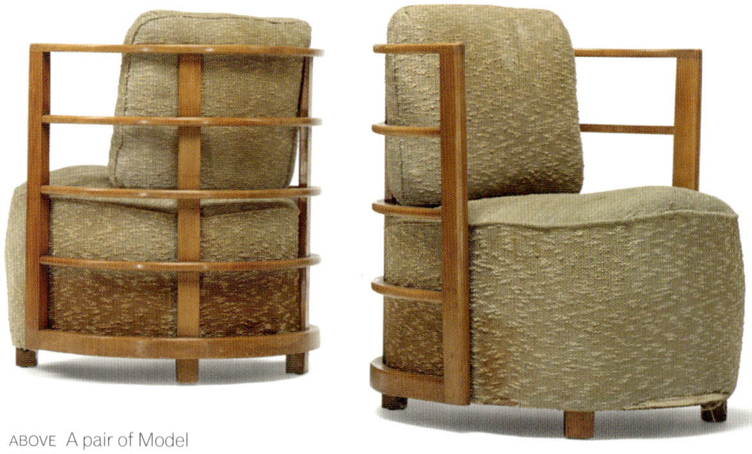

ABOVE A pair of Model 3451 bentwood armchairs, produced by Herman Miller, 1934, in beech with upholstery.

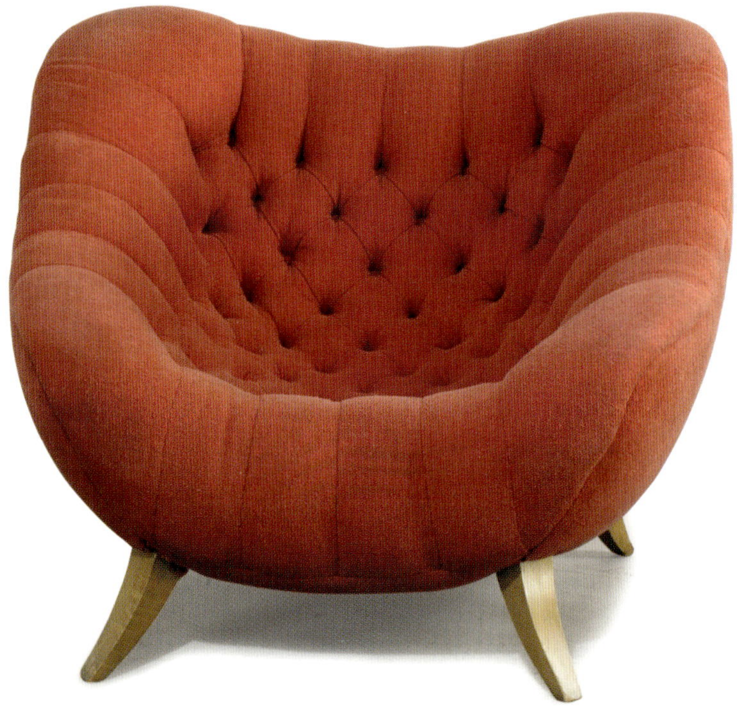

ABOVE A Gilbert Rohde lounge chair, made by Herman Miller, *c.* 1935, in bleached mahogany with upholstery.

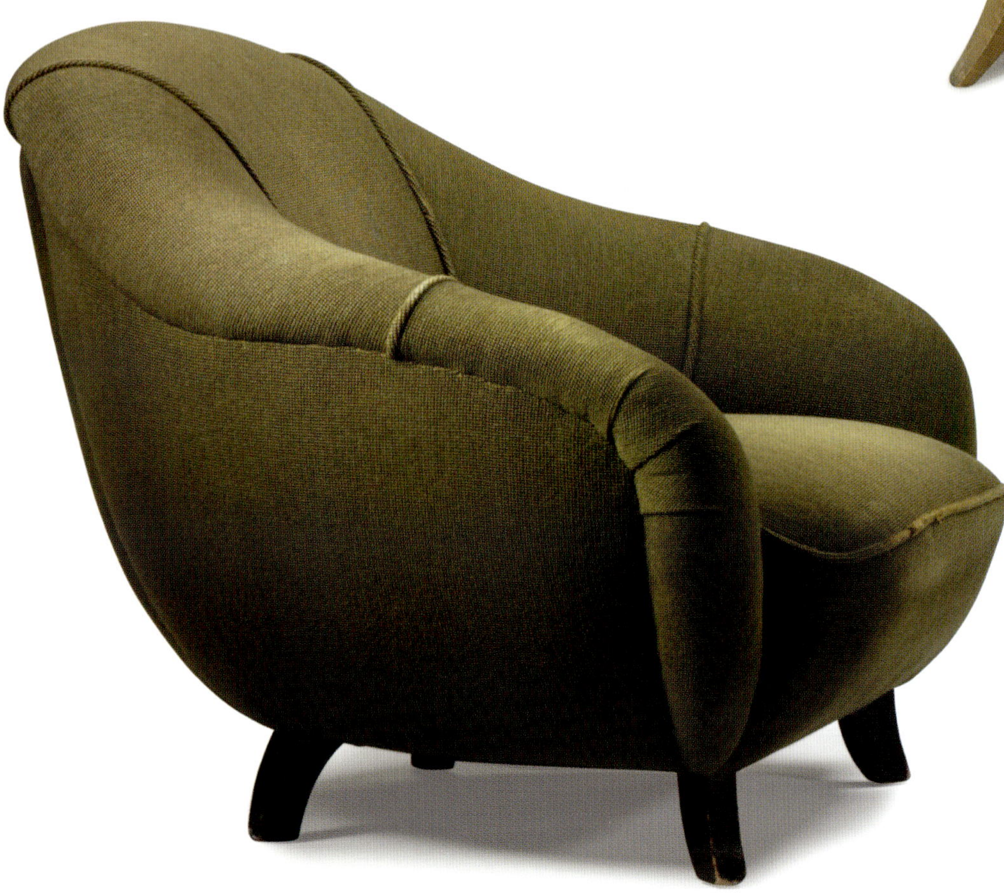

LEFT A Model 3950 easy chair, produced by Herman Miller, *c.* 1933, in walnut and upholstery.

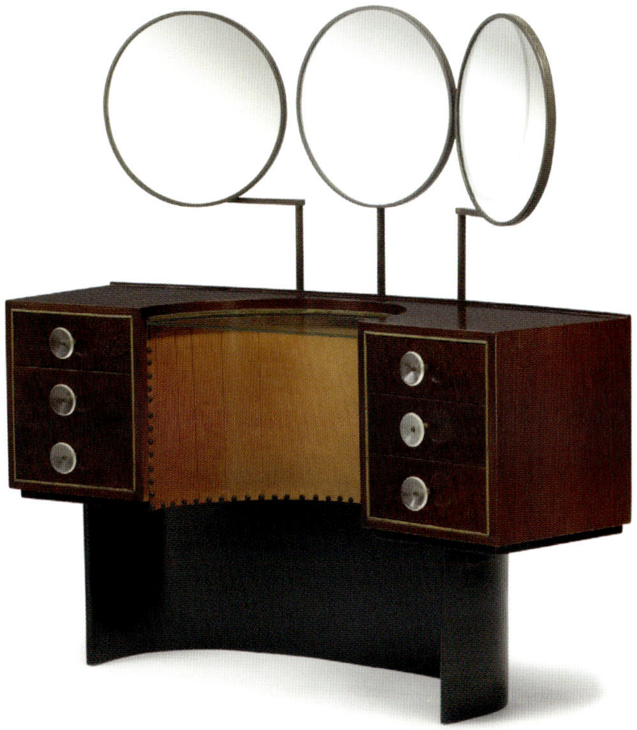

ABOVE A vanity table, manufactured by Herman Miller, 1939, made with East Indian laurel, sequoia burl, oak, acrylic, brass, steel, leather and mirror glass.

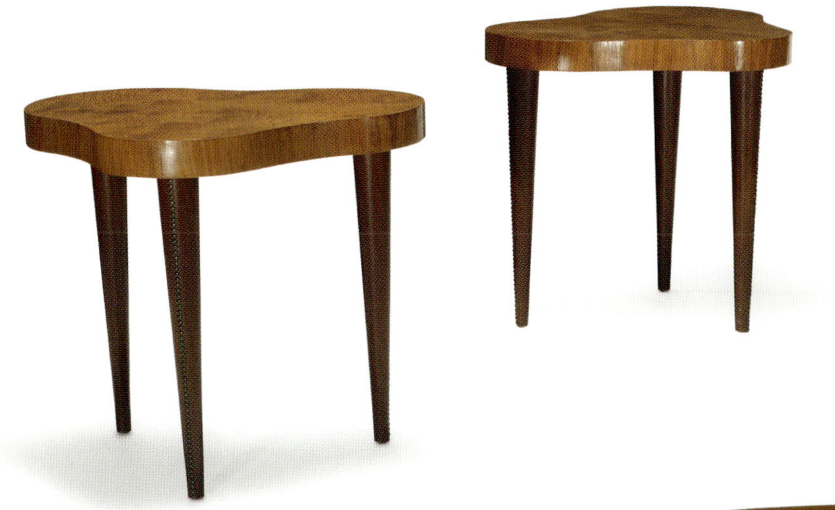

ABOVE Model 4187 occasional tables, made by Herman Miller, *c.* 1940, in acacia burl, leather and brass.

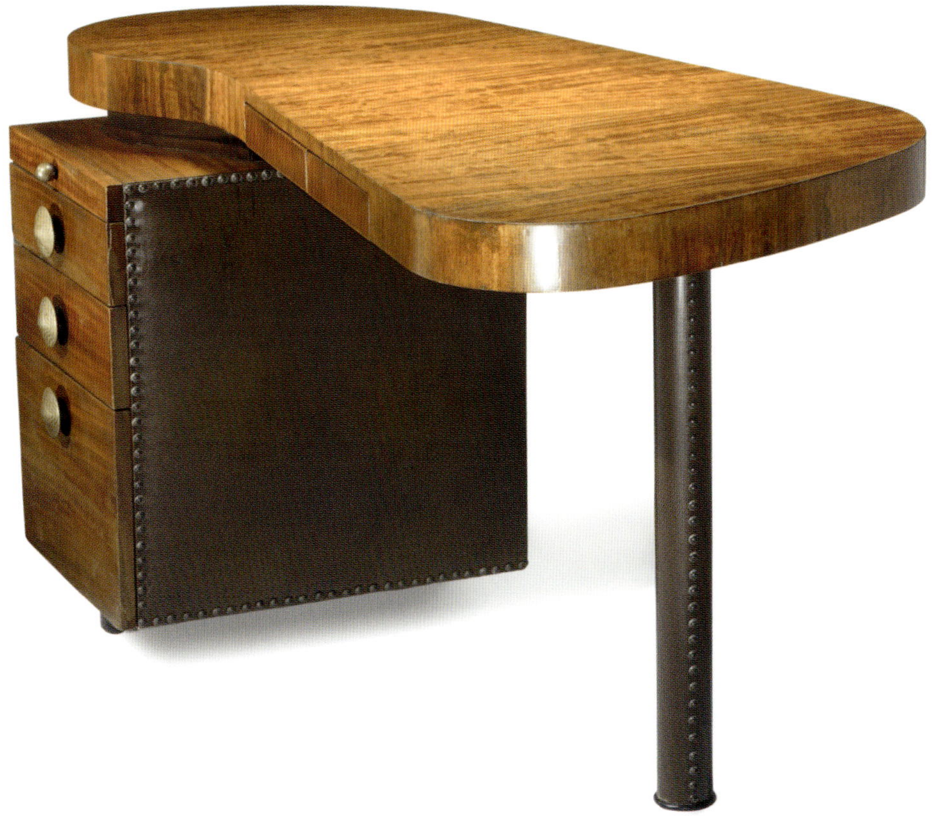

RIGHT A Model 4106 desk, from the Paldao collection, made by Herman Miller, *c.* 1940, in paldao, vinyl, brass and glass.

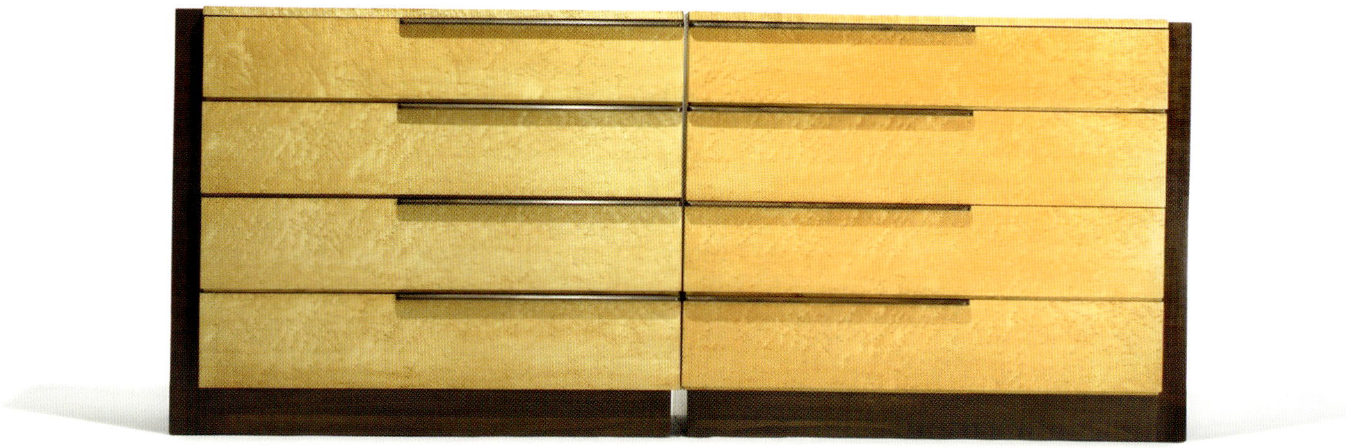

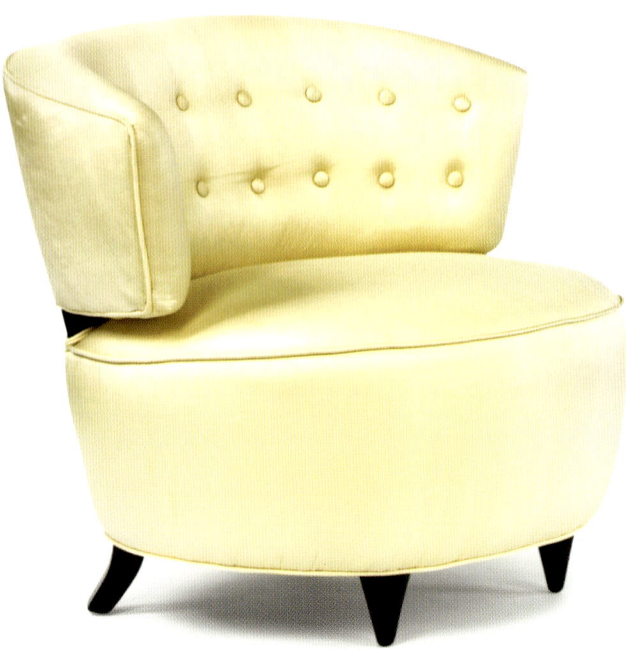

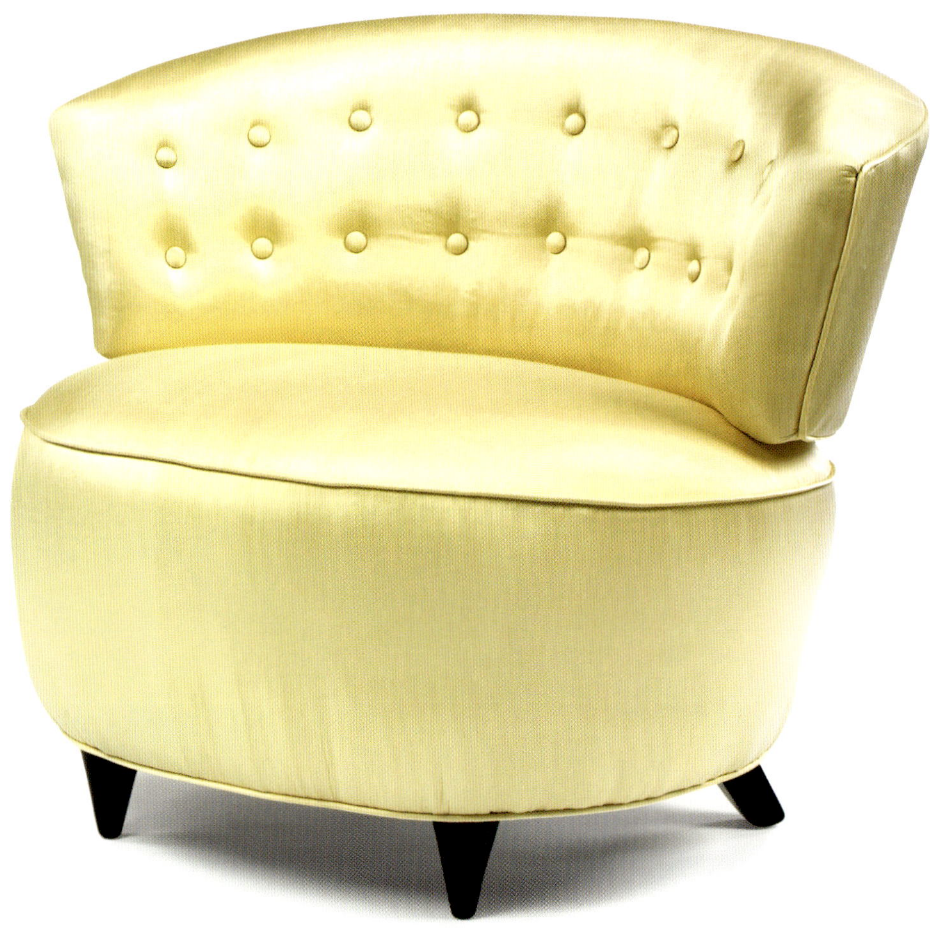

TOP Model 3626 and 3627 twin cabinets, produced by Herman Miller, *c.* 1940, made with burled white ash veneer and walnut.

ABOVE A pair of lounge chairs, made by Herman Miller, *c.* 1940, in lacquered wood and upholstery.

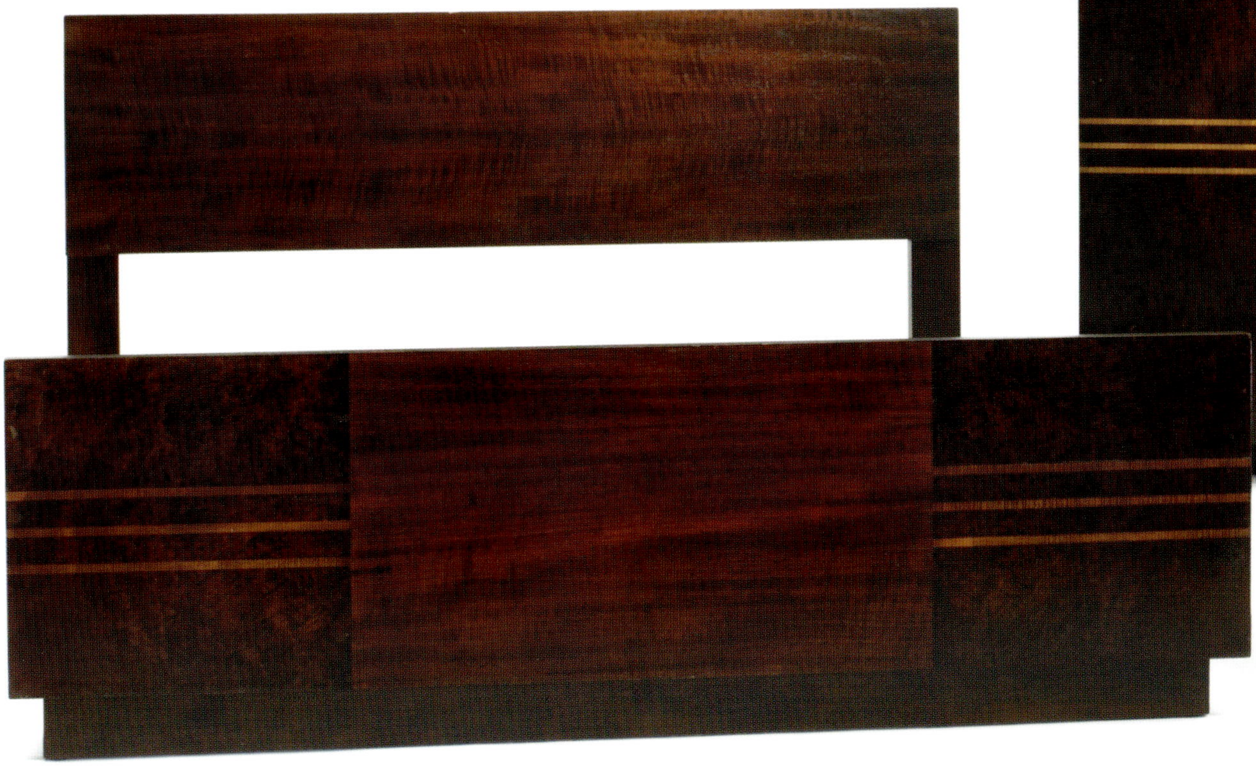

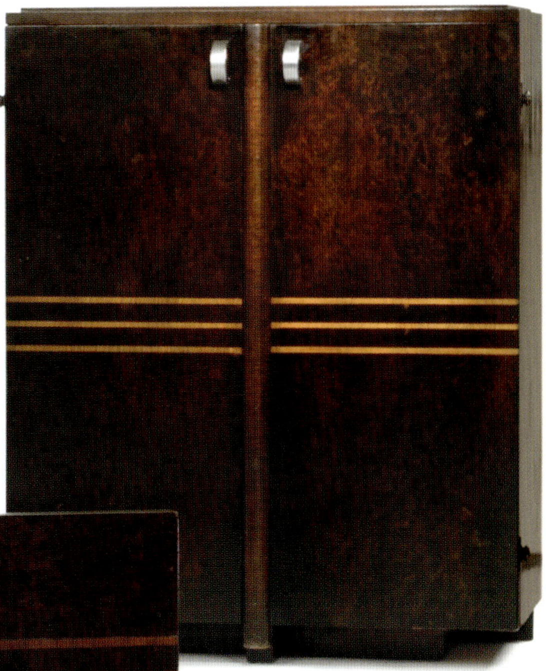

LEFT & ABOVE A matching bed and dresser from the 3317 series, manufactured by Herman Miller, 1933, in sequoia burl, castano and harewood.

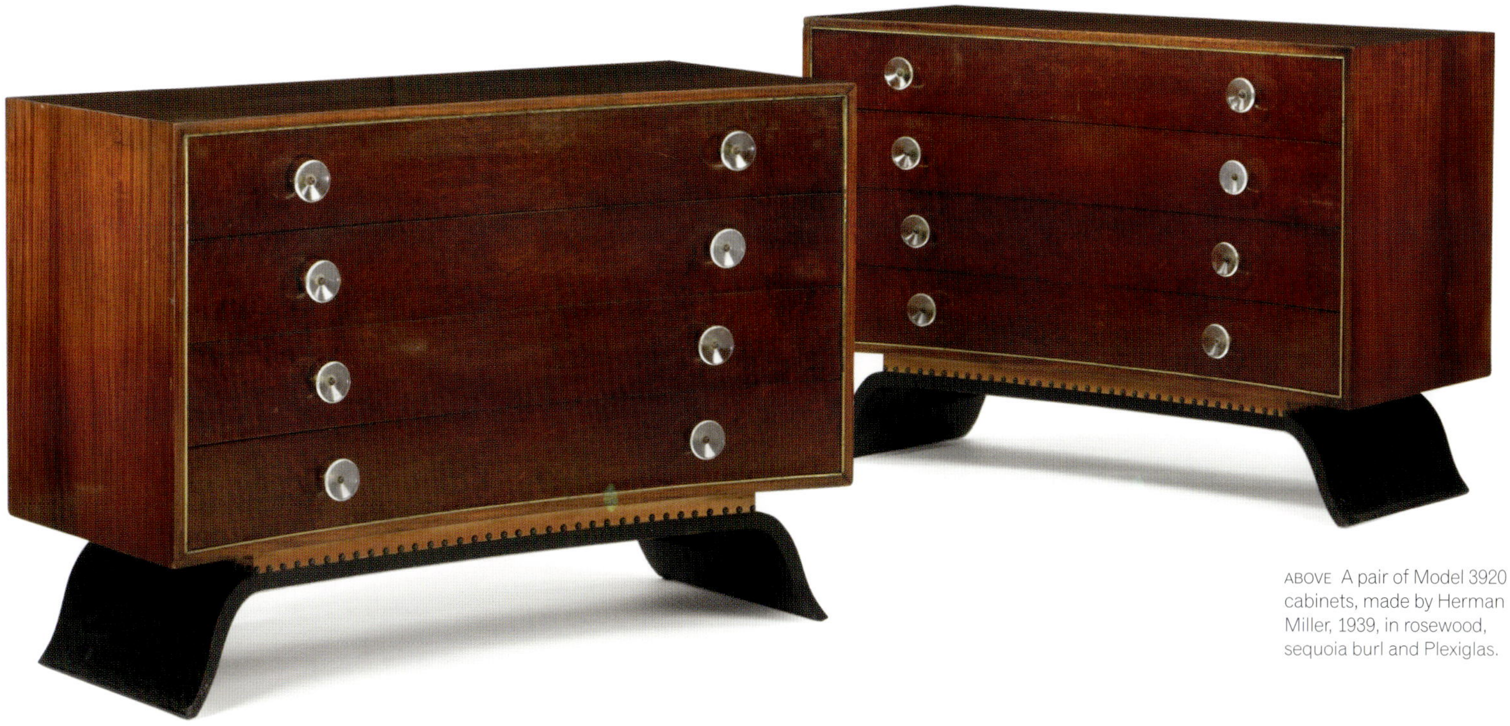

ABOVE A pair of Model 3920 cabinets, made by Herman Miller, 1939, in rosewood, sequoia burl and Plexiglas.

KEM WEBER
(1889–1963)

The German-born designer Kem Weber planned to visit California for less than a year but ended up staying a lifetime. Asked in 1914 by his mentor, the architect Bruno Paul, to supervise the installation of the German pavilion at the Panama-Pacific International Exposition in San Francisco, Weber boarded the liner *George Washington* and crossed the Atlantic. With the onset of World War I and new restrictions on travel, he soon found himself trapped in America. After a difficult few years, he embraced the opportunity to begin a new life, becoming one of a number of early émigrés – such as Rudolph Schindler and Richard Neutra – who helped to pioneer and promote modernist architecture and design on the West Coast. Weber designed houses, interiors and products but is best known for his innovative furniture designs of the Thirties, including his flat-pack, self-assembly Airline chair of 1934 (see also p. 18).

Karl Emanuel Martin Weber was born in Berlin and came from a family of soap manufacturers. After a rebellious period at school he was asked to leave and was found an apprenticeship with cabinetmaker Eduard Schulz in Potsdam, which Weber embraced. He then enrolled at the Unterrichtsanstalt des königlichen Kunstgewerbemuseums, or School of the Royal Museum of Applied Arts, in Berlin, where Bruno Paul was the director. He caught the attention of Paul, who offered him work with his own atelier while he continued with his studies; in 1910, Paul asked Weber to help complete the German pavilion at the International Exposition in Brussels. After Weber finished his studies in 1913, Paul invited him to take the trip to America that was to change the course of his life.

In California, where Weber combined his initials to create the simpler and less Germanic name of Kem, the war years were not easy and Weber made ends meet with a range of jobs, including running a chicken farm. In 1918, he and his new wife settled in Santa Barbara, where Weber began teaching and also opened a workshop to produce his early designs.

Weber's fortunes changed when he was offered a design role in 1921 with the Los Angeles furniture company Barker Brothers, one of the largest manufacturers and retailers on the West Coast, run by Charles Horace and William Barker. Within just a few months Weber was made the head of the company design studio and gradually began pressing for a modern collection that would sit alongside Barker Brothers' more traditional output. Weber finally managed to persuade the company to launch a modern line after an inspirational visit to Europe and particularly Paris in late 1924.

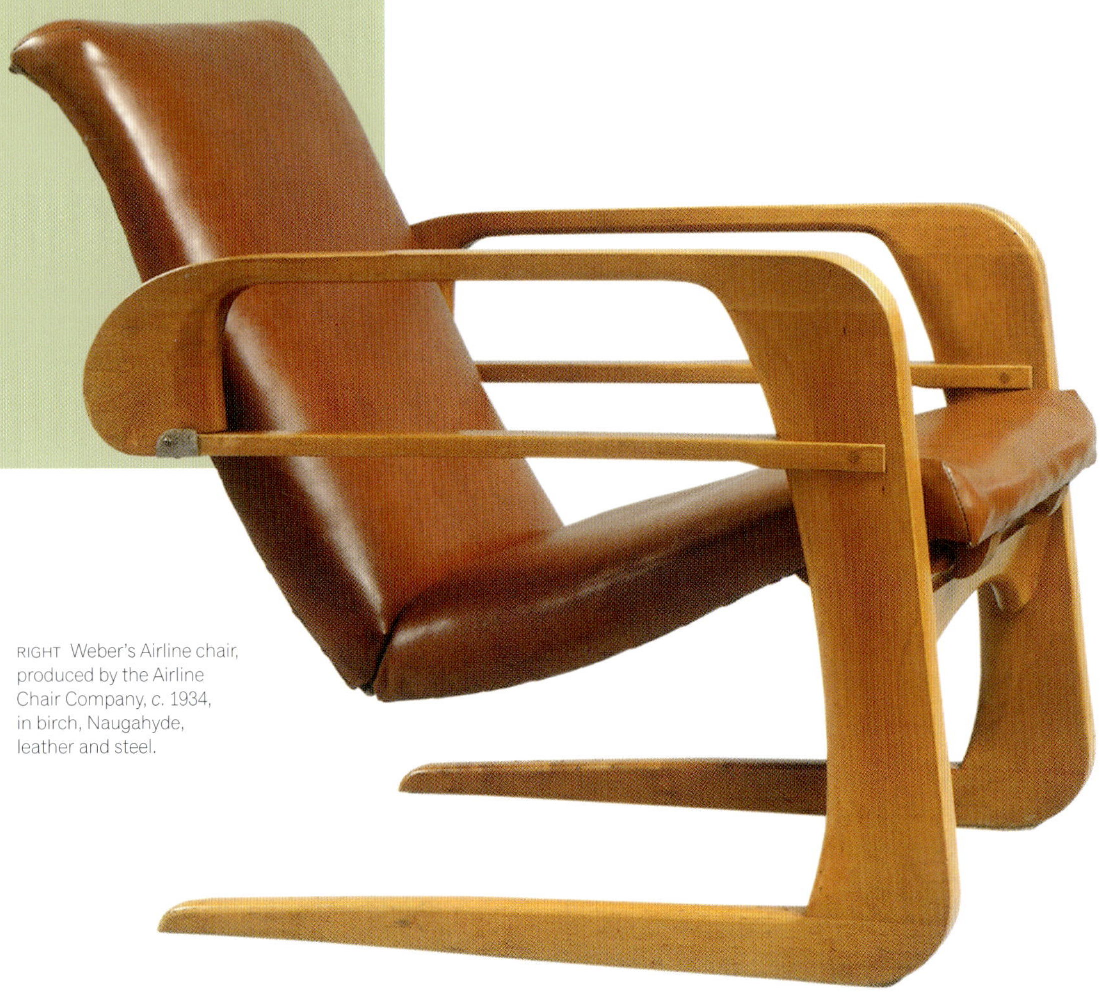

RIGHT Weber's Airline chair, produced by the Airline Chair Company, c. 1934, in birch, Naugahyde, leather and steel.

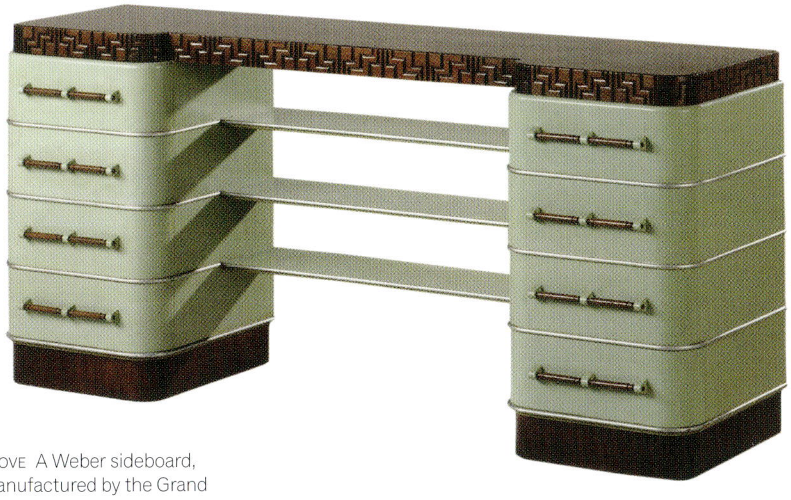

ABOVE A Weber sideboard, manufactured by the Grand Rapids Chair Company, Michigan, 1928, in lacquered metal with a carved walnut top.

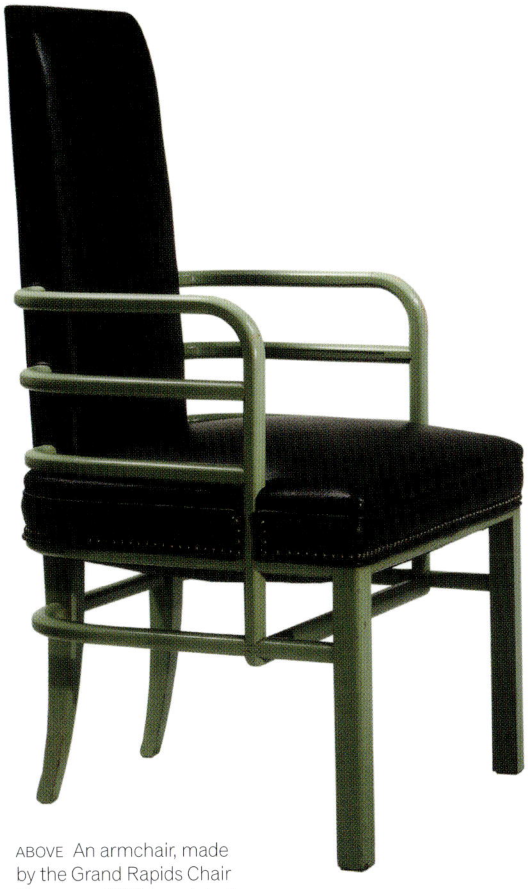

ABOVE An armchair, made by the Grand Rapids Chair Company, 1928, in painted beech with black leather upholstery.

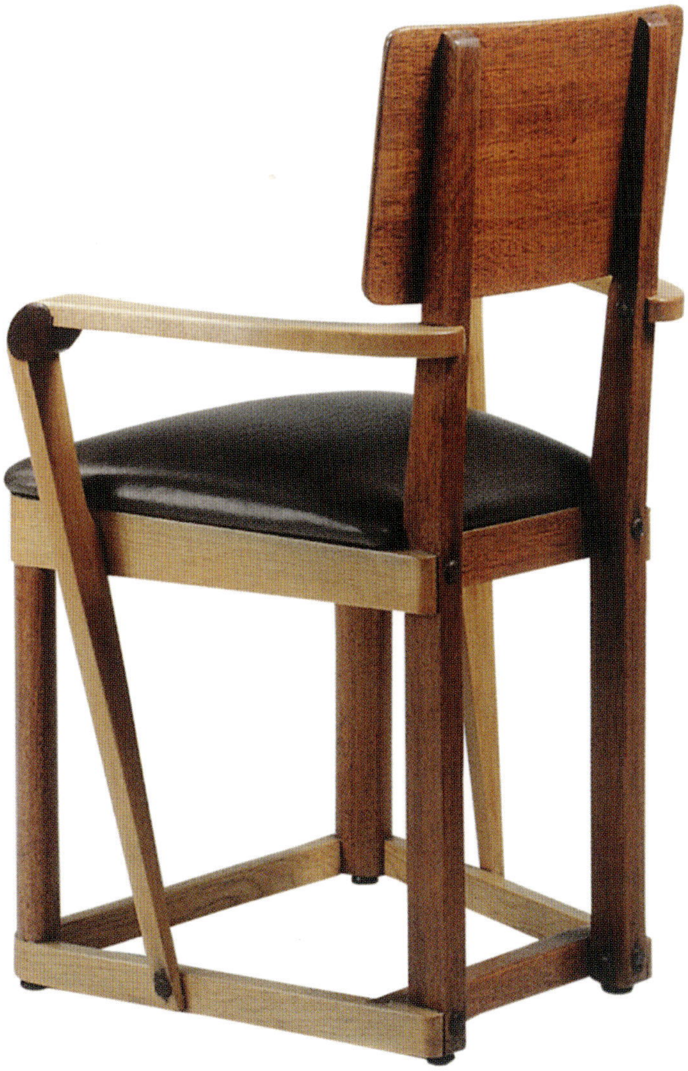

ABOVE A Bentlock armchair, 1931, in ash and mahogany with leather upholstery.

In 1925, Weber designed and built a house for his family and another for a friend, but more importantly he now had the green light for Modes & Manners, a 'shop in shop' within the Barker Brothers stores, dedicated to a new collection designed by himself, drawing on the interest and fascination that the seminal Exposition des Arts Décoratifs et Industriels Modernes in Paris the previous year had encouraged. The legendary Hollywood actress Louise Brooks was pictured with two of Weber's Deco-inspired chairs in 1926. The success of Modes & Manners brought in fresh clients and commissions, including the interiors of the Mayfair Hotel in Los Angeles (1927).

Encouraged by these successes and anxious to explore new opportunities, Weber resigned from Barker Brothers and launched his own design studio towards the end of 1927. In the early Thirties, Weber began to actively explore new methods of making and assembling furniture. In particular, there was a series of Bentlock pieces – lightweight modern timber furniture made using a 'self-locking woodcorner', or small timber pin that was glued in place to help hold the piece together. The most striking of these designs was Weber's cantilevered Bentlock chair of 1931, a dynamic and curvaceous piece somewhat reminiscent of ALVAR AALTO and MARCEL BREUER's sculptural plywood furniture. The wooden pins were clear to see, with the construction technique forming part of the essential character of the piece.

Despite the promise of the Bentlock collection, it proved difficult to produce in volume. But the cantilevered chair, in particular, was a key step in the evolution of Weber's most famous design, the Airline chair. Again, the piece was cantilevered and streamlined, drawing inspiration from the dynamism of the new passenger planes and their interiors on the one hand, but offering a reworking of the generic club chair on the other. The Airline chair featured a lightweight but highly tensile birch frame that supported a floating seat coated in leather or Naugahyde. Importantly, the seat flexed and adjusted to the weight of the user, offering an ergonomic level of comfort. And crucially the chair could be delivered flat-packed and easily assembled at home for a price of just under $25.

The Airline chair was full of ingenuity and style, offering an early forerunner of the affordable flat-pack furniture that has become ubiquitous today. Unfortunately for Weber, once again he was ahead of his time and the seat did not take off. Many of the 300 chairs originally produced ended up in the new studio complex that Weber designed for the filmmaker Walt Disney in Burbank, completed in 1940.

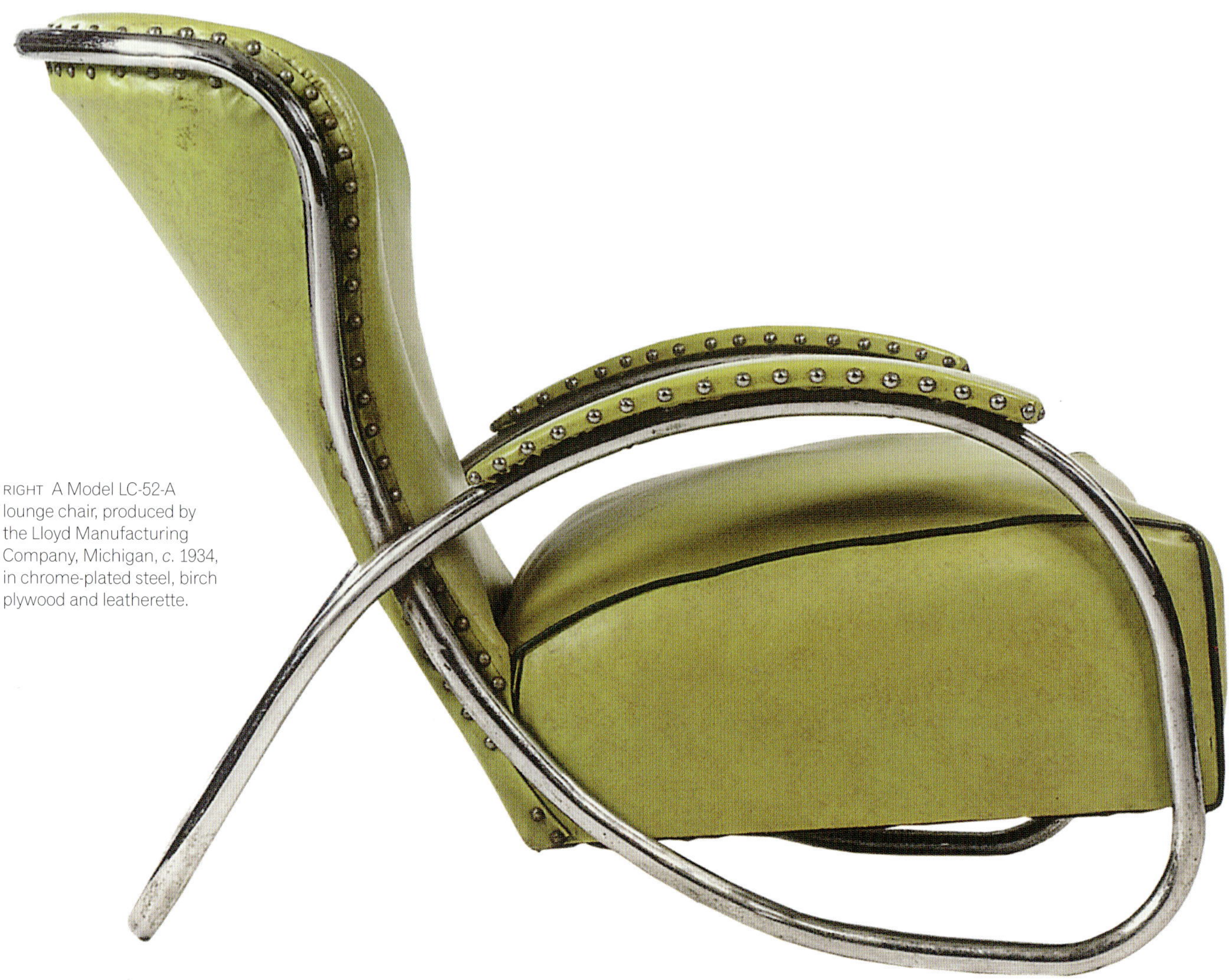

RIGHT A Model LC-52-A lounge chair, produced by the Lloyd Manufacturing Company, Michigan, *c.* 1934, in chrome-plated steel, birch plywood and leatherette.

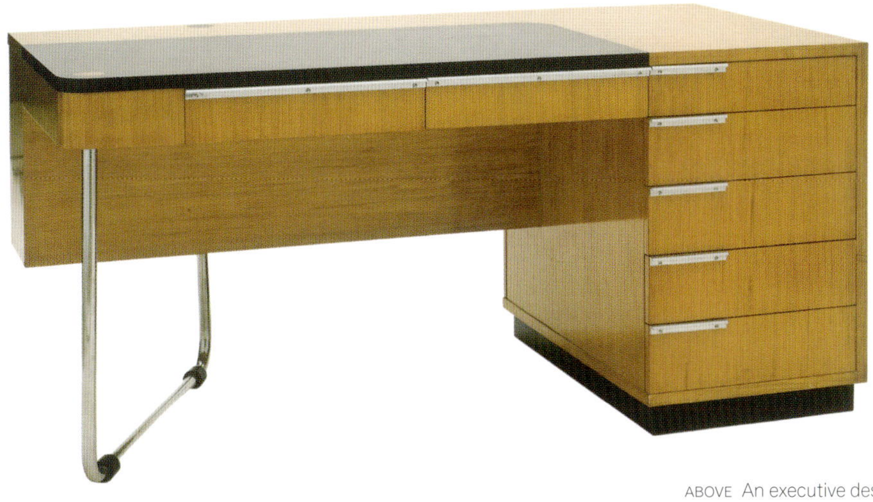

ABOVE An executive desk shown at the Golden Gate Exposition in San Francisco, 1939, made of chrome-plated tubular steel, satinwood, Bakelite and aluminium.

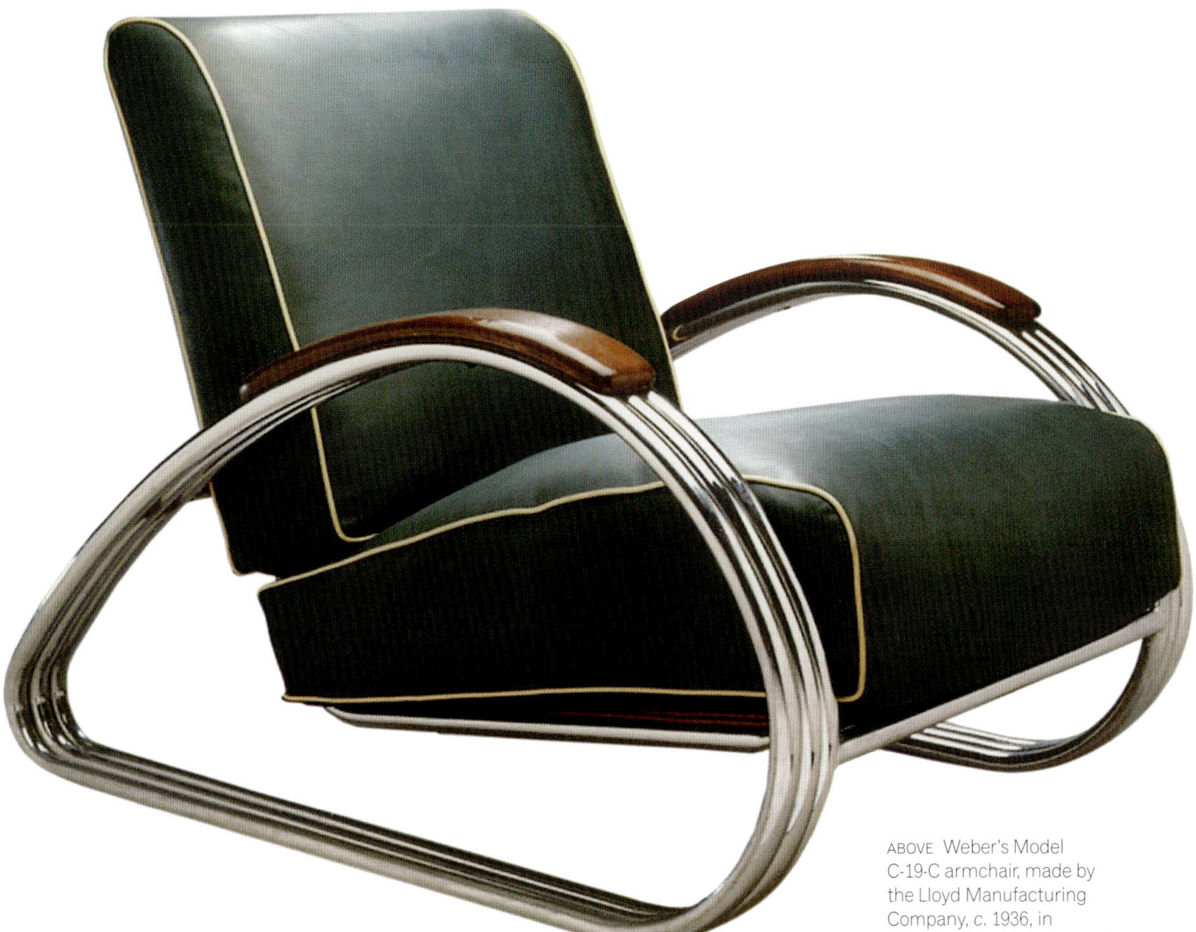

ABOVE Weber's Model C-19-C armchair, made by the Lloyd Manufacturing Company, *c.* 1936, in chrome-plated tubular steel with Naugahyde upholstery.

LIGHTING

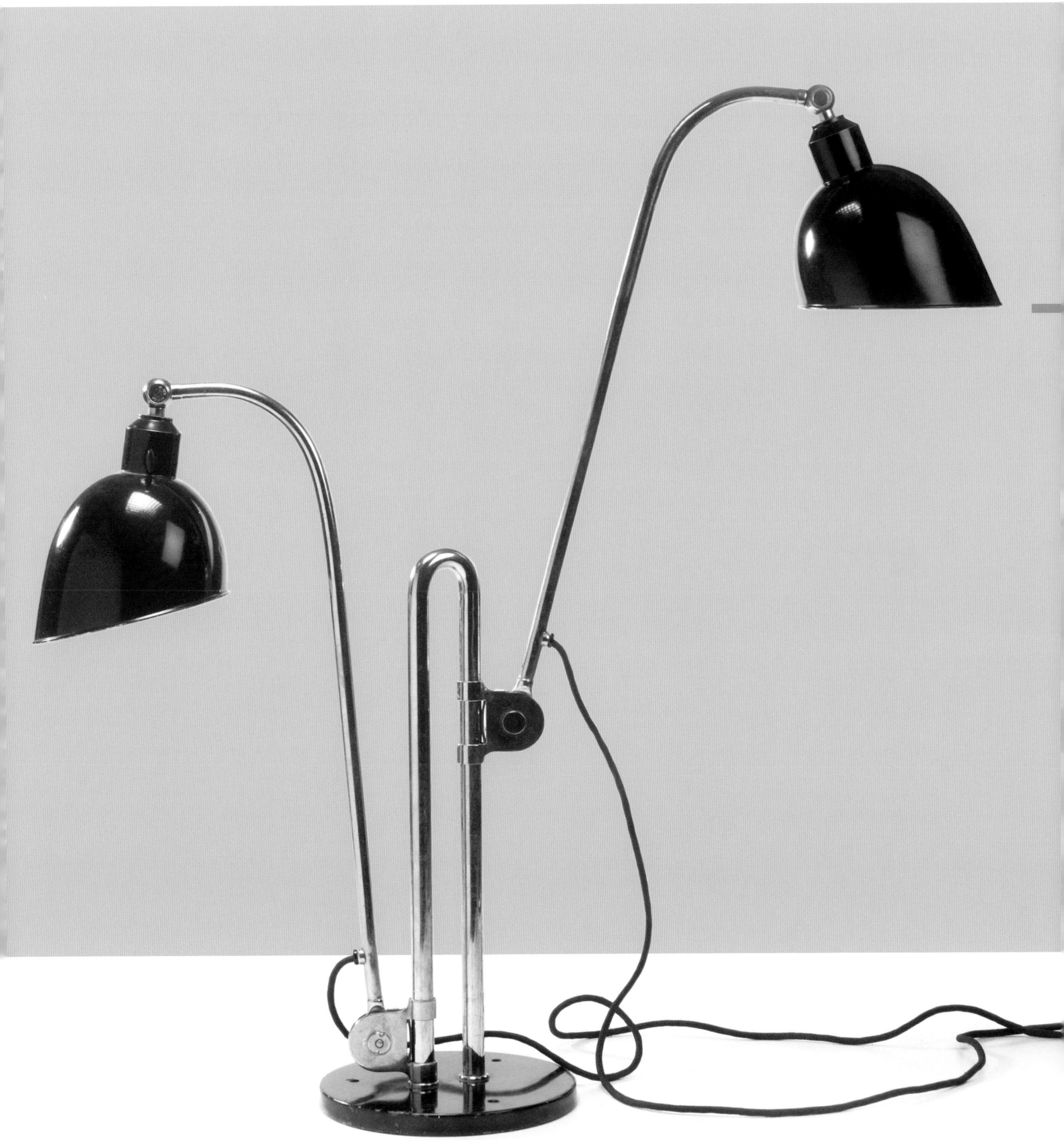

In an age when we take 'flick of the switch' electric lighting for granted, along with electricity itself, it is easy to forget that in the early years of the 20th century it was seen as almost miraculous. Commercial manufacturing of incandescent light bulbs – along with the infrastructure needed to generate and produce electric lighting – only began in the late 1870s. Despite the rapid evolution of lighting and power generation in the years that followed, electric lighting took many decades to become a familiar part of daily life and begin to surpass gaslight. The vast majority of designers working in the early modernist period, therefore, had spent their childhoods in homes, schools and colleges illuminated by the soft glow of gas-powered lighting or – in more rural and remote parts of the world – by oil lamps. The design of electric lighting evolved in parallel with the birth of the 20th-century machine age itself, forming a new medium that advanced at a rapid and extraordinary pace.

Early experimentation with electric lighting focused on arc lighting – led by Humphry Davy, William Edwards Staite, Jean-Bernard-Léon Foucault, Charles Brush and others – in which electricity jumped between two carbon rods placed close to one another, with the glowing current forming brilliant illumination. By the late 19th century, arc lighting was developed to a point where it became commercially viable, yet the light was so intense that it was only suitable for open public spaces or factory settings.

Incandescent bulbs developed almost in parallel to arc lighting, but again it was only in the last decades of the 19th century that the science and technology of the electric light bulb reached a point where it was ready to step into the home. Joseph Swan in Britain and Thomas Edison in America were at the forefront of attempts to perfect the light bulb, in which an electrical current passed through a filament contained within the vacuum of a glass bulb. The vacuum slowed and controlled the process of illumination, preventing the glowing filament from burning out too quickly. Edison developed a screw fitting for his incandescent bulb, and Swan chose the bayonet.

The 1880s marked the beginning of the true revolution of light. As well as working on perfecting the light bulb and its manufacture, Edison and Swan and their disciples began working on the infrastructure that was necessary to produce the electricity required to power their lighting systems, including generators and dynamos. In 1880, Edison installed electric lighting on the ocean liner SS *Columbia*; just two years later his company was ready to offer domestic lighting to the residents of New York City. On both sides of the Atlantic and beyond, there was an explosion of electrical activity as an entirely new industry developed around the possibilities offered by electrical light and the generation of electrical power. Companies such as Philips in the Netherlands and AEG in Germany – which commissioned architect and designer PETER BEHRENS to design not just new electrical lamps and products but also giant turbine-making factory halls – grew quickly in the early years of the 20th century, becoming powerful industrial concerns.

Designers responded to this new field of interest in different ways, with some focused on the utility and function of lighting and others more concerned with ways in which they could domesticate table lamps, ceiling lights and sconces and thus persuade consumers to embrace the technology. In the early years of lighting production, many lamps and lighting products were essentially attempts to disguise the electric bulb while treating the character of the light in much the same way as traditional gas lamp illumination. At the same time, preoccupations with form and artistry mirrored the popularity of aesthetic styles such as Arts and Crafts or Art Nouveau.

Lighting design of the Art Nouveau period, in the early years of the 20th century, took on a particularly flamboyant character. Lamps created by Louis Majorelle & Daum Frères in France, for instance, or Émile Gallé, adopted exotic, sinuous

INTRODUCTION

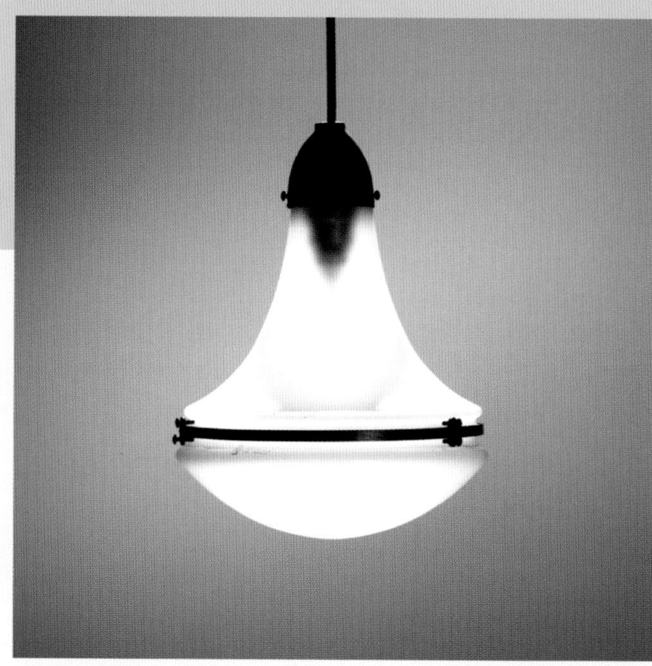

PAGE 111 A table lamp designed by Christian Dell, produced by Berlag, *c.* 1930, in enamelled steel, chrome-plated brass and Bakelite.

ABOVE A pendant lamp designed by Peter Behrens, Germany, *c.* 1910, in copper, brass and glass.

forms with a combination of metalwork and glass shades that copied natural shapes, from magnolias to mushrooms. In the States, Tiffany Studios explored similar preoccupations, with exquisitely crafted lighting inspired by nature, including their Lily, Lotus, Dragonfly and Apple Blossom lights.

Arts and Crafts-inspired lighting by Gustav Stickley, Greene & Greene, Charles Rennie Mackintosh and Frank Lloyd Wright embraced a more geometrical or linear approach and made use of crafted timber as well as glass and metalwork. The designs took on the aesthetics of simpler lanterns and sconces, but once again there was the intention of hiding the bulb and softening the light while often harking back to an era of candlelight.

Yet at the same time other influences and sources of inspiration were making themselves known. The theatre was one key area of innovation, with theatrical designers such as Norman Bel Geddes and Henry Dreyfuss involving themselves in theatre lighting, as well as set design, before stepping into the world of industrial and product design. In Italy, Mariano Fortuny* initially developed his iconic Projector floor lights (1907) for use in the theatre, with a directional spotlight contained within a concave linen backed shade sitting upon a tripod base; later, the floor lamp was adopted for residential use, while Fortuny also developed table lamps and other lighting designs.

Photography, too, helped inform the evolution of lighting design, with photographic studios looking at fresh ways to manipulate light – both directly and indirectly – to achieve different visual effects in a comparable way to the theatre. Pietro Chiesa and others noted the impact of indirect lighting within photography and the way light could be reflected and bounced off ceilings and other surfaces; his famous Luminator light of 1933 evolved, in part, from such thinking.

These investigations into the creative and atmospheric implications of various kinds of lighting overlapped with the

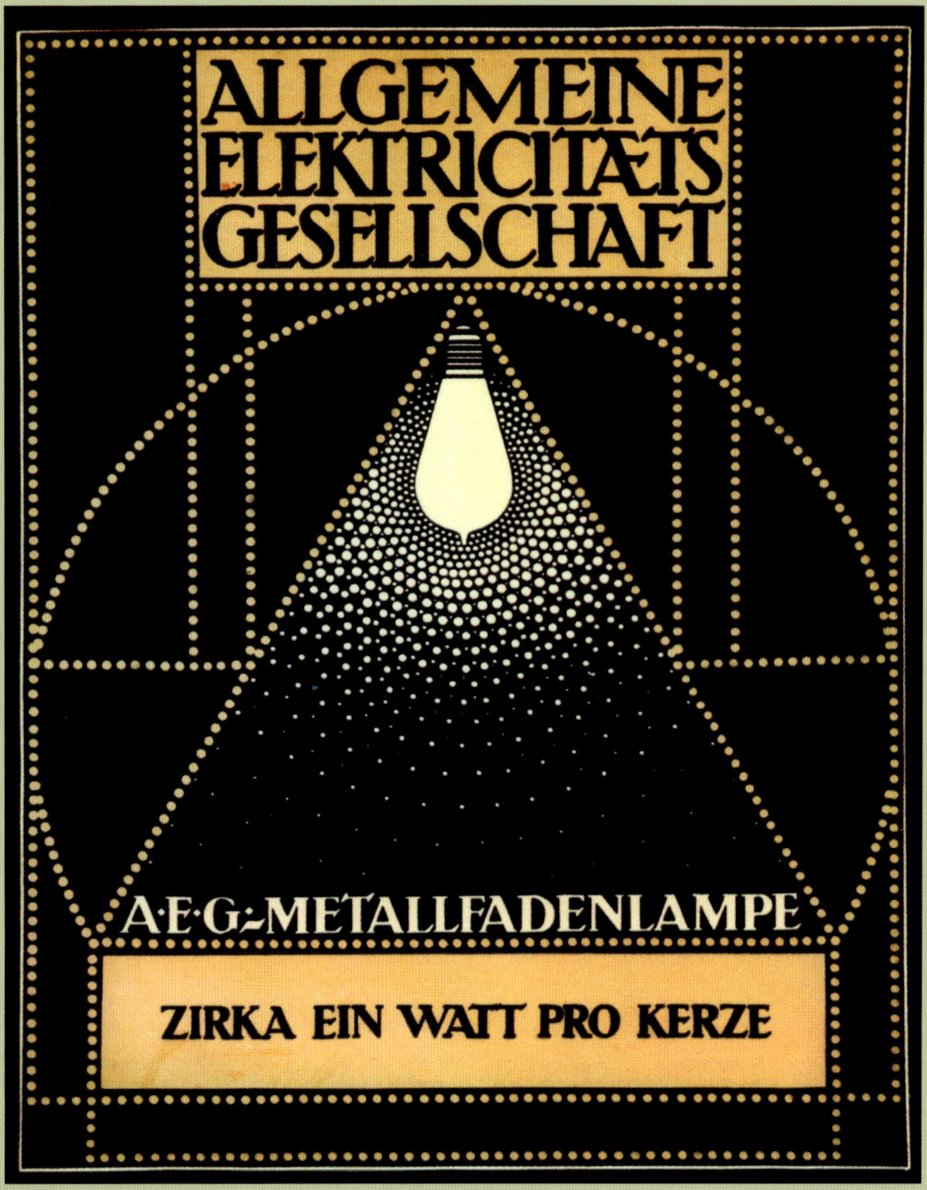

ABOVE An advertising poster, designed by Peter Behrens, 1907, promoting AEG's electric light bulbs.

focus on function that formed the preoccupation of many early modernist lighting designers. Here, the emphasis was not upon disguising the source of the light or creating a sculptural or decorative centrepiece, as was so often the case with Art Nouveau craftsmen and designers. Pioneering proto-modernists such as Josef Hoffmann* and Otto Wagner* adopted a semi-scientific attitude to lighting, with relatively little importance attached to embellishment or decoration. Some of Hoffmann and Wagner's lighting designs from the early years of the 20th century – when the age of electricity was still in its infancy – have a surprisingly 'modern' aesthetic. Hoffmann's M109 table light of 1903, for instance, with its nickel-plated brass base and half-moon frosted glass shade, feels like a highly sophisticated foretaste of work that would appear thirty or forty years later. Wagner's wall light of 1902, originally designed for the telegraph office of the *Die Zeit* newspaper office in Vienna, is a stripped-down, functional design with a simple nickel-plated mount holding a bare ceramic bulb with no shade.

Increasingly, as one plots the progress of early 20th-century and early modernist lighting design, one sees a twin-track approach developing by the Twenties and early Thirties. On the one hand, Art Deco* designers such as René Lalique, Jean Perzel*, Albert Cheuret and Jacques-Émile Ruhlmann* continued to explore sculptural shapes, intricate craftsmanship and fine materials within their lighting designs, although their work was distinctly different in character from the colourful exoticism of their Art Nouveau forebears, reflecting machine age concerns with more geometrical forms and a focus on metallic finishes and frosted or etched glass. But on the other hand, the pioneering modernists were increasingly focused on function and industrial means of production, even if they shared some common interests concerning the impact of various different or complementary forms of lighting, such as table lamps, uplighters, wall lights and pendant ceiling lights.

Modernist lighting designs by Marianne Brandt, Wilhelm Wagenfeld and Christian Dell were influenced by close association with the teachings of the Bauhaus and its emphasis on both functionalism and 'democratic' design, meaning well-designed but affordable products that could be factory-made with a kit of standard components. Many of Brandt's designs, in particular, were employed at the Bauhaus campus itself, either in the masters' houses designed by Walter Gropius or the workshops.

The preoccupation with function in lighting design at the Bauhaus and elsewhere led to a broad collection of task lighting during the late Twenties and Thirties, including desk lights and table lamps, as well as specialist lights such as the Model 404 Giso piano lamp designed by J. J. P. Oud and manufactured by W. H. Gispen (1927). A key element of many of these task light designs was the ability to control and manipulate light through flexible systems, such as pivoting spotlight heads, extending arms and the use of counter-weights to vary the height and reach of a lamp. The Anglepoise desk lamp developed by George Carwardine, using a spotlight mounted upon a spring-loaded

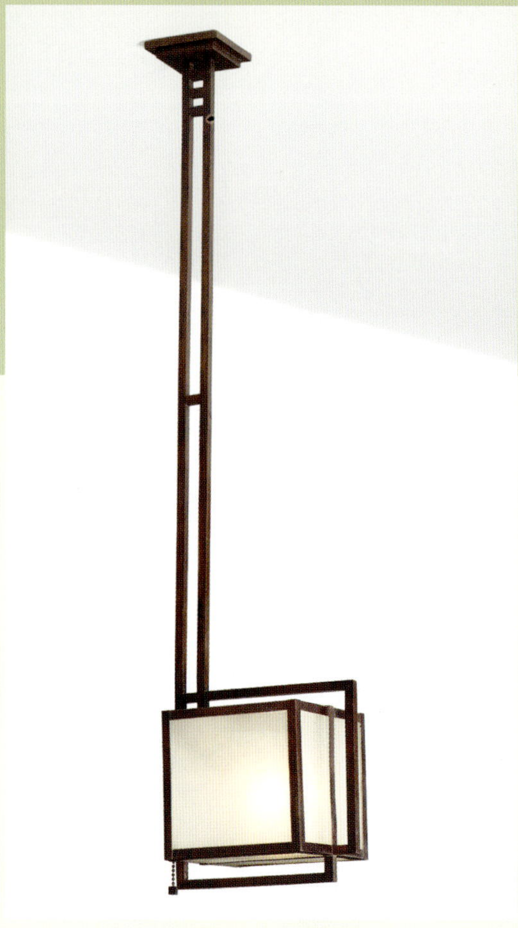

RIGHT A hanging lamp designed by Frank Lloyd Wright for the John Store House in Hollywood, 1923.

arm, is a classic example of the early modernist push towards adaptable task lighting.

Certain designers such as Poul Henningsen became specialists in the field of lighting, concentrating their thinking and energy for experimentation upon this new and vibrant medium. But many fine examples of Twenties and Thirties lighting came from multi-talented architects and designers who worked across a spectrum of different disciplines. These included architect and designer Gerrit Rietveld, whose famous Hanging lamp of 1922 was originally designed for a doctor's consulting room; a later, highly minimal design of 1924, featuring a simple glass disc reflector hovering over a naked bulb, was first created for the Muller House in Utrecht. Others who designed seminal lighting as one strand of a wider portfolio of interests included Pierre Chareau, Eileen Gray and Jean Prouvé, whose famous Jib wall light of *c.* 1942 was another stripped-down, minimalist design with a bare bulb mounted on a tubular steel arm fixed to a wall-mounted fixing plate that allowed the arm to swing through 180 degrees.

The experiments of the pioneering modernists and their particular fascination with functionality and task lighting helped to lay the groundwork for a very different approach to lighting in the mid-century period. By the Fifties, so much more was understood about function and various different forms of lighting, offering a powerful foundation to a new generation of designers who were able to experiment with new materials – from paper to plastics – and sculptural, organic forms. Mid-century lighting was free to become more playful, colourful and exciting, liberated to some extent by the work of the early modernists and their many innovations with methods of manufacture, materiality and the complex science of light, with its profound importance not only in facilitating work, education and production but also in establishing different kinds of moods, atmospheres and emotions, even if – after everything – the effect might be a return to the feeling of candlelight, or oil lamps, and an age of simplicity.

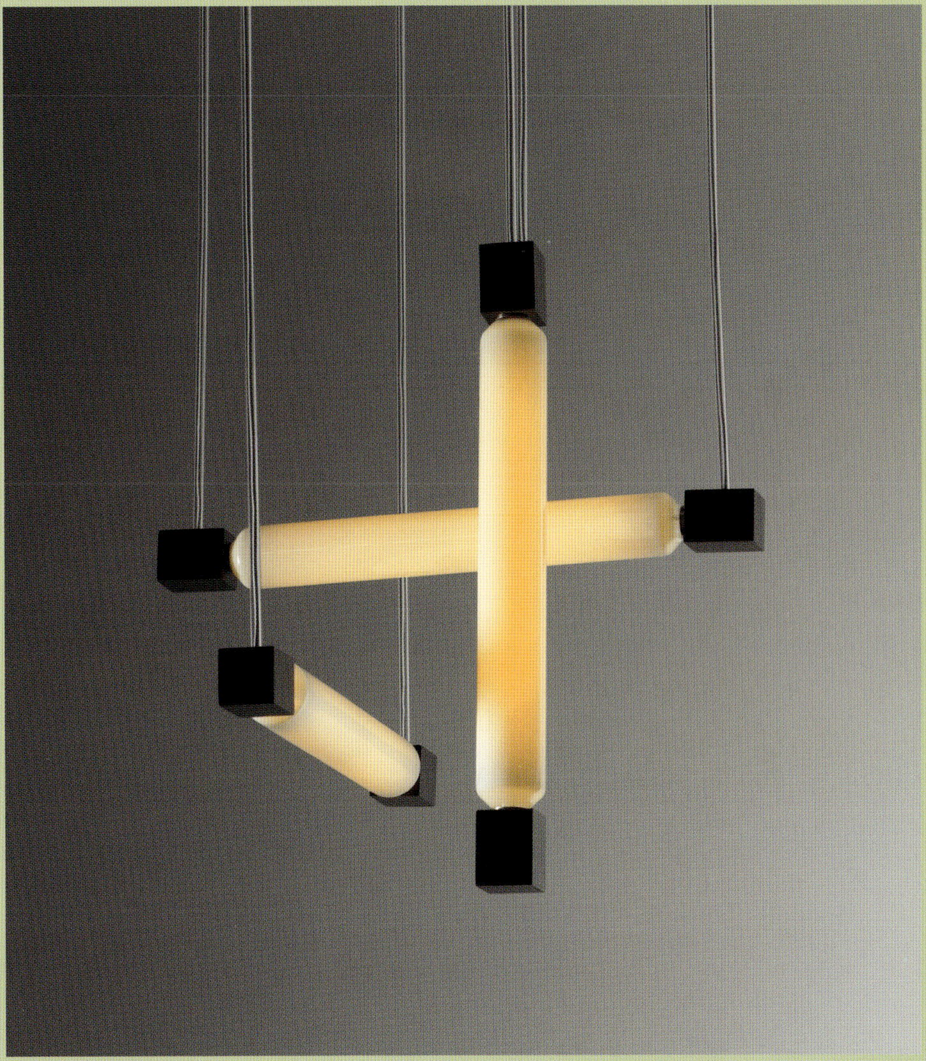

ABOVE Gerrit Rietveld's hanging light, originally designed *c.* 1920 for the consulting rooms of Dr Hartog in Maarssen, the Netherlands, manufactured by Gerard van de Groenekan in glass and stained oak.

In the years between the wars, a revolution took place in the world of design, as the style of luxury, epitomized by the artist/patron model, transitioned towards a more rationalist and populist view of design as an instrument of change. Emphasis shifted from production to consumption as skilled craftsmen, artists and designers embraced machine age methods to create a new vernacular in decorative arts. Amongst designs, lighting forms were some of the most experimental and accessible as new materials and technologies became more readily available. Further, as electric lighting was increasingly used for signage, skyscrapers and theatre, designers too found inspiration in the dramatic effects that lighting could bring to an interior.

The design revolution was not a directly linear progression: designers all over the world were challenging expectations and redefining what it meant to be modern. In Europe, and France in particular, Art Deco*, a radical new aesthetic that reached its peak in the roaring Twenties, rose to international prominence as an alternative to post-war austerity. Ornate and decadent, the masterworks of this movement were created for the wealthy, painstakingly crafted from rare woods with inlays of precious materials.

Deco lighting set itself apart from the biomorphic exoticism of Art Nouveau lighting – exemplified by the work of Jules Jouant, Louis Majorelle, Émile Gallé and Tiffany Studios – where illumination spilled from florid flower petals or shining domes held aloft by sculpted branches. Deco lighting was generally more in tune with the character of the electric age, yet still inclined towards ornament, with a resilient emphasis on craft, artistry and sculptural forms, even if these were streamlined and spoke – more or less – of a kind of modernity.

One thinks of René Lalique, Daum Frères or Simonet Frères in France, using etched or moulded glass to create ethereal lamps and statement chandeliers. Lalique's glass lighting, in particular, played with the idea of a cascading glass fountain of light or a glowing glass star. Albert Cheuret retained a love of natural forms and points of inspiration even as he sought to weave these into a more 'contemporary' Deco aesthetic in his sconces and floor lamps. Jean Perzel* took a more restrained approach to ornamentation, yet his work was still highly sculptural and drew upon a palette of rich materials, including chrome, nickel-plated steel and sandblasted glass. Edgar Brandt's Cobra lamp of 1925 helps to illustrate the dominant Deco aesthetic with the use of exquisitely crafted bronze and glass shades by the premier producer of art glass in France, Daum Nancy. Stylistically, this lamp was fit for a queen, and was one of a series of lights – including La Tentation and Le Serpent Naja – in which Brandt used the idea of a rising snake supporting a bowl of fire.

Yet at this same time Bauhaus designers were exploring form and function in the absence of ornamentation, forming a stark contrast with the work of the French Deco masters. WILHELM WAGENFELD's table lamp from 1924, comprised of simplistic geometric forms and constructed of industrial glass and metal, proudly expresses its function as the necessary components, the switch and wiring become the decorative elements. Le Corbusier's dictum from 1923, 'The house is a machine for living in', is a breathtakingly radical proposition that found its best proponents in the German Bauhaus. Founded in 1919 and based on the idea of the German guild system, the Bauhaus had influences that were far-reaching. To be free of clutter, to be efficient, were signs of being modern. Embodied in the Bauhaus aesthetic was a radical challenge to the cultural norm of society. Modernism strove for a utopian vision, and the elements of the home were part of that manifesto.

THE AGE OF ENLIGHTENMENT: THE EVOLUTION OF LIGHTING FROM ART DECO TO MODERNISM

RICHARD WRIGHT

Designs emanating from the lighting workshop were the most commercially successful forms to come out of the Bauhaus. CHRISTIAN DELL, foreman of the metal workshop at the Bauhaus from 1922 to 1925, designed some of the most popular pieces. Dell's creations are variations on a basic design vocabulary utilizing the fundamental function to drive the form, but his lamps also express a technical interest; they are designed to offer variability of light as their shades tilt and raise and lower. The constraints of the electrical wire are often used as a visual element, with its subtle curved movement working in opposition to the central vertical metal rod.

Like Dell's, MARIANNE BRANDT's lighting designs have become emblematic of this time in history. Brandt was the head of the Bauhaus metal workshop in 1928 and is credited with negotiating the collaboration between the school and the progressive manufacturer Kandem Lighting, thus introducing hundreds of designs to the market. Her Kandem table lamp, with its clean form, adjustable light source and simple push-button on/off, has become one of the most replicated designs. The Kandem/Bauhaus collaboration would continue until 1933 and resulted in the production of more than 50,000 lamps.

By 1929 a group of designers and architects in France had formed the Union des Artistes Modernes (UAM)*, headed by ROBERT MALLET-STEVENS, with notable members including LE CORBUSIER, RENÉ HERBST and EILEEN GRAY, among many others. In opposition to designing for the elite and wealthy, the group aimed to make design more widely accessible. Emphasizing design over decoration, the lighting produced was markedly different from its Art Deco predecessors. René Herbst's suspended ceiling lamp has a clean, industrial aesthetic composed of nickel-plated steel discs or diffusers suspended by tubular rods.

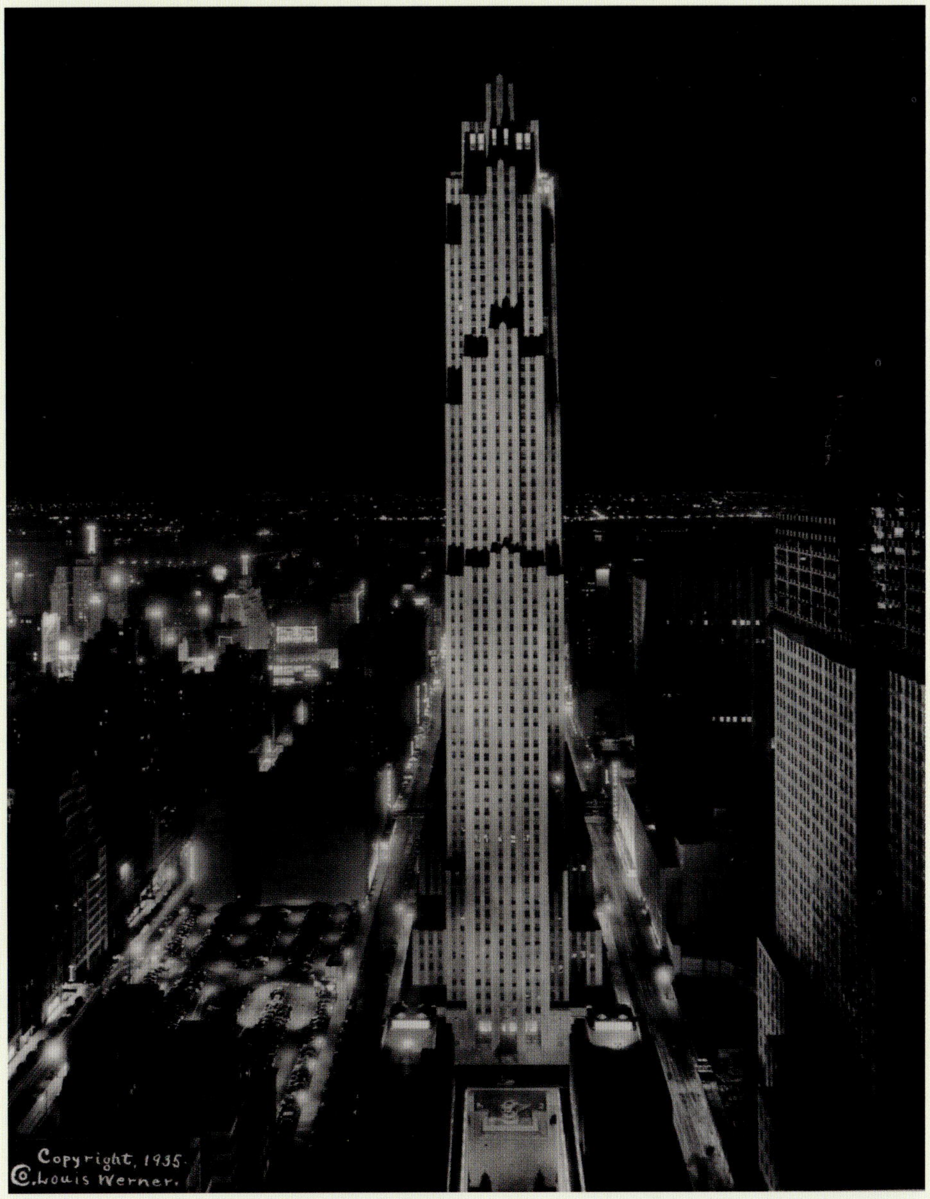

ABOVE A night view of the illuminated RCA Building at the Rockefeller Center, New York, c. 1935.

Yet the break from the decadence that was Art Deco was not always so neatly drawn. Several designers and companies created lighting designs that shared the rich materiality of the Art Deco movement with the more pure, geometric functionalism of the modernists. For instance, Pierre Chareau bridged the worlds of Art Deco and modernism with designs such as the LP180 table lamp composed of alabaster and iron. Similarly, Maison Desny*, a French design firm formed by Desnet and René Nauny and active from 1927 to 1933, was widely known for its innovative lighting, accessories and furniture featuring geometric shapes and pure surfaces. Modern in style, Maison Desny's functional works have a richness of materiality – from chrome-plated and nickel-plated metals and glass to silver and exotic woods. Table lamps in nickel-plated brass with repeated layers of thick glass and simplicity of form reflect admiration for machine-made qualities, while also presenting a refined elegance and sophistication.

In the United States, Donald Deskey created several lighting designs that were at the same time sophisticated yet streamlined. His table lamp with its smoothly curved arm and half-dome fixture captured an entirely different take on modern – one that was more scientific, based on motion and speed, resulting in a rich, aerodynamic aesthetic. Taking these concepts even further, Walter Dorwin Teague designed the Polaroid lamp in 1939. Made of Bakelite and aluminium, the Polaroid lamp was functional as well as affordable.

Examples of forward-thinking and modern lighting designs were found around the world during the late Twenties and Thirties. The designs of Poul Henningsen for Louis Poulsen in Denmark embraced industry and mass production with a line of PH table lamps and ceiling lights made with various-sized glass and metal shades; the materials were selected for their reflective qualities and illumination. Geometric and mathematical in composition, Henningsen's lighting designs embody the modern tenet of 'design before decoration', though some works, such as his Piano lamp with its curled stem, include playful flourishes.

In Italy, Pietro Chiesa – invited by Gio Ponti in 1932 to head FontanaArte – played an important role in the canon of modern lighting design. His Luminator floor lamp stands out for its indirect illumination, its tall slender form, and its cone-shaped opening that both hides and releases the light source. Stripped of decoration, the simple form is derived from its function, but the elegant styling belies the hand of the artist.

Through the work of the modernist, the aesthetics of function triumphed in the field of lighting design. It was during the Twenties and Thirties that lighting moved away from the luxury and decorative styles of Art Deco, with underlying forms derived from the historical tradition of candle- or oil lamp light. Technical innovation and engineering quickly evolved along with the science of light, and pioneering designers created a new visual language of illumination. The resulting designs still look radical today.

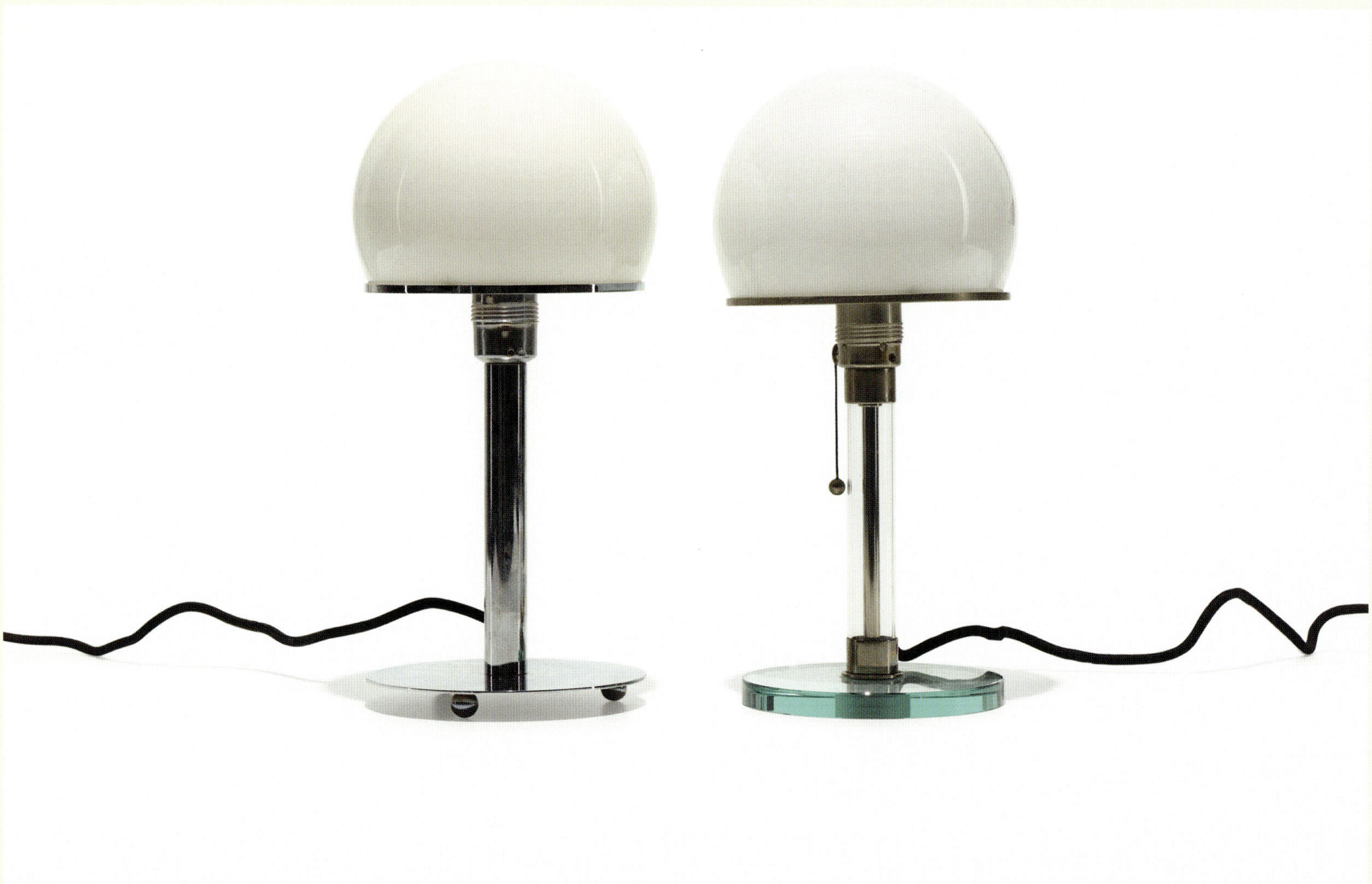

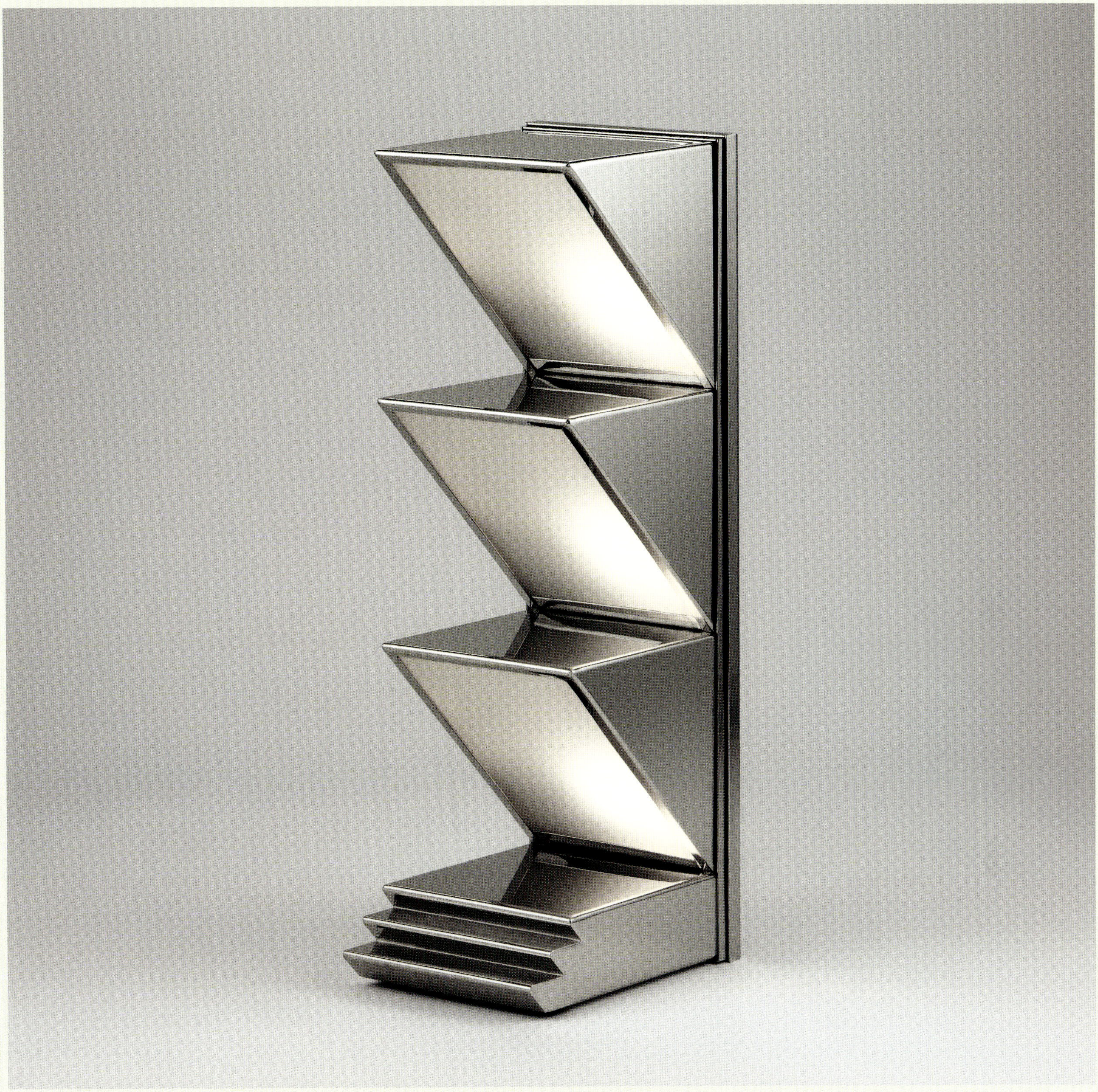

OPPOSITE Table lamps, designed by Wilhelm Wagenfeld, 1924.

ABOVE A replica (courtesy of Terry Tynan Lighting) of a Vollmer lamp, designed by Donald Deskey, 1928.

MARIANNE BRANDT
(1893–1983)

A designer of lighting and industrial products, as well as a photographer, painter and sculptor, Marianne Brandt was one of the multi-talented and influential figures closely associated with the evolution of the Bauhaus. Among her many designs, she produced lighting for the Bauhaus buildings, including the new masters' houses, designed by WALTER GROPIUS in the mid-Twenties. Encouraged by Gropius and others, she also created a number of key lighting designs intended for mass production, and was particularly innovative in her work with functional task lighting.

Marianne Liebe was born in the East German city of Chemnitz. In 1919, she married the Norwegian painter Erik Brandt but the couple only lived together for a few years and divorced in 1935. She studied painting and sculpture at the Hochschule für Bildende Kunst Weimar, after some time at a private art school. She then spent some years in Norway and France before enrolling at the Bauhaus in Weimar in 1924, where she was taught by JOSEF ALBERS, LÁSZLÓ MOHOLY-NAGY, Paul Klee* and Wassily Kandinsky* during her preliminary studies before joining the metal workshop, headed by Moholy-Nagy.

Brandt soon distinguished herself in the metalworking department, despite being in a minority as a female student. Her early Bauhaus designs included a teapot, as well as ashtrays with similar rounded bodies and a discreet X-shaped base. Brandt was conscious of the ambition expressed by Gropius to develop well-designed, democratic products within a 'laboratory of mass production' provided by the Bauhaus.

In 1926, Brandt moved with the Bauhaus to Dessau and a year later took charge of lighting design with the metal workshop, before becoming its director from 1928 to 1929. Much of Brandt's energy was directed into her lighting designs, including collaborations with a small number of Bauhaus colleagues and students. One of her early projects was the ME78B hanging lamp (1926), designed in conjunction with Hans Przyrembel. This elegant pendant light made of aluminium featured a simple saucer shade combined with an innovative pulley system and counter-weight, which allowed the height of the lamp to be adjusted with ease; the pendant was used in a number of locations upon the Dessau campus, including the metal, weaving and architectural departments, as well as the dining room of Gropius's own house.

Brandt's designs suggested a powerful interest in the science and mechanics of lighting, with the aim of creating flexibility for the user in terms of directing and manipulating the source of illumination. This was also true of one of Brandt's most successful designs: the Kandem light of 1928, designed in conjunction with Hin Bredendieck. The Kandem desk or bedside lamp featured a spotlight-style head to direct the light in a specific beam, along with a pivot where the head met the slim supporting arm, which in turn was anchored within a sturdy aluminium base, complete with a push-button on/off switch.

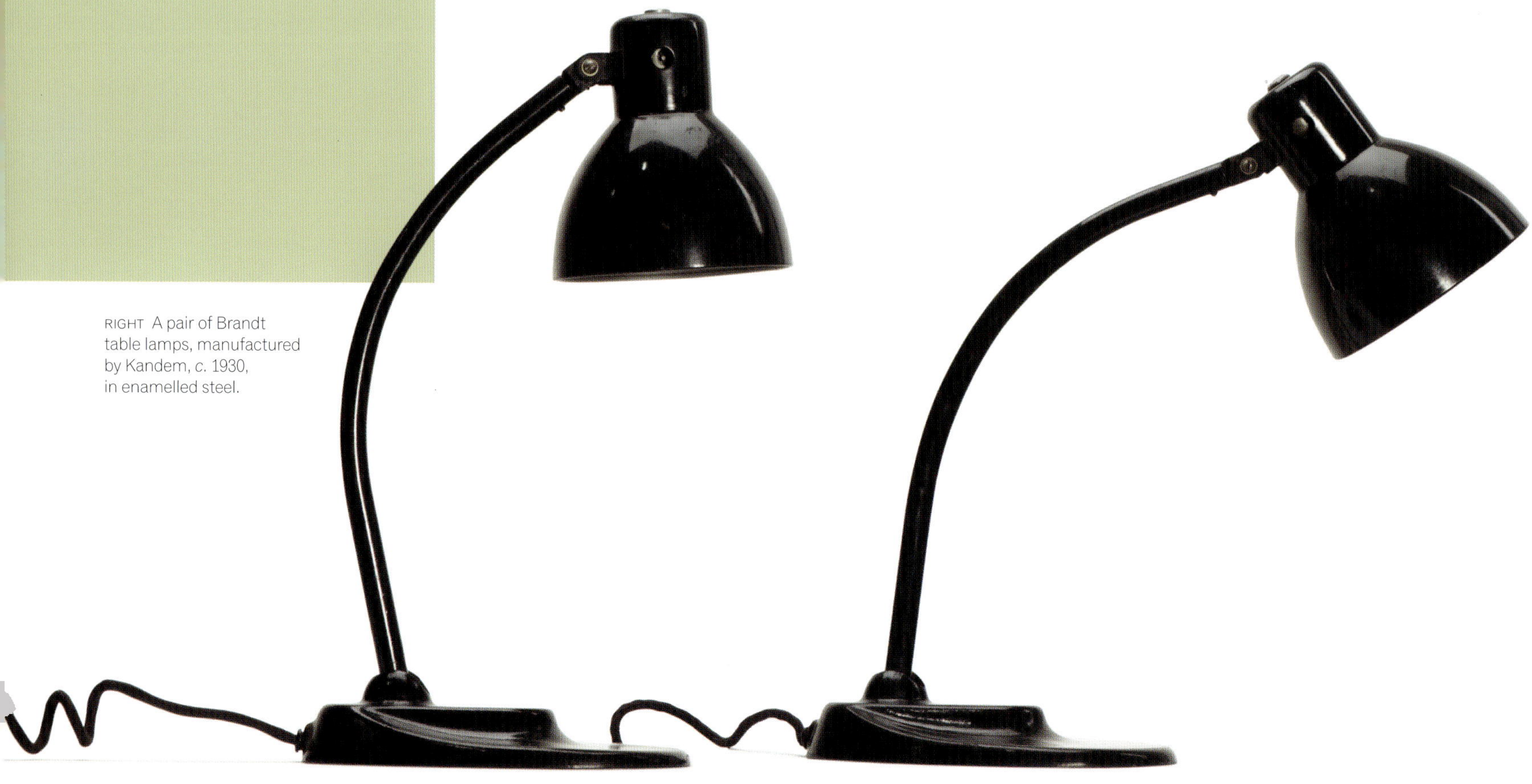

RIGHT A pair of Brandt table lamps, manufactured by Kandem, *c.* 1930, in enamelled steel.

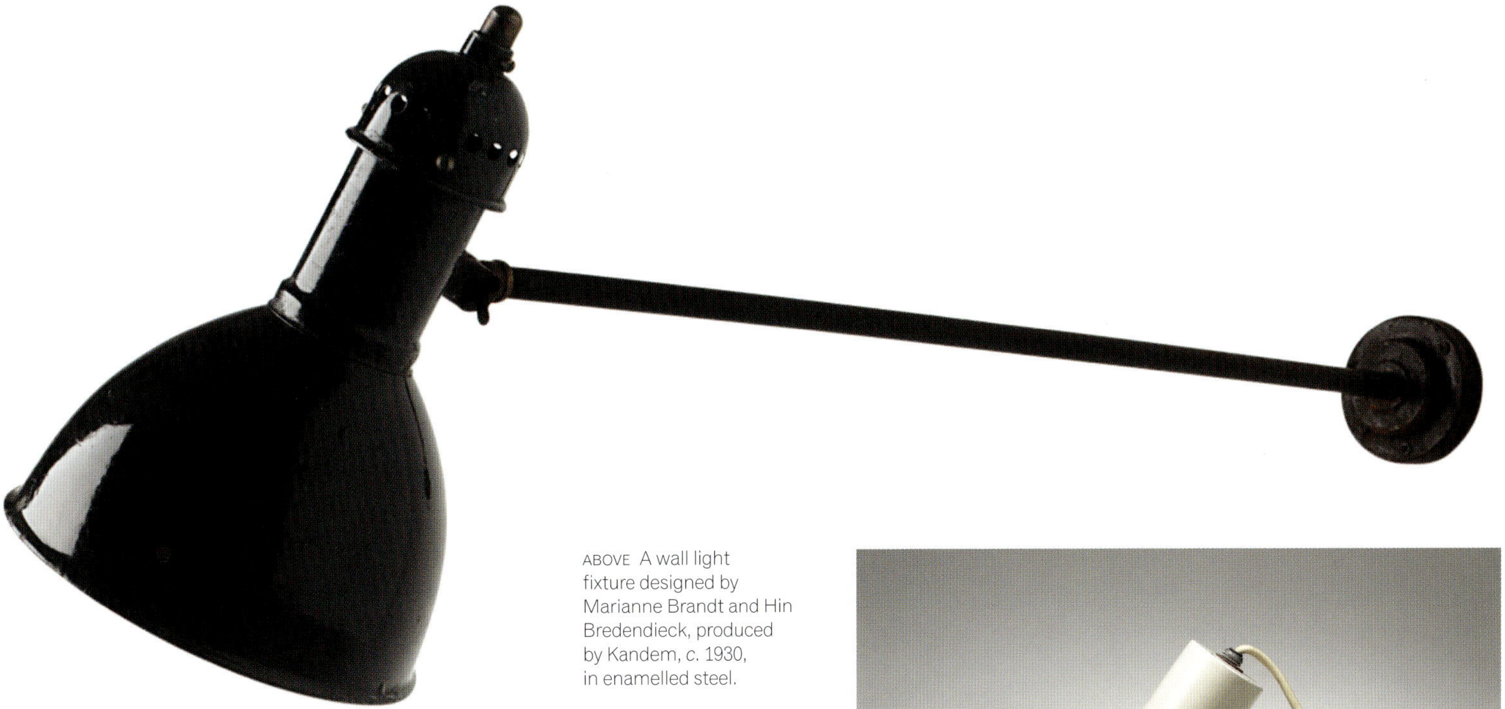

ABOVE A wall light fixture designed by Marianne Brandt and Hin Bredendieck, produced by Kandem, *c*. 1930, in enamelled steel.

RIGHT A Brandt table lamp, manufactured by Kandem, 1930, in enamelled steel.

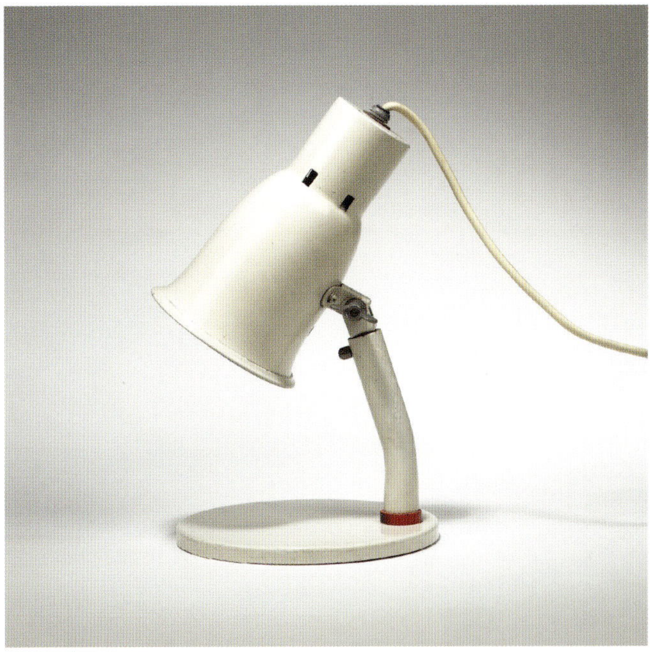

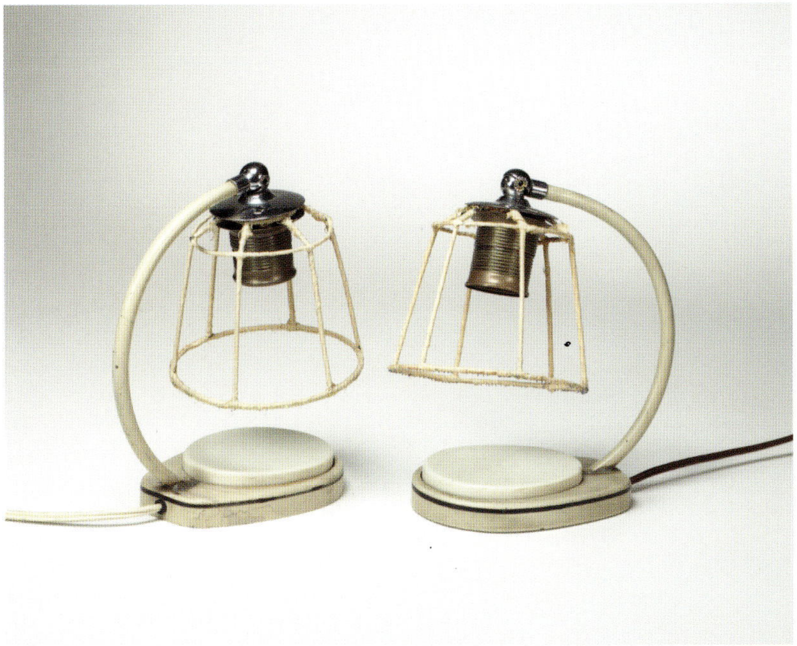

ABOVE A pair of Brandt task lights, produced by Ruppelwerke, 1930s, in enamelled metal and plastic.

RIGHT A table lamp manufactured by Ruppelwerke, 1930s, in enamelled metal and glass, with a large plastic switch.

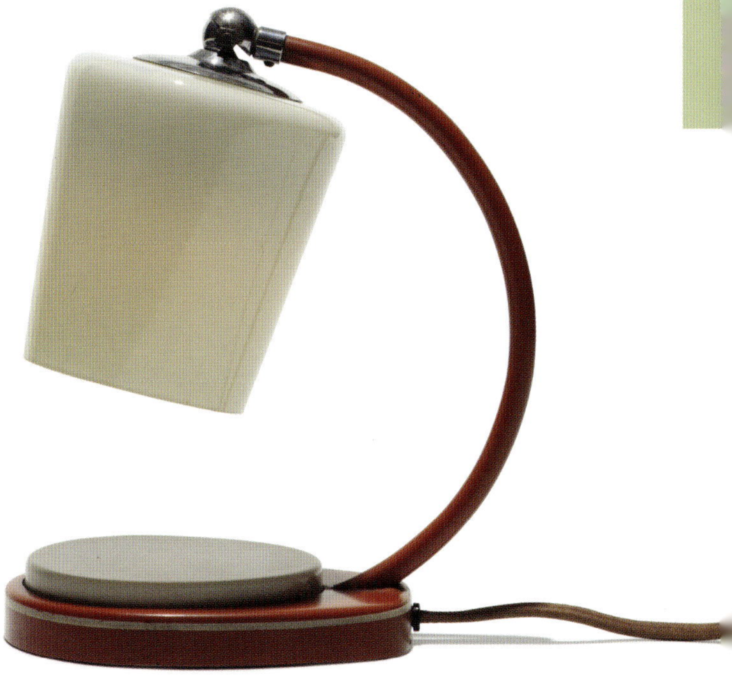

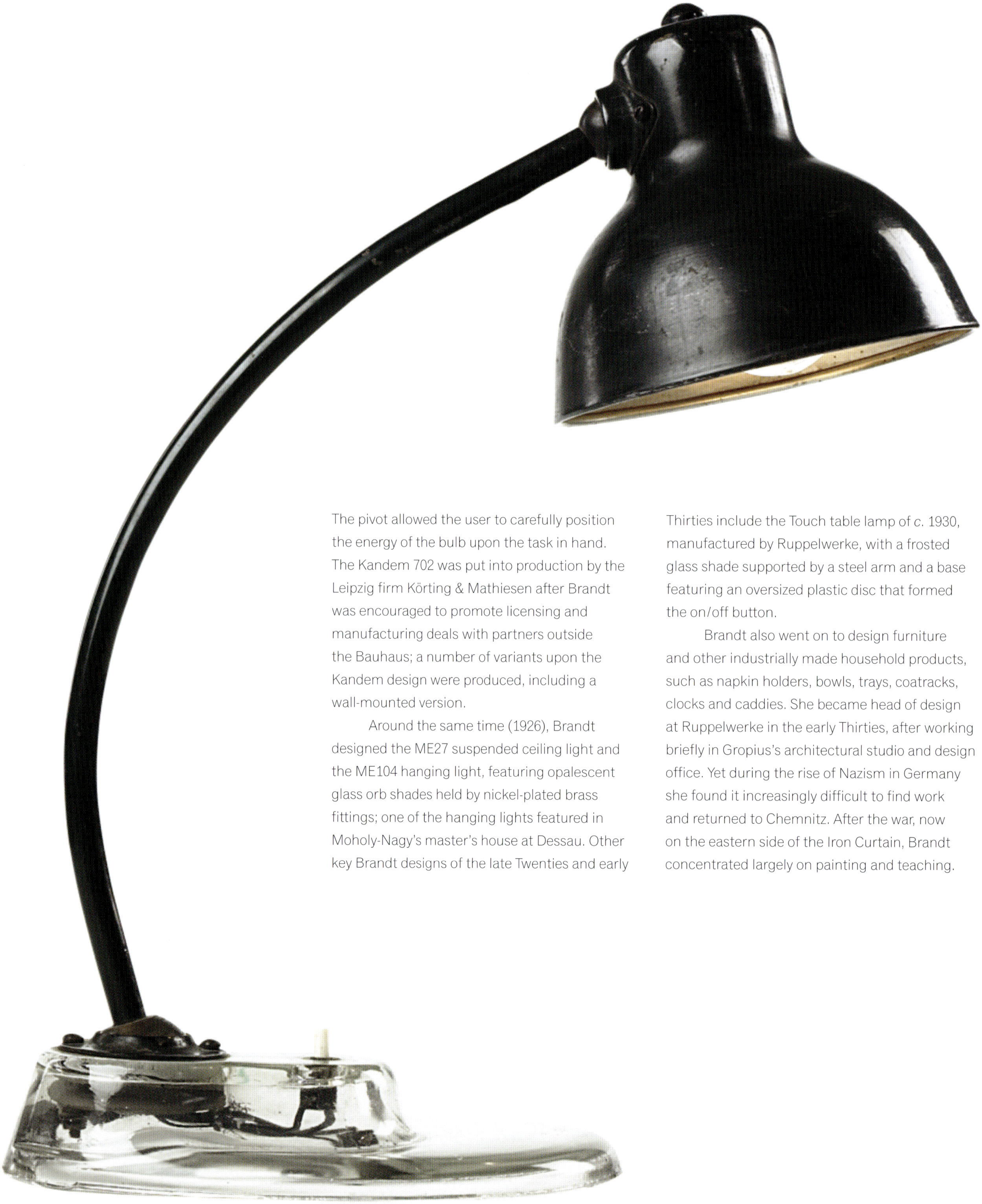

The pivot allowed the user to carefully position the energy of the bulb upon the task in hand. The Kandem 702 was put into production by the Leipzig firm Körting & Mathiesen after Brandt was encouraged to promote licensing and manufacturing deals with partners outside the Bauhaus; a number of variants upon the Kandem design were produced, including a wall-mounted version.

Around the same time (1926), Brandt designed the ME27 suspended ceiling light and the ME104 hanging light, featuring opalescent glass orb shades held by nickel-plated brass fittings; one of the hanging lights featured in Moholy-Nagy's master's house at Dessau. Other key Brandt designs of the late Twenties and early Thirties include the Touch table lamp of *c.* 1930, manufactured by Ruppelwerke, with a frosted glass shade supported by a steel arm and a base featuring an oversized plastic disc that formed the on/off button.

Brandt also went on to design furniture and other industrially made household products, such as napkin holders, bowls, trays, coatracks, clocks and caddies. She became head of design at Ruppelwerke in the early Thirties, after working briefly in Gropius's architectural studio and design office. Yet during the rise of Nazism in Germany she found it increasingly difficult to find work and returned to Chemnitz. After the war, now on the eastern side of the Iron Curtain, Brandt concentrated largely on painting and teaching.

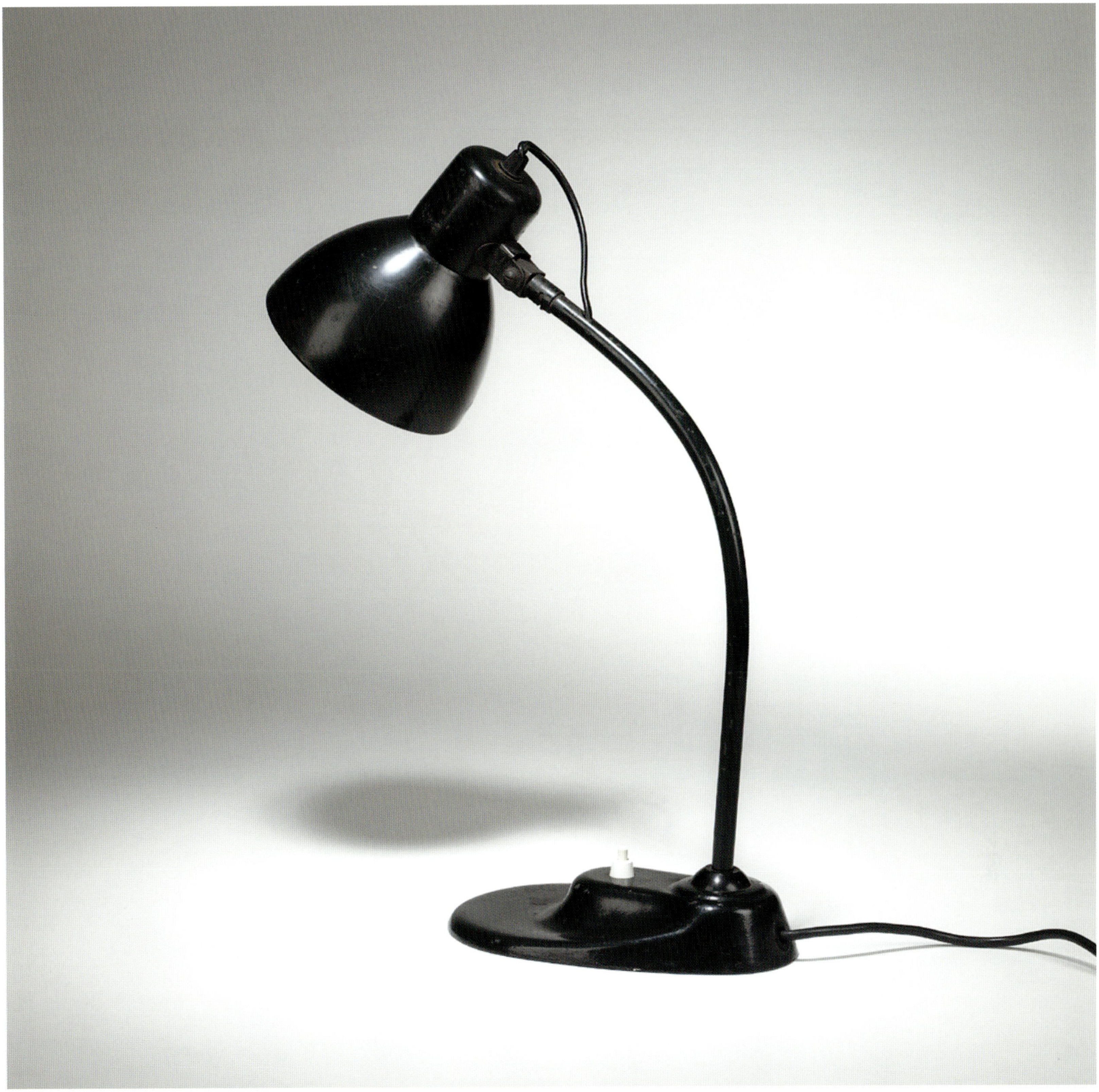

OPPOSITE A Brandt table lamp, produced by Kandem, *c.* 1930, in enamelled steel, glass and Bakelite.

ABOVE A desk light, made by Kandem, 1930s, in enamelled metal and plastic.

GEORGE CARWARDINE
(1887–1947)

A number of key lighting designs of the early modernist period emerged from dedicated experiments with function, focused upon the manipulation, reflection and direction of artificial light. This was particularly true of a new generation of task lighting, which emerged largely from a need for efficient, practical illumination in workshops and factories, yet spread to more domestic contexts and office applications. In 1927, the French designer Edouard-Wilfrid Buquet came up with a functional but elegant desk or table lamp made of nickel-plated brass and aluminium that featured two interconnected, pivoting arms floating upon a base; the lower arm held the counter-weight and the upper held a spotlight lamp. Around the same time, the English designer George Carwardine was thinking in a similar way and began experimenting with a spring-tensioned light. This was to become the Anglepoise – a ubiquitous presence, even today, upon countless desks.

Carwardine was an engineer, who worked in the English car industry and specialized in suspension systems. He worked for the Horstman car company in Bath, founded by Sidney Horstmann, the son of a German clockmaker (the second 'n' was dropped from the company name during World War I).

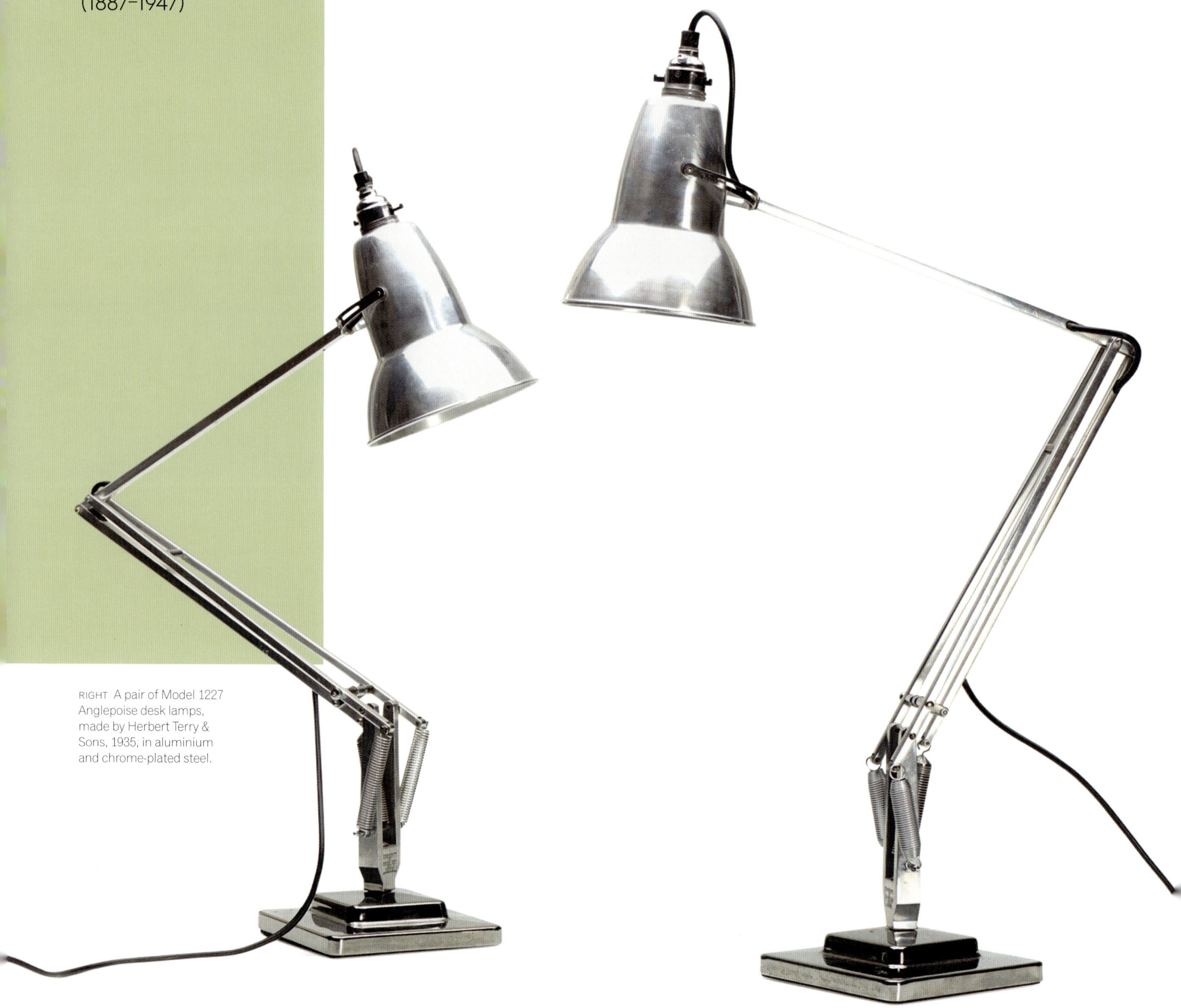

RIGHT A pair of Model 1227 Anglepoise desk lamps, made by Herbert Terry & Sons, 1935, in aluminium and chrome-plated steel.

Horstman produced around 3,000 cars from 1915 to 1929, when the worldwide economic recession provoked a rapid downturn in car-making, but the company continued to be respected for its work on innovative suspension systems.

One of the victims of the Depression was Carwardine himself, who left his job as chief designer with Horstman to found his own engineering company, also based in Bath, known as Cardine Accessories. From a small workshop, Carwardine began applying his knowledge of suspension to lighting, developing early prototypes of a task light that used a combination of coiled

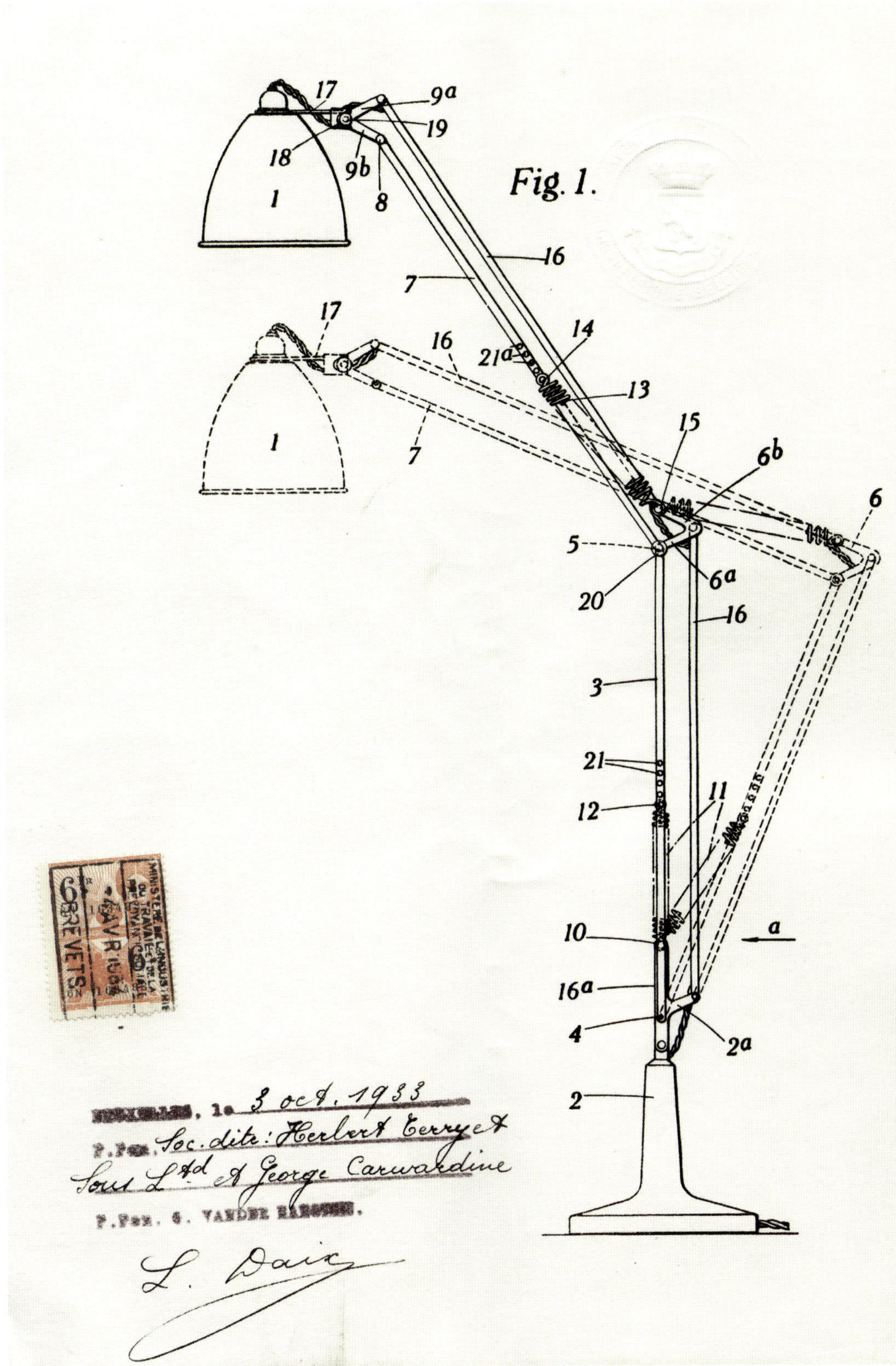

ABOVE The original patent document (dated 1933) for Carwardine's Model 1208 desk lamp, manufactured by Herbert Terry & Sons.

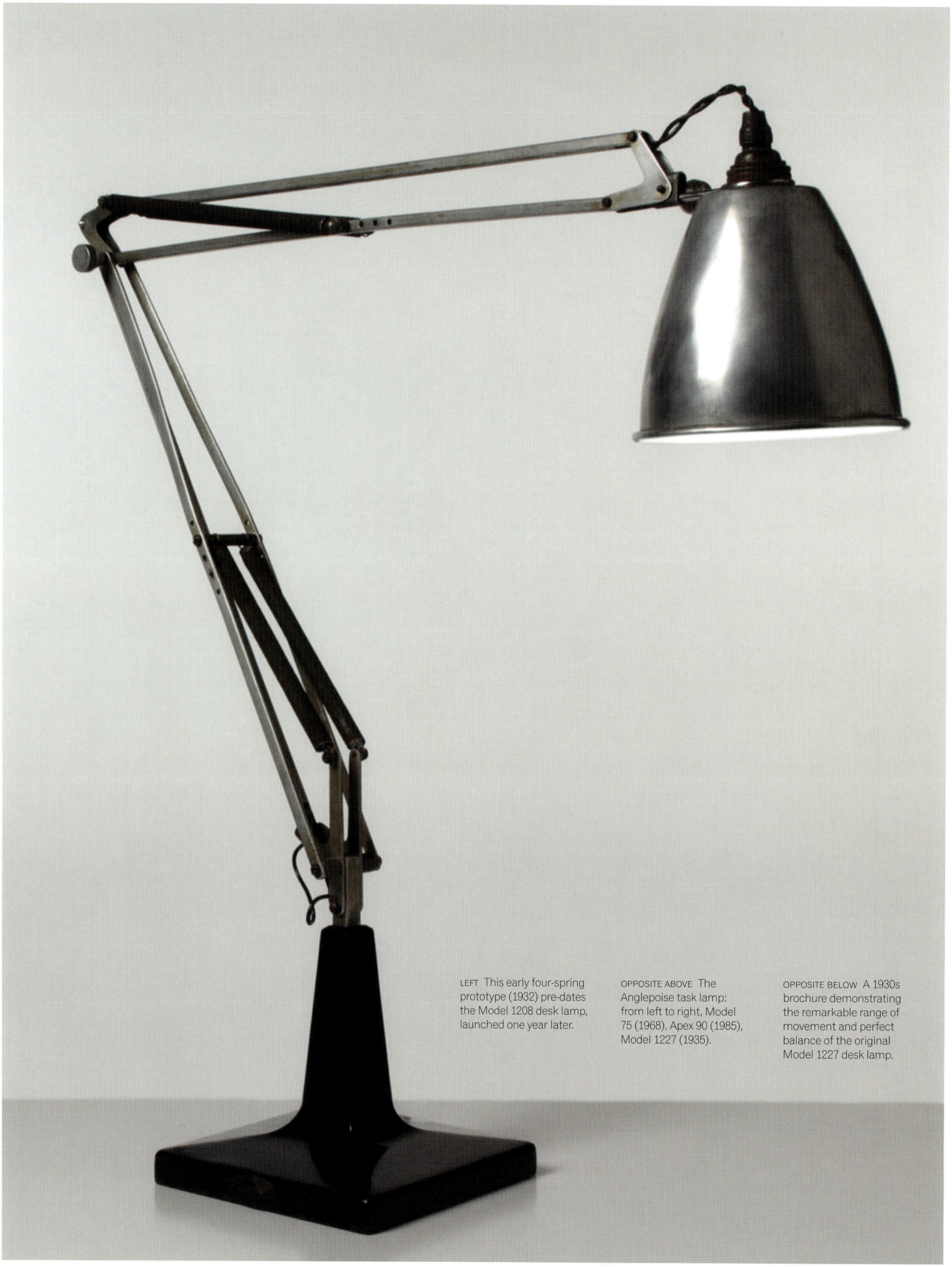

LEFT This early four-spring prototype (1932) pre-dates the Model 1208 desk lamp, launched one year later.

OPPOSITE ABOVE The Anglepoise task lamp: from left to right, Model 75 (1968), Apex 90 (1985), Model 1227 (1935).

OPPOSITE BELOW A 1930s brochure demonstrating the remarkable range of movement and perfect balance of the original Model 1227 desk lamp.

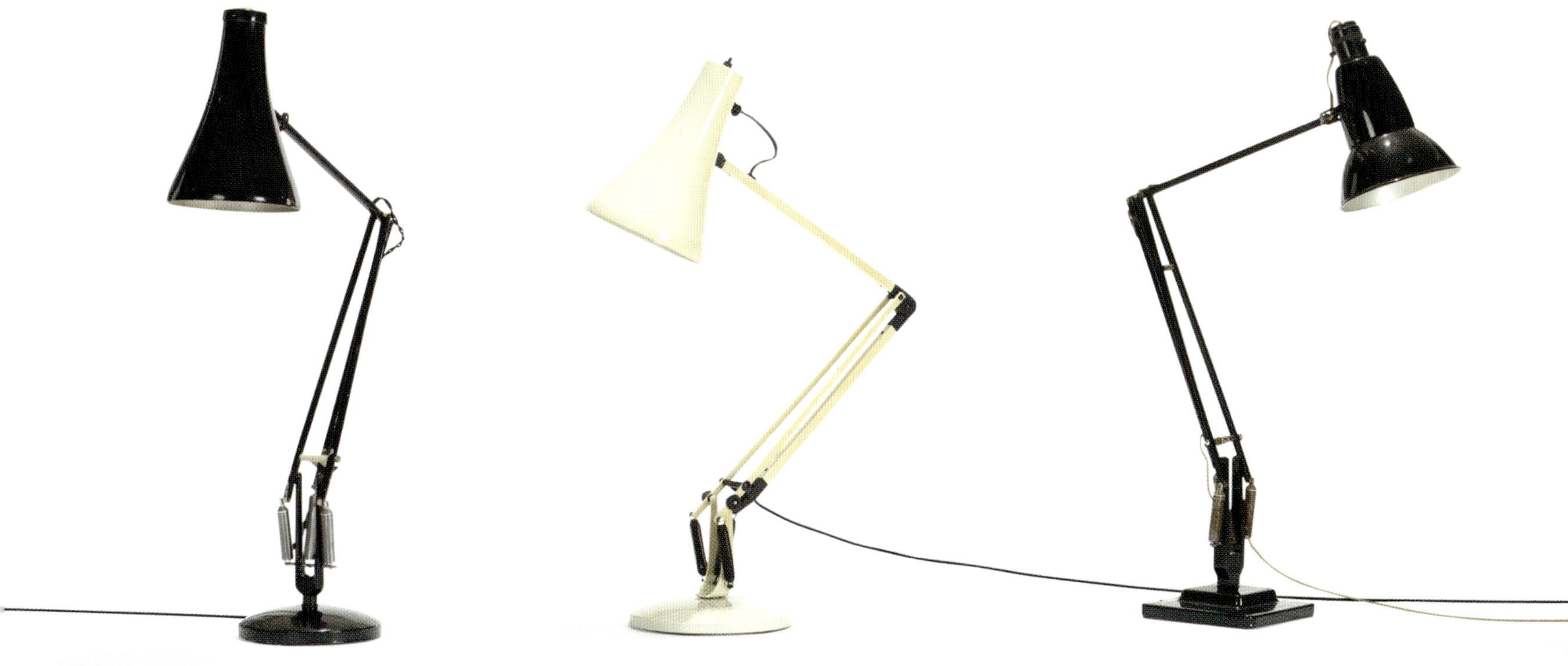

springs and levered joints or pivots. The results were highly effective, adaptable and relatively simple, yet the ingenuity of the thinking behind them was completely fresh, using the idea of spring-loaded tension to control and hold the position of the lights rather than counter-weights and balances.

In 1932, Carwardine launched his first design incorporating the patented concept of 'equipoising'. The light has often been compared to a human arm, with the upper and lower portions linked by a pivoting elbow and an adjustable aluminium spotlight in place of the hand; the arm was held in place by a solid metal base, and four springs were used to tension the light and hold the arm in position.

Carwardine began producing early batches of the lamp, aimed at commercial use, but soon found that he could not keep up with demand. So in 1933 he joined forces with Herbert Terry & Sons, a spring manufacturer based in Redditch, which had been supplying the coiled metal springs for the first lamps. Additional Anglepoise lights soon followed, including the four-spring 1209, with longer arms, and then in 1935 a more refined, three-spring version known as the Model 1227, marketed for residential and office use. The Anglepoise became hugely successful in many different contexts which demanded a light source that could be easily controlled and directed, with variations used in hospitals and factories. The Anglepoise company continues to make the lamp today.

A few years after the launch of the 1227, a Norwegian lighting designer and manufacturer Jacob Jacobsen acquired a licence to produce a version of the Anglepoise. Jacobsen adjusted the design again, using a rounded base rather than Carwardine's linear, ziggurat format; the shape of the spotlight was also amended, along with other refinements. The Luxo L-1 lamp, first launched in 1937, adopted a parallel trajectory to the success of its Anglepoise siblings, also selling in the millions.

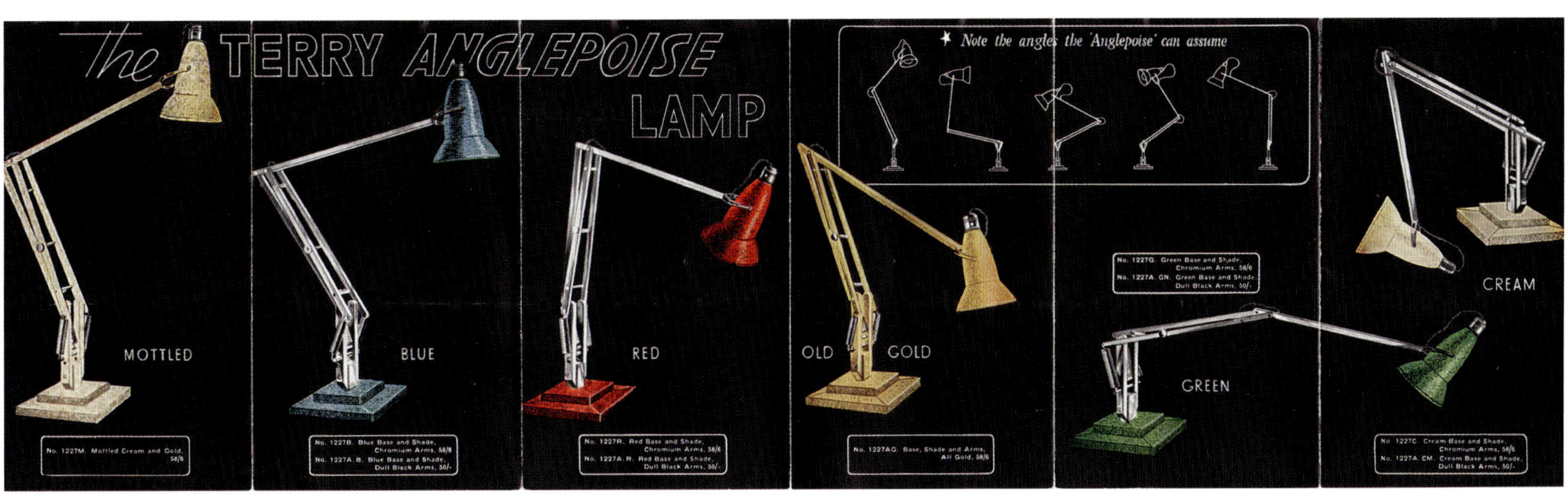

CHRISTIAN DELL
(1893–1974)

Along with MARIANNE BRANDT and WILHELM WAGENFELD, Christian Dell was one of the key figures in the sphere of lighting design to emerge from the Bauhaus. His work was led by an emphasis on function, with Dell creating a striking portfolio of desk lamps and task lights, some of which are still in production today, while also adhering to the Bauhaus tenets of creating affordable, factory-made products with kits of standardized parts. Yet, at the same time, there was a powerful sculptural quality to Dell's work, which helps to explain its ongoing popularity and relevance. His parabolic shades and saucer-like light fittings anticipated some of the more sculptural and organic shapes that emerged in post-war and mid-century lighting created by Serge Mouille, Greta Grossman and others.

Dell was born in the German city of Offenbach am Main, close to Frankfurt. He went to nearby Hanau to study at the Königlich Preussische Zeichenakademie (Royal Prussian Drawing Academy) and to take an apprenticeship in silversmithing at Schleissner & Söhne. Dell went on to work briefly as a silversmith in Dresden before continuing his studies under Henry van de Velde* at the Kunstgewerbeschule (School of Arts and Crafts) in Weimar, the school that eventually became the Weimar Bauhaus.

After World War I, Dell worked as a silversmith in both Munich and Berlin, before establishing his own atelier in Hanau. In 1922, he joined the metalworking department at the Bauhaus school in Weimar, led by LÁSZLÓ MOHOLY-NAGY. Dell served as a tutor at the Bauhaus up until 1925 before moving on to the Frankfurter Kunstschule (Frankfurt Art School) rather than making the move to the new Bauhaus campus at Dessau.

Many of Dell's most famous lighting designs date from this period, in the late Twenties and early Thirties, when he was working in Frankfurt while continuing to embrace the functionalist ethos of the Bauhaus. One of Dell's most successful early lights was the Rondella desk or table lamp of 1927. The Rondella was a flexible task light with a parabolic metal shade fitted to a slim tubular steel arm with a ball and socket joint; both the spotlight and the arm – which could slide up and down on a companion column mounted on a round base – were easily adjustable. The Rondella was manufactured by a company of the same name based in the town of Oberursel, and later by another local firm, Bünte & Remmler.

The Rondella was among a series of similar designs for task lamps, defined by the combination of function and flexibility plus the distinctive parabolic spotlight shade. The series included the Rondella Polo desk light of 1928 and the Polo Populär desk light of 1929. Another close variant, the Dell-Lamp Type K (1930), was produced in the thousands for the social housing projects of New Frankfurt; a wall-mounted version of the light was also manufactured, along with a design that could be clamped to a table.

A second collection of lights followed in the early Thirties, produced by Kaiser & Co. This was the Idell collection, which grew into a catalogue of designs that included not only desk lamps and table lights but also wall-mounted lights and floor lamps. The unifying factors that distinguished the

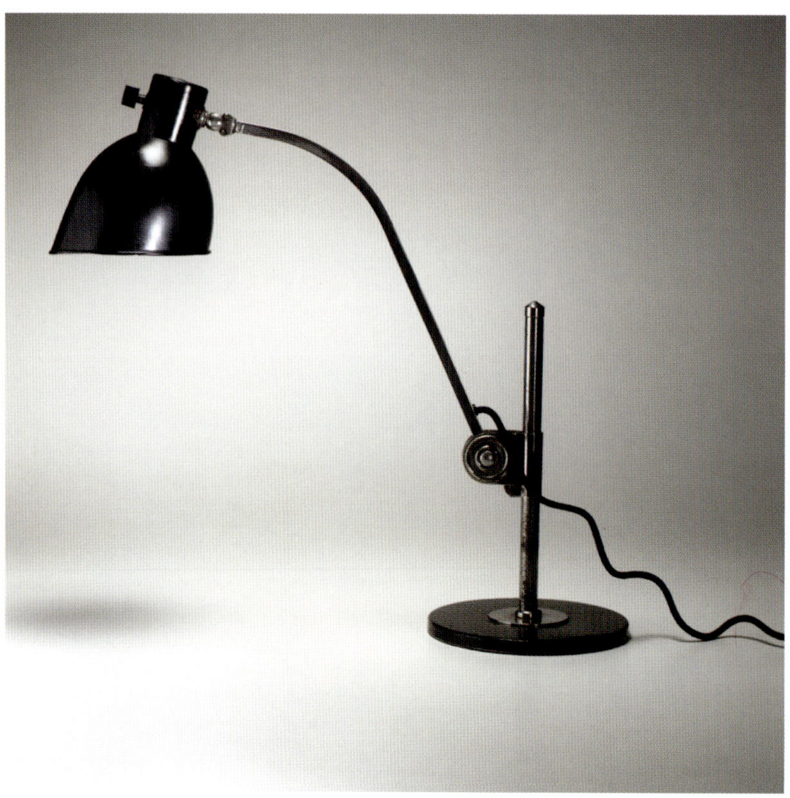

RIGHT Christian Dell's Type K lamp, produced by Kaiser & Co., *c.* 1932, in enamelled metal and steel.

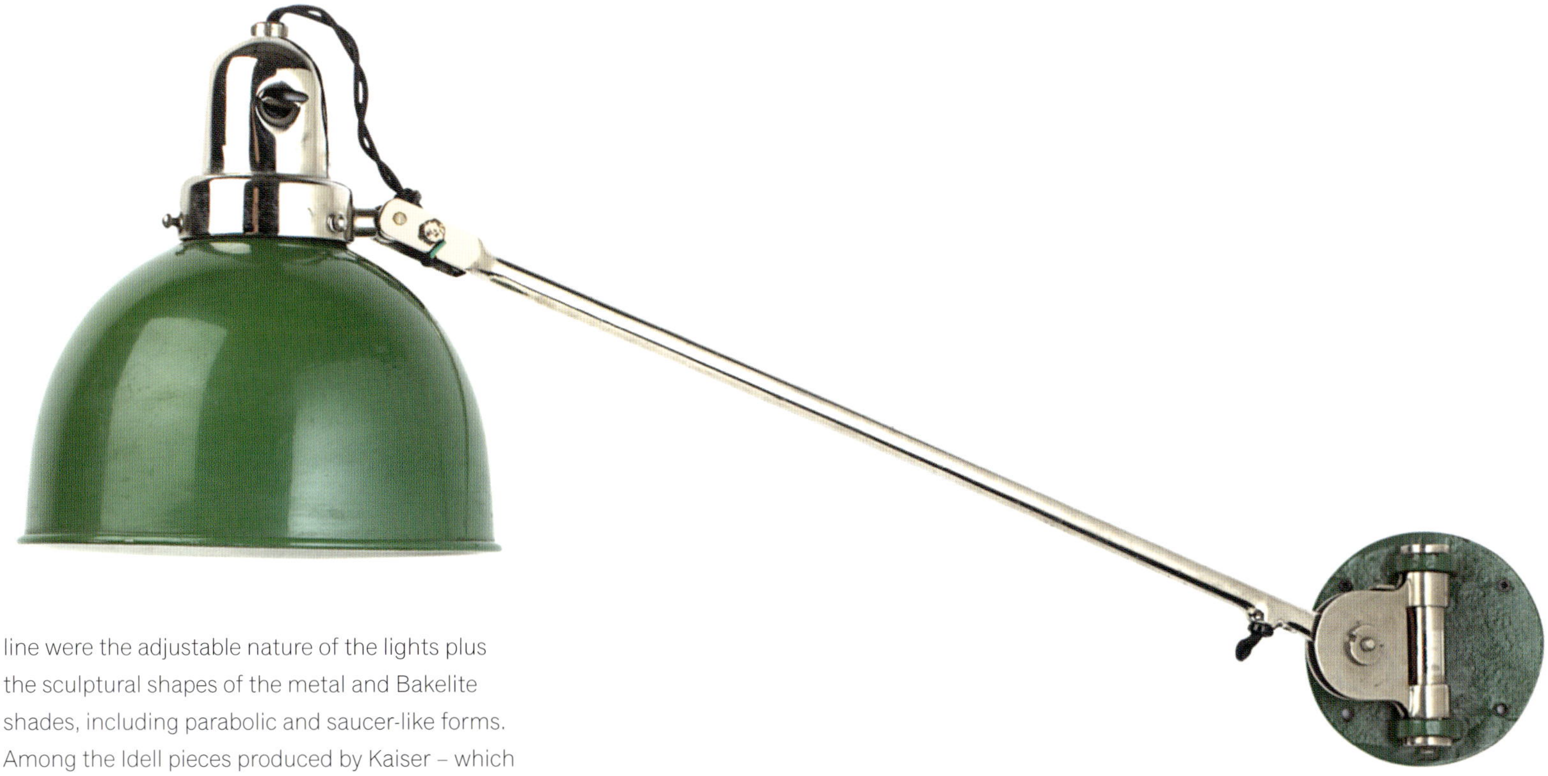

line were the adjustable nature of the lights plus the sculptural shapes of the metal and Bakelite shades, including parabolic and saucer-like forms. Among the Idell pieces produced by Kaiser – which shared many common components – the stand-out designs include the 6580 'Double Dell' table lamp (*c.* 1932), with twin adjustable arms and shades embedded in a single round base, and a wall-mounted Idell light with an extending scissor-action arm.

Along with many other modernist architects and designers in Germany, Dell found his ability to work increasingly frustrated by the Nazi authorities during the latter half of the Thirties. In the post-war years, he returned to his work as a silversmith and later opened a jewelry business in Wiesbaden. Yet his pre-war lighting designs still manage to feel current.

ABOVE A wall lamp designed by Christian Dell, *c.* 1930, in enamelled metal and chrome-plated steel.

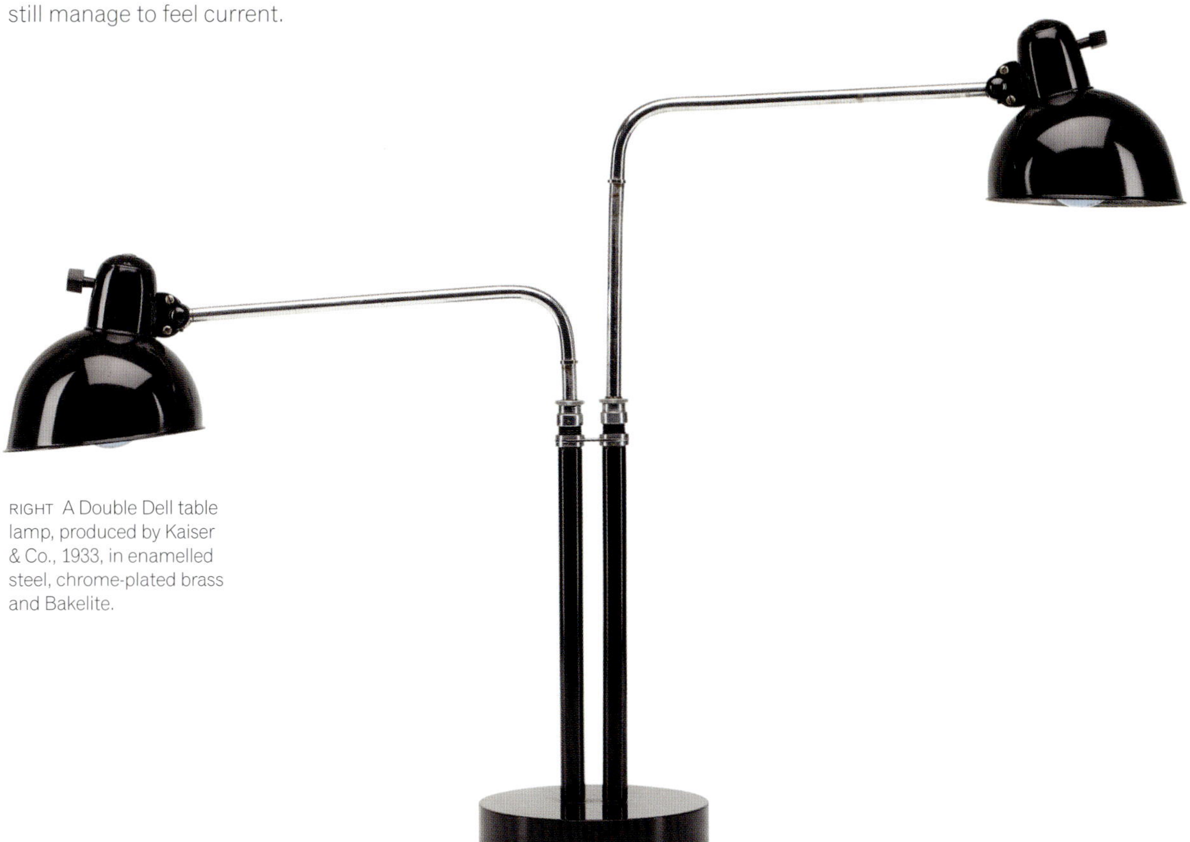

RIGHT A Double Dell table lamp, produced by Kaiser & Co., 1933, in enamelled steel, chrome-plated brass and Bakelite.

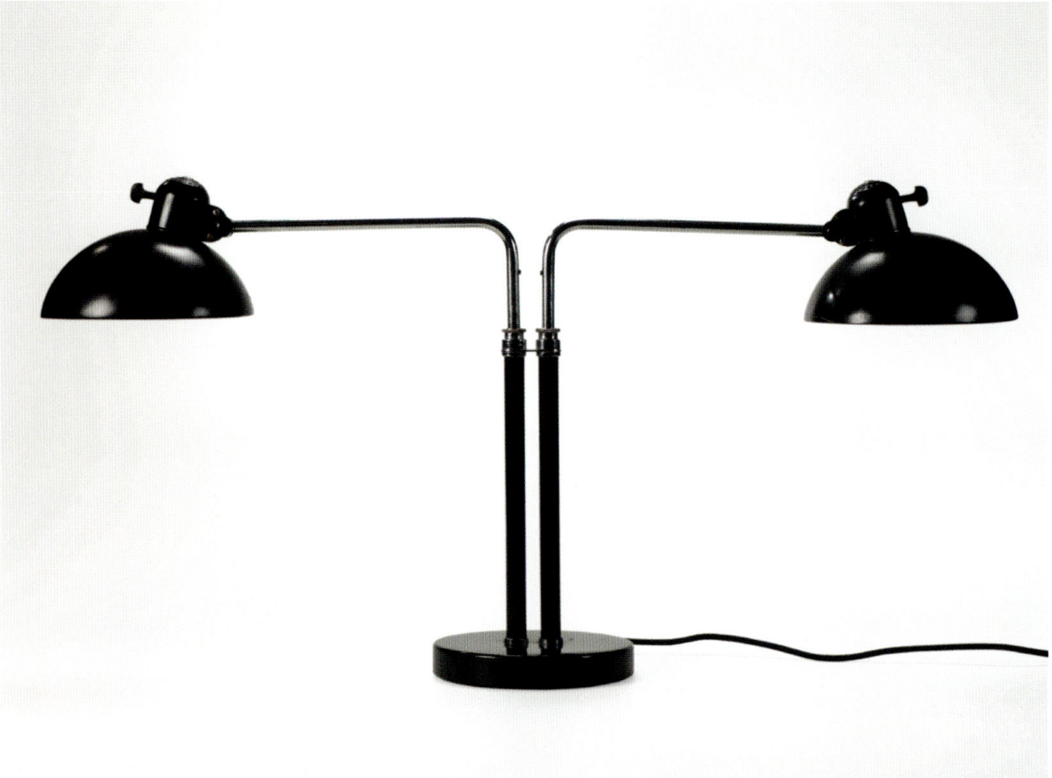

ABOVE The adjustable Model 6580 double table lamp, made by Kaiser & Co., 1933, in enamelled and chrome-plated brass and Bakelite.

RIGHT An early Dell table lamp, produced by Kaiser & Co., 1920s, in enamelled metal.

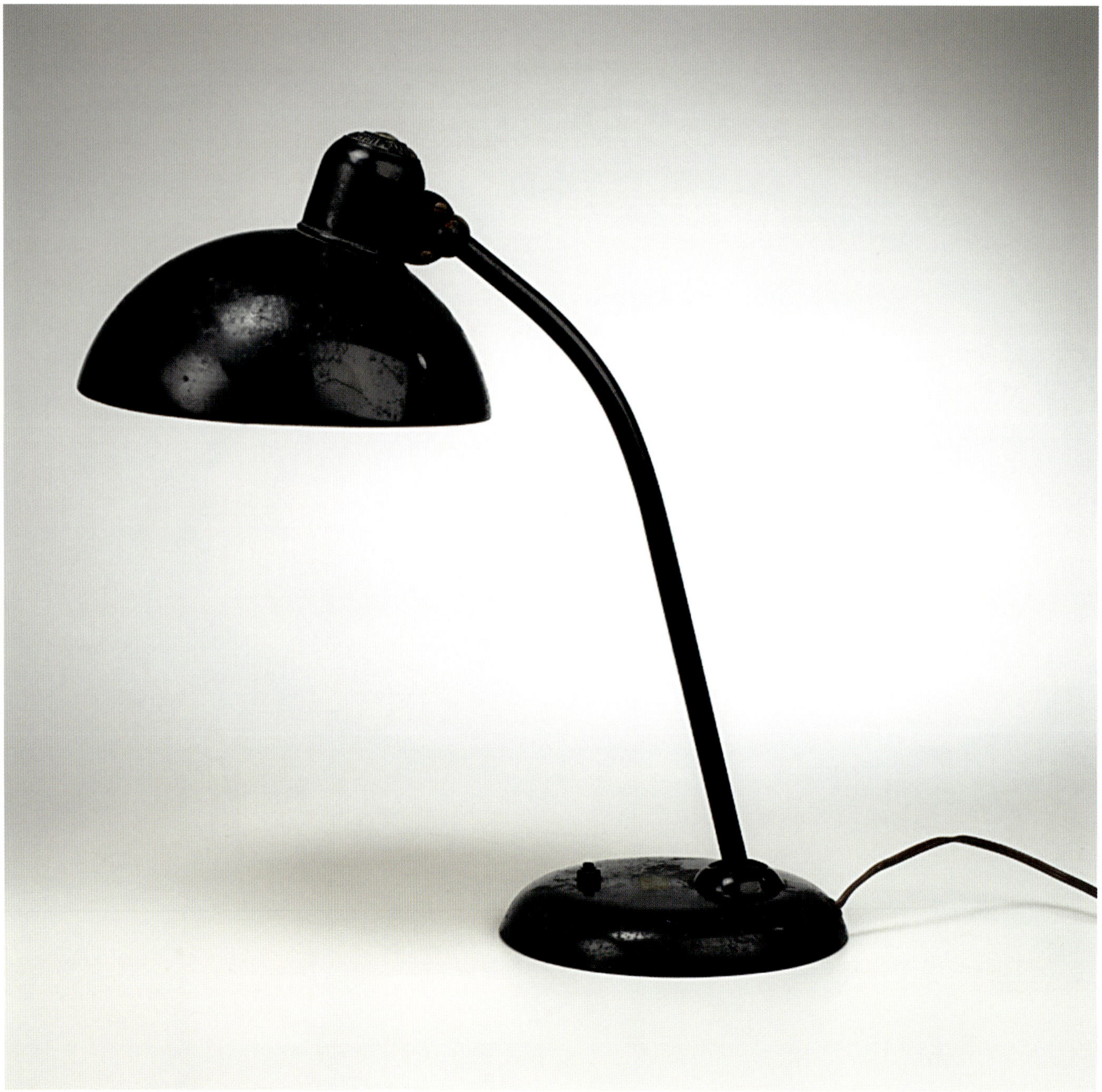

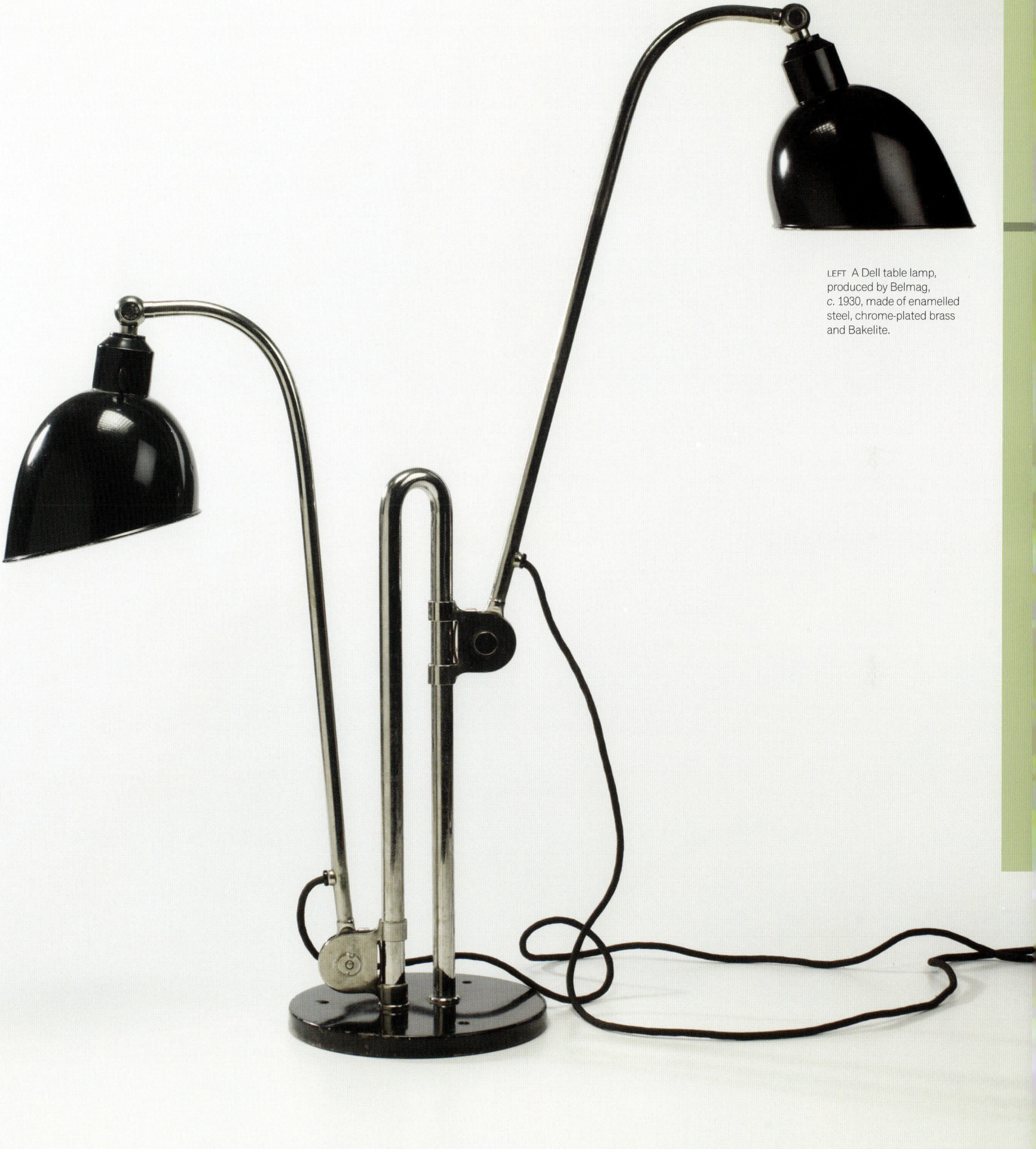

LEFT A Dell table lamp, produced by Belmag, *c.* 1930, made of enamelled steel, chrome-plated brass and Bakelite.

POUL HENNINGSEN
(1894–1967)

During the early years of the 20th century, designers tended to approach lighting as an artistic medium, concentrating on sculptural forms, sinuous shapes, and a flamboyant use of materials and colour. There was surprisingly little thought given to the science of electric lighting and its manipulation, despite the fact that artificial light was born from scientific process. While much early lighting seemed concerned with disguising the electric bulb and replicating the effect of a candelabra or gas light, early modernist designers such as the Danish pioneer Poul Henningsen began to adopt an approach that was initially led by a focus on function and the careful manipulation in search of 'harmony in lighting'.

Henningsen grew up in a home illuminated by oil lamps, which cast a warm, subtle glow through a space. He was born in Ordrup, in Denmark, and was the son of two writers. He studied architecture in Copenhagen and went on to work as a journalist, editor and critic, while also writing poetry and film scripts. However, he remained committed to design and became fascinated with lighting.

Unsettled by the intensity and glare of electric light bulbs, Henningsen began to analyze the impact of various shades and systems upon the diffusion of light throughout a room. 'Electricity has made it possible to gorge on light, which is a new phenomenon,' he said. 'The golden rule that illumination ... must be pleasant to look at has been forgotten in the lighting high.'[1]

Inspired by the inviting interiors of his childhood, Henningsen began to work on lighting designs that softened electric light. He examined, with scientific rigour, the way in which the use of different materials and colours impacted upon the intensity and warmth of the light produced. In the early Twenties he formed an alliance with Sophus Kaastrup-Olsen, the new head of the Danish manufacturers Louis Poulsen & Co., which was to become one of the great creative and productive partnerships in the history of 20th-century lighting design.

The Poulsen company began as a family concern in Copenhagen in the 1870s and, under the leadership of Louis Poulsen himself, focused on producing and selling tools, machinery and

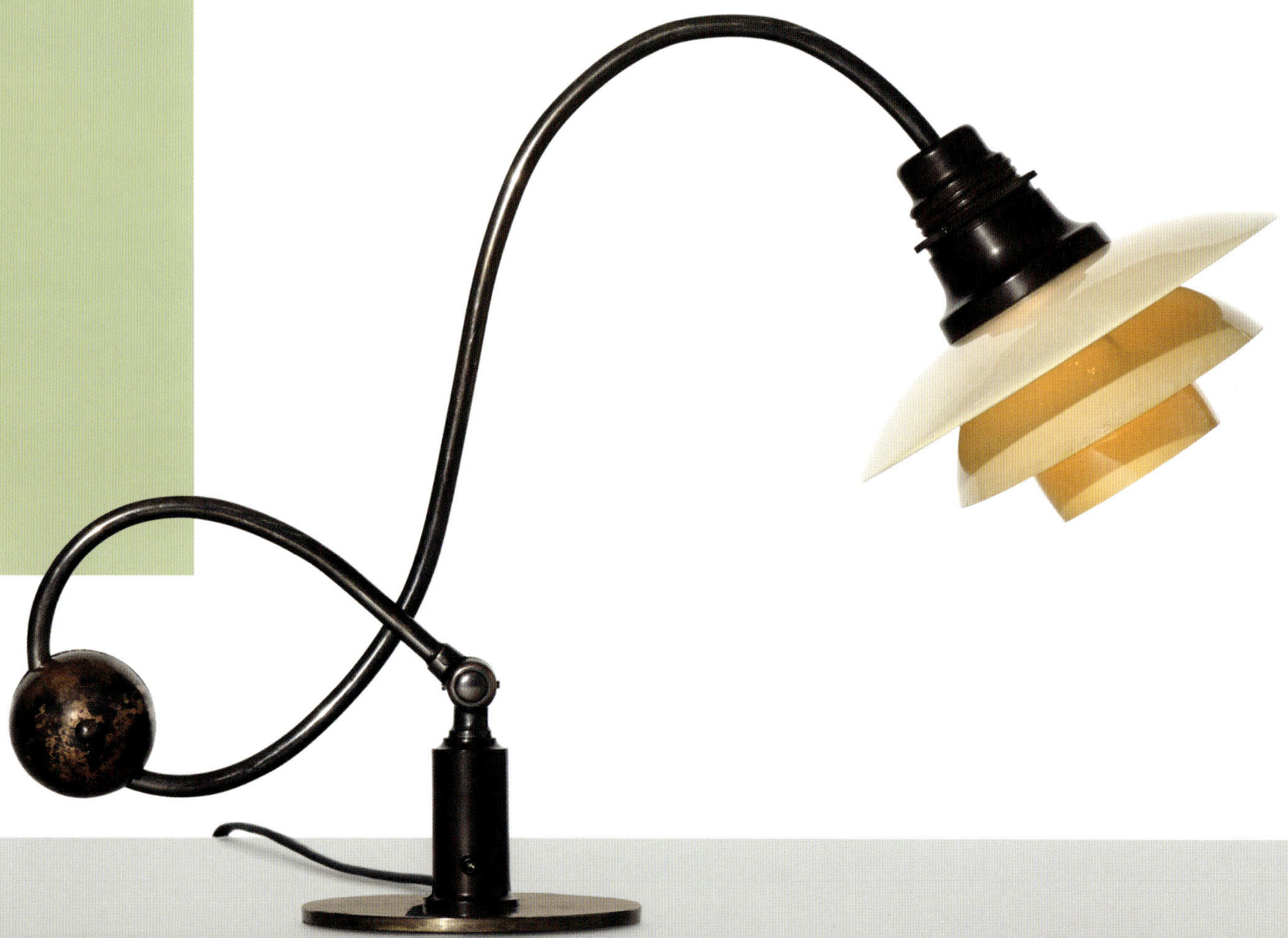

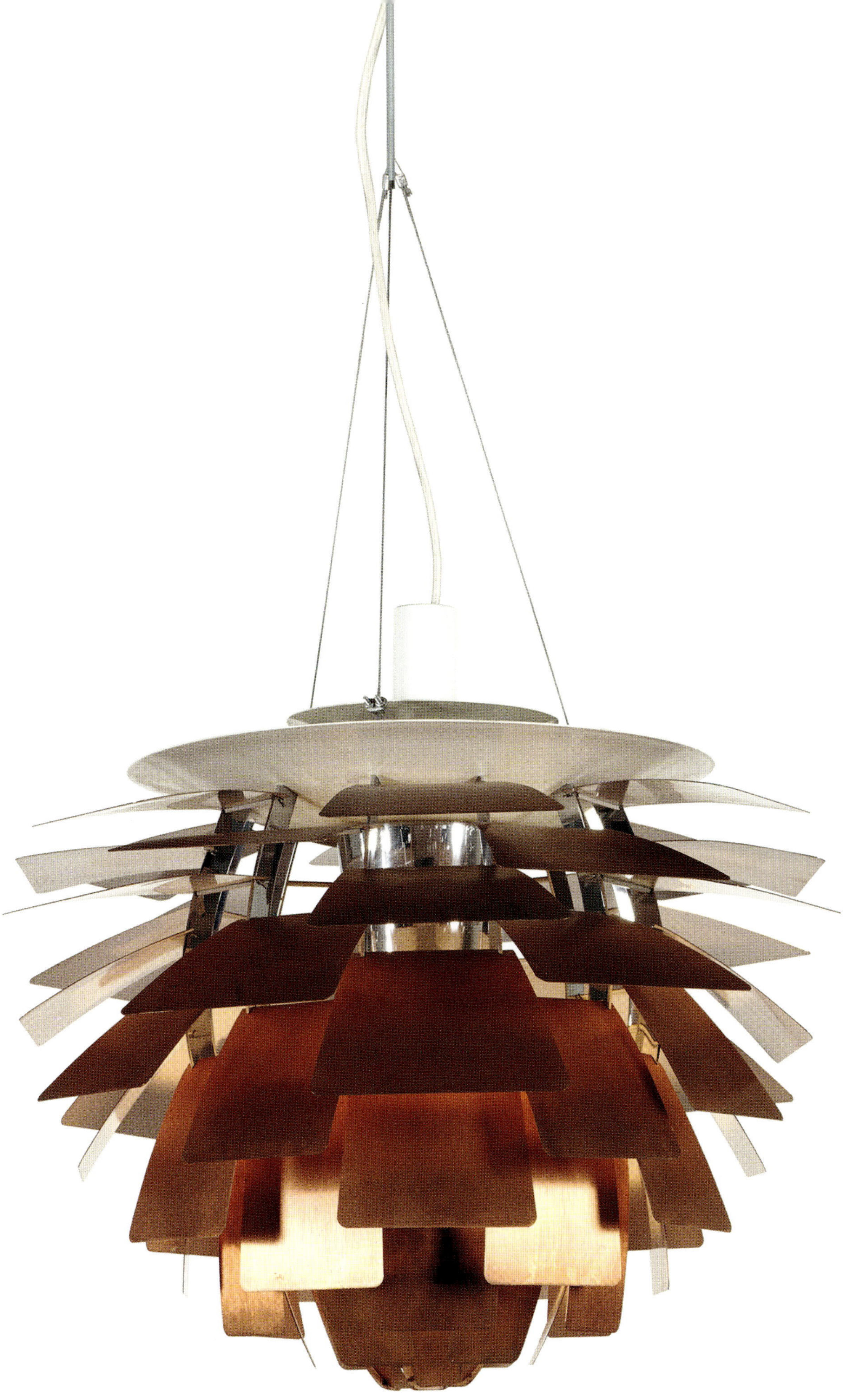

OPPOSITE A PH 22 piano lamp, produced by Louis Poulsen, 1931, made with bronze, Bakelite and glass.

ABOVE Poul Henningsen's iconic Artichoke lamp, manufactured by Louis Poulsen, 1958, in copper, aluminium and stainless steel.

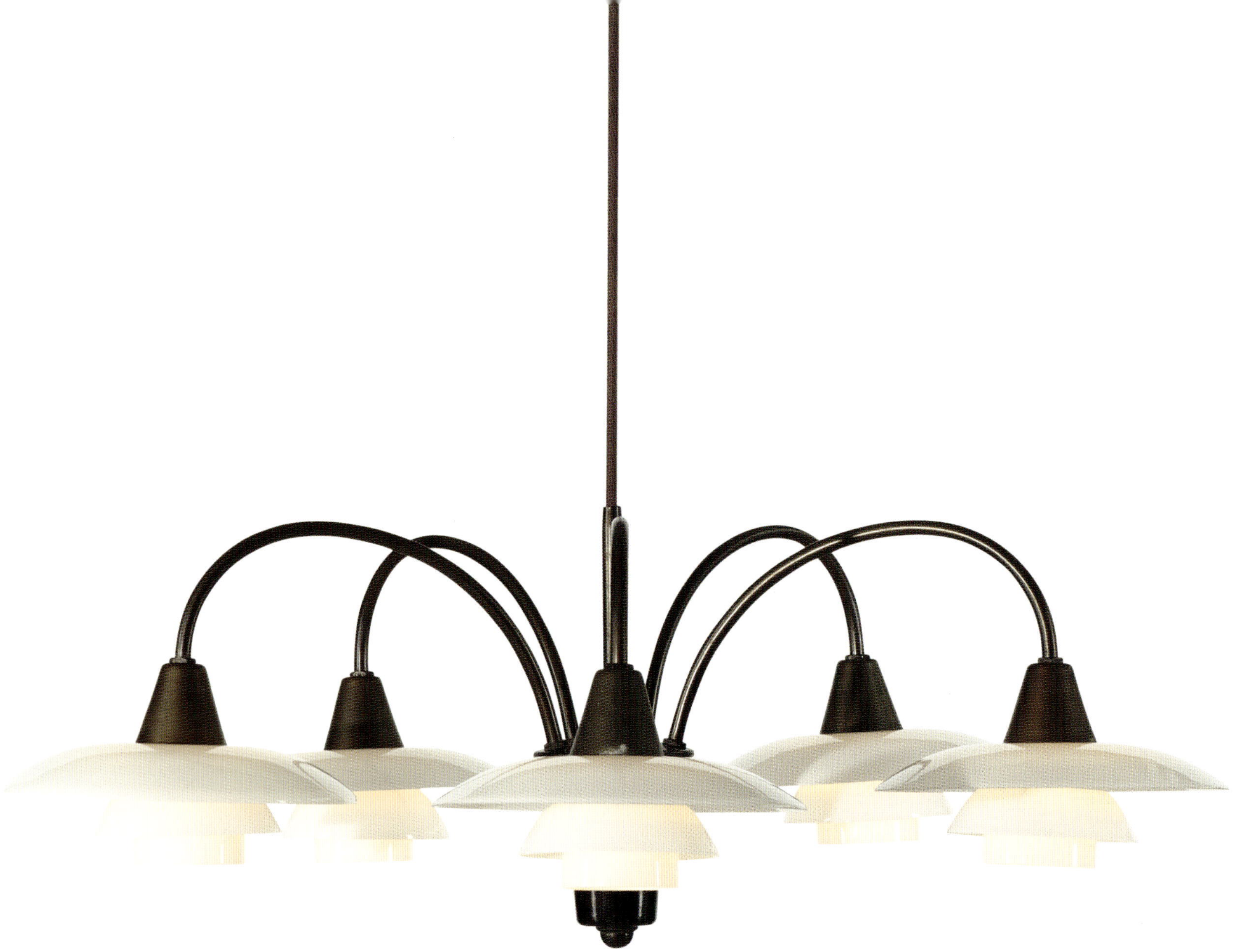

electrical supplies. Kaastrup-Olsen took over the company in 1917, concentrating on tapping into the growing market for electrical services and products. He embraced Henningsen's approach to lighting design and commissioned the designer to develop a range of 'PH' lights, with the aim of presenting the first designs at the Exposition Internationale des Arts Décoratifs et Industriels Modernes of 1924–25 in Paris.

These early PH table lamp designs used a triptych of concentric shades, either in metal or glass, to soften the glare of the bulb and produce a seductive, ambient light. The designs were well received both in Paris and Denmark, and Henningsen produced a range of variants using the same stacked system of complementary shades, including ceiling lights and wall lamps.

In 1927, Henningsen produced his famous Septima hanging light, using seven concentric, circular glass shades; more than ever, the lamp resembled the layered petals of an upturned flower, suggesting the designer's talent for fusing science with sculptural artistry. By the time of the Exposición Internacional de Barcelona of 1929, where Louis Poulsen had a stand, the PH product line included an extensive collection of Poul Henningsen's designs.

During the early Thirties, Henningsen continued to develop new lighting designs, while also experimenting with tubular steel furniture, which included his spring-like Snake chair (or 'Slange-stolen') of 1931. Fresh PH designs of this period include floor lights and the PH 3 chandelier, using a collection of triptych-style shades arranged upon spokes embedded in a central brass mount; a variation on the design, known as the Cascade chandelier, appeared a few years later (*c.* 1934).

Henningsen's work was admired internationally and noted in Germany, in particular. Here, Otto Müller became the distributor for the Louis Poulsen collection. In the early Thirties, Müller produced his own range of 'Sistrah' lamps, largely inspired by Henningsen's work.

The highly successful collaboration with Louis Poulsen continued in the post-war period. The Fifties saw the appearance of Henningsen's iconic PH Artichoke ceiling light (1958), which used a series of metal leaves, or petals, arranged around a light, steel frame.

OPPOSITE The Cascade chandelier, manufactured by Louis Poulsen, 1933, made of bronze, brass, frosted glass and Bakelite.

BELOW The PH 2/1 Kippe table lamp (left), 1933, in brass, frosted glass and Bakelite, along with the Model 2/1 table lamp (right), 1931, in bronze, enamelled steel and frosted glass, both manufactured by Louis Poulsen.

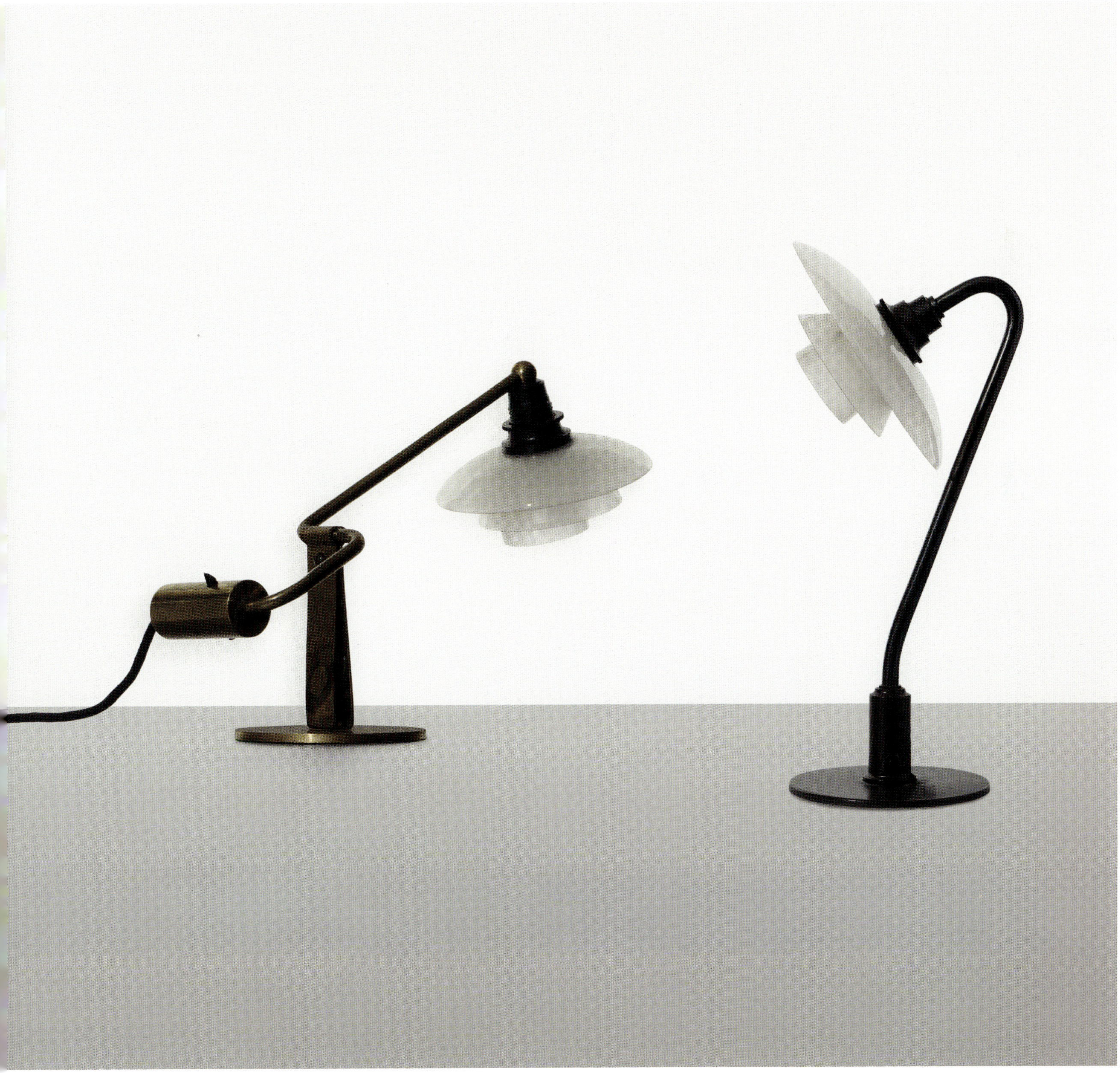

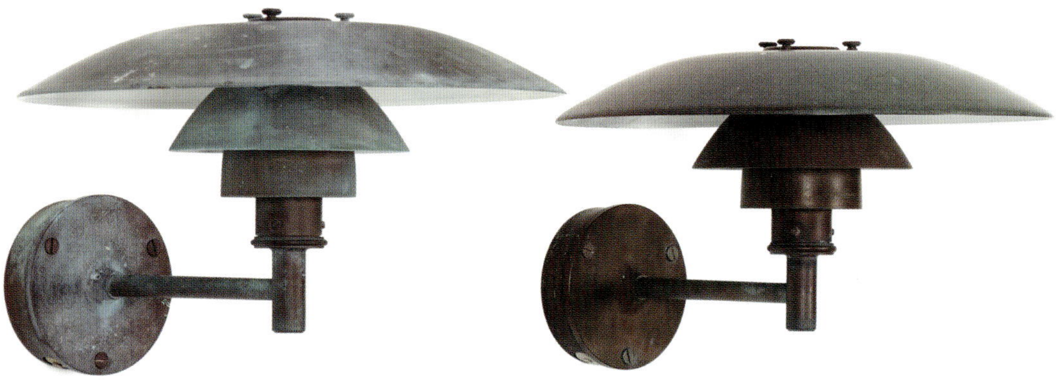

ABOVE A pair of PH 4/3 wall lamps, produced by Louis Poulsen, 1927, in copper.

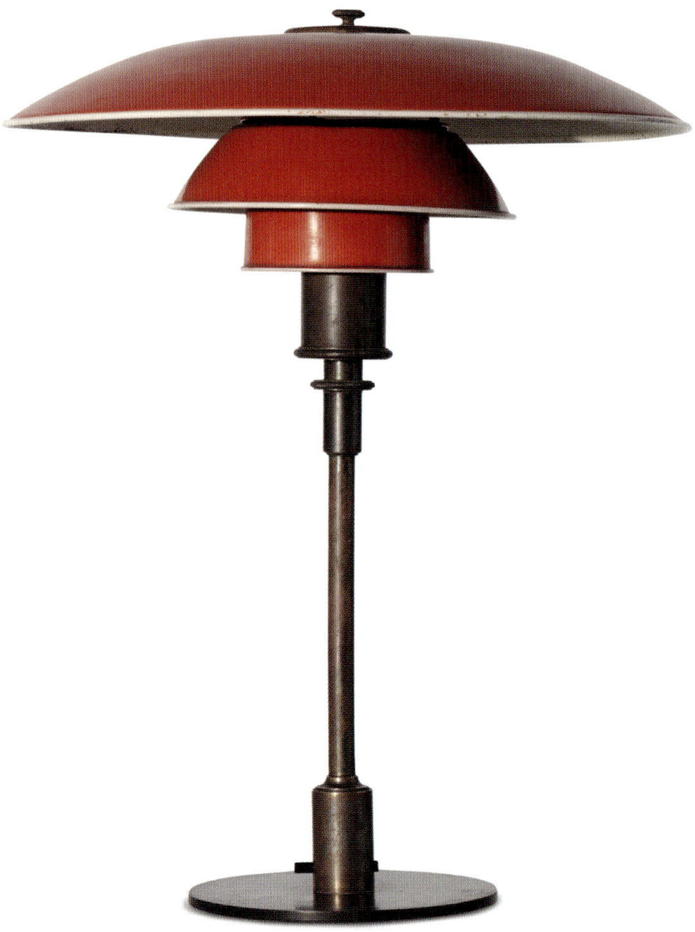

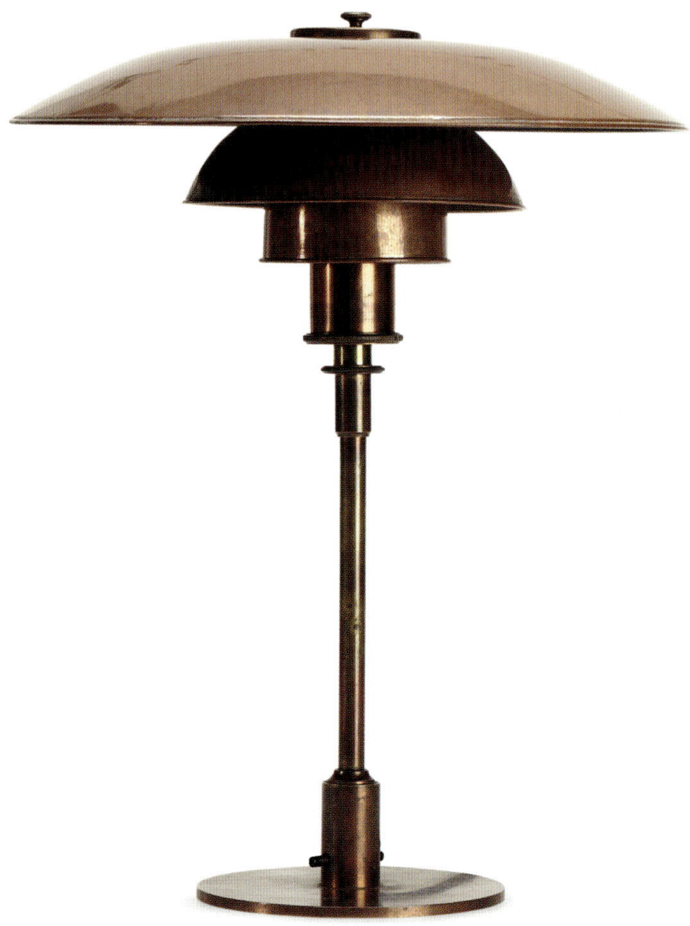

ABOVE & RIGHT Two versions of the PH 4/3 table lamp, designed in 1927, in red enamelled brass and bronze (above), and copper and brass (right), both made by Louis Poulsen.

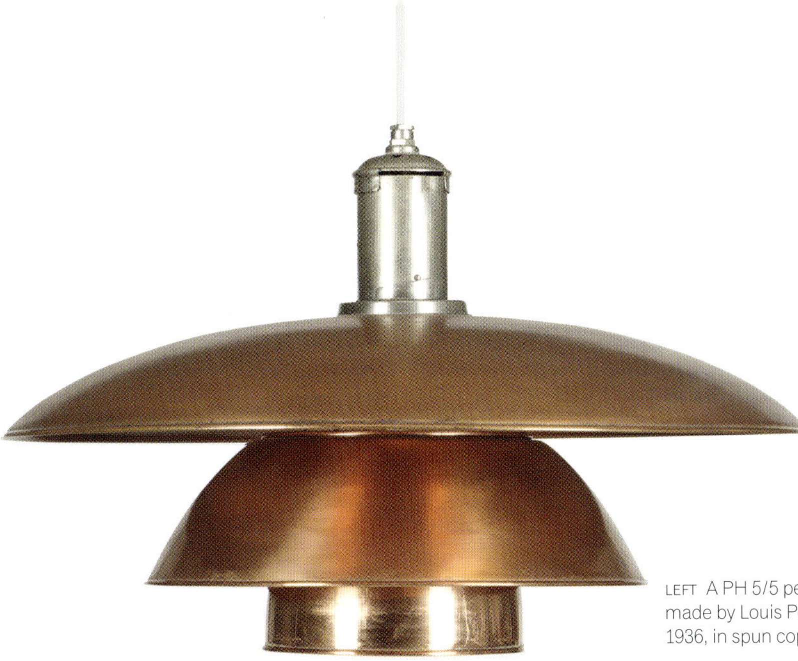

LEFT A PH 5/5 pendant light, made by Louis Poulsen, 1936, in spun copper and tin.

BELOW A PH 3.6/2.5 table lamp (left), designed in 1927, made of nickel-plated and enamelled metal, Bakelite and glass, with a PH 1/1 table lamp (right), 1936, in enamelled brass, Bakelite and glass, both produced by Louis Poulsen.

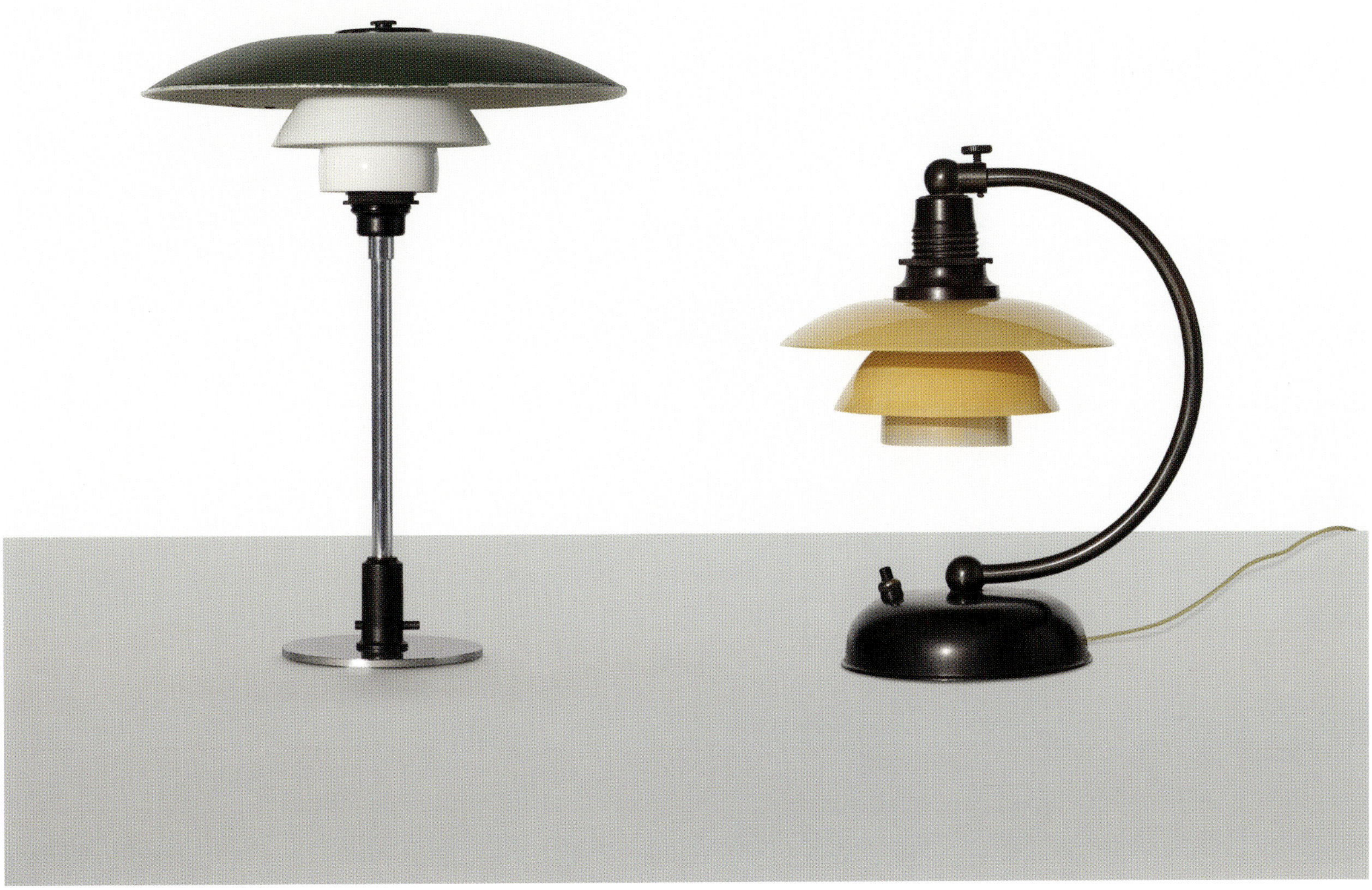

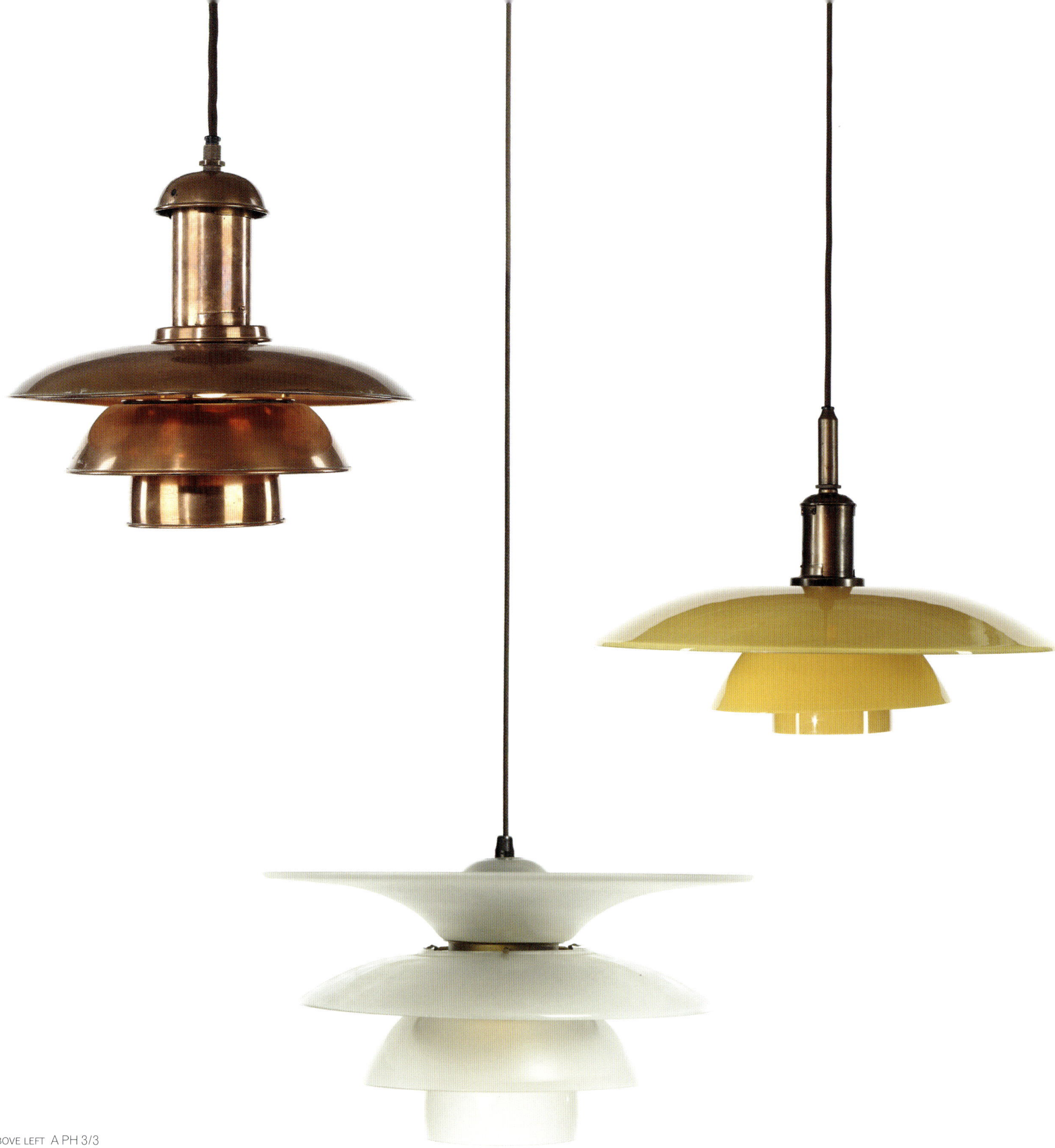

ABOVE LEFT A PH 3/3 hanging lamp, manufactured by Louis Poulsen, *c.* 1926, in copper.

ABOVE CENTRE The PH 4 chandelier, produced by Louis Poulsen, 1932, in frosted glass, brass and Bakelite.

ABOVE RIGHT A PH 5/4 pendant lamp, made by Louis Poulsen, designed 1929, in glass and brass.

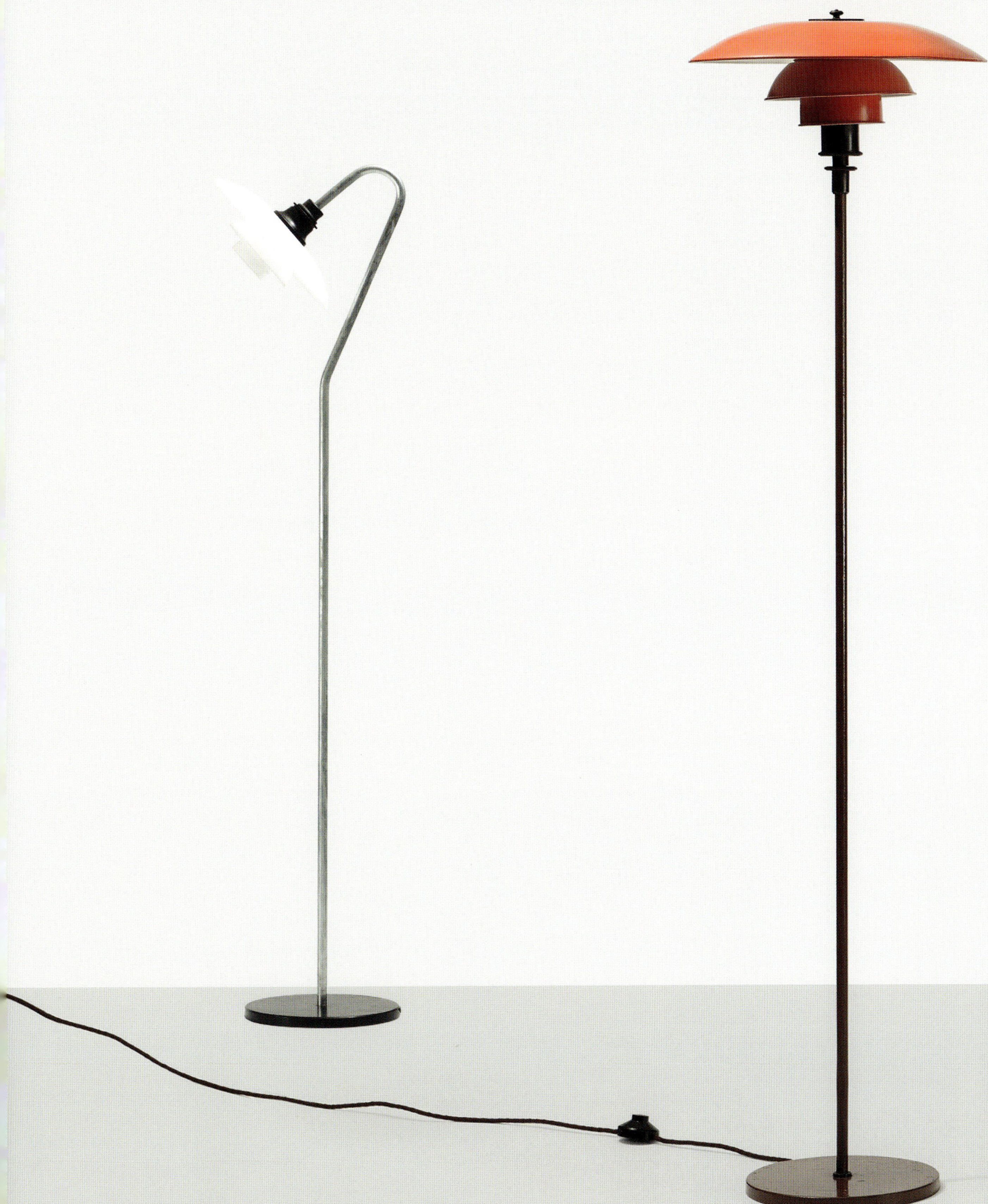

FAR LEFT The PH 3/2 floor lamp, made by Louis Poulsen, 1931, in enamelled and chrome-plated steel, brass, Bakelite and opal glass.

LEFT Poul Henningsen's PH 4/2 floor lamp, produced by Louis Poulsen, 1933, made of enamelled copper, enamelled steel and brass.

WALTER VON NESSEN
(1889–1943)

The German-born and educated designer Walter von Nessen was one of a highly significant wave of émigrés who helped to shape American design during the Thirties and beyond. Von Nessen moved to New York in the mid-Twenties, with his Swedish wife, Greta (or Margaretta), and established himself as one of the most important and influential East Coast lighting designers of the period. While guided by a devotion to functionalism, von Nessen was particularly fascinated with indirect sources of lighting and flexibility. These two interests combined in his series of adjustable table and floor lamps (*c.* 1927 onwards), which featured a pivoting arm (and shade) that could rotate upon its axis. This swing-arm series created an elegant and sophisticated solution to the need for adaptable lamps and lighting, with a number of these designs still in production today.

Von Nessen was born in the city of Iserlohn, in western Germany, and studied at the Kunstgewerbeschule in Charlottenburg, near Berlin. Here, he was taught by the architect, furniture designer and educator Bruno Paul, one of the co-founders of the Deutscher Werkbund*. Von Nessen contributed to the design of subway interiors and taught at Charlottenburg himself before moving to Stockholm in 1919, where he spent a number of years – concentrating mostly on furniture design – until he and Greta (1900–1978) decided to emigrate to the United States in 1925.

Initially, von Nessen continued to focus on furniture and interior design, but within a few years he decided to devote the majority of his time to modern, functional lighting, sensing the lack of serious competition within the field. With his wife Greta, he co-founded Nessen Studio in 1927, based in New York's Murray Hill district. The company prototyped and manufactured von Nessen's innovative designs, and Nessen Studio's swing-arm lights began to appear that same year.

It was the versatility of von Nessen's designs that made them stand out in the market place. His NT922 table lamp and NF987 floor light of 1927 featured similar circular, chrome-plated metal bases and slender, pole-like stems supporting the carefully engineered swing arms. These were capped with simple, almost innocuous linen shades, lending the lights a subtlety that made them appropriate for many different aesthetic contexts and interiors, from Art Deco* apartments to more traditional, period environments. A number of other von Nessen designs from the early Thirties combined striking chrome-plated bases with these relatively simple shades.

The success of these early designs, which featured in spaces designed by Donald Deskey, Eliel Saarinen and others, gave the von Nessens the confidence to expand the studio, even in the midst of the Great Depression. Von Nessen began to experiment with other materials, such as Bakelite and Vitrolite, and to explore new and

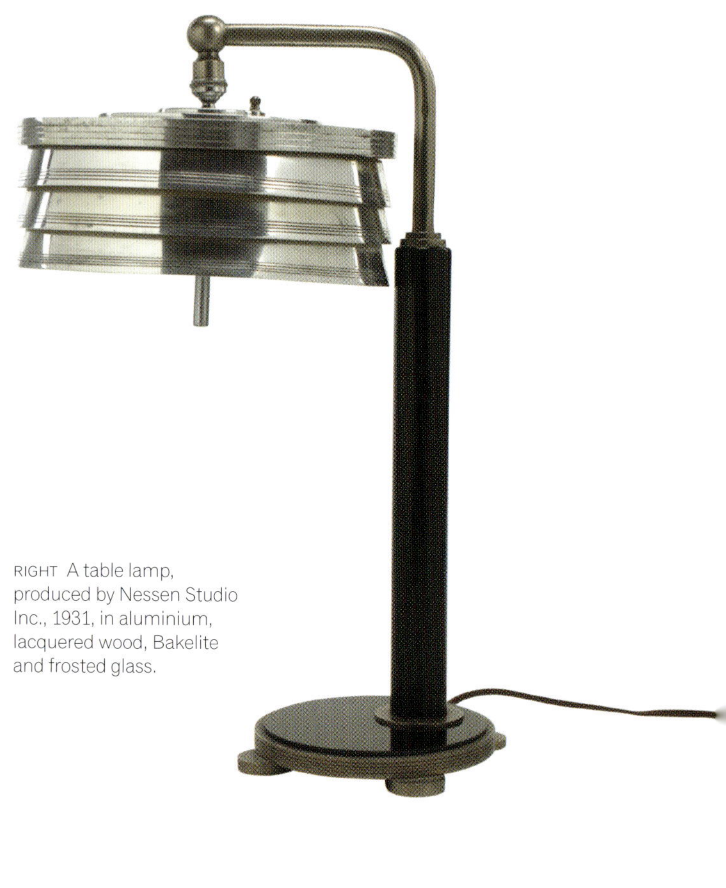

RIGHT A table lamp, produced by Nessen Studio Inc., 1931, in aluminium, lacquered wood, Bakelite and frosted glass.

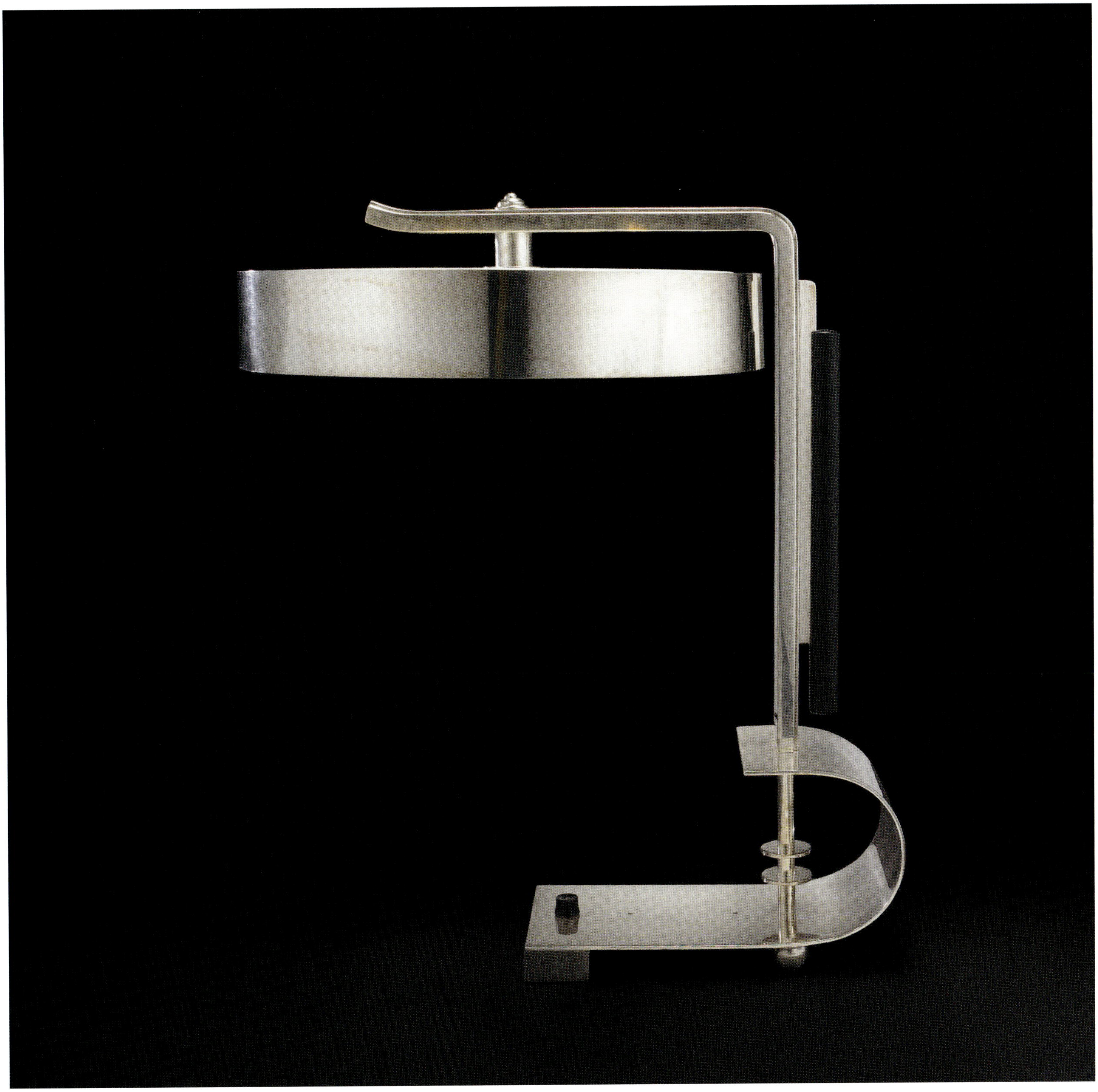

ABOVE A Walter von Nessen table lamp, *c.* 1930, made with silver plate, glass and ebony.

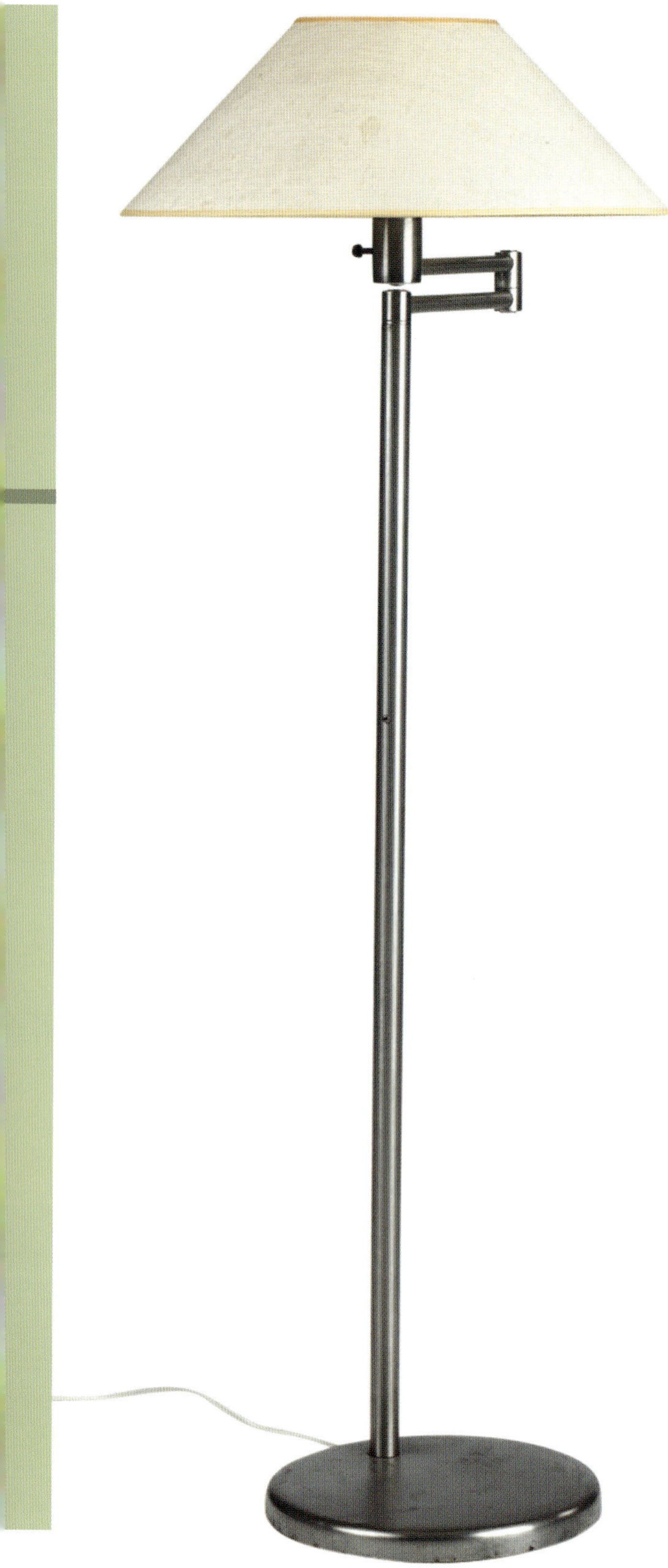

more theatrical and flamboyant forms. There was a striking aluminium and chrome-plated steel floor lamp (*c.* 1928) with a conical spotlight, or uplighter, developed with Saarinen for the dining hall at the Kingswood School in Bloomfield Hills, Michigan. There were also occasional designs for other producers, such as an intriguing brass lamp of 1932, made by the Miller Lamp Co., with a series of concentric metal plates, resembling a large cog or piece of industrial machinery on the one hand and a maritime beacon on the other. Around the same time, von Nessen was also active designing furniture, tableware and other products.

The Nessen Studio was temporarily closed following von Nessen's death in 1943. But it was re-opened and revived two years later by Greta, who continued producing her husband's designs, while adding complementary work by other designers. Nessen Studio's designs of the mid-century period included Greta von Nessen's own Anywhere lamp of 1951 – a light, portable and playful piece that fitted in beautifully with the spirit and aesthetic preoccupations of the post-war period.

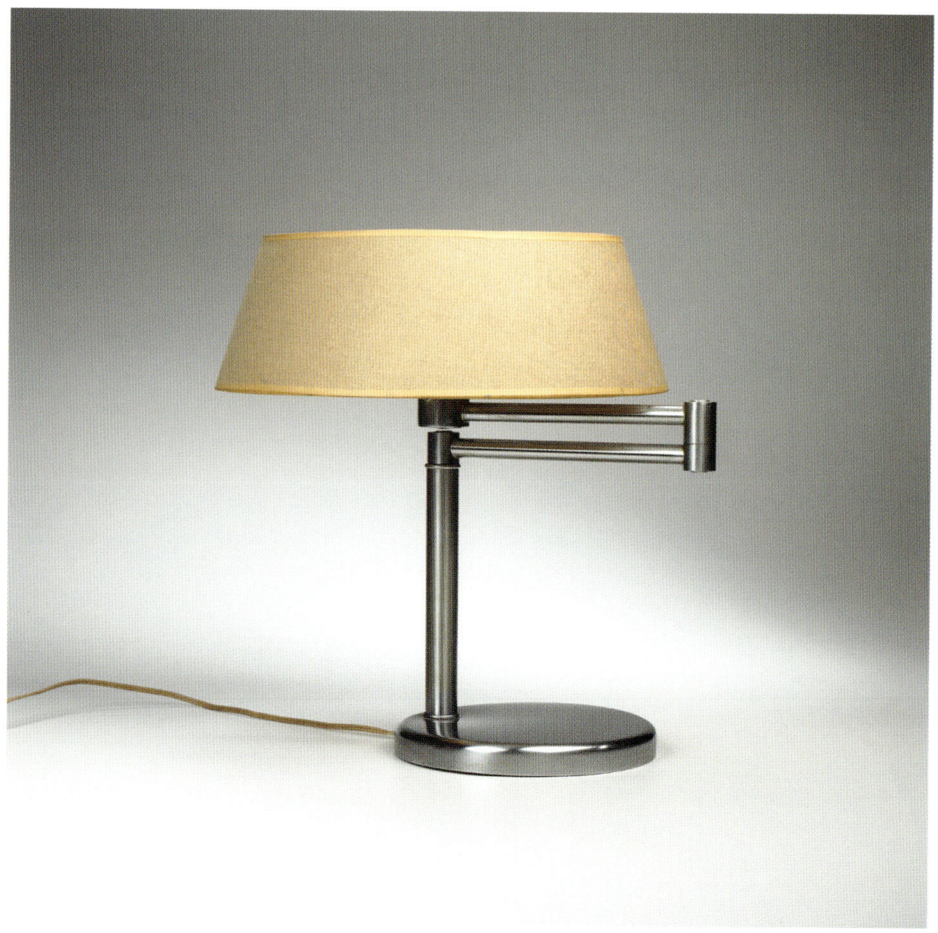

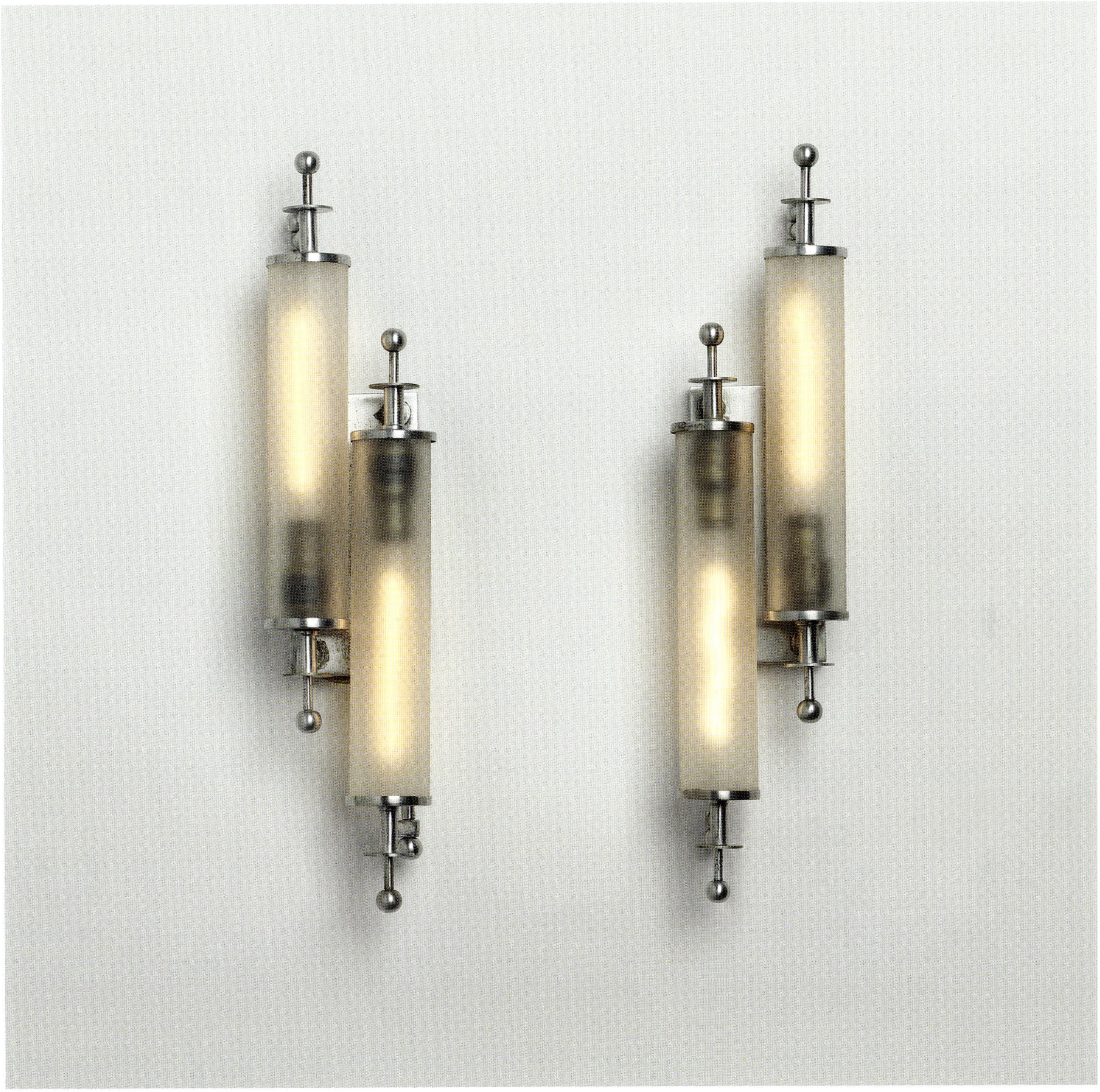

OPPOSITE An adjustable floor lamp and table lamp, made by Nessen Studios, manufactured *c.* 1950, in satin chrome-plated steel and glass with fabric shades.

ABOVE A pair of lamp sconces, produced by Nessen Studio, *c.* 1930, in nickel-plated brass and frosted glass.

BELOW A pair of adjustable floor lamps from the Morton D. May House in Ladue, Missouri, made by Nessen Studio, 1940, in aluminium with silk shades.

OPPOSITE A pair of table lamps, produced by Nessen Studio, *c.* 1935, in chrome-plated steel with linen shades.

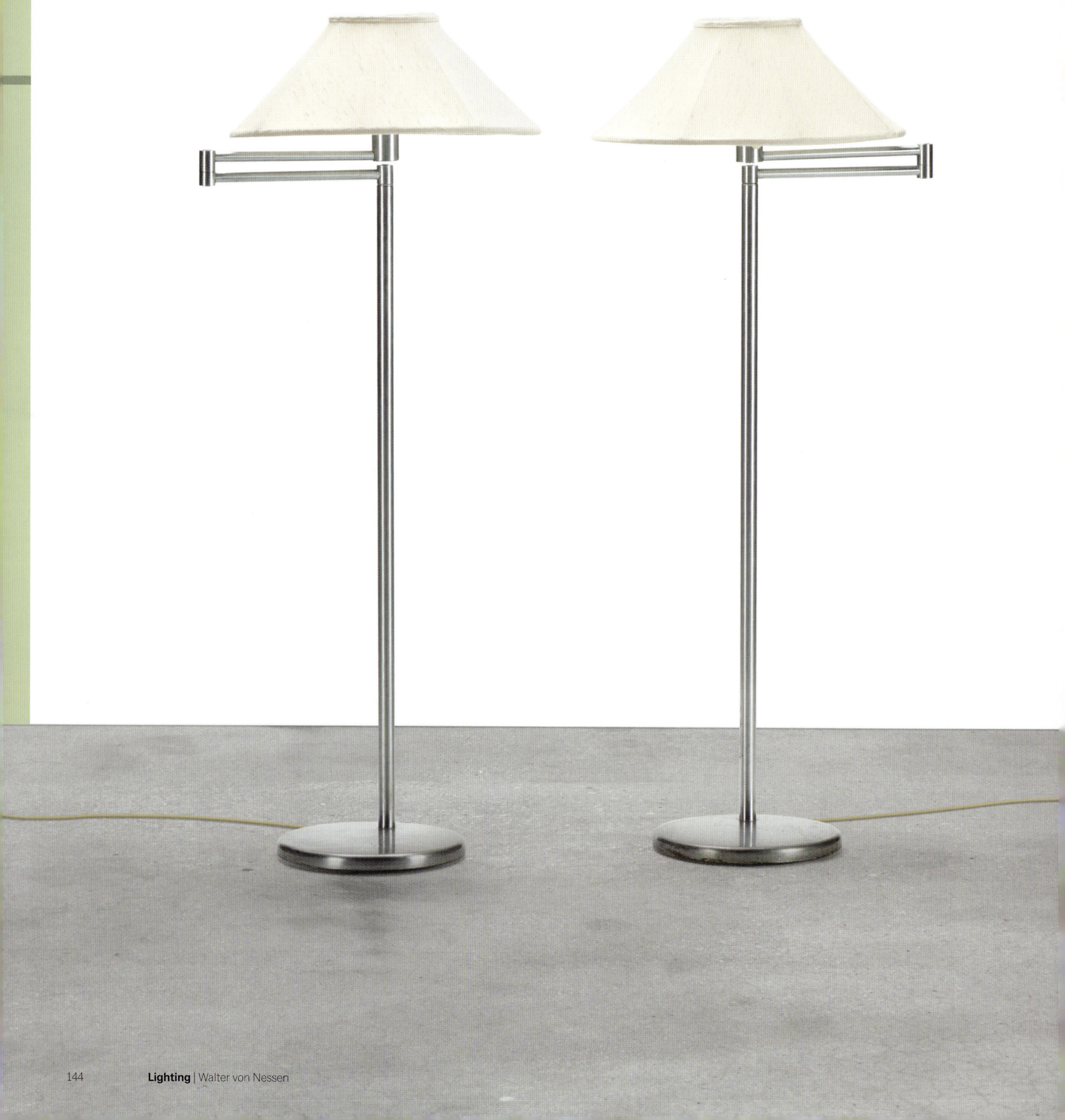

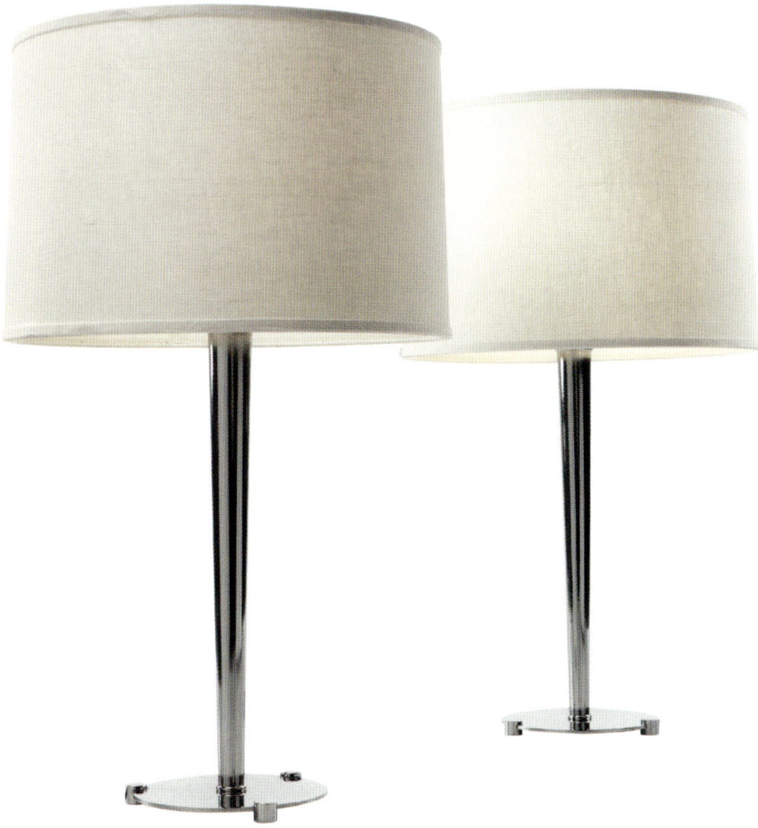

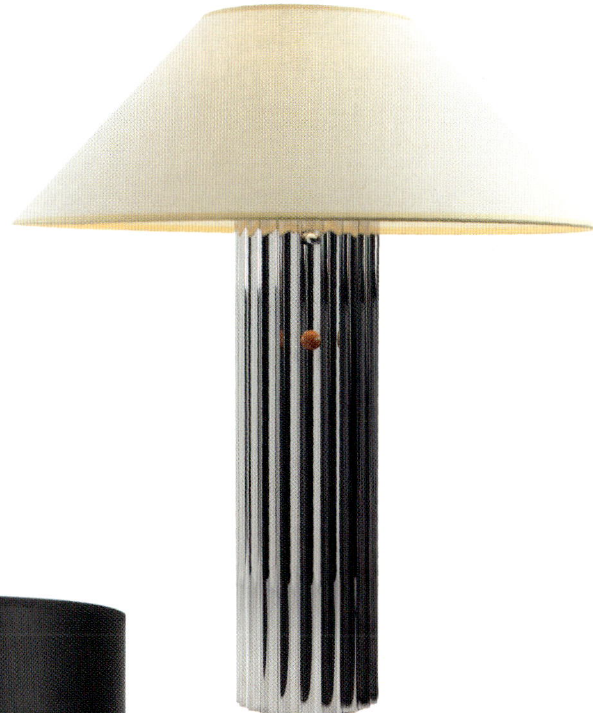

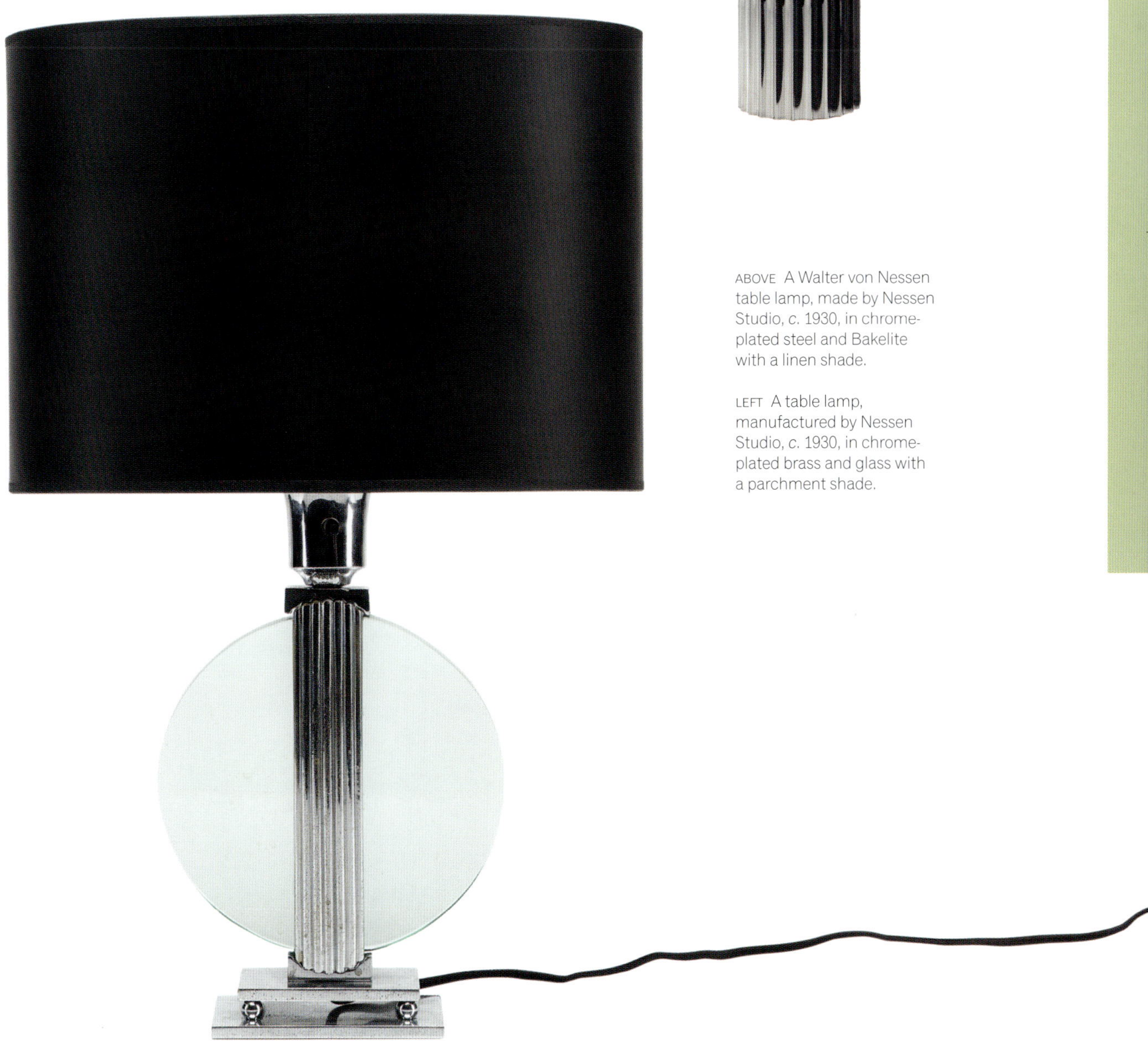

ABOVE A Walter von Nessen table lamp, made by Nessen Studio, *c.* 1930, in chrome-plated steel and Bakelite with a linen shade.

LEFT A table lamp, manufactured by Nessen Studio, *c.* 1930, in chrome-plated brass and glass with a parchment shade.

GIO PONTI
(1891–1979)

PIETRO CHIESA
(1892–1948)

The Italian architect, designer, teacher and writer Gio Ponti has been described as the godfather of modern Italian design. Certainly, the breadth and depth of his portfolio over a long and illustrious career is startling, embracing houses and towers, furniture, ceramics, and industrial and product design. Ponti's early success as a lighting designer was also highly important. As the co-founder of FontanaArte in 1932, Ponti created a number of pioneering modernist designs and collaborated with the company's artistic director, Pietro Chiesa, who created one of the most famous Italian pre-war lamps: the Luminator of 1933.

Giovanni Ponti was born in Milan and remained intimately connected with the city throughout his working life, even during post-war periods of international success and global commissions. He studied architecture at the Polytechnic of Milan and served as a captain of the pioneer corps in the Italian army during World War I. In the early Twenties, he completed his studies and opened his first architecture and design office in Milan.

The early years of Ponti's career suggest an energetic dialectic between neoclassical influences and the pull of the modern, also encapsulated within the Italian Novecento* movement, of which Ponti was – for a time – a leading figure. His early buildings explore this tension before moving decisively in the direction of modernity. As the artistic director of the ceramics manufacturer Richard-Ginori between 1923 and 1930, he explored a similar conversation between past, present and future in his playful and expressive ceramic designs. In 1927, Ponti designed silverware for Christofle and glassware for Paolo Venini, and he launched the design journal *Domus* in 1928 while balancing his growing workload with architectural projects.

FontanaArte was founded in 1932 to serve as the progressive arm of Luigi Fontana's glass factory, creating lighting and furniture primarily in glass. Ponti invited Pietro Chiesa to become artistic director of the company. Chiesa was already a respected glassware and furniture designer, having launched his own Milanese atelier – Botega di Pietro Chiesa – in 1921. In the

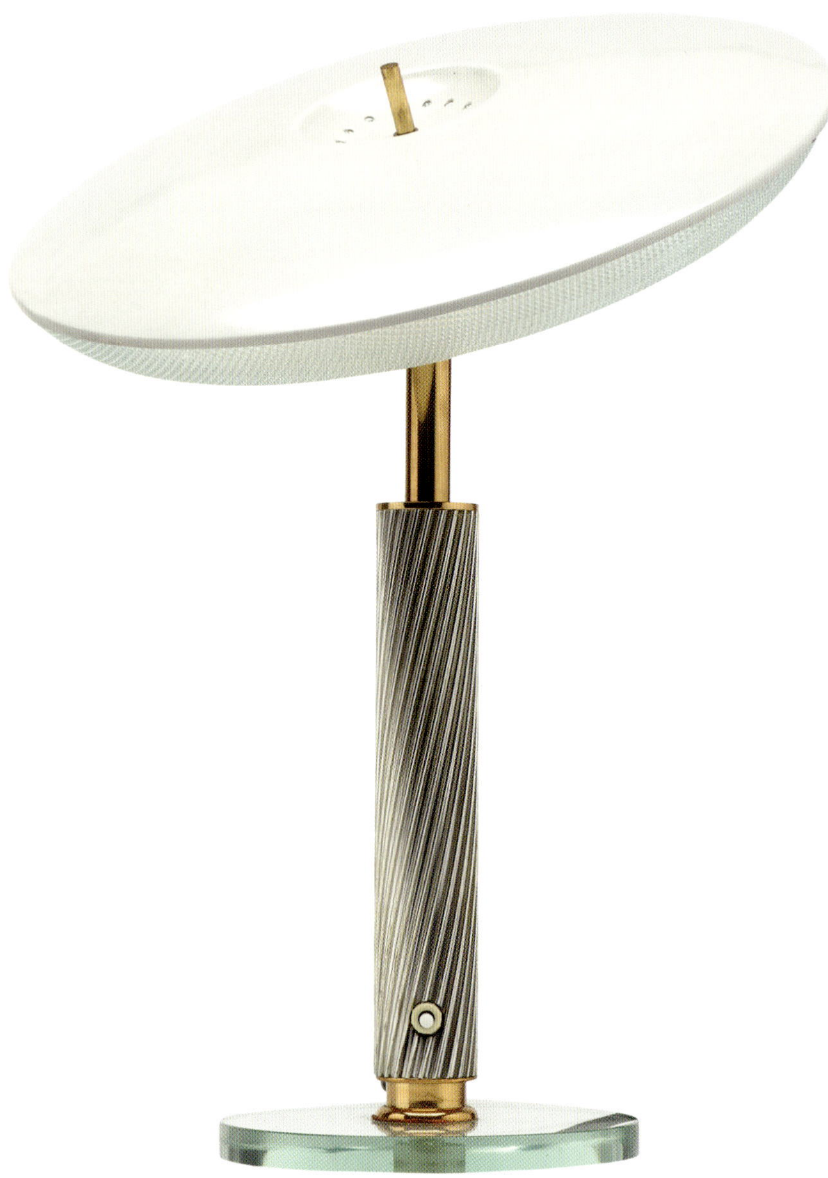

RIGHT A table lamp with a pivoting top designed by Pietro Chiesa, produced by FontanaArte, *c.* 1940, in aluminium, brass, crystal and cut glass.

late Twenties, both Ponti and Chiesa were among the contributors to Il Labirinto, which produced a range of furniture with a Novecento influence.

Ponti's most famous and innovative lighting design for FontanaArte was the Model 0024 hanging light of 1931 – a futuristic ceiling light, which used a series of concentric glass discs arranged around the central light source, fitted together with chrome mounts. The result was a floating, illuminated satellite with the light percolating through the glass plates to create an intriguing geometrical pattern; in many ways, Ponti's design was a forerunner of the space age lamps of the post-war period, but it also offered parallels with the work of the pioneering Danish designer Poul Henningsen. Another key design from the same year was Ponti's Bilia table lamp, forming another graphic, geometric composition, with a round, frosted glass bubble perched upon the point of a metal cone.

Pietro Chiesa was also highly productive and experimented with a range of technical and formal lighting solutions. These included a series of floor

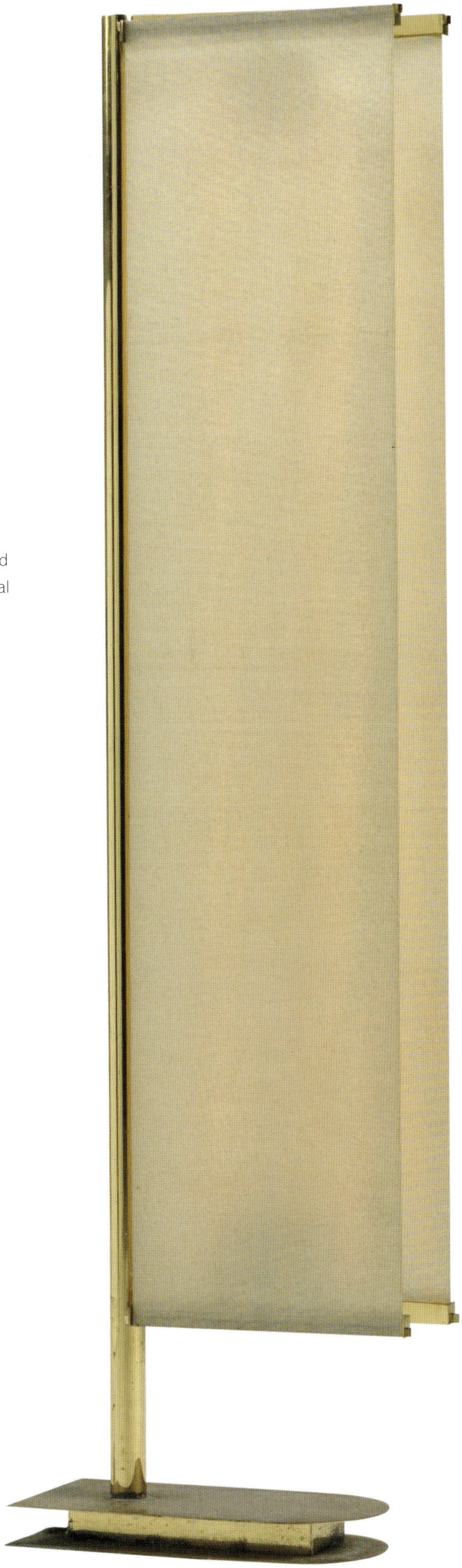

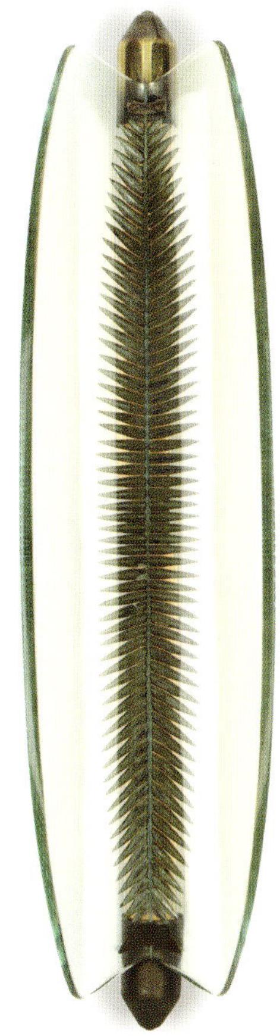

ABOVE A sconce lamp by Pietro Chiesa, manufactured by FontanaArte, *c.* 1945, in carved and polished crystal and brass.

LEFT A floor lamp by Pietro Chiesa, manufactured by FontanaArte, *c.* 1933, in brass, brass-plated steel and linen.

lights, dating from 1933, which resulted from Chiesa's interest in indirect lighting, as opposed to directed task lighting. One of FontanaArte's most successful designs of the period was Chiesa's Luminator floor lamp (1933), which had an air of ambiguous, machine-like modernity. The form was relatively simple: a modest round base supporting a tall, slim, brass stem topped with a fluted spotlight throwing the light up to the ceiling and bouncing it around the room. The tube and fluted spotlight were seamless, resembling a kind of funnel or vertical exhaust pipe; at first glance, without a shining bulb, it was not at all obvious that the Luminator was a light at all. Much later, in 1954, Achille and Pier Giacomo Castiglioni produced their famous B9, partly inspired by Chiesa's earlier model.

Chiesa also designed one of FontanaArte's most successful pieces of furniture, the Fontana table of 1932, which has remained in production ever since. Here, Chiesa was experimenting with the structural possibilities offered by sheet glass and created the table with one plane of glass curved at either end to form symmetrical supports. Beautifully simple and structurally innovative, the Fontana table still speaks – like many of Chiesa's lighting designs, and those of his fellow pioneer Gio Ponti – of a powerful combination of functionalism and futurism.

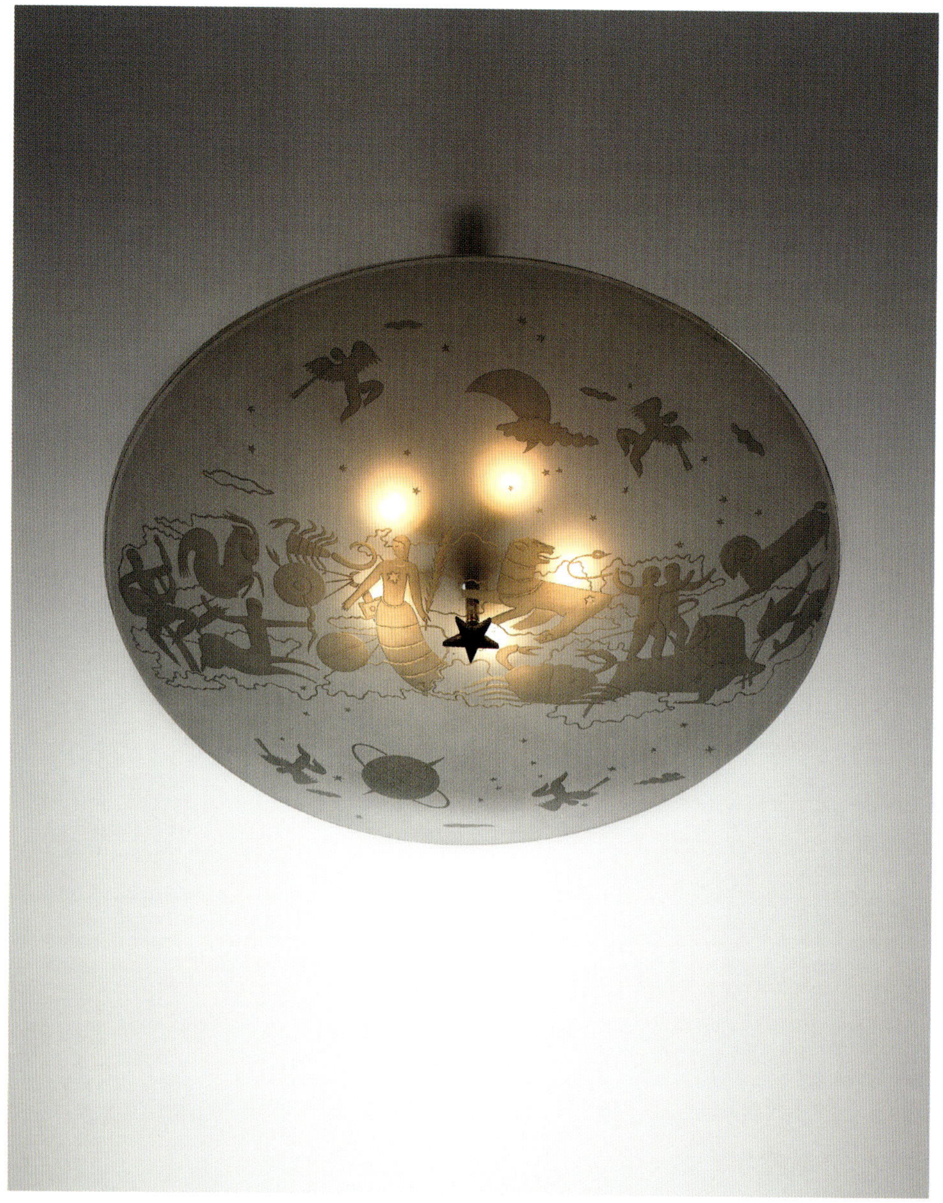

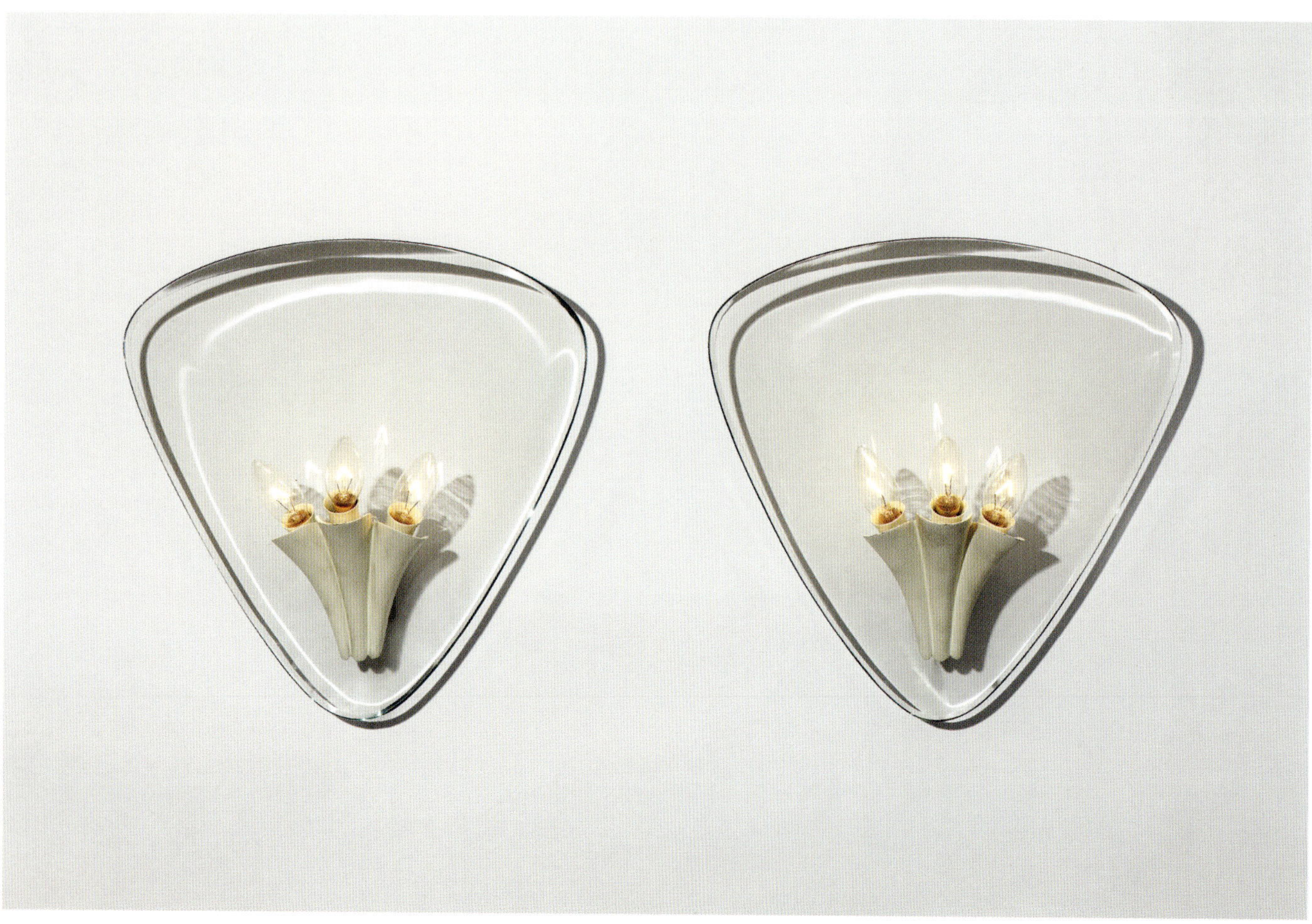

ABOVE Gio Ponti's Zodiac chandelier, produced by FontanaArte, 1936, in carved crystal, frosted glass and aluminium; the light was produced for the 1937 International Exhibition in Paris.

LEFT A pair of sconce lights by Pietro Chiesa, made by FontanaArte, *c.* 1940, in enamelled aluminium, brass and crystal.

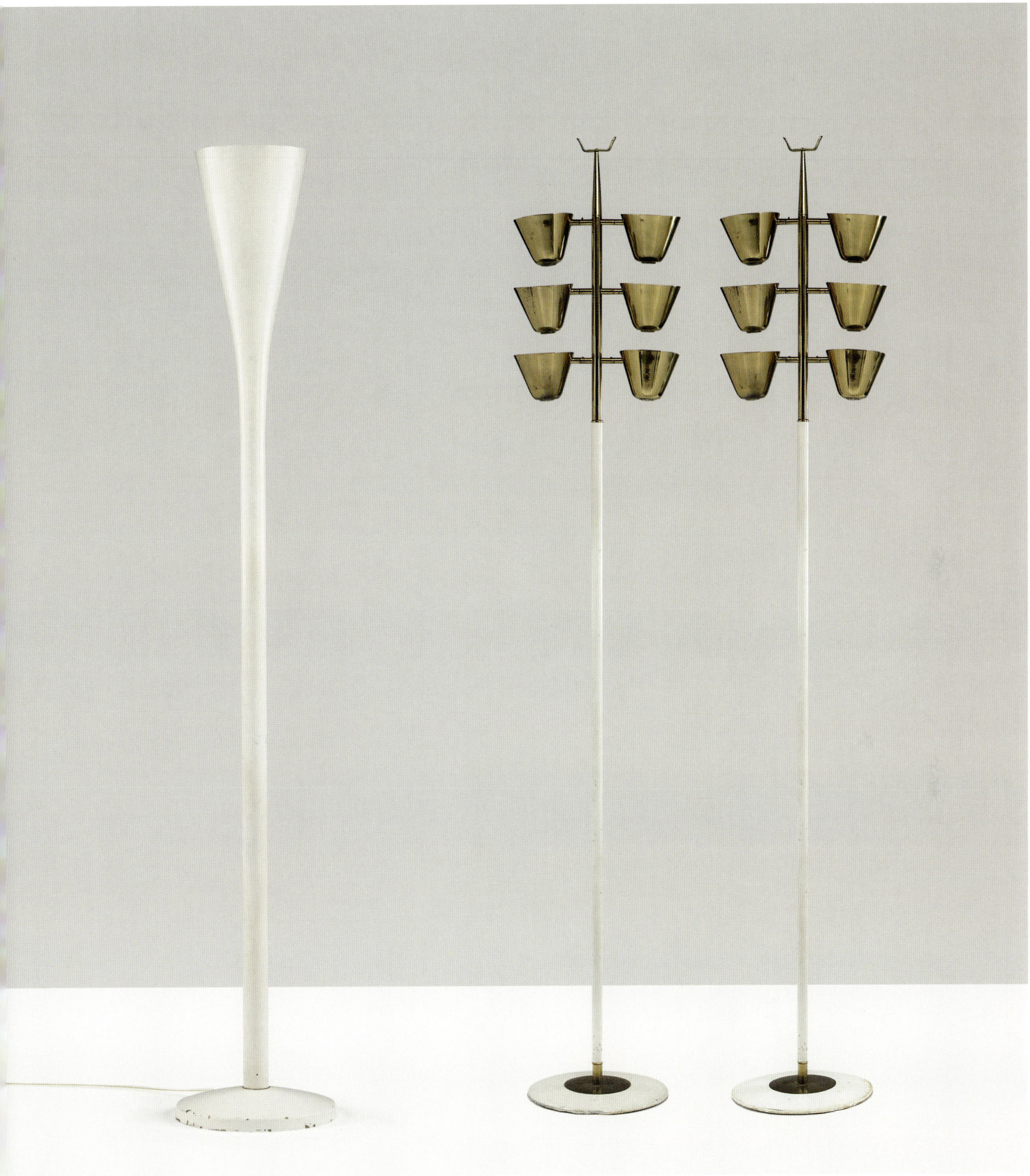

ABOVE LEFT The Luminator floor lamp designed by Pietro Chiesa, made by FontanaArte, *c.* 1936, in enamelled metal and aluminium.

ABOVE RIGHT A pair of floor lamps by Chiesa, made by FontanaArte, 1938, in brass, enamelled aluminium, iron and glass.

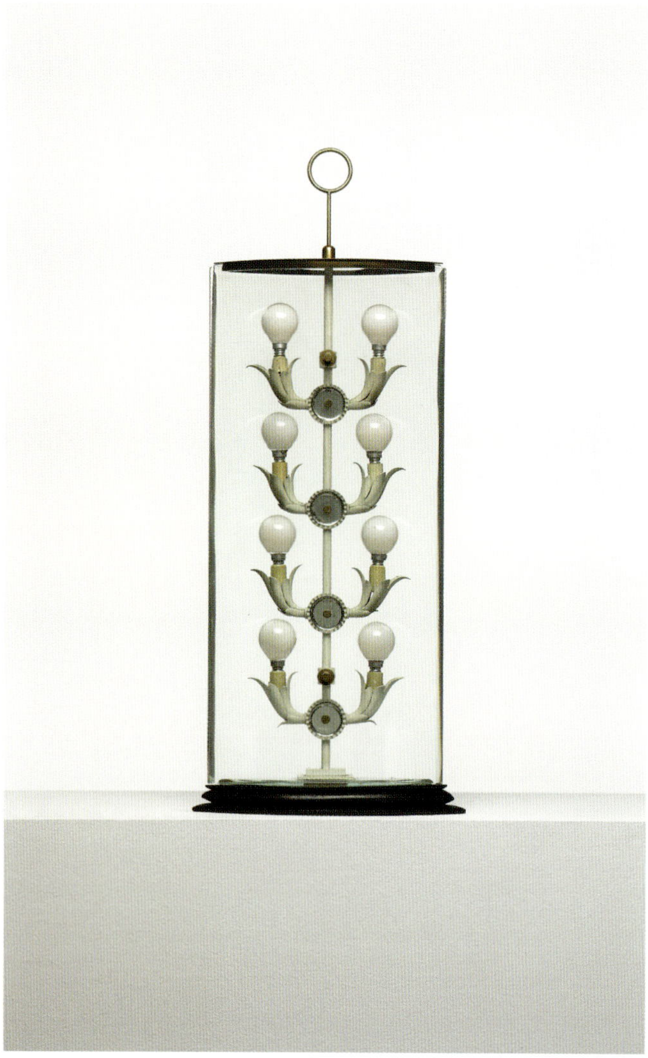

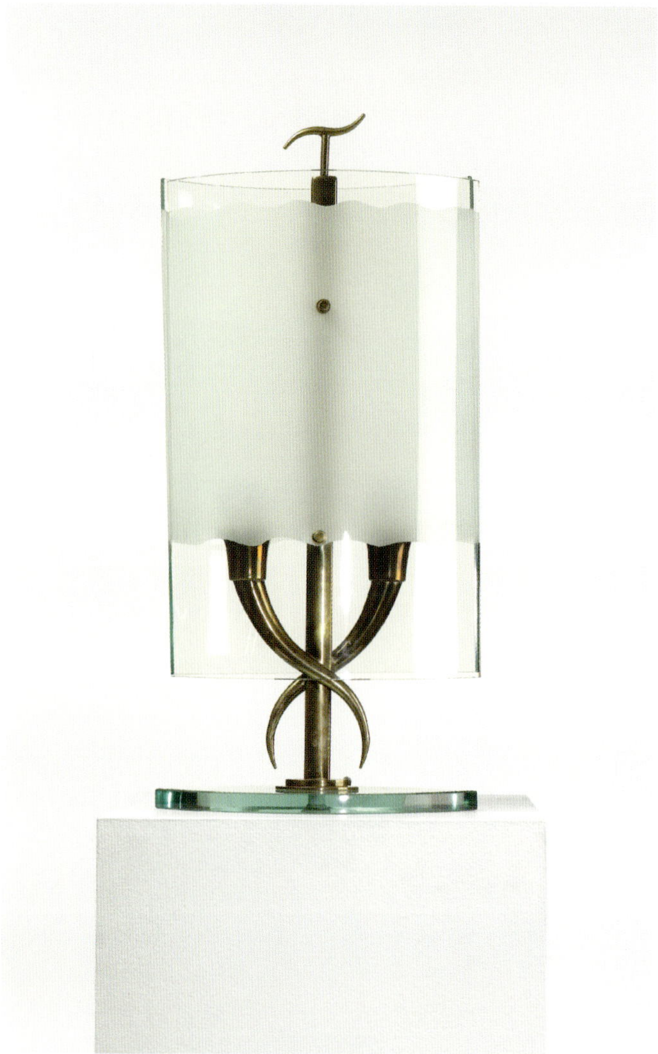

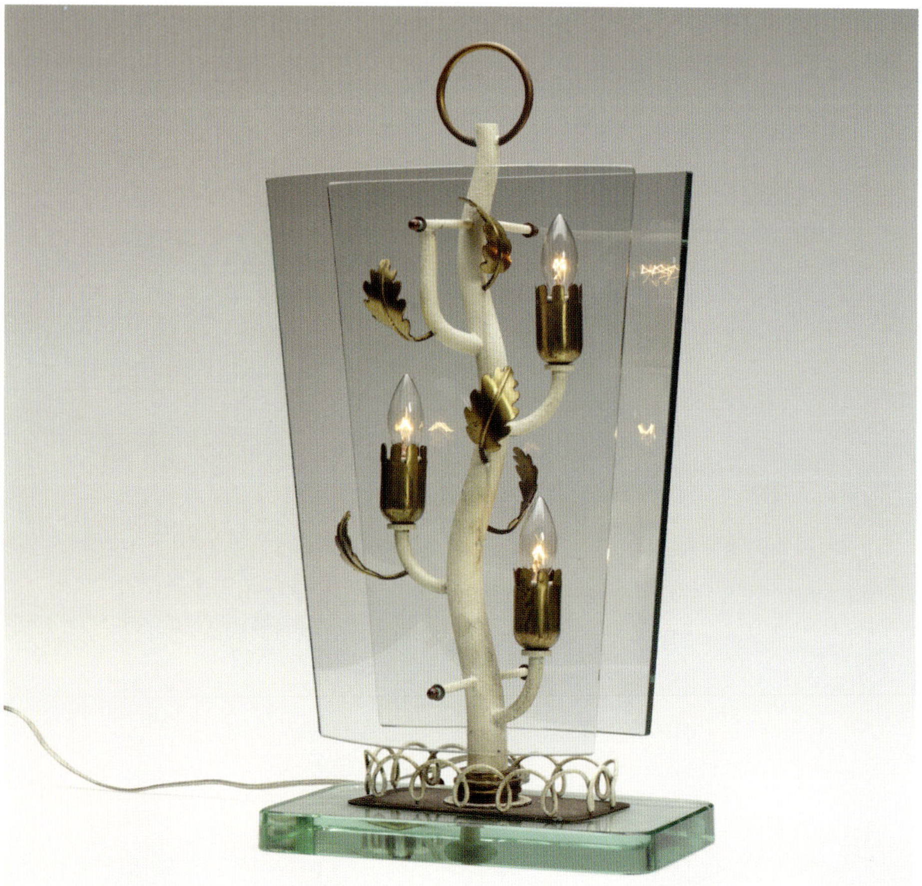

ABOVE LEFT A table lamp by Pietro Chiesa, manufactured by FontanaArte, 1936, in crystal, glass, lacquered wood and brass.

LEFT A table lamp designed by Chiesa, produced by FontanaArte, c. 1930, in glass and brass.

ABOVE A table lamp by Gio Ponti for FontanaArte, c. 1940, in crystal, brass and etched glass.

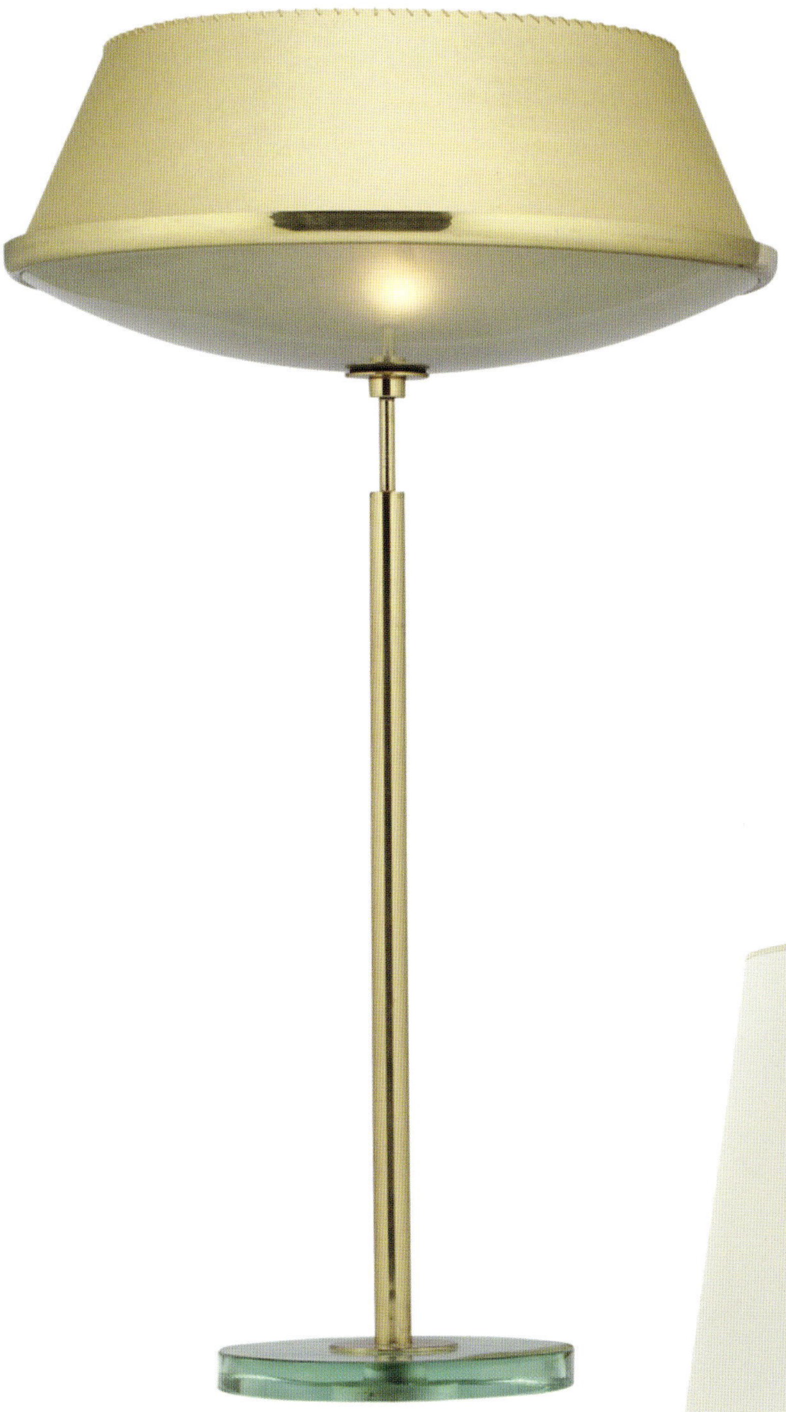

ABOVE A table lamp designed by Pietro Chiesa, made by FontanaArte, 1939, in crystal, glass, brass and parchment.

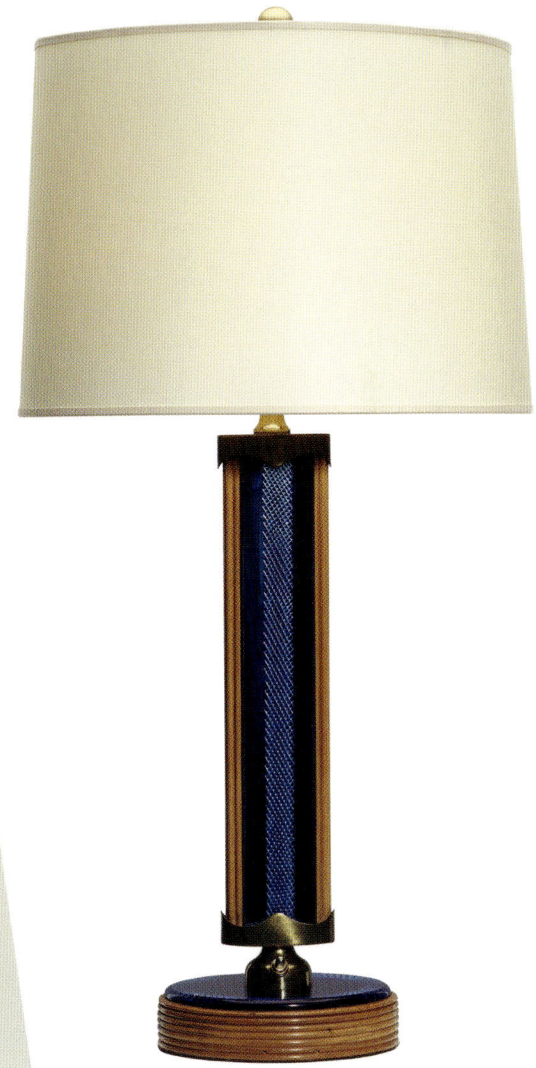

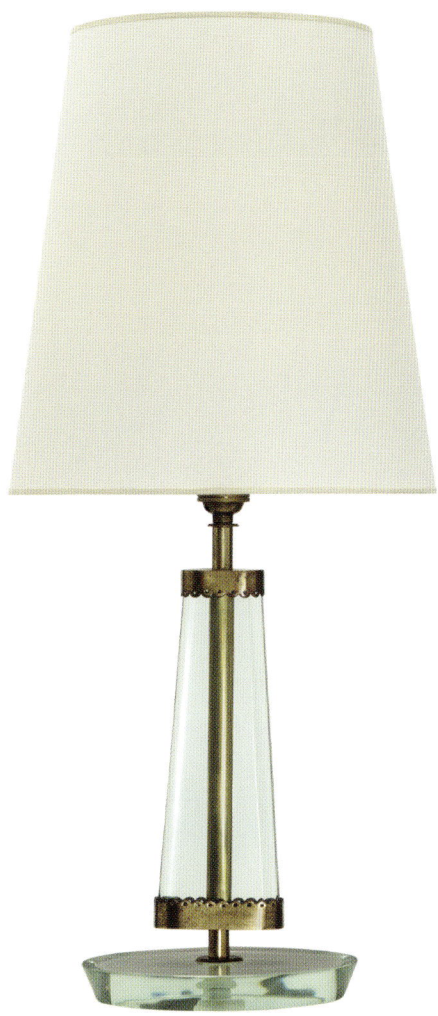

ABOVE A Chiesa table lamp, made by FontanaArte, *c.* 1943, in cut glass and cherry wood with a parchment shade.

LEFT A table lamp by Chiesa, produced by FontanaArte, *c.* 1940, in crystal and brass with a silk shade.

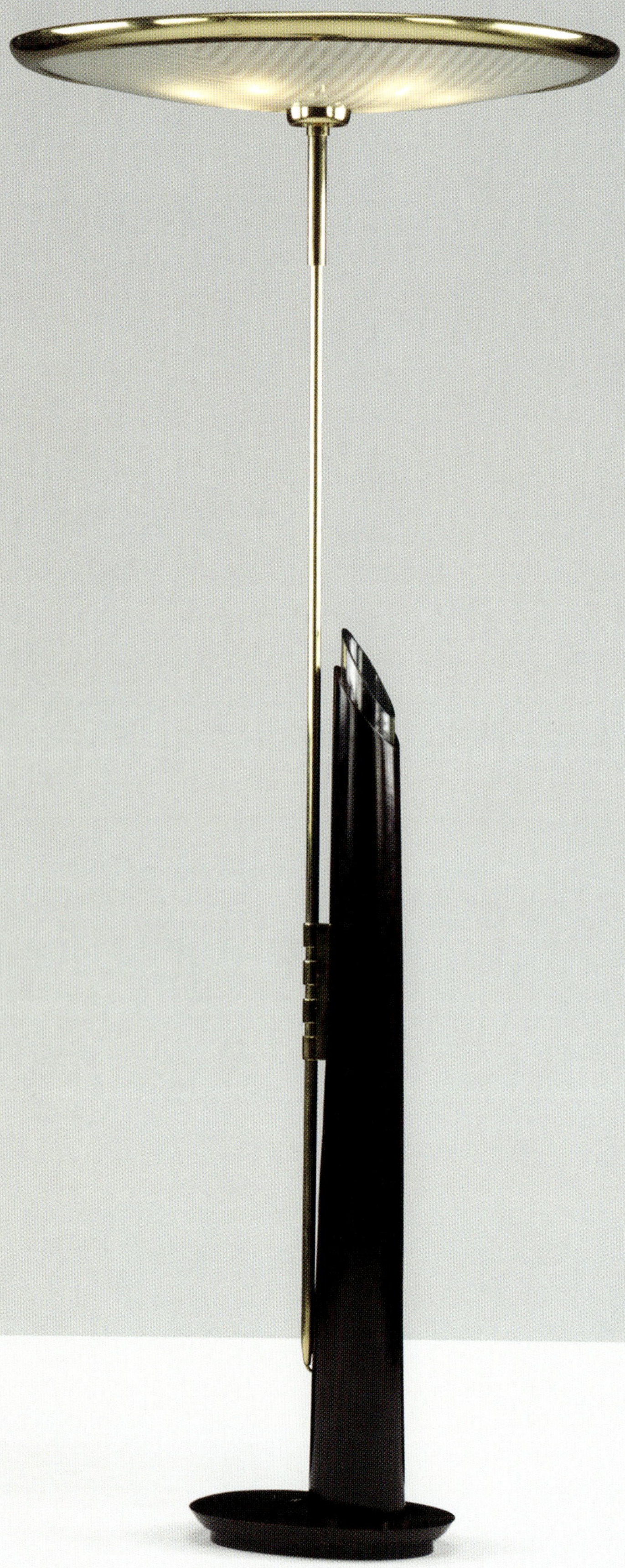

LEFT A floor lamp by Pietro Chiesa, produced by FontanaArte, *c.* 1940, in lacquered wood, brass and glass.

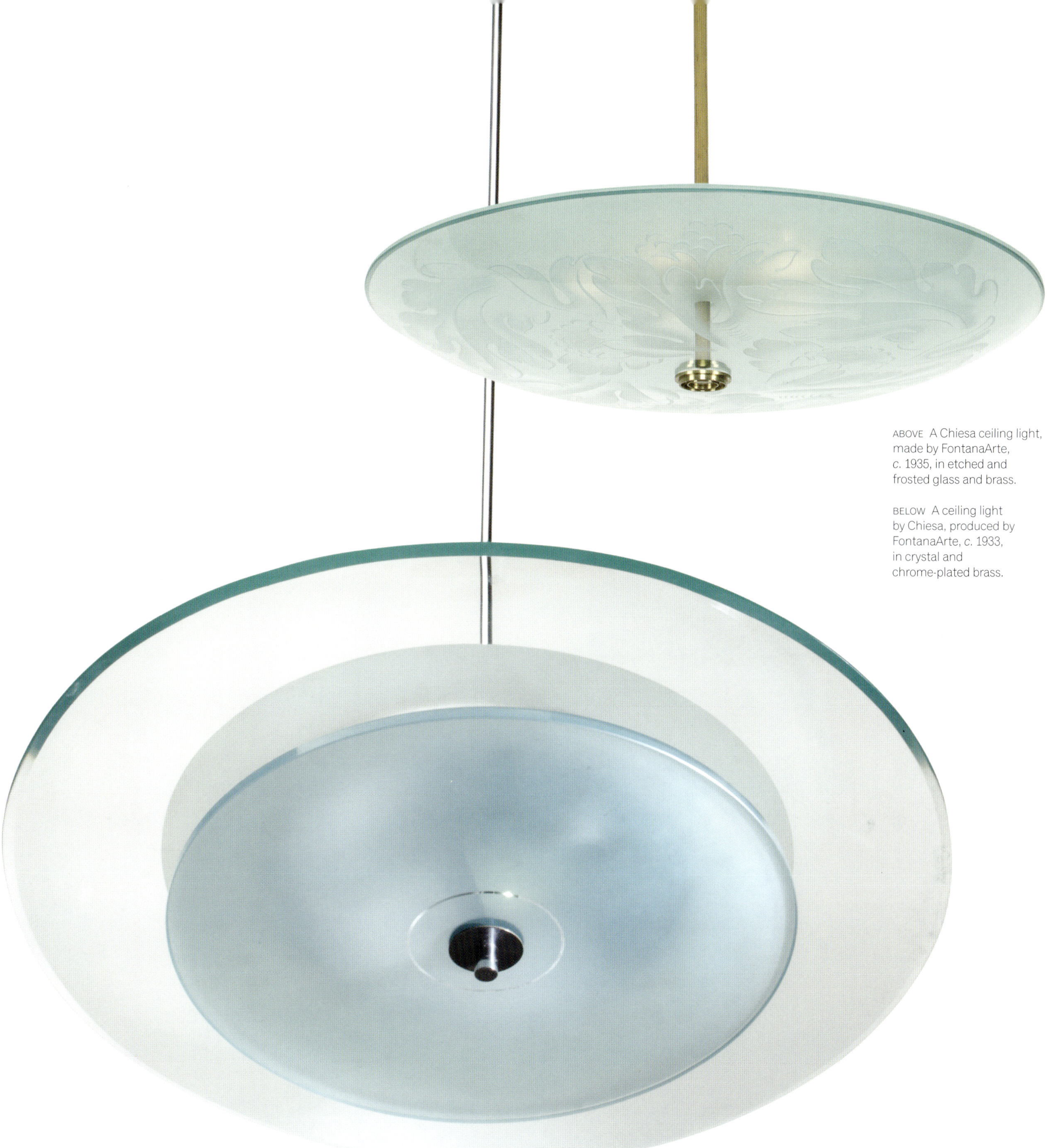

ABOVE A Chiesa ceiling light, made by FontanaArte, *c.* 1935, in etched and frosted glass and brass.

BELOW A ceiling light by Chiesa, produced by FontanaArte, *c.* 1933, in crystal and chrome-plated brass.

CERAMICS & GLASS

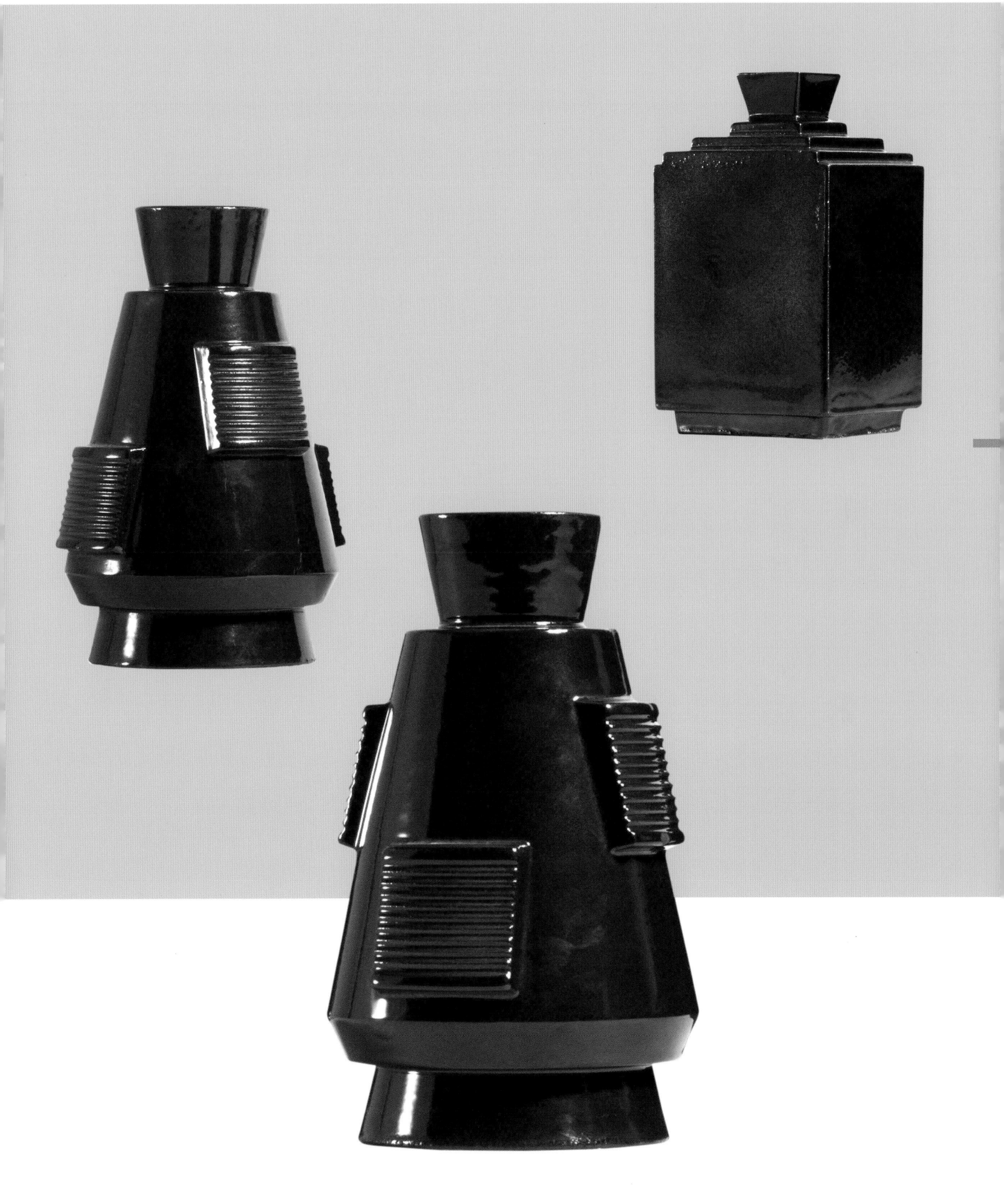

INTRODUCTION

The dialectical tension between industry and craft was nothing new but it came to a head during the modernist period, particularly within the fields of ceramics and glass. The rise of the machine and factory production – along with a perceived and related decline in handmade, crafted design – had been of great concern to supporters of the Arts and Crafts movement back in the late 19th century. The tension only increased as the 20th century wore on, and reached a climax during the 'machine age' of the Twenties and Thirties, when production-line technology was perfected by Henry Ford* and other industrialists, while modernist designers argued for a more democratic approach to design in which factory machinery made well-conceived and well-constructed products available to a mass audience.

This complex and sometimes troubled relationship between industrial manufacturing and craft lies at the heart of the modernist project. For some, the relationship was symbiotic and natural. But for others the rise of machine-age production was deeply troubling and led to intense concern about deteriorating standards in craftsmanship and artistry. Among those who shared such worries were many designers and commentators regarded as modernists or, at least, sympathetic to modernism. They included the sculptor, engraver and typographer Eric Gill*. As he wrote in 1931, in his book on typography: 'The conflict between industrialism and the ancient methods of handicraftsmen which resulted in the muddle of the 19th century is now coming to its term. But tho' industrialism has now won an almost complete victory, the handicrafts are not killed, and they cannot be quite killed because they meet an inherent, indestructible, permanent need in human nature.... The two worlds can see one another distinctly and without recrimination, both recognizing what is good in the other – the power of industrialism, the humanity of craftsmanship.'

One of the most powerful voices in the world of ceramics was Gill's friend Bernard Leach, who argued passionately in favour of the authenticity and character of studio pottery and against mass production. Leach believed in the power of craft, in learning and in pottery's contextual roots within vernacular tradition. Yet his approach and aesthetic were also avant-garde and – arguably – modern. He and his greatest student-turned-master potter, Michael Cardew, developed 'standard' collections of ceramics, alongside their art pottery, which they attempted to produce and fire in large enough numbers to make them more widely available and affordable. In their own way, Leach and Cardew were seeking a kind of accommodation between the means of production and authentic craftsmanship.[1]

Within the highly influential but relatively short-lived ceramics workshop of the Weimar Bauhaus there was both a master of form (Gerhard Marcks*) and a master of craft (Max Krehan*). Within the studio – whose students included Marguerite Wildenhain and Otto Lindig – there was an emphasis on craftsmanship, artistry, learning and technique as well as a growing pressure (from Bauhaus director Walter Gropius and others) to develop ways in which the production of new designs could be scaled up to an industrial or semi-industrial level.

A number of Bauhaus ceramicists went on to design pieces suited to factory production. Wildenhain, who emigrated to the Netherlands and then America, was passionate – like Leach – about themes such as authenticity and character, while seeking to pass on her knowledge of craft and technique in her teaching and her books. But she also recognized – like her

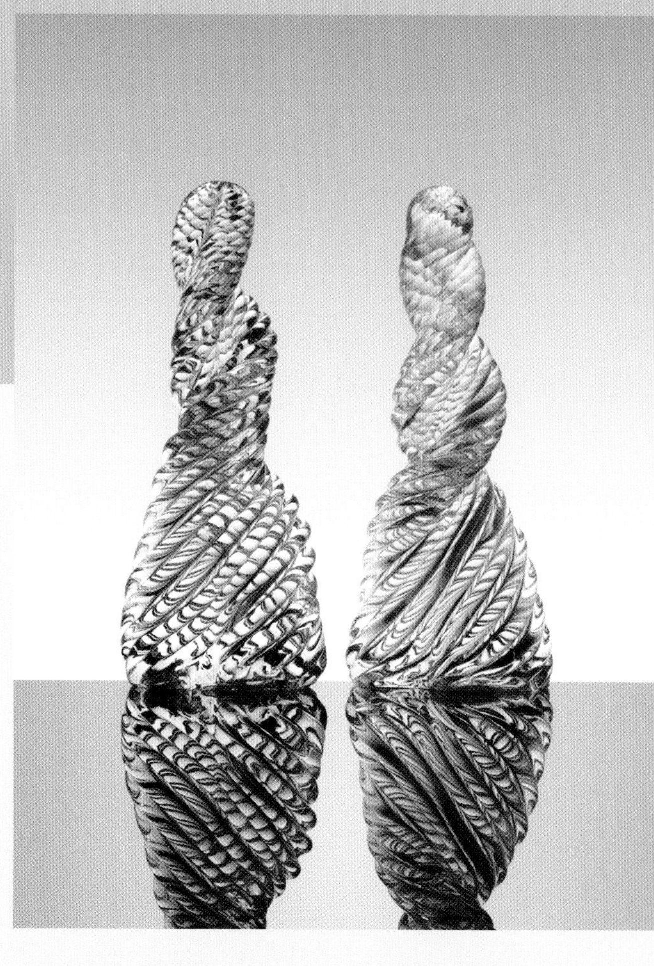

instructors at the Bauhaus – that this could provide a foundation for a ceramicist to apply their work to factory production: 'I have said before, and want to stress again, that there will always be need for the craftsman and the individual worker, not because of their production (which is small, anyway) but because of the human development of the man himself; still I do want to remark, on the other hand, that there is nothing anomalous in a well-trained potter's making models for mass production. It is a much cleaner solution than to try to repeat by hand an unlimited number of identical pieces; that can only result in killing any original talent, imagination – and, in the end, the whole man – in the process.'[2]

The mantra coined by American designer Russel Wright that 'good design is for everyone' could have come from the lips of Gropius and his Bauhaus philosophers. Wright designed the American Modern range of dinnerware, launched in 1939, which became one of the best-selling ceramic collections in history. He considered himself one of the new breed of industrial designers and created a self-made brand around his work, which also included glassware and furniture. Wright was not a fully trained ceramicist, but – like many of his contemporaries – he had studied art and sculpture; there was an undeniable sculptural beauty and fluidity to the American Modern collection.

Fine artists and sculptors were to have a profound effect on the evolution of both ceramics and glassware during the modernist period. While they may not have been fully immersed in the artisanal crafts and techniques of these disciplines, they brought a profound sense of artistry to their work and many of them sought close collaboration with technicians within the field itself.

There were some artists – such as Eric Ravilious and Raoul Dufy – who saw the surface of ceramics as another 'canvas' suited to their work, which was applied as a form of decoration. But many others were concerned with every aspect of the product, particularly the form as well as the pattern, texture and colour. They included the Swedish painter Wilhelm Kåge*, who became the art director of the Gustavsberg Pottery in 1917. Early pieces, such as his Argenta range, combined classical and Art Deco* references, but increasingly Kåge became interested in the sculptural language of his work, as seen in his Vaga porcelain collection of 1939. These cream-coloured pots featured rippled, clam-like necks and rims: their organic character was a foretaste of the extraordinary forms and textures explored by the artist and his protégés during the mid-century period, when Scandinavian ceramics was at the height of its power.

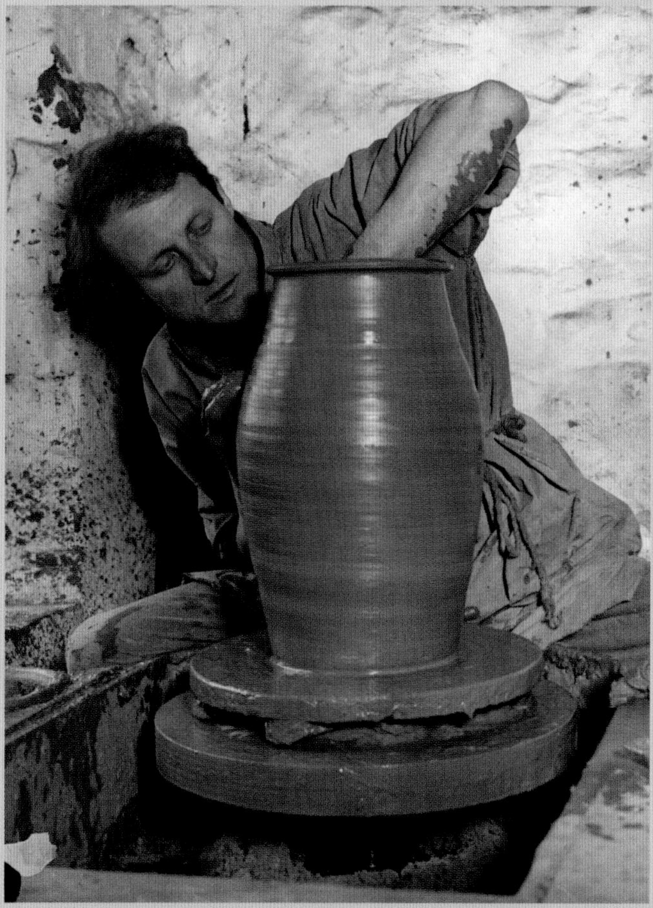

Along with Kåge, modernist ceramicists and glassware designers of the late Twenties and Thirties laboured under the shadow, to some extent, of the Art Deco movement, which – in these mediums, in particular – was so rich and distinctive in character, form and decoration. It was a challenge to forge a new aesthetic in the face of such powerful work by René Lalique, Daum Frères, Clarice Cliff, Émile Decoeur* and others. A number of ceramic designers – including Robert Lallemant* and Keith Murray* – have been placed in the Art Deco camp but have also been described as early modernists; similarly, in the

PAGE 155 A collection of three glazed earthenware vases by Robert Lallemant, France, c. 1930.

LEFT A portrait of potter Michael Cardew at work on the potter's wheel, taken in Winchcombe, Gloucestershire, UK, in 1936.

OPPOSITE A pair of Diamante Obelischi Model 9034 glass vases by Paolo Venini and Carlo Scarpa, made by Venini in Murano, Italy, 1934.

world of glass, a number of key figures emerged from the Deco style and sought experimentation with pioneering modernist forms and a more restrained approach to decoration.

In the world of glassware, the relationship between art and industry also became increasingly important. Swedish glass-maker ORREFORS benefited from a series of vital relationships with sculptors and artists during the Twenties and Thirties, including Edward Hald and Edvin Öhrström. Rival glass-maker Kosta – which struggled during the Twenties but revived in the Thirties and thrived in the post-war period with the help of Vicke Lindstrand – also worked with a number of painters and fine artists, including Ewald Dahlskog, Sven Erixson and Tyra Lundgren.

In Italy, too, where glass-making was centred primarily upon the Venetian island of Murano, a series of new alliances between factory technicians and art directors or consultants brought in from the worlds of art and architecture helped spur on the creation of new forms, ideas and production techniques. During the Twenties and Thirties, PAOLO VENINI established his new glassworks with the help of sculptor Napoleone Martinuzzi, painter Tomaso Buzzi and architect Carlo Scarpa.

Drawing upon thinking about form, colour, pattern and texture from beyond the confines of the industry itself brought in not just fresh blood but also an open, liberated attitude, with artists and artisans free to experiment and, in doing so, create a fresh aesthetic that took Venini into a new realm. The creative impetus within the design and manufacturing of ornamental glass began to shift from France, which had been a world leader in Art Deco glassware, to Italy and Scandinavia.

Germany, meanwhile, stepped up production with the help of designers such as the Bauhaus-trained WILHELM WAGENFELD, who worked with producers Schott & Gen. and Vereinigte Lausitzer Glaswerke (VLG). Wagenfeld designed a heat-resistant glass teapot and tea set (1931) for Schott, and then in 1935 was appointed art director at VLG, where he headed an 'artistic design laboratory' populated by designers and technicians. Wagenfeld himself designed a series of press-moulded bowls for VLG and his famous set of Kubus storage containers (1938) – a stackable collection of rectangular, pressed glass pieces for kitchen and table use.

In the fields of both glass and ceramics, the early modernist period saw a level of innovation and new thinking that brought ingenious, functional designs to a growing band of machine-age consumers – pieces largely defined by a relative simplicity and lack of ornamentation, but with a pleasing sense of artistry in their conception and form. At the same time, factory-made art ceramics and ornamental art glass were given fresh impetus and new direction by the many alliances forged between artists, sculptors, technicians and entrepreneurs. There was also a place for studio potters and small-scale producers such as Leach and Cardew.

Although the dialectical tension could be somewhat fierce on both sides, industry and craft reached some kind of accommodation during the modernist period to the benefit of all. Sculptors worked with factory engineers, artists inspired new concepts of glass-making, and architects and industrial designers were also welcome collaborators. There was a vibrant process of cross-pollination, echoed in other disciplines, particularly graphic design and typography. It was a process that continued and accelerated during the mid-century period, when ceramics and glassware became more expressive, sculptural, organic and playful, with the industries fed by the imaginations of artists, architects and creative thinkers of all kinds.

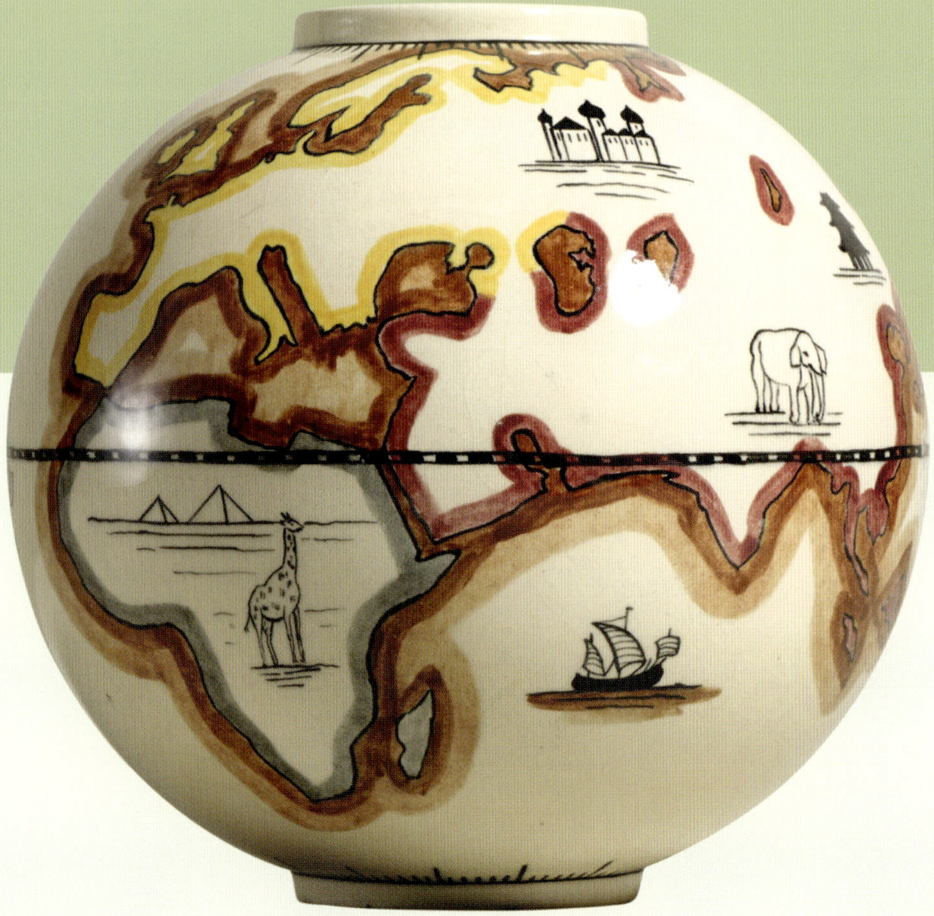

LEFT A globe vase in glazed earthenware by Robert Lallemant, made in France c. 1930 for Macy's.

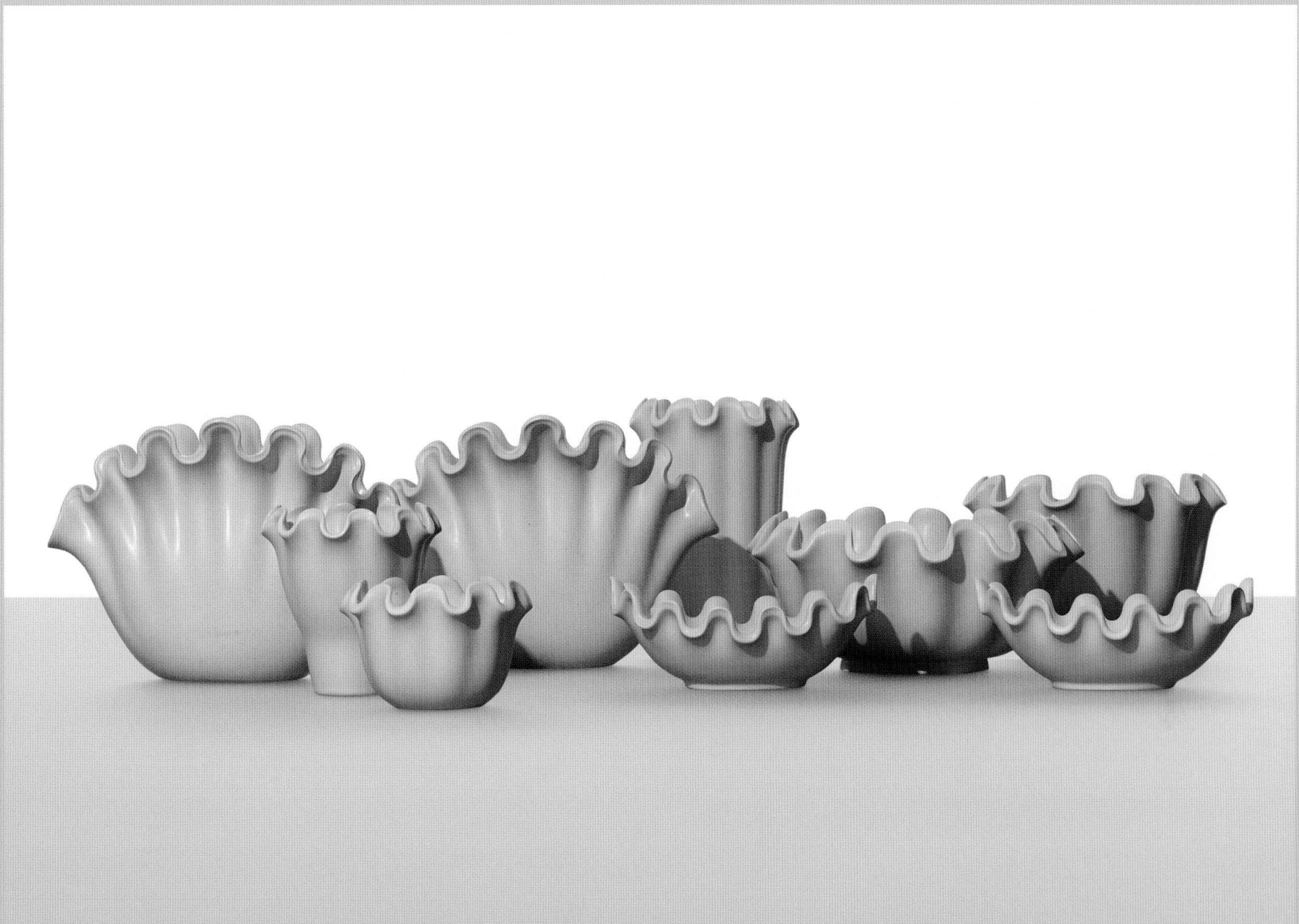

ABOVE A selection of porcelain pieces from Wilhelm Kåge's Vaga collection, made by Gustavsberg in Sweden, *c.* 1939.

THE NEW AESTHETIC: THE EVOLUTION OF MODERNIST CERAMICS IN THE INTER-WAR YEARS

MEL BUCHANAN

Ceramics rode the wave of change in design and craft during the Twenties and Thirties, reflecting greater society's vacillation between extreme conditions in economics, politics and culture. Between the cataclysmic First and Second World Wars there was a worldwide economic depression and the rise of fascism, but also great advancements in science, technology and the arts. In the bumpy transition into the modern era, exciting ceramic ideas percolated as ceramicists searched for both national and personal definition, some reflecting on a romanticized craft history and others anticipating a utopian industrialized future.

Forecasting the erosion of artistic hierarchies in the coming 20th century, clay artists were at once part of craft, industrial and fine art communities. They explored different aesthetics, from updated versions of neoclassicism to bold modern abstraction. In the Twenties, ceramicists both innovated with industrial production and re-embraced ancient hand-craft techniques. Most pointedly, inter-war ceramic designers showed a wide variety of intent behind their art making, some being part of progressive social movements while others offered luxury objects to select markets. In this array we see vibrant kernels of individuality that would mature into the sophisticated expressiveness of mid-century studio pottery, and also modern designers embracing thoughtful mechanization for unprecedented results in industrial ceramics.

In 1920, Bernard Leach, a British potter with ties to Japan, and Shoji Hamada, a potter central to Japanese folk arts, together founded a pottery in St Ives, England. The Leach Pottery extolled the fusion of Eastern and Western ceramic traditions and became enormously influential to 20th-century studio potters. Furthering the 19th-century Arts and Crafts movement's reverence for hand-craft, Leach brought to England the Japanese *Mingei* revival of traditional ceramics. Leach was a prolific promoter of his philosophy, a teacher who directly influenced Lucie Rie and Hans Coper, fellow Brit Michael Cardew, and American Warren MacKenzie. Leach's aesthetic was predominantly hand-thrown traditional shapes, like wide-body vases and single-handled pitchers, together with Eastern-origin forms like tea bowls. His decoration nodded to British historic techniques such as slip clay mixed with Japanese glaze and firing techniques, notably raku.

Bernard Leach's legacy is the elevation of the individual artist-potter and of small-scale ceramic production, and his philosophy can be understood as oppositional to the growing industrialization of the Twenties. In the inter-war era this long-simmering adulation of the East flourished into a mania by a particular artistic set searching for an alternate belief system to the one that led to the horrors of World War I.

Progressive modernists shared this desire for revolution amid the inter-war era's palpable social anxiety, but instead had an artistic response that looked to the future. These avant-garde art circles were centred in post-revolution Russia, the Wiener Werkstätte in Austria, and the Deutscher Werkbund* and Bauhaus in Germany. Some of the most fascinating ceramic designs of the Twenties come out of Russia. In the abstract shapes and elemental colours of seminal Russian art figures Wassily Kandinsky* and Kazimir Malevich*, teapots and platters made at the State Porcelain Factory used a bold modernist art language to carry socialist political messages to the Russian proletariat and beyond. These propaganda plates are among the most innovative ornament found on ceramics in the 20th century, though notably production techniques and forms remained essentially traditional.

The German Bauhaus opened in 1919 as a state-funded school, with lofty goals to transform society through the development of a new type of designer. This modernist knew hand-craft fabrication, embraced industry, eschewed ornament, and understood art psychology. The school's ceramics workshop in Dornburg was at a geographical remove from the Bauhaus proper in Weimar, but it also started at a pedagogical remove as its masters Gerhard Marcks* and Max Krehan* emphasized hand-craft before manufacture. In April 1923, Bauhaus director

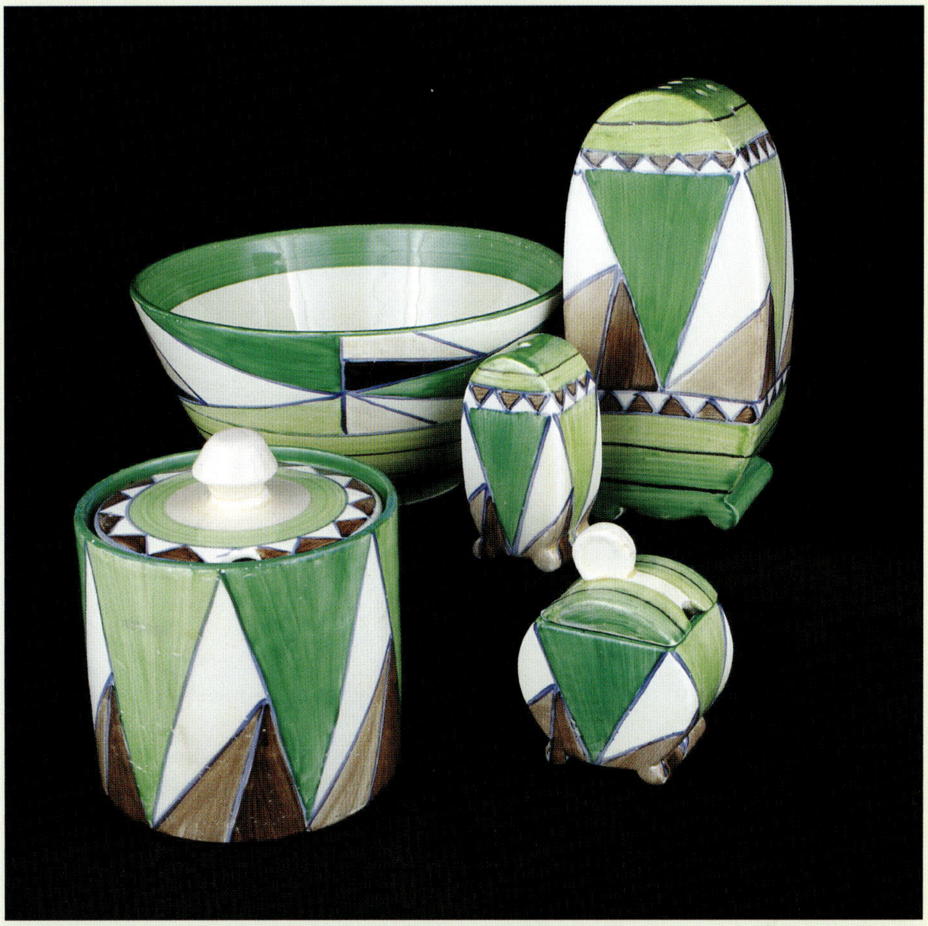

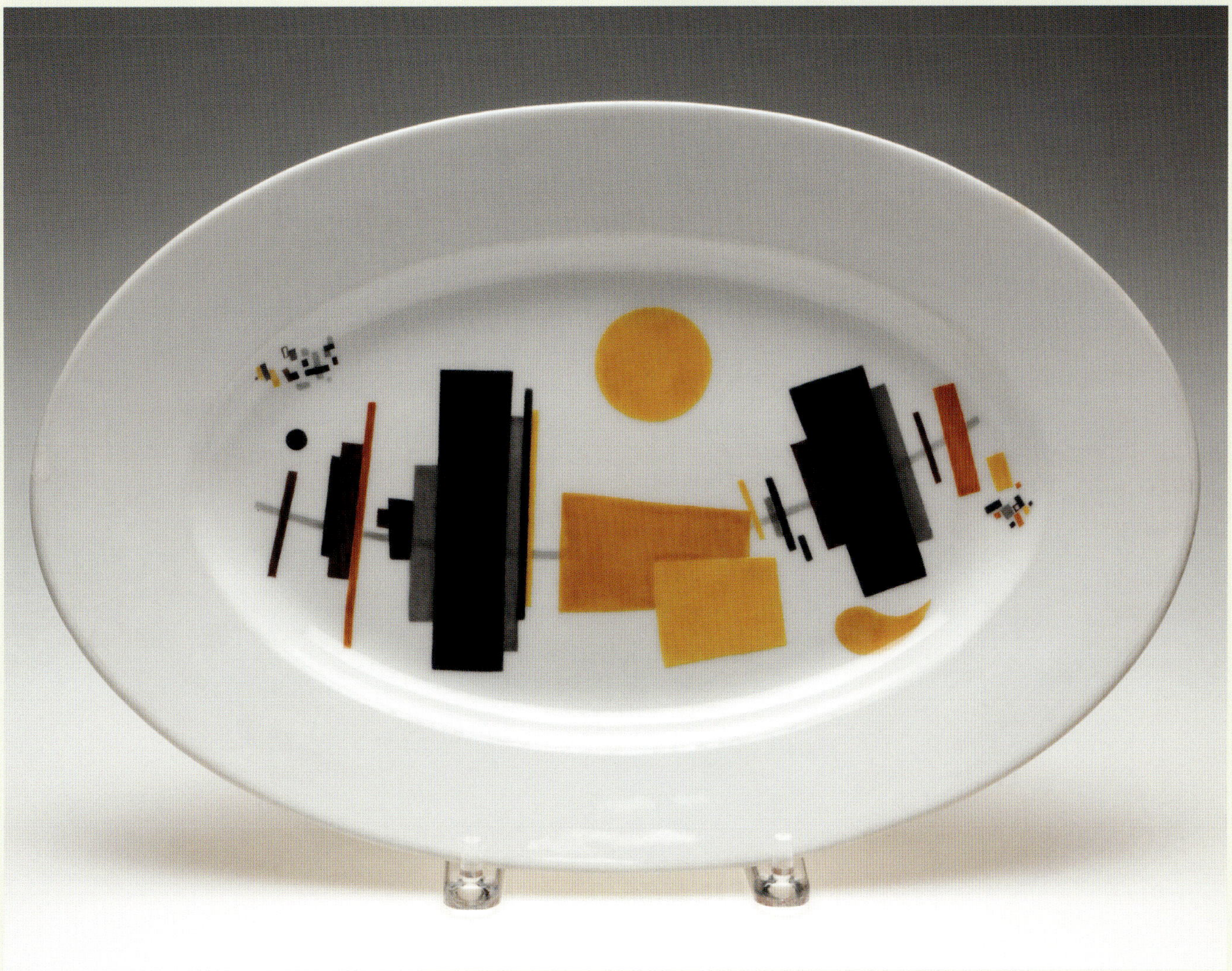

ABOVE Clarice Cliff's Killarney series, including a bowl, sugar sifter and shaker, honey pot and mustard pot, decorated in a 'Bizarre' pattern, 1920s.

LEFT A Suprematist platter in porcelain, with overglaze decoration by K. Marebura, *c.* 1920, after Kazimir Malevich; manufactured by the Russian State Porcelain Factory.

Walter Gropius complained of this, writing to Marcks, 'I had a look at your many new pots. Almost all of them are unique, unrepeatable; it would be positively wrong not to look for ways of making the hard work that has gone into them accessible to large numbers of people.... We must find ways of duplicating some of the articles with the help of machines.'[3] By 1924 the Bauhaus pottery had Otto Lindig and Theodor Bogler, directors who introduced cast ceramics to the curriculum. Their industry-friendly designs reflected Bauhaus ideals of basic geometry, looked machine-made, and could (in theory) be affordably mass-produced.

At the same moment that the tiny Bauhaus was experimenting in ways that would have a magnified influence on later 20th-century design, the eyes of the world were on Paris for the 1925 World's Fair – the International Exposition of Modern Decorative [and Industrial] Arts, or in French the *arts décoratifs* that gave 'Art Deco'* its name. The fair included only what organizers termed '*Art Moderne*', ushering in the popularity of an unabashedly luxurious decorative style that revitalized motifs from Ancient Egypt to Neoclassical Louis XVI, from Cubism to Japanese art. Unlike other contemporary modern art movements, with idealistic social manifestos, *Art Moderne* was purely and almost hedonistically decorative.

Italian architect and designer Gio Ponti is not usually thought of as a ceramicist, but in the Twenties he was the artistic director of the esteemed Milanese Richard-Ginori porcelain company. Ponti's ceramic designs – playing to Italian nationalism with ornament pulled from classical mythology and Etruscan frescoes – were awarded 'Grand Prix' in porcelain at the 1925 fair. Notably, he achieved the supreme elegance prized by French taste-makers, but also worked within the industrial production facilities of the Richard-Ginori factory. Ponti was aware of his moment in design history, specifically noting that: 'Industry is the manner of the twentieth century, its method of creation.'[4]

The sumptuous Art Deco of the Twenties also appealed to popular tastes. The style became especially affordable when executed in ceramics. English designers Clarice Cliff and Susie Cooper offered on-trend plates and teapots in inexpensive earthenware. Their joyful expression is generally applied as a colourful ornament over traditional forms (usually a ceramic factory 'blank'), marking the limitations of Staffordshire's embrace of modernist design. The English Potteries' traditional method of divided production had separate designers and craftsmen overseeing different stages of form and ornament. Taking root in avant-garde inter-war Europe was the idea that good design was holistic through all processes of conception, production and decoration. In the Thirties, forward-looking designers fleeing Nazi Germany, such as ceramicist Grete Marks, encountered the English Potteries' opposition in executing their ideals of industrial marriage of form and ornament. Keith Murray's* architectonic designs with Wedgwood were an important exception. In those elegant works an architectural form is perfectly matched by a restrained, non-ornamented surface.

Luxury market Art Deco, forward-looking industrial modernism and romantic revivalist tendencies all had trickle-

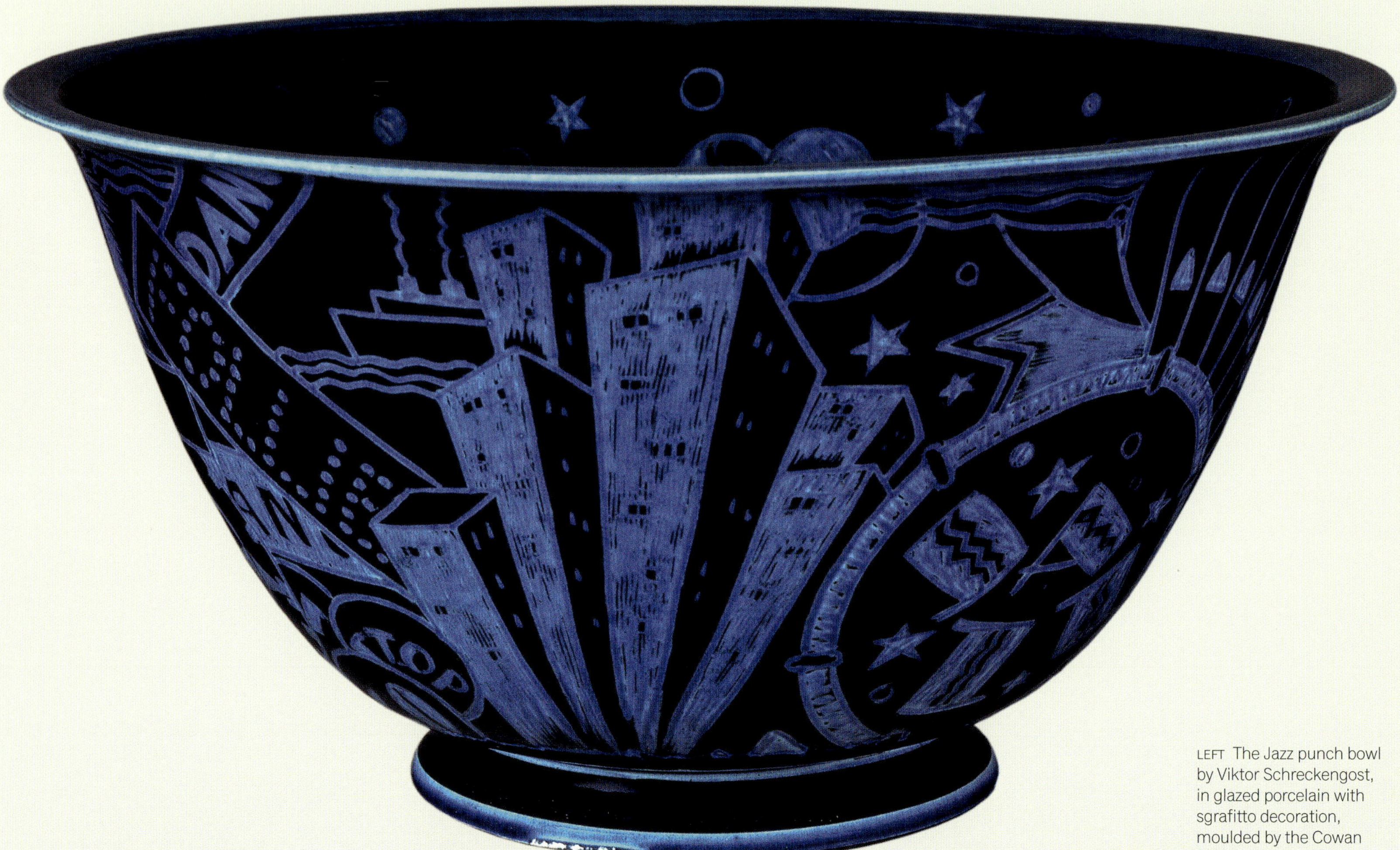

Left The Jazz punch bowl by Viktor Schreckengost, in glazed porcelain with sgrafitto decoration, moulded by the Cowan Pottery Studio, Ohio.

down counterparts in American ceramics. Art Deco received a uniquely American interpretation, coming to be the style associated with 1920s' urbanization. In 1931 Eleanor Roosevelt ordered a ceramic design that would come to be an icon. She wanted something 'New Yorkish' from Ohio's Cowan Pottery, and their designer Viktor Schreckengost delivered the 'Jazz' punch bowl. The famous bowl is uniquely American in production and theme, an homage to New York City's bright lights at night, but with its bold black and deep blue graphic quality it showed that Schreckengost had absorbed his ceramic study with Viennese cutting-edge designers of the Wiener Werkstätte.

The inter-war period in the United States was marked by the Art Deco 'roaring twenties', followed by the 1929 economic crash that begat the dramatic economic downturn of the Thirties. The Welfare Art Program of the Works Progress Administration included ceramicists in its efforts to maintain professional artists during the Depression, resulting in public art commissions that pushed ceramics into large, architectural scales. Waylande Gregory developed a honeycomb method of ceramic construction, allowing for monumental displays like the 'Fountain of the Atoms' he presented at the 1939 New York World's Fair. RUSSEL WRIGHT's restrained 1937 design, American Modern dinnerware, brought a softer 'biomorphic' form of modernism into American households. Between 1939 and 1959 Wright's American Modern became the best-selling ceramic ever manufactured.

We see in the inter-war years the patterns of personal expression and increasingly thoughtful design for manufacture that would result in the American mid-century's dominant period of both experimental ceramics and industrial production.[5] After World War II, California emerged as a centre of both expressive art clay and mass-market tableware. American craft communities flourished under the support of patrons such as Aileen Osborn Webb, who founded the American Craft Council in 1943 and what is now New York's Museum of Arts and Design in 1956. The arrivals into the United States of Otto and Gertrud Natzler in 1938 and the Bauhaus-trained MARGUERITE WILDENHAIN in 1940, each fleeing the Nazi invasions of Vienna and Holland respectively, mirror the shift to the United States for world leadership in both ceramic culture and industry.

The Twenties and Thirties ceramicist, having crossed boundaries between fine arts, craft and industrial production, was a bellwether for the 20th century's erosion and tension between those categories. Ceramics scholar Garth Clark points out that this tension between the individuality of the studio ceramics movement, or 'craft', and the modernism of industrial ceramics, or 'design', was often upheld by institutions as early as the Twenties and Thirties. The Museum of Modern Art in New York, for instance, enforced the division through their inter-war 'Machine-Art' and 'Good Design' exhibitions that included only industrially produced ceramics, not hand-crafted ones. Clark notes that: '[MoMA] conceived its curious policy that functional ceramics made by industry could be art, whereas vessels made by a potter with a wealth of intuition were craft and, by implication, of lesser importance.'[6] This continued tension between craft, fine arts and design, with clay having a vital interconnection to all three, would be the subject of a fervent conversation for the next chapter of design history.

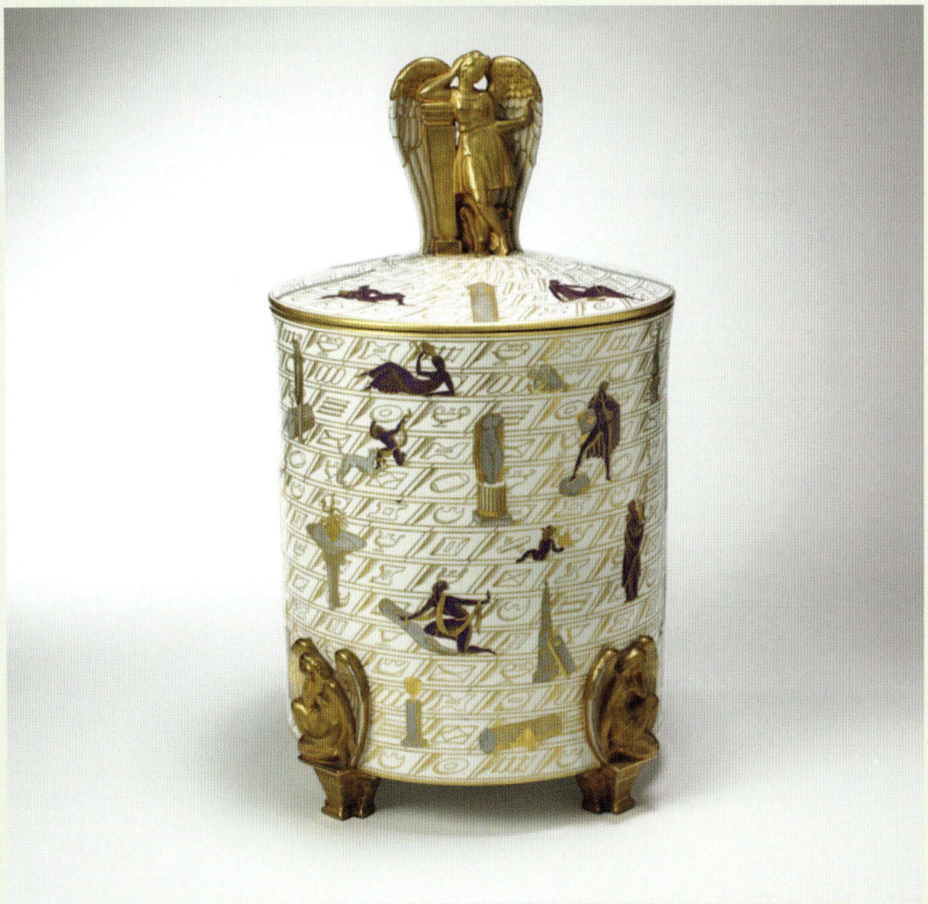

ABOVE A ceramic by Gio Ponti, from the 'La passeggiata archaeologica' (An Archaeological Stroll) series.

BERNARD LEACH
(1887–1979)

MICHAEL CARDEW
(1901–1983)

To describe the pioneering British studio potter Bernard Leach, or his close disciples, as 'modernist' would be controversial and tenuous. Yet, there were aspects of Leach's avant-garde approach and outlook that overlapped with modernist preoccupations, including an emphasis on form, a respect for functionality, and a stripping away of excess ornament and decoration. In some ways, Leach's work – particularly his 'standard ware' collections – could be seen as modern in character, although his passion for craft and vernacular techniques, both Eastern and Western, also allies him with an Arts and Crafts philosophy.

The greatest point of difference between Leach and the modernist mainstream was in the means of production. During the Thirties, in particular, there was a rapid push towards industrial, factory-made collections of standardized ceramics that were affordable and functional. This was anathema to Leach, who was passionate about craftsmanship, the art of making and the character of the materials used. 'My frequent criticism of mass-produced wares should not be regarded as an attack upon the machine so much as an exposure of the false standards of beauty, whether of commercial origin or of debased court taste, which have accompanied the rise of industrialism,' Leach wrote.[7]

The notion of a product intimately connected to the maker and rooted in both place and history was vital to Leach, with this philosophy playing an integral part within the character of his studio pottery, as practised by Leach himself and key students, such as Michael Cardew. Their work, and teachings, became powerfully influential not just in Britain but internationally and helped shape the work of younger generations of studio potters.

Leach himself was something of an East-West fusion. He was born in Hong Kong to English parents, although his mother died in childbirth; his father was a judge and his grandparents, who looked after him as an infant, were missionaries in Japan. At the age of ten, Leach went to school in England and then enrolled at the Slade School of Fine Art. He turned his back on a very brief career in banking to pursue his interests in art and painting, studying at the London School of Art.

In 1908, Leach returned to Japan and spent much of the next decade there and in China. In 1911, he became fascinated by Japanese raku ceramics, or traditional tea ware, and began to study with the master ceramicist Urano Shigekichi. During the following years in Asia, Leach devoted himself to perfecting his knowledge of making and firing, as well as developing his own style of slipware (using simple slip coatings made with a mix of clay and water). Around 1919, he met the Japanese potter Hamada Shoji, who became a close friend and collaborator; a year later, in 1920, the two made their way to England.

Settling in St Ives, Cornwall, Leach and Hamada established the Leach Pottery, building the first kilns and producing the first pieces of stoneware and earthenware, while experimenting with local clays and timber fuels for the kilns. Hamada went back to Japan in 1923 and that

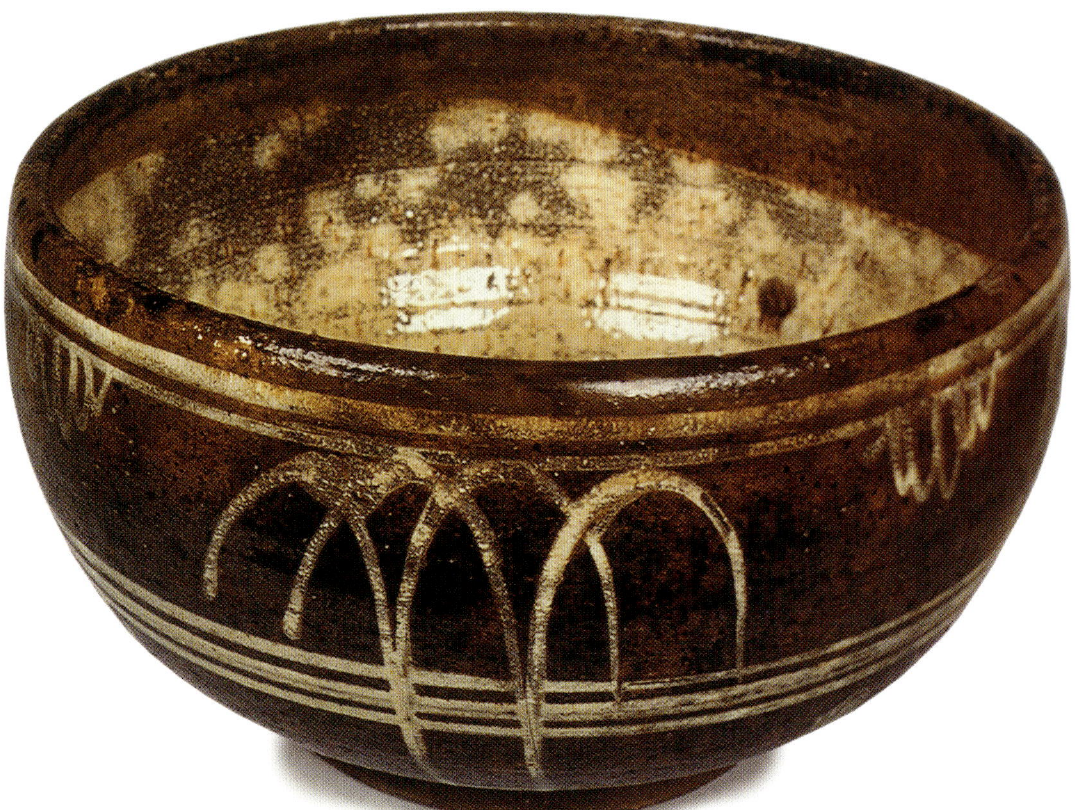

RIGHT An earthenware bowl made by Michael Cardew, *c.* 1937.

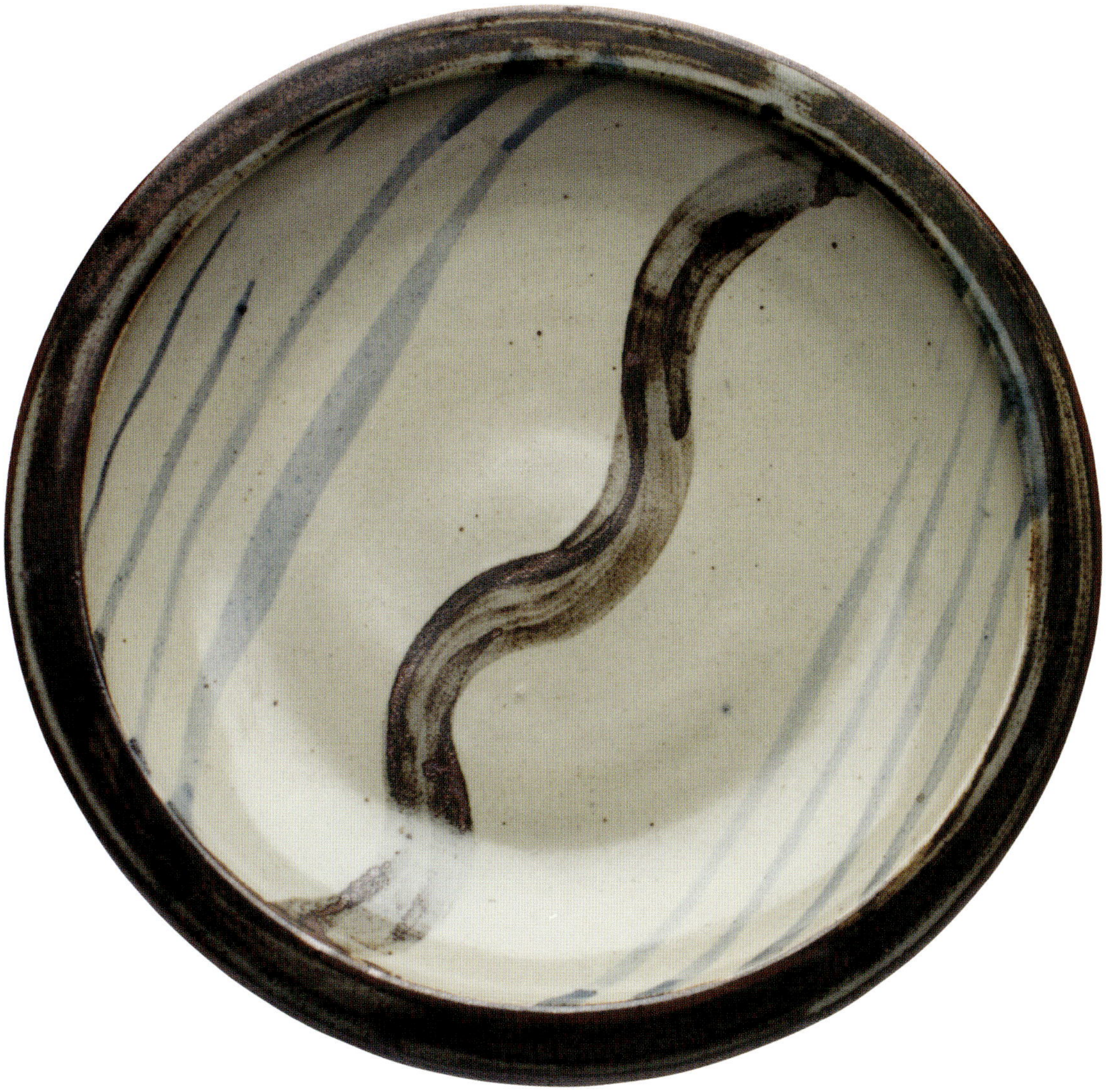

ABOVE A decorated cake dish from the Standard Ware collection designed by Bernard Leach, first produced *c.* 1946.

RIGHT A Celadon tea plate from the Standard Ware range, made of Cornish stoneware clay and produced at the Leach Pottery.

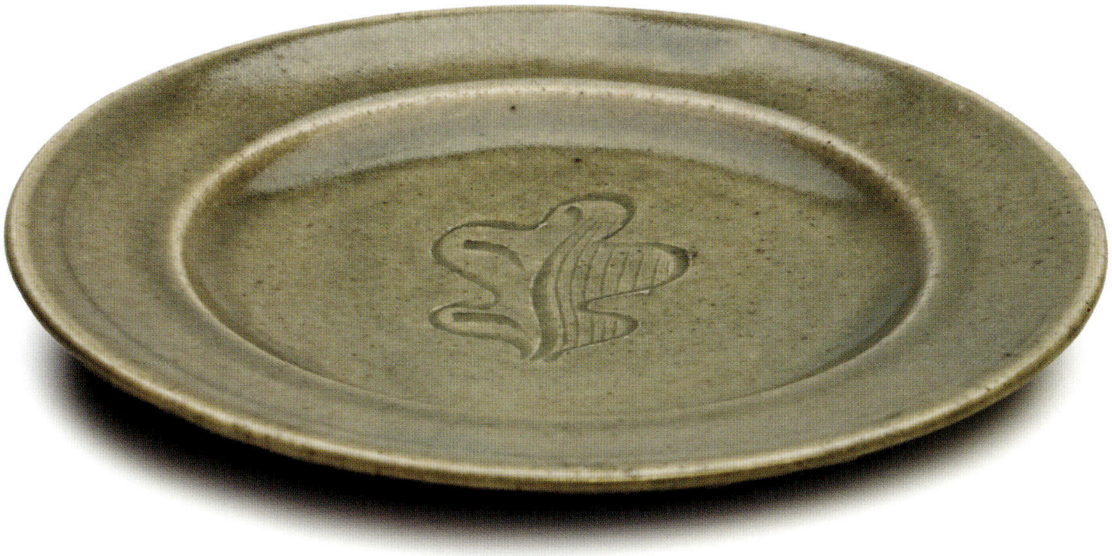

same year Michael Cardew, a young Oxford graduate, arrived and became Leach's most accomplished and trusted student. Leach began to exhibit and publish his thoughts on the techniques and philosophy of ceramics; his son David also joined the pottery and became an important part of its evolution.

Leach believed that the majority of his studio-made pieces should have a utilitarian and functional use, suggesting that such hand-crafted work should be available and affordable for everyday users. Yet, within the context of the austerity years of the early Thirties especially, the Leach Pottery generally attracted informed and relatively affluent collectors, while struggling to produce tableware in volumes that would make it both profitable and more affordable. Leach began to differentiate between his art pottery pieces and his 'standard ware', which was sold more widely and listed in catalogue form by 1946.

Michael Cardew also believed that his work should have utilitarian use within the home. Having left the Leach Pottery in 1926, he founded the Winchcombe Pottery in Gloucestershire, eventually building a kiln that could hold as many as 4,000 pieces in the hope of producing on a scale large enough to reach wider audiences of consumers while still retaining a focus on craft and authenticity. Later, Cardew moved into teaching and spent many years working and instructing in Nigeria during the post-war era.

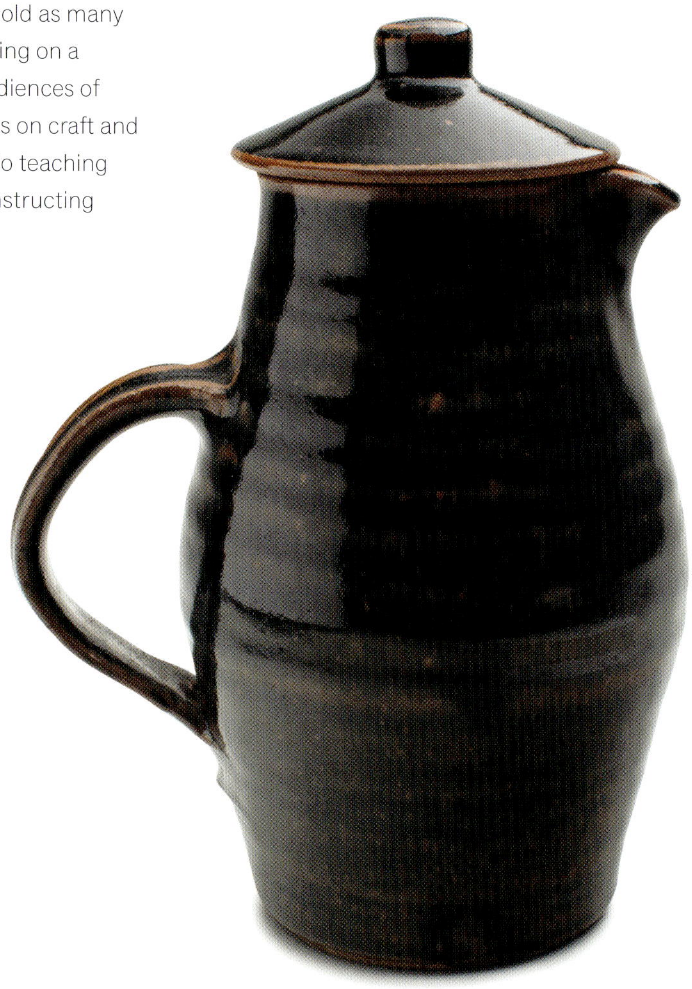

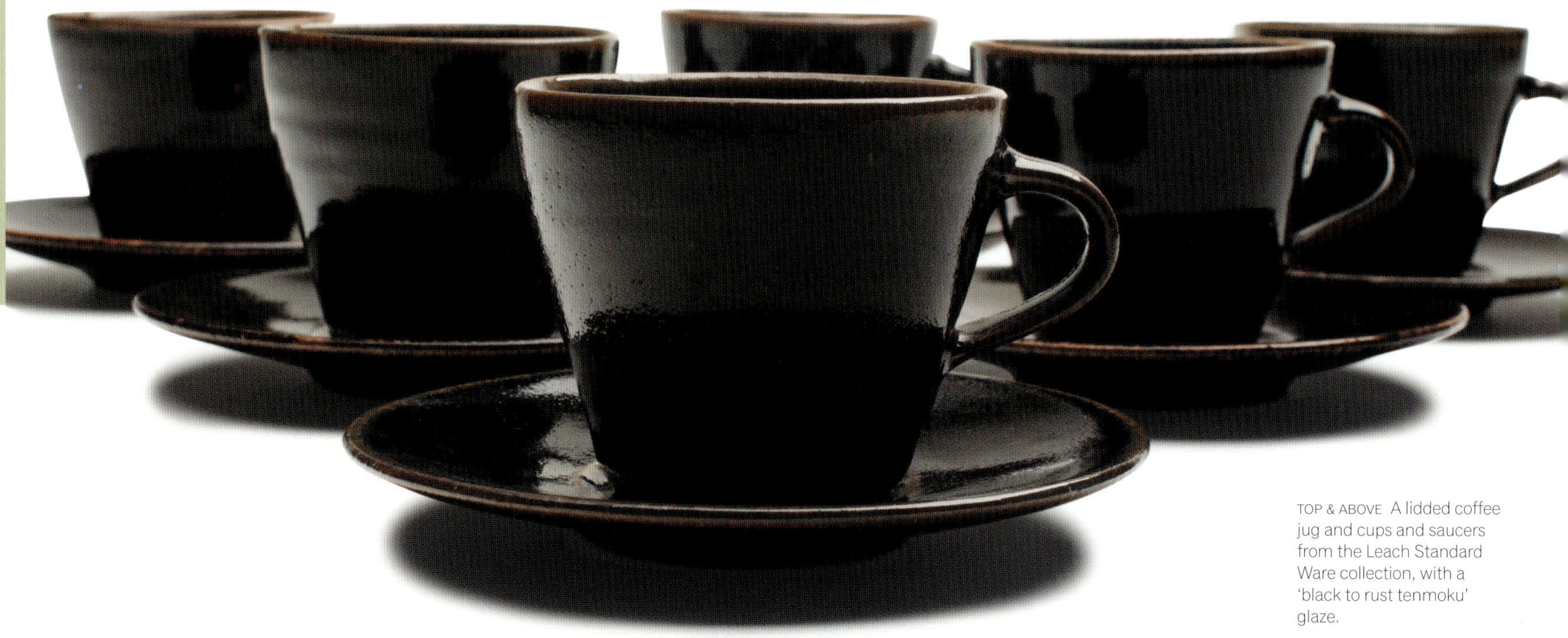

TOP & ABOVE A lidded coffee jug and cups and saucers from the Leach Standard Ware collection, with a 'black to rust tenmoku' glaze.

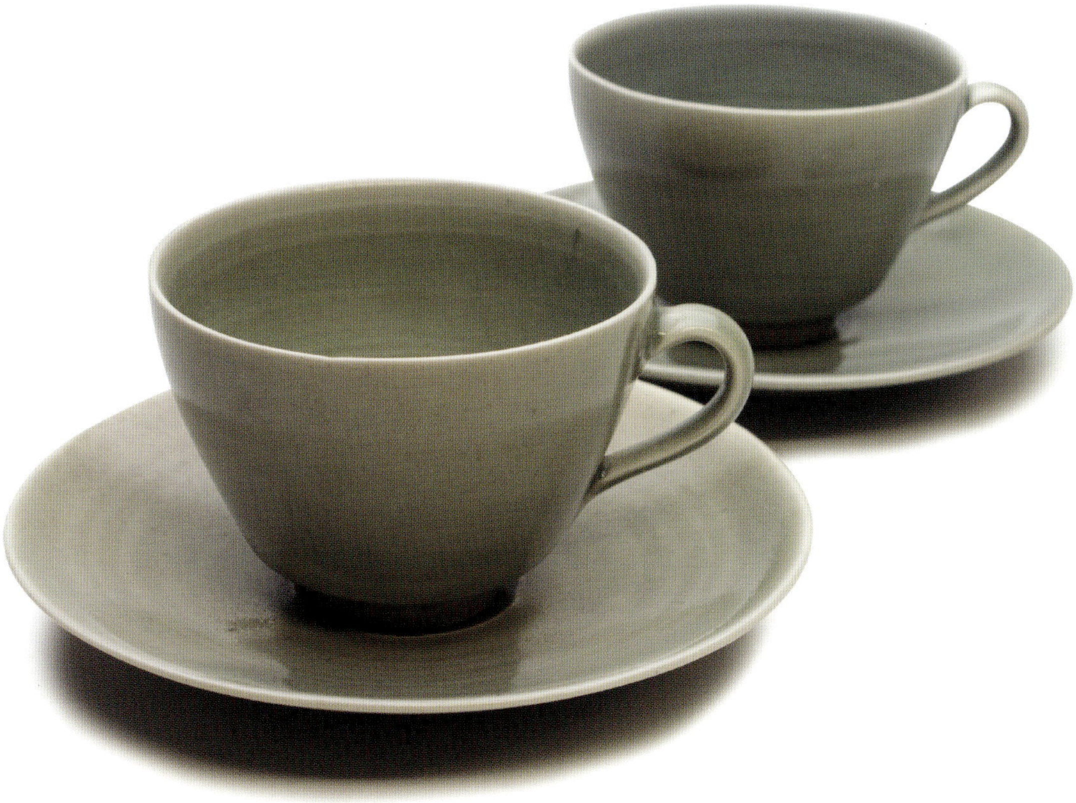

ABOVE Celadon tea cups and saucers from Bernard Leach's Standard Ware collection.

BELOW An open sugar bowl, with a 'black to rust tenmoku' glaze, from the Standard Ware range.

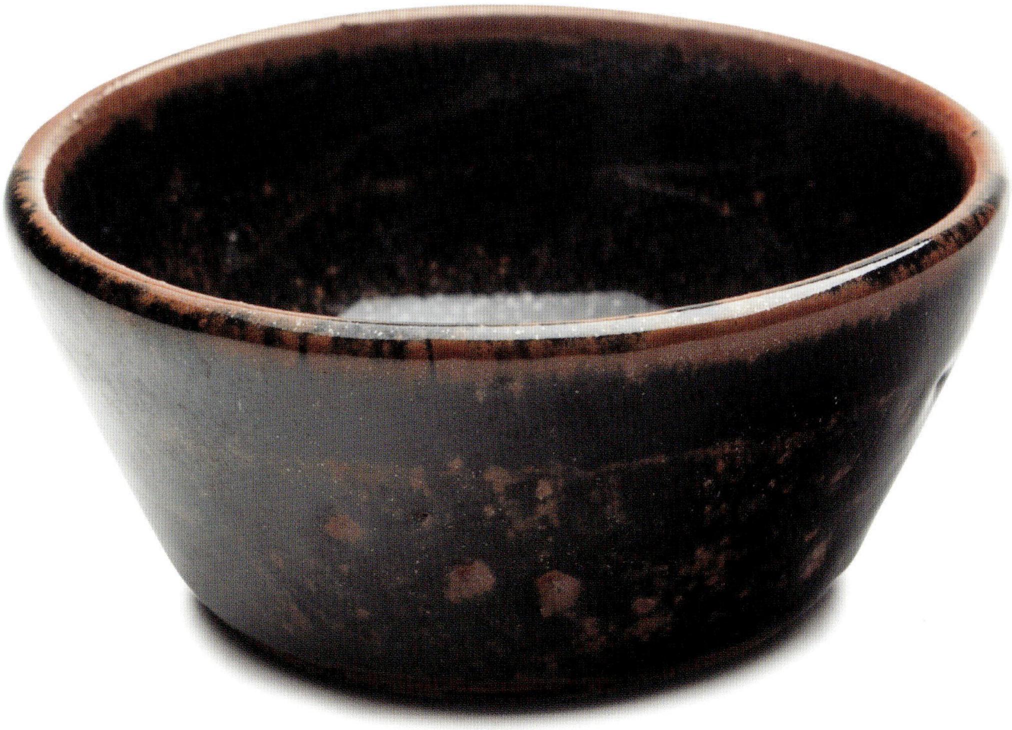

OTTO LINDIG
(1895–1966)

THEODOR BOGLER
(1897–1968)

The pottery workshop at the Bauhaus was short-lived but highly influential. In the space of just a few years, from 1919 to 1925, the Weimar Bauhaus department trained a number of talented ceramicists and developed new techniques aimed at expanding production on an industrial scale. Under pressure from Bauhaus director WALTER GROPIUS, the ceramics workshop sought to step away from the potter's wheel and use ceramics casts to produce new earthenware, characterized by a combination of sculptural artistry and utility, with a lack of ornamentation and a simple palette of glazes. Two of the key figures in the short life of the workshop – and its subsequent legacy – were Otto Lindig and his brother-in-law, Theodor Bogler.

Otto Lindig was born in Thuringia and initially studied art at the College of Drawing and Carving in Lichte before moving on to serve an apprenticeship with sculptor Max Bechstein. In 1913, Lindig moved to Weimar where he attended both of the institutions that would soon combine to form the Weimar Bauhaus, beginning with a course in ceramics with Henry van de Velde* at the Kunstgewerbeschule (School of Arts and Crafts), followed by studies in sculpture at the Grossherzoglich Sächsische Hochschule für Bildende Kunst (Grand Ducal Saxon College of Fine Arts) under Richard Engelmann. Following the creation of the Bauhaus in 1919, under the direction of Gropius, Lindig joined the school to pursue his interests as a sculptor, before enrolling in the nascent ceramics studio as an apprentice in 1920.

The ceramics workshop was run by a master of form, Gerhard Marcks*, and a master of craft, Max Krehan*. Marcks, like Lindig, possessed a love for both sculpture and ceramics, while Krehan came from a family of master potters. Together, they established the ceramics department within Krehan's existing studio at Dornburg, around twenty miles away from Weimar itself. The location of the workshop gave the productive, working studio a degree of independence, but Gropius soon began applying pressure on Marcks and Krehan to move away from unique, one-off designs and devote themselves to pioneering techniques for mass production: 'It would be positively wrong not to look at ways of making the hard work that has gone into them accessible to large numbers of people.... We must find ways of duplicating some of the articles with the help of machines.'[8]

Lindig became the most successful of the Bauhaus ceramicists in applying Gropius's philosophy, developing a number of designs during the Twenties that were put into production. Lindig adopted the Bauhaus aesthetic, focusing on simplicity of line and geometrical forms, while largely using cream, off-white and organic glazes.

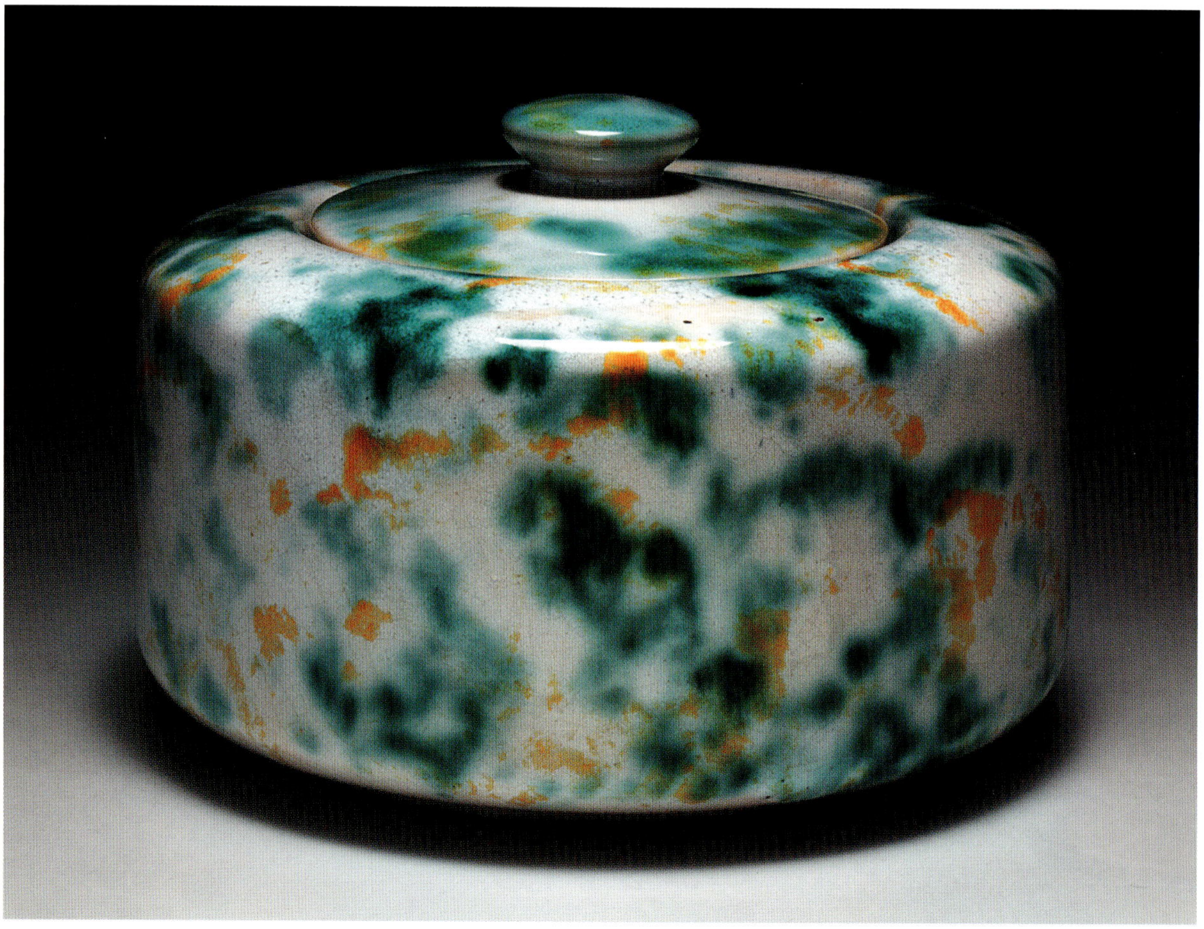

RIGHT A lidded jar by Theodor Bogler, made by Velten-Vordamm Keramik, *c.* 1925, in stoneware with faience shards and a cloudy pewter glaze.

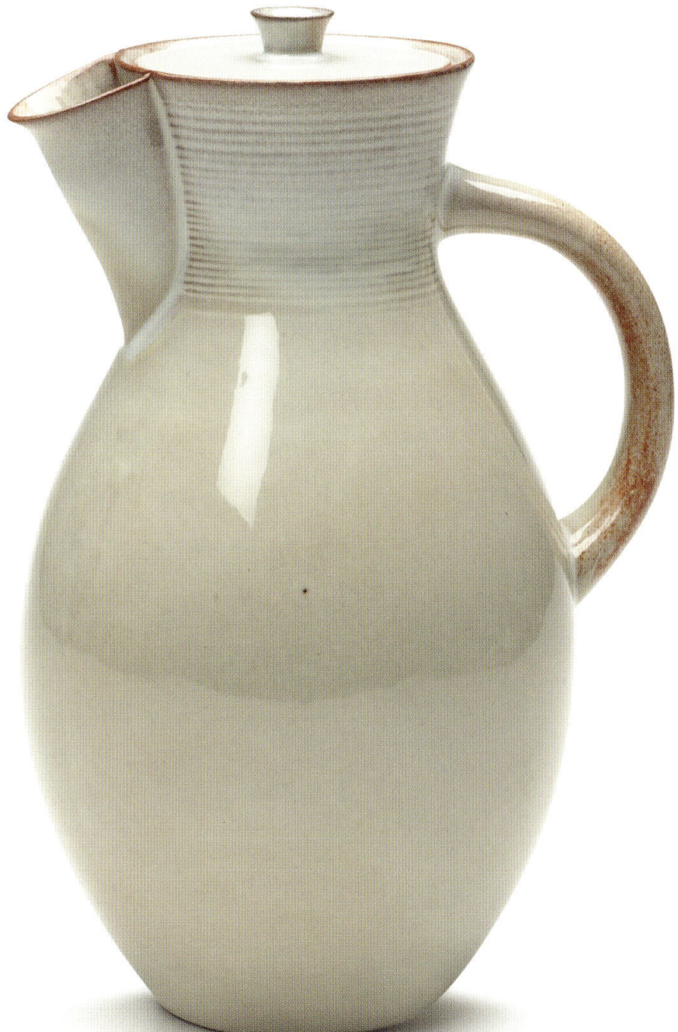

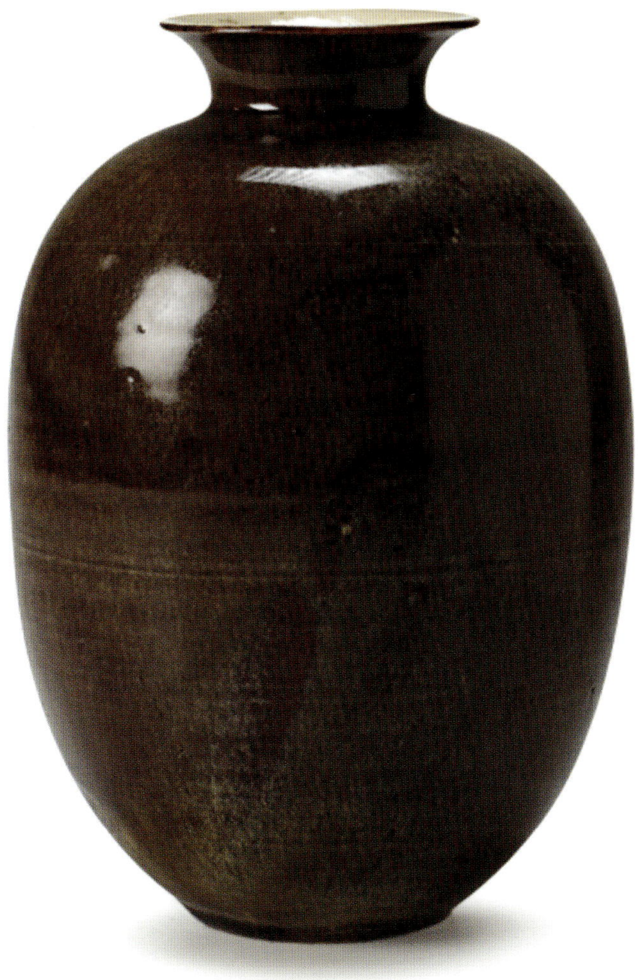

ABOVE A cocoa pot designed by Otto Lindig, made by Karlsruhe Majolika, first designed in 1923 and manufactured in 1929, in glazed earthenware.

LEFT A vase by Otto Lindig, made at his studio in Dornburg, *c.* 1927.

Yet he also brought a sculptor's eye to his work, as seen in his tall, lidded coffee and cocoa pots of the early Twenties, with their distinctive spouts and – in some cases – striking, elongated bodies or necks. By 1923, Karlsruhe Majolika had begun producing coffee and tea services to Lindig's design, featuring an extensive range of cups, saucers, pouring pots and jugs. In 1924, Lindig took over the technical and then the commercial management of the Dornburg workshops (Marcks left in 1924 and Krehan died in 1925). After the Bauhaus moved to Dessau in 1925, Lindig became director of the Dornburg studio, which continued under the jurisdiction of the Weimar College of Crafts and Architecture until 1930, and then became independent (in the post-war period Marcks invited Lindig to join him at the State College of Art in Hamburg).

Theodor Bogler took a foundation course at the Bauhaus in 1919, before joining the Dornburg workshop in the same year as Lindig, serving an apprenticeship before becoming a journeyman and then taking over the business administration of the studio for a short time. Bogler was responsible for one of the most innovative and famous designs to emerge from the Weimar Bauhaus ceramics workshop during its short life. This was the Combination teapot of 1923 – a pioneering earthenware pot consisting of a handful of separate components that could be used in different combinations, either as a teapot, pouring bowl, or bowl and cover.

After leaving the Bauhaus in late 1924, Bogler was ordained as a monk and then a priest, but continued working as a ceramicist before and after the war years. Other key students who attended the Bauhaus ceramics workshop included Grete Marks and Marguerite Wildenhain.

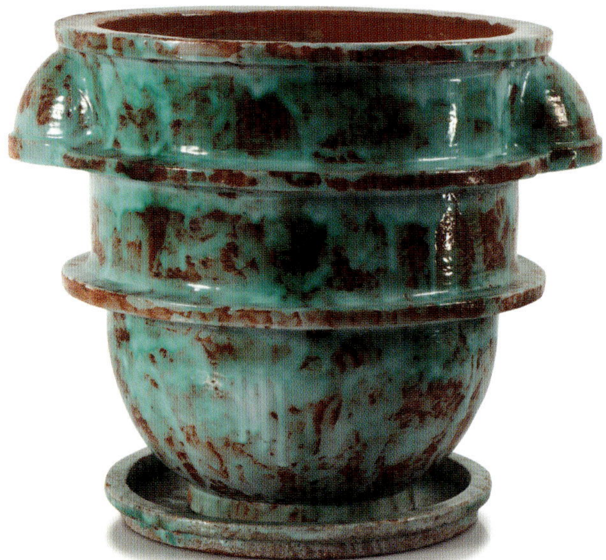

ABOVE A Theodor Bogler cachepot with saucer, made by Velten-Vordamm Keramik, *c.* 1925, in glazed majolica.

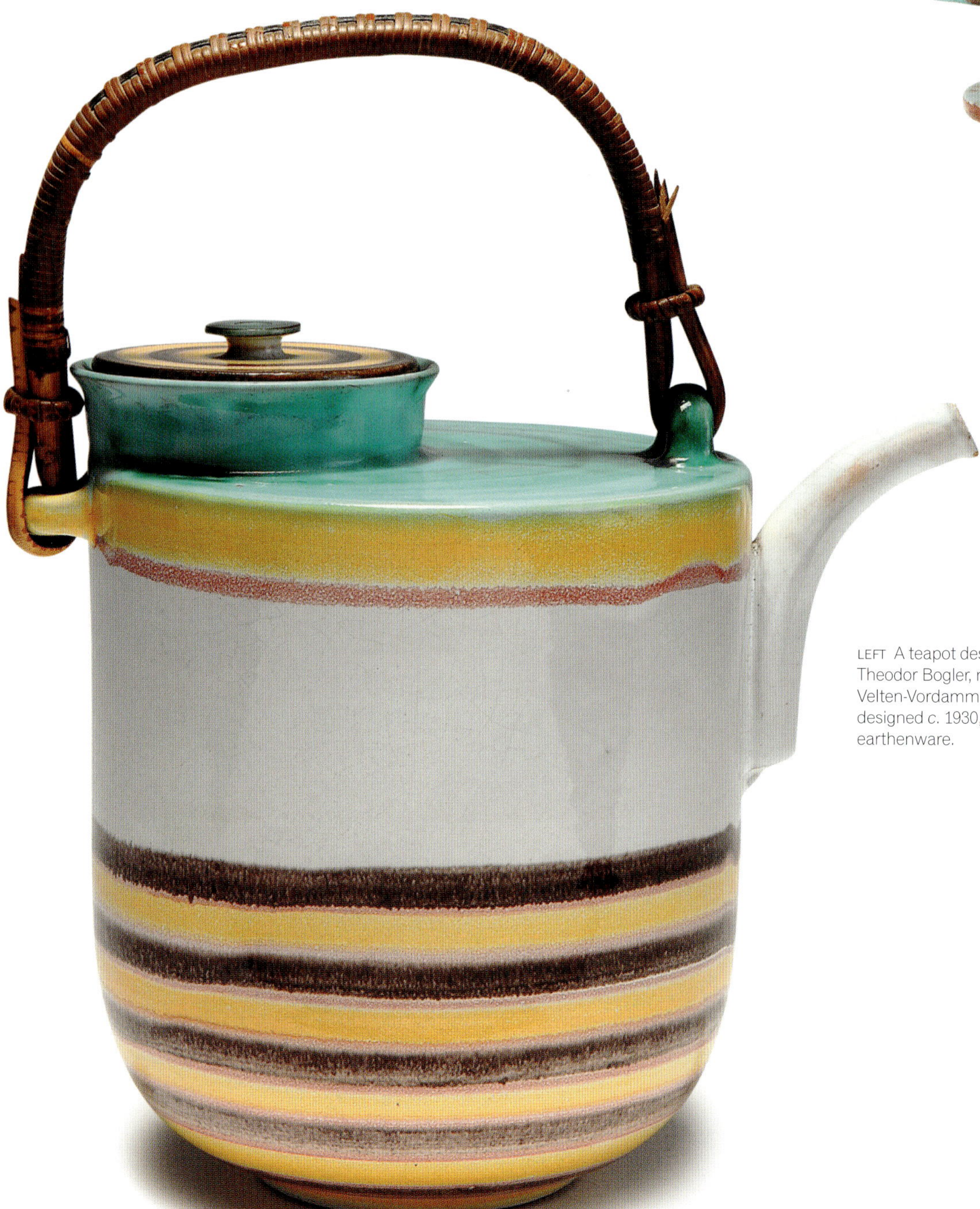

LEFT A teapot designed by Theodor Bogler, made by Velten-Vordamm Keramik, designed *c.* 1930, in glazed earthenware.

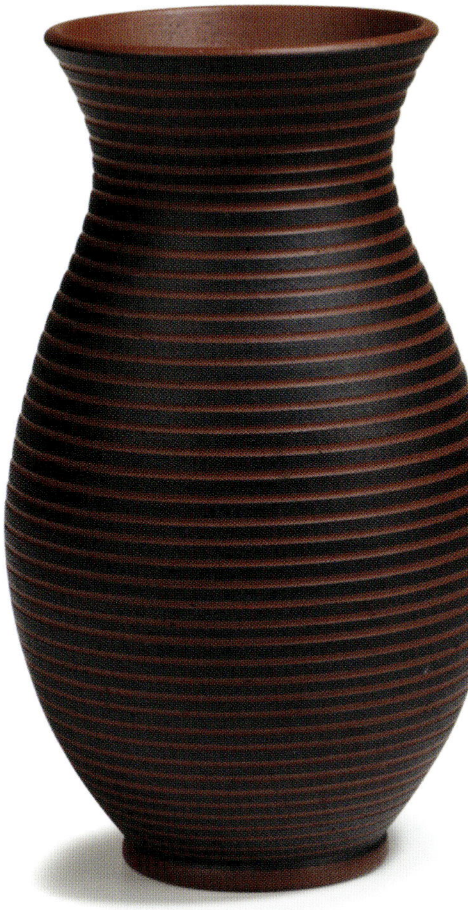

LEFT A vase by Theodor Bogler, designed in 1928, produced in Höhr-Grenzhausen, in glazed and incised red terracotta.

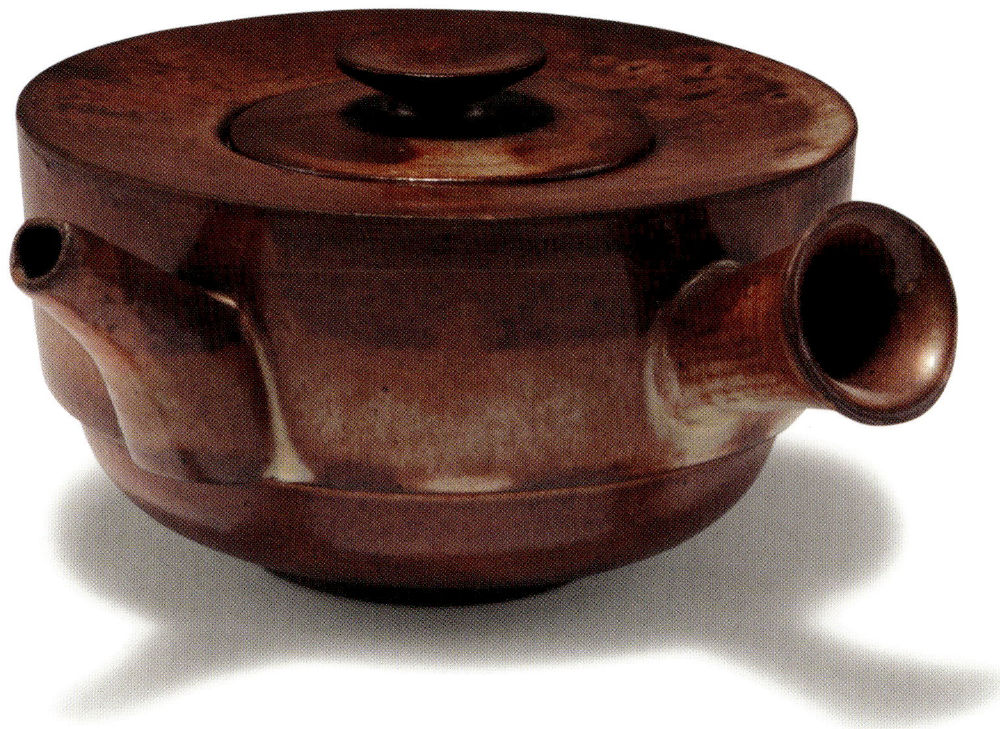

ABOVE A tea jug by Theodor Bogler, made at the Dornburg ceramic workshop, *c.* 1924, in glazed stoneware.

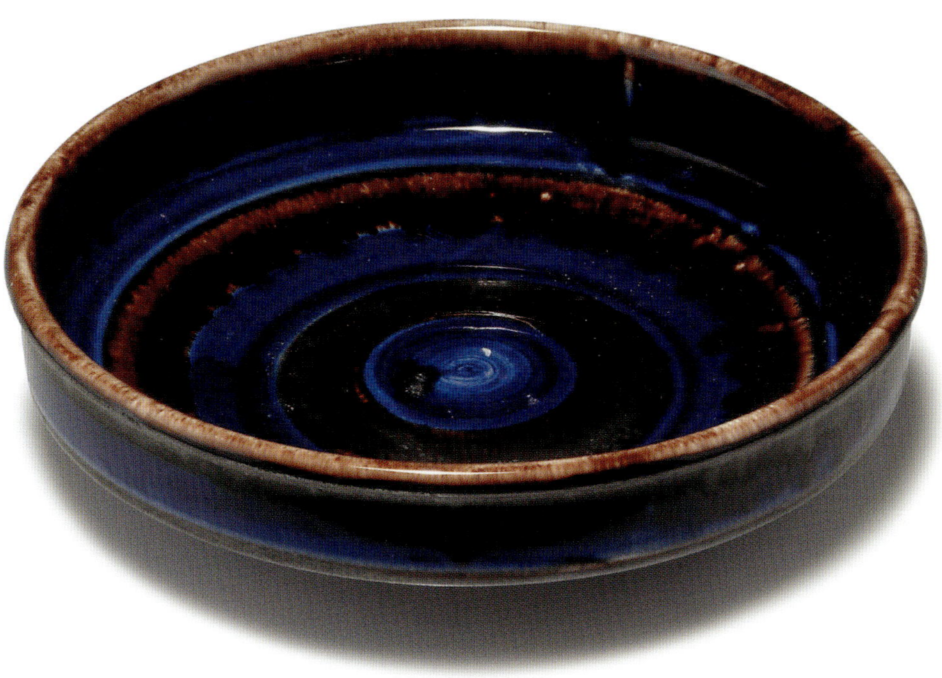

LEFT A bowl by Theodor Bogler, made by Velten-Vordamm Keramik, in glazed stoneware.

FOLLOWING PAGES A tea set for six people by Otto Lindig, produced by Karlsruhe Majolika, *c.* 1931, in glazed earthenware.

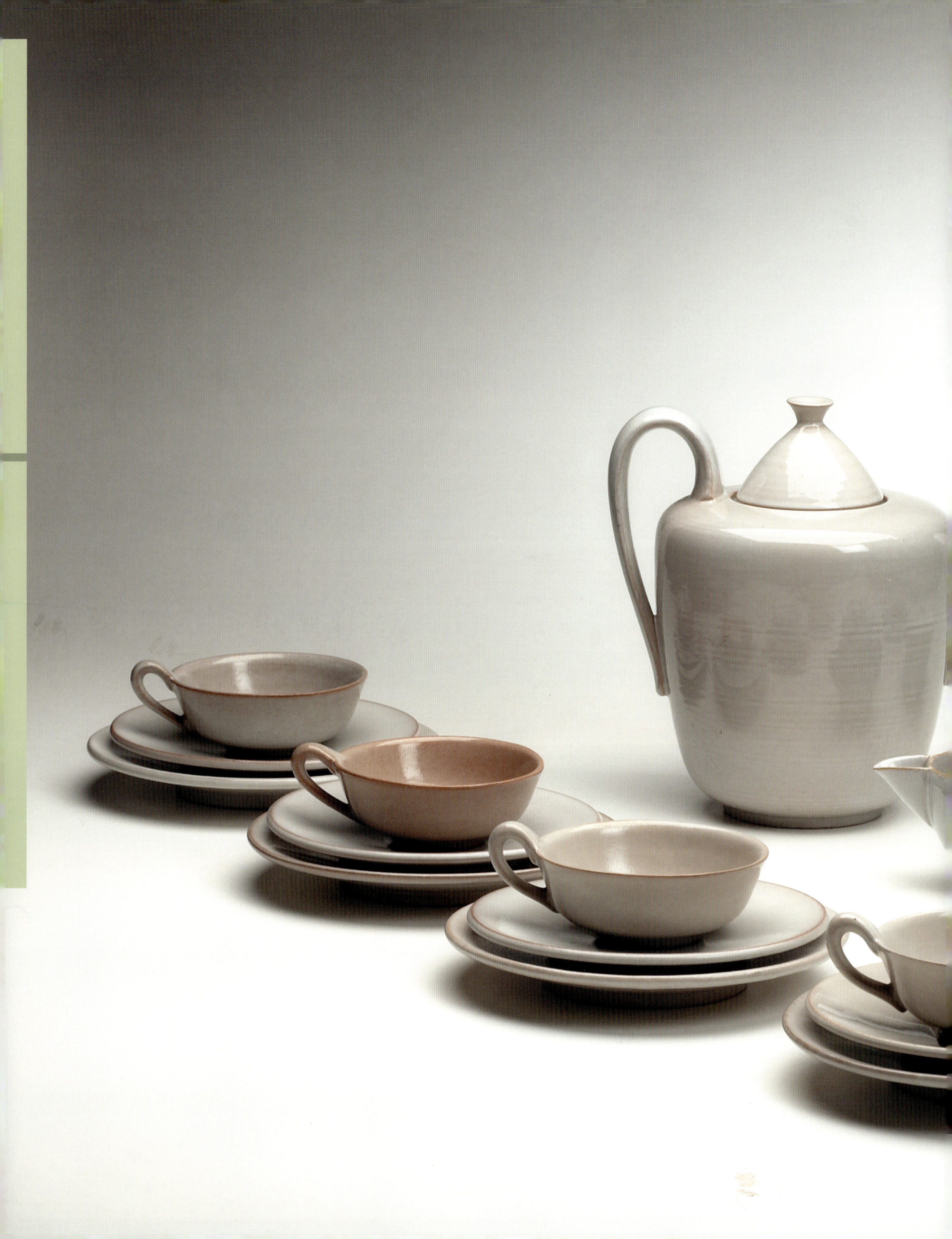

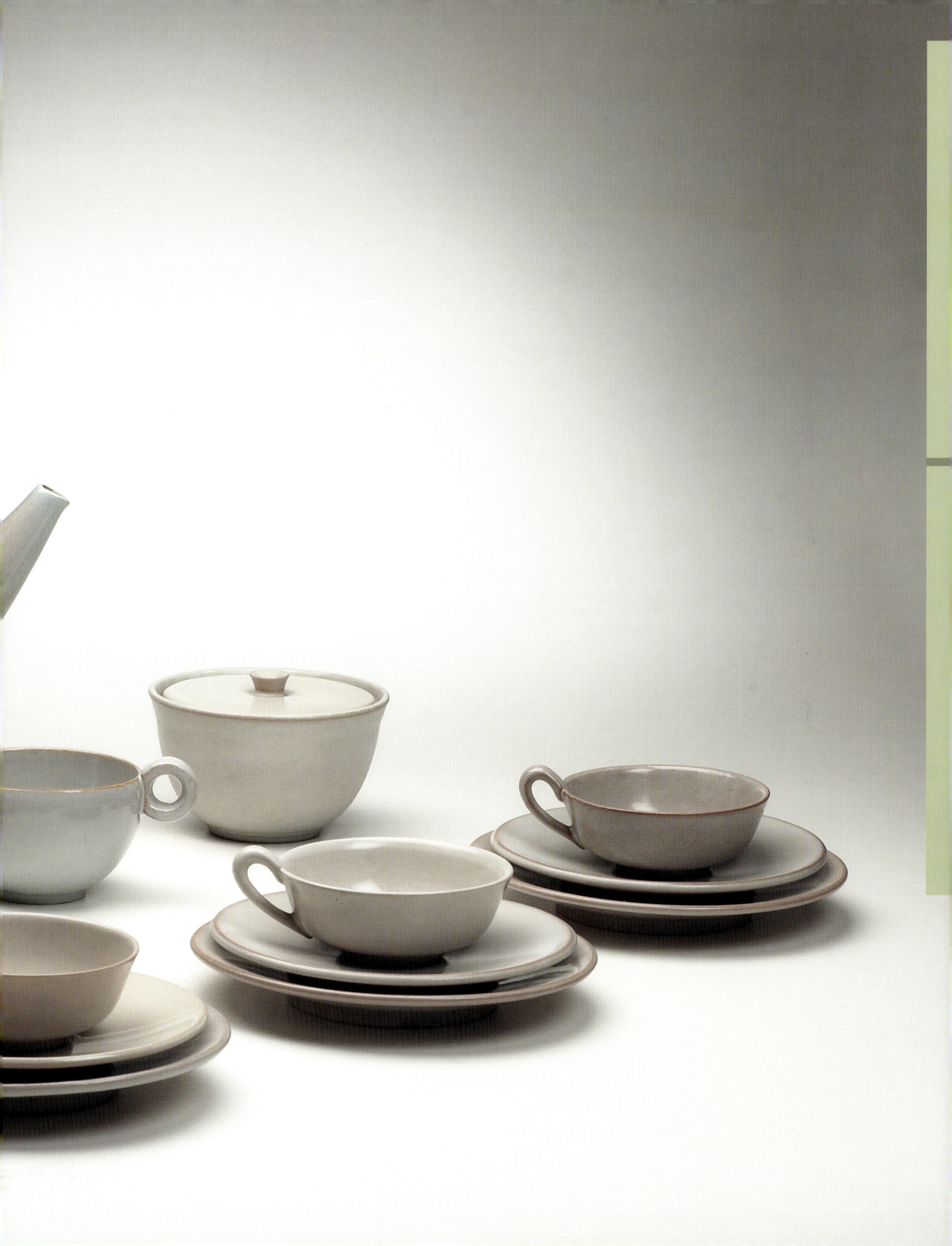

GRETE MARKS
(1899–1990)

The creative careers of many German architects, designers and artisans were damaged, destroyed or interrupted by the rise of the Nazi regime in the late Thirties. Hitler's government was profoundly hostile to modernist design in general, and to work created or produced by Jewish designers or manufacturers in particular. Ceramicist Grete Marks was a modernist pioneer from an artistic Jewish family, so perhaps it is no surprise that her work was labelled as 'degenerate' within the pages of propaganda minister Joseph Goebbels's newspaper *Der Angriff* and that her pottery business was shut down through the authority of the German Reich. It was a blow from which Marks never fully recovered.

Margarete – or Grete – Heymann was born in Cologne in the final year of the 19th century. She trained as a painter at art schools in both Cologne and Düsseldorf before enrolling in 1920 at the new Bauhaus school established in Weimar. She studied initially with Johannes Itten* before joining the Bauhaus pottery workshop in nearby Dornburg, led by Gerhard Marcks* and Max Krehan*. The pottery studio was a male-dominated environment, with students including Otto Lindig and Theodor Bogler. There was just one other woman student in the department, Marguerite Wildenhain, and Marks had to persuade her tutors to allow her to join the studio.

A strong and independent character, with clear ambitions of her own, Marks seemingly clashed with her Dornburg tutors and left the Bauhaus in 1921. Two years later, she married the industrialist Gustav Löbenstein and the two of them soon established their own pottery in Marwitz, known as Haël-Werkstätten. Some of Marks's work of the Twenties and Thirties embraced a similar kind of aesthetic to that being explored by fellow ceramicists from the Bauhaus, with an emphasis on geometrical forms, limited ornamentation and a restrained use of glazes. Yet there was also a willingness to experiment, with certain designs exploring more organic and unusual, sculptural shapes, and Marks became increasingly interested in using pattern and colour. Other pieces played with notions of abstraction and deconstruction, as with a striking black vase from the period embellished with a small collection of circular and diamond-shaped handles.

The pottery became a success, employing over one hundred workers and selling its collections – particularly tableware and tea services – not just in Germany but also abroad; one notable supporter was Ambrose Heal*, who sold Marks's ceramics at his store in London. But this period of happiness and success was relatively short-lived. In 1928, Löbenstein was killed in a car accident, leaving Marks to run the business alone

BELOW & OPPOSITE Two examples of tea services by Grete Marks, both *c.* 1930.

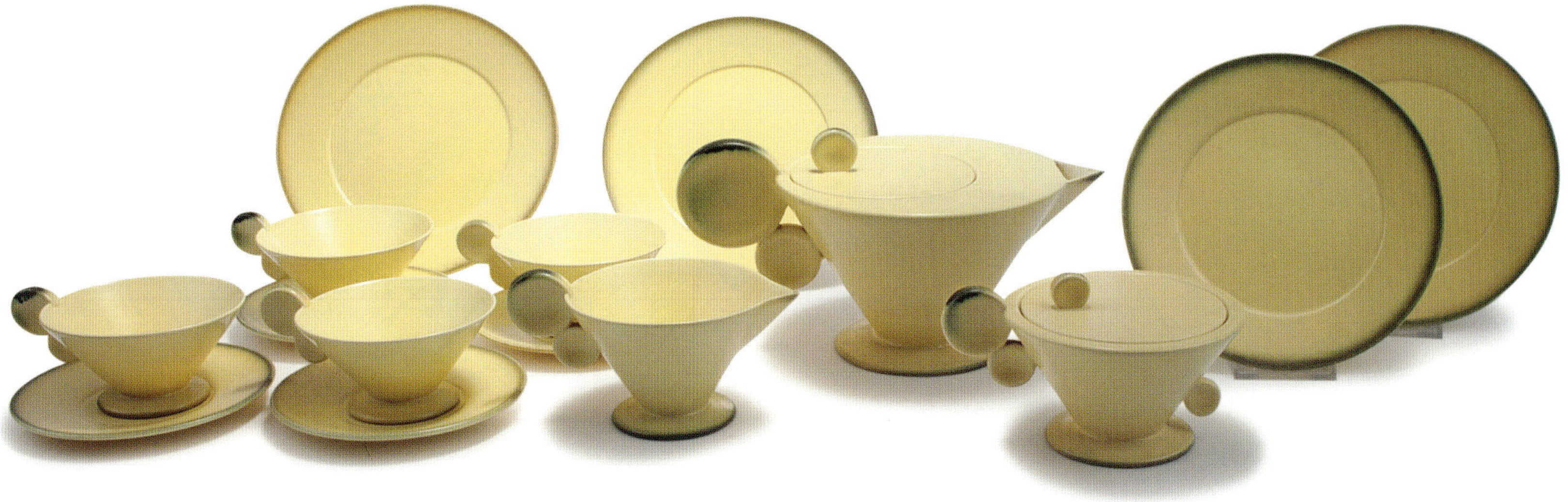

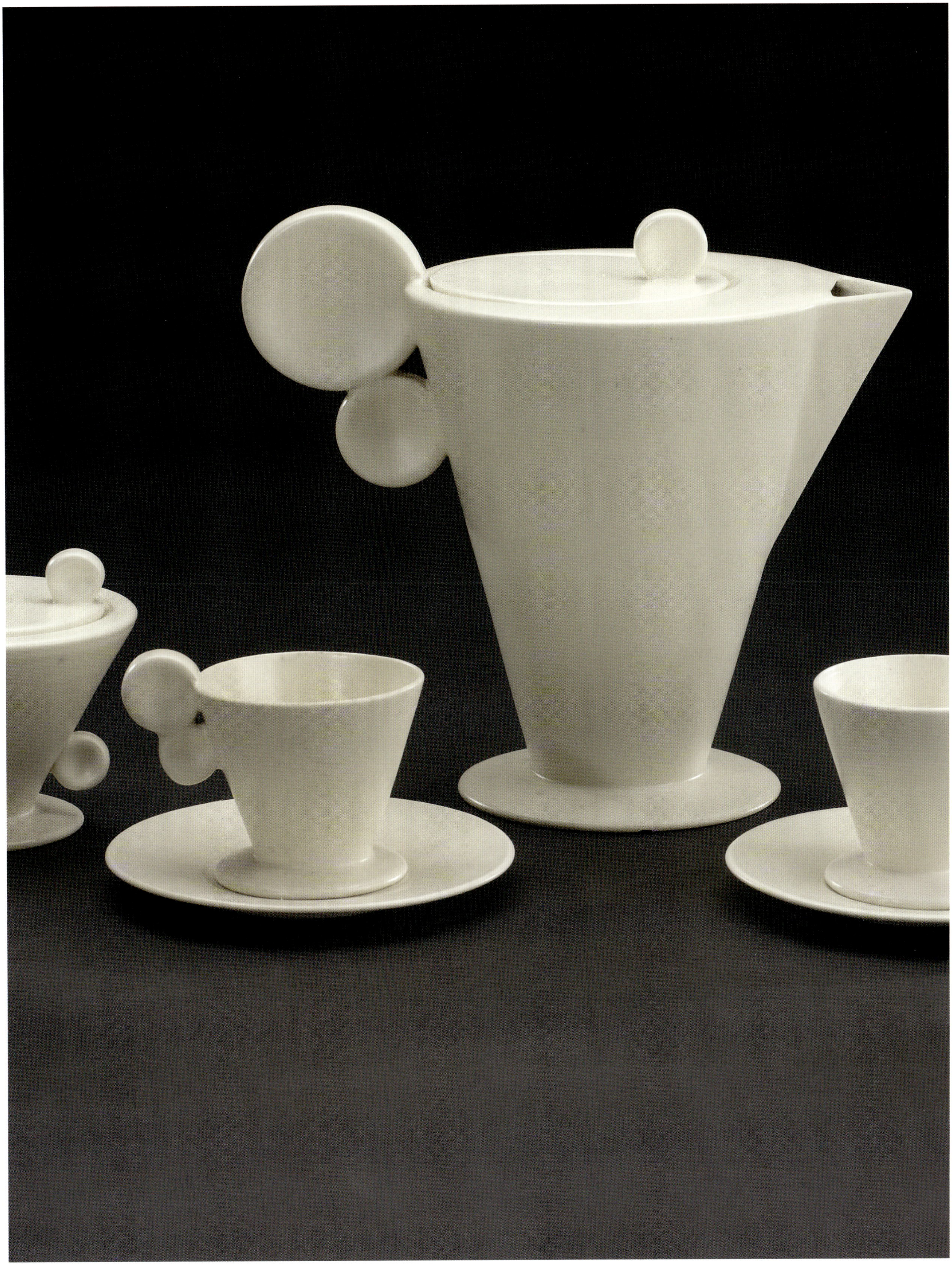

and look after their two young sons. The effects of the Great Depression then made themselves felt on the business, followed soon after by growing oppression from the Nazi authorities. Factory production halted in 1932, and the business was confiscated via compulsory purchase in 1935.

Ambrose Heal helped Marks and her family emigrate to England, where she found work in the Potteries, centred upon Stoke-on-Trent. After a short period teaching at Burslem School of Art, Marks found support from the manufacturer Mintons, who offered space and encouragement to establish her own brand – Greta Pottery – produced by the Mintons factory. The work was carried by Heal's, Fortnum & Mason and others, but the relationship with Mintons came to an end after Marks fell out with the company's art director.

In 1938, Marks rented space of her own in Stoke and continued Greta Pottery on a small scale, largely relying on buying undecorated earthenware from other producers and hand-painting the pieces with her own abstract designs; there was also freelance design work for other local potteries, some of which drew upon her German portfolio. In 1938 she also married her second husband, Harold Marks, and in 1945, at the end of World War II, the family moved to London. In later years, Grete Marks returned to painting, but with relatively little success, and there were only occasional experiments in ceramics.

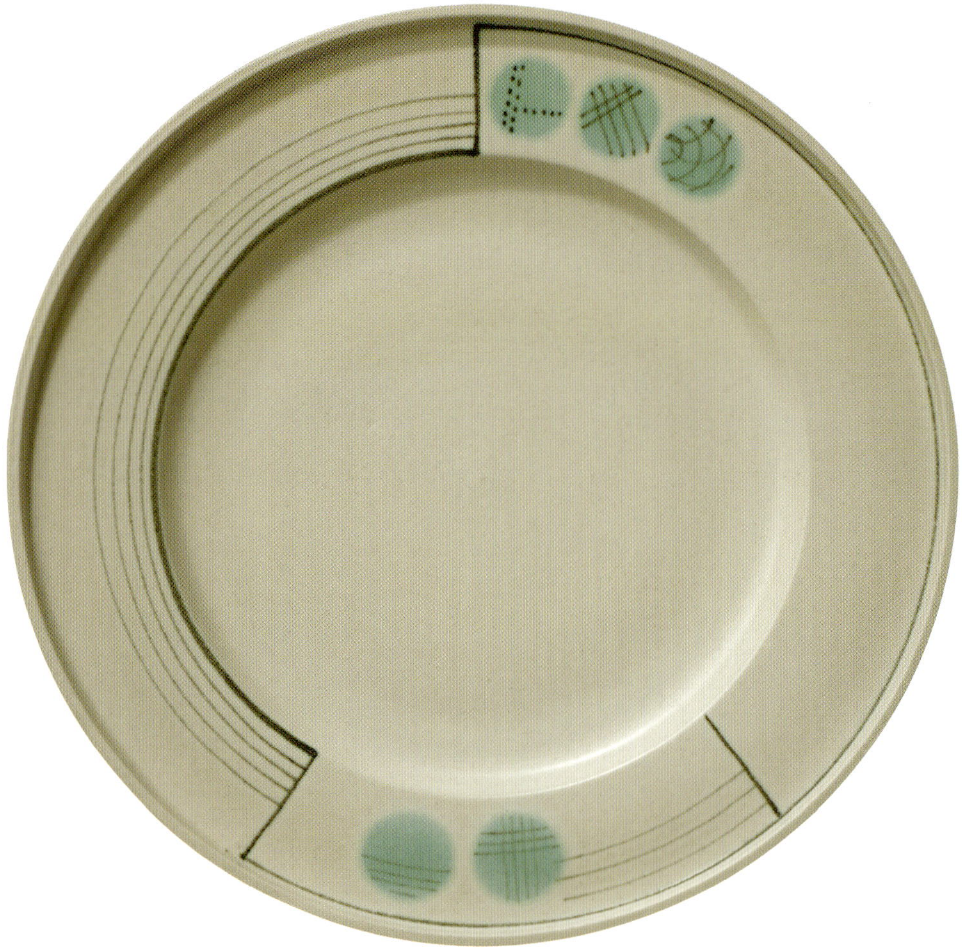

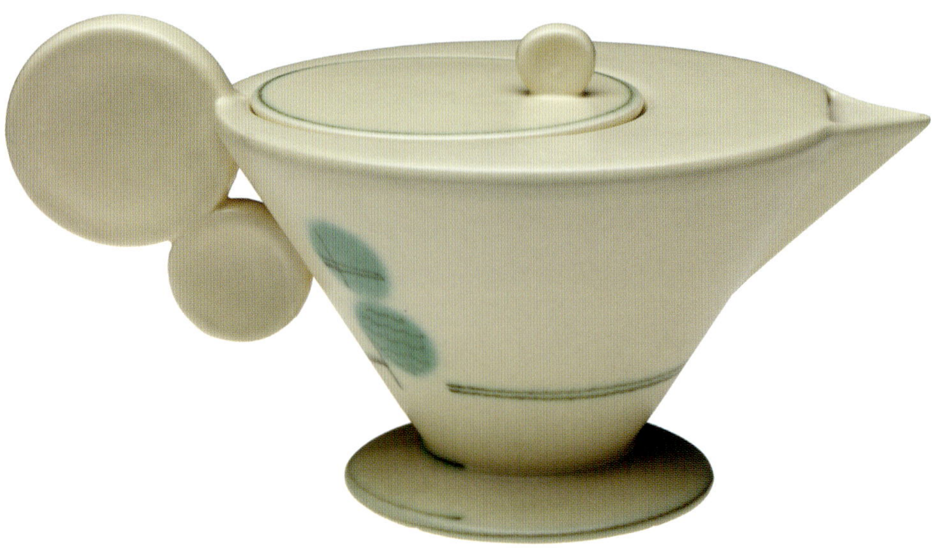

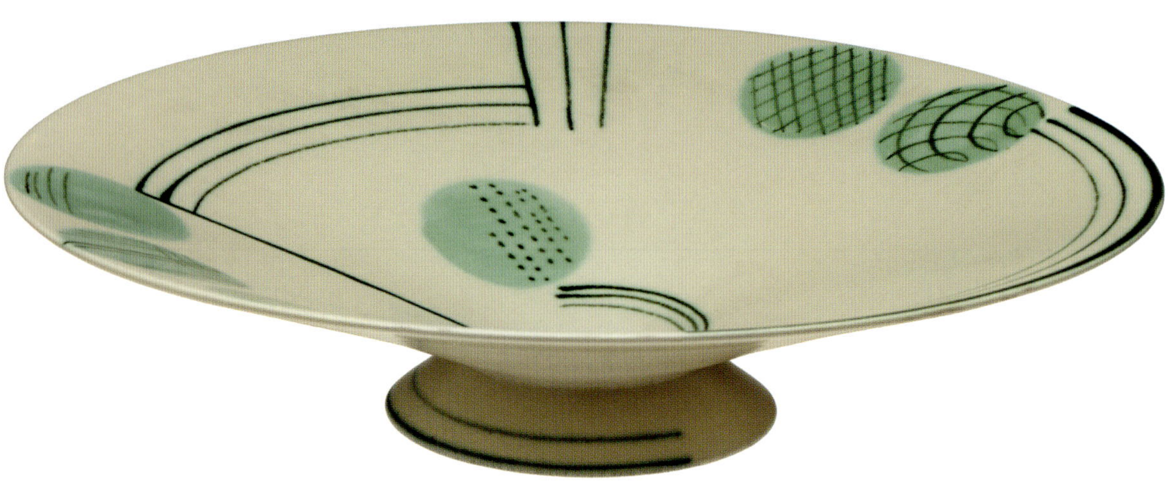

ABOVE & BELOW Three designs in a similar style by Grete Marks, manufactured by Haël-Werkstätten in Marwitz, *c.* 1930, in glazed earthenware.

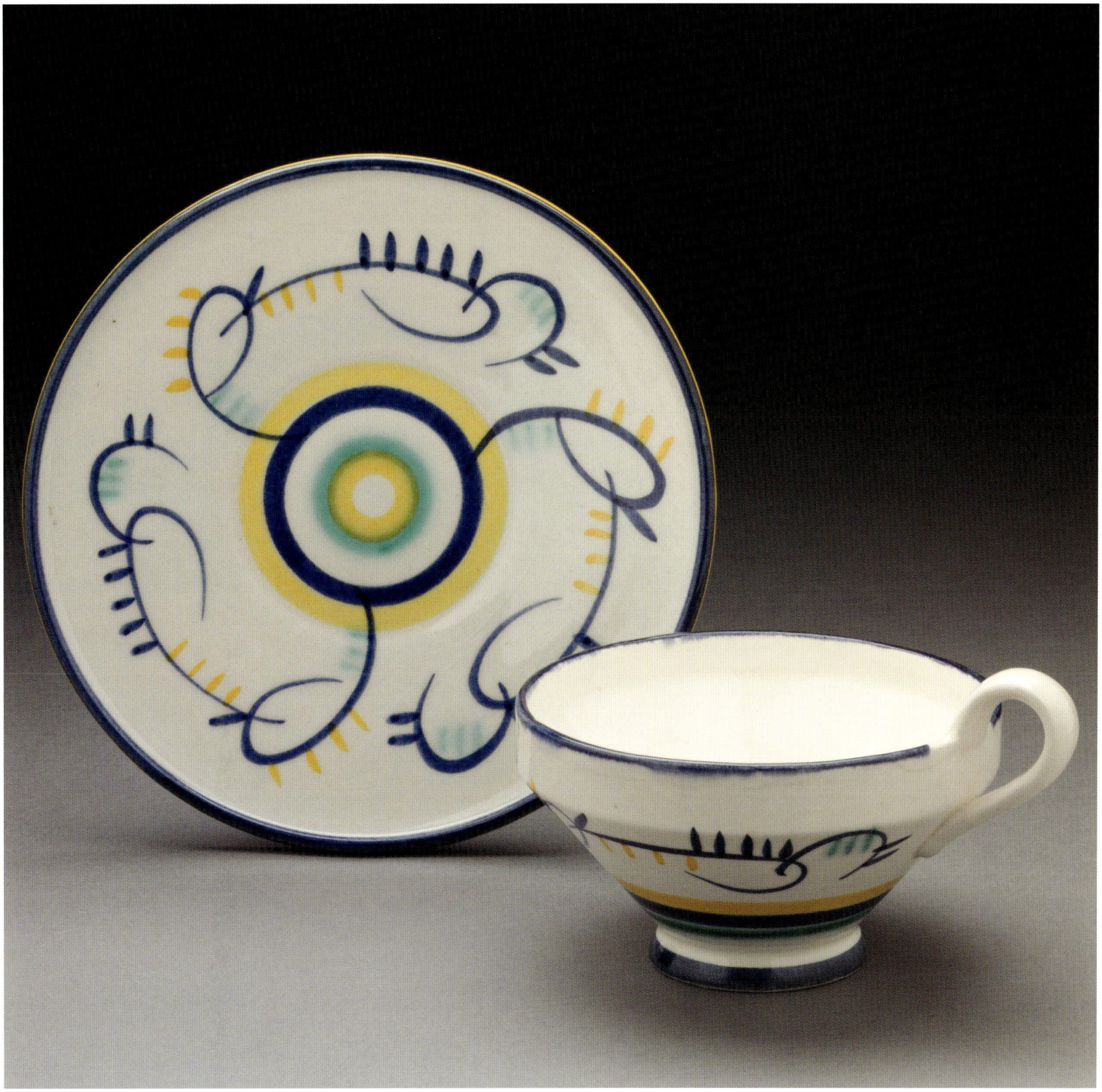

ABOVE A teacup and saucer by Grete Marks inspired by the work of Wassily Kandinsky, manufactured by Haël-Werkstätten, 1929, faience with hand-painted decoration.

ORREFORS

EDWARD HALD
(1883–1980)

EDVIN ÖHRSTRÖM
(1906–1994)

VICKE LINDSTRAND
(1904–1983)

The pre-eminent Swedish glassware company Orrefors, based in Småland, first began life in the late 19th century as a factory producing bottles and window glass. The company moved into the field of art glass after being taken over by Consul Johan Ekman in 1913. From this point onwards, Orrefors developed a tradition of collaborating with artists and sculptors on the design of its collections, forging a series of highly productive relationships that introduced fresh thinking and new techniques throughout the course of the 20th century. The Twenties and Thirties, in particular, formed a vibrant period, during which the combination of experimentation, technical expertise and artistry took the reputation of the company to a new level and helped establish Sweden as a key focal point in the evolution of modernist glassware.

Edward Hald was a Swedish painter, who trained in Stockholm, Copenhagen, Vienna and Paris, and was married to a studio potter. In 1917, he started designing ceramics (for Rörstrand) and also began working for Orrefors, around the same time as fellow artist Simon Gate. During the Twenties Hald designed a number of pieces of engraved clear glass, including his Fireworks bowl of 1921 and his African vase of 1926; many of Hald's engraved pieces were highly pictorial, with a narrative thread, and suggested an Art Deco* influence.

In 1933, Hald became both managing director and artistic director of Orrefors. As well as engraving, for which Orrefors became renowned, Hald and his colleagues began exploring other techniques, particularly 'graal', in which a piece of glass was engraved or acid-etched and then reheated and encased in another outer layer, leaving the pattern floating behind this new wall. The technique was first developed at Orrefors around 1916 and evolved over the decades that followed, with Hald making the most of the graal method in more colourful vases of the Thirties, decorated with abstract geometrical patterns, as well as his famous Fish Graal designs, where a pattern made of seaweed and ethereal fronds – with silhouetted fish swimming among them – appeared to float within the layers of glass, as though captured and encased.

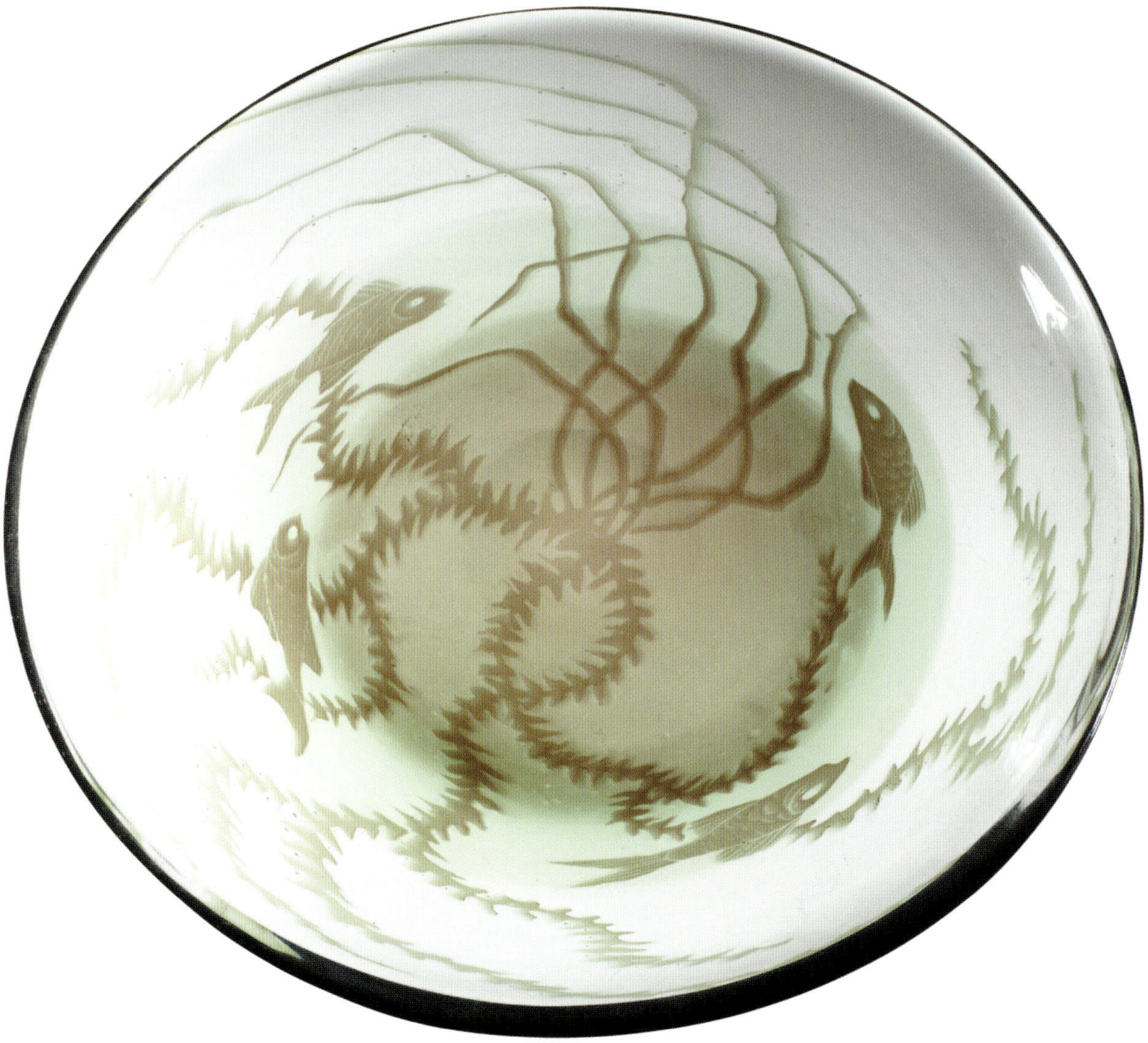

RIGHT A Fish Graal glass plate by Edward Hald, produced by Orrefors, 1939.

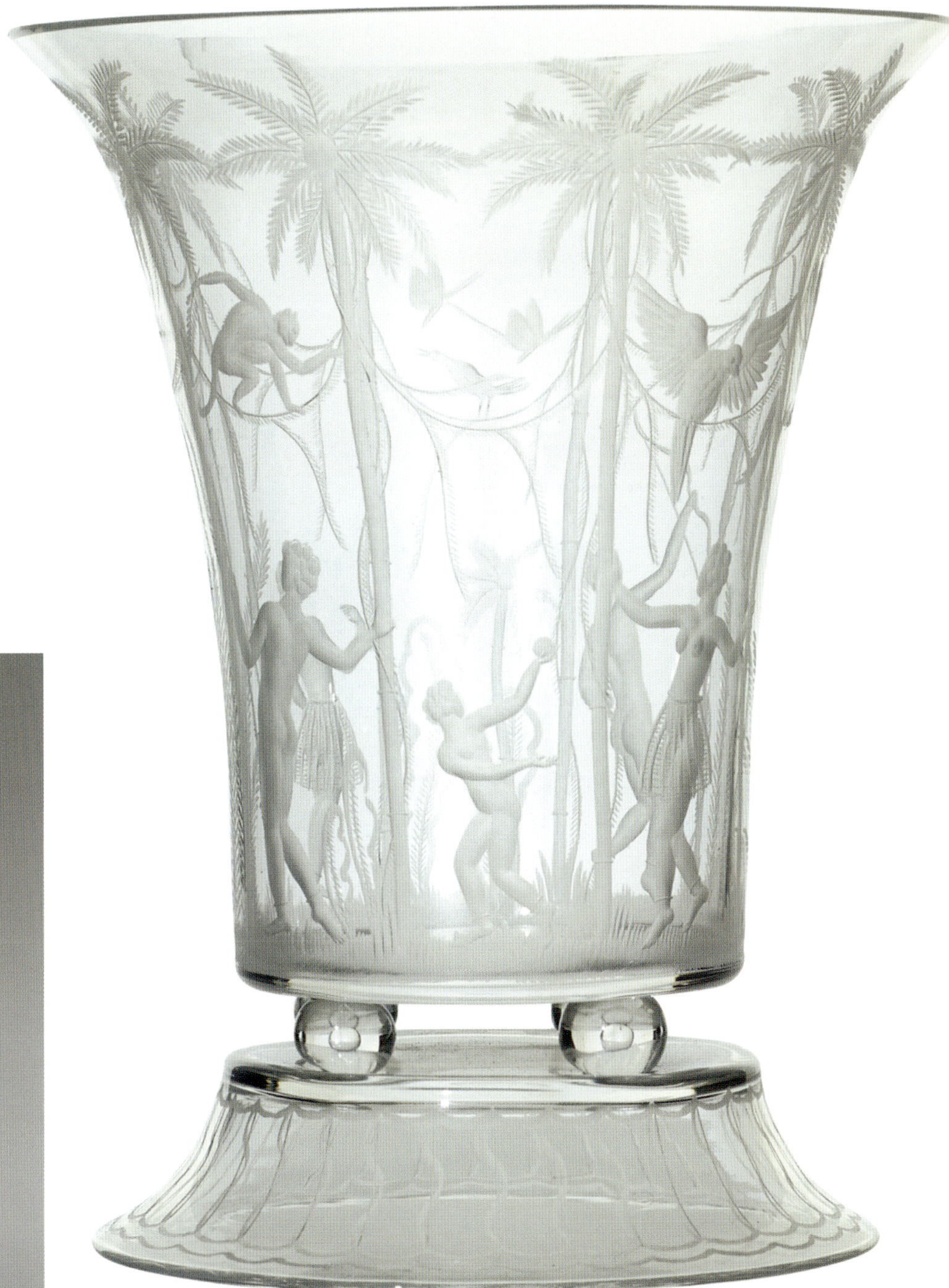

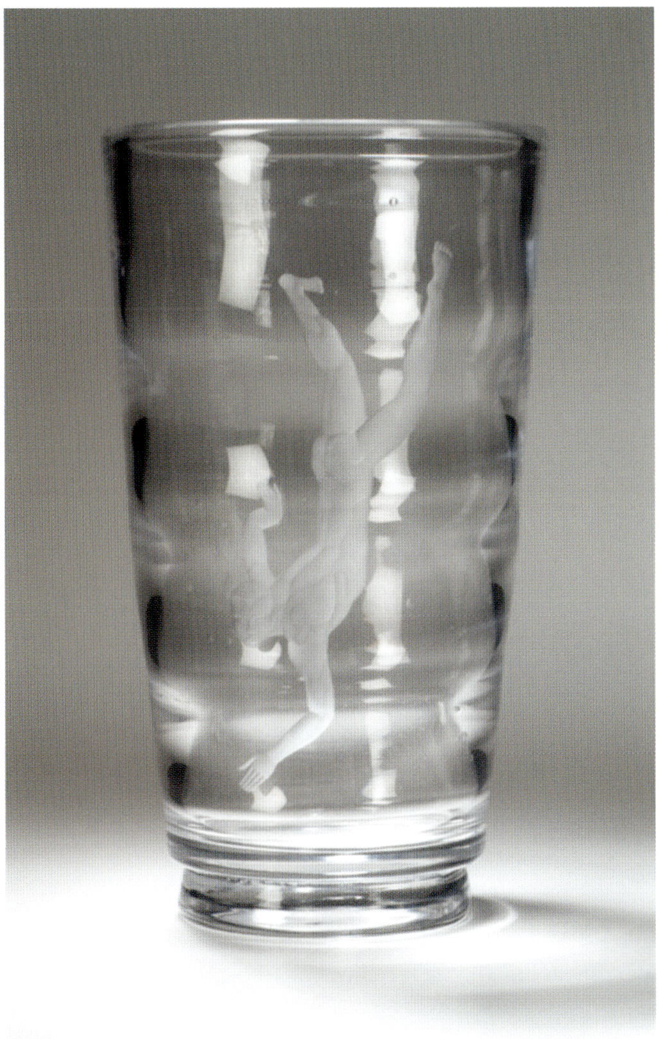

ABOVE An African vase designed by Edward Hald, made by Orrefors, 1926, in engraved glass.

LEFT A Pearl Diver vase by Vicke Lindstrand for Orrefors, 1934, in etched glass.

The sculptor Edvin Öhrström also started working with Orrefors during the Thirties. Having trained at the Royal University College of Fine Arts in Stockholm, Öhrström began contributing ideas to Orrefors from 1932 onwards, helping to develop a new technique – sometimes known simply as 'Edvin' – which marked a new stage in the evolution of the graal method, with more fluid, abstract patterns captured alongside air trails within the encased layers of glass. During the late Thirties and early Forties, Öhrström experimented with coloured glass and patterns populated with favoured motifs, including shells, sea urchins, doves and human figures. The work was mostly in the form of vases, but later Öhrström also designed plates and bowls.

A third key figure of the period was Vicke (Viktor) Lindstrand, who worked alongside Öhrström and Orrefors glass blowers to perfect collections of layered glassware, which became known as the 'Ariel' series (named after the spirit character in Shakespeare's *The Tempest*). Lindstrand studied graphic design in Gothenburg, Sweden, before working as an illustrator and joining Orrefors in 1928. Early Lindstrand designs featuring wheel-engraved figures or motifs on clear or semi-clear glass had a Deco flavour, but Lindstrand adopted the Ariel technique with enthusiasm and rigour, developing a highly successful and influential series of vases employing colour, sculptural forms and fluid patterns.

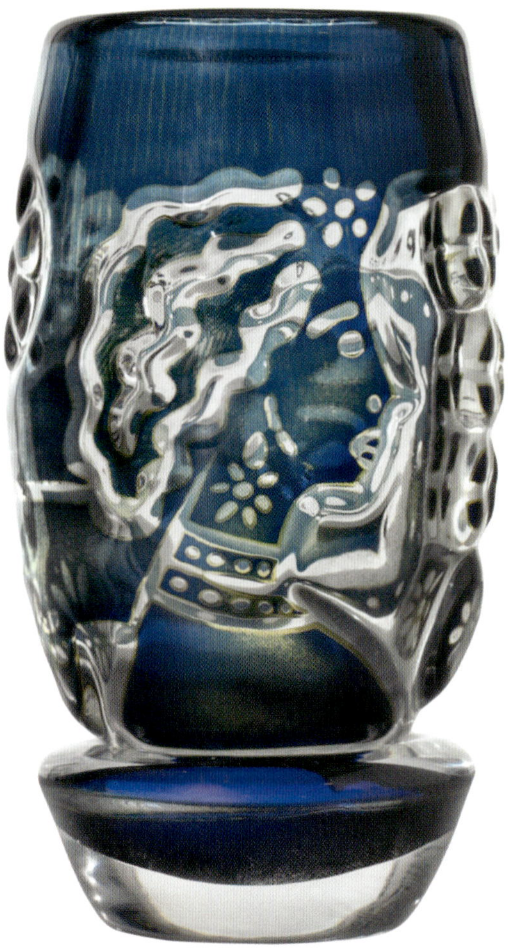

ABOVE An Edvin glass vase by Edvin Öhrström for Orrefors, 1944.

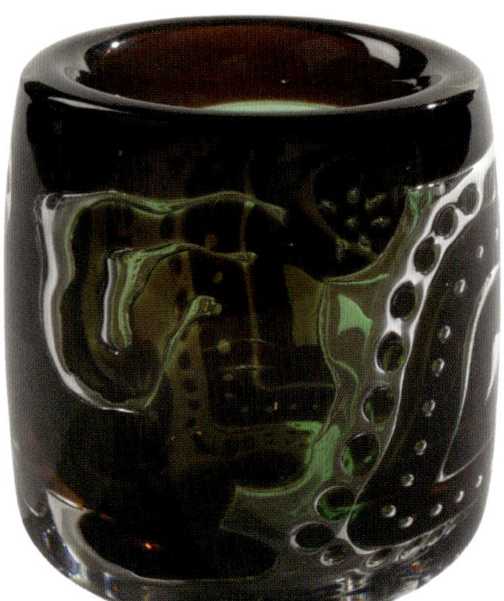

ABOVE An example of an Ariel vase by Edvin Öhrström, Orrefors, 1937.

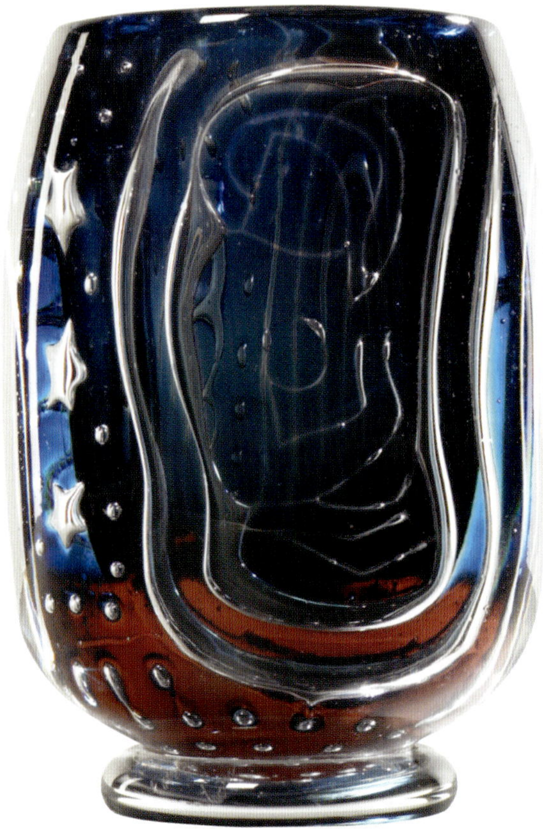

ABOVE An Ariel vase by Edvin Öhrström, made by Orrefors, 1939.

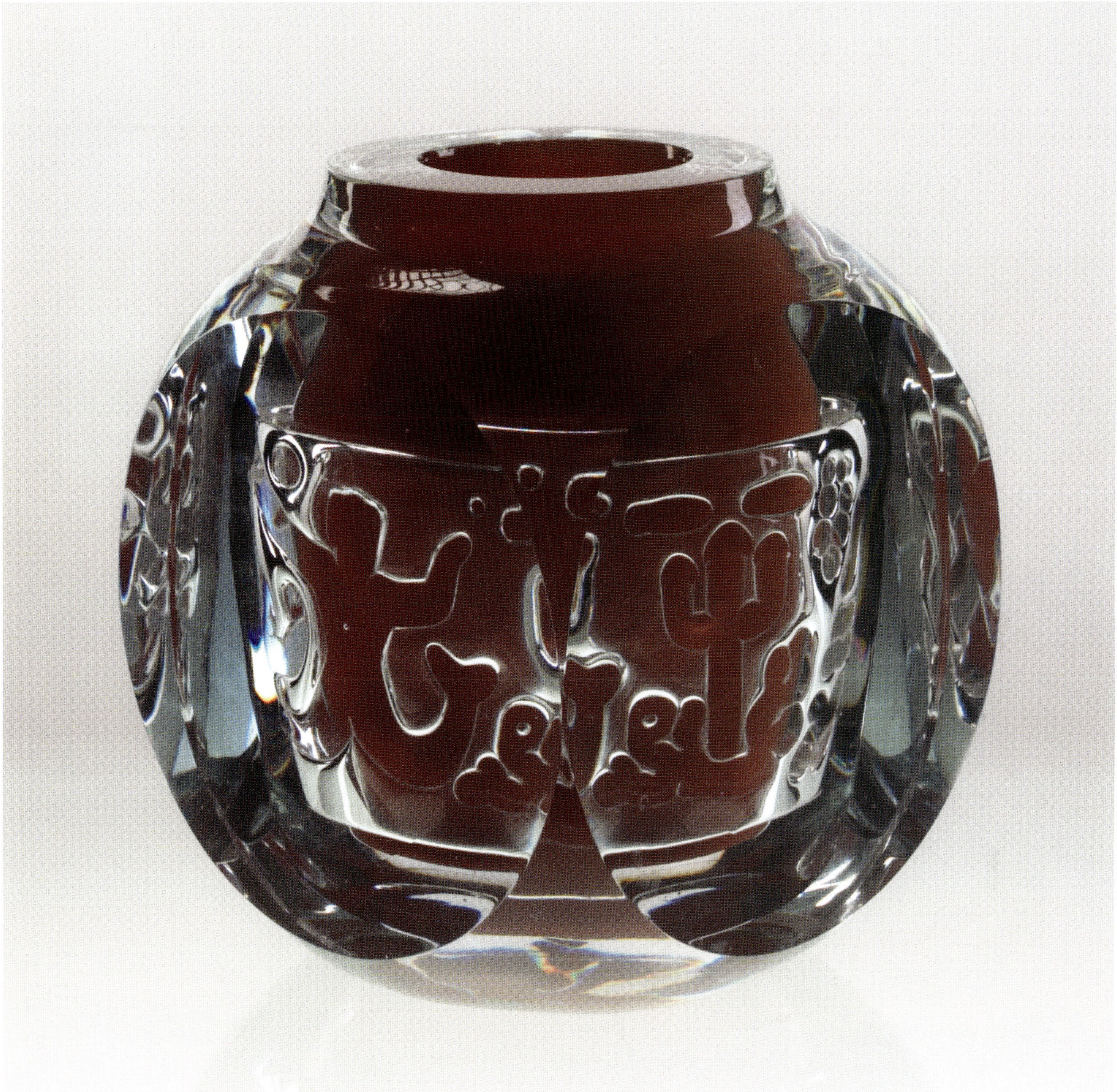

ABOVE An Ariel vase designed by Edvin Öhrström, made by Orrefors, 1939, with internally decorated glass.

PAOLO VENINI
(1895–1959)

CARLO SCARPA
(1906–1978)

The Italian entrepreneur Paolo Venini helped to revolutionize the Venetian glass industry not only by developing new forms and glass-making techniques, but also by engaging in a series of highly productive and creative collaborations. As well as employing some of the most talented glass blowers and technicians for his factory on the Venetian island of Murano, Venini sought out talented designers who brought fresh vision and modern thinking to the factory's collections. One of the most fruitful and respected collaborations of the Thirties was Venini's association with architect and designer Carlo Scarpa, whose own practice was based in Venice.

Paolo Venini himself trained as a lawyer in Milan but was part of a family that had been involved in glass-making near Lake Como. In 1921 he co-founded a new glass-making company in Murano with antiques dealer Giacomo Cappellin, introducing a fresh approach and ethos to this long-established hub of Italian glass manufacture. A few years later, Cappellin and Venini separated, and both went on to establish their own independent glassware enterprises.

From the Twenties onwards, Venini combined technical innovation and key relationships with artistic directors and designers. Early Venini art directors included sculptor Napoleone Martinuzzi (1925–31) – among whose key designs were the Soffiati (*c.* 1925–27) and Zanfirico series (1929) of vases – followed by painter and architect Tomaso Buzzi (1932–34). Martinuzzi and Buzzi helped establish Venini's reputation for fresh modernity, with Venini himself playing an increasingly important role in the evolution of new techniques and the encouragement of experimentation.

Another key innovator of the Thirties was the Swedish designer Tyra Lundgren, who initially worked in the field of ceramics before applying herself to glassware from 1933 onwards. Lundgren created a number of strikingly delicate and

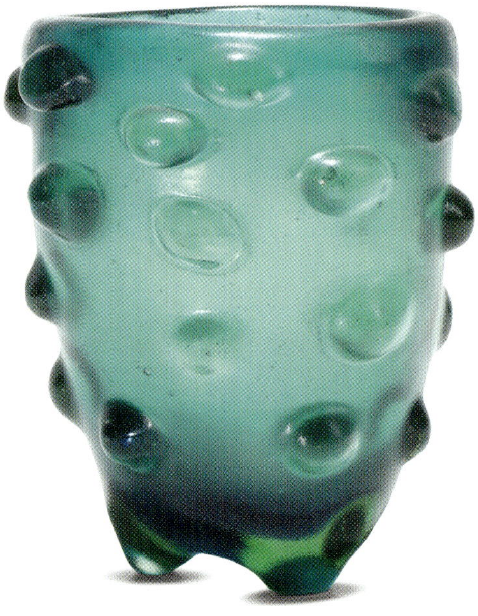

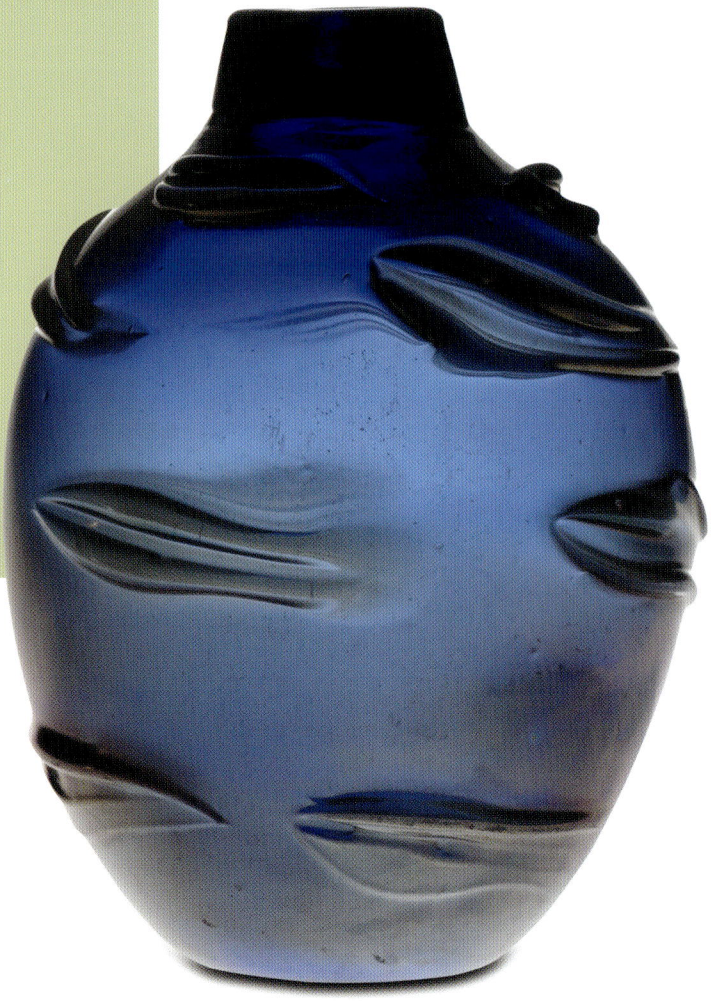

ABOVE Carlo Scarpa's Bugne vase, produced by Venini in Murano, *c.* 1936, in iridized corroso glass with decoration in relief.

LEFT A Model 3659 Rilievi vase designed by Scarpa, produced by Venini, *c.* 1935, in iridized blue and amber glass with raised decoration.

sculptural designs for Venini during the late Thirties, most notably her Filigrana leaf bowl series (*c.* 1938).

The Venetian architect and designer Carlo Scarpa first began designing glassware for Giacomo Cappellin's company, but, after the firm closed in 1932, Scarpa was drawn to Venini. Scarpa combined his work as an architect and exhibition designer with his work for Venini, who became a close friend. Scarpa began by contributing design ideas on a freelance basis before assuming the position of artistic director from 1934 to 1947.

This period of collaboration during the Thirties and early Forties became a golden era for the firm, with a powerful level of technical brilliance fused with modern forms. The range of pieces was extensive and included lighting designs, as well as vases and bowls, with these Scarpa-Venini pieces infused with a vibrant spirit and fresh character defined not only by modern forms but also by the use of colour, texture and decorative

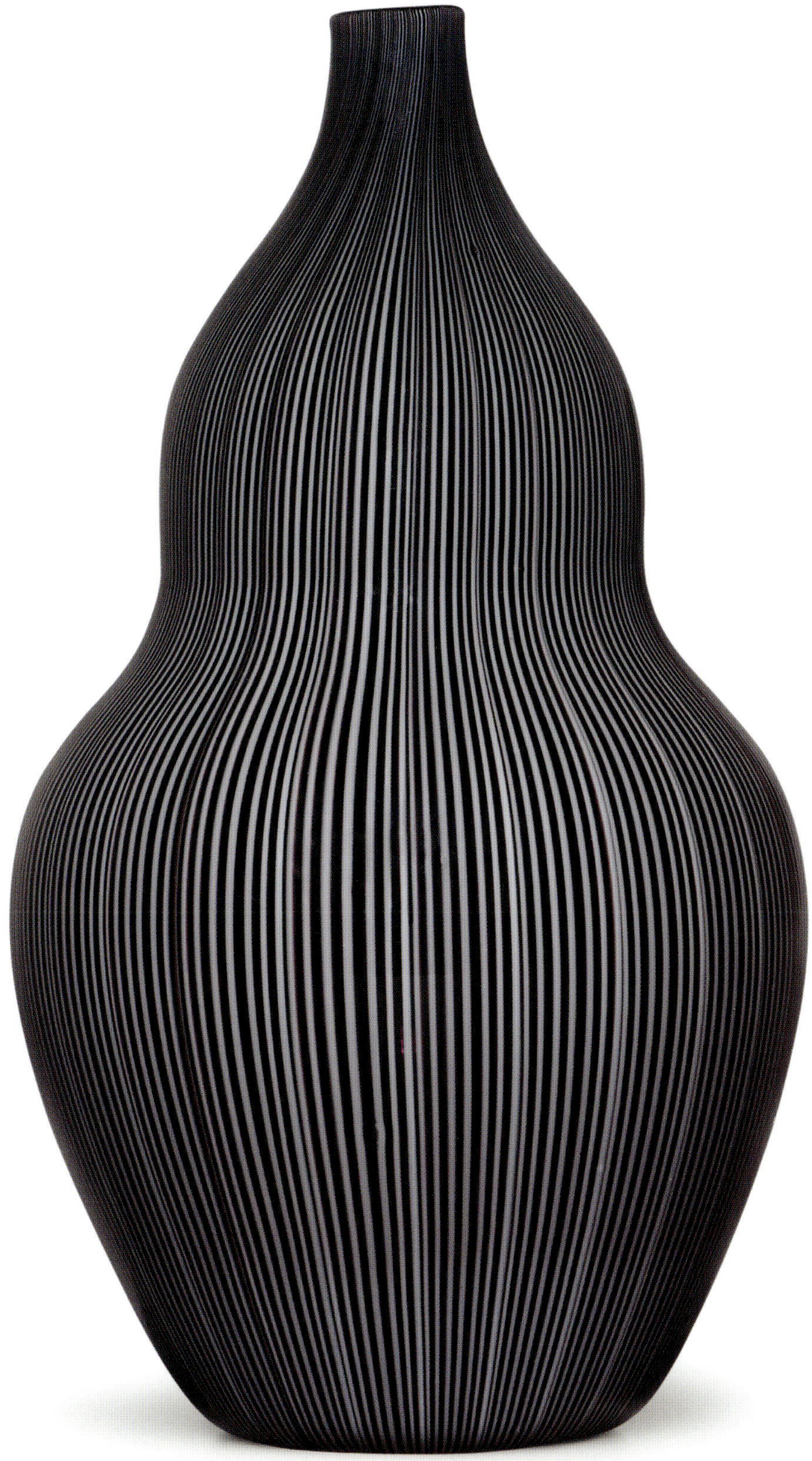

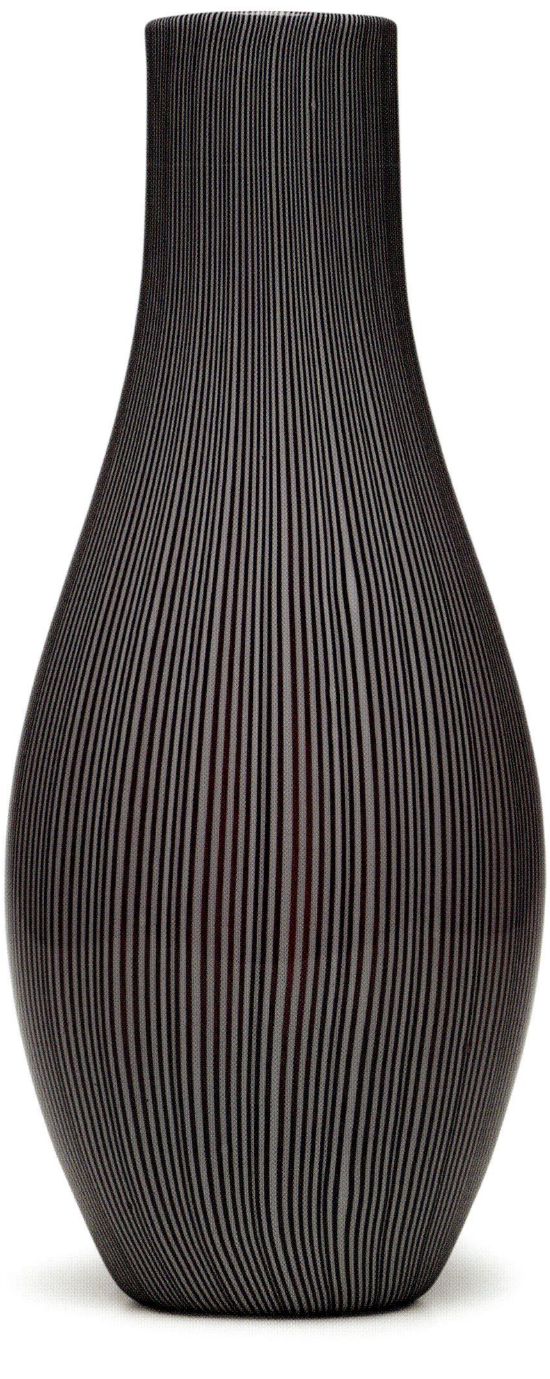

ABOVE An early Tessuto vase by Carlo Scarpa, produced by Venini, *c.* 1940, made with fused polychrome glass strands.

LEFT A Model 3547 Tessuto vase designed by Scarpa, made by Venini, *c.* 1945, with fused polychrome glass canes.

pattern, including stripes, bands, folds and waves. The level of artistry and innovation expressed within little more than a decade was extraordinary.

In terms of form, among Scarpa's most striking designs were his Diamante and Diamante Obelischi vases (*c.* 1935), which looked like twisted pieces of glacial ice, and his shell-shaped Conchiglie vase of *c.* 1942. In terms of texture, the Bugne series (1936) was a particular delight, with the surface of the vases coated with irregular lumps and bumps, like a pearlescent seed pod; the Battuto collection (1940) featured a dimpled surface made by wheel-carving the glass.

Colour and pattern were fully explored in designs such as the Tessuto range (*c.* 1945), with its fine candy cane stripe, and the Sommerso series (*c.* 1934), which featured fragments of gold leaf contained within the glass itself. Scarpa and Venini's Murrine collection (*c.* 1936) used rectangular cross-sections of coloured glass cane (known as 'murrine') in ordered geometrical patterns.

After World War II, Venini continued to thrive and innovate, enjoying another marked period of success during the mid-century period, with designs by Gio Ponti, Fulvio Bianconi and others. Scarpa's architectural reputation also flourished during the post-war period, and he added furniture design to his portfolio during the Sixties and Seventies.

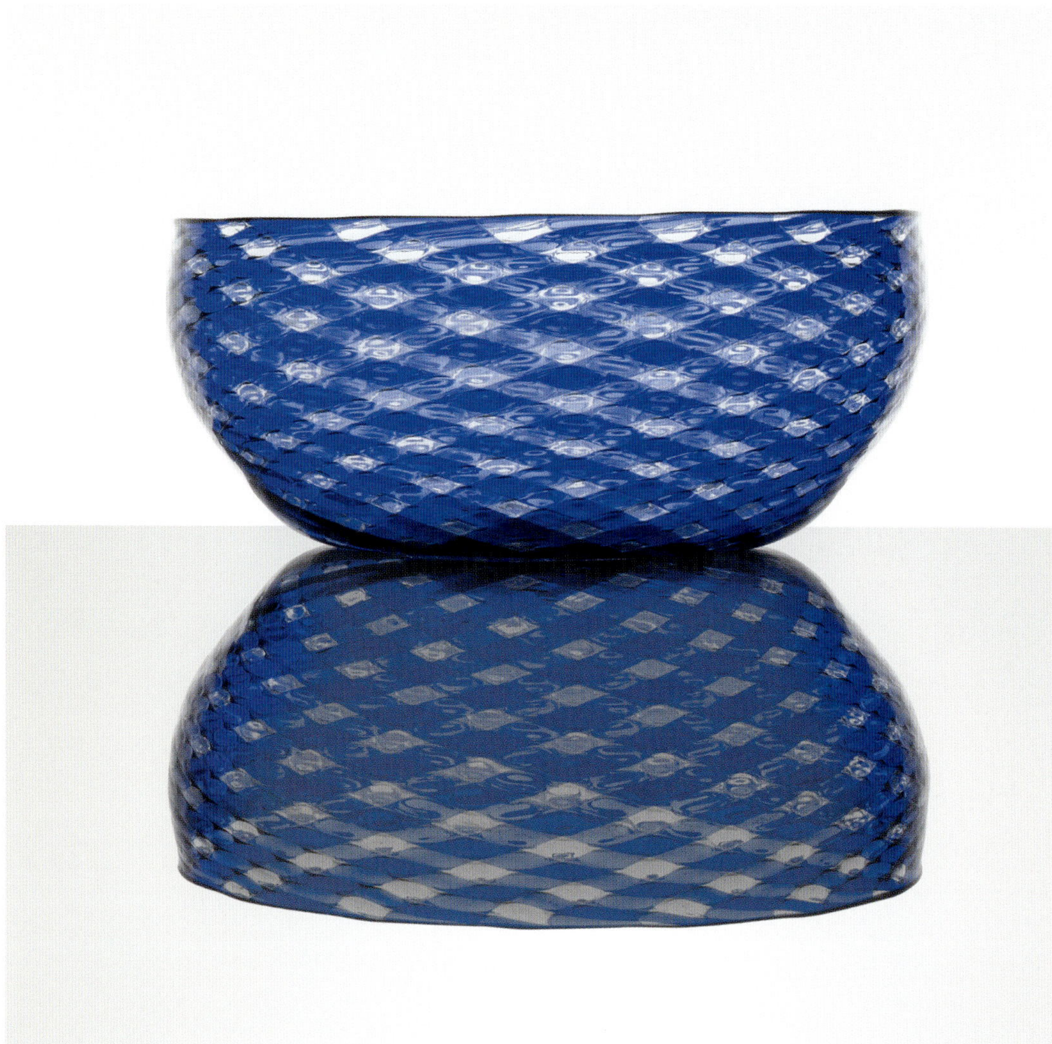

ABOVE Carlo Scarpa's Reticello bowl, produced by MVM Cappellin, 1927.

OPPOSITE A Tessuto vase by Carlo Scarpa, made by Venini, *c.* 1940.

BELOW A Model 4003 Murrine Romane bowl designed by Paolo Venini and Carlo Scarpa, made by Venini, *c.* 1936, with rectangular coloured glass murrines.

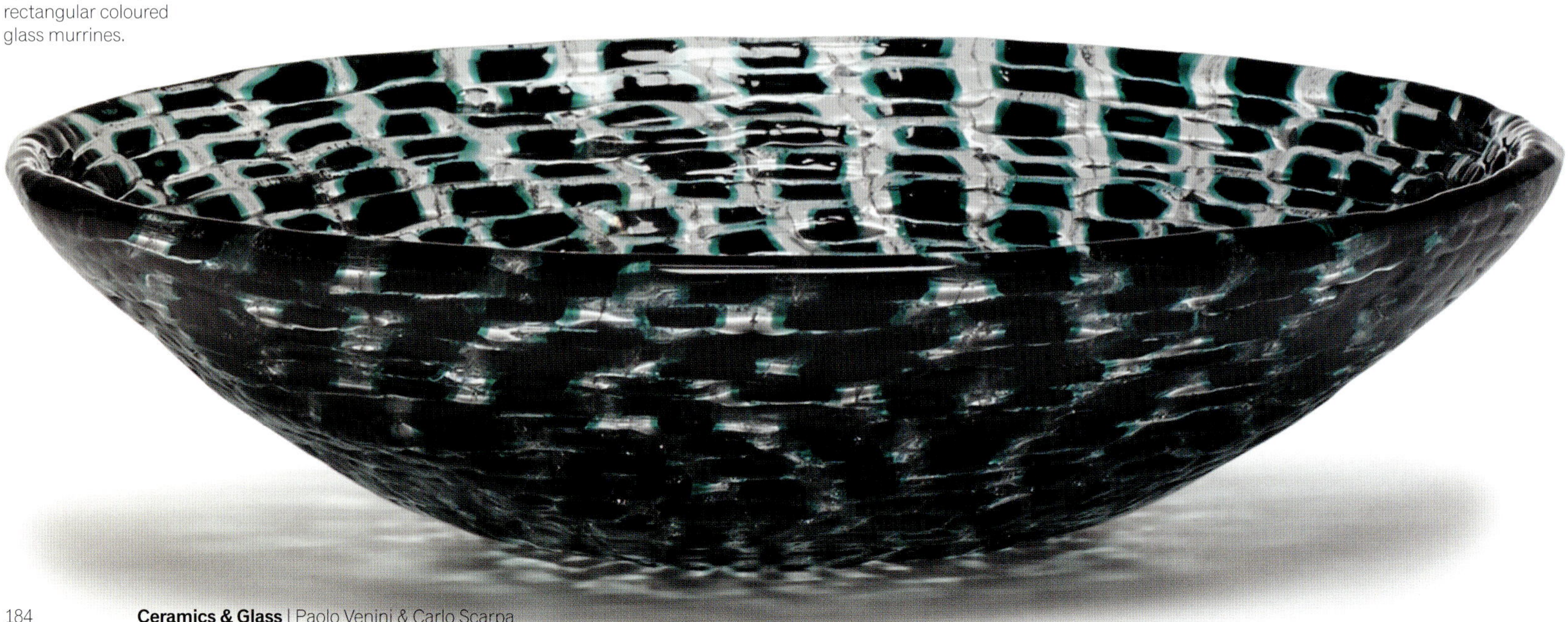

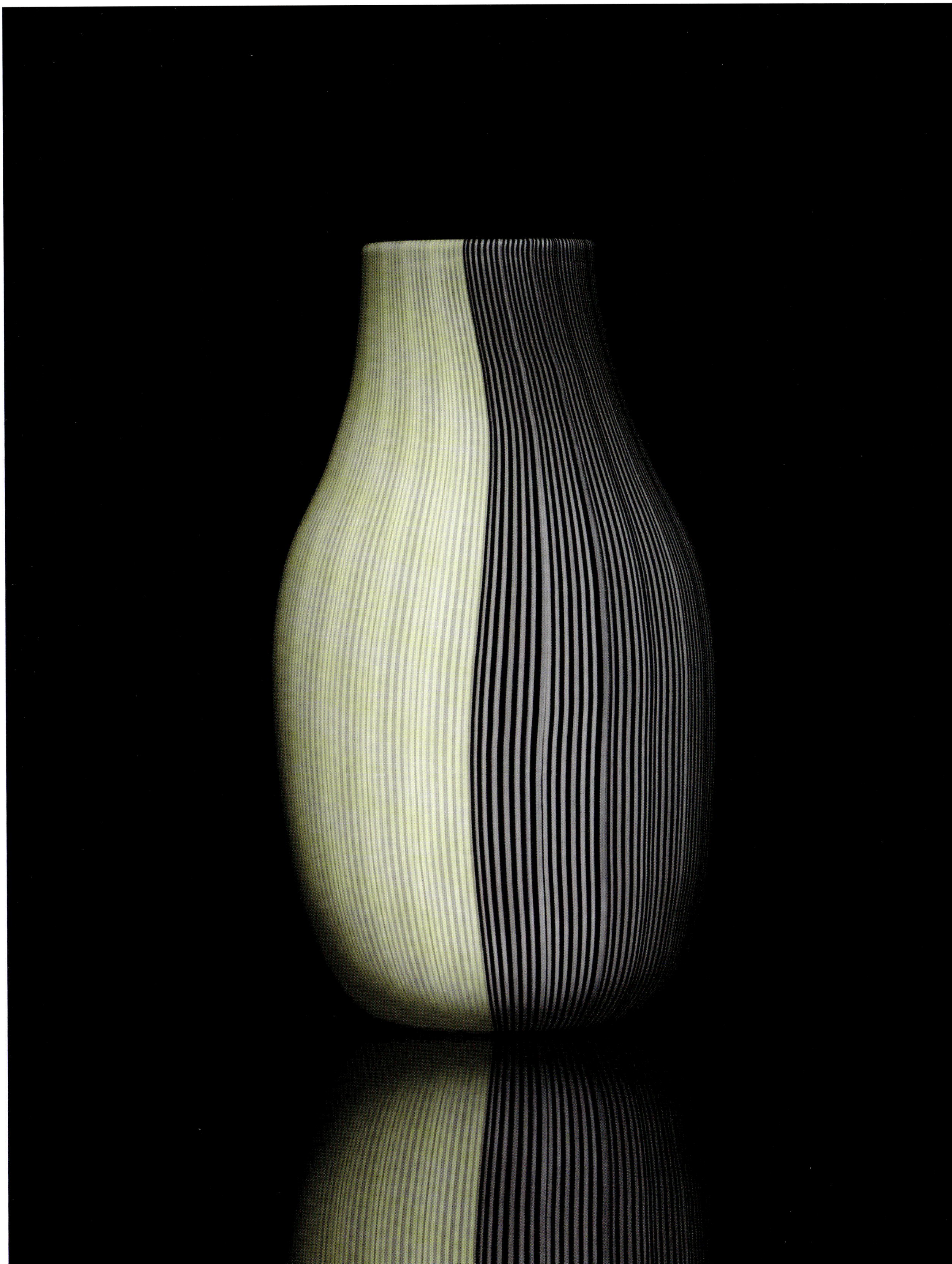

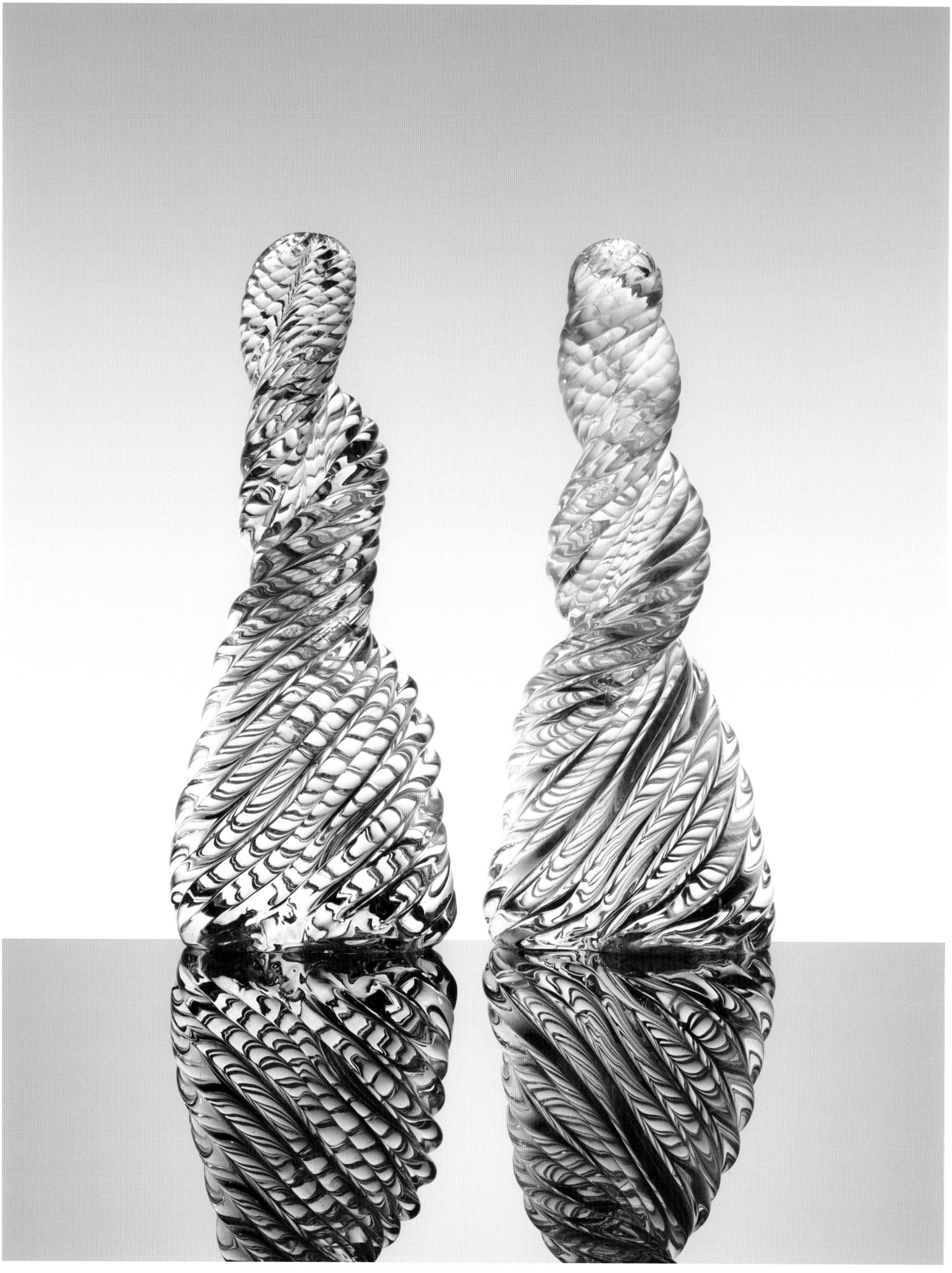

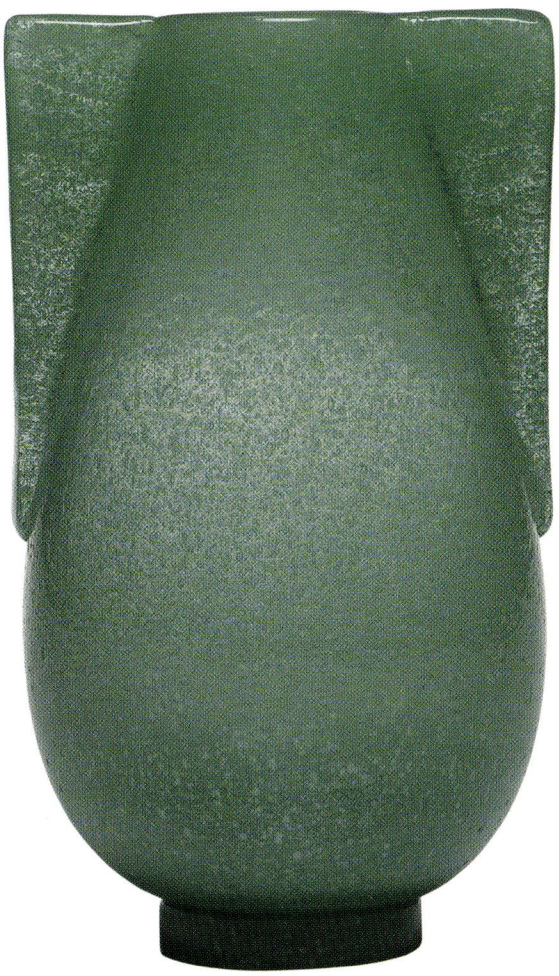

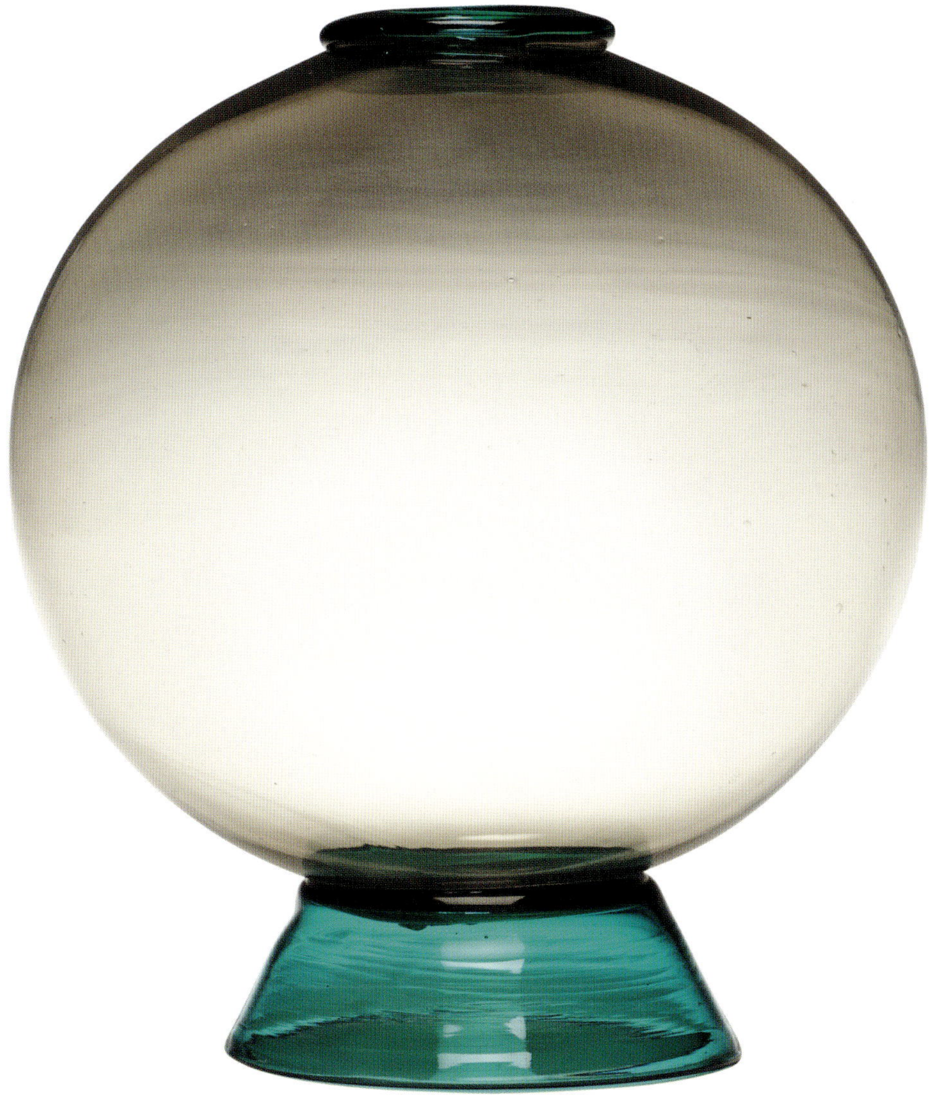

ABOVE Carlo Scarpa's Transparenti vase, produced by MVM Cappellin, 1926, in soffiato glass.

BELOW A Model 1351 Conchiglie vase, designed by Carlo Scarpa, made by Venini, *c.* 1942, in iridized glass.

ABOVE A Model 11016 Bollicine vase by Carlo Scarpa, produced by Venini, *c.* 1932.

OPPOSITE A pair of Model 9034 Diamante Obelischi pieces by Paolo Venini and Carlo Scarpa, made by Venini, 1934.

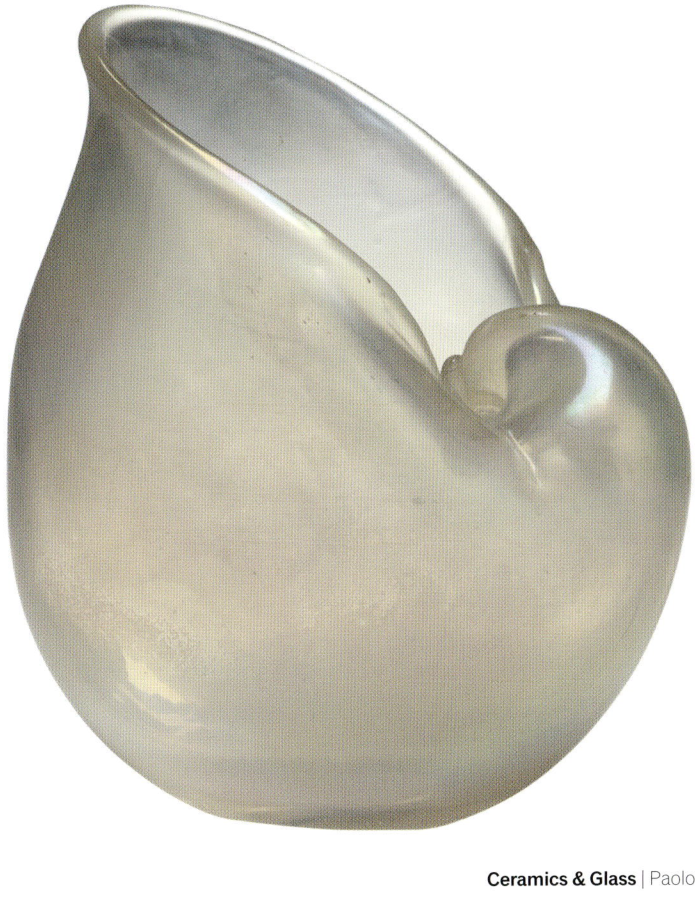

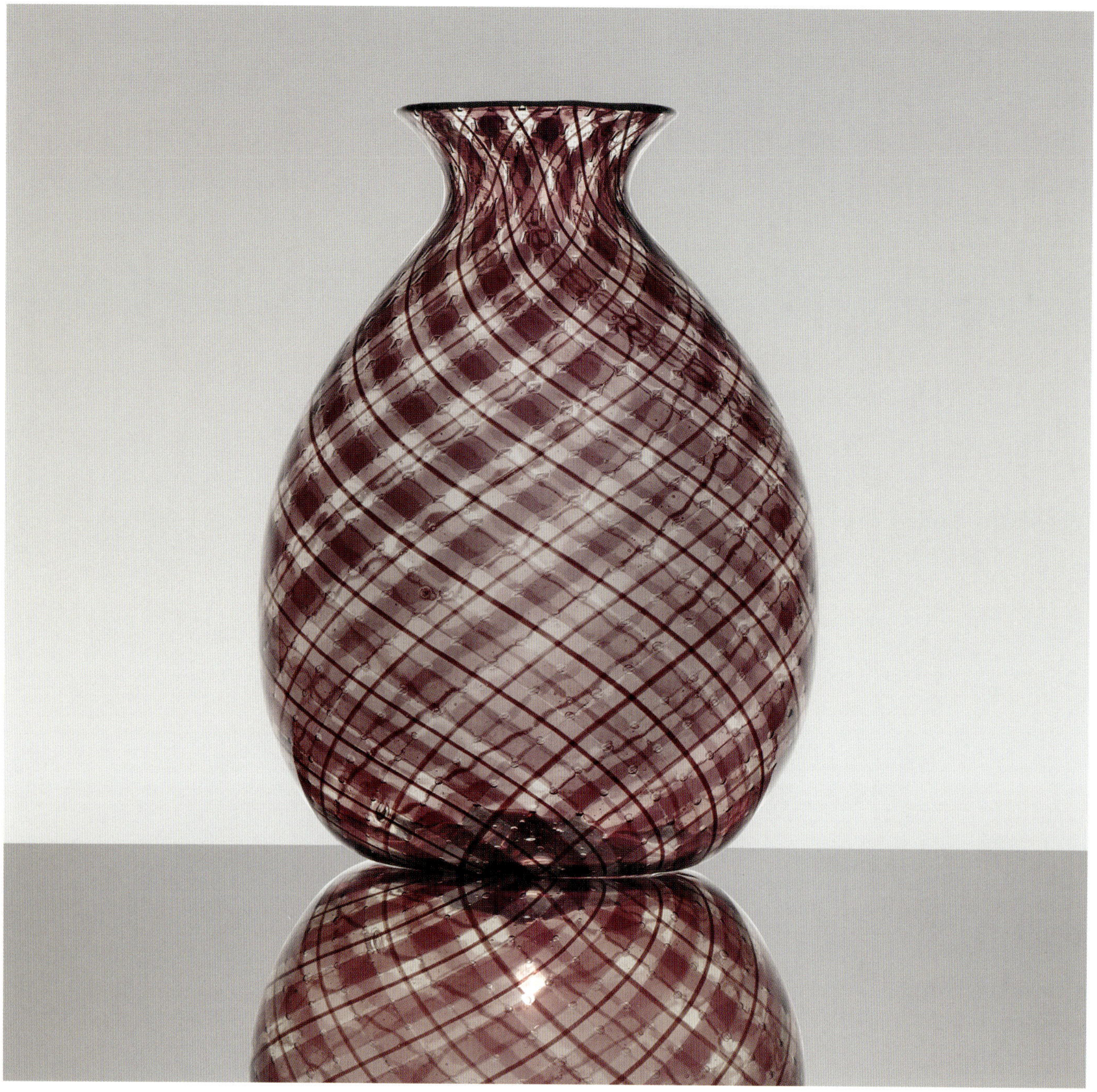

ABOVE A Reticello vase by Carlo Scarpa, produced by MVM Cappellin, *c.* 1927, with alternating glass canes.

RIGHT A Model 2938 Battuto bowl by Scarpa, made by Venini, 1940.

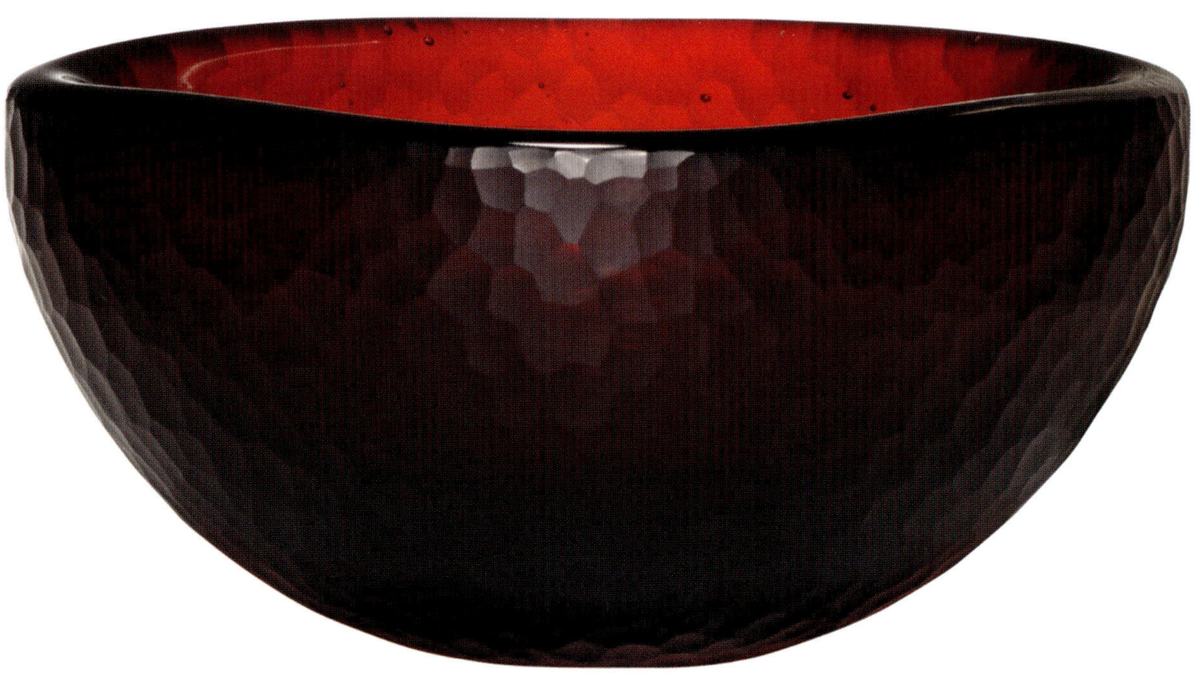

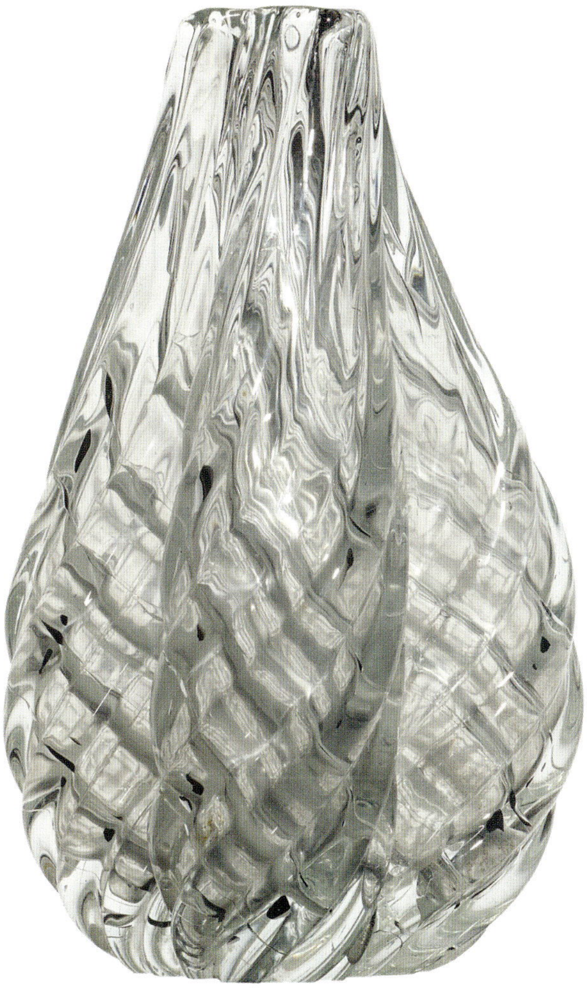

ABOVE A Diamante vase by Paolo Venini and Carlo Scarpa, made by Venini, *c.* 1935.

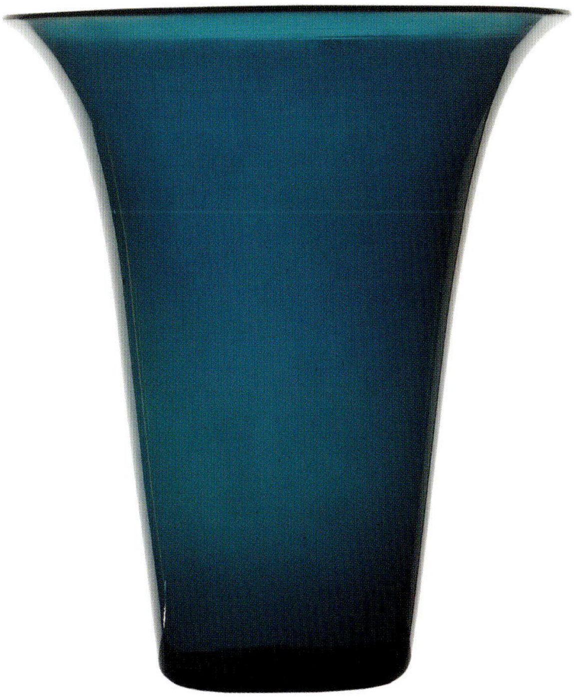

ABOVE A Model 3781 Iridato vase, designed by Carlo Scarpa, produced by Venini, 1940.

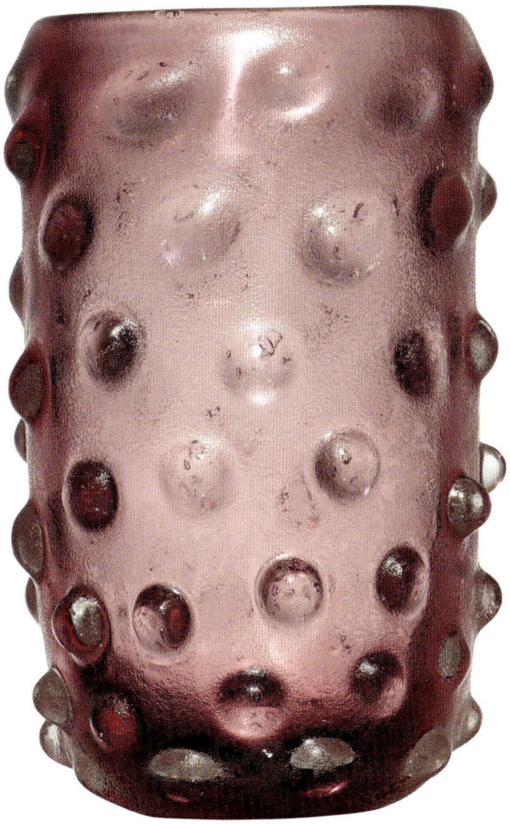

LEFT Scarpa's Bugne vase, produced by Venini, 1936, in iridized corroso glass.

MARGUERITE WILDENHAIN
(1896–1985)

A number of former Bauhaus masters and students forged successful and highly influential second careers in the United States across a range of disciplines. However, in the field of ceramics, Marguerite Wildenhain was unique in transferring her talents and teaching methods to an American context, inspiring a new generation of students and studio potters in the post-war period.

Wildenhain's early life was somewhat peripatetic. Born Marguerite Friedlaender, she had a German father and an English mother, while her place of birth was Lyon in France and she took French citizenship. She was educated in England and Germany before studying sculpture at the University of the Arts in Berlin. After finding work in a porcelain factory, Wildenhain decided to enrol at the Bauhaus in Weimar in 1919, becoming one of the new school's first students.

Following her foundation course at the Bauhaus, Wildenhain progressed to the ceramics workshop, which was situated in the nearby town of Dornburg. Here, her master of form ('Formmeister') was Gerhard Marcks* and her master of craft ('Lehrmeister') was Max Krehan*. The years that Wildenhain spent with Marcks and Krehan were to shape her approach to ceramics; her first book, *Pottery: Form and Expression*, was dedicated to her two Bauhaus masters. Fellow students in the ceramics workshop included Otto Lindig, Theodor Bogler, Grete Marks and her husband-to-be, Frans Wildenhain (1905–1980). She graduated in 1926 with the title of 'master potter'.

Following her graduation, Wildenhain was made head of ceramics at the Burg Giebichenstein School of Fine and Applied Art in the nearby city of Halle. Here, she designed work that would be suitable for factory production, including a dinner service and a Halle tea service collection (*c.* 1930). During the early Thirties, Wildenhain also worked for the Königliche Porzellan-Manufaktur, with her best-known pieces including her white porcelain vase of *c.* 1931, with its over-sized funnel sprouting from a modest egg-like body; the form was sculptural but also played with geometry.

After Adolf Hitler's National Socialist government took power in Germany in 1933, Wildenhain was in a vulnerable position due to her family's Jewish heritage. She and her husband moved to the Netherlands in 1937, founding a new pottery together, known as Little Jug. After World War II began, Wildenhain arranged to emigrate to America, arriving in 1940; her husband, a German citizen, was not so fortunate and remained behind, eventually being drafted into the German army.

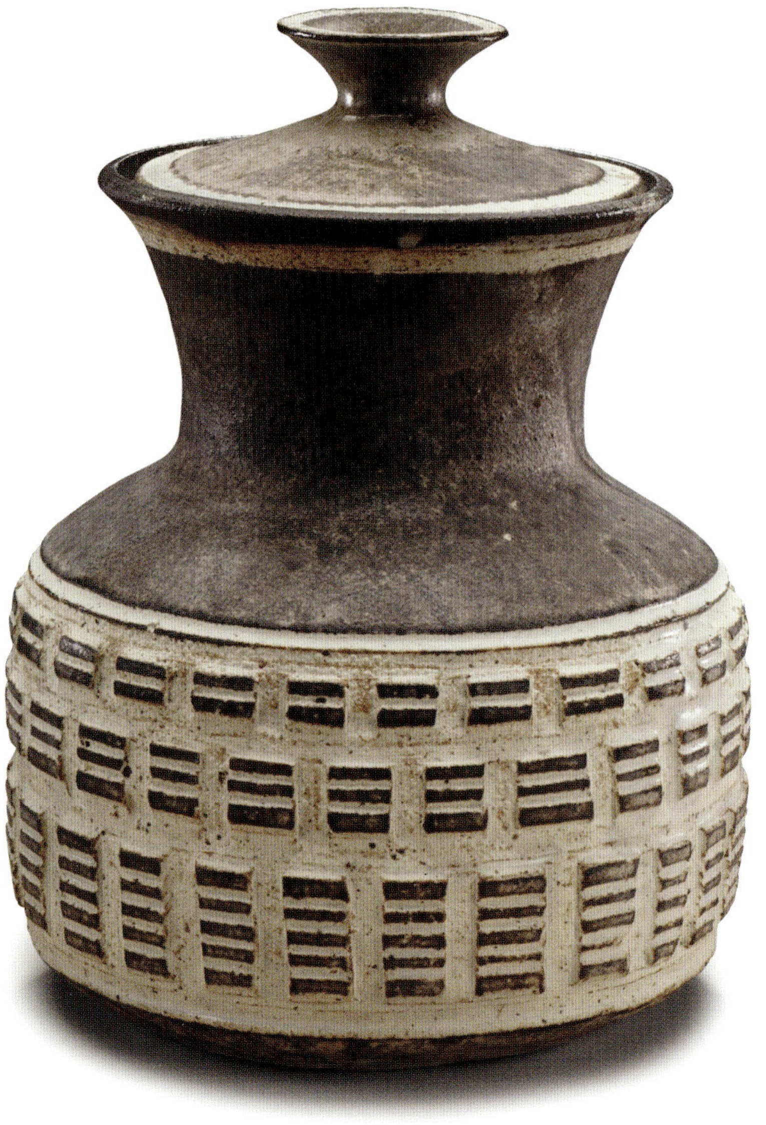

RIGHT A large earthenware jar by Marguerite Wildenhain, *c.* 1950.

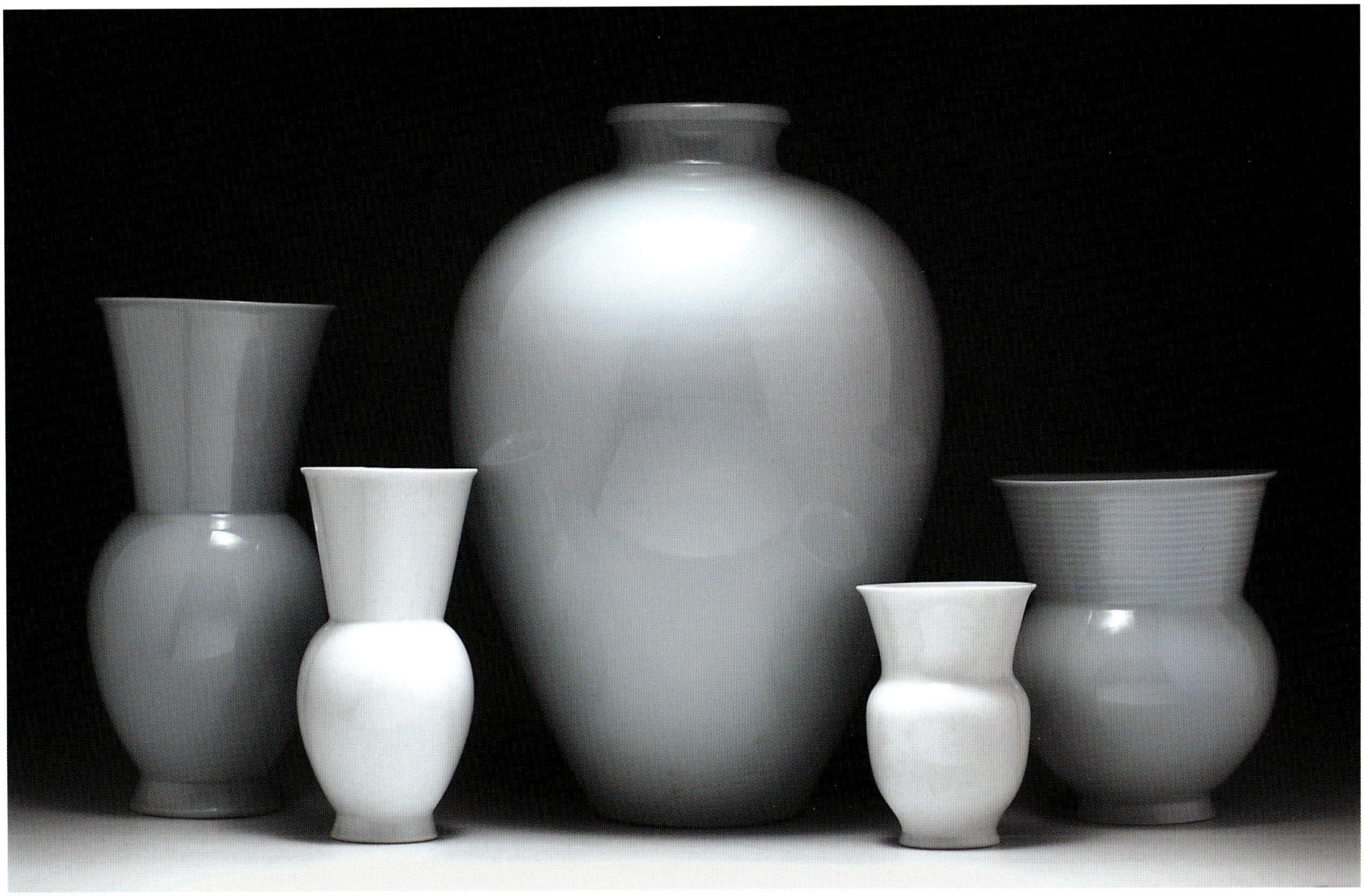

ABOVE Hallesche Form vases by Marguerite Wildenhain, manufactured in Berlin, 1931.

BELOW A stoneware vase by Wildenhain, manufactured by Burg Giebichenstein, *c.* 1922.

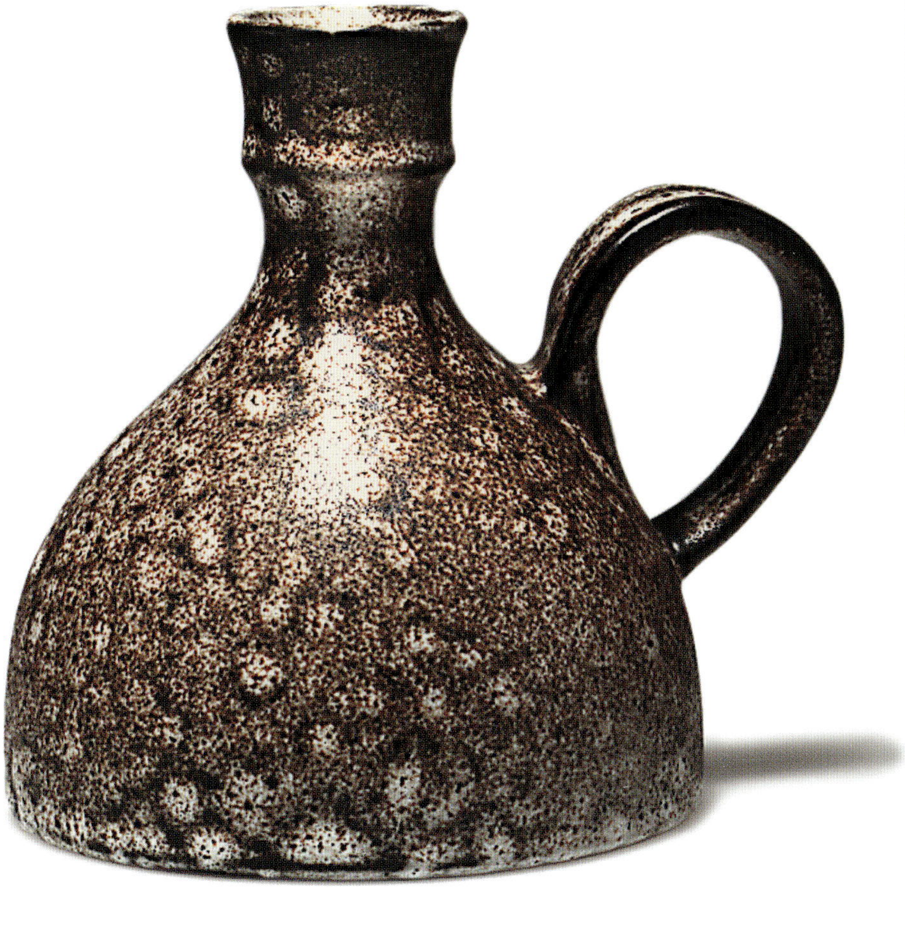

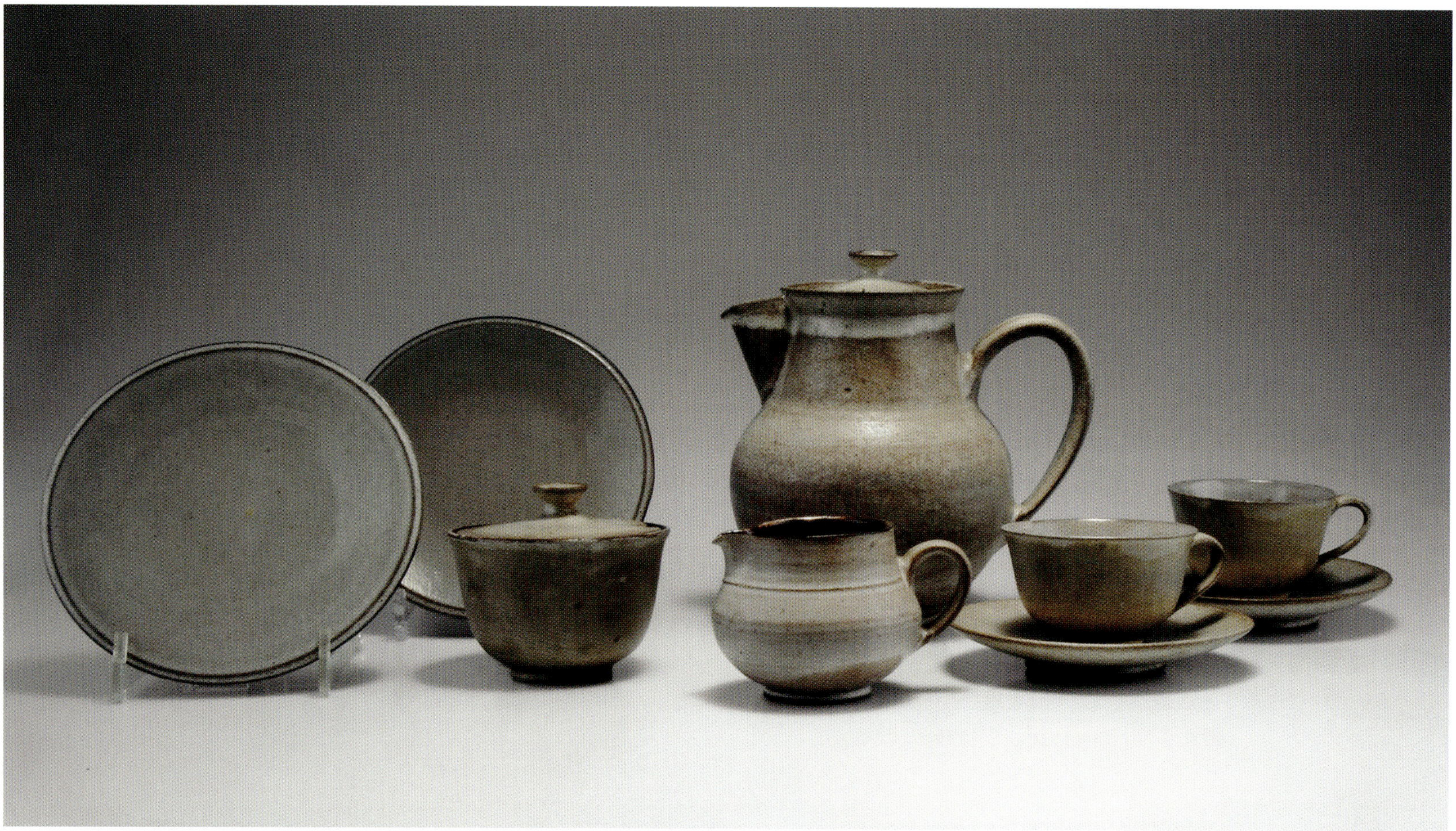

ABOVE A seven-piece coffee set by Marguerite Wildenhain, made by Burg Giebichenstein, *c.* 1930, in stoneware.

In the United States, Wildenhain took short-lived teaching posts before moving to Sonoma County, north of San Francisco, California, in 1942. Here, she settled at the artists' colony of Pond Farm at the invitation of its wealthy patrons and founders, architect Gordon Herr and his wife Jane, a writer. Situated on ranchland, which now forms part of a state park, Pond Farm included gardens and orchards run on sustainable principles, but also hosted a new house for Wildenhain – designed by Gordon Herr – and a restored barn used as a pottery studio, along with other buildings, new and old.

Under the direction of the Herr family, Pond Farm hosted a series of workshops across a number of fields of art and design. Gordon Herr ran classes in architecture, Frans Wildenhain (who arrived in 1947) taught sculpture, and Wildenhain herself ran classes in ceramics, focused on craft and technical studio skills, while other resident and visiting artists ran other courses. By 1952, the work of Pond Farm had ground to a halt after a number of family tragedies – including the death of Jane Herr – and the collapse of Marguerite and Frans Wildenhain's marriage; subsequently Frans moved on to the Rochester Institute of Technology and remarried.

The school and workshop of Pond Farm Pottery, however, survived, with Marguerite Wildenhain continuing to live, work and teach there. Her classes became increasingly well known and Wildenhain also produced three books on ceramics. The Farm became something of a place of pilgrimage for young students attending her summer school; she also became an ardent supporter and defender of American studio pottery, arguing with BERNARD LEACH, who criticized the lack of a 'tap root' within the country's art ceramics.

Much of Wildenhain's later work was hand-produced studio pottery, mostly fired with earth-coloured glazes, some with playful decoration. Yet she also continued to recognize the merits of factory production, advocated by Gropius and others at the Bauhaus.

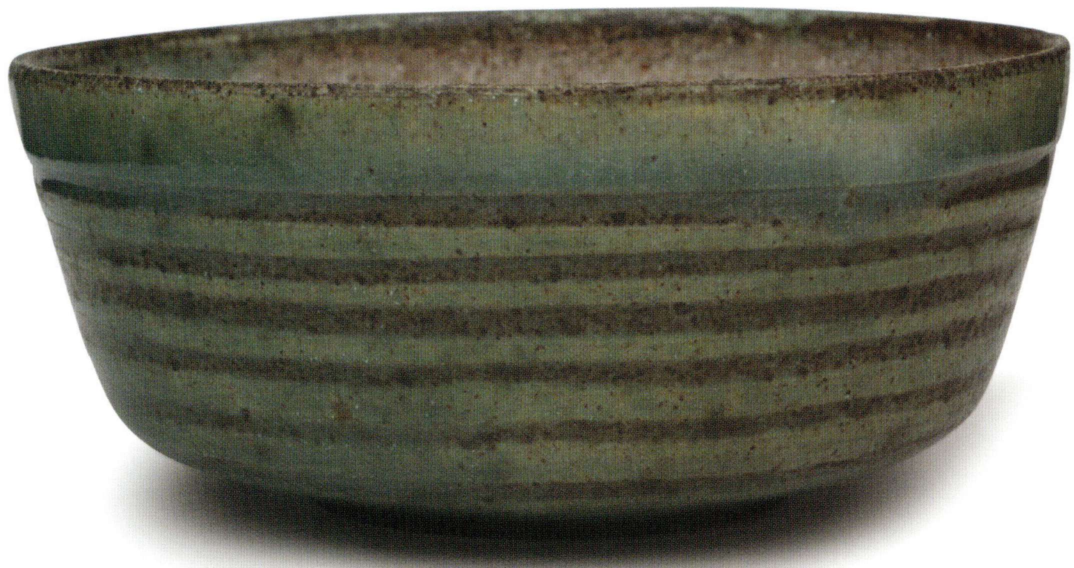

ABOVE & BELOW Two glazed bowls by Marguerite Wildenhain, 1946.

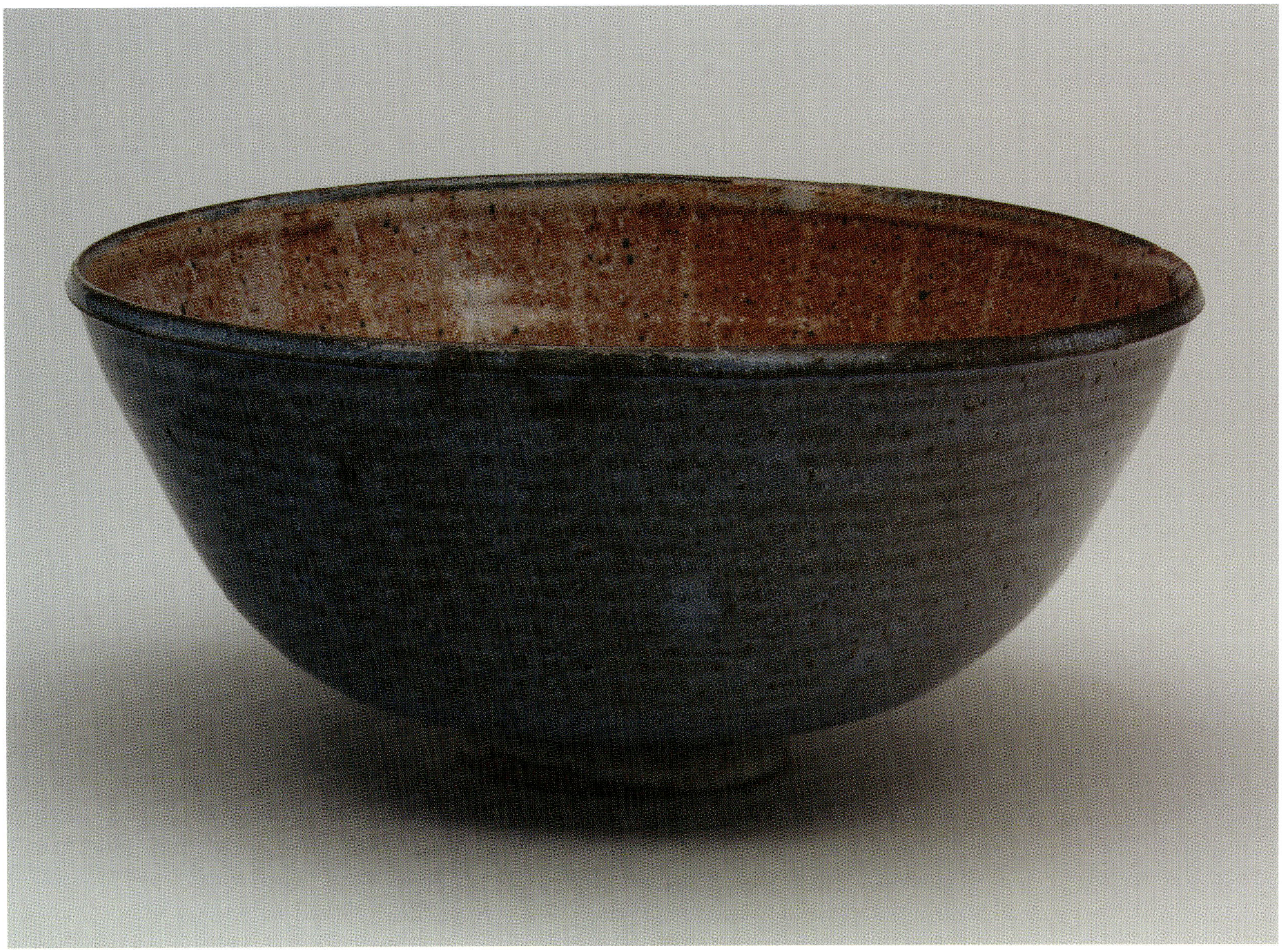

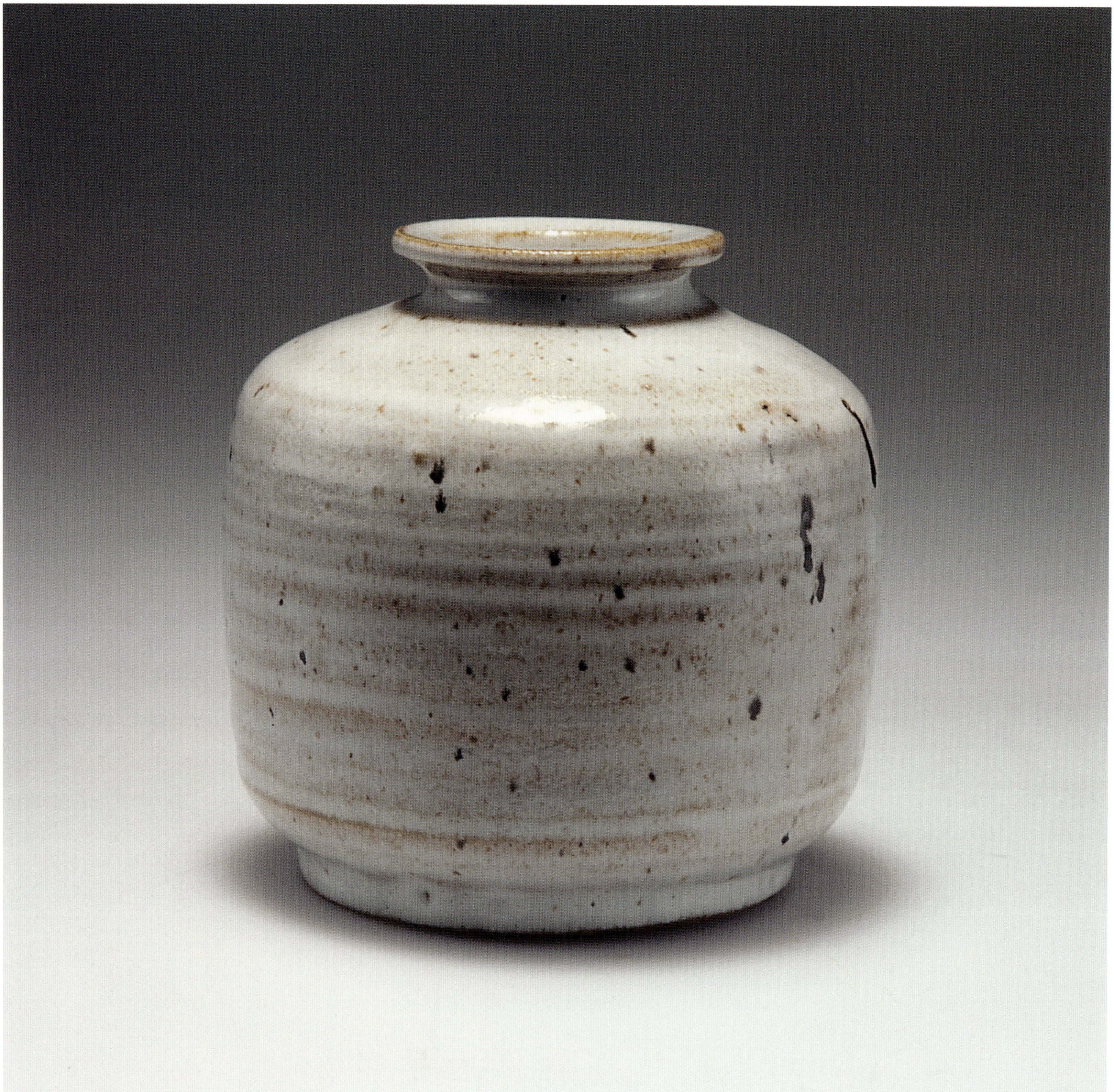

ABOVE A small vase by Marguerite Wildenhain, manufactured by Burg Giebichenstein, *c.* 1926.

OPPOSITE An earthenware pitcher designed by Wildenhain, *c.* 1950.

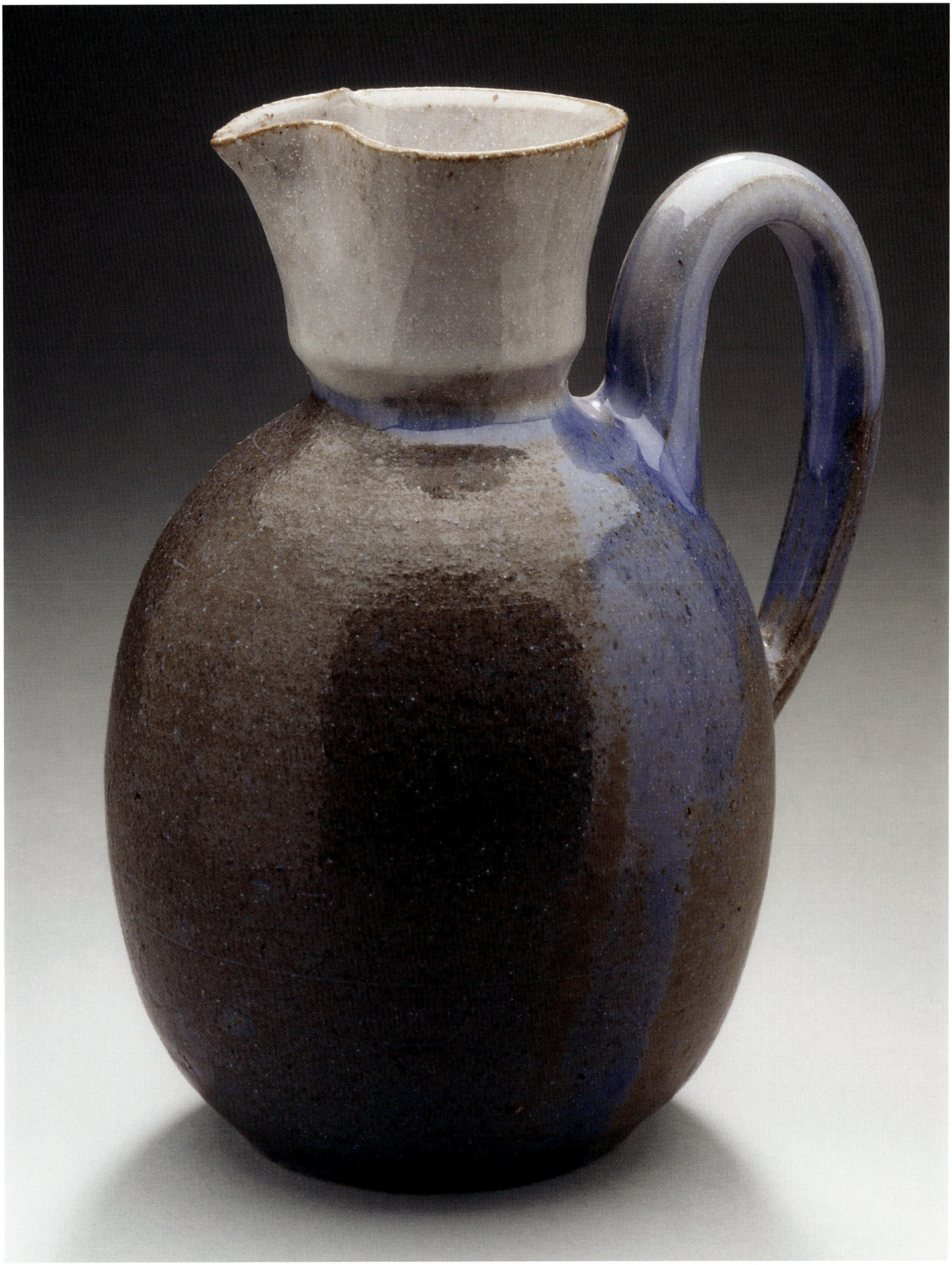

RUSSEL WRIGHT
(1904–1976)

The Wall Street Crash of 1929 and the Great Depression that followed, damaging the economy worldwide, inevitably affected the design community. Within this age of austerity, the excesses of the Jazz Age seemed inappropriate, while the rising competition within the market place saw producers and manufacturers look to make products that were not only functional and utilitarian, but also aesthetically pleasing and affordable. The American industrial designer Russel Wright embraced this challenge with intelligence and creativity, adopting the motto 'good design is for everyone' and promoting a democratic approach to the invention of new lines of furniture and homeware, particularly in the field of ceramics but also adopting glass and, later on, plastics for his collections.

Born in Lebanon, Ohio, Wright was the son of a judge, and he himself was encouraged to pursue a legal career. Yet even at high school, Wright was interested in art and design, taking classes at the Art Academy of Cincinnati followed by time at the Art Students League of New York. He took a degree in law at Princeton University, but also pursued work as a sculptor at the same time. After graduating, Wright went to work as a theatre set designer with NORMAN BEL GEDDES.

In 1927, Wright established his own design studio in collaboration with Mary Small Einstein, a sculptor and designer who he married that same year. Mary Wright played a vital part in the evolution of the business, helping to instigate new ideas but also developing the notion of a Wright design brand that could be applied across many different disciplines; she was also the co-author of the couple's highly influential book, *Guide to Easier Living*, which was published in 1950.

Designs from the late Twenties and early Thirties included vessels, vases and cocktail sets in spun aluminium, and silver-plated flatware; from relatively early on his working life, much of Wright's work focused on the idea of the dining table as the focal point of the home. In

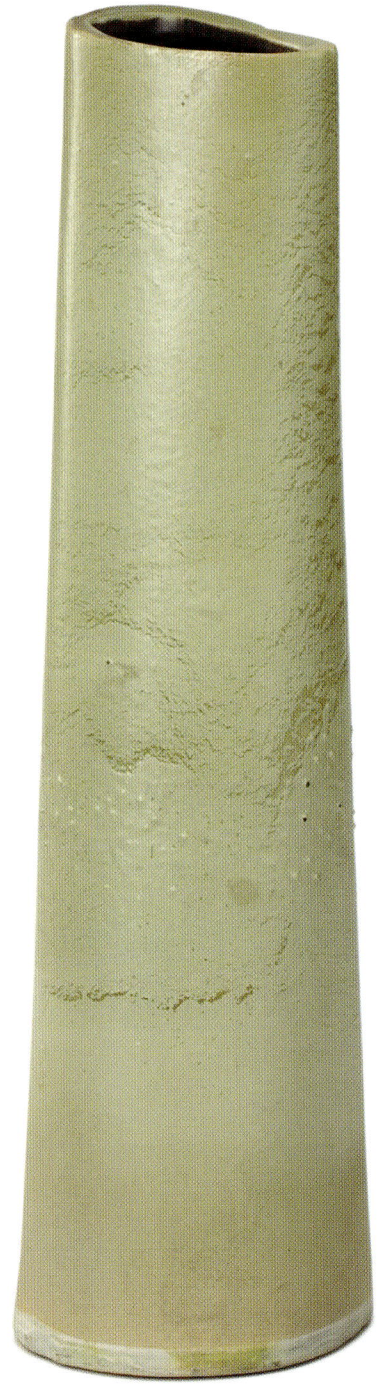

RIGHT A floor vase, designed by Russel Wright and made by Bauer, *c.* 1946, in glazed stoneware.

BELOW An American Modern salt and pepper set, produced by Steubenville, *c.* 1939, in glazed stoneware.

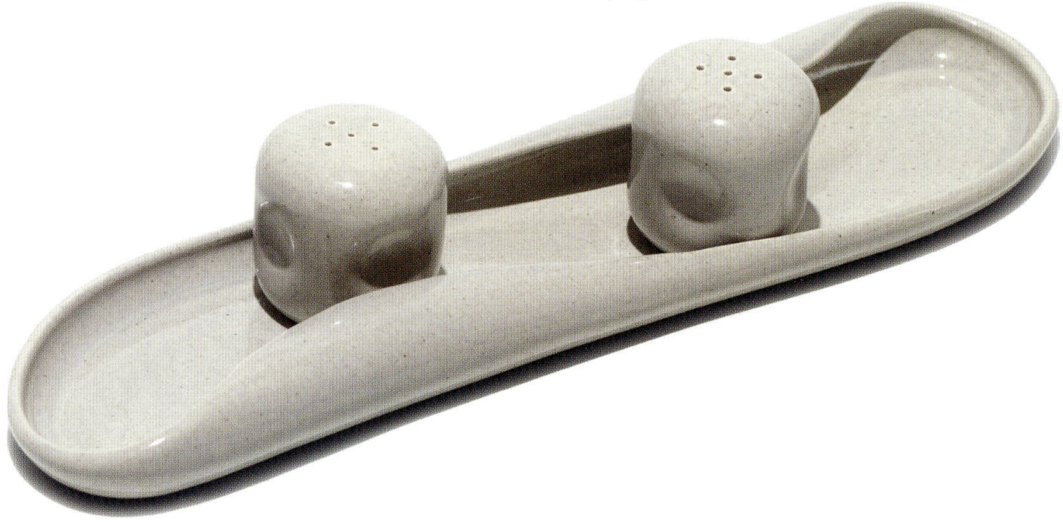

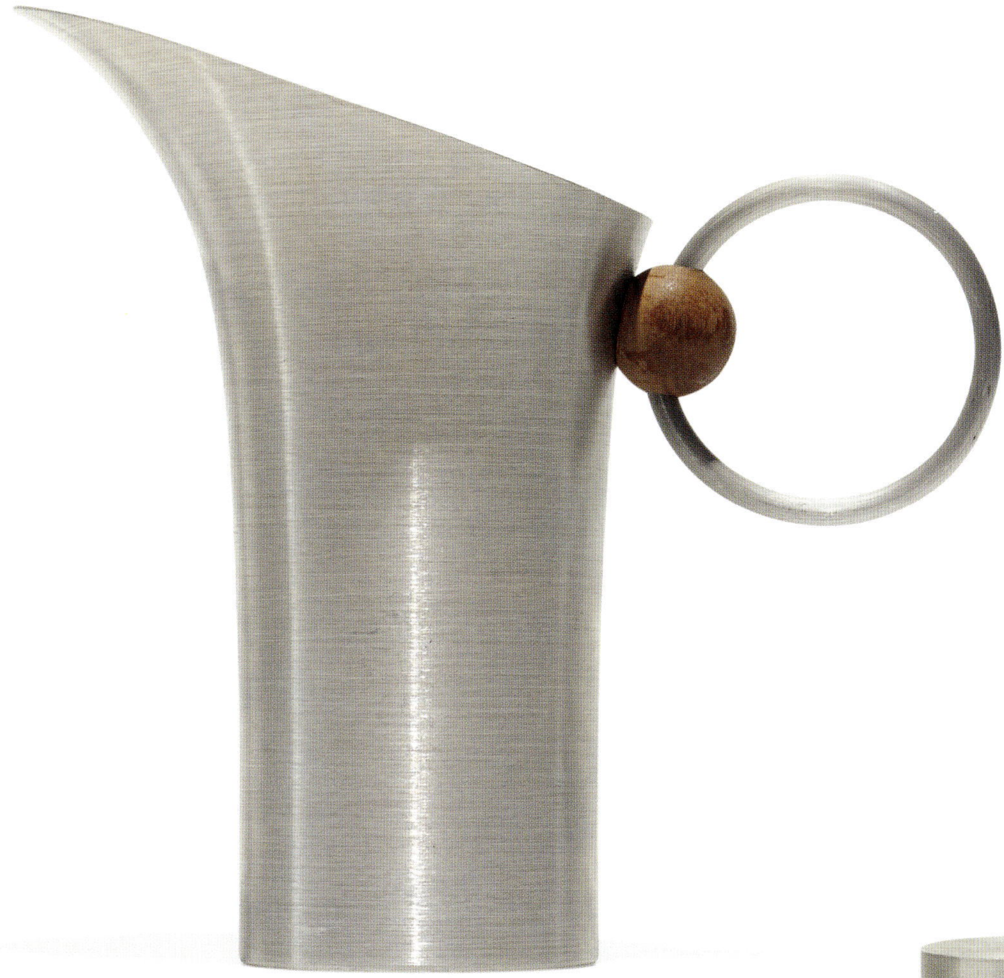

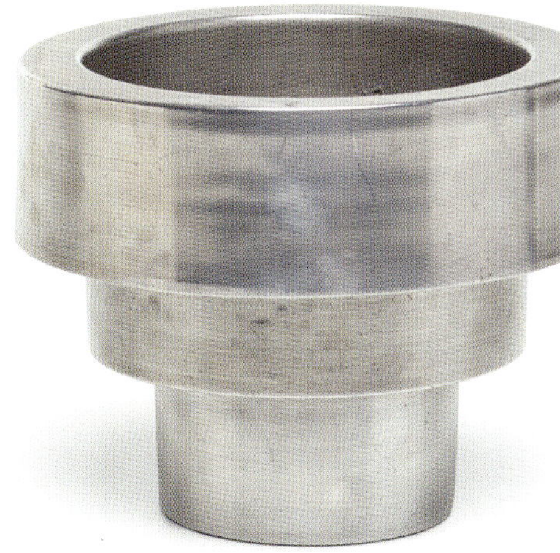

ABOVE A vase by Russel Wright, *c.* 1930, in satin-finished spun aluminium.

LEFT A Russel Wright pitcher, made by Russel Wright Inc., *c.* 1932, in spun aluminium and beech wood.

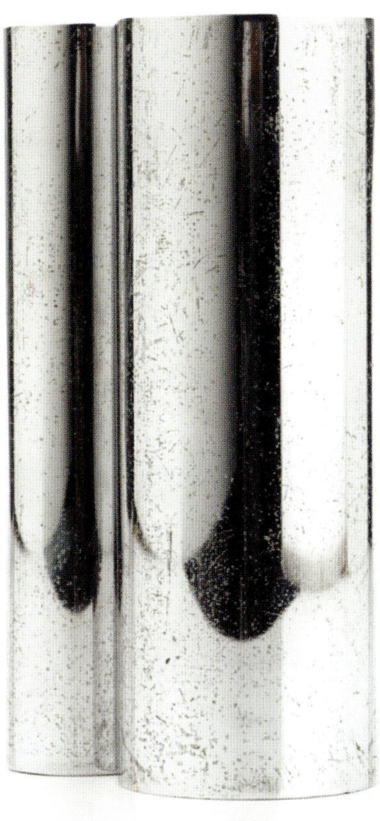

LEFT A double cylinder vase by Russel Wright, 1930, in chrome-plated steel.

RIGHT A Model 326 cocktail set by Russel Wright, made by Russel Wright Inc., 1930, in spun aluminium and cork.

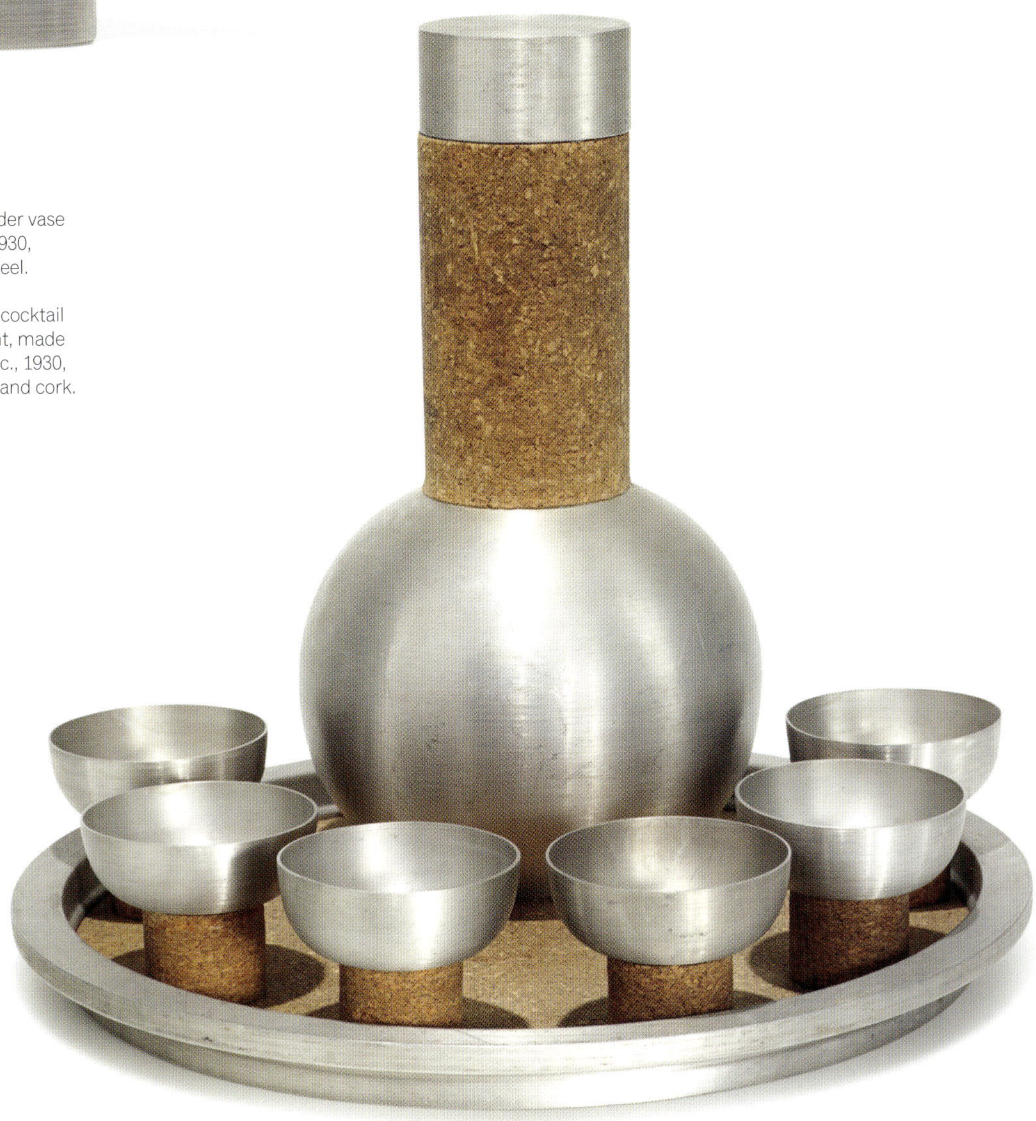

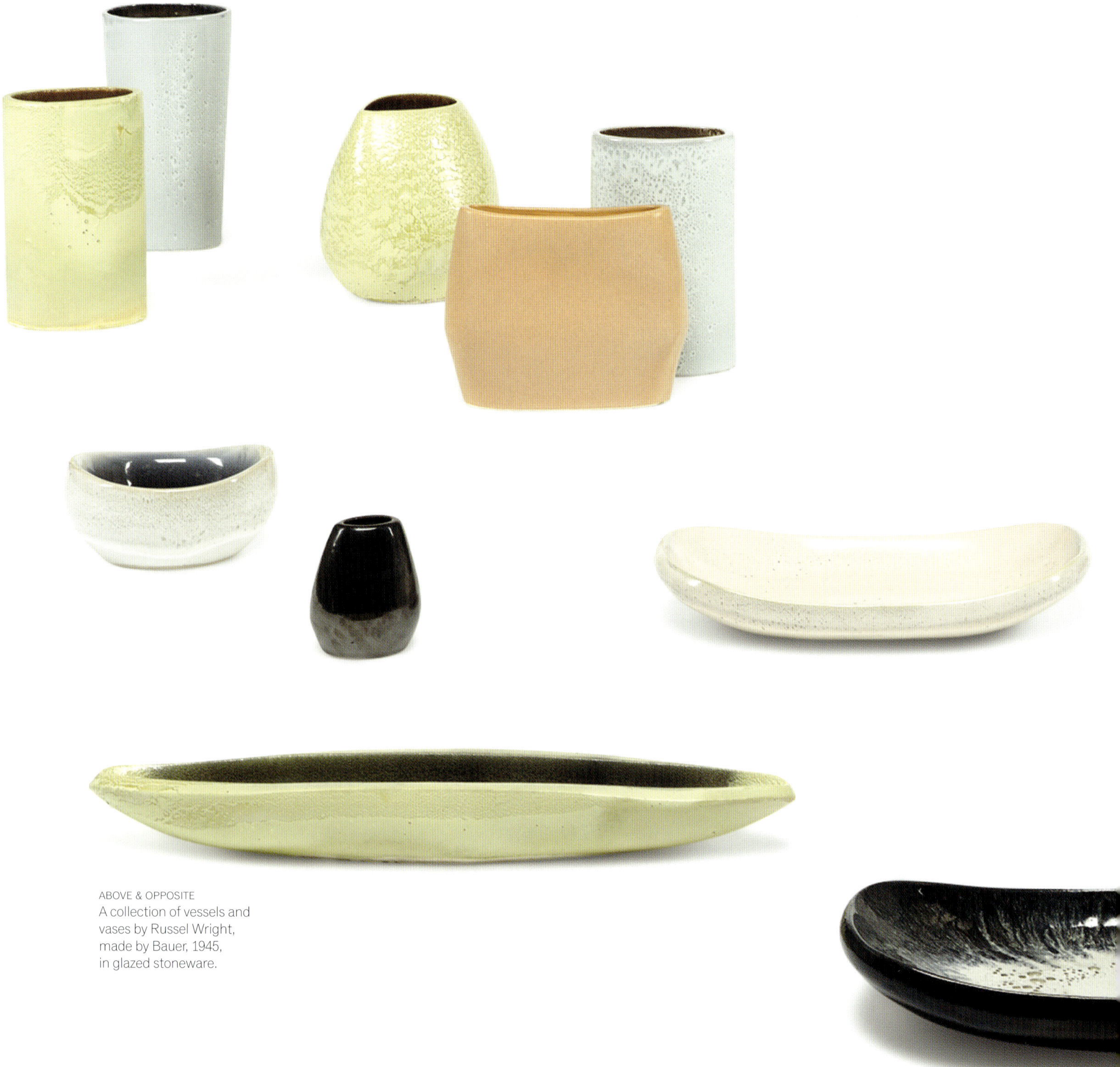

ABOVE & OPPOSITE
A collection of vessels and vases by Russel Wright, made by Bauer, 1945, in glazed stoneware.

the mid-Thirties, Wright designed his American Modern line of maple and blonde timber furniture, which was manufactured by Conant-Ball in Massachusetts and branded with his name.

Four years later, in 1939, Steubenville began producing Wright's American Modern collection of ceramic dinner- and tableware. The coordinated collection was factory-produced in large volumes, with rounded and inviting shapes that were pleasant to the touch and to the eye. The pieces were undecorated, with a simple glazed finish, but were available in an original range of evocative and enticing colours, including sea foam, coral, cedar green and chartreuse. Wright's American Modern tableware provided a pleasing and vibrant addition to the home, while being affordable; importantly, pieces could be bought one by one, so that a full range could be assembled over time.

'In considering the desirable general character of the shapes for dishware, I thought that since most foods are amorphous in shape that the dinnerware should be of simple or geometric form yet without sharp angles which would attract more attention than the food, and I found that there should be no decorations on the surface, and thus I designed the American Modern line of dinnerware.'[9]

During the key years of production, from 1939 to 1959, around 250 million pieces of American Modern dinnerware were sold, making it one of the most successful ceramic collections of all time. In more recent times, the range has been reissued by the Bauer pottery company, who worked with Wright in the Forties and Fifties on more sculptural ceramics, such as vases and bowls, produced in smaller volumes.

In 1945, Wright designed his first glassware set, for Century Metalcraft, followed by pieces for Fostoria, Appleman Art Glass and American Crystal. During the Fifties, Wright continued to design new collections of both ceramics and glass while expanding his work as an industrial designer, adding home appliances, accessories, lighting and other products to his portfolio.

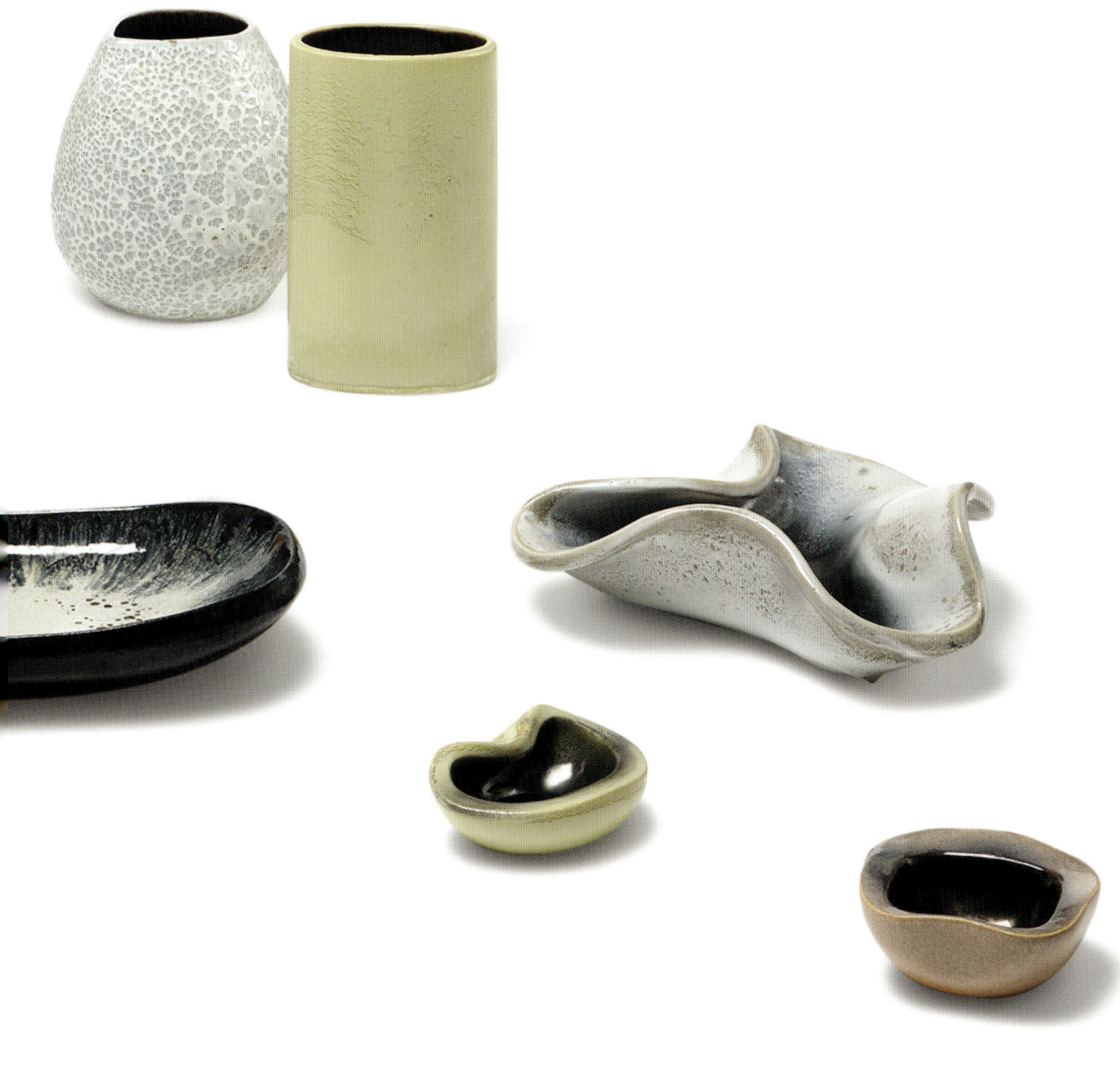

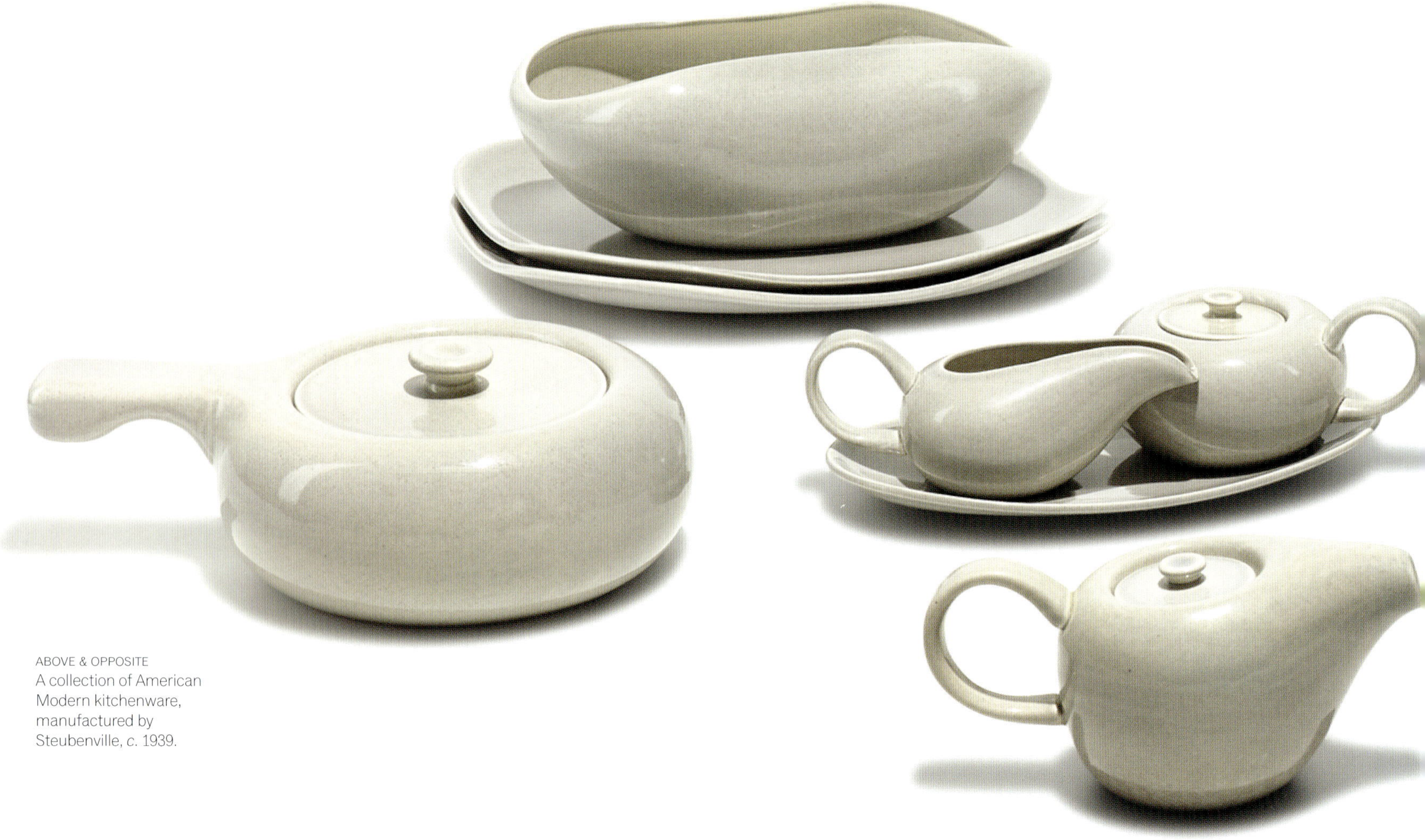

ABOVE & OPPOSITE
A collection of American Modern kitchenware, manufactured by Steubenville, *c.* 1939.

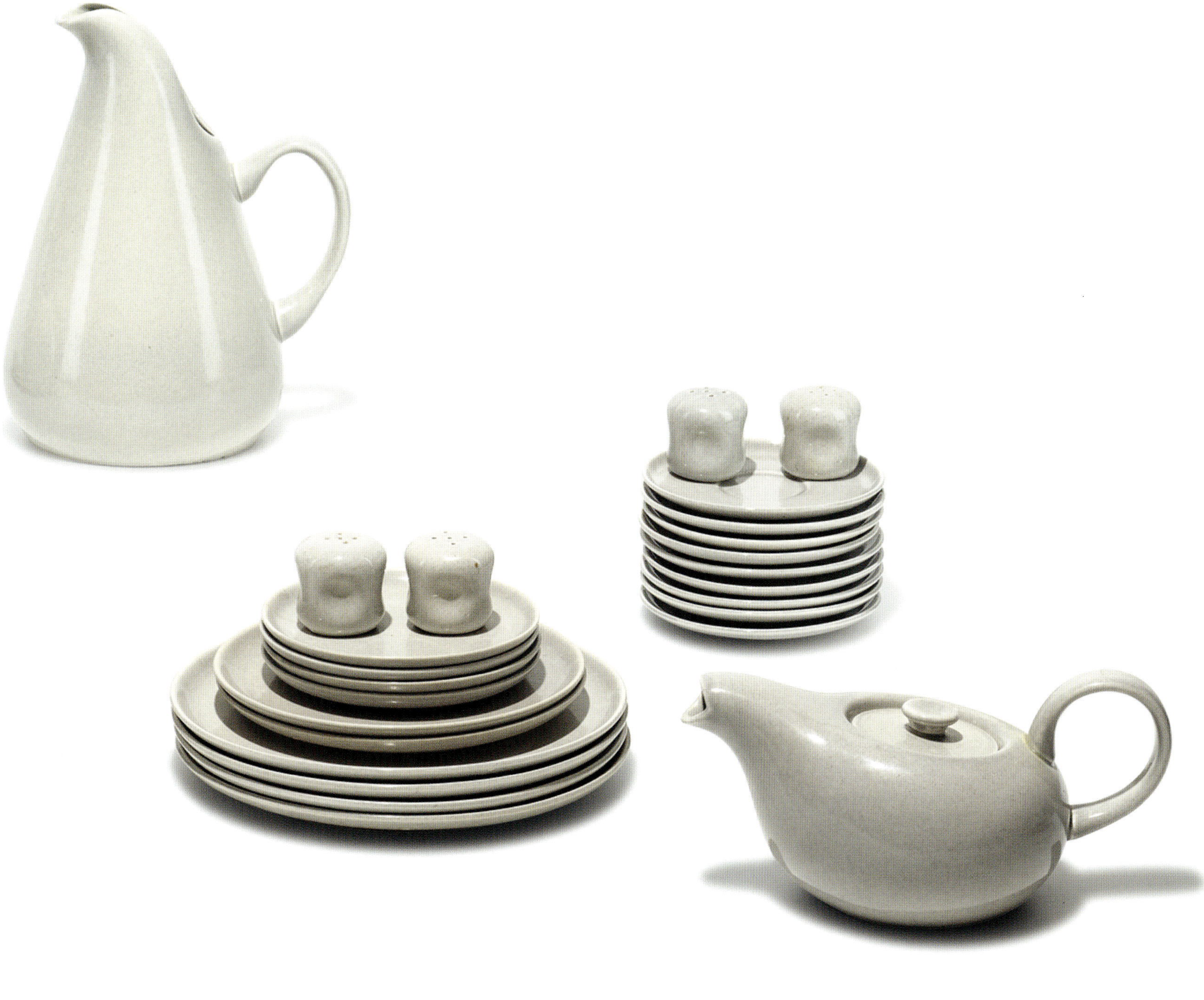

INDUSTRIAL & PRODUCT DESIGN

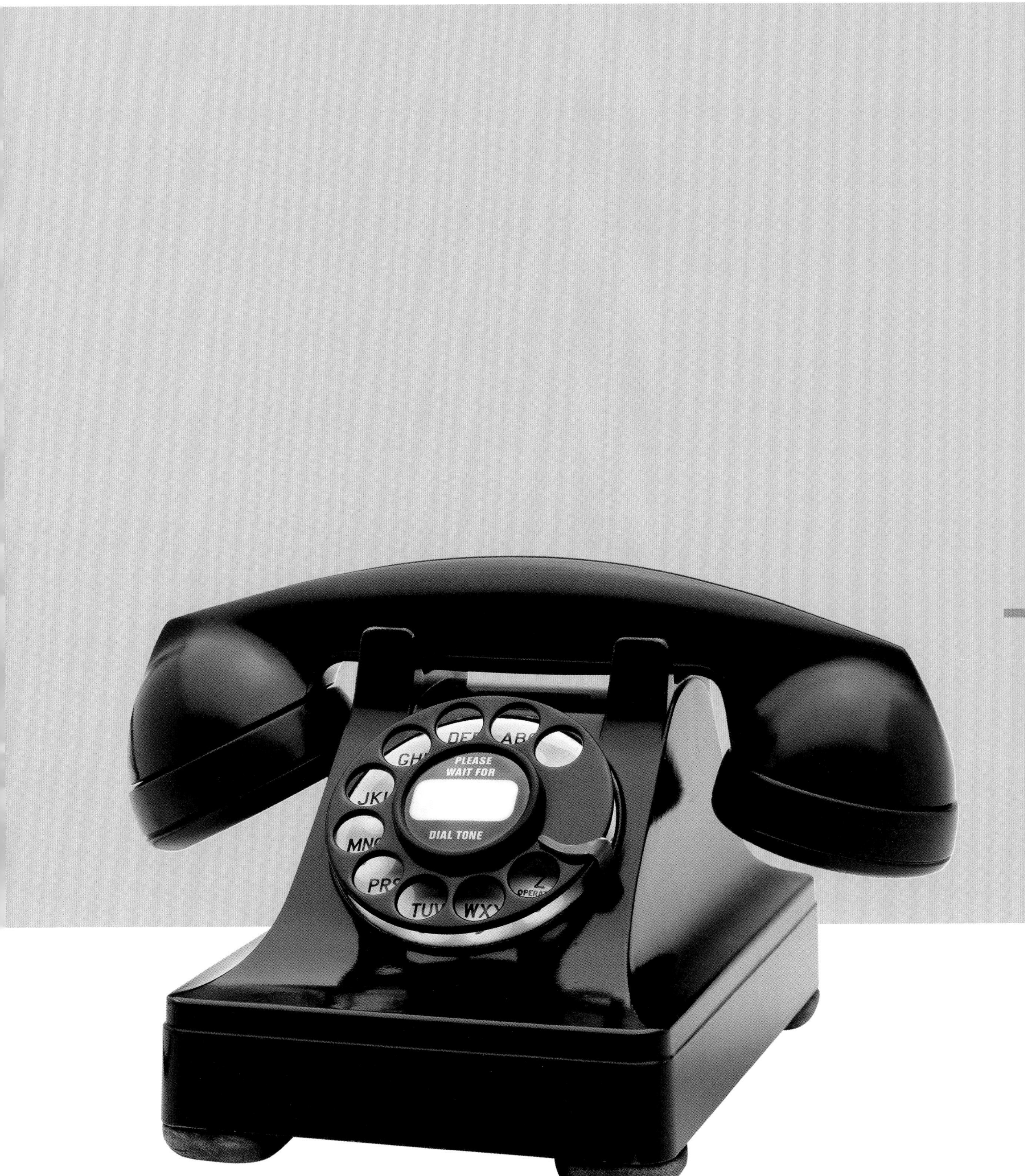
PLEASE
WAIT FOR
DIAL TONE

INTRODUCTION

The art and craft of industrial design was born during the early modernist period between the wars. The 'new profession' was staffed by pioneers from other disciplines drawn to the challenge of bringing order and a degree of beauty to categories of machine-age products and appliances that had evolved through engineering, led by the necessities of function and practicality but with little thought given to styling and aesthetics. These pioneers stepped out of many different careers.

Some, like WILHELM WAGENFELD and FERDINAND PORSCHE, emerged from practical fields with apprenticeships in metalworking or engineering. But others came from more artistic backgrounds, with training in painting or illustration. PETER BEHRENS, RAYMOND LOEWY and WALTER DORWIN TEAGUE worked as illustrators before moving into product and industrial design, while NORMAN BEL GEDDES and his protégé, HENRY DREYFUSS, worked in set design and brought an element of theatre to their new vocations. There was a sense in which this disparate group of innovators had to define themselves and their new careers as they went along, with HAROLD VAN DOREN, Teague and others writing practical guidebooks and manuals to this fresh and fast-changing discipline.

'Looking back to the early days ... I can say that we proceeded with considerable naïveté,' said the Franco-American designer Raymond Loewy, talking in the Seventies. 'We didn't realize what the repercussions might be decades later. That a new profession was beginning around us! Aside from wanting to make a living, we were all nice fellows, pure at heart and simple enough to believe that by improving a product functionally, safely, qualitatively, and visually, we were contributing something valuable to the consumer, his sense of aesthetics, and to the country. But we also proved that good appearance was a highly saleable commodity, opening the floodgates to all sorts of later unethical merchants and phony "designers"....'[1]

Some early industrial designers worked closely alongside engineers and technicians, labouring to fully understand the workings and mechanics of a product before attempting to improve upon it. Others were essentially concerned only with the outward shell and its appearance, with little concern about what 'lay under the hood'. Yet each and every one of them understood that the consumer of the Twenties and Thirties had a free choice about what they decided to buy with the money in their wallet or purse. Faced with the option of deciding between a primitive product and a beautifully conceived, elegantly detailed, streamlined alternative, they would always choose the latter. Faced with this undeniable truth, many progressive entrepreneurs and businesses came – in increasing numbers – to the realization that the new profession could make a real difference to the desirability of their products and, therefore, the bottom line.

In manuals such as Van Doren's *Industrial Design: A Practical Guide* (1940) and Teague's *Design This Day: The Technique of Order in the Machine Age* (also 1940), the pioneers sought to bring discipline to their discipline, explaining the precepts, principles and practicalities to which they adhered. A number of American leaders – including Loewy and Dreyfuss – also banded together and formed the Society of Industrial Designers in 1944, electing Teague as the first president.

Within the initial pages of his book, Van Doren offered this useful definition:

'Industrial design is concerned with three-dimensional products or machines, made only by modern production methods as distinguished from traditional hand-craft methods. Its purpose is to enhance their desirability in the eyes of

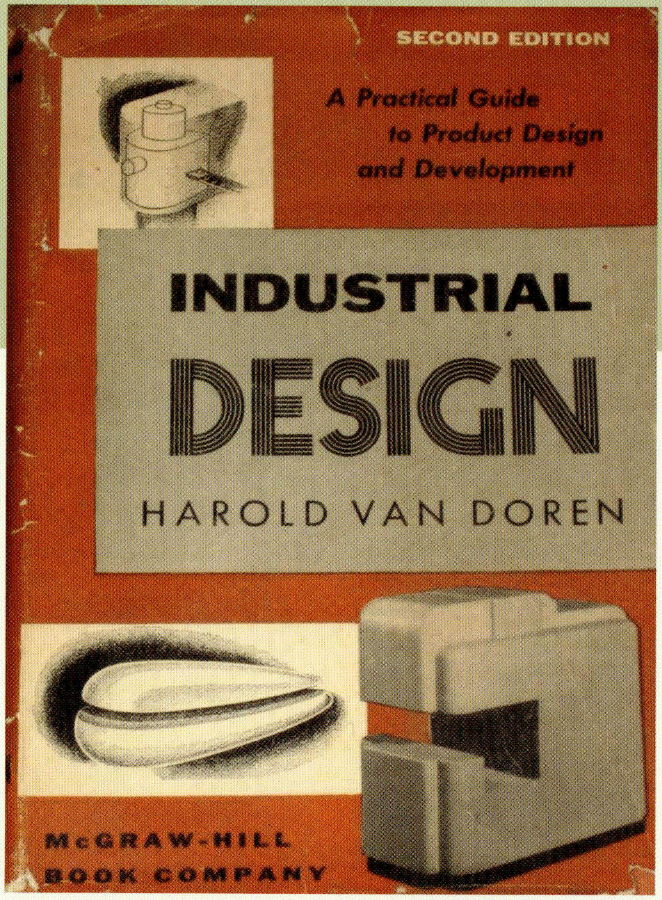

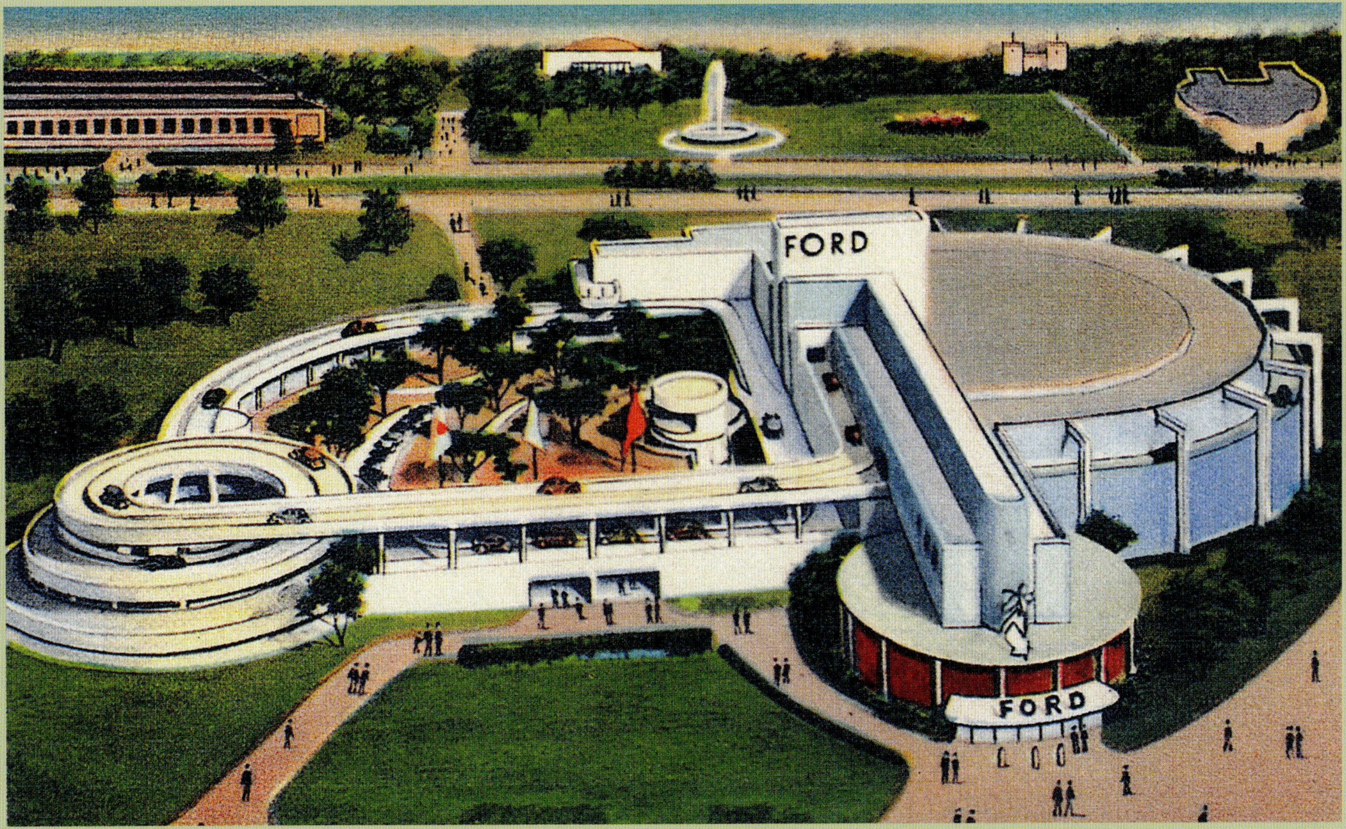

PAGE 203 A Model 302 telephone designed by Henry Dreyfuss, *c.* 1937.

LEFT The Ford Motor Company building at the New York World's Fair, 1939–40, as illustrated on a promotional postcard; the architect was Albert Kahn, and the designer of the Ford exhibition Walter Dorwin Teague.

the purchaser through increased convenience and better adaptability of form to function; through a shrewd knowledge of consumer psychology; and through the aesthetic appeal of form, color, and texture.'[12]

In many respects, industrial design was a product, in itself, of the machine age, as Van Doren suggests. The focus of the profession was upon factory-produced products, which sometimes lacked attention to aesthetics, along with a new generation of domestic appliances and labour-saving machines that had come from the hands of inventors and engineers who were not always concerned with how desirable their work might appear to the end user.

The machine-age model of production – together with a key catalyst for the business of industrial design – came in part from the automobile industry. Industrialist and engineer Henry Ford* perfected the idea of moving assembly lines to produce his Model T, which was first introduced in 1908. By 1927 Ford had sold over 15 million of these cars, using a standardized set of components pieced together on his super-efficient production lines. As efficiency increased so the price of the cars fell, driving more and more sales; fewer workers were needed, but those that found a job with Ford earned a good wage. This was the ethos of 'Fordism'.

Production-line techniques – also informed by time and motion studies developed by Frederick Winslow Taylor – were applied to a growing range of products and other kinds of consumer-oriented machines. In Germany, Peter Behrens designed a range of electrical products for AEG, but he also designed their factories, equipped with assembly lines that were sometimes on a vast scale, as with their Turbine Hall in Berlin (1909).

Wolf Dohrn, the executive director of the Deutscher Werkbund*, described a visit to the AEG factory: 'Like trees in an avenue, to the right and left of the passageways ... were the work stations with their machine tools, and the dynamos, turbines and other machines in course of being built. Internal transportation

OPPOSITE Harold Van Doren's guidebook to a new profession, *Industrial Design*, first published in 1940.

through the hall is carried out from above in the roof space, by travelling cranes running on tracks. They bring the finished machines to testing rigs under the gallery on which we stand.... Throughout the entire space, no transmission [chains or belts] interrupts the free movement. There are only individual power units – that is to say, each machine tool has its own motor. In this way it can be brought to the work being undertaken.'[13]

Industrial assembly-line techniques made possible a wave of new and cheaper products, from cars to vacuum cleaners and even entire kitchens (as in the case of Margarete Schütte-Lihotzky's* Frankfurt Kitchen). This fitted in neatly with the modernist preoccupation – expressed in the teachings of the Bauhaus masters, in particular – of making good design democratic and affordable to all, based upon a set of standardized, factory-made components.

Increasingly, during the Twenties and Thirties, car producers and other industrially led manufacturers looked beyond engineering to the styling of their products. The key expression of this movement in the car industry was streamlining, the introduction of rounded, aerodynamic bodies and shells that spoke of speed and energy (see also Anne Hoy's essay on streamlining, p. 208). In America, General Motors president Alfred P. Sloan and his head of design, Harley Earl*, did much to promote the concept of streamlining and – later on – the more dubious concepts of 'dynamic obsolescence' and the annual upgrade. In Germany, Ferdinand Porsche first produced the streamlined Beetle in the Thirties. Many other industrial designers of the period, such as Loewy, worked for the car industry and for the railroads, which also embraced streamlining in the late Twenties and early Thirties. Planes, trains and automobiles – along with airships and even liners – spoke of machine-age dynamism and an increasing obsession with speed.

While streamlining had its roots in transportation of all kinds, the aesthetic soon began appearing in many other fields of product design, from tractors to drinks bottles to radios. Industrial designers were commissioned to undertake a process of seduction, tempting consumers to adopt a new wave of domestic appliances that had begun life as clunky, clumsy and even intimidating collections of component parts; this process accelerated dramatically in the late Twenties and Thirties, as more and more appliances were powered by electricity. This was especially true of the washing machine, for example, with the earliest electrical models resembling an unsightly combination of tubs and motor. Reinvention with streamlining techniques encased the tubs in a hygienic, rounded shell that looked as though it might receive an invitation into the home. The same was true of many other home appliances, such as the vacuum cleaner and the electric refrigerator, which evolved from the idea of the ice box.

The growing audience for home entertainment and communication was also served by an alliance of manufacturers, engineers and industrial designers. The radio was reduced down

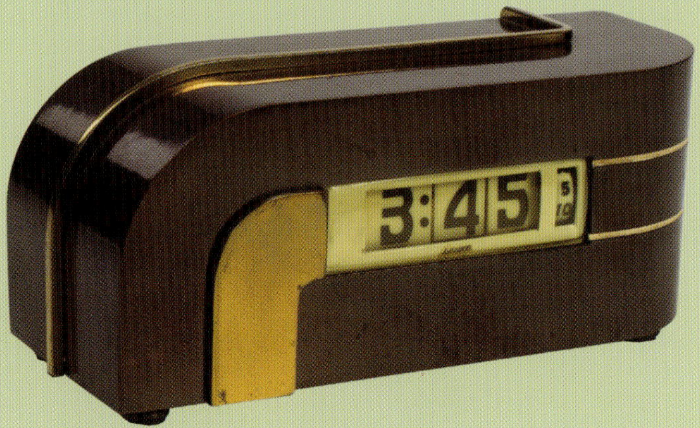

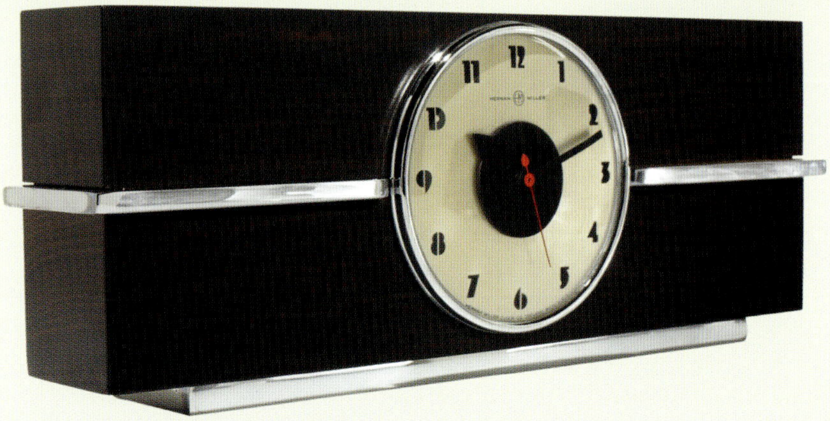

LEFT A Model 6366 mantle clock designed by Gilbert Rohde, made by Herman Miller, *c.* 1932, in macassar ebony, chrome-plated steel, aluminium and glass.

ABOVE Kem Weber's Zephyr table clock, produced by Lawson Time Inc., 1933, in copper-plated brass and enamelled steel.

from a heavy piece of 'furniture' to a crafted, rounded object, such as Teague's Nocturne radio of 1935 or Bang & Olufsen's futuristic Hyperbo 5RG radio of 1934; gramophones and television sets were not far behind. Telephones, cameras and typewriters were all redesigned in more enticing packages, aided by the evolution of a new palette of materials such as Bakelite, the earliest synthetic plastic, which was introduced in 1909 and widely used during the Thirties. These were innovative products that marked a mid-way point between early, prototypical inventions and the miniaturization process that gathered pace in the Fifties and Sixties.

The process of seduction often extended into the packaging and presentation of products, with a growing overlap between industrial design and graphic design. Indeed, the nascent nature of the profession and its undefined character gave industrial designers an extraordinary freedom to expand into many different fields of design, spreading their wings into corporate identity, architecture, lighting, furniture, glassware, ceramics and many other sub-disciplines. There was a glorious opportunity for a kind of design 'land-grab', with ateliers such as Loewy's and Teague's growing rapidly before and after World War II.

This land-grab even extended into the future, with Bel Geddes and others coming up with concepts for extraordinary 'dreamlined' ships, titanic seaplanes driven by a great battery of engines, as well as flying cars; his 'Futurama' exhibition at the 1939–40 New York World's Fair presented an entire vision of an imagined world contained in a futuristic, streamlined pavilion. RICHARD BUCKMINSTER FULLER was another true visionary, applying principles of modular, factory construction to the home itself, producing a highly influential though commercially flawed prefabricated home in the form of the futuristic Dymaxion House. Always ahead of the times, Fuller ultimately found his greatest success in the post-war period with his patented geodesic dome building system. In defining and shaping their own profession, such modernist pioneers also established the solid foundations for the future of industrial design itself.

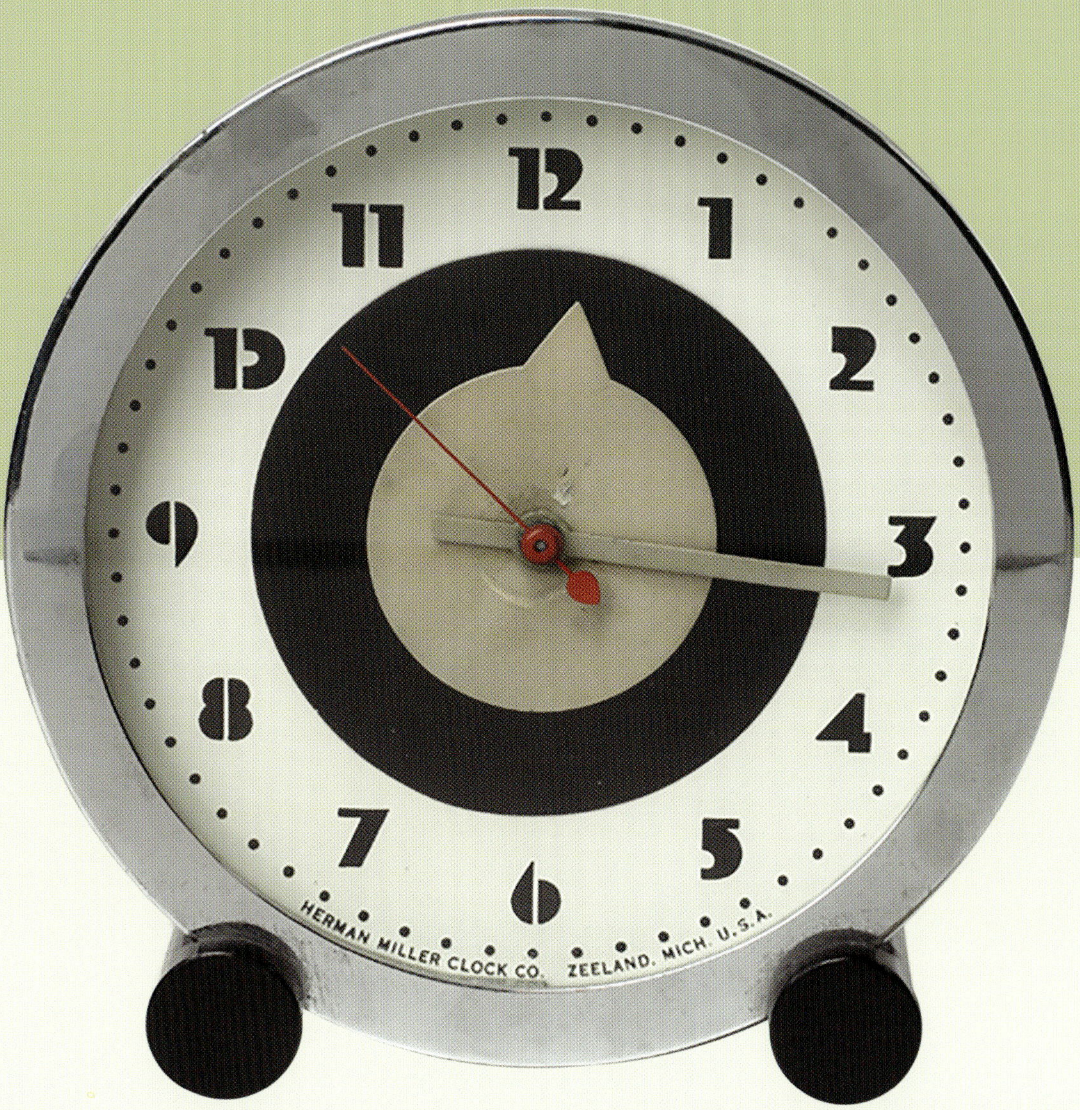

RIGHT A table clock designed by Gilbert Rohde, manufactured by Herman Miller, c. 1933, in chrome-plated steel and enamelled metals.

During the Depression, streamlined design captured America's popular imagination, in products and in pictures. It had 'eye-appeal', wrote its champions, who identified it with machine-age modernity, progress and speed. Streamlining originated in engineering research, in aerodynamic studies to facilitate transportation, which found that smooth, simplified, often parabolic or teardrop forms offer the least resistance to water and wind. This form language shaped steamships and aircraft, and lent its dynamism to automobiles by the Twenties and locomotives by the Thirties. Industrial designers heralded it, and their rise as a profession coincided with its wide acceptance. Their new streamlined consumer goods evoked America's preeminence in manufacturing, flight and auto-mobility: to buy them was seen as almost patriotic while the nation worked to recover from the Wall Street Crash of 1929.

The heyday of US streamlining can be dated from 1932 through 1940, key years for NORMAN BEL GEDDES, one of the most influential among the decade's visionary and publicity-driven designers. His widely read book *Horizons* (1932) illustrated and explained how he would apply streamlining to manufactures from planes, trains and buses to future homes. Opening on the eve of World War II, his 'Futurama' exhibit for General Motors was the most popular at the 1939–40 New York World's Fair, *The World of Tomorrow*. Bel Geddes's thrilling display envisioned America in 1960 prospering through rapid yet friction-free growth, in quarter-of-a-mile-high skyscrapers connected to suburbia, farms and open spaces through tiers of transportation, crowned by a national system of interstate highways. Detroit's continued dominance of world car sales since the Twenties was assumed. The triumphalist American narrative of Crash and recovery via mass production and mass consumption (with some federal support) starred Bel Geddes and 'streamliners' like RAYMOND LOEWY, WALTER DORWIN TEAGUE, HENRY DREYFUSS, KEM WEBER, and scores of others;[4] captains of major industries; and supporters (and detractors) among US cultural critics. The story differed from Thirties' Hollywood yarns mainly in that much of it came true.

Critics and Consumers Support Streamlining

'The automobile and the aeroplane are not the weakest but the best of our machine products, a distinction which they share with American kitchen equipment and bathroom fixtures,' wrote Lewis Mumford in 1928. He lauded 'the unified mass and the slick stream-lines of the modern car', and found autos and aircraft 'both gaining in beauty as they were adapted more carefully to the mechanical requirements...'.[5]

The idea that vehicles and appliances were evolving functionally and aesthetically with their 'fitness to purpose' came from European and American defences of modern architecture. Erich Mendelsohn* and LE CORBUSIER illustrated US industrial structures and skyscrapers in their books on modern building;[6] and these German and French architects visited America in 1924 and 1935 respectively. In 1936 Sheldon Cheney and Martha Candler Cheney applied the evolutionary theme to streamlined goods in *Art and the Machine*:

> STREAMLINED...: As an aesthetic style mark, and a symbol of twentieth-century machine-age speed, precision, and efficiency, it has been borrowed from the airplane and made to compel the eye anew, with the same flash-and-gleam beauty re-embodied in all travel and transportation machines intended for fast-going.[7]

'FLASH-AND-GLEAM BEAUTY': THE STREAMLINED DECADE IN AMERICA

ANNE H. HOY

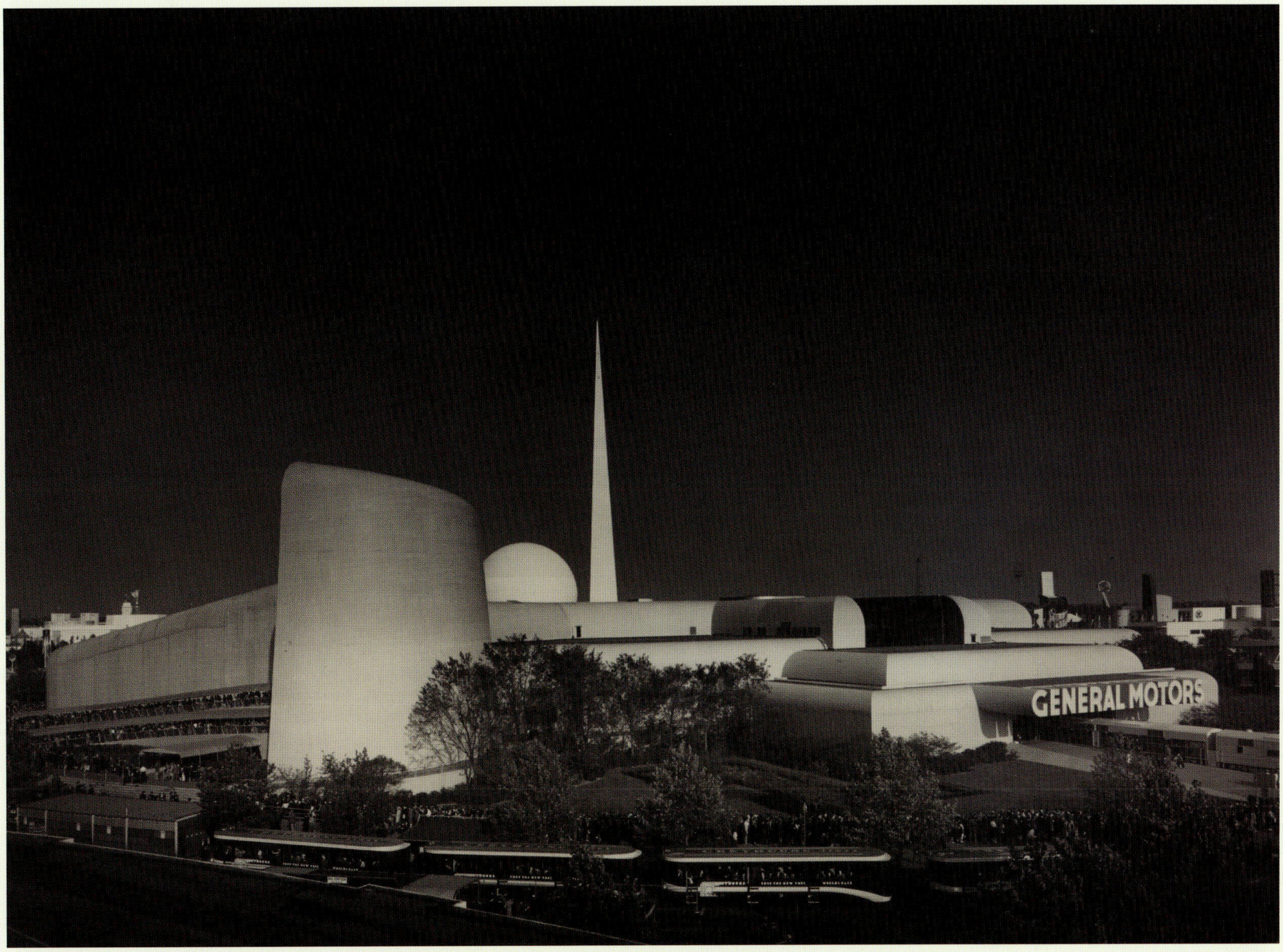

ABOVE The General Motors Highways and Horizons building by Norman Bel Geddes, which housed his 'Futurama' exhibit at the 1939–40 New York World's Fair; the iconic Perisphere and Trylon modernist structures can be seen in the background. At the far left, visitors queue on twin ramps with streamlined crowd control.

Firsts in flight aroused American pride and world fascination, from the Wright brothers motoring aloft in 1903 to Charles Lindbergh's flying the Atlantic nonstop in 1927. In mass media, early Thirties photographs of aircraft joined Twenties close-ups of machinery and factories to honour precision manufacturing. Streamlining – like that shown in Margaret Bourke-White's *Goodyear Zeppelin*, 1931 – won the technological beauty contest in American magazines and newspapers.

Streamlining's marketing and styling success is based on that of America's premier industries of the Twenties, and auto manufacturing was foremost in creating the country's high standard of living. During his production of Model Ts, 1908–27, Henry Ford* sold 15 million cars while reducing the price by more than a third. The workers on his assembly lines could afford to buy what they produced. Period surveys announced that rural Americans valued their cars and trucks over indoor plumbing; in fact, 93 percent of Iowa farmers owned motor vehicles by 1926.[8] Installment buying put cars within reach of all but the poverty-stricken, and competition among Ford, General Motors and Chrysler made cars faster yet safer and more comfortable. But the market was nearing saturation by the late Twenties. By the early Thirties, the Big Three had adopted streamlining, for both improved performance and allure.

GM's brilliant automotive design and salesmanship under Harley Earl*[9] and 'buying on time' became paradigms for household durables and appliances as the US economy recovered from its nadir in 1934, when one in four adults was unemployed. For consumers, annual aerodynamic style changes created better looks that were read as meaning better quality; for producers, sleek metal housings were cheaper to change than mechanical parts.[10] In car culture at the time, 'planned obsolescence' did not mean wasted money and manipulated buyers but shinier status symbols and ever-higher velocity. At the 1933–34 Century of Progress International Exposition in Chicago, visitors could admire Chrysler's Airflow car, RICHARD BUCKMINSTER FULLER's three-wheeled, teardrop-shaped

Dymaxion No. 3 auto prototype, the Goodyear blimp, and the first streamlined train, the Burlington Zephyr, which swept into the fair's opening day in 1934 in a record-breaking Denver–Chicago run of fourteen hours.[11]

In 1934, advertising executive Egmont Arens telegrammed President Franklin D. Roosevelt offering to lecture across the country on 'Streamlining for Recovery'. Arens headed the first styling and design department in an advertising agency, and his and Roy Sheldon's book *Consumer Engineering: A New Technique for Prosperity* (1932) saw consumer sales spurred by psychologically astute ads as vital to a capitalist renaissance. FDR's response is not recorded. But federal support for home building and improvement, like the National Housing Act of 1934, fostered popular consumption of appliances and products from large to small.

The Style's Artistic Rivals

The year 1934 marked elite as well as mass acceptance of streamlining. At its annual design exhibition, New York's Metropolitan Museum of Art mounted *Contemporary American Industrial Art*, where Raymond Loewy and Lee Simonson's 'Model Office and Studio for an Industrial Designer' outshone its rivals. These latter were dismissed as 'Zigzag Moderne' and 'modernistic' by the tastemakers of the city's Museum of Modern Art, who upheld a third current. Promulgated by Philip Johnson*, Henry-Russell Hitchcock and Alfred Barr of MoMA was the rationalist functionalism of modern European and Russian architecture. This is better known as the 'International Style'*, so-called from the museum's 1932 exhibition and catalogue, and shown extending into product design in MoMA's *Machine Art* exhibition of 1934. There, Bauhaus designs met ball bearings and laboratory beakers, stressing the utility and resulting abstract beauty they shared.

Since the Sixties the Zigzag Moderne has been termed Art Deco* after the simplified Cubo-Futurist style seen at the 1925 Paris Exposition Internationale des Arts Décoratifs et Industriels Modernes. The US had declined to contribute any designs, and its decorative arts industry spent the next half-dozen years cultivating the high art-derived language. Certain emigré and American-born designers favoured hand-crafted limited-edition or one-off designs of complex faceted forms rendered in costly, sometimes exotic materials. But the Crash stifled sales and commissions by the previously active DONALD DESKEY, Kem Weber and Paul Theodore Frankl*, among others.[12] When business began reviving around 1932, they had learned to streamline.

Perhaps surprisingly, the machine art upheld by MoMA was closer to streamlined design than Art Deco. Both approaches embraced mass production for its high volume and low cost; modern materials like tubular steel, chromium, aluminium, heat- and impact-resistant glass, and plastics (such as Bakelite) for their durability and 'hygiene'; and a technologically inspired, ornament-free form language. But the sides warred over aerodynamic principles applied to stationary products, form *not* following function but concealing it. The difference was not only visual but moral. The platonic solids of cube, cylinder and sphere in MoMA's modern language came from ideal geometry and served socialist and utopian design. Streamlining's shining bullet-, rocket- and cockpit-like forms, by contrast, appealed to impulse spenders escaping economic and psychological depression, and exploited their designers' experience in the arts of illusion.

Few of the designers were engineers. Rather, theatre sets and design and lighting united Bel Geddes, Loewy, Arens and Dreyfuss. Art, advertising graphics and illustration were among the skills of Loewy, Teague, HAROLD VAN DOREN, GILBERT ROHDE and Lurelle Guild.

After Bel Geddes, Loewy was best known in the Thirties. Photographed for the Metropolitan Museum's *Industrial Art* exhibit, the dapper French-born designer presided over an interior where streamlining unified the walls, ceiling, elongated window and tubular steel furnishings. The continuous curves were accented by horizontal metal bands, like the triple 'speed lines' of car grilles and vents: the total resembled the control room of a spotless ocean-going yacht.

Such metaphors of easy mobility offended MoMA. Loewy's prototype for a pencil sharpener, never realized,

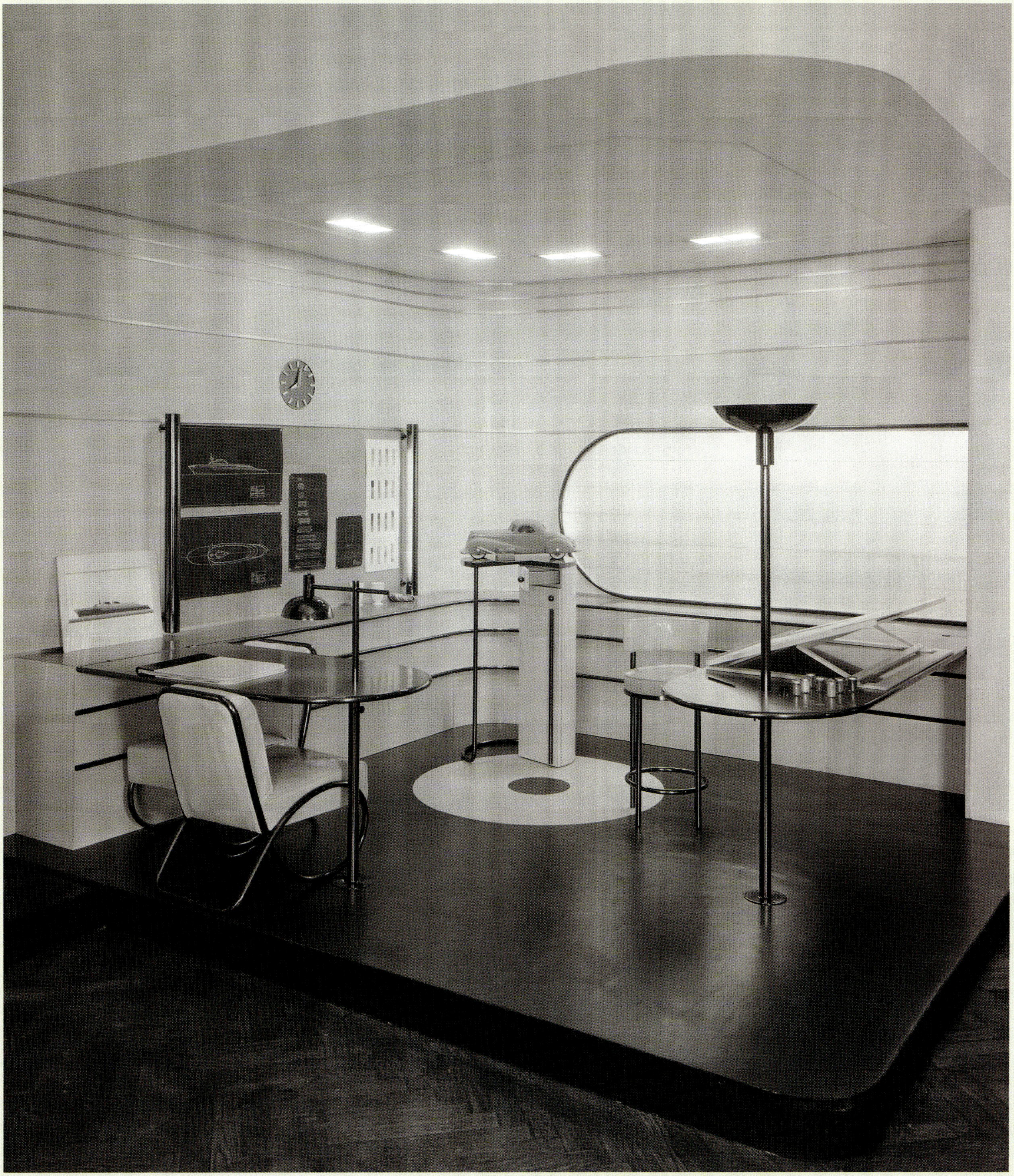

ABOVE Raymond Loewy and Lee Simonson's 'Model Office and Studio for an Industrial Designer', at the *Thirteenth Annual Exhibition of Contemporary American Industrial Art*, Metropolitan Museum of Art, New York, 1934–35.

aroused Alfred Barr to damn its resemblance to a fighter plane; and his colleagues condemned small electronics that looked capable of hurtling through space. Only in the Eighties, when postmodernism introduced irony and retro style quotations to the arts, was it widely accepted that form could follow fun, and that one function of design could be futurist fantasy.

Choice Appliances and Architecture

More than Raymond Loewy's pencil sharpener, his Coldspot refrigerator redesign for Sears, Roebuck & Company challenged functionalists. It was the first appliance promoted primarily for its style. A few rows of moulded speed lines and a chromium 'feather touch' handle alone accented its unitary sculptural form. Loewy's three subsequent models varied the sleek formula only slightly, yet succeeded in increasing sales in five years, from 15,000 units to eight times that many.[13] In 1949 Loewy was the first designer to appear on the cover of *Time* magazine ('he streamlines the sales curve').

As for house cleaning, Lurelle Guild redesigned the Electrolux vacuum in 1937 to resemble a rocket crossed with a sled. Moving on low metal gliders, the elongated canister with hose promised to reach under any furniture and save the housewife effort and time. Henry Dreyfuss called the look 'cleanlining'. The Guild model was hawked door to door, and was bought by nearly a million households in one year. In the decades in US culture known for their war against germs and slums (both identified with 'the unwashed', aka immigrants), such appliances and their advertising played on homeowners' anxieties about family health and social standing. 'Hygiene' had several meanings before World War II.[14]

In American architecture streamlining was choice. William Lescaze* and George Howe's Philadelphia Savings Fund Society Building (1929–32) may be America's first fully modernist skyscraper, and its streamlining, in its corner entrance and banking floor, was an 'enriching variation in ... straight-line Functionalism', Lescaze stated.[15]

In his multi-level 1933–35 residence outside New York City for Richard H. Mandel, Edward Durell Stone synthesized Bel Geddes's 'House of Tomorrow' concept of 1931 with the principles of the International Style as defined by MoMA. More famously, Frank Lloyd Wright's S. C. Johnson Administrative Complex (1936–39) in Racine, Wisconsin, folded allusions to streamlining, Mendelsohn's expressionist buildings and Dutch modernism into his 'organic' idiom. Wright and Bel Geddes had both met Mendelsohn in 1924, and they knew his German department stores with their façades of repeated curving bands of floors. On the brick exterior of Wright's Administration building are thin parallel bands of lighting, like speed lines, which emphasize its continuous horizontals. These curves and the recessed entrance also recall Amsterdam School architecture such as Michel de Klerk's Het Schip complex (1917) in the Dutch seaport.

BELOW Electrolux Model 30 vacuum cleaner, 1937–38, in steel, aluminium, rubber and vinyl, designed by Lurelle Guild.

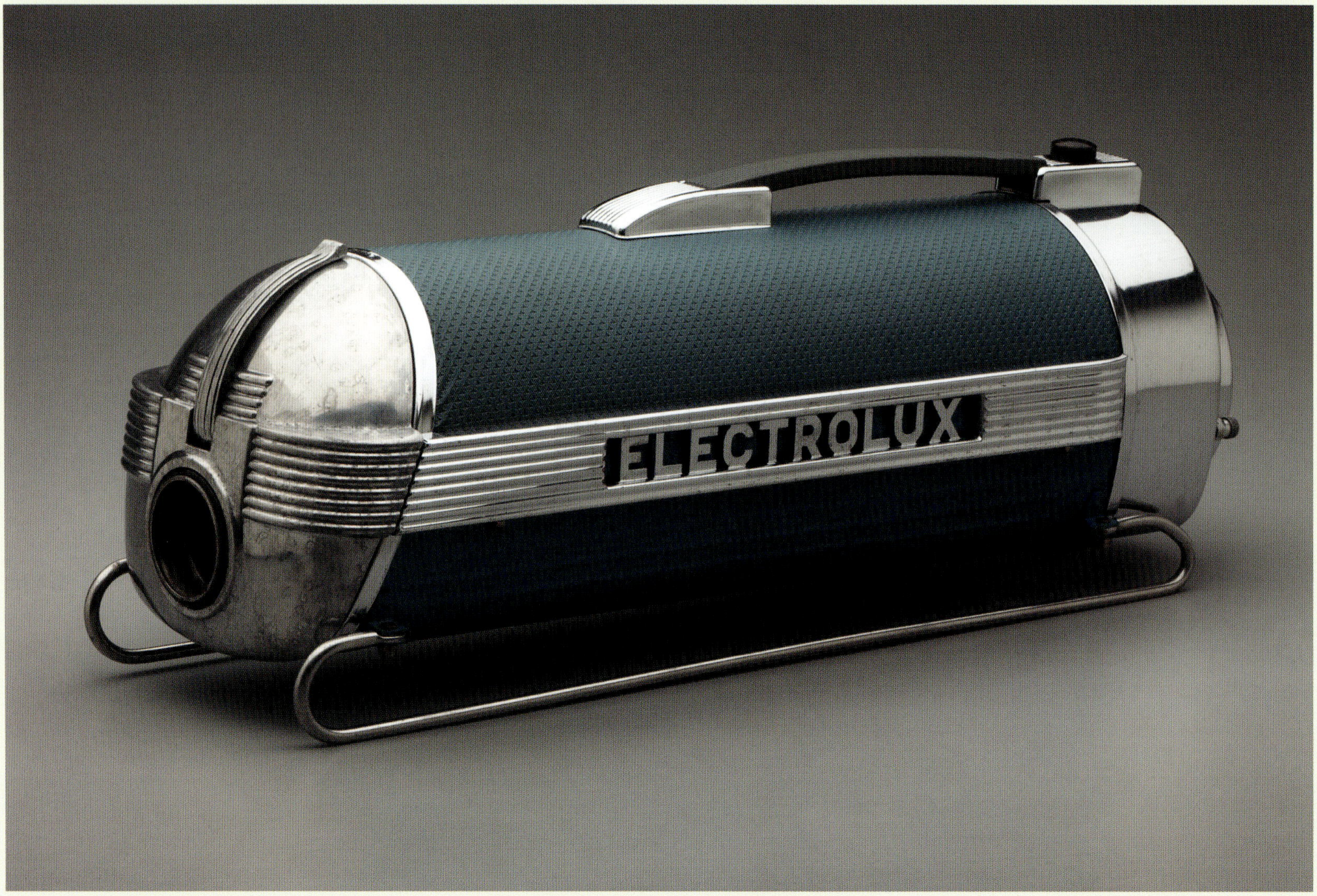

Outside America, nautical references abound in the De La Warr Pavilion (1935), a public building at Bexhill-on-Sea, England. There Mendelsohn and SERGE CHERMAYEFF, now émigré architects, responded to the beach site with a stack of semi-circular ferro-concrete bays offering an observation deck, a café and a flat roof for sunbathing. A cantilevered helical staircase, visible from outside, is part of the Pavilion's suave appeal.[16]

Of all these designs, the most theatrical was Bel Geddes's 'Futurama' in his GM Building for the New York World's Fair. Two long pedestrian ramps rose to the entrance near the highest point of a single curving roofline. Like a streamlined appliance, this smooth monocoque exterior gave no hint of the highly detailed scale model of America within or the machinery powering the visitors' seated ride across it. The Bel Geddes vision of transcontinental transport climaxed the identification of streamlining with prosperity – the free enterprise kind, linking corporate production and individual consumption. Here the highway supplanted the skyscraper as the emblem of America. In the GM building with its showroom of gleaming cars, streamlining triumphed, 'with the same flash-and-gleam beauty re-embodied in all travel and transportation machines'.

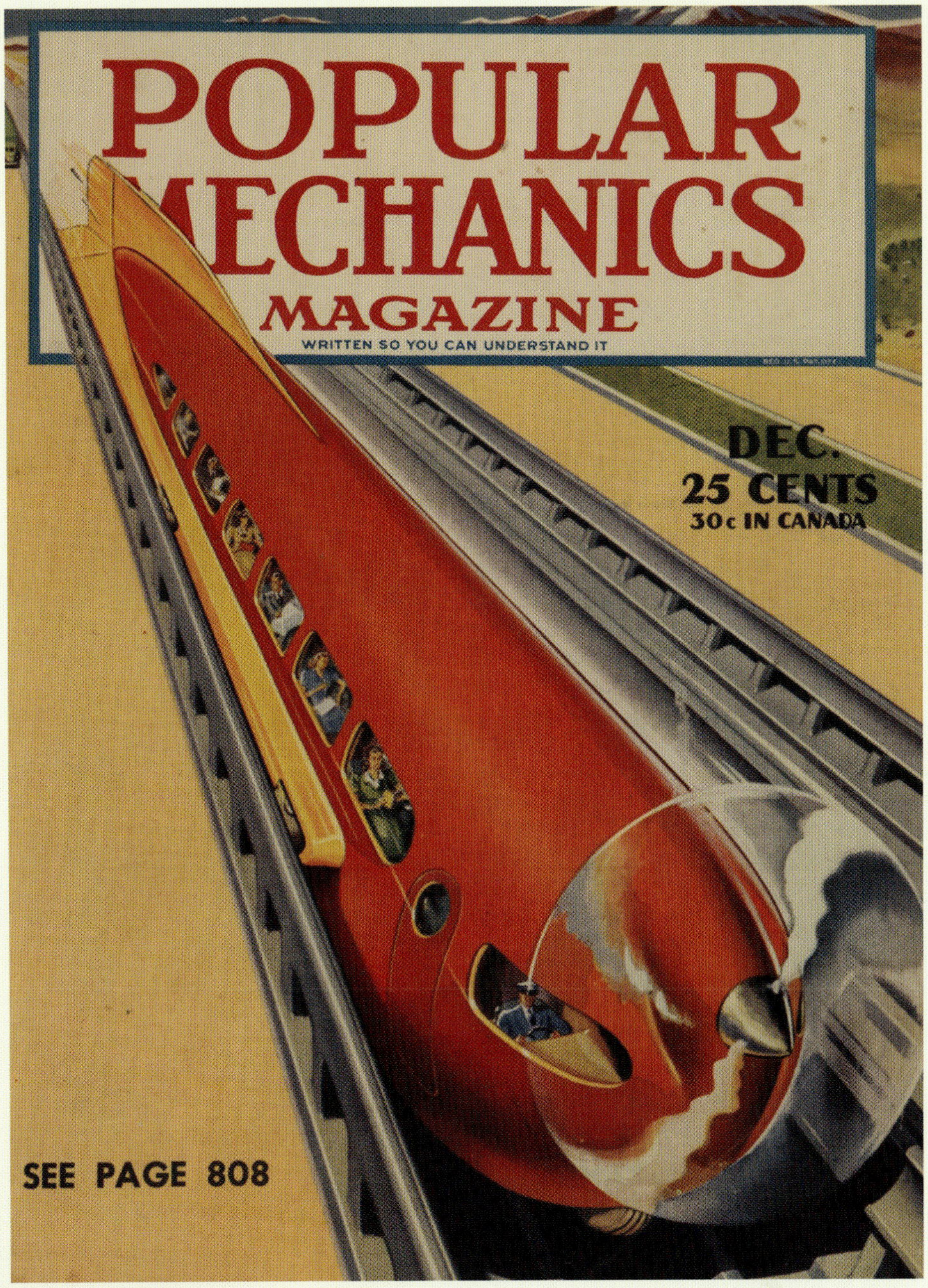

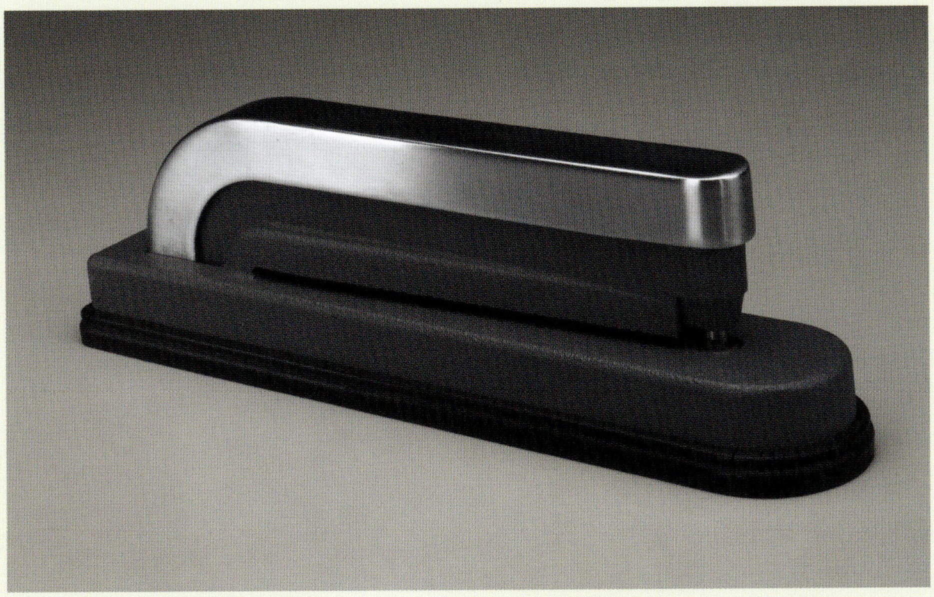

LEFT Computer card hole punch, or scoring stencil punch, made by an unknown designer for IBM (International Business Machines Corporation, New York), c. 1939, in painted steel, aluminium and rubber.

ABOVE Front cover of the American magazine *Popular Mechanics*, December 1940, featuring a monorail by Walter Dorwin Teague.

PETER BEHRENS
(1868–1940)

As an architect and multi-talented designer – who worked across the fields of industrial, product and graphic design, as well as furniture, glass and ceramics – Peter Behrens earned the right to recognition as one of the founding fathers of modernism. In his own work and in his teaching, Behrens promoted the idea of a pragmatic alliance between art and utility, advocating a pared-down aesthetic in which ornament was treated sparingly but crafted sophistication remained a priority. He became a co-founder of the influential Deutscher Werkbund*, which also argued for a creative collaboration between the worlds of art and industry. As he once wrote: 'The practical object does not seem to us to be any longer entirely subservient to mere utility, but combines therewith a certain degree of pleasure.'[17]

The significance of this alliance between artistic beauty and utilitarian function became progressively more critical to Behrens as his career progressed and expanded, taking him well beyond the sphere of architecture alone. Later, he also spoke of the importance of industrial design as a process of 'creating forms that accord with the character of the object and that show new technologies to advantage'. Yet this still allowed room, within Behrens's work, for buildings and products that were beautifully conceived and detailed, even within industrial processes of production.

Behrens began his career as an artist. Born in Hamburg, he studied fine art in Karlsruhe and Düsseldorf before moving to Munich, where he started working as a painter, illustrator and graphic designer. In 1899, he moved to the Darmstadt Artists' Colony, where he designed a house in 1901 for himself and his wife, which was much respected as a rounded and cohesive expression of Jugendstil*, the German equivalent of Art Nouveau.

For four years from 1903, Behrens served as director of the Düsseldorf School of Arts and Crafts, before embarking on a new career path. In 1907, he established his own architectural practice in Potsdam and co-founded the Deutscher Werkbund. In that same year, he also began a long relationship with the company AEG, which was to play a large part in defining his legacy and in promoting the birth of industrial design as a discipline in itself.

AEG (Allgemeine Elektricitäts-Gesellschaft) was founded in the 1880s by the entrepreneur and industrialist Emil Rathenau, who acquired German patent licences for the production of Thomas Edison's light bulb. The company grew quickly and by 1907 was one of the leading European producers of not just light bulbs but also generators, motors, transformers and electrical products.

Emil Rathenau and his son Walter maintained many links with writers, artists and intellectuals, and as a company AEG had a proud record of association and collaboration with a number of respected designers, architects and typographers. Behrens was appointed as an artistic consultant, a freelance arrangement which allowed him to pursue other interests even

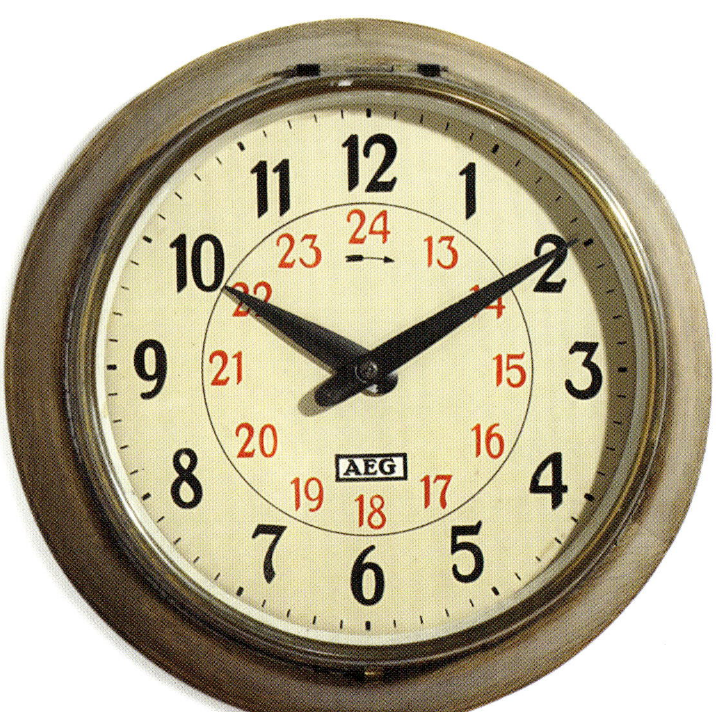

RIGHT A wall clock designed by Peter Behrens, made by AEG, *c.* 1915, in oak, brass, enamelled metal and glass.

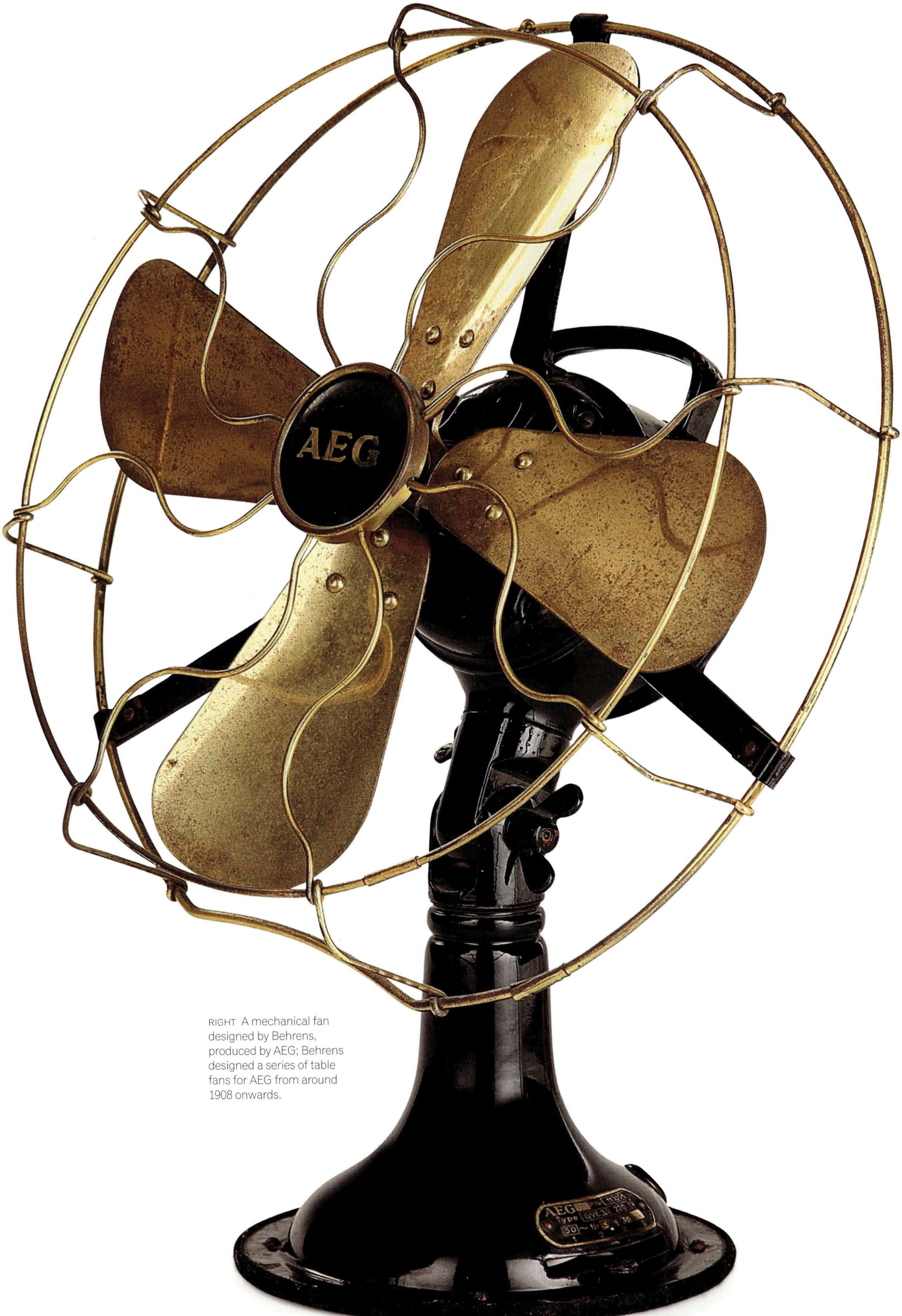

RIGHT A mechanical fan designed by Behrens, produced by AEG; Behrens designed a series of table fans for AEG from around 1908 onwards.

as his influence over AEG grew to a point where he has been credited with the creation of a full 'corporate identity' and – in doing so – establishing a comprehensive precedent for the term itself.

Behrens's remit at AEG was wide-ranging. It embraced architecture, beginning with exhibition stands and pavilions, and then expanded to workshops, power houses and his landmark Turbine Hall in Berlin of 1909. Described as a cathedral of industry, this vast steel, concrete and glass supershed – designed for the production of turbine engines and large enough to accommodate internal lifting cranes – is one of the key markers within the early evolution of modern architecture, with its combination of dramatic open space and a stripped-down industrial aesthetic. Over the coming years Behrens designed a number of other large-scale AEG factory buildings, along with housing schemes for company workers.

He was also involved in many other aspects of design for AEG. Drawing upon his skills as a graphic designer and illustrator, he designed the corporate logo (1907) and exhibition posters, and standardized typography for use in company literature and advertisements. By 1916, he was using an elegant sans-serif design, which reinforced the message of modernity.

At the same time, Behrens also began working on a range of products for AEG, beginning with arc lamps and lighting but also including electric clocks, kettles, clocking-in machines, desk fans and coffee grinders. Collectively, they formed a new generation of products for the machine-age consumer, establishing a fresh portfolio of domestic appliances and labour-saving goods.

Behrens's influence upon architecture and design was profound. Le Corbusier, Walter Gropius and Ludwig Mies van der Rohe were among the younger architects who passed through his atelier. Yet his impact on the nascent profession of industrial design proved to be just as important, if not more so.

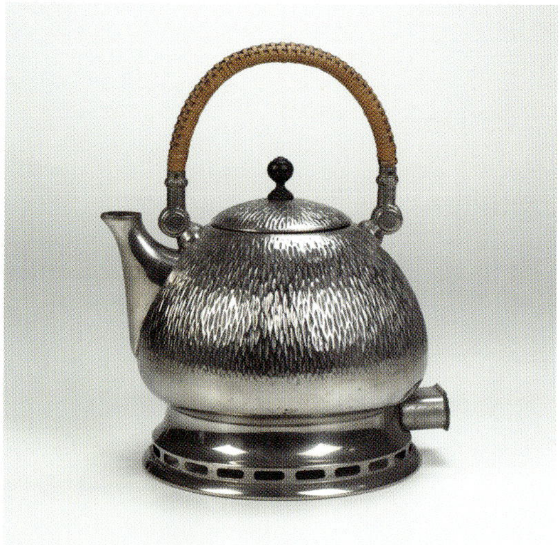

ABOVE A Peter Behrens tea kettle, made by AEG, c. 1909, in nickel-plated brass with a rattan handle.

OPPOSITE Cutlery sets by Peter Behrens, designed for the Wertheim department store in Berlin and for the Behrens House in Darmstadt, 1901–2, in silver, stainless steel and gilded silver.

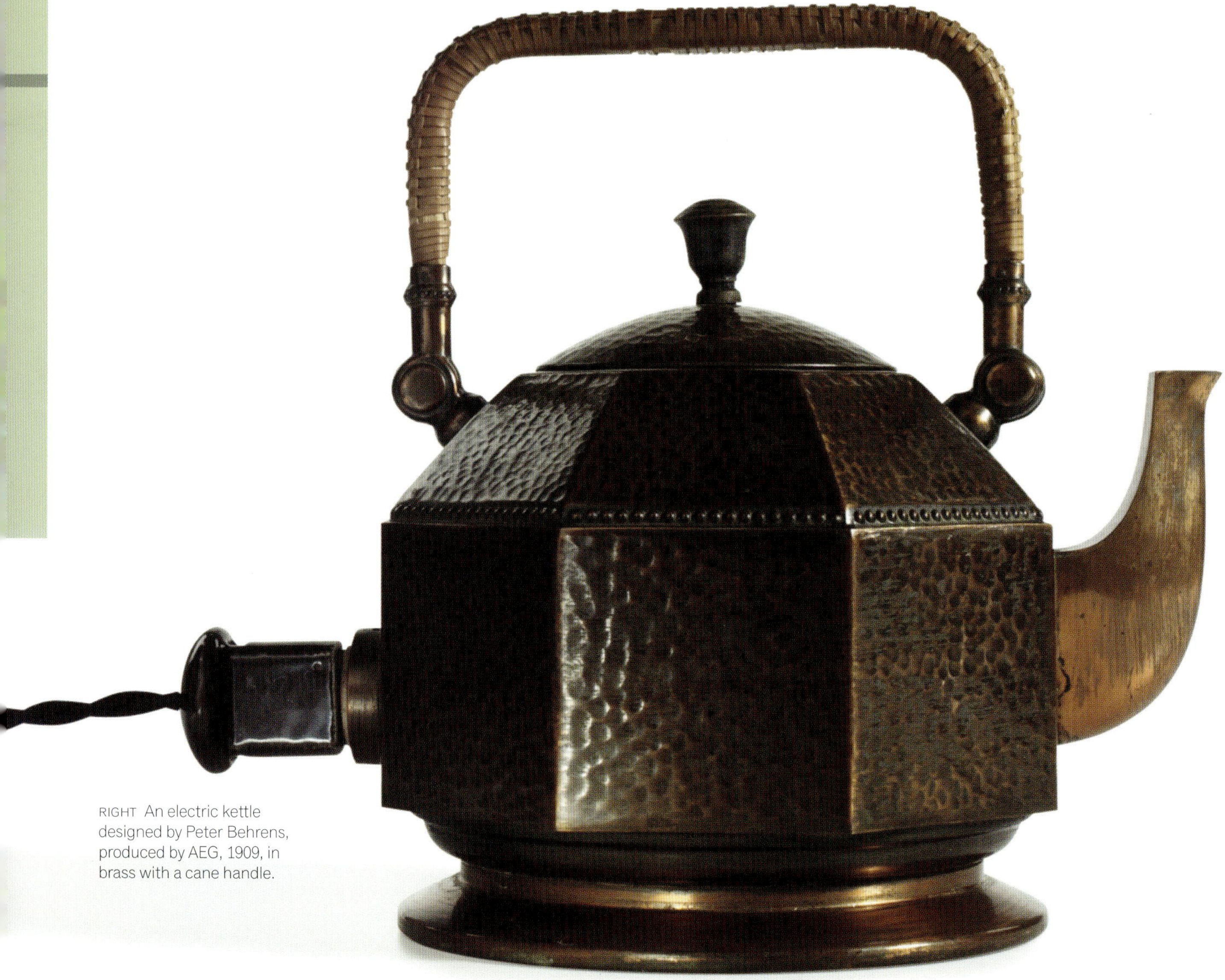

RIGHT An electric kettle designed by Peter Behrens, produced by AEG, 1909, in brass with a cane handle.

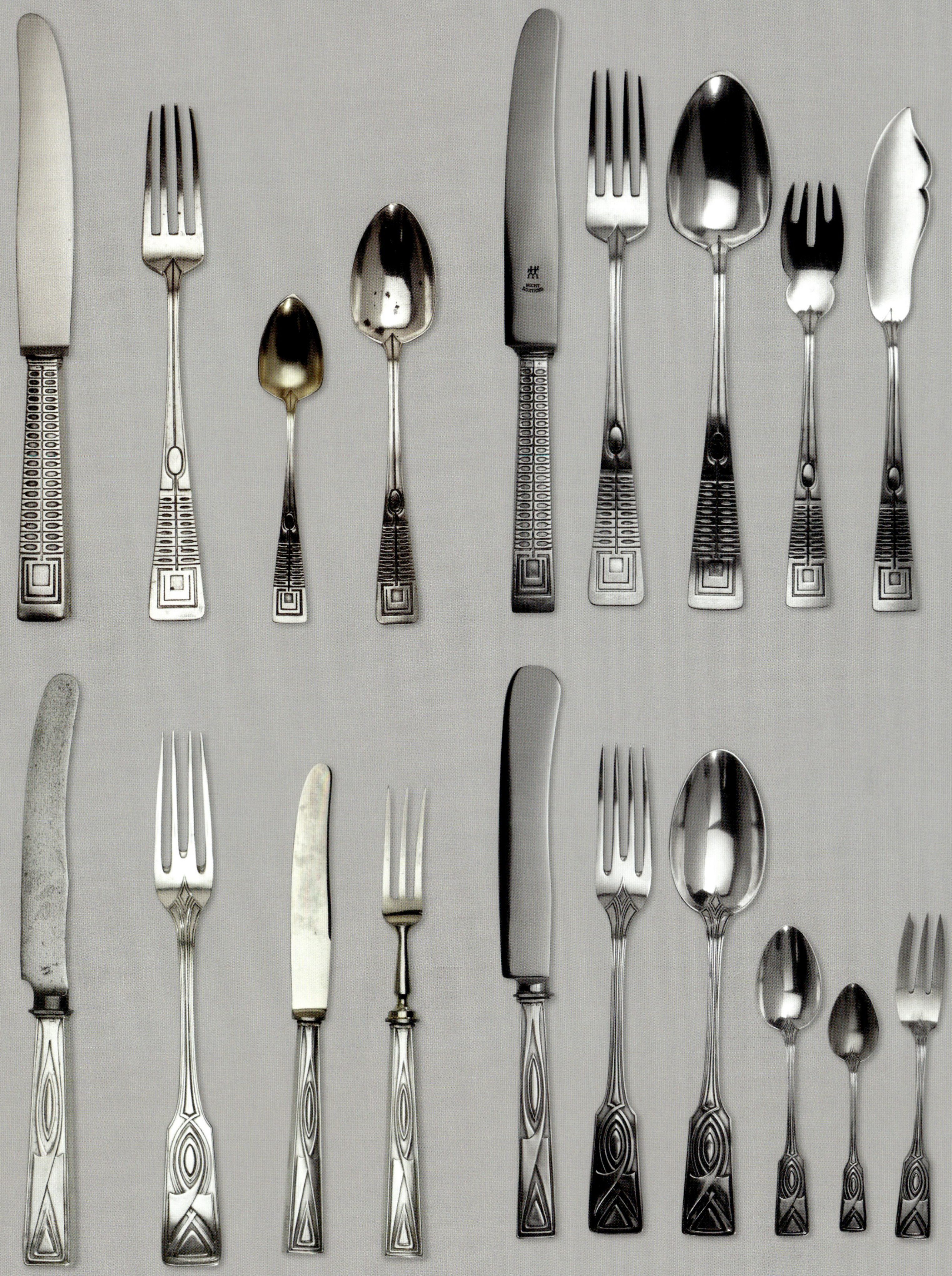

HAROLD VAN DOREN
(1895–1957)

In the year 1940, Harold Van Doren published one of the first handbooks devoted to his young profession. *Industrial Design: A Practical Guide* was a comprehensive manual addressing the many challenges of machine-age design styling, manufacturing and presentation. Van Doren divided his book into four clear parts, beginning with an attempt to define and describe the profession itself, followed by a survey of fundamental design principles, progressing to an analysis of technique and, finally, practice. He included a number of informative case studies: a 'streamlined sled', a meat grinder and a four-pocket dough divider.

'The job of an industrial designer is to interpret the function of useful things in terms of appeal to the eye; to endow them with beauty of form and color; above all to create in the consumer the desire to possess,' wrote Van Doren.[18] This notion of the 'desire to possess' has become a vital ingredient within industrial design, offering the essential point of difference between a mundane object and an enticing product. Van Doren saw the power of design to create everyday objects of desire within a process of aesthetic seduction. This was particularly true of the new generation of domestic appliances: functional, industrially produced machines that had a utilitarian purpose as labour-saving devices, yet also needed to be pleasing enough in appearance to tempt the consumer. Seeking to persuade customers of the desirability of a whole portfolio of goods, Van Doren turned his attention to the design, function and styling of kitchen scales and washing machines, cooking ranges and refrigerators.

As with his contemporaries Walter Dorwin Teague and Norman Bel Geddes, Van Doren's formative years were largely focused on fine art. Following his graduation from Williams College in Chicago and service in an ambulance unit during World War I, Van Doren studied at the Art Students League in New York, then won a fellowship to study art history in Paris. During his years in France, he worked as a lecturer (at the Louvre) and as a translator. From 1927 to 1930, he served as assistant director of the Minneapolis Institute of Arts before a dramatic change of course.

In 1931, Van Doren formed a partnership with John Gordon Rideout and opened an office in Toledo, Ohio, devoted to industrial design. Early commissions included the Air King radio (1933) made of synthetic Plaskon, with a silhouette inspired by the wave of new skyscrapers being built in New York and Chicago. The partners were also invited to design a series of products and appliances for the Toledo Scale Company.

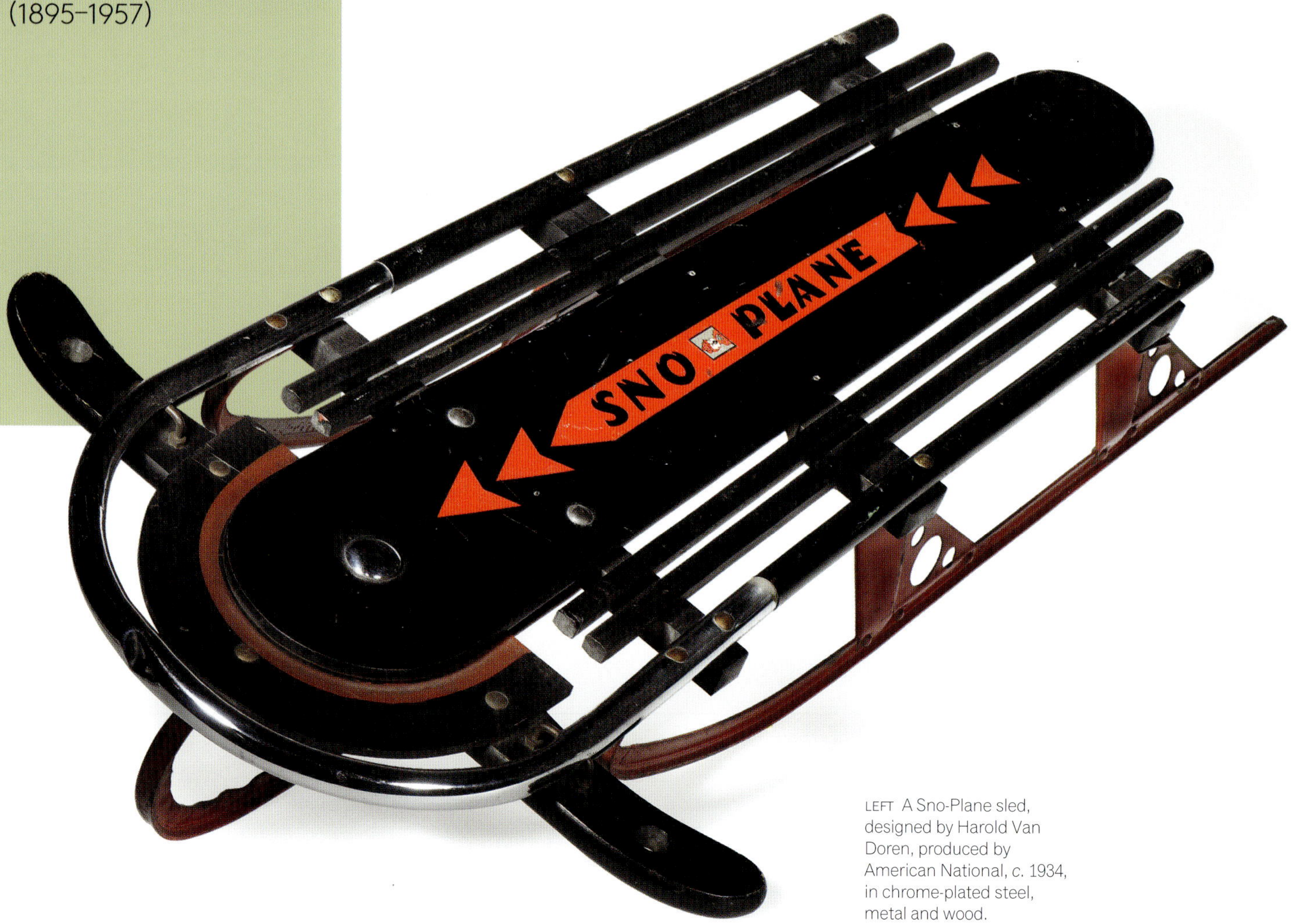

LEFT A Sno-Plane sled, designed by Harold Van Doren, produced by American National, *c.* 1934, in chrome-plated steel, metal and wood.

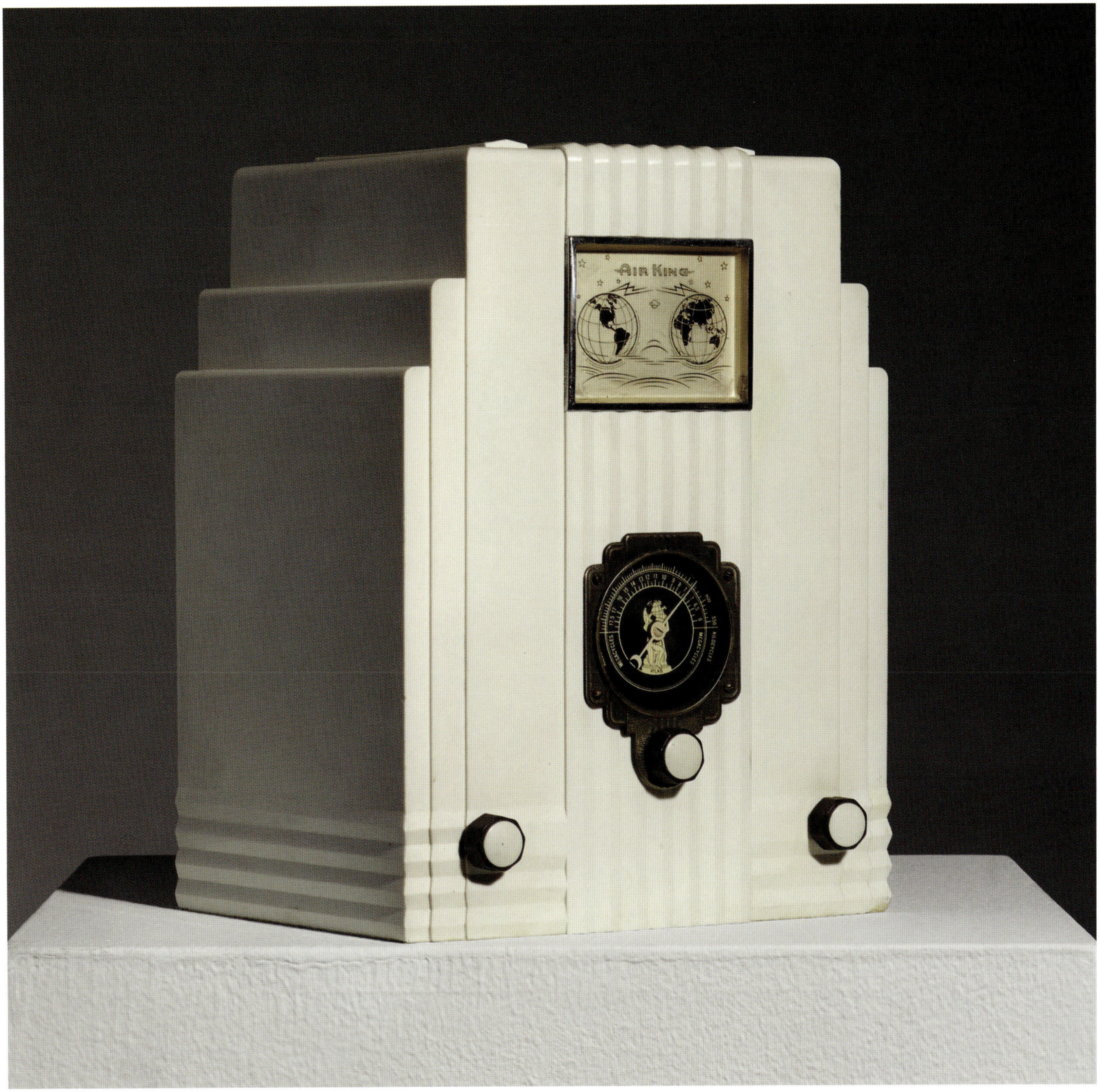

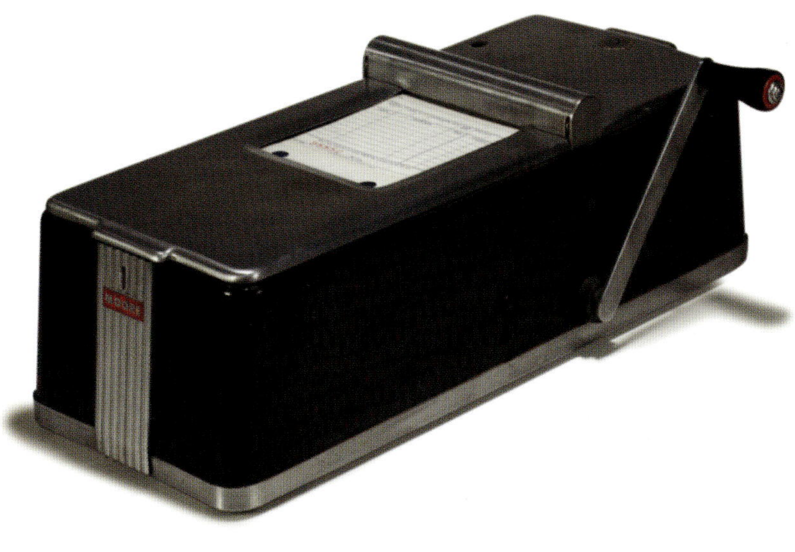

LEFT A Harold Van Doren receipt machine, made by Moore, 1930s, in aluminium and Bakelite.

ABOVE A Model 66 Air King 'Skyscraper' radio by Harold Van Doren, produced by Air King Products, 1933, in plastic, brass and glass.

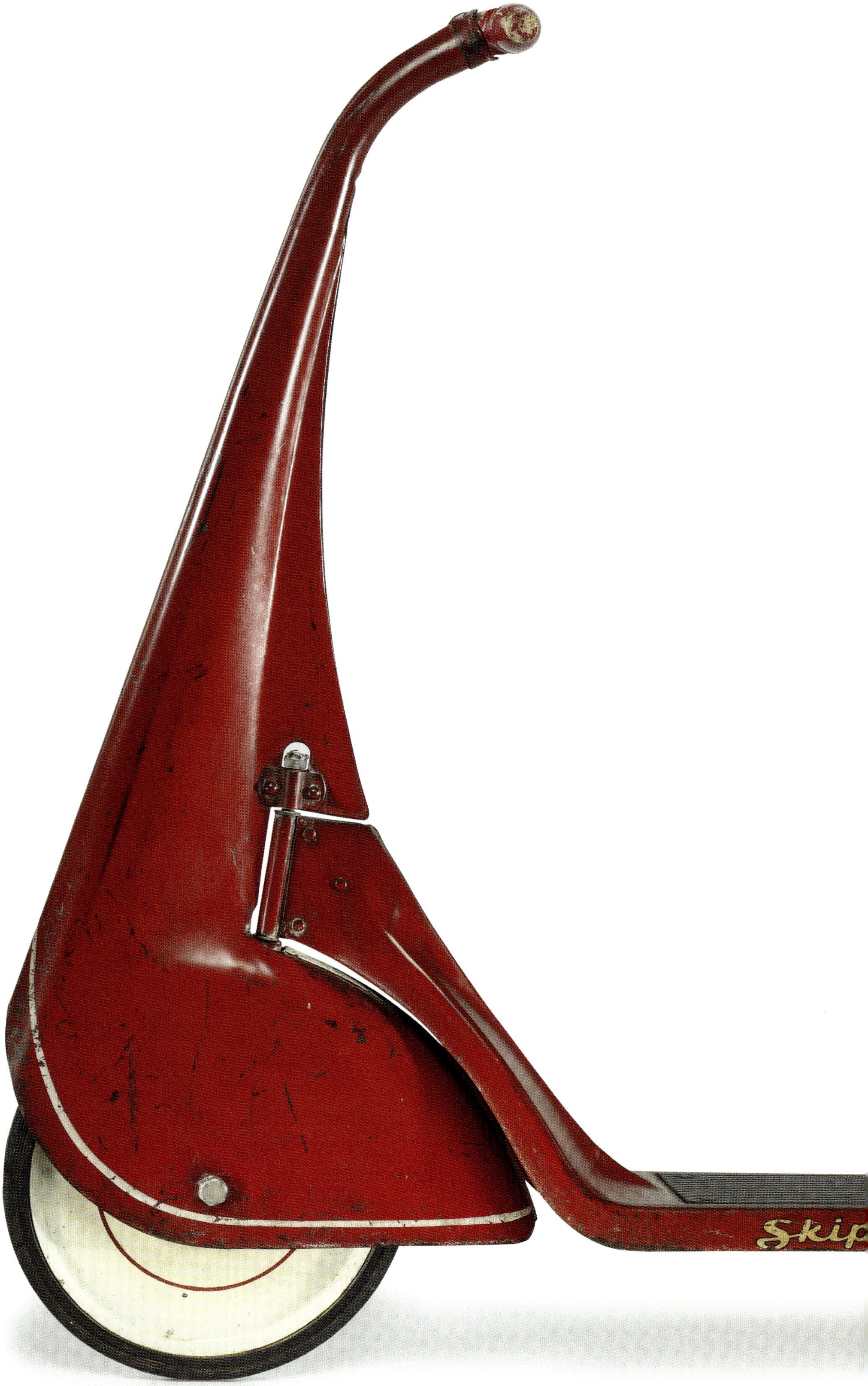

RIGHT Harold Van Doren's streamlined Skippy-Racer scooter, manufactured by American National, *c.* 1933, in steel and rubber.

Within just a few years, Van Doren had earned a reputation as one of the leading industrial designers in America, winning a number of awards.

He was a skilled pioneer of the streamlined aesthetic and devoted a chapter of his own book to the theme. 'Streamlining has taken the modern world by storm,' he wrote, with tongue in cheek. 'We live in a maelstrom of streamlined trains, refrigerators, and furnaces; streamlined bathing beauties, soda crackers, and facial massages.'[19] Noting the origins of streamlining within the automobile industry, as well as aerodynamic trains and planes, Van Doren embraced 'nonfunctional' streamlining, applying the approach to all kinds of domestic products, which were given rounded edges and sinuous casings.

Among Van Doren's most endearing and successful designs were a series of children's scooters, cycles and sleds that used the language of streamlining to maximum effect. In the late Twenties, Van Doren and Rideout designed a child's ride-on scooter consisting of little more than two wheels, a footplate and a handle, produced in an engaging vivid red colour; variants on the design were patented in the Thirties. American National's Sno-Plane used similar techniques, transforming the ride-on sled into a sophisticated speeding arrow, complete with dorsal fins. As Van Doren put it, 'children love streamlining'.

In 1939, Van Doren designed a 'Master Washer' for Maytag, finished in a pure white rather than the ubiquitous grey-greens and browns used for many pre-war appliances; over time such finishes, along with the term 'white goods', were to become standard. Van Doren opened a second design office in Philadelphia in 1941, followed by a New York office in 1944; that same year he co-founded the Society of Industrial Designers.

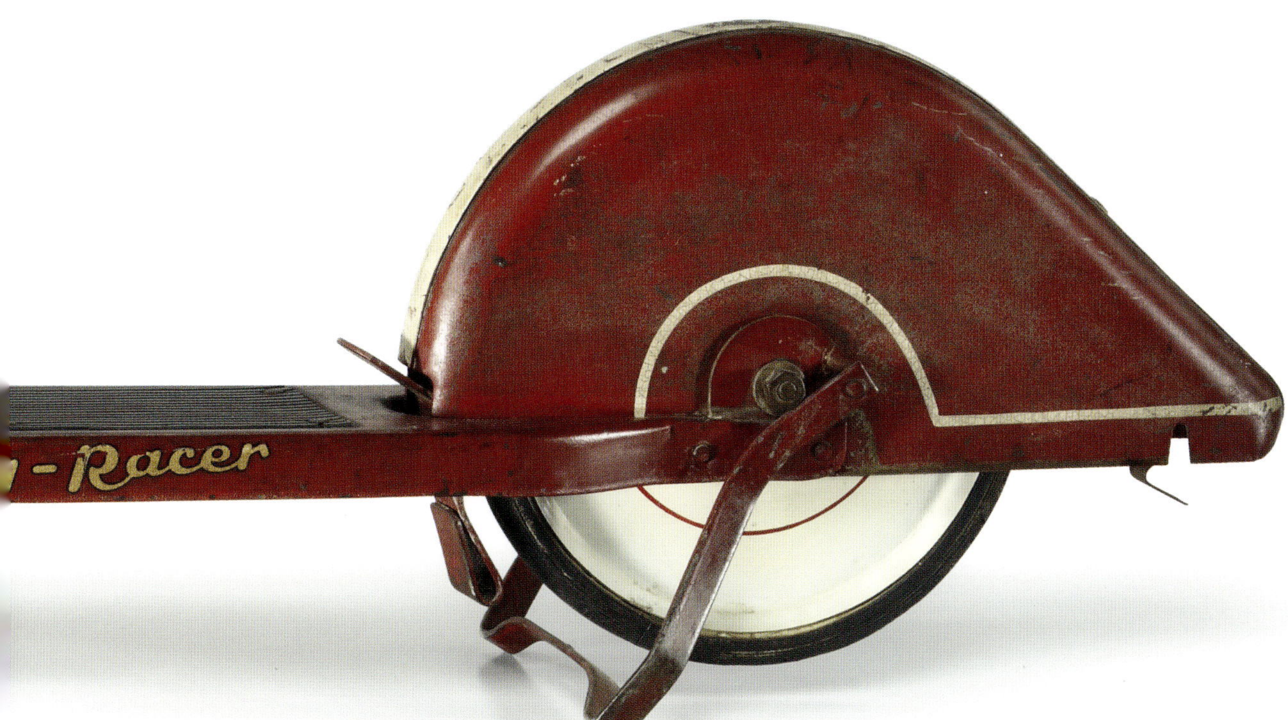

HENRY DREYFUSS
(1904–1972)

The J-3a Hudson locomotive, designed by Henry Dreyfuss in 1938, has become one of the most iconic products of the early modernist era. Commissioned by New York Central Railroad, these sleek engines – with their bullet-shaped noses, punctuated by a central spotlight – offered a definitive image of futuristic, machine-age dynamism. The design itself spoke of speed and efficiency, offering an early foretaste of the aerodynamic bullet trains that came to prominence later in the 20th century. With a maximum speed of 103mph, the new-generation steam locomotives became the workhorses of the Central Railroad's '20th Century Limited', a flagship service between New York and Chicago.

The streamlined Hudson featured a rounded boiler shell and cab, yet the sequence of drive wheels below was left exposed and even illuminated at night, emphasizing the power and raw energy of the train. Famously, Dreyfuss turned his attention to the totality of the design, working not only on the locomotive but also on the carriages, the dining car and observation car, including the interiors and graphic identity of the service. 'In designing a train we not only design the train outside and in, but also every detail and accessory including the porters' uniforms, matches, magazine covers, china, silver, etc. We feel that if we are designing a product, any detail related to that product in the way of cartons, display, lettering, etc, is part of the job.'[20]

By claiming such a broad remit, Dreyfuss helped to shape the character of his own profession, which was still in its infancy during the Thirties, and helped to develop the concept of a corporate identity. Dreyfuss's personal ability to combine professional integrity with such a commanding eye for every element won him many clients and admirers.

Dreyfuss himself came from a modest background. His father was a tailor of German heritage who died in his late thirties, leaving the family in some hardship. But Dreyfuss was awarded a two-year scholarship to the New York Society for Ethical Culture High School, which played a key role in encouraging his interests and future direction. Here, he received a general grounding in the arts, but was also introduced to the applied arts and crafts, including ceramics, textiles, typography and metalwork.

Having graduated from school, Dreyfuss went on to take some classes with NORMAN BEL GEDDES, who was working as a stage set designer. Once the classes had finished, Dreyfuss stayed on in Bel Geddes's office until 1924, working as an assistant. He then worked in the field of theatre and stage design in his own right, creating the sets for a series of productions, including Broadway shows. This

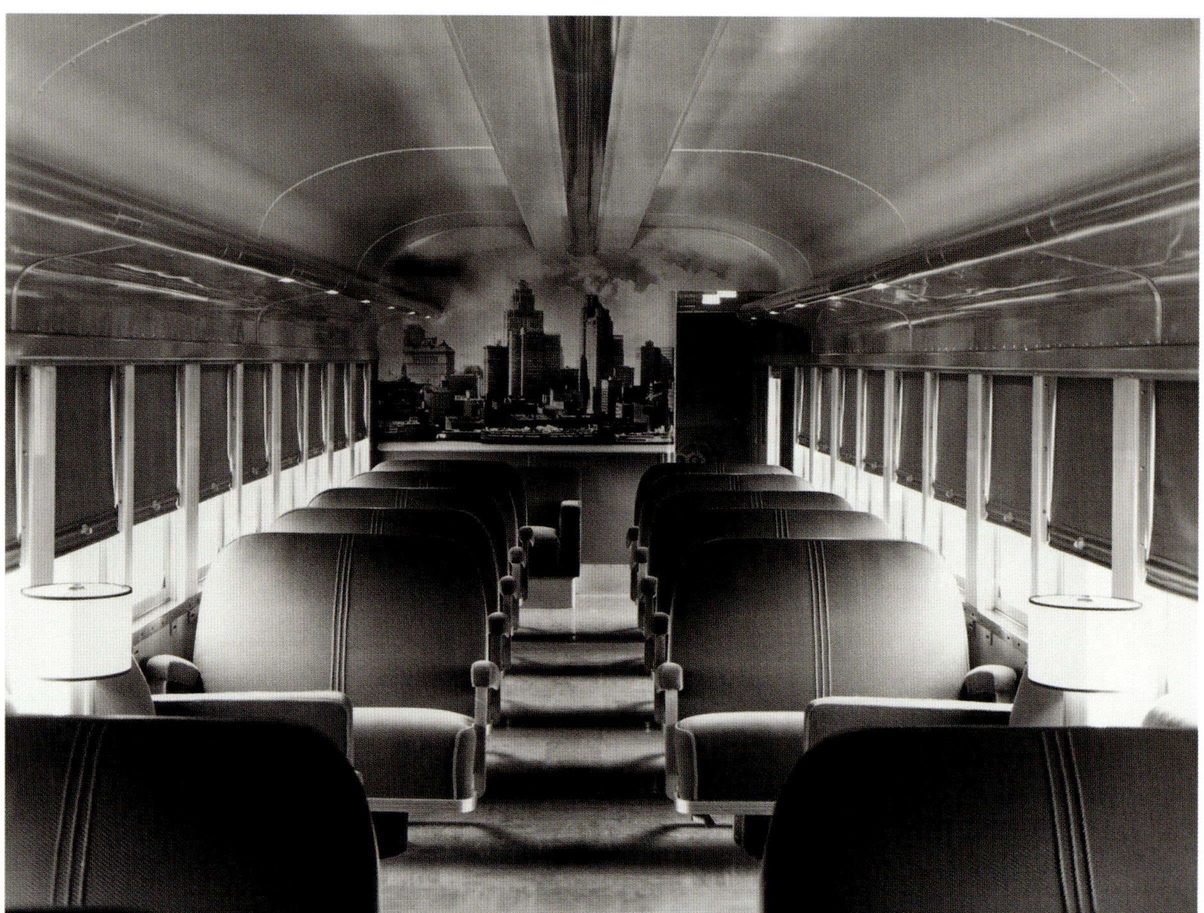

RIGHT The Dreyfuss-designed interior of a passenger coach on the New York Central Railroad's Mercury train, 1936.

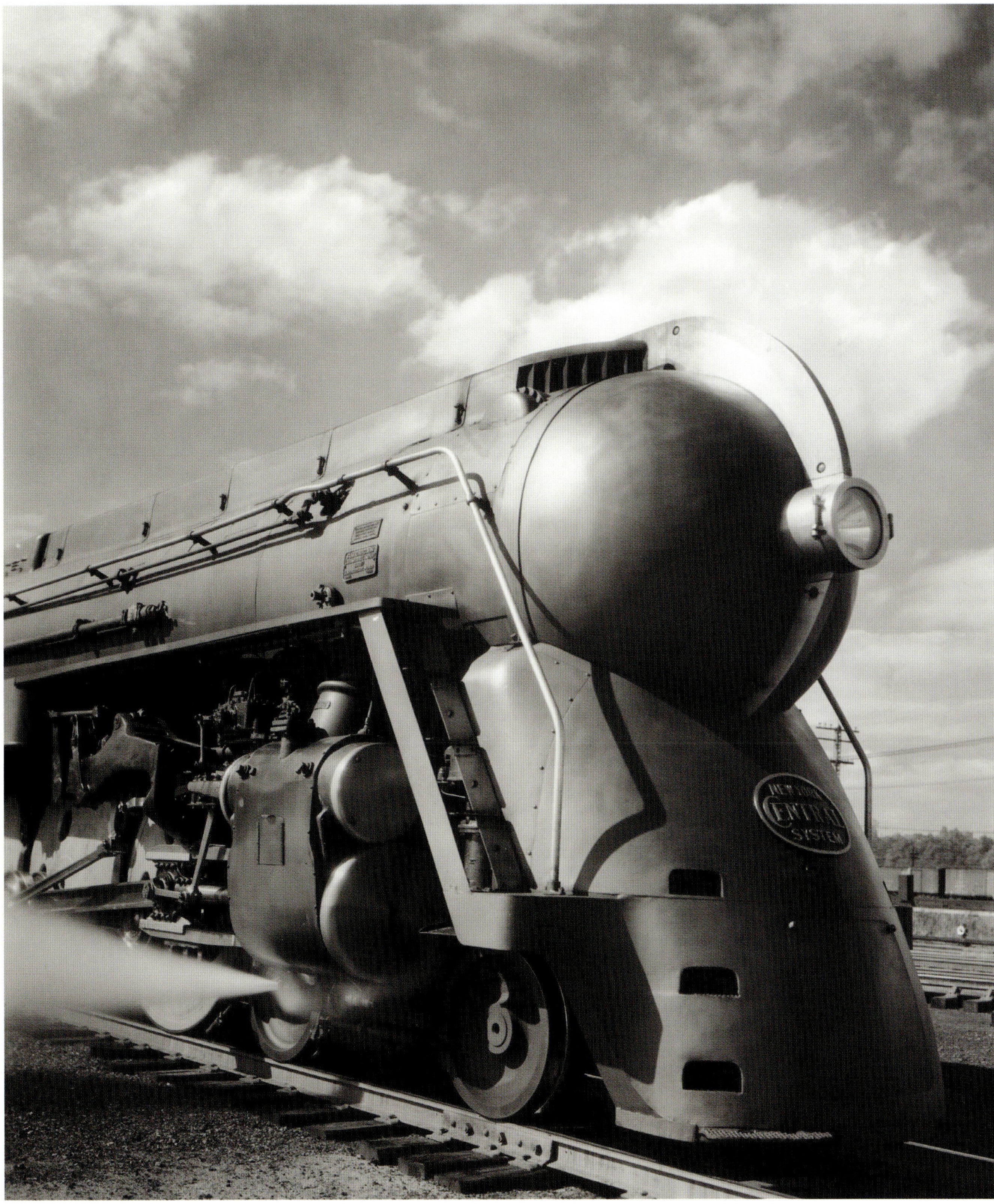

ABOVE A 4-6-4 Hudson steam locomotive designed by Henry Dreyfuss, built by the American Locomotive Company and operated by the New York Central Railroad, *c.* 1938.

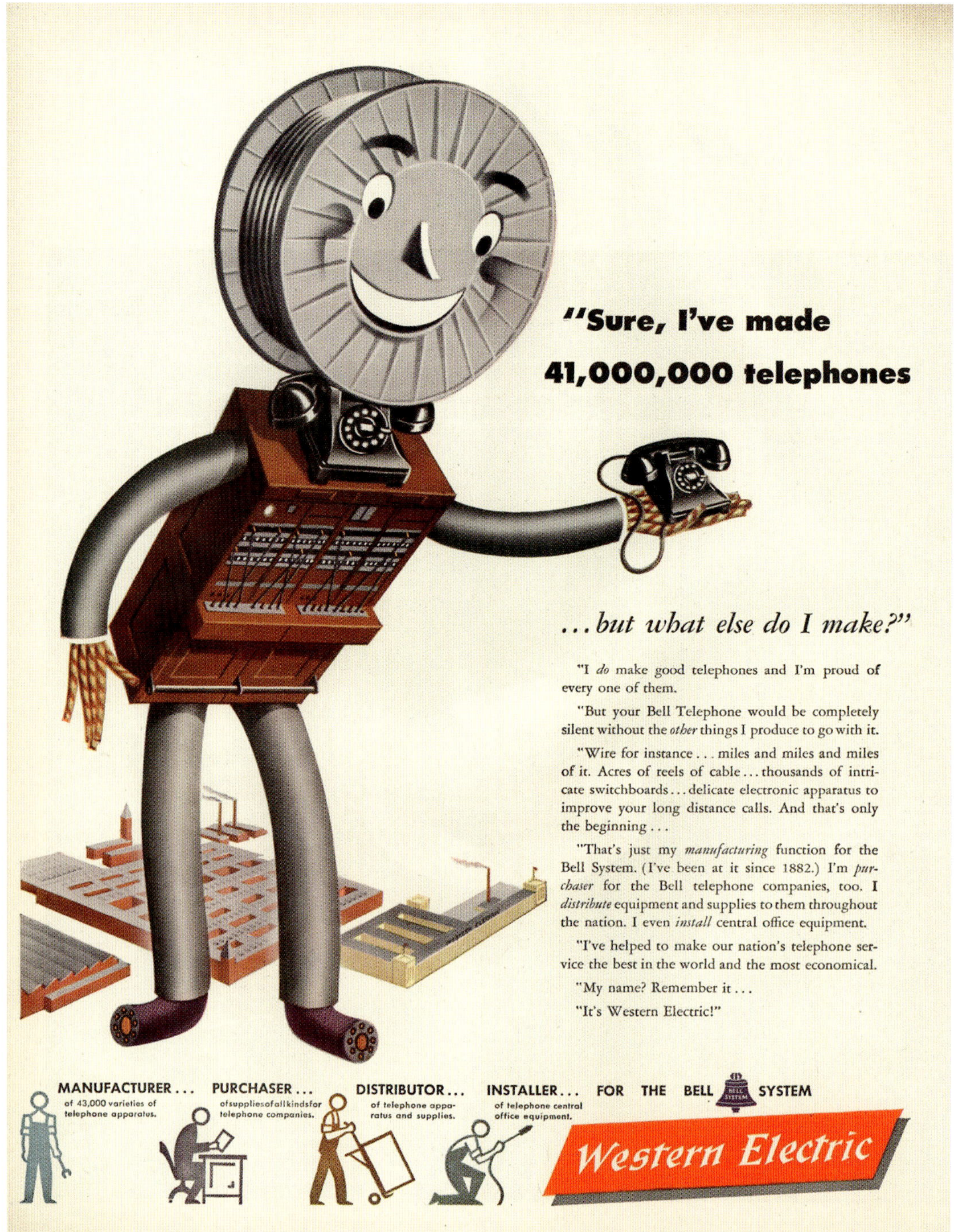

ABOVE A Western Electric advertisement for 300 Series telephones designed by Henry Dreyfuss.

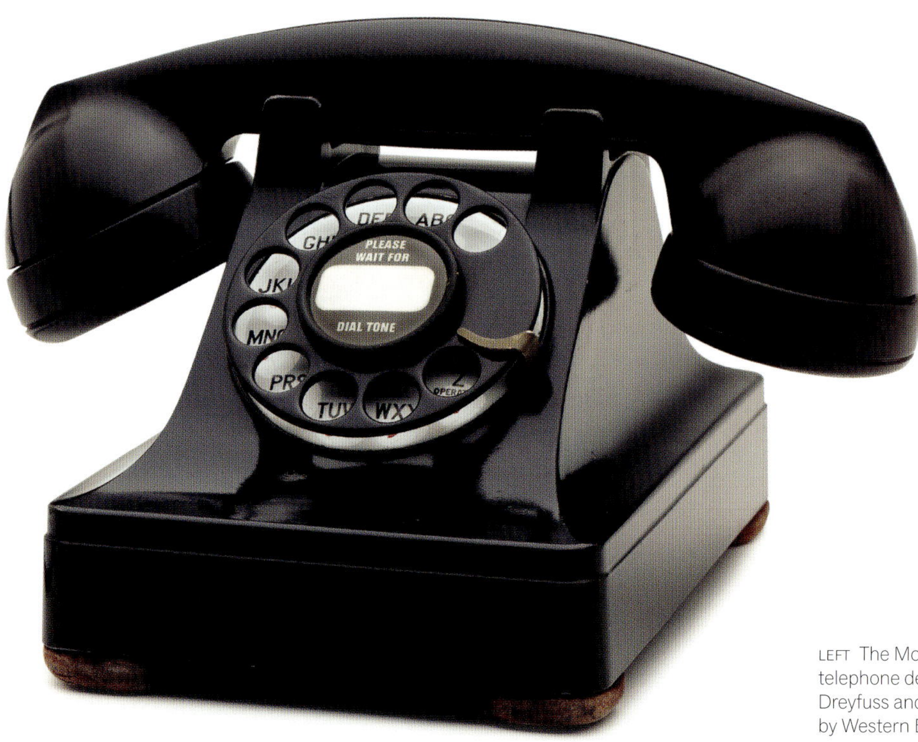

LEFT The Model 302 telephone designed by Dreyfuss and manufactured by Western Electric, *c.* 1937.

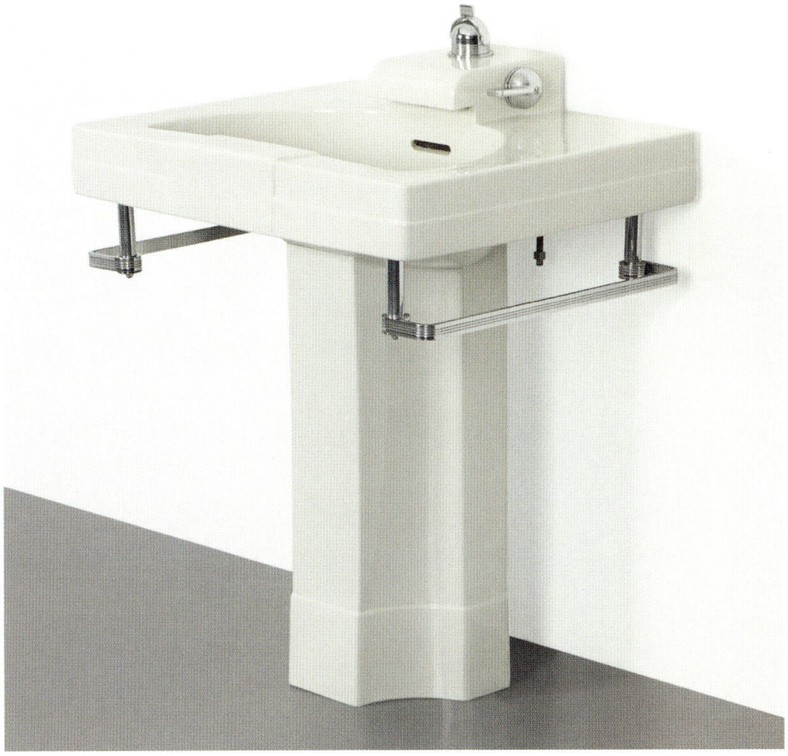

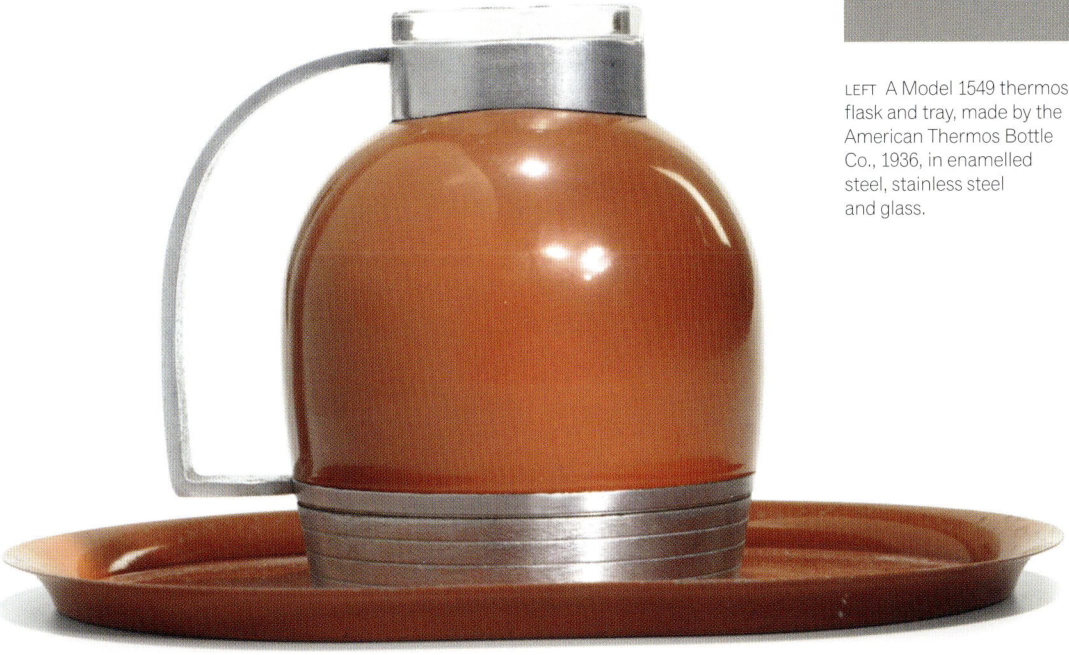

LEFT A Model 1549 thermos flask and tray, made by the American Thermos Bottle Co., 1936, in enamelled steel, stainless steel and glass.

ABOVE Dreyfuss's Crane Neuvogue pedestal sink and faucet, made by the Trenton Potteries Company for Crane Co., *c.* 1939, in porcelain, aluminium and chrome.

was followed by the interior design of a number of theatres for RKO, before he opened an office for industrial design in 1928; his well-connected wife, Doris Marks, helped establish the office and promote useful links with potential clients.

Early commissions included packaging and kitchen utensils, which were followed by designs for a series of household appliances, including refrigerators for General Electric and household heating thermostats for Honeywell. The workload grew quickly, with Dreyfuss determined to immerse himself in the design process of each product, analysing the needs of the consumer on the one hand and working in close collaboration with engineers on the other.

Two of Dreyfuss's most famous projects found their way into countless homes. The first was a new vacuum cleaner for Hoover, the company first founded in Ohio by William Henry Hoover in 1908, having bought a patent for an 'electric suction sweeper' from inventor James Murray Spangler. The early upright machines were basic, clumsy and a little intimidating, in a primitive mechanical way. Dreyfuss, with his Model 150 cleaner of 1936, introduced a more streamlined design with a Bakelite housing for the motor. The refreshed machine – which helped to pioneer a new verb, 'to hoover' – was efficient and user-friendly, while appealing to the Thirties obsession with domestic cleanliness and familial health.

Similarly, Dreyfuss reinvented another technological staple: the telephone. The Bell Telephone Company launched an invited competition to redesign their telephones, which commonly took a 'candlestick' form in the early Thirties, with a separate mouthpiece and receiver. By threatening to walk away from the project, Dreyfuss persuaded Bell to let him have direct access to their engineers on an equal footing as he developed his design for the Model 300 telephone (1937), with its ergonomic, combined ear- and mouthpiece resting upon a contoured cradle made of shiny black phenolic resin with a numbered central dial. This telephone continued to be produced up until 1950. In the Sixties, Dreyfuss pioneered a new phone for the Western Electric Company, the Trimline (1964), which was also to have a profound influence upon the evolution of modern communications.

RICHARD BUCKMINSTER FULLER
(1895–1983)

In so many ways, Richard Buckminster Fuller was ahead of his time. Back in the Thirties, the architectural designer, inventor, theorist and entrepreneur pioneered design concepts that would only begin to seep into the mainstream decades later. These included, above all, concepts for prefabricated houses, though Fuller also pioneered new approaches to transport and off-the-grid living. Most prominent among his building products of the pre-war era was his futurist Dymaxion House, perhaps the closest thing to a 'machine for living in' ever produced during the early modernist period.

Fuller was preoccupied with the modernist ideal of making good design available to the masses through industrialized, factory-production processes, yet at the same time he was famously driven by the idea of 'less for more', emphasizing the need to employ resources, energy and materials in a sensitive and sustainable way for the good of 'spaceship earth'. As one of the most original thinkers, theorists and teachers of the 20th century, Fuller was to have a profound influence upon younger generations of designers and architects.

He was born in Massachusetts to a well-heeled New England family. His maternal grandmother bought Bear Island in Penobscot Bay, Maine, where the family spent a great deal of their summers, with the young Fuller gaining here a love of the ocean and sailing. His father died young, just as Fuller was graduating from Milton Academy. He went on to study at Harvard University but was expelled twice over.

After working in a number of manual jobs, Fuller enlisted in the US Navy Reserve and went on to train at the Naval Academy in Annapolis before joining the Navy itself. This was a formative period, with service as a naval communications officer and rescue boat commander. Having left the Navy in 1919, Fuller co-founded a construction company called Stockade with his father-in-law in 1922, specializing in lightweight building methods. In late 1927, the company ran into financial problems and Fuller was ousted.

Following this period of crisis, Fuller rallied and began to apply his construction and engineering knowledge to the idea of prefabrication. Early concepts for 4D housing – with the fourth dimension defined as 'time' – included a prefabricated apartment tower originated in 1928. Around the same time Fuller began working on initial designs for a 4D, factory-produced home, which was christened the Dymaxion House in 1929 (a combination of the words 'dynamic', 'maximum' and 'tension').

Heavily influenced by Fuller's preoccupation with boats and naval design, the first Dymaxion House concept was a compact hexagonal model with the roof and floor plates radiating from a central supporting pillar, while the floor plan was divided into a series of spaces according to function and fitted with utilitarian pods containing the bathroom, kitchen and services. Fuller exhibited various drawings and models of the house, looking to secure investors, but the Wall Street Crash of 1929 made the immediate evolution of the project more than challenging.

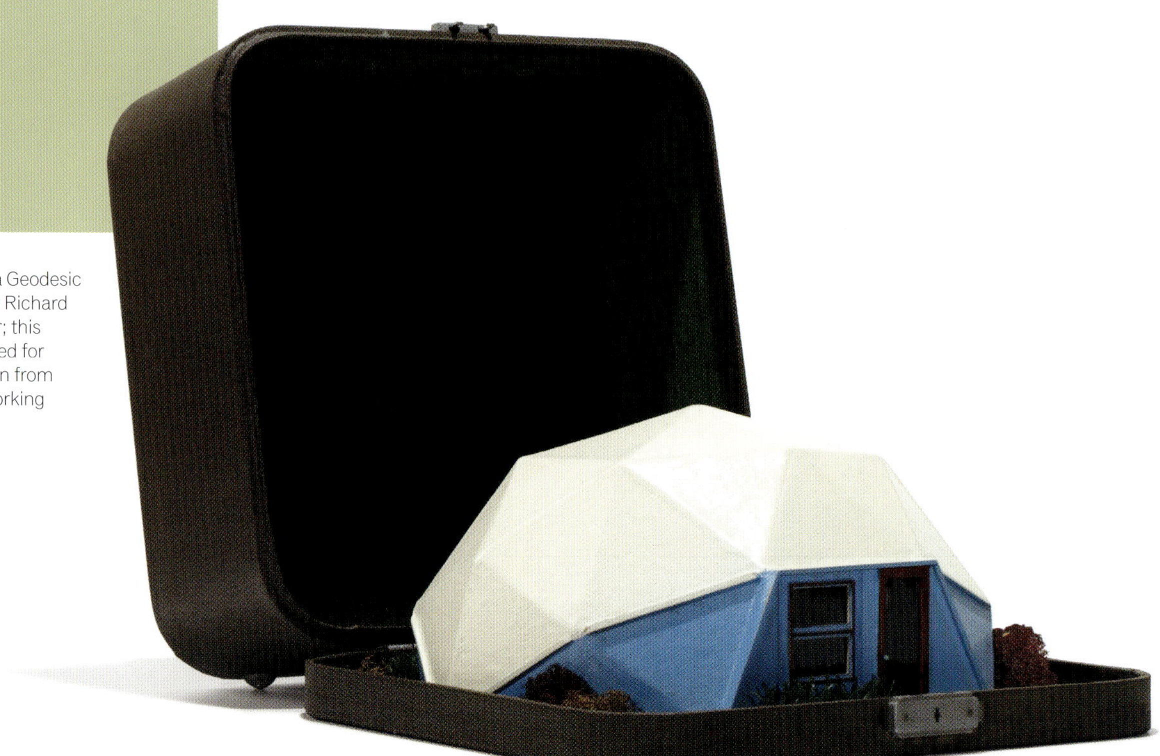

RIGHT A model of a Geodesic Home designed by Richard Buckminster Fuller; this model was produced for the use of salesmen from the Pease Woodworking Company, c. 1960.

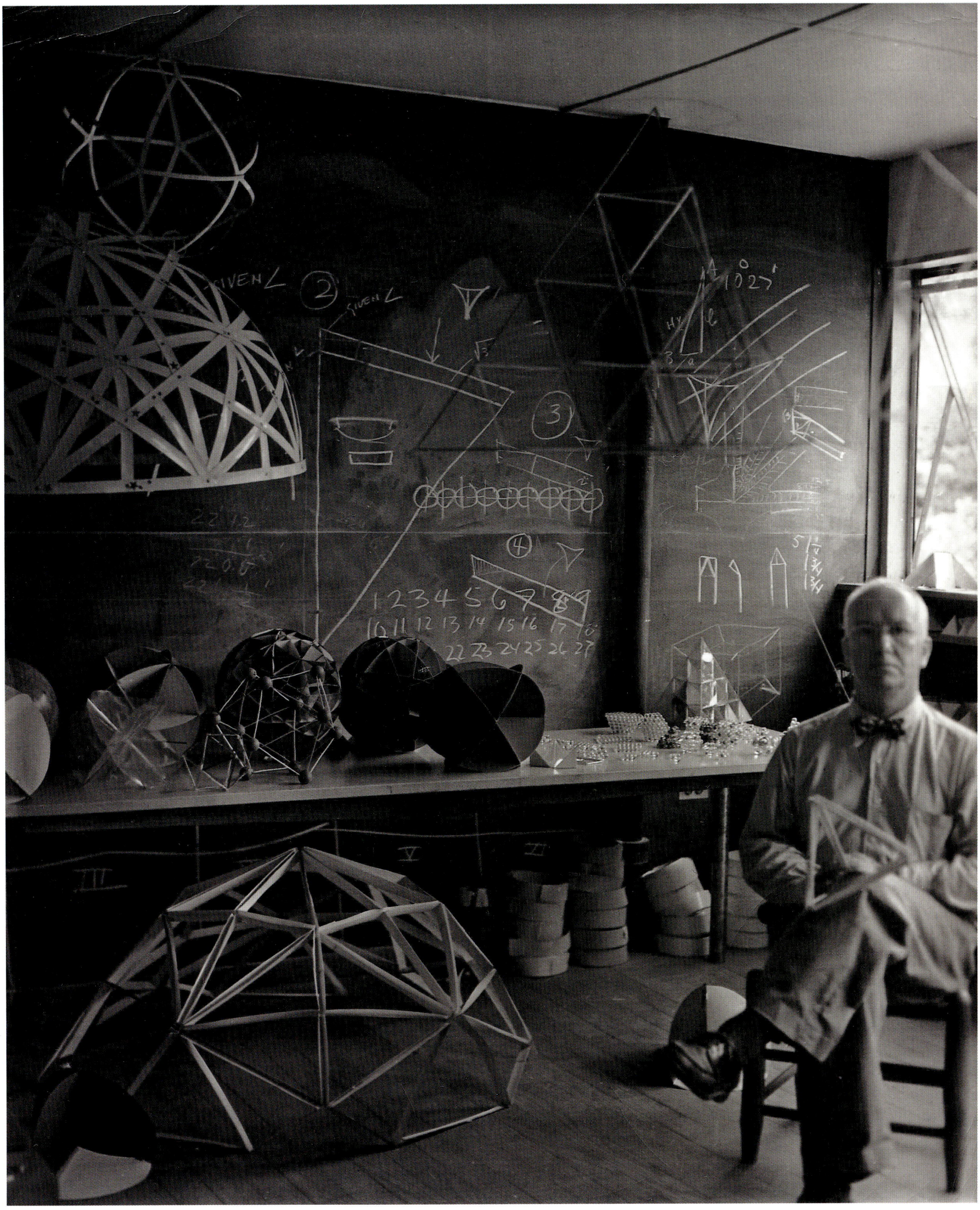

ABOVE A portrait of Richard Buckminster Fuller taken in his classroom at Black Mountain College, North Carolina, 1948.

He, however, continued to develop his ideas and during the early years of World War II developed Dymaxion Deployment Units – flexible, lightweight structures for military use in remote locations. Towards the end of the war, Fuller realized that there would soon be a golden opportunity to create mass-produced civilian housing to help during the reconstruction effort in Europe and elsewhere. By 1945, he had completed the first prototype of what was now known as the Dymaxion Dwelling Machine (also known as the 'Wichita House').

The refined Dymaxion Dwelling featured a rounded, streamlined body made of aluminium panels supported by a lightweight metal frame. The entire house was designed in such a way that it could be shipped to site in a large tube, and easily and rapidly assembled. Fuller Houses Inc. developed two prototype buildings and began to seduce potential clients and investors, only for Fuller to falter and delay at the crucial moment, looking to refine the design one step further before production. Investors began to walk away – an inglorious end for a project that ultimately helped to shape a push towards prefabricated housing concepts in the final quarter of the 20th century.

Fuller also developed a second project in parallel with his prefabricated home concept. The Dymaxion Car, like the house, looked like a prop from a science-fiction movie, with a streamlined fuselage and prominent tailfin. The car had a sequence of elongated, linear windows and the engine was situated to the rear. It ran on three wheels, with the front two providing propulsion and the third, back wheel used for steering. Fuller produced two prototypes in 1933, one of which was exhibited at the Century of Progress International Exposition in Chicago that October, only for the reputation of the vehicle to be marred by an accident in which another car drove into the Dymaxion, killing its driver. Fuller used part of his inheritance from his mother to help fund a third prototype in 1934, but – despite initial interest from a number of automotive manufacturers, including Chrysler – he ran out of money and was forced to close down his factory. The Dymaxion Car has become something of a cult design icon in more recent years and an endless source of fascination.

Despite his setbacks and disappointments, Fuller continued exploring new ideas with admirable determination and enthusiasm. In the post-war years he began a second career as an educator, working with students at the Massachusetts Institute of Technology and elsewhere. During these teaching programmes, Fuller enlisted the help of his students in developing his most successful and enduring design initiative from 1948 onwards: the geodesic dome.

Light, strong and adaptable, Fuller's patented dome can be seen as the product of a natural evolution of many of the ideas first explored in the Dymaxion House. By the Fifties, the domes were capturing the attention of government bodies including the US military, who began using them as emergency shelters. In total, over 300,000 geodesic domes have been built around the world, including the American Pavilion at the Montreal Expo of 1967 and a central portion of the Amundsen-Scott South Pole Station, used between 1975 and 2003.

BELOW The third prototype of Richard Buckminster Fuller's Dymaxion Car at the Chicago World's Fair, 1934.

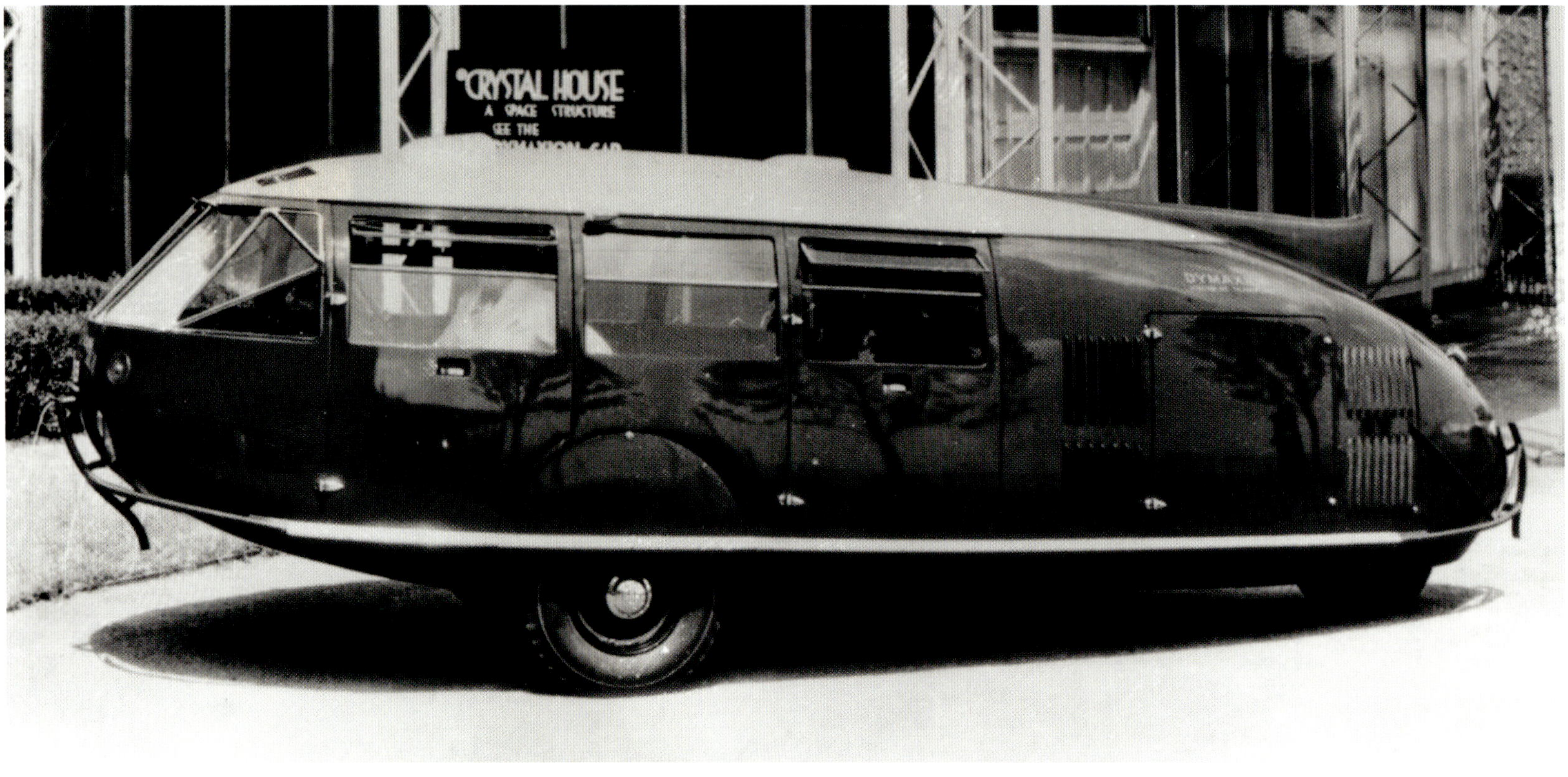

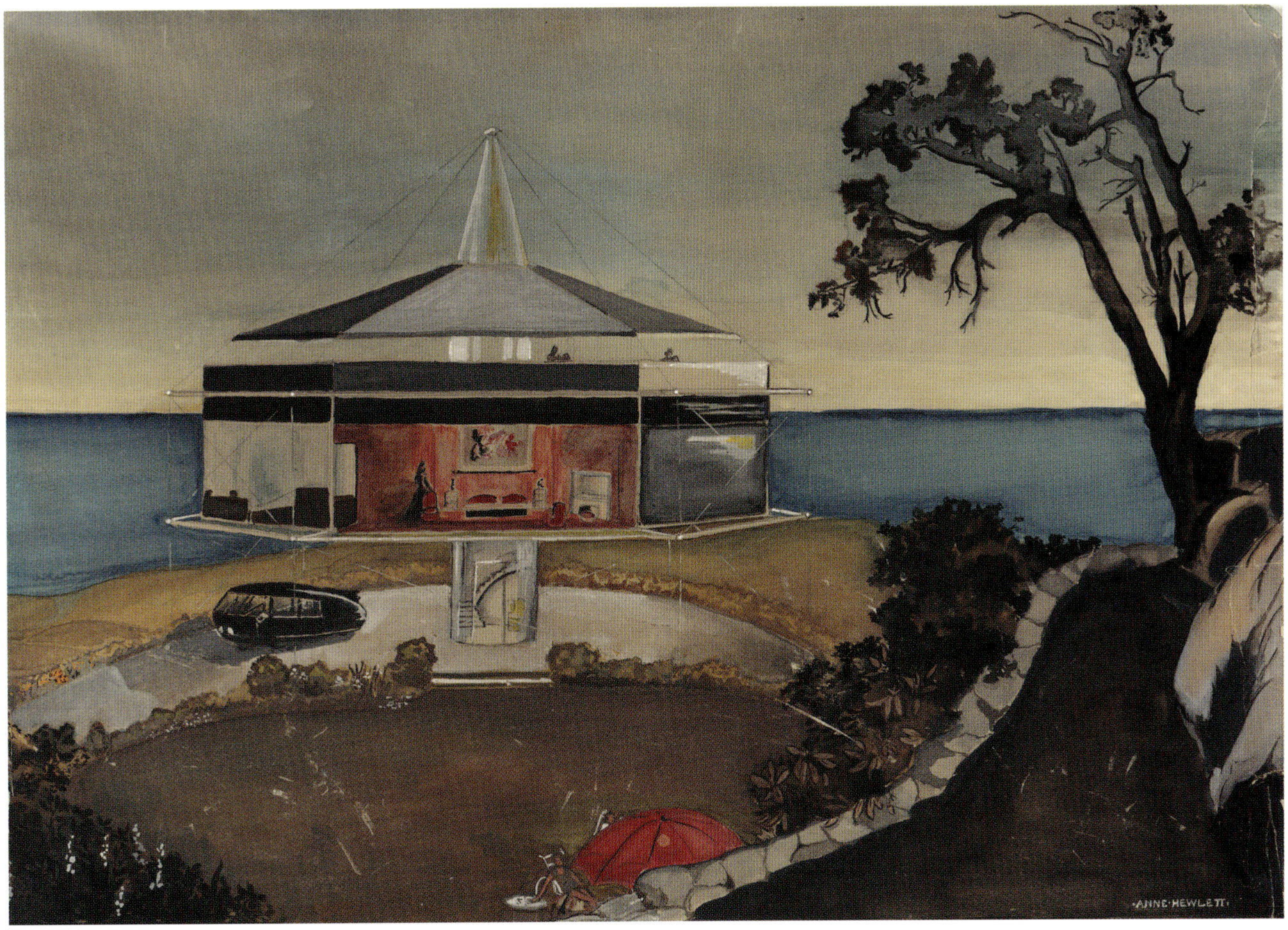

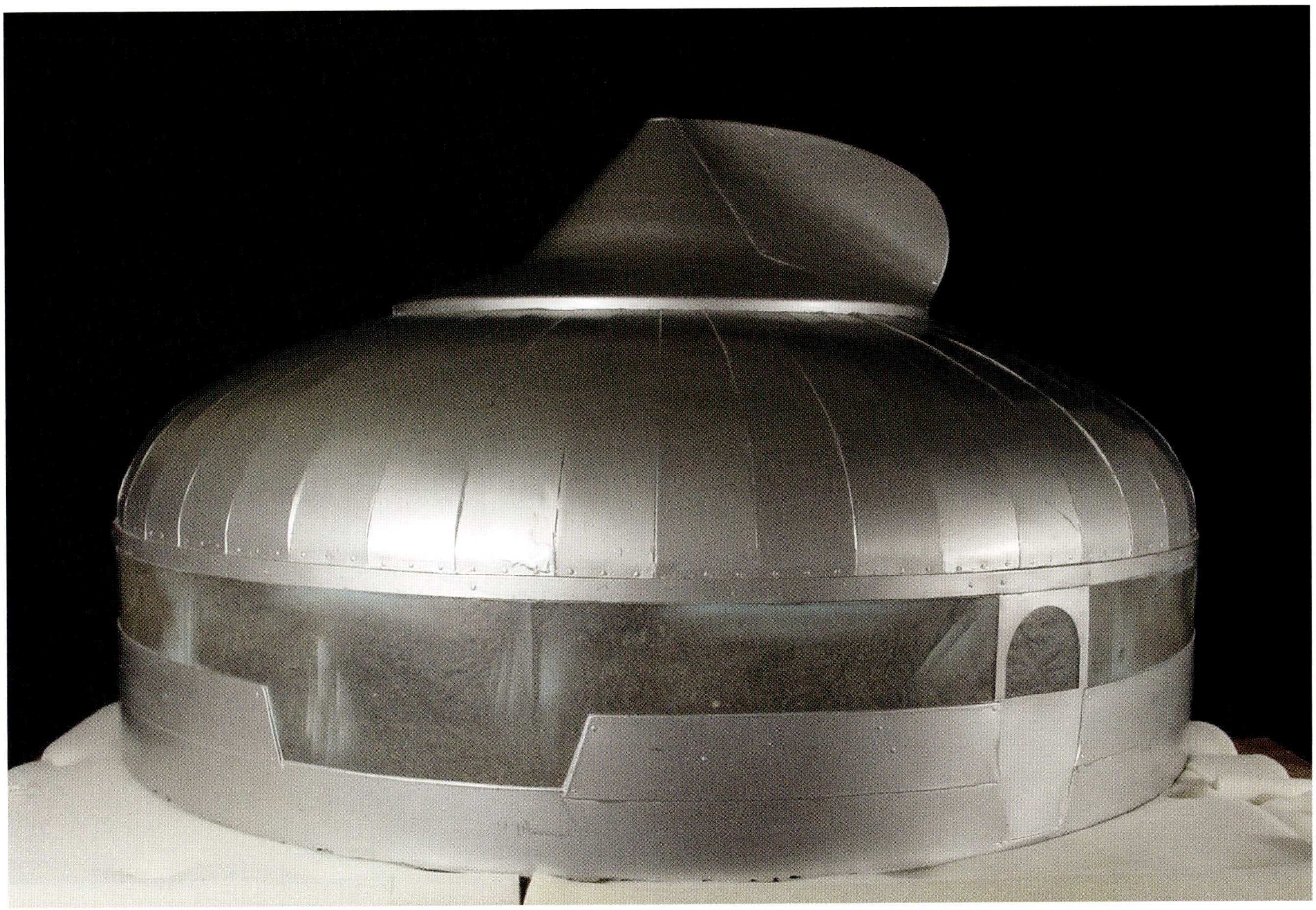

ABOVE A watercolour of Fuller's Dymaxion House, painted by his wife Anne Hewlett Fuller, 1932.

LEFT A model of Fuller's Dymaxion Dwelling, or Wichita House, 1946.

NORMAN BEL GEDDES
(1893–1958)

For Norman Bel Geddes, theatre and design were inextricably linked. A consummate showman and publicist, Bel Geddes was brilliant at creating a sense of dramatic interest around himself, his work and his many clients, for whom publicity and press relations became part of his service offer. Yet, more than this, he fused theatre and design in his visual and aesthetic approach. As one of the first and most prominent industrial designers of the Thirties, he was more concerned with the outward appearance and cinematic dynamism of a product rather than its mechanics. Indeed, this passing interest in the engineering of everything from household appliances to super-yachts and futuristic automobiles was, perhaps, Bel Geddes's Achilles heel, which undermined the value of some of his own concepts, theories and presentiments, many of which were way ahead of his own time.

Bel Geddes enjoyed playing the role of the visionary futurist, portraying and promoting a streamlined world of speed and connectivity, with bubble cars sweeping across slender bridges and along super-highways until they reached pinnacle skyscrapers or airports, served by a new generation of airliners. In many ways, Bel Geddes offered a vision of a dream world, created by an alliance of technology and design. There was even a special word for this talent: 'dreamlining'.

This particular role as a visionary was best expressed in Bel Geddes's striking 'Futurama' exhibit at the 1939 New York World's Fair, funded by General Motors. The exhibition – housed in an extraordinary curvilinear pavilion, accessed by a long, winding, processional, ramped walkway – offered a fusion of theatrical spectacle and dreamline design. Audiences were treated to the future in miniature in the form of a vast model, complete with motorways and bridges, dams and air terminals, skyscrapers, parks and cathedrals. Spectators were offered badges, or buttons, on departure, reading: 'I Have Seen the Future'.

Norman Geddes was born in Adrian, Michigan, in 1893. His father died young, leaving the family struggling. The young Norman showed a talent for drawing and illustration and took classes at the art institutes of Chicago and Cleveland before taking on work as an illustrator. He adopted the 'Bel' in tribute to his first wife, Helen Belle Sneider, whom he married in Toledo, Ohio, in 1916. That same year, the couple moved to Los Angeles.

In California, Bel Geddes embraced a career in theatre set design, which was to continue throughout much of his working life to varying degrees. As a stage designer he was credited with many innovations, such as the first combined focus and flood lamp and the quick-change pivot stage. In 1917, Bel Geddes moved to New York, where he began working for the Metropolitan Opera Company and Broadway theatres. Within the space of just a few years he had won a reputation as a leading theatre designer.

In 1927, he began designing window displays for New York stores, and then a year later opened an office devoted to industrial design. Over the coming years he designed a range of products, including furniture for the manufacturers Simmons, scales and appliances for the Toledo Scale Company, kitchen ranges for the Standard

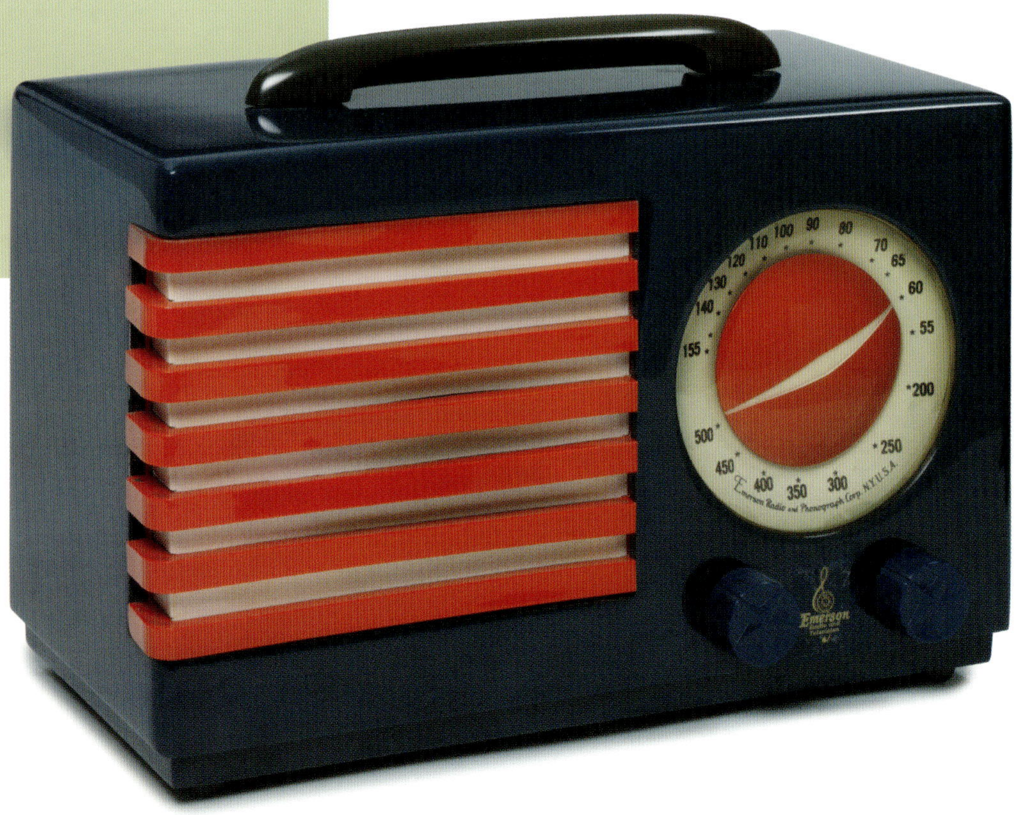

LEFT Norman Bel Geddes's Emerson 'Patriot' radio, c. 1940.

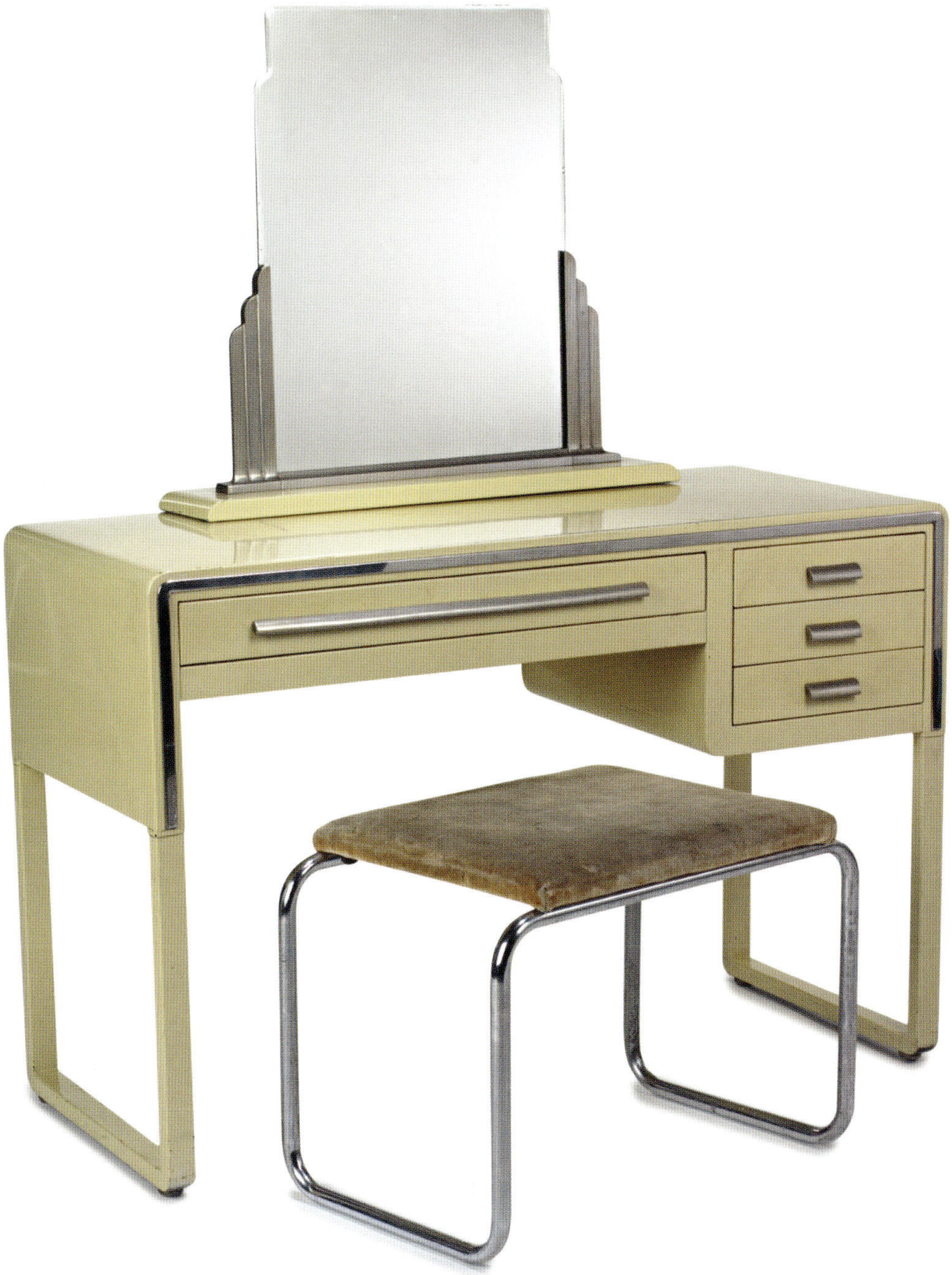

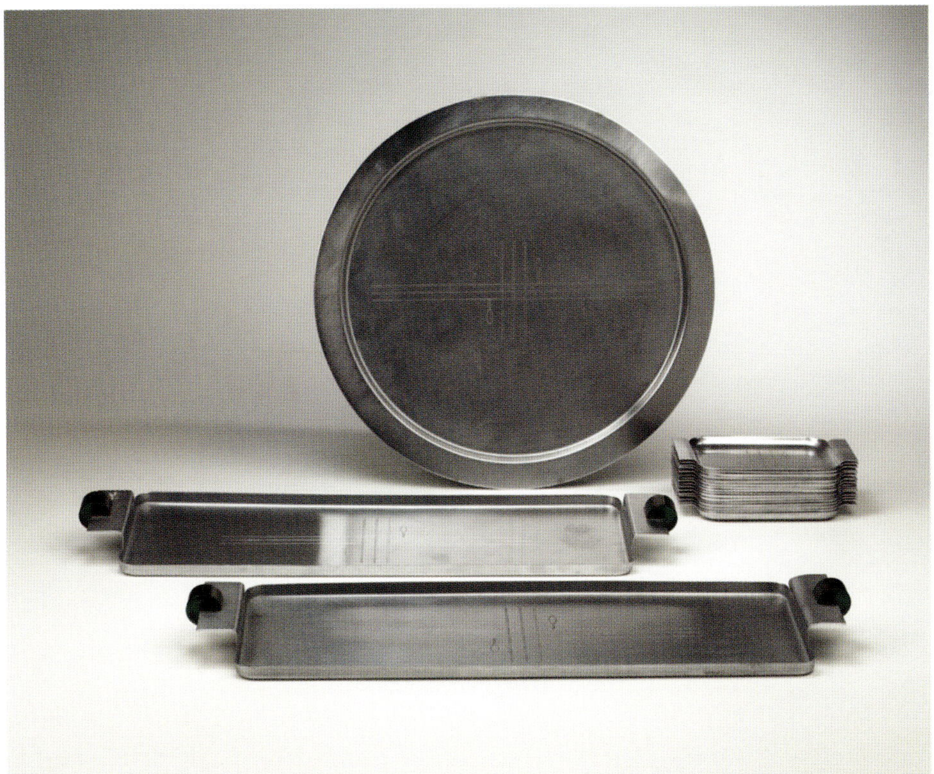

ABOVE A Model 16085 vanity, mirror and stool designed by Norman Bel Geddes, made by Simmons, c. 1935, in lacquered wood, aluminium, upholstery and glass.

LEFT Serving trays designed by Norman Bel Geddes, made by Revere, 1930s, in aluminium and Bakelite.

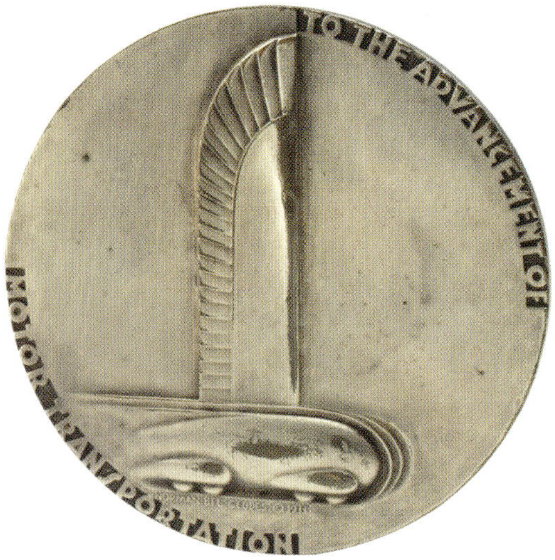

ABOVE A presentation medal designed by Norman Bel Geddes to commemorate the 25th anniversary of General Motors, 1933.

Gas Equipment Corporation, refrigerators for the General Electric Company, vacuum cleaners for Electrolux, and homeware for Revere Copper and Brass, including a cocktail set with shapes and styling inspired by the new-generation skyscrapers of Manhattan.

In all of his work, Bel Geddes was a leading advocate of streamlining. This approach was applied not just to product design, but also to a range of ideas for transportation by road, rail, ocean and air. In 1933, Bel Geddes was commissioned by Pan Am to design the interiors of its Clipper fleet of seaplanes, which flew between Miami and destinations in Latin America. The hybrid notion of the seaplane was explored around the same time in a number of concepts for super-sized airliners that could take off from the water. Later there was an idea for a flying car, another futuristic hybrid, which was also explored by Bel Geddes's former employee, Henry Dreyfuss.

Bel Geddes's work in the automobile industry reflected an ongoing interest not only in cars, but also in infrastructure and the implications for cities and nations of growing car use and ownership. In 1933, he was commissioned by Chrysler to work on the styling of its Airflow model, a car that the company hoped would be as popular as Henry Ford's* Model T. The ambition was for a car that would be relatively inexpensive and adaptable, with a rear-mounted engine, but sales were modest and the car that truly met the challenge was Ferdinand Porsche's Volkswagen Beetle.

Around the same time, 1931–33, Bel Geddes explored designs for a dreamline bus, a streamlined locomotive, a rounded liner (which almost resembled a submarine with windows) and a teardrop-shaped automobile with an elongated sequence of curved windows, a design that echoed in some respects Richard Buckminster Fuller's Dymaxion Car of 1933.

Many of Bel Geddes's concept designs were never put into production, but production was not always the point. A large number of his sketches were intended for promotional purposes. His highly influential ideas meant that his name came to be seen as enhancing the value of the products he created or styled, like the signature of an artist or craftsman.

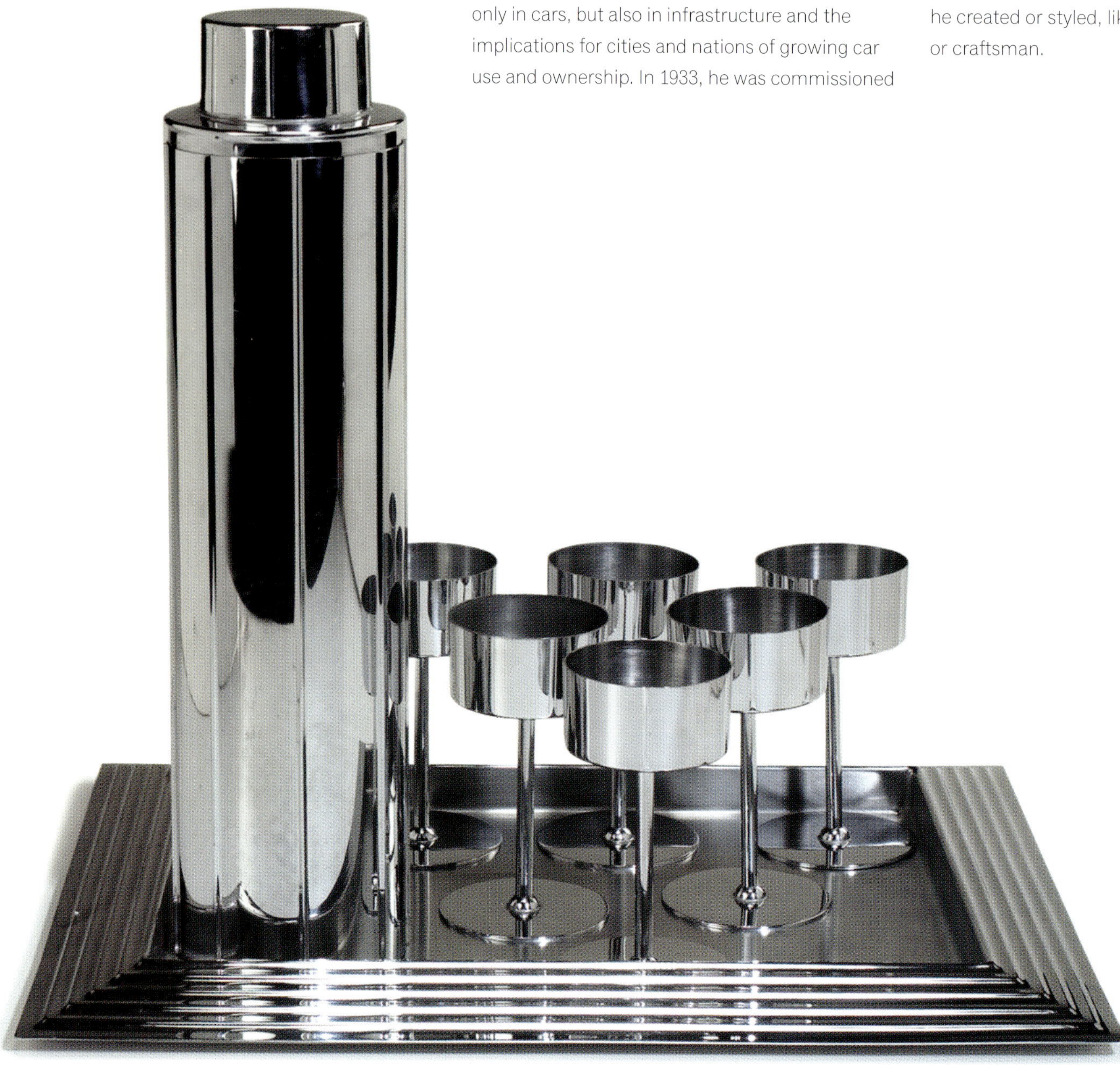

LEFT A Manhattan cocktail set, made by the Revere Copper & Brass Co., 1937, in chrome-plated brass.

OPPOSITE A cigarette card depicting a concept for a streamlined transatlantic vessel designed by Norman Bel Geddes, *c.* 1935.

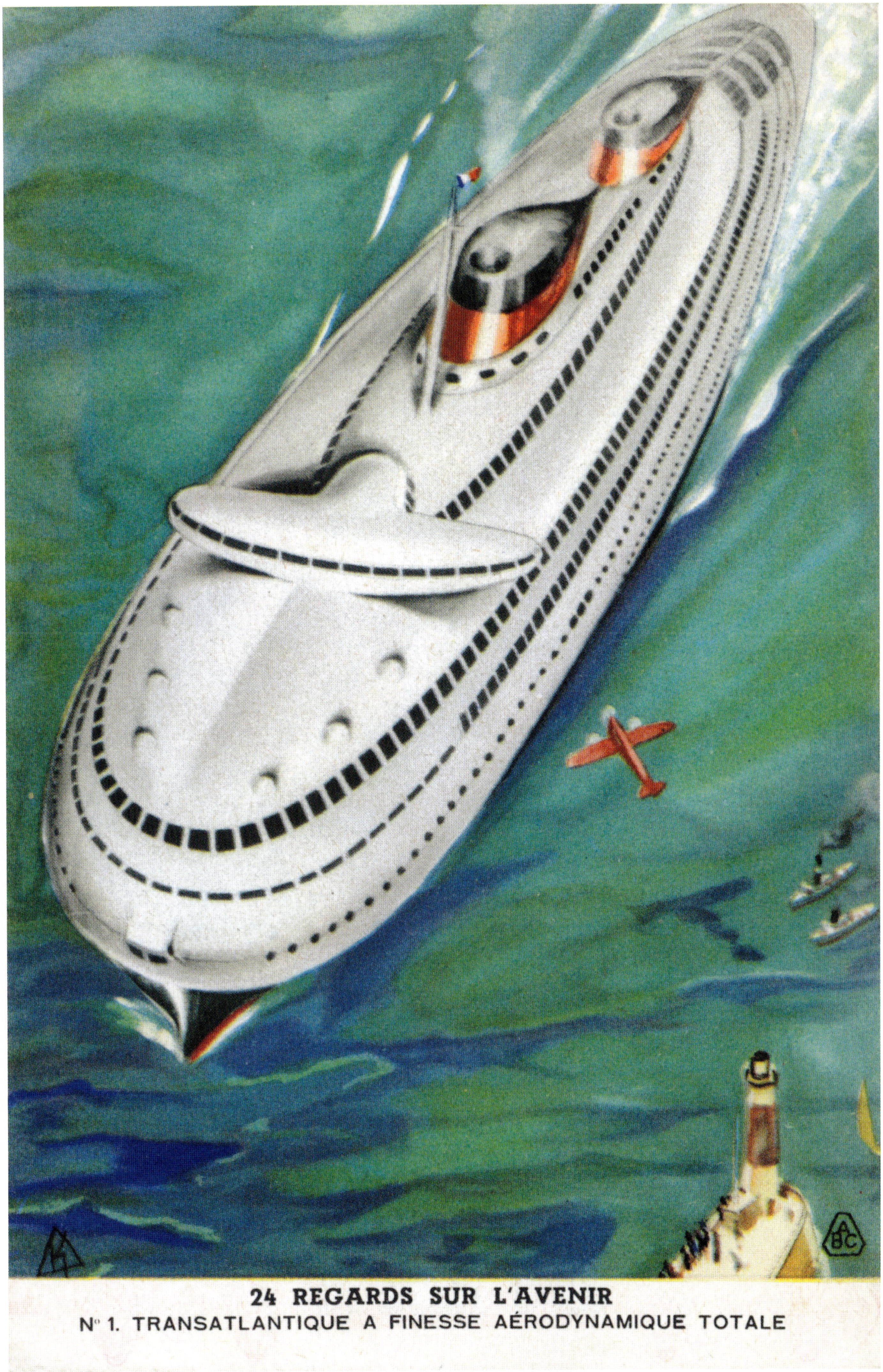
24 REGARDS SUR L'AVENIR
N° 1. TRANSATLANTIQUE A FINESSE AÉRODYNAMIQUE TOTALE

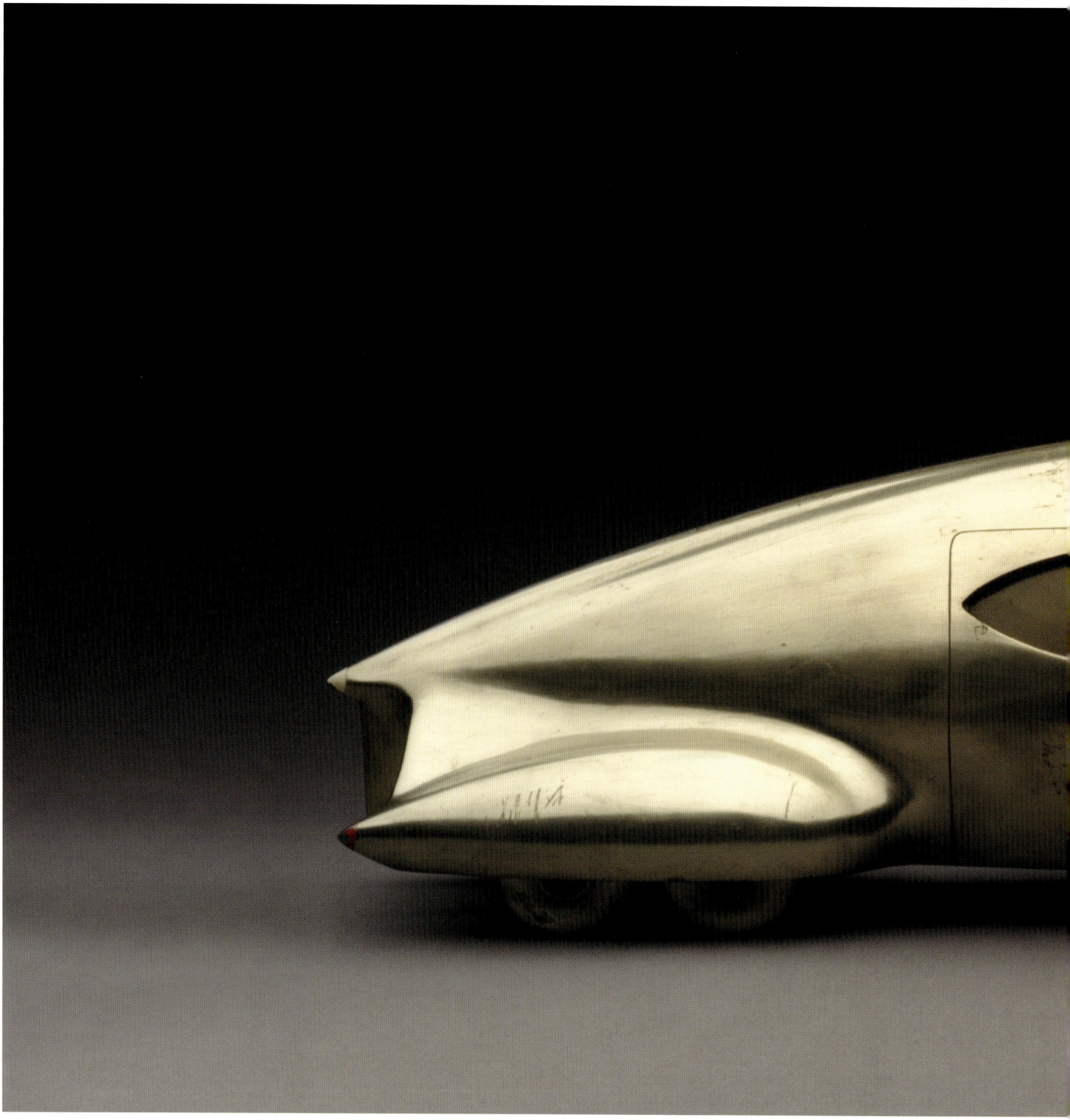

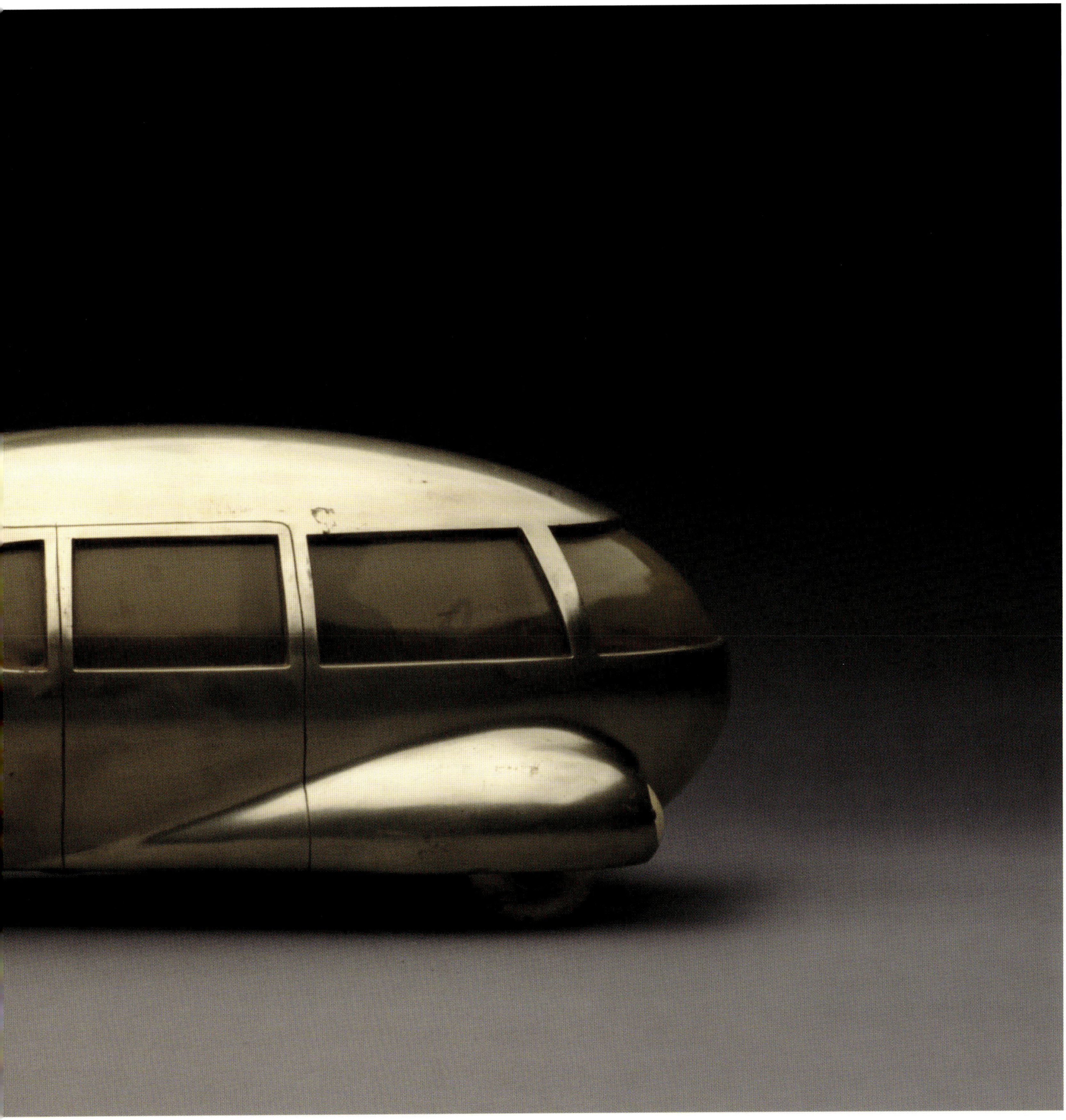

ABOVE A model of a finless vehicle, Motor Car Number 9, designed by Norman Bel Geddes, *c.* 1933, in brass and plastic.

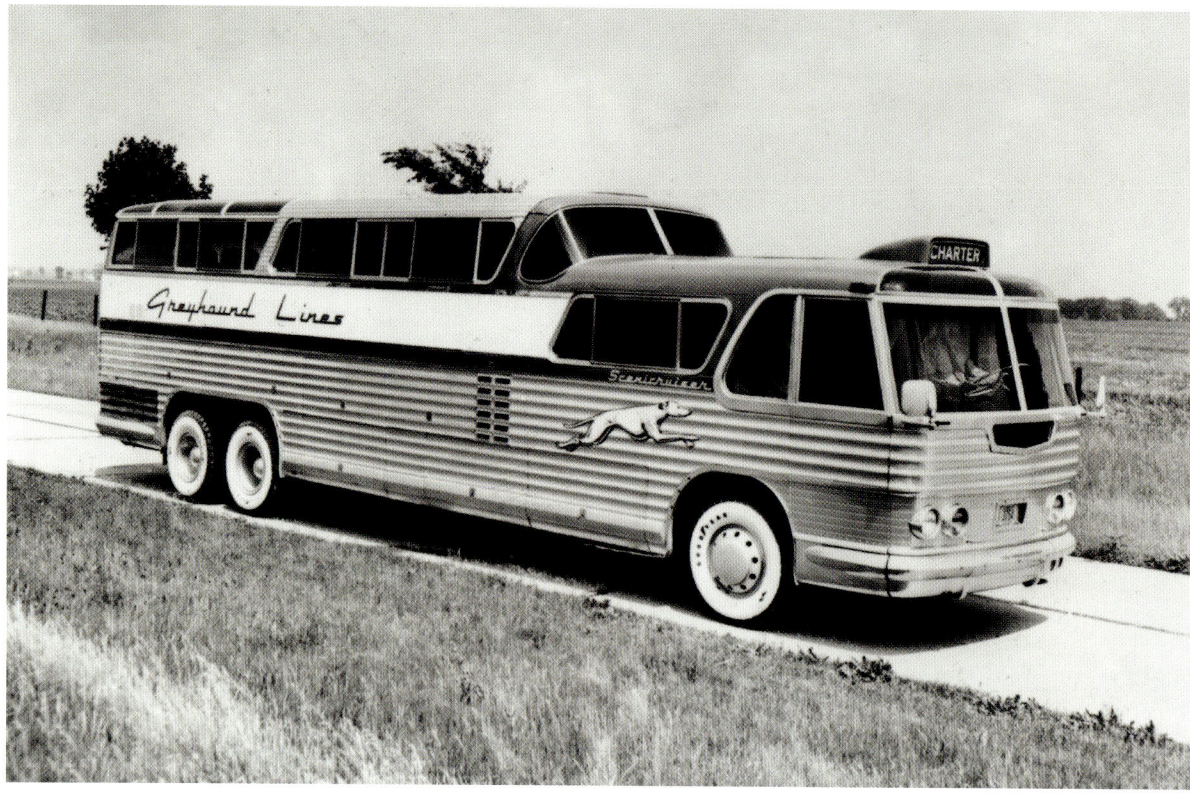

RAYMOND LOEWY
(1893–1986)

One of the first and one of the most respected industrial designers, Raymond Loewy was devoted to the concept of speed. This was expressed, in a personal way, within his own love of racing cars and motor boats; Loewy attended Le Mans every year and took a high-performance driving course at the age of 70. Yet the fascination with speed was also explored to the full within his lengthy professional career, with a wealth of commissions for the automobile industry, railroads and airlines, as well as for passenger ships and – ultimately – even spacecraft.

'I have a secret amour with transportation of every sort even today,' said Loewy, speaking in the late 1970s. 'Every three or four years, you know, I still design a new car, whether or not it gets produced.... Urban transport now, high-speed planes, speed boats, buses, all that kind of thing, has always had its appeal. I think speed and mobility are the touchstone of our age.'[21]

From the early Thirties onwards, Loewy indulged his love of speed – and the aerodynamic aesthetic – within a range of innovative machine-age designs, informed by his own preoccupation with streamlining. In 1938, he began an association with Studebaker, which lasted until 1962, with cars including the Champion (1939), the Starliner (1953) and Avanti (1962). The relationship with Detroit was long and fruitful, although Loewy grew critical of the excessive 'tinsel and trash' styling that crept into the industry in the post-war period. However, he fully appreciated the importance of the seductive elegance and finesse of the car's outward appearance and the huge value of an emotive reaction from consumers to the automobile's sensual curves. At the same time, he believed that form and function should be united: 'good design is not an applied veneer; it must be integral to the project'.[22]

Loewy's relationship with the railway industry also helped to define his career. In 1932, the president of the Pennsylvania Railroad commissioned him – rather dismissively – to design a station trash can. Along with the bin, Loewy provided a speculative rendering of a new 'triplex' locomotive, a striking design with a rounded nose that almost resembled a finless whale; it was enough to launch a long and productive relationship. Loewy went on to develop one of the first electric locomotives in the form of the GG1 (1934). This was highly innovative in every sense, from its shape and styling to its construction methods and power system. The T1 (1937) and S1 (1938) offered further refinement of the streamlined dynamic of the engines.

Loewy also worked with the Greyhound bus company, and designed speed boats as well as a passenger ferry for the Virginia Ferry Corporation

OPPOSITE Raymond Loewy's Greyhound 'Scenicruiser' bus, c. 1946.

BELOW Studebaker's 'State' President sedan, designed by Raymond Loewy, 1939.

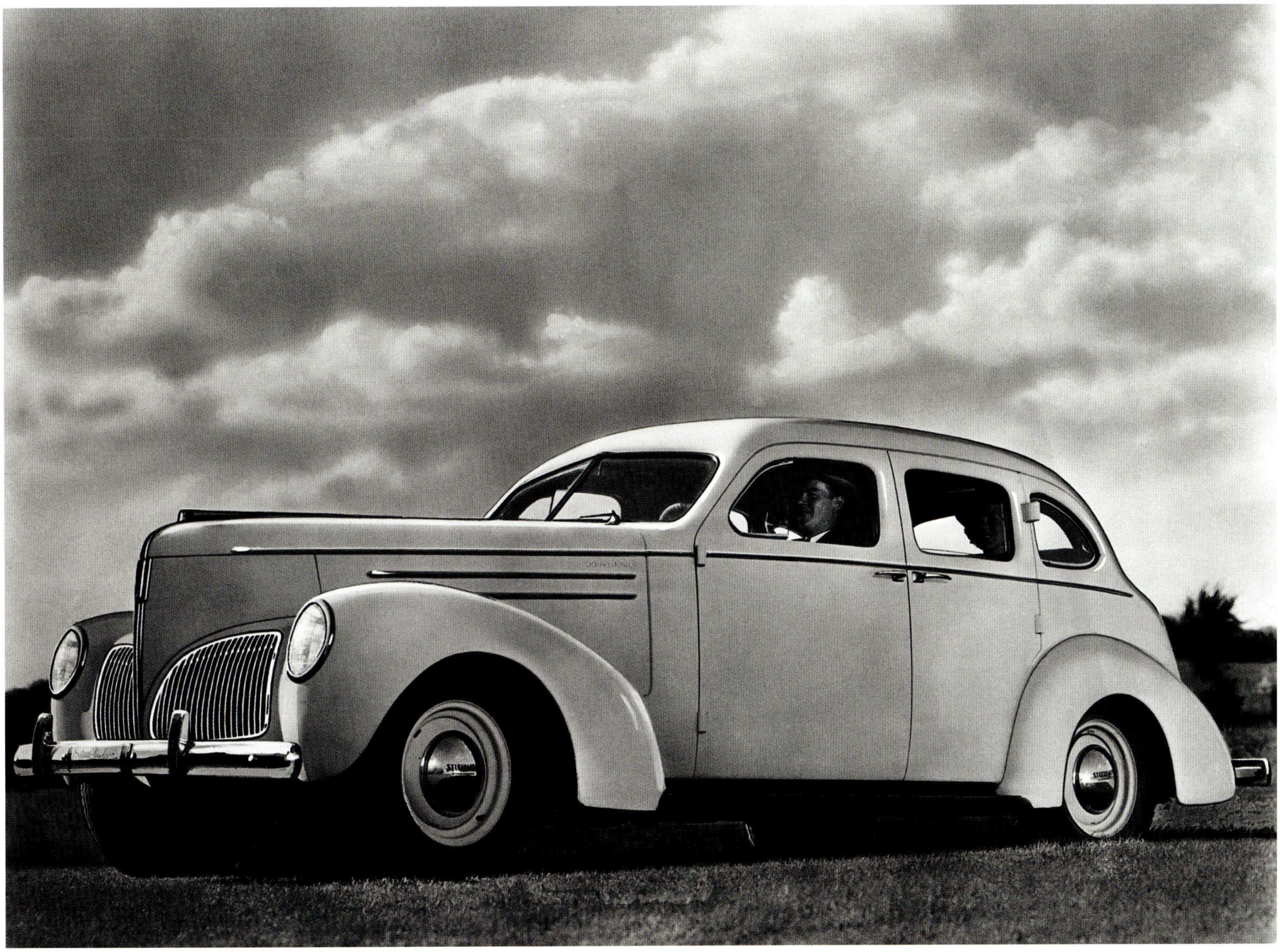

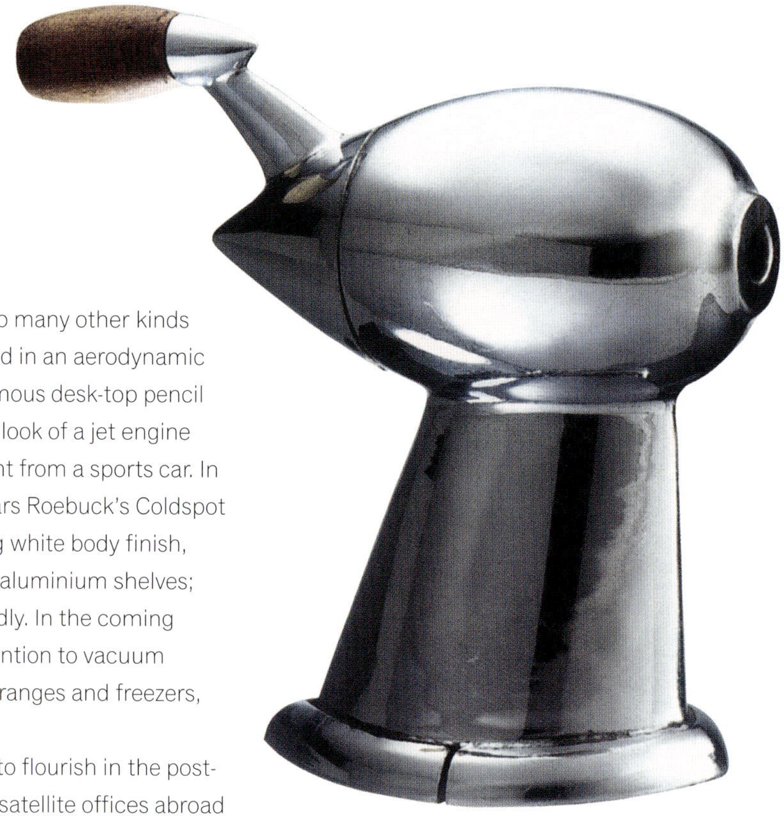

known as the *Princess Anne* (1933), this too with an attention-grabbing, streamlined silhouette. In aviation, Loewy created interiors for the Lockheed Constellation in 1945 and, in the 1960s, he designed the interiors and livery for President Kennedy's Air Force One. In the late Sixties and Seventies, Loewy's team worked with NASA, most notably on the Skylab project.

This long and involved association with all kinds of transportation reinforced the sense of glamour and mystique that surrounded Loewy throughout his working life, despite his relatively humble beginnings. Born in France, he served in the French Army during World War I before arriving in New York in 1919 with just fifty dollars. He found work as a window dresser with Macy's department store, followed by freelance commissions as a fashion illustrator for retailers and magazines such as *Vanity Fair* and *Harper's Bazaar*.

In the early Thirties, Loewy helped to invent the 'new profession' of industrial design. As well as his many successes grounded in machine-age transport, he brought the notion of speed – via streamlining techniques – to many other kinds of products that were refined in an aerodynamic style. These included his famous desk-top pencil sharpener of 1933, with the look of a jet engine casing or the hood ornament from a sports car. In 1934, Loewy redesigned Sears Roebuck's Coldspot refrigerator, with a gleaming white body finish, rounded edges and shining aluminium shelves; Coldspot sales climbed rapidly. In the coming years, Loewy turned his attention to vacuum cleaners for Singer, kitchen ranges and freezers, radios and television sets.

His career continued to flourish in the post-war period, with successful satellite offices abroad and commissions from Coca-Cola, Lucky Strike and other household-name companies. There were houses in Palm Springs, Mexico and France, and a transatlantic social life. In 1949, Loewy even made it onto the cover of *Time* magazine (see also p. 212). Arguably, he did more than anyone to establish industrial design as a discipline in itself and to lend it structure and substance.

ABOVE Raymond Loewy's Teardrop pencil sharpener, 1934.

BELOW An experimental radio designed by Loewy, made by Hallicrafters, c. 1946, in plastic and Bakelite.

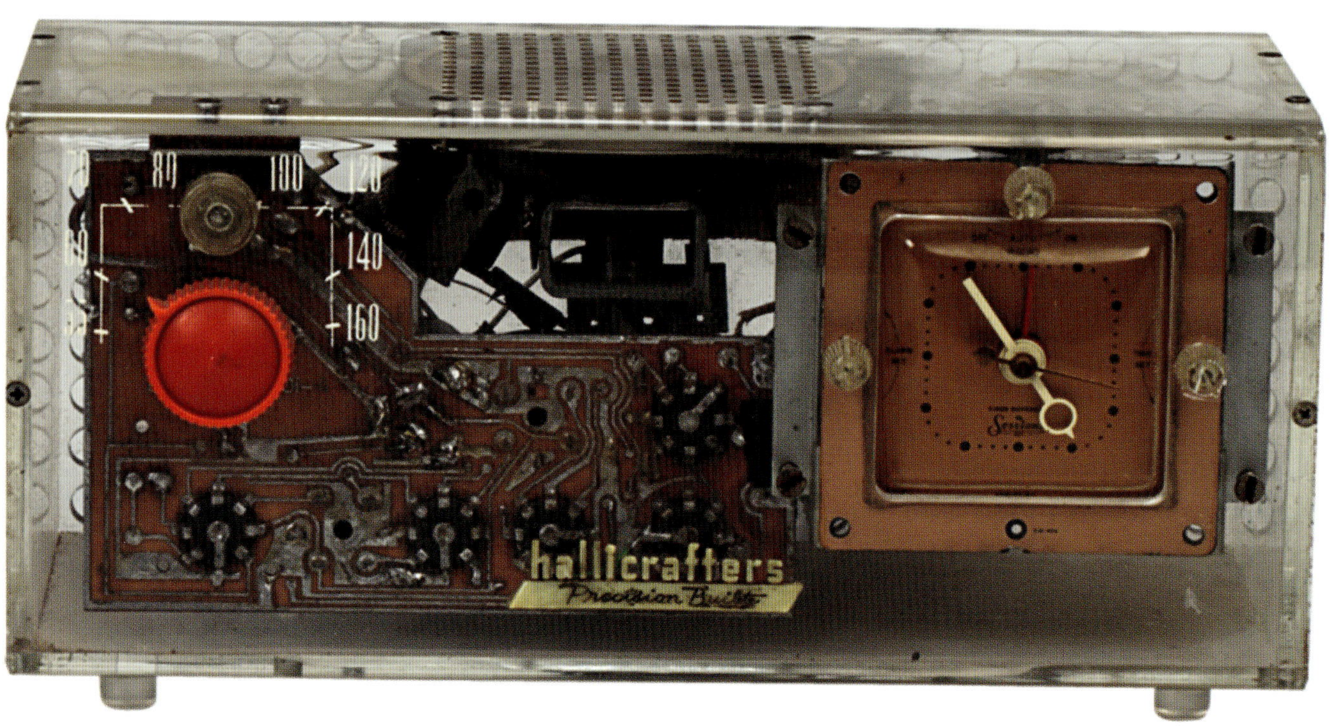

RIGHT Loewy's PRR S1 steam locomotive, known as 'The Big Engine', at the New York World's Fair, 1939.

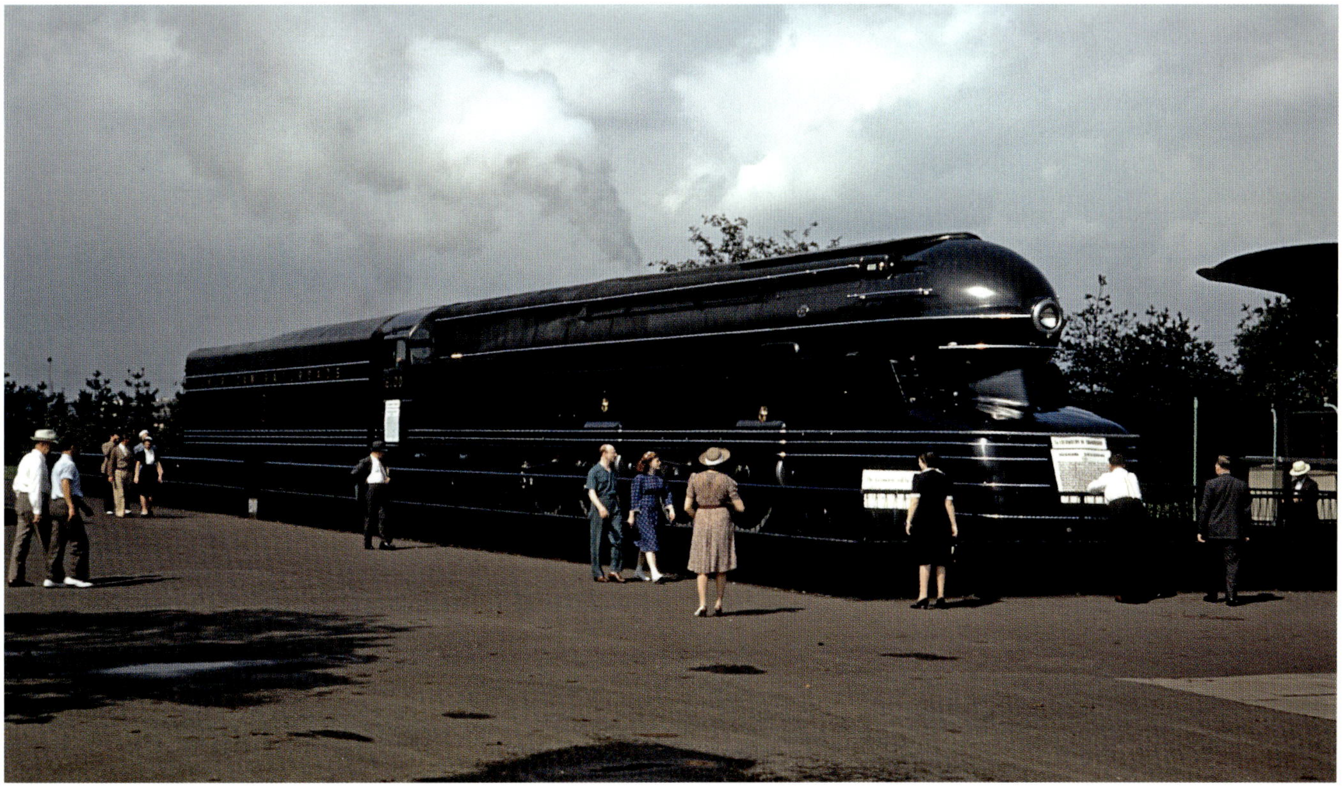

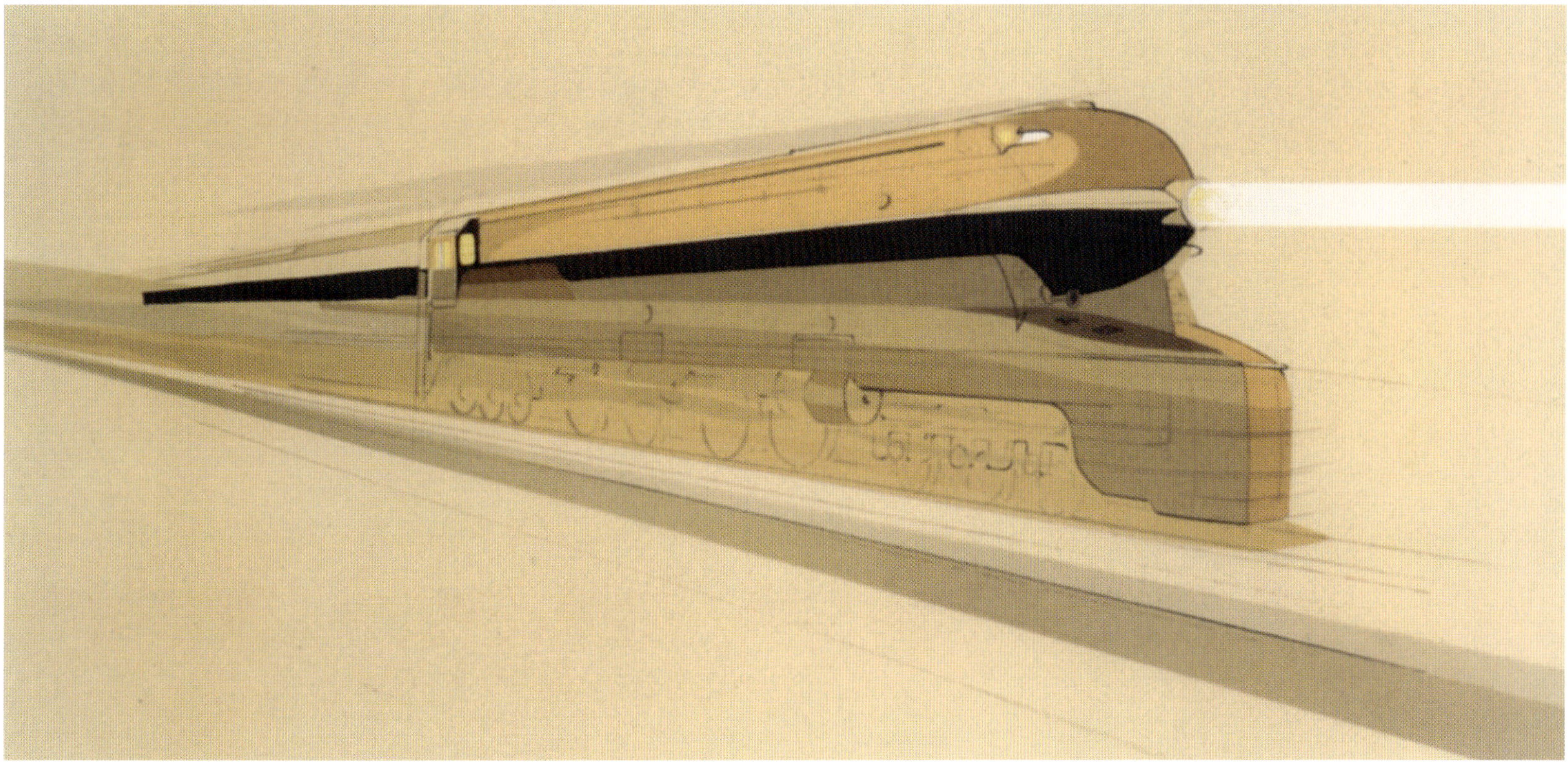

ABOVE A drawing of an S1 locomotive for the Pennsylvania Railroad Co., 1937.

FERDINAND PORSCHE
(1875–1951)

The success of the Volkswagen Beetle is extraordinary by any standards. Over 20 million of these cars have been sold worldwide, making it the best-selling and longest serving car in history. Over its extensive life, the Beetle has become beloved, starring in feature films and becoming a 'character' in its own right. It was the first truly modern, affordable, compact family vehicle – produced well in advance of competitors such as the Mini, Fiat 500 and Citroën 2CV – and an updated version remains popular today. Somehow, this magical car also managed to avoid the significant taint of its intimate association with Adolf Hitler. Despite a connection to fascism that would usually be highly damaging for a company, brand or designer, the Beetle's creator, Ferdinand Porsche, is still lauded as one of the most visionary figures in the automotive world.

Porsche was born in Bohemia, still at the time part of Austro-Hungary but later Czech; eventually he would take German citizenship. His father was a panel beater and Porsche studied at the Imperial Technical School in Reichenberg. From an early age, he showed a talent for engineering, and he began working for the electrical company Béla Egger & Co. in Vienna at the age of eighteen, stimulating a profound interest in electrical power and propulsion.

After joining Jacob Lohner & Company, a coach-building firm also based in Vienna, Porsche developed his first car designs using battery power. The Egger-Lohner – or 'P1' – was launched in 1898 with a design that resembled a horseless carriage. The innovation lay more in its propulsion, with batteries driving electric motors that turned the wheels. Three years later, Porsche produced the world's first hybrid car, using a petrol engine in combination with a battery pack. With Porsche in the driving seat, the Lohner-Porsche hybrid won the Exelberg Rally of 1901.

Following his military service, Porsche went on to work with Austro-Daimler as chief engineer, developing a number of new designs and also becoming a leading figure in motor sport. In 1923, he moved on to the Stuttgart company that became Mercedes-Benz, although the Great Depression of 1929 saw him temporarily out of work.

In 1931, Porsche laid the foundations of the company that would evolve into his eponymous car brand. Based in Stuttgart, he established his own engineering business with the help of former colleagues and major backers such as his son-in-law Anton Piëch, who was to become a key member of the family enterprise, along with daughter Louise and son Ferry. Porsche worked on new car design commissions and developed a 16-cylinder racing car for Auto-Union in 1933.

That same year, Hitler – the German Chancellor – raised the idea of a people's car or 'Volkswagen'. This was to be an accessible, reliable, economical, versatile small car that could be made available to all via a state-supported savings scheme. Porsche was well placed to bid for the contract to design the car, having already researched the idea of a modestly scaled automobile suitable for mass production.

In June 1934, Porsche began working on the car that became the Beetle after winning the job from the Automobile Industry Association

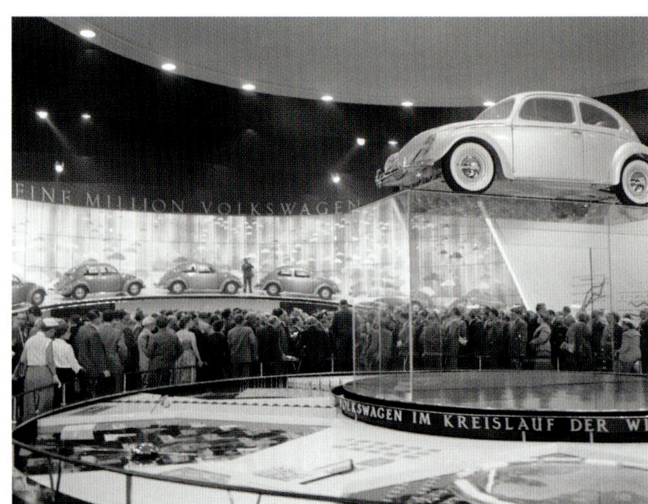

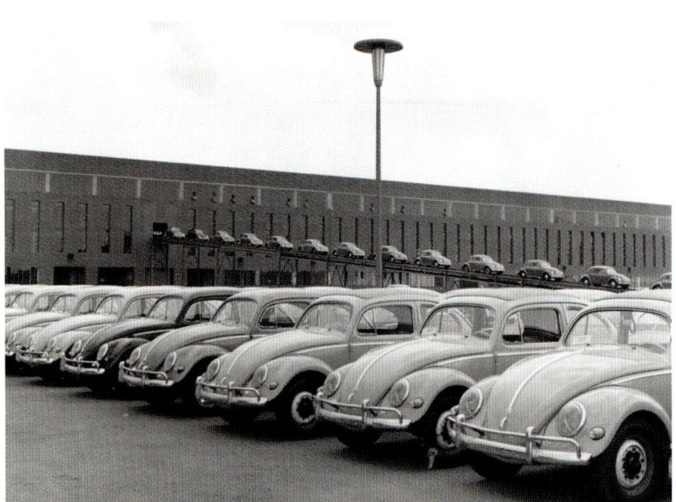

ABOVE A presentation at the Frankfurt Motor Show, 1955, celebrating a million Volkswagen Beetles.

LEFT Volkswagen Beetles lined up at the VW factory in Wolfsburg, Germany, *c.* 1956.

OPPOSITE A Volkswagen advertisement, *c.* 1936; the caption reads 'The KdF Car', with KdF standing for 'Kraft durch Freude', or 'strength through joy'.

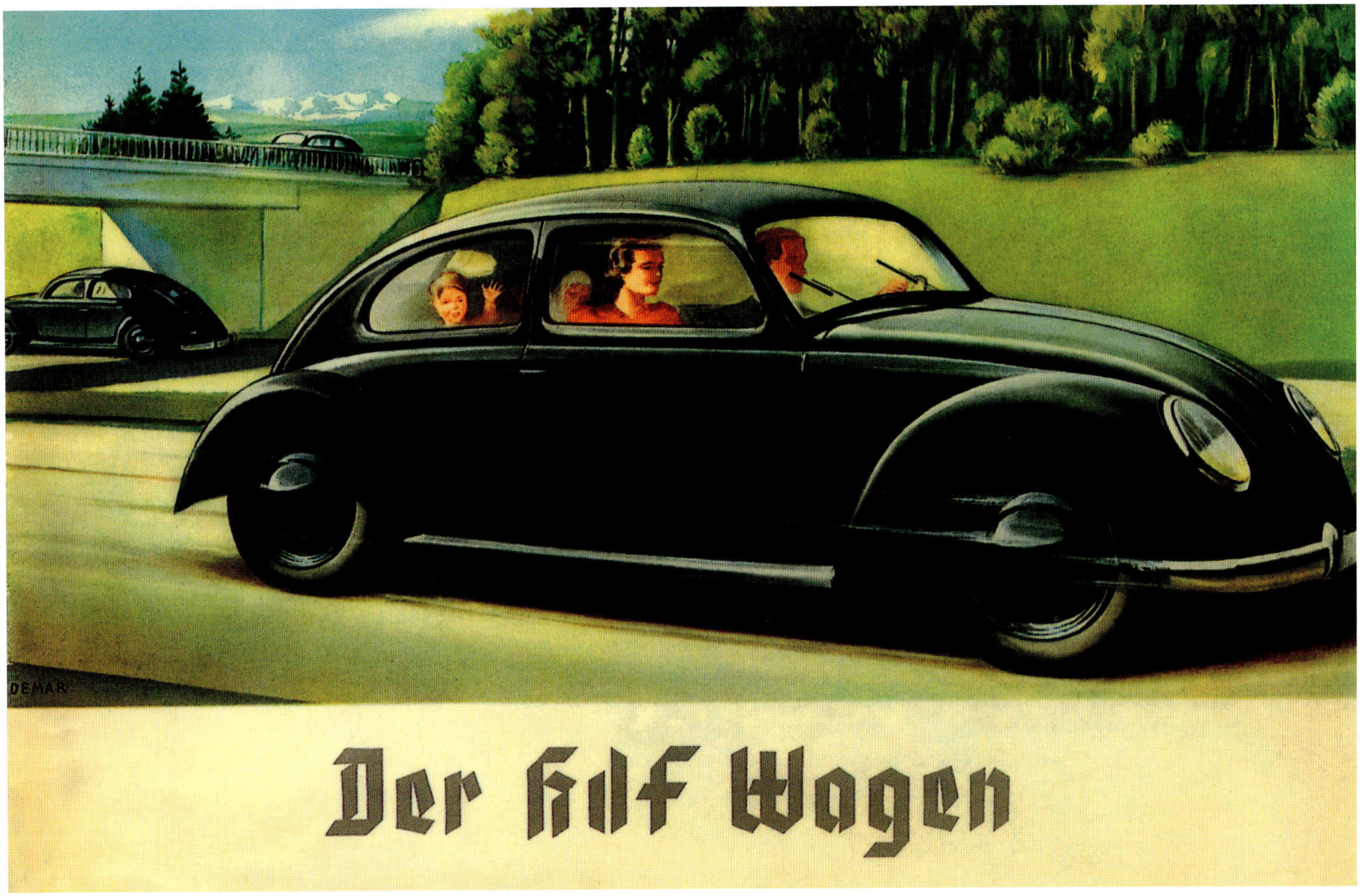

of the Reich. Early prototypes were produced in Stuttgart in 1935, with Ferry supervising the first test drives. Porsche's colleague Erwin Komenda contributed to the aerodynamic styling of the car, which featured a rear-mounted, air-cooled engine. In 1936, Daimler-Benz produced thirty W30 prototype cars for Porsche, already recognizable as the 'Beetle'. Over the following few years, further batches of prototypes were manufactured and rigorously tested, and in 1937 Porsche was appointed managing director of the Volkswagen company; a year later construction work began on a new factory in Fallersleben, later folded into the neighbouring city of Wolfsburg. Hitler wanted the car to be known as the 'Kraft-durch-Freude-Wagen' or 'Strength Through Joy Car'. Reporters – noting the neat, rounded, shell-like body of the vehicle – called it the 'Beetle', with the French opting for 'Coccinelle' (or 'Ladybird').

By the autumn of 1939, when World War II began, the Wolfsburg factory had only just begun producing Beetles and production was soon switched to support the German war effort. The same platform – or chassis – was used to produce vehicles for the military, including the Kübelwagen. Porsche, a member of the National Socialist German Workers' (or Nazi) Party, devoted himself to a number of military projects during the war years and is credited with the design of the Tiger II tank, one of the fearsome icons of the wartime design and production 'machine' overseen by Albert Speer*, Hitler's architect and designer-in-chief.

In the post-war years, the Beetle was seen as a key tool in the bid to restart German manufacturing and begin reconstruction. Supported by a British Army initiative, the damaged VW factory was reopened and production of the Beetle resumed. In 1972, with just over 15 million Beetles made, the car overtook the Ford Model T as the world's most widely produced car.

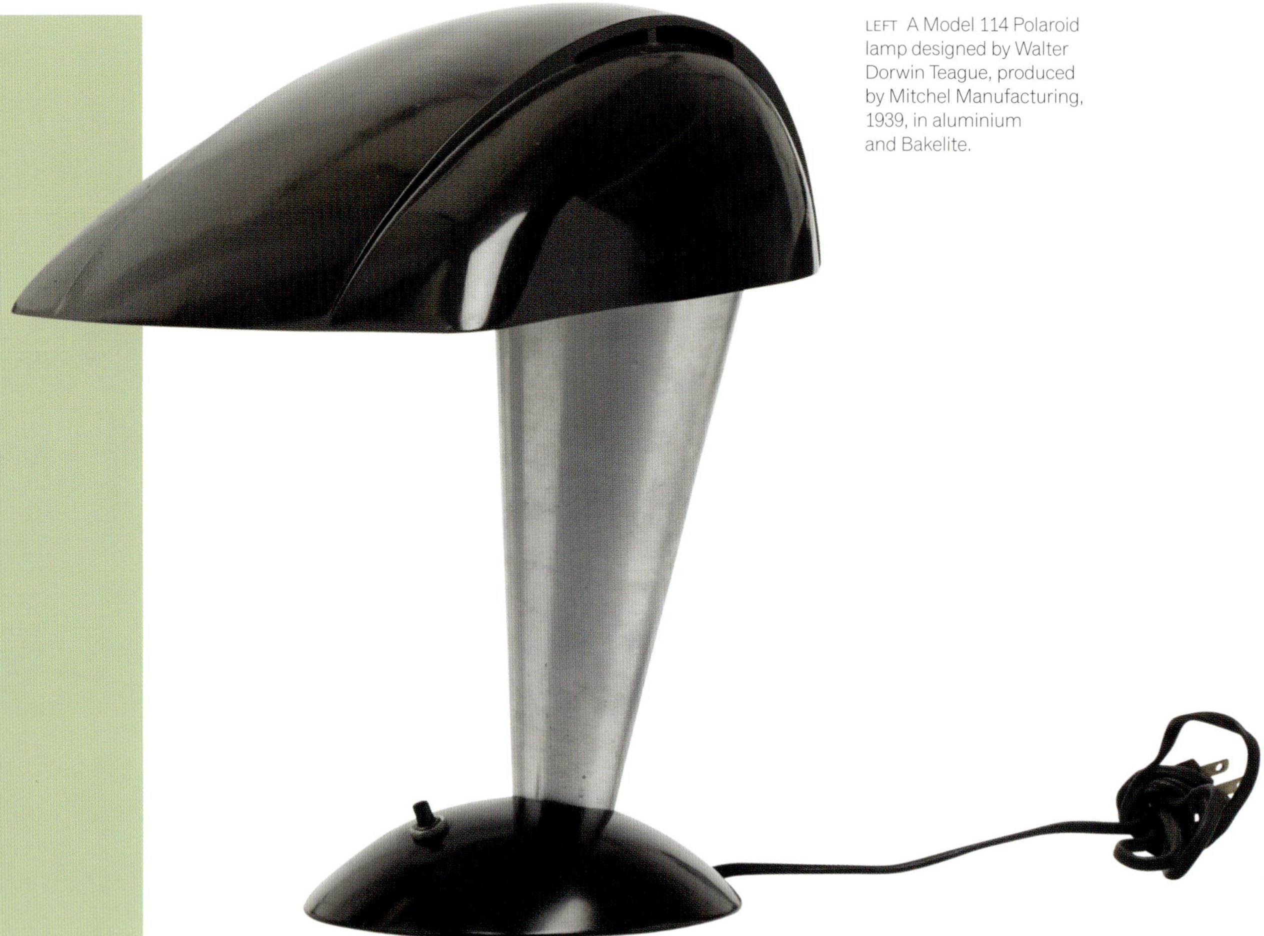

LEFT A Model 114 Polaroid lamp designed by Walter Dorwin Teague, produced by Mitchel Manufacturing, 1939, in aluminium and Bakelite.

WALTER DORWIN TEAGUE
(1883–1960)

The Twenties and Thirties saw the introduction of a whole new wave of consumer appliances and domestic machines. These were factory-made products with a semi-industrial or commercial provenance, including vacuum cleaners, refrigerators and washing machines – labour-saving, utilitarian devices that gradually made their way into the home. At the same time, there was another collection of devices associated with pleasure, communication and entertainment, which also became more accessible and affordable as the modernist era progressed. Radio sets were particularly popular, followed by early television sets, and there were telephones and cameras. This was the beginning of a revolution in consumer technology that was to help define the 20th and 21st centuries. One of the first and leading designers to lend shape, style and identity to such products was the American designer Walter Dorwin Teague.

Teague railed against the ugliness of early factory-produced appliances, which had first evolved according to function and practicality, with little thought given to craft or elegance. He saw an opportunity to introduce aesthetics into the equation, which – in itself – would seduce a fresh and increasingly discerning audience. 'Everything made by the hand of man is art and has within itself a potentiality of beauty,' he argued. 'There are but two categories of art: if a thing – anything – is made well enough, it is fine art, and beautiful; if it is made badly, crudely, clumsily, it is vile art, and ugly.'[23]

Among the consumer products that Teague is most intimately identified with stands the camera. In the early years of the 20th century, camera technology advanced rapidly to a point where a large piece of professional equipment was reduced down to a simple, easy-to-use, hand-held device. George Eastman introduced celluloid film in the late 1880s and his first Kodak box camera at around the same time. In 1900, Kodak launched the Brownie box camera, which became increasingly popular in various incarnations over the following decades.

In 1927, Eastman Kodak contacted Teague. The company wanted to make their products more aesthetically appealing to buyers and so Teague began working with their engineers, immersing himself in the technology and styling of the cameras but also examining the packaging and presentation of the product lines. Among Teague's most successful designs were the Baby Brownie of 1934 and the Brownie Hawkeye of 1950, along with the compact, folding 1A Gift Kodak of 1930 – an artwork in itself, with a streamlined leatherette case emblazoned with a Cubist pattern that had a strong Art Deco* influence, which was echoed upon the lid of the presentation box. This highly

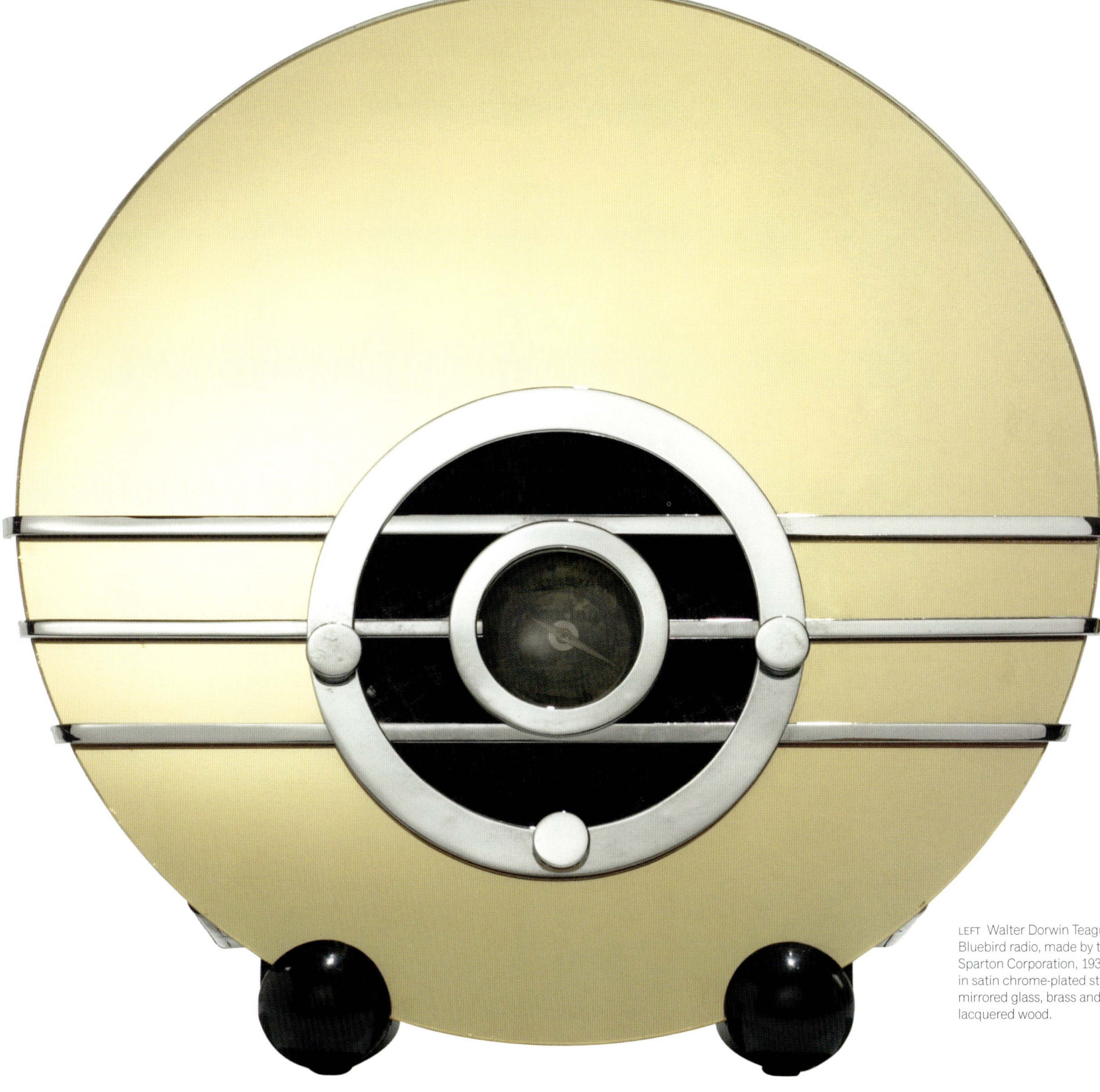

LEFT Walter Dorwin Teague's Bluebird radio, made by the Sparton Corporation, 1934, in satin chrome-plated steel, mirrored glass, brass and lacquered wood.

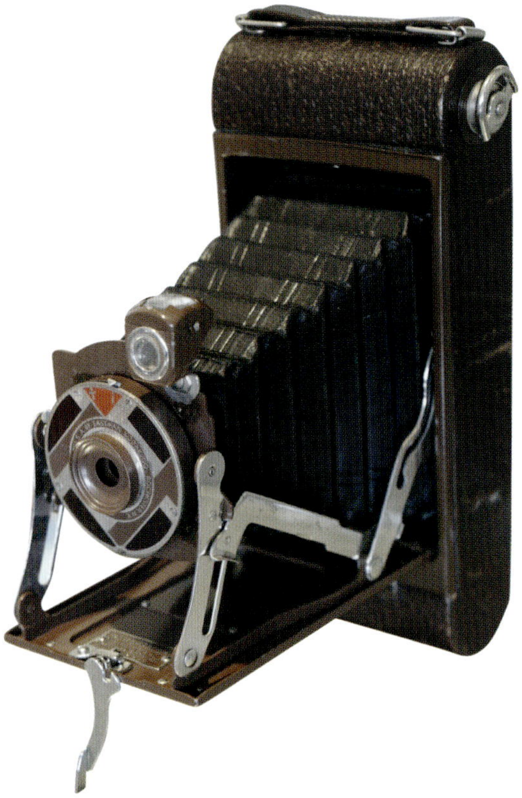

desirable camera was a joy to look at, and invited touch and interaction. Later, Teague also designed the Kodak Bantam Special of 1936, one of the first consumer cameras to use colour film, housed in an aluminium body with a fold-out lens. The long relationship with Kodak extended into display material and corporate identity, and was a catalyst for Teague's rapid progress in the new career of industrial design; later, he also worked with Polaroid, designing the Land camera of 1948.

Like his colleagues and competitors Raymond Loewy and Norman Bel Geddes, Teague began his working life as an illustrator. Born in Indiana, his first ambition was to be an artist and he studied painting at the Art Students League in New York. He went on to provide illustrations for catalogues and advertising, taking posts at two New York advertising agencies before moving into freelance graphic design, packaging and typography. In the late Twenties, he took a European tour and became familiar with the Bauhaus and the work of Le Corbusier. Upon his return, he stepped into a new vocation and became one of the first and most influential industrial and product designers, ultimately building a studio whose success was only comparable with that of Raymond Loewy.

Beyond the Kodak relationship, Teague turned his attention to another consumer favourite: the radio. Among his most striking designs was the Nocturne of 1935 for the Sparton Corporation: a crafted moon with a blue face, including a speaker grill and a neat set of dials. The sinuous receiver was elegantly conceived and enigmatic, offering few immediate indicators as to its function and providing a conversation piece in itself.

There were also cash registers and lamps (including the Polaroid 114 desk light of 1939), as well as refrigerators and pianos, such as the Centennial for Steinway. Teague expanded into transportation, with commissions such as the Marmon Model 16 automobile of 1930 and train interiors for the Hartford Railroad. Later, in the post-war period, Teague and his growing office – which was continued after his death by his son – began a long relationship with Boeing, designing interiors for a series of the company's airliners. Along with Loewy and Henry Dreyfuss, Teague was a founding member of the Society of Industrial Designers, and he became its first president in 1944.

ABOVE A Model 1A Gift camera, made by Eastman Kodak, *c.* 1930, in enamelled metal with a lacquered cedar box.

BELOW Walter Dorwin Teague's Beau Brownie camera, produced by Kodak, 1930, in enamelled metal and plastic.

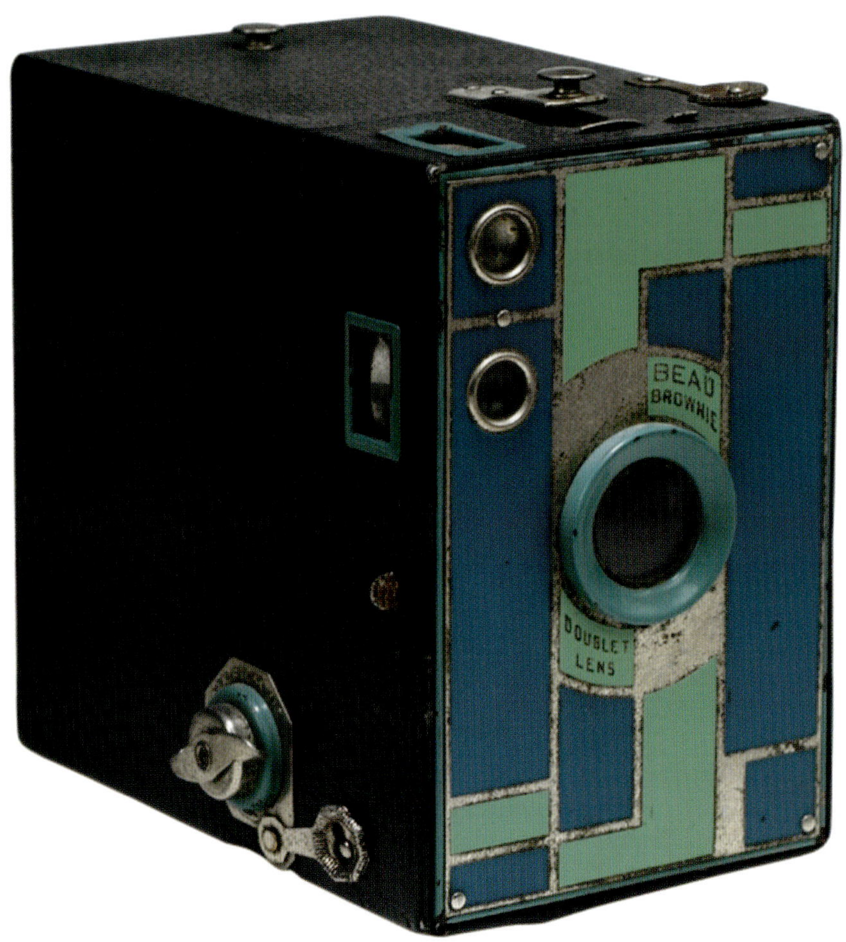

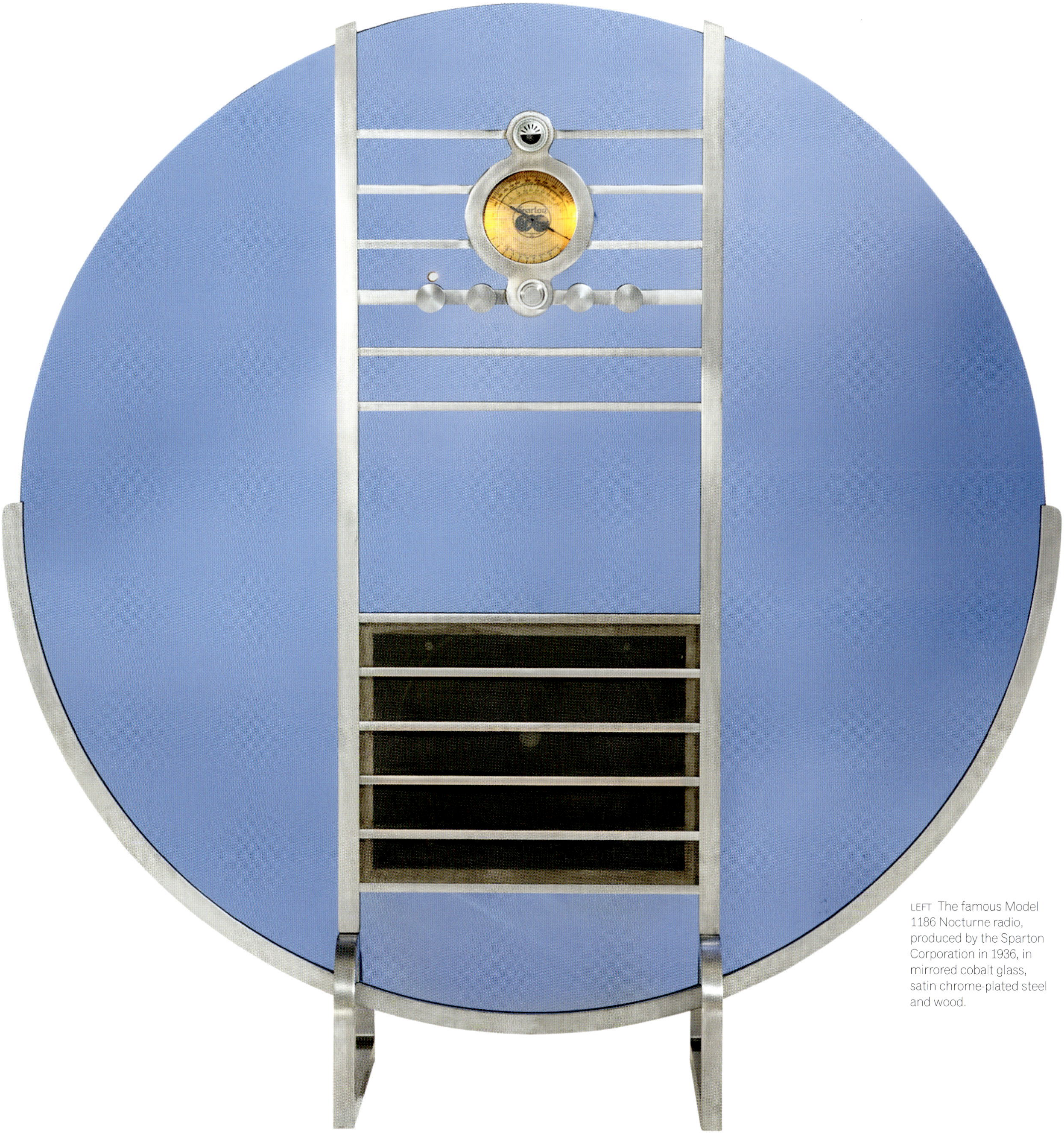

LEFT The famous Model 1186 Nocturne radio, produced by the Sparton Corporation in 1936, in mirrored cobalt glass, satin chrome-plated steel and wood.

WILHELM WAGENFELD
(1900–1990)

The multi-talented German designer Wilhelm Wagenfeld worked across many different design disciplines. For some he is best known as a lighting designer, but for others his work in glassware makes him unique. Yet Wagenfeld also designed door handles and typewriters, cutlery and silverware. As such, he became one of the most influential European industrial designers of the early modernist period. A number of his designs from the Twenties and Thirties – defined by a combination of ingenuity and clean lines – are still in production today, suggesting the depth of Wagenfeld's prescient talent and the enduring relevance of his work.

Wagenfeld was born in Bremen in northwestern Germany. He studied in the city, as well as serving an apprenticeship with the silverware makers Koch & Bergfeld, before moving to Hanau where he won a scholarship to study at the State Academy, graduating in 1922. In 1923, Wagenfeld decided to continue his studies at the Bauhaus in Weimar and enrolled in the metal workshop, supervised by László Moholy-Nagy. Here, he developed one of his most iconic designs: the MT8 – or 'Bauhaus' – table lamp (also known as the WG24).

With inspiration and input provided by silversmith and industrial designer Carl Jacob Jucker, Wagenfeld perfected a simple, graphic lamp with a cylindrical metal shaft rising from a round metal base and topped with an opaque, half-moon glass dome. Early prototypes were produced in the Bauhaus workshops, as Wagenfeld and his colleagues sought ways to gear the lamp to mass production – one of the key ambitions of the Bauhaus ethos – and to develop a glass shade from the Schott & Gen. Jenaer Glas company that would be robust enough to resist the heat of the bulb without cracking. Later, the light was manufactured by Schwintzer & Gräff; it is still produced under licence by Tecnolumen.

After the Bauhaus moved from Weimar to Dessau in 1925, Wagenfeld stayed on, lecturing at the State Academy of Crafts and Architecture (Staatliche Bauhochschule Weimar) and working under Richard Winkelmayer in the metal workshop, becoming director of the department himself from 1928 to 1930. During this period Wagenfeld designed a number of products, including the WD28 door handle for the hardware company S.A. Loevy. He also began working with Schott & Gen., creating a number of designs, including his famous glass teapot and service of 1931.

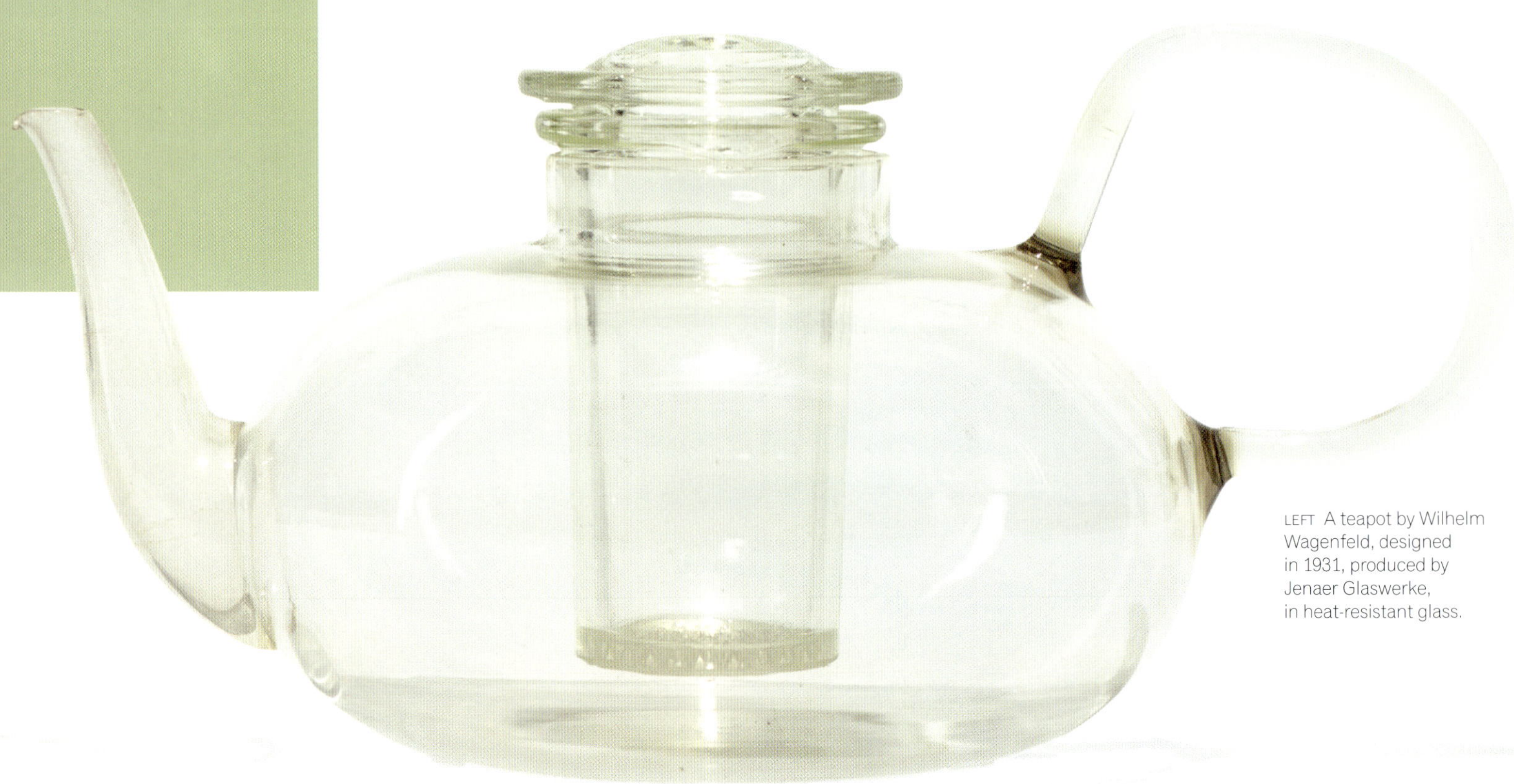

LEFT A teapot by Wilhelm Wagenfeld, designed in 1931, produced by Jenaer Glaswerke, in heat-resistant glass.

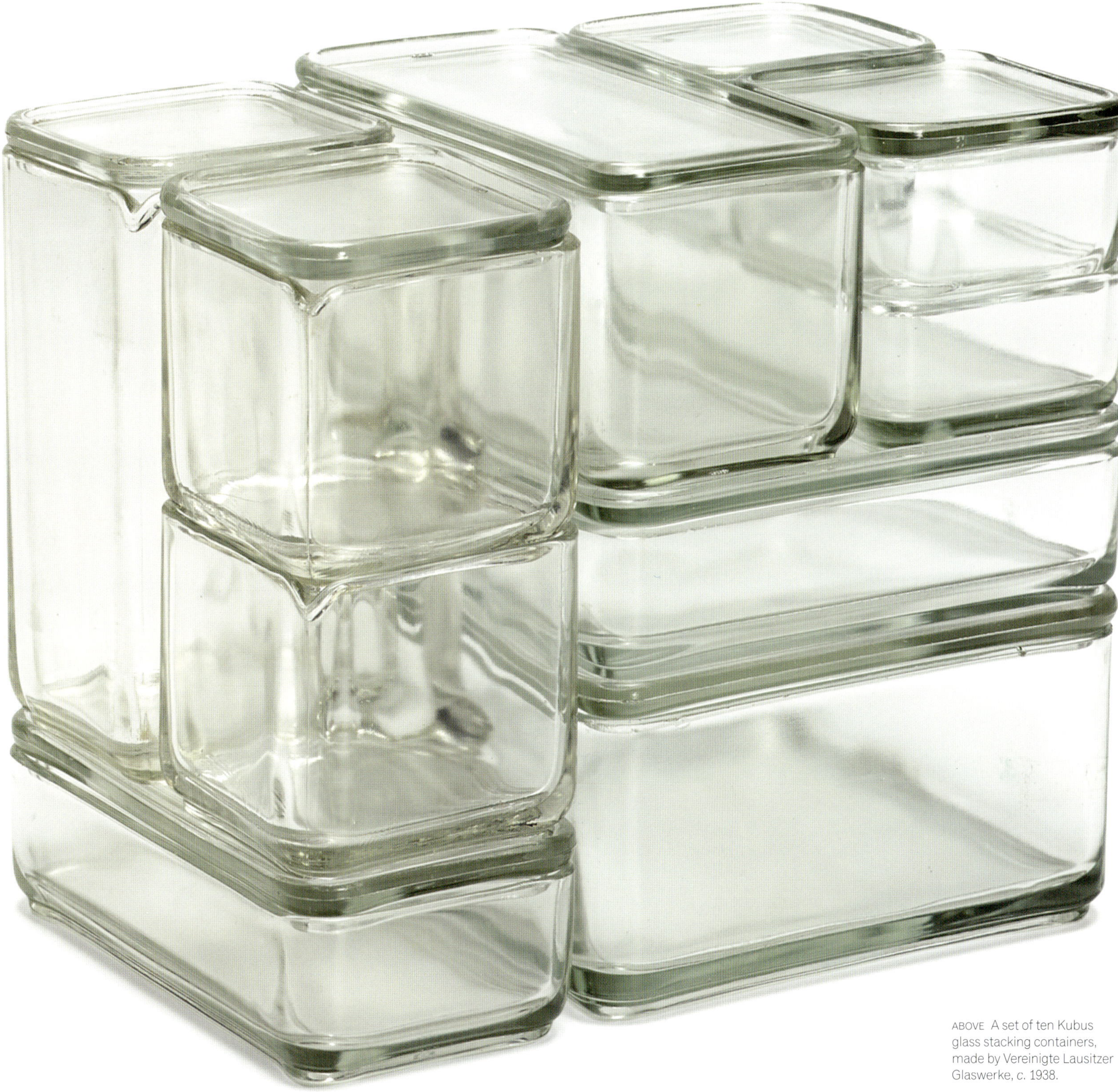

ABOVE A set of ten Kubus glass stacking containers, made by Vereinigte Lausitzer Glaswerke, *c.* 1938.

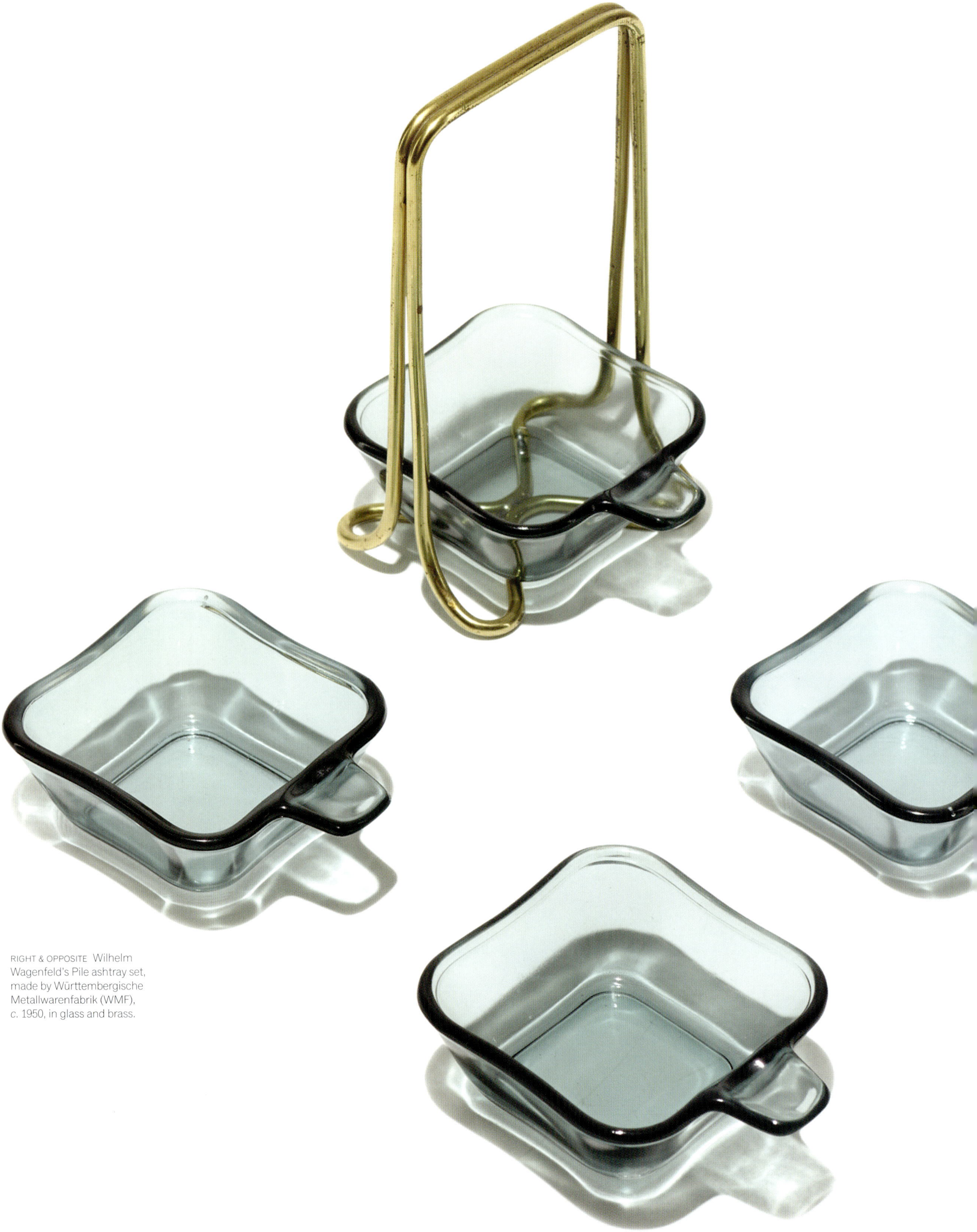

RIGHT & OPPOSITE Wilhelm Wagenfeld's Pile ashtray set, made by Württembergische Metallwarenfabrik (WMF), c. 1950, in glass and brass.

The head of the glassworks, Erich Schott, had begun experimenting with borosilicate glass, which was hard-wearing and heat-resistant. Working alongside the company's engineers, Wagenfeld looked at ways of applying the new materials to an innovative range of functional and engaging homeware. The most striking result was the clear glass teapot, complete with a conical glass combined infuser and lid, which held the tea immersed in the hot water of the pot. In this way, the infusion process became visible, allowing the user to judge its efficacy but also providing a miniature drama upon the tea table. The teapot was complemented by glass cups, plates and a tea tray; a few years later, in 1935, Wagenfeld added a glass coffee percolator to the collection. Here, again, the process of making was laid bare and explicit.

In 1938, Wagenfeld designed his Kubus storage containers, again in glass. Originally made by the Vereinigte Lausitzer Glaswerke, the lidded jars came in a range of sizes and could be stacked together. The Kubus pieces were hygienic, practical and affordable, yet elegant enough to make the journey from the kitchen to the dining table, where they could be used for serving. Here, again, the transparency was not only pleasing but useful, offering the opportunity to check contents and quantity at the glance of an eye. During the Thirties, Wagenfeld also designed ceramics for Rosenthal and Fürstenberg.

During the war, Wagenfeld was drafted into the army and sent to the Russian Front, where he was captured and imprisoned. In the late Forties, he began teaching again in Berlin and restarted his design work, opening his own atelier in Stuttgart in 1954. Many of Wagenfeld's best known post-war products were manufactured by the Württembergische Metallwarenfabrik, better known as WMF. These included his Max & Moritz salt and pepper set (1952), along with tableware collections; there were also lighting designs for Peill & Putzler and others, and a combined radio and record player for Braun.

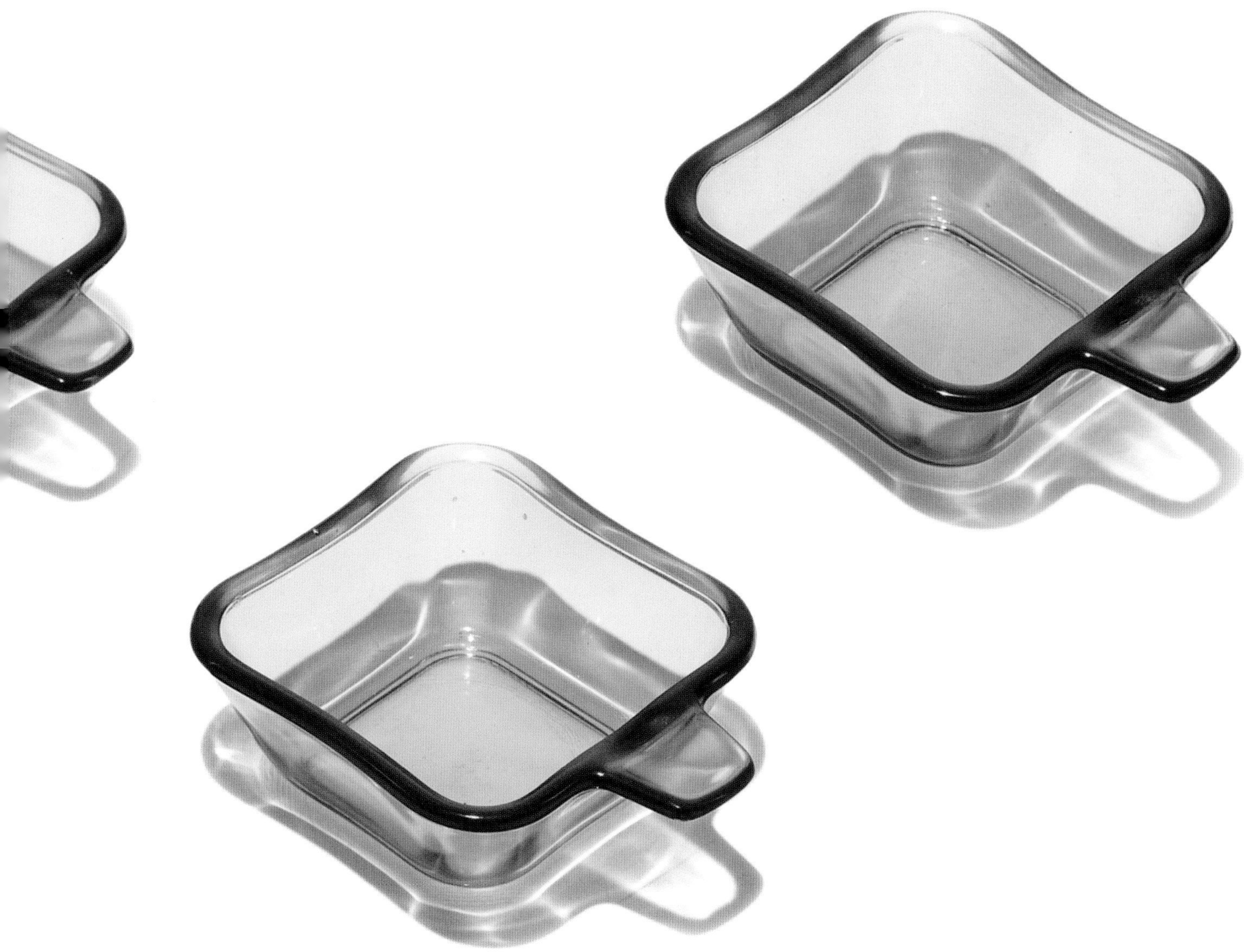

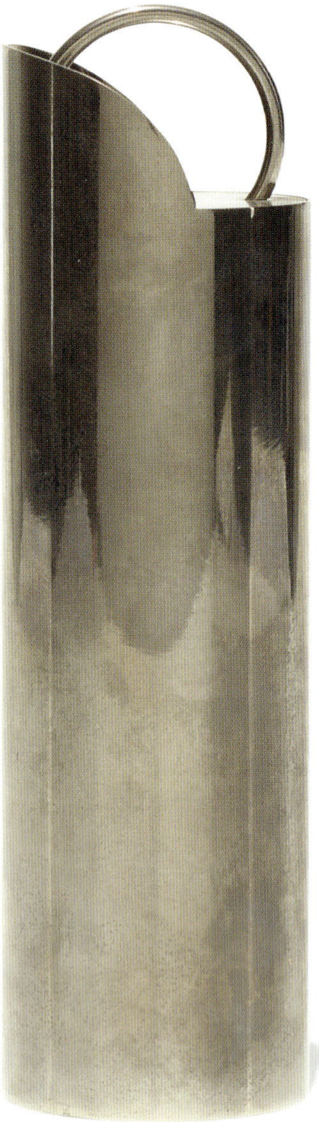

ABOVE A WT24 tea canister designed by Wilhelm Wagenfeld in 1924, made by Tecnolumen, in silver-plated brass.

RIGHT A tea caddy by Wagenfeld, made by Württembergische Metallwarenfabrik (WMF), *c.* 1950, in silver and brass-plated steel.

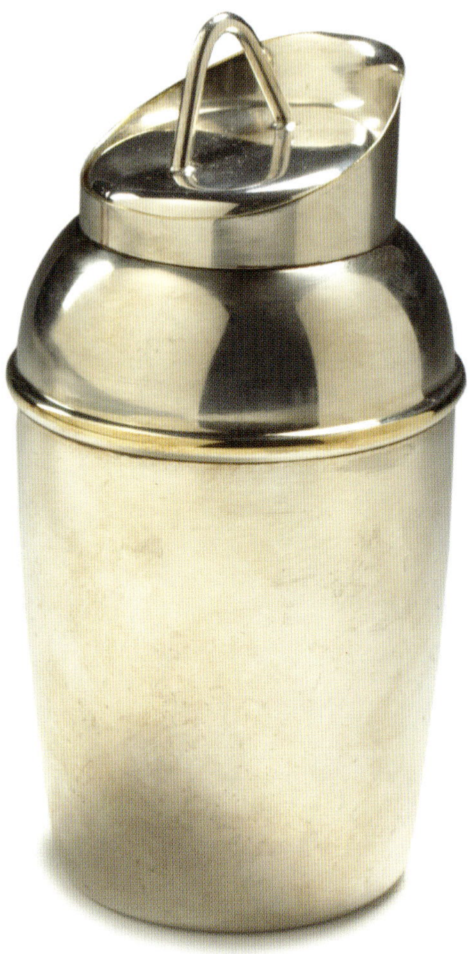

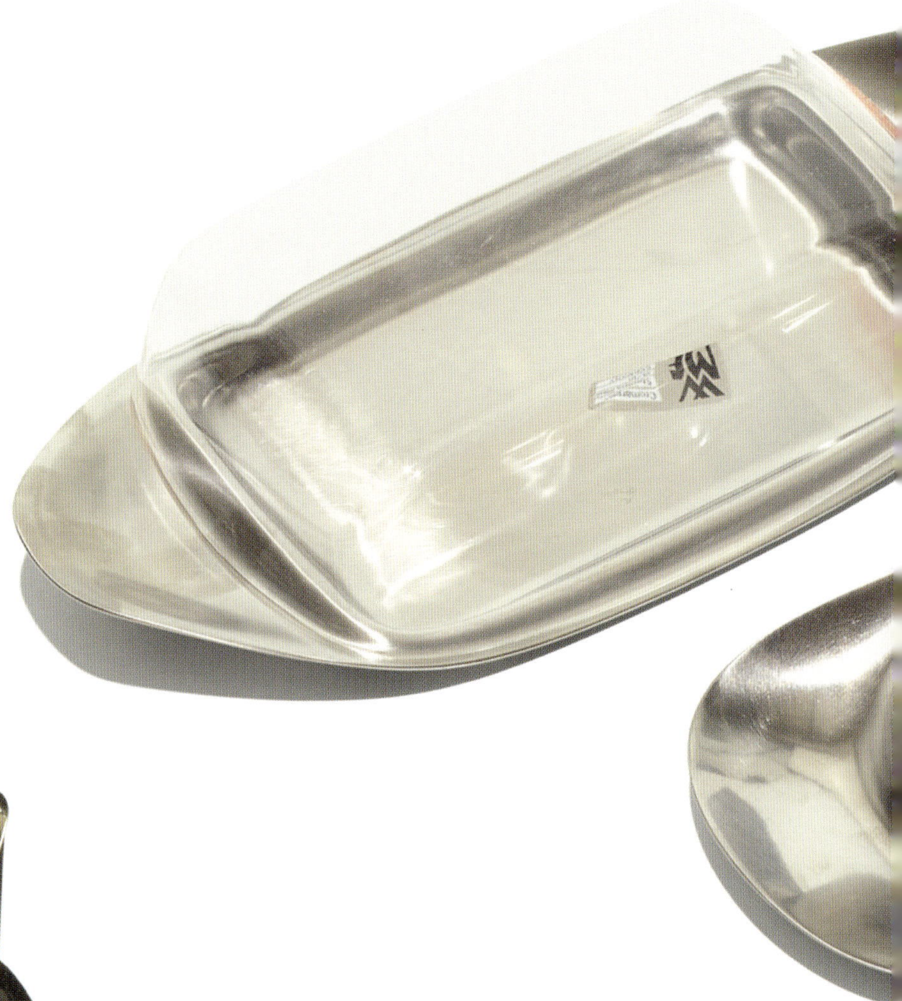

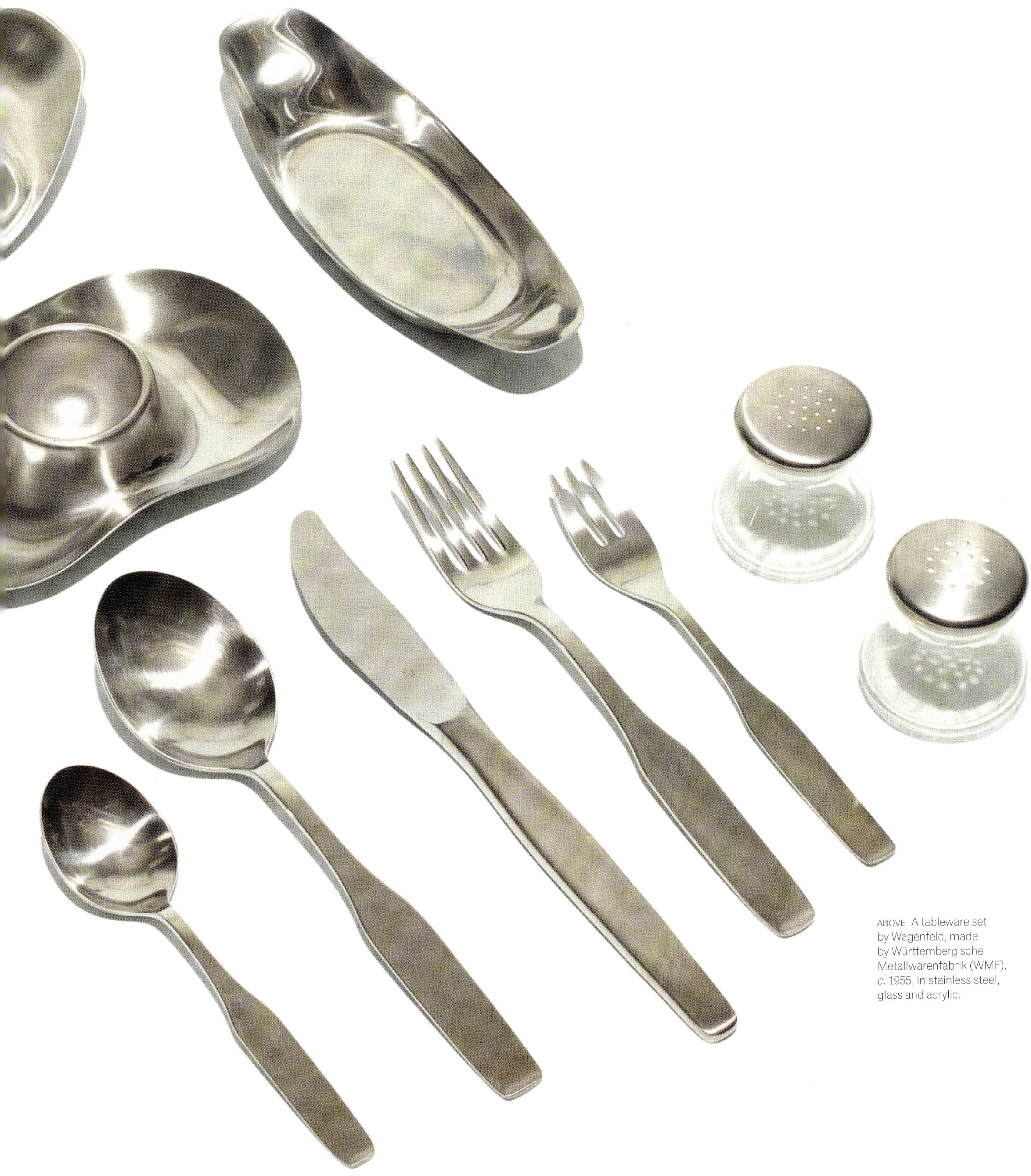

ABOVE A tableware set by Wagenfeld, made by Württembergische Metallwarenfabrik (WMF), *c.* 1955, in stainless steel, glass and acrylic.

GRAPHICS & POSTERS

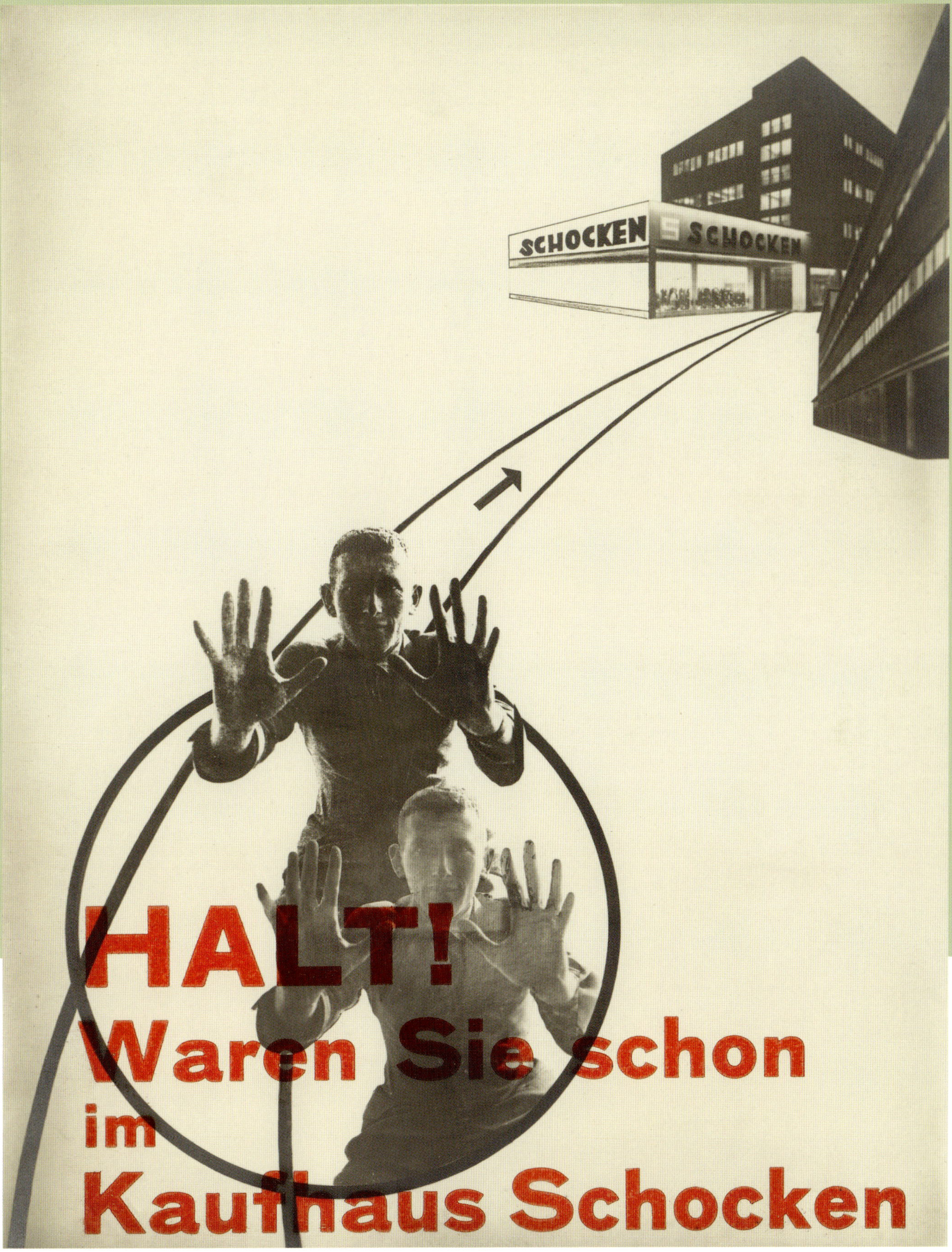
SCHOCKEN
SCHOCKEN
HALT!
Waren Sie schon
im
Kaufhaus Schocken

In his landmark book of 1928, entitled *The New Typography*, JAN TSCHICHOLD declared war on tradition. He presented the choice between old and new, between archaic conventions and modernity, as a battle. He began his book with a quote from the abstract artist Piet Mondrian, one of the founders of the De Stijl* movement in Holland: 'We must never forget that we are now at a turning-point of civilisation, at the end of everything old. This parting of the ways is absolute and final.'[1]

Tschichold's manifesto – which was partly inspired by the pioneering work he had seen at the Bauhaus, as well as the art and graphics of De Stijl – set out the principles and methodology of a modern approach to type and graphic design in the machine age, arguing in favour of standardization and geometrical precision. He compared typographical design to architecture, talking of the 'new and splendid' housing and buildings designed by WALTER GROPIUS and LE CORBUSIER. Tschichold offered a choice between two ways of living, while damning the 'absurdity' of Rococo palaces and Gothic castles.

'In the battle between the old and the new, it is not a question of creating a new style for its own sake. But new needs and new contents create new forms which look utterly unlike the old. And it is just as impossible to argue away these new needs as it is to deny the need for a truly contemporary style of typography. That is why printers today have a duty to concern themselves with these questions. Some have forged ahead with energy and creative success: for the rest however, it seems that there is still almost EVERYTHING to do.'[2]

The battle lines between old and new were, for modernists such as Tschichold, clear and stark. 'The New Typography' – a phrase initially coined by artist, designer and Bauhaus tutor LÁSZLÓ MOHOLY-NAGY – argued in favour of sans serif fonts, abstract purity, geometrical precision, and a degree of dynamism introduced by a bold, mathematical use of primary colours and asymmetrical compositions that made much of the way that diagonal lines and golden triangles relate to the square or rectangle formed by the page or poster. In the other corner of the ring stood the florid example of decades of excess and aesthetic confusion.

In the Twenties, graphic design as a dedicated and singular profession was – like industrial design – still in its infancy. Up until that point, the business of what we now think of as graphic design was undertaken by a whole range of specialists, including commercial artists, layout artists, illustrators, typographers and compositors. During the late 19th century, the evolution of new printing techniques coincided with the rapid growth of advertising, as well as fresh and corresponding expansion in the magazine and journal publishing industry. The period was exemplified, in some respects, by the rise of the 'art poster', which served the worlds of theatre, entertainment and – increasingly – travel, as well as expanding into the field of consumer product advertising. These poster images were accomplished artworks in themselves, as seen in the work of Jules Chéret, Toulouse-Lautrec, Aubrey Beardsley and others. With the rise of Art Nouveau and Jugendstil*, these advertising posters became increasingly exuberant, sensual and exotic, as can be seen in the flamboyant work of Alphonse Mucha, Théophile Steinlen and others, where the lettering often followed the character of the images, becoming sinuous and organic as it took on a biomorphic quality of its own.

World War I saw the rise of the propaganda poster – a much more direct form of communication with a corresponding sense of severity and austerity. The Russian Revolution of 1917

INTRODUCTION

and the rise of the new Communist state saw another kind of propaganda evolve, and the evolution of Constructivist art and graphic design. Designers such as ALEXANDER RODCHENKO placed their talents at the service of the state and state enterprises, creating a completely fresh approach influenced by abstract art movements such as Cubism. Constructivist posters used abstraction combined with bold blocks of colour, particularly red, white and black, as well as new techniques such as photomontage.

The importance of Cubism and avant-garde abstraction also impacted upon the evolution of De Stijl in Holland. The movement itself was named after the eponymous journal, co-founded and edited by THEO VAN DOESBURG, and was preoccupied with geometrical forms, abstraction and the use of vivid blocks of primary colour against a neutral backdrop. These preoccupations crossed the borders of art, graphic design, interiors and architecture, with De Stijl counting GERRIT RIETVELD and Robert van 't Hoff* among its champions.

Both the work of De Stijl and the Russian Constructivists was to have an impact on a number of key figures at the Bauhaus – such as László Moholy-Nagy, HERBERT BAYER and JOSEF ALBERS – as well as the 'new typography'. At the Bauhaus, under the direction of Walter Gropius, there was a rich, multi-disciplinary process of cross-pollination between the fields of art, typography, architecture and product design, which served as a powerful foundation for the experimental approach to graphics and sans serif fonts explored by Moholy-Nagy, Bayer and others. This was to have – in turn – a profound impact upon Tschichold and the birth of his seminal text.

Within the Bauhaus and its philosophy, there was also a strong emphasis on function and clarity. Modernist graphic

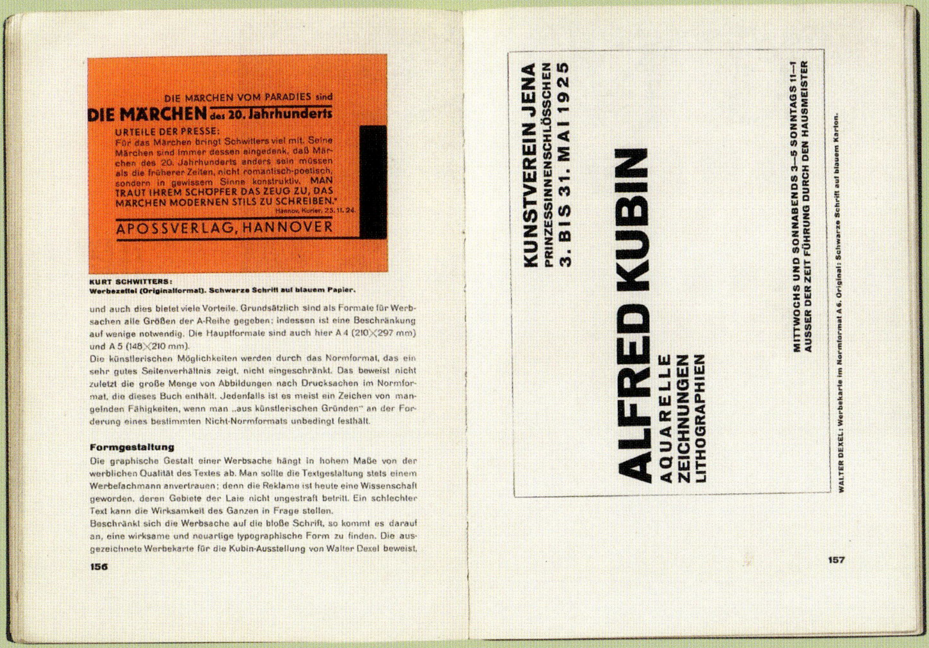

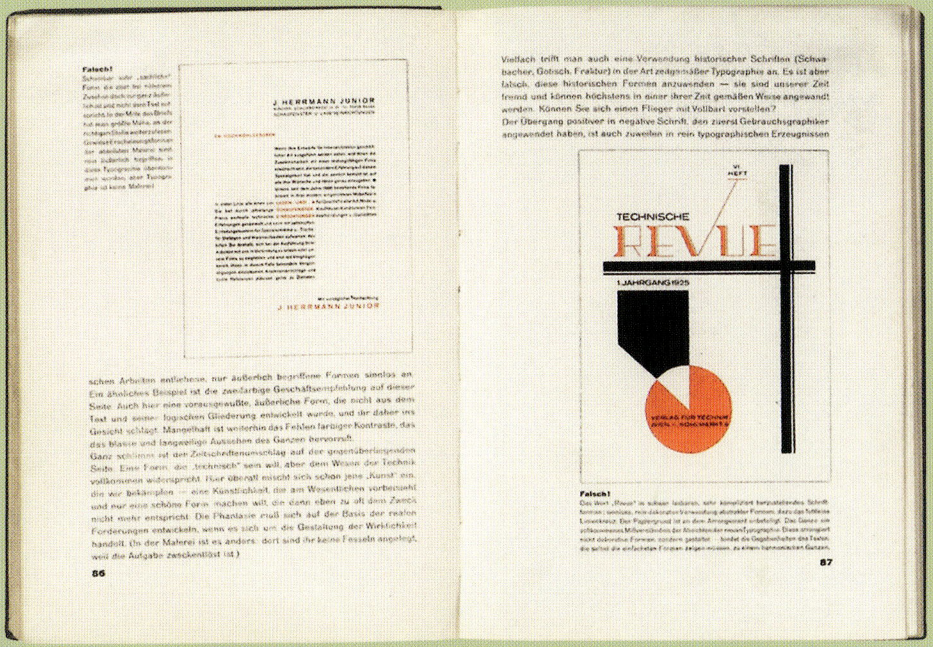

PAGE 253 A poster design by László Moholy-Nagy, 1927.

ABOVE Two spreads from Jan Tschichold's *The New Typography* (*Die neue Typographie*), first published by Bildungsverband der Deutschen Buchdrucker, Berlin, 1928.

design and typography, along with other forms of design, sought to strip away ornament and excess in favour of a pared-down, pure aesthetic suited to the machine age. Fonts were sans serif and easily accessible and legible, while modern techniques – particularly photomontage – became part of the readily available pool of resources. While many influential figures in the world of early modernist graphic design had emerged from the world of fine art and often kept one foot either side of the border, figureheads such as Bayer and Tschichold established themselves as specialists immersed in the fast-growing profession of graphic design.

Tschichold's contention that the modern world demanded a modern approach to type and graphic design was, arguably, in tune with the increasingly complex nature of everyday life. This was exemplified by the London Underground map of 1933 drawn by Henry Beck, along with the sans serif type designed by Edward Johnston (assisted by Eric Gill*, whose Gill Sans type was adopted by the London and North Eastern Railway). The map, famously, took an abstract approach to the various lines of the system rather than basing it upon a geographical or topographical template. This helped to simplify the diagram and, along with colour coding for the lines and the type font, provide clarity and simplicity for the end user. This approach was an inspiration both to designers and transportation companies around the world, helping to further the development of 'information design'.

Railway companies, along with shipping lines, were also key clients for a fresh wave of travel posters in the Twenties and Thirties. Just as industrial designers such as HENRY DREYFUSS and RAYMOND LOEWY benefited from the machine-age patronage of the railway companies, so too did graphic designers and poster artists. The growth of the travel industry was a spur to some of the greatest designers of the period, such as A. M. CASSANDRE, whose iconic posters of transatlantic liners such as the *Normandie* helped to sum up a brief but golden era of streamlined escapism. For younger designers, too, such as Herbert Matter*, who produced a striking series of Swiss ski travel posters using photomontage, the travel industry was a vital source of both income and progressive patronage.

Matter was one of the many European designers who made the journey to America during the course of the Thirties, going on to work with Knoll and others in the post-war period. Others who made the journey included Albers and Moholy-Nagy, who brought Bauhaus-inspired curriculums to American schools and colleges. Bayer, too, moved to the States, while Cassandre

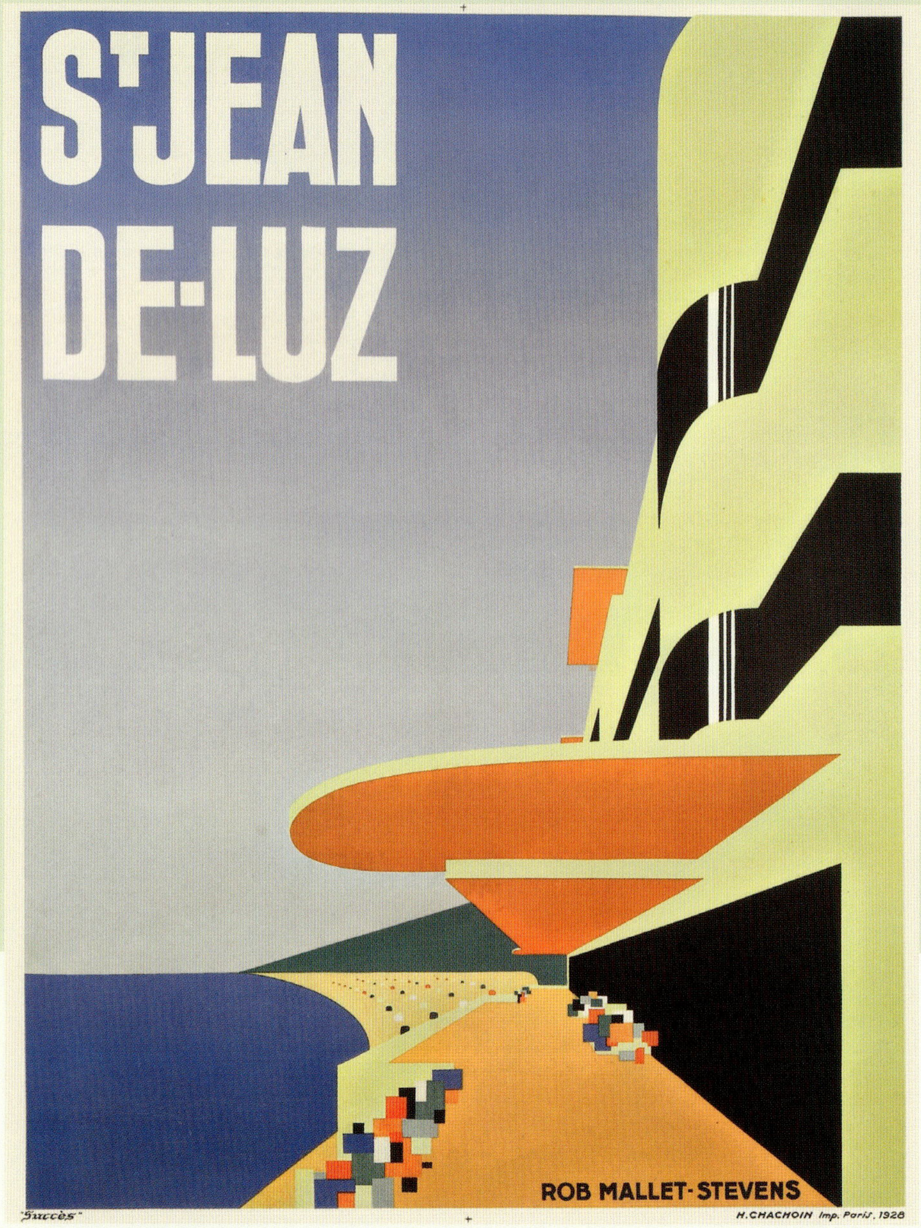

ABOVE A travel and tourism poster promoting Saint-Jean-de-Luz, designed by Robert Mallet-Stevens and printed by H. Chachoin, Paris, 1928.

spent a period of time in New York working with *Harper's Bazaar*, *Fortune* magazine and others.

The nascent American graphic design industry was undoubtedly strengthened and revitalized by this influx of talent, which also included Czech designer Ladislav Sutnar and legendary *Harper's Bazaar* art director Alexey Brodovitch*, who moved from Russia to Paris and then on to New York. Although born in America, Edward McKnight Kauffer worked largely in Paris and London until he, too, returned to America. Collectively, they created an extraordinary pool of original talent, adding to home-grown creatives such as Lester Beall* and Paul Rand, all contributing to the powerful surge of American advertising and graphic design in the mid-century period. A number of pioneering industrial designers, such as Raymond Loewy, branched out into graphic design as they extended their reach into branding, logos, packaging and corporate identities; Loewy famously redesigned the Lucky Strike cigarette packet in 1942.

Tschichold and *The New Typography* were also to have a particular impact on the evolution of the Swiss School of typography and design in the post-war period, also known as the International Graphic Style. Along with Adrian Frutiger, Max Miedinger and Otl Aicher, one of the key figures of the Swiss School was Max Bill*, who had trained at the Bauhaus. Bill was the co-founder of the Ulm School of Design in Germany, which lasted from 1953 to 1968, and followed a Bauhaus-inspired model; Josef Albers served as a guest tutor. Bill and Tschichold, who began to doubt some of the crucial tenets of his own book in later years, had a very public rift. But there is no doubting the ongoing impact and influence of the 'new typography' and early modernist thinking upon the evolution of post-war graphic design.

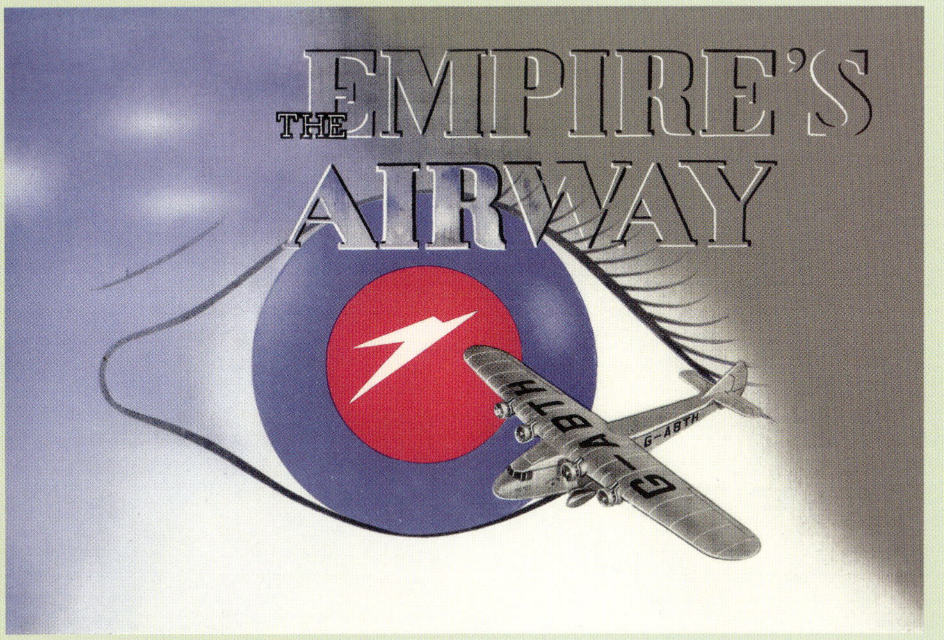

ABOVE A booklet cover for Imperial Airways, designed by László Moholy-Nagy, 1930s.

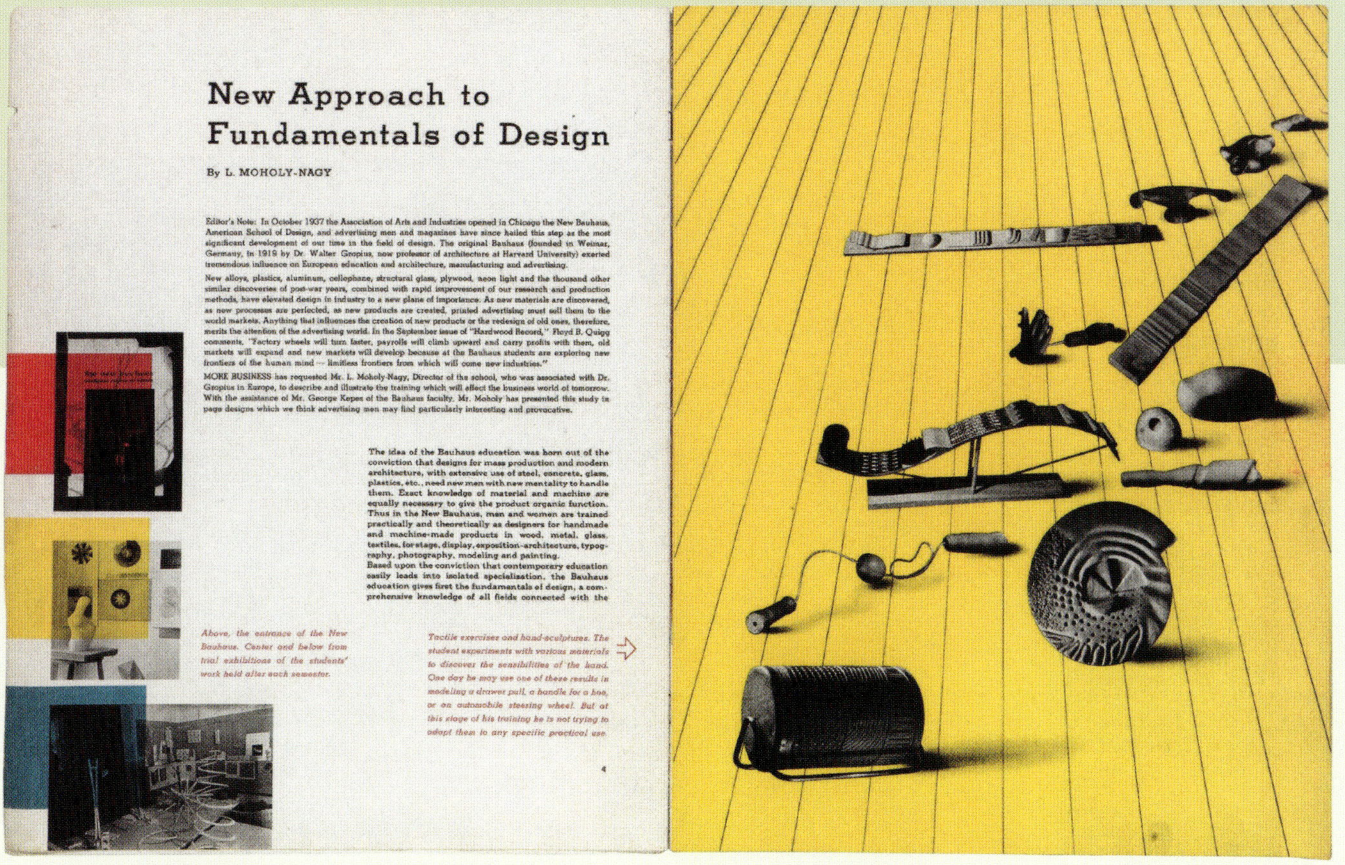

New Approach to Fundamentals of Design

By L. MOHOLY-NAGY

Editor's Note: In October 1937 the Association of Arts and Industries opened in Chicago the New Bauhaus, American School of Design, and advertising men and magazines have since hailed this step as the most significant development of our time in the field of design. The original Bauhaus (founded in Weimar, Germany, in 1919 by Dr. Walter Gropius, now professor of architecture at Harvard University) exerted tremendous influence on European education and architecture, manufacturing and advertising.

New alloys, plastics, aluminum, cellophane, structural glass, plywood, neon light and the thousand other similar discoveries of post-war years, combined with rapid improvement of our research and production methods, have elevated design in industry to a new plane of importance. As new materials are discovered, as new processes are perfected, as new products are created, printed advertising must sell them to the world markets. Anything that influences the creation of new products or the redesign of old ones, therefore, merits the attention of the advertising world. In the September issue of "Hardwood Record," Floyd B. Quigg comments, "Factory wheels will turn faster, payrolls will climb upward and carry profits with them, old markets will expand and new markets will develop because at the Bauhaus students are exploring new frontiers of the human mind — limitless frontiers from which will come new industries."

MORE BUSINESS has requested Mr. L. Moholy-Nagy, Director of the school, who was associated with Dr. Gropius in Europe, to describe and illustrate the training which will affect the business world of tomorrow. With the assistance of Mr. George Kepes of the Bauhaus faculty, Mr. Moholy has presented this study in page designs which we think advertising men may find particularly interesting and provocative.

The idea of the Bauhaus education was born out of the conviction that designs for mass production and modern architecture, with extensive use of steel, concrete, glass, plastics, etc., need new men with new mentality to handle them. Exact knowledge of material and machine are equally necessary to give the product organic function. Thus in the New Bauhaus, men and women are trained practically and theoretically as designers for handmade and machine-made products in wood, metal, glass, textiles, for stage, display, exposition-architecture, typography, photography, modeling and painting.

Based upon the conviction that contemporary education easily leads into isolated specialization, the Bauhaus education gives first the fundamentals of design, a comprehensive knowledge of all fields connected with the

Above, the entrance of the New Bauhaus. Center and below from trial exhibitions of the students' work held after each semester.

Tactile exercises and hand-sculptures. The student experiments with various materials to discover the sensibilities of the hand. One day he may use one of these results in modeling a drawer pull, a handle for a hoe, or an automobile steering wheel. But at this stage of his training he is not trying to adapt them to any specific practical use.

4

LEFT A spread from László Moholy-Nagy's essay 'New Approach to Fundamentals of Design', in a 1938 issue of *More Business: The Voice of Letterpress Printing and Photo-Engraving* devoted to the Bauhaus.

THE BIRTH OF MODERN TYPE: TYPOGRAPHICAL REVOLUTION AND THE ADVENT OF MODERNIST GRAPHIC DESIGN

RICK POYNOR

The formulation of the 'new typography' in the 1920s was a moment of profound transformation in the practice of visual communication. During these years, the industrial activity of typography became fully modern. This European movement established benchmarks and ideals that inform graphic and typographic design to this day. The individuals who developed these principles and campaigned for their implementation remain leading figures in the history of graphic design. JAN TSCHICHOLD, perhaps the greatest in terms of his contribution and influence, would famously later renounce his earlier commitment to the new typography, but its tenets had taken hold, achieving international acceptance after World War II, and his achievement as a typographic reformer could not be undone.

LÁSZLÓ MOHOLY-NAGY, a Hungarian artist, teacher and theorist, was the first to use the term 'new typography', in 1923, as the title of an essay in the book *Staatliches Bauhaus in Weimar, 1919–1923*. Typography, he argued, should be communication of the most intense kind, and it must aim for total clarity. Legibility should never be impaired by a predetermined aesthetic, and letters should not be forced to conform to a pre-existing framework. Type in any size, geometric form or colour should run not only horizontally, but in all linear directions. 'We want to create a new language of typography whose elasticity, variability and freshness of typographical composition is exclusively dictated by the inner law of expression and the optical effect.' The viewer would be confronted by the 'shock effect' of new typefaces, as well as other visual devices, and the new typography would offer an experience that was simultaneously visual and communicative.

A chaotic early version of this new typography was by that time flourishing among artists and poets in avant-garde art movements and publications. In a manifesto published in 1913 titled 'Destruction of Syntax – Imagination without Strings – Words-in-Freedom', the rambunctious Italian Futurist* Filippo Tommaso Marinetti proclaimed the need for a typographic revolution that would disrupt and dynamically remake the 'so-called typographical harmony of the page, which is contrary to the flux and reflux, the leaps and bursts of style that run through the page'. If necessary, within a single page, the Futurist writer would employ 20 different typefaces and a mixture of type styles, jumping from bold to italic, to amplify the expressive impact of the language. In Marinetti's *Futurist Words-in-Freedom*, published in 1919, the letters burst free from the parallel lines that normally contain and discipline them and explode in a blast of ink in all directions across the paper.

A similar liberation could be seen in the angular geometrical type arrangements and page architecture of Russian Constructivists* such as ALEXANDER RODCHENKO and El Lissitzky*, author of 'Topography of Typography' (1923), another key statement, and in the way the Dadaist anti-art provocateurs collaged together words from different sources in their visual broadsides. In the *Kleine Dada Soirée* poster (1922) by THEO VAN DOESBURG and Kurt Schwitters, interlocking letters drawn by hand form a dense barrage of alphabetical noise superimposed on the word 'DADA' tumbling behind them.

This period of the new typography led by artist-typographers has been characterized as its visionary phase. What was needed next was a more systematic articulation of principles so that the new way of designing could be brought into the realm of ordinary printing. In October 1925, *Typographische Mitteilungen* published a special issue devoted to 'Elemental Typography' guest-edited by Tschichold,

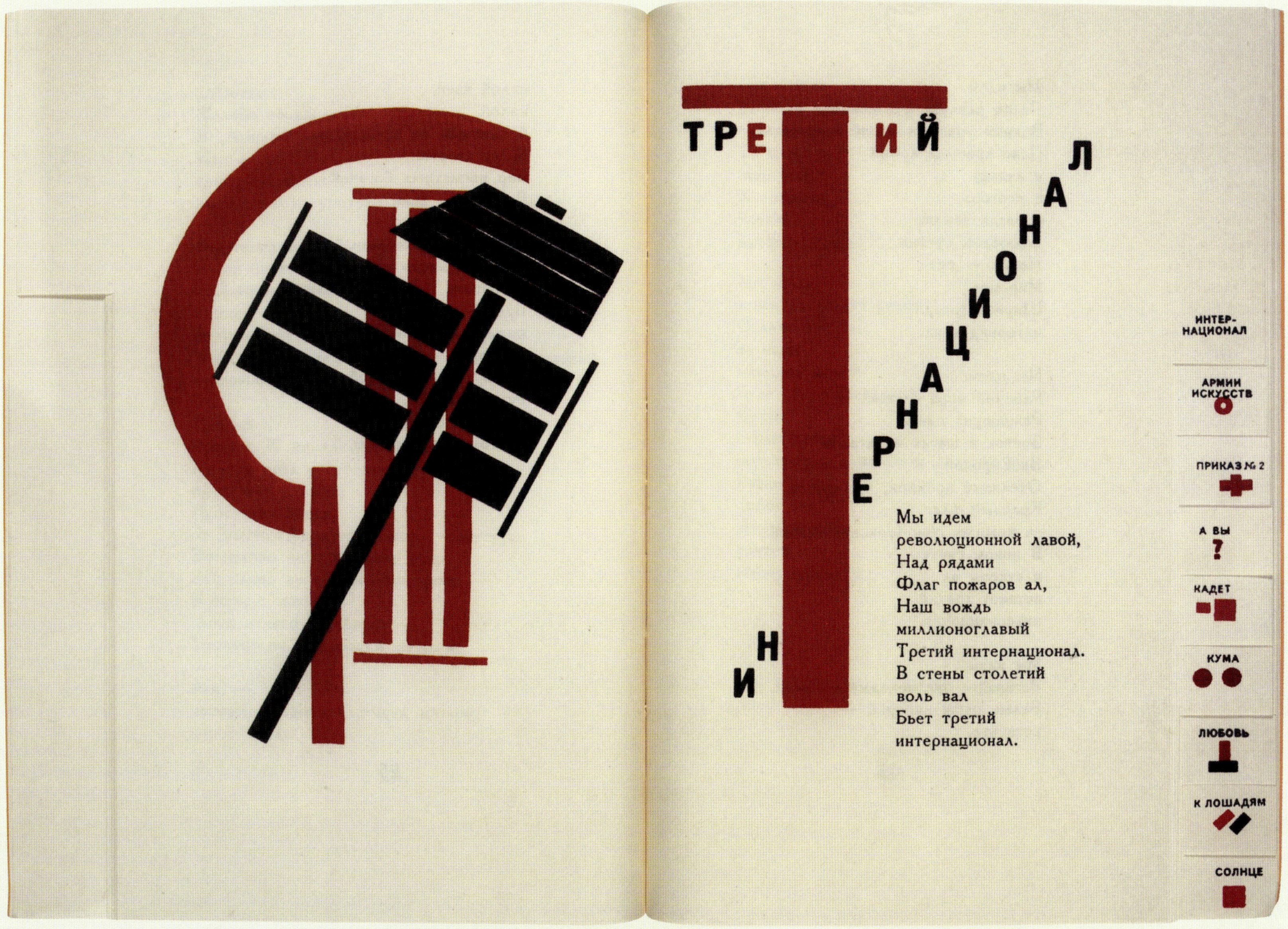

ABOVE A spread from *For the Voice*, showing a layout for 'The Third International' (a poem by Vladimir Mayakovsky), designed by El Lissitzky, 1923.

a professional typographer conversant with every aspect of print. The magazine included Tschichold's list of 10 precepts, beginning with the statement 'The new typography is oriented towards purpose' and followed by other essential outcomes of the new design: simplicity, ordering, organization, economy, differentiation, legibility and urgency of communication. The typography of pre-modern times had been burdened, often gratuitously, with ornamental devices used to decorate the page. Elemental typography rejected these fanciful effects, although 'inherently elemental' forms such as squares, circles and triangles were permitted as part of a fully integrated construction.

Unlike earlier advocates of the new typography, Tschichold focused on practicalities of typeface selection and the attributes of letterforms, including the unsuitability of those with an overly national character. The most elemental letterform, he advised, was the sans serif, which could be used in all its variations of weight, although he acknowledged that serif types were at that point more legible for text setting. HERBERT BAYER, director of printing at the Bauhaus, gave his geometrical sans serif alphabet the name 'Universal' (1925), but it was not put into production as a typeface. Paul Renner's similarly forward-looking sans serif Futura (1927–30), which employs a consistent stroke width throughout the character set, became one of the most enduring typefaces of the modernist era.

In 1928, Tschichold developed his thinking at book length in *The New Typography: A Handbook for Modern Designers*, in which he urges printers to engage with the need for a form of typography that corresponds to the demands of a new technical age (surprisingly, an English translation did not appear until 1995). Tschichold devotes a section to 'old typography' from 1440 to 1914, concluding that in recent history there has been a 'senseless outpouring' of new types offering poor variations of historical themes, while printers obliged to submit to the whims of book artists' private styles have succumbed to decadence. Instead, he proposes a rigorous typography based on 'logical, "non-personal" work, using impersonal materials, which alone will make possible a free, impersonal creativity'. The application of impersonal laws and elements would assure the emergence of a 'collective culture' based on pure design.

The principal difference between old and new typography lay in the shift from the symmetrical organization of the page to an asymmetrical structure. In the traditional page, all the elements – titles, subheadings and text – are positioned around a central axis. In the new typographic page, the elements could be placed to the left or right in dynamic arrangements that broke away from the stiffness of this backbone. Text columns aligned on the left were typeset with even word spacing and a ragged edge on the right. This treatment of longer texts remained controversial for many years and when Tschichold's second book (1935) was finally published in English in 1967, under the title *Asymmetric Typography*, Tschichold followed convention and justified the text to the full column width.

The delays in publishing English versions of these seminal texts are signs that the new typography was not embraced everywhere in the Twenties and Thirties, despite its aspirations to universalism. Its centre was Germany until 1933 when Adolf Hitler became Chancellor and the Nazi Party came to power. Tschichold was arrested and later that year he fled the country. Although the new typography would have a lasting impact, it was the creation of a relatively small band of designers, such as Bayer and Walter Dexel, author of the essay 'What is New Typography?' in Germany, Piet Zwart* and Paul Schuitema in the Netherlands, Karel Teige and Ladislav Sutnar in Czechoslovakia, and the Hungarian Lajos Kassák, as well as the figures already mentioned.

In these years, the movement's influence was mostly felt in Central and Eastern Europe and Scandinavia. In

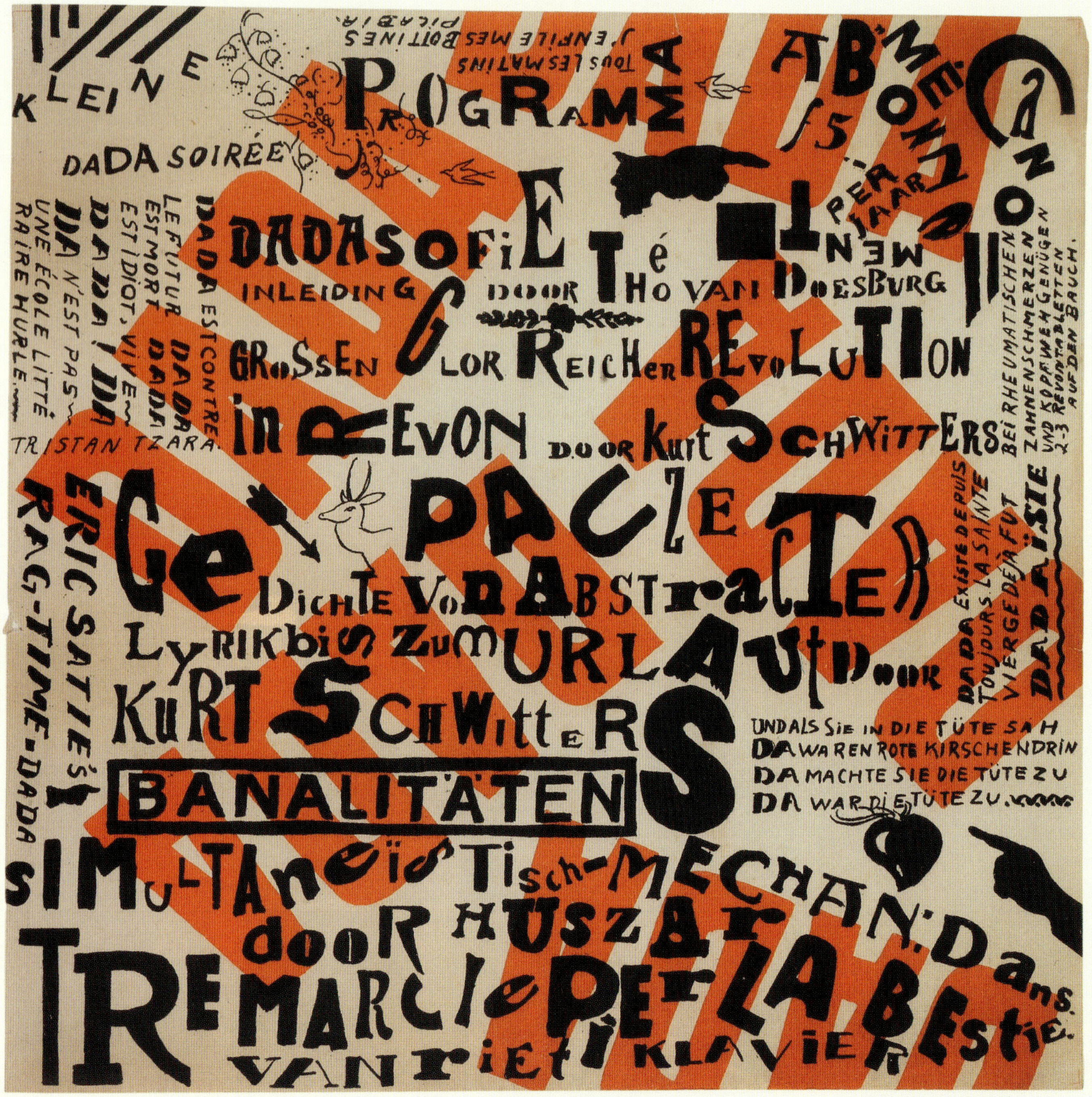

LEFT Theo van Doesburg and Kurt Schwitters's poster for a 'Small Dada Soirée', Dada Tour of the Netherlands, 10 January–14 February 1923.

Thirties France, the Art Deco* style was far more pervasive in advertising and everyday communication; the design manual *Mise en Page: The Theory and Practice of Lay-out* (1931) by Alfred Tolmer, a printer in Paris, was a bravura application of Art Deco mannerisms to the page. Tschichold had an exhibition in London in 1935 and wrote a few articles about the new typography for English magazines such as *Commercial Art* and *Typography*, which eclectically blended both modernist and traditional subjects. In 'What is this "Functional" Typography?', a profile of Tschichold for *Printing* in 1936, Robert Harling, *Typography*'s editor, crowned him as 'the dominant voice in Continental typography' – 'deadly serious, quiet, insidious and *logical*'. For the most part, the new typography was too clinical for traditional British taste and its lessons were still being absorbed in the late Fifties.

In Switzerland, however, modernist typography found highly fertile soil. Early exponents such as Dexel had stressed the need for 'an objective and impersonal presentation free of individuality'. The aim of this functionalist design was clarity: to communicate the essence of a message in words and images without aesthetic interference. The Swiss architect, artist, designer and theorist Max Bill* exemplified this approach. After studying at the Bauhaus, Bill returned to Zurich in 1929 and worked for the next decade as a graphic designer. In a leaflet designed in 1932 for the Wohnbedarf furniture company, he used the sans serif typeface Akzidenz Grotesk, set entirely in lowercase, while typographic panels and photographs illustrating the products alternated in a vertical flow that was both precisely descriptive and graphically expressive of a lifestyle based on rationality and simplicity.

During the Fifties, Swiss typography would begin to emerge as an 'international style' founded on the precepts of the new typography and Bill, his authority cemented by his participation in the pre-war phase, would be a pivotal influence on its development. By 1958, the Neue Grafik (new graphic design) was widely recognized and a magazine with that title disseminated the cause of functionalism, and an adherence to clarity, asymmetry and the sans serif, around the world. Absorbed into 'Swiss Style', the new typography remains an essential point of reference and a source of inspiration for graphic designers everywhere.

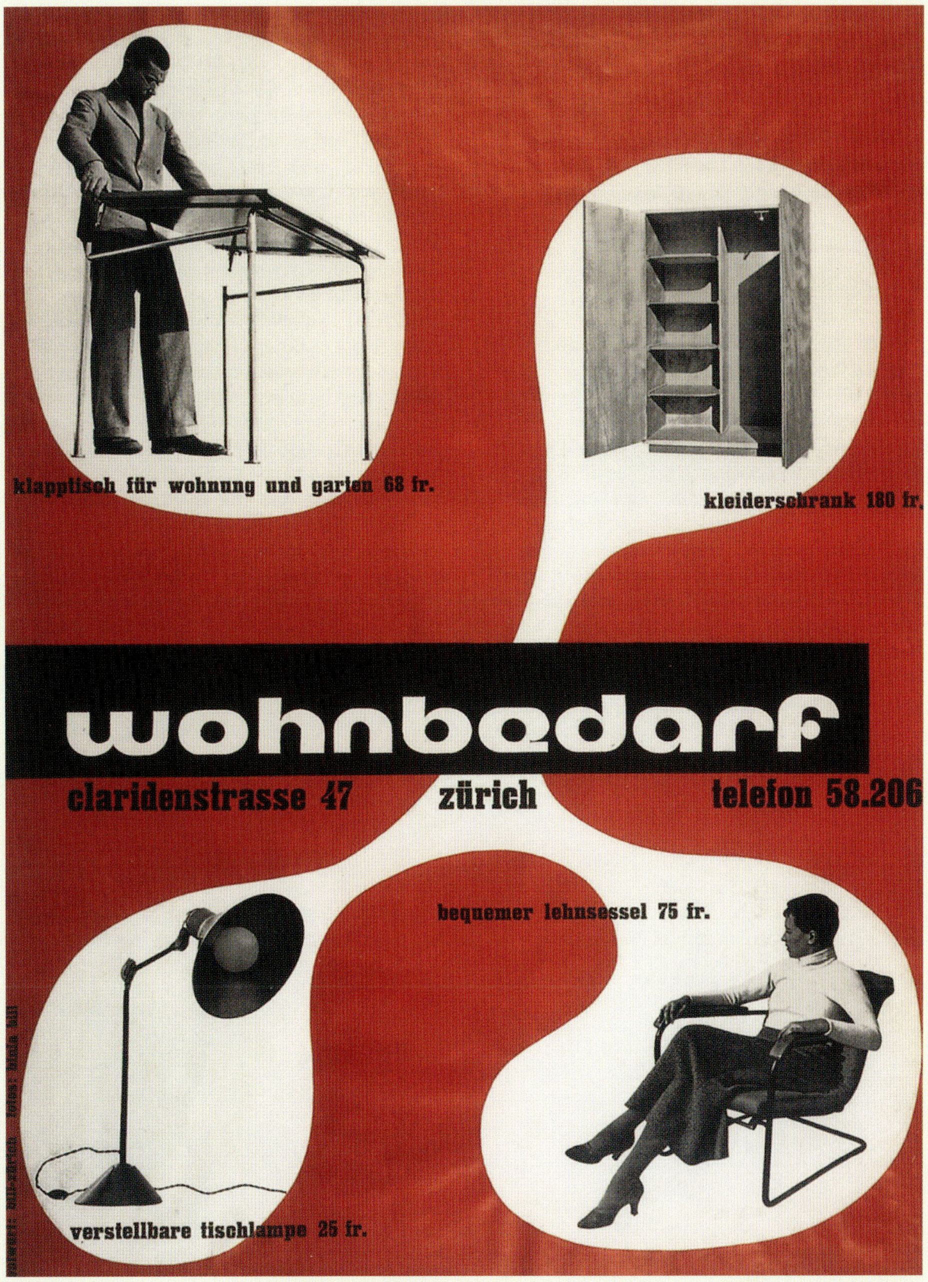

ABOVE A letterpress leaflet for the furniture company Wohnbedarf, designed by Max Bill, 1932.

JOSEF ALBERS
(1888–1976)

An artist, designer, typographer and educator, Josef Albers was a key figure in the evolution of a fresh approach to art, abstraction and colour theory. Through both his own work and his teaching – which was profoundly influenced by his years at the Bauhaus – Albers encouraged his students and followers to embrace a process of artistic self-discovery and growing awareness of colour, form, texture and line. The key ambition was 'to open eyes' rather than impose a set of strict rules or teach specific techniques. For Albers himself, the process led towards an increasing emphasis upon geometric abstraction and purity, famously explored in a long sequence of paintings and screenprints created over many years, known as 'Homage to the Square', which began in 1949. These pieces, along with seminal illustrated texts such as *Interaction of Color* (1963), were to have a profound impact upon a younger generation of post-war Abstract Expressionist artists and pioneers of the Op Art movement.

Born in the German town of Bottrop in Westphalia, Albers worked as a primary school teacher for five years. He then retrained as an art teacher in Berlin, followed by further studies in Munich and, from 1920, at the Bauhaus in Weimar where he took the foundation course (or 'Vorkurs') run by Johannes Itten*. In 1922 he met fellow student Anni Fleischmann at the Bauhaus, and she later became his wife. Along with Marcel Breuer, Albers was one of a small and select group of talented Bauhaus students who were then invited back as teachers – or 'young masters' – by director Walter Gropius.

As a Bauhaus master from 1923 onwards, Albers taught on the foundation course alongside László Moholy-Nagy. But he was also appointed as a master in the Weimar glass workshop, with his own work from this period including stained glass, glassware and furniture. When the Bauhaus moved to Dessau in 1925, Josef and Anni Albers* moved with the school and took up residence in one of the new masters' houses designed by Gropius. Josef

ABOVE An abstract oil painting by Josef Albers, 'Ochre Around Red & Browns', 1947.

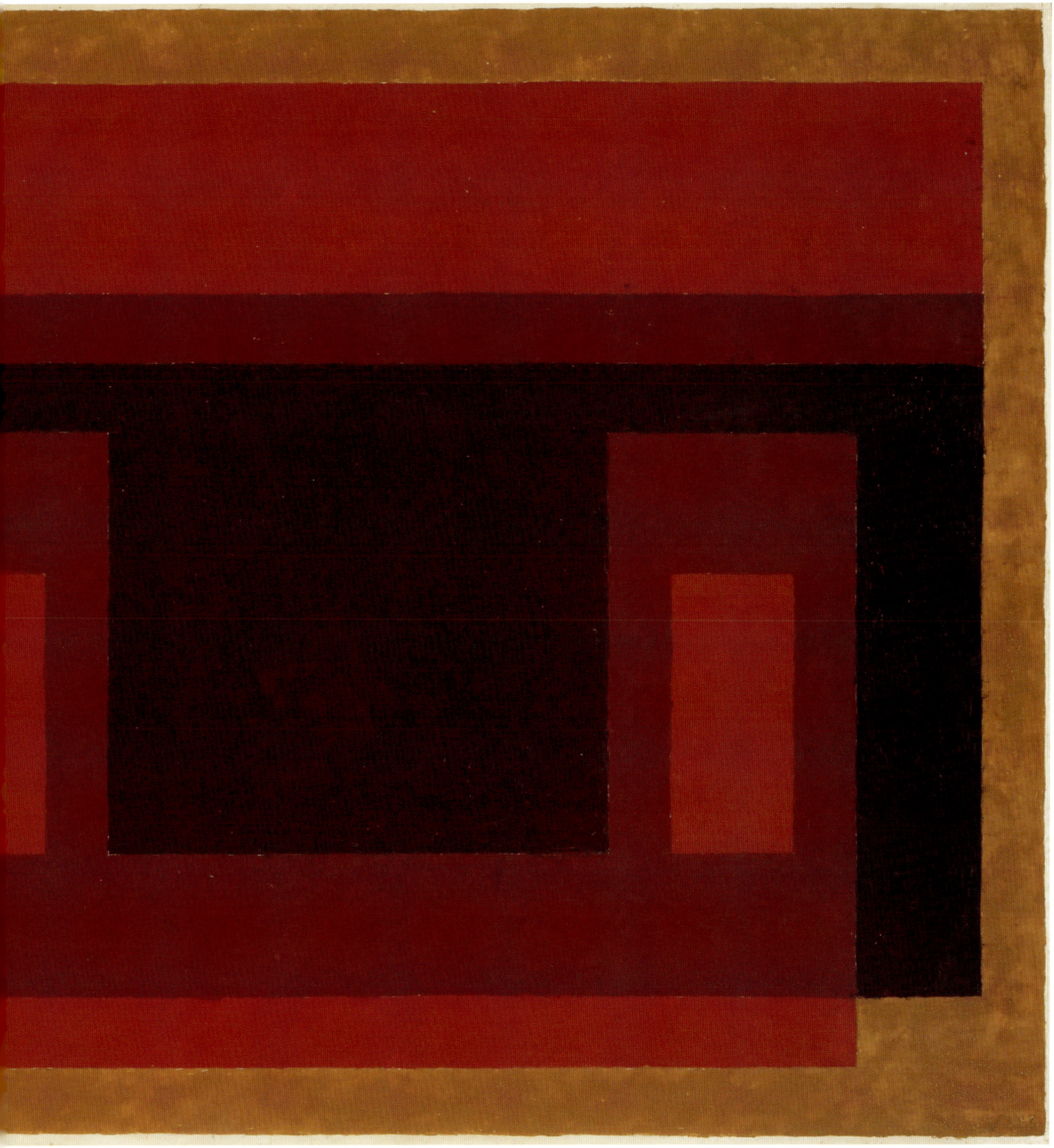

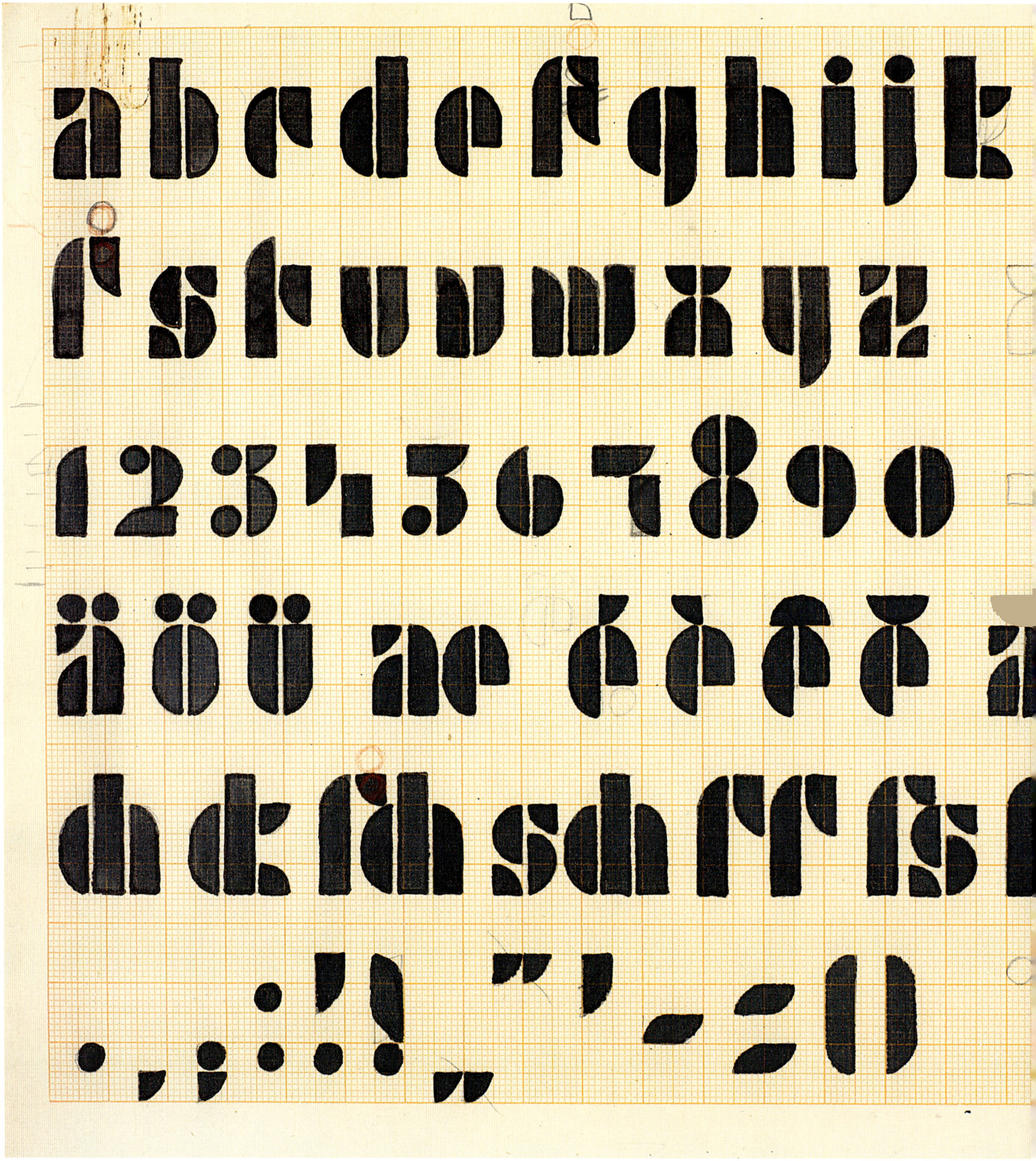

ABOVE A design study for a universal typeface in ink and pencil on graph paper by Josef Albers, *c.* 1926.

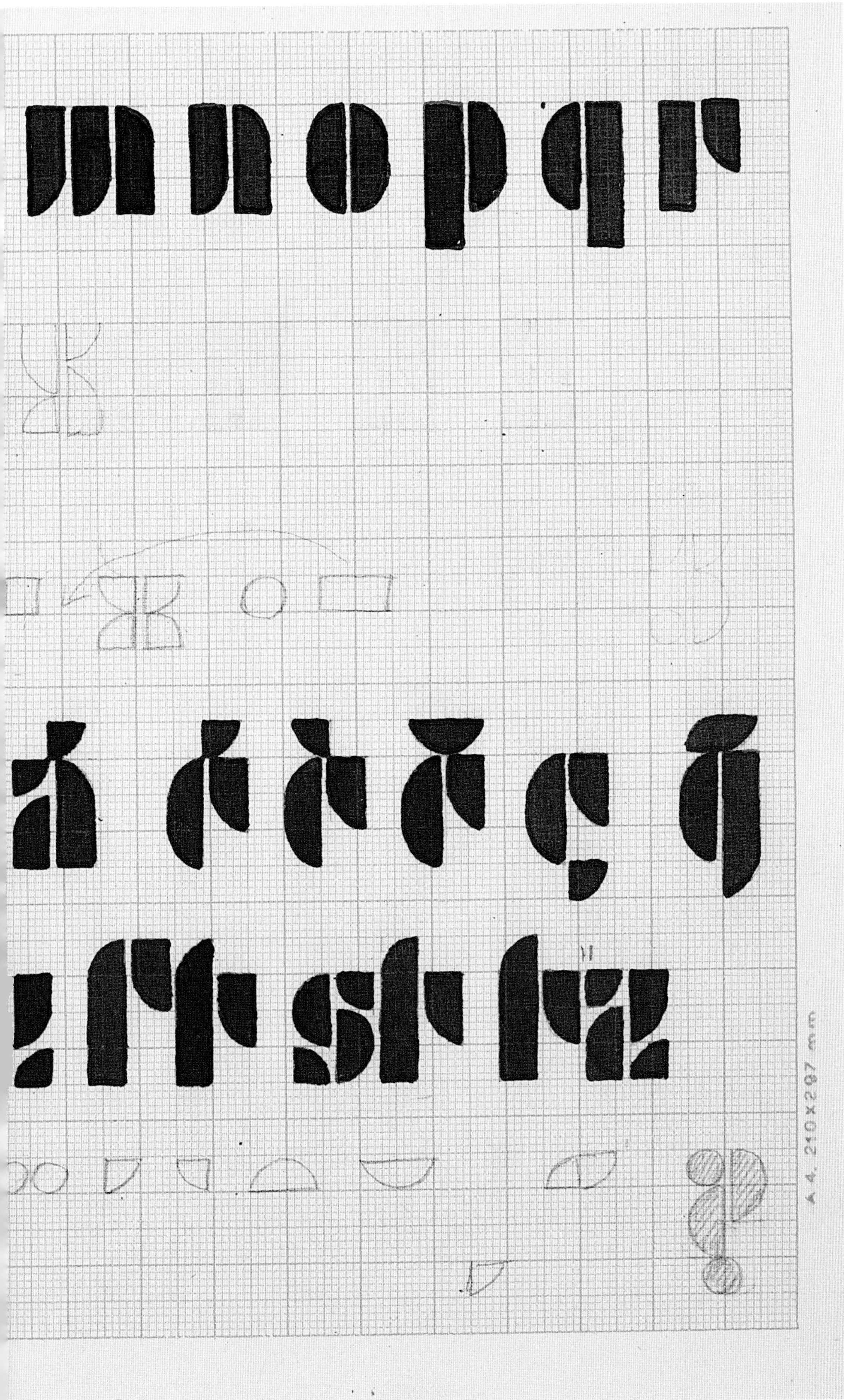

Albers continued to teach on the foundation course but was also involved in other aspects of the school curriculum. Other colleagues from this period included artists Paul Klee* and Wassily Kandinsky*.

In 1925, while at the Bauhaus, Albers designed his striking and original 'Stencil' typeface. This was a sans serif font, made up of a small and thoughtful palette of repeated shapes. It was distinctive and beautiful, with something of an Art Deco* flavour, yet it was also rather impractical as the semi-abstracted letters could be difficult to read.

In 1933, with the forced closure of the Bauhaus under pressure from the Nazi authorities, Josef and Anni Albers emigrated to America. Josef became head of the art department at the Black Mountain College in North Carolina, while continuing with his own work as a painter. After sixteen years at Black Mountain, Albers was then invited to assume the headship of the design department at Yale University, where he rewrote the curriculum on Bauhaus-based principles; he also served as a visiting professor at the Ulm School of Design (Hochschule für Gestaltung) co-founded by another ex-Bauhaus pupil, Max Bill*.

Albers's work on colour theory, which was also explored extensively in his own geometric paintings and artworks, was to prove particularly influential. One of his key preoccupations concerned the 'relativity' of colour and the way that the perception of colours could be affected by their relationship or juxtaposition with other colours, as well as by the perspective of the viewer and the nature of the light and surrounding conditions. Albers's paintings explored these relationships and the borders between colours.

Among the many students that Albers taught were Robert Rauschenberg and Cy Twombly; graphic designer Victor Moscoso, who experimented with 'vibrating colour' in his psychedelic posters of the Sixties, was also a student. Albers continued to explore other avenues of art and design, as well as painting and teaching, including large-scale murals for locations such as the Pan Am Building in New York (1963), co-designed by his former Bauhaus colleague, Walter Gropius, and he continued to work in the field of stained glass.

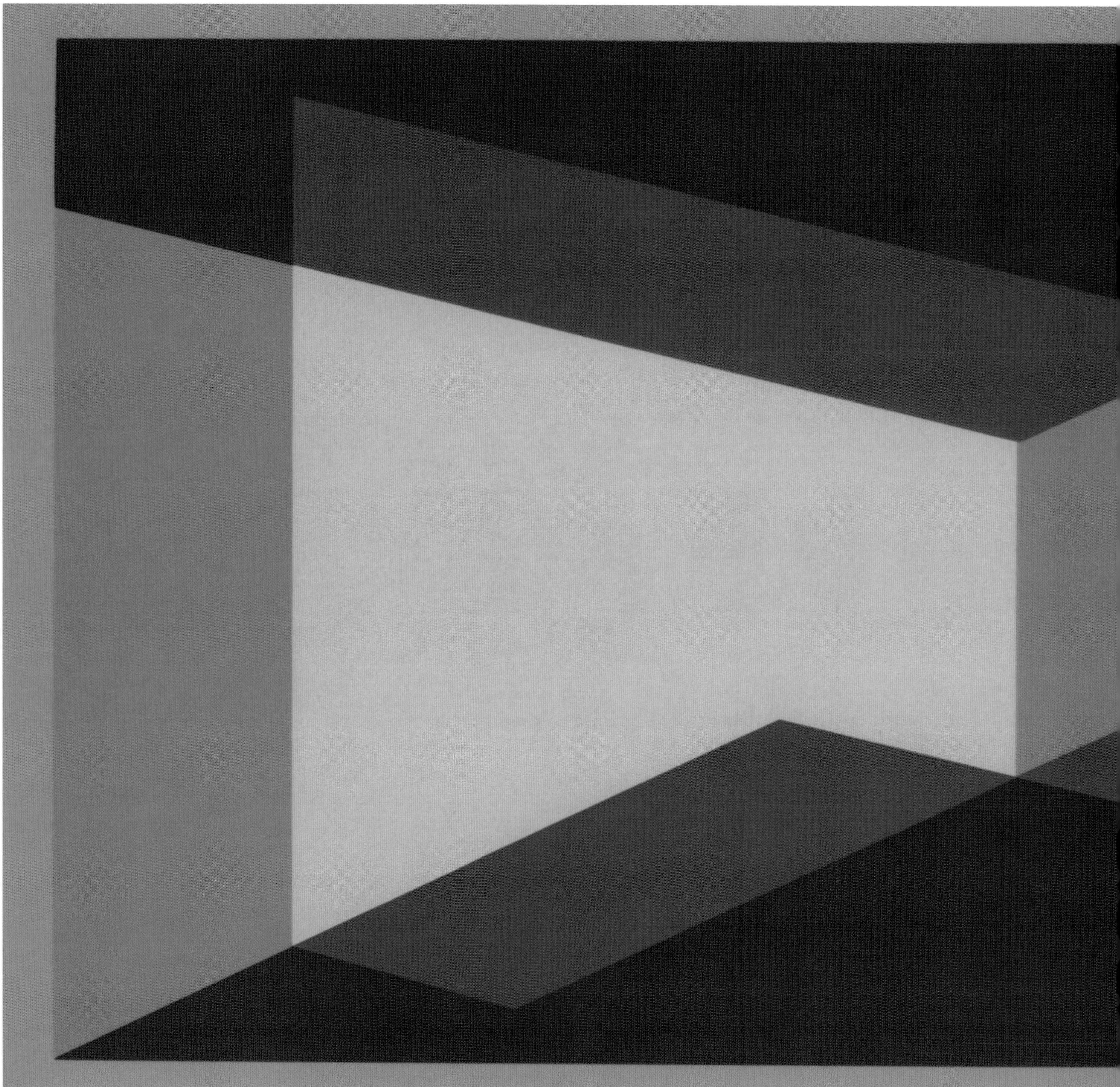

BELOW A post-war screenprint by Josef Albers, 'I-S 1 VV II', part of a limited edition of a hundred prints.

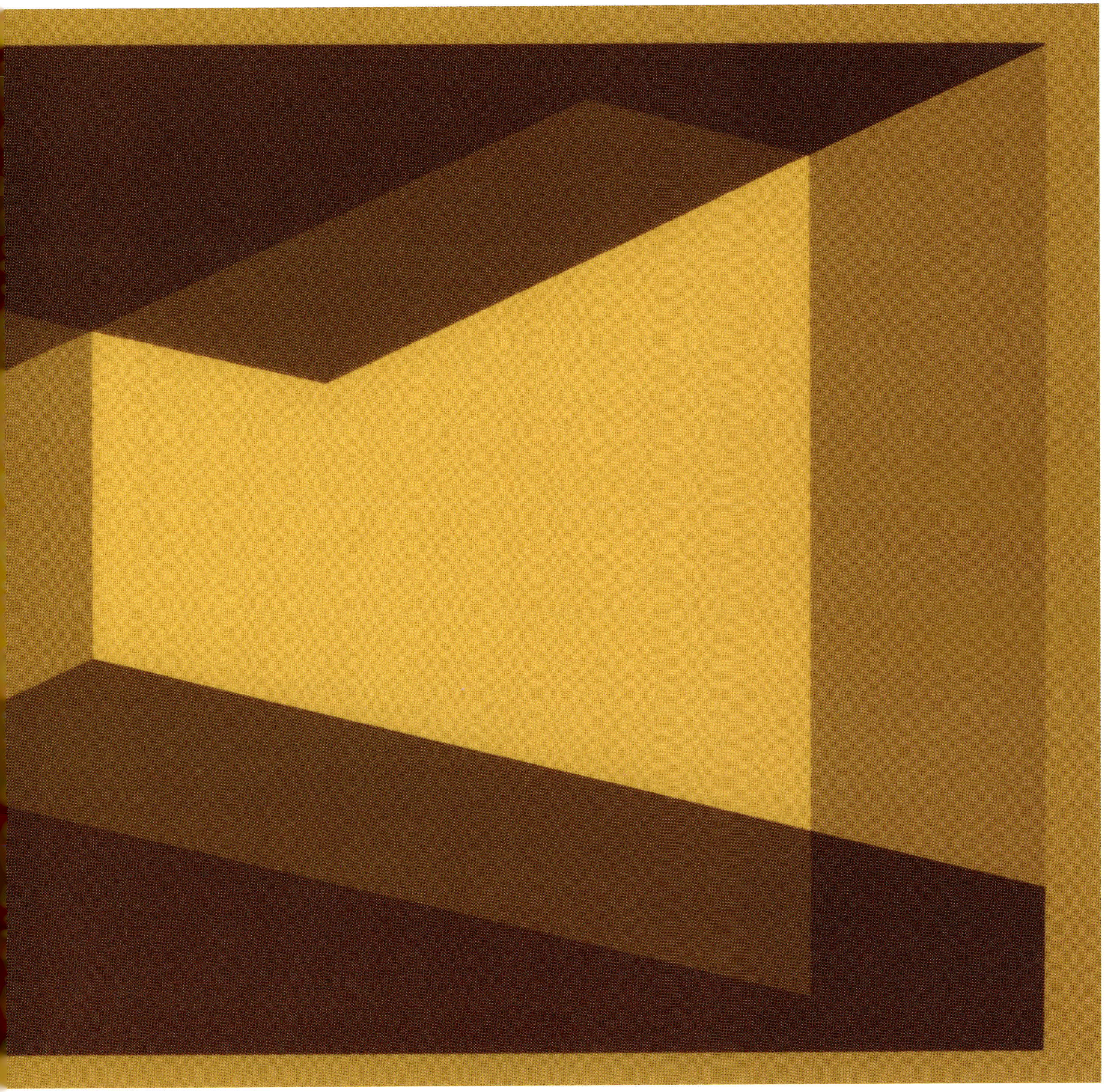

HERBERT BAYER
(1900–1985)

The graphic designer and typographer Herbert Bayer played a vital part in the presentation of the Bauhaus's visual identity. Along with his colleague László Moholy-Nagy, Bayer evolved what was effectively a brand identity conveyed through Bauhaus posters, magazines and exhibitions, partly defined by Bayer's use of progressive typography and a palette of geometric forms and bold colours. His experimental 'Universal' typeface of 1925, along with other work from his Bauhaus period, was a key influence upon the development of modernist graphic art in the years that followed.

Born in the Austrian town of Haag, Bayer served an apprenticeship in the Linz office of architect and designer Georg Schmidthammer, followed by a period at the artists' colony in Darmstadt, Germany. In 1921, he enrolled as a student at the Weimar Bauhaus, taking the foundation course led by Johannes Itten*, and then joining the mural painting department, where he was taught by Oskar Schlemmer* and Wassily Kandinsky*.

In 1925, along with former students Marcel Breuer and Josef Albers, Bayer was invited to join the teaching staff at the new Bauhaus campus in Dessau. Here, he became a young master within the print and advertising workshop, where he worked until 1928.

Bayer embraced the opportunity to experiment with both typography and graphic art. In the field of typographic design, he sought to address the challenge of creating a new approach in line with the Bauhaus's emphasis on machine-age design suited to factory production, with an emphasis on clear, legible lettering that could be easily standardized and reproduced. Most famously, he created the 'Universal' typeface. This was a sans serif 'mono-alphabet' with no upper-case capital letters: 'Why should we write and print in two alphabets? We do not speak in a capital "a" and a small "a". A single alphabet gives us practically the same result as the mixture of upper- and lower-case letters, and at the same time is less of a burden on all who write.'[3]

The Universal typeface was used for most publications and posters produced by the Bauhaus in Dessau; a variation known as 'Bayer-type' (1935) was later produced by the Berlin-based Berthold type foundry. The brave dismissal of the entire upper case represented, in itself, a radical break with tradition and spoke of typographic revolution. In this way, Bayer's work had a particular impact on the growth of the 'new typography' developed by Jan Tschichold and others, who were inspired by the Bauhaus model.

In addition to the radical use of typography, Bayer employed a distinctive use of colour and geometric form in his poster and journal designs for the Bauhaus, which suggested the influence of the Russian Constructivists* and other avant-garde modernists, such as followers of De Stijl*. Triangular forms, cones, spheres, squares and rectangles appear again and again, suggesting mathematical precision, but also expressive of volume and scale.

Commentators have pointed to the way in which Bayer sought to find a method of echoing or mirroring – in abstract, graphic form – the

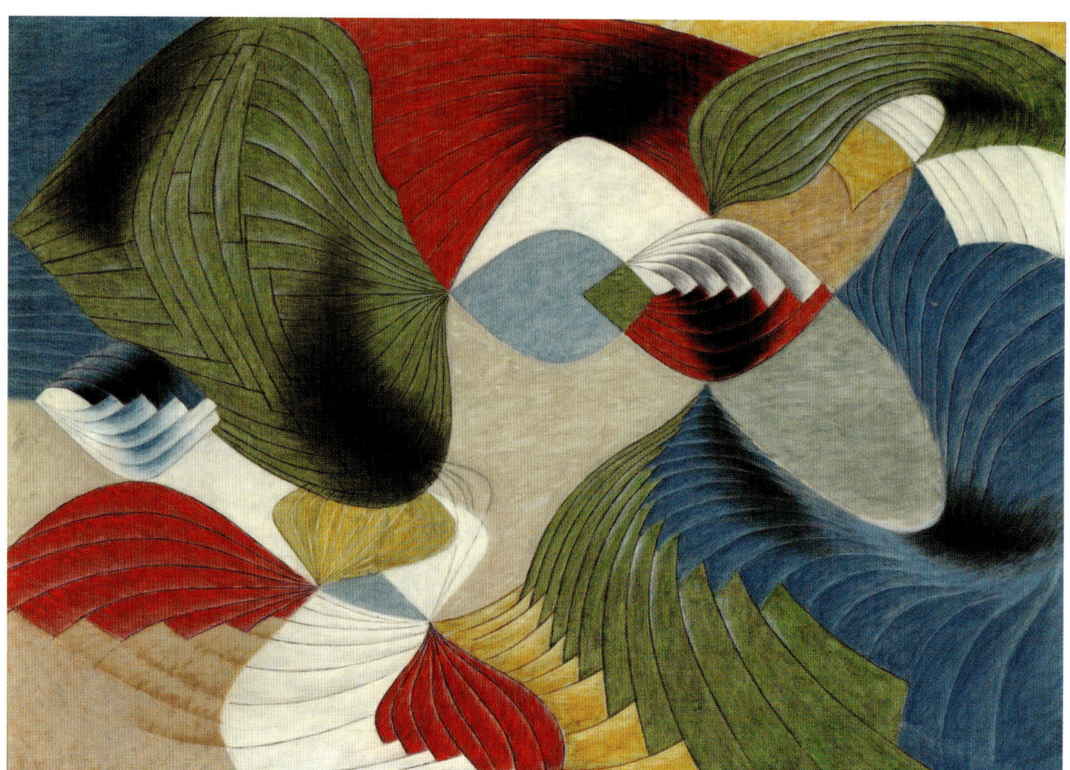

RIGHT A 'Formations' oil painting on canvas by Herbert Bayer, 1949.

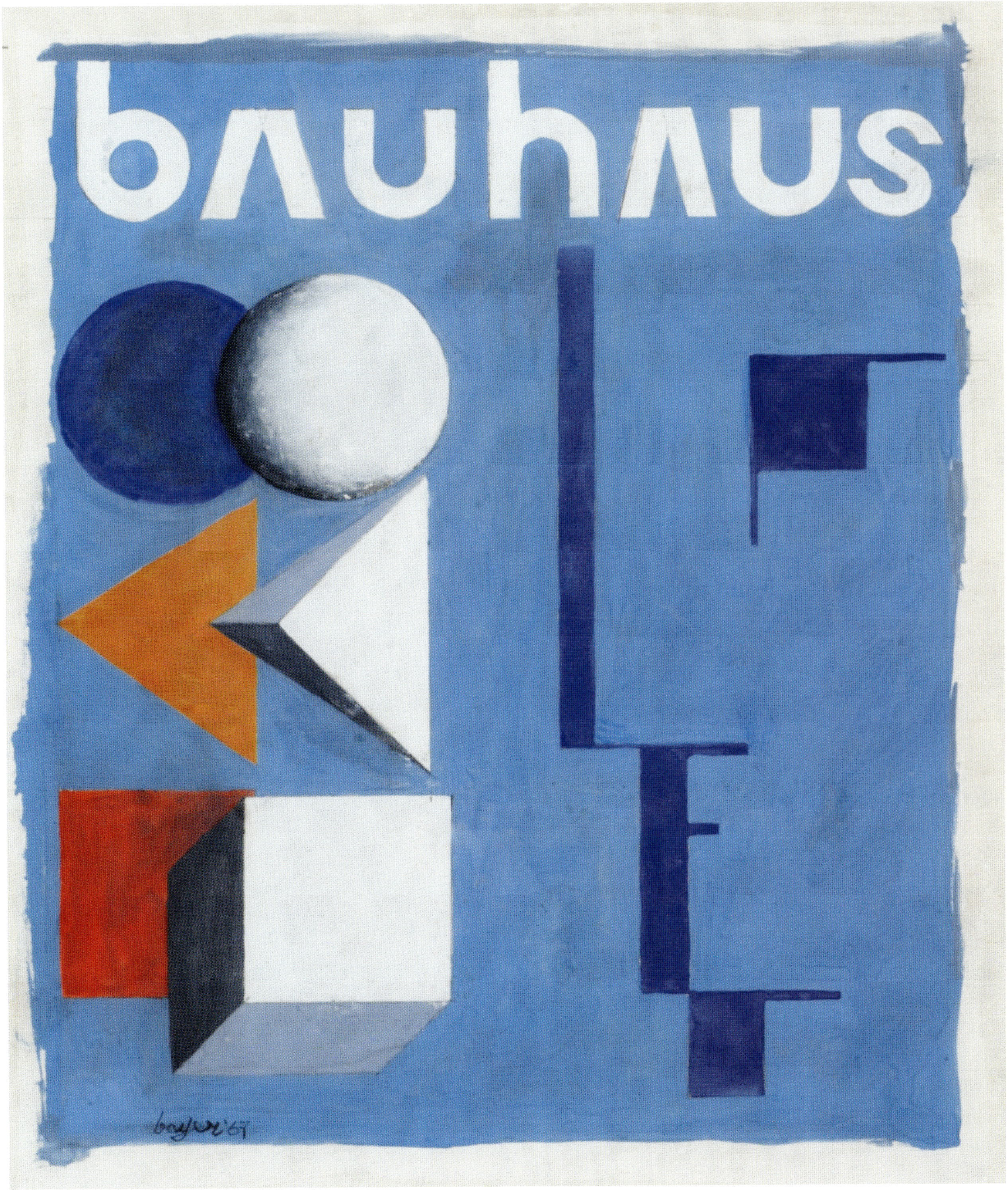

ABOVE A study for a Bauhaus poster by Herbert Bayer for the travelling exhibition '50 Jahre Bauhaus', 1968.

modernist architecture of the new Bauhaus campus, as well as its machine-age ethos. The vivid use of colour is also striking, suggestive again of the love of primary colours seen in – for instance – Dutch De Stijl.

After Bayer left the Bauhaus, in 1928, he moved to Berlin, working with the advertising agency Studio Dorland as director and serving as an art director for a number of magazines. He also designed posters for several exhibitions, including a 1930 exposition in Paris organized by the Deutscher Werkbund*.

In 1938, Bayer followed many of his Bauhaus contemporaries and emigrated to America; he designed the catalogue and poster for a major Bauhaus exhibition held at the Museum of Modern Art in New York that same year, curated by former director WALTER GROPIUS (fifty years later, in 1968, Bayer designed artwork for another Bauhaus exhibition, held in Stuttgart).

Bayer's American design career carried him across a number of different fields, including graphic design, architecture and landscape design. In 1946 he moved to Aspen, where he became a consultant to the packaging company Container Corporation of America (CCA) and worked for a number of other clients. Projects for CCA included the 'Great Ideas of Western Man' branding campaign, in which different designers – including Bayer himself – illustrated landmark quotes and ideas by writers and philosophers; the campaign, using an oblique and indirect method of communication, ran for thirty years.

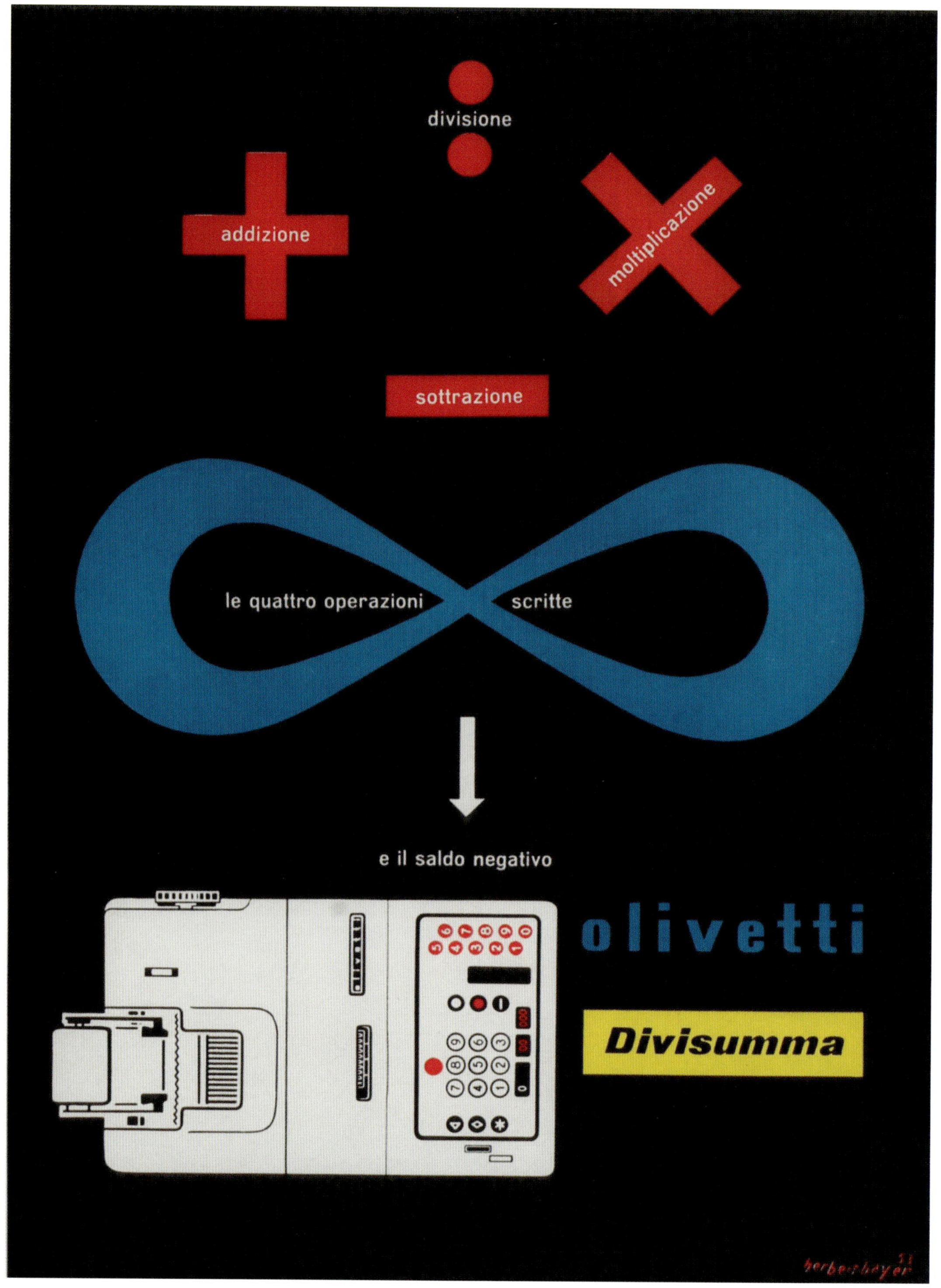

ABOVE Advertising graphics by Herbert Bayer for Olivetti, 1942.

OPPOSITE Bayer's 'Signs & Signals', in watercolour and gouache on paper, 1939.

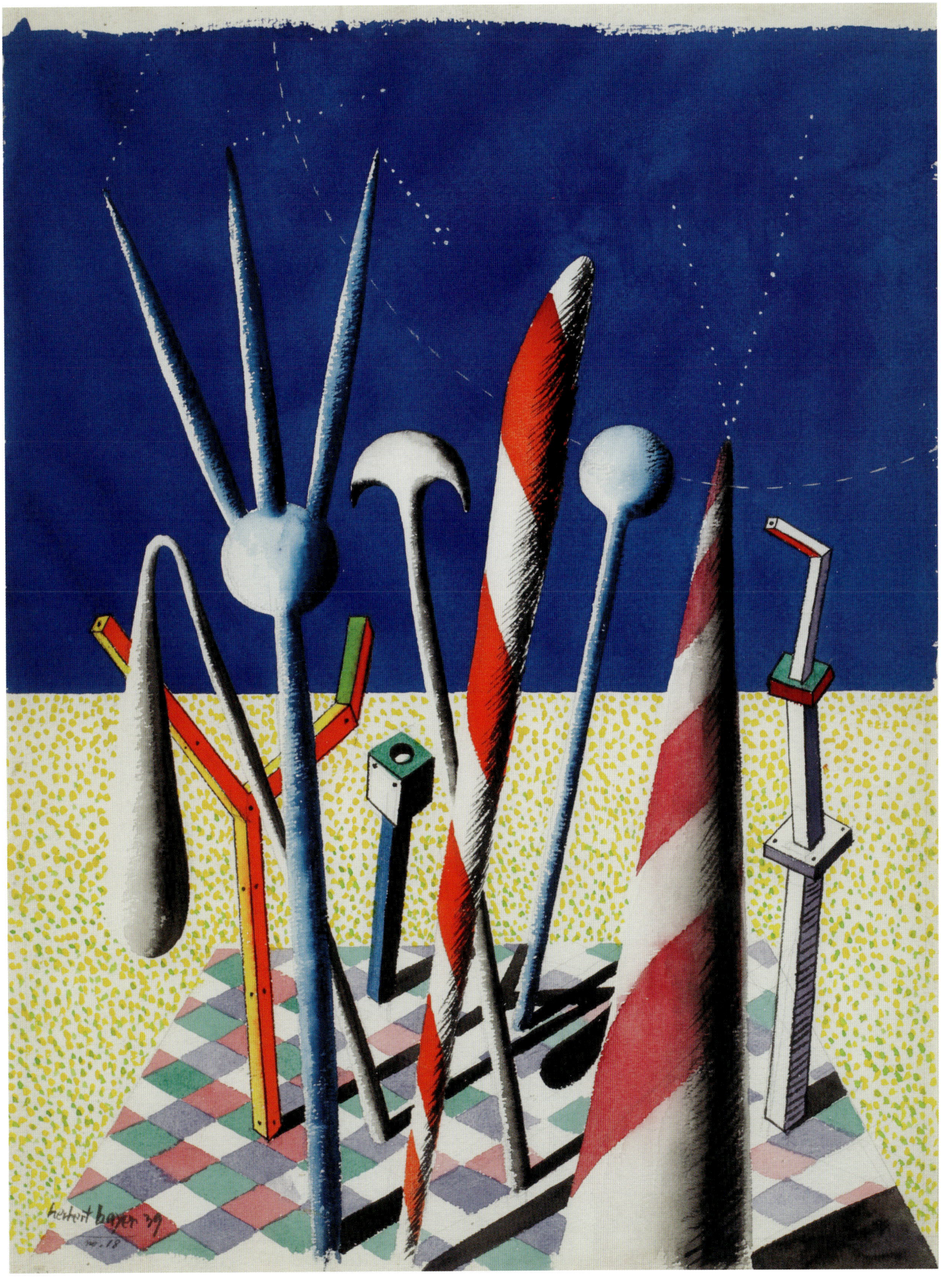

A. M. CASSANDRE
(1901–1968)

The iconic Twenties posters designed by French graphic designer and typographer A. M. Cassandre speak of a glamorous world of travel, speed and Jazz Age indulgence. They evoke a carefree and outward-looking world before the Great Depression, a brief interlude during which anything seemed possible. But more than this, the posters speak of a fresh spirit in design itself, with the vivid deployment of colour, shape and form enhanced by a bold but limited use of typography and the application of innovative techniques such as triangulation to introduce a sense of dynamism and kinetic energy. The artistry of the images, too, is striking, with Cassandre's liners and night trains simplified, reduced and abstracted into powerfully expressive representations of machine-age exoticism. Imbued with a strong Art Deco* aesthetic, these evocative posters were key markers in the evolution of modernist graphic design.

Adolphe Mouron was born to French parents in the Ukraine, only settling in France from 1915 onwards. He studied art and painting at the Académie Julian in Paris, before establishing his first atelier in 1922 and adopting the name 'Cassandre'. One of his early posters for the furniture store Au Bûcheron (1923) sealed his reputation, with its trademark sense of intelligent dynamism and graphic strength, combining a striking use of type and triangular forms like shining sunbeams. The large posters appeared on hoardings throughout Paris and won Cassandre an award at the Exposition Internationale des Arts Décoratifs et Industriels Modernes in 1925.

The work that followed embraced similar principles, distilling ideas, brands and images down to a powerful essence that could be appreciated in a moment and within the busy motion of the morning commute. Key commissions focused upon transport, including the 'Nord Express' poster of 1927 for the Chemin de Fer du Nord, where an abstract black locomotive becomes a futuristic speeding machine hurtling into an open land of possibilities. Similarly, Cassandre's 'L'Atlantique' poster of 1931 provided a vision of escapism, with the streamlined bulk of a liner promising a new beginning. His famous 1935 poster 'Normandie' remains a defining image of transatlantic travel, encapsulating one of the era's most familiar liners.

These images fused many different influences with sophisticated originality, with aspects of Cassandre's work suggesting avant-garde ideas and techniques developed by Cubist, Surrealist and Constructivist* artists. Cassandre also explored frame-by-frame narratives, as seen in his 'Dubo Dubon Dubonnet' poster of 1932, in which a semi-completed image of a figure drinking at a café table completes itself in three stages like a work of animation.

Yet Cassandre always remained intensely aware of the intent and purpose of his work, describing his posters as 'a means of communication between the seller and the public' – a vivid, concise burst of visual information like an illustrated telegram. At the same time, he believed that hoarding posters could bring life and colour to the streetscape, offering an opportunity to 'enliven' rather than depress the urban context.

Cassandre soon found his career developing at a rapid pace, with brands such as Bugatti, Pernod and Pathé looking to benefit from the thoughtful energy of his work. In 1930, he co-founded the Alliance Graphique agency in Paris, along with Charles Loupot and Maurice Moyrand. He also applied himself to typography, designing a series of influential new typefaces, including the Deco-influenced 'Bifur' type of 1929 and the sans serif 'Peignot' font of 1937, which has more in common with the 'new typography' being developed in Germany and Switzerland around this time.

From 1936 to 1939, he spent time in New York, where the Museum of Modern Art mounted an exhibition of his work and he designed magazine covers for *Harper's Bazaar*. After a brief period serving in the French army during World War II, he began to focus increasingly upon theatre design and painting, with occasional commercial projects, including the monogram logo for Yves Saint Laurent (1963). Suffering from depression following several years of setbacks and disappointments, he committed suicide in 1968.

RIGHT A. M. Cassandre's iconic poster promoting French Line's *Normandie* transatlantic liner, 1935.

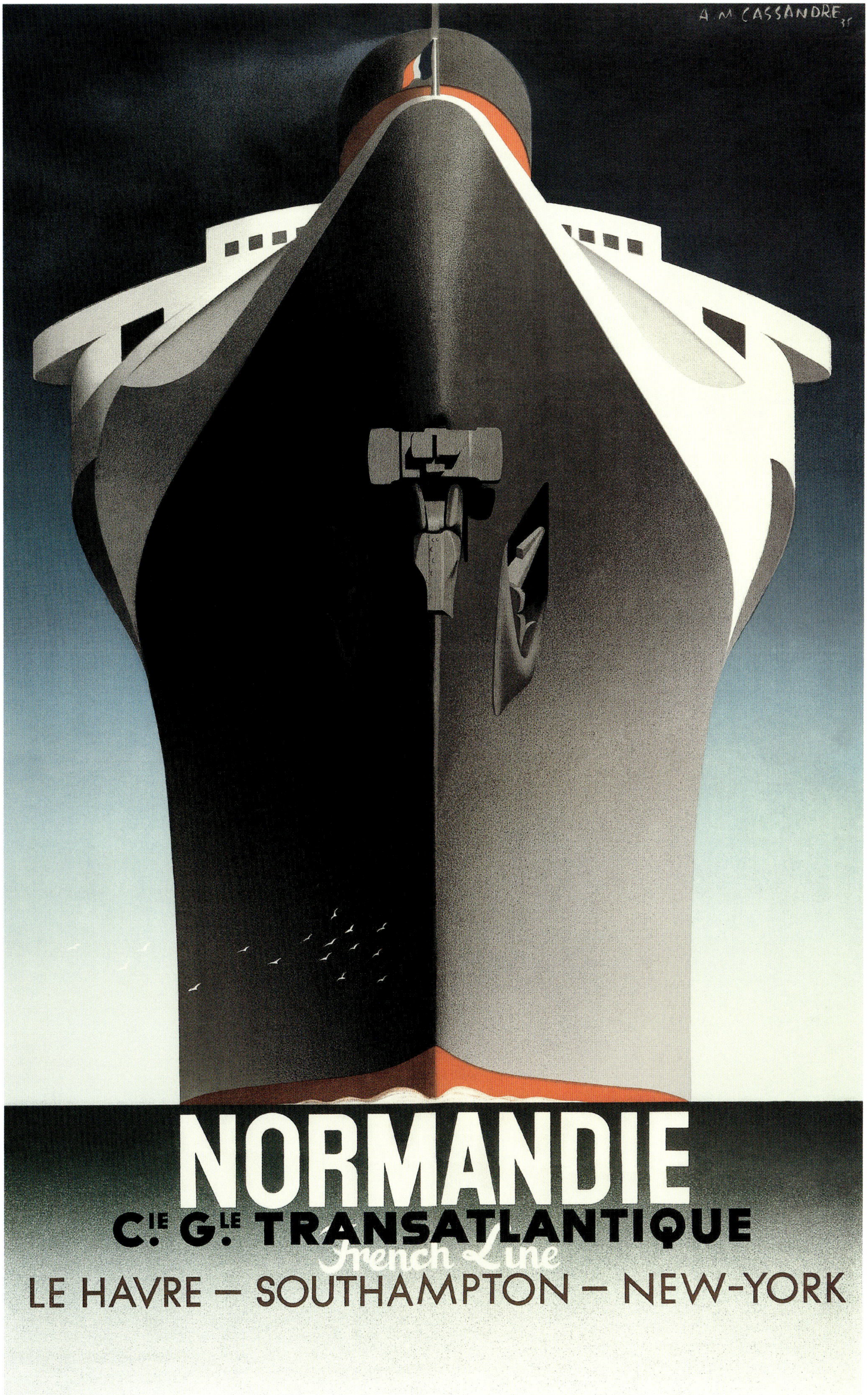

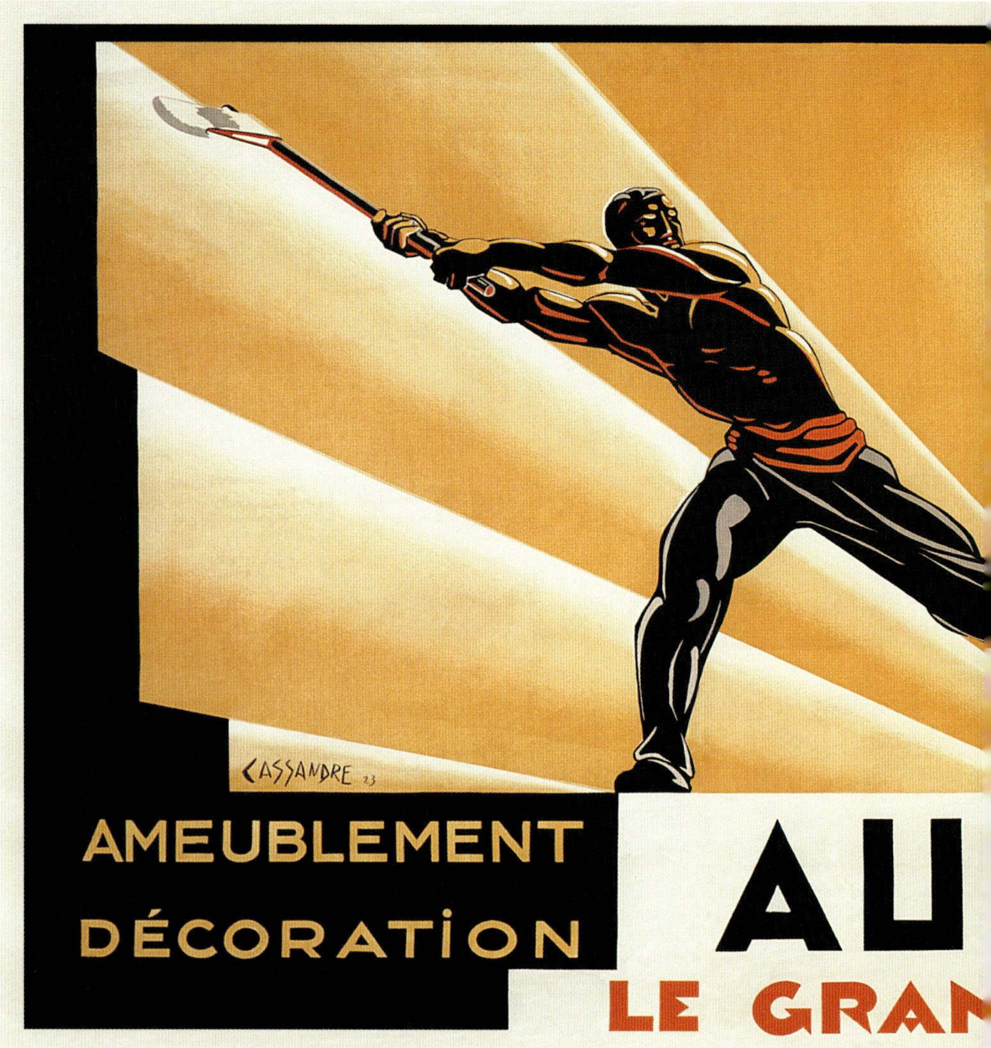
CASSANDRE 23
AMEUBLEMENT
DÉCORATION
AU
LE GRAN

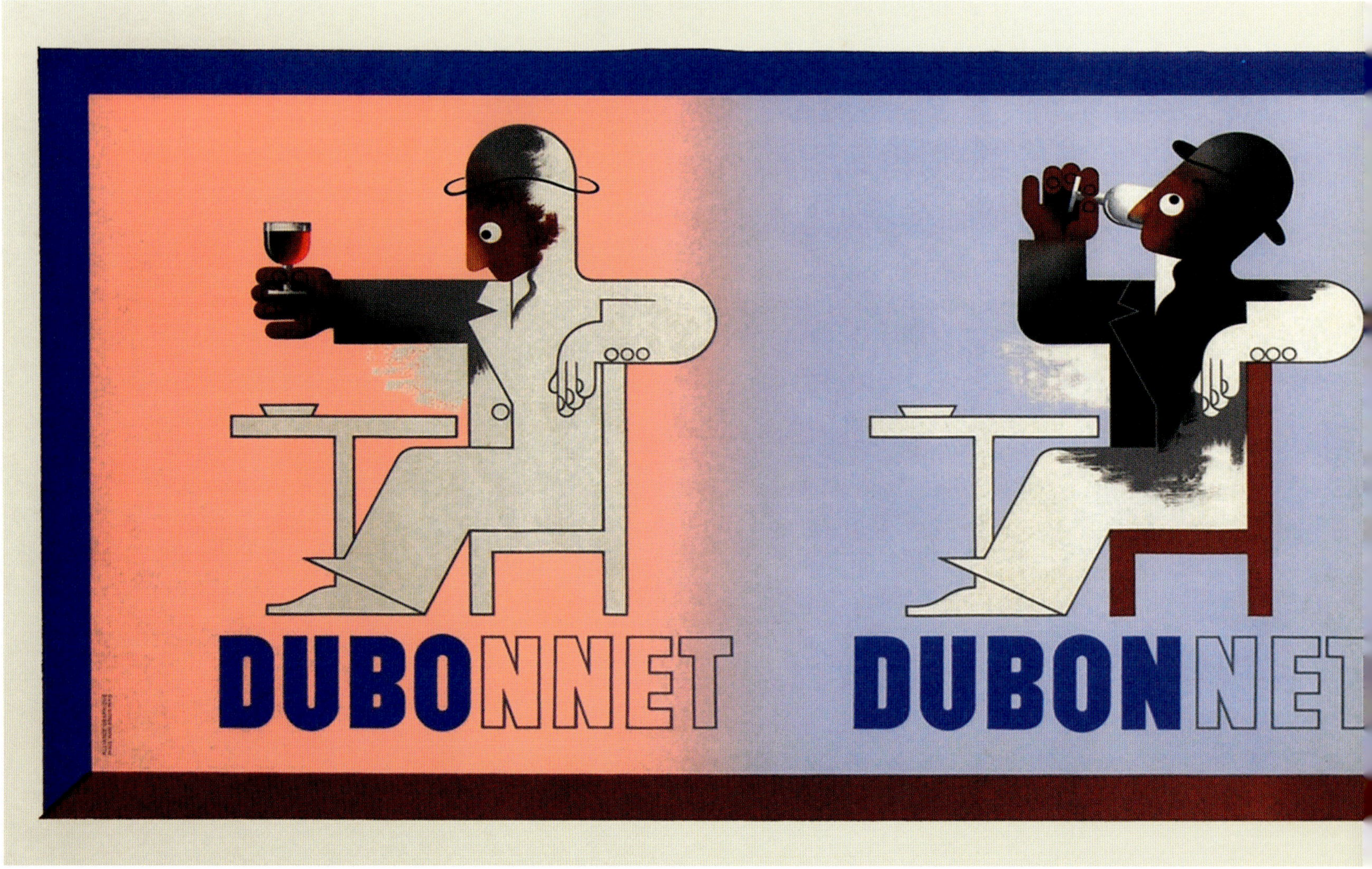
DUBONNET
DUBONNET

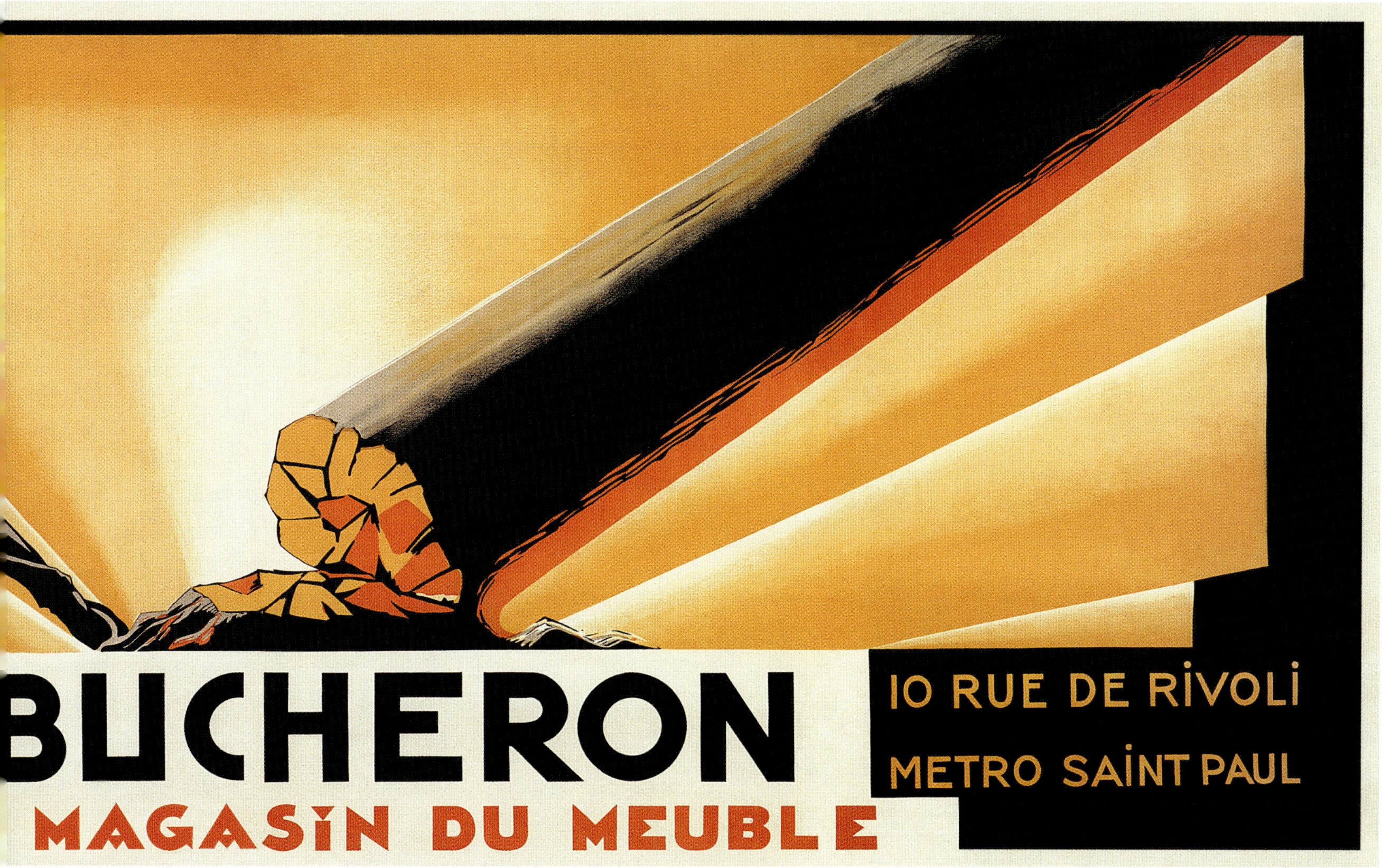

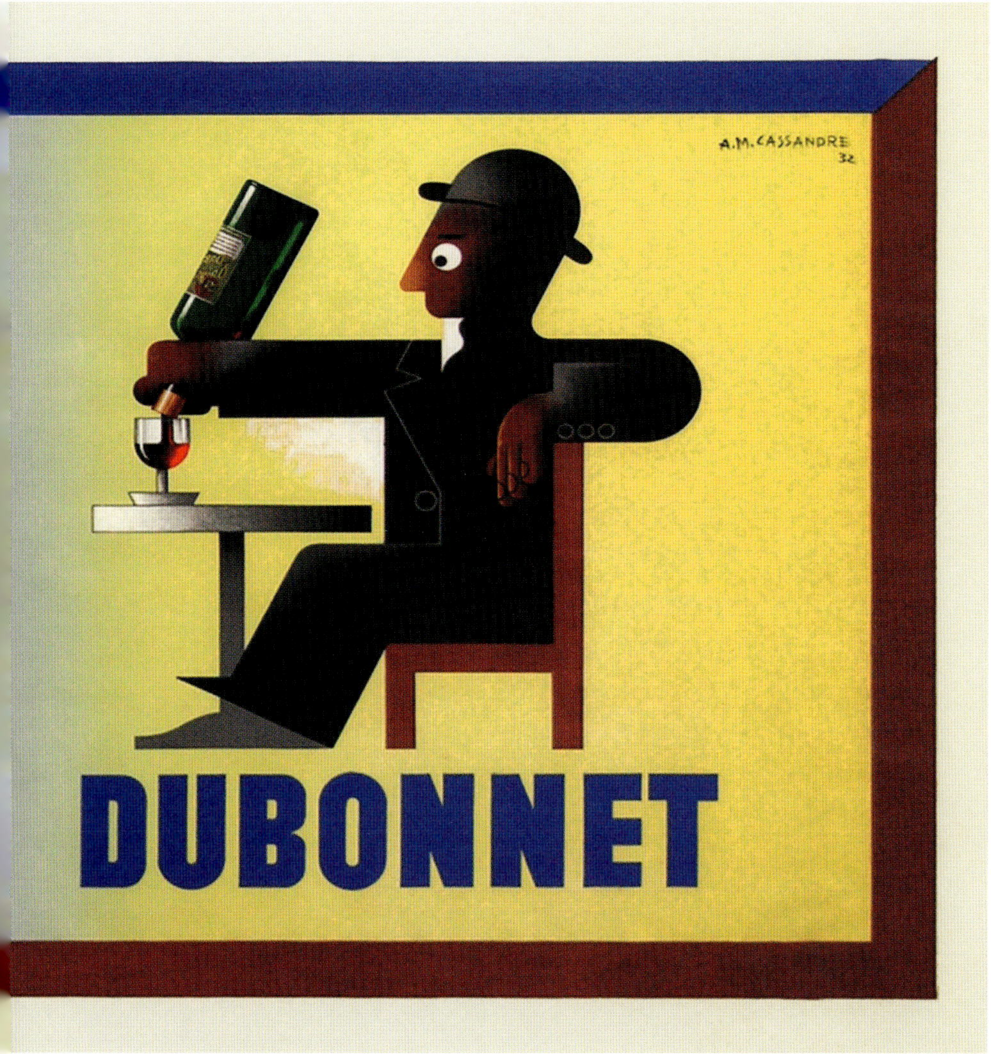

LEFT An advertising poster for Dubonnet, known as 'Dubo Dubon Dubonnet', 1932.

ABOVE A. M. Cassandre's poster for the Au Bûcheron furniture store in Paris, 1923.

THEO VAN DOESBURG
(1883–1931)

The Dutch artist, designer and typographer Theo van Doesburg is often described as the 'ambassador' of De Stijl*. As the co-founder of this influential modernist sub-movement and editor of the journal that gave the group its name, van Doesburg was certainly a key figure in the evolution of De Stijl. Sometimes he is also called a 'theorist' yet, more than this, he believed in a holistic approach to art, architecture and design, in which the principles of De Stijl – such as abstraction, geometric and linear forms, and the use of bold blocks of primary colour within a neutral backdrop – were integrated and expressed upon the 'living' canvas of a home or a building, so that they became a part of everyday life.

Born in Utrecht as Christian Küpper, van Doesburg adopted his first name and surname in honour of his stepfather and added the 'van' between them. He initially studied drama in Amsterdam before teaching himself to paint. Around 1916–17, he established contacts with artists and designers who were to form the core of the De Stijl movement, including artist Piet Mondrian, designer and architect Gerrit Rietveld, designer Vilmos Huszár, and architect and designer J. J. P. Oud*. In collaboration with Oud, van Doesburg worked on a number of house projects and co-founded an artists' collective known as De Sphinx.

In 1917, van Doesburg launched the journal *De Stijl* and co-founded the movement. Working with Huszár, who designed a number of covers for the journal and contributed articles, van Doesburg helped to shape the visual identity of De Stijl and give voice to the theories and ideas that underpinned it. In 1919, he also created an experimental new typeface known as 'Alphabet' or 'Square Alphabet', in which each letter sits within a box (forming an early version of the grid system employed by post-war designers). The typeface – later relaunched as 'Architype Van Doesburg' – has a curiously digital quality, anticipating in some ways the ordered reduction of early computer-generated type.

As time went on, van Doesburg increasingly experimented with the design of the journal, creating a fresh aesthetic in 1921 with a new logo that used a bespoke sans serif typeface that he created together with Mondrian. Using the pseudonym I. K. Bonset, he also included his own 'visual poetry' in the magazine, creating dynamic, experimental collages of words and type. In addition, he published work by the Russian Constructivist* designer El Lissitzky* and began to experiment increasingly with diagonal triangulation and asymmetry. Van Doesburg's De Stijl colleague Piet Zwart* took a more commercial direction within the field of graphic design,

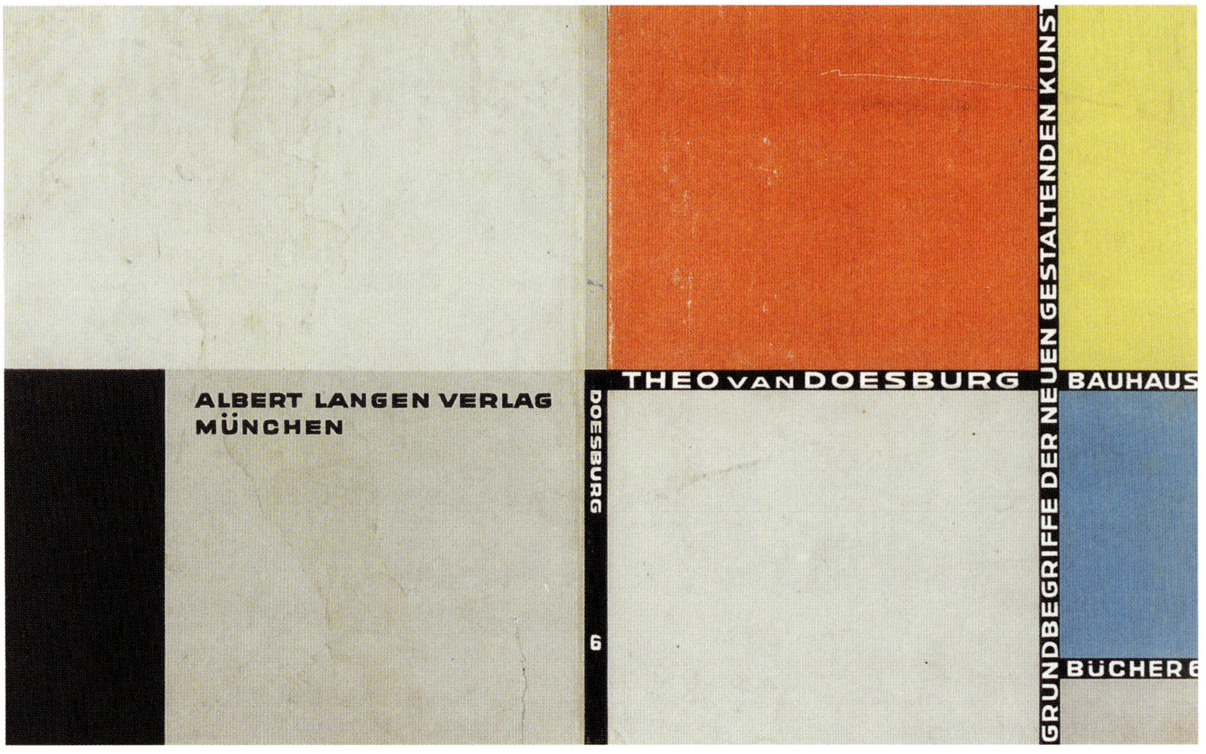

RIGHT The cover design for Theo van Doesburg's book *Principles of Neo-Plastic Art*, published by Albert Langen, Munich, 1925.

OPPOSITE Van Doesburg's cover design for *Mécano* magazine, issue 3, 1922.

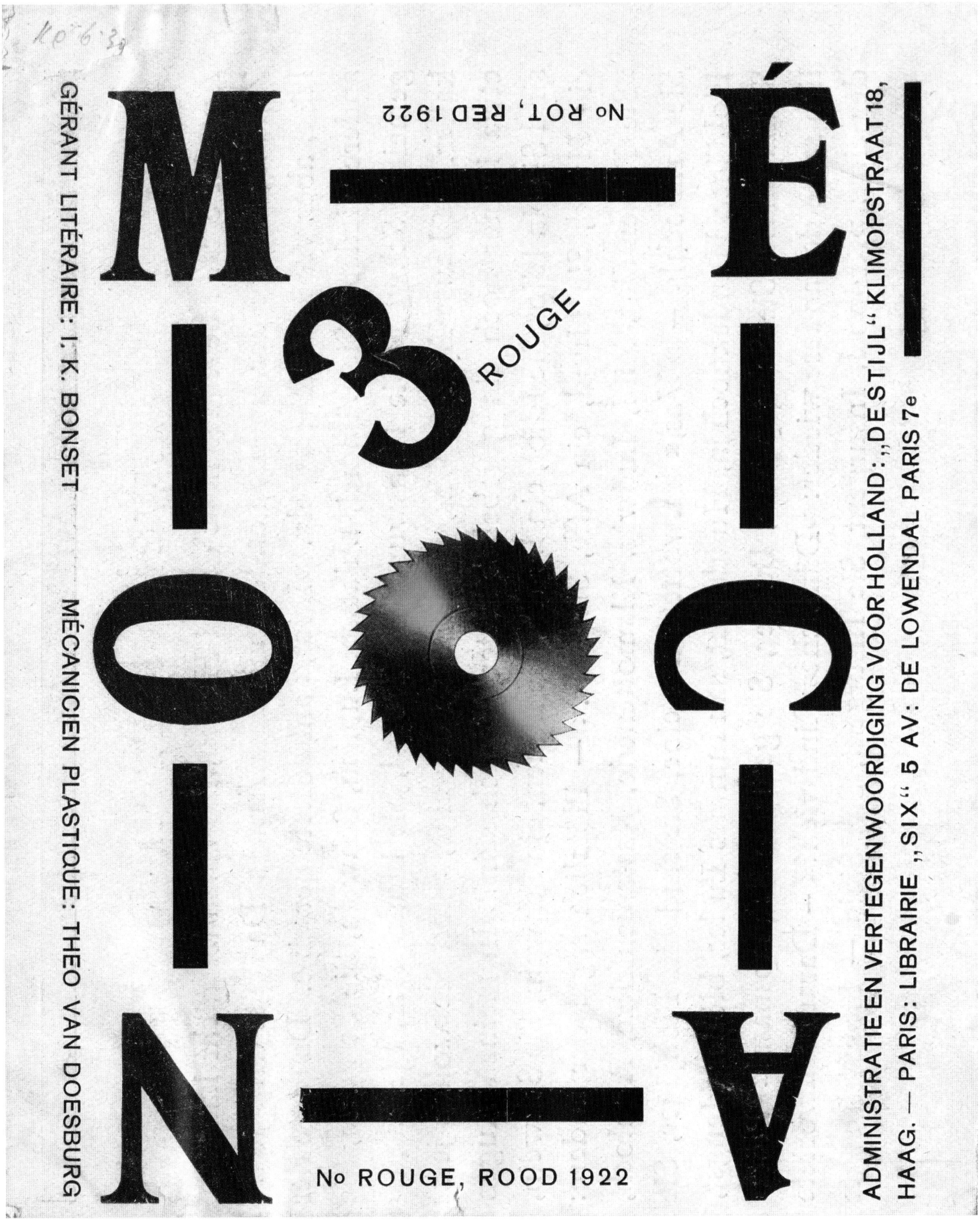
MÉCANO
3 ROUGE
No ROT, RED 1922
No ROUGE, ROOD 1922
GÉRANT LITÉRAIRE: I. K. BONSET
MÉCANICIEN PLASTIQUE: THEO VAN DOESBURG
ADMINISTRATIE EN VERTEGENWOORDIGING VOOR HOLLAND: „DE STIJL" KLIMOPSTRAAT 18,
HAAG. — PARIS: LIBRAIRIE „SIX" 5 AV: DE LOWENDAL PARIS 7e

applying his ideas to advertisements and posters, as well as typography.

In addition to editing the De Stijl journal, van Doesburg worked as a book designer, creating a series of children's books during the Twenties in association with artist Kate Steinitz. Increasingly fascinated by Dadaism, he went on to launch a second magazine, *Mécano*, exploring Dada-inspired ideas and thinking (which he became preoccupied with for a brief period); a third title, *Art Concret*, was launched in 1929.

In 1922, van Doesburg moved to Weimar in Germany with the hope of gaining employment at the Bauhaus. Director WALTER GROPIUS decided that van Doesburg, who had a reputation as a possessive De Stijl dogmatist, was not best suited to a teaching post. Instead, he settled nearby for a time, setting up a 'satellite' bureau to spread the ethos of his movement. Around 1924, there was a split with Mondrian over the direction and detail of De Stijl, essentially over the use of triangulation; the two men were reconciled in 1929.

Beyond van Doesburg's work within the field of magazines, graphics and typography, he pursued his career as a painter. He also continued to collaborate on interiors for houses and other projects, including a café in Strasbourg and the 'flower room' at the Villa Noailles, the house in the South of France designed by ROBERT MALLET-STEVENS (1925). After a period in Paris, he moved to Davos, Switzerland, where he died of a heart attack at the age of just 47.

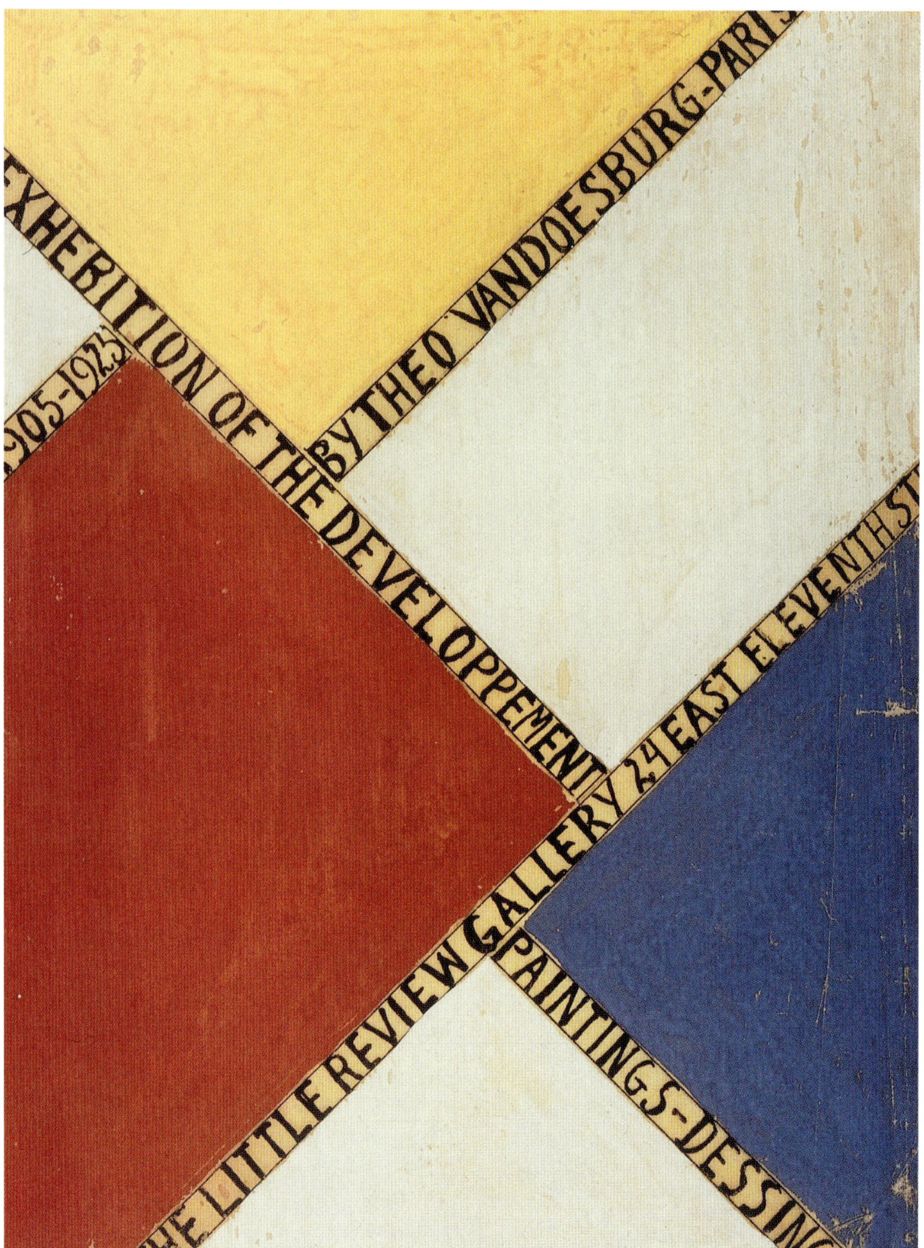

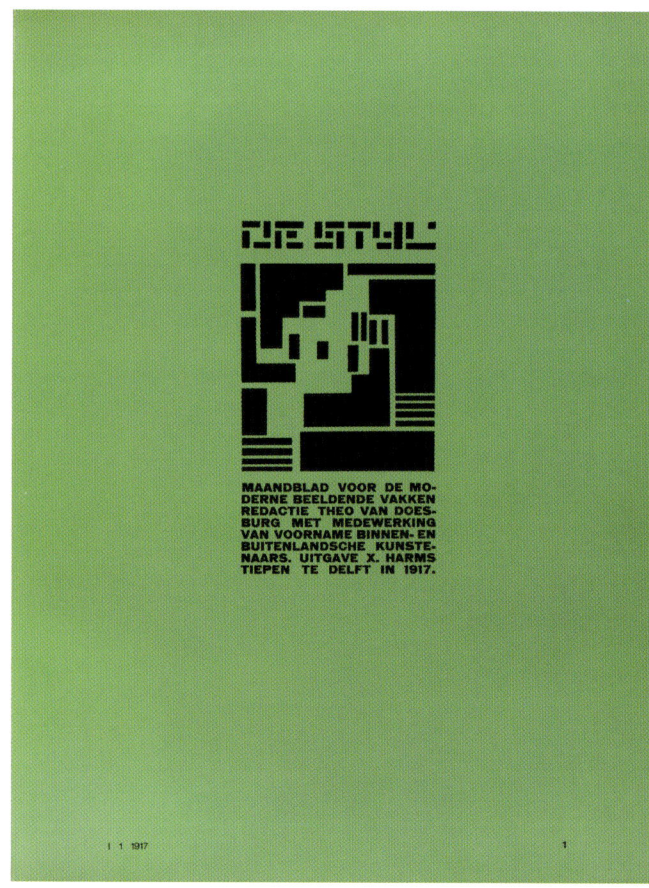

ABOVE Graphics for an early De Stijl publication, 1917, by Theo van Doesburg and Vilmos Huszár.

LEFT An exhibition poster design for a Theo van Doesburg show at The Little Review Gallery in New York, 1923.

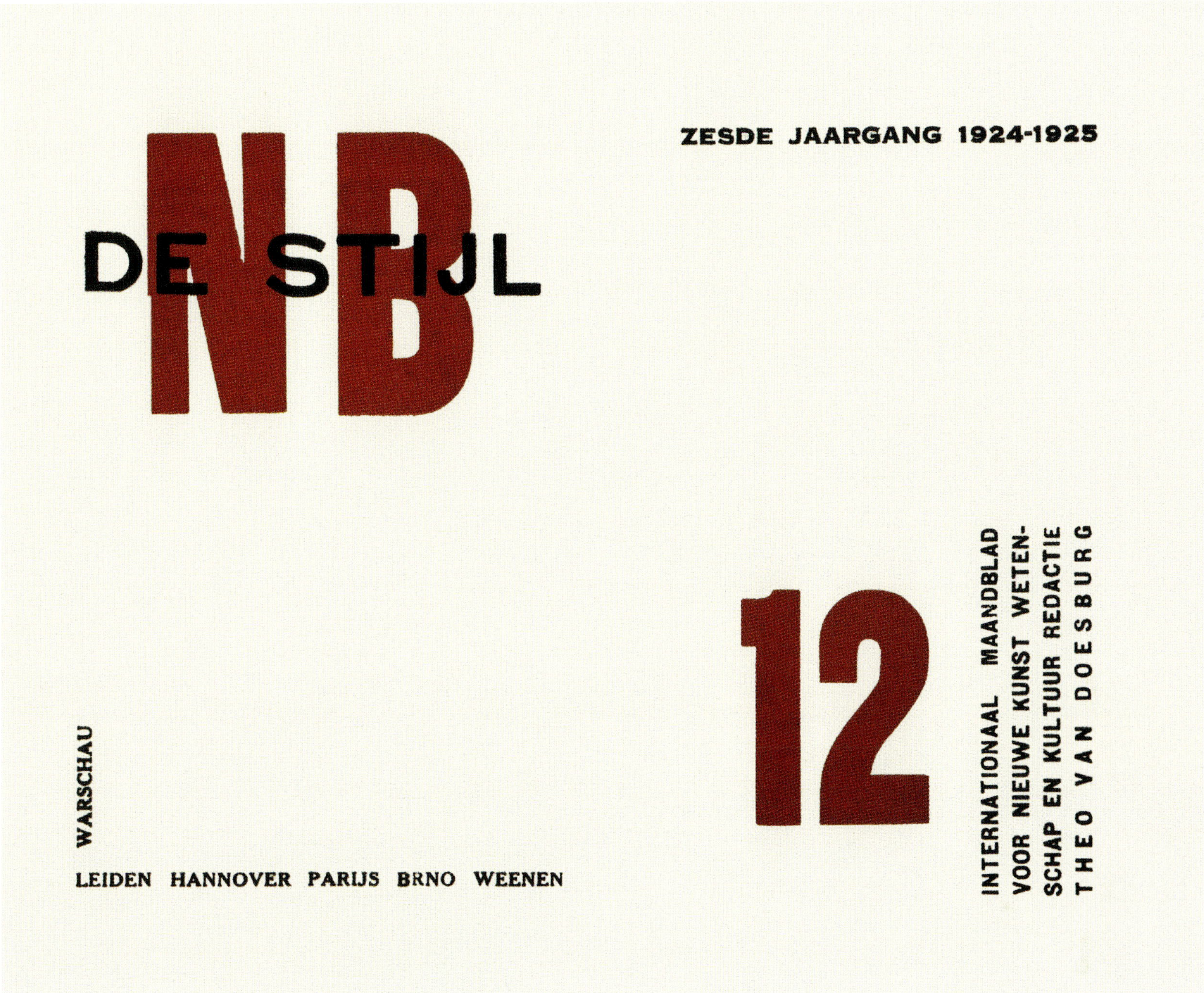

ABOVE Cover art for issue 12 of the *De Stijl* magazine, designed by Theo van Doesburg, 1924–25.

LÁSZLÓ MOHOLY-NAGY
(1895–1946)

A painter, artist, photographer, designer, typographer, writer and educator, László Moholy-Nagy spent his working life pushing boundaries and experimenting with new ideas. As an intellectual polymath, he was well suited to a role at the Bauhaus, where he taught the foundation course alongside Josef Albers and helped instil fresh rigour to a wide-ranging curriculum. Moholy-Nagy also played a key role in developing the Bauhaus's own publishing department, designing the visual identity of the books and formulating suitable typography, as well as writing essays and texts. Along with colleague Herbert Bayer, he helped to formulate the graphic presentation of the school as a whole, both through its publishing programme and through key exhibitions. In later years, he sought to build upon the legacy of the Bauhaus by opening a school in Chicago, known as the New Bauhaus.

Born László Weisz in southern Hungary, he adopted the name of his uncle and guardian – Gusztáv Nagy – after his father abandoned the family; he later added 'Moholy' after the name of the town closest to where he grew up. In 1913, Moholy-Nagy enrolled as a law student at the University of Budapest. However, his studies were interrupted by World War I, in which he served as an artillery officer and was sent to the Russian Front, where he was wounded.

Following the war years, Moholy-Nagy switched direction and decided to pursue his growing interest in art. He exhibited his watercolours in Budapest and moved to Vienna in 1919, and then on to Berlin a year later. Over the following years, Moholy-Nagy was increasingly influenced by the work of the Russian Constructivists*, by the ethos of De Stijl*, and by the experimental art and theatre of the Dadaists, contributing to the Dada magazine, *Ma*. Increasingly, he began to experiment with new mediums, including sculpture, film and photography; his first wife, Lucia, was a photographer and helped her husband develop his technical knowledge of the craft.

He first met Bauhaus director Walter Gropius in 1922 and a year later was invited to become a tutor, joining the school in Weimar as one of its youngest members of staff. He took over the foundation course – or 'Vorkurs' – from the increasingly eccentric Johannes Itten* and began teaching alongside Josef Albers. Together, they refocused the preliminary course in a way that was

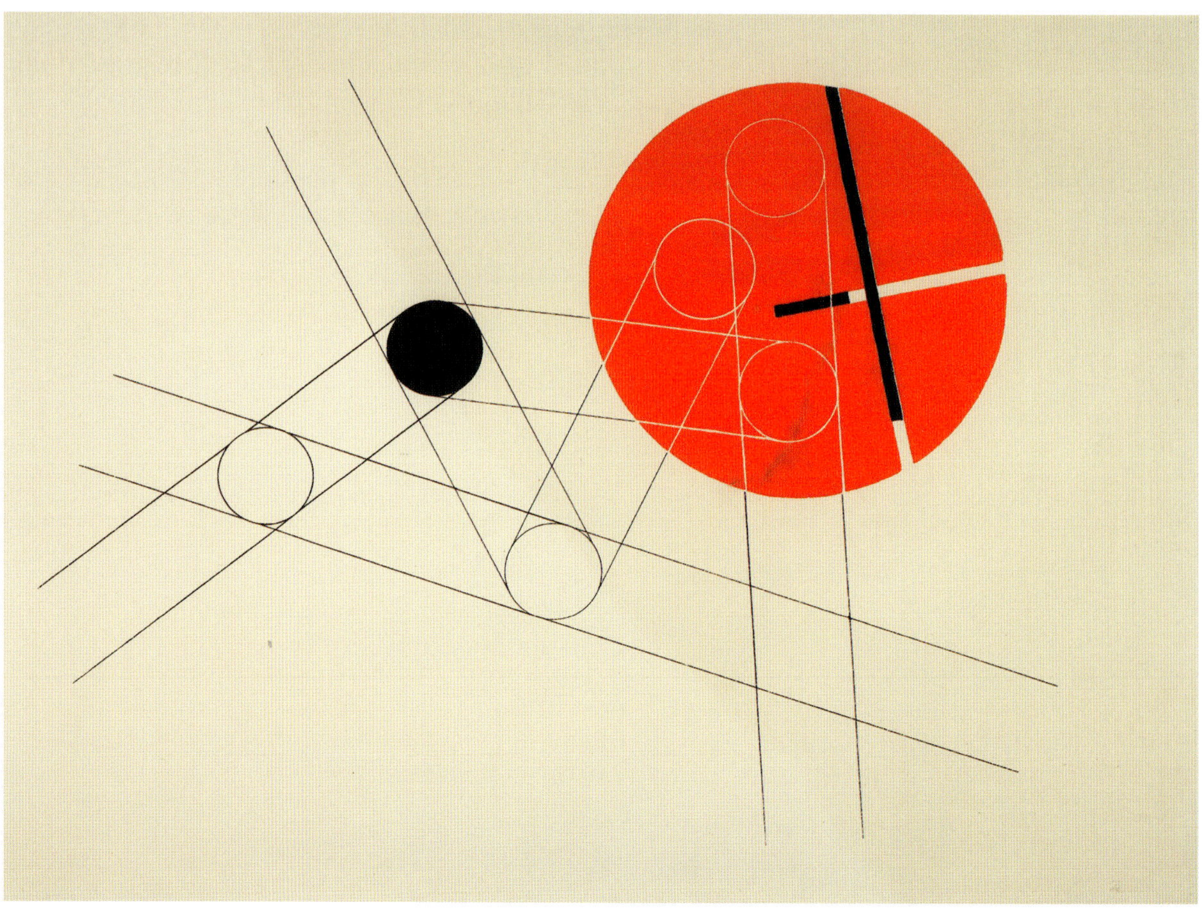

RIGHT An untitled screenprint by László Moholy-Nagy, posthumously printed from an original design.

more in tune with Gropius's concentration upon the integration of the applied arts, architecture and design within a new age of mechanical reproduction and industrialization. Moholy-Nagy, who also taught in the metal workshop, explored the use of new, industrially made materials and advocated technical drawing within a push towards a more ordered approach to form and geometry. He and Albers sought to create a fusion curriculum, combining the imagination and creativity of the artist with the technical ability of the engineer.

Moholy-Nagy helped to plan a Bauhaus exhibition in 1923, held in Weimar, designing the exhibition catalogue and other material. Around the same time he began to develop initial ideas for the Bauhaus Press, designing a striking logo for the imprint. The first of fourteen Bauhaus titles appeared in 1925, after the move to Dessau, with Moholy-Nagy designing all but three of the books and developing a complementary typographical approach, including collaborations with Herbert Bayer. He also contributed essays and texts, with some of the titles – including volume eight on painting, photography and film – written by Moholy-Nagy alone.

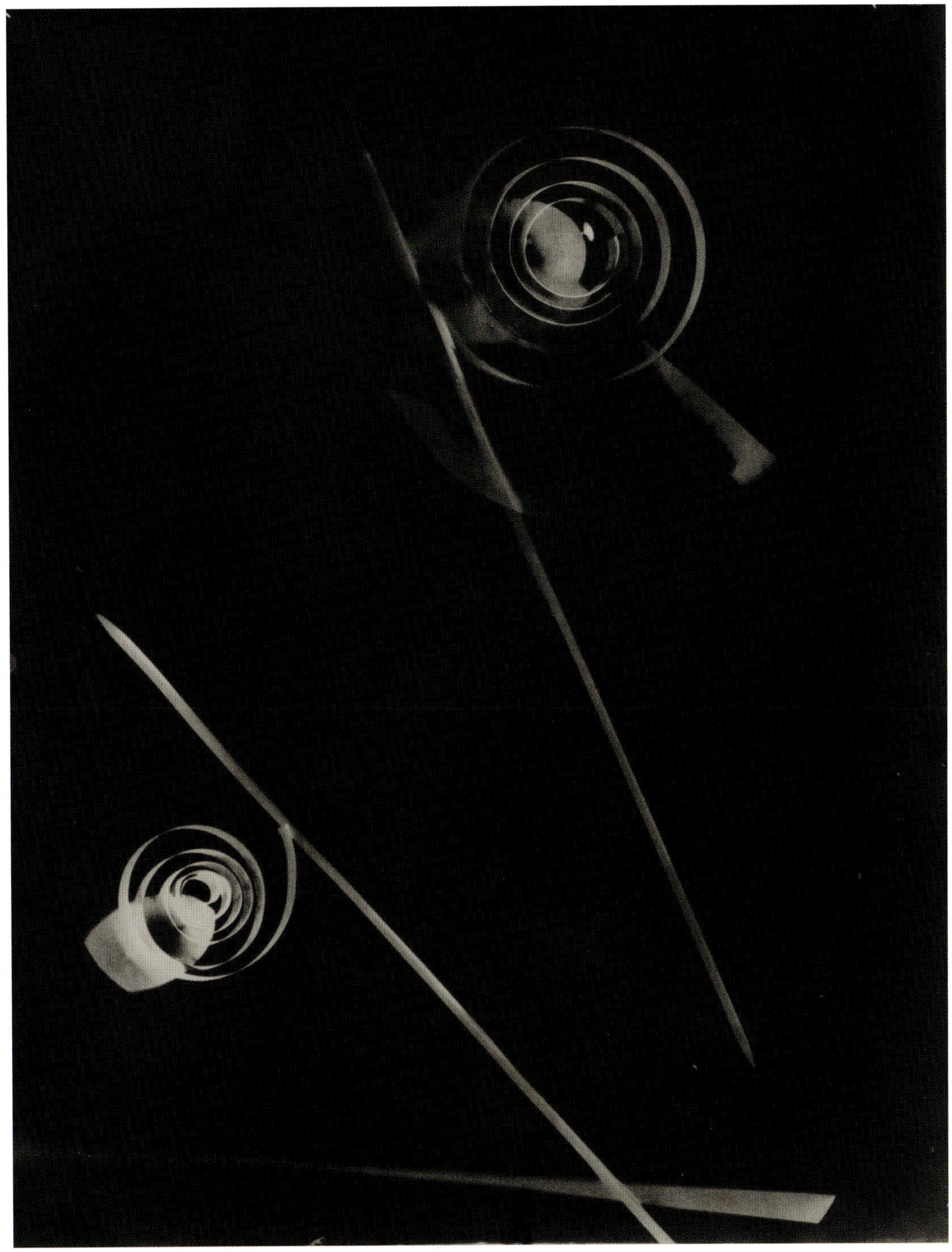

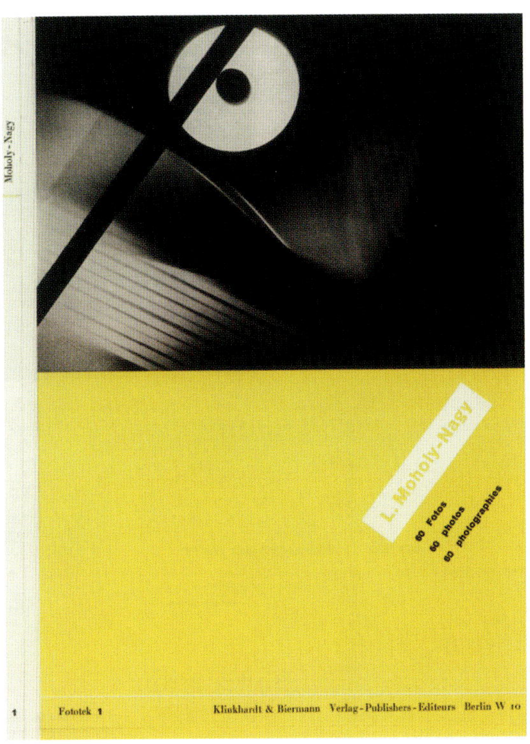

ABOVE A gelatin silver print of a photogram by Moholy-Nagy, 1922.

LEFT Cover art for Moholy-Nagy's publication *Fototek 1*, 1930.

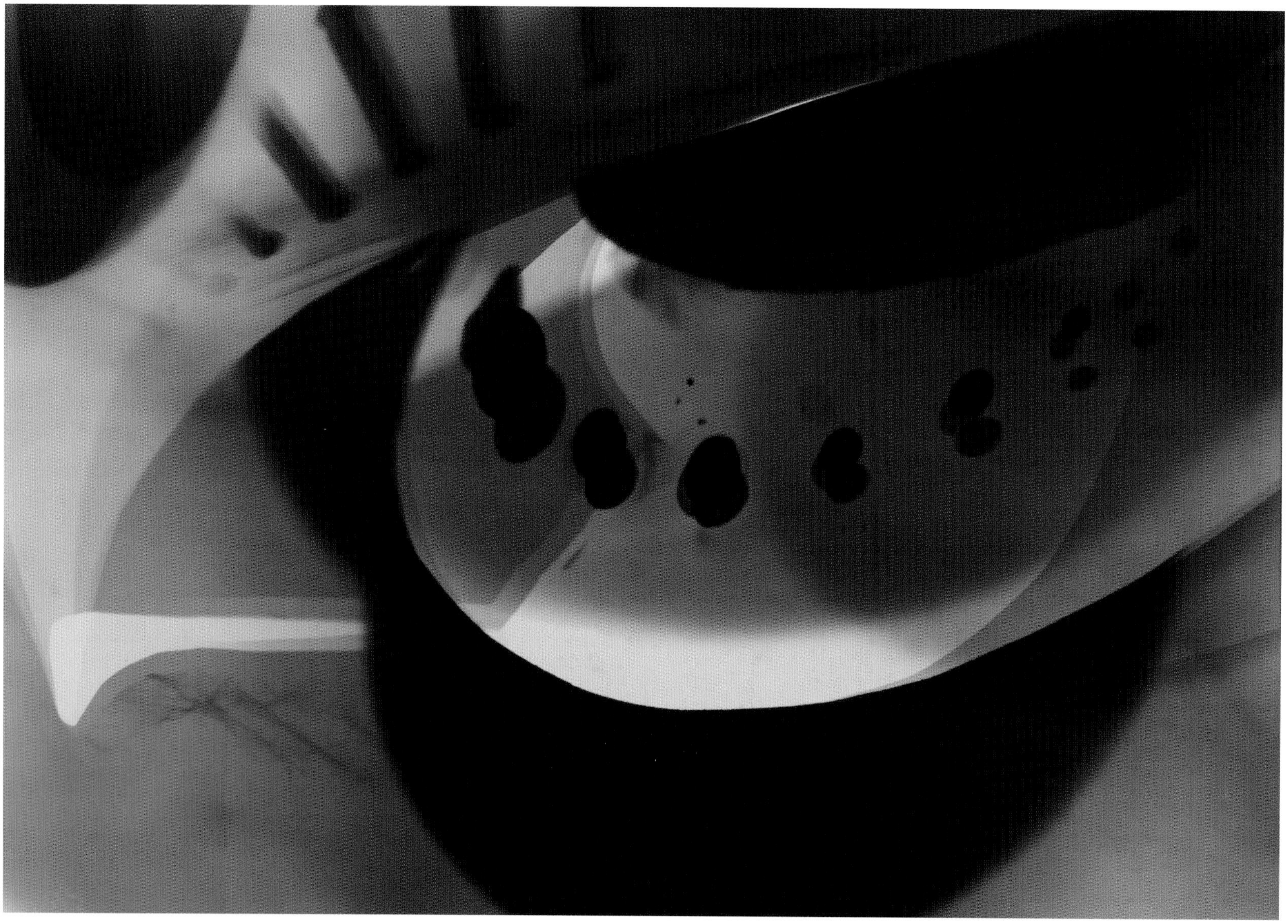

While at the Bauhaus, Moholy-Nagy continued to explore new techniques in art, photography and graphic design, while sometimes fusing different mediums. His photograms of this period – created by placing objects over photographic paper and exposing them to light – often took on the look of graphic collages.

In 1928 Moholy-Nagy resigned from the Bauhaus and moved to Berlin, where he concentrated on graphic design and typography, as well as continuing his work as an artist. He served as an editor at the magazine *i10*, which he helped to establish, and in 1929 signed a contract to design covers for the fashion magazine *Die Neue Linie*. In 1931 he collaborated with Bayer again on designs for the National German Exhibition (Deutsche Bundesausstellung) held in Berlin.

Given the worsening political situation in Germany, Moholy-Nagy moved to Amsterdam in 1934 and then to London a year later, where he lived and worked for two years, reuniting with a number of Bauhaus exiles. He designed posters for Imperial Airways and London Transport, as well as promotional material for the Isokon* furniture company.

With the help of Gropius, Moholy-Nagy moved on to America in 1937. In Chicago he found sponsors to help him launch the New Bauhaus design school, only for the institution to close again a year later due to lack of funding. However, in 1939 he successfully launched the School of Design, again in Chicago, which later became the Institute of Design.

ABOVE An untitled gelatin silver print by László Moholy-Nagy and George Barford, 1939.

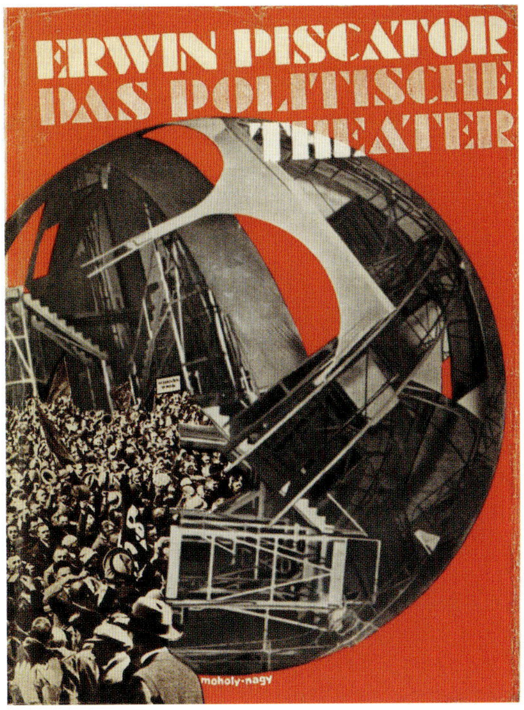

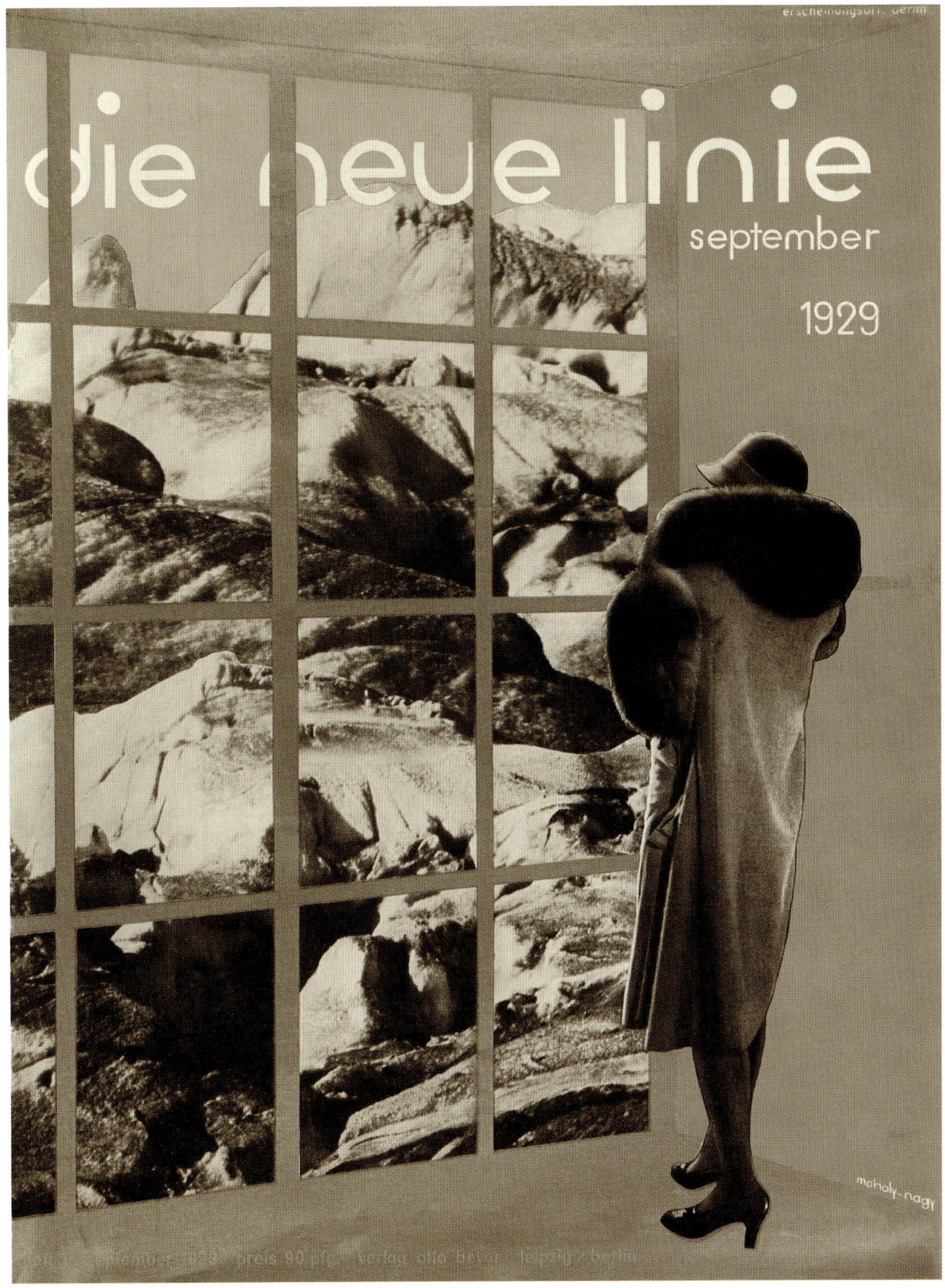

ABOVE Moholy-Nagy's cover design for Erwin Piscator's *Das Politische Theater*, 1929.

RIGHT The September 1929 edition of *Die Neue Linie* magazine, with cover art by Moholy-Nagy.

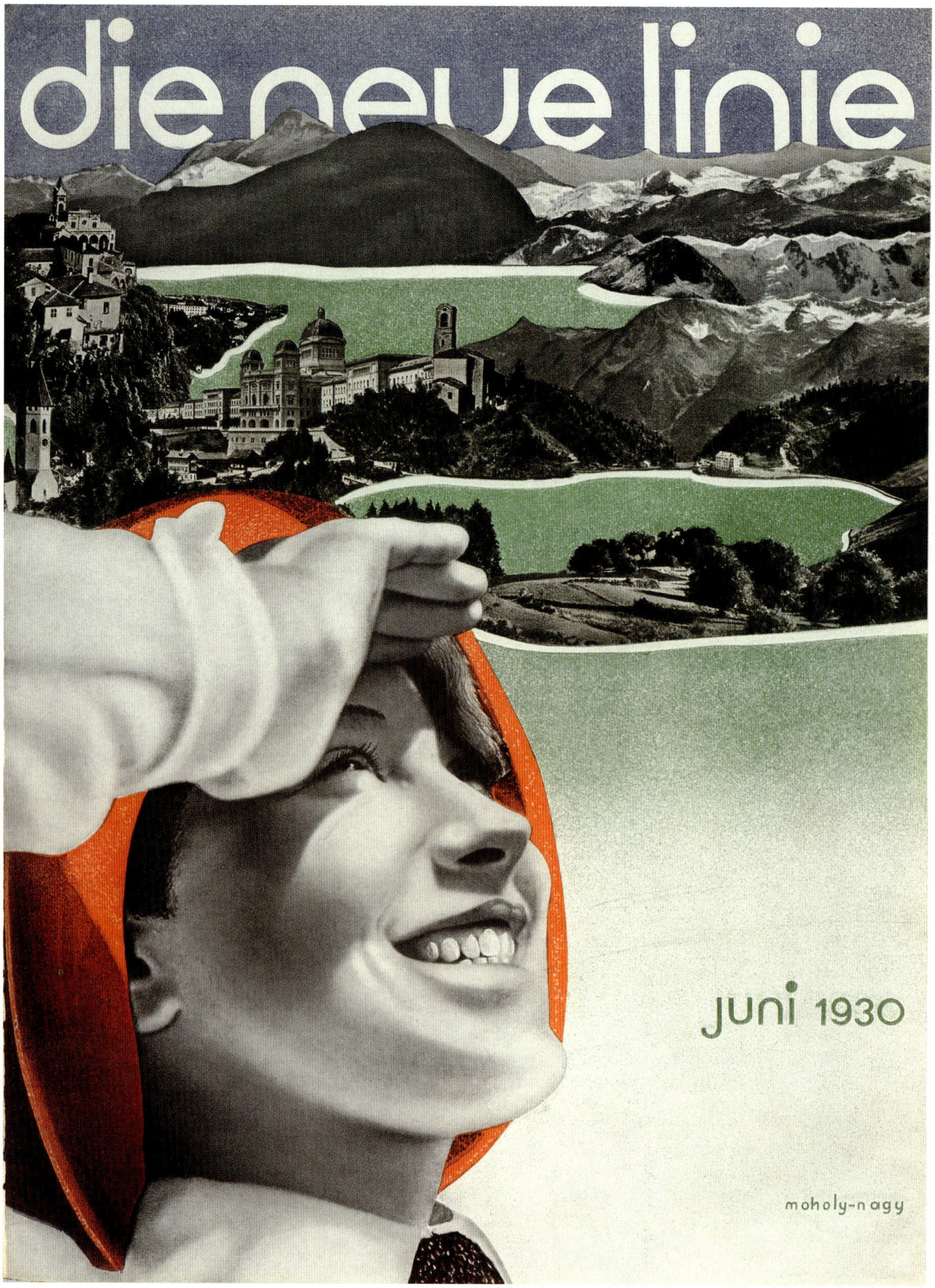

LEFT László Moholy-Nagy's cover design for *Die Neue Linie* magazine, June 1930.

RIGHT An advertisement for London Underground, 1936.

BELOW Moholy-Nagy's artwork for Isokon's Long chair, designed by Marcel Breuer, 1936.

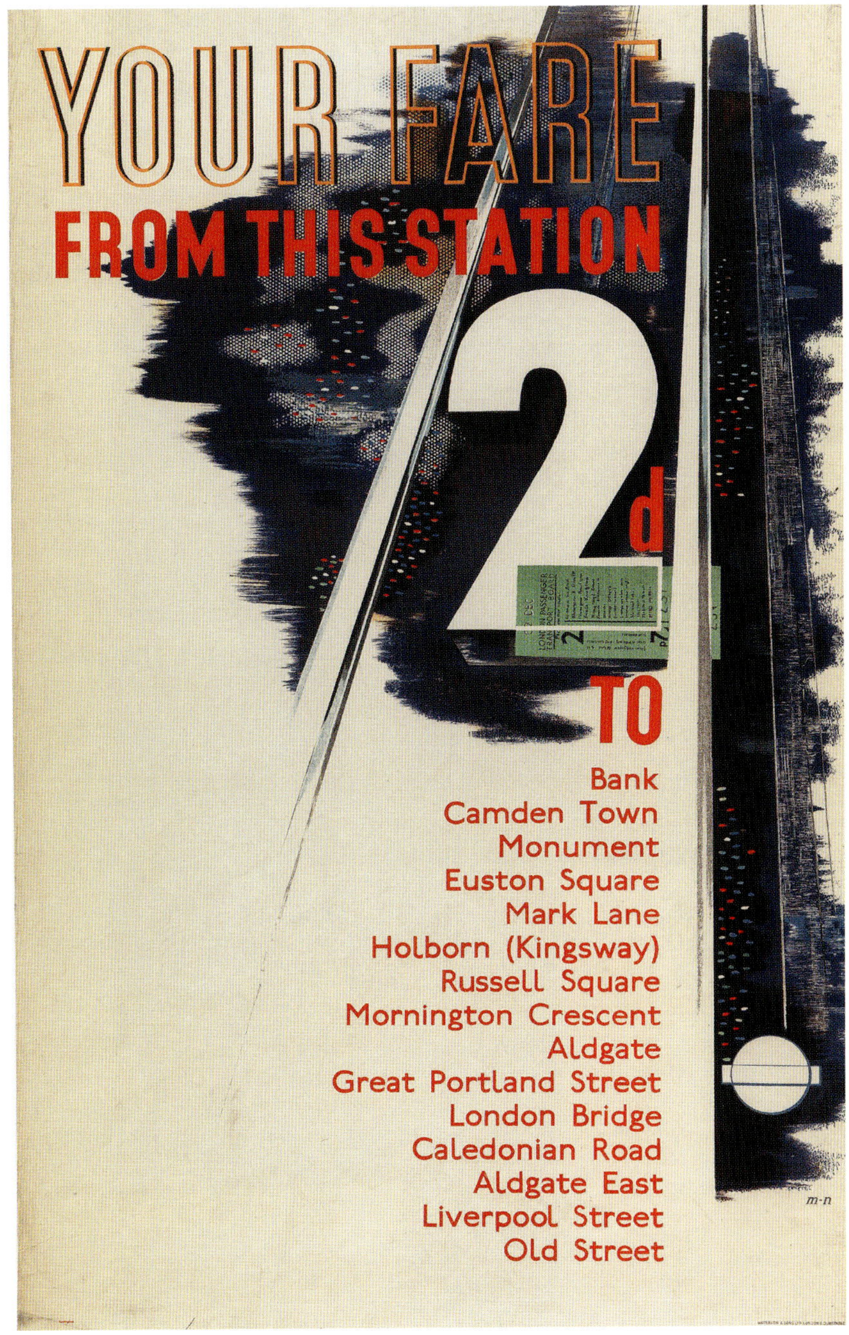

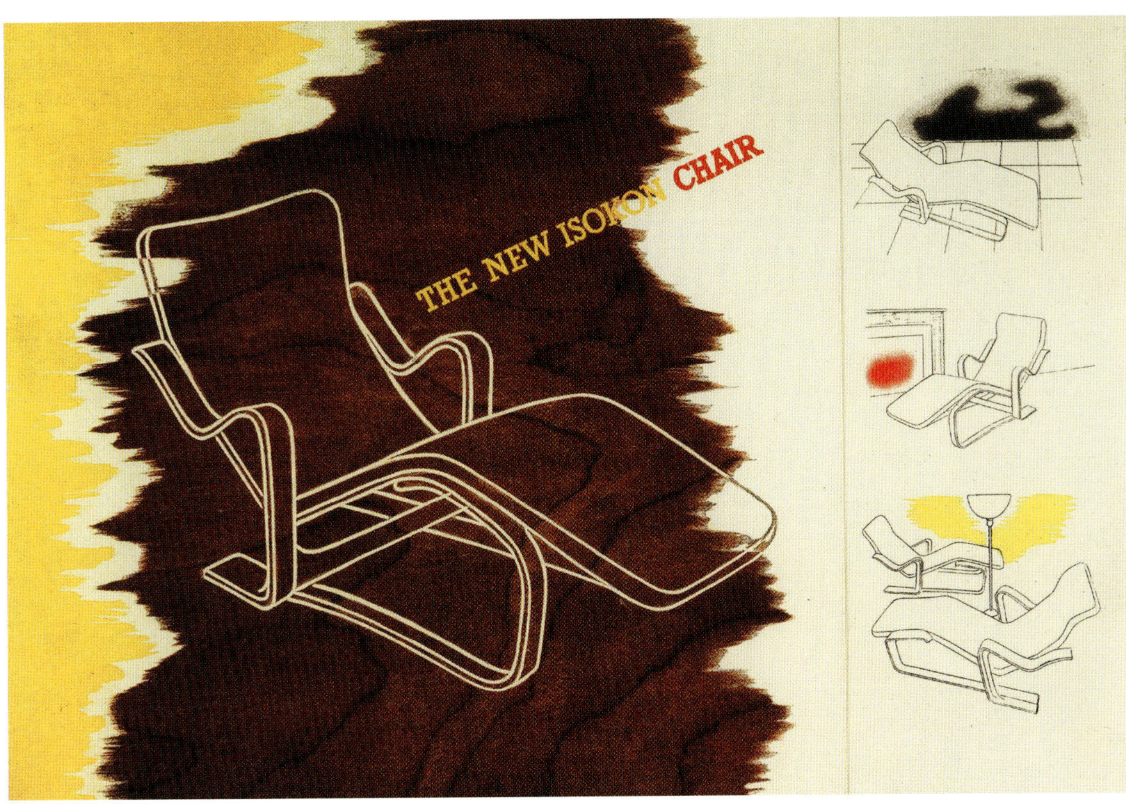

ALEXANDER RODCHENKO
(1891–1956)

The Russian graphic designer, photographer and artist Alexander Rodchenko was one of the key pioneers of a Constructivist* approach to visual design, particularly the advertising poster. Embracing the opportunity to place his talent in the service of the freshly installed Communist state – which was seen by many, in its early years, as a means to invent a new utopian society – Rodchenko created a powerful series of images for state-sponsored companies and institutions which spliced propaganda elements with original and seductive use of imagery, as well as a bold use of Cyrillic typography. Through his work, he helped develop a number of fresh techniques – including triangulation, photomontage and 'loud typography' – that were to have a profound impact on artists and designers not just in Russia and Europe, but all over the world.

Born in St Petersburg, Rodchenko came from a working-class background and found his initial interest in art through the pages of magazines and journals. He studied at the Kazan Art School and later in Moscow. Initially he concentrated on painting but became increasingly interested in Constructivism and switched direction. He was particularly influenced by the work of sculptor Vladimir Tatlin, while his wife – artist Varvara Stepanova – was also a key figure in the evolution of the movement.

Influenced by Cubism and avant-garde abstract art, the Constructivists argued for geometric purity and a highly ordered and linear approach to form and composition. Rodchenko himself expressed this drive towards abstraction in his triptych painting of 1921 entitled 'Pure Red Colour, Pure Blue Colour and Pure Yellow Colour', in which conventional representation was abandoned and the work became pure blocks of vivid colour. Rodchenko was also influenced by the work of artist Kazimir Malevich* and followers of Suprematism*, such as El Lissitzky*, who took a dogmatic and extreme approach to the elimination of representation and expressionism in art.

The Constructivists, in particular, argued that their work should also serve a practical purpose within the state, seeking to apply their creativity and ally themselves with industry, architecture and design in the service of the new Communist state. Rodchenko and others used the word 'Productivism' to sum up this pragmatic 'artist as engineer' approach, which overlapped – in some ways – with the Bauhaus philosophy of fusing art, creativity and design with the ambition of improving day-to-day living through a new generation of industrially made, affordable and intelligently conceived products.

Abandoning painting, Rodchenko began working for a number of state-owned companies and causes. His posters of the Twenties tended to feature a bold use of red, white and black, echoing the colours of the Bolshevik revolution itself and suggesting a propagandist subtext. One of his most famous posters of this period is 'Books' (1925), designed for the Leningrad branch of the state publishing house. The poster encapsulates many of the techniques for which Rodchenko is best known, within a collage that features a

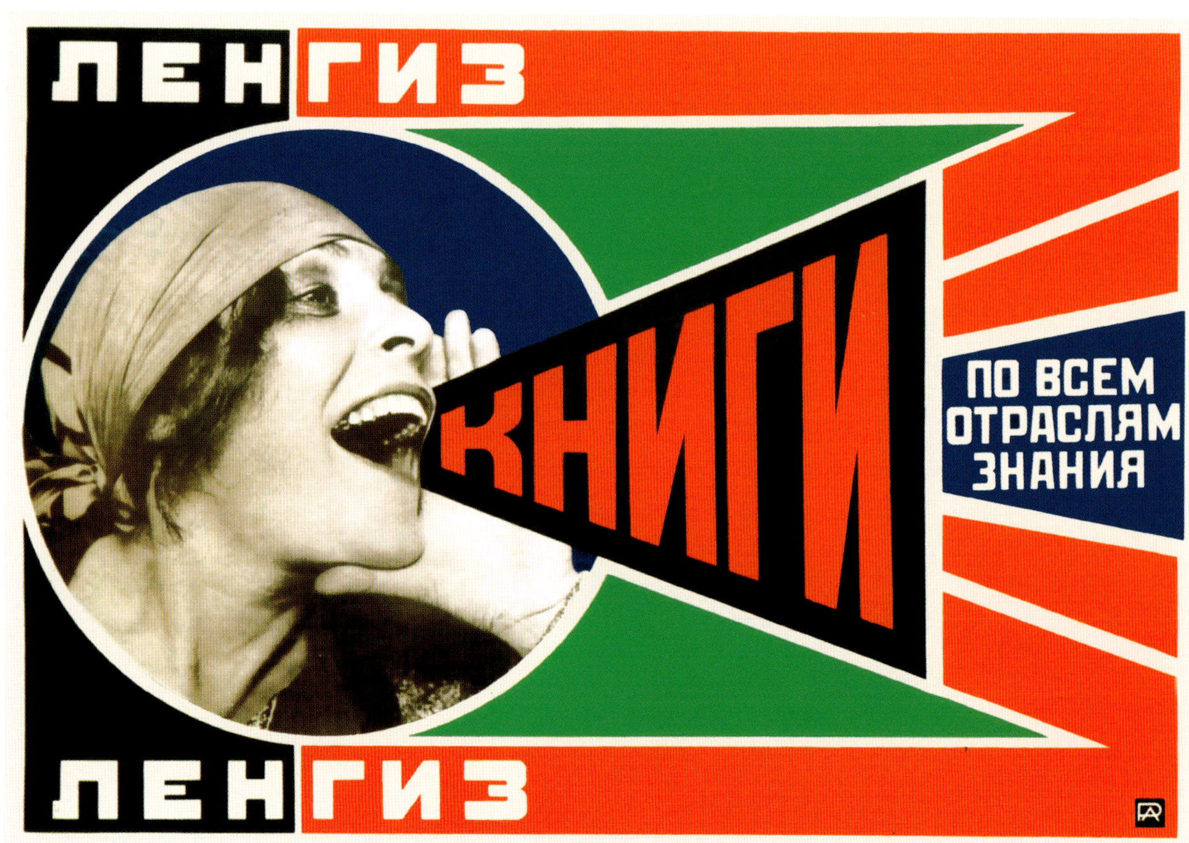

RIGHT Alexander Rodchenko's 'Books' advertising poster, featuring actress Lilya Brik, for the publishers Gosizdat, 1925.

OPPOSITE A Rodchenko advertising poster for pacifiers produced by Rezinotrest, a Soviet rubber industry trust, 1923.

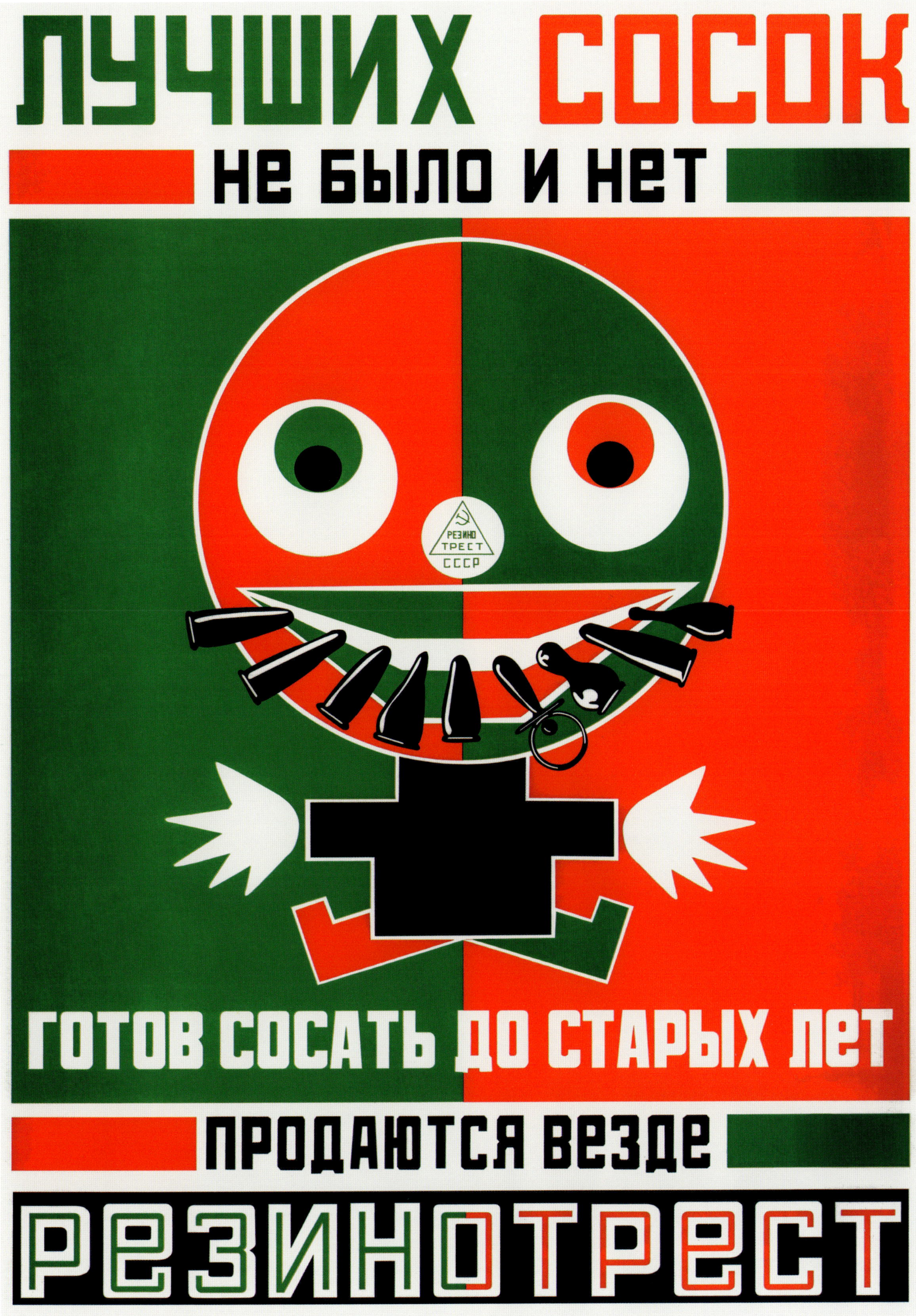
ЛУЧШИХ СОСОК
НЕ БЫЛО И НЕТ
РЕЗИНО ТРЕСТ СССР
ГОТОВ СОСАТЬ ДО СТАРЫХ ЛЕТ
ПРОДАЮТСЯ ВЕЗДЕ
РЕЗИНОТРЕСТ

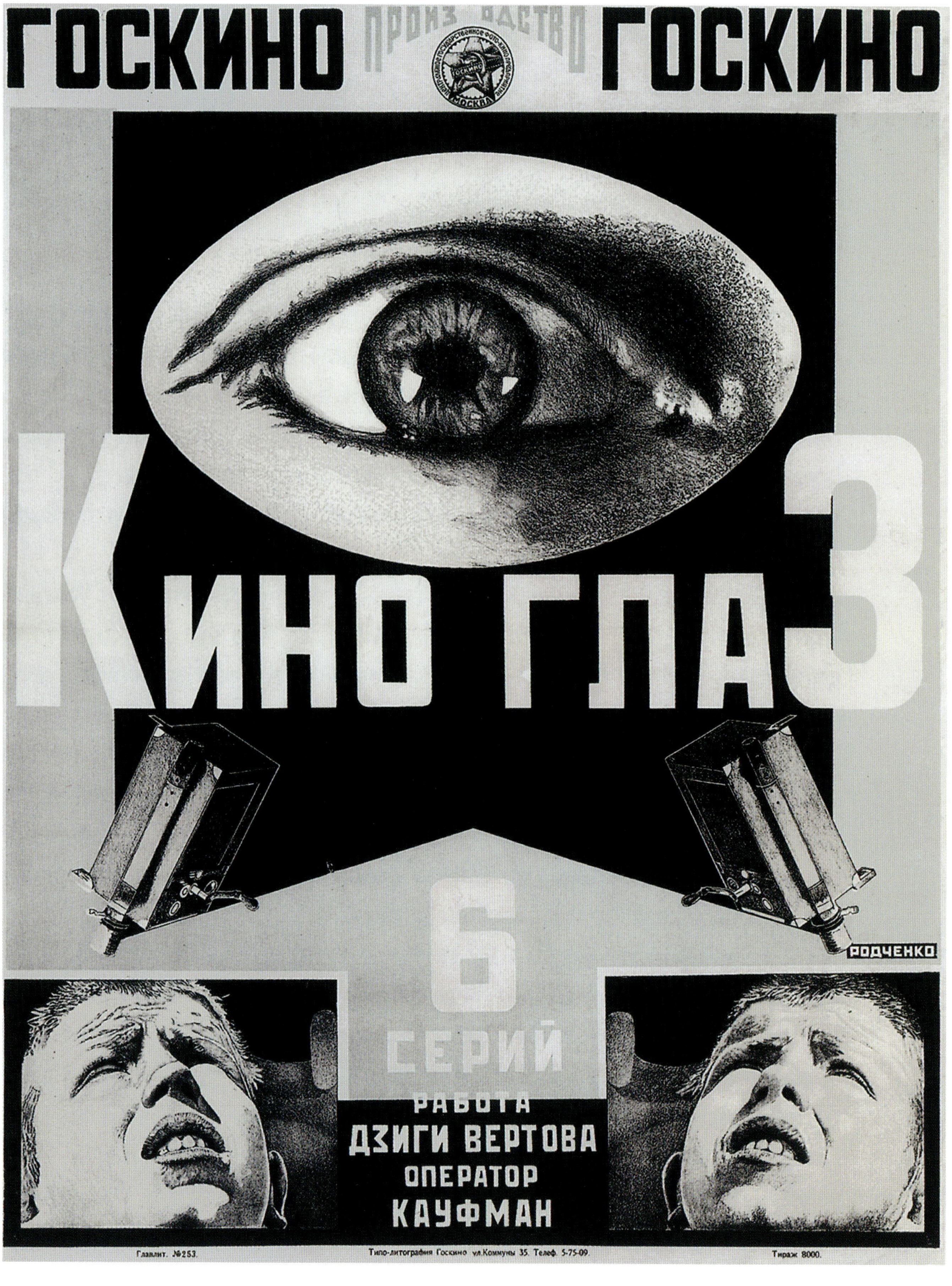

LEFT A poster promoting the Kino Glaz ('Cine Eye') newsreel, designed by Alexander Rodchenko, 1924.

OPPOSITE A poster design for the Soviet state airline Dobrolet, *c.* 1923.

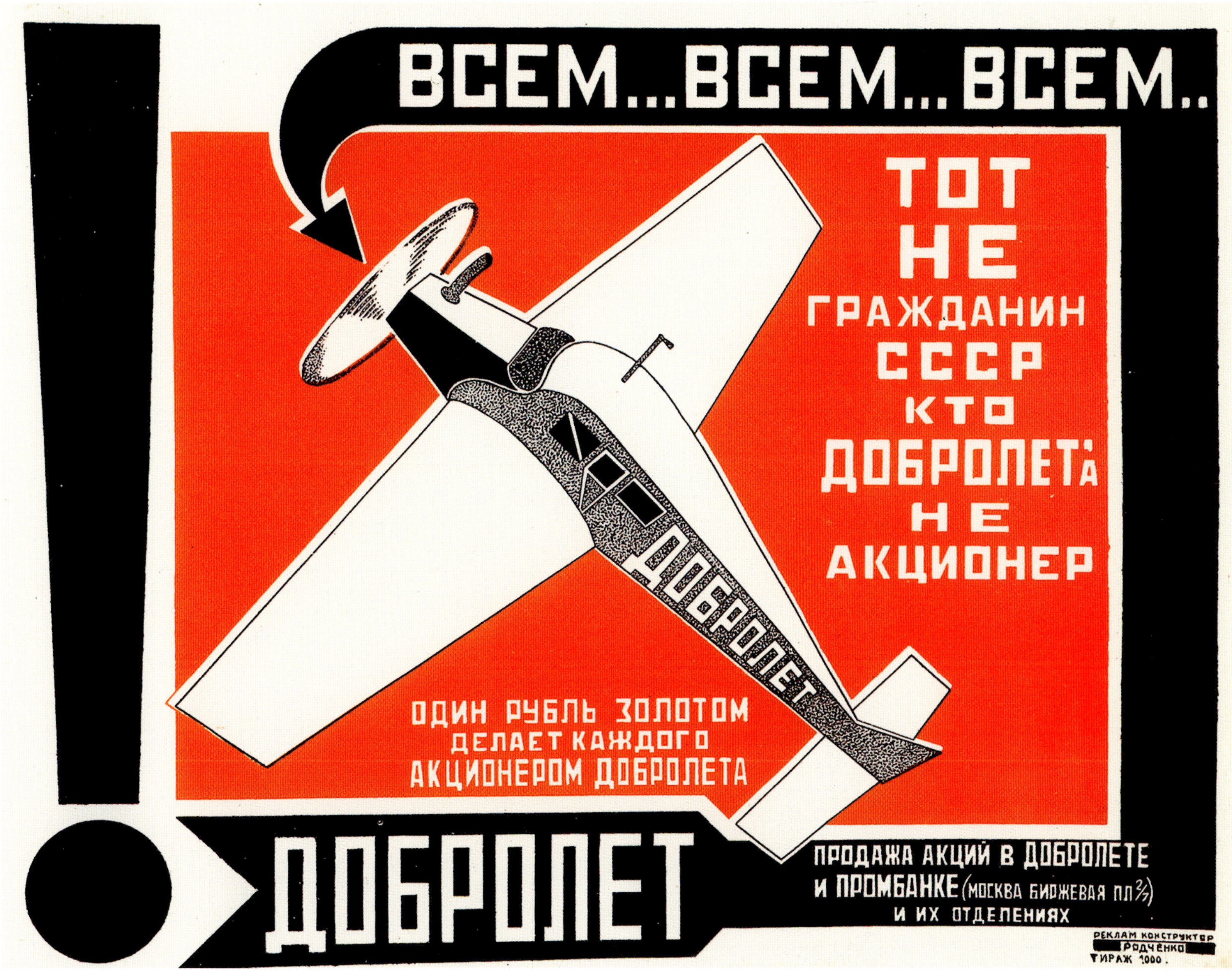

photograph of a woman (Lilya Brik, lover of the poet Vladimir Mayakovsky) with a triangular megaphone speech bubble emerging from her open mouth, its bold – or 'loud' – type shouting 'books in all branches of knowledge'. The poster is, at the same time, almost like a flag or banner, with its abstract use of colour blocks and geometry. It was a poster that spoke, literally, to the viewer with a clear, vibrant and arresting message.

Other vivid Rodchenko works of this period include a 1923 poster, designed with his wife, for babies' pacifiers, in which an abstract, geometrical baby becomes an extraordinary and machine-like mascot. He also designed advertisements for the state air service, Dobrolet, featuring – again – a vivid use of red, white and black, along with a climbing Junkers aeroplane set upon the diagonal within the linear rectangle of the poster (*c.* 1923).

In 1923 Rodchenko formed an advertising and graphic design agency with Vladimir Mayakovsky, their clients including the state-owned food store chain, Mosselprom. These advertising posters feature dynamic employment of colour and type sitting within powerful geometric and symmetrical combinations of triangular forms floating within rectilinear frameworks.

Around this same period, which was marked by high productivity, Rodchenko became increasingly interested in photography and photomontage techniques. He provided some surreal photomontage illustrations for Mayakovsky's poetry in the mid-Twenties, while also designing film posters – such as 'Kino Glaz', or 'Cine Eye' (1924) – that used the technique to great effect. Around the same time, Rodchenko provided graphic elements for the interiors of the Russian 'Workers' Club' exhibited in Paris at the Exposition Internationale des Arts Décoratifs et Industriels Modernes of 1925.

Rodchenko also designed layouts, typography and visuals for journals, and increasingly involved himself in education, teaching at a number of applied art schools and becoming a professor at the Vkhutemas technical college in Moscow. Unlike some of his contemporaries, such as El Lissitzky, Rodchenko seldom ventured or worked outside of Russia, yet his influence was to travel broadly over the coming decades.

JAN TSCHICHOLD
(1902–1974)

The beginning of modernism was marked by an explosion of manifestos. A whole library of statements and declarations sought to explain the practice, theory and thought behind radical new approaches to architecture, art and design. It was a way of trying to make sense of what was, in effect, the start of a cultural revolution. In the field of graphic design, one of the key manifestos of the period was Jan Tschichold's 'The New Typography' ('Die neue Typographie'), published in 1928.

'The essence of the New Typography is clarity,' wrote Tschichold. 'This puts it into deliberate opposition to the old typography whose aim was "beauty" and whose clarity did not attain the high level we require today. The utmost clarity is necessary today because of the manifold claims for our attention made by the extraordinary amount of print, which demands the greatest economy of expression.'[4]

Tschichold's clear and elegant volume laid out the key principles, as he saw them, for a new era of graphic design. These included the use of standardized and functional sans serif fonts, geometrical precision and dynamic asymmetrical layouts, given that Tschichold regarded symmetrical design as part of the old order. The book was highly influential and helped to establish a set of guidelines soon taken up by the Swiss School (also described as the International Graphic Style), with its own emphasis on pure, functional fonts and a grid-based approach to layouts.

Interestingly, Tschichold was one of the few graphic designers and typographers to emerge from a 'type' background, rather than being trained in the world of art and sculpture. Born in Leipzig, Germany, he was the son of a sign painter and studied at the Leipzig Academy for Graphic Arts, followed by work as a calligrapher.

In 1923 Tschichold attended the Bauhaus exhibition in Weimar organized by, among others, László Moholy-Nagy. There he was particularly taken by the typographic work he saw by Moholy-Nagy, the latter's Bauhaus colleague Herbert Bayer and others. The term 'The New Typography' was itself first coined by Moholy-Nagy.

Around the same time Tschichold became increasingly aware of the work of De Stijl* in Holland and the dynamic aesthetic and visual approach formulated by the Russian Constructivists*, including Alexander Rodchenko and El Lissitzky*. These various influences impacted upon his own work in the fields of typography, book design and poster design.

One of Tschichold's posters from around this time, promoting the film *The Woman Without a Name* (1927), exhibits a strong Constructivist influence, with its use of a vivid red, white and black colour palette, asymmetrical composition, triangulation and a bold but limited use of sans serif type. Other Tschichold film posters of the same period, such as 'Die Hose' (1927) and 'Napoleon' (1927), employ similar techniques.

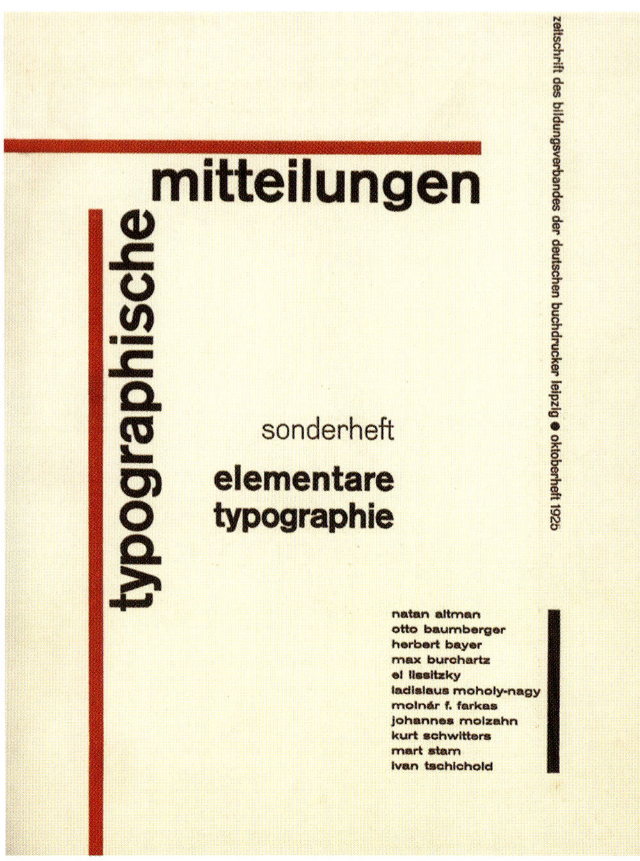

RIGHT Jan Tschichold's cover design for 'elemental typography' (*elementare typographie*), a special insert for the publication *Typographische Mitteilungen*, 1925.

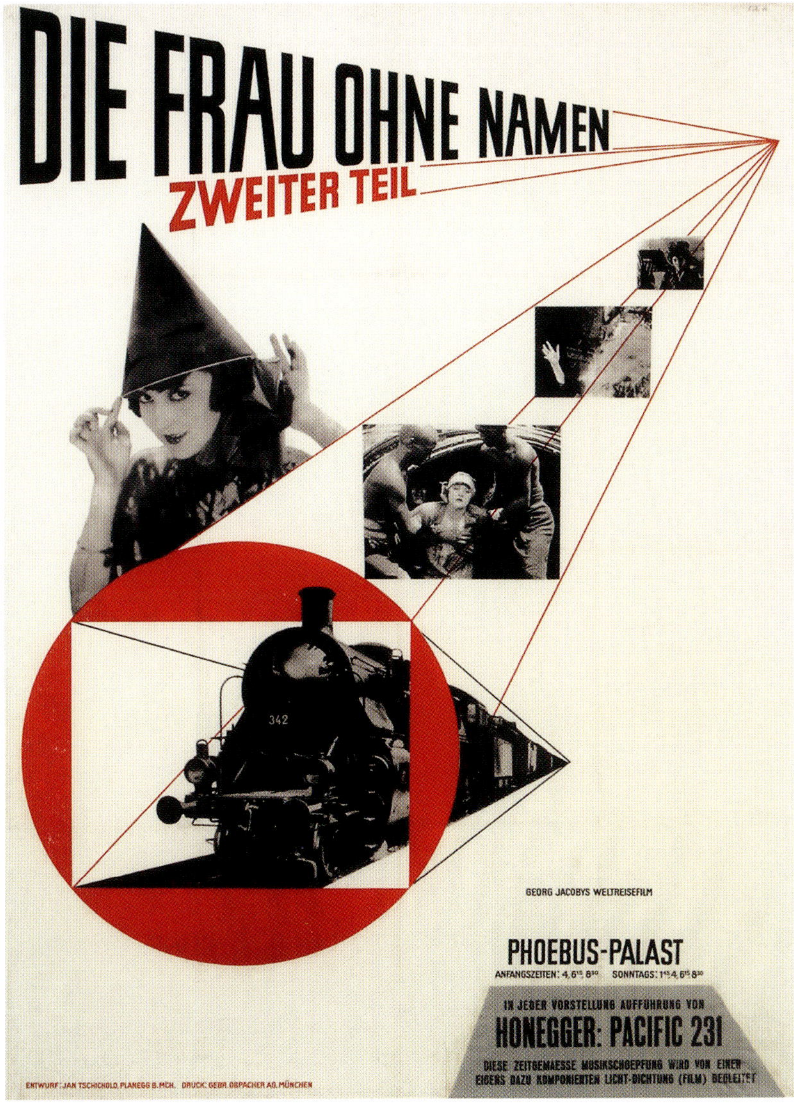

LEFT A film advertising poster for *The Woman Without a Name, Part Two* (*Die Frau ohne Namen, Zweiter Teil*), 1927.

BELOW Tschichold's poster for the film *The Pair of Trousers* (*Die Hose*), 1927.

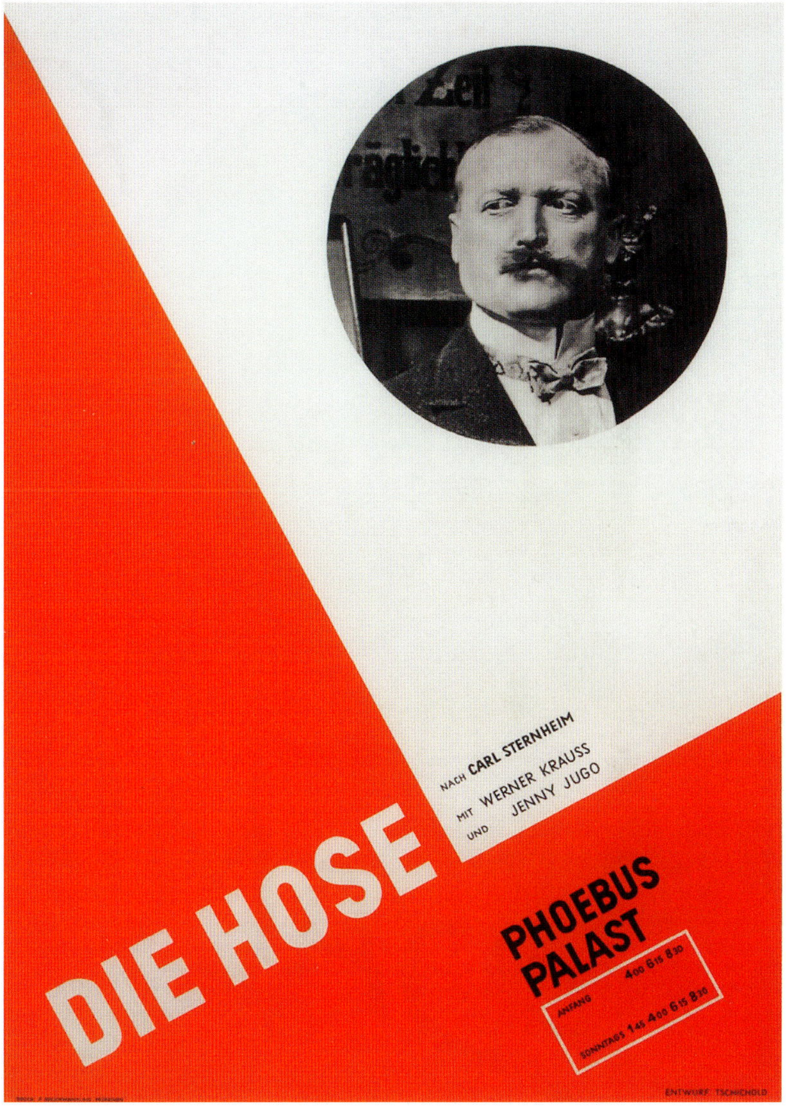

In 1925, Tschichold edited an issue of a Leipzig journal called *Typographic News* (Typographische Mitteilungen), in which he began to set out the key principles of the 'new typography'. Within a few years, his thinking had evolved further and he set out his ideas in detail within *The New Typography*. Key to the success of the book was that it followed its own principles within its design and was heavily illustrated with examples of practical work by Tschichold and his contemporaries. The book served, then, as both a manual and a manifesto, printed in a convenient size so that it could sit in the pocket like a notebook. The emphasis throughout was on functionality, clarity and making it new; the book was both pragmatic and didactic.

'The New Typography is distinguished from the old by the fact that its first objective is to develop its visible form out of the functions of the text,' Tschichold wrote. 'It is essential to give pure and direct expression to the contents of whatever is printed; just as in the works of technology and nature, "form" must be created out of function. Only then can we achieve a typography which expresses the spirit of modern man.'[5]

In the years that followed, Tschichold attempted to follow the line of his party. He designed a number of new fonts, including Transit (1931), Zeus (1931) and Saskia (1932); much later, in 1967, he created Sabon. Along with many other modernist colleagues and contemporaries in Germany, however, he fell under the suspicion of the Nazi state, and in 1933 he and his wife were arrested. Helped by a sympathetic policeman, they were given tickets to flee to Switzerland, where Tschichold continued to work in the field of book design.

Appalled by the authoritarian nature of the Nazi regime, Tschichold began to worry about the strict dogma of his own master work (even if this did not square at all well with the state's hostility to modernism). Just as Tschichold's text was being taken up by followers of the Swiss School, its author began to question it within a revisionist process, prompting a hostile reaction from Swiss advocates such as Max Bill*.

Yet Tschichold himself continued to work and develop his thinking. Among his key projects of the post-war period was his appointment as art director for Penguin Books in London, from 1947 to 1949. Here, he kept the Gill Sans typeface developed by Eric Gill*, but helped to perfect the familiar Penguin paperback look and logo.

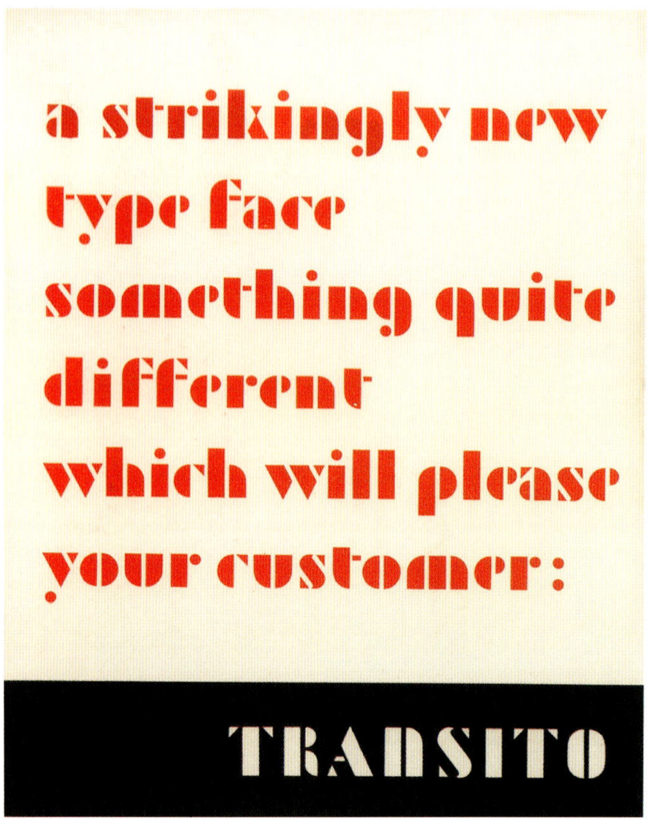

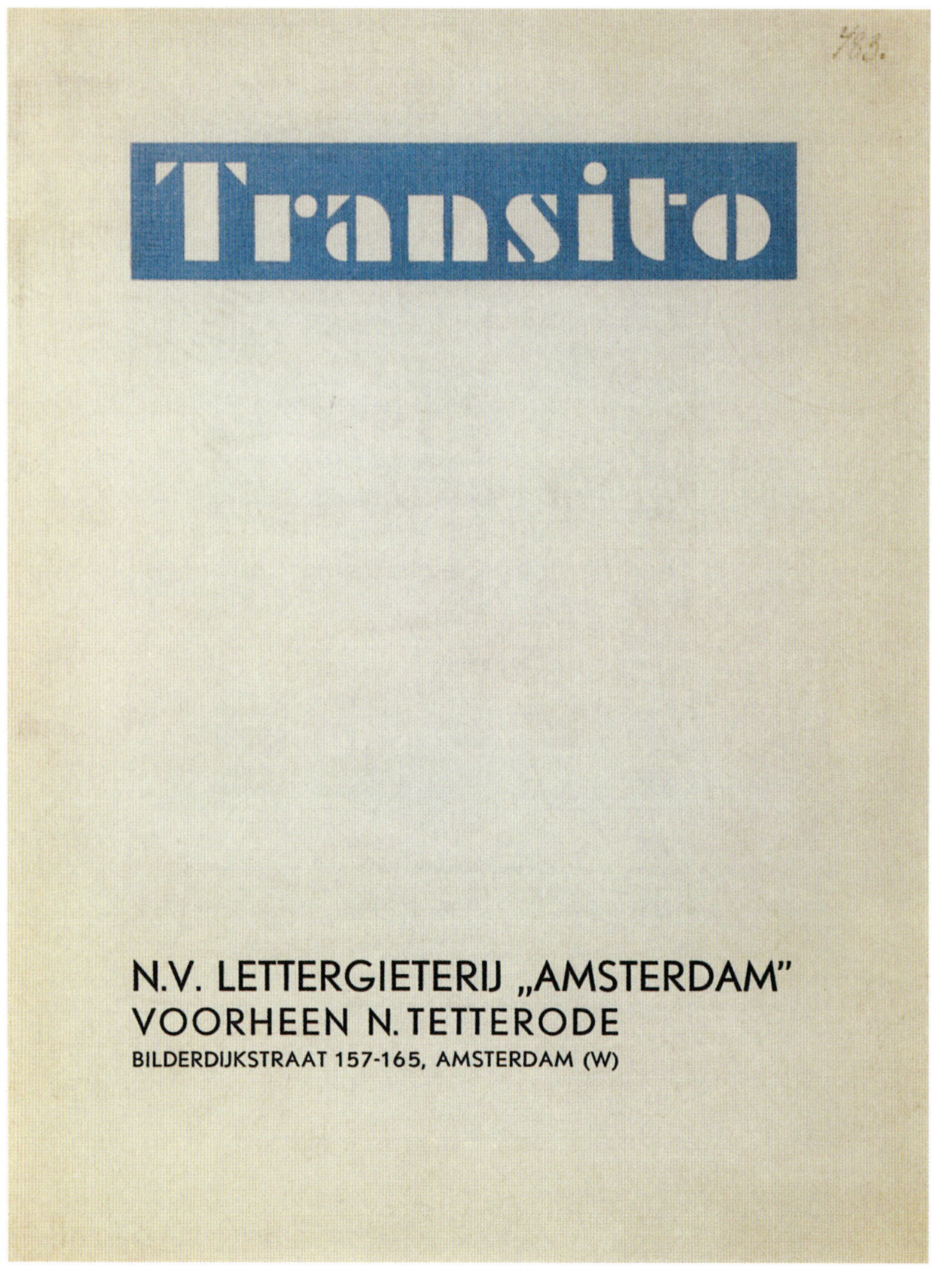

ABOVE & RIGHT A page and cover art for *Transito*, a type specimen book published by Lettergieterij, Amsterdam, 1931.

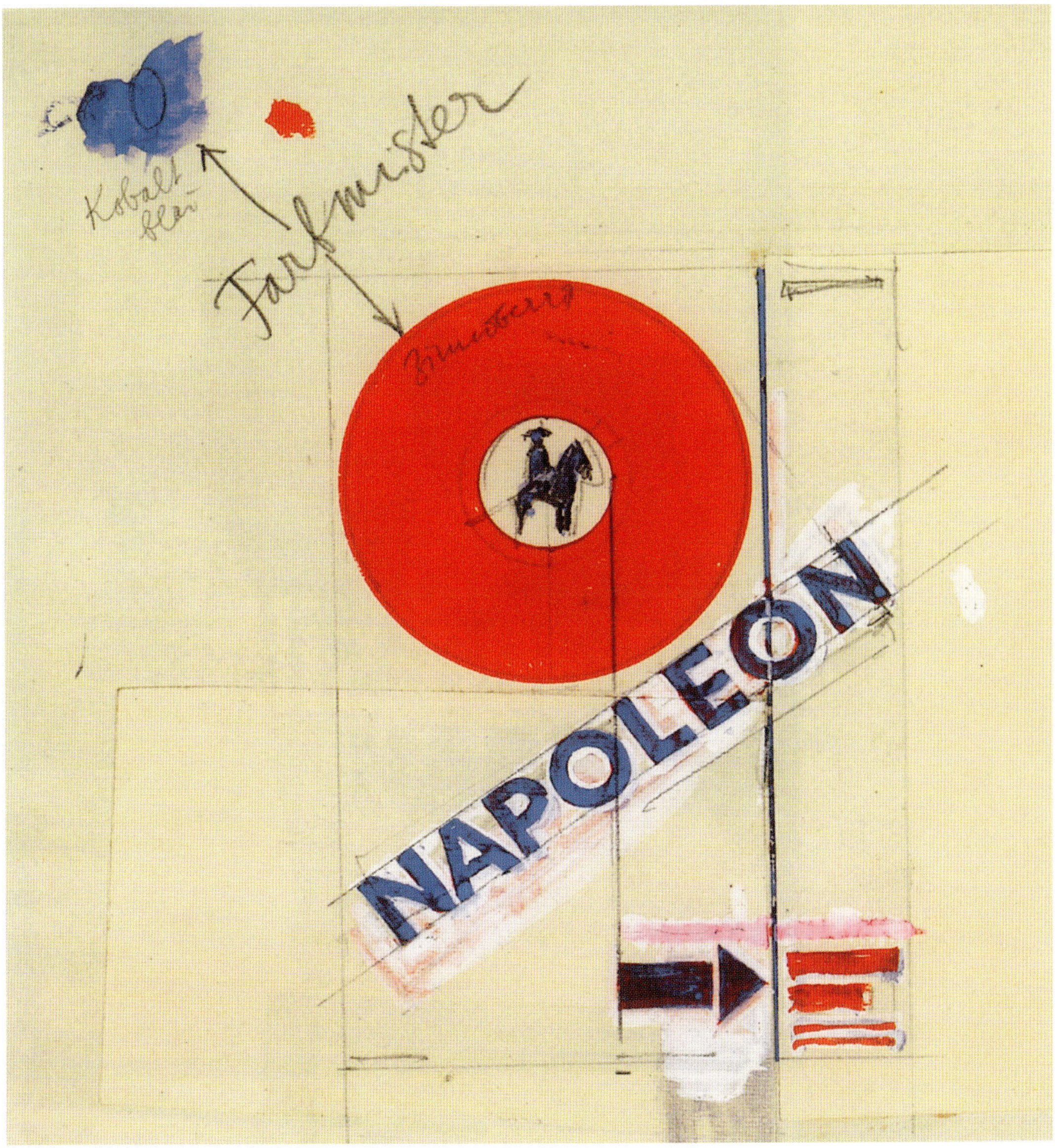

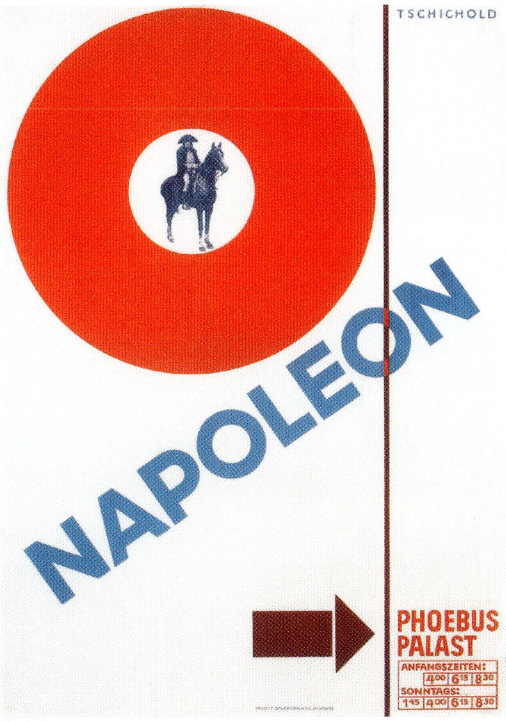

ABOVE & RIGHT A draft sketch by Jan Tschichold for the film *Napoleon*, and the published poster, 1927.

HOUSES & INTERIORS

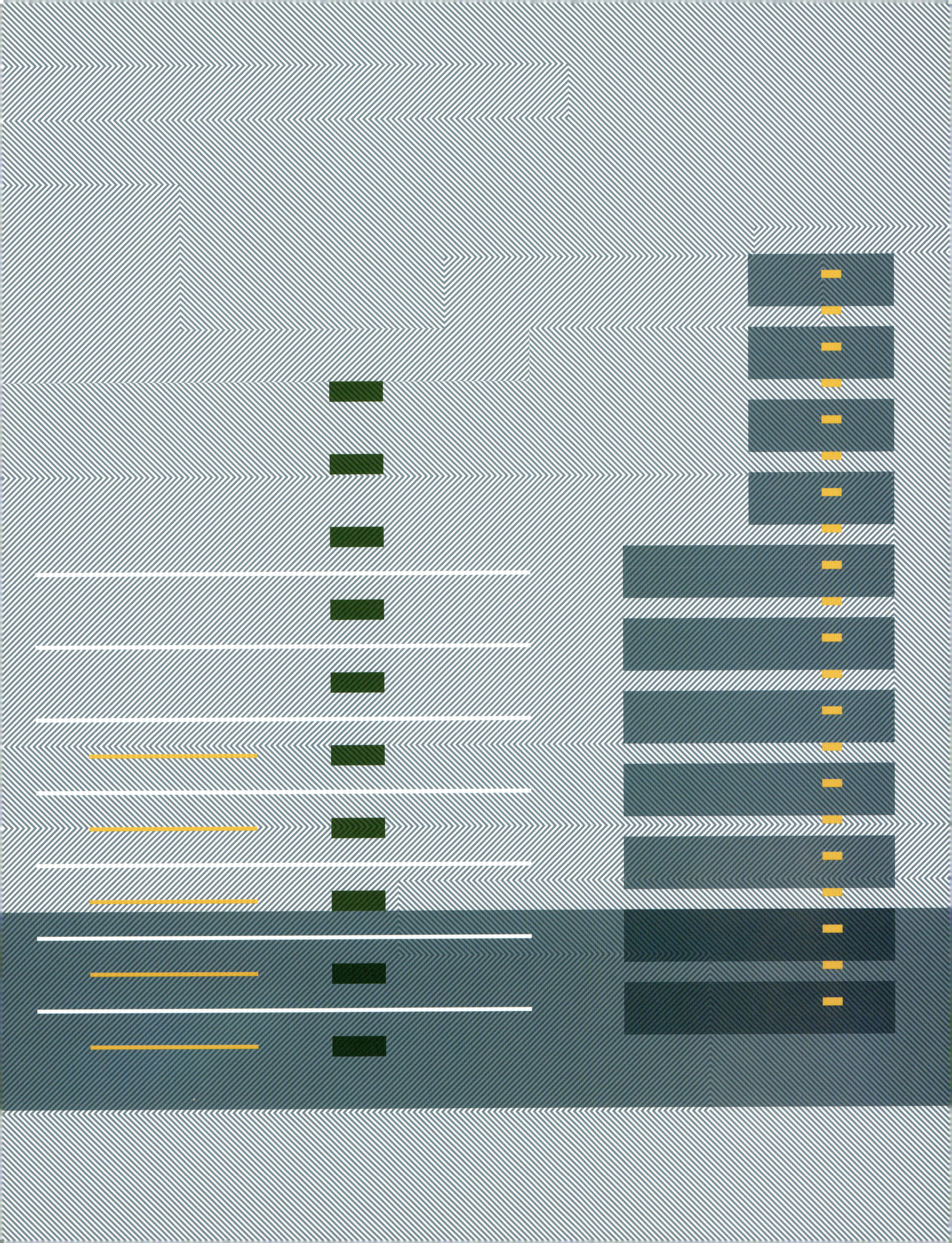

INTRODUCTION

On the 9th of February, 1932, a show opened at the Museum of Modern Art in New York called *Modern Architecture: International Exhibition*. It was curated by the architectural historian Henry-Russell Hitchcock and curator Philip Johnson*, who also wrote an associated book together entitled 'The International Style'. The exhibition later travelled to a number of other American cities.

Both exhibition and accompanying book helped to publicize and define a modernist approach to architecture and codify the ingredients of this fresh spirit. 'Today a single new style has come into existence,' Hitchcock and Johnson argued. 'This contemporary style, which exists throughout the world, is unified and inclusive.'[1]

The exhibition featured work by influential European modernists, including Le Corbusier, Walter Gropius, Alvar Aalto and Ludwig Mies van der Rohe, as well as a number of American architects including Richard Neutra. Hitchcock and Johnson defined several common principles related to the structure, order and aesthetic of the houses and buildings of the International Style*, closely related to Le Corbusier's 'Five Points of Architecture', a key modernist manifesto statement.

Le Corbusier's Five Points were most clearly expressed in the Villa Savoye (see p. 370) and encompassed the notion of a free façade, liberated from a load-bearing function by a structural framework of reinforced concrete or steelwork. The liberated façade became a curtain wall that could be infilled with glass rather than stone or brick, introducing sunlight and creating a sense of transparency; for Le Corbusier the windows became horizontal ribbons of light (or 'fenêtres en longueur'). At the same time, the structural lattice of the house liberated the internal floor plan, creating a fluid interior layout rather than a series of cellular rooms, with the universal or open plan becoming another key characteristic of the modernist house. Le Corbusier added two other points, namely the use of piloti (or supporting pillars) to raise the building above the ground, and the creation of integrated terraces and outdoor rooms on the flat roof.

If one takes the Five Points and the aesthetic offered by Hitchcock and Johnson's exhibition, one has the 'classic' characteristics of the modernist home: a linear, rectangular form with elevated and free-flowing living spaces, a flat roof line, and integrated outdoor areas contained within the floor plan and structure of the building itself. Ramps were another common feature, emphasizing the processional nature of buildings such as the Villa Savoye or Giuseppe Terragni's Villa Bianca (see p. 420), which offered a progressive journey through the home, moving through a variety of internal volumes as well as weaving between inside and outside spaces. The boundaries between interior and exterior were broken down with the extensive use of terraces, decks and balconies, while the unglazed surface of the house tended to be rendered with cement and painted a gleaming white.

As well as embracing this vivid, crisp, linear aesthetic, the modernist approach of the International Style embraced a fresh way of living within the home. There was a radical shift from the traditional, cellular room pattern of period 18th- or 19th-century homes in favour of a more informal and fluid layout with inter-connected living and dining spaces, although many modernist homes still featured separate kitchens and often staff quarters as well.

The modernists of the International Style famously embraced Mies van der Rohe's dictum of 'less is more', sweeping away ornament and excess in favour of a more functional and ordered approach, both within and without. Materiality and finish, having been laid bare, became more important and expressive than ever, with Mies's Villa Tugendhat, for instance (see p. 380), featuring statement walls of marble and macassar ebony. Form may have followed function, but this is not to say that the International Style home was devoid of expressive materials, colour or texture, even if these elements were edited down in favour of an almost minimalist approach.

Le Corbusier himself, as an artist and painter of some renown, was never afraid of colour and was fond of decorating buildings with murals of his own design. A number of other modernists outside the strict circle of the International Style, such as FRANK LLOYD WRIGHT, had a more complicated relationship with decoration and ornament, and only periodically embraced pattern and motif. Wright and Scandinavian modernists such as Alvar Aalto were strongly immersed in a craft tradition and particularly passionate about the warmth of natural materials such as wood and stone, which they then allied with new methods of building in concrete and steel.

Le Corbusier famously talked of the house as a 'machine for living in', declaring that 'our own epoch is determining, day by day, its own style'.[2] The godfather and prophet of modernism, who loved a grand statement, declared that 'there is a new spirit: it is a spirit of construction and of synthesis guided by a clear conception'.[3] Yet for Le Corbusier and many other modernists, a machine was not only functional but beautiful. A machine was engineered and utilitarian, but it was also crafted and expressive.

For many architects and designers, a key point of inspiration was the ocean liner. The great transatlantic ships of the first half of the 20th century were emblematic of both new technology and engineering, but also of connectivity, travel, internationalism and progress itself. The liners were glamorous and evocative, but also designed with great intelligence and thought. They combined open, spacious and dramatic communal spaces and intimate, crafted cabin retreats, complete with integrated furniture, space-saving devices and modern services.

'A seriously minded architect, looking at it as an architect (i.e. a creator of organisms), will find in a steamship his freedom from an age-long but contemptible enslavement to the past,' wrote Le Corbusier. 'The steamship is the first stage in the realization of a world organized according to the new spirit.'[4]

PAGES 294–5 The spiral staircase and ramp within the Villa Savoye, Poissy, by Le Corbusier, 1931.

BELOW LEFT Le Corbusier pictured in 1935 with a model of the Villa Savoye.

BELOW Catalogue for the *Modern Architecture* exhibition at the Museum of Modern Art in New York, 1932, which included Le Corbusier.

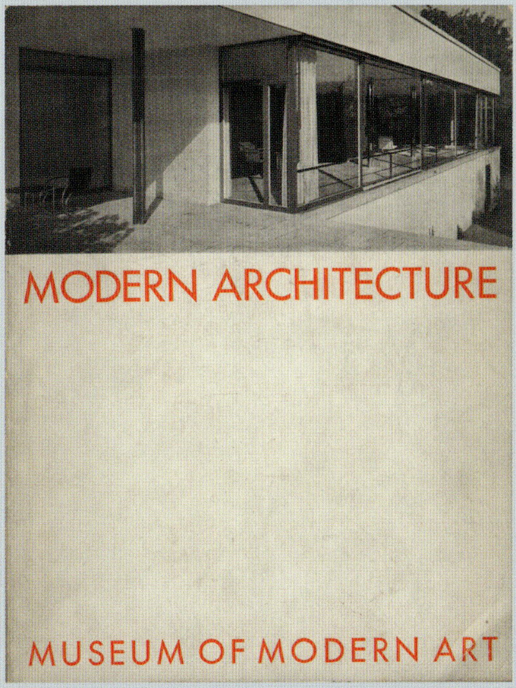

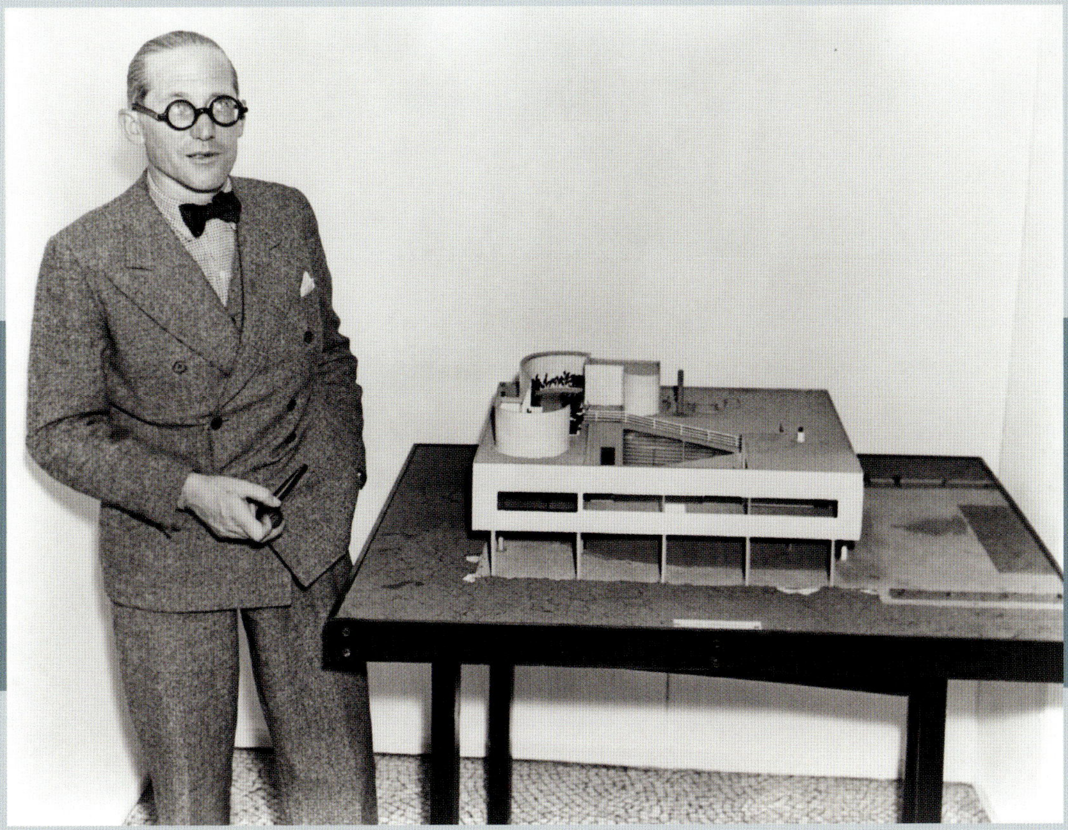

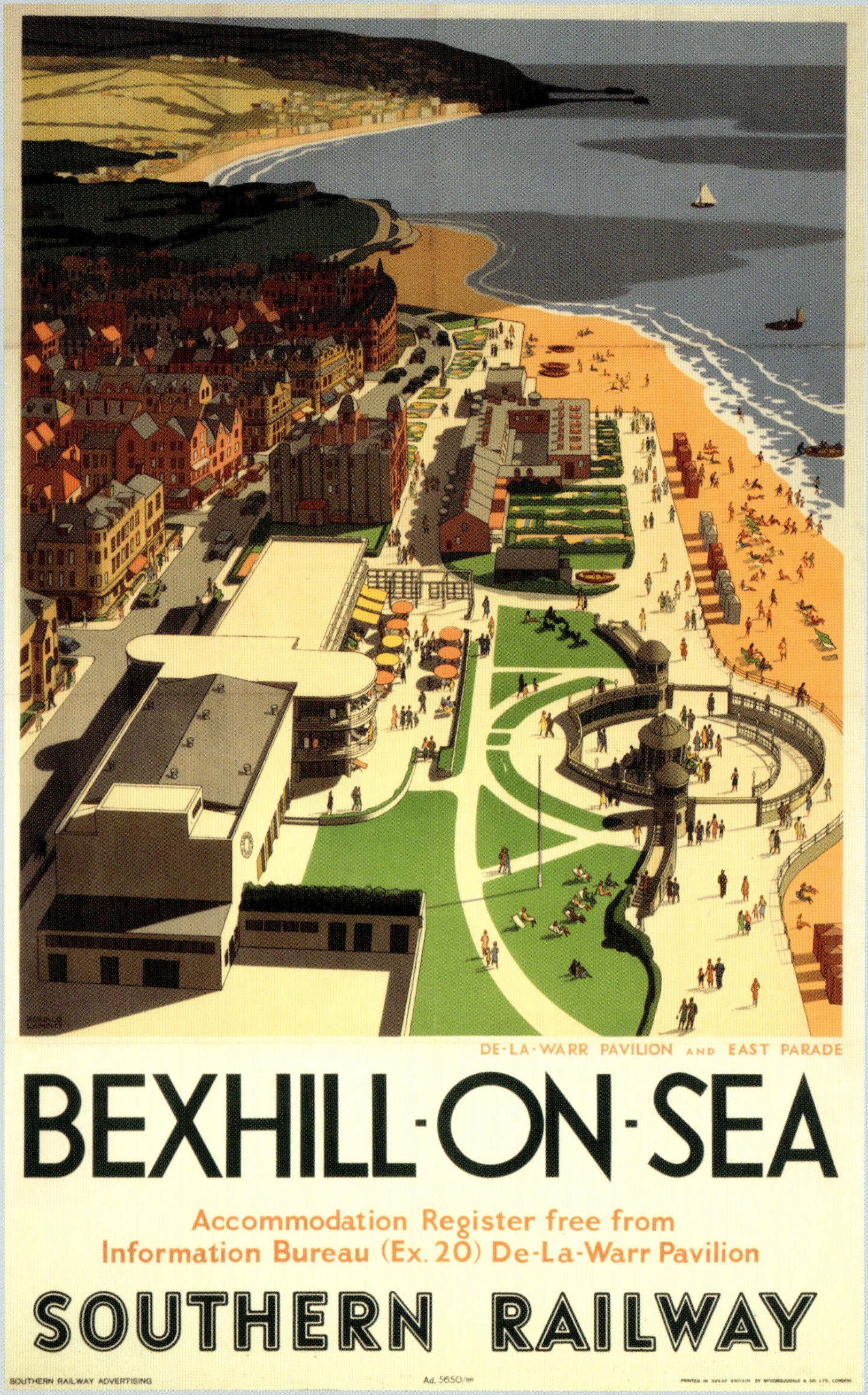

Many pioneering houses and apartment buildings of the machine age have been compared to grounded ocean liners. They include Gray and Badovici's E-1027 but also Hans Scharoun's Haus Schminke (see p. 408), Richard Neutra's Lovell Health House (see p. 384) and Wells Coates's* Isokon building in Belsize Park, North London (see p. 444).

At the same time, the car was another great influence. As with the ocean liner, the car was a point of inspiration on the one hand and a finely tuned example of form follows function on the other. But it was also a symbol of modernity that deserved integration within the architecture of the home itself. Villa Savoye, for example, had garaging and chauffeur's quarters tucked under the elevated body of the house. Similarly, at Mies's Villa Tugendhat, a small pavilion alongside the main house holds a garage and staff accommodation, forming a satellite structure to the master building.

This preoccupation with technology was all part of the spirit of the age, as Le Corbusier put it, and the ambition to 'make it new' among machine-age modernists. It should not be forgotten that modernist architects were themselves pushing through a technological boundary, and that new materials and advanced engineering were a key part – a foundation stone – of the modernist home encompassed by the Five Points and the International Style. Architects such as Auguste Perret and Rudolph Schindler helped pioneer construction methods using concrete, while Neutra and Pierre Chareau experimented with steel frameworks and glass walls.

The early modernists were, essentially, testing out new ways of building as well as new ways of living. It is no wonder, perhaps, that so many narratives of the great modernist houses and apartment buildings are littered with battles over leaking roofs, damp rooms, structural defects or poor finishes (see the Villa Savoye, for example, or the Isokon building, or Wright's Fallingwater).

As one surveys the stories of the early modernist homes, they do start to become metaphorical battlegrounds. Very few make for happy or easy experiences for architect and clients. Houses such as Ernő Goldfinger's Willow Road building (see p. 336) or Connell, Ward & Lucas's* 66 Frognal (see p. 444) were only built after overcoming fierce objections from conservative neighbours and planning authorities. Politics, anti-semitism and extremism also took a terrible toll on a number of houses of the Thirties, as well as upon their architects and owners: Haus Schminke and Villa Tugendhat are poignant examples of this. A number of modernist architects, including Wells Coates, SERGE CHERMAYEFF and even Frank Lloyd Wright, experienced occasional or perpetual money worries. There is a sense in which the process of making it new, of striving to be different, was never easy.

Yet the legacy of the pioneering modernists and their buildings is profound. The modernists literally prepared the ground for a new way of living. The revolution they started gathered pace in the mid-century period, when such ideas were developed further and progressed until, eventually, they became staples of the contemporary home. The more organic approach advocated by Wright and Aalto also gained prominence in the post-war period, when the soft warmth of Scandinavian and Californian modernism became so popular.

Philip Johnson, the young curator of the International Style exhibition, was one of those who advanced the lessons of modernism in the decades that followed. Having turned to the practice of architecture himself, he designed his famous Glass House in New Canaan (1949), where notions of the curtain wall, transparency, universal space and open-plan living were taken to a new extreme. The building itself owes a very great deal to the ideas of Mies van der Rohe, who became an occasional working partner of Johnson's, while the interior was populated with Mies's furniture. In such ways and others, the ideas, principles and dreams of the early modernists were pushed out into the wider world.

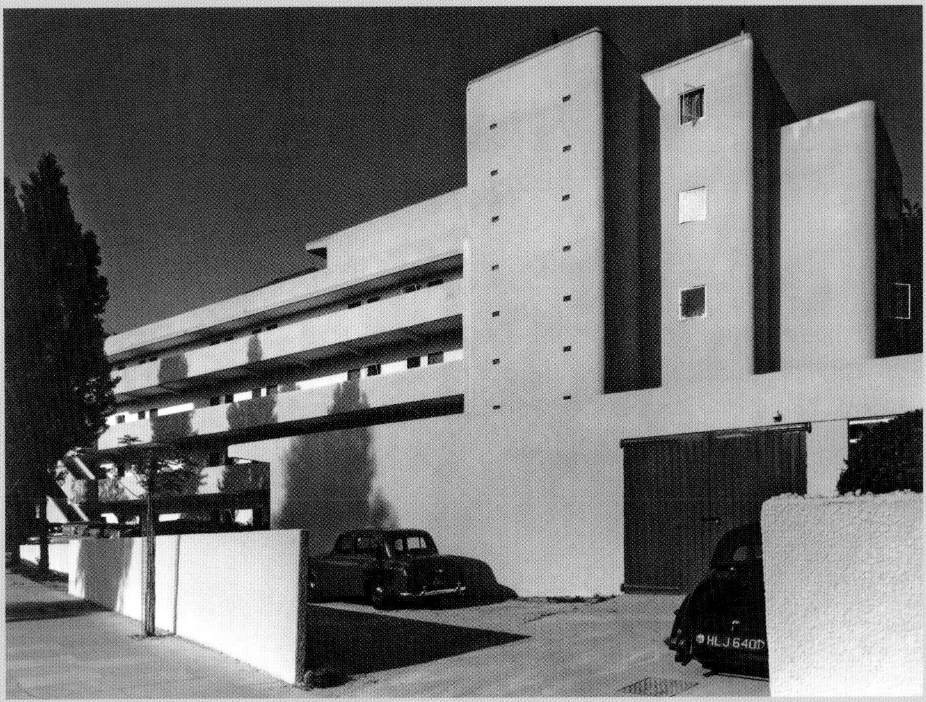

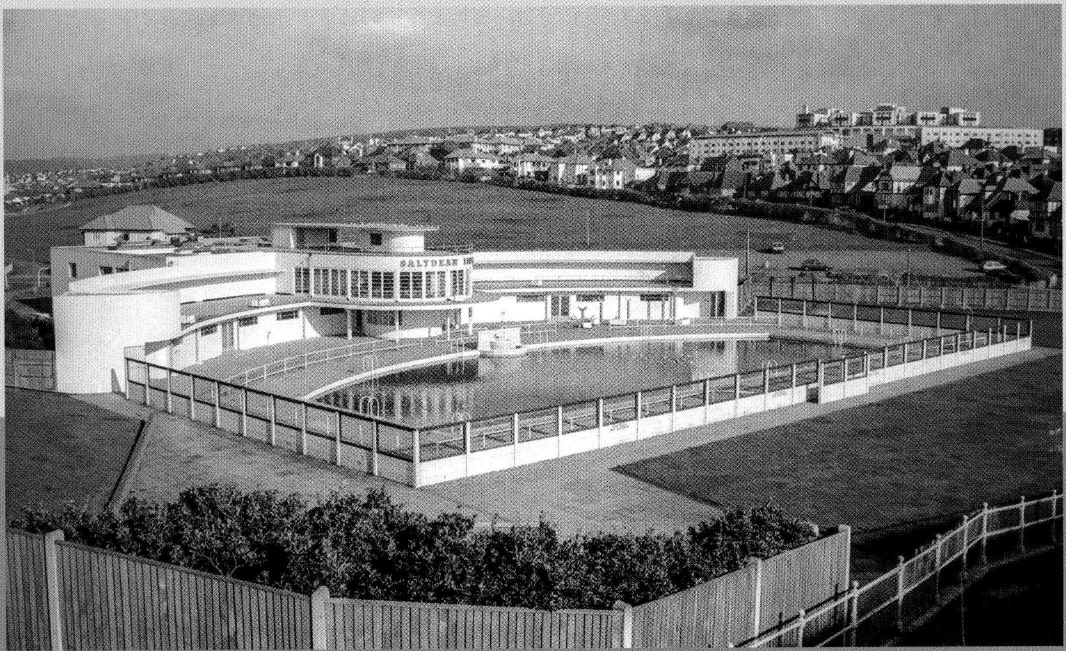

OPPOSITE A Southern Railway advertising poster for Bexhill-on-Sea, with artwork by Ronald Lampitt featuring the De La Warr Pavilion designed by Serge Chermayeff and Erich Mendelsohn, completed in 1935.

ABOVE The Lawn Road Flats, or Isokon Building, in Belsize Park, London, designed by Wells Coates and completed in 1934.

LEFT Saltdean Lido, in Brighton & Hove, designed by Richard Jones and completed in 1938.

Advocates of the 'Modern Movement' – an appropriate term coined by English writers to describe the advent of modernism in architecture and design in the Twenties – generally adopted a fetishist view of the machine. That is, they pledged allegiance to the view that industrialization, mechanization and standardization were the principal causes of change in architecture and design, but they usually stopped short of actually adopting industrialized solutions to practical problems of design. Take the tubular steel chair. Almost every architect in Europe who wished to assert his or her 'modern' credentials, designed tubular steel chairs between 1925 and 1937. Industrial productions were frequently cited as inspiration for these designs. For example, the impressive exhibition curated by Adolf Schneck in Stuttgart in 1928, and its catalogue *Der Stuhl*, compared steel and wooden furniture, and explicitly included car seats and bicycle frames as prototypes. And yet these chairs were invariably made by hand and at great expense. It was not until the Thirties that some tubular steel chairs were industrially produced at competitive prices: for example, the RP6, designed by the Austrian designer Bruno Pollak between 1931 and 1932 and manufactured by PEL in the Accles and Pollock Paddock Works in Oldbury from 1934 – a chair that continued to be used into the Fifties in British school and church halls. Many of the pioneers of tubular steel furniture, such as Gerrit Rietveld, Charlotte Perriand, Eileen Gray, Marcel Breuer and Alvar Aalto, quickly turned away from tubular steel to design in wood and other materials in the Thirties. Consumers preferred the warmth and comfort of wooden chairs and armchairs, even among modern designs.

The example of the tubular steel chair is symptomatic of the problem, but we must dig deeper to understand the causes, which were not merely practical. The underlying reasoning for revolutionizing architecture and design depended on the Zeitgeist argument. The grounds for abandoning traditional methods and the adaptation of historic styles were summed up in the formula: the world has been revolutionized by machines, so architecture and design must follow suit. The 'spirit of the age' was thought to pick up automatically the fundamental rules of each new age. However, while fashion had been changing with the times, architecture and design continued to imitate the past. No-one dressed like Louis XV's courtiers in the 1920s, but most public buildings in France continued to imitate classical and Rococo styles into the Twenties. The new forms of transportation were particularly evocative of these changes. They not only stated the problem – distances have been shortened – but they also pointed the way to the future.

Le Corbusier famously wrote three chapters in his book *Vers une architecture* (1923) which he called 'Eyes that do not see'. The chapters dealt with ocean liners, airplanes and automobiles. His argument is more complex t han often assumed. Although he often picked visual elements from photographs of airplanes and ocean liners – wing struts became concrete pilotis, promenade decks were evoked by the long horizontal windows of his concrete buildings – it was not his aim to imitate the forms of modern machines. The argument was one of method. If the problem of the airplane had been posed in the way architects think of buildings, Le Corbusier argued, they would have remained at the stage of imitating birds. If the problem of the house was stated correctly, given modern conditions, the modernist house would automatically ensue, complete with pilotis, long windows, cantilevers and roof gardens. Thus, Le Corbusier called his design for a mass-produced concrete house 'Citrohan', in imitation of the Citroën automobile, and tried to involve the car industrialists to invest in his projects. A housing project for Peugeot at Audincourt

WELCOME TO THE MACHINE: THE POWER OF THE AUTOMOBILE AND NEW TECHNOLOGY IN SHAPING THE MODERN HOME

TIM BENTON

(1924) was unsuccessful, but he managed to persuade Gabriel Voisin to invest in his Esprit Nouveau pavilion at the 1925 Exposition Internationale des Arts Décoratifs et Industriels Modernes in Paris, and he persuaded Voisin's managing director Mongermon to let him have his first second-hand Voisin car in exchange for a house project which remained on the drawing board. Le Corbusier owned three Voisin cars in succession and travelled all over France and as far as North Africa in them. But he was not a 'petrol head'; it was his cousin Pierre Jeanneret and his friend Amédée Ozenfant who knew about cars.

Le Corbusier crossed the Atlantic in the *Normandie* in 1935 and took the *Graf Zeppelin* to fly to Rio in 1936, but in neither case did he describe the experience. He took a few shots of the zeppelin at Recife with his cine camera and took hundreds of photographs of the SS *Conte Biancamano* on his return to France, but in general his enthusiasm for mechanical form had dwindled by the Thirties.

The sin of direct imitation of industrial products was explicitly raised by Bruno Taut in his book *Modern Architecture* (published in English and German in 1930). Illustrating a cropped view of an ocean liner, he drew a parody of a modern house in which the marine details could clearly be identified. And he mocked the details of modern architecture which imitated ocean liners – the little balconies, the balcony railings, the porthole windows.

Nor were the productions of the new industrial revolution necessarily better than the work of the past. Le Corbusier, in particular, always insisted that buildings such as the Parthenon and the apse of St Peters in Rome set the standard for everything which followed. The problem was to try to achieve, with the modern means of production determined by industrialization, work of formal and spiritual value comparable to the great work of the past. Thus, the famous double-page spread in *Vers une architecture* which compared two Greek temples with two motor cars was not meant to suggest that the cars were as good as antique architecture; rather, the argument takes the form 'a is to b as c is to d'. In other words, within the techniques and traditions of the Greek temple, the perfection achieved in the Parthenon compared to the temple of Hera at Paestum built a century earlier can be compared to the improvement in design of the Delage sports car compared to the clumsy Humber a mere 14 years previously. The machine dictates the means but not the ends.

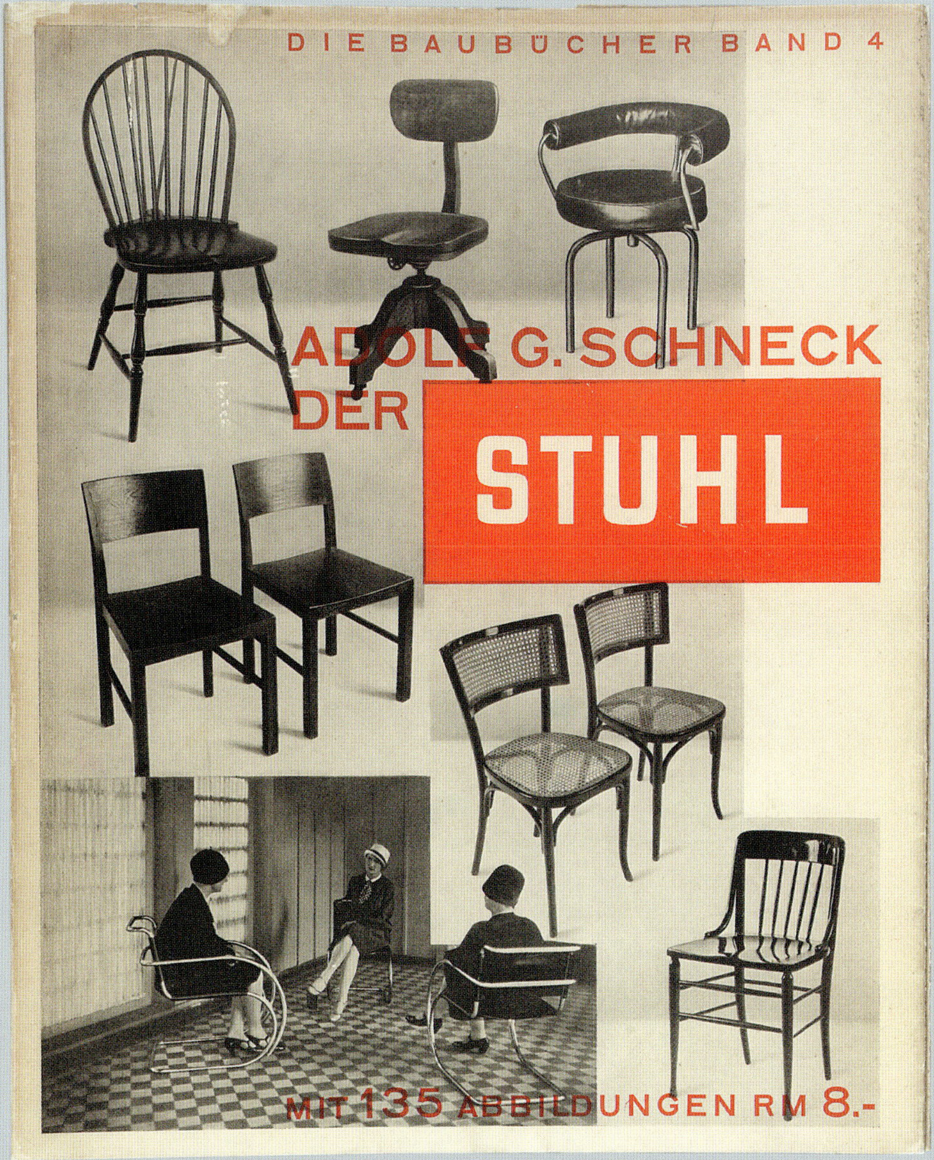

ABOVE *Der Stuhl* catalogue, by Adolf G. Schneck, published in Stuttgart, 1928.

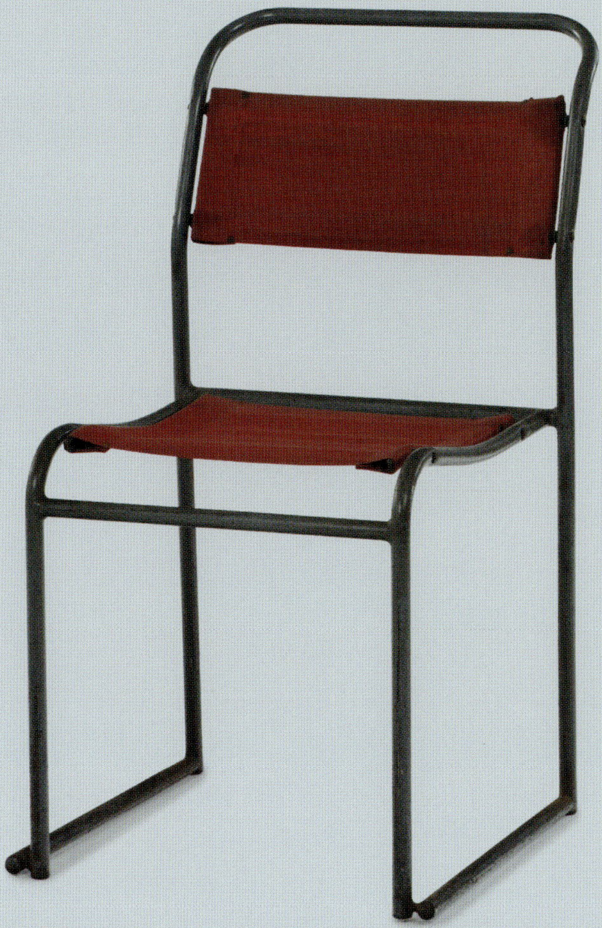

LEFT The RP6 stacking chair, in bent tubular steel with stretched canvas seat and back, designed by Bruno Pollak, 1931–32.

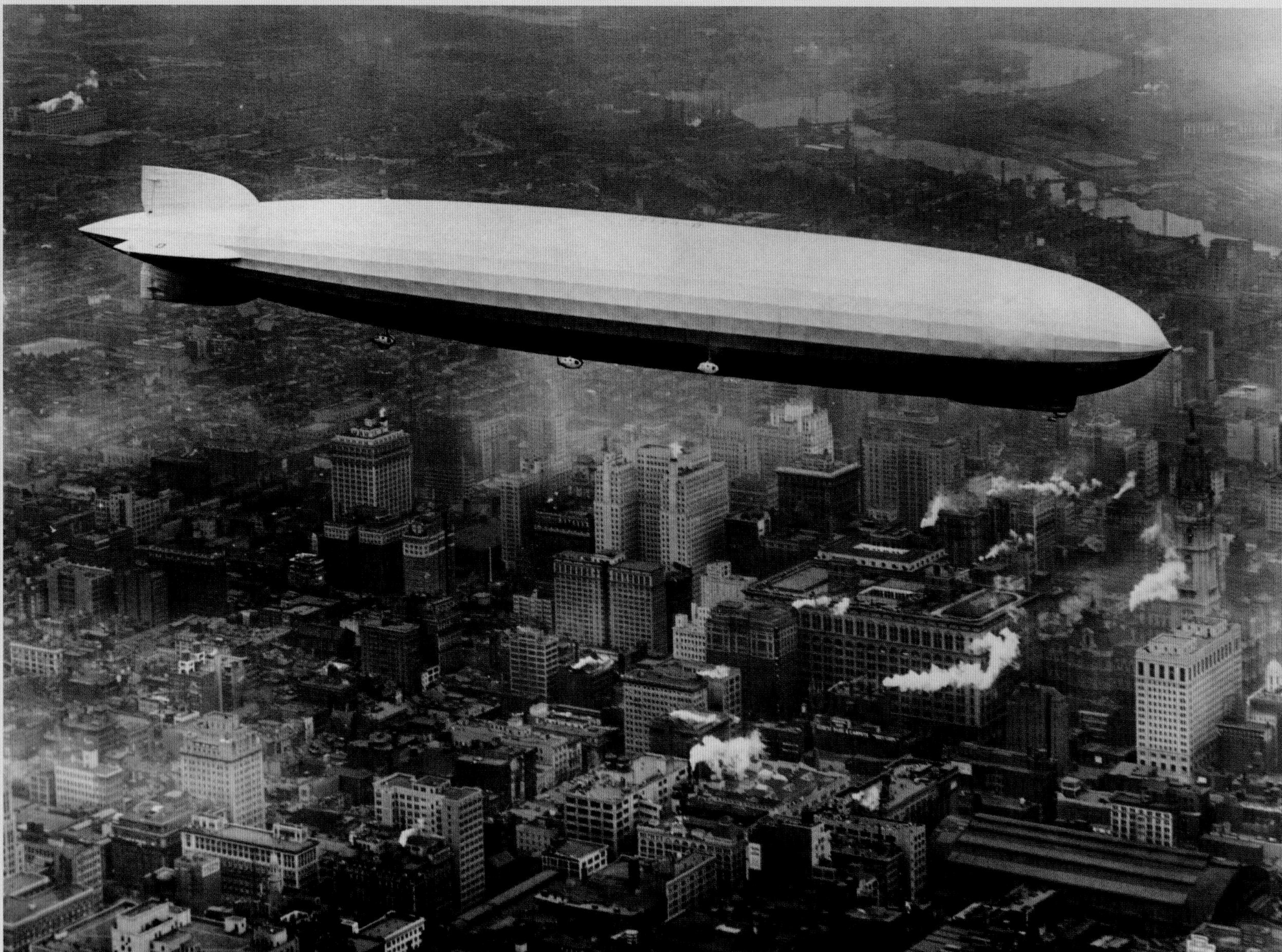

ABOVE The LZ 129 *Graf Zeppelin*, over Philadelphia, Pennsylvania, 16 October 1928.

The most important impact of the machine, and in particular motor cars and airplanes, was evocative rather than practical. Filippo Tommaso Marinetti's founding and manifesto of Futurism*, published under the title 'Le Futurisme' on 20 February 1909 in *Le Figaro* in Paris, captured the essence of the spirit of the machine which galvanized a generation. He describes a group of aesthetes in a Milanese salon, drinking late into the night and wallowing in the narcissistic self-indulgence of Symbolist poetry, who are roused by the sound of the electric tram and jump into their motor cars, to tear through the city, causing havoc as they go, until the inevitable accident, when their expensive, hand-built limousine turns over into a factory ditch. As the car is winched out, the coachwork comes away and only the functional chassis remains. This was a vivid metaphor for the casting away of tradition and comfort to embrace a new reality and, in the Futurist example, violent political demonstrations which would see them identified as prototypes and pioneers of Fascism. Marinetti was an enthusiastic militarist and explicitly cited the warplane as an expression of virile sexual aggression. But, even in the case of the Futurist artists, the car and the airplane were less central to their iconography than the principles of dynamic movement and simultaneity of feeling. Most of Umberto Boccioni's paintings and sculptures were of men, women and crowd scenes, perceived as if in constant flux and movement. It was not until well into the Fascist era that a second generation of Futurists took up the depiction of the world from airplanes – the *Aeropittura futurista* school, which flourished in the Thirties with state patronage.

So the machine was a stimulant, a moral example, a source of inspiration, but not primarily an object of imitation. When Le Corbusier was asked to edit a picture book of airplanes by the English publisher Studio, he turned away a number of the photographs he was sent to select from. He was less interested in the machines themselves – those with speed or altitude records, or those with outstanding technical specifications. In fact, like the Futurists, he was more interested in what they could enable you to do. He had had his first experience of flight in Argentina in 1929 and had been mesmerized by this god-like view of the world as a living entity. From then on, he flew whenever he could and he described his impressions of flying over the M'zab valley in Algeria in 1933 in the book *Aircraft*. Paradoxically, it was this experience of mechanized transportation that contributed to his fundamental move away from admiring mechanization to seeking to understand the forces that shaped the natural world. It was the 'cosmic forces' that he perceived at work in the South American jungle and the North African desert which inspired him. By 1930, he was seeing industrialization as a menace, an evil which replaced honest craftsmanship with mass-produced kitsch. He wrote about this movingly after a road trip in Spain in his beloved Voisin automobile, following the new motorway. He observed that the arrival of new transportation created a brief miracle, whereby the visitor could easily see unspoiled countryside and an untouched peasant culture. But, just as the railway lines had swollen the cities and created a false culture of industrialized consumption and shoddy housing, so the road would have the same effect. 'Civilization' – product of industrialization – had become a bad word for Le Corbusier, and this affected his architecture. He rediscovered the pleasures of rough stone and wood and threw away the 'Five Points of a New Architecture' that had been the slogan for his concrete modernist buildings.

Alvar Aalto, the great Finnish modern architect, was one of the first to abandon the hard, functionalist look in his buildings, using natural materials and organic planning to have his buildings fit into the landscape as well as possible. Some German architects, such as Hans Scharoun, followed suit. But most modernist architects continued to evoke the spirit of the machine as they understood it, throughout the Thirties. To understand why this is so requires an extended explanation. But two factors are significant. The first is that, for many German, Dutch and Czech architects, modern architecture was primarily motivated by political ends. The aim was not to make beautiful buildings, necessarily, but to create cheap housing and solve the problems of the modern city. Although these had always been given as arguments for the Modern movement, many architects, including 'leaders' of the movement such as Le Corbusier, Ludwig Mies van der Rohe and Marcel Breuer – were reluctant to adopt a fully functionalist approach. For the functionalists, industrial production of buildings was the only solution to bringing down prices, and they went as far as they could to include standardization and mass production of components, in imitation of the machine. The other factor is more difficult to explain. For Le Corbusier, architecture was 'the masterly, correct and magnificent play of volumes brought together in light'. His was always a sculptural aesthetic, influenced by the engineering of concrete grain silos. For many Dutch and Central European architects, the miracle of modern engineering pushed in a different direction, towards transparency, stimulated by the filigree effects one finds in plunging views of the Eiffel Tower or the Transporter Bridge at Marseilles (two structures constantly photographed by modernist architects). In this utopian vision, the aim of modern architecture, like the state, was to wither away and become insubstantial. An extensive literature exists about this notion of transparency, which was extremely influential. The idea that the structure supporting a chair or enclosing a space could become invisible, or nearly so, haunted many designers. In this way, the machine achieved its apotheosis as a transcendence of the physical.

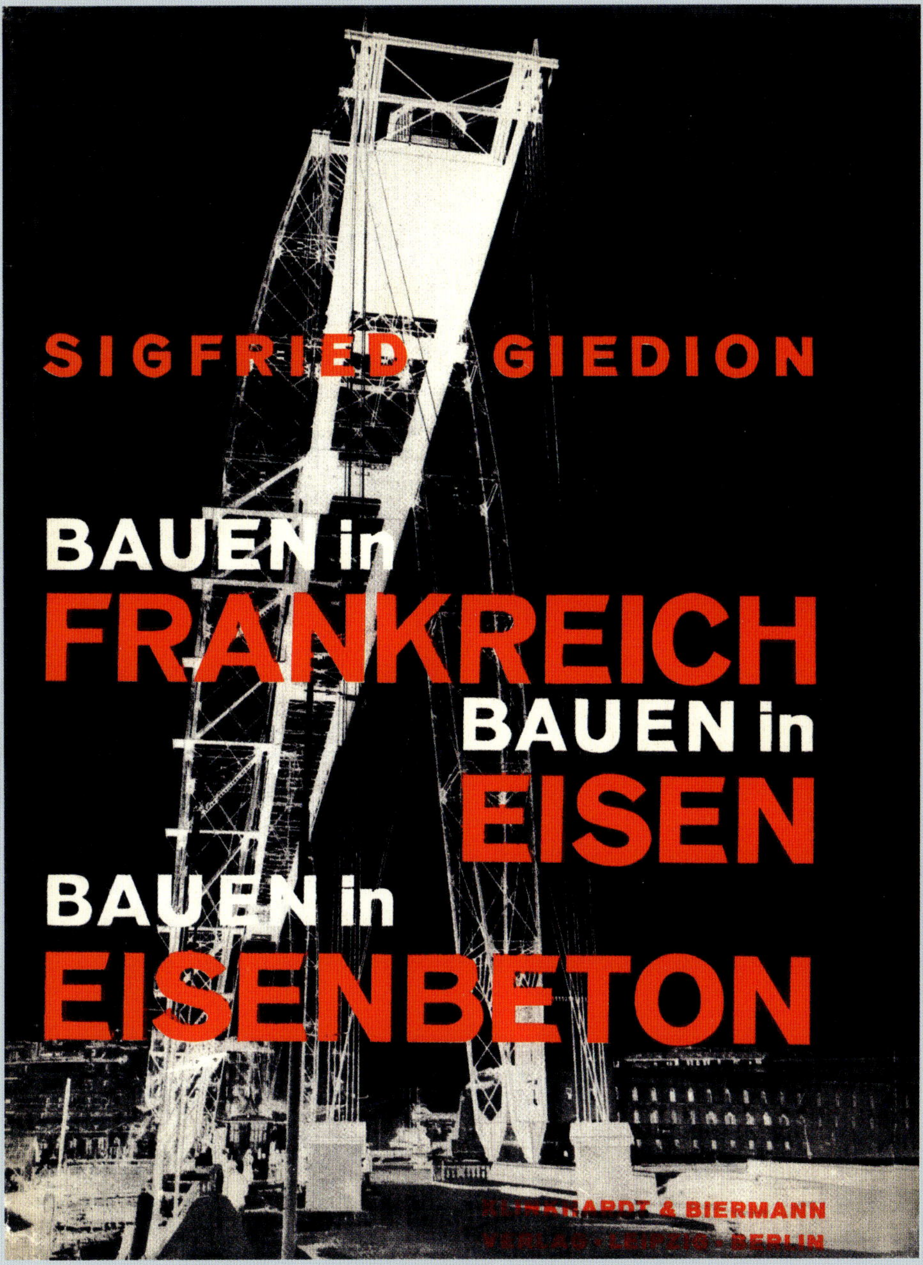

BELOW Front cover of the book *Bauen in Frankreich, Bauen in Eisen, Bauen in Eisenbeton*, with a design by László Moholy-Nagy, based on the negative of a photograph taken by author Sigfried Giedion of the Transporter Bridge at Marseilles.

THE BAUHAUS IN AMERICA: THE MODERNIST ÉMIGRÉS AND THEIR INFLUENCE ON AMERICAN ARCHITECTURE, INTERIORS AND DESIGN, 1920–1940

MARGRET KENTGENS-CRAIG

'The Bauhaus is dead.' These are the words that one of the school's most distinguished designers addressed to the American architect Harwell Hamilton Harris at their first encounter in 1943. In MARCEL BREUER's opinion, the Bauhaus had 'served a very useful purpose' at one time but 'that time had passed'.[5]

Whatever the grounds for this assessment, it was not gained from a historic perspective. The Bauhaus as a revolutionary institution of higher learning had been closed only ten years prior. The unconventional name indicated the primary goal: 'a house to be built'. And not just any house. WALTER GROPIUS's founding manifesto – with its bold woodcut illustration by Lyonel Feininger – anticipates nothing less than a soaring glass structure to be created in unity by visual artists and sculptors under the roof of architecture: 'Let us then create a new guild of craftsmen without the class distinctions that raise an arrogant barrier between craftsman and artist! Together let us desire, conceive, and create the new structure of the future, which will embrace architecture and sculpture and painting in one unity and which will one day rise toward heaven from the hands of a million workers like the crystal symbol of a new faith.'[6] This call signals a break with the 'dead styles' of the past and with traditional academic education, which no longer met the needs of design and production in the post-World War I industrial age. In early Bauhaus education the distinction between the fine arts and the crafts was bridged, so that every student would first become a certified craftsman before turning their attention to higher goals. Gropius's intent was not to 'propagate any "style", system, dogma, formula, or vogue but simply to exert a revitalizing influence on design'. He sought no less than 'the vital spark of life behind life's ever-changing form'.[7]

During its short life span of 14 years, the Bauhaus experienced two more directors, Hannes Meyer* and LUDWIG MIES VAN DER ROHE. It moved from Weimar to Dessau and finally to Berlin, changing its institutional status and faculty members along the way. Starting as a revolutionary, holistic design school, it evolved into an experimental design laboratory with 'workshops for modernity',[8] then into a pragmatic architecture school. The constant evolution caused tension. Towards the end, implosion within the school coincided with the rise of the Nazis and their scorn for modernism, which eventually led to the closing of the Bauhaus. From an immediate aftermath of dissolution and displacement, persecution or at times collaboration, Bauhaus masters and students who survived the Third Reich initiated a post-World War II revival of ideas in Germany and abroad, especially in the United States.

As great an impact as the Bauhaus would have – in the decades to come – on design, architecture and education in the USA, it was preceded by European and American modernists in the United States before 1933. Albert Kahn's River Rouge Glass Plant for the Ford Company in Dearborn, Michigan (1924–28), Richard Neutra's Lovell Health House in Los Angeles (1927–29) and the Philadelphia Savings Fund Society Building by George Howe and William Lescaze* (1929–32) are only a few examples. The machine aesthetic pursued in the Bauhaus workshops after 1922 reverberated in the New York *Machine-Age Exposition* of 1927, which confronted the American public for the first time with the concept that the machine could provide a formal aesthetic code.[9]

The transfer of ideas was not a one-way street. When the work of FRANK LLOYD WRIGHT was published in the famous 1910 Berlin *Wasmuth Portfolio*, it left a permanent impression on Mies and Gropius. Early in the Bauhaus existence, publications, exhibitions, correspondence, lectures, travel and visits in both directions ensured a remarkable exchange between both sides of the Atlantic.

In America, interest in the Bauhaus awoke almost immediately after its foundation in 1919. Throughout the Twenties much of the attention focused on architecture. In light of Gropius's founding manifesto, and his first footprint as an architect on American soil with an entry in the 1923 *Chicago Tribune* competition, this did not come as a surprise. Outside attention was further channelled to architecture when, in 1925, faculty and students designed and built the physical manifestation of their programme: the well-publicized Dessau Bauhaus Building. All three directors of the Bauhaus – Gropius, Meyer and Mies – were architects, the first and third already with stellar reputations before their appointments. American books and periodicals also featured their work.

The dominance of architecture within the professional discourse prevailed into the early Thirties even if, years earlier, American correspondents in Berlin had reported on the German art scene and introduced visual artists of the Bauhaus, notably Paul Klee*, Wassily Kandinsky* and Lyonel Feininger, to American audiences. The collector Katherine S. Dreier, who co-founded the Société Anonyme art organization, organized solo exhibitions in America for Klee and Kandinsky as early as 1923 and 1924 respectively, and around the same time the art dealer Emmy Esther (Galka) Scheyer began promoting the artists' group 'The Blue Four', which included Bauhaus painters. However, they were mainly discussed as artists in their own right rather than as Bauhaus faculty. A change in perception occurred with the very first Bauhaus exhibition in America when the Harvard Society of Contemporary Art, in 1930, presented the Bauhaus as an integrated design school, with works by Feininger, Johannes Itten*, Kandinsky, Klee, Oskar Schlemmer* and Lothar Schreyer. The Germanic Museum at Harvard continued showing Bauhaus protagonists, including Josef Albers (1936), Kandinsky (1937) and Feininger (1938), as well as Klee and Mies (1940).[10]

This more authentic picture of the Bauhaus came further to the surface when, in 1933, Josef and Anni Albers were the first Bauhaus faculty members to emigrate to the United States to teach at the newly established Black Mountain College in North Carolina. Josef Albers was one of the most versatile and experienced protagonists of the school – not an architect but an artist, photographer, designer, poet and theorist. As a pedagogue he had re-defined the so-called *Vorkurs* – a mandatory preliminary course at the Bauhaus that would later be adopted worldwide.

By the end of the Thirties, the couple were able to present a remarkable oeuvre 'made in America' and became prominently exhibited. Anni Albers's work included wall hangings, tapestries, a series of necklaces made of industrial materials, and bobby pins strung on whatever-at-hand ribbons (*c.* 1940). Josef Albers's resume in the visual arts extended to a leading role at Black Mountain College. He designed its seal and library bookplate (1935), as well as a desk and a chair in leather

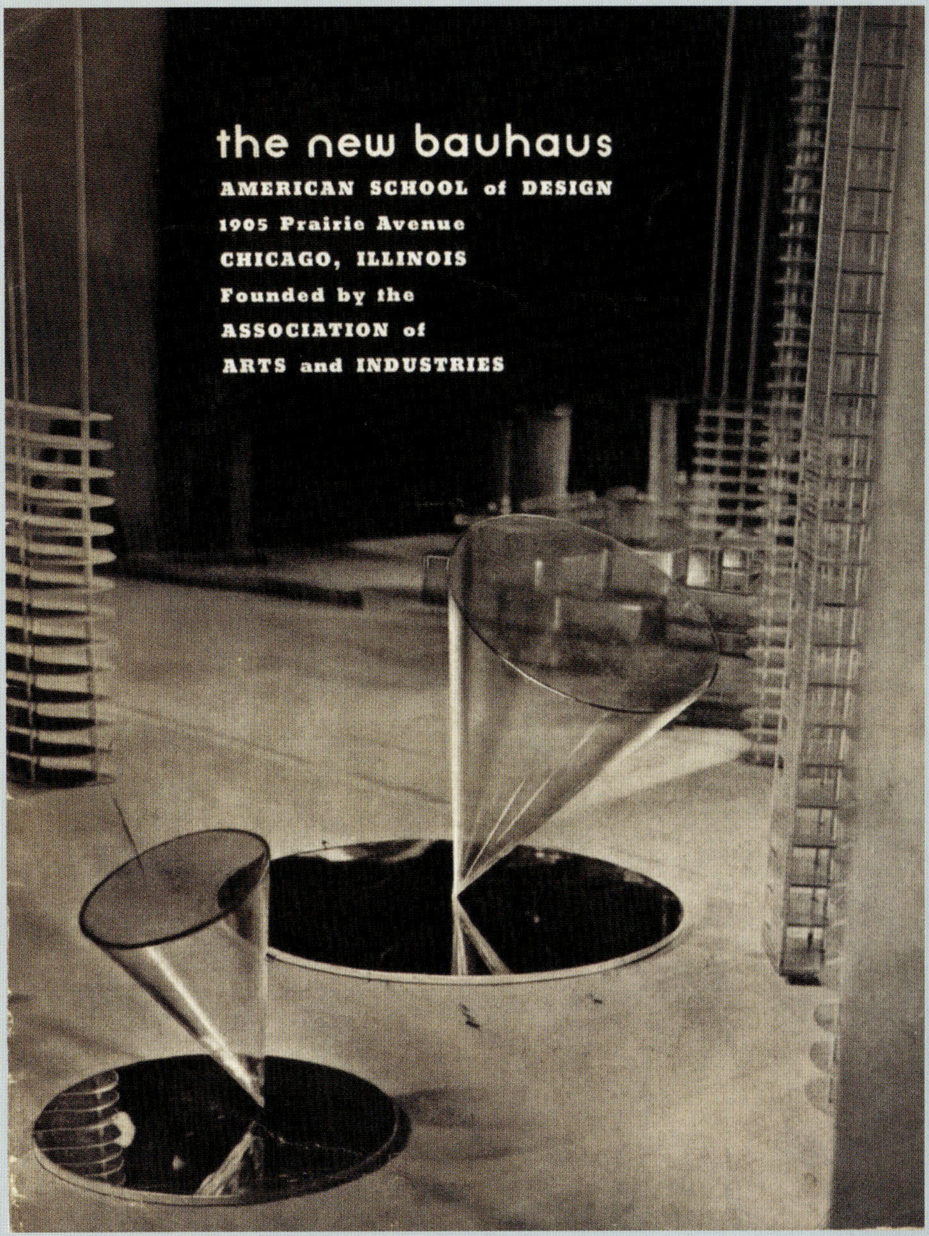

ABOVE The cover of the first school prospectus for the New Bauhaus, designed by László Moholy-Nagy, 1937.

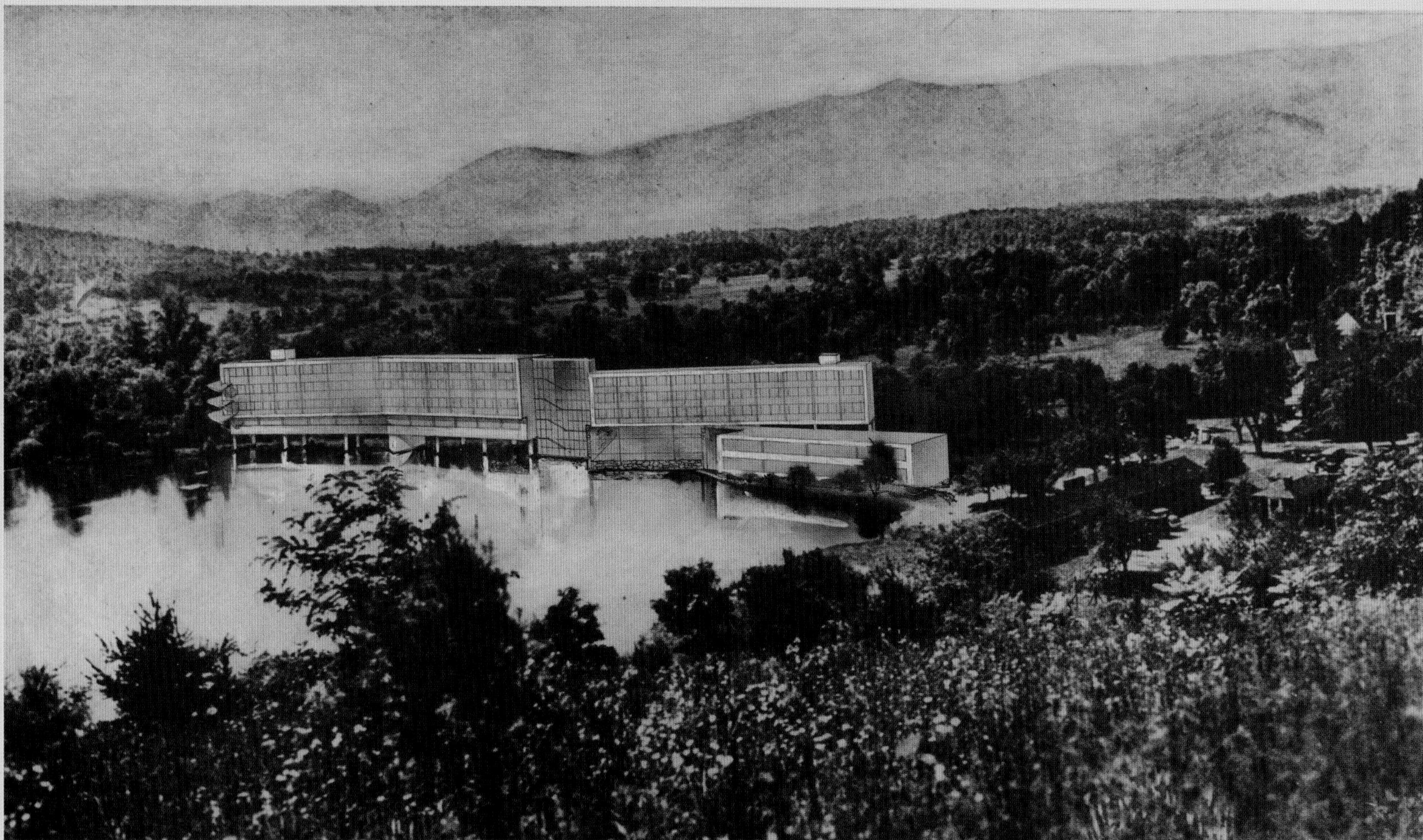

ABOVE A drawing of Black Mountain College, Lake Eden, North Carolina, 1938–39, with the school seen across the valley (photographic reproduction with collage, black ink, white gouache and graphite on white wove paper).

and wood (*c.* 1940). Through his artwork he investigated the nature and behaviour of colour and geometric linear constructs, foreshadowing his life-long methodological studies into the relativity of human visual perception.

Albers invited other Bauhaus artists to the college. In 1939, Gropius and his business partner Marcel Breuer were commissioned to design its new campus. However, under the cloud of World War II and German-American tensions, fundraising efforts fell short. The project did not materialize; only building plans and a model reflect what could have become a most spectacular manifestation of Bauhaus architecture and design in the South.

In the American reception throughout the Twenties and Thirties, the Museum of Modern Art in New York played a key role. When founded in 1929, its director, Alfred Barr Jr., established the remit to study, collect and exhibit modern art. A department of architecture was added and led by his Harvard friend Philip Johnson*. Barr and Johnson travelled to Europe and visited the Bauhaus. As early as 1927 they had explored the Weissenhof* exhibition in Stuttgart, a display of avant-garde architecture, planned and built under Mies's artistic direction.

In 1932, the two men, along with the historian Henry-Russell Hitchcock, distilled their observations of modern architecture in the formula of the 'International Style'*. The *Modern Architecture: International Exhibition* brought the Bauhaus and other manifestations of avant-garde architecture to the public stage, along with an accompanying catalogue and book. During its six-week run in New York, followed by a two-year tour, the *International Exhibition* reached enough of an audience to have a lasting impact.

For the aftermath of the Bauhaus this meant on one hand a critical introduction to the North American professional world, but on the other a fateful reduction of its ideas to a 'Style' that described commonalities among the architectural avant-gardes at the cost of their individual aesthetic principles and social concerns.

The Bauhaus as a comprehensive new type of art and design school made it to centre stage when in 1938 the Museum of Modern Art displayed a Bauhaus exhibition in its initial quarters in Rockefeller Center. While the 1930 Harvard exhibition had travelled to New York and Chicago, the MoMA show reached a much wider audience, which meant a major boost in the awareness of the Bauhaus. The exhibition and catalogue were both supervised by Walter Gropius and showed representations of the Bauhaus workshops, typography, architecture, painting, photographs and models of the Dessau Bauhaus building, tubular steel furniture, metalwork, and ceramics.[11] Unfortunately, the event also laid the foundation for long-lasting misconceptions about the school by limiting its scope to the Gropius years.

By the late Thirties, an increasing American presence of Bauhaus art, design, and architecture reflected the exodus of creative power from Nazi Germany. The country's blood loss of its scientific, artistic and intellectual elite meant a gain for the United States, even if only few refugees and émigrés were lucky enough to pick up the threads of their lives where they had left off. Gropius was one of them. In 1937, he was appointed chair of architecture at Harvard University. He brought his protégé Marcel Breuer and both were soon given the opportunity to build their own modern houses in Lincoln, Massachusetts. Gropius sought work and positions for other Bauhaus colleagues as well, among them HERBERT BAYER and LÁSZLÓ MOHOLY-NAGY.

In 1937, Moholy-Nagy came to Chicago to direct the New Bauhaus school, anticipating a collaboration between art and industry, and thus closely following the spirit of the original Bauhaus. In what was considered the most authentic attempt to revive the Bauhaus as an institution, the faculty sought new frontiers, in particular in the development of photography and the methodology of related education. Moholy-Nagy believed that photography could convey a 'new vision' of seeing objects beyond the physical capacities of the human eye. Former Bauhaus students Hin Bredendieck and Marli Ehrman were among the teachers.

The same year, through the initiative of Alfred Barr Jr., Mies also entered the scene of the Bauhaus and America. While Barr failed in his attempt to get Mies the commission for the new quarters of the Museum of Modern Art, he suggested the architect for a summer house in Jackson Hole, Wyoming, commissioned by a MoMA trustee, Mrs Stanley Resor. This project was Mies's first design for an American site. In 1938, Mies was appointed dean of the Armour Institute, now Illinois Institute of Technology. Unlike Gropius and Moholy-Nagy, Mies did not intend an institutional revival of the Bauhaus but rather focused on his work as an architect and investigations 'into the tectonic expression of modern technology that had been initiated by the Chicago School'.[12] Early into his tenure, the Art Institute of Chicago showed a small selection of his works, photographs and drawings, and models of three early houses. His first large-scale design opportunity came in 1939 with the master plan for the entire Armour Institute campus.

Mies's first realized work in North America came about in New York when Philip Johnson hired him to design the interior of his one-bedroom apartment in midtown Manhattan, aspiring not only to bring the architect's exquisite design to his own living environment but also to use the apartment as a showcase of the aesthetics and thought as practised during Mies's directorship at the Bauhaus. In collaboration with his partner LILLY REICH, Mies created a multifunctional space with supports of chrome steel, luxurious materials such as raw silk for floor-to-ceiling curtains, exotic wood for equally high bookcases, white vellum and tubular steel for chairs.[13]

The fact that the Bauhaus was the only school worldwide to successfully translate avant-garde visions into education helped it find fertile ground in the United States. With its advanced technology and the rational, sober, geometric language of the machine ingrained in its built environment, the country was a natural harbor for the Bauhaus. Unlike the German writers and playwrights who lost their creative medium in foreign-speaking lands, the leading architects and designers of the Bauhaus found receptive communities in which their ideas and careers could thrive.

The death of the Bauhaus as a beacon of modernism was without doubt the intent of the National Socialists when they harassed the school to the point where the faculty finally called it quits. What then looked like the closing chapter and defeat for the Bauhaus ideal ironically turned around in the revival of its ideas throughout the world, particularly in America.

At the time of Breuer's statement of last rites, it was hardly conceivable that the Bauhaus would arguably become the most influential art, design and architecture school of the 20th century. In 1948, Charles Kuhn, then the director of Harvard's Busch-Reisinger Museum, stated: 'The Bauhaus today is still a living force and has incalculable effect on the architecture, industrial design, and artistic education in America.'[14] Four years later, *The New York Times* confirmed his assessment: 'The Bauhaus never died. Its principles have been disseminated in other centers of learning.'[15] How is this possible if, as a school, the original Bauhaus could not be replicated? In Mies's words, 'The Bauhaus was not an institution with a clear program – it was an idea, and Gropius formulated this idea with great precision.... The fact that it was an idea, I think, is the cause of this enormous influence the Bauhaus had on every progressive school around the globe. You cannot do that with organization, you cannot do that with propaganda. Only an idea spreads so far.'[16]

BELOW The study in Philip Johnsons's apartment at Southgate 424 East 52nd Street, New York, c. 1930, designed by Ludwig Mies van der Rohe and Lilly Reich.

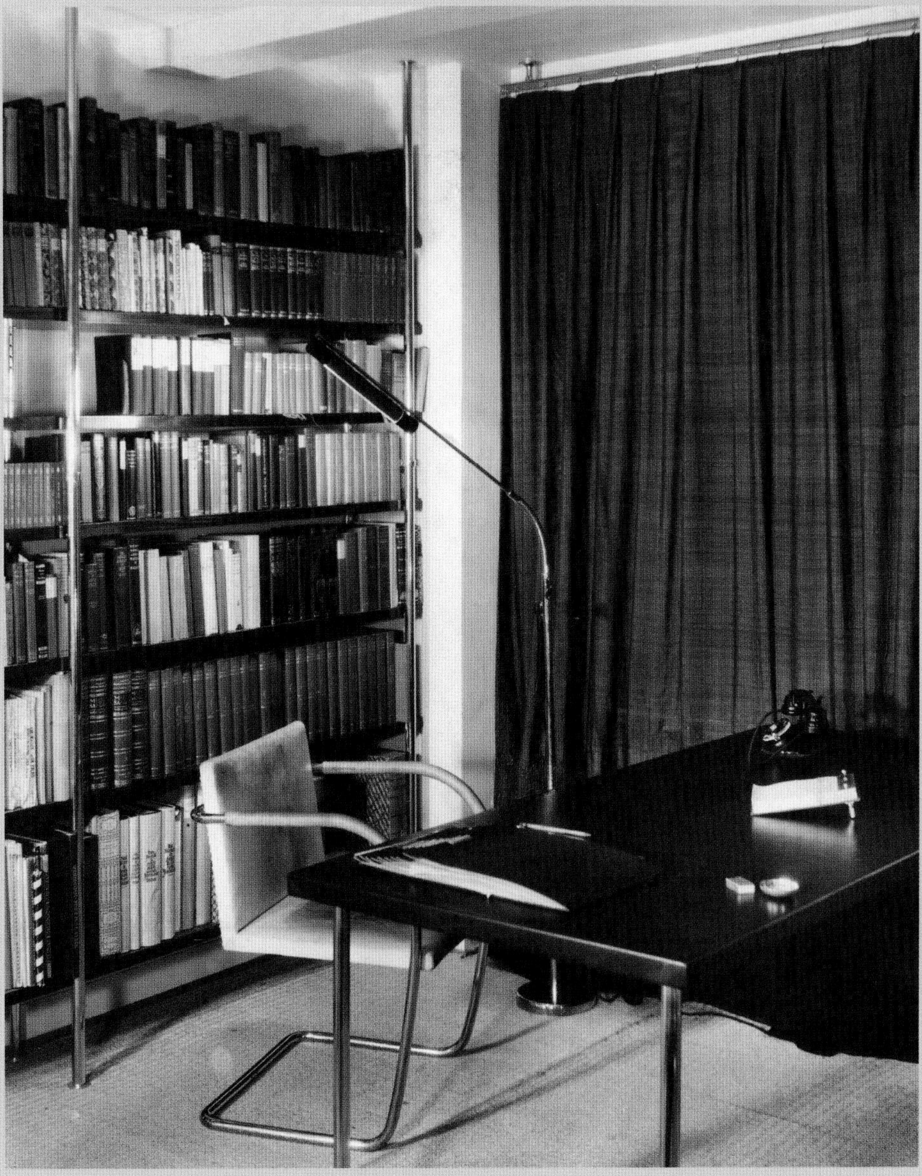

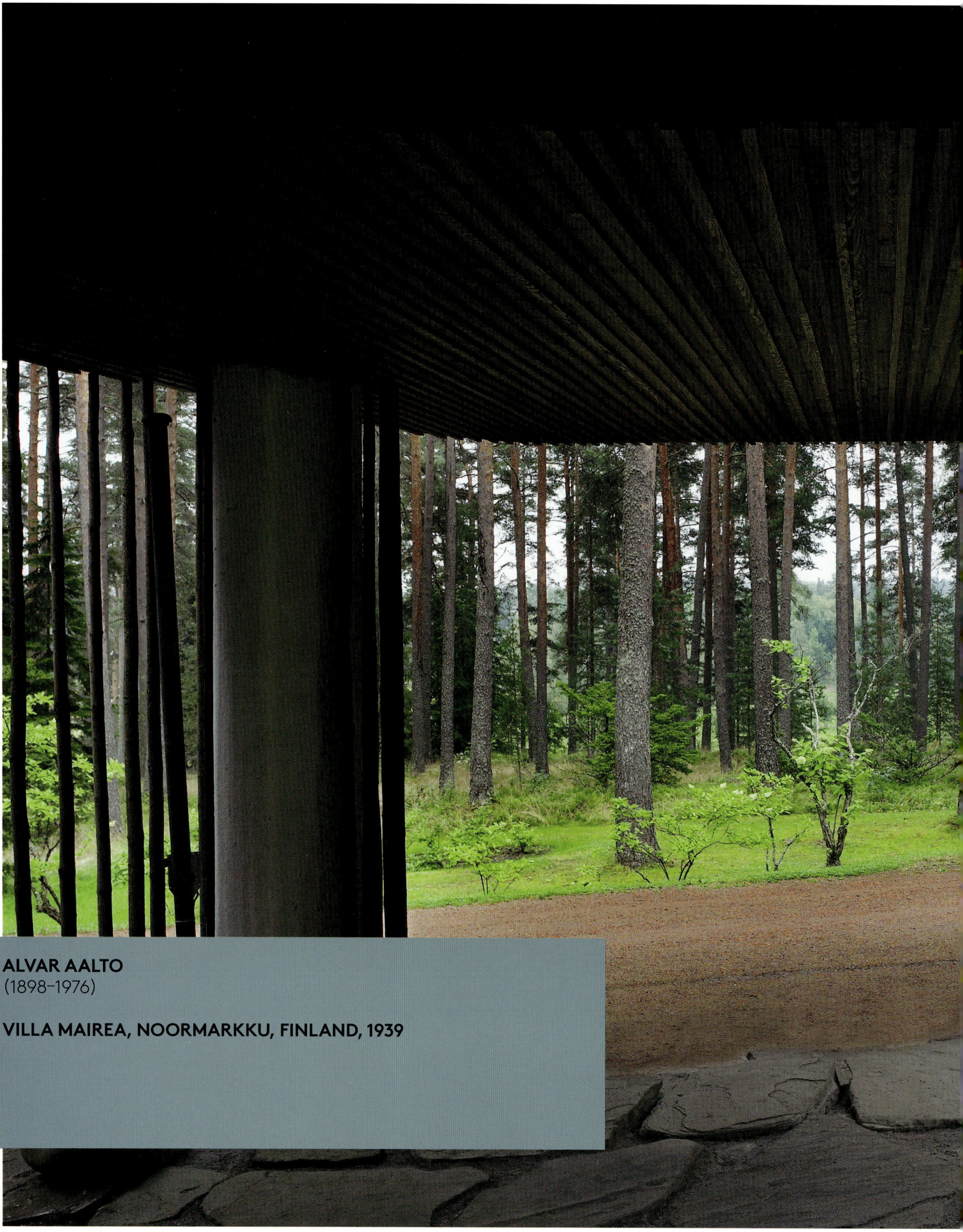

ALVAR AALTO
(1898–1976)

VILLA MAIREA, NOORMARKKU, FINLAND, 1939

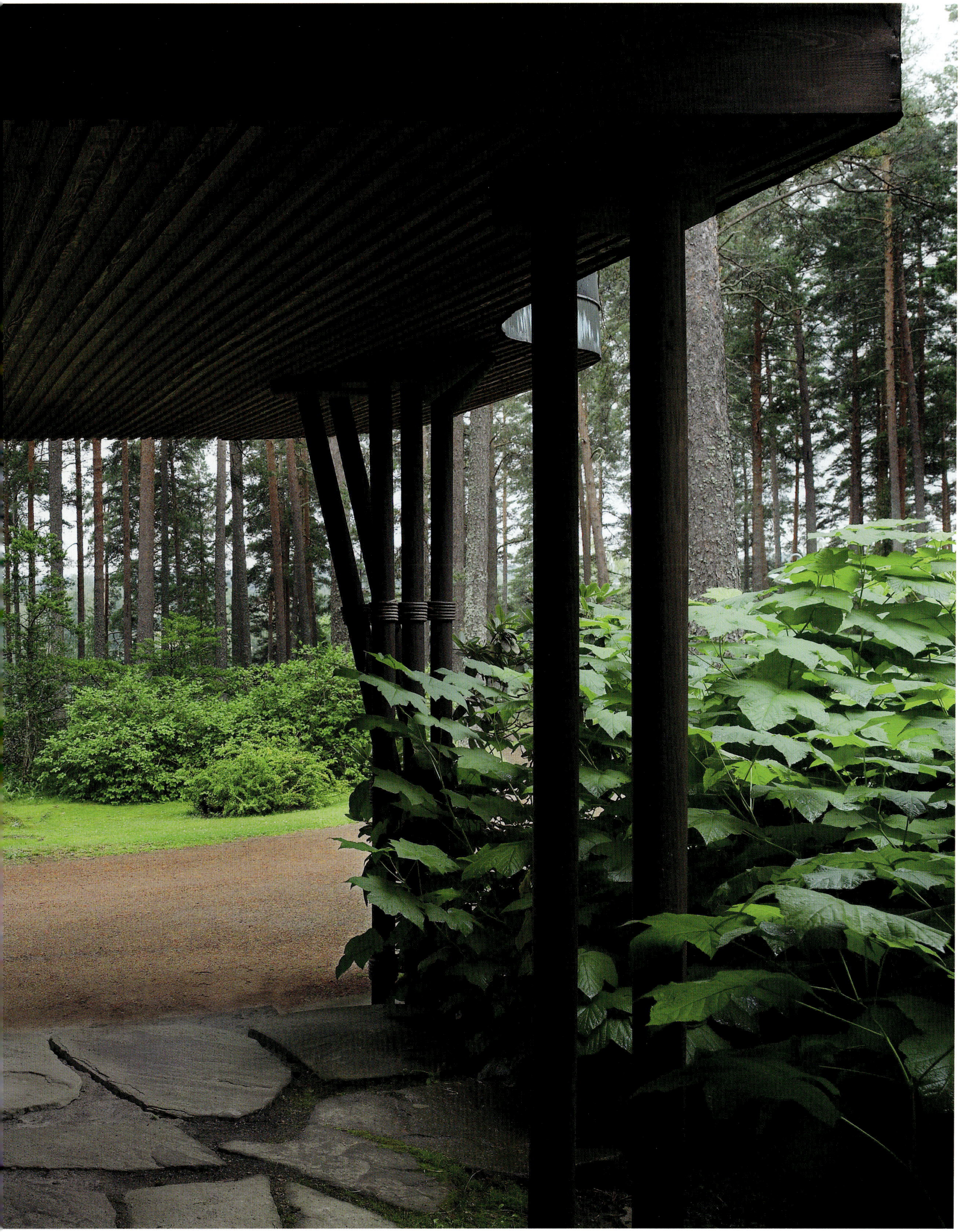

Finnish architect and designer Alvar Aalto was one of the greatest and most innovative exponents of 'soft modernism'. His work fused a pioneering approach to the tenets of modernism and its fresh ways of living with a deep-rooted love of organic and expressive natural materials. This fusion can be seen at its best in the form of Villa Mairea, one of Aalto's most rounded and delightful achievements. It is, typically, a total work of art, showing Aalto's hand in the architecture, interiors, furniture and landscaping.

Alvar Aalto was born in the small Finnish town of Kuortane in the rural Southern Ostrobothnia region. His father was a land surveyor, and the young Aalto often accompanied him on his working trips during the summer months. After the death of Aalto's mother, when Aalto was only five, his father remarried and the family moved to the city of Jyväskylä. Having shown promise as an artist at an early age, Aalto studied architecture at Helsinki's Institute of Technology, before establishing his own practice in the city in 1923. He married fellow architect Aino Marsio, who collaborated on many of Aalto's commissions, a year later.

Early projects were influenced by Scandinavian neoclassicism, while Finnish vernacular was also an important source of inspiration. But by the late Twenties Aalto was also looking much further afield, travelling throughout Europe and becoming familiar with the work of the early modernists. One of his key early projects was the Paimio Sanitorium, completed in 1933. This exhibited Aalto's unique version of warm modernism, which placed humanism at its heart, within a design approach that was all-encompassing.

Villa Mairea, completed six years later, also exhibited Aalto's trademark consideration, from the sinuous door handles to the kidney-shaped swimming pool. The house was the special fruit of a highly productive friendship with two of Aalto's most important and enlightened clients: Maire Gullichsen, after whom the house is named, and her husband Harry. Maire was the daughter of the Finnish industrialist Walter Ahlström, who had built up a vast timber and forestry business, and had a particular love of art, painting and architecture. Her husband became head of the Ahlström company in his late twenties after the untimely death of Maire's father. The Gullichsens shared progressive ideas and a love of culture, and they became key patrons for Aalto.

In 1935, together with the art critic Nils-Gustav Hahl, Maire Gullichsen and Alvar and Aino Aalto co-founded Artek, the furniture company that began producing the designs – mostly in timber and plywood – that had emerged from Aalto's various commissions, including Paimio. Not long after, they began discussing the idea of a summer house on the Ahlström estate at Noormarkku, which was already home to a saw mill and two earlier family villas sitting among the woods.

Aalto was given a generous level of freedom by his clients, although their requests for an art gallery, library and other key spaces led to a floor plan that was initially compartmentalized and relatively traditional in character. This more formal arrangement was soon abandoned in favour of a large, open-plan living space at the heart of the house, where artworks would be displayed more openly. Harry Gullichsen's insistence upon a separate library, where he could hold private meetings, resulted in a distinct space to one side of the living area where the bookshelves serve as partition walls. The large dining room lies beyond the living area, connecting to a service wing, while Maire Gullichsen had a studio of her own at first-floor level, floating above a winter garden room.

For such a substantial building, Villa Mairea manages to retain a level of modesty, partly due to the way that the house is sensitively sited within the forest. Here, context is all-important and the strong relationship with the surroundings is maximized throughout. This is expressed not only in the way that windows and apertures frame views of the forest, but in repeated echoes of the woods within the house itself. Timber (pine, hornbeam, teak) is used extensively for the interiors, along with slate, brick and ceramic tiles, while supporting pillars are wrapped in rattan mirroring the tree trunks in the woods beyond.

Most striking throughout is the crafted, organic quality of the spaces, which softens the modernity of the building, together with the more sinuous elements that subvert the linear plan. There are echoes of Wright's Fallingwater here, but the Villa Mairea is a true original, infused with Aalto's experimental vision and the spirit of the Scandinavian countryside.

PAGES 308–9 The projecting porch that shelters the entrance to Villa Mairea is a transitional zone between inside and out, which held particular importance for Alvar Aalto.

OPPOSITE The collection of slim columns that support the sinuous, floating roof canopy of the porch echo the trunks of the pine trees in the surrounding woodlands.

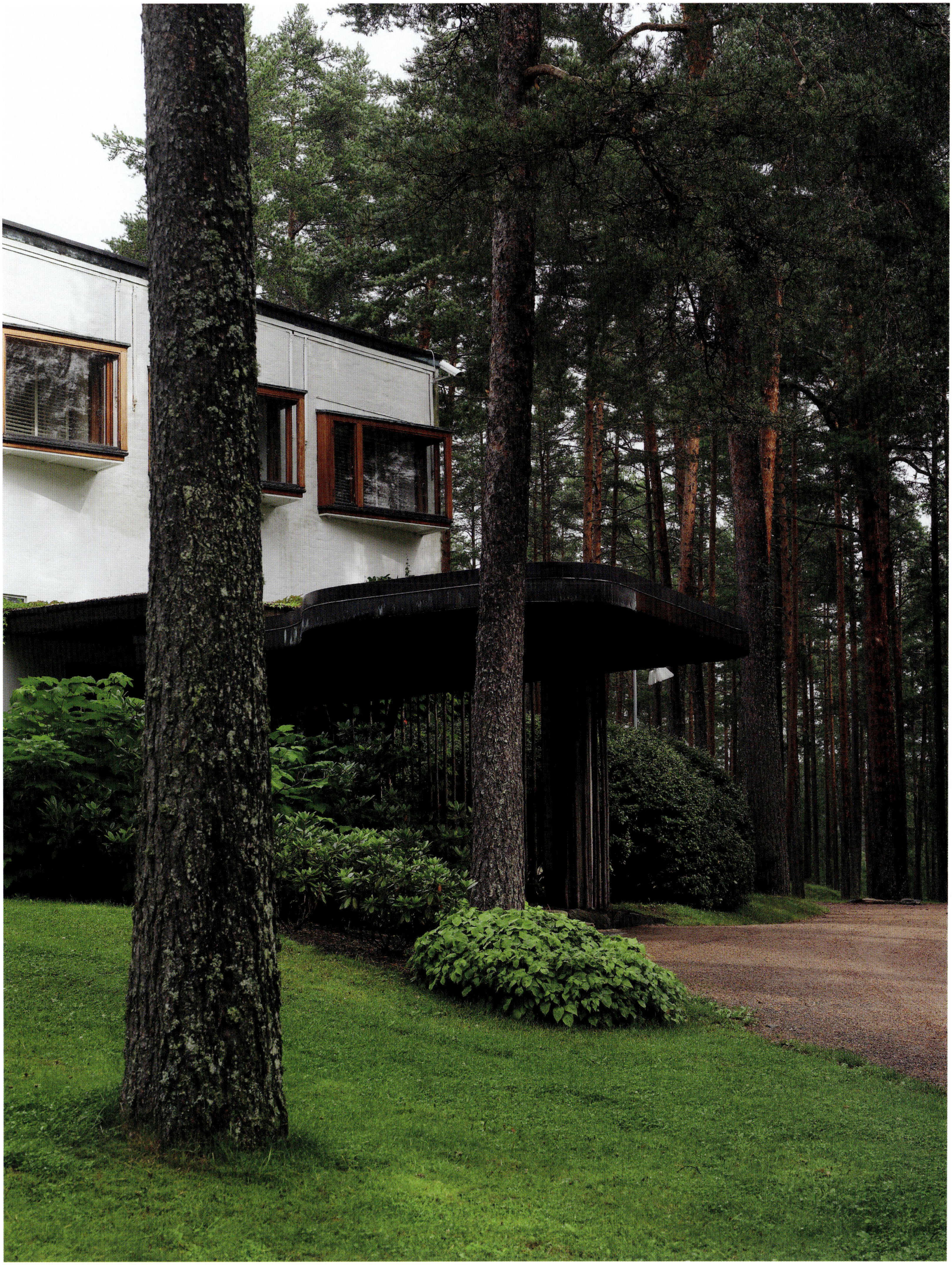

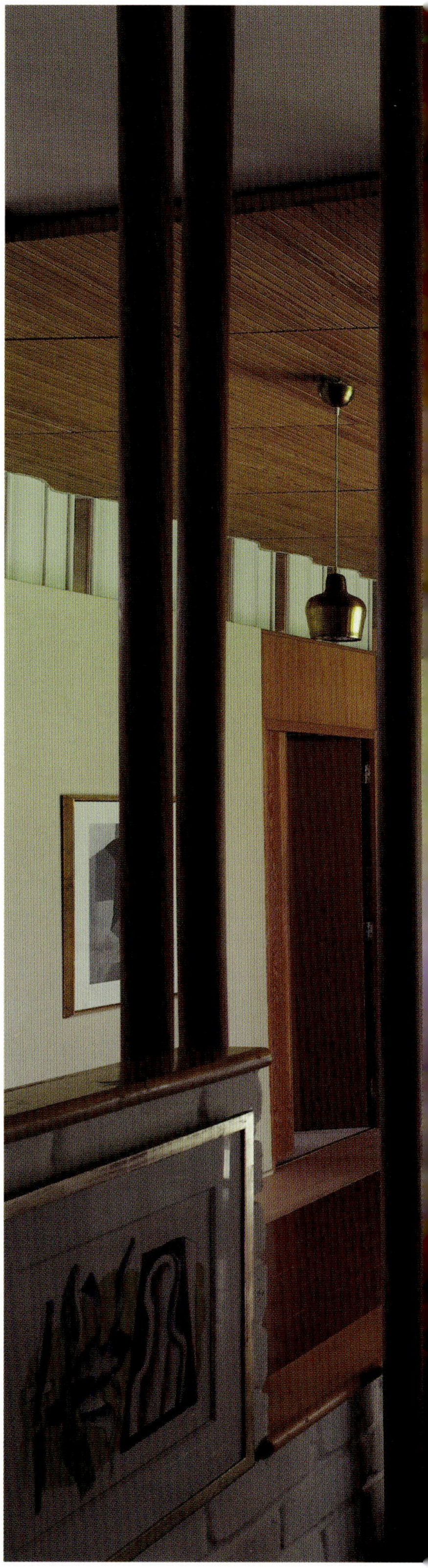

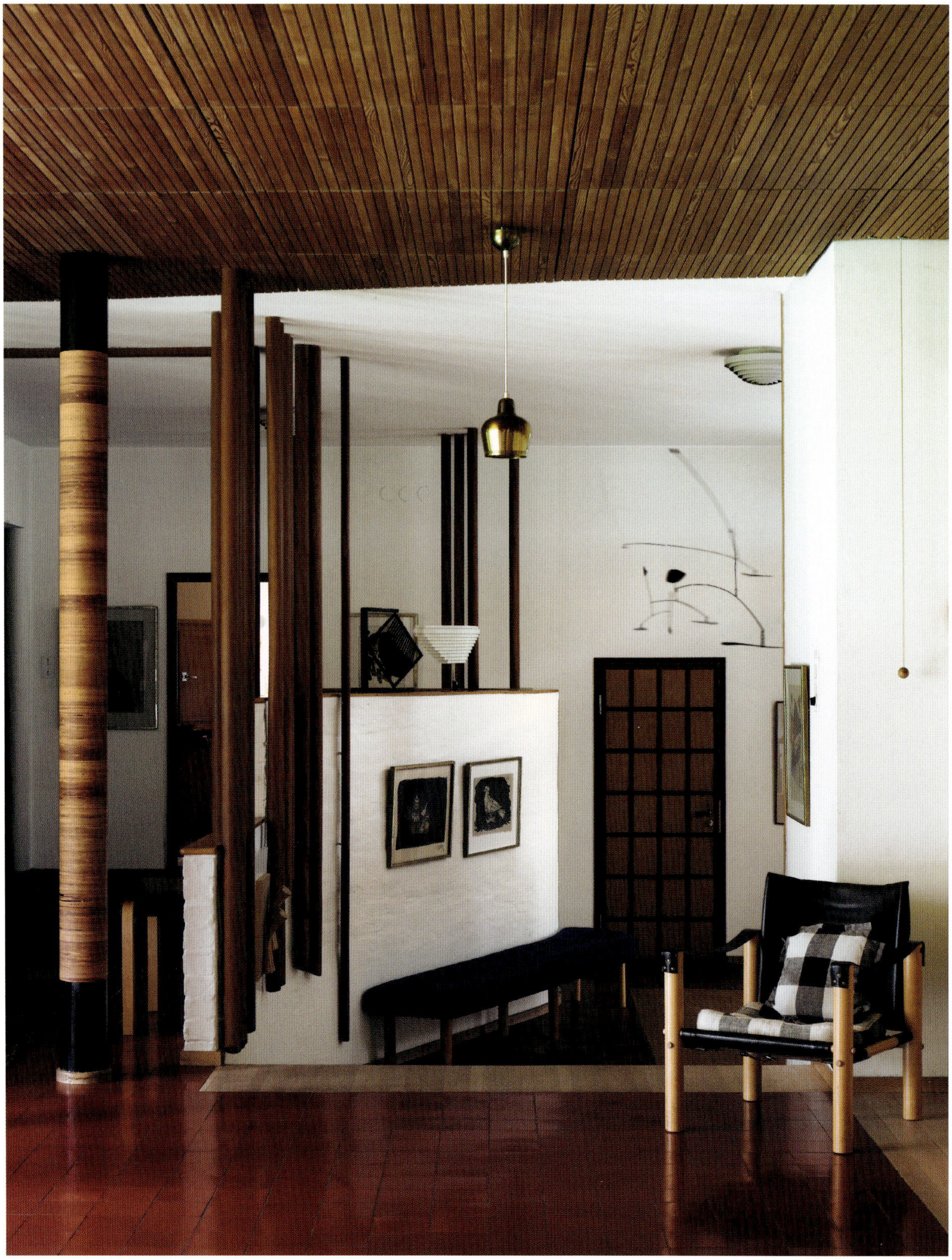

ABOVE The symbolism of the forest continues inside the house, where the slender steel columns used as both supports and screens are wrapped in raffia or birch slats; the ceiling is composed of thin pinewood slats.

LEFT The organic qualities of the house are evident in the entrance hallway, which leads directly into the spacious central sitting room, with a corner fireplace.

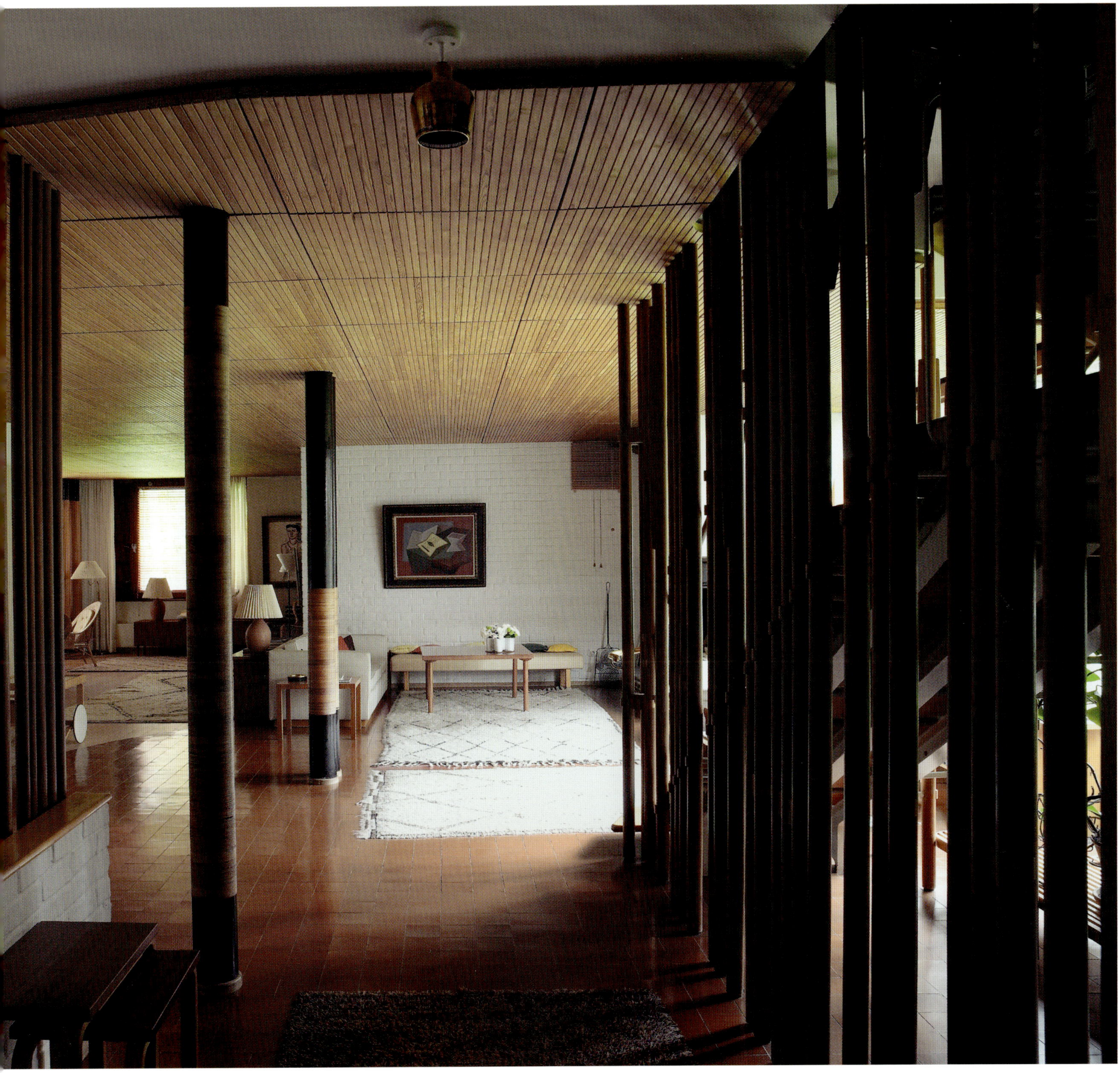

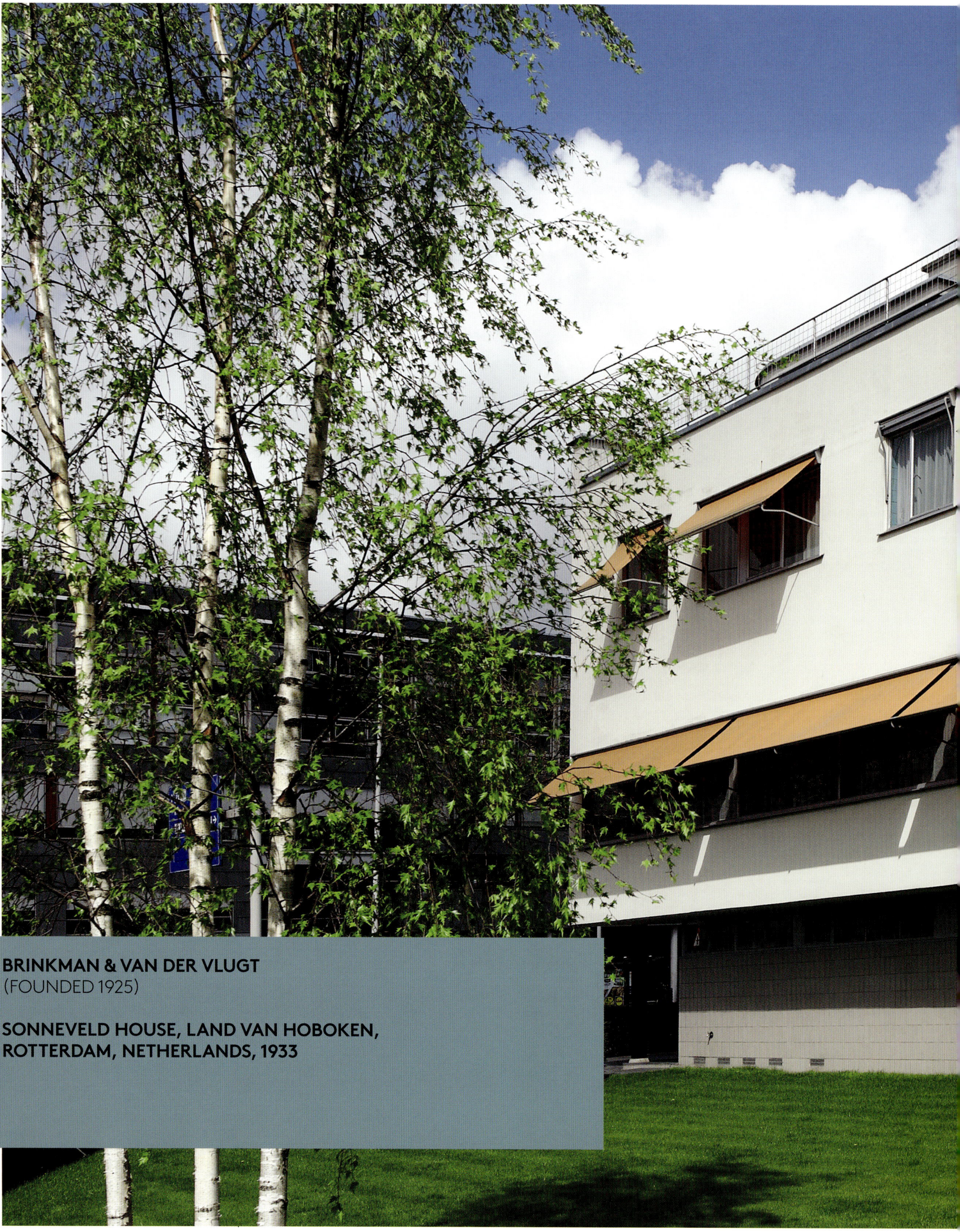

BRINKMAN & VAN DER VLUGT
(FOUNDED 1925)

SONNEVELD HOUSE, LAND VAN HOBOKEN, ROTTERDAM, NETHERLANDS, 1933

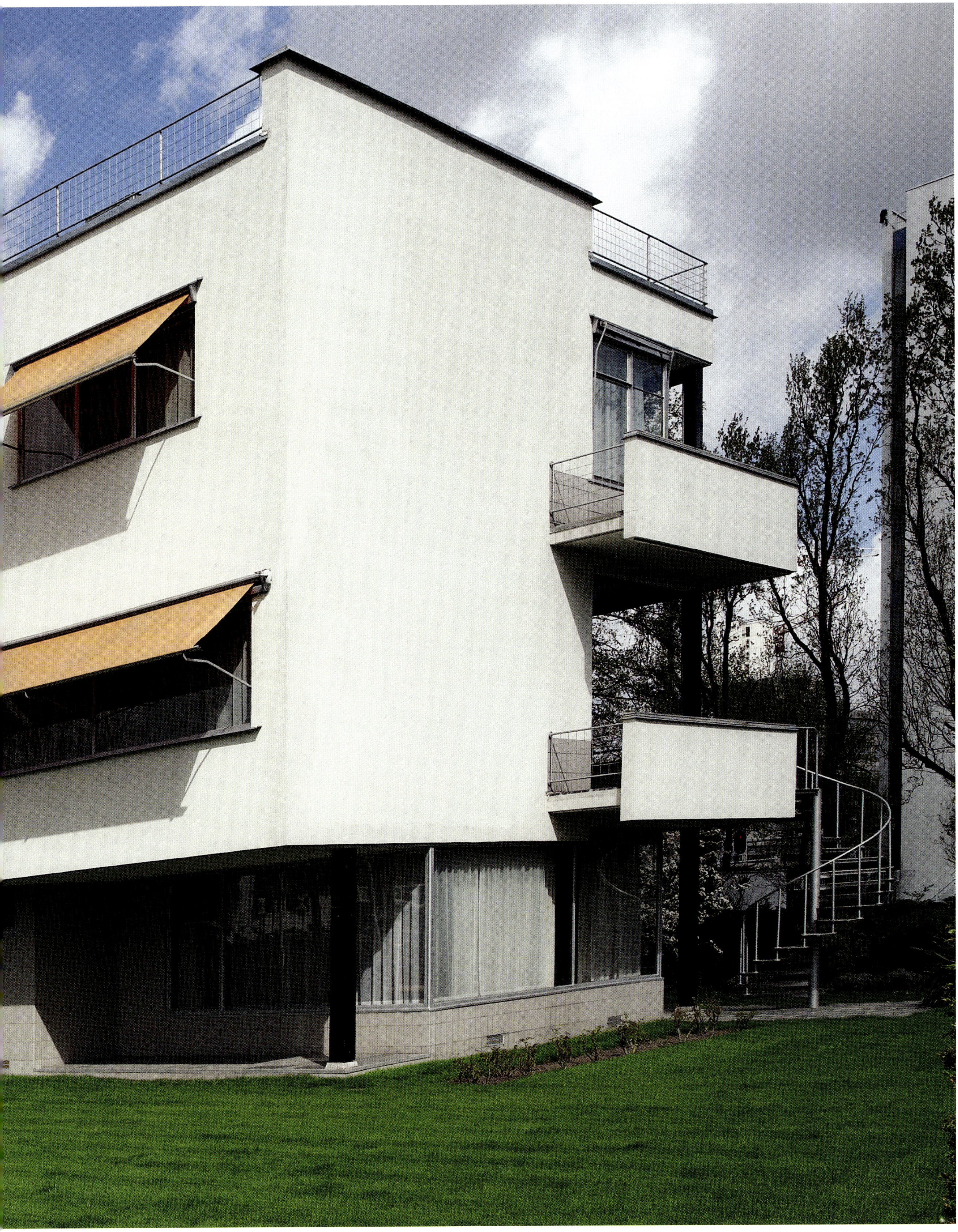

In an age before air travel, the great ocean liners offered the fastest and most graceful mode of international passage. During the inter-war years, liners became ever grander and more luxurious as the appetite for both tourism and trade increased. These vast ocean-going ships became symbols of modernity in themselves and emblematic of a new era of adventure. It is no wonder, then, that the visual design language of the liners slowly began to infuse the world of residential and hotel design. Countless marine motifs infiltrated the modernist home, from porthole windows, and decks and terraces, to cabin-like bedrooms with a wealth of fitted furniture.

The Dutch businessman Albertus Sonneveld was one of a new generation of self-made men who were regular passengers on the transatlantic liners. As the director of the tobacco division of the Van Nelle conglomerate, Sonneveld made trips to America each year, and he grew to appreciate the design and luxury of the ships and hotels that he experienced first-hand.

His interests were shared by his wife, Gésine Sonneveld-Bos. Together, they decided to commission a house for themselves and their two daughters, Puck and Gé (Magdalena and Gésine), on a leafy plot of land devoted to a new development of villas on the edge of the Land van Hoboken estate in Rotterdam. They turned to architect Leendert van der Vlugt (1894–1936) of Brinkman & Van der Vlugt to design a bespoke and distinctly modern home in the Nieuwe Bouwen, or Dutch Modernist, style.

Van der Vlugt studied architecture at the Academy of Arts and Technical Sciences in Rotterdam, before joining Michiel Brinkman's practice in 1920. After his father's death Johannes Brinkman invited Van der Vlugt to become his business partner, and the firm worked on a

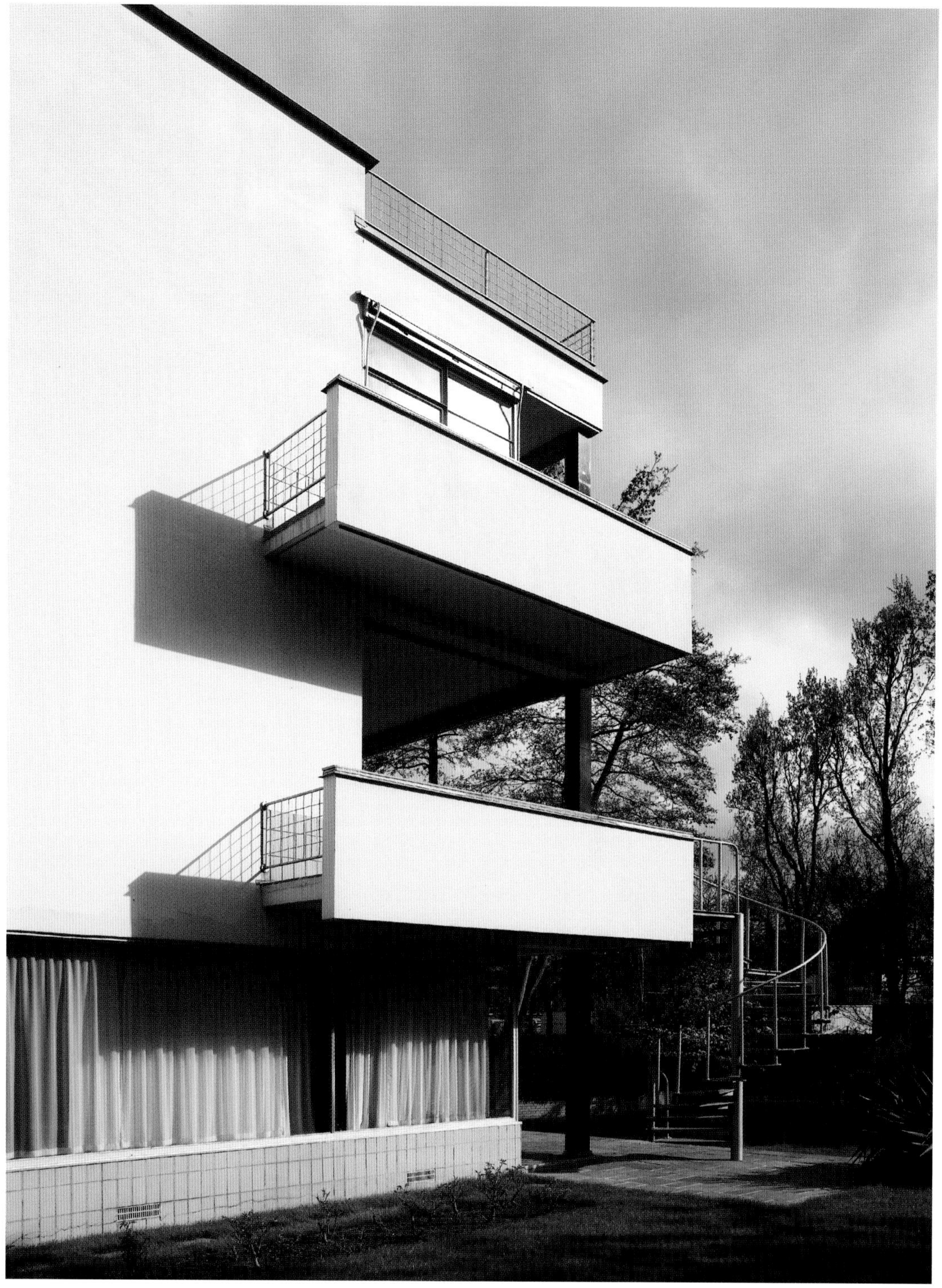

PAGES 314–15 The ground floor of this crisply rendered, three-storey house is largely devoted to the entrance area and service spaces; the main living spaces are at mid-level.

ABOVE & OPPOSITE The upper two storeys feature substantial recessed balconies, connected to the living room (mid-level) and master suite (upper level); an outdoor spiral staircase connects the lower of the two balconies to the garden.

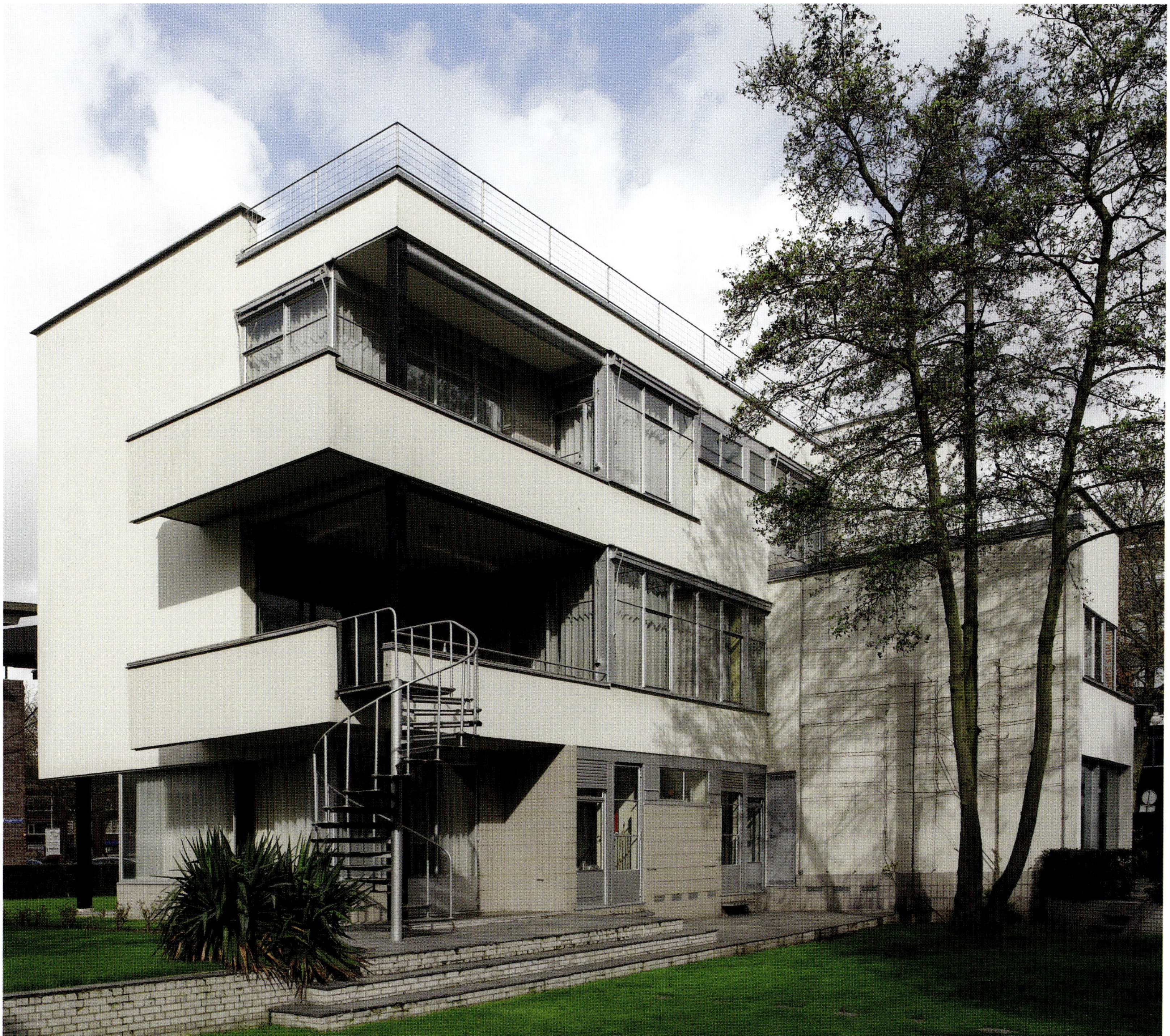

number of projects related to Van Nelle, including the Van Nelle Factory in Rotterdam (1930) and the Van der Leeuw House (1929) for another senior company executive.

Van der Vlugt designed the three-storey Sonneveld House according to the particular needs and desires of his clients, creating a bespoke home that embraced modernist thinking and new domestic technology of all kinds. The ground floor was designed with a large entrance hall and a studio for the children, as well as accommodation for two household servants. The principal living spaces were placed on the floor above, which featured an integrated balcony, with an external spiral staircase leading directly down to the landscaped garden. On the other side of the house there was a continuous ribbon window running almost the entire length of the building. The living area included a lounge, with a Steinway piano, a library and a dining room in a flexible open-plan arrangement, where a sliding leatherette wall could be used to separate library and lounge and curtains could be drawn across to lightly section off the dining zone. The kitchen was also at mid-level, while the family bedrooms and bathrooms were above, on the top floor.

The house of steel and concrete embraced many of the tenets of international modernist design: open-plan living, curtain walls, integrated decks and terraces, extensive horizontal windows and more. The Sonnevelds also wanted the latest home technology, including a functional modern kitchen, internal telephones and a system of call lights for attracting the attention of the servants. Much of the furniture, particularly in the bedrooms, is integrated, lending cohesion to the design but also saving space and lending the bedrooms a cabin-like quality.

The vivid use of colour throughout the interiors is also striking, contrasting with the snow-white exterior. Artist Bart van der Leck was consulted on colour choices, collaborating with Metz & Co. on the selection of curtain fabrics and upholstery. Much of the furniture – including the tubular steel chairs and beds – was designed by Willem Hendrik Gispen*, who was also responsible for much of the lighting, and another leading figure in the Nieuwe Bouwen movement.

The result is a house of great warmth and character on the one hand but a highly functional, practical and adaptable family home on the other. A careful restoration programme of the house, which is now open to the public, reveals both to the full.

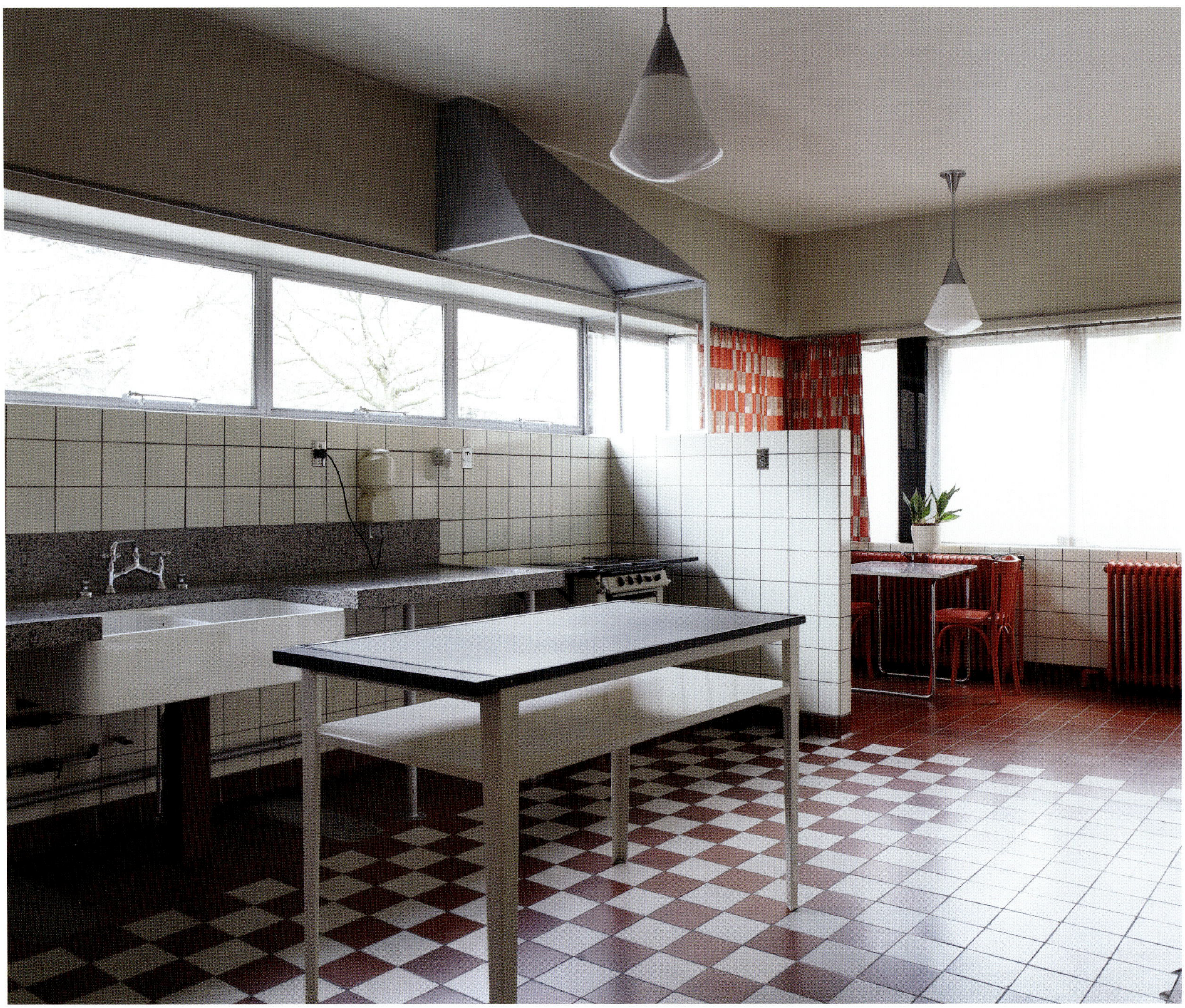

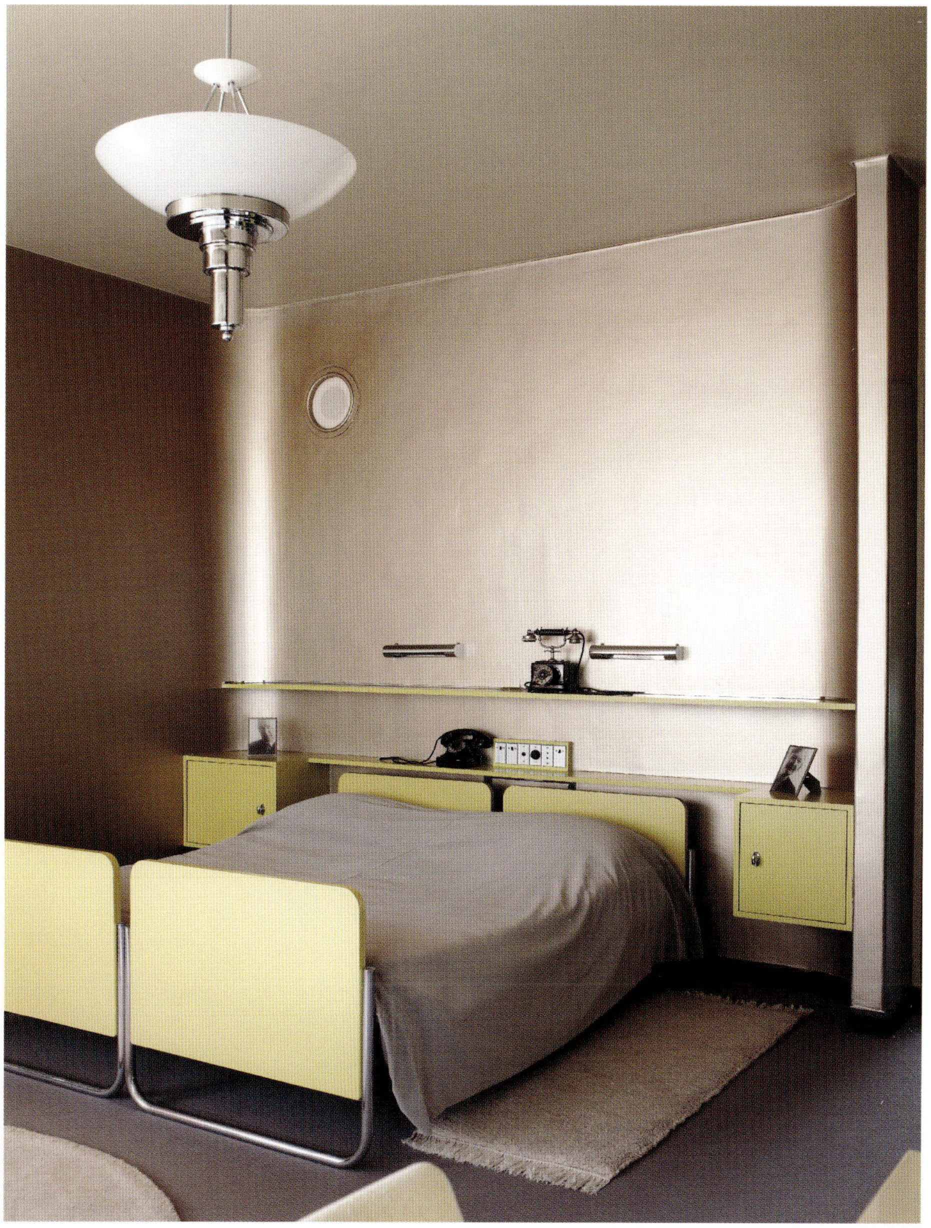

ABOVE The master suite features both a telephone with an outside line and an intercom telephone system; the bedroom is part of a substantial suite that also features a bathroom and dressing room.

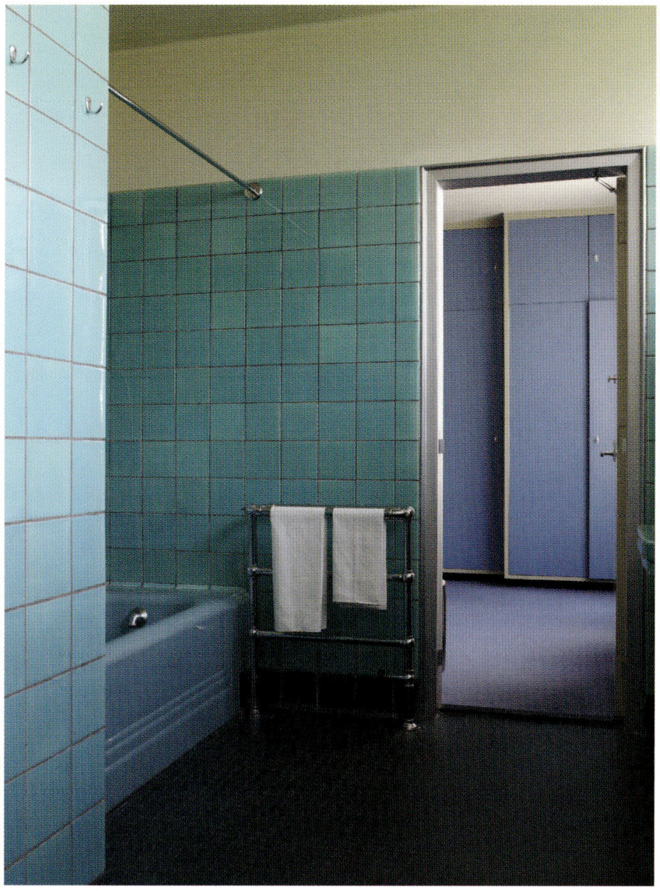

OPPOSITE The interiors of the house feature the extensive use of colour, in contrast with the crisp white façade; the kitchen is dominated by a combination of red and white, with the curtains around the breakfast area by Metz & Co., which supplied most of the fabrics used in the house.

ABOVE The bathrooms are both functional and colourful, with a suitably maritime turquoise blue used for the matching wall tiles and sanitary ware, which was made in Pittsburgh by the Standard Sanitary Manufacturing Company.

RIGHT The first-floor library sits at one end of a flexible living space, where a 'harmonica' wall in leatherette can be used to create separation from the main living room alongside; the library furniture is by Gispen.

BELOW A detail of the ground-floor studio, which featured an integrated bank of furniture holding a secret kitchenette, storage and a fitted sofa between them.

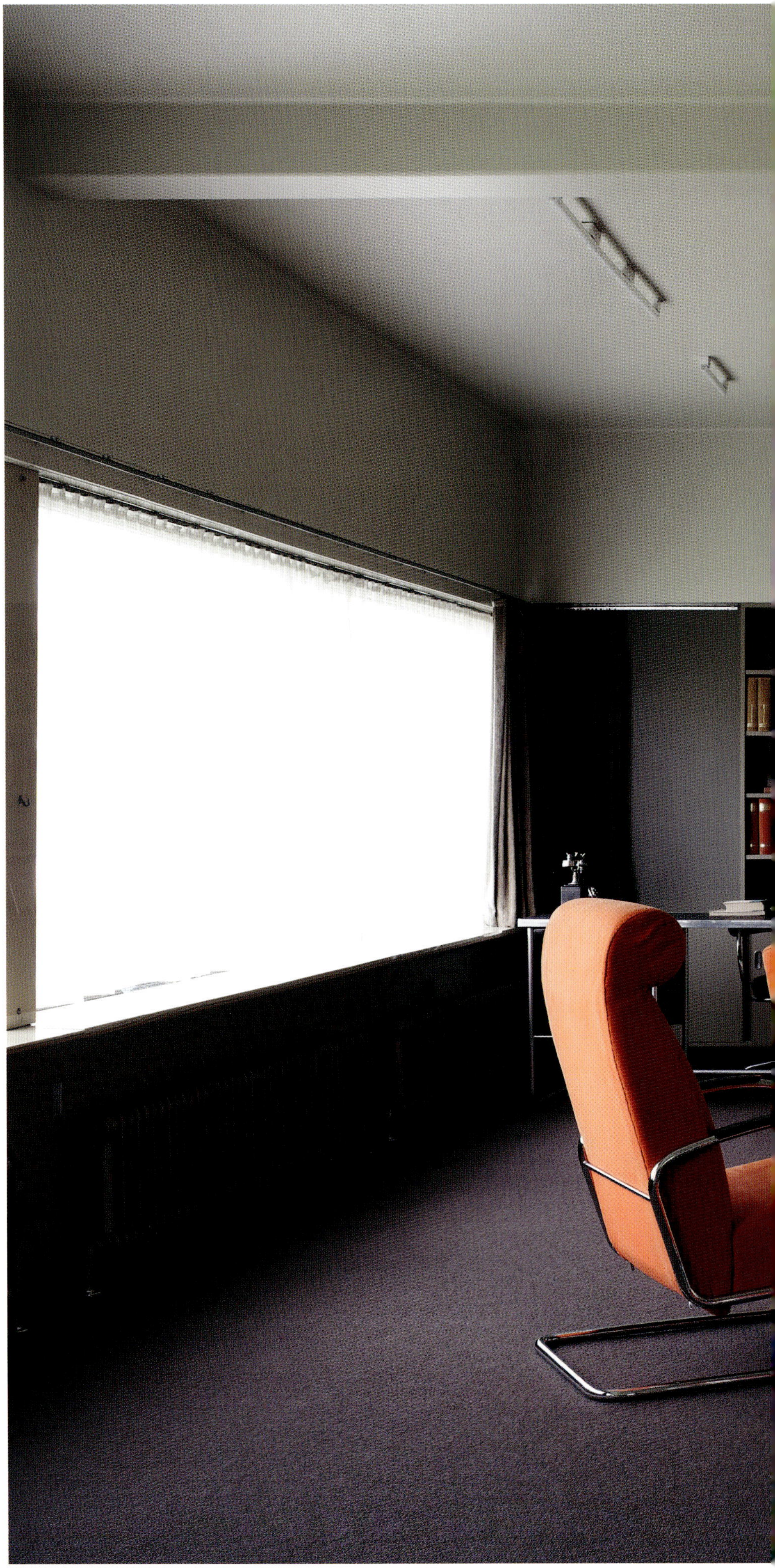

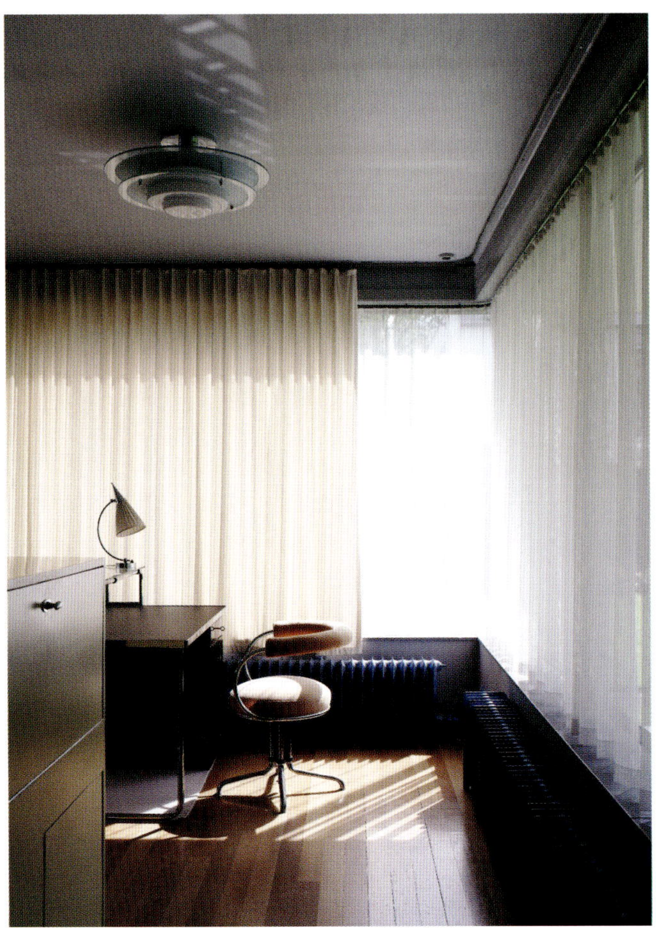

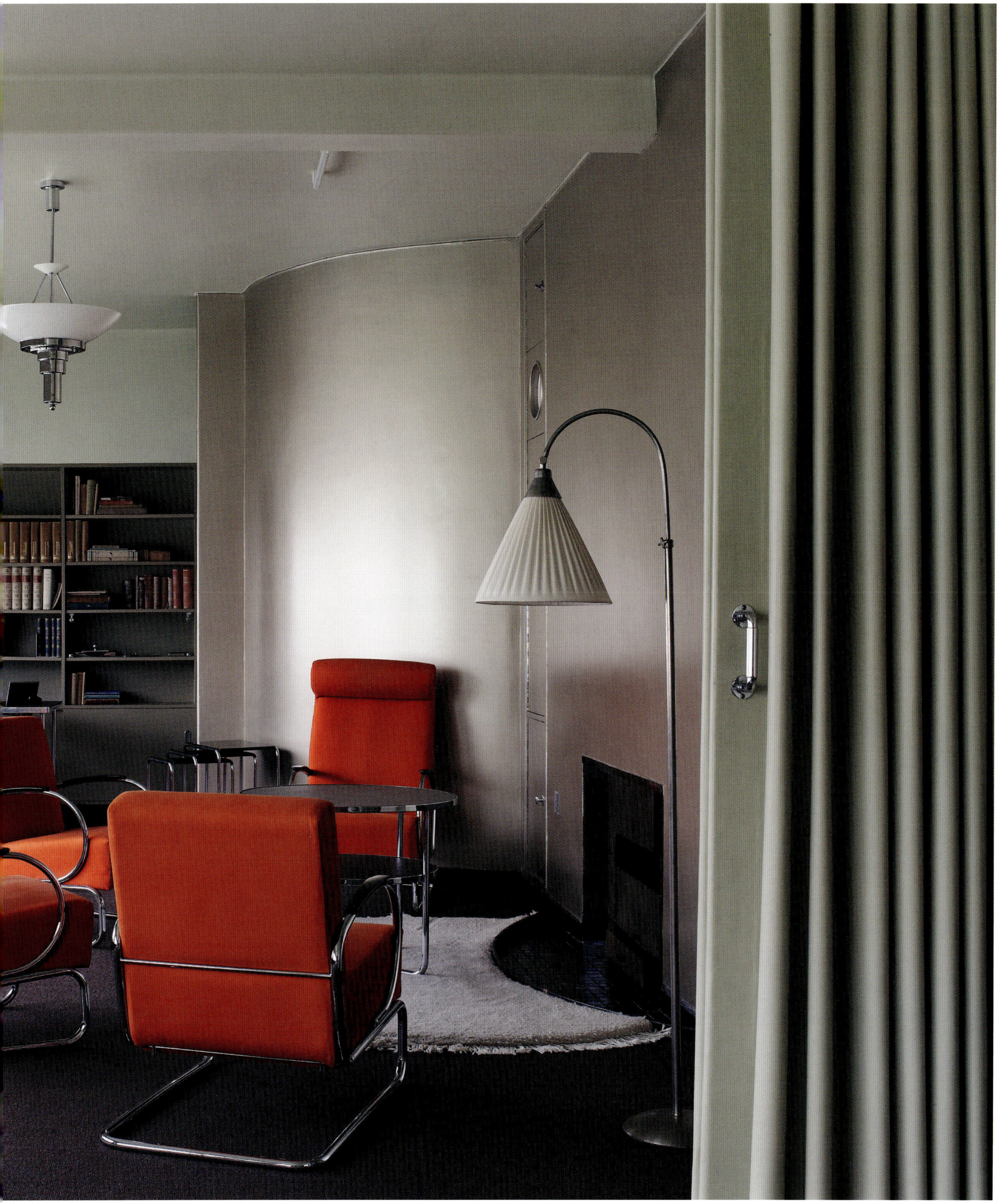

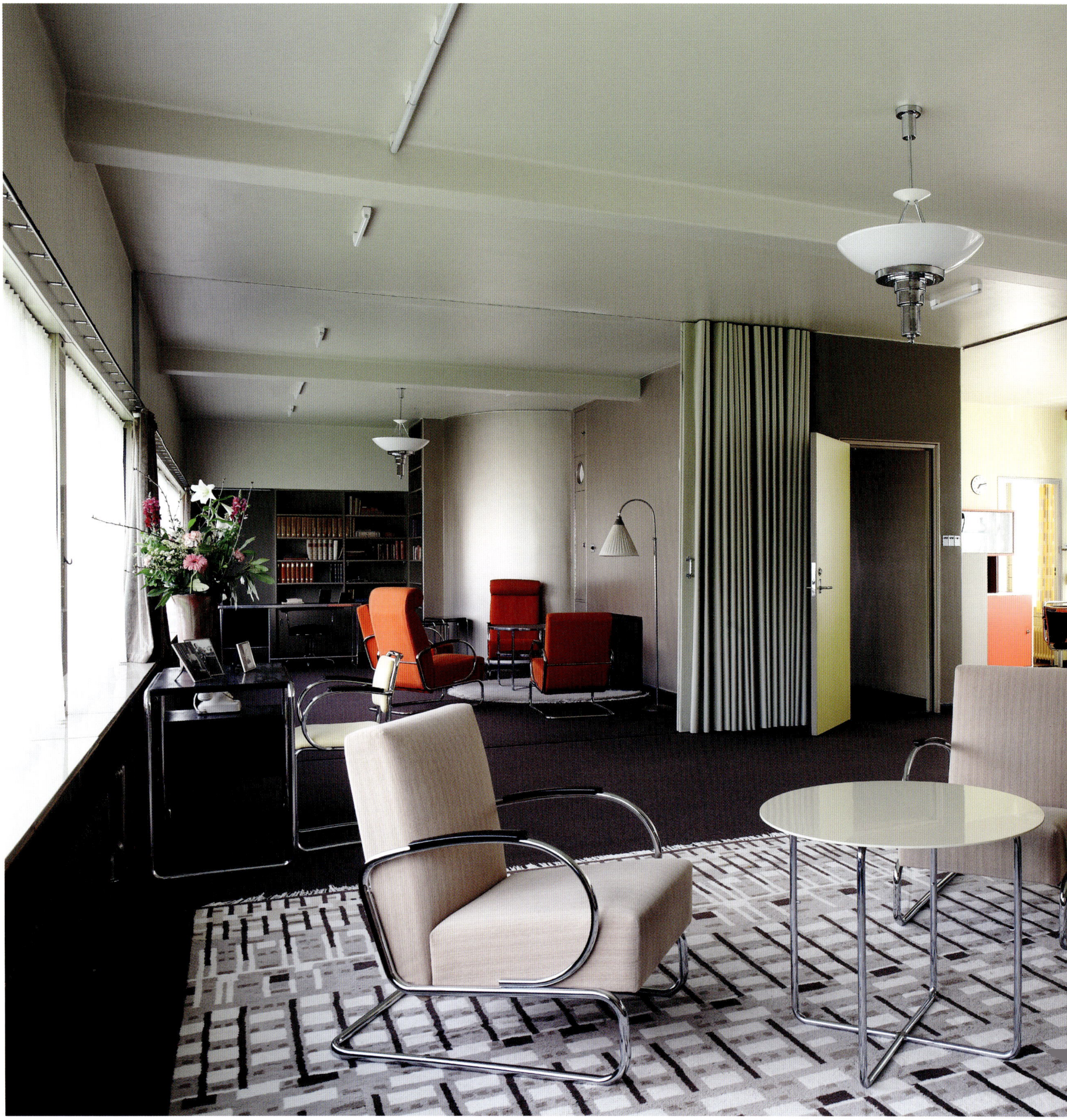

ABOVE A long view of the combined sitting room and library, with a glimpse through to the dining room; the geometric rug was designed by Elise Djo-Bourgeois.

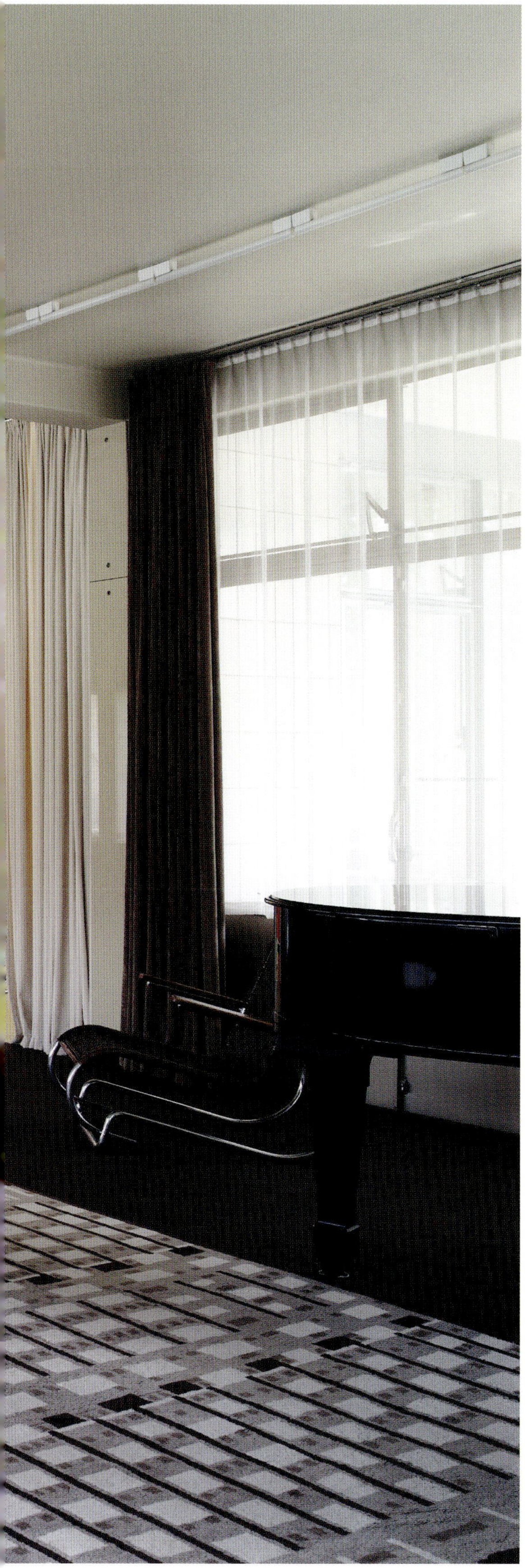

RIGHT This colourful dressing room forms part of the master suite, with a wealth of fitted cupboards and integrated storage.

BELOW The dining room sits at one remove from the sitting room alongside; the tubular steel furniture is by Gispen, while an intercom telephone has been wall-mounted.

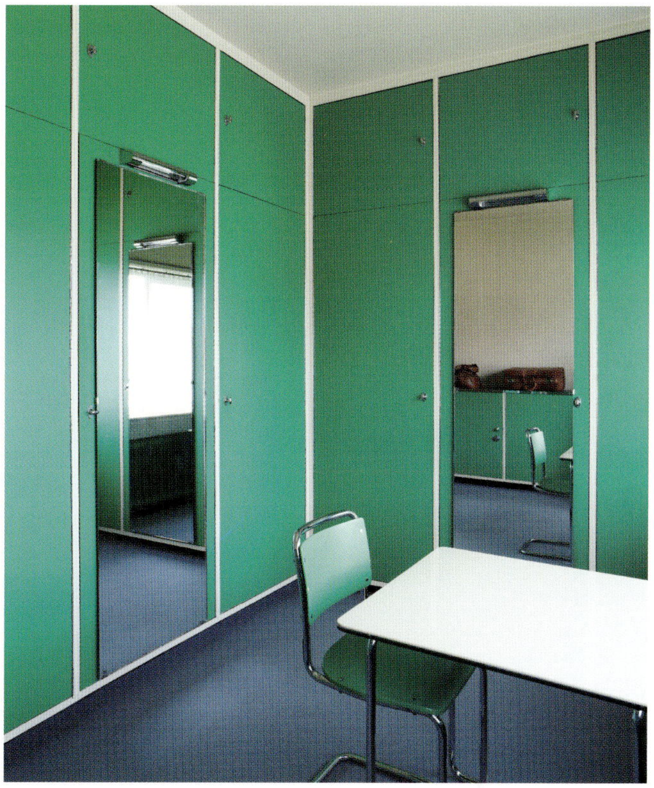

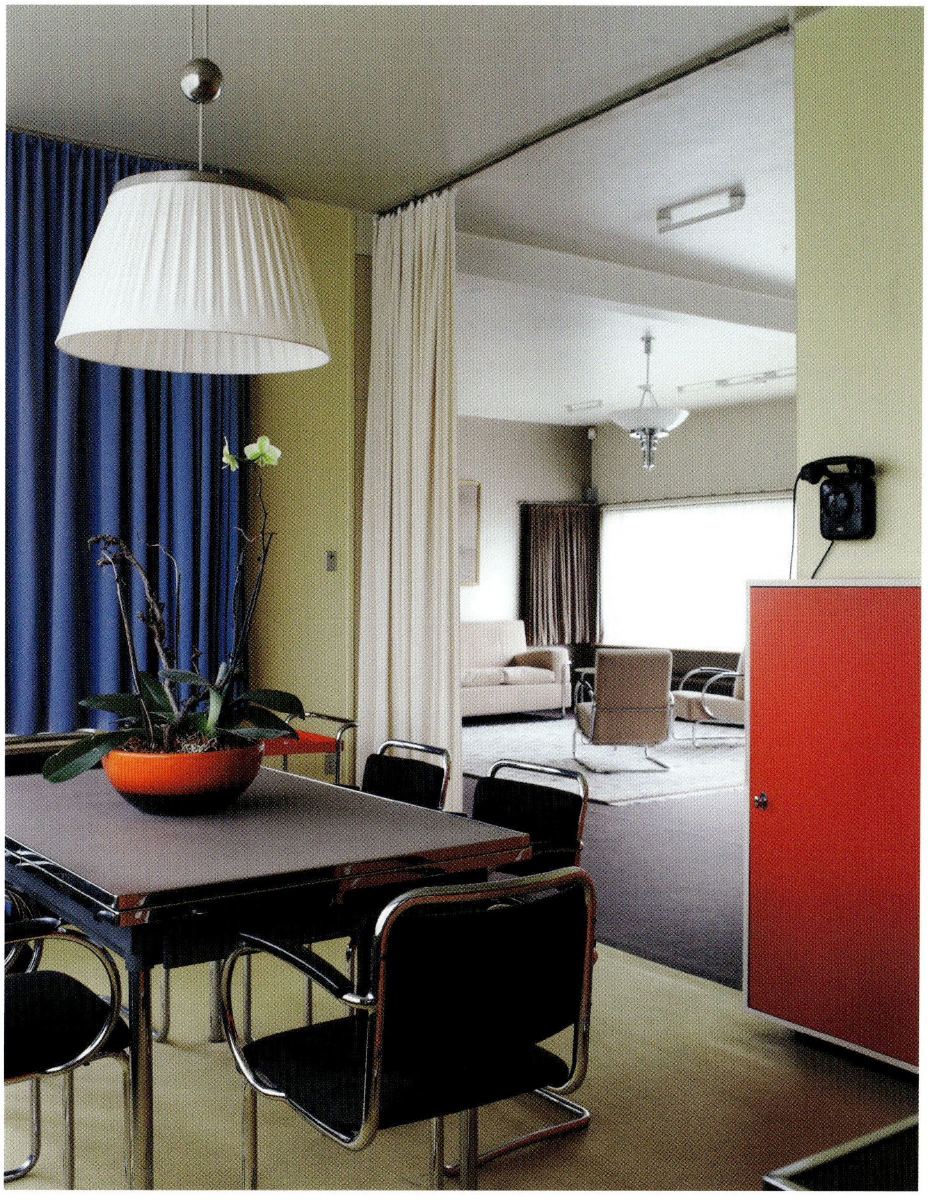

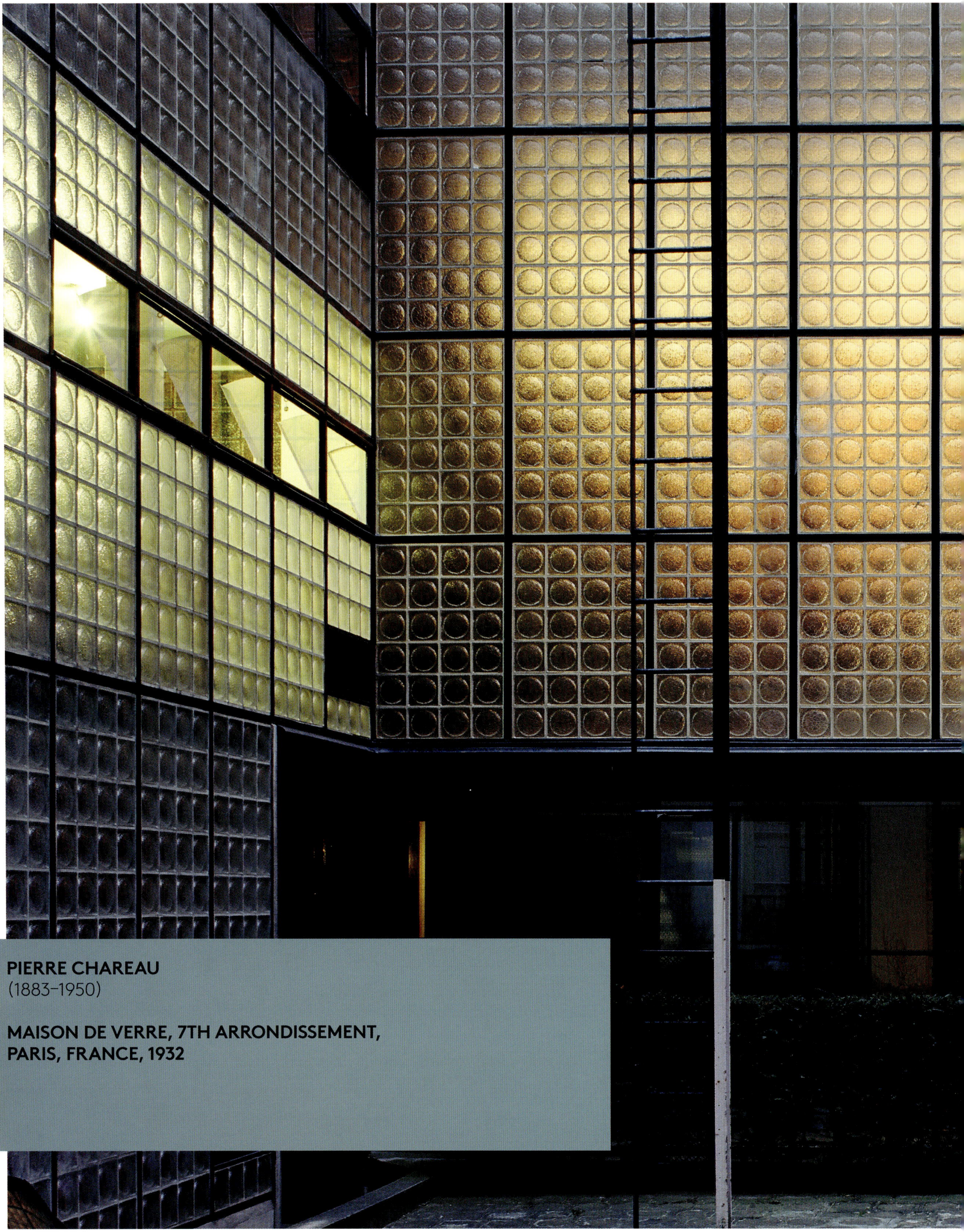

PIERRE CHAREAU
(1883–1950)

MAISON DE VERRE, 7TH ARRONDISSEMENT, PARIS, FRANCE, 1932

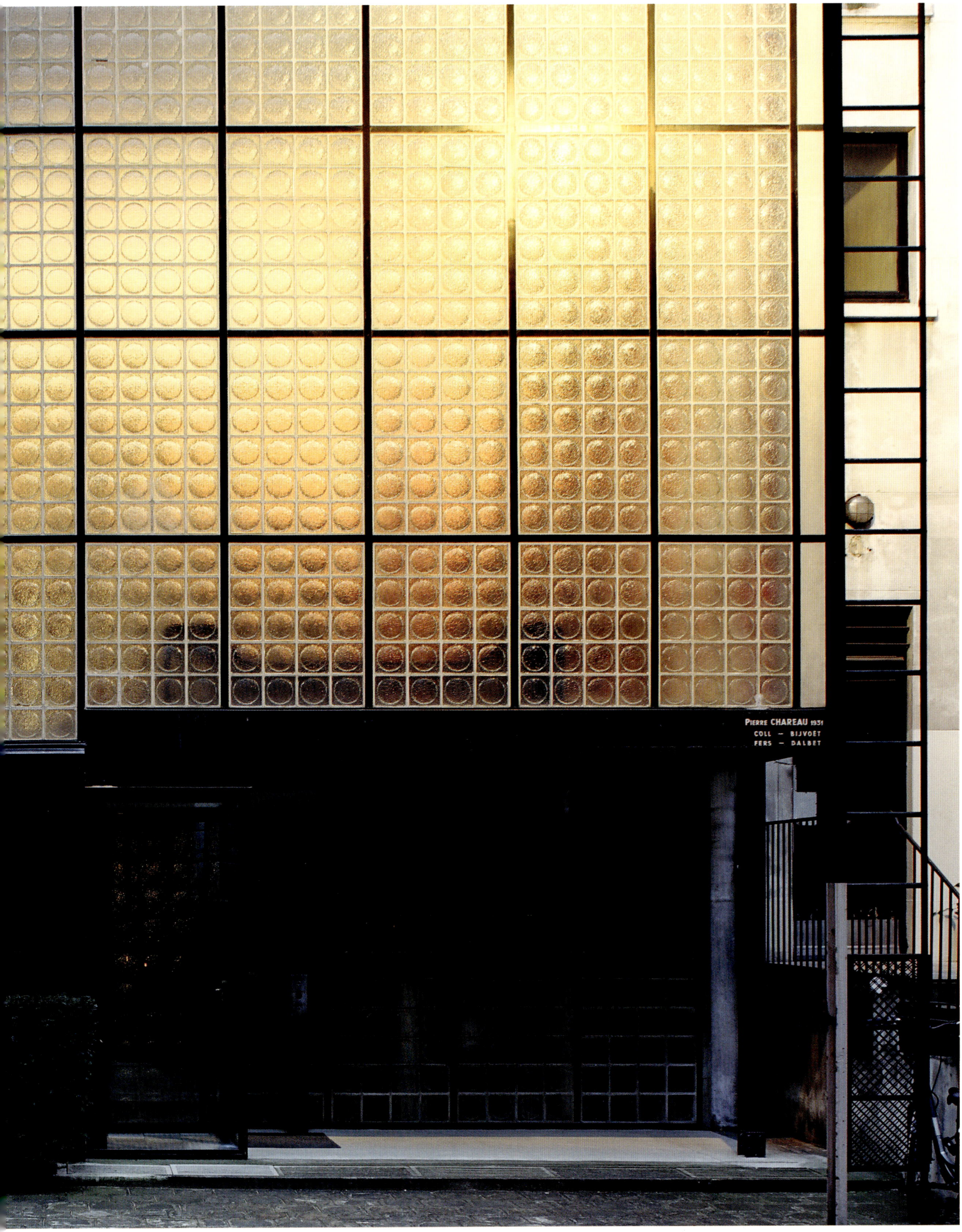
PIERRE CHAREAU 1931
COLL — BIJVOET
FERS — DALBET

The French designer and architect Pierre Chareau worked in the fascinating hinterland between Art Deco* and modernism. Much of his early work in furniture design and interiors has been described as essentially Deco in character, with its crafted quality, sinuous lines and an emphasis on materiality. Yet, like EILEEN GRAY and JEAN-MICHEL FRANK, Chareau criss-crossed the complex borders between the crafted luxury of Deco and the machine-age functionalism of the early modernists. A man of both wood and steel, both a craftsman and an engineer, Chareau managed to fuse a range of influences in spaces that were avant-garde, innovative, and rich in bespoke detailing and ergonomic delight. His greatest and most rounded achievement – combining his love of architecture, interiors and furniture – is the Maison de Verre.

Chareau was born in Bordeaux and studied architecture at the École Nationale Supérieure des Beaux-Arts in Paris. He went on to work for five years with the furniture company Waring & Gillow and also married an Englishwoman, Louise Dyte, who was to help run his furniture and interiors business. In 1924, he established his own furniture shop in Paris and began collaborating with Louis Dalbert, a gifted metalworker. In the late Twenties, there was a commission for a golf clubhouse at Beauvallon near Saint-Tropez and also a hotel in Tours.

It was Chareau's wife who first introduced him to Annie Dalsace, who – along with her husband, Dr Jean Dalsace – went on to commission the Maison de Verre. The Dalsaces asked Chareau to design the interiors of their apartment in Boulevard Saint-Germain, as well as a consulting room for Dr Dalsace; Chareau also designed interiors and furniture for Annie Dalsace's parents' country house, and her uncle helped Chareau win the Beauvallon commission.

Annie Dalsace confirmed her position as Chareau's most important patron when she persuaded her father to help fund the Maison de Verre. The project was unique in so many respects, involving the partial replacement of an 18th-century town house within a hidden courtyard. A sitting tenant on the top floor of the house could not be moved, so Chareau suggested a startling solution, offering to resupport the top-floor apartment on a series of steel columns while slotting in an entirely new three-storey house within the space below. The use of the supporting pillars freed up the façade of the building, which became a non-structural curtain wall, where Chareau created his famous façade of glass bricks. These translucent bricks, which had previously only been used on small-scale or semi-industrial projects, offered both natural light and privacy, and formed an extraordinary glowing lantern in the dusk.

Collaborating with Dalbert and architect Bernard Bijvoët, Chareau turned his hand to every possible detail. On the ground floor he created a series of spaces for the use of Dr Dalsace, including consultation and waiting rooms, as well as a reception area and surgery. Here, the aesthetic was functional but also highly considered, with every element thought through and every modern convenience provided. A curved screen helped to shroud the main staircase to the upper levels from patients and visitors, serving to maintain the privacy of the accommodation on the two floors above.

Ascending the staircase, one was drawn into the 'great room': a double-height living space illuminated by the glass brick façade, while a vast library wall sat to one side, complete with a bespoke rolling bookcase ladder. The stairwell was guarded by a balustrade made up of low bookshelves, while to the rear a sliding wall could be used to separate – or connect with – an adjoining study.

'The great room is a beating heart,' wrote Dominique Vellay, Annie Dalsace's grand-daughter. 'It is like a modern cathedral, where eleven orange and black columns studded with rivets and bolts impose a rhythm and form the framework of the house.... Wherever one looks in this house, something is happening. The more one looks, the more one discovers the volumes, the different materials, the meticulous details.'[17]

Completed in 1932, the house survived the war years unscathed, as did its contents, which were stored away for safekeeping and later reinstated; the Maison de Verre remained in the hands of the family until a private sale in 2007. Chareau moved to London with his wife during the war and then on to the United States. One of his few completed post-war projects – a studio on Long Island for the painter Robert Motherwell – was demolished in the Eighties. The Maison de Verre, however, is in safe hands and remains a powerful testament to the virtuosity of one of modernism's most imaginative pioneers.

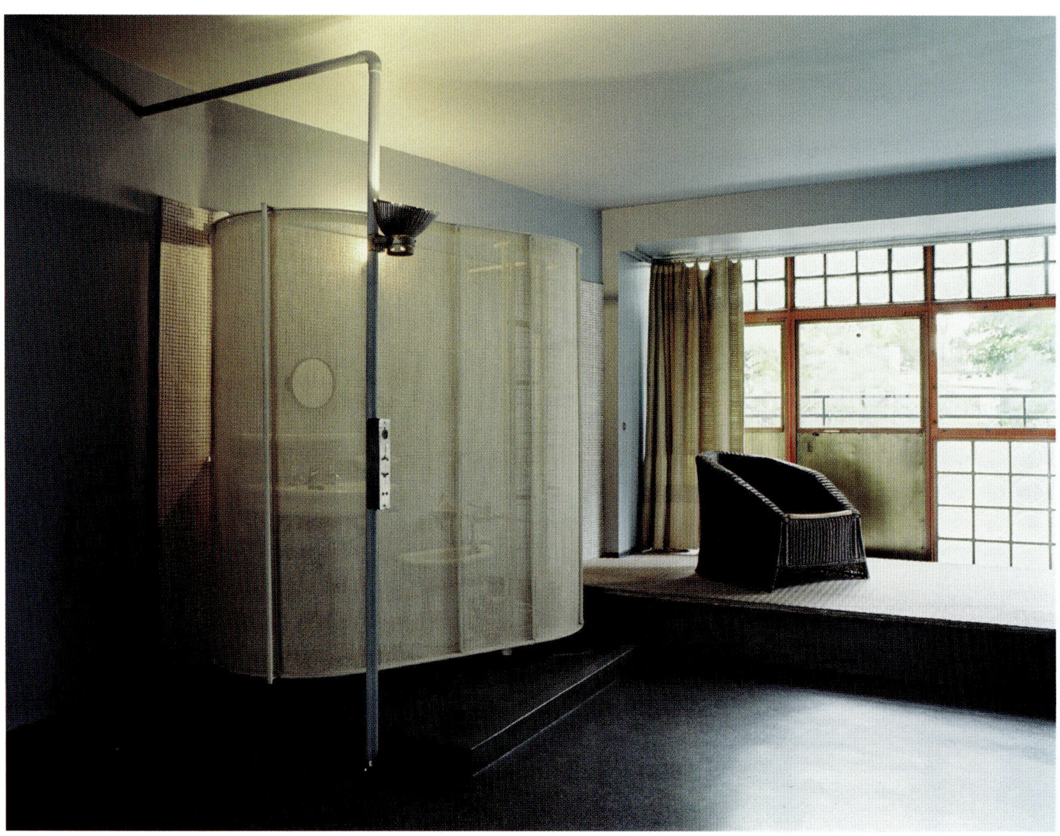

PAGES 324–5 The iconic façade of the Maison de Verre, composed of a geometric lattice of glass bricks made by Saint-Gobain; the entrance zone flows into the 'public' surgery on the ground floor.

LEFT Dr Dalsace's surgery was located on the ground floor, with private living spaces arranged on the two levels above; the surgery is, in itself, extensive, with reception and waiting areas, as well as Dr Dalsace's consulting room and a separate space for examination and procedures.

OPPOSITE Ground-floor circulation spaces feature functional rubber floor tiles, while turning wheels can be used to open operable sequences of windows for ventilation; the back of the building flows out to a garden.

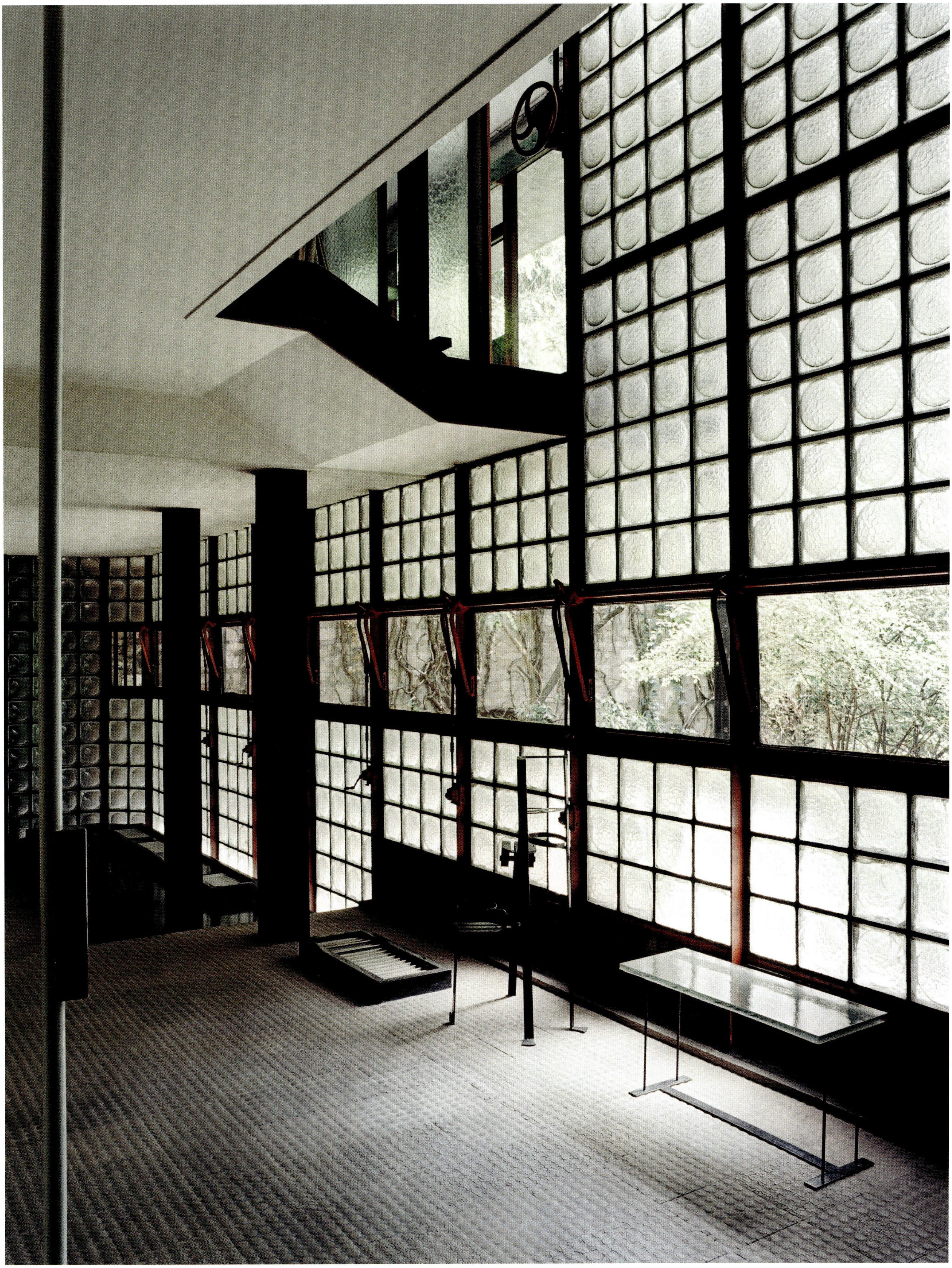

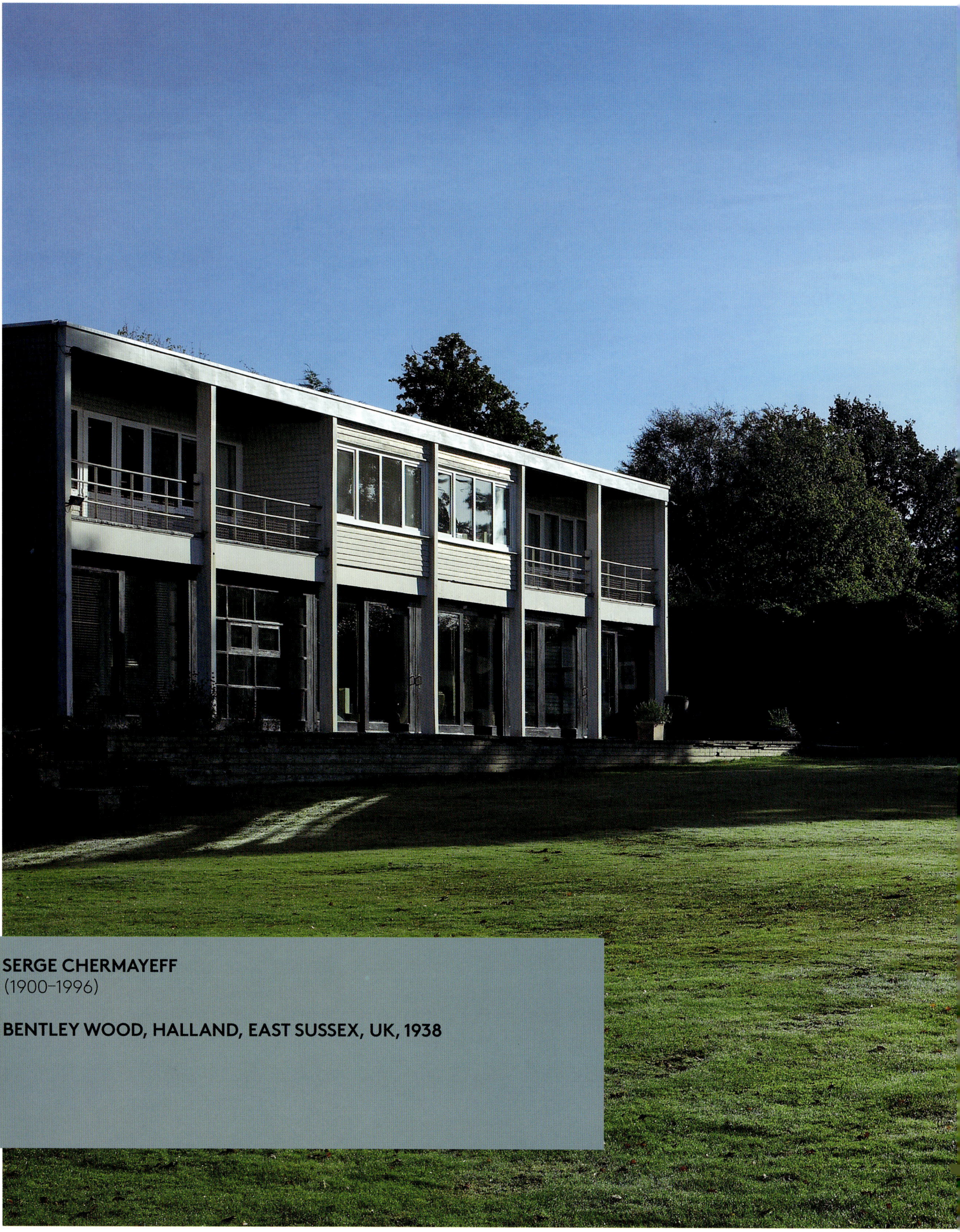

SERGE CHERMAYEFF
(1900–1996)

BENTLEY WOOD, HALLAND, EAST SUSSEX, UK, 1938

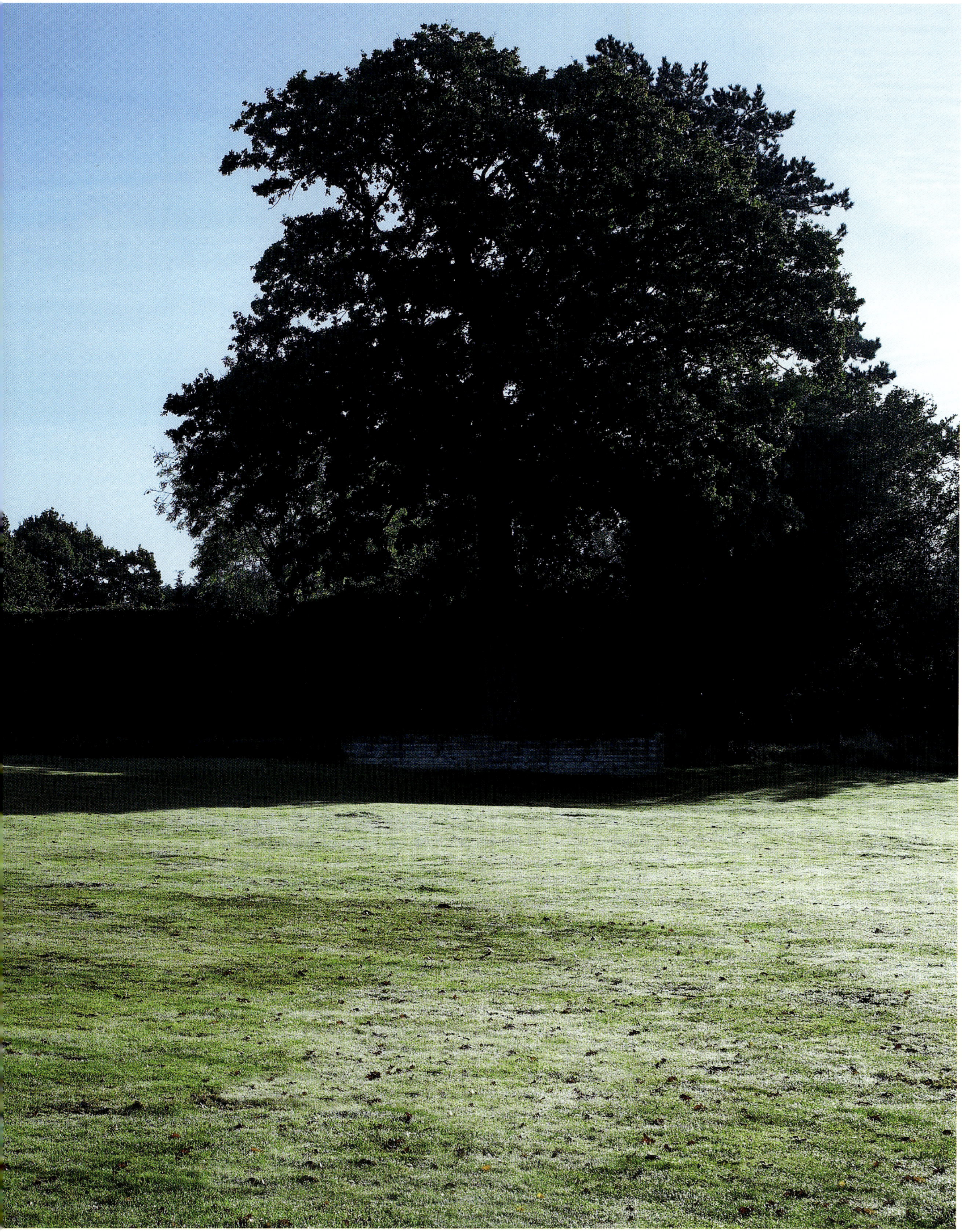

The architectural designer Serge Chermayeff was a colourful and flamboyant character for a golden age of architecture and design. He was at the height of his career during the Thirties, when he completed some of his most famous buildings, including the De La Warr Pavilion in Bexhill-on-Sea, designed with Erich Mendelsohn* and opened in 1935. His financial collapse and his decision to emigrate to America, along with his family, in 1939 seemed to mirror the implosion of Europe itself as it descended into war.

'Chermayeff' was an assumed name for a man who spent his early years constantly reinventing himself. He was born in Chechnya with the name Serge Issakovitch and his family grew wealthy when land that they owned was found to be sitting on an oil field. Young Serge was sent to boarding school in England (Peterborough Lodge, followed by Harrow) but his hopes of going to Cambridge were ended by the Russian Revolution, which saw his family lose their property and wealth.

Chermayeff worked as a reporter and illustrator, while also painting and indulging a love of dance, winning many competitions and enjoying the life of a London dandy of the early Twenties. He married in 1928 and took a position around the same time as head of the Modern Art Department of the interiors and furniture firm Waring & Gillow. Chermayeff was able to indulge and develop his interest in modernist design within a secure environment, supported by society patrons.

Although he had no formal training as an architect, Chermayeff decided to step up his ambition by combining his interest in interiors with more substantial architectural projects. A modest cubist house in Rugby for a biology master at Rugby School was completed in 1933, just before Chermayeff joined forces with Erich Mendelsohn, who had emigrated from Germany with his wife and daughter, leaving behind a highly successful career as a pioneering modernist architect, including commissions for department stores and villas, such as the Mendelsohn House in Berlin, finished just a few years before he and his Jewish family fled Nazi Germany.

Chermayeff and Mendelsohn's partnership was brief but productive. There was a house project in Old Church Street in Chelsea (1935) and, most notably, the De La Warr Pavilion, a project won by competition and still one of the most elegant and seductive modernist buildings in Britain, with Mendelsohn taking the architectural lead and Chermayeff masterminding the interiors. Soon after the completion of their seaside masterpiece, the two men fell out and dissolved their partnership, with Mendelsohn moving on to the United States in 1941.

Bentley Wood marked the crowning achievement of Chermayeff's solo career as an architectural designer. He bought 52 hectares of land in East Sussex around the time that the De La Warr Pavilion was being completed, with the aim of building a house for himself and his family. The finished house was to use a jarrah wood timber frame and red cedar cladding, together with expanses of glass to connect inside and out.

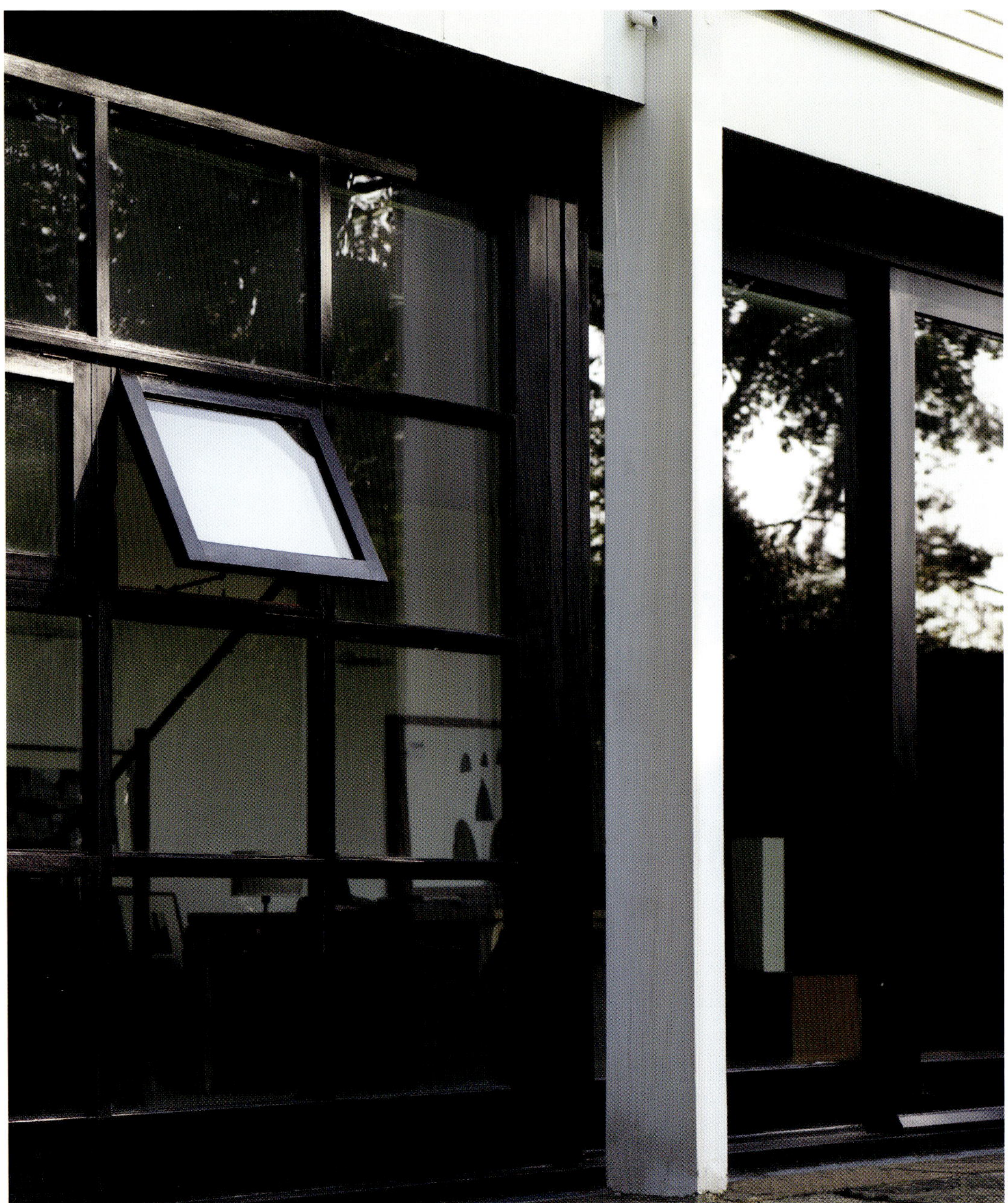

PAGES 328–9 The substantial gardens and grounds around Bentley Wood were designed in collaboration with landscape architect Christopher Tunnard.

LEFT & OPPOSITE The design of the building promoted a vivid and constant relationship between indoors and out, with long sequences of glazing along the façade on both levels.

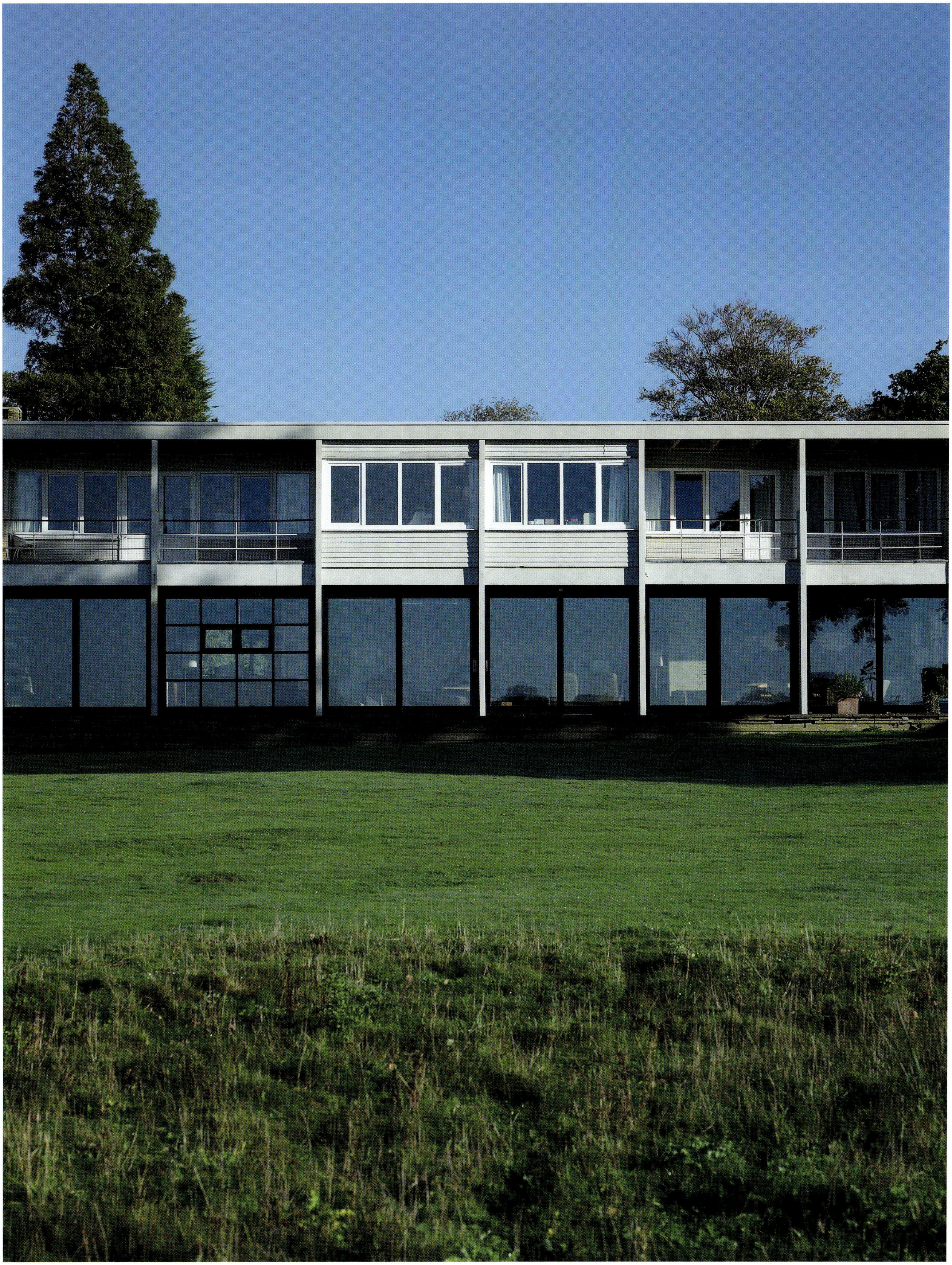

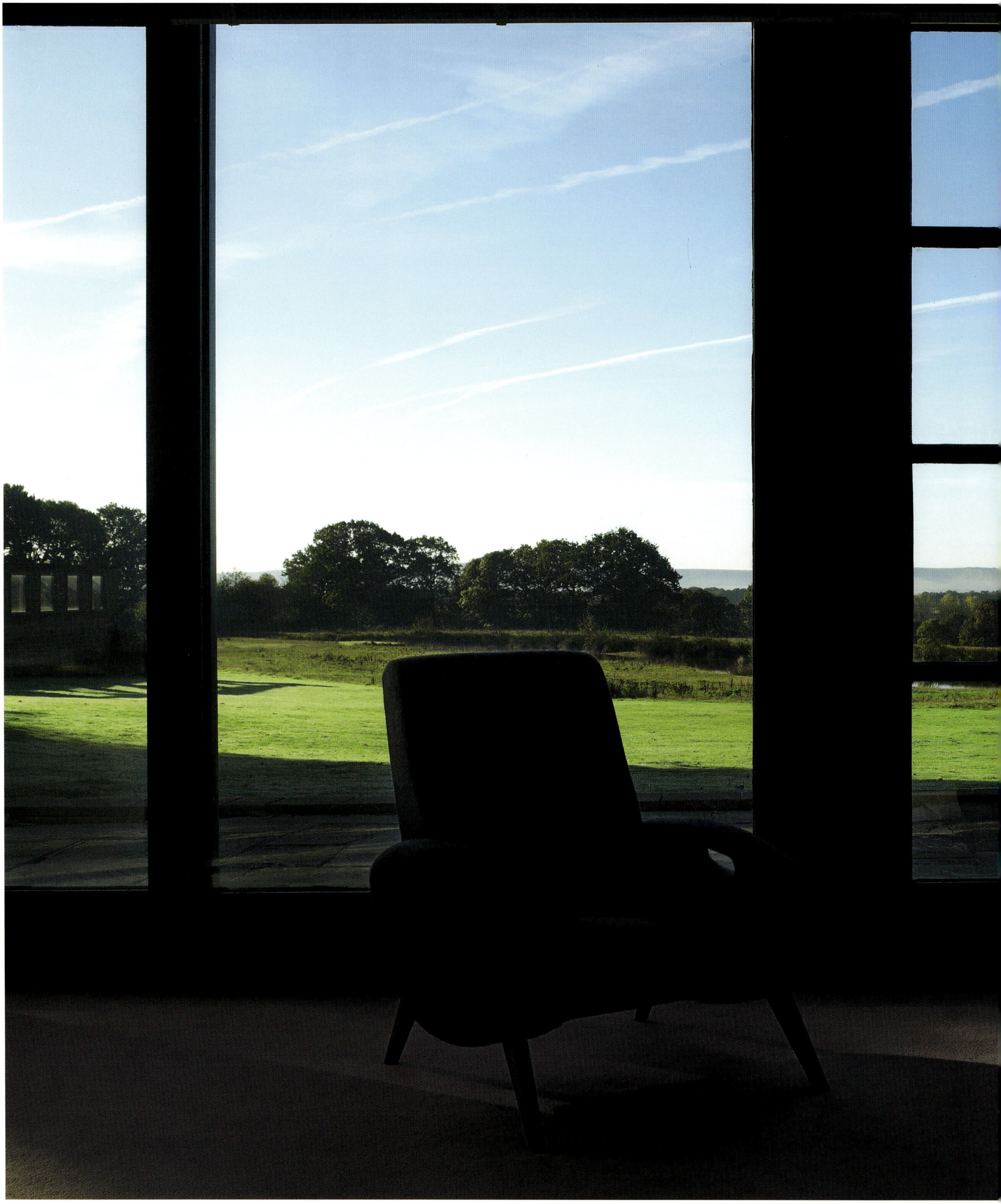

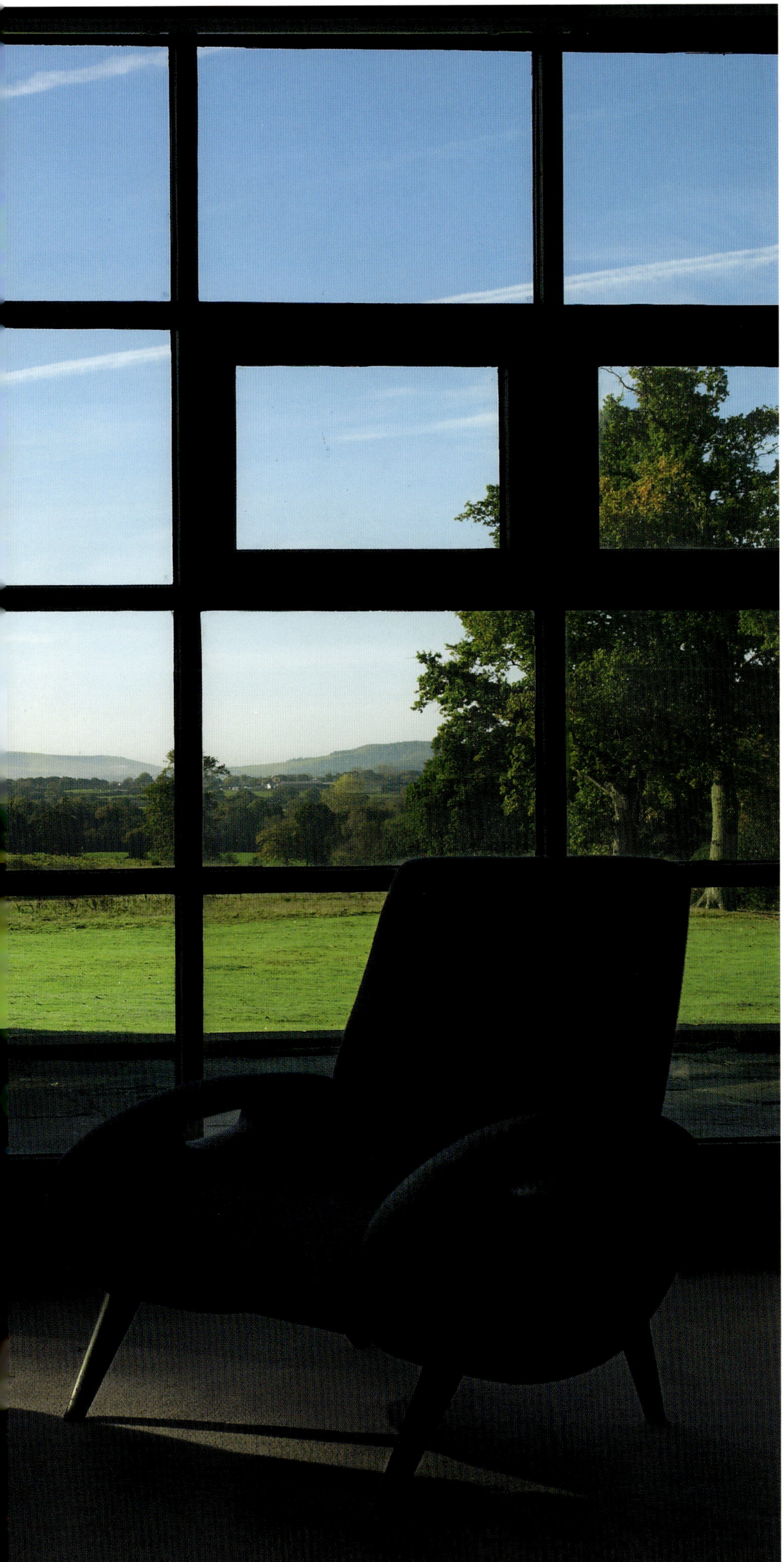

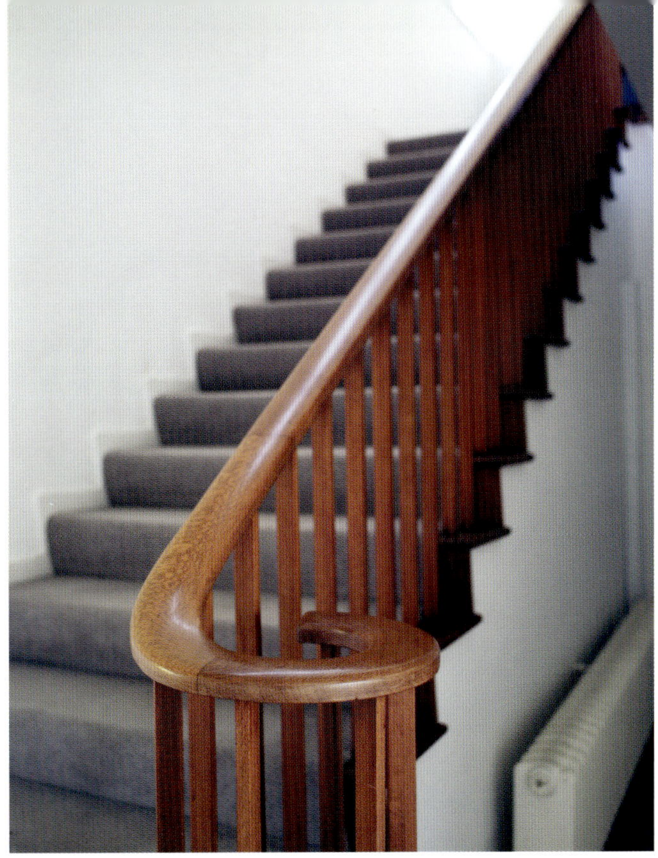

ABOVE Bentley Wood was described by one commentator as a 'Rolls Royce of a house', with luxurious materials, crafted elements (such as the staircase) and much in the way of new technology.

LEFT The main living spaces offer extensive views over the Sussex landscape, with expanses of woodland beyond a modest boating pond.

In many ways, the two-storey house anticipated post-war, mid-century modern houses of California and other parts of the States, with its emphasis on connecting to the landscape and dissolving boundaries between inside and outside living. Walls of glass on the lower level rolled back to allow a fluid connection between the principal living spaces, the adjoining terrace and the landscaped gardens beyond. Upstairs, a sequence of four bedrooms arranged over six bays was complemented by partially sheltered balconies tucked under the flat roof. Chermayeff worked on the landscaping with Christopher Tunnard and commissioned a Henry Moore statue for the garden.

Within, Chermayeff's ambition was equally advanced, aiming for a combination of character, luxury, refinement and technology. There were artworks by John Piper, Ben Nicholson, Barbara Hepworth, Alexander Calder* and Pablo Picasso. Bespoke furniture designs by Chermayeff himself mixed with pieces from Thonet* and elsewhere. The house featured concealed heating, piped music, and push-button bell pushes and integrated telephones in the bedrooms.

Frank Lloyd Wright and Ernő Goldfinger visited the house, and it was widely praised. Yet within months the project had bankrupted Chermayeff. Like his former partner, he emigrated to America, reinventing himself once again as a teacher and writer, as well as continuing to paint. He left behind one of the most accomplished modernist houses in Thirties England.

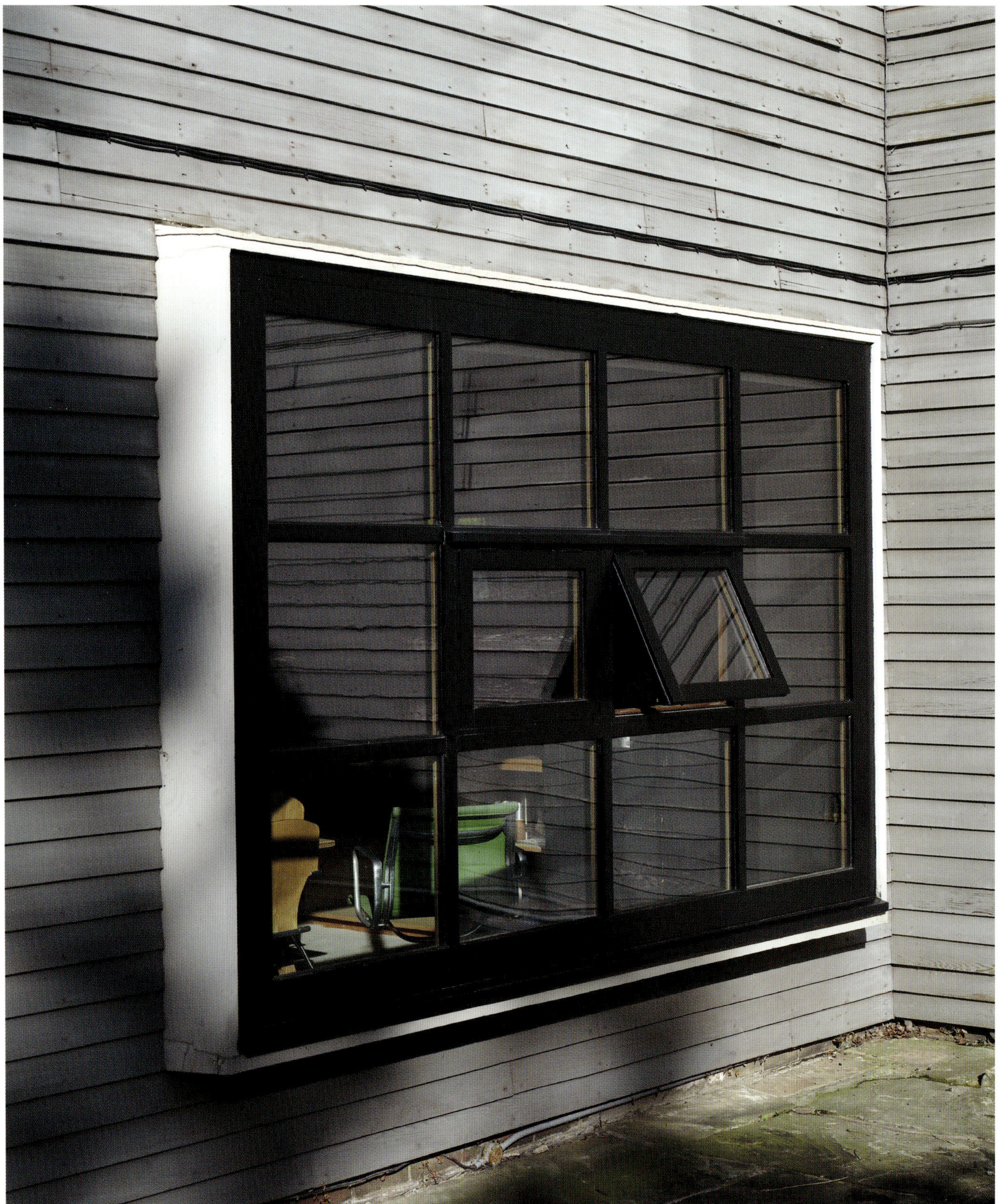

Left & opposite Glazing at the lower level opens up to the adjoining terrace at the front of the house; originally each of the five bedrooms on the upper level, including the master suite, featured access to a balcony but in later years the two central bays were filled in.

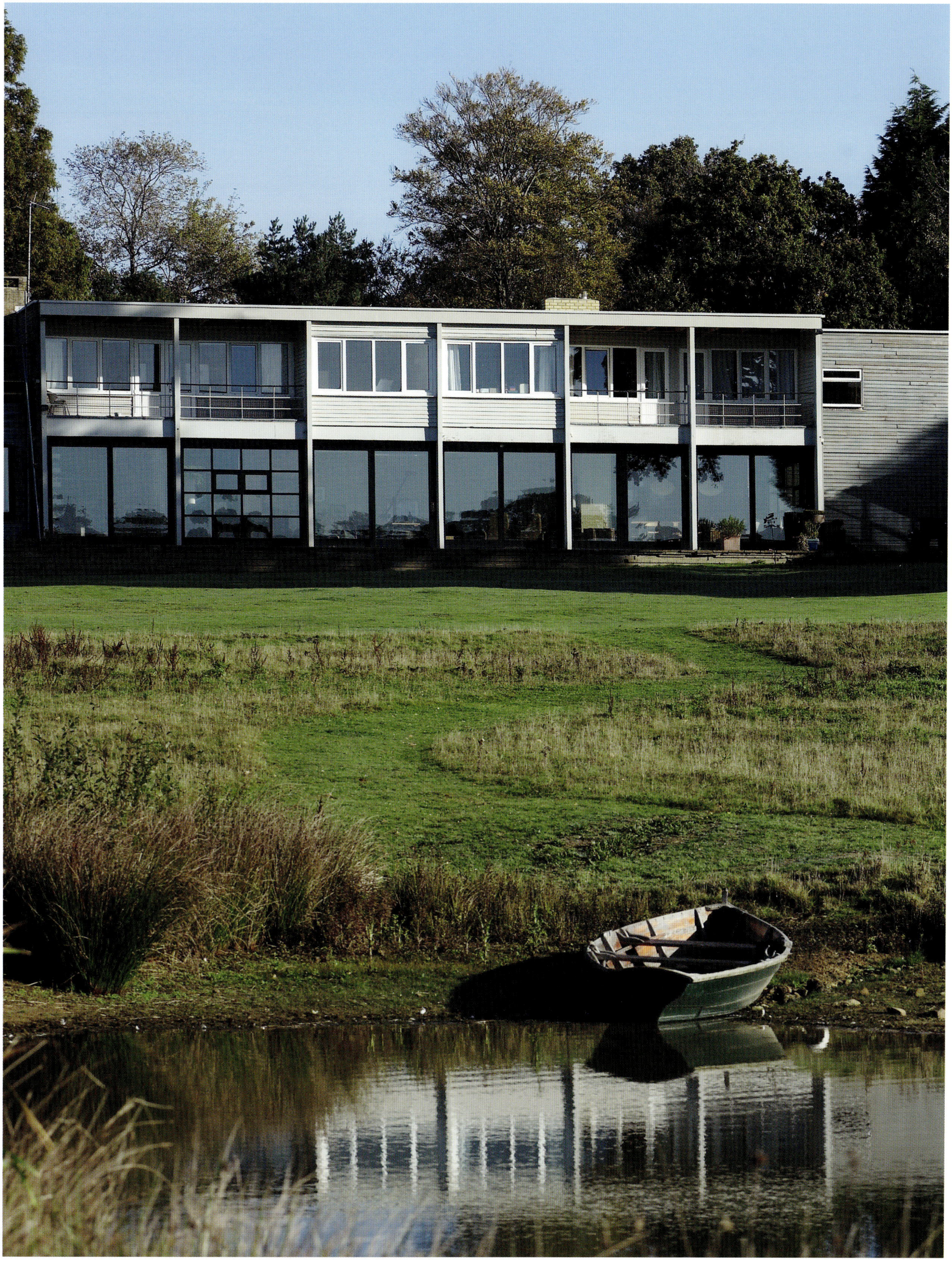

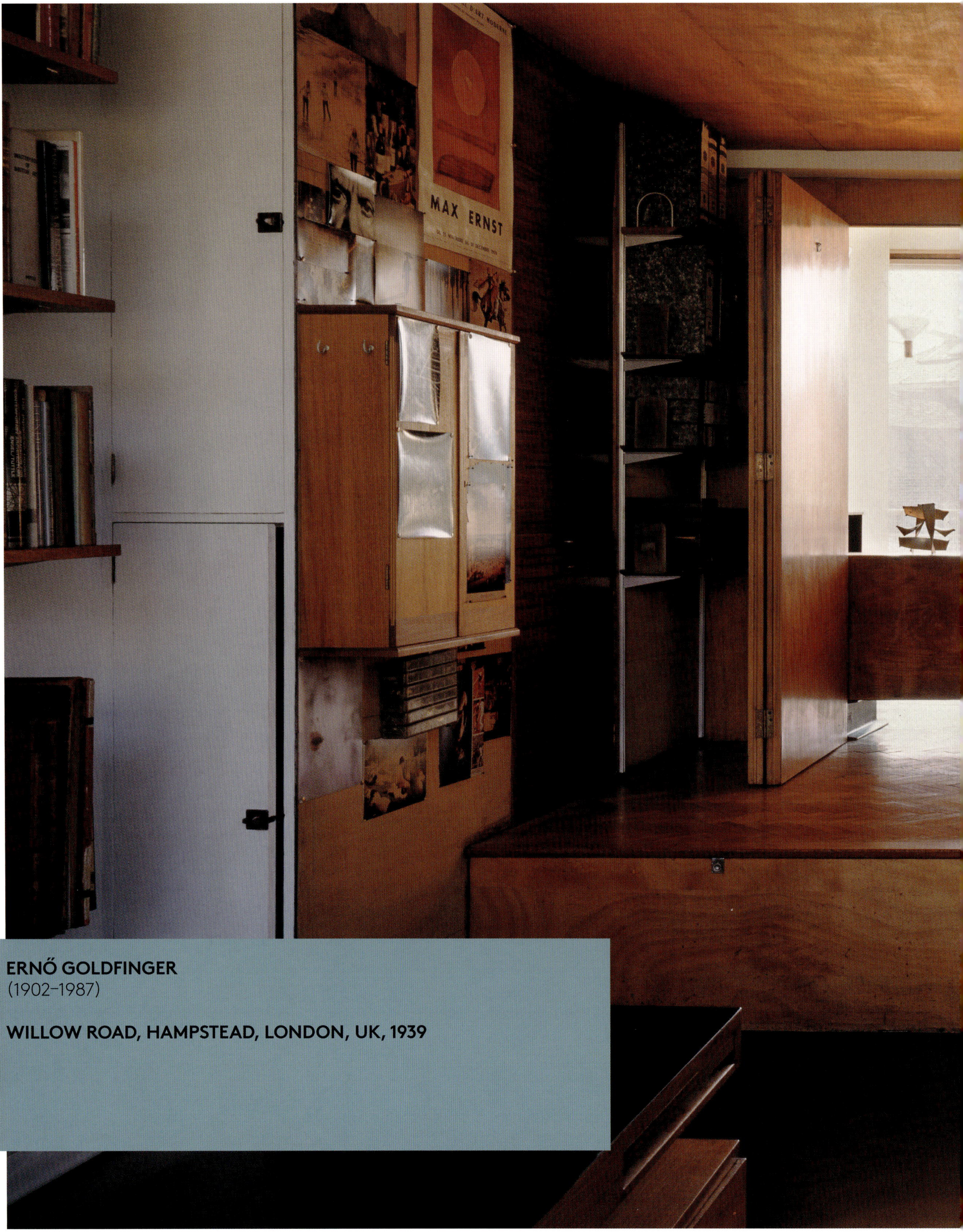

ERNŐ GOLDFINGER
(1902–1987)

WILLOW ROAD, HAMPSTEAD, LONDON, UK, 1939

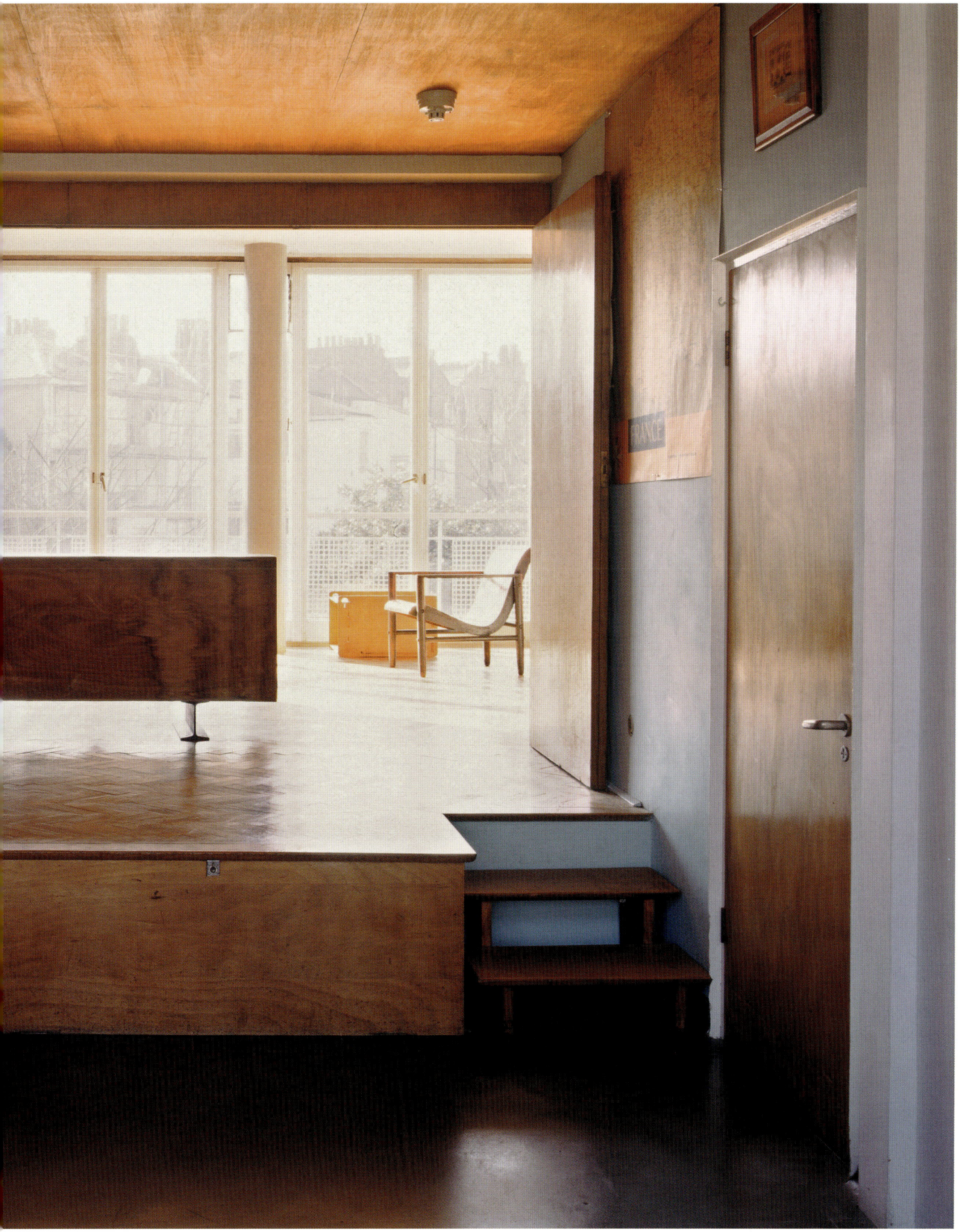
FRANCE

The Hungarian-born, Paris-educated, Anglophile architect Ernő Goldfinger was not an admirer of what he described as the 'kasbah' school of modernist architecture. By this, he meant the familiar rectilinear houses of the inter-war years, resembling cubist ships, crisply rendered and painted a gleaming white. Goldfinger preferred a more expressive approach, where the materiality and texture of a building became a chief part of its natural character, while the tectonic dynamics of the architecture were also clear for all to see and enjoy. The building that best encapsulates his approach during the Thirties is his own house at 2 Willow Road, Hampstead, conceived in conjunction with its neighbours to either side, forming a prototypical modernist terrace.

'A young architect ought to be made to build his own house first,' declared Goldfinger. 'It is the only way to learn. And at his own expense. I don't know in all cases, but he ought to have the chance to show what his ideas really are.'[18] In Goldfinger's case, the house was actually funded by his wife, Ursula Blackwell, who came from the wealthy family behind the Crosse & Blackwell soup and food company. But it was – apart from a number of extensions, interiors commissions and smaller projects – his first major completed architectural project and a dramatic statement of intent, with a family home for himself, Ursula and their children, along with a neighbouring house for sale and another for rental.

Goldfinger was himself born to a wealthy family in Budapest. His father was a lawyer who got involved in the management of the family's forestry concerns in Transylvania, where the Goldfingers would often spend their summers. Goldfinger settled in Paris, where he studied and worked for 14 years. He joined the atelier of architect Léon Jaussely and from there gained a place at the prestigious École Nationale Supérieure des Beaux-Arts.

Goldfinger formed a brief partnership with another Hungarian émigré, André Sive, and they worked on a number of minor commissions, as well as developing a tubular steel chair together. He attended conferences of the Congrès International d'Architecture Moderne (CIAM)* and first met Ursula Blackwell in Paris, where she was studying art. Their circle included many respected artists and photographers: Lee Miller, Man Ray, Max Ernst and Fernand Léger.

In 1934, Goldfinger moved to London and married the same year. He and his wife settled into an apartment at Highpoint I, designed by Berthold Lubetkin*; one of their neighbours was another émigré architect, Erich Mendelsohn*. Using money from Ursula's trust fund, Goldfinger acquired a site in Hampstead occupied by four small Georgian cottages just across the road from the open spaces of Hampstead Heath.

The three houses in the replacement terrace were designed as a triptych, resembling one large dwelling, with a series of piloti framing the entrances and service spaces on the ground floor, the living spaces on the piano nobile above lifted by a long ribbon of windows, and the bedrooms sitting on the top floor, defined by a neat line of seven square windows punctuating the brick façade that disguised the concrete structure within.

The interiors of the house were as rounded and as pioneering as the exteriors. The Goldfingers' house was the largest of the triptych and sat at the centre, with a concrete spiral staircase almost in the middle of the plan. The stairs doubled as an additional structural support but also helped save on circulation space. The first-floor living areas were highly fluid in nature, with a large studio space to the rear complemented by living and dining areas partly defined by a change in floor level and sliding screens. Bathrooms and service spaces throughout were modest and functional, but the family did make room for a maid's quarters alongside two integrated garages on the ground floor.

Goldfinger resumed his career after World War II and is best known for his large-scale housing projects of the Fifties, including Balfron Tower (1967) and Trellick Tower (1972). Goldfinger himself took an apartment at the top of Balfron and would host parties for tenants: he is often seen as the inspiration for the sinister architect who inhabits the penthouse in J. G. Ballard's novel, *High-Rise* (1975). The Willow Road residence, complete with its art collection and library, was the first modernist home acquired by the National Trust after the deaths of Ernő and Ursula Goldfinger, and was opened to the public in 1996.

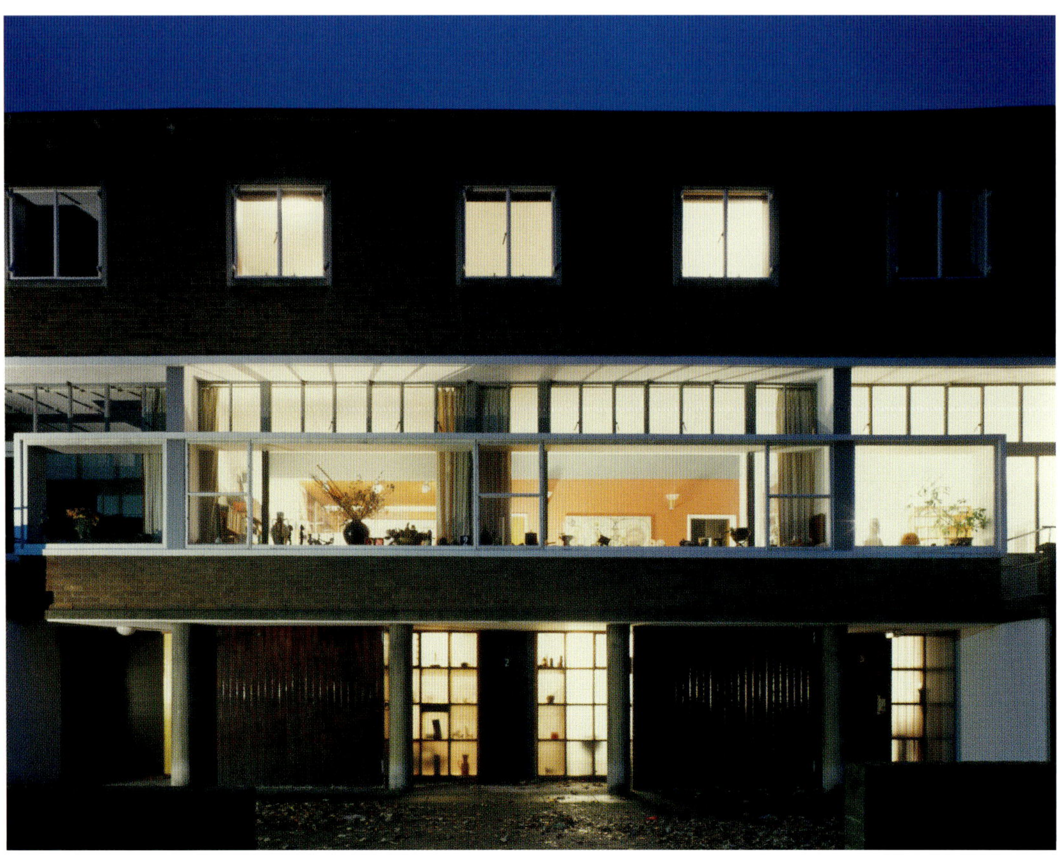

PAGES 336–7 The main living spaces are on the piano nobile, with a fluid, semi-open-plan arrangement that sees a change in floor level (plus sliding screen doors) help to demarcate the living room to the front from the studio space to the rear.

LEFT The Goldfinger house, which is now in the care of the National Trust, is the central residence of a terrace of three modern houses designed by the architect.

OPPOSITE The spiral staircase sits at the very centre of the house, with concrete steps, a brass handrail and a rope balustrade; a circular skylight above the stairwell draws in natural light.

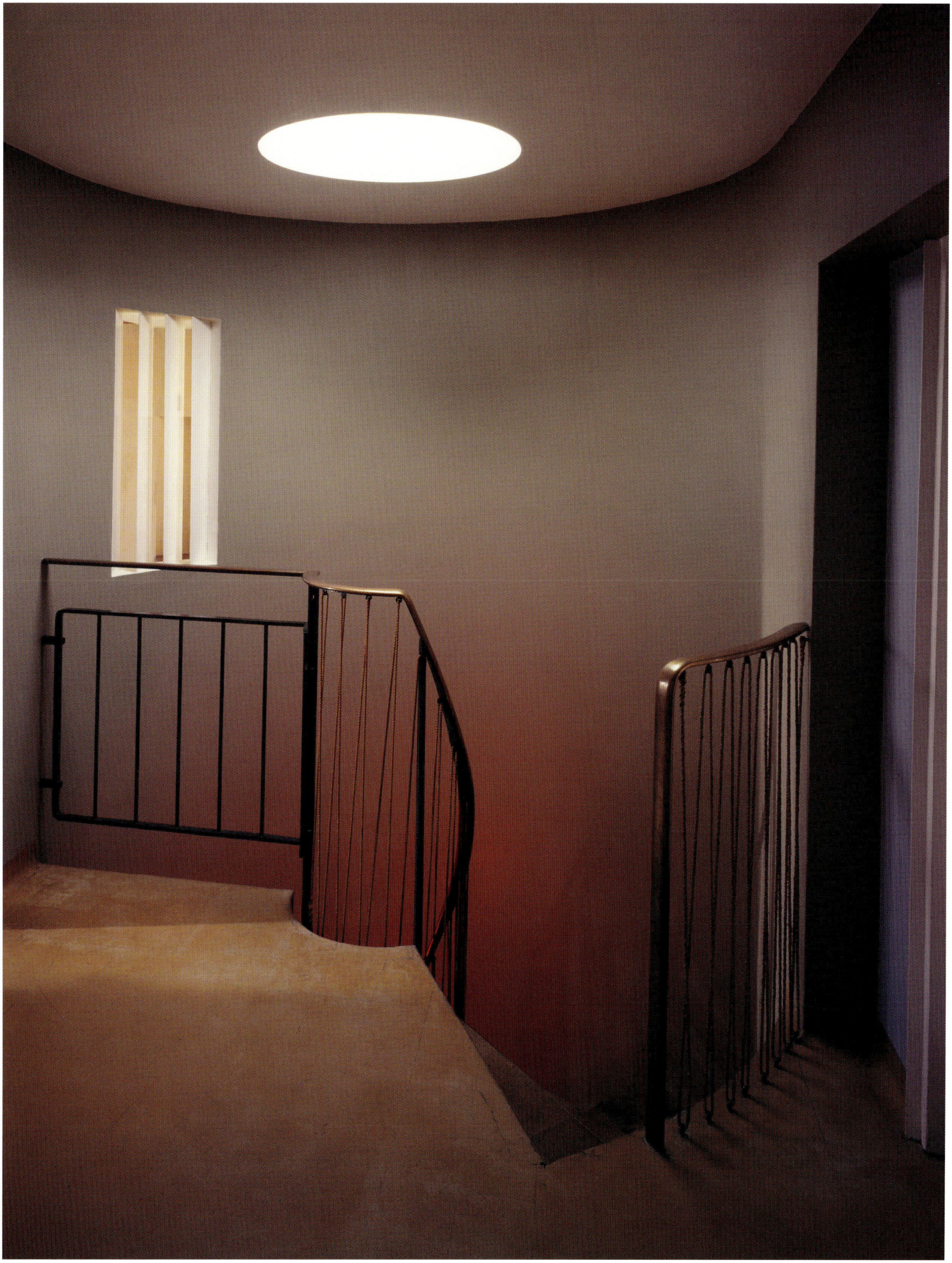

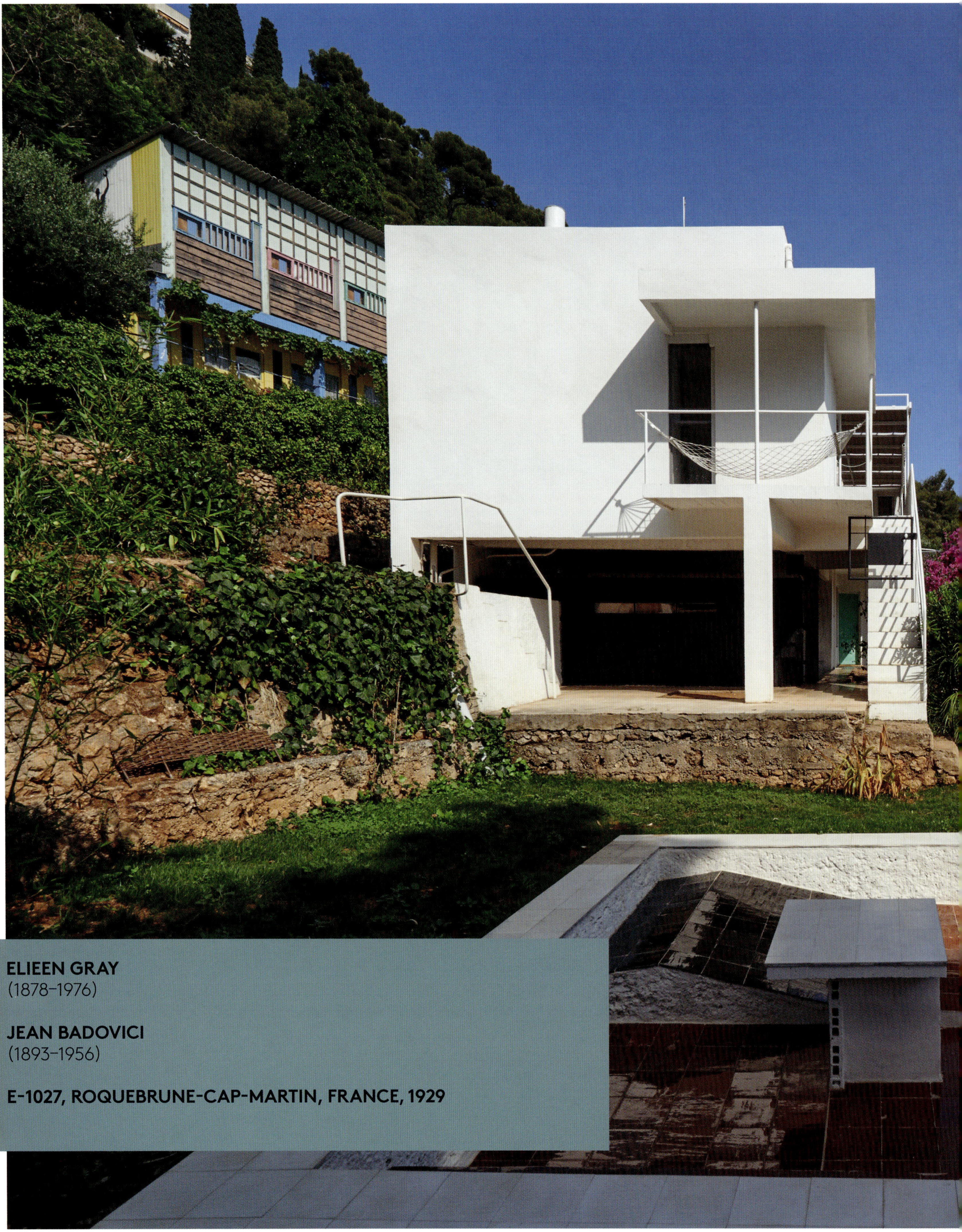

ELIEEN GRAY
(1878–1976)

JEAN BADOVICI
(1893–1956)

E-1027, ROQUEBRUNE-CAP-MARTIN, FRANCE, 1929

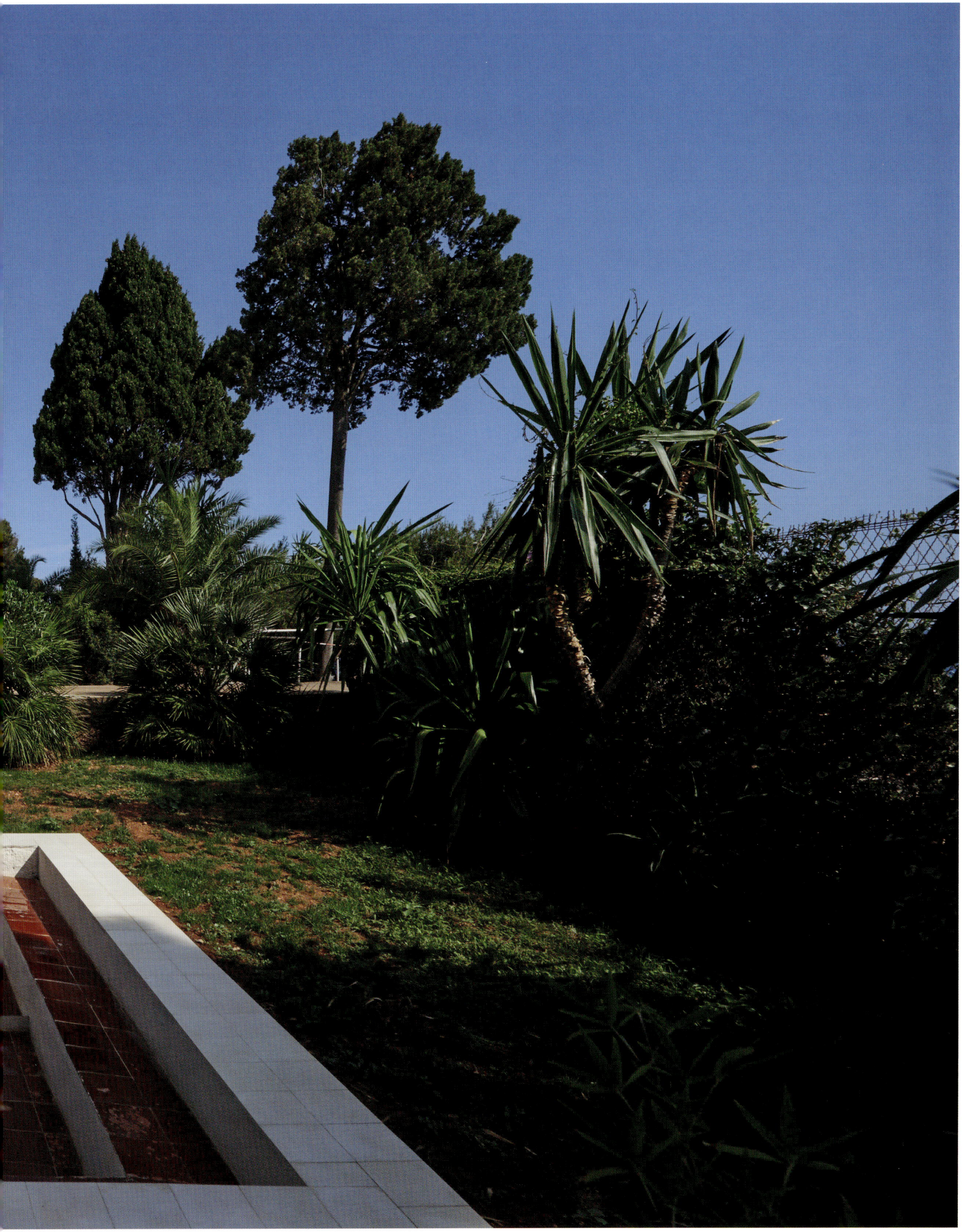

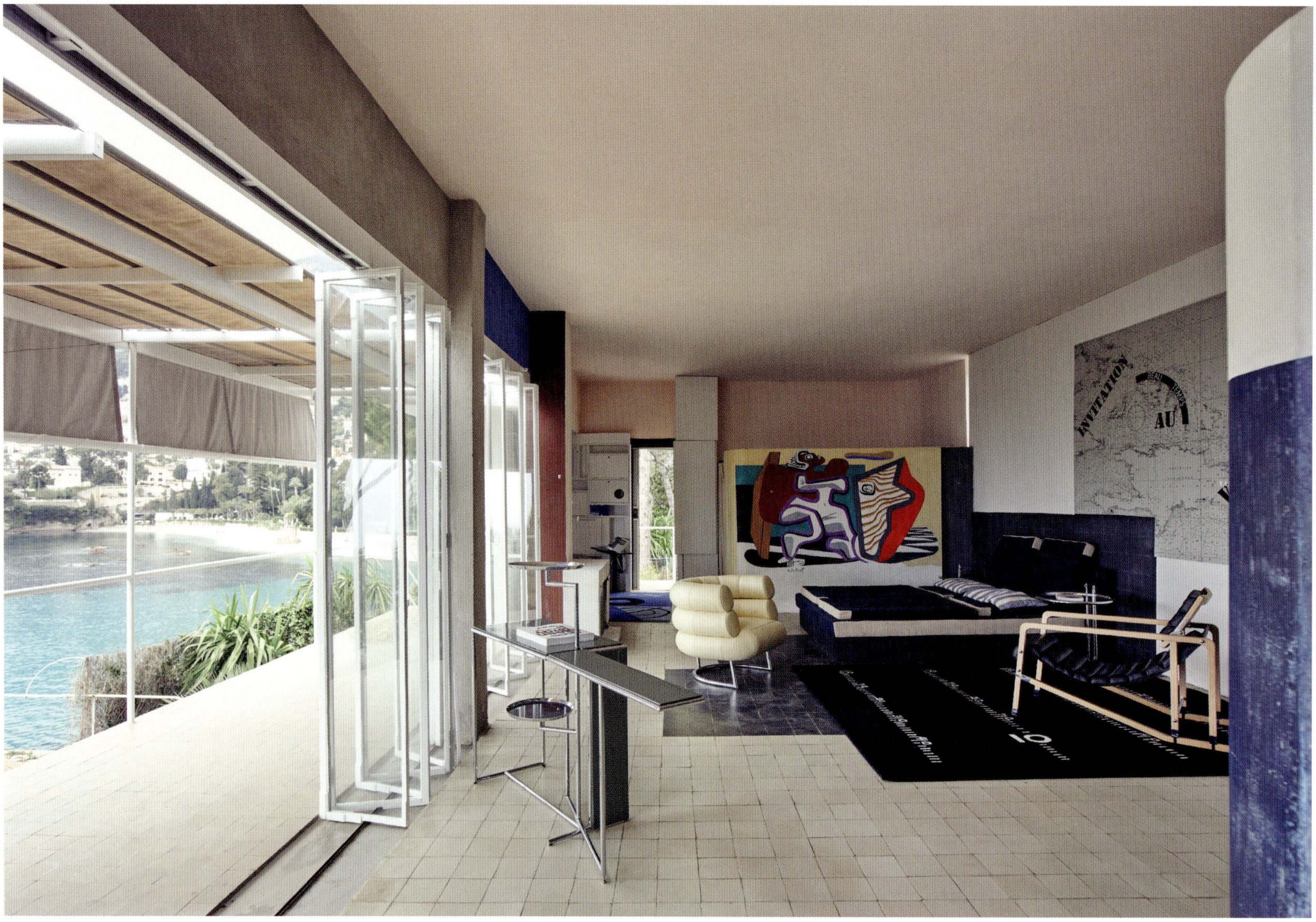

PAGES 340–41 The E-1027 house is tucked gently into the rugged slope of a hill, which features a series of stone-walled terraces, formerly used for cultivation.

ABOVE The main living room on the upper level is a multi-functional space, with a daybed in one corner plus seating and other furniture designed by Eileen Gray; the sliding doors (specified by Jean Badovici) open onto the elevated balcony.

RIGHT The exteriors of the house, with its upper-level 'decks', suggest a nautical influence; much of the ground floor is taken up by a sheltered terrace, or verandah, that sits underneath the main living spaces above.

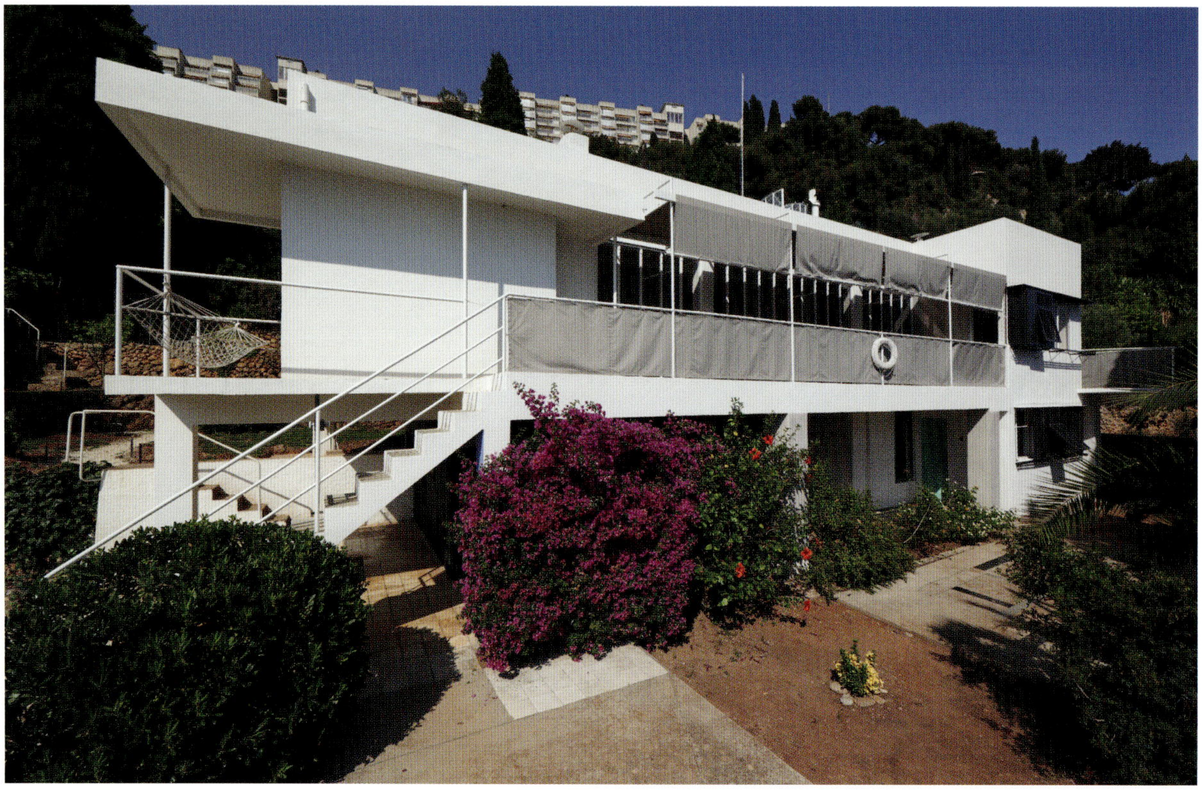

The villa by the sea that Eileen Gray designed for herself and her lover, Jean Badovici, sits in a small enclave only accessible via a slim path that runs alongside the rail line carrying the Metrazur between Marseilles and Ventimiglia, just across the Italian border. Voices carry upwards from the stony beach in the cradle of the bay below, along with the purring engines of the elegant pleasure boats cruising the green-blue waters. It is the picture-book setting for a modernist adventure peopled by large personalities, with Eileen Gray herself firmly at its heart.

Gray was born into an aristocratic Irish family, but eventually settled in Paris where she concentrated for many years upon her work as a furniture designer (see p. 48). Here she met Jean Badovici, a Romanian-born architect and journalist who, in Paris in 1923, launched the magazine *L'Architecture Vivante*, which soon became the bible of French modernism, championing the work of – among others – Le Corbusier, who became a friend of Badovici's.

Intrigued by her characteristic independence and determination, and impressed by her talents, Badovici encouraged Gray to broaden her work into architectural projects. Despite an age gap of 14 years, they also became lovers, although Gray retained a defensive reserve about this and other relationships – with men and women – across her unmarried life.

Badovici commissioned Gray to build a house for the two of them by the sea, and E-1027 became her first major architectural design beyond a series of interior and renovation projects. In 1925, she found the site at Roquebrune and bought the land in Badovici's name. The design of the house was largely Gray's conception, but included a number of suggestions and amendments from Badovici. From 1926 to 1929 Gray lived alone in a flat in Roquebrune as the house was built, with every piece of material transported along the narrow path by wheelbarrow.

The ship-like building was raised up on piloti, with decks, integrated terraces, ribbon windows and open-plan spaces. But this was no Corbusian 'machine for living in'. Purposefully, the house was a more organic and emotional creation that responded to the site, the views across the sea, and Badovici and Gray's own desires and ambitions. 'A house is not a machine to live in,' she declared. 'It is the shell of man, his extension, his release, his spiritual emanation.'[19]

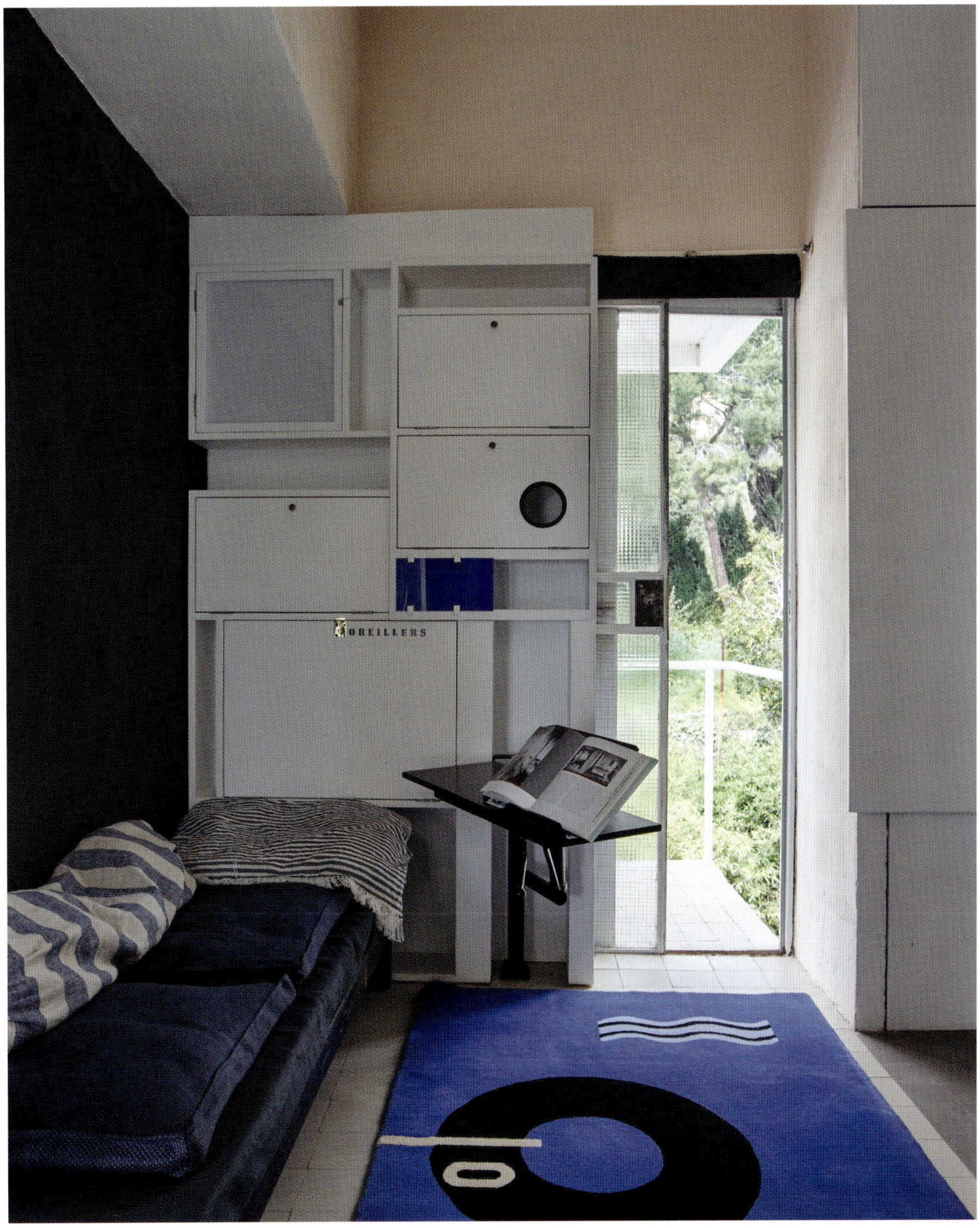

ABOVE The house features many integrated pieces of furniture designed by Gray, as well as free-standing pieces; this alcove sits just beyond the sitting room and opposite a small shower room.

Upstairs there was a large, open-plan and multi-functional living area with a sequence of folding, floor-to-ceiling glass windows (designed by Badovici) to the balcony. There was space enough here for a dining area and a sleeping zone for guests, while integrated shelving and furniture designed by Gray made the space both more cohesive and practical. The upper storey, which enjoyed the best of the views, included the master bedroom and bathroom, as well as the kitchen, situated close to the entrance at the rear of the house.

The design integrated terraces, courtyards and balconies, oriented towards the sea, while canvas awnings were tied to the metal balustrades of the balcony for shade. Crucially, the interiors

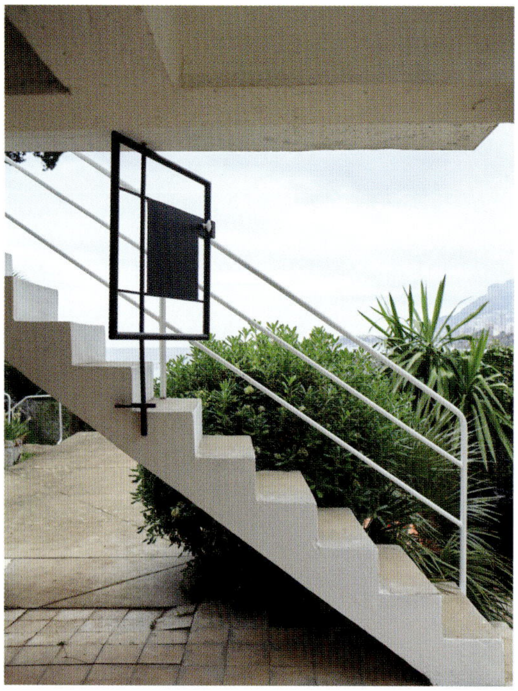

ABOVE Two outdoor staircases at opposite ends of the house lead to the main living spaces on the upper level; the ground floor is largely devoted to the maid's quarters, a guest bedroom and the sheltered terrace.

BELOW A discreet shower room sits to one side of the living room; the daybed is a flexible design that can also serve as a sleeping couch.

OPPOSITE Colour has been used extensively, as seen in the master bathroom.

of the house were conceived as an essential part of the whole, with Gray designing some of her most celebrated pieces of furniture specifically for E-1027, including her circular glass E-1027 table, Transat deck chair and Bibendum armchair.

Badovici was delighted, and both he and Gray spent a number of seasons at the house. Le Corbusier came to visit in 1937 and was instantly mesmerized by the building, later writing in a letter to Gray of his appreciation of 'the rare spirit which dictates all of its organization, both inside and outside'.

But all began to unravel as Badovici's womanizing prompted Gray to keep her distance, building a new house for herself, known as Tempe a Pailla, further down the coast near Menton. Badovici allowed Le Corbusier, who had become increasingly obsessed with Gray's achievement, to paint a series of murals at E-1027, which infuriated its architect, who saw the murals as vandalism.

Le Corbusier's proprietorial attitude to both the house and the site never left him. He had fallen in love with the area and the sea and could not shake off his obsession. During World War I, while Gray stayed in France under supervision as a resident alien, E-1027 was occupied by German and Italian soldiers, while Gray's own home near Menton and her apartment in Saint-Tropez were looted or badly damaged. After the war, Gray's architectural output was very limited, but included another house for herself at Chapelle-Sainte-Anne, known as Lou Pérou (1961).

Eventually Le Corbusier managed to obtain a small plot of land a stone's throw away from E-1027 and built a modest wooden cabin for himself and his wife, completed in 1952 and almost the total antithesis to E-1027 – and much of his own work – in terms of its sheer simplicity. From here he kept an eye on Gray's creation right up to his death on the rocks below in 1965 from a heart attack following a swim in the ocean.

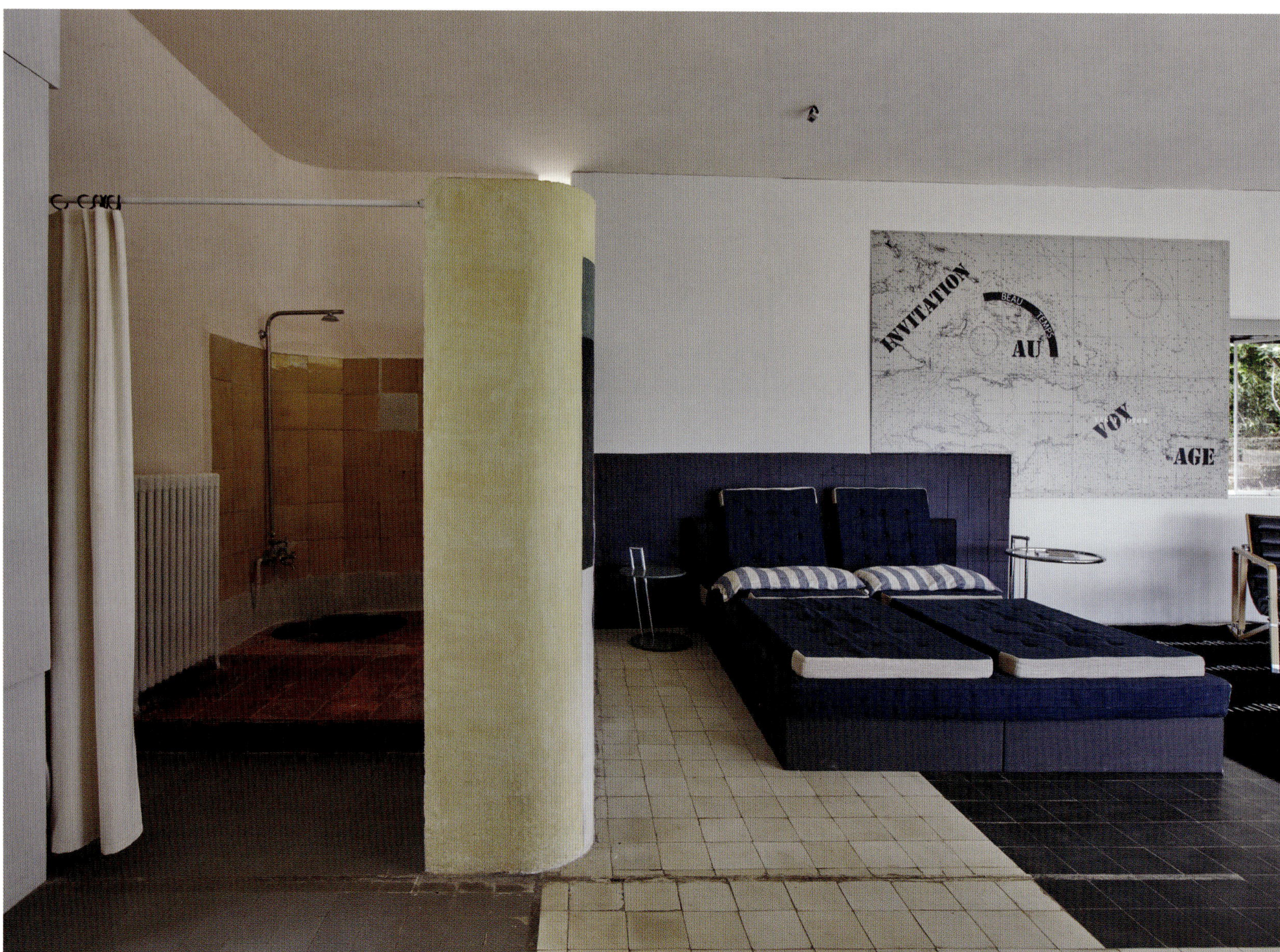

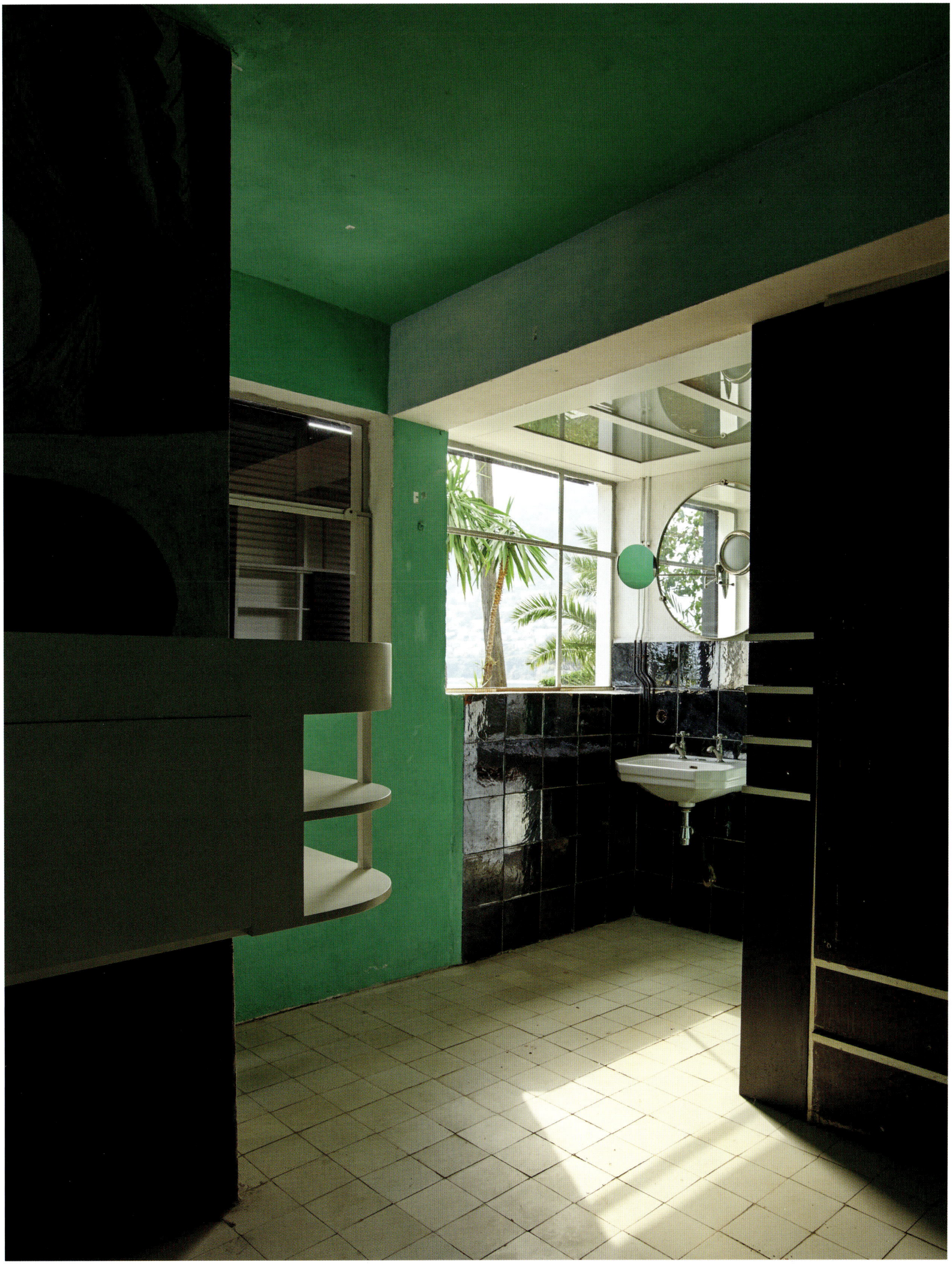

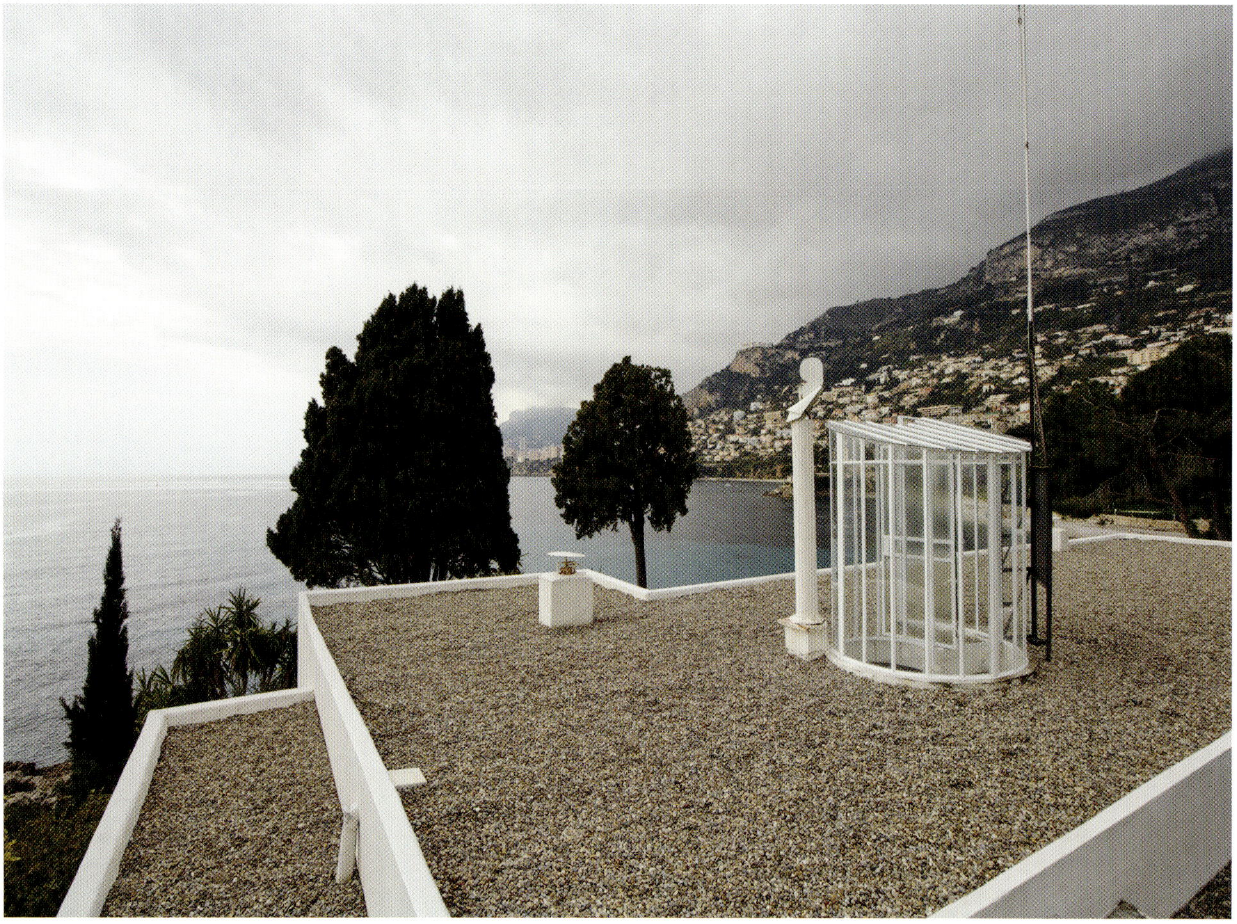

LEFT & OPPOSITE From its elevated position, the house enjoys spectacular views of the Mediterranean, best seen from the upper level and its balconies.

RIGHT The dining area sits to one side of the main living room; an indoor spiral staircase sits beyond the painted wall.

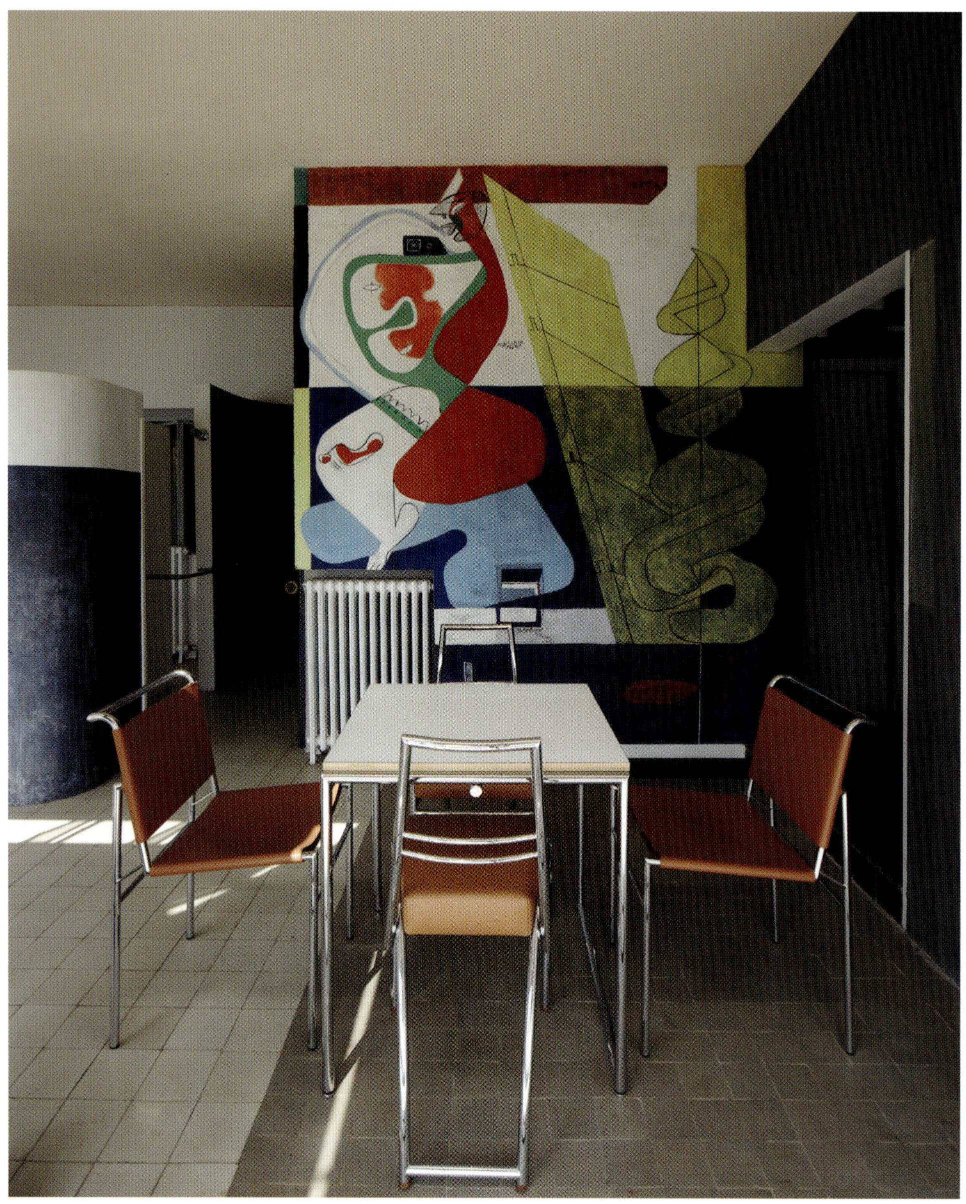

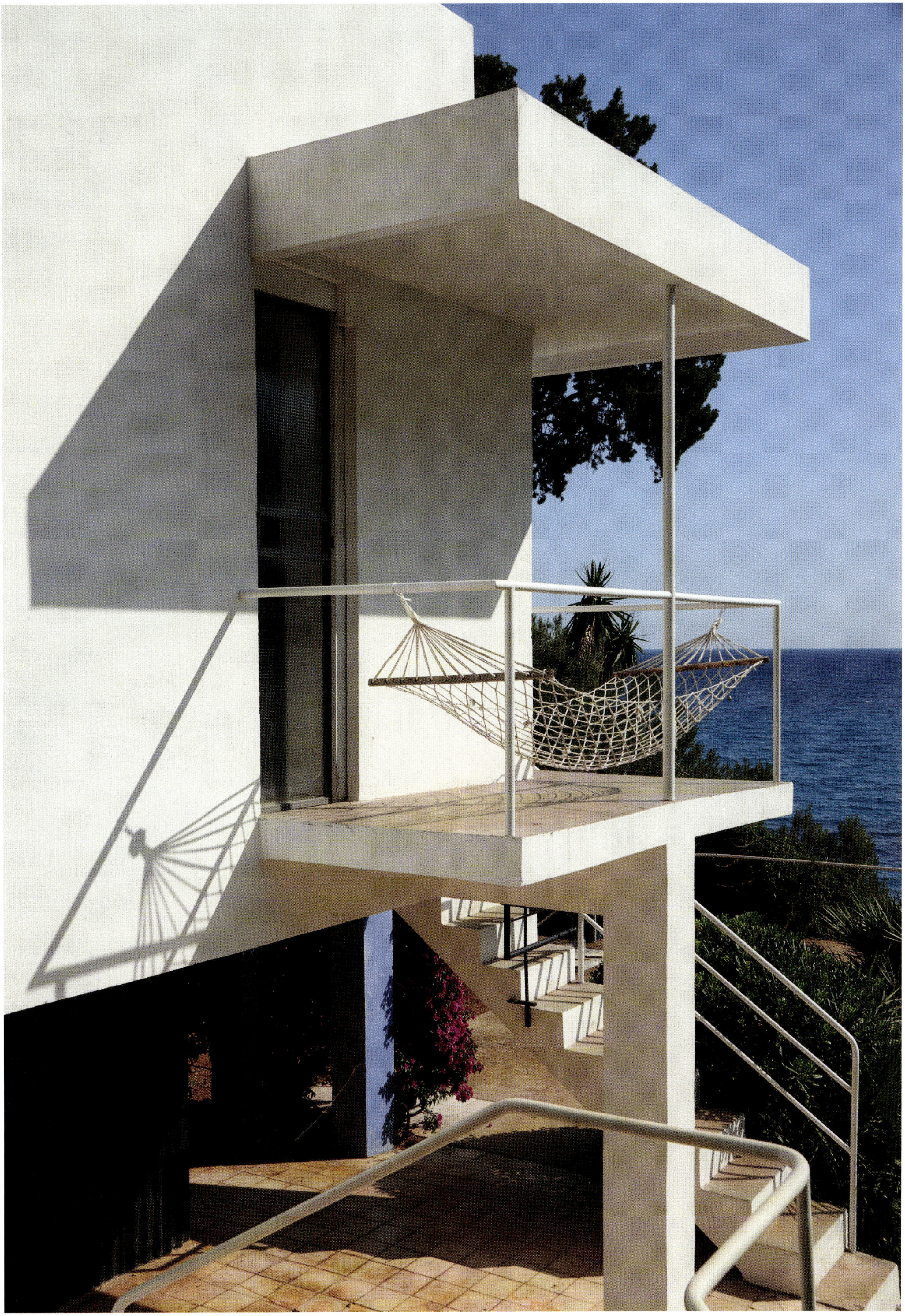

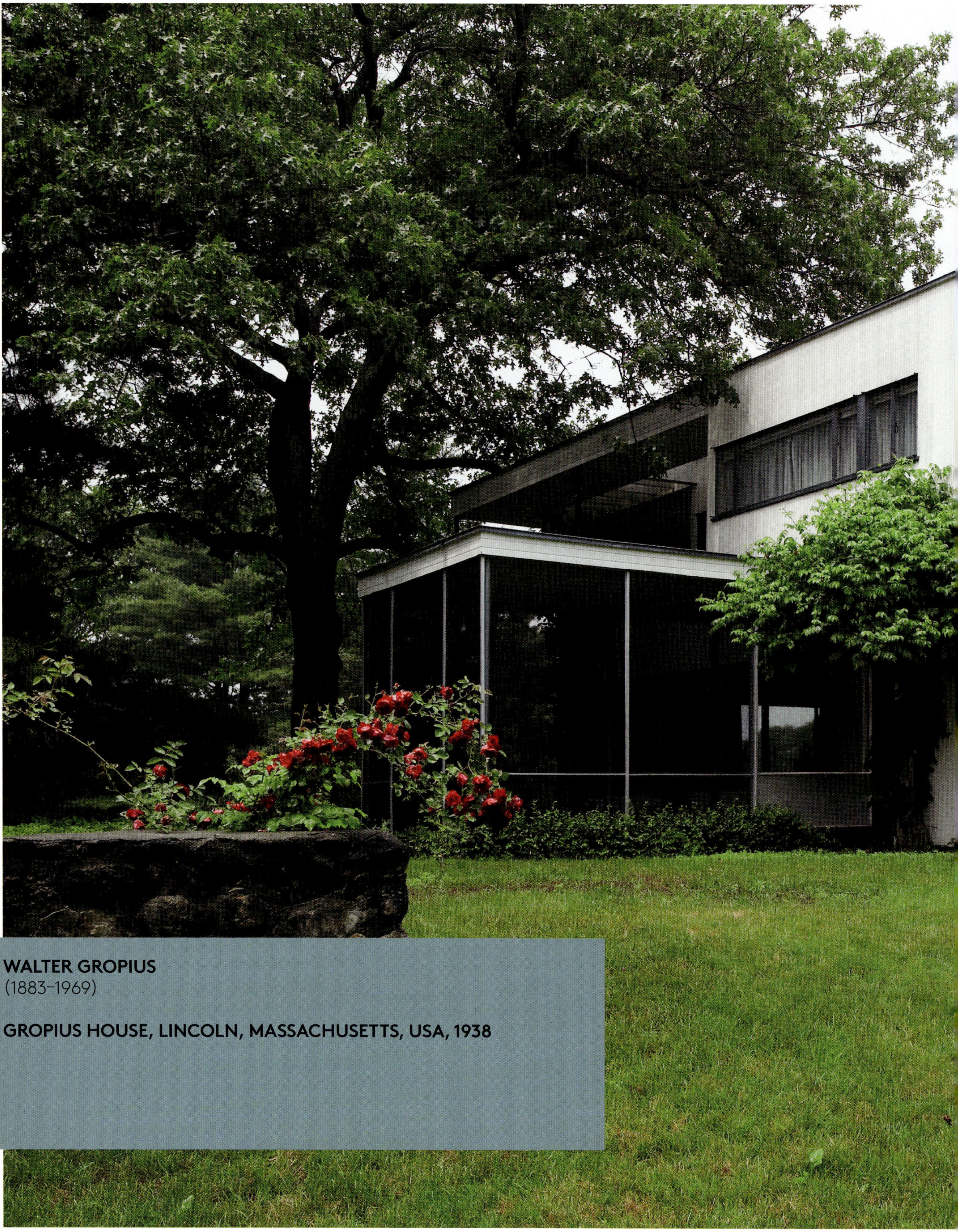

WALTER GROPIUS
(1883–1969)

GROPIUS HOUSE, LINCOLN, MASSACHUSETTS, USA, 1938

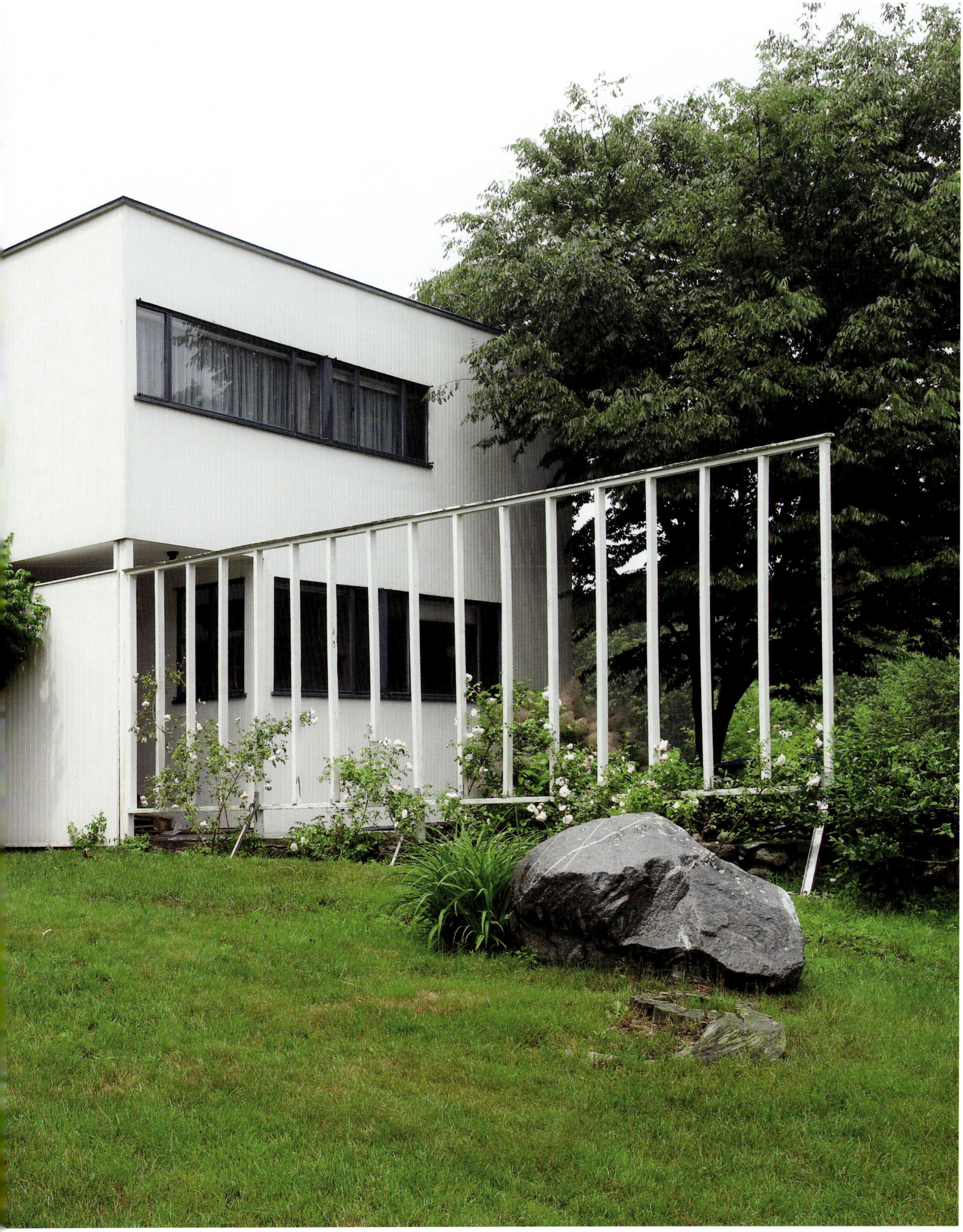

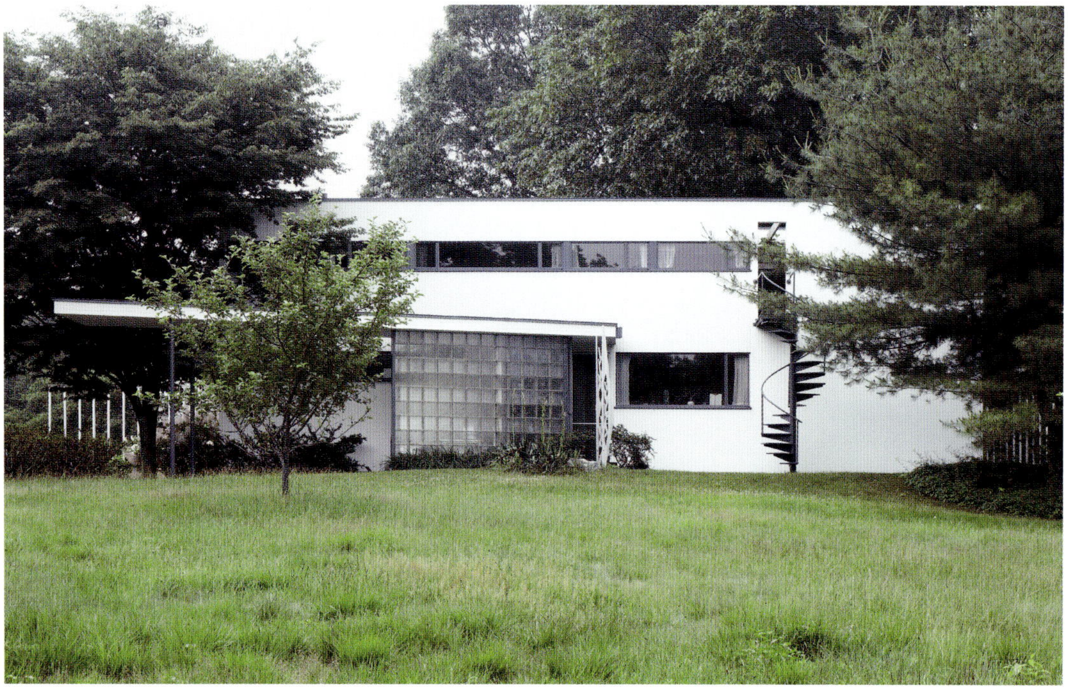

PAGES 348–9 The timber-framed house updates and reinterprets many traditional elements in a highly modern form, including the screened porch that sits at the rear of the building.

LEFT & BELOW A number of axial sight lines are continued in the garden, as seen in the linear gravel garden (below) which continues outwards from the body of the house in the direction established by the rear porch.

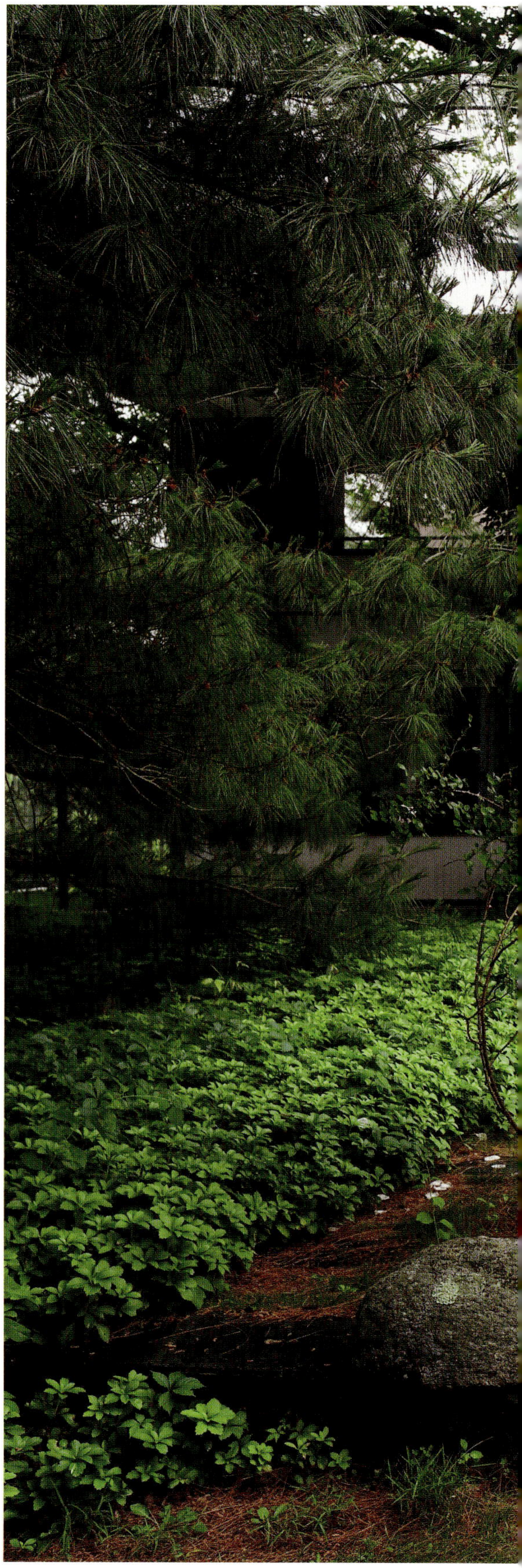

As both the father of the Bauhaus and a pioneering and original architect, Walter Gropius was one of the most influential figures of the modernist movement. Having worked with PETER BEHRENS in Berlin and then run his own practice in Potsdam, Gropius was named the director of the College of Fine Arts and the School of Arts and Crafts in Weimar, in the year 1919. Gropius combined the two institutions – and the worlds of art, craft and design – in the form of the Bauhaus, which became a byword for avant-garde modernity.

Gropius created a unique teaching approach and curriculum at the Bauhaus, with students beginning with a general introductory course in art and design before moving on to specialize in specific workshops run by dedicated masters, including studios devoted to cabinet making, typography, textiles, pottery and other disciplines. The Bauhaus model was innovative and progressive, with tutors including Paul Klee*, JOSEF ALBERS, MARCEL BREUER and Wassily Kandinsky*, but at the same time Gropius was determined to ensure connections to the real world and to encourage students to design for function, as well as respecting art and craft, while developing ideas that could translate into mass-produced, industrial products.

In 1925, the Bauhaus relocated to the town of Dessau, encouraged by the local council and a synergy with local industry and production, which included the Junkers airplane factory. Here, Gropius designed the Bauhaus school buildings and a series of houses for rent by the senior masters. The campus became a modernist portfolio in itself, consisting of a series of crisp, rendered buildings with fluid floor plans, curtain walls and picture windows. These were spaces full of light and drama, with the masters' houses (1926) forming a cohesive collection of white boxes surrounded by pine trees, with integrated decks and balconies.

The glory days of the Bauhaus were, of course, cut short by politics and the rise of the National Socialist Party, which was suspicious of the school's outlook and teaching methods. Gropius stepped down as director in 1928, succeeded briefly by Hannes Meyer* and then LUDWIG MIES VAN DER ROHE, before the school was shut down altogether in 1932. Many of the Bauhaus masters, including Gropius, emigrated to America and often via England, where Gropius lived and worked (in partnership with Maxwell Fry*) for a short period.

In 1937, Gropius and his family arrived in the States, where he became head of the architecture department at Harvard, supported by Breuer with whom he also formed an architectural partnership. The first building that Gropius completed in America was his own house in Lincoln, Massachusetts, on land provided for him by a generous patron. The site consisted of four acres not far from Walden Pond and around half an hour's journey from the Harvard campus.

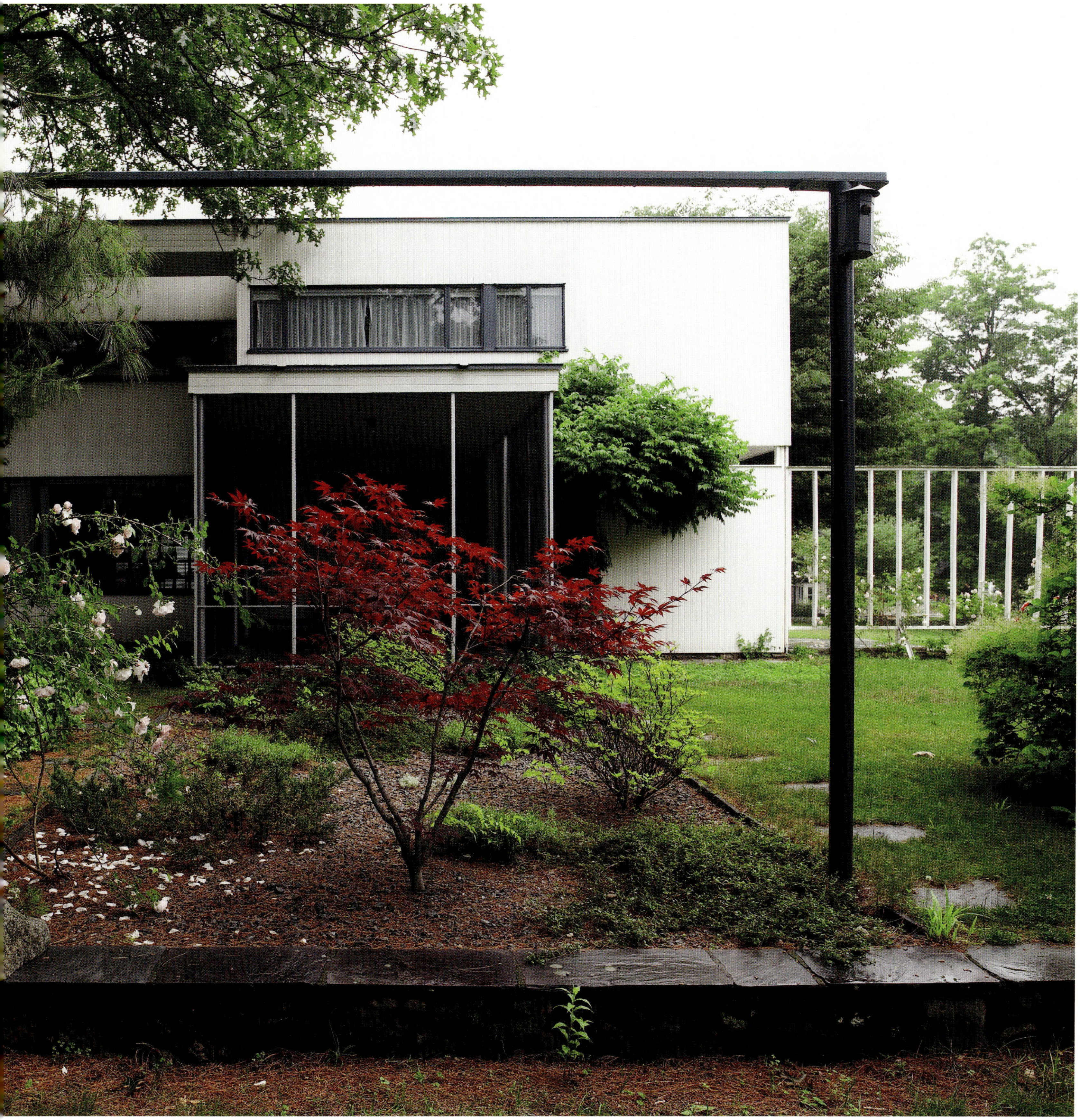

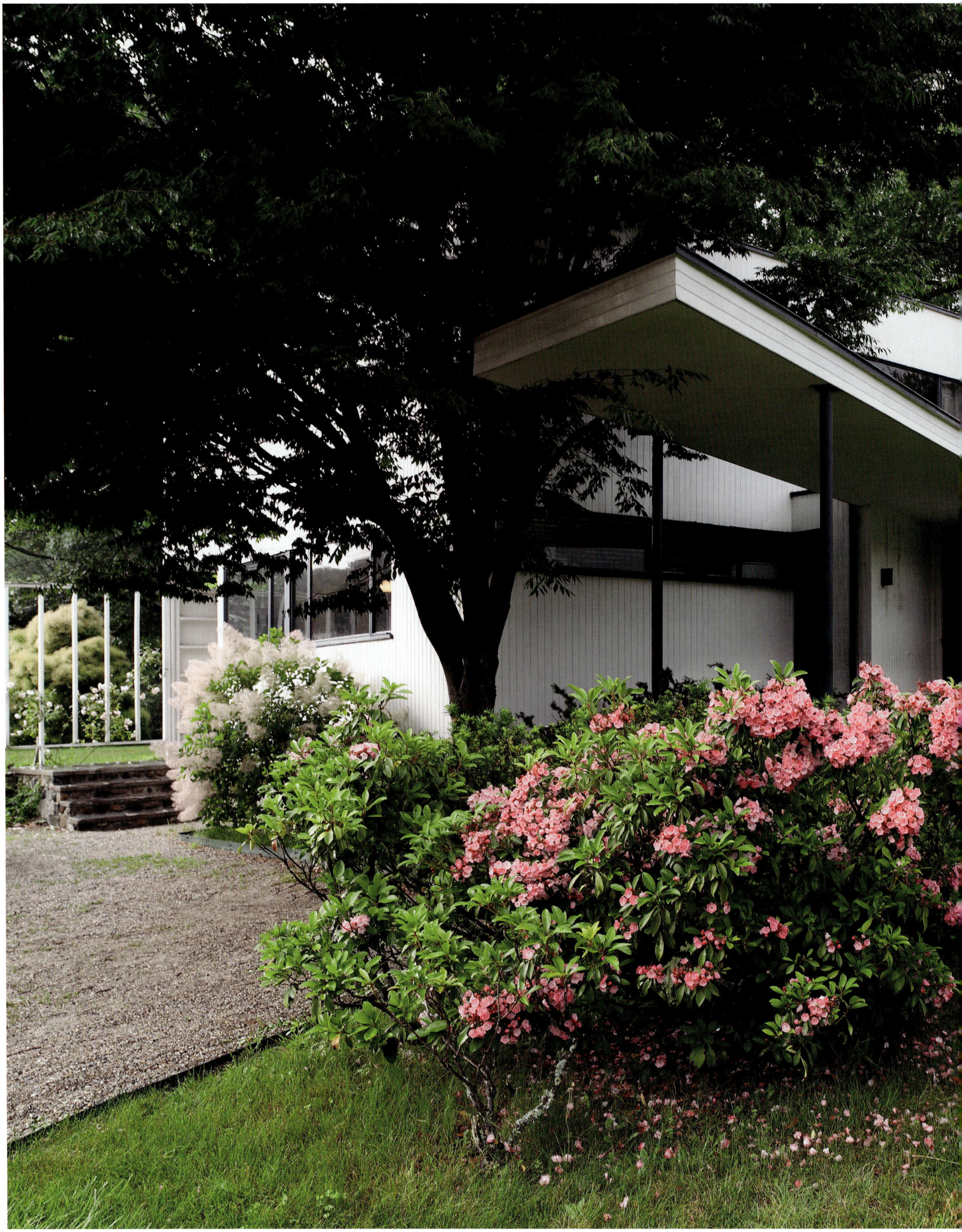

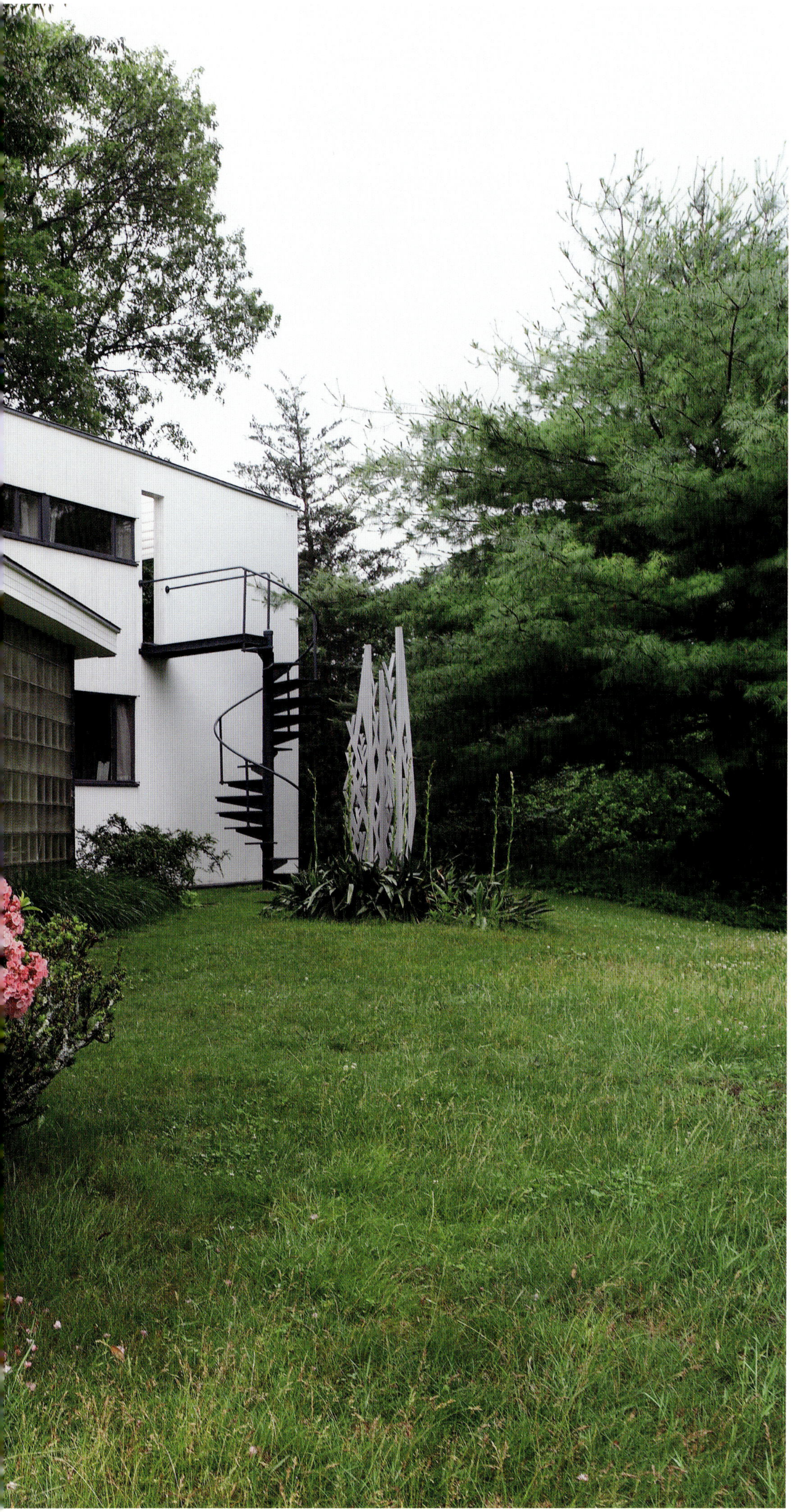

The house exemplifies many key Bauhaus and modernist principles, whereby form and function are intertwined and intelligent engineering and planning allow freedom in the floor plan, a fluid journey from one space to another as well as openness to the landscape via large expanses of glazing. The building has a two-storey cubic form, with ribbon windows, subtly subverted by an angular entrance canopy that projects like a wing from one side of the building and a screened porch at the other. On the ground floor, the living and dining areas are arranged in a largely open-plan formation, with a separate study alongside; the kitchen, service spaces and a maid's room are at the opposite end of the house. Upstairs, there are three bedrooms and an integrated roof terrace, which offers a sun deck and viewing platform looking out across the countryside.

The Gropius House is, of course, distinctly modern in form and design. It has been seen as a key exemplar of the International Style*, although Gropius himself disliked the term. For Gropius the house was highly contextual and rooted in a sense of place and the needs of his wife and daughter, whom he consulted extensively throughout the design process.

'[W]hen I built my first house in the USA ... I made it a point to absorb into my own conception those features of the New England architectural tradition that I found still alive and adequate,' Gropius wrote. 'This fusion of the regional spirit with a contemporary approach to design produced a house that I would never have built in Europe with its entirely different climatic, technical and psychological background.'[20]

The Gropius-Breuer architectural partnership lasted just a few years, with Gropius forming a new practice, The Architects Collaborative, in 1945. Key projects of the post-war period include the Graduate Center at Harvard University (1950) and the landmark Pan Am (or MetLife) Building in New York (1963), which remains one of the most visible and famous statements by the founding father of the Bauhaus.

LEFT The main entrance is protected by a projecting canopy, which extends outwards at an angle and lightly subverts the essentially box-like form of the house itself; the exterior spiral staircase leads to a roof deck on the upper level.

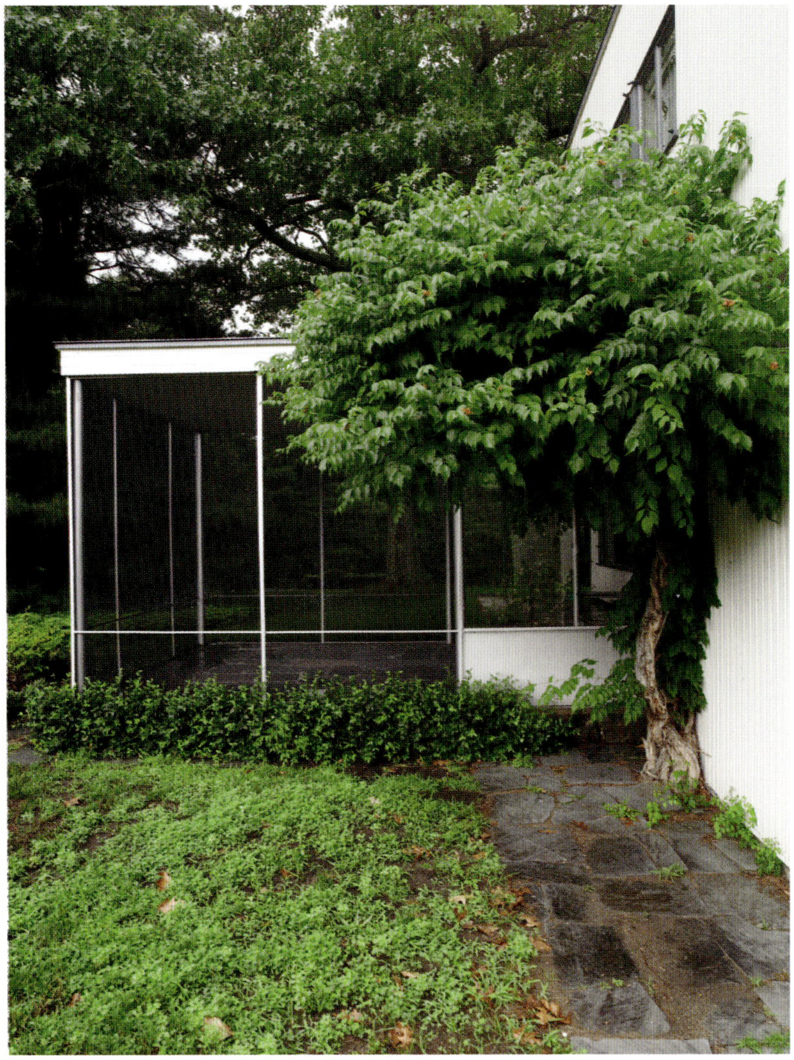

RIGHT The ground-floor living and dining room is a fluid, open space; the study to one side is separated from the living space by an internal wall of glass bricks.

BELOW Looking through the front door to the entry porch, which is sheltered by both a canopy and a partial wall of glass bricks.

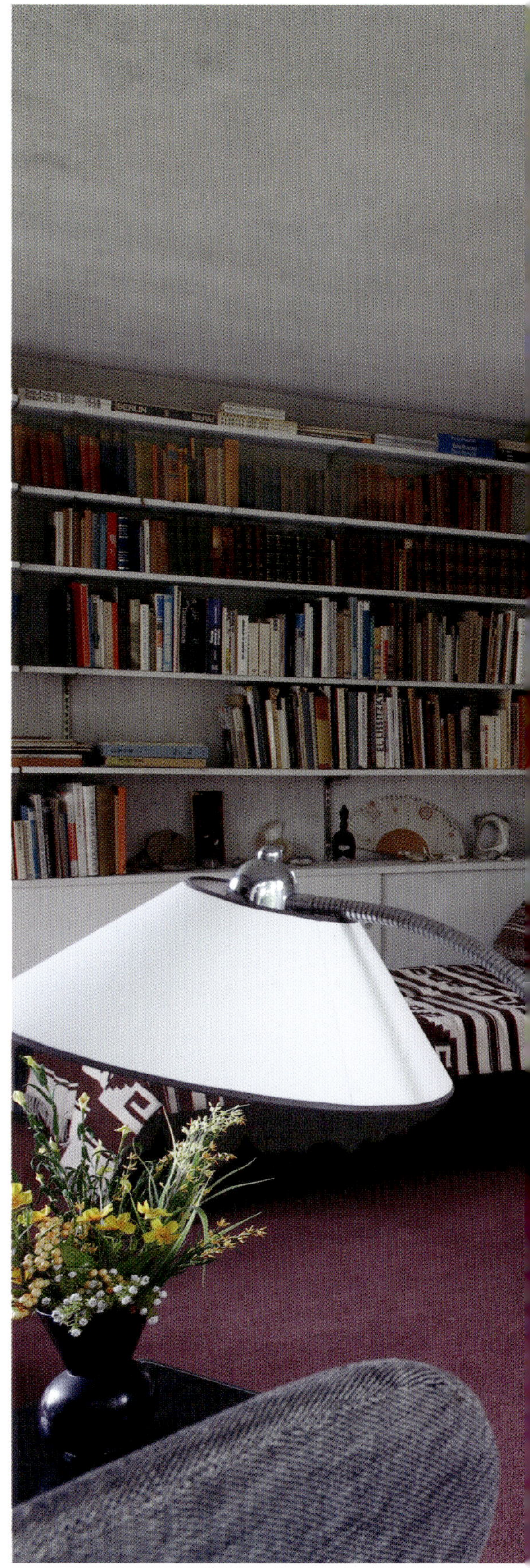

ABOVE A view of the screened porch at the rear of the house, with the trees beyond; the site was formerly an orchard.

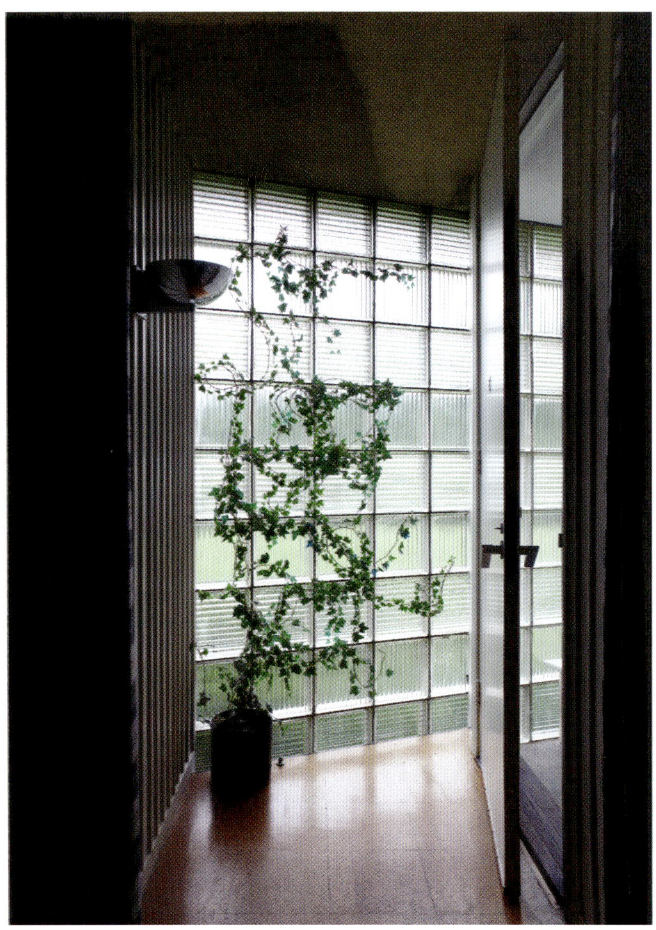

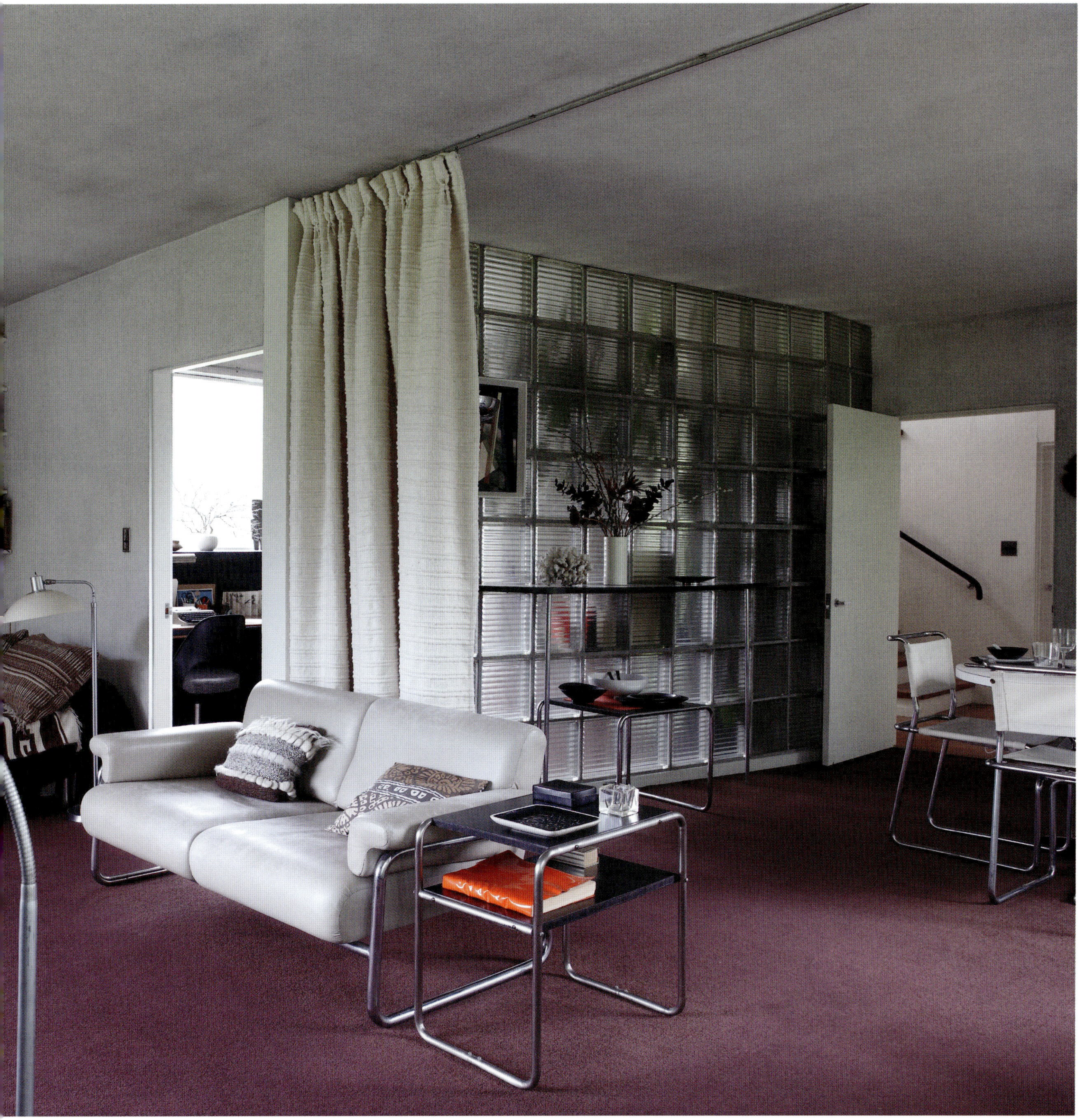

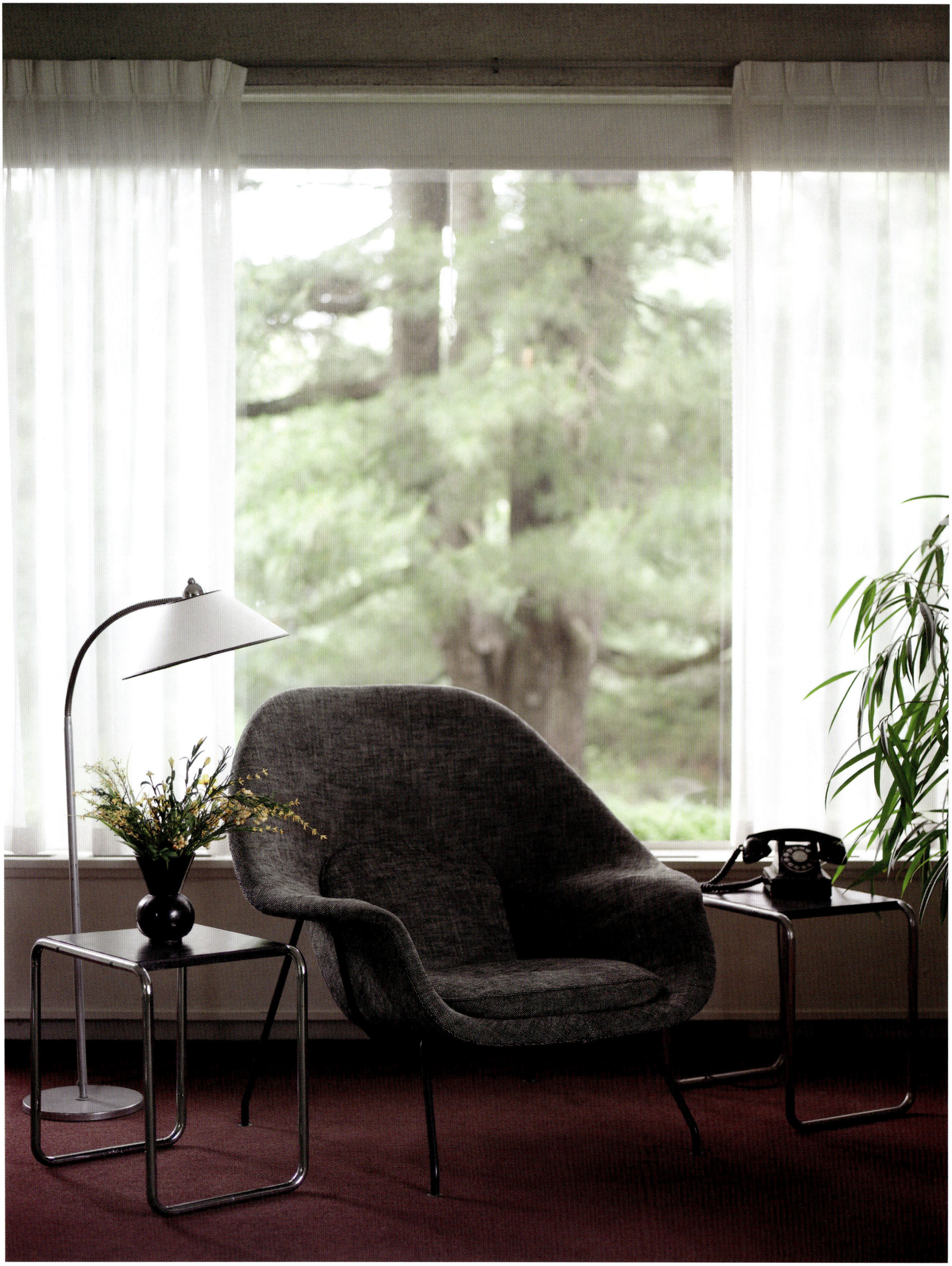

OPPOSITE Walter Gropius's choice of furniture included many pieces designed by his Bauhaus colleague Marcel Breuer, as well as Eero Saarinen's Womb chair (1946).

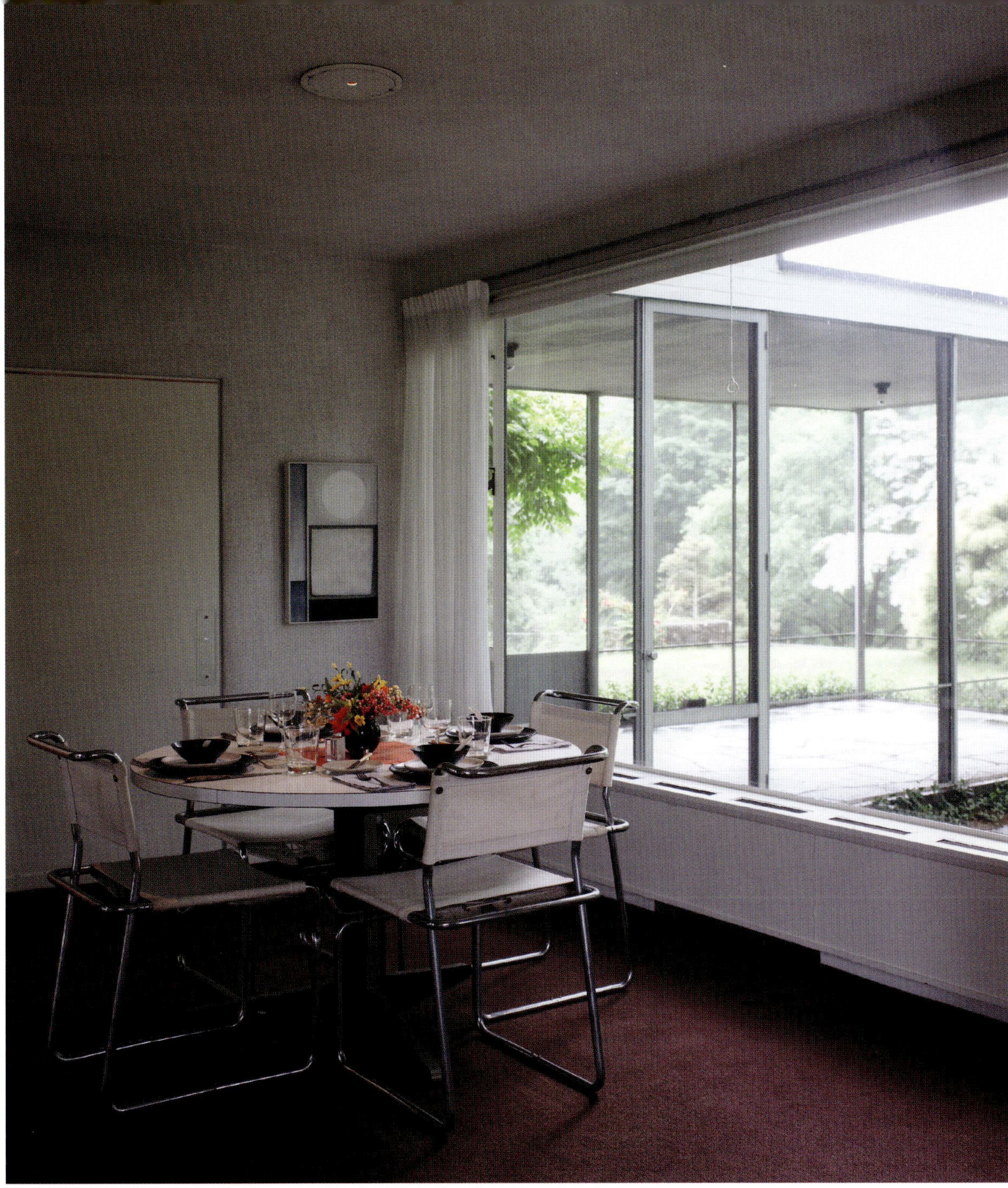

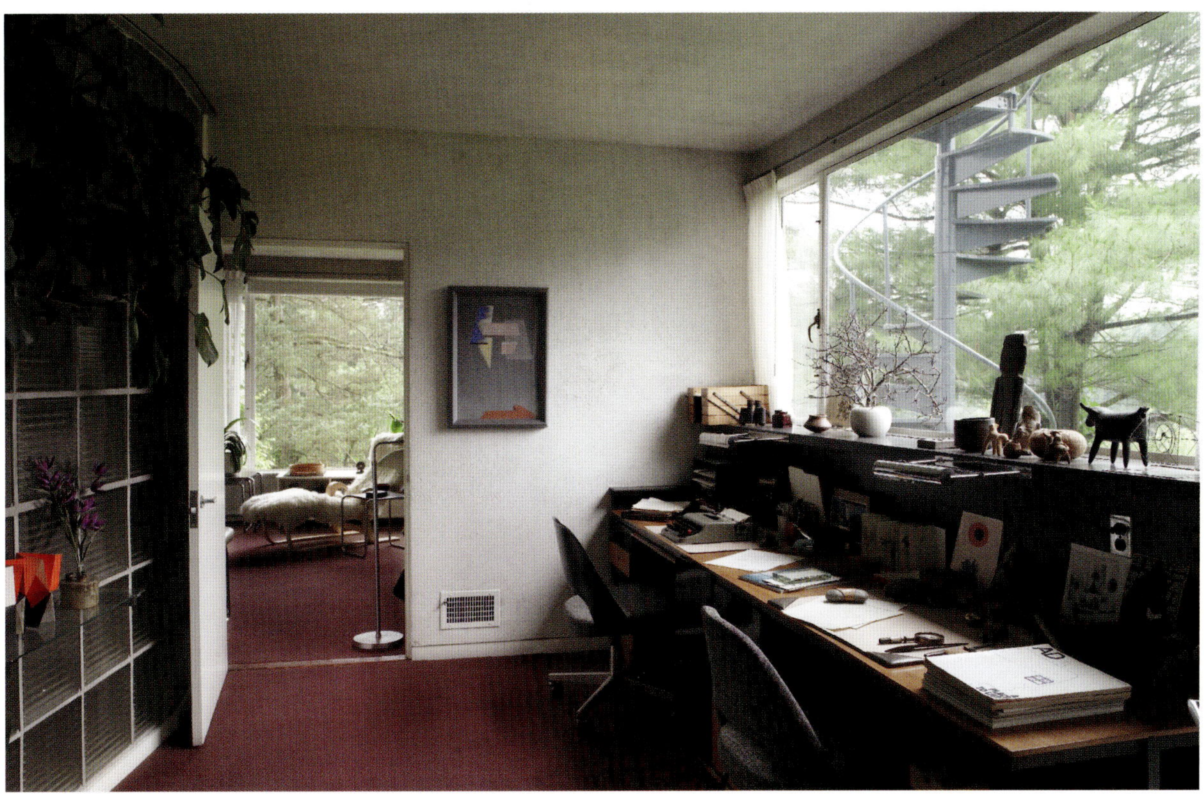

ABOVE The dining area sits to one side of the main living space; the dining chairs are Marcel Breuer designs.

LEFT A view of the study, which is situated at the front of the house close to the main entrance, offering a view of any approaching visitors.

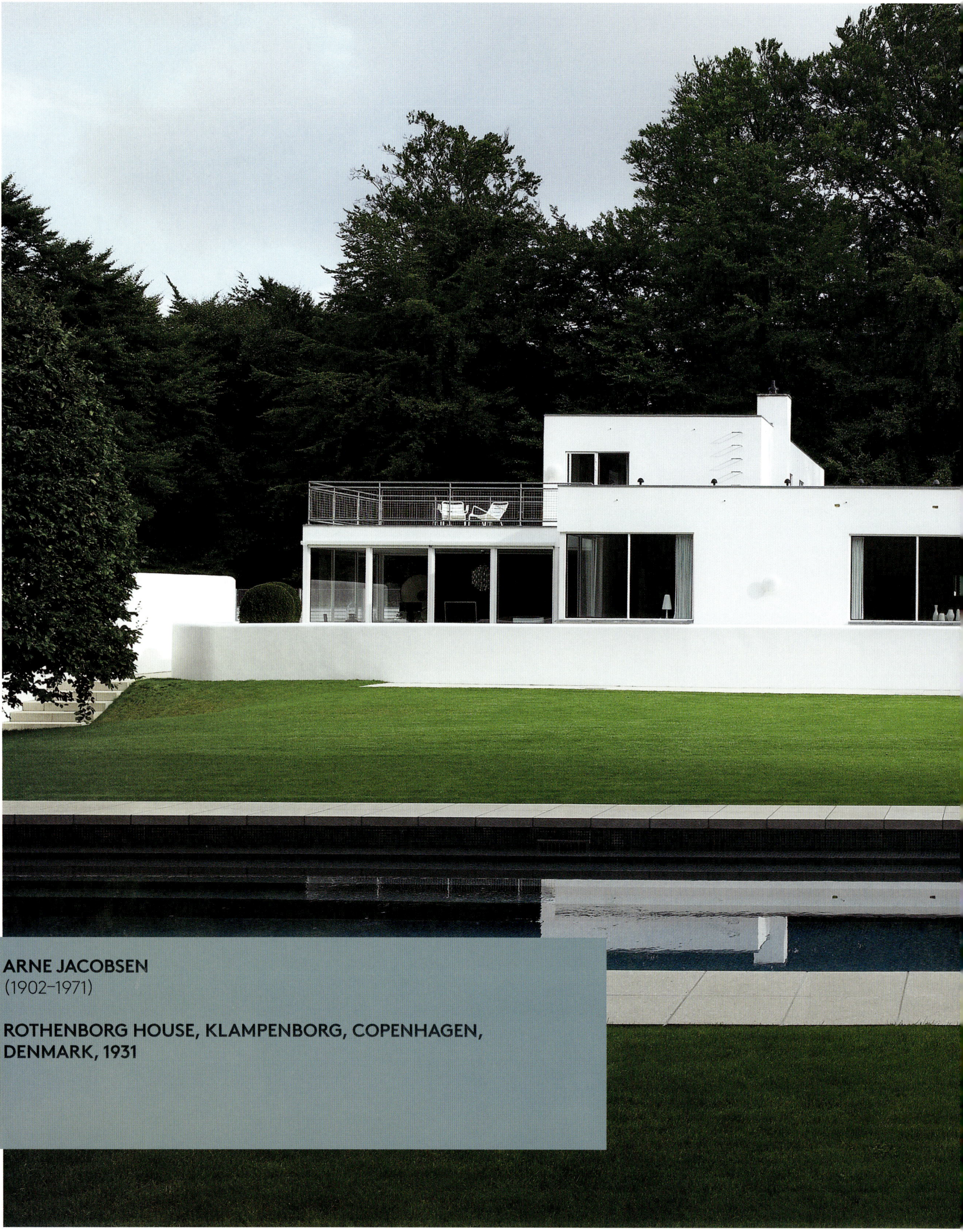

ARNE JACOBSEN
(1902–1971)

ROTHENBORG HOUSE, KLAMPENBORG, COPENHAGEN, DENMARK, 1931

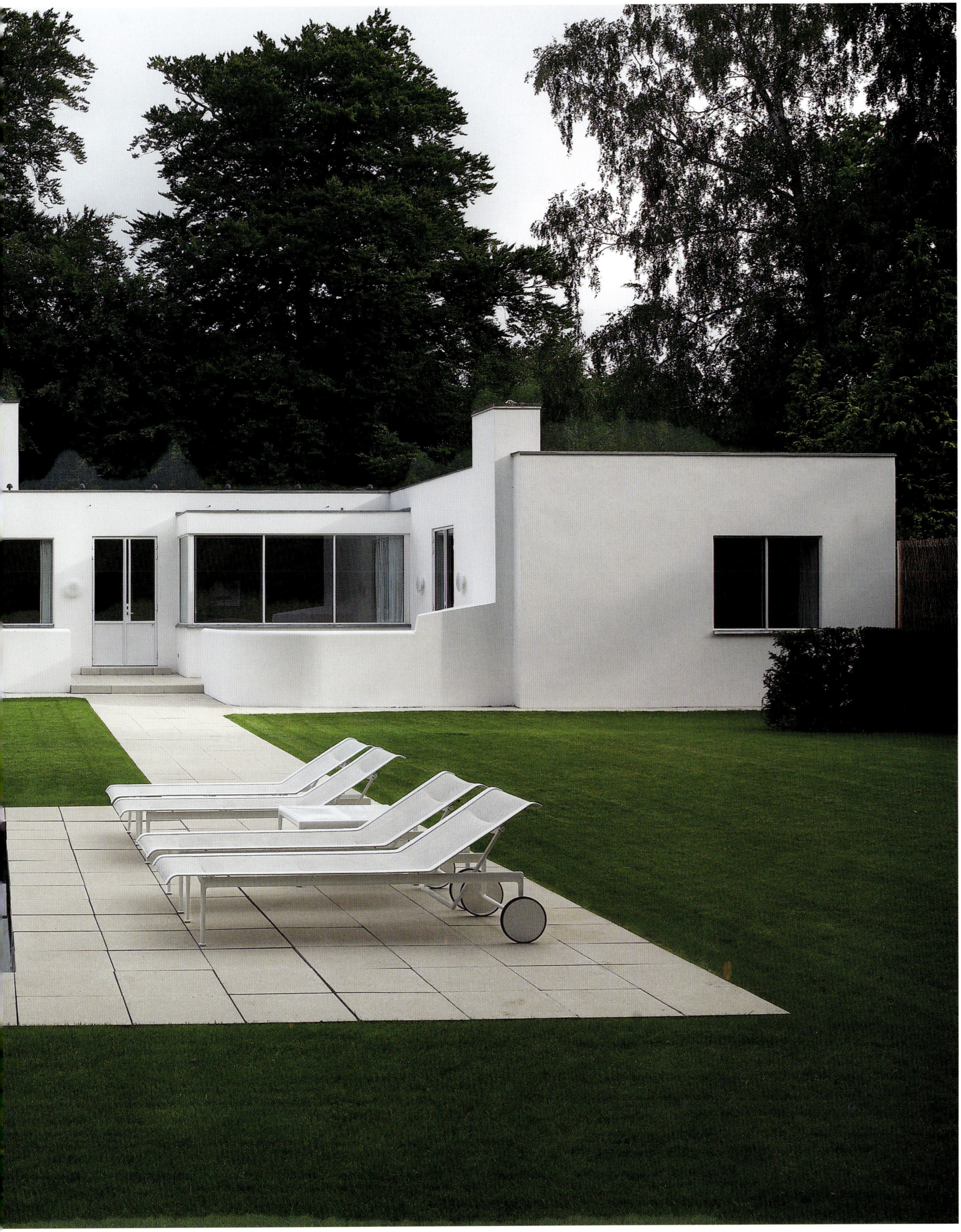

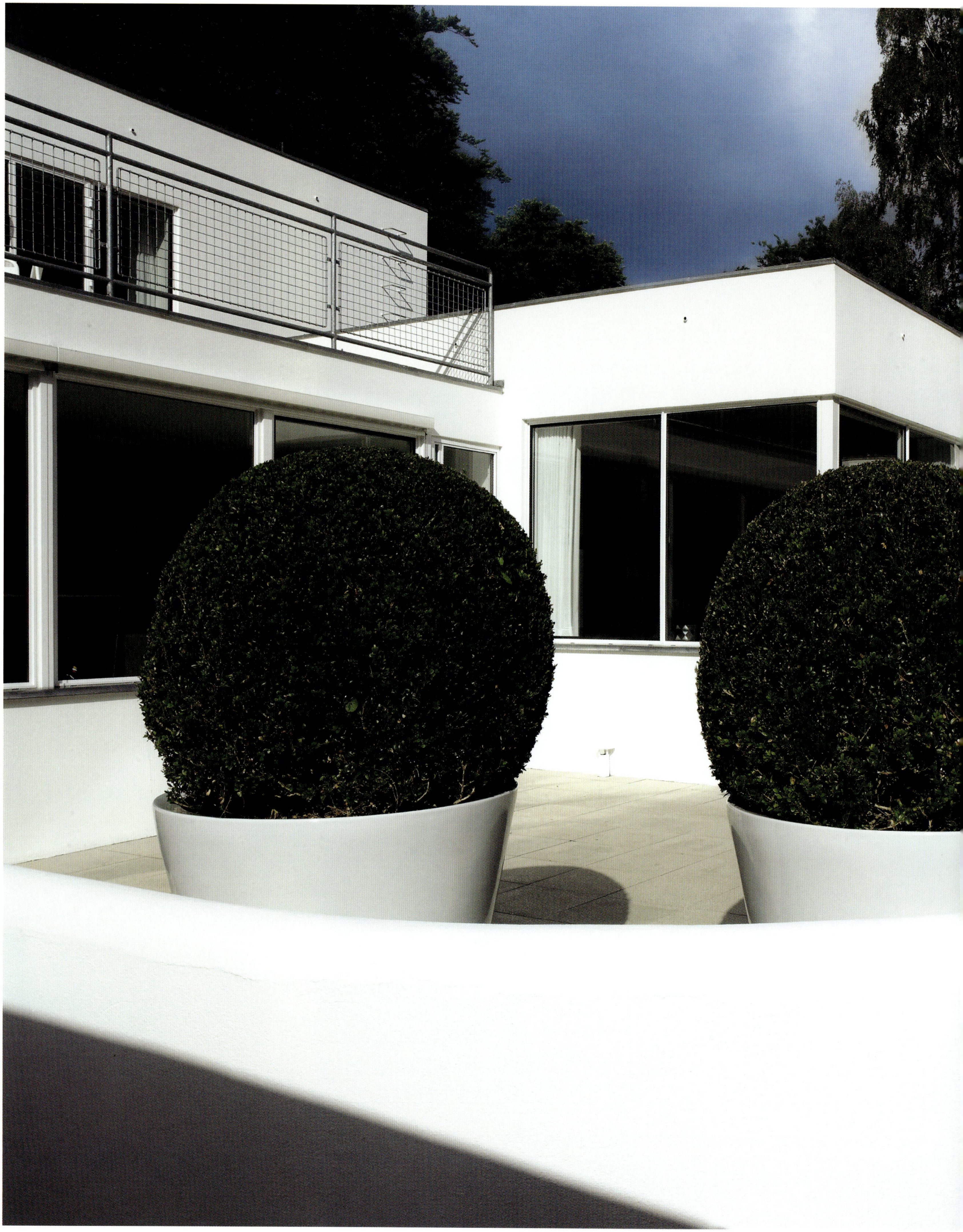

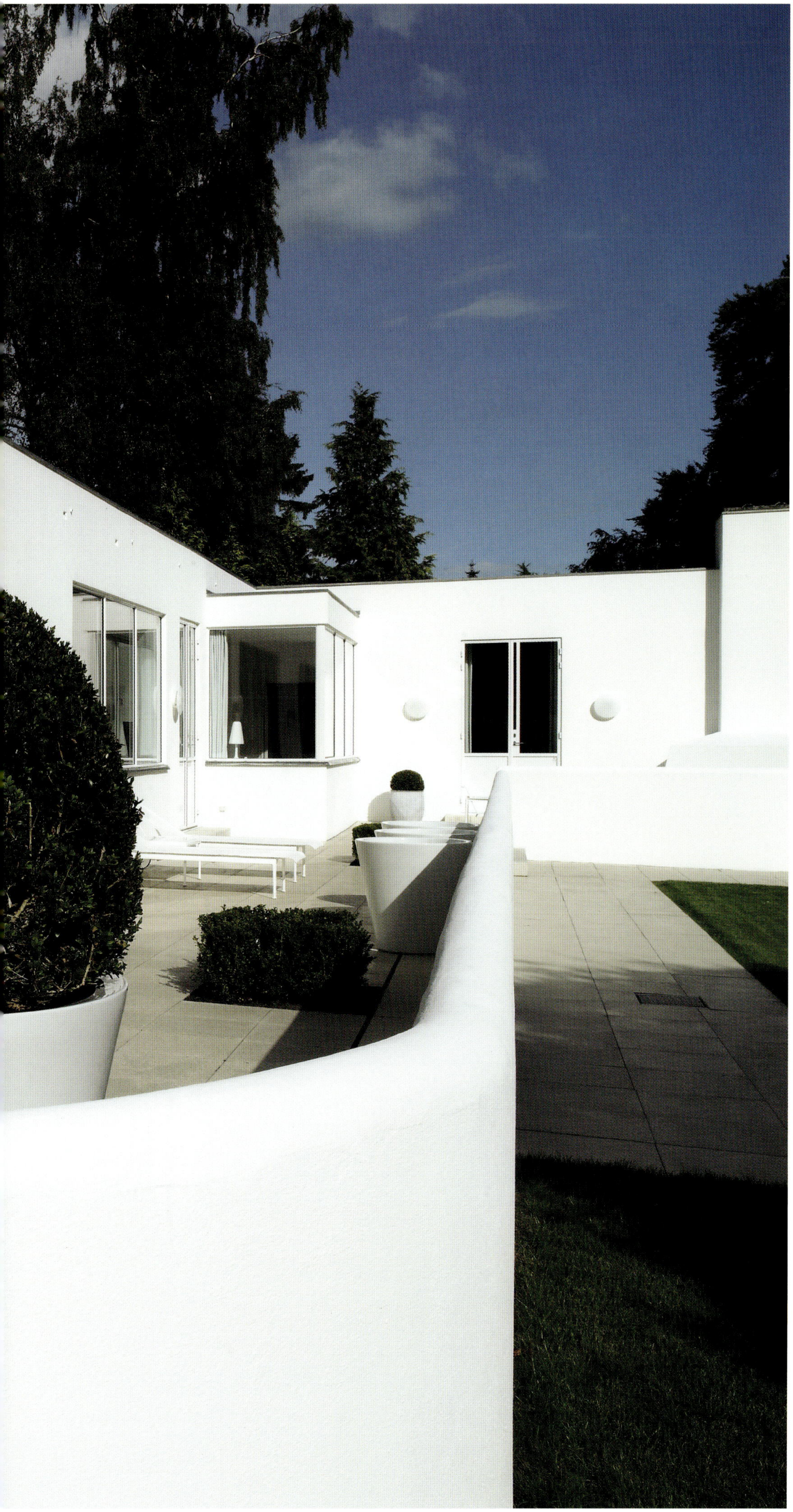

In 1929, Arne Jacobsen designed the 'House of the Future'. Created in collaboration with Flemming Lassen* for an exhibition in Copenhagen, the house was a two-storey circular building in concrete and glass, which featured roof decks and a helicopter pad, as well as wind-down windows and a boathouse. This prototype dwelling marked a turning point in Jacobsen's career, with the rejection of the neoclassical influences that had dominated Nordic design in favour of something more forward-thinking, influenced by the growing force of modernism in Europe and beyond.

The House of the Future also marked Jacobsen out as a polymath designer. Like his Finnish contemporary ALVAR AALTO, Jacobsen was much more than an architect and ready to look at a building in its totality, turning his attention to the interiors, furnishings, landscaping and detailing. His projects, famously, tended to be 'Gesamtkunstwerks' – total works of art. Many of his most famous creations emerged from key commissions for houses, hotels and buildings that resulted not only in a fully crafted space but also in some of the most familiar and resilient furniture and lighting designs of the 20th century.

Having shown talent as an artist as a child, Jacobsen trained as a stonemason initially, which encouraged a love of both craft and materials. He went on to study architecture at the Royal Academy of Arts in Copenhagen, and as a student travelled to Paris and to Berlin, where he discovered the work of the Bauhaus masters. In his early projects, including a house for himself in Charlottenlund (1929), Jacobsen veered towards modernist and functionalist forms, with cubist outlines, white rendered façades and large windows framing the views of the landscape beyond.

PAGES 358–9 & LEFT The crisp white outline of the house stands out vividly against the woodland backdrop, while terraces and a roof deck provide outdoor rooms connected to the landscape.

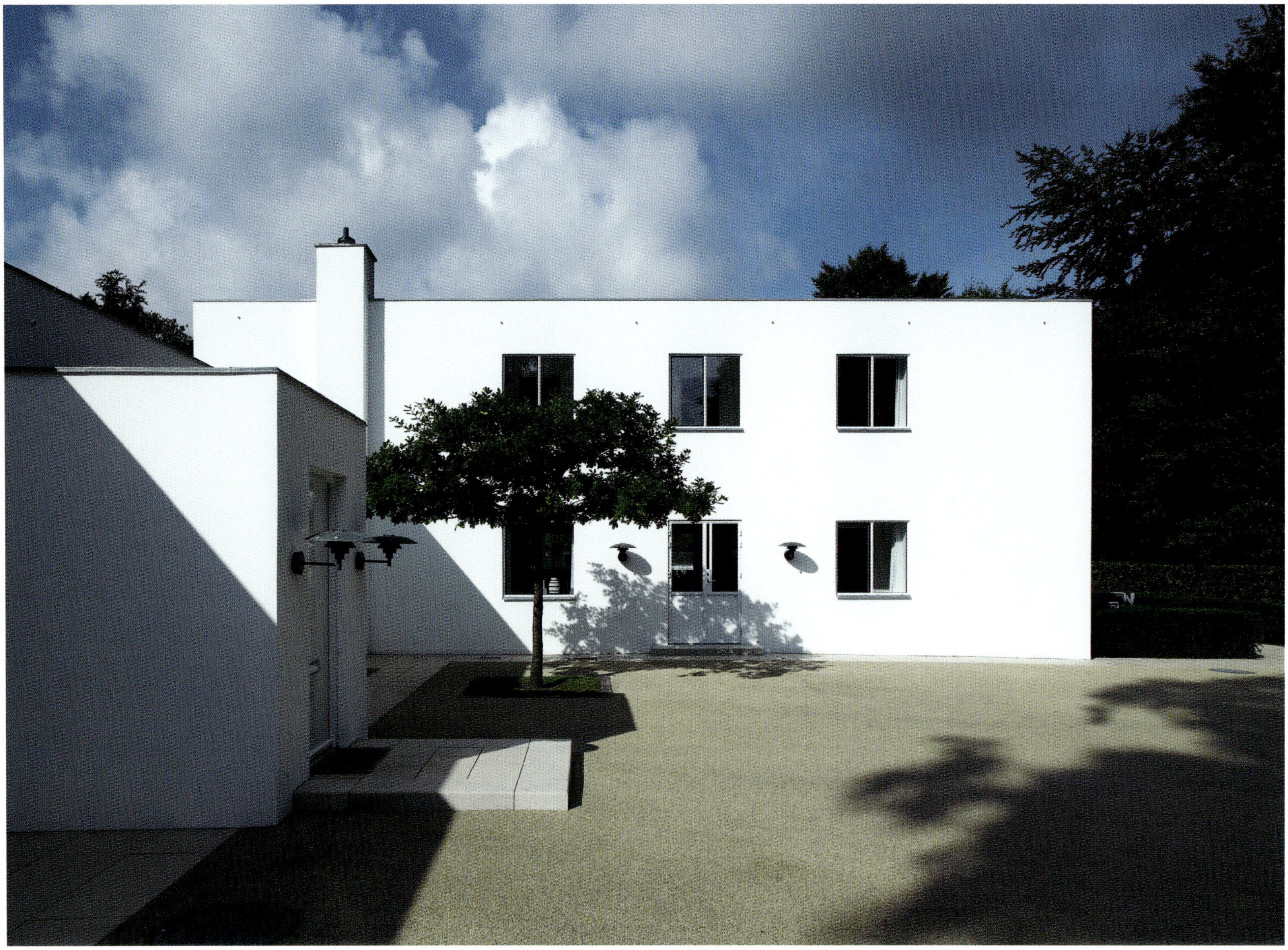

The same was true of the Rothenborg House in Klampenborg completed just a few years later. A Copenhagen suburb by the sea, overlooking the Øresund strait, Klampenborg is home to a number of projects by Jacobsen, including the Bellevue beach resort (1932), the Bellavista housing estate (1934), and the Bellevue Theatre and Restaurant (1935). The house itself was commissioned by lawyer Max Rothenborg and his wife, who had a particular interest in modern architecture and interiors. They asked Jacobsen for a large villa in the modern style, granting him a considerable degree of freedom to further his ideas and create a rounded but progressive home.

The house is arranged in a U-shaped plan, with an entrance courtyard flanked by two wings, the smaller wing holding a garage and the larger, two-storey wing holding a kitchen and service zone on the ground floor with bedrooms above. The heart of the house holds a broad hallway leading to a large living room, which in turn feeds out onto a generous terrace overlooking the gardens and surrounding beech trees. The sitting room is flanked by the dining room plus winter garden to one side, while the master suite sits on the other. As well as the long terrace, complete with an outdoor fireplace, there is a substantial roof terrace. Together, these complementary outdoor spaces, combined with the crisp, sleek outline of the white rendered building, suggest the decks of an ocean liner looking out into the landscape.

With its large windows, linear outline and essential purity, the house arguably has more in common with the work of the Bauhaus modernists than fellow Scandinavian pioneers

ABOVE The main entrance sits within a U-shaped courtyard, sheltered by the formation of the house and garaging.

OPPOSITE Key parts of the house are single-storey, but there is also a two-storey wing and a basement level largely devoted to service spaces.

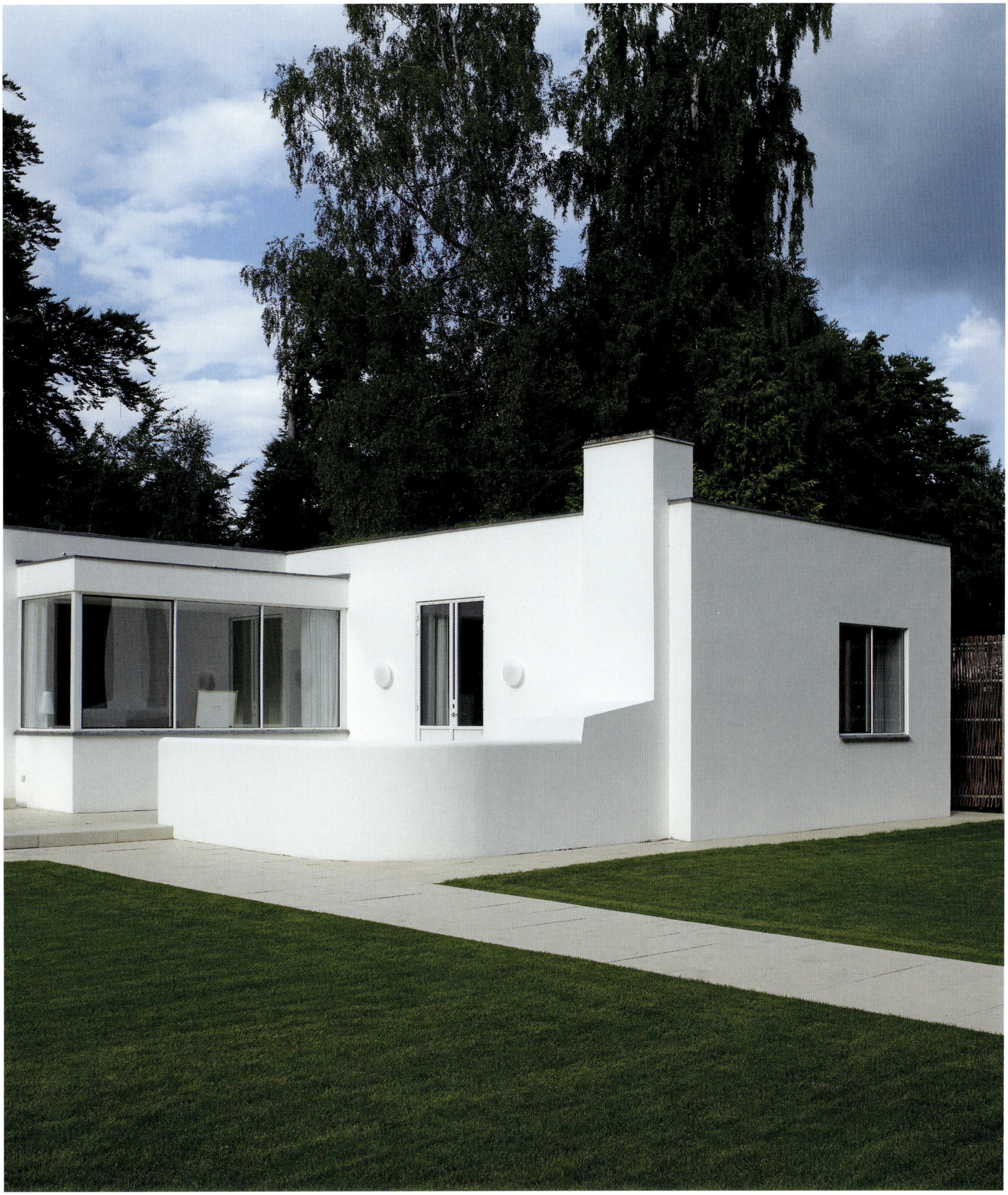

RIGHT The master bedroom features a corner fireplace, one of a number of integrated hearths inside and out.

BELOW The main living spaces look out over the gardens and connect with the adjoining terrace.

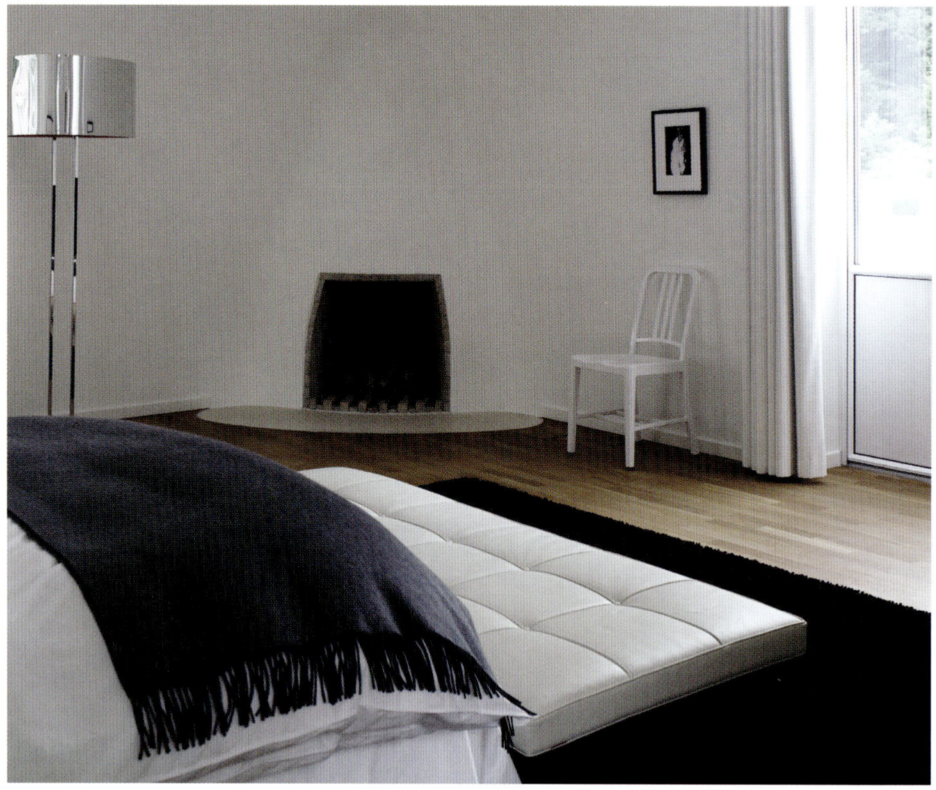

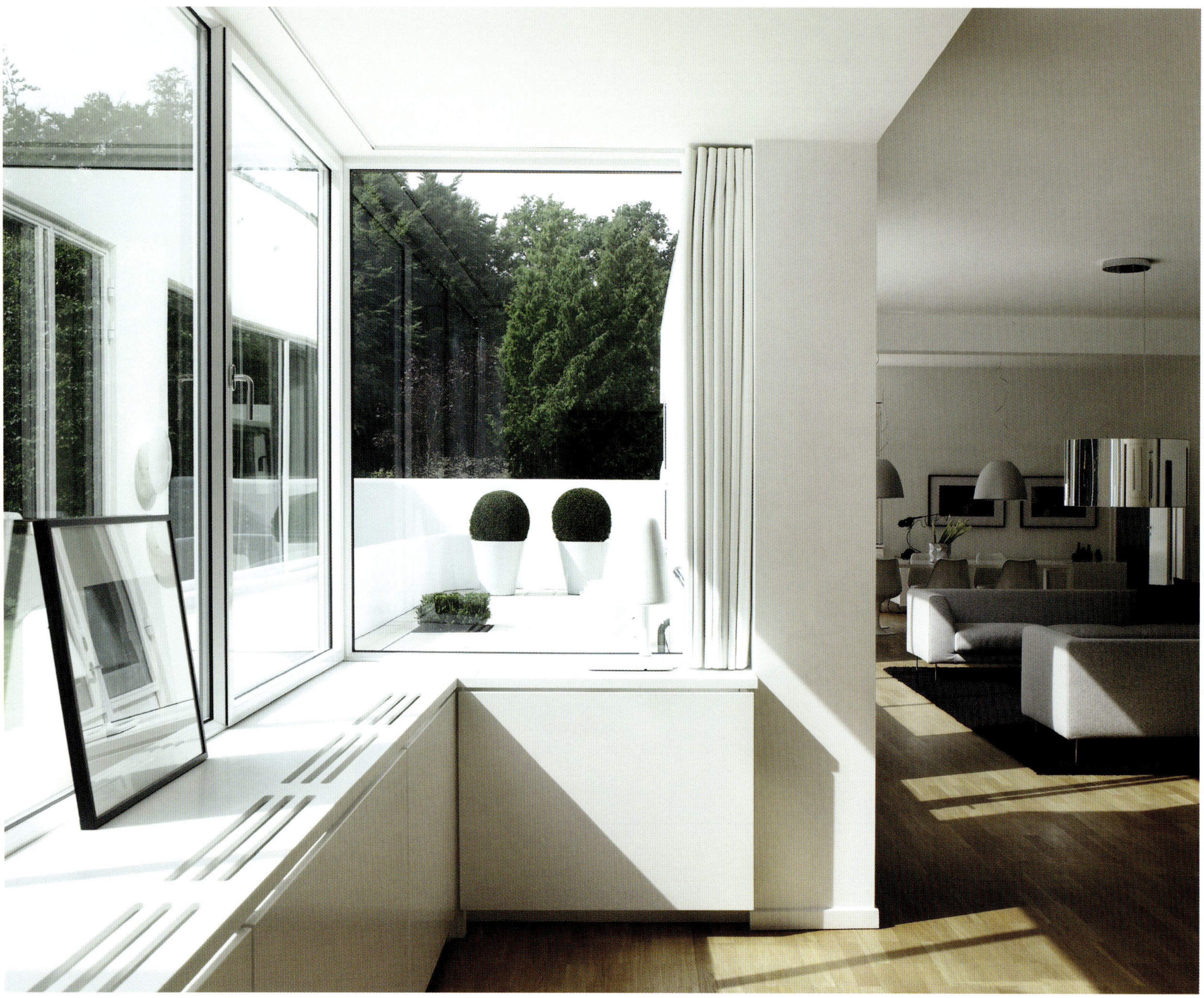

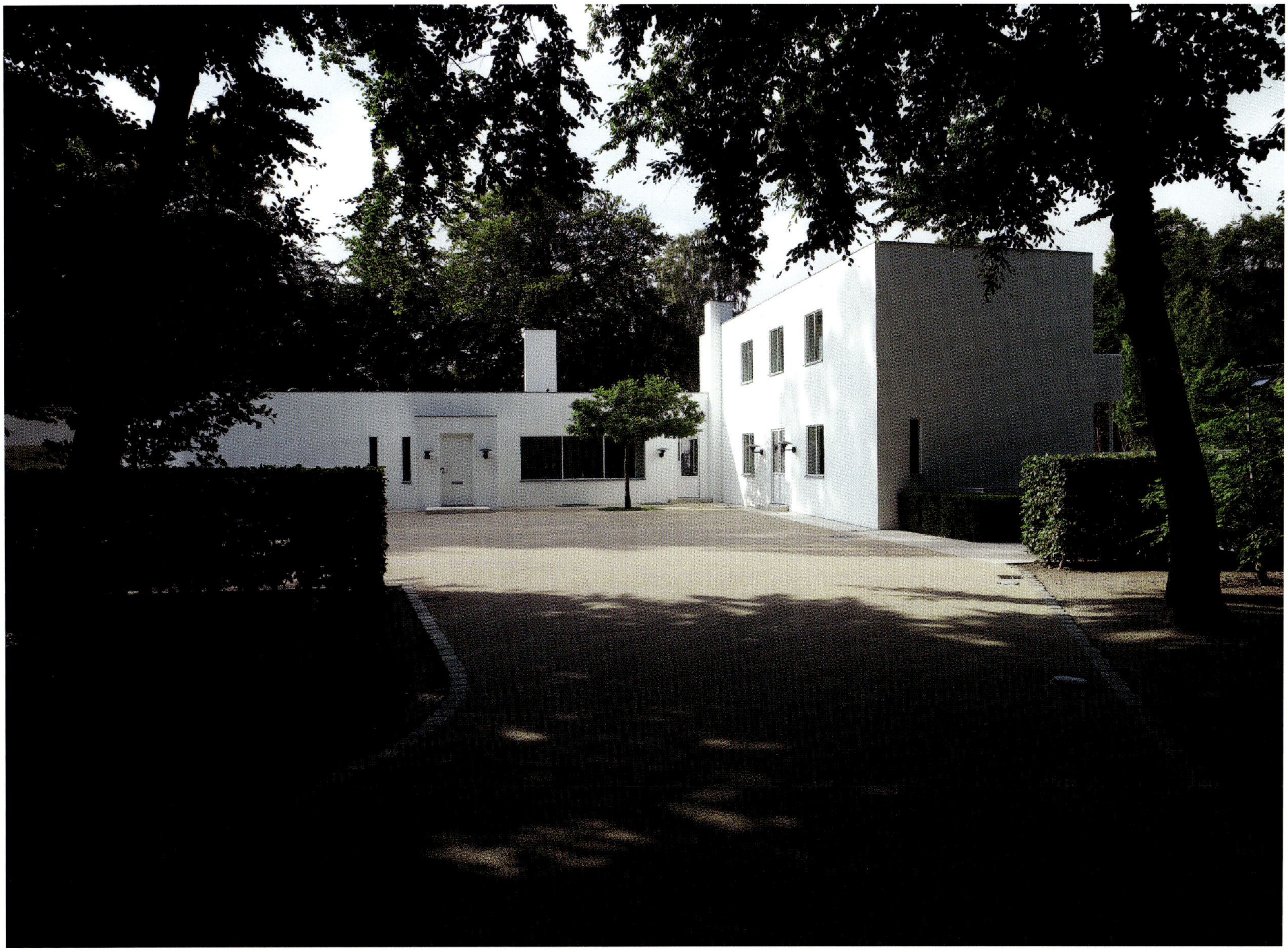

such as Aalto. Its strong relationship between inside-outside living and its cohesive interiors – significantly altered in later years – mean that the Rothenborg House forms a key marker in the evolution of Nordic modernism as well as Jacobsen's own design philosophy.

Post-war projects saw Jacobsen take his work to a larger scale while also exploring commissions abroad. The SAS Royal Hotel & Air Terminal project in Copenhagen, completed in 1960, was one of Jacobsen's most accomplished and integrated projects including both architecture and landmark furniture designs, featuring the Egg, the 3300 and Swan chairs (manufactured by Fritz Hansen), as well as the AJ lamp. Similarly, the commission for St Catherine's College, Oxford (1966), was all-pervasive, with Jacobsen involved in every detail down to the trees in the courtyard and the fish in the ornamental pond. Jacobsen's famous Oxford desk chair and other pieces emerged from the project as well.

Jacobsen's experiments in plywood furniture, partly influenced by the work of Charles & Ray Eames* in the States, also emerged from a specific context. The famous Ant chair, with its three legs and tucked ant-shaped silhouette, was first developed for the canteen of the Novo Nordisk healthcare company (1952). The Series 7 plywood chair of 1955 – the best-selling chair produced by Fritz Hansen and an icon of mid-century design – was an evolution of the shape and production techniques used for the Ant.

ABOVE Mature trees help soften the approach to the house; seen from the rear, the notion of juxtaposing different volumes and scales in various parts of the building becomes more apparent.

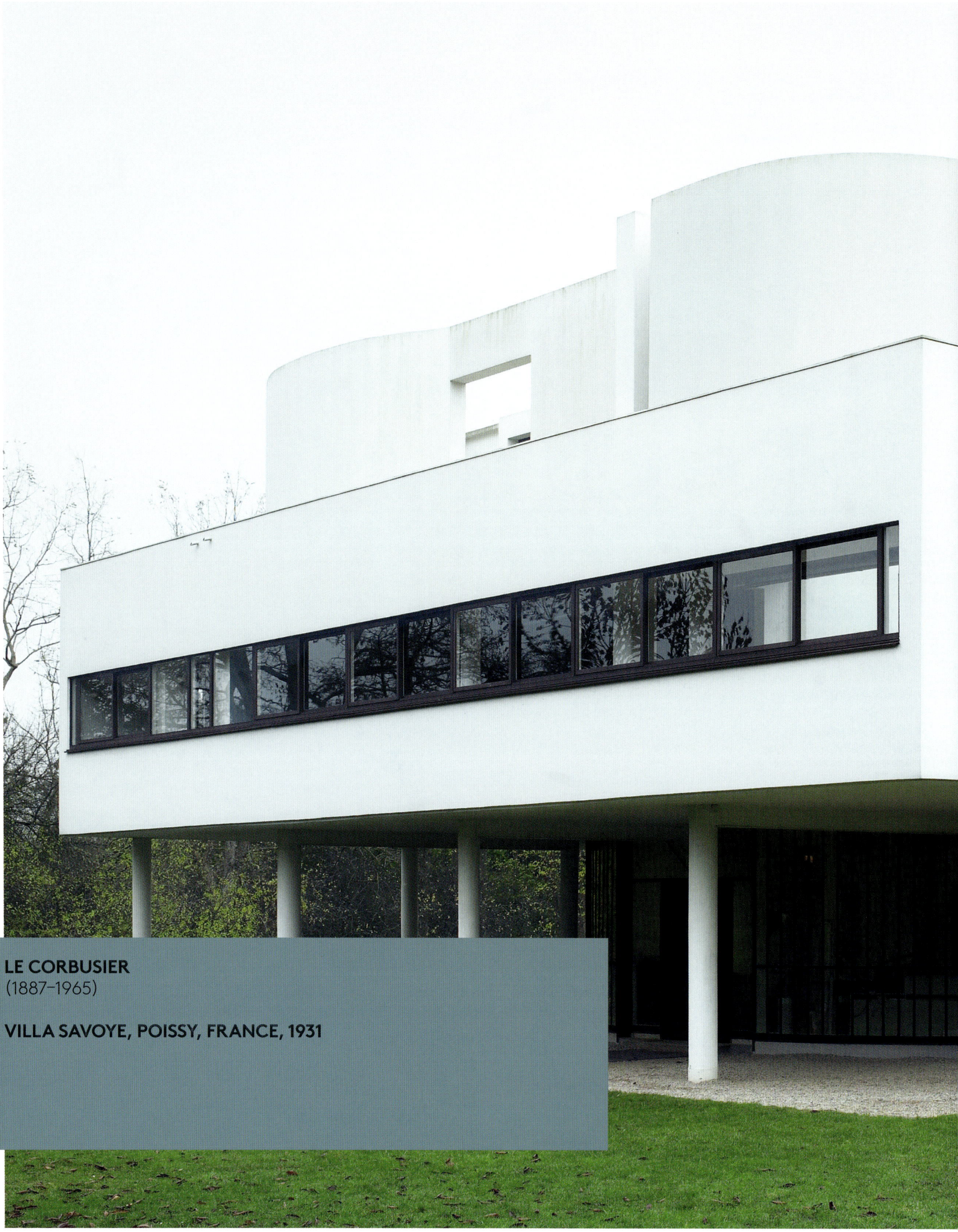

LE CORBUSIER
(1887–1965)

VILLA SAVOYE, POISSY, FRANCE, 1931

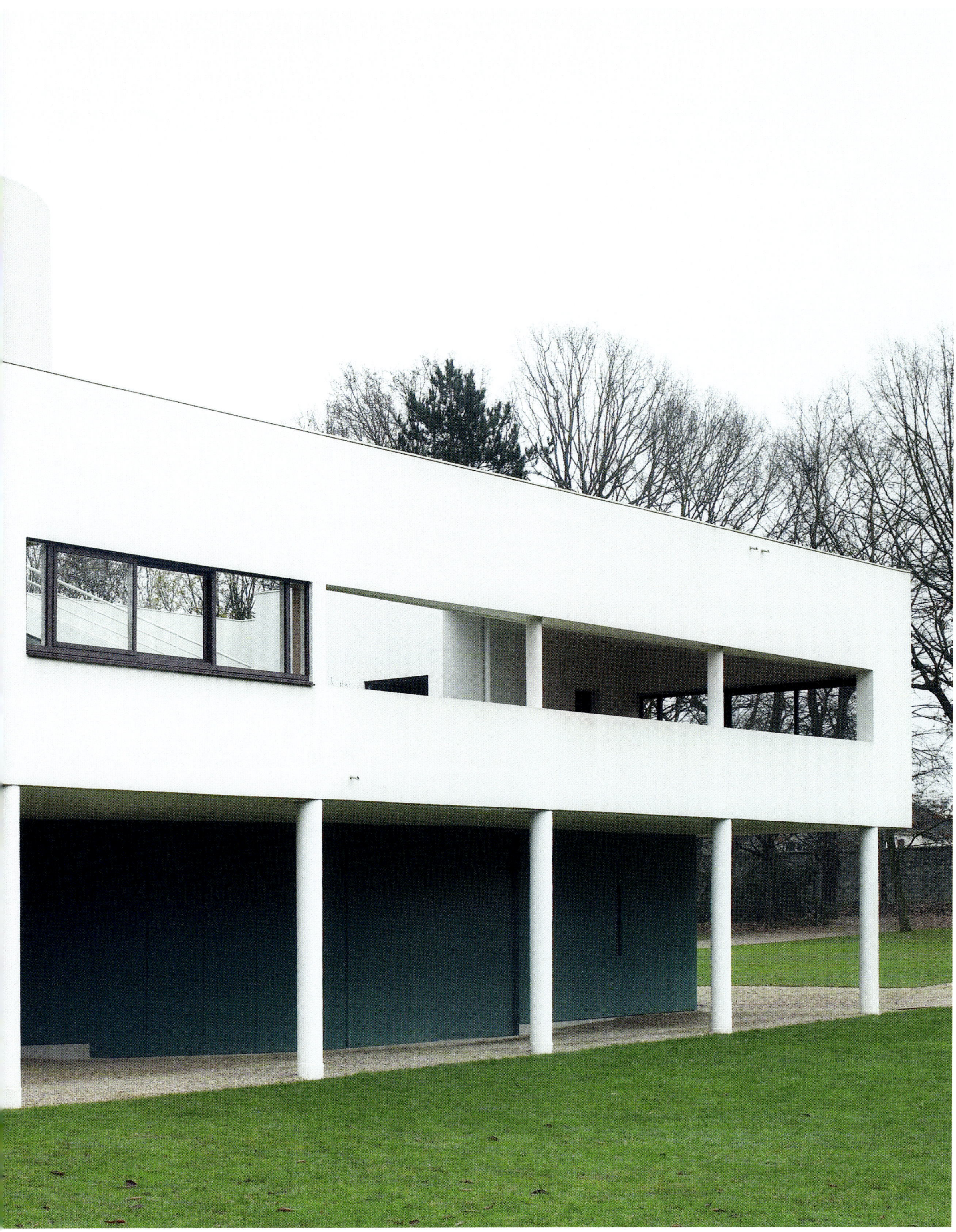

For the great father figure of modernist architecture, the house was more than 'a machine for living in'. Le Corbusier was certainly obsessed by notions of architectural purity and functionality, as well as the influence of machine-age technologies on modern ways of living, yet he also saw the house as an emotional and spiritual focal point for daily life.

'One concern has been uppermost in my mind,' he wrote, towards the end of his life. 'To make the family sacred, to make a temple of the family home.'[21] And writing in 1925 he talked of the two principal roles of the home: 'First, it is a machine for living, which means a machine designed to make our daily work as rapid and simple as possible, and to look after our bodily needs attentively, providing comfort. It also provides surroundings where meditation can take place, and a place in which beauty brings the repose of spirit which is so indispensable.'[22]

Le Corbusier's critics and detractors tend to stress the more didactic nature of his buildings and his writings; the zeal with which he presses for a new way of living, whether in a family villa or a mass housing project. Yet there is, for all the controversies that complicate his life and work, an ongoing belief in the power of modern architecture to facilitate an easier, more comfortable and more enjoyable way of living. It is the power of this belief, more than anything, that has helped to make Le Corbusier such an influential figurehead for the modernist movement. And this fusion of the functional and the spiritual within house and home can be seen at its height in the Villa Savoye.

Today, the villa sits among suburban sprawl on the edge of Paris, but back in the Twenties and Thirties the area still had a semi-rural quality. Seven hectares of pasture and woodland were

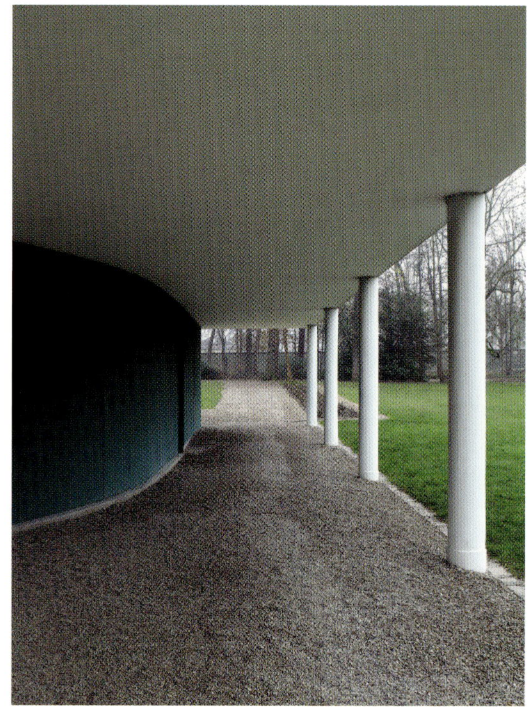

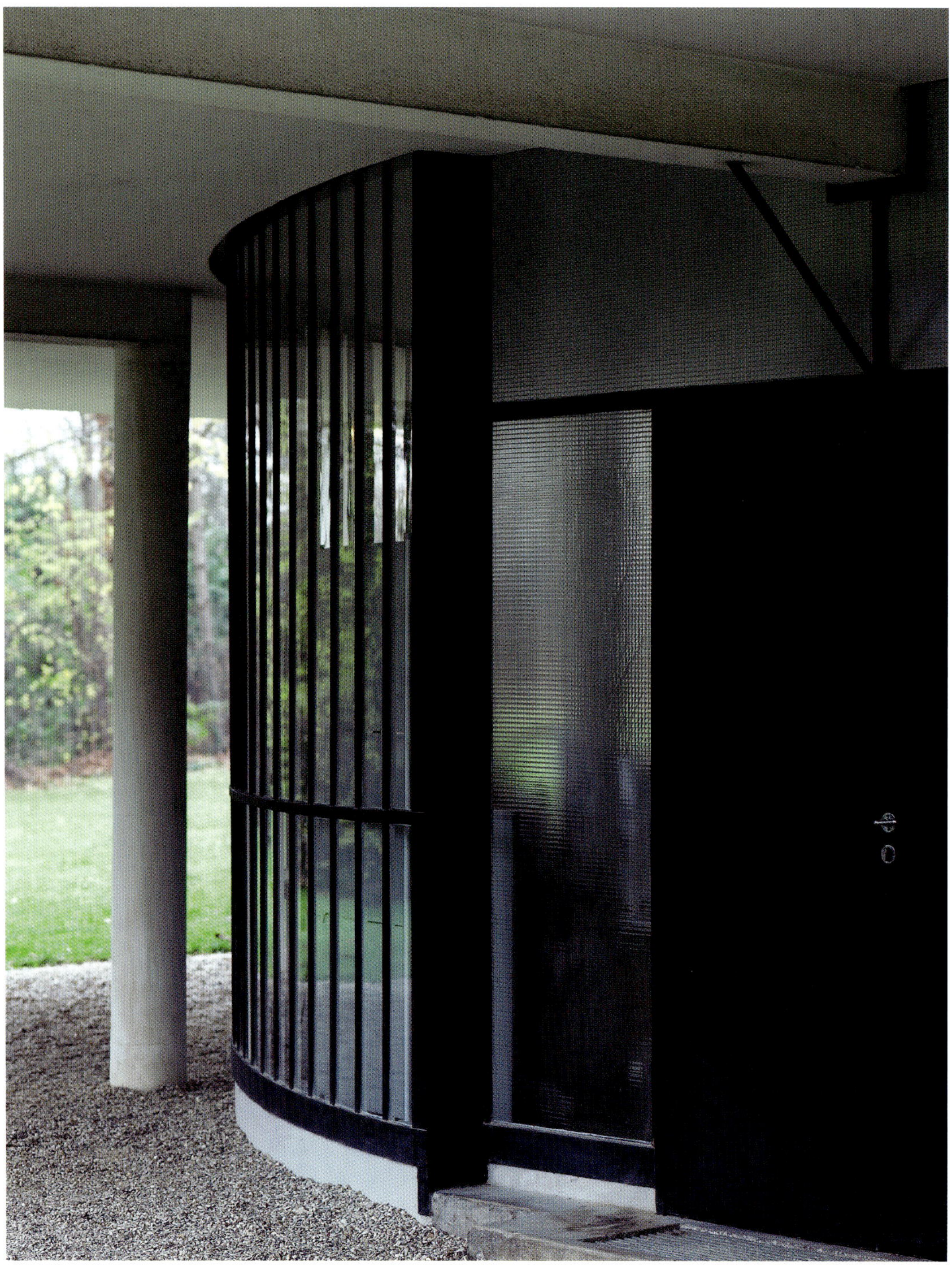

PAGES 366–7 This iconic 'purist' villa is enhanced by its setting, with the geometric structure floating upon an open green lawn, bordered by trees.

LEFT & ABOVE The main body of the building hovers upon a collection of pilotis, helping to shelter the entrance routes on the ground floor, which also holds staff accommodation and a garage.

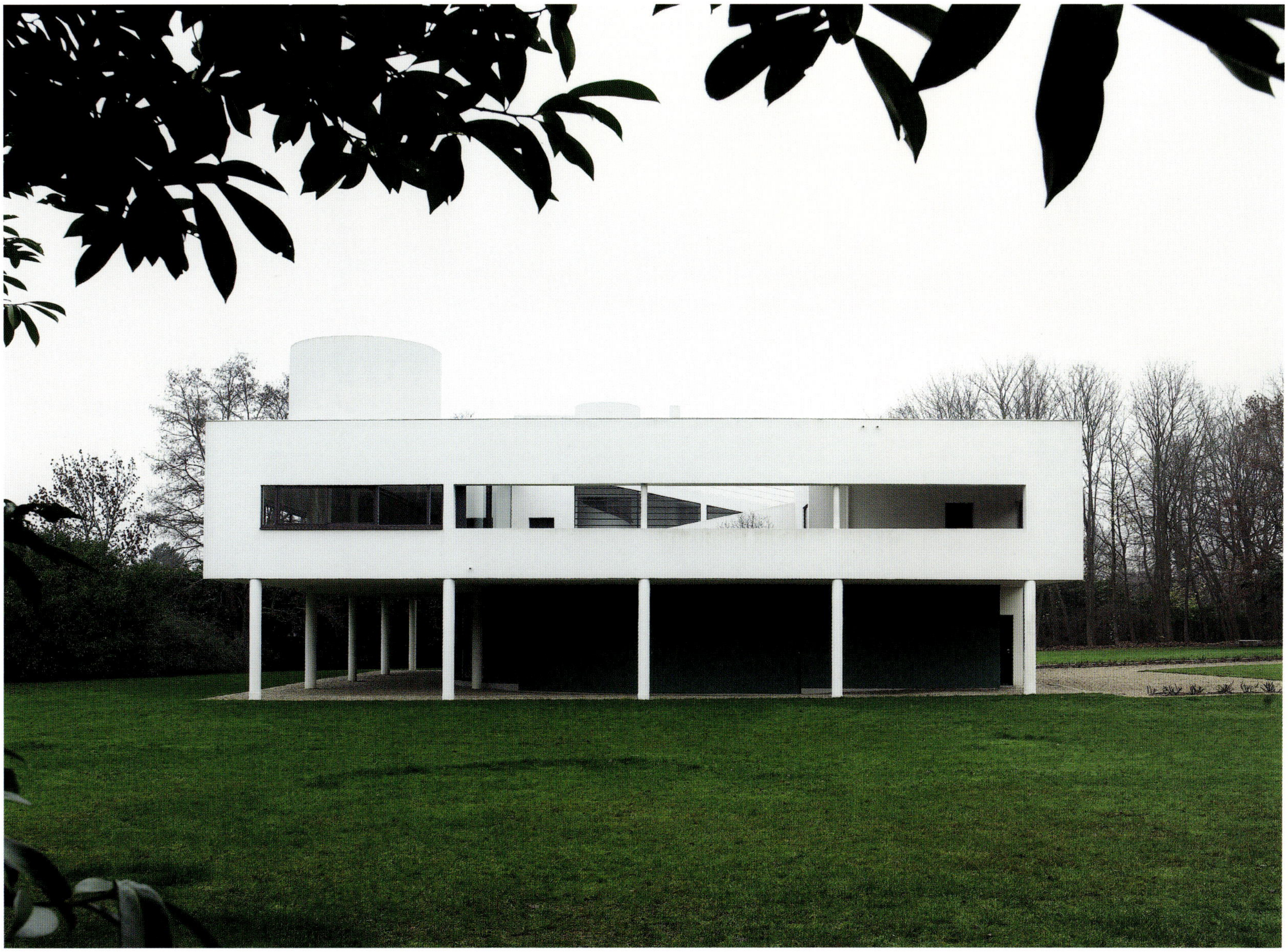

acquired by Pierre Savoye, an insurance company director, and his wife. They asked Le Corbusier to create a weekend retreat, a modest drive away from their Parisian home, for themselves and their son. They first got in touch with the architect in 1928, having become familiar with his work through friends; Le Corbusier described them as open-minded and without any preconceived ideas.

The architect was born Charles-Edouard Jeanneret in the Swiss watch-making town of La Chaux-de-Fonds. Later he would assume both French citizenship and his pseudonym, thought to derive from the French word 'corbeau', meaning crow or raven. He began his studies at the town's art school, but took up travelling at an early age, immersing himself in architecture, art and design in Tuscany, Vienna and Paris. He went to work, in Paris, with AUGUSTE PERRET, one of the early pioneers of reinforced concrete construction, who became something of a mentor. He also worked with PETER BEHRENS in Berlin, and travelled in Eastern Europe, before settling in Paris in 1917.

Le Corbusier's earliest house projects in Switzerland appear almost as works in progress – markers upon the pathway towards his vision of a truly modern architecture. By the time he began working on a sequence of 'purist' Parisian villas in the Twenties – including Villa Savoye – he was a different man and much more than an architect. He was now a writer, theorist, painter, interior designer, and – along with his cousin and collaborator PIERRE JEANNERET – a furniture designer as well (see p. 64).

ABOVE The geometric outline, horizontal windows, slim pillars and giant 'funnels' on the roof (which shelter the sun terrace) reinforce the notion of a machine-age masterpiece.

His 'Five Points' were issued in 1927, forming a brief modernist manifesto. The Five Points embraced a series of key characteristics that Le Corbusier saw as vital to the success of modernist architecture: pilotis, or supporting pillars, which elevated the building above the ground plane; a free or universal floor plan and a free façade, made possible by a structural frame of reinforced concrete rather than supporting walls; horizontal ribbon windows; and lastly integrated roof decks and terraces, providing outdoor rooms.

Sitting upon an open meadow, surrounded by woodland, Villa Savoye neatly encapsulates these principles in one powerful building, with the majority of its exteriors painted snowy white. Supported by a series of piloti, the rectangular body of the house seems almost to float above the ground floor. The undercroft of the villa – designed in association with Jeanneret – is devoted to an inset, U-shaped entrance area, which also holds garaging for three cars and accommodation for a chauffeur and servants. A series of long ribbon windows and apertures form distinct banners at first-floor level, while the reinforced concrete structure allows for a free and fluid floor plan, with a choice of outdoor rooms contained within the plan and upon the roof.

A dynamic architectural 'promenade' through the building is offered via a series of

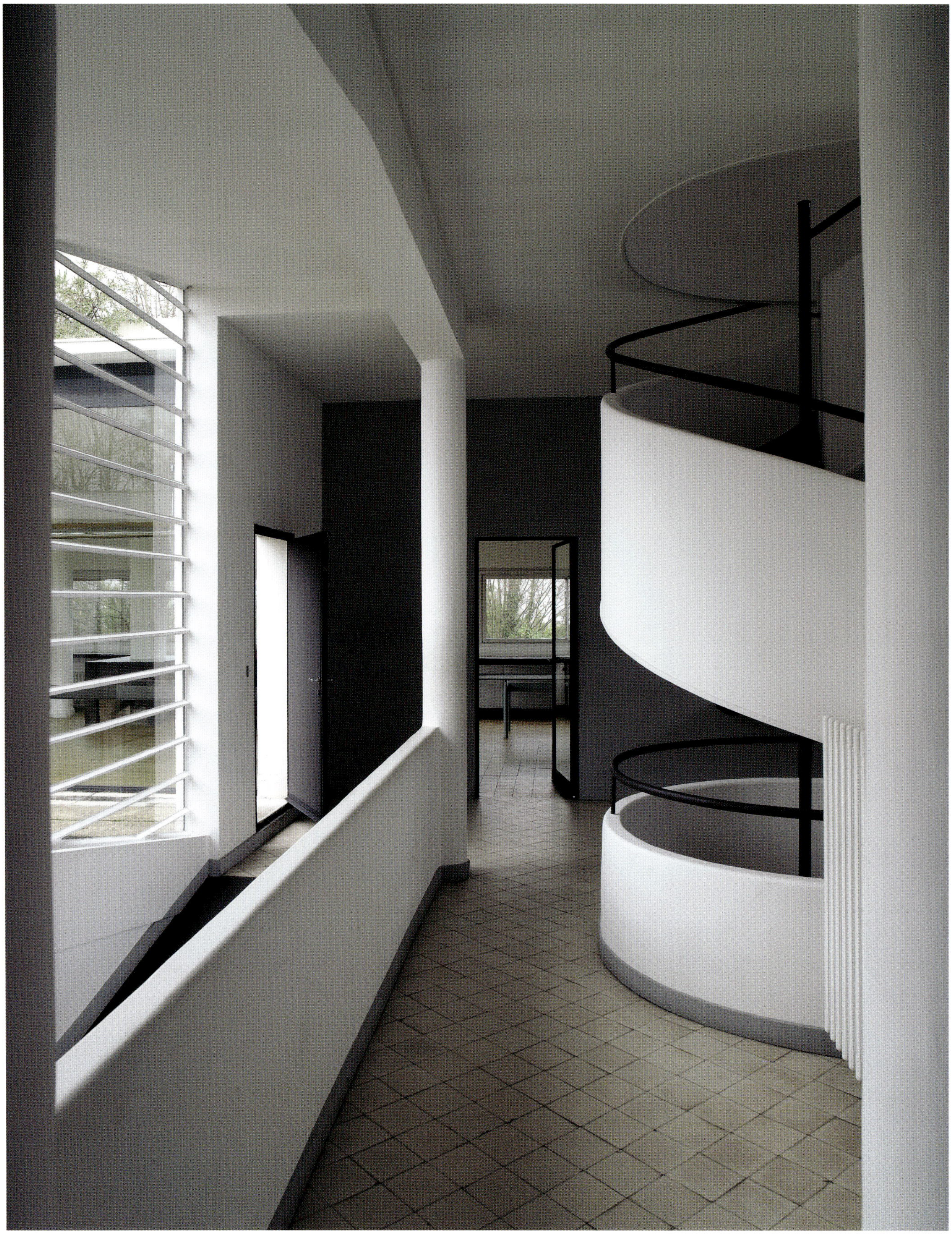

LEFT The house features alternative circulation routes, including a central spiral staircase plus a series of ramps that provide a more drawn-out process of discovery as one progresses through the building.

OPPOSITE Colour is used in places, standing out vividly against the dominant whites and neutrals; blue mosaic tiles add character to the master bathroom, which features an integrated couch by the bath.

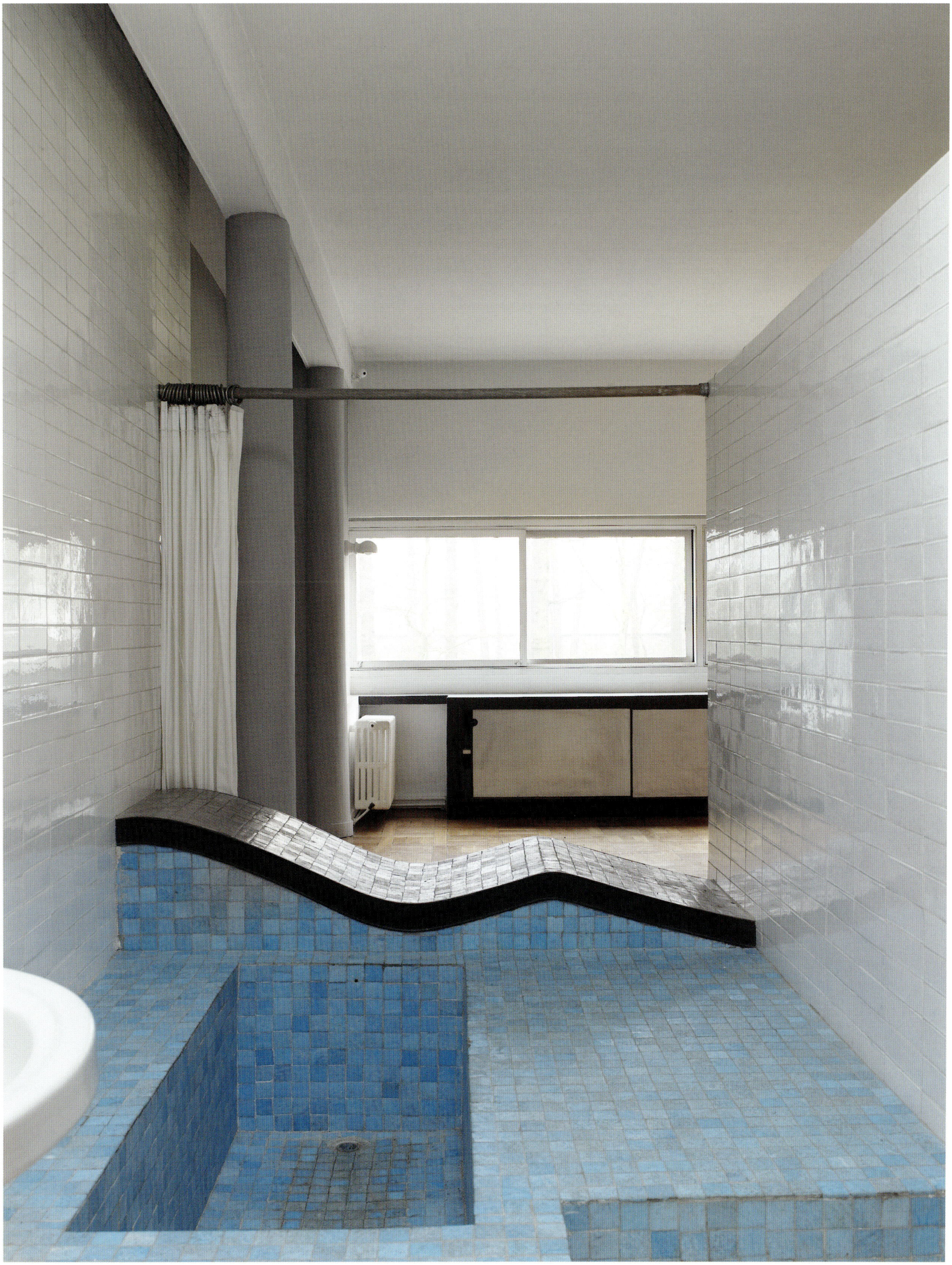

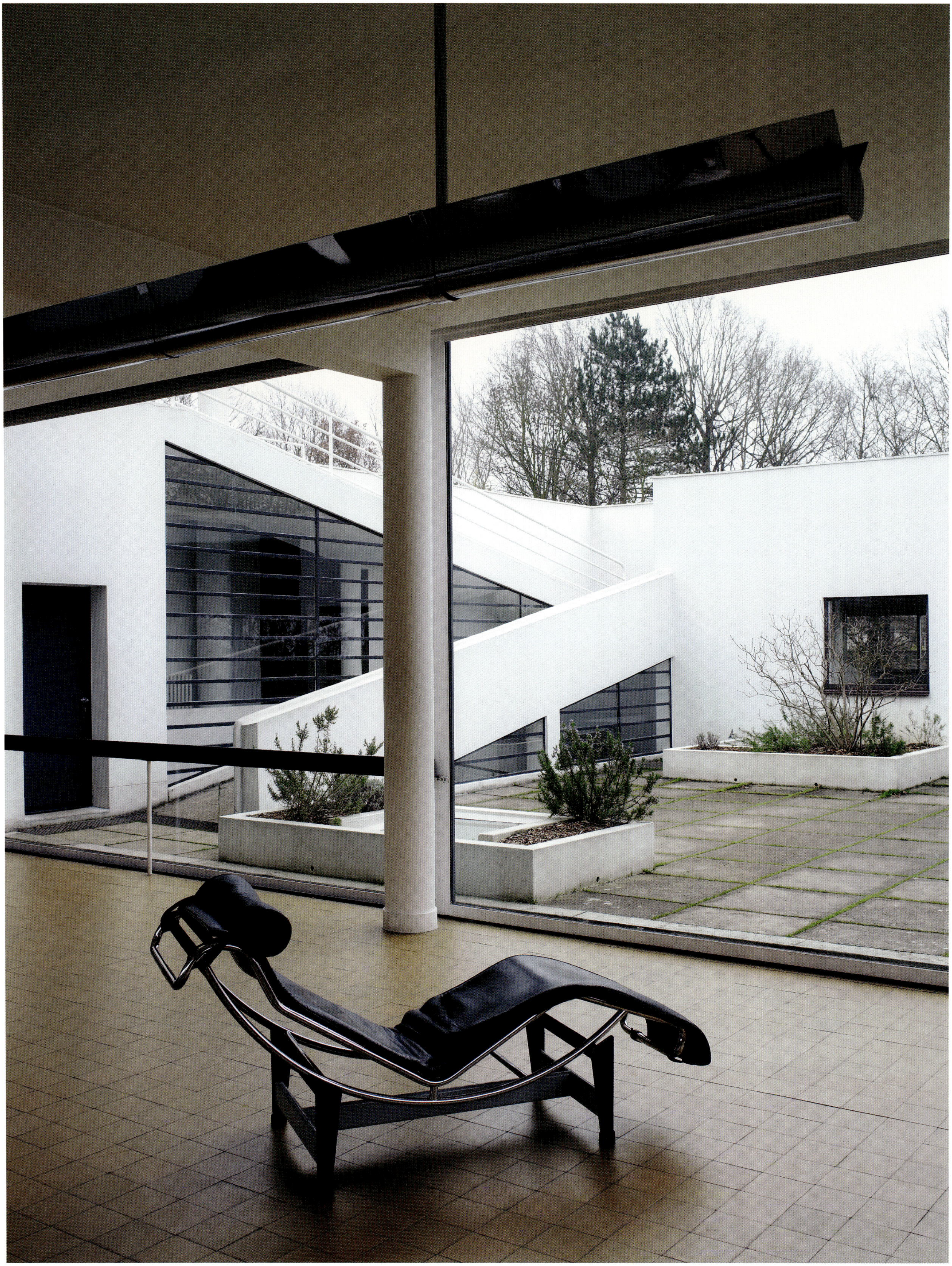

ramps, complemented by a spiral staircase. At first-floor level, there are three bedrooms, a large bathroom, a kitchen, and a spacious living room which connects to a large terrace – integrated within the outline of the building – via a wall of sliding glass. From this mid-level terrace, a ramp leads up to a solarium on the roof – an additional outdoor room, partially shaded by protective curved walls which have the look of vast funnels when the house is viewed from a distance. As well as the Five Points, the interiors explore colour, movement, transparency, and connections between indoor and outdoor space.

For the Savoye family, however, the house was not a complete success. Problems during construction caused a number of ongoing leaks and a series of complaints, not helped by the fact that the Savoyes' son suffered from health problems that could only be made worse by living in a house that was cold and damp. These issues were never fully resolved and letters were still passing between architect and client in 1936–37, shortly before the family abandoned the house.

The building was requisitioned by the German army during the war and later used by American forces and only listed as a historic building in the mid-Sixties, when an initial restoration programme was carried out. It is now one of the iconic emblems of early modernist architecture and design.

Le Corbusier went on to reinvent himself many times over. There were the ambitious and profoundly influential brutalist housing schemes of the post-war years, including Unité d'Habitation in Marseilles (1952), and the sculpted beauty of the Church of Notre Dame du Haut (1955). There were projects in Tokyo, India, Brazil and America, including his contribution to the United Nations Headquarters in New York (1952). It is fascinating to note that Le Corbusier declared himself to be happiest when living and working in his small timber 'Cabanon' (1952) by the sea at Roquebrune-Cap-Martin. Here, within this simple retreat, the great modernist found a combination of function and spiritual repose. This is where he died of a heart attack, in 1965, after going for a final swim in the sea below.

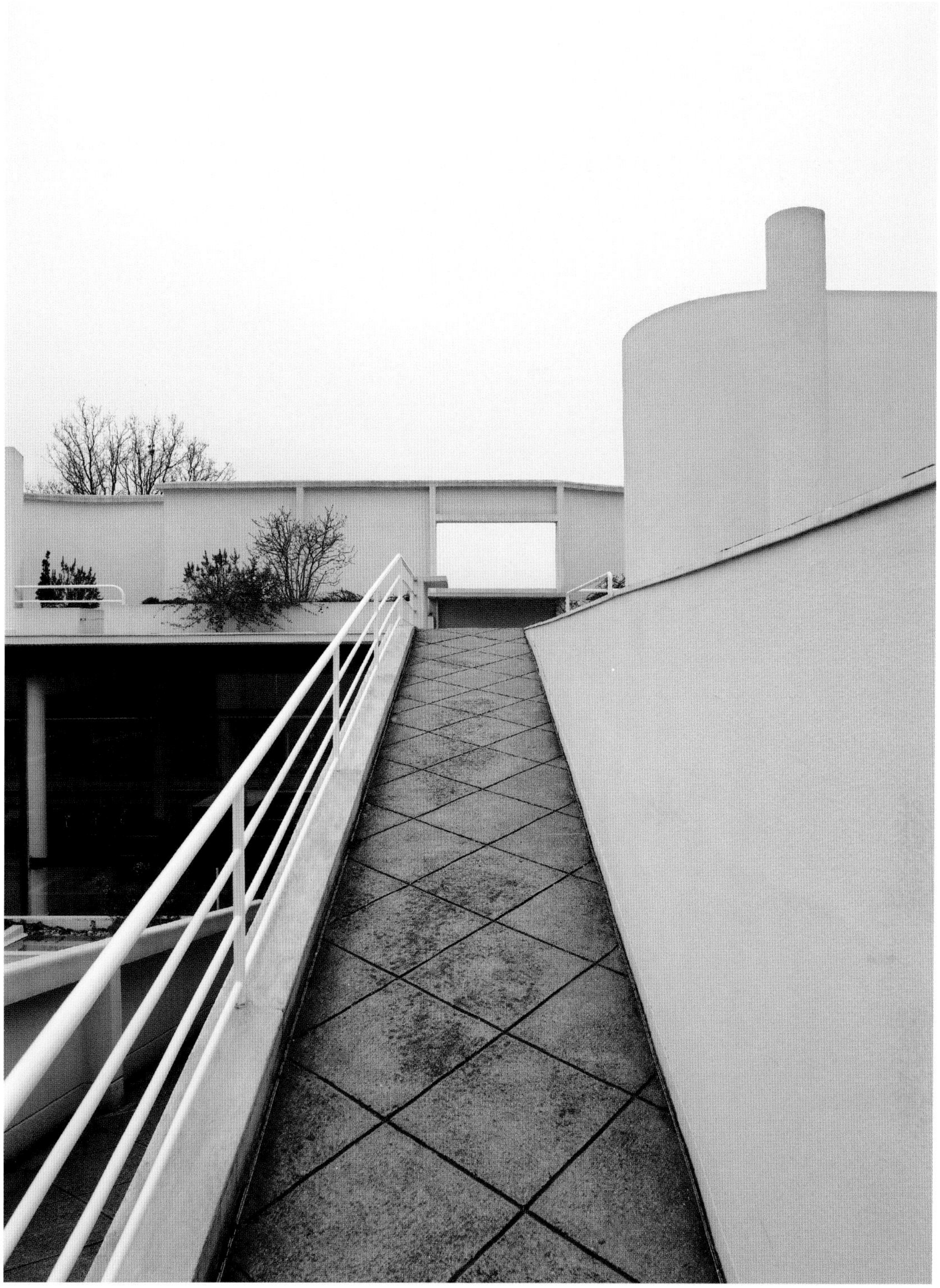

OPPOSITE The principal living spaces are situated at the mid-level of the house, arranged around an elevated terrace.

ABOVE From this terrace, the processional ramp leads further upwards to a sun deck at the very top of the building.

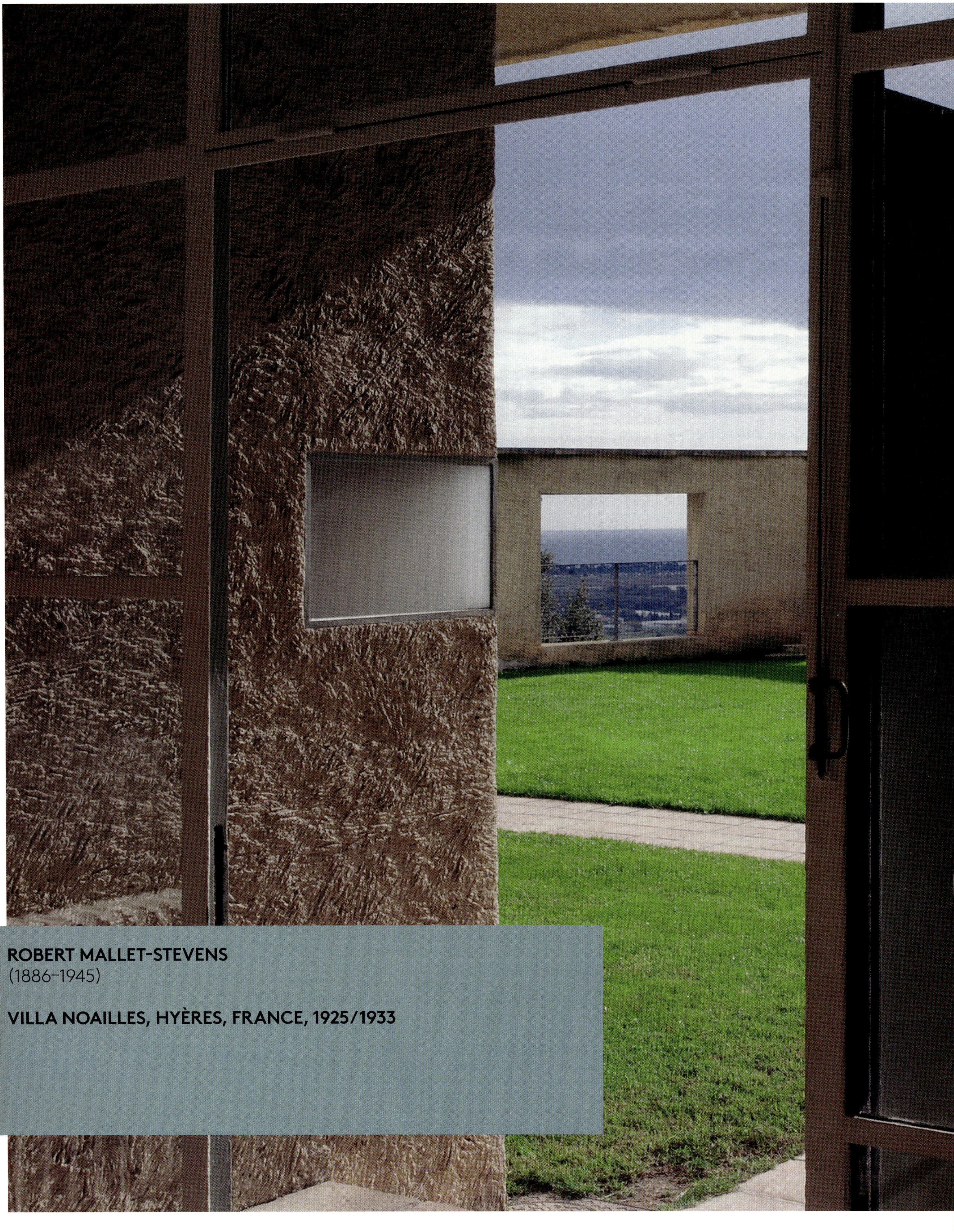

ROBERT MALLET-STEVENS
(1886–1945)

VILLA NOAILLES, HYÈRES, FRANCE, 1925/1933

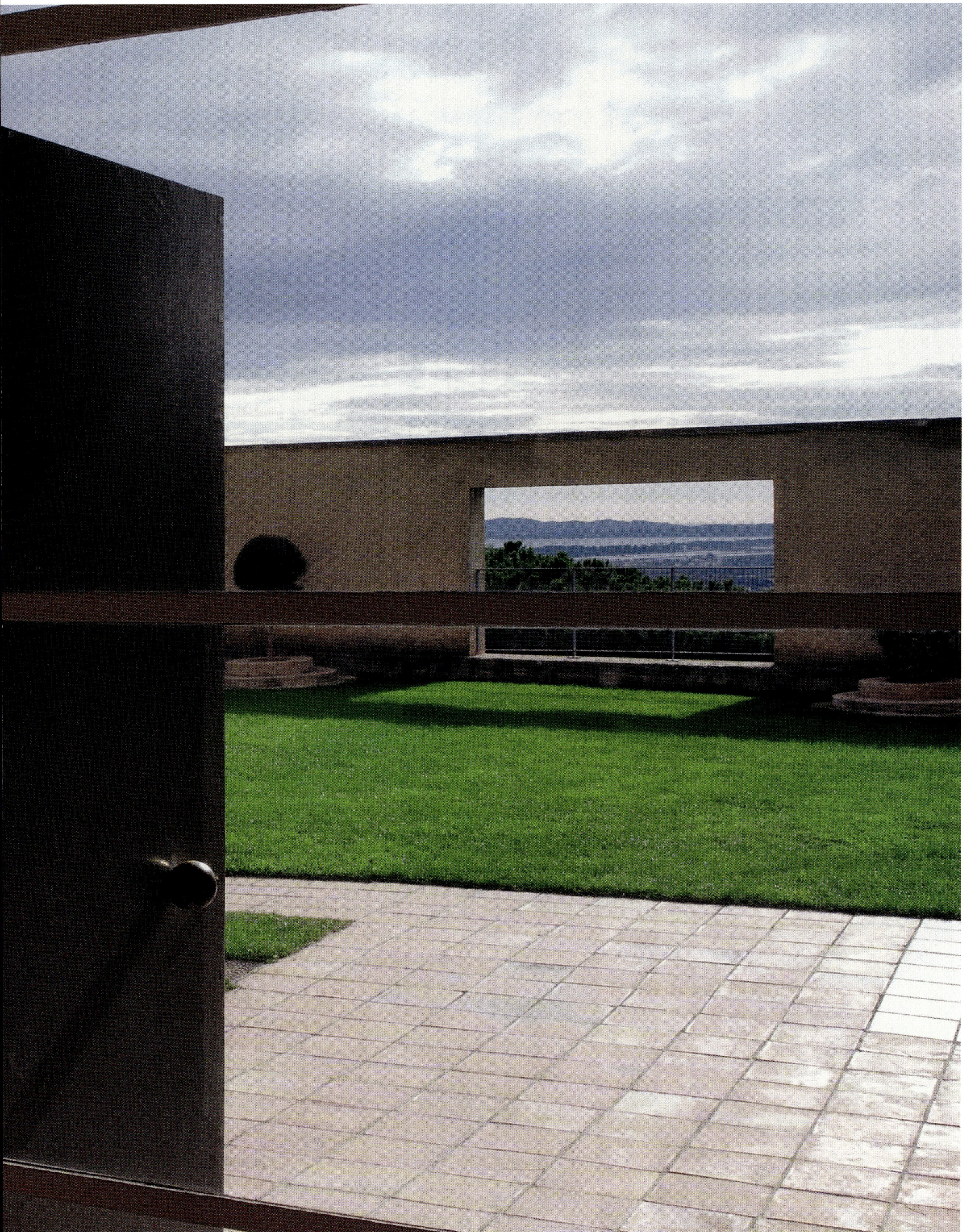

The Viscount de Noailles and his wife, Marie-Laure, were great patrons of both the arts and design during a brief but golden age of creativity and excess between the wars. The de Noailles were collectors of modern art but they were also film enthusiasts, who financed three key films of the period, including Jean Cocteau's first film (*Le Sang d'un Poète*) and Luis Buñuel and Salvador Dalí's highly controversial masterwork, *L'Age d'Or*. The third film was Man Ray's *Les Mystères du Château du Dé*, which was partly filmed at the Noailles' extraordinary early modernist villa in Hyères, designed by French architect and designer Robert Mallet-Stevens and widely regarded as the architect's own masterpiece.

The house was the particular passion of Charles de Noailles, who had a strong interest in both architecture and landscape design. His Parisian mansion featured interiors remodelled by Jean-Michel Frank, and de Noailles was well versed in the work of the early modernists. His mother gave him a substantial parcel of land in Hyères on the Provençal coast, complete with the ruins of an old château, and the couple initially approached Ludwig Mies van der Rohe in Berlin, asking him to build a new house on the site. When Mies told them that he was too busy to accept the commission, they considered Le Corbusier and visited the Bauhaus masters and their students, without making a decision.

'We then asked the opinion of Mr Metman, director of the Musée des Arts Décoratifs and an old friend of both our families, concerning the construction,' Charles de Noailles wrote of the project. 'He did not hesitate, saying that when the museum had given exhibitions of modern architecture, the only one who seemed to him to have any taste or imagination was Mallet-Stevens.'[23]

Charles de Noailles visited the architect at the Parisian site of five new villas in the 16th arrondissement, on a street now known as rue Mallet-Stevens, which included a home for himself and another for sculptors Joël and Jan Martel. The two men got on well, although, according to the Viscount: 'To be quite frank, I found he had great imaginative qualities but a little too much taste for the unexpected.'[24]

Up until the early Twenties, Mallet-Stevens had in fact built very little. He was born in Paris and his father was an art expert and appraiser, who helped to champion the work of the French Impressionists. Mallet-Stevens studied architecture at the École Spéciale d'Architecture and grounded himself in the work and philosophy of the early modernists, guided by his interest in new materials, functional plans and forms, logical and hygienic modern living, and machine-age technology.

During World War I Mallet-Stevens volunteered in the French air force, gaining a profound respect for the power of the machine and an enduring interest in aviation. In the post-war years he became interested in set design and contributed to ten films between 1919 and 1929, lending his work a sensitivity to perspective, as well as a somewhat theatrical quality.

The original house that Mallet-Stevens designed for the de Noailles family at Hyères, completed in 1925, was relatively modest, with five bedrooms, a living room and a drawing room in a cubist arrangement of ascending

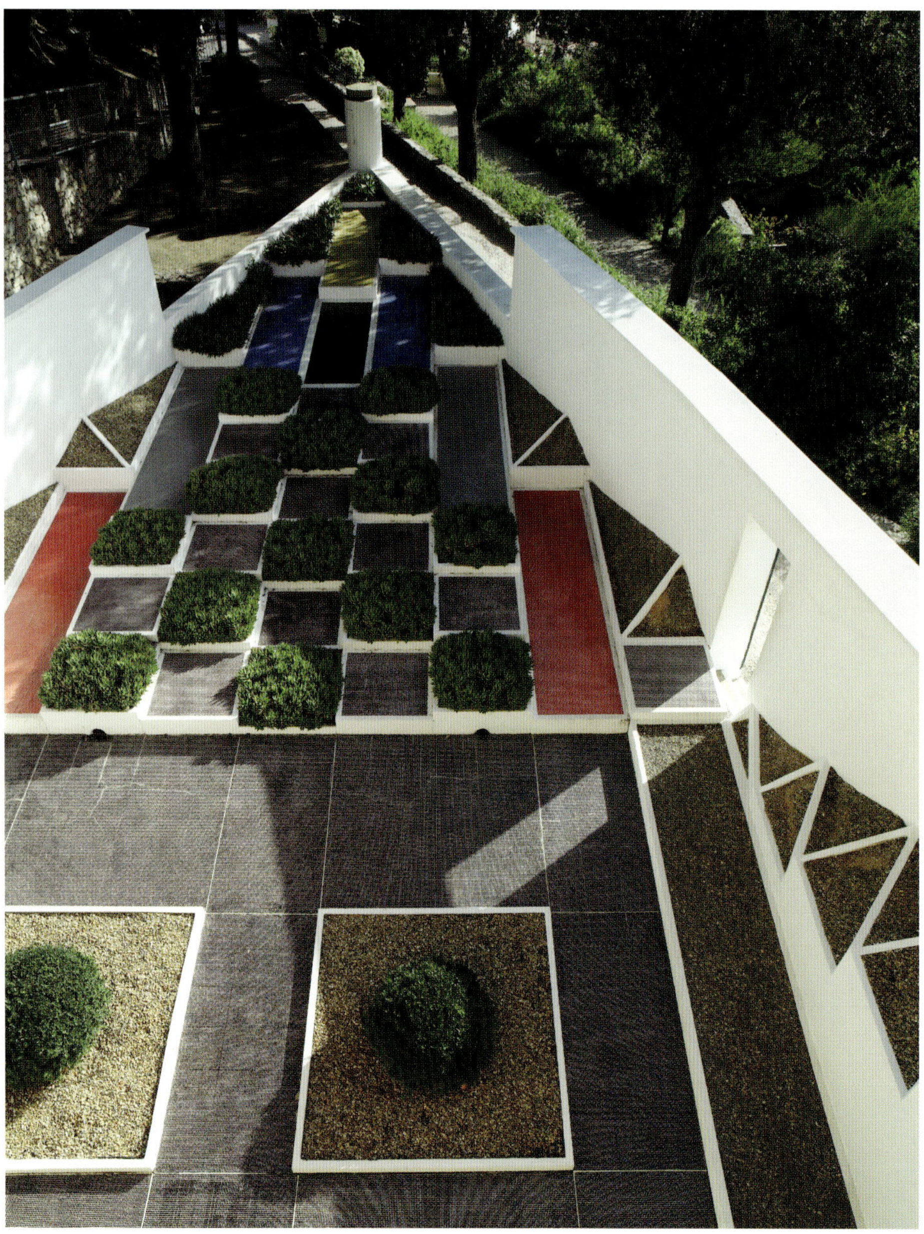

PAGES 374–5 & OPPOSITE The house enjoys a prominent position in the landscape, with a walled garden pierced by 'windows' that frame edited views of the countryside.

LEFT The Cubist-inspired gardens around the house were designed by Gabriel Guévrékian, one of a number of artists and designers who contributed to the residential complex.

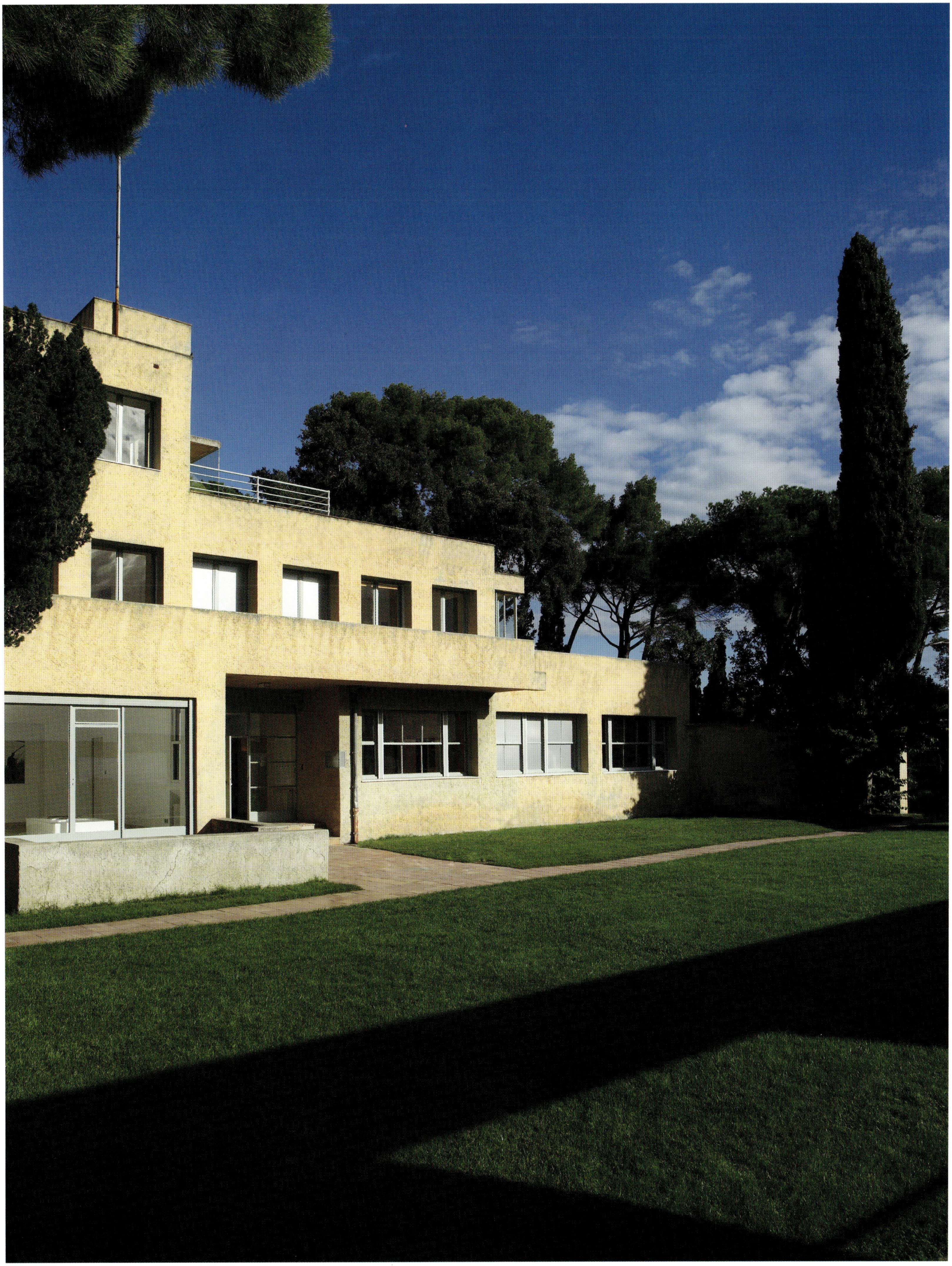

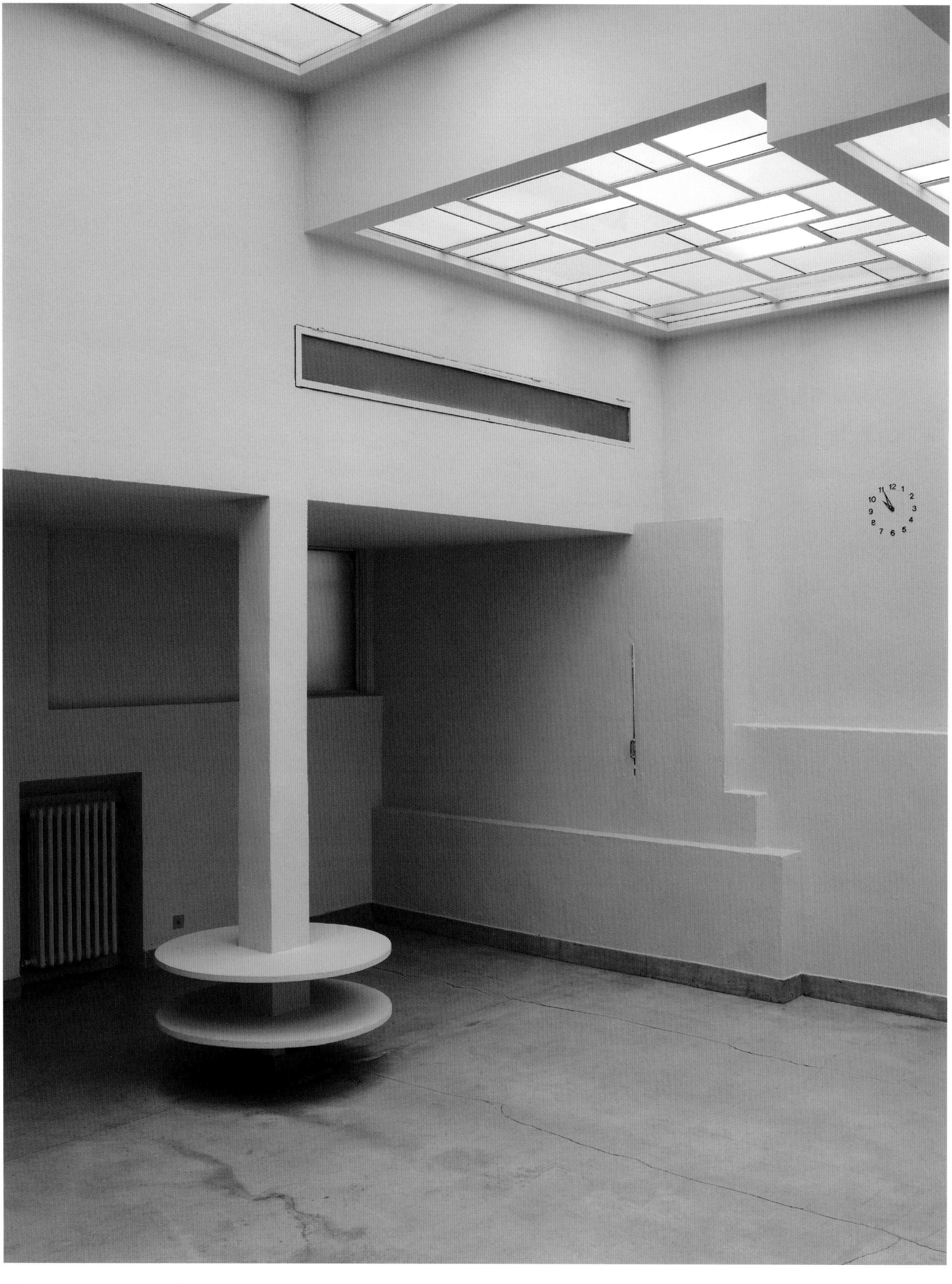

rectangular forms, creating the impression of a complex, geometrical ziggurat, made of reinforced and rendered concrete and pierced by large windows. But almost as soon as the initial house was completed the family began adding extra elements to the project. There was an annex with additional bedrooms, a squash court, a gym, an indoor swimming pool and service wings, as well as terraces and a large parvis or patio.

Ultimately the villa grew to around sixty rooms, attended to by nineteen servants. The family extended their patronage to include other designers, artisans and artists, with a triangular cubist-inspired garden by Gabriel Guévrékian, furniture by Mallet-Stevens and PIERRE CHAREAU (who also designed a glass room for the house), and artworks by Alberto Giacometti*, Jacques Lipchitz, Constantin Brancusi, Henri Laurens and others.

Charles and Marie-Laure de Noailles spent a good deal of time living apart, given that they were spectacularly ill suited, but both made use of the villa. During World War II the house was requisitioned by the Italian army, yet continued to be used by the family until the early Seventies, when it passed to the local municipality for safekeeping. Villa Noailles has since been restored twice over and opened to the public.

Mallet-Stevens fled to the Lot-et-Garonne region of France with his wife during the Occupation and went into hiding; he died in 1945 after a kidney operation. After decades of neglect, his work has been rediscovered and lauded in more recent years. There are other important projects of the period, including the Villa Poiret for couturier Paul Poiret (1923), the Villa Cavrois (1932) in the city of Croix, and the rue Mallet-Stevens, but the Villa Noailles represents his most rounded and cohesive achievement.

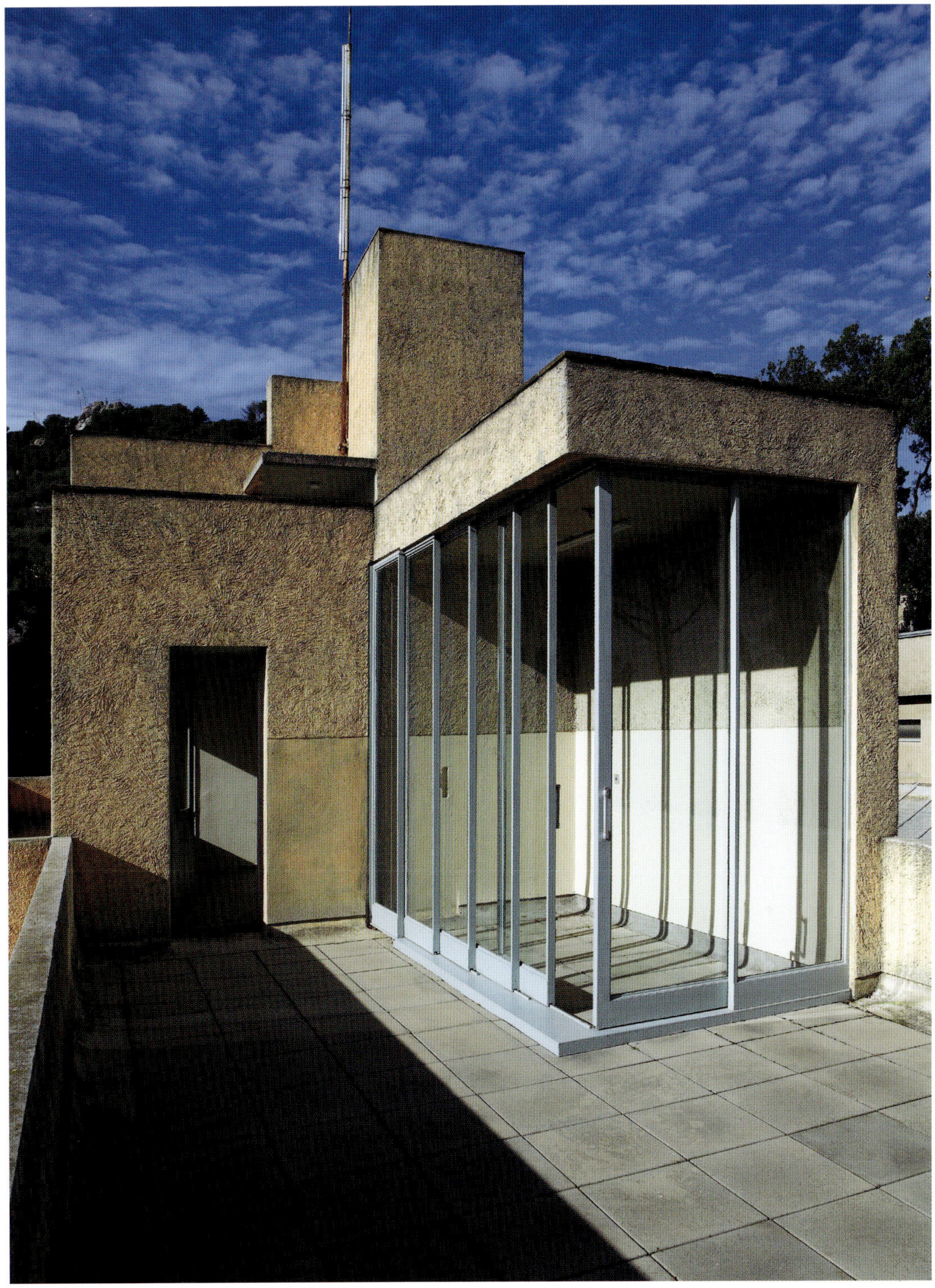

OPPOSITE The 'Pink Salon' features a series of geometric skylights, with patterns echoing the Cubist gardens outside.

ABOVE This 'chambre en plein air' up on the roof terrace originally featured a daybed designed by Pierre Chareau.

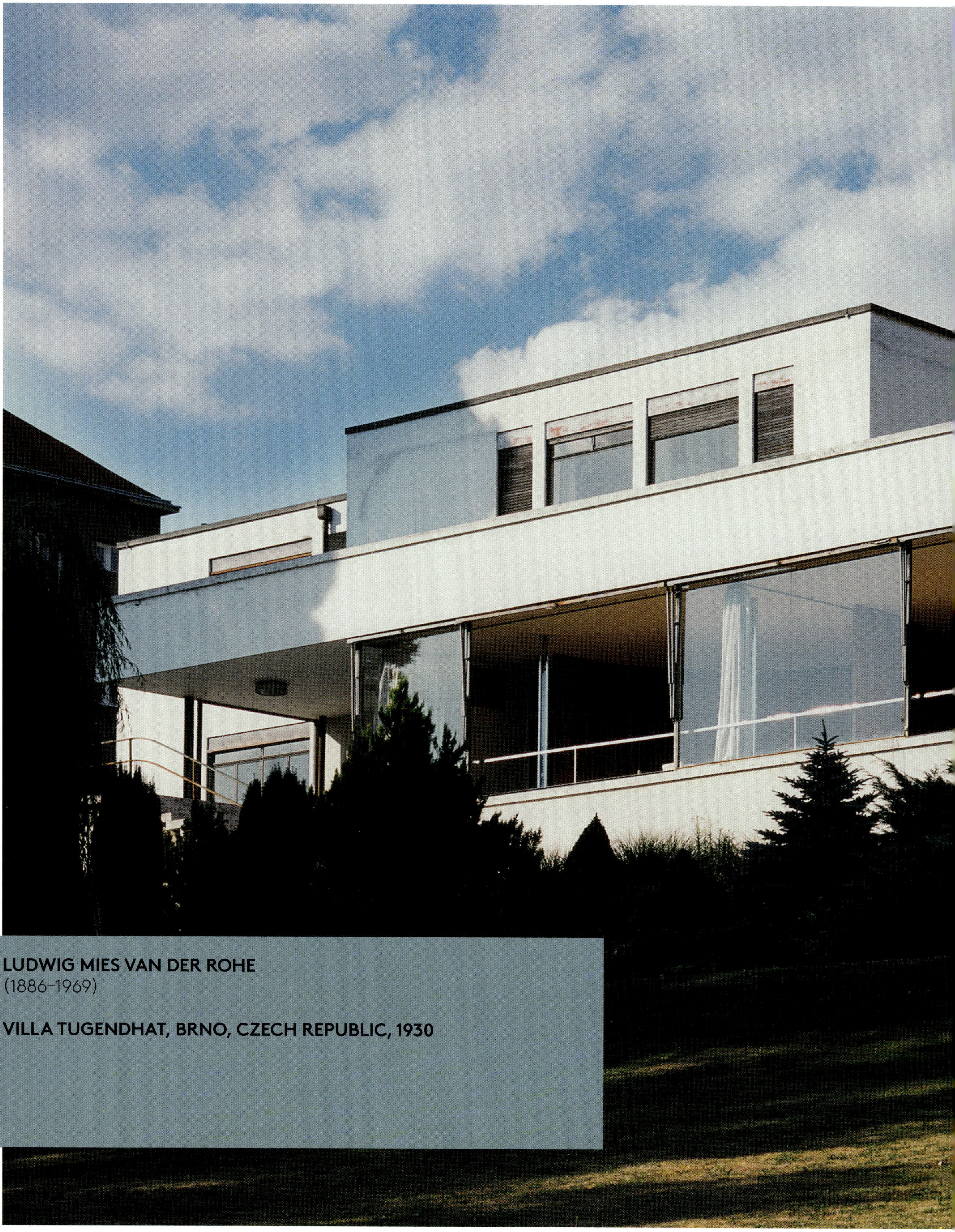

LUDWIG MIES VAN DER ROHE
(1886–1969)

VILLA TUGENDHAT, BRNO, CZECH REPUBLIC, 1930

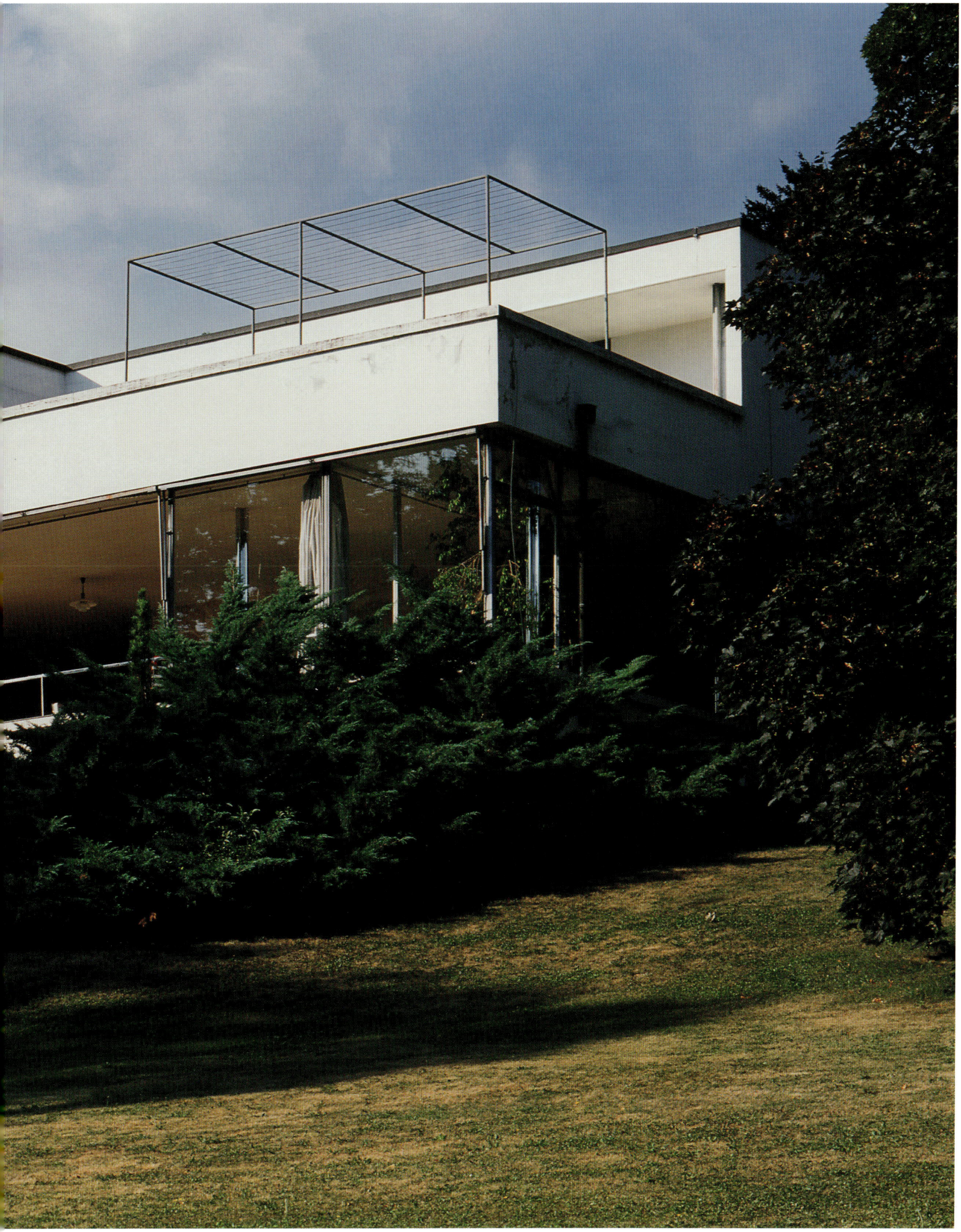

The story of Villa Tugendhat mirrors, in many ways, the uneasy birth of European modernism itself. The house was conceived in a spirit of optimism by a pioneering architect and his enlightened clients, looking to create something fresh, innovative and even revolutionary. Villa Tugendhat is a building full of sophisticated and powerful ideas, as well as state-of-the-art technology and engineering, but at the same time it was also a family home, tailored to a particular way of living and lifted by expressive materials and a strong connection to the gardens and city beyond. Yet the glory years of the house were very few, as it became a victim of extreme politics and violent upheaval, occupied first by the Nazis and then by the Russians, who caused great damage to the building while its owners and architect went into exile.

For Ludwig Mies van der Rohe, the Villa Tugendhat represents the finest of his European houses and – along with the Barcelona Pavilion of 1929 – one of his most formative and influential projects of the inter-war period. The Brno house came at a time when Mies and his work were much in demand. In 1927, he was placed in charge of the Weissenhof Estate* in Stuttgart – a showcase exhibition of the Deutscher Werkbund*, which was itself a collective formed by a number of leading German modernist architects.

Soon afterwards the German Weimar Republic commissioned Mies to design the pavilion at the International Exhibition in Barcelona. This pavilion was intended as a temporary structure (although it has since been rebuilt) for ceremonial functions and to serve as an introduction to the German section of the exhibition. With its steel framework, curtain walls and expanses of glass – as well as the use of finely crafted natural materials – the pavilion has many features in common with Villa Tugendhat, although the programmes of the two projects were entirely different. In 1930, Mies was appointed head of the Bauhaus in Dessau.

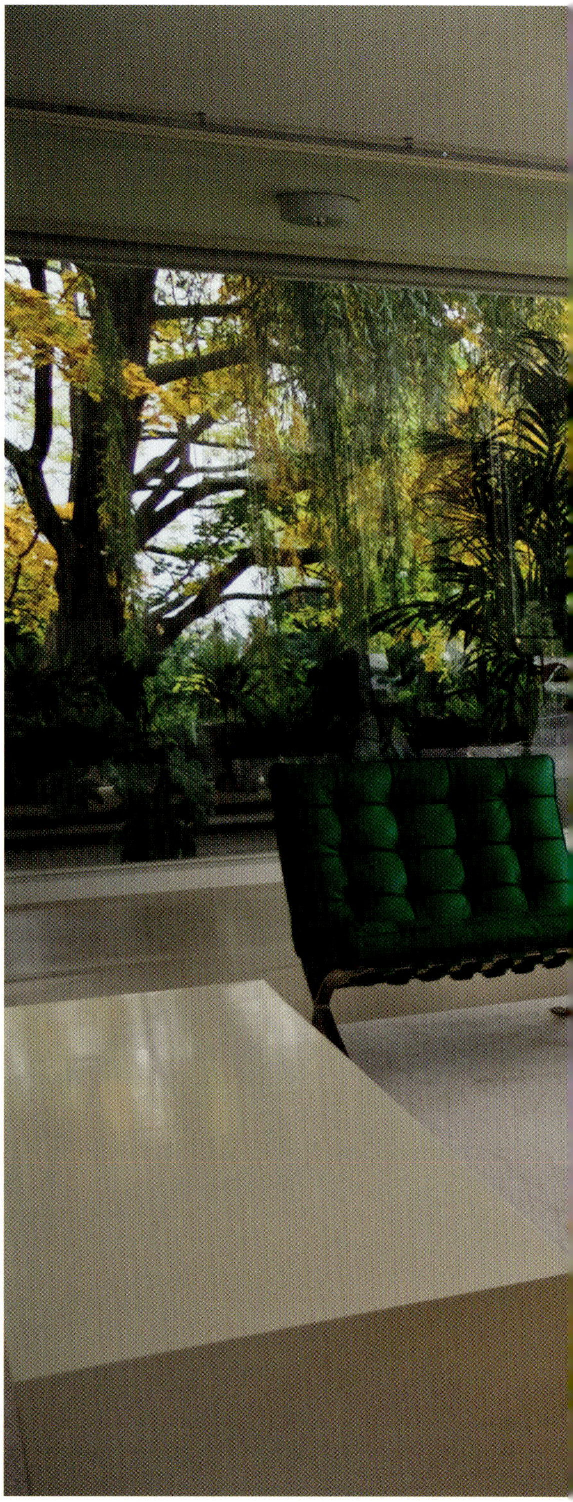

Grete Löw-Beer became familiar with Mies's work when she was living with her first husband in Germany; she knew both the Weissenhof project and some of the architects' early houses. When her unhappy marriage broke up, she returned to Czechoslovakia, where her wealthy Jewish-German family were industrialists with interests in textile manufacturing. Here she married Fritz Tugendhat, whom she had known since childhood; he too was involved in the textiles business in Brno.

Grete Tugendhat's father gifted her a parcel of land next to the family's own Art Nouveau villa on a hillside overlooking the city. Both Grete and Fritz Tugendhat wanted a new beginning, so her thoughts turned to Mies, who designed a contextual structure that responded to the hillside setting and the open vista. From the access street behind, the house resembles an enigmatic bungalow, punctuated by a gap between the main building and a small pavilion nearby housing a garage and staff quarters. The entrance hallway, five bedrooms and a generous roof terrace inhabit what is, in fact, an upper level. It is only as one descends to the storey below – pushed into the hillside itself with a concrete retaining wall – that the house truly reveals itself.

Here, the benefits of using a structural steel framework for the house become very clear. A similar approach had been used for the Barcelona Pavilion but in Brno the system forms a very early example of a residential application, with the absence of load-bearing walls allowing Mies to create a vast ribbon of floor-to-ceiling glass that dominates the largely open-plan living space. Two sections of this glass wall drop down into the floor mechanically, further dissolving the junction between inside and outside, and allowing free access to an adjoining terrace.

The living area is lightly zoned in subtle ways, without interrupting the flow of light or the horizontal ribbon of glass, while the steel columns are disguised in chrome-plated shrouds, forming glimmering reflective pillars. An onyx screen helps to lightly separate a seating area and a study/library, while a curving semi-circular enclosure made of crafted macassar ebony helps define a dining area and also demarcates the 'servery' beyond. Working in collaboration with LILLY REICH, Mies also turned his attention to the interiors and furnishings of the house (see p. 80).

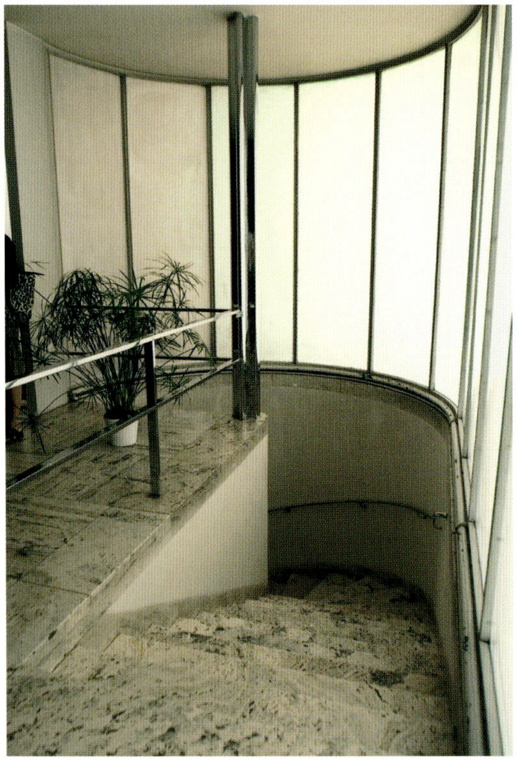

The project was begun in 1928 and completed in 1930. Yet in just a few years the political and cultural landscape of Central Europe changed dramatically. In 1933 the new Nazi government forced the closure of the Bauhaus, and in 1938 Mies left for America, beginning a second important phase of his career. Back in Brno, the Jewish heritage of the Tugendhat family, along with their support for German political refugees, made them vulnerable. In May 1938 the family moved to Switzerland, and they later emigrated to South America. A restoration programme began in the Eighties, and the revived house is now open to the public.

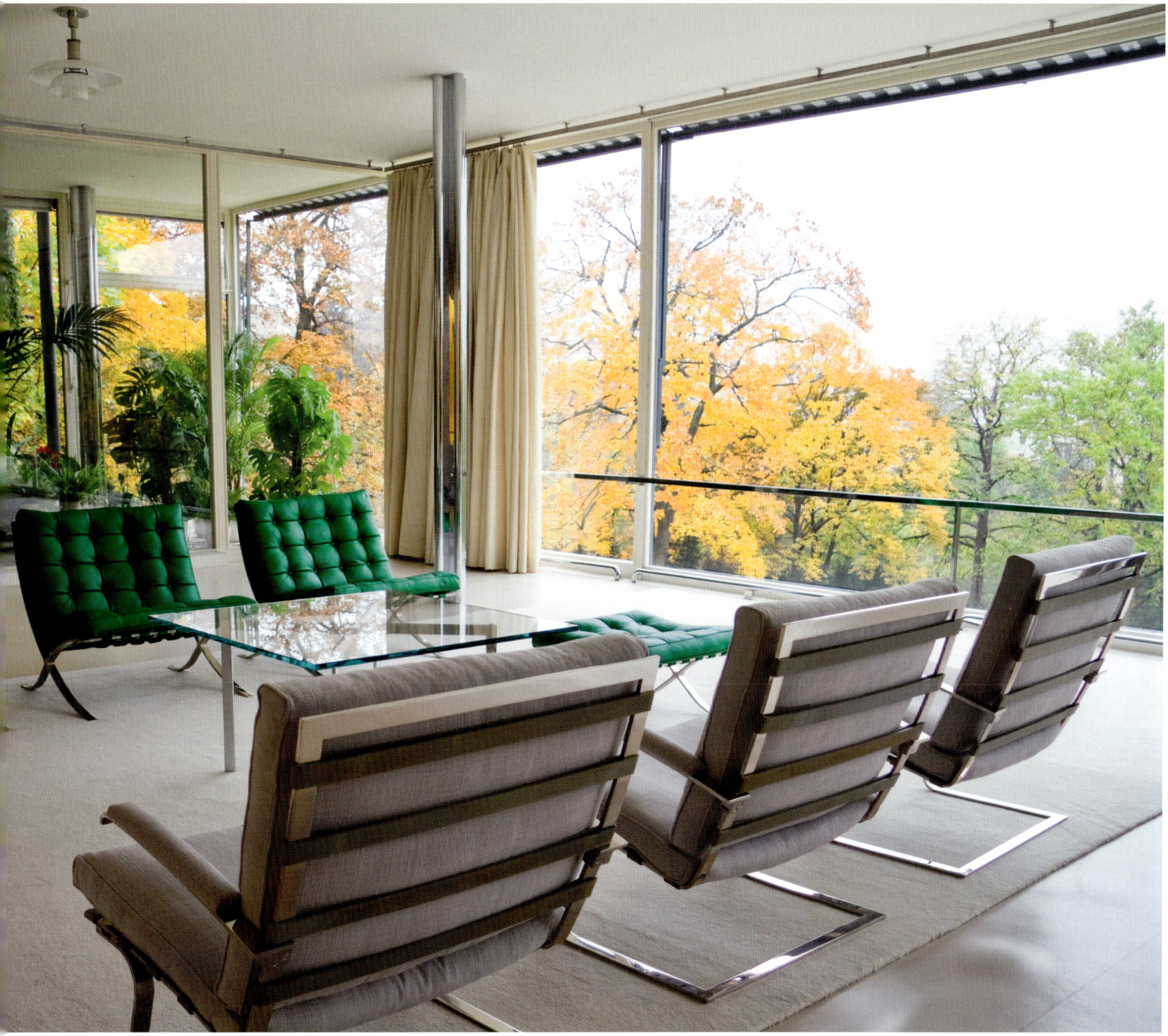

PAGES 380–81 & OPPOSITE Tucked into the hillside, the house is arranged on two levels with the main living spaces on the lower storey and bedrooms and a roof terrace above; the entrance is to the rear, on the upper floor, with stairs leading directly from the entry hall to the living room below.

ABOVE A view of the main seating area within the fluid sequence of living spaces on the floor level, with the winter garden beyond; the floor-to-ceiling windows to the exterior sink down into the floor, creating an open connection to the landscape.

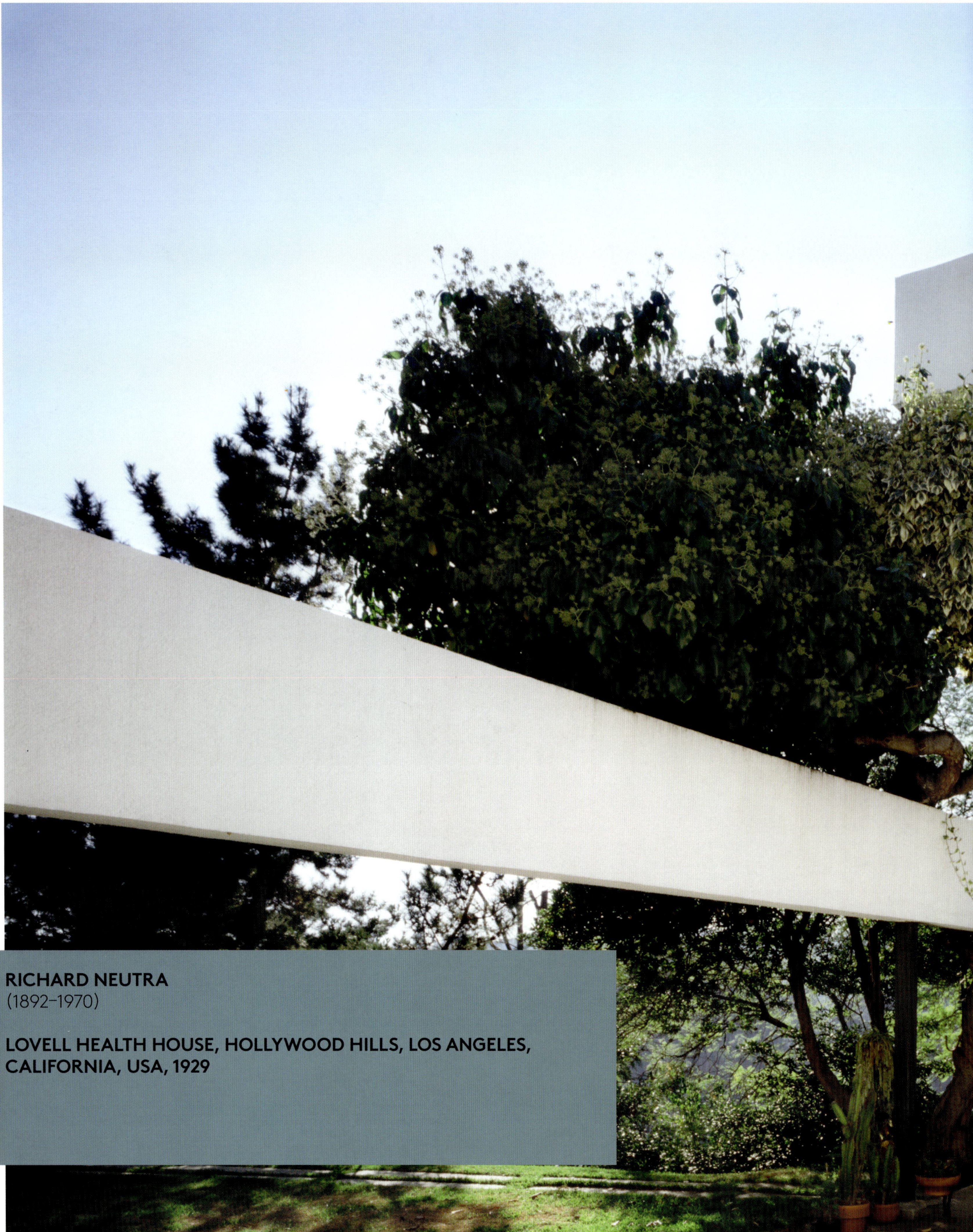

RICHARD NEUTRA
(1892–1970)

LOVELL HEALTH HOUSE, HOLLYWOOD HILLS, LOS ANGELES, CALIFORNIA, USA, 1929

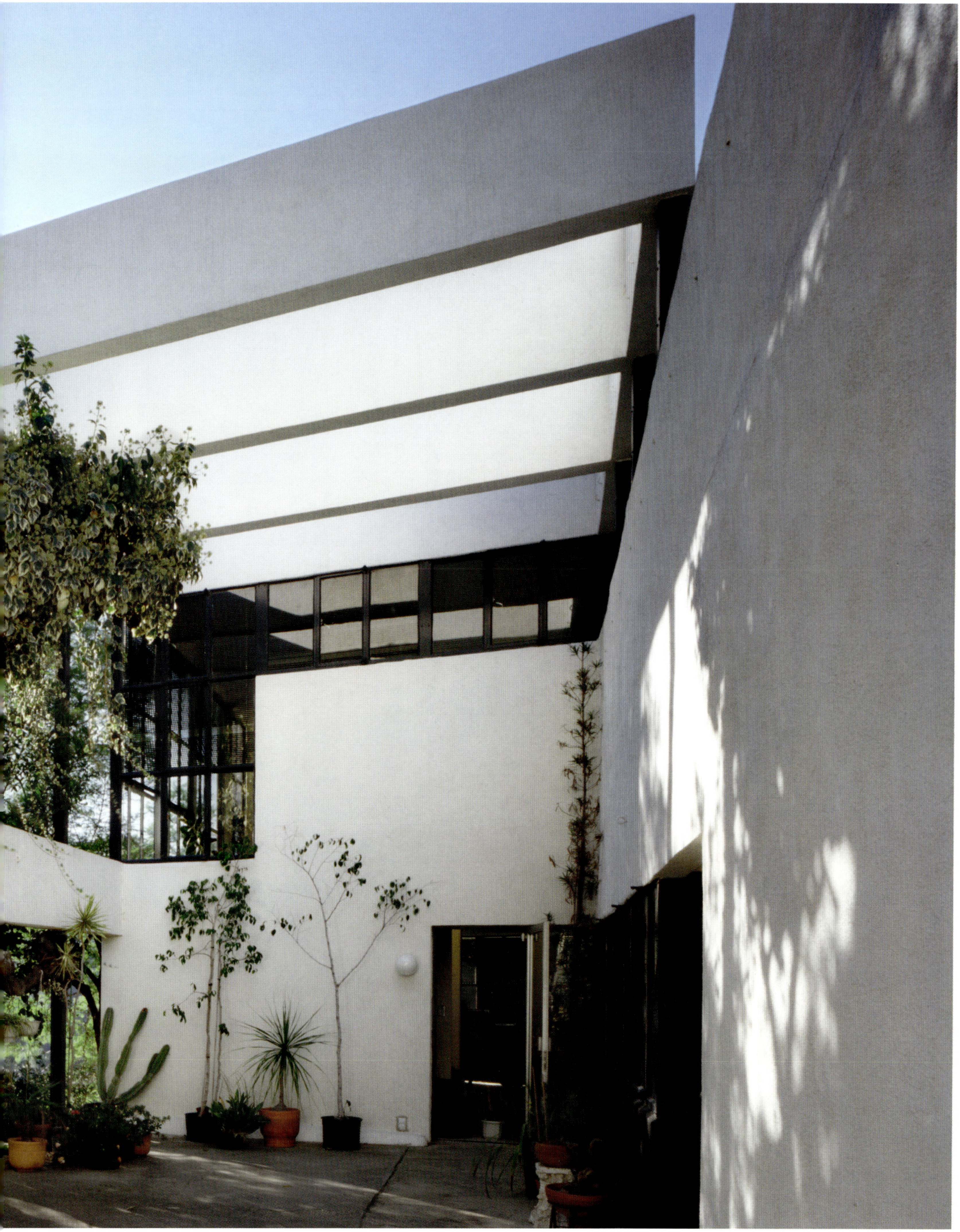

Among many modernist architects of the Twenties and Thirties there was a sincere belief in the impact of architecture and design upon well-being. Modernist architecture was not only rational, functional, logical and practical but also health-promoting and hygienic, standing in contrast to the sprawling slums and tenement buildings that blighted many cities around the world. A thoughtful, considered, modern building had the power to enhance the daily lives of its occupants in a multitude of ways, adding to a sense of physical and mental ease.

Richard Neutra was a famous believer in this power of architecture and design. The architect caught malaria during his military service with the Austrian Imperial Army during World War I and spent time recovering in hospitals and rest homes in Vienna, Slovakia and Switzerland. Neutra saw his war years as a tragic waste, creating a terrible gap in his life and career that took many years to fill. Before the war he had begun his studies in architecture at the Technical University in Vienna, studying with Adolf Loos* and influenced by Otto Wagner*; here he also met Rudolph Schindler, a friend and colleague who shared his fascination with America and the work of Frank Lloyd Wright. Schindler emigrated to America in 1914, but Neutra was only able to graduate in 1918, after his military service.

In the early Twenties he worked with the modernist architect Erich Mendelsohn* in Berlin, and he married Dione Niedermann. In 1923, the Neutras finally left for America. There, Neutra worked as a draughtsman with a Chicago firm, Holabird & Roche, and spent some months at Frank Lloyd Wright's Taliesen East residence and studio, after meeting Wright at the funeral of architectural pioneer and godfather of the modern skyscraper Louis Sullivan*.

In 1925, the Neutras moved on to Los Angeles and settled into the pair of vacant studio rooms at the pioneering house that Schindler had built at North Kings Road (see p. 412). The two friends collaborated on a number of projects before concentrating upon their own individual activities and clients. One client that they held in common was Philip Lovell.

Lovell was a 'naturopath'. As a physician with a considerable following, plus a newspaper column, he was a believer in drawing upon nature itself to help in the healing process and in maintaining good health, promoting regular exercise and calisthenics, a vegetarian diet, nude sunbathing and fresh air. Lovell and his wife Leah commissioned Schindler to design a beach house in Los Angeles (1926), as well as a mountain cabin and a desert house. After the collapse of the cabin's roof under winter snow and issues with the balconies at the beach house, Lovell turned to Neutra to design his principal residence. It was a decision that placed strains upon the relationship with Schindler, but one that also launched Neutra's career.

Both Lovell and Neutra shared this belief that architecture and the way that one lives can greatly affect health and well-being. Neutra developed a philosophy of what he called 'biorealism', centred upon the need for a strong and vital relationship between indoor and outdoor living space. In this respect and many others, Lovell was the dream client. The Lovells wanted terraces and outdoor spaces for exercise and enjoying the climate, as well as exterior sleeping porches alongside the bedrooms.

The steep canyon-side site on Dundee Drive also steered Neutra's design, with the house pushed into the hillside. Neutra developed a pre-fabricated steel framework, which was assembled in just 40 hours on site and fixed to the concrete foundations; it has been described as the first all-steel-framed house in America. There were also great expanses of glass, enhancing the views and the connections with the outdoors.

The rectangular form of the building had a crisp cubist quality and a lightness of touch, with piloti at ground level supporting the 'floating' house and partially framing the semi-sheltered swimming pool, while the principal living spaces – including a dramatic open-plan living room – were at mid-level, and bedrooms and entry at the upper level.

Lovell was delighted with the house: 'I did not have a single complaint,' he wrote. 'Everything I represented you produced for me – and it seemed that it was your philosophy as well as mine. Not only did you create a masterpiece, but you did it so economically that I was surprised and pleased.'[25]

The Lovell Health House opened a whole series of doors for Neutra. In later projects, which included apartment buildings and many landmark houses, he continued to explore the combination of spatial and structural innovation that helped make the Lovell project so special. Key commissions included the Miller House (1937) in Palm Springs for another devotee of the healthy Californian lifestyle and 'functional exercise', Grace Lewis Miller. Neutra also produced one of the greatest houses of the mid-century period – the Kaufmann House (1947), again in Palm Springs – which took the theme of connectivity between inside and outside living to a new level.

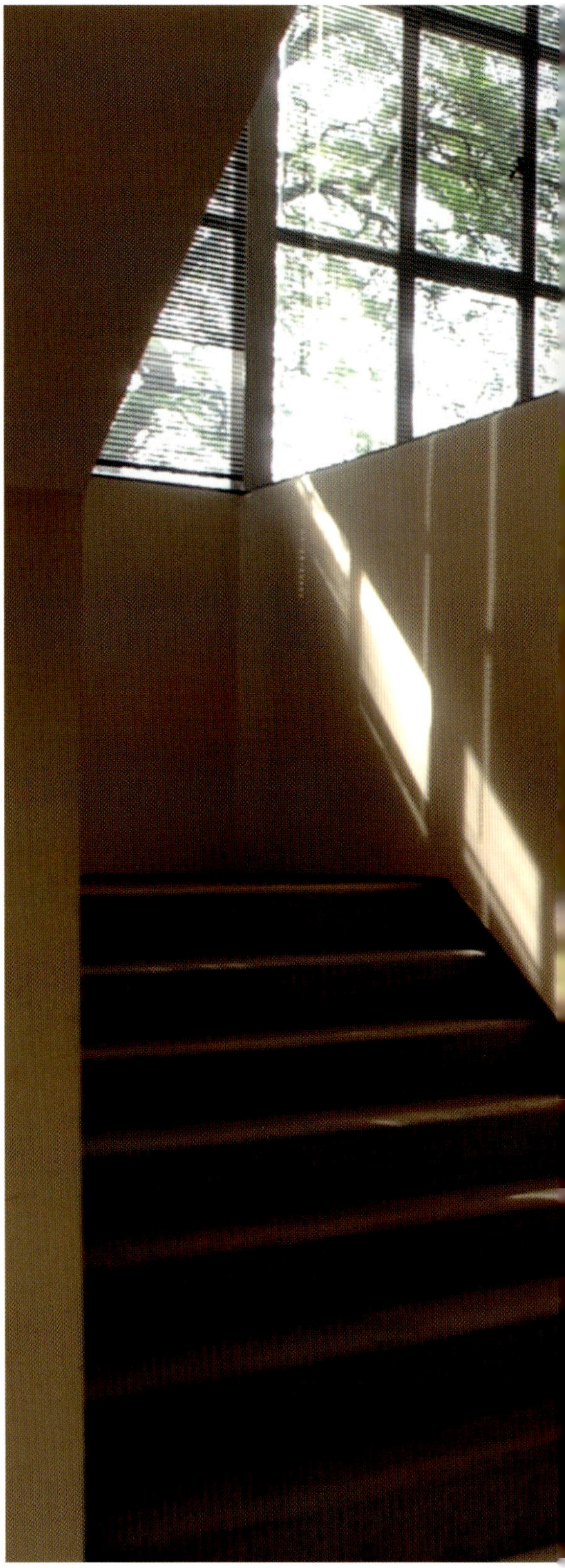

PAGES 384–5 & RIGHT
Pushing outwards from the crest of the ravine, the house looks out into the open canyon, complemented by a series of terraces, porches and balconies, including the swimming-pool terrace at the base of the building.

BELOW The library, seating area (with a fireplace) and dining area form an open sequence of spaces, with the staircase to one side sitting within an atrium framed by a double-height wall of glass leading to the uppermost storey, which holds the master suite and connects with the street-level entry sequence.

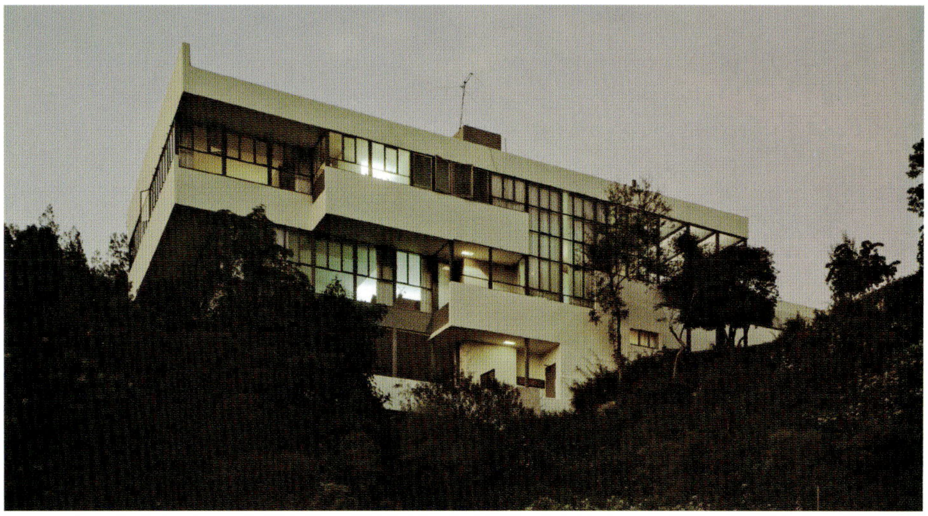

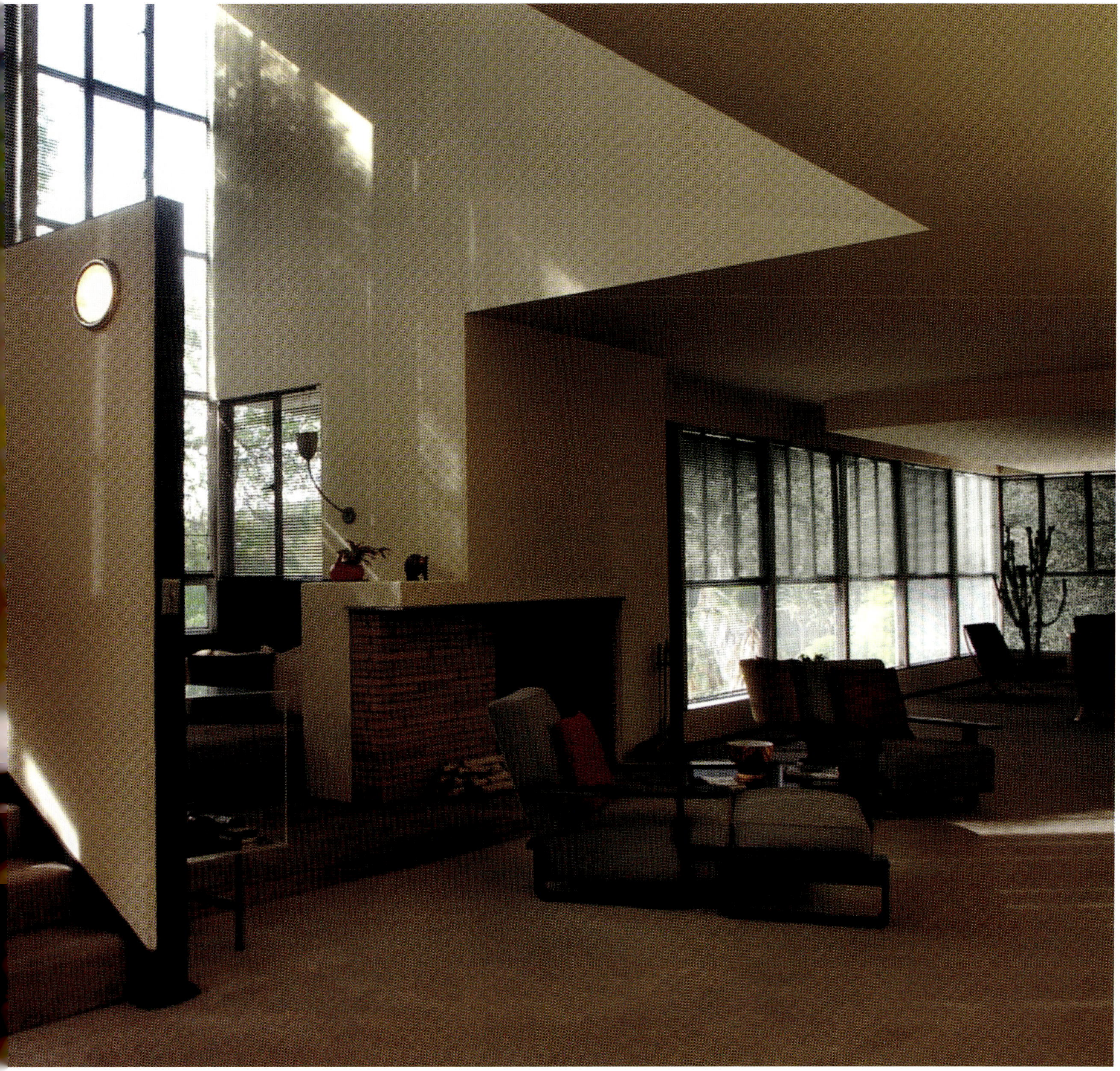

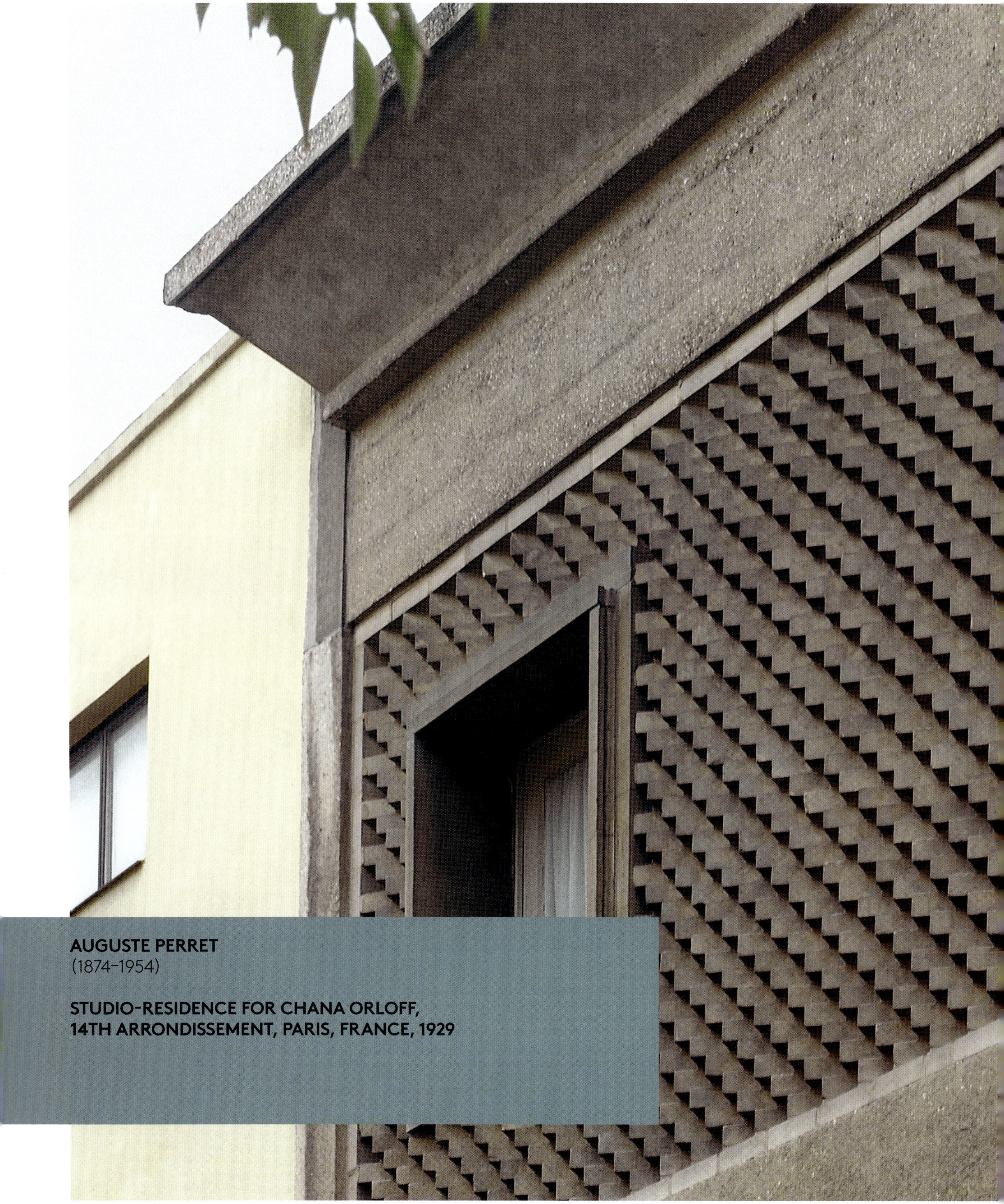

AUGUSTE PERRET
(1874–1954)

STUDIO-RESIDENCE FOR CHANA ORLOFF, 14TH ARRONDISSEMENT, PARIS, FRANCE, 1929

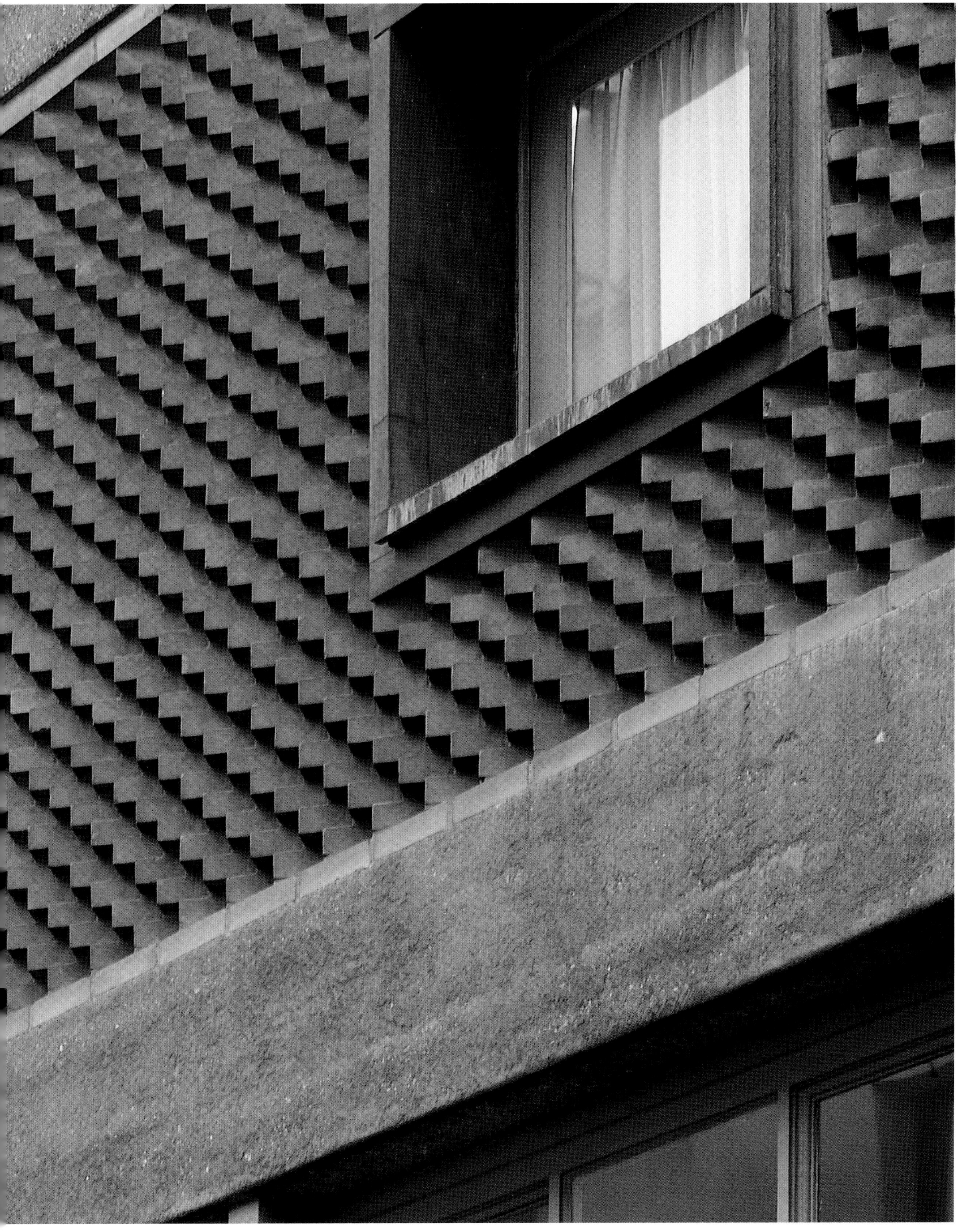

The French architect Auguste Perret was a pivotal figure who operated at the junction points of various different worlds. He was a pioneer and an innovator, who played a vital role in transforming reinforced concrete from a marginal, industrial material into one of the key building tools of modernism itself. Yet, at the same time, Perret's work was rooted in the past, with an enduring respect for neoclassical architecture and historical precedents. While his love of engineering and construction may have been revolutionary, he was not the kind of avant-garde hero who delights in breaking with the lessons of history and tradition. 'Perret was a firm and solid mooring in the swell that divided the nineteenth century from the twentieth,' observed LE CORBUSIER,[26] who served a fourteen-month apprenticeship in the studio of his mentor.

Perret also worked in an engaging hinterland between engineering and art. He called himself a '*constructeur*', grounded in the practicalities of the building site and craft. Yet he was also a fine draughtsman and had studied – along with his brother Gustave – under architect Julien Guadet at the École des Beaux-Arts in Paris. Perret himself played a key role as an educator, serving as a mentor not just to Le Corbusier but also to other modernists, such as Berthold Lubetkin* and ERNŐ GOLDFINGER.

Perret was the son of a stone mason and was one of three brothers. Along with Gustave and Claude, he worked in his father's busy atelier from a young age, with 'Perret et Fils' enjoying success with a series of apartment building commissions. After their father's death in 1905, Claude Perret focused on running the finances and business administration of the company, while Auguste concentrated chiefly upon design and Gustave took more responsibility for construction. The

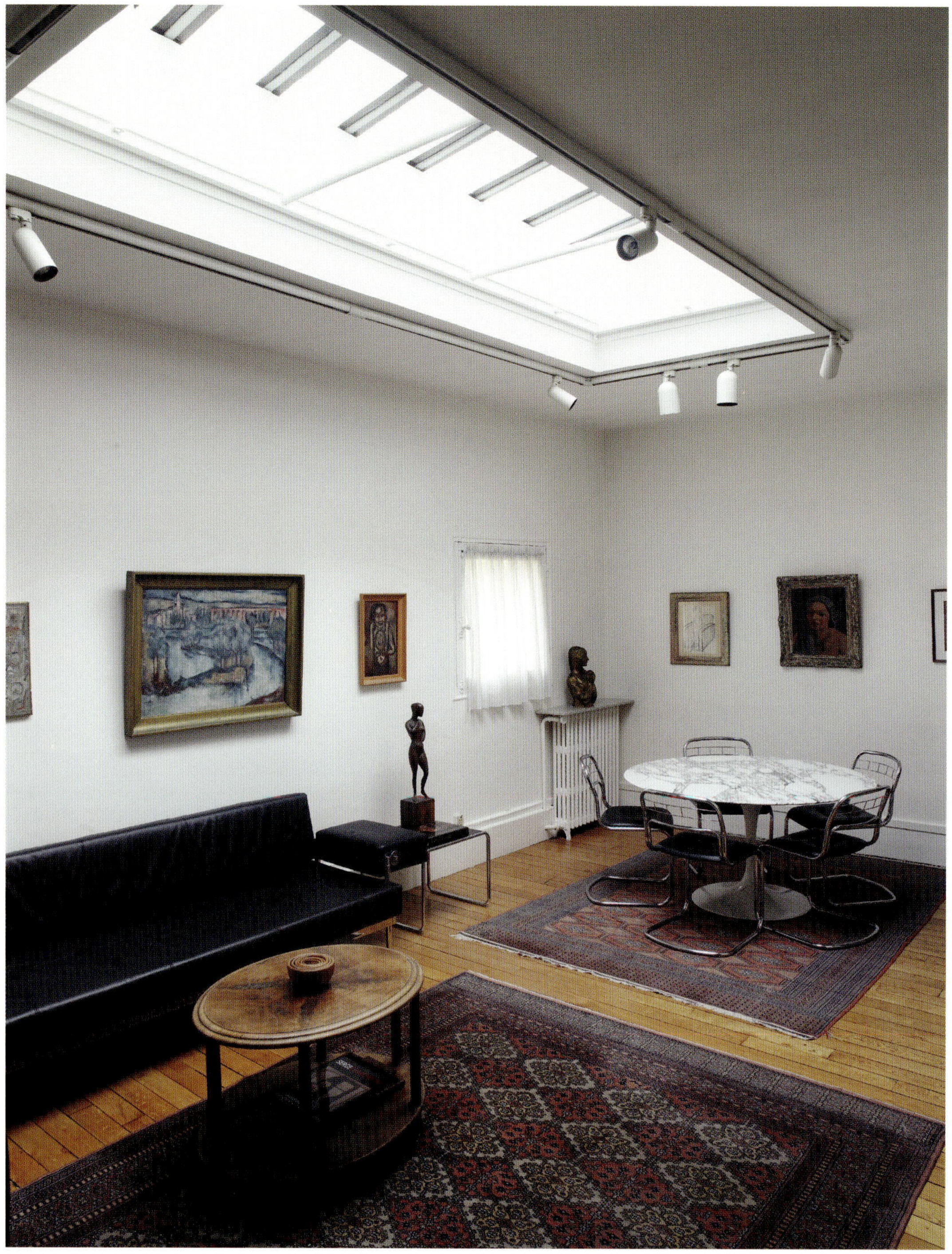

PAGES 388–9 & OPPOSITE The concrete-framed building features a cubist-style façade, with a sequence of extensive glazing introducing light to the art studio on the lower levels and an infill of brick adding texture to the upper band of the building.

ABOVE A private apartment for Chana Orloff and her son was created at the top of the studio-residence, featuring a large skylight that serves as the principal source of sunlight in the living/dining area.

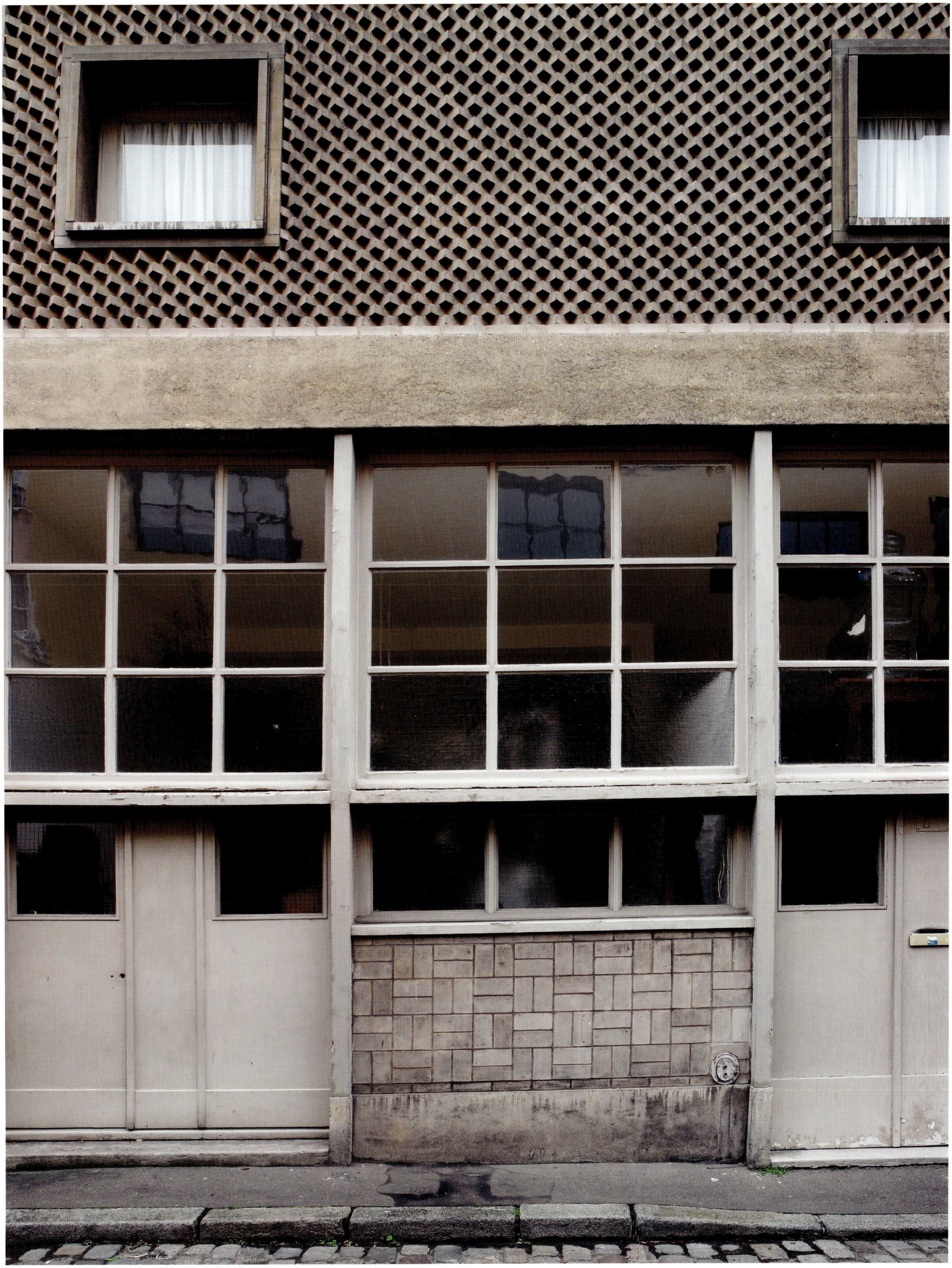

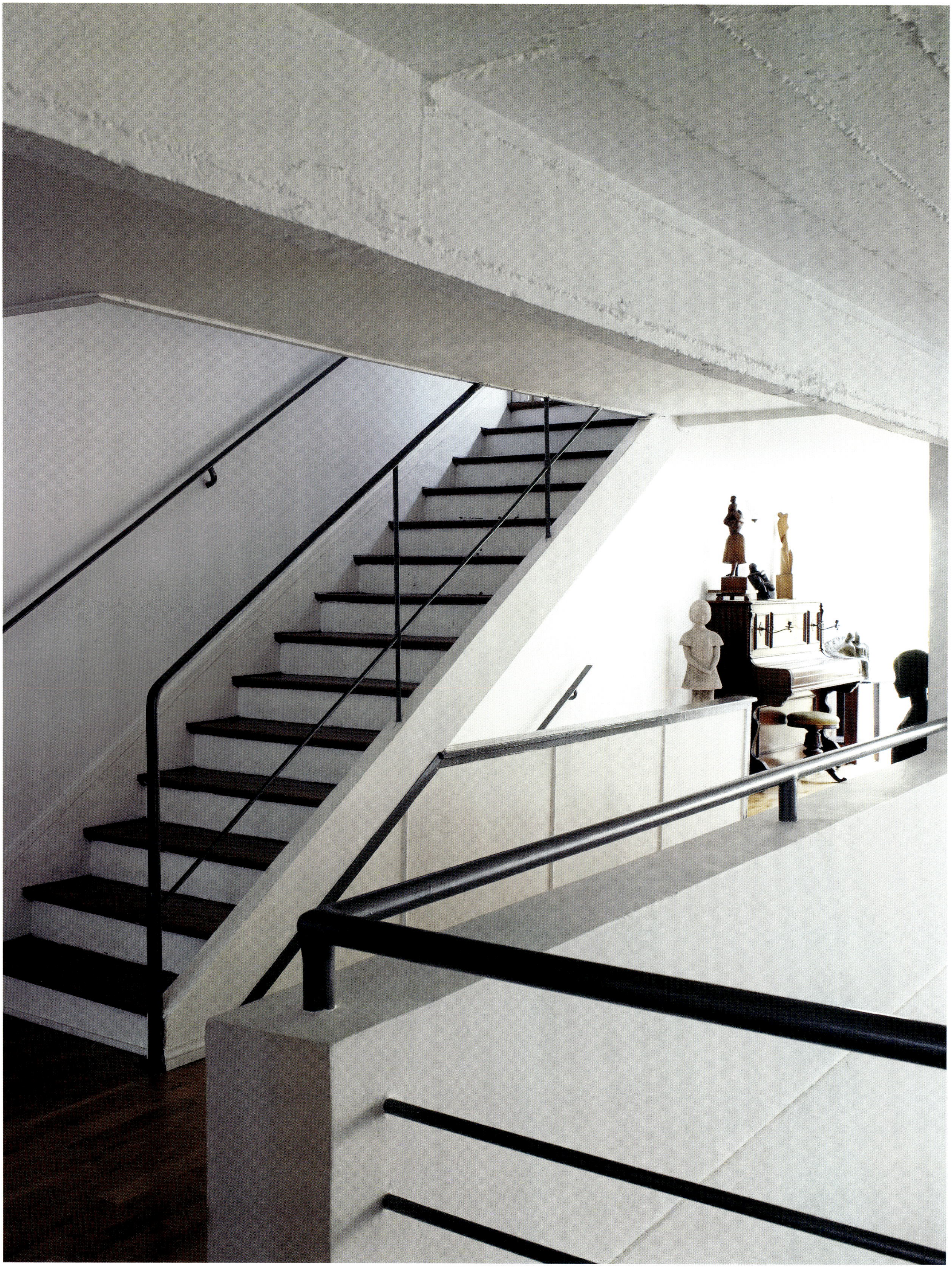

atelier decided to specialize in reinforced concrete buildings, with architecture credited to 'A. & G. Perret' and construction and engineering to 'Perret Frères'.

During the Twenties there were a number of key projects for artists and collectors, including a home in Versailles (1924) for the graphic designer A. M. Cassandre and a studio-residence in Paris for painter Georges Braque (1927), as well as another for the painter Mela Muter (1928), which was rented by the French sculptor and artist Jean Dubuffet, who eloquently praised Perret's design of the building. The notion of a studio-residence had become fashionable in Paris, as in other parts of Europe, and the Perrets designed a series of such buildings over the inter-war period. One of these was a commission for Chana Orloff, completed in 1929.

Orloff was a fascinating figure. She was one of nine children born in the Ukraine to a farm worker, with the family emigrating to Palestine in 1905. Her talents as a seamstress were noted at school and she was encouraged to pursue her studies in Paris, where she combined an apprenticeship with evening courses. She went on to study drawing and art history at the École Nationale Supérieure des Arts Décoratifs and sculpture at the Académie Russe. She married the poet Ary Justman in 1916, but he died three years later, leaving her with a young son to support. Despite these challenges, Orloff continued to forge an independent and successful career and became well known for her cubist woodcut portraits and sculpted busts of many celebrated artists and musicians.

Orloff asked Auguste Perret to design a studio-residence in the Villa Seurat to serve as a gallery-showroom, a workshop and a home. These studio-residences, as a new typology, invited an experimental approach in every respect. They

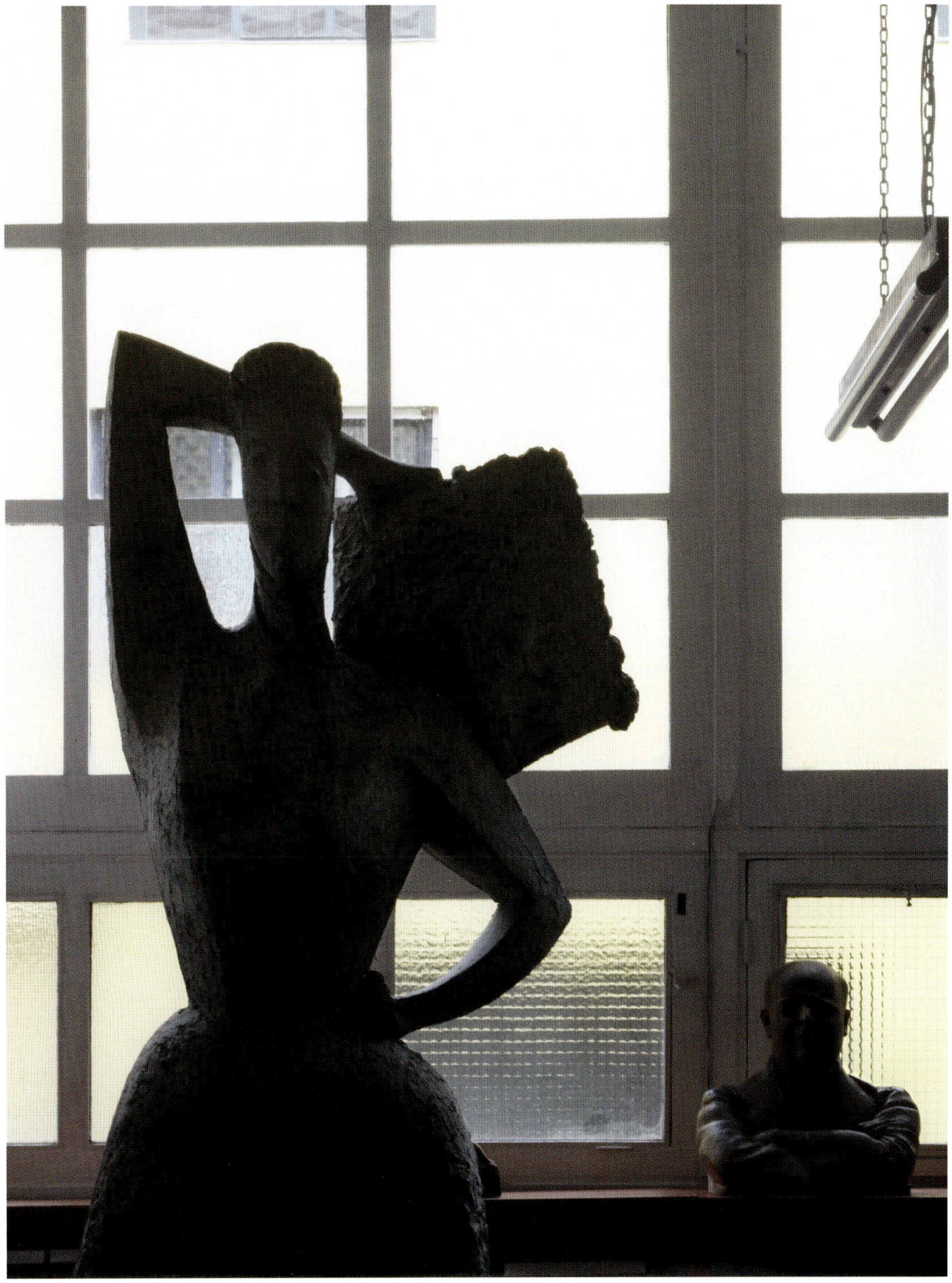

OPPOSITE The double-height studio is overlooked by a mezzanine level – as well as a circulation landing around the stairs – which forms a more intimately scaled space for both working and displaying smaller sculptures.

ABOVE Taller pieces of work, such as this standing figure, sit within the main body of the studio space, which draws light from the large sequence of windows to the front of the building.

demanded light and open space, which made them highly suited to reinforced concrete-framed construction that allowed for fluid space within and large expanses of glazing on the façade. The frontage of the house has a semi-industrial quality, with garage-style double doorways at street level, a large band of glazing above and then a chevron brick infill on the upper level, punctuated by two windows and topped with a cornice. Texture is provided by the materials, with the only 'ornamentation' provided by the pattern of the brickwork infill between the bands of concrete.

The philosophy of simplicity and reduction continues within. The ground floor features a double-height, timber-floored gallery to the front and a workshop to the rear, with an open mezzanine level forming a partial mid-level to the building. A simple staircase to one side reaches up to the apartment at the top, with two bedrooms, a bathroom and an open-plan living space facing the street and illuminated by a large skylight.

The pared-down nature of the space and its semi-domestic status lend it a 'modernity' that contrasts with some of Perret's other residential projects. The modest scale also contrasts with some later commissions, including the striking dynamism of the Museé des Travaux Publiques in Paris (1948) and buildings created as part of the reconstruction of Le Havre, following heavy devastation during World War II.

Orloff and her son fled to Switzerland during the war. Upon her return, she found that the studio-residence had been broken into and many of the contents destroyed. The studio was, however, restored, and in the later years of her life Orloff travelled and worked between her Paris base and Israel.

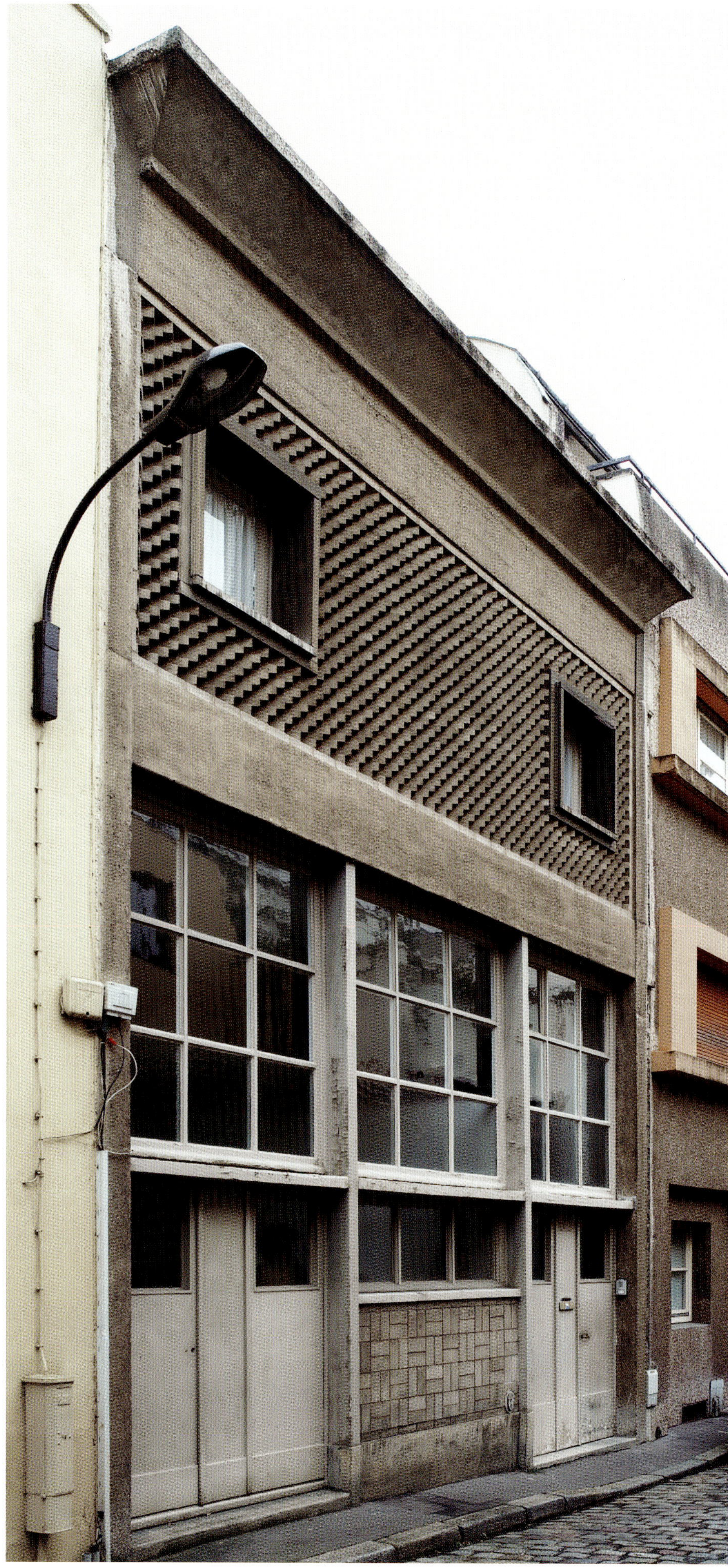

RIGHT & OPPOSITE The majority of the ground floor is devoted to Orloff's double-height studio space, with its simple white walls and practical wooden floors; the mezzanine level offers a kind of 'bridge' overlooking the activity below.

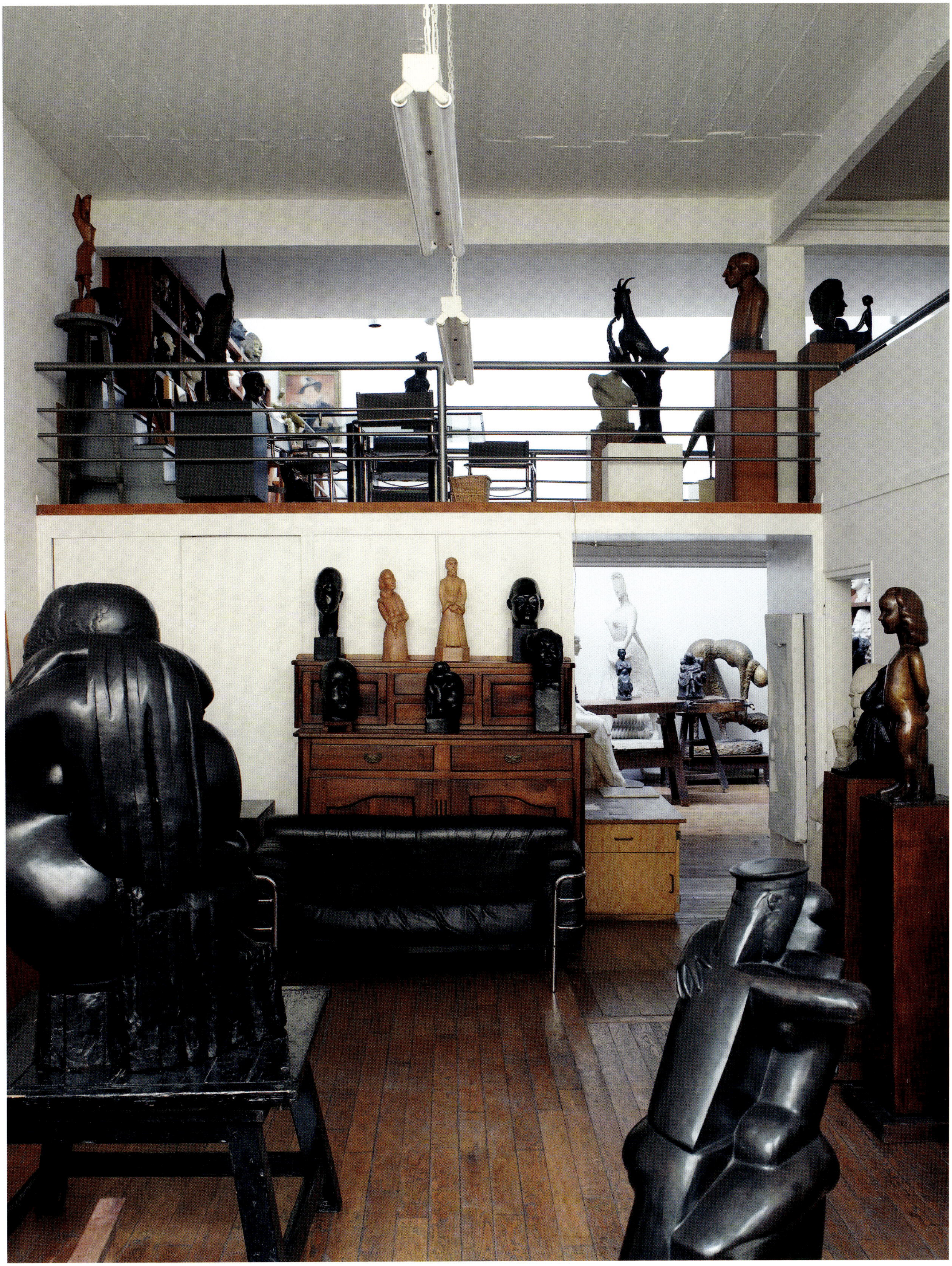

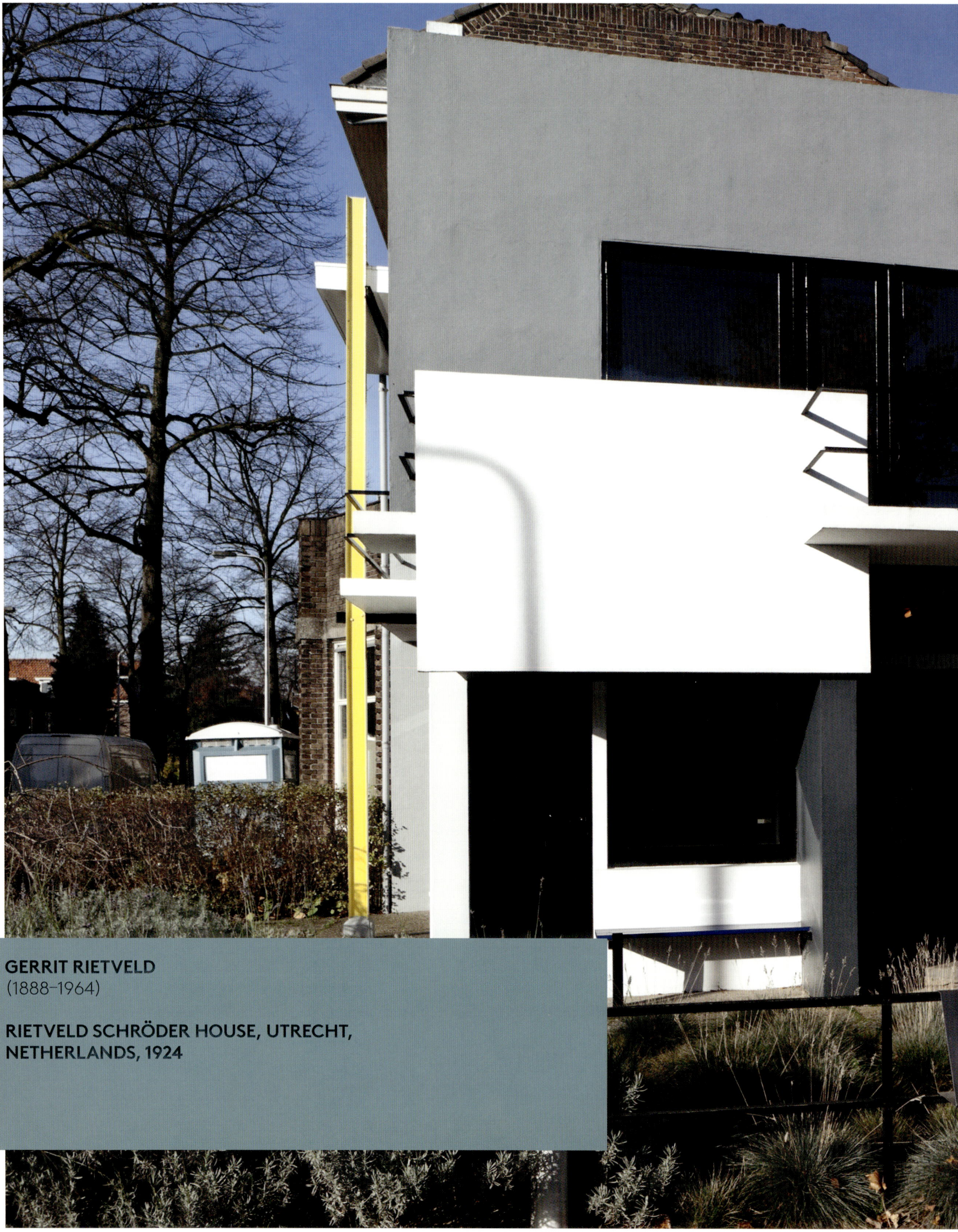

GERRIT RIETVELD
(1888–1964)

RIETVELD SCHRÖDER HOUSE, UTRECHT, NETHERLANDS, 1924

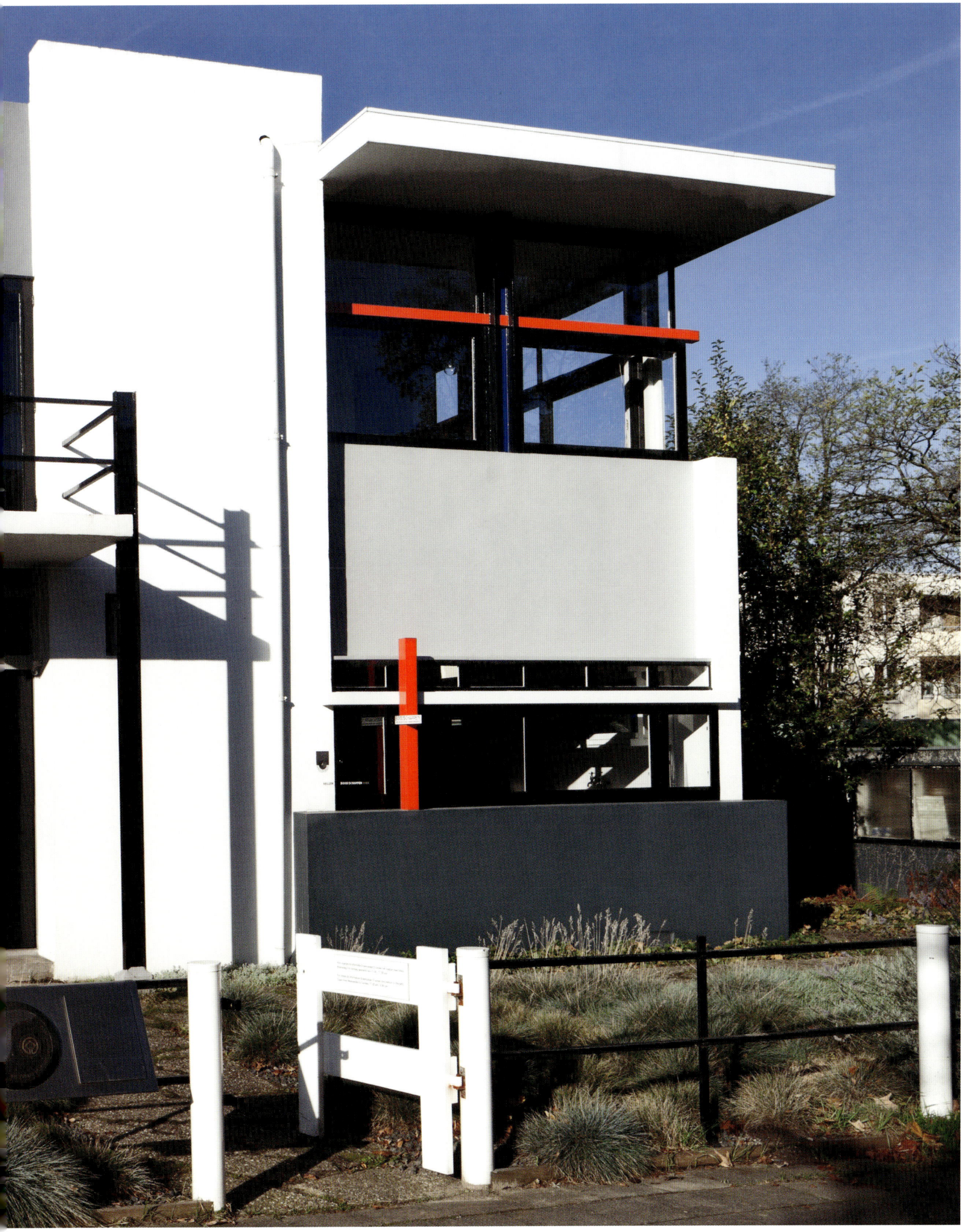

Truus Schröder was an unusual and imaginative client, and the kind many architects can only dream of. A wealthy patron, she was also progressive and liberated, and pushed her architect, Gerrit Rietveld, in new and unexpected directions. She shone within a collaborative approach, concentrating upon the interiors of the house in particular, to the point of contributing to future Rietveld projects.

A young widow with three children, Schröder first met Rietveld in 1921, not long after he had established his architectural practice. Initially, she asked him to work on interior projects at her 19th-century home; then on the design of her new house, also in Utrecht.

Up until this commission Rietveld was best known as a furniture designer and leading light in the Dutch De Stijl* movement, which promoted a fresh modernity based upon geometric abstract form (see p. 92). His Red/Blue chair had already made him an influential figure; the Schröder House became the most accomplished and complete structural expression of the movement, with its powerful geometry and rich use of colour.

'The building of this house is an attempt to break free of the humdrum excesses, which around 1920, after the honest styles of Berlage and other innovators, still influenced the architecture,' Rietveld said. 'We used solely primary forms, shapes and colours because these are so elementary and because they are free of association.'[27]

The house was also spatially radical and experimental. Schröder set certain criteria, particularly for the ground floor – dominated by an open-plan, multi-functional living space, plus a separate master bedroom – where each room was to have direct access to the exterior. These conditions stretched Rietveld into ever more original thought.

Planning codes meant that the upper level was officially designated as an attic, and here Rietveld positioned the children's bedrooms. Yet his client suggested that all solid partitions be removed upstairs, creating a dynamic mixed-use space. Rietveld established a flexible system of sliding partitions in response. In designing a building where the resident was constantly asked to make choices and actively engage with the space, Rietveld addressed his displeasure at the passive nature of the early 20th-century home. At the same time, he expressed his skills as a furniture designer in a number of bespoke and integrated designs.

A studio on the ground floor became the base for Rietveld's atelier up until 1932. Much later, in the Fifties, following the death of his wife, Rietveld returned to the house and made it his home until his own death in 1964. Over the years a number of changes were made to the building, which were accommodated by the adaptable design.

Bequeathed by Schröder to a dedicated foundation and later opened to the public, the house stands at the heart of an extraordinary and long-lived creative exchange. At the same time, it is one of the most powerful manifesto statements of not just De Stijl but the modernist movement itself.

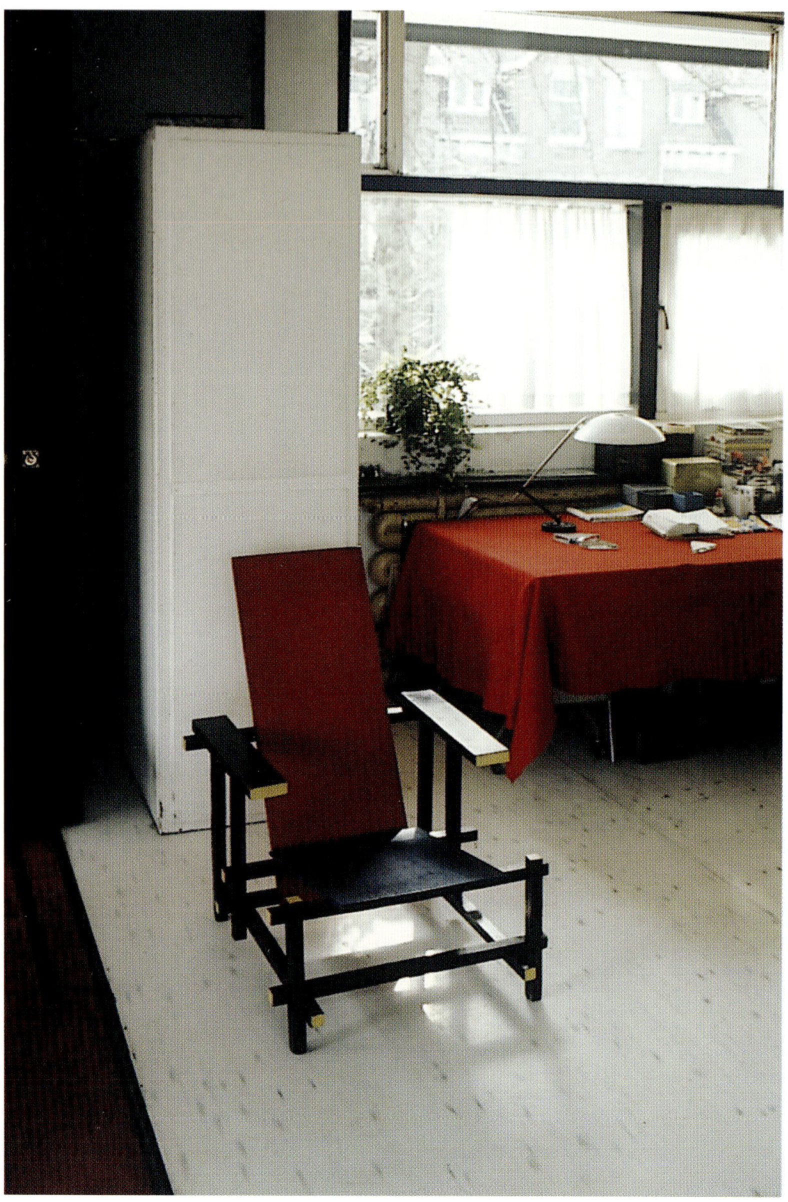

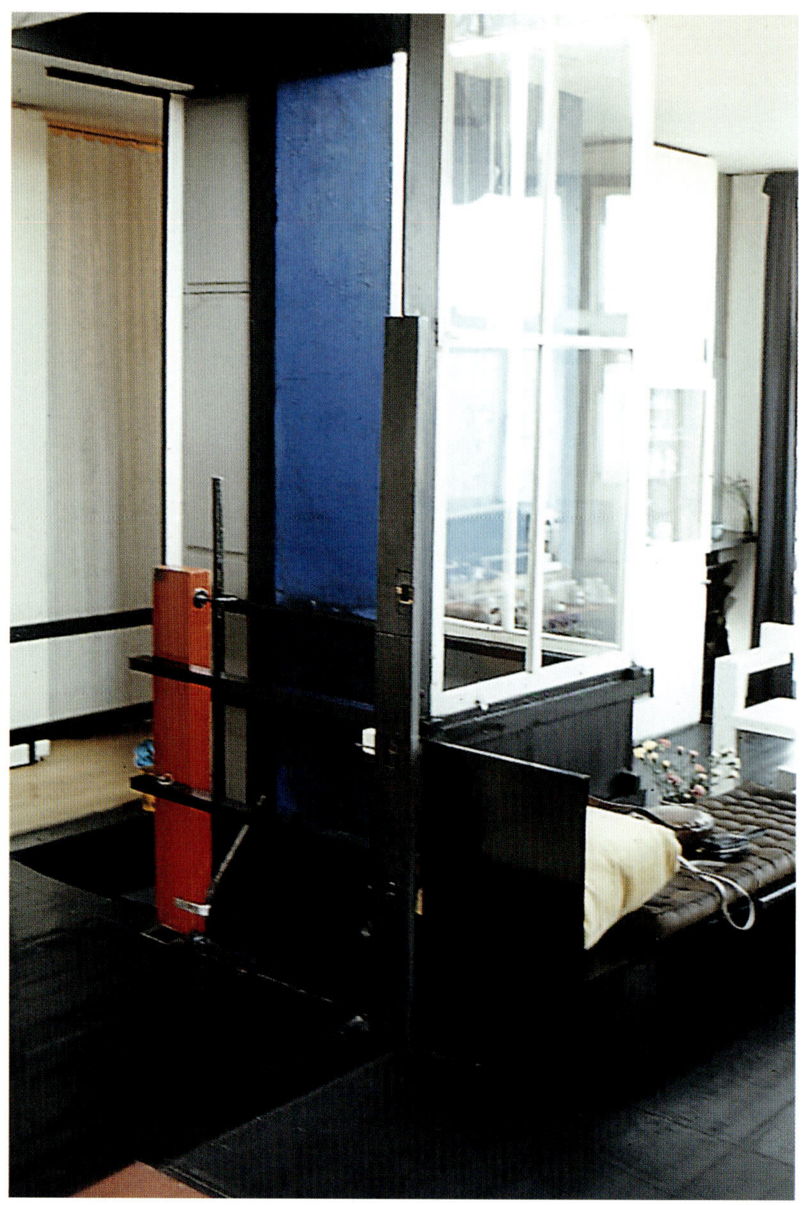

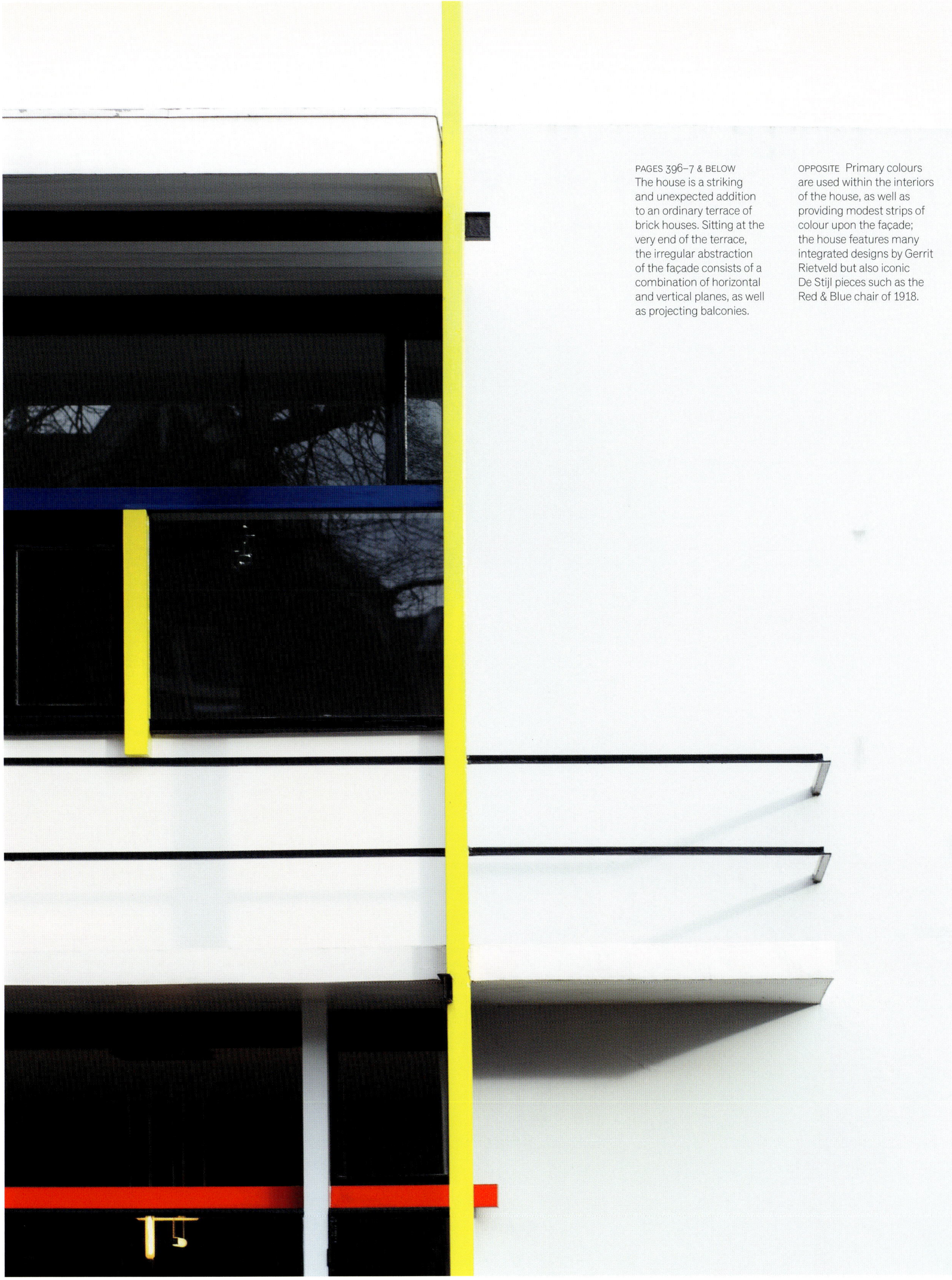

PAGES 396–7 & BELOW The house is a striking and unexpected addition to an ordinary terrace of brick houses. Sitting at the very end of the terrace, the irregular abstraction of the façade consists of a combination of horizontal and vertical planes, as well as projecting balconies.

OPPOSITE Primary colours are used within the interiors of the house, as well as providing modest strips of colour upon the façade; the house features many integrated designs by Gerrit Rietveld but also iconic De Stijl pieces such as the Red & Blue chair of 1918.

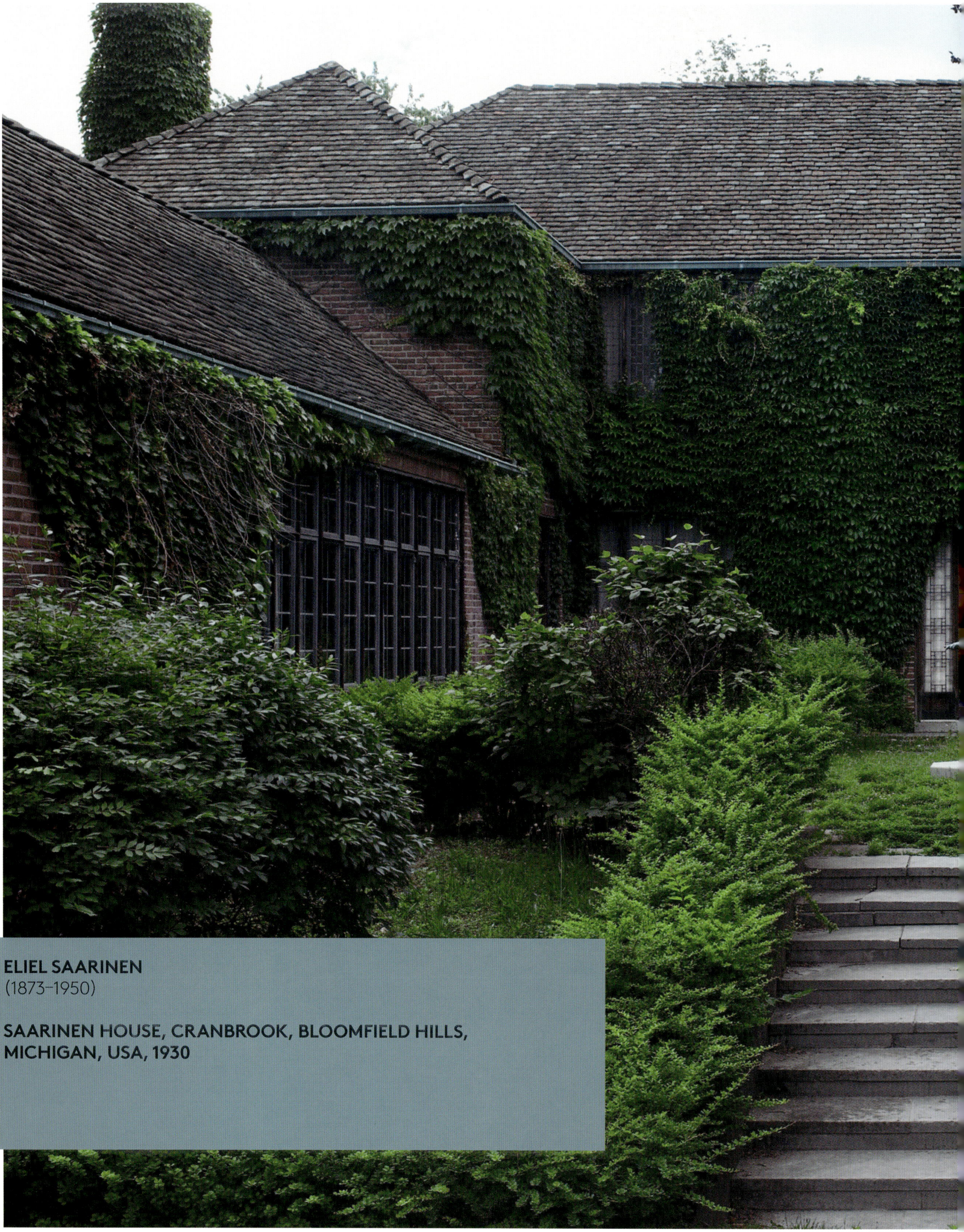

ELIEL SAARINEN
(1873–1950)

SAARINEN HOUSE, CRANBROOK, BLOOMFIELD HILLS, MICHIGAN, USA, 1930

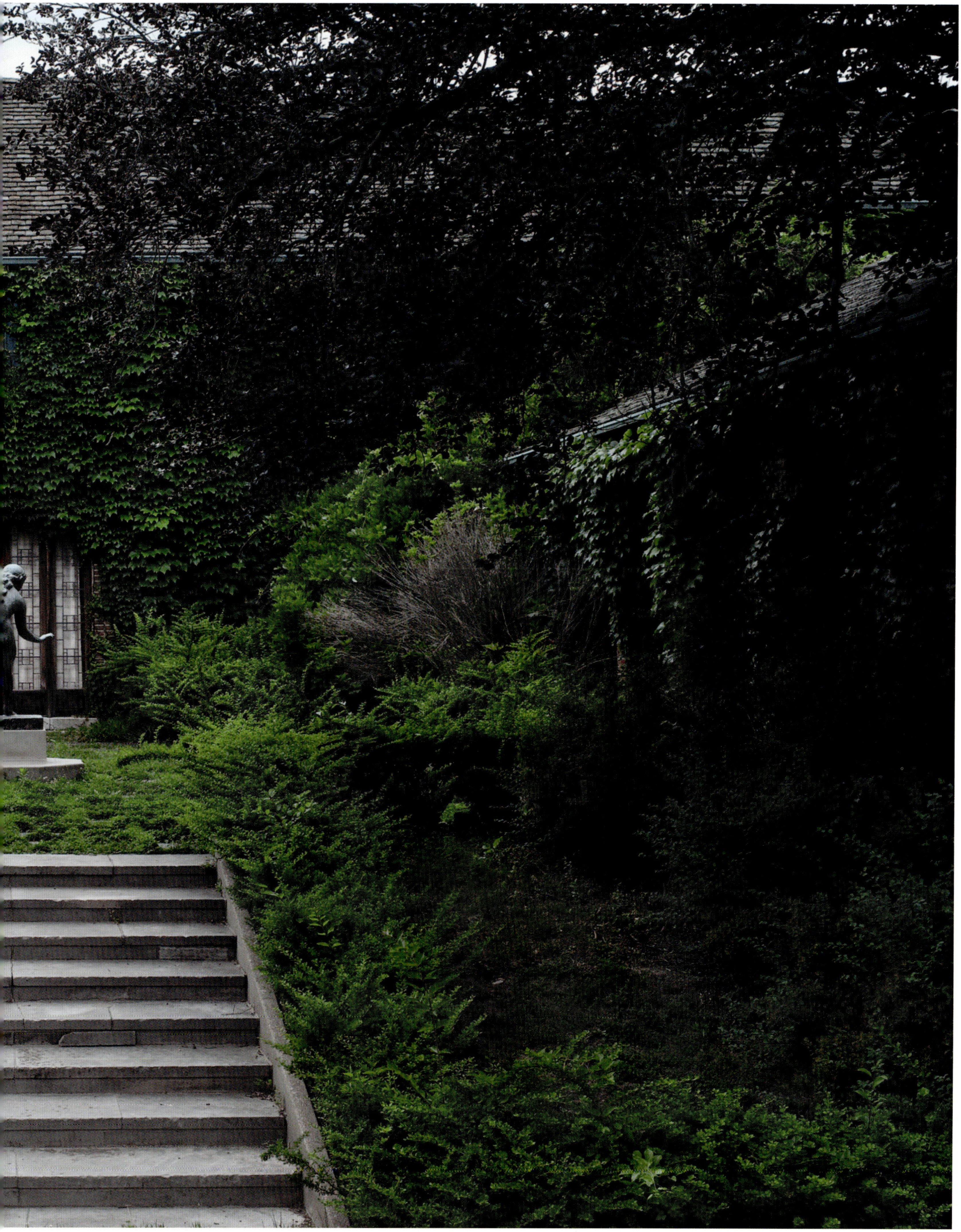

The Finnish architect Eliel Saarinen was described as a 'bridge builder' by his contemporary, ALVAR AALTO. Saarinen began working in Finland in a largely Arts and Crafts manner, initially in partnership with Herman Gesellius and Armas Lindgren, with key buildings including the Finnish National Museum of 1912 and the Helsinki Railway Station of 1919. But Saarinen's open-minded approach fused many different influences, and after he arrived in America in the early Twenties his work became more modern in character, with a combination of Art Deco* and more experimental elements. The house that he designed for himself and his family on the campus of the Cranbrook Academy of Art in 1930 splices these different threads into one cohesive whole, forming a 'bridge' in itself between craft and modernity.

Saarinen studied architecture and fine art in Helsinki before forming his first practice with university classmates Gesellius and Lindgren. One of their first projects was the Finnish Pavilion for the Paris Exposition in 1900, and soon after the partners completed a country villa for themselves known as Hvitträsk (1903), where they combined architecture, interiors and furniture design. It was an all-encompassing approach to design without borders that Saarinen applied to many of his own commissions and that informed his philosophy as an educator.

In 1922, Saarinen responded to an international competition to design a new skyscraper for the *Chicago Tribune* newspaper. Saarinen came second but his proposal was widely praised and earned him a runner-up prize of $20,000. This award helped fund Saarinen's move to America along with his wife Loja, an accomplished textile designer, and their two children, Pipsan and Eero.

One of Saarinen's new admirers was the wealthy newspaper publisher and philanthropist George Booth. Before moving into the newspaper business, Booth had taken an apprenticeship as an architect himself and was an enlightened supporter of architecture and the arts. He invited Saarinen to develop ambitious plans for an educational community at Cranbrook, twenty miles north of Detroit, that would soon include schools for both girls and boys, a science institute, a museum and library, plus the Cranbrook Academy of Art.

As well as serving as Cranbrook's chief architectural officer, Saarinen was asked to be the Academy of Art's founding president, a post that he held from 1932 until 1946, while Loja Saarinen headed the weaving department. The Academy soon became a multi-disciplinary centre of design excellence, with some similarities to the Bauhaus model; alumni include Florence Knoll*, Harry Bertoia, and Charles & Ray Eames*.

Echoing the idea of the masters' houses at the Bauhaus in Dessau, designed by WALTER GROPIUS, the Academy campus was created with a series of faculty houses, including Saarinen's own home, completed in 1930. The Saarinen House manages to combine aspects of both traditionalism and modernity without a jolt. Designed at the same time as its neighbouring faculty house, the Saarinen building has a discreet presence to the street, and most of the windows on that side of the house – within a very public avenue of dormitories and academic buildings – look into hallways and thoroughfares to preserve a sense of privacy. The larger windows and French doors are to the back of the building, where the house opens out to form a U-shaped plan around a large courtyard, partially protected to the sides by a covered porch to one aspect and a studio wing to the other.

Within, the Saarinens created a series of rich, warm spaces with a wealth of colour, texture and natural light. Loja Saarinen designed curtains, rugs and wall hangings while also collaborating closely with her husband. Saarinen involved himself in each and every aspect of the interiors, from the finely detailed panelling and furniture in the dining room to the 'raisin and silver' tiled fireplace in the living room.

Undoubtedly it was the studio that was the best used and most hard-working part of the house, as well as the most modern in design and feel, forming a flexible and open space with a barrel-vaulted roof. The studio was large enough for three drafting tables at its centre, as well as an office for Saarinen at one end and a seating alcove at the other, closest to the doorway through to the living room. Saarinen designed a number of pieces of furniture for the studio, including a series of tubular steel chairs in a modernist style. With its sense of scale and large window looking onto the courtyard, this was a welcoming space as well as a working environment.

PAGES 400–1 A covered porch to one side and the Saarinens' studio to the other form projecting wings that help shelter a verdant courtyard garden at the rear of the house.

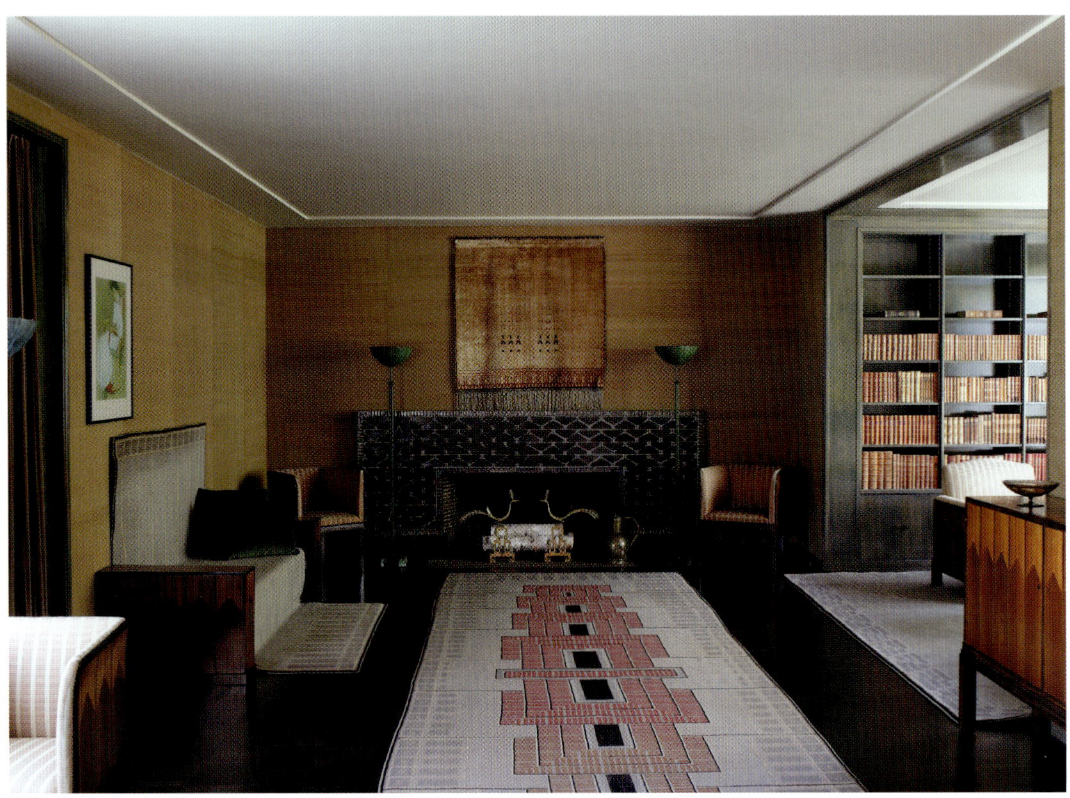

RIGHT & OPPOSITE The rugs in the living room and dining room are by Loja Saarinen, while the majority of the furniture was designed by Eliel Saarinen and made by Tor Berglund.

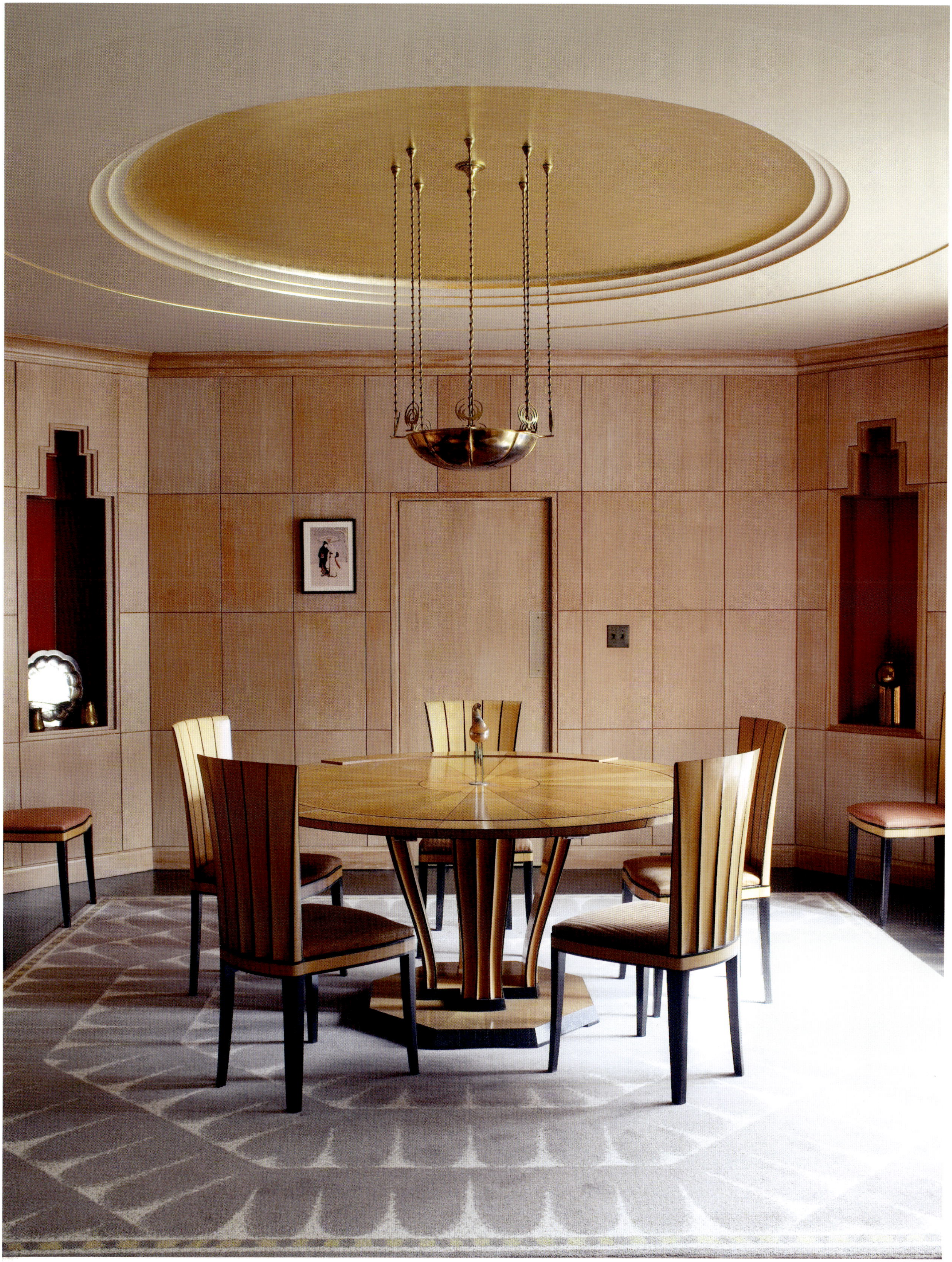

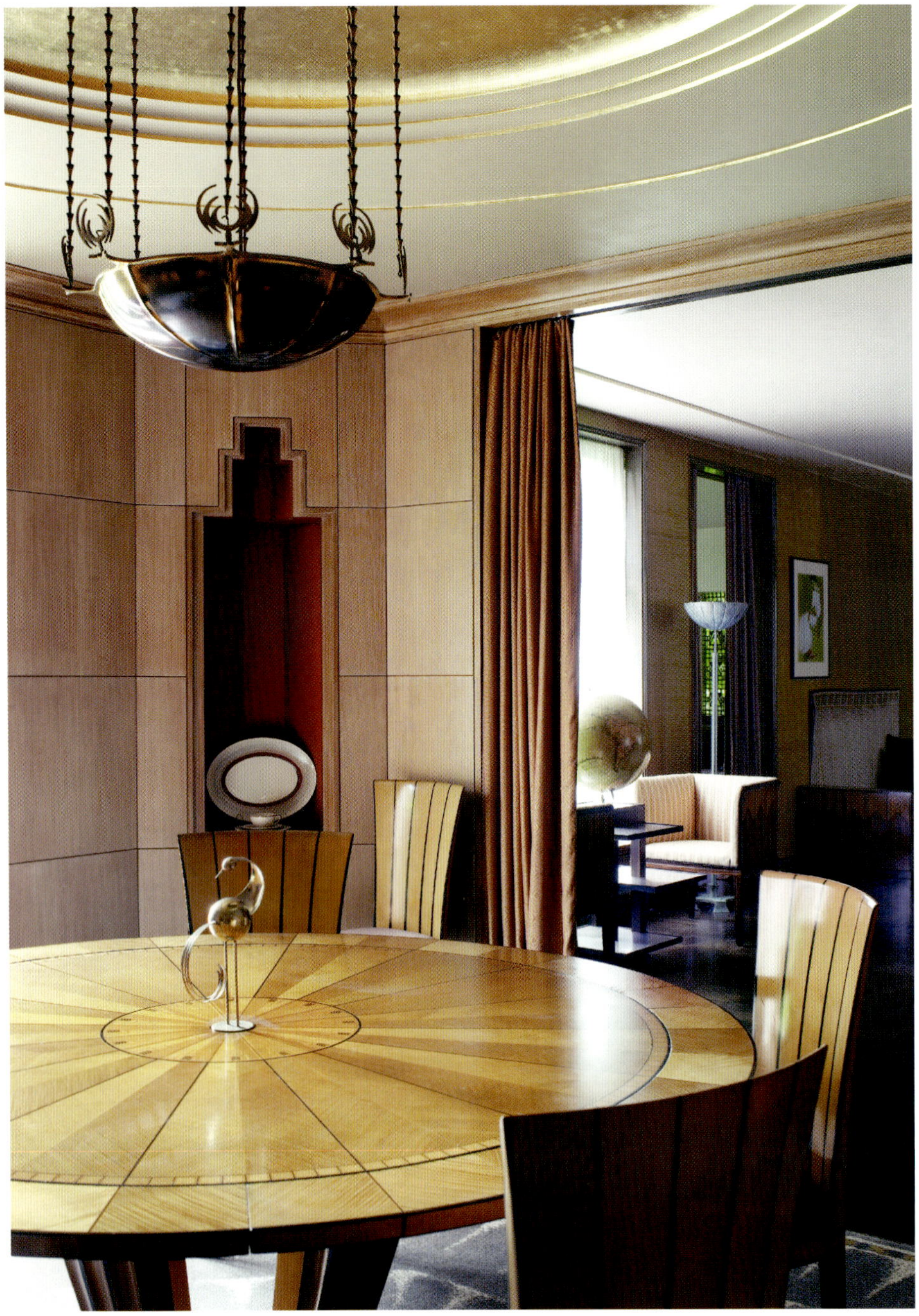

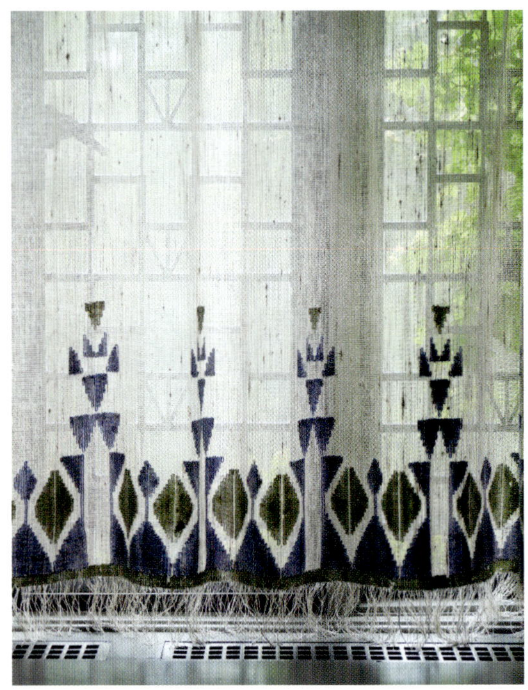

THIS PAGE & OPPOSITE
The house is a fusion of Arts & Crafts, Art Deco and early modernist influences, with the sophistication of the interiors evidenced by the glasswork, tiles, textiles and the timber panelling in the dining room.

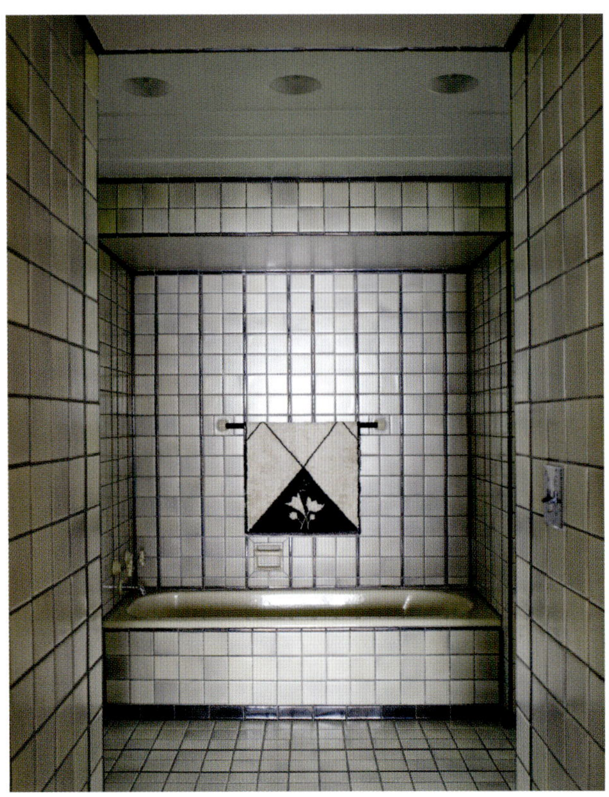

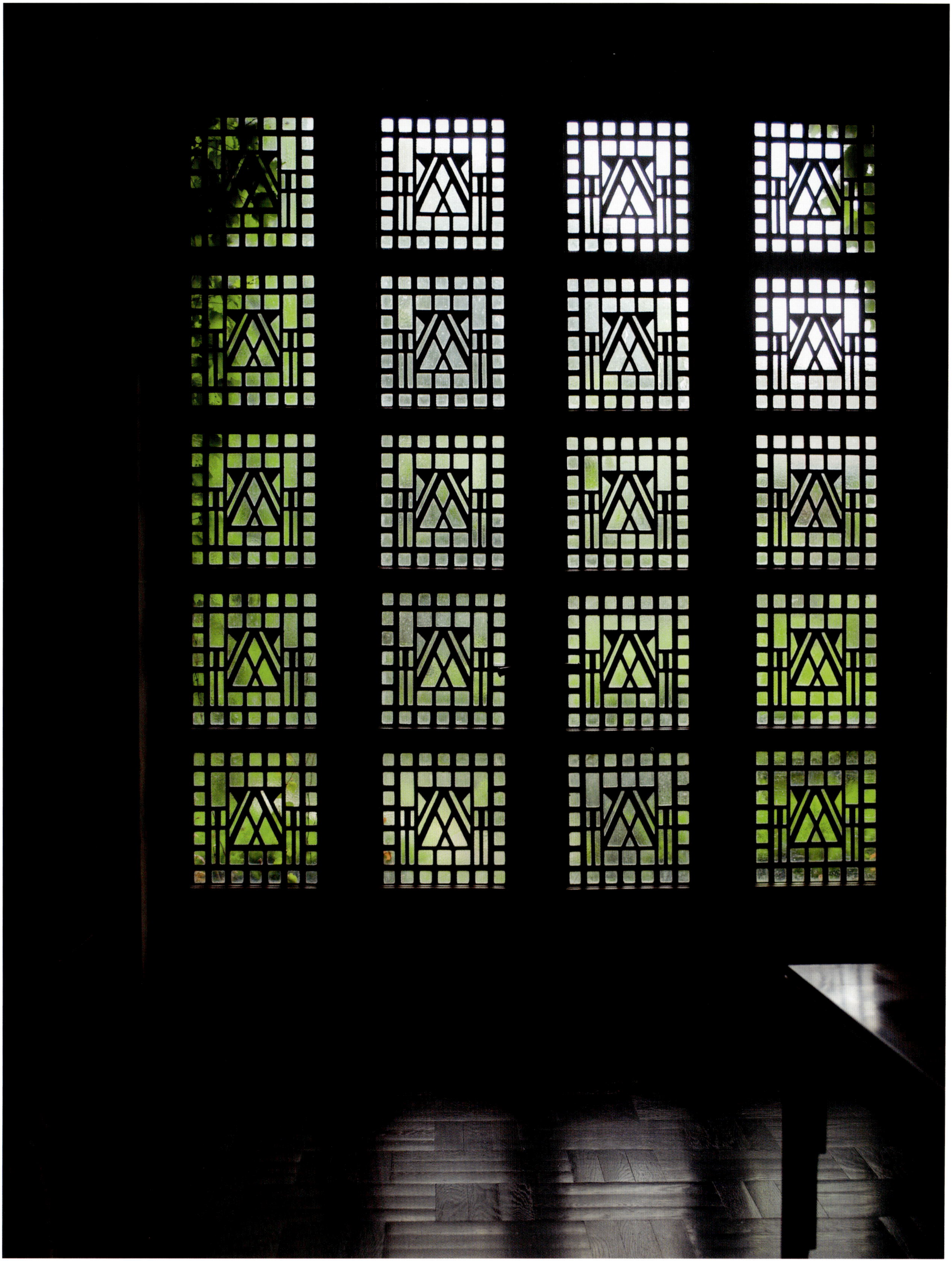

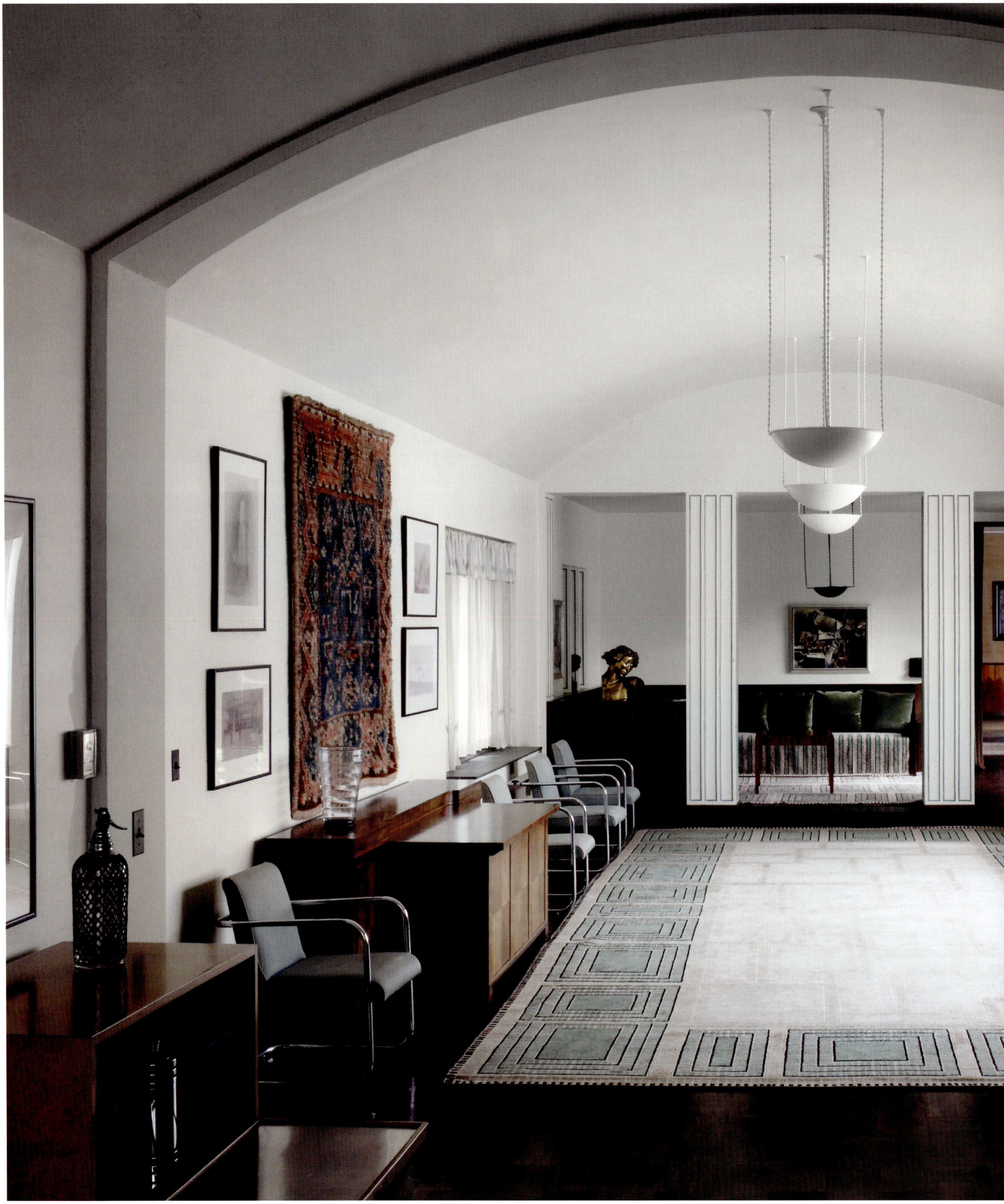

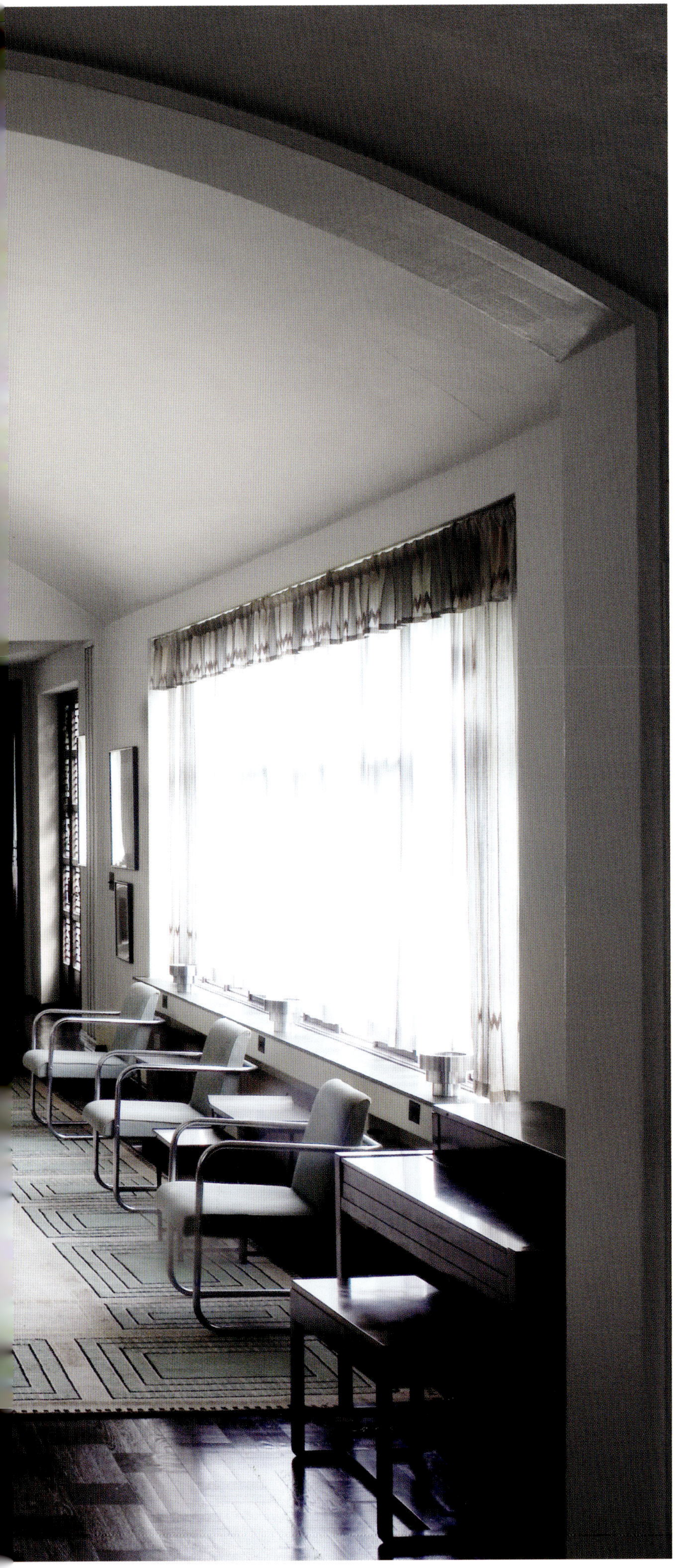

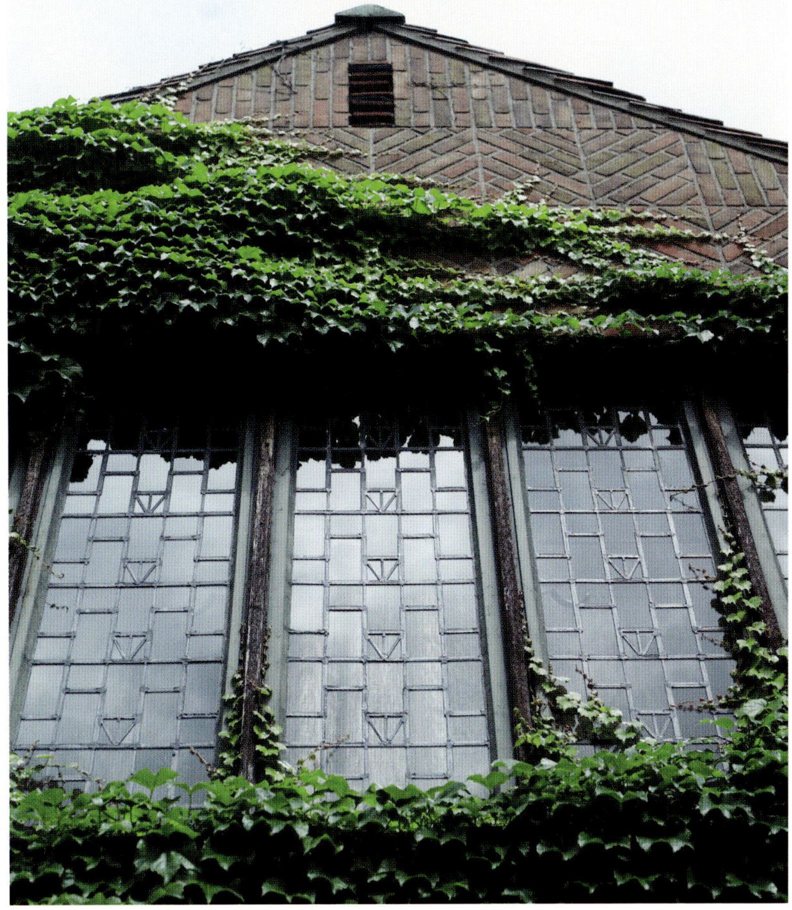

ABOVE The glasswork and brickwork provide texture and pattern, softened by the advancing greenery.

LEFT The vaulted studio was a key part of the house, used by both Loja and Eliel Saarinen; the space was also used for functions and entertaining, when the furniture could be pushed back to the sides of the space.

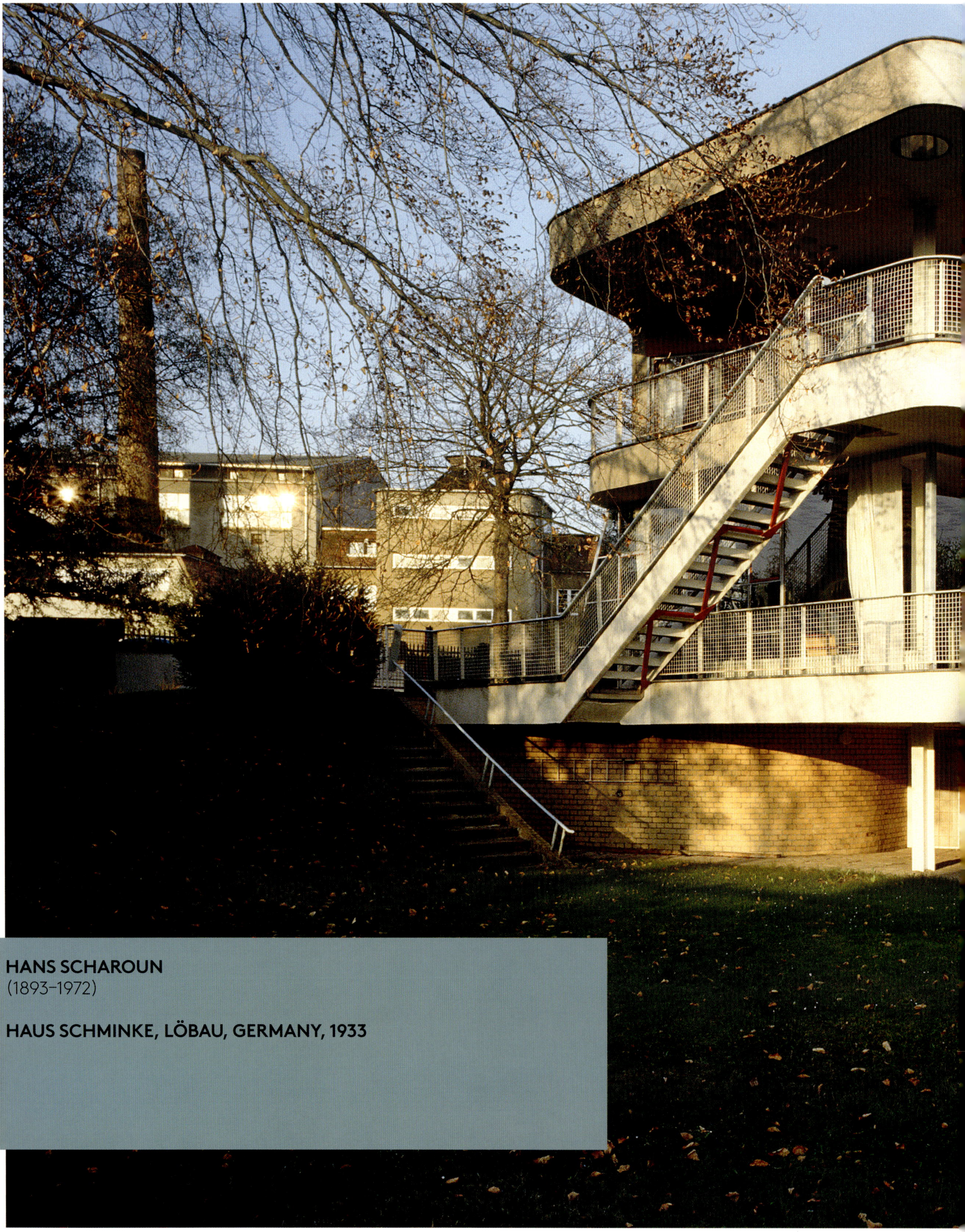

HANS SCHAROUN
(1893–1972)

HAUS SCHMINKE, LÖBAU, GERMANY, 1933

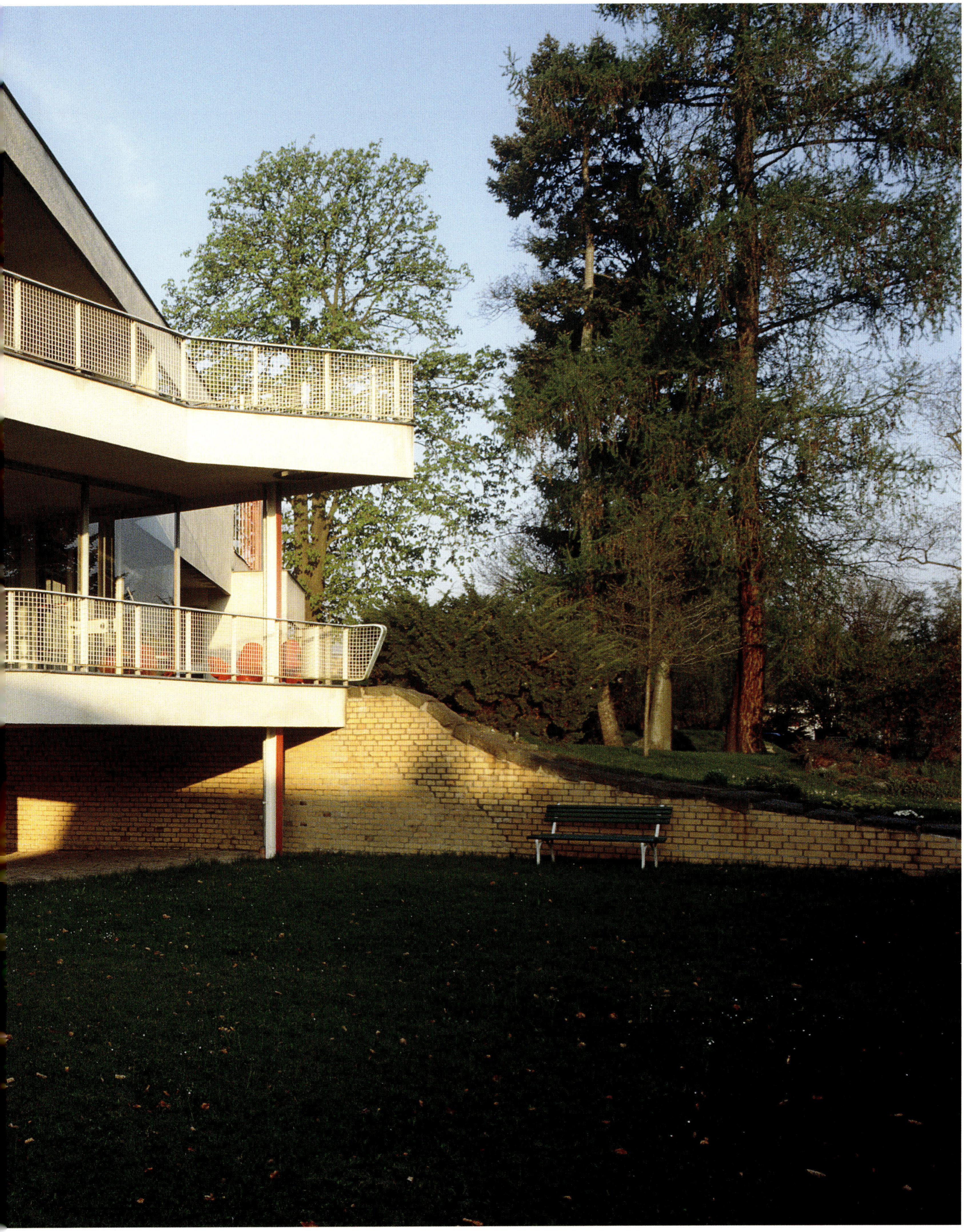

With their decks, terraces and smooth, white-painted exteriors, many early modernist houses and apartment buildings have been compared to ships. There was an obvious maritime influence playing upon the work of a significant number of modernist architects and designers, and a more general fascination with a new generation of glamorous ocean-going traffic. The ships themselves were symbols of optimistic modernity, encompassing a streamlined aesthetic approach.

German architect Hans Scharoun was more explicit than most about the influence of naval design upon his work. He grew up in the ship-building town of Bremerhaven, which was also a busy sea-port for trade and passengers; as a boy he would watch the liners leaving for New York. 'One wishes to see something of the boldness of modern ship structures transferred to the design of houses and thus hope to overcome the pettiness and restrictedness of today's housing,' he wrote in 1928.[28]

This influence made itself known in a number of ways. There was a dynamism of form to many of Scharoun's buildings, with sinuous lines and soaring decks subverting the rectangular blocks of modernism. At the same time, there was a strong emphasis on functionality and ergonomics, with the strict rationality and space-saving mentality of ship design informing some of his residential projects. Prime examples are Scharoun's House 33 (1927), designed for the Deutscher Werkbund*'s Weissenhof* exhibition in Stuttgart, directed by LUDWIG MIES VAN DER ROHE.

Fritz and Charlotte Schminke first became aware of Scharoun's work through the Werkbund exhibitions, which they attended together. The couple were devoted to the idea of creating a modern, bespoke house for themselves and their four children on the green edge of Löbau, a town near Dresden. The Schminke family owned a pasta factory in the town, called Loeser & Richter, which produced the well-known Anker brand of noodles. After his father's early death, Fritz Schminke took over the running of the business and contacted Scharoun to commission a new family home.

The greatest challenge lay with the site itself. The gardens and the best of the views were to the north, but the architect and clients also wanted to make the most of the southern light, while avoiding a vista of the factory and the street. The complex positioning and formation of the house were a contextual response to these concerns, with the villa angled on the sloping site accordingly. Substantial balconied decks on both floors thrust outwards into the landscape like the bow of a ship, while large expanses of glazing maximize the natural light and help to reinforce the vivid sense of connection with the gardens.

Within, the house was also highly bespoke, tailored to the specific needs of the family, who wanted a functional, ordered, low-maintenance home that could be easily managed with the help of one maid. On the ground floor, the maid's quarters, service spaces and a Frankfurt Kitchen (the compact, fitted design created by Margarete Schütte-Lihotzky*) were positioned at the rear of the house, alongside the entrance hall and a playroom. The stairs in the hallway were placed on a diagonal, helping to steer visitors – via sliding doors – to the main living area beyond. This is a fluid and generous space leading down to an integrated winter garden (complete with miniature pond) and a glass-walled conservatory, which feeds out in turn to the elevated, cantilevered terrace projecting out into the garden. The upper level of the house hosts a guest suite at one end and the master suite at the other. Between them, there are two simple bedrooms for the children; storage was provided by a line of fitted cupboards in the hallway.

Rather like Mies's Villa Tugendhat (see p. 380), with which Scharoun's residential masterpiece is often compared, Haus Schminke soon became a victim of war and politics. The house was requisitioned by the Soviet Red Army and Fritz Schminke was held for a time as a war criminal given that the family pasta business had supplied the German army. The factory was also confiscated, and Charlotte and Fritz Schminke left in 1951 and moved to Lower Saxony. The house later served as a youth club, but was finally restored in the Nineties and opened to the public.

Unlike many of his contemporaries, Scharoun stayed on in Germany throughout the Thirties. He was banned from bidding for public contracts under Nazi rule, but did design a small number of private houses in the mid-Thirties. During the war years he surveyed bomb damage in Berlin, and he became involved in reconstruction efforts in the post-war years. There were school projects and apartment buildings, and then a late flowering of his reputation following the success of his Berlin Philharmonic building in 1963.

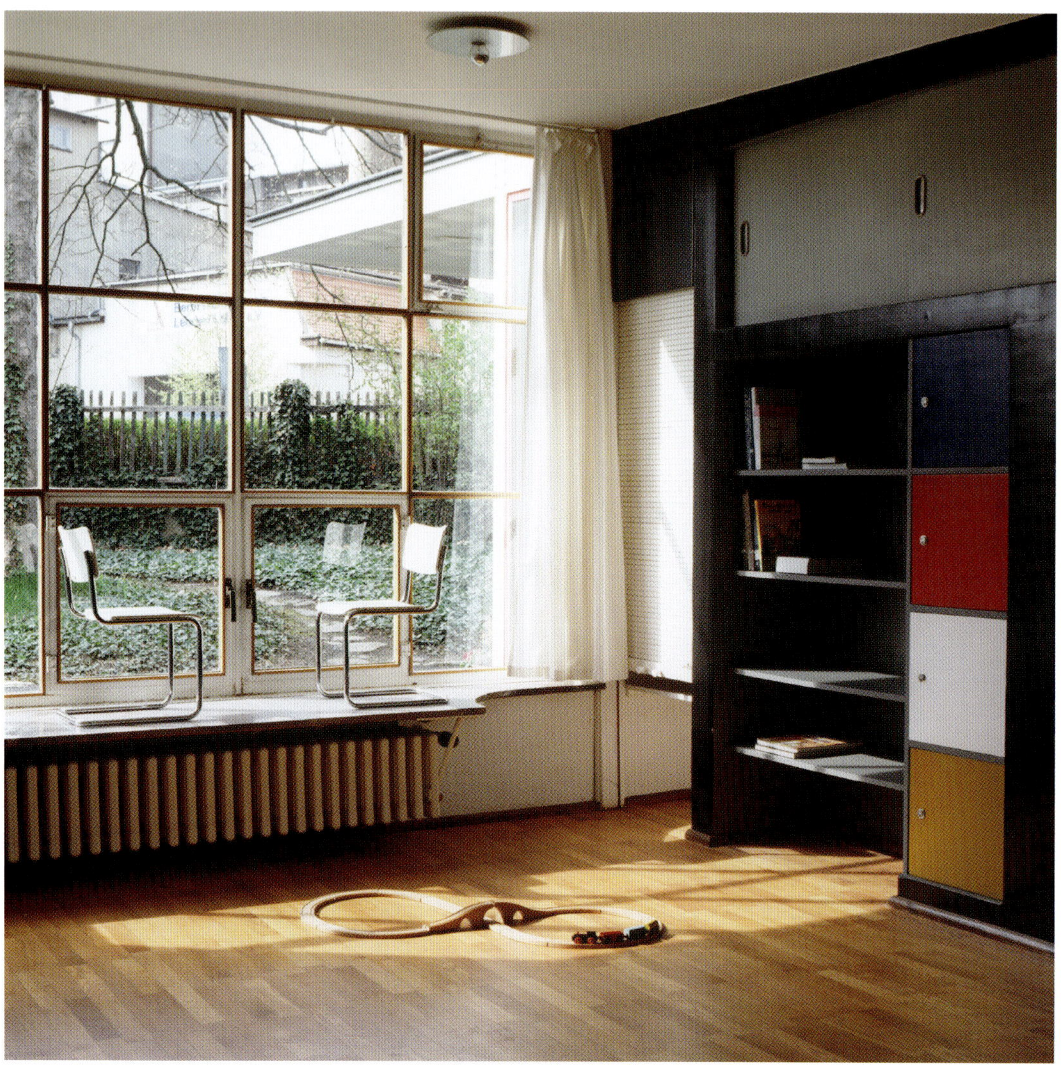

PAGES 408–9 & RIGHT The exteriors of the house have a nautical quality, with the bows of the 'ship' formed by elevated terraces pushing out into the garden.

OPPOSITE The playroom sits to one side of the entrance hall and can be partitioned with a sliding curtain; the lockers for the children are colour-coded.

BELOW The ground-floor seating area is part of an open, fluid space that includes space for a grand piano and a dining area beyond.

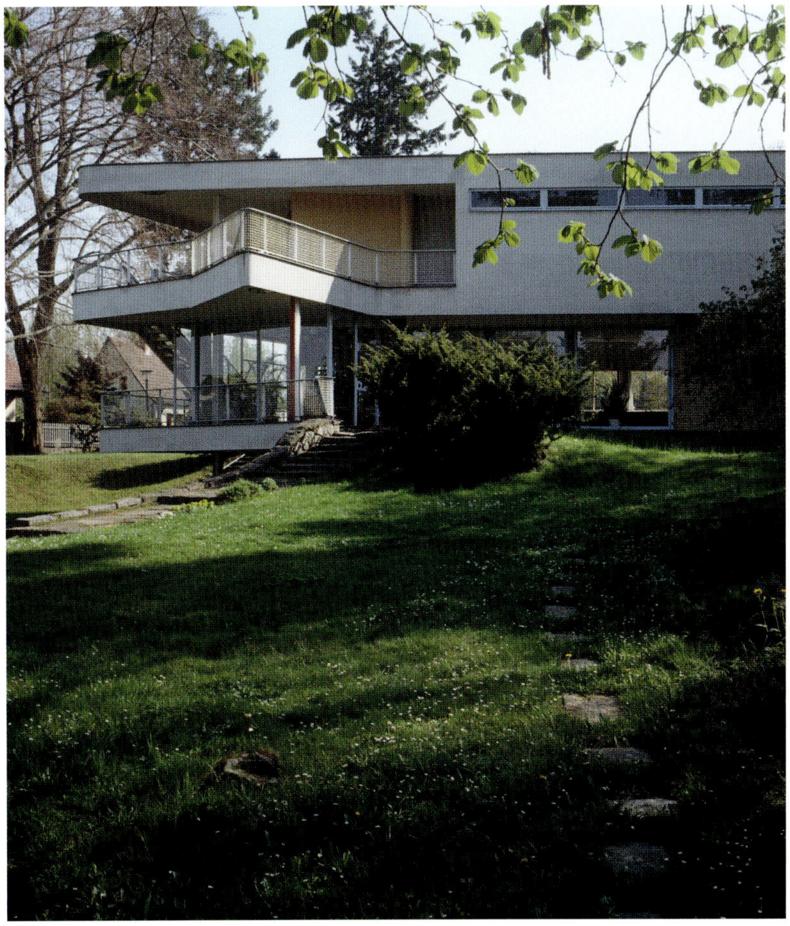

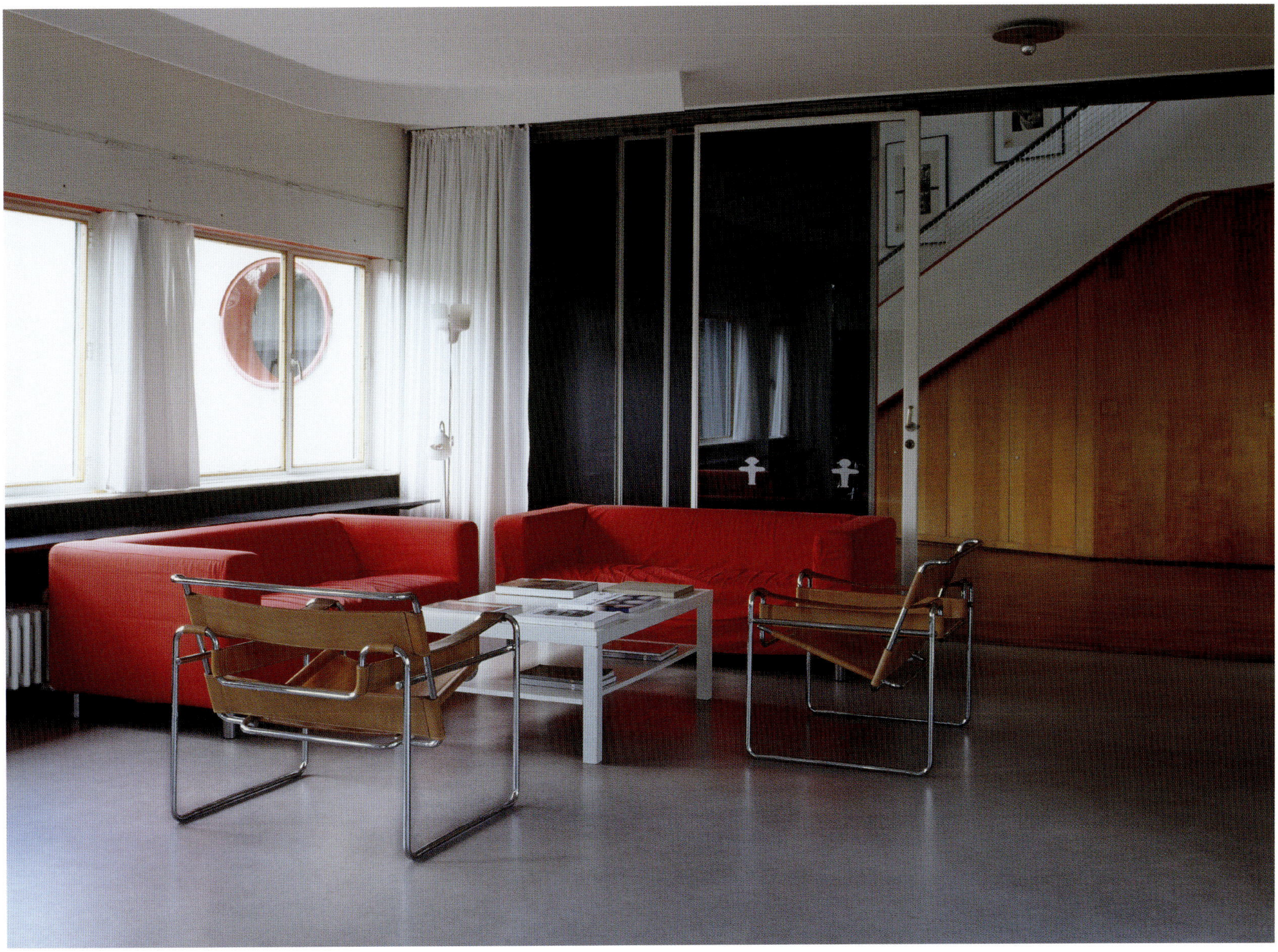

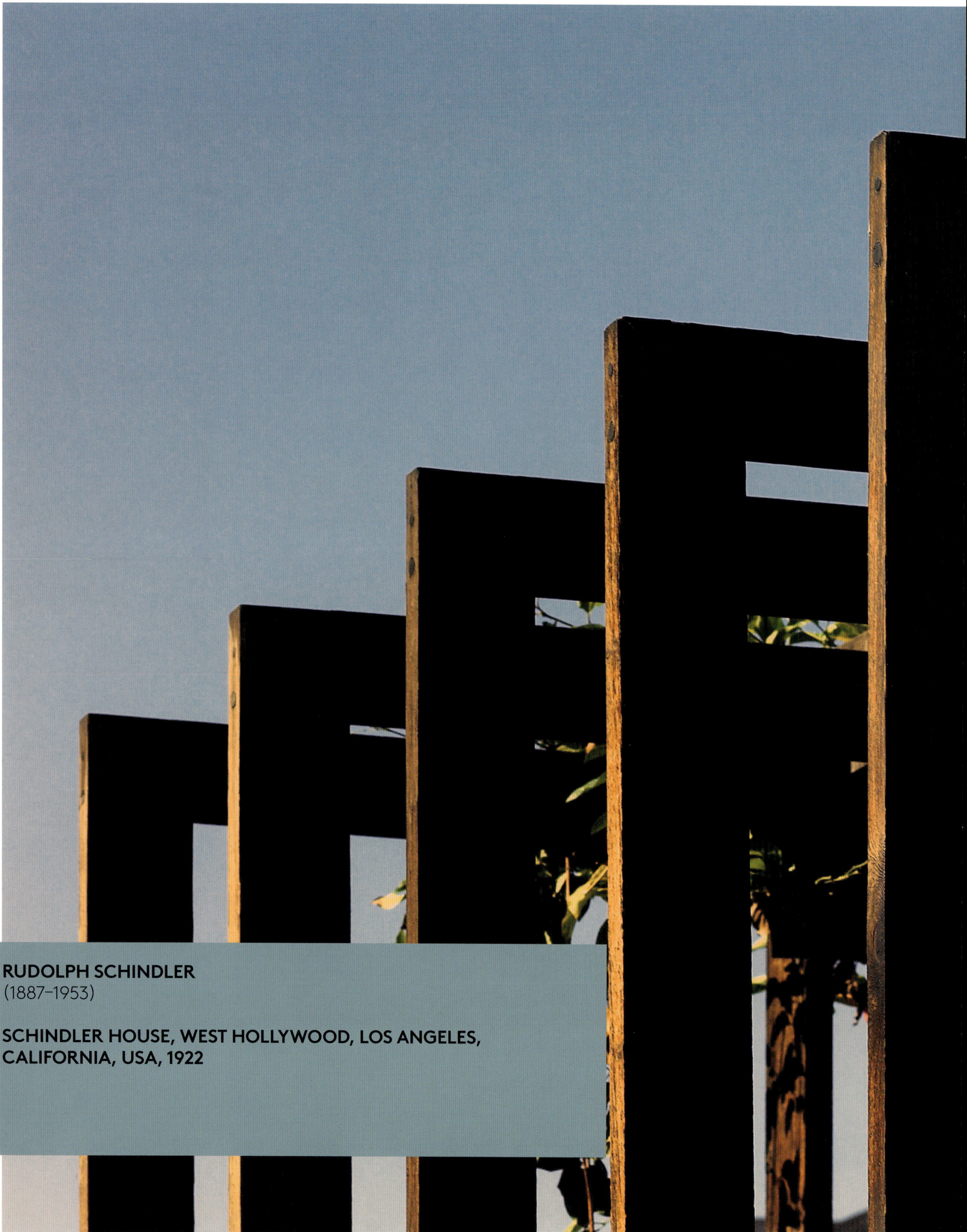

RUDOLPH SCHINDLER
(1887–1953)

SCHINDLER HOUSE, WEST HOLLYWOOD, LOS ANGELES, CALIFORNIA, USA, 1922

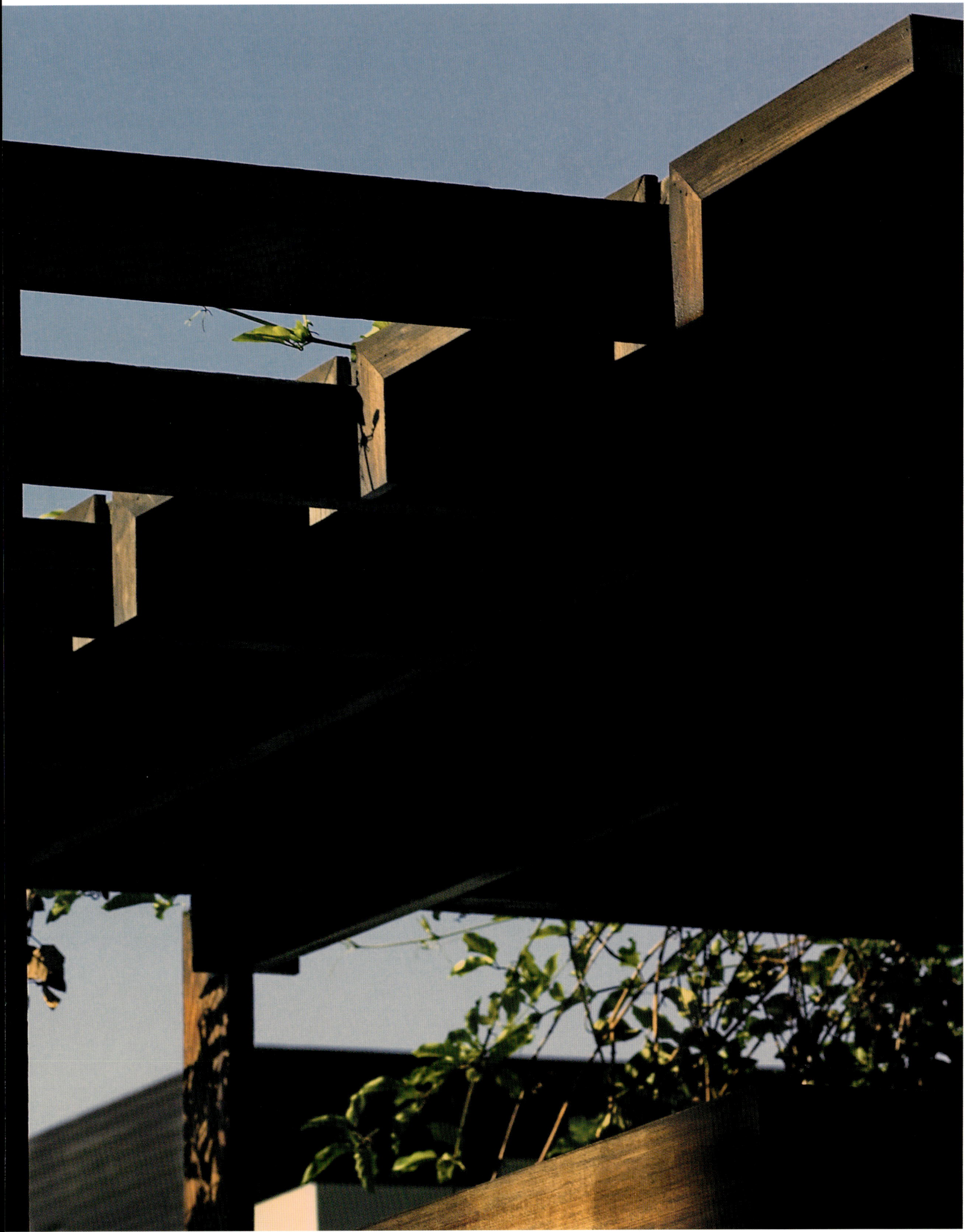

The house that Rudolph Schindler built for himself on North Kings Road in Los Angeles is one of the earliest exemplars of a style that has become known as 'Californian Modern'. It has also been described as one of the world's truly modernist homes, revolutionary in its structure, its floor plan and the way that the house forges an intimate sense of connection with the surrounding terraces and gardens. The thoughtfulness of this inside-outside relationship and the spatial dynamic of the house became defining traits of Schindler's work in general, and of the Californian Modern way of living itself.

Schindler was born in Vienna and studied at the Technical University followed by the Academy of Fine Arts, where he was taught by Otto Wagner* and attended lectures by Adolf Loos*; both were formative figures in the early evolution of what became known as modernist architecture and design. Schindler went on to work in the Viennese office of architects Mayr & Mayer before taking Loos's advice and emigrating to America in 1914.

One of the chief reasons that Schindler decided on America was the influence of Frank Lloyd Wright. Schindler hoped to gain a position with Wright and wrote to him accordingly, although he had to wait until 1917 before Wright finally took him on. Wright had just accepted the commission to design the Imperial Hotel in Tokyo – a project that would take many years to complete and kept Wright in Japan for long periods of time. Schindler played a key part in maintaining his office in Chicago during these years, contributing to the hotel project itself and helping to develop other new commissions. Later, he was asked to work on Wright's Hollyhock House in Los Angeles, and in 1920 Schindler and his new wife, Sophie Pauline Gibling (known as Pauline), who was an art and music teacher, moved to California to supervise the construction of Hollyhock for client Aline Barnsdall.

Schindler and his wife decided to build a new home for themselves in Los Angeles, to serve as both a base and as a calling card for Schindler's own architectural studio. They joined forces with one of Pauline's college friends, Marian Chace, and her husband Clyde, who was a skilled builder and engineer with experience of tilt slab construction, whereby concrete is poured into flat timber moulds on site and the 'cured' walls are then levered into place. The Schindlers and Chaces agreed to share the house and make best use of the tilt slab system.

The floor plan that Schindler developed was highly informal and, for the times, revolutionary. Each of the quartet was given their own open-plan studio, with each couple's pair of rooms linked by a hallway and combining to create two 'legs' of a single-storey pinwheel; the third and final leg held a communal kitchen, a guest bedroom and a garage. There was no communal living room or dining room.

'The basic idea was to give each person his own room – instead of the usual distribution,' said Schindler, 'and to do most of the cooking right on the table, making it more a social "campfire" affair than the disagreeable burden to one member of the family.'[29]

The notion of a roofed camp site was continued with the extensive use of glazed walls, framed in redwood, that would slide away to connect with the adjoining terraces and courtyard gardens. There was an outdoor fireplace and up on the roof Schindler designed timber-framed 'sleeping baskets' with canvas awnings that could serve as summer bedrooms. Within, the finishes were unornamented and raw: exposed concrete floors, redwood ceilings and woodwork, with timber furniture designed by Schindler.

Conceived in a bohemian spirit, with the idea of recreating a perpetual vacation spirit, the house was tested by reality. The Chaces moved to Florida in 1924 after the birth of two children and were replaced for a time by architect Richard Neutra and his wife and son (Schindler had first met Neutra in Vienna and they formed a brief working partnership before heading their separate ways). The Schindlers' marriage became strained and Pauline moved out before finally returning to the house in the late Thirties and living, as part of an estranged couple, within her own section of the pinwheel.

Schindler continued to live and work in the house right up until his death, formulating plans for a whole series of Californian projects, including apartment buildings and individual houses, among them the Lovell Beach House (1926).

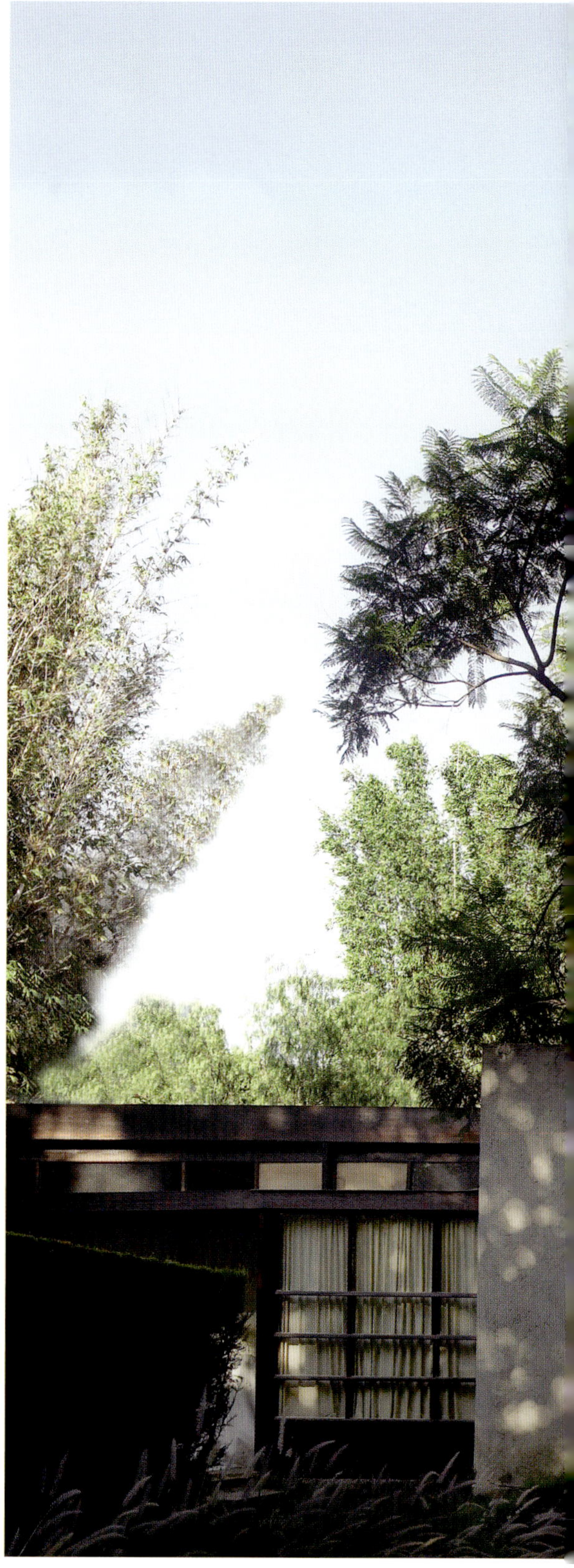

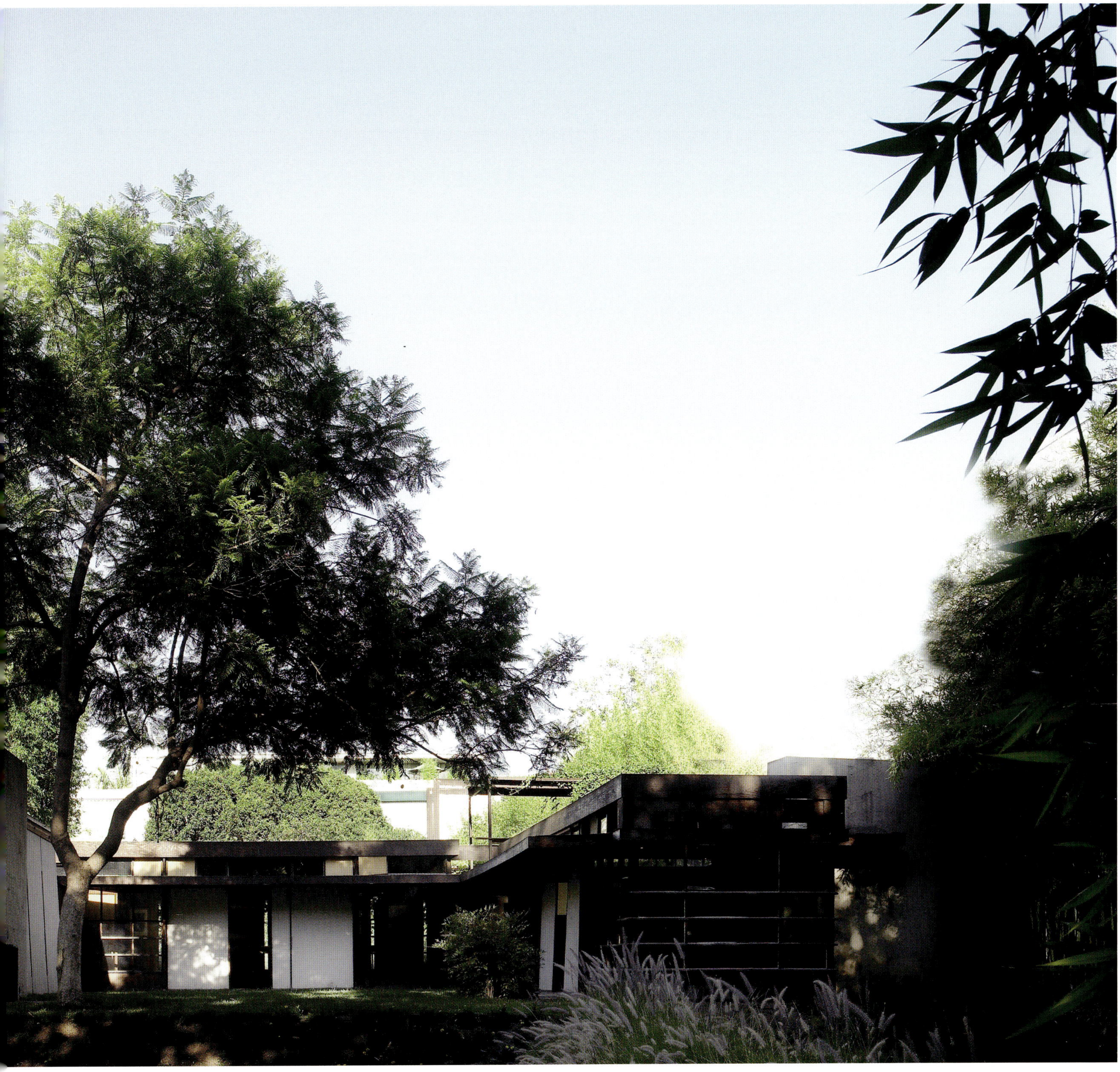

PAGES 412–13 & ABOVE
Fluid and flexible, the Schindler House forges a strong relationship between inside and outside living spaces, featuring patios front and back plus fresh air sleeping 'baskets' up on the roof deck.

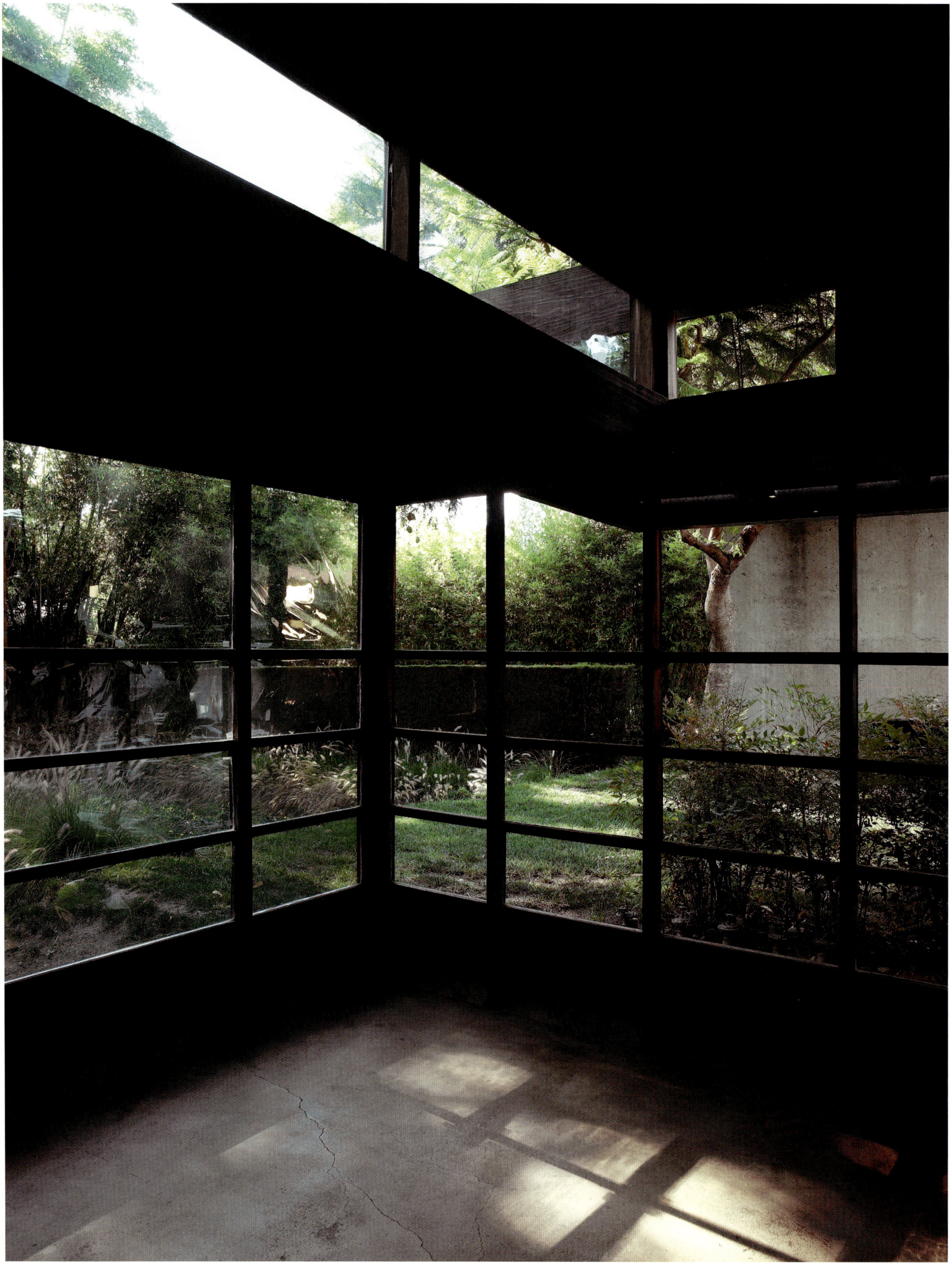

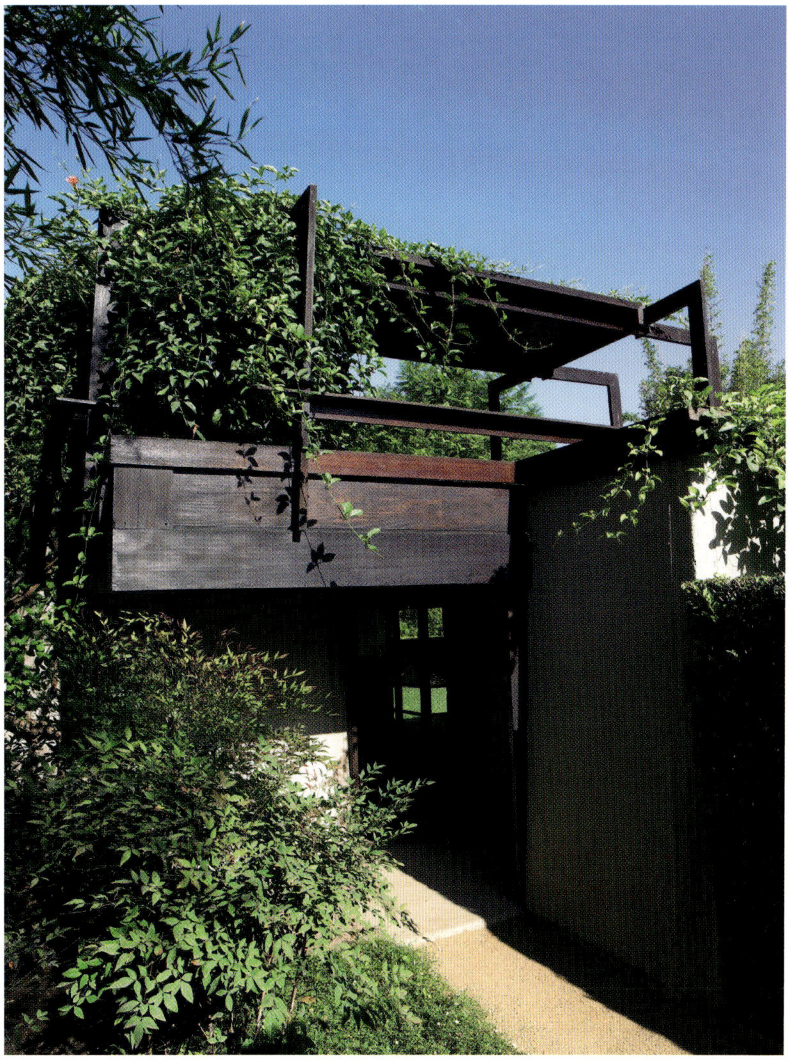

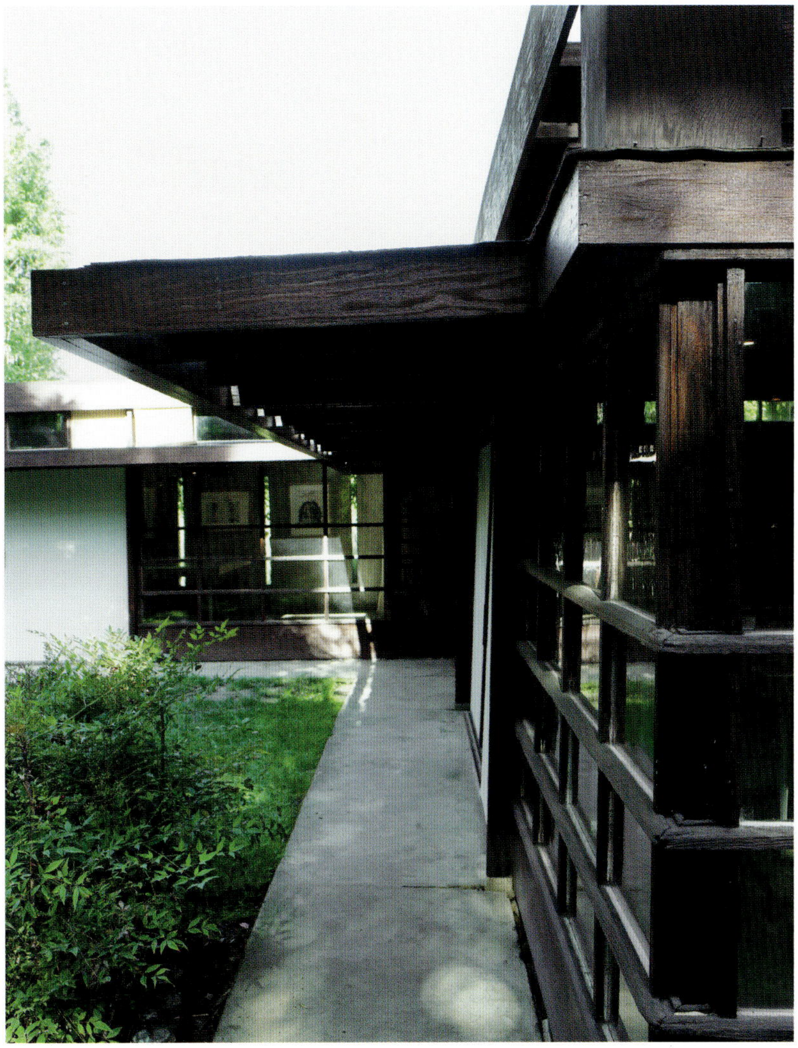

OPPOSITE & THIS PAGE
The 'cooperative dwelling' features a series of studio spaces radiating around a central fireplace; these studios are adaptable, offering open living spaces to be filled as desired, with extensive glazing in each to the gardens and patios.

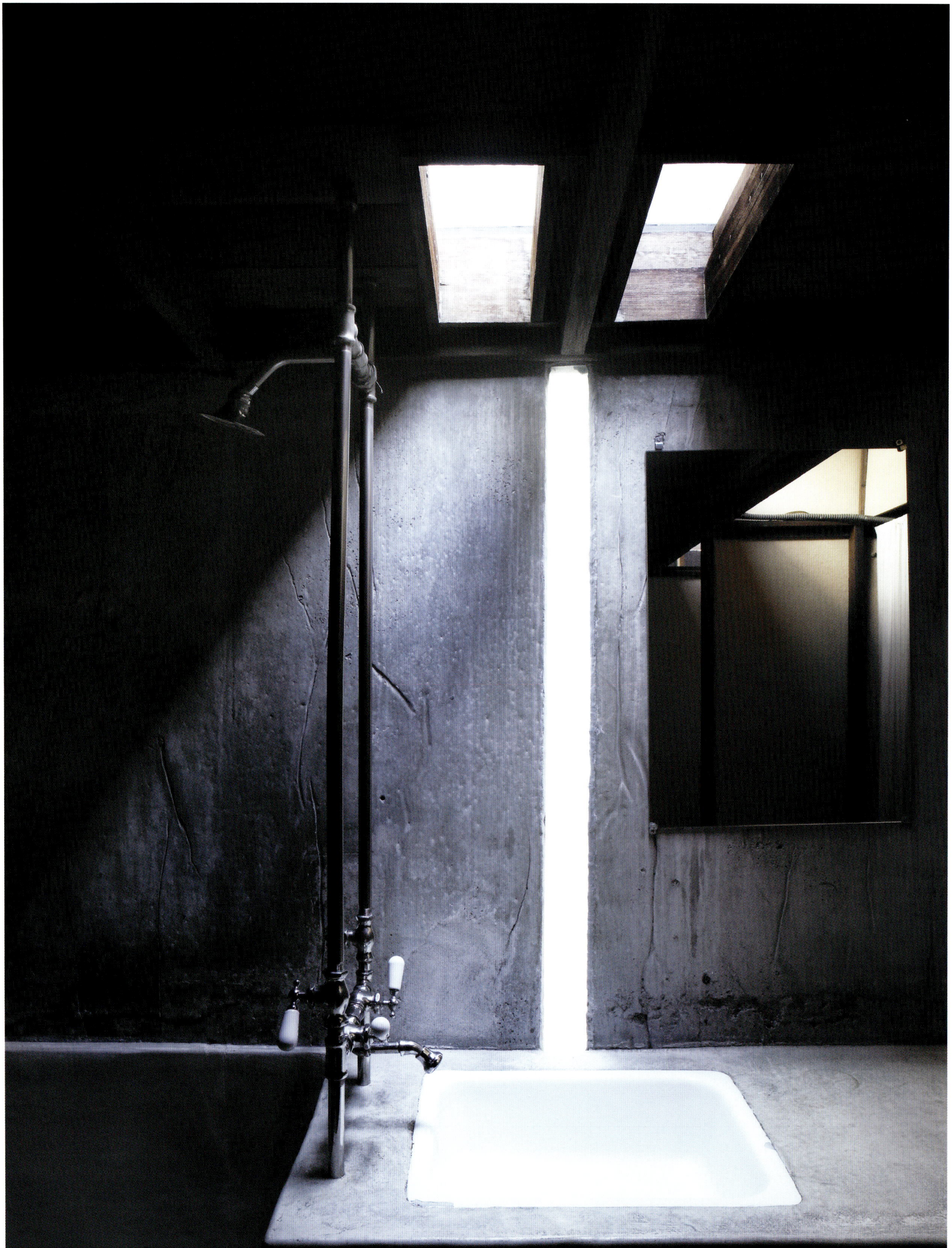

OPPOSITE & THIS PAGE
The Schindler House plays with light in many different ways, using skylights over private spaces such as the bathrooms, slot windows within the slab-tilt walls, and clerestory windows to complement the banks of glazing overlooking the garden; much of the original furniture was also designed by Schindler in redwood.

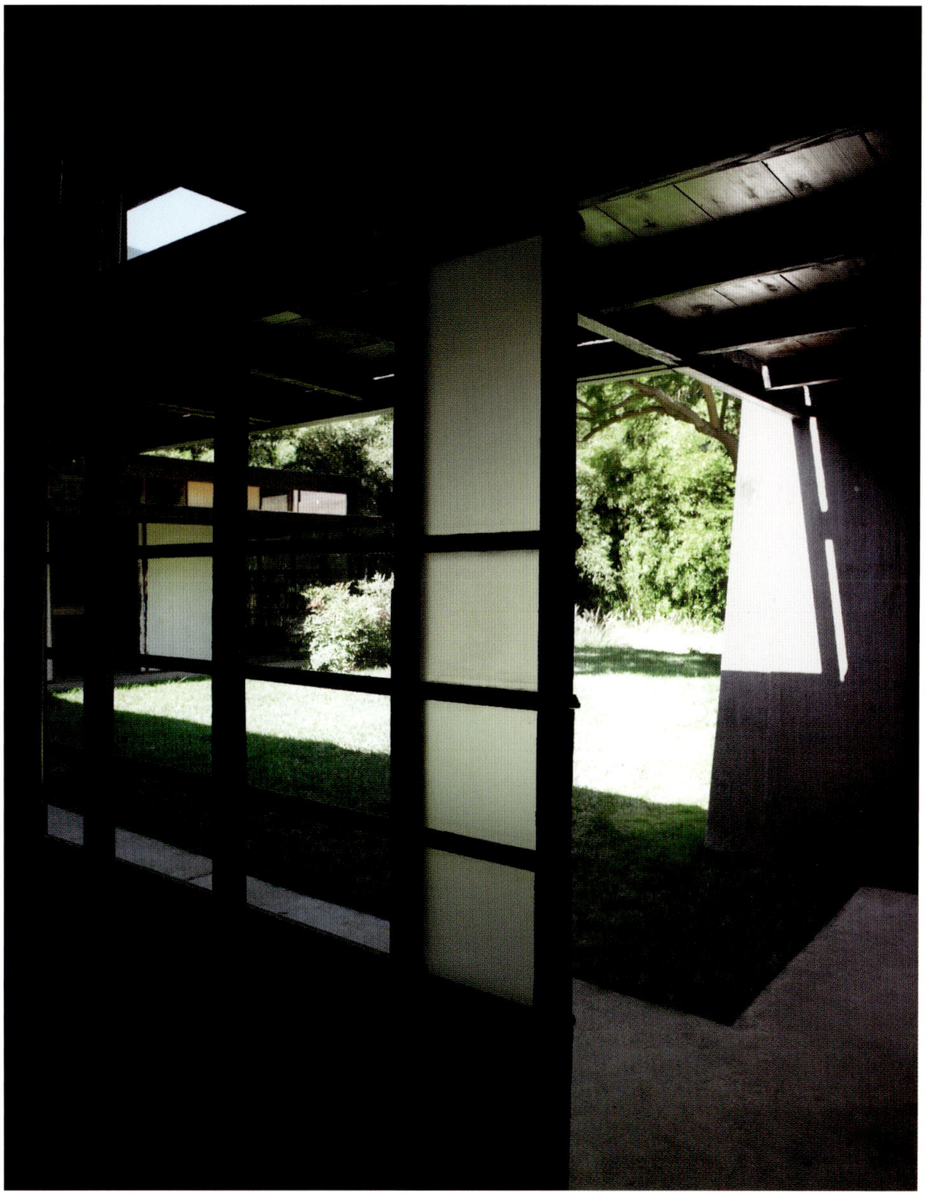

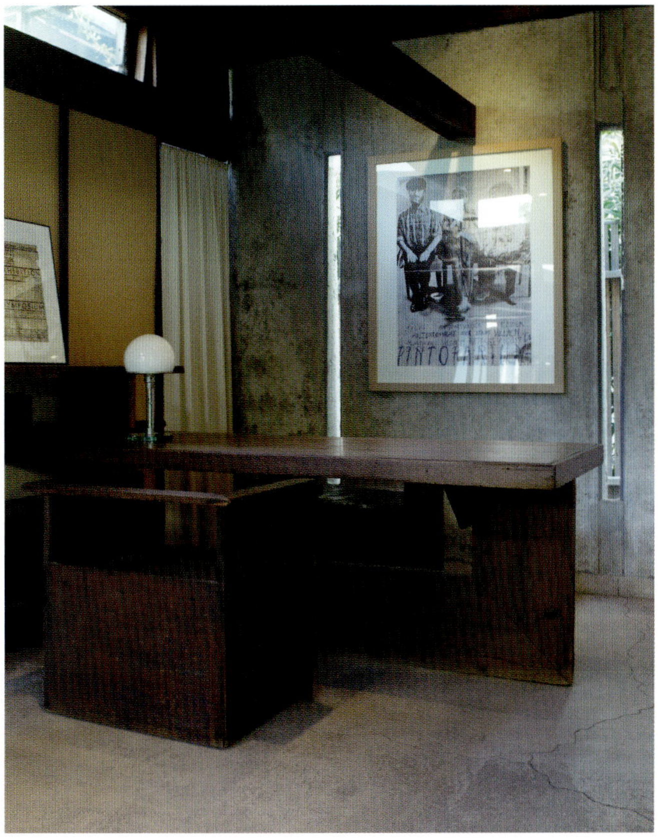

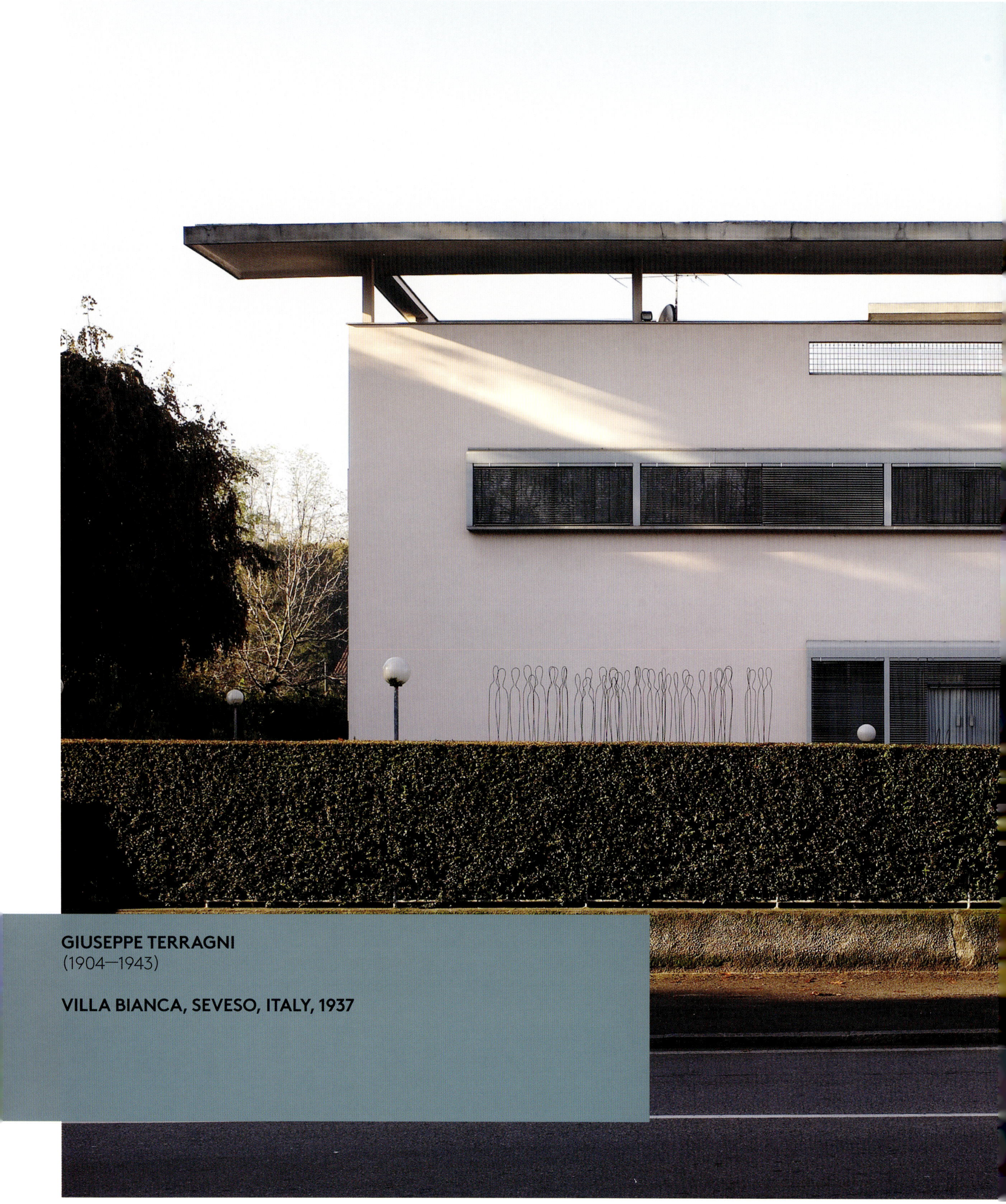

GIUSEPPE TERRAGNI
(1904–1943)

VILLA BIANCA, SEVESO, ITALY, 1937

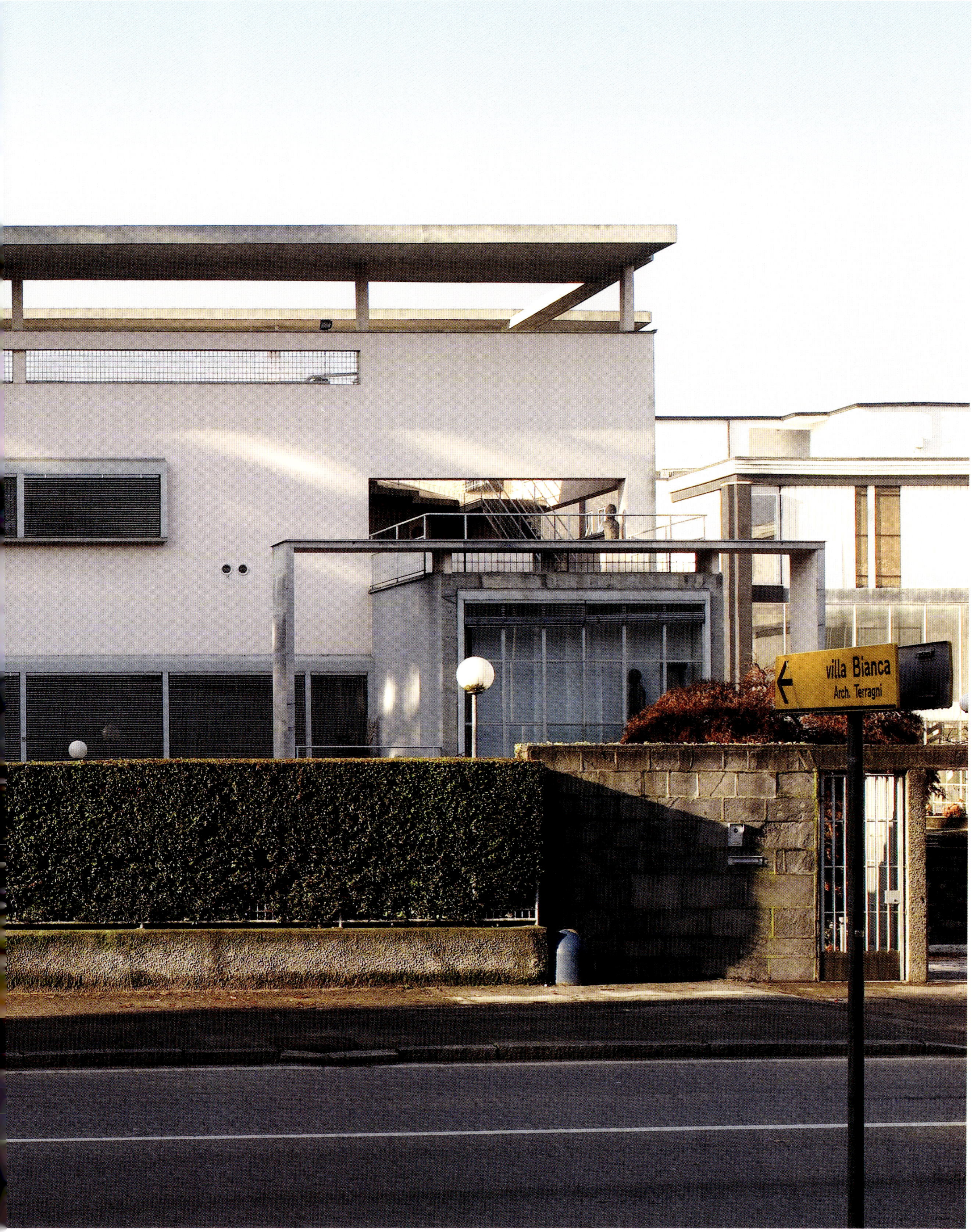
villa Bianca
Arch. Terragni

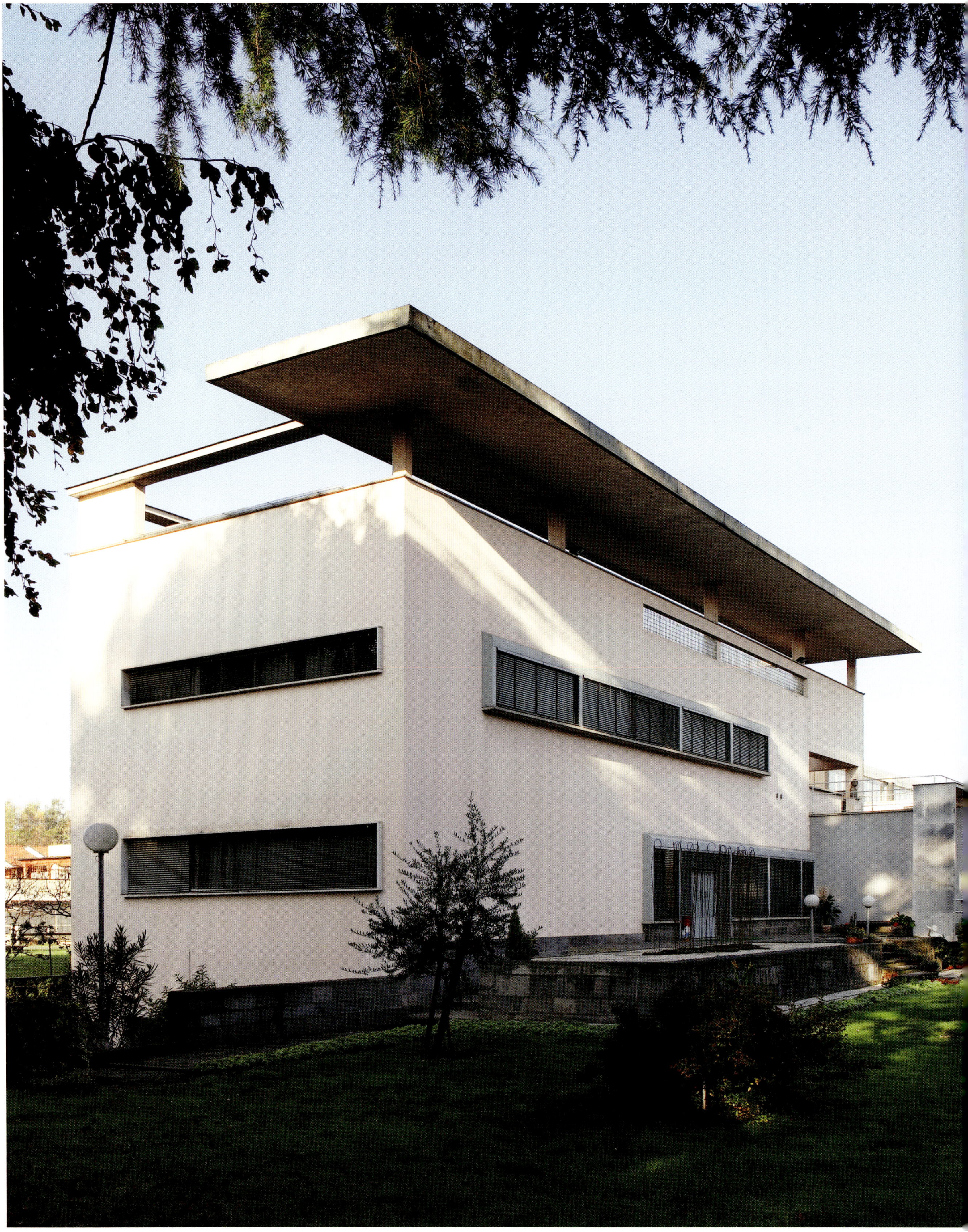

After the collapse of the Weimar Republic and the rise of the Nazi party in Germany in the Thirties, the state turned its back on modernist architecture. Hitler and his fascist dictatorship turned to a grandiose and intimidating version of neoclassicism, epitomized by the work of Albert Speer*, while modernism was largely dismissed as subversive and unpatriotic.

In Italy, the relationship between Mussolini's fascist state and modernist architecture was far more complicated. The Italian fascists saw architecture as a way of expressing the revival and rebirth of Italy itself, and largely embraced modernity in all its forms, seeing itself as progressive and technocratic, and commissioned buildings from a number of Italian Rationalist* architects. They included Giuseppe Pagano, Marcello Nizzoli*, Adalberto Libera* and Giuseppe Terragni, who was a founding member of Gruppo 7, a group of rationalist architects that sought a third way between neoclassicism and Futurism*, creating modern buildings of logical precision and order with a pared-down aesthetic.

Terragni was born in the town of Meda near Milan, where his father owned a construction company. The latter encouraged his son's interest in architecture and he studied at Como's Technical College and then at the Politecnico di Milano. He graduated in 1926 and co-founded Gruppo 7 the same year. Influenced by the work of LE CORBUSIER in France and WALTER GROPIUS in Germany, Terragni and Gruppo 7 participated in the 1930 Milan Triennale, exhibiting the Casa Elettrica – an eye-catching house in a distinctly modernist style, with extensive glazing and a substantial roof terrace.

In 1927, soon after graduating in architecture, Terragni opened his own architectural office with his brother Attilio, an engineer. He participated in a number of exhibitions, sometimes in association with Gruppo 7, and also collaborated

PAGES 420–21 & LEFT Villa Bianca makes use of integrated decks, terraces and balconies at every level, some of which offer private outdoor rooms within a relatively urban setting.

RIGHT A ramp connects the house to the rear garden and offers an alternative circulation route to the formal entrance of the front façade.

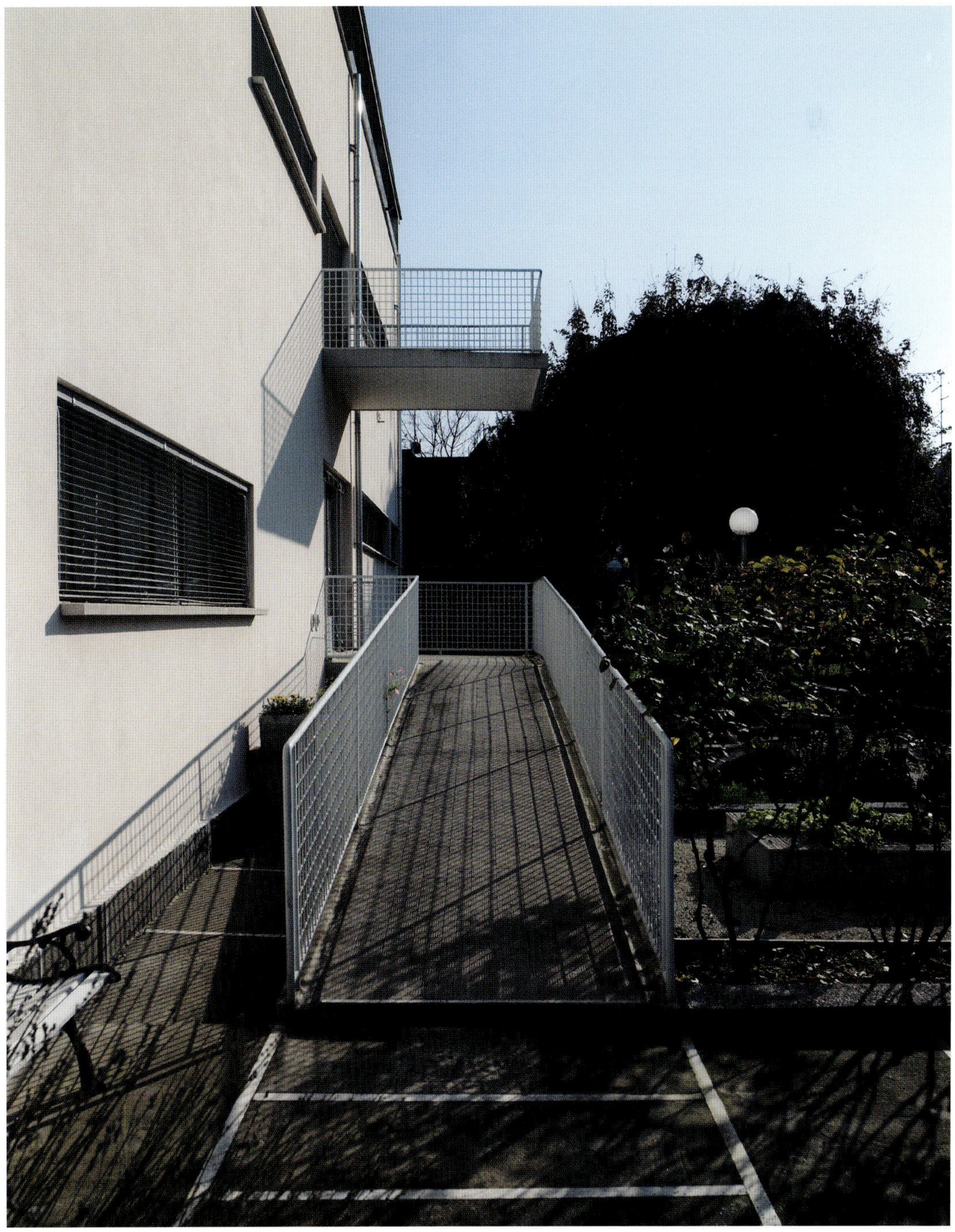

with Pietro Lingeri on a number of projects, including a series of apartment buildings. One of his key commissions of the Thirties was the Casa del Fascio in Como of 1936, forming the branch headquarters of the National Fascist Party; it was later converted into a police station. A year later, Terragni completed the Sant'Elia Nursery School, also in Como, which is seen as one of his masterpieces, followed by a second Casa del Fascio, in Lissone, completed in 1939.

Two of Terragni's key houses of the period are Casa del Floricoltore (1937) and Villa Bianca (1937). The Casa del Floricoltore in Rebbio for the flower grower Amedeo Bianchi conforms to all of Le Corbusier's Five Points of New Architecture, with its free façade, flat roof, ramped entrance and integrated roof terraces; it was even raised up above the landscape on a series of piloti.

The project for the Villa Bianca overlapped with the flower grower's house almost exactly, although slightly larger in scale and more ambitious in its programme. The house was built in the town of Seveso, between Como and Milan, for Terragni's cousin, Angelo Terragni. The building is, again, essentially rectangular, but the form is broken down and subverted in a number of ways. A lower-ground level holds garaging and service spaces, helping to raise up the main body of the house, with the principal living areas floating slightly above the site and flowing outwards to a terrace to one side; a ramp walkway on the opposite side of the house provides an alternative to the main entrance.

The floor above holds the bedrooms and a study, as well as a double-height courtyard, complete with a stairway that ascends to the upper roof terrace, all contained within the high walls that form the cubist outline of the house and topped by floating, cantilevered canopies. The dynamic progression through the house and the shift of spatial volumes as one journeys from internal to external space and up to the layered roof terraces invites comparison with Le Corbusier's Villa Savoye (see p. 366).

In 1939, Terragni was drafted into the Italian army and in 1940, after Italy entered World War II, he was sent for service in Russia as a captain in an artillery regiment. By 1943, he was suffering from a nervous breakdown and was evacuated home on a hospital train. He died in Como in July of that year, just before the coup that generated the fall of Mussolini and his fascist government.

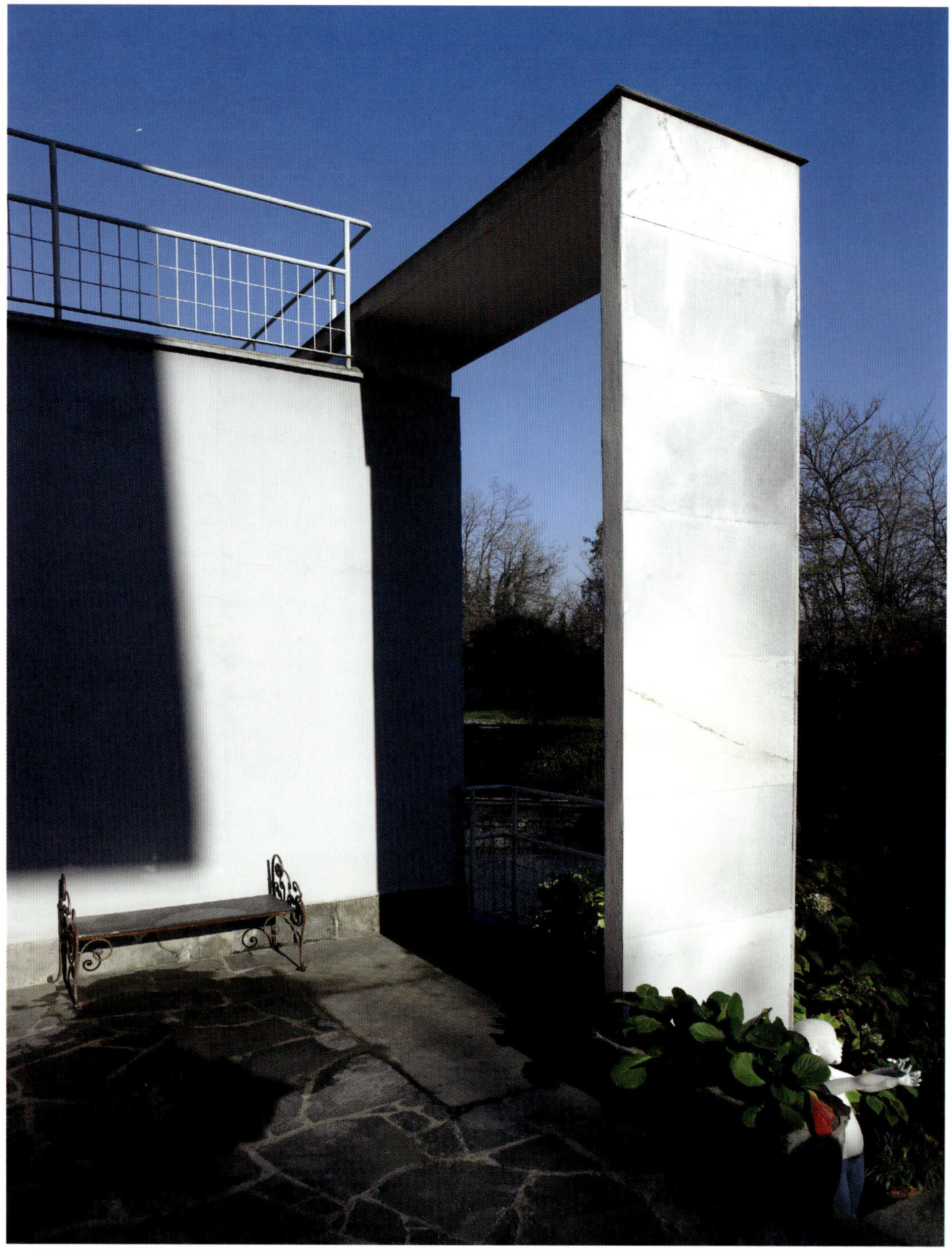

ABOVE This seating area sits upon the terrace around the main entrance, with the projecting structure behind it enclosing the living room with a roof terrace above it.

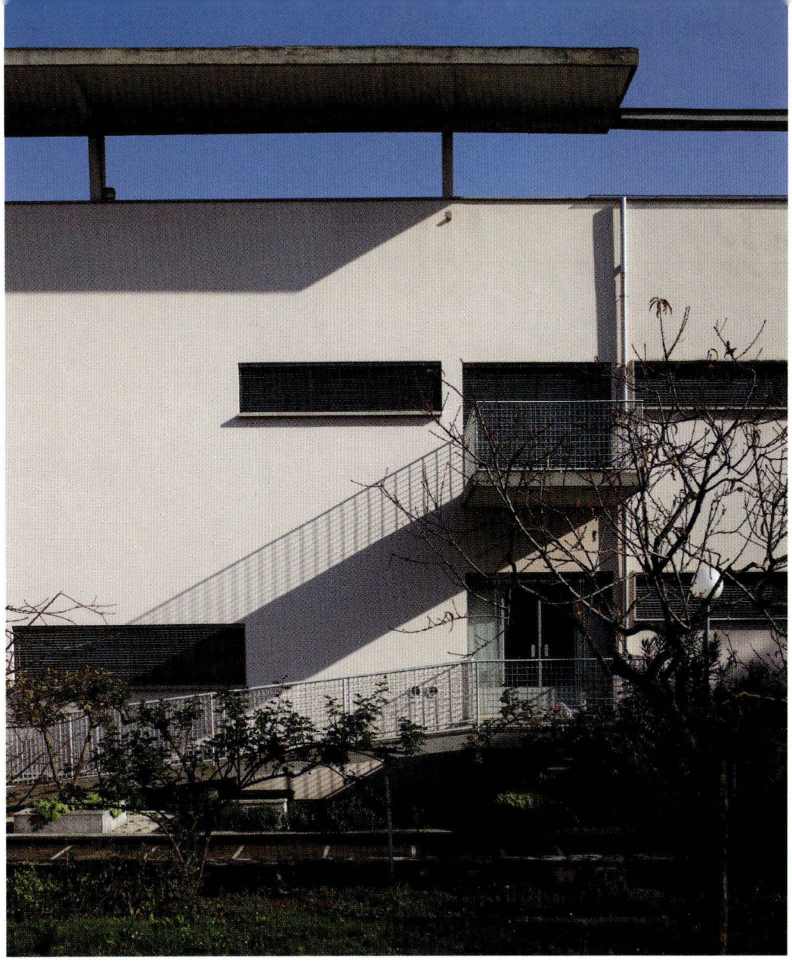

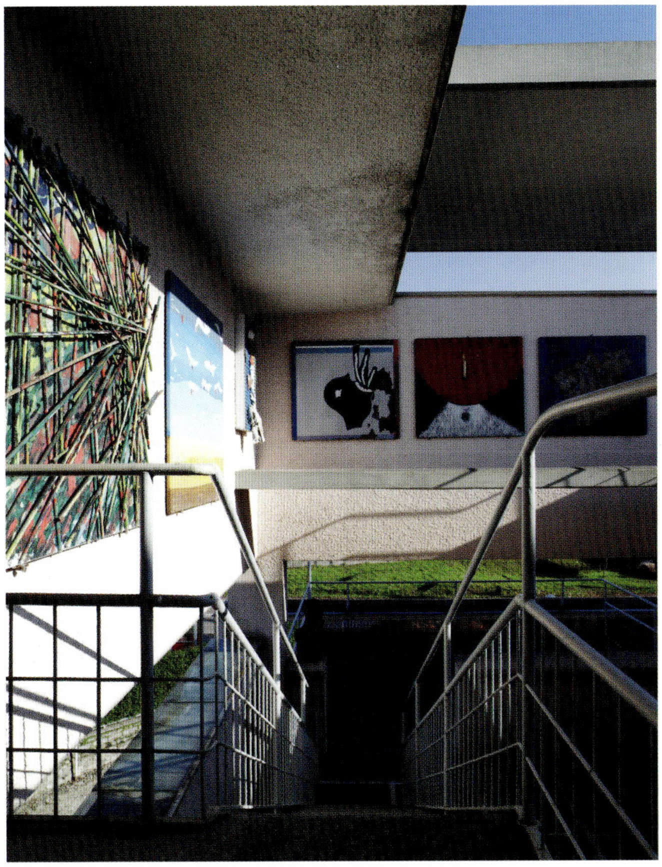

THIS PAGE The house offers a choice of circulation routes, with this staircase (above & right) from the upper storey to the roof deck contained within a sheltered terrace; there is also an internal stairway at the heart of the floor plan.

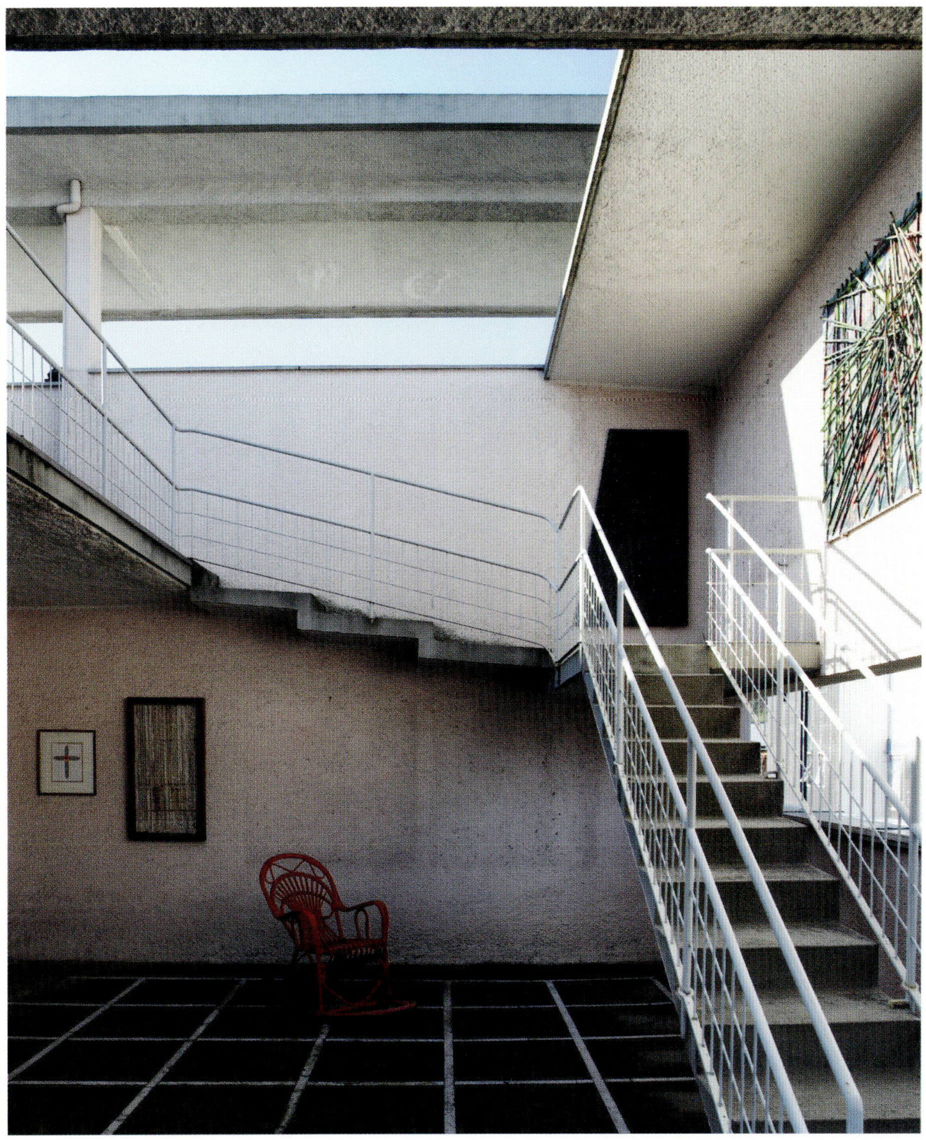

RIGHT This extraordinary floating canopy helps provide a degree of shading on the roof deck, but also offers some level of protection for the terrace on the floor below.

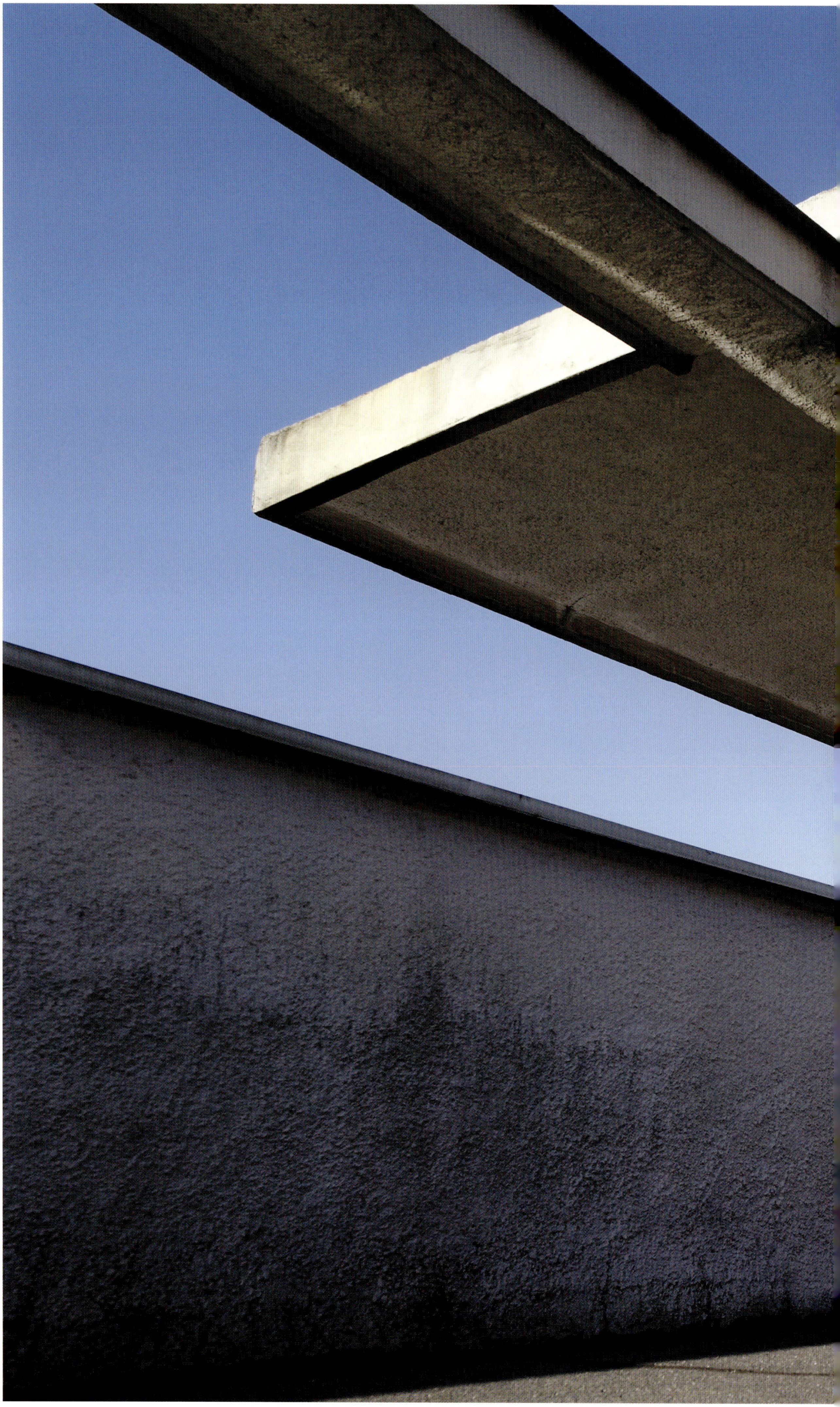

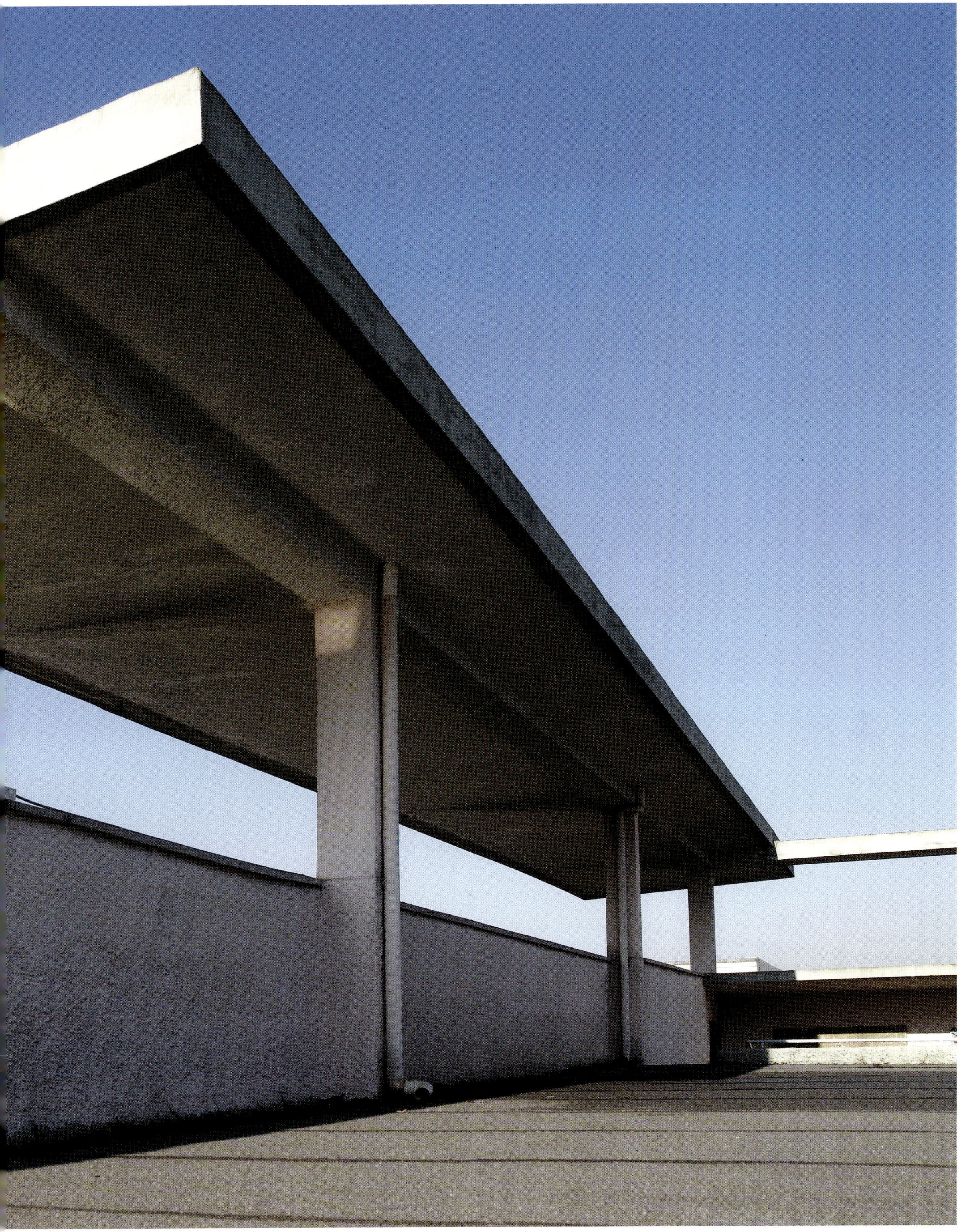

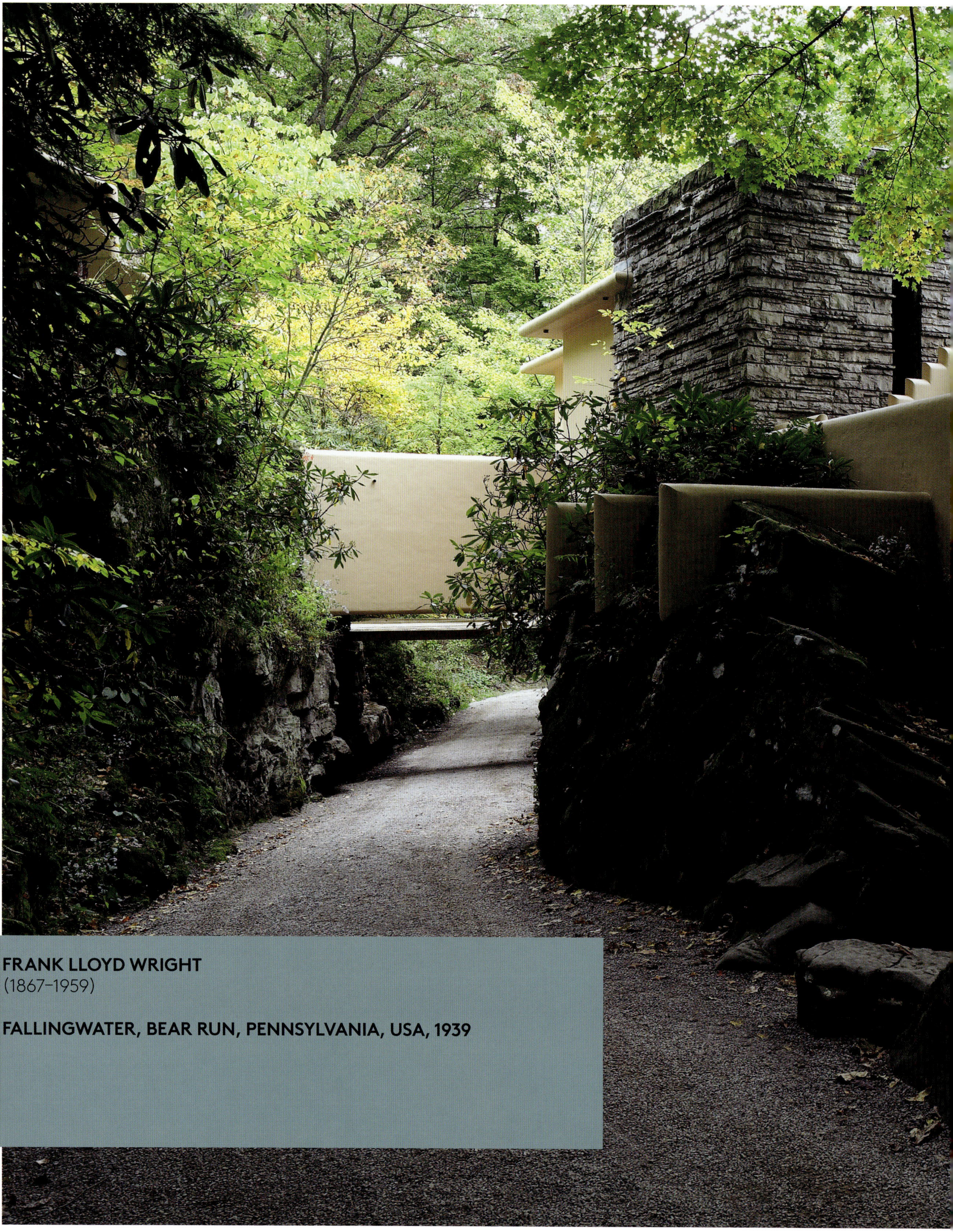

FRANK LLOYD WRIGHT
(1867–1959)

FALLINGWATER, BEAR RUN, PENNSYLVANIA, USA, 1939

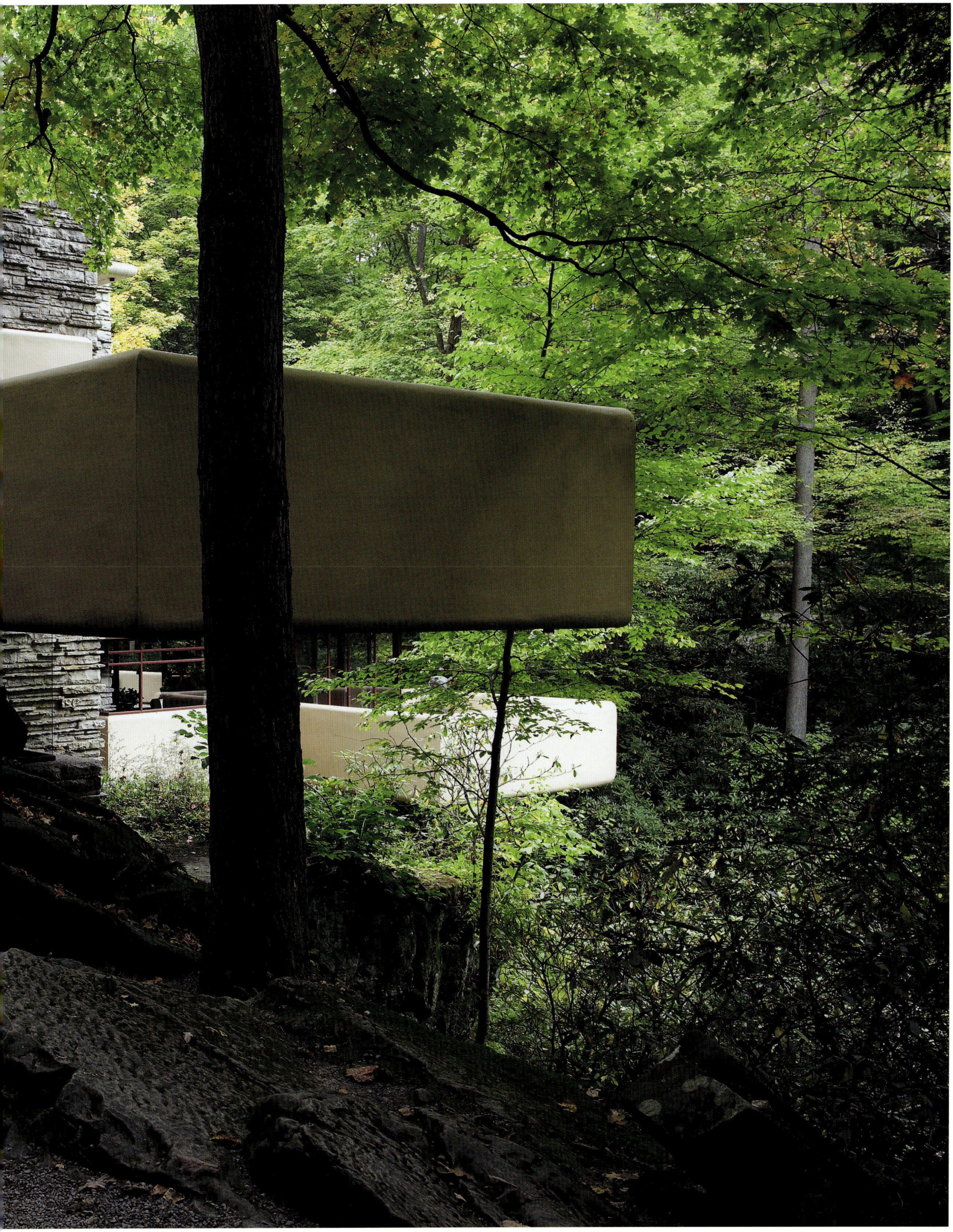

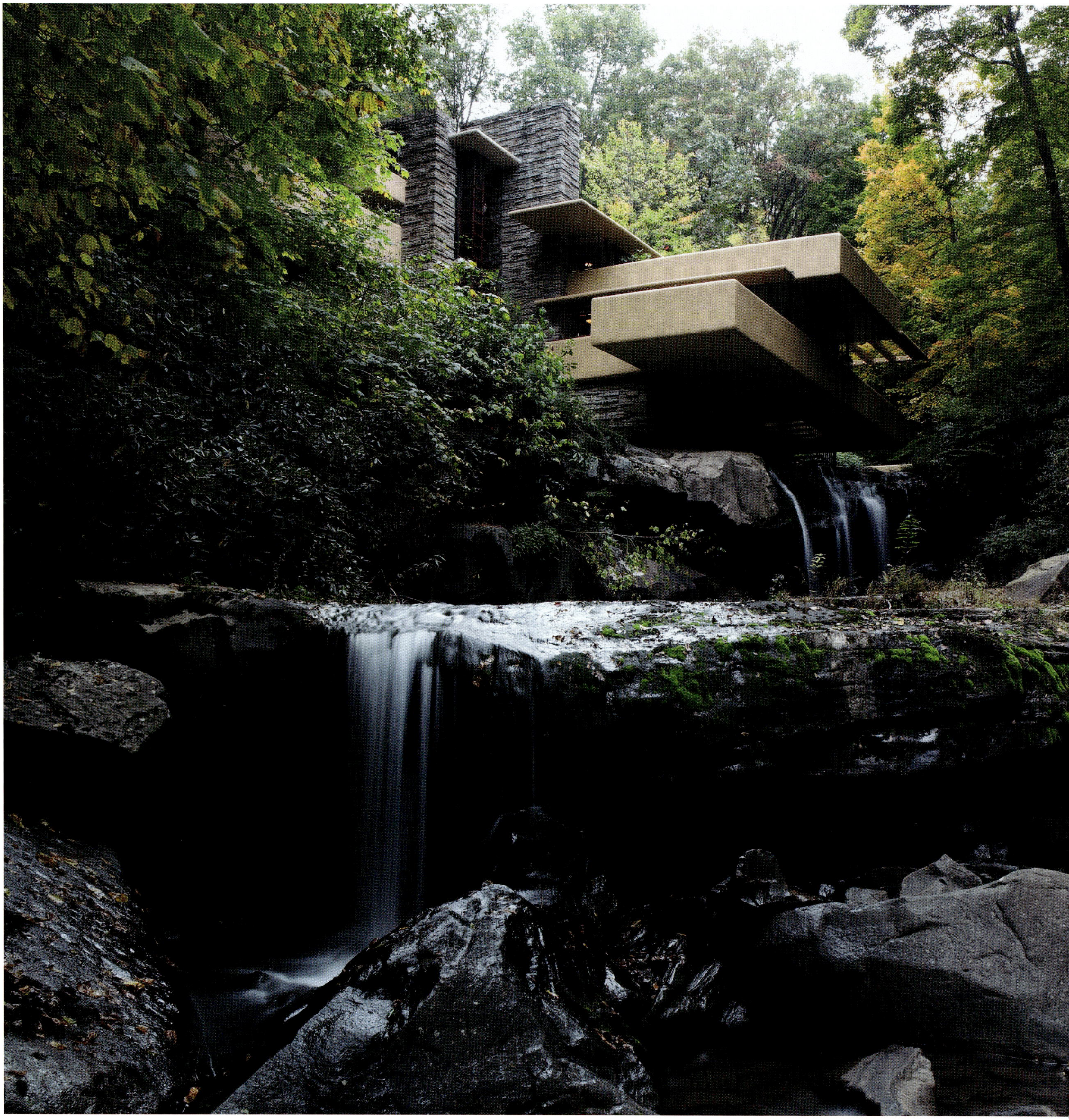

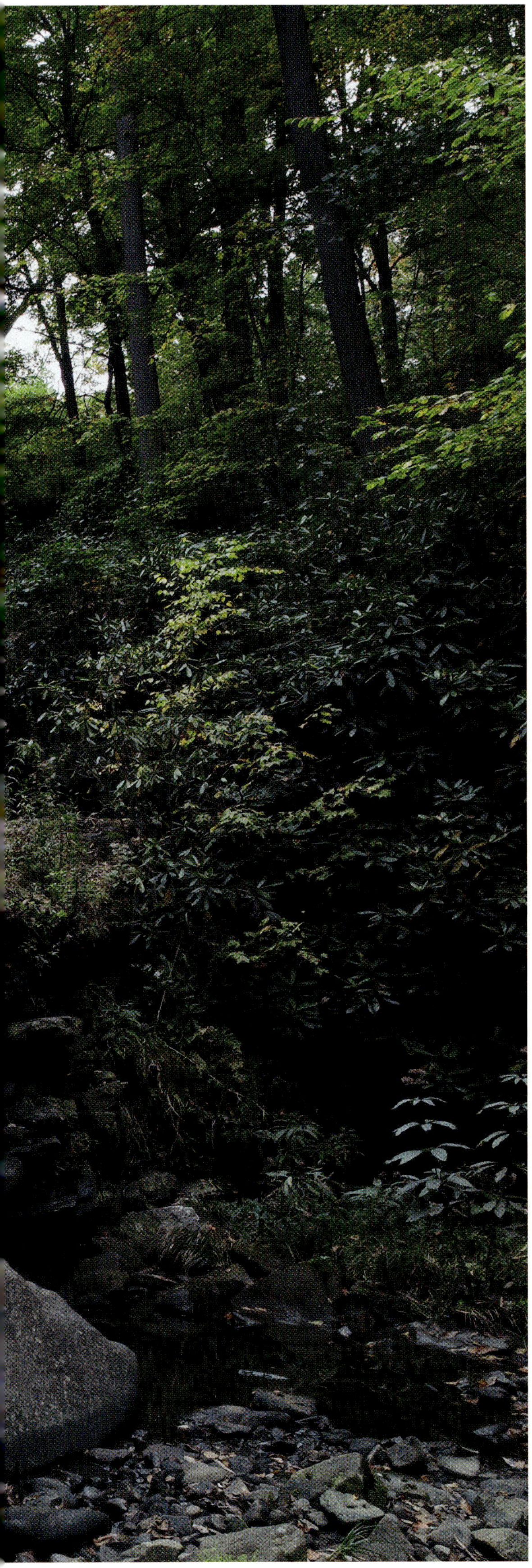

The American architect and designer Frank Lloyd Wright was the great pioneer and promoter of organic modernism. He advocated a highly contextual approach to architecture in every sense, with a building responding to a specific setting within a considered and tailored solution. He also argued for an architecture of integrity that placed the individual at its heart, suggesting the transformative power of design to shape our lives, sustain us, and connect us to nature and the land.

'It is in the nature of any organic building to grow from its site, come out of the ground into the light – the ground itself held always as a component basic part of the building itself,' he wrote. 'And then we have primarily the new ideal of building as organic. A building dignified as a tree in the midst of nature.'[30]

Wright used the term 'organic' to speak of a 'natural' building, integrated with its site, the environment and its residents. But he also spoke of the honest character of materials, believing they should be allowed to express themselves freely and without disguise. There was a strong Arts and Crafts influence to many of his earlier projects and much of his crafted timber furniture.

Yet Wright was never a traditionalist. He was open to historical influences, but dismissive of the predominance of the neoclassical style in American architecture, arguing for an architectural approach and language rooted in America itself.

As well as using natural materials and emphasizing craft, Wright experimented with fresher components such as reinforced concrete and concrete blocks. Cast cement blocks, also known as 'textile blocks', were used to build a number of his houses in the Twenties, such as the Alice Millard House in Pasadena (1924) and the Charles Ennis House in Los Angeles (1924).

Yet, in many ways Wright was most comfortable in rural settings, where his love of landscape and a truly organic approach could be more fully indulged. He grew up in a rural part of Wisconsin, where his father was a minister and his mother a teacher. Wright spent many childhood summers working on his uncle's farm nearby and spoke of them as being highly formative. He studied engineering at the University of Wisconsin but already knew that his future lay in architecture, so decided to leave for Chicago.

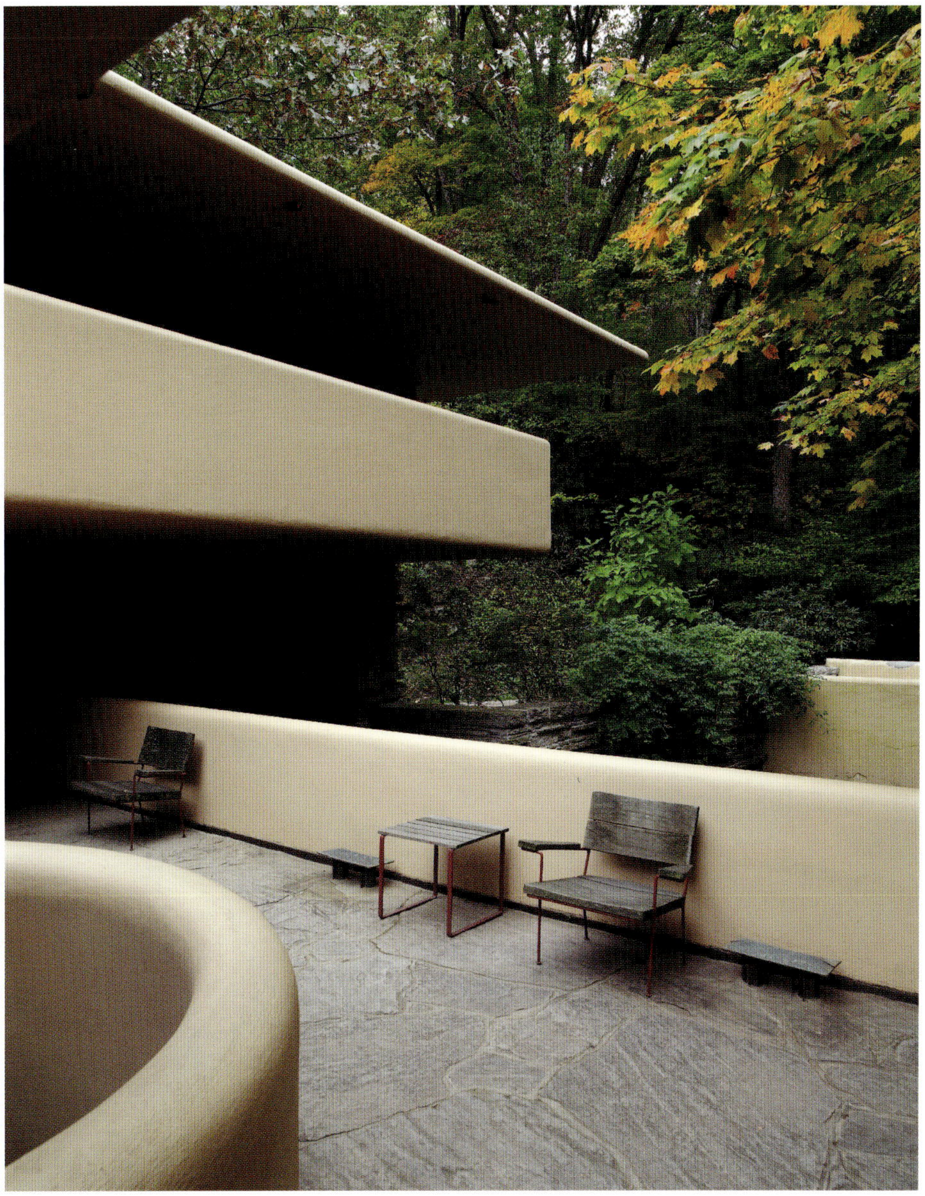

PAGES 428–9 & ABOVE Fallingwater is the foremost example of 'organic' architecture, with the house forming a considered and constant relationship with the topography and its surroundings.

RIGHT The house offers a series of terraces and integrated outdoor spaces at every level, promoting and enhancing the sense of connection with the landscape.

Wright eventually found a position with Louis Sullivan's* firm and rose quickly. During the five years that Wright spent with Sullivan he was placed in charge of the firm's residential work, but was ultimately asked to leave after spending too much time on private commissions. In 1893, Wright established his own practice in Chicago.

In 1911, Wright built his first rural home and studio at Taliesen, near Spring Green in Wisconsin, close to where he grew up. Taliesen was a constant presence in Wright's turbulent personal life and was twice rebuilt after devastating fires. Taliesen and its sister project, Taliesen West (1937) in Arizona, also played a key role in Wright's professional career, serving as studio retreats and hosting fellowships for apprentices and acolytes.

One of Wright's Taliesen students was Edgar Kaufmann Jr., the son of a wealthy department store owner, based in Pittsburgh. Over a number of years Wright and the Kaufmann family got to know one another until Kaufmann Sr. asked the architect to look at a site that the family owned in the Allegheny Mountains, Pennsylvania, where they had a small winter cabin. Here there was a stream called Bear Run where the family enjoyed picnics and sometimes swam in the water pools.

'He loved the site where the house was built and liked to listen to the waterfall,' said Wright. 'So that was a prime motive in the design. I think you can hear the waterfall when you look at the design. At least it's there and he lives intimately with the thing he loved.'[31]

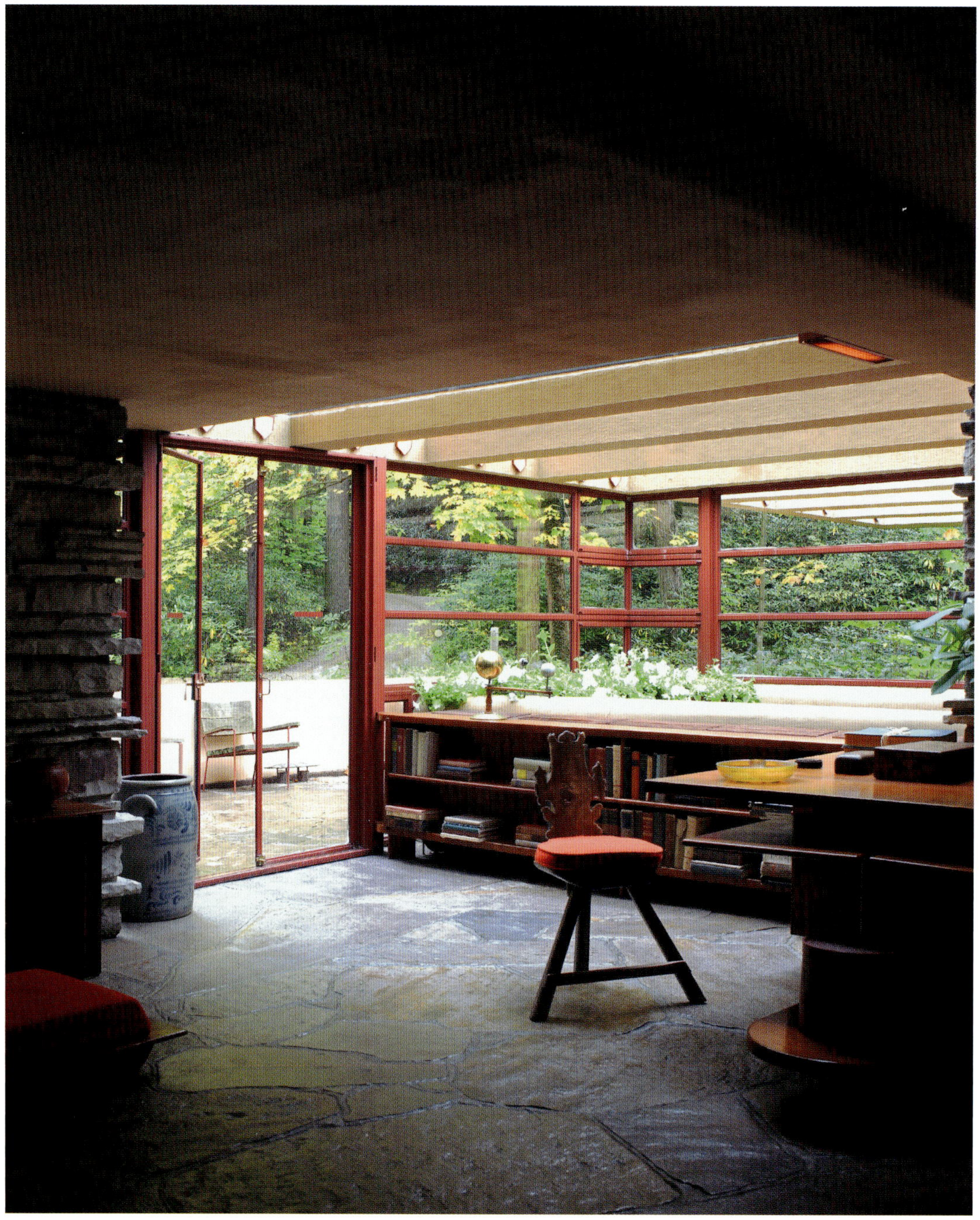

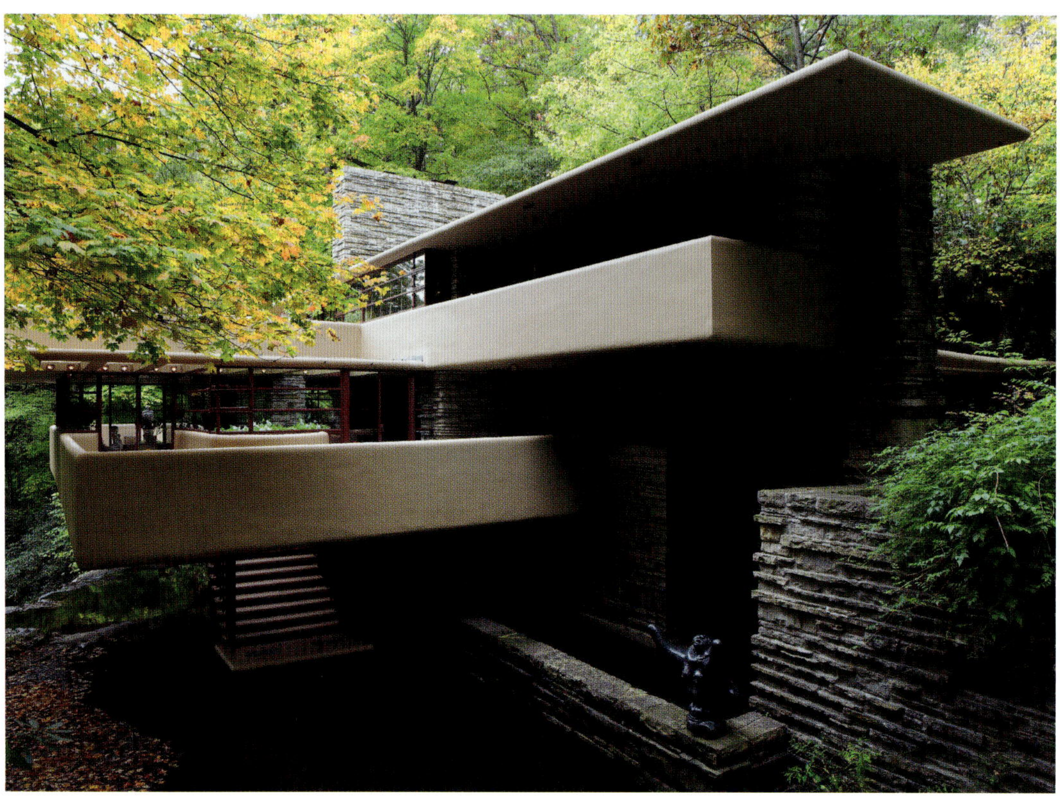

ABOVE Internal spaces such as the study, on the top level of the building, seek vivid connections with the verdant setting beyond the windows and the adjoining terraces.

LEFT & OPPOSITE The artistic composition and tectonics of the house juxtapose the horizontal elements of the projecting concrete terraces and floor plates with the vertical elements in locally quarried stone.

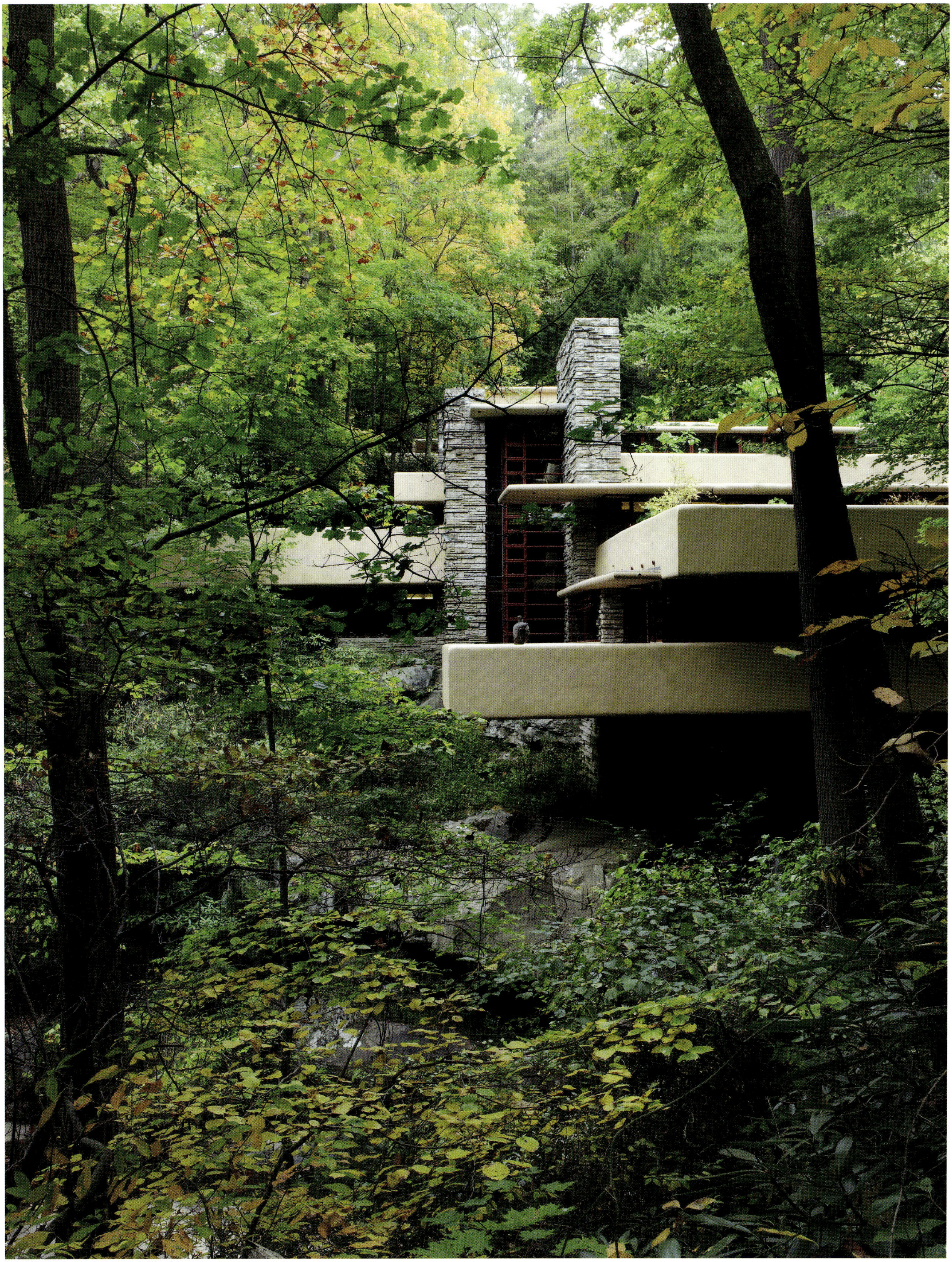

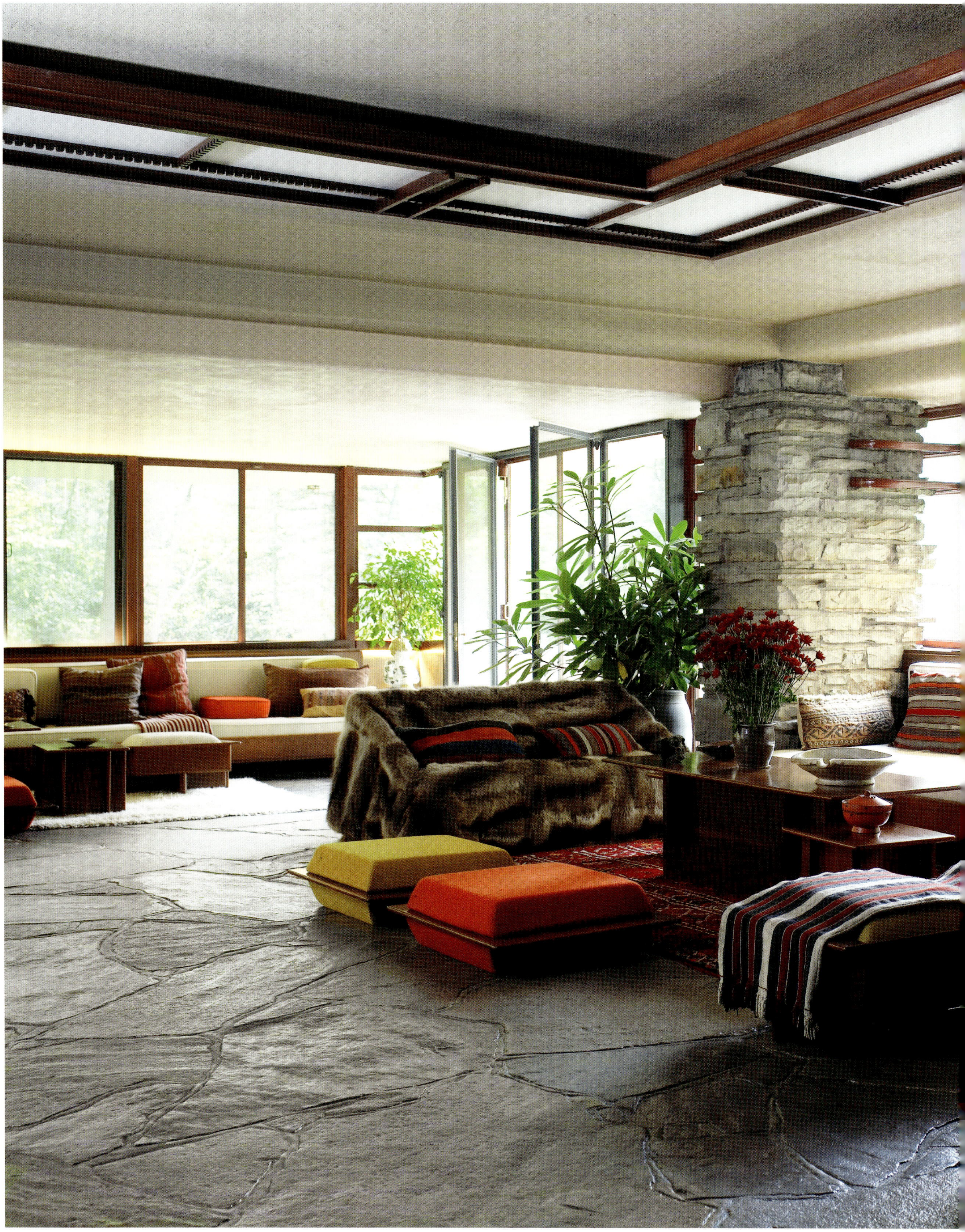

The Kaufmann House is the fullest expression of Wright's version of organic architecture. Reinforced concrete slabs were used to cantilever the three-storey house out over the waterfalls of Bear Run itself, while vertical stack walls of local stone helped anchor the building to the site. The reinforced concrete framework of the house took the structural load from the façade, allowing Wright to create horizontal bands of glass that connected the house to the surrounding woodlands. The relationship between inside and outside space, which was so crucial to Wright, was further enhanced by the provision of terraces, platforms and plunge pools. The integration of house and setting was complete.

Wright had many phases over a long and distinguished career, culminating in the radical dynamism of the Guggenheim Museum in New York (1959). Yet Fallingwater – despite subsequent structural issues – has become one of the best beloved houses of the 20th century and the great exemplar of warm modernity, comparable with the soft modernism of the great Scandinavians such as Alvar Aalto. It represented a clear alternative to the crisp, linear, white residences of the International Style*, which Wright criticized for their sterility, and became a seminal house for mid-century architects of the post-war period who embraced the organic approach to the full.

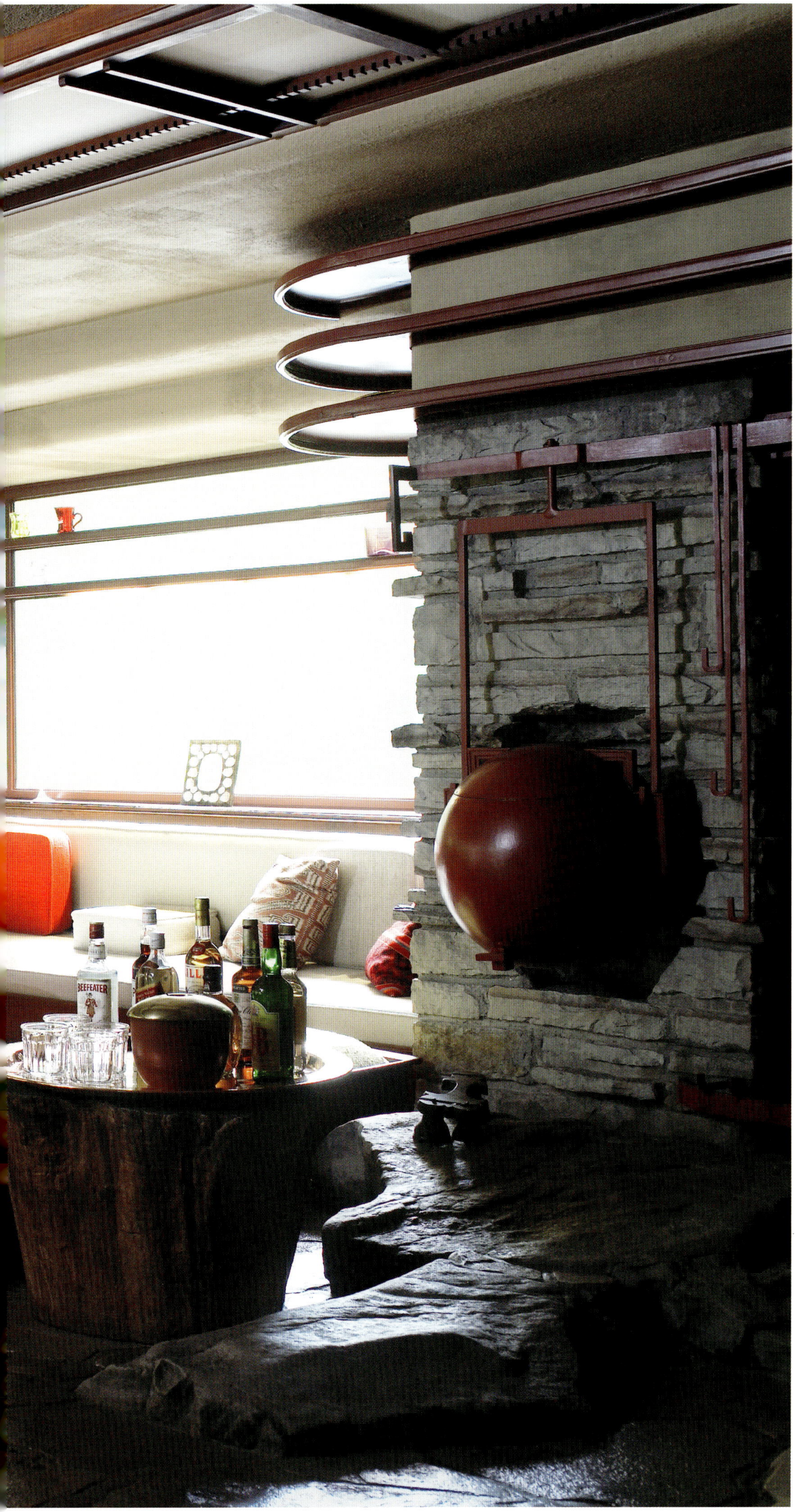

LEFT The living room on the ground floor forms part of an open sequence of generously scaled and inter-connected spaces, including the dining area and adjoining terraces; the floors are paved in stone, while the fireplace sits alongside pieces of bedrock pushing through the flagstones.

RIGHT A covered walkway connects the main building to a guest house nearby; the fluid character of the canopy acts as a foil to the strong horizontal and vertical elements seen elsewhere at Fallingwater.

BELOW A detail of Edgar Kaufmann Sr.'s bedroom, where the lightness of the glazing meets the monumentality of the stone walls.

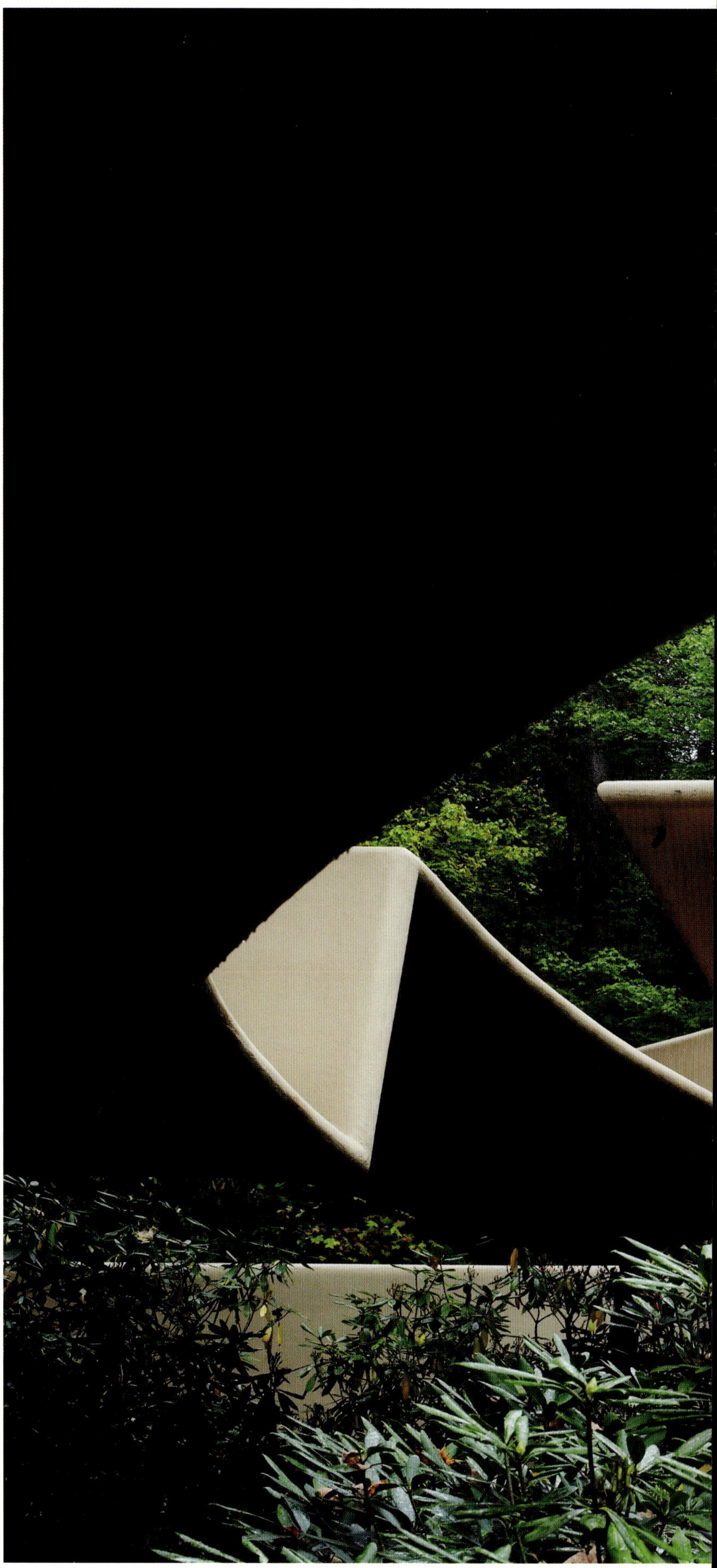

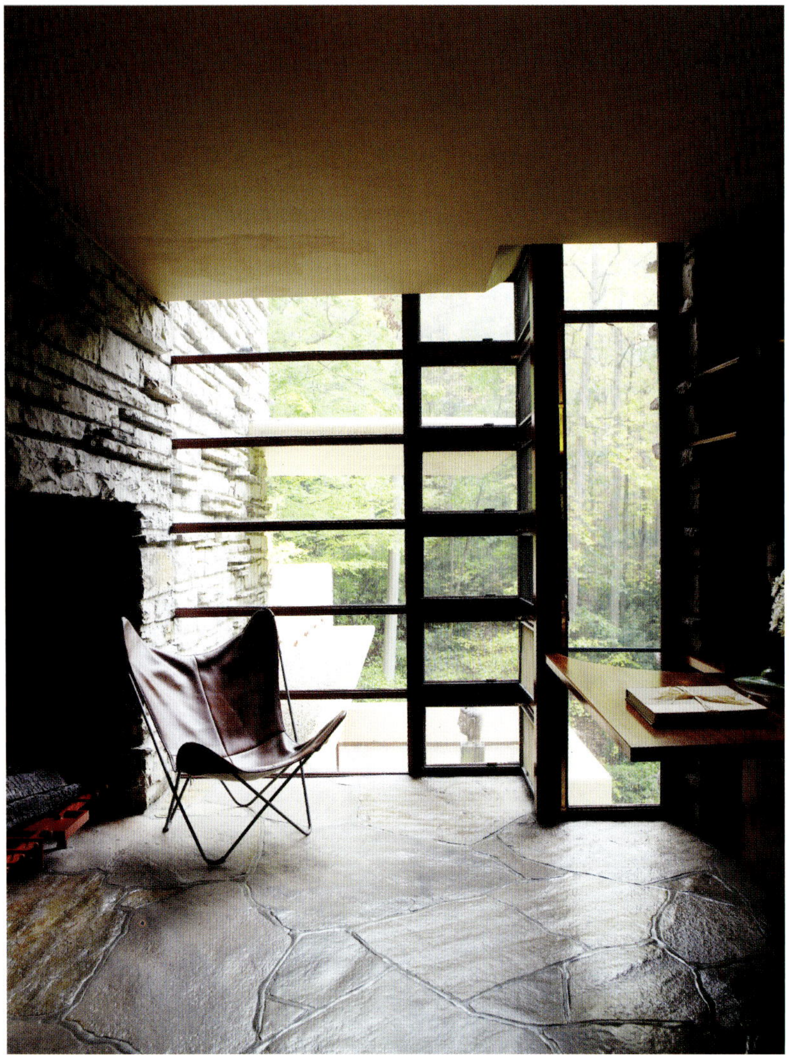

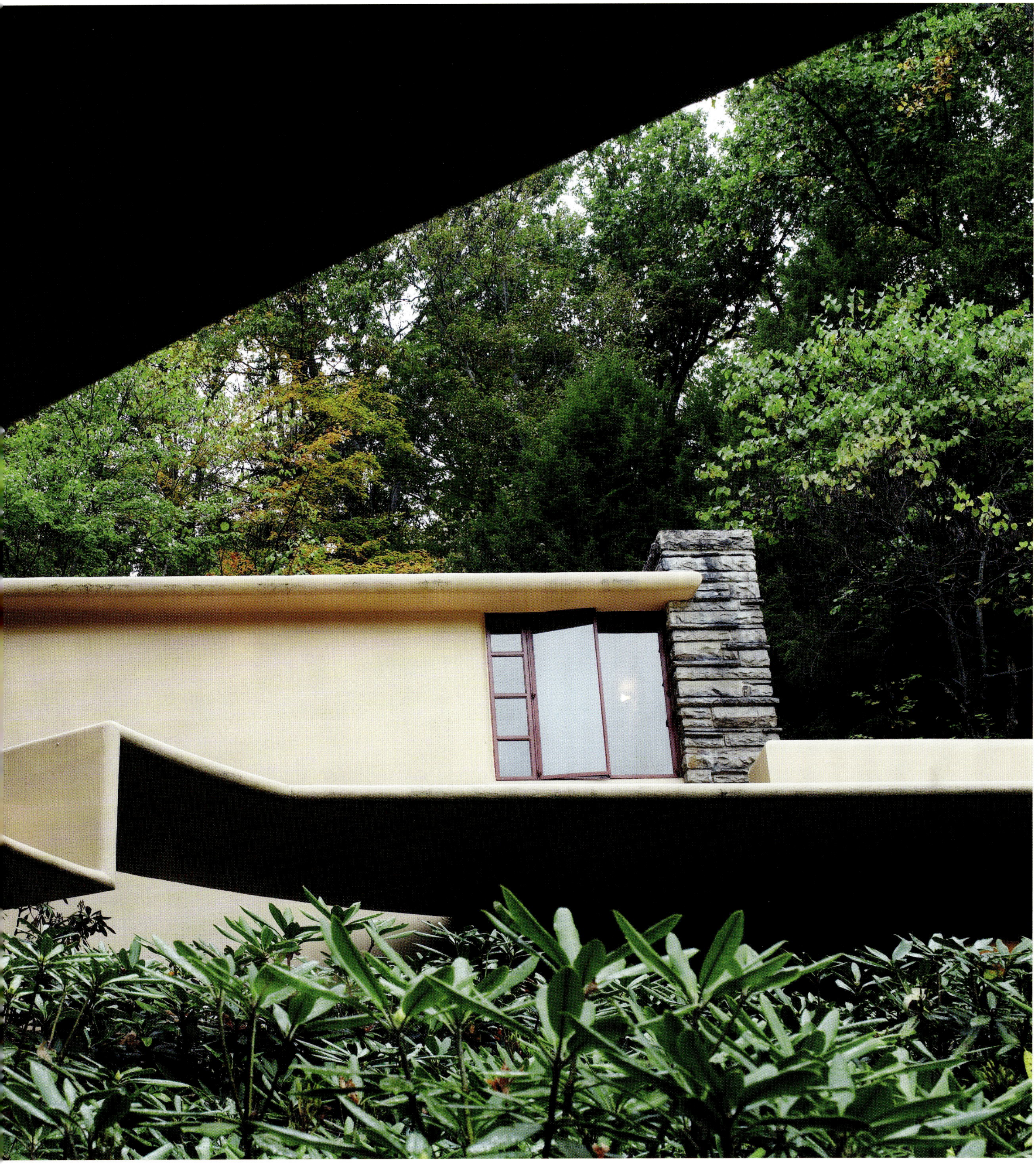

A–Z OF DESIGNERS, ARCHITECTS & MANUFACTURERS

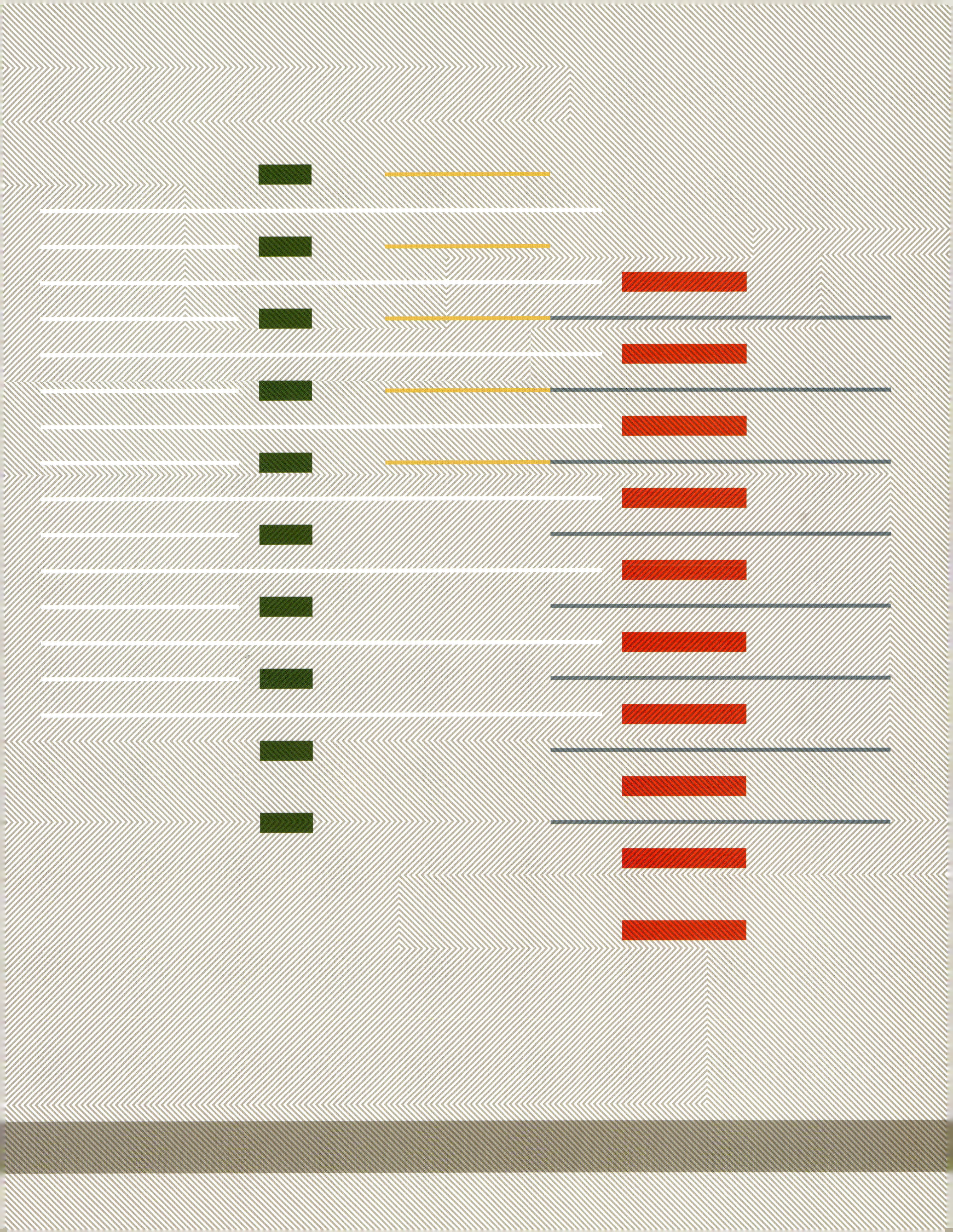

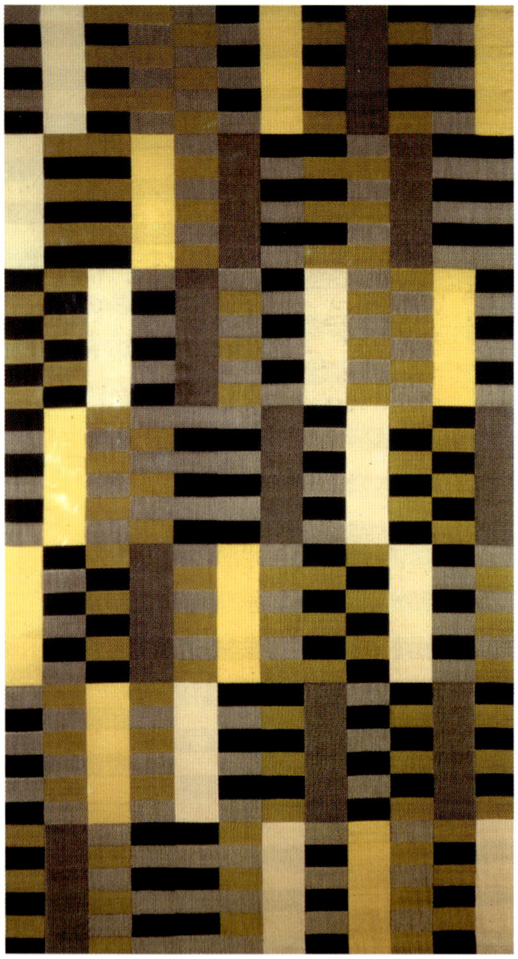

LEFT Anni Albers's wall hanging 'Black White Yellow', designed in 1926, in silk and rayon.

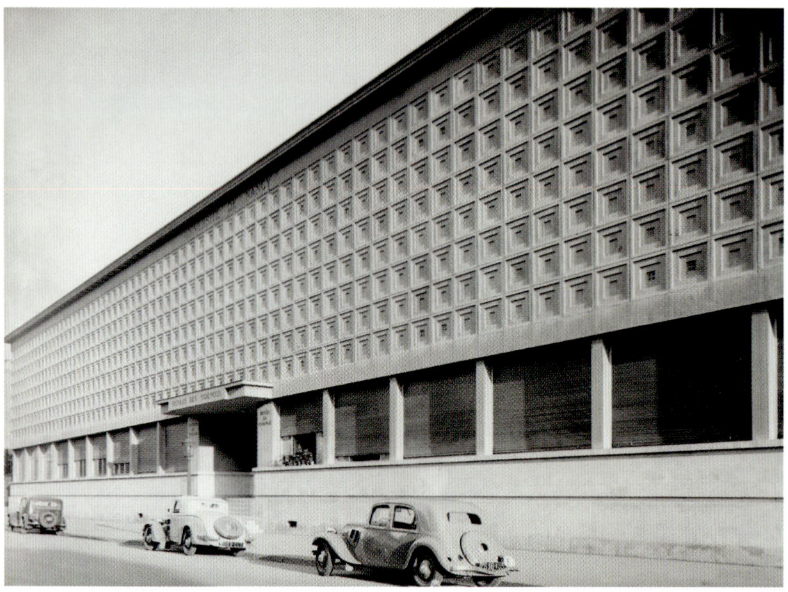

ABOVE The Institut de Zoologie in Nancy, France, 1933, designed by Jacques and Michel André.

AALTO, ALVAR (1898–1976)
See p. 308.

ALBERS, ANNI (1899–1994)
Influential German textile artist Anni Albers studied at the Bauhaus, where she eventually became a master in the weaving department, working alongside Gunta Stölzl*. On moving with her husband Josef Albers to America, she taught at Black Mountain College, where she developed successful ranges of textile designs. She began working with Knoll*, among others, and in 1949 the Museum of Modern Art in New York exhibited her work. She wrote a key text about her craft, *On Weaving*, which was published in 1965.

ALBERS, JOSEF (1888–1976)
See p. 262.

ANDRÉ, JACQUES (1904–1985)
The son of French architect Émile André, a founding member of the École de Nancy, Jacques André took over the family practice with his brother Michel in the Thirties. He collaborated on a number of projects with Jean Prouvé, who was also based in Nancy, including their Garden chair of 1936, made of perforated steel and acrylic glass. In the post-war period, André concentrated on the design of public buildings.

ART DECO
Named after the Exposition Internationale des Arts Décoratifs et Industriels Modernes held in Paris in 1925, Art Deco was a stylistic movement of the Jazz Age, at its zenith in the late Twenties and early Thirties. With its focus on sophisticated materials, luxurious finishes and fine craftsmanship, Art Deco was largely a movement of wealth, yet also influenced architecture and transport, particularly in its connection to 'streamlined' aesthetics. The work of a number of early modernist designers and architects – such as Eileen Gray, Pierre Chareau, Donald Deskey and Jacques-Émile Ruhlmann* – suggests a significant overlap between the Art Deco style and the beginnings of the modern movement, which was partly defined by the paring down of superfluous decorative elements, the use of more 'democratic' materials and the push towards factory production.

ARUP, OVE (1895–1988)
Anglo-Danish structural engineer Ove Arup studied in both the UK and Denmark, becoming a specialist in reinforced concrete construction. After working in Hamburg he settled in London and founded his own consultancy – now a global business – in 1938. During the Thirties he worked closely on building projects with Berthold Lubetkin's* Tecton* architectural partnership and also collaborated with, among others, Ernő Goldfinger and Wells Coates*.

ASPLUND, ERIK GUNNAR (1885–1940)
See p. 28.

BADOVICI, JEAN (1893–1956)
See p. 340.

BAWDEN, EDWARD (1903–1989)
English illustrator, graphic designer, printmaker and painter Edward Bawden studied at the Cambridge School of Art, and subsequently won a scholarship to the Royal College of Art. As well as book covers and illustrations, he produced a series of advertising posters during the Thirties, including designs for London Underground and Imperial Airways. He served as an official war artist during World War II before resuming his career from his Essex base. His work, often compared with that of colleague Eric Ravilious, is distinctly English – modern and fresh in style but also whimsical, playful and colourful.

BAYER, HERBERT (1900–1985)
See p. 268.

BAZELEY, GEOFFREY (1906–1989)
After studying architecture at Cambridge University and the Architectural Association, English architect Geoffrey Bazeley began working with SERGE CHERMAYEFF'S practice in 1933. He founded his own architectural office in Penzance, Cornwall, in 1936, largely on the basis of a commission for Tregannick, an influential modernist house in nearby Sancreed.

BEALL, LESTER (1903–1969)
American graphic designer Lester Beall studied history of art at the University of Chicago, where he was influenced by a number of modern artistic movements such as Constructivism*. He began his career in Chicago before moving to New York in 1935. In addition to his magazine work, he was well known for his advertising posters, including a series of Thirties posters for the Rural Electrification Administration, which combined a vivid use of abstract images, colour and typography.

BECK, HARRY (1902–1974)
The schematic map of the London Underground produced by draughtsman and designer Henry ('Harry') Beck in 1931 was one of the first non-geographical transit diagrams, in which metro lines and stations were ordered and abstracted rather than represented in terms of true scales or distances. With each line rendered in a different colour and with clarity of type, a vast amount of complex information was quickly and simply conveyed. Seen as an early example of grid-based graphic design, and sometimes compared with the work of abstract modern artists such as Piet Mondrian, Beck's approach has played a key role in the evolution of transit maps worldwide.

BEHRENS, PETER (1868–1940)
See p. 214.

BERNHARD, LUCIAN (1883–1972)
Born in Stuttgart, Lucian Bernhard studied in Munich before settling in Berlin in 1901. He established himself as a leading proponent of the 'Sachplakat', or 'object poster', in the first decades of the 20th century, producing bold advertising artworks for companies such as Adler, Bosch and Manoli, with bright colours, abstract representations of products and large, vivid typography. Bernhard also developed a series of new typefaces over the course of the Twenties and Thirties, continuing his career in New York after emigrating in 1923.

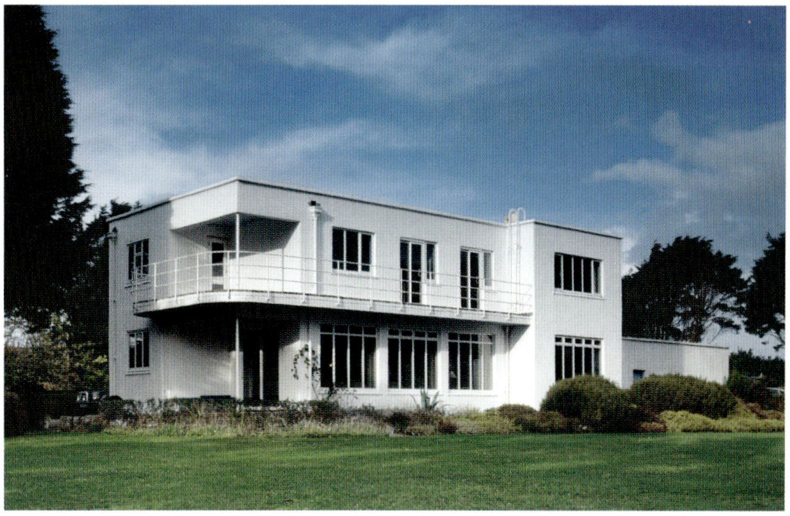

ABOVE An exterior view of Tregannick, a modernist residence in Sancreed, Cornwall, UK, designed by Geoffrey Bazeley, 1936.

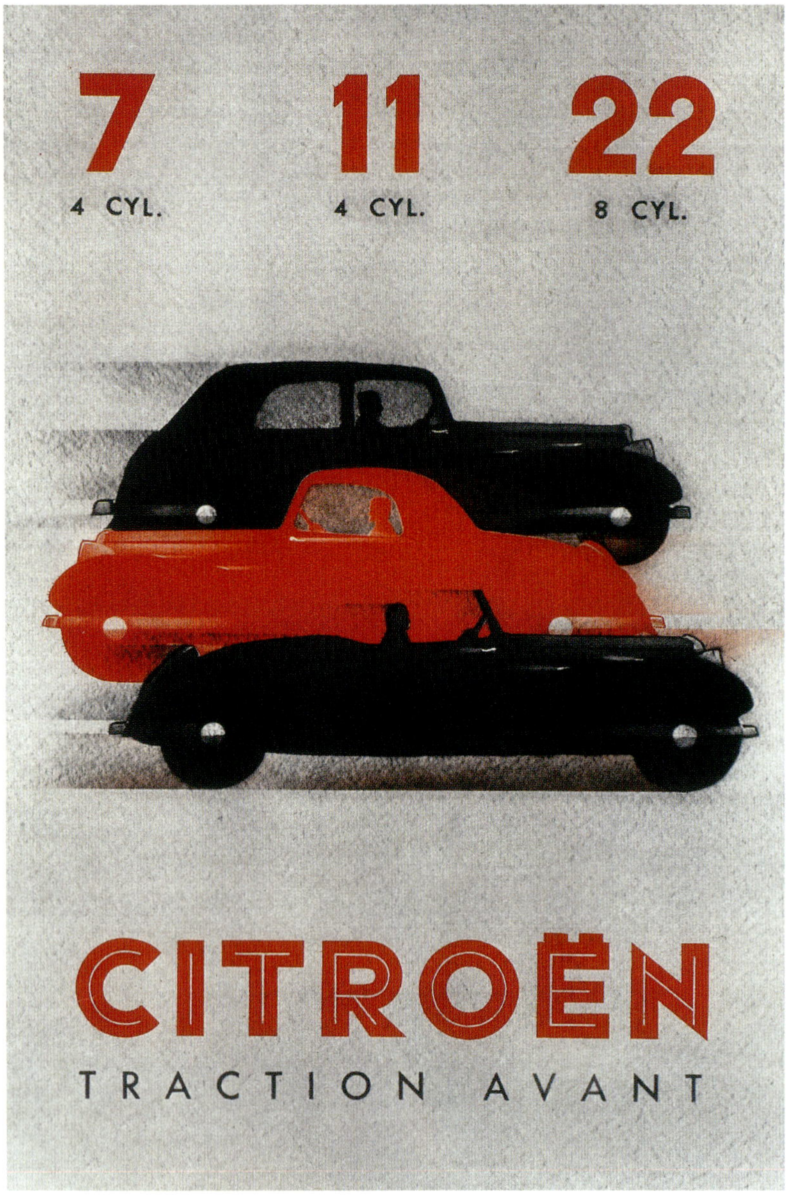

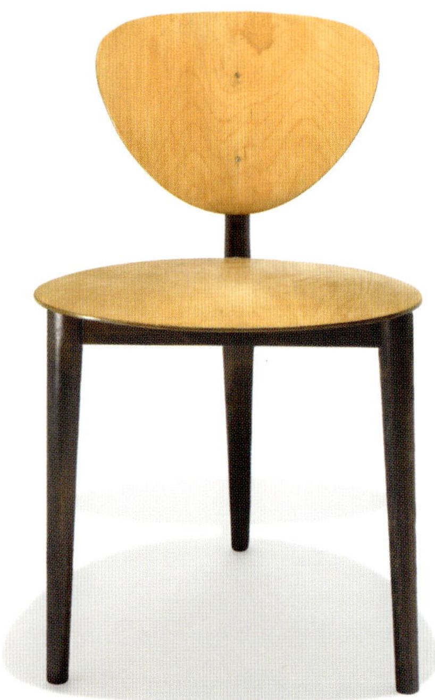

ABOVE A 1934 Citroën advertising poster designed by Flaminio Bertoni, promoting the motor manufacturer's Traction Avant, or front wheel drive, collection of cars.

LEFT Max Bill's Three-Legged chair, produced by Horgenglarus in Switzerland, 1949, in beech and stained wood.

BERTONI, FLAMINIO (1903–1964)
Italian designer Flaminio Bertoni is best known for his work for Citroën from 1932 onwards. Projects included the Traction Avant (1934) and the 2CV, which was designed in the late Thirties but not produced until 1949; he also designed the DS, which was launched in 1955.

BILL, MAX (1908–1994)
After studying at the Bauhaus, Max Bill went on to achieve success in a range of disciplines – architecture, industrial design, furniture design and graphic design. During the Thirties, while based in Zurich, his focus was mainly on architecture and graphics. As an educator, he began his teaching career in Zurich, before co-founding the Ulm School of Design in 1953, based upon the Bauhaus model.

BOGLER, THEODOR (1897–1968)
See p. 168.

BONET, ANTONIO (1913–1989), **JORGE FERRARI-HARDOY** (1914–1977) **& JUAN KURCHAN** (1913–1972)
The Spanish architect and designer Antonio Bonet and his Argentinian colleagues Jorge Ferrari-Hardoy and Juan Kurchan, all worked in LE CORBUSIER's Parisian office in the late Thirties. In 1938, the three moved from France to Argentina, where they developed one of their most famous works: the Butterfly chair. A lightweight, modern re-working of the folding campaign chair, the piece featured a tubular steel frame and a leather or canvas sling seat. Manufactured by Artek-Pascoe and then Knoll* and others, the chair has been widely copied and imitated ever since.

BRANDT, MARIANNE (1893–1983)
See p. 120.

BREUER, MARCEL (1902–1981)
See p. 32.

BRINKMAN & VAN DER VLUGT
See p. 314.

BRODOVITCH, ALEXEY (1898–1971)
Graphic designer and art director Alexey Brodovitch was born in Russia and settled in Paris in 1920. Here, he began his career as a designer and applied artist, winning a number of medals at the Exposition Internationale des Arts Décoratifs et Industriels Modernes of 1925. Launching his own studio, he designed advertising posters for Athélia, Cunard and others, as well as working in the field of book illustration. In 1930 he moved to America and began teaching, as well as freelancing. From 1934 to 1959 he served, with distinction, as the art director of *Harper's Bazaar*, where he established collaborations with leading photographers and artists, while creating powerful page layouts with an emphasis on white space and 'floating' images.

BURNET, TAIT & LORNE

The architectural practice founded by John James Burnet (1857–1938), Thomas Smith Tait (1882–1954) and Francis Lorne (1889–1963) had offices in both Glasgow and London, initially focused on buildings in a neoclassical style. Over the course of the late Twenties and Thirties, the practice embraced modernism, seen both in individual house projects (such as the series at Silver End near Braintree, Essex, 1927) and more substantial commissions, including the Royal Masonic Hospital in Hammersmith, London (1933), and Burlington Danes School, also in Hammersmith (1936).

CALDER, ALEXANDER (1898–1976)

Having studied both mechanical engineering and art in the United States, American artist and sculptor Alexander Calder moved to Paris in 1926, where he became friends with a number of avant-garde creatives, such as Jean Arp, Marcel Duchamp and Joan Miró; he also met Theo van Doesburg and visited Piet Mondrian's studio. He created a kinetic, sculptural circus ('Cirque Calder') in the late Twenties and went on to design a series of distinctive, colourful 'mobiles' – suspended sculptures that bridged the worlds of art and design. Calder settled in Connecticut in 1933, developing an original sequence of mobiles and standing sculptures, as well as painting murals, including panels at a number of New England houses designed by Marcel Breuer.

CARDEW, MICHAEL (1901–1983)

See p. 164.

CARWARDINE, GEORGE (1887–1947)

See p. 124.

CASSANDRE, A. M. (1901–1968)

See p. 272.

CHAREAU, PIERRE (1883–1950)

See p. 324.

CHERMAYEFF, SERGE (1900–1996)

See p. 328.

CHIESA, PIETRO (1892–1948)

See p. 146.

CIAM (INTERNATIONAL CONGRESS OF MODERN ARCHITECTURE)

The CIAM – or Congrès International d'Architecture Moderne – was one of the most influential, organized groupings established by modernist architects to promote their ideas and provide a forum for debate and discussion. The body was founded in 1928 in Switzerland by 28 European architects, including Le Corbusier, Pierre Jeanneret, Pierre Chareau and Gerrit Rietveld; later members included Walter Gropius and Alvar Aalto. Participants gathered at a number of key conferences from 1929 onwards, with each one devoted to a particular theme, such as 'The Minimum Dwelling' (Frankfurt, 1929) and 'The Functional City' (Athens, 1933). These gatherings were interrupted by World War II then resumed in 1947, but CIAM began to splinter and finally broke up in 1959.

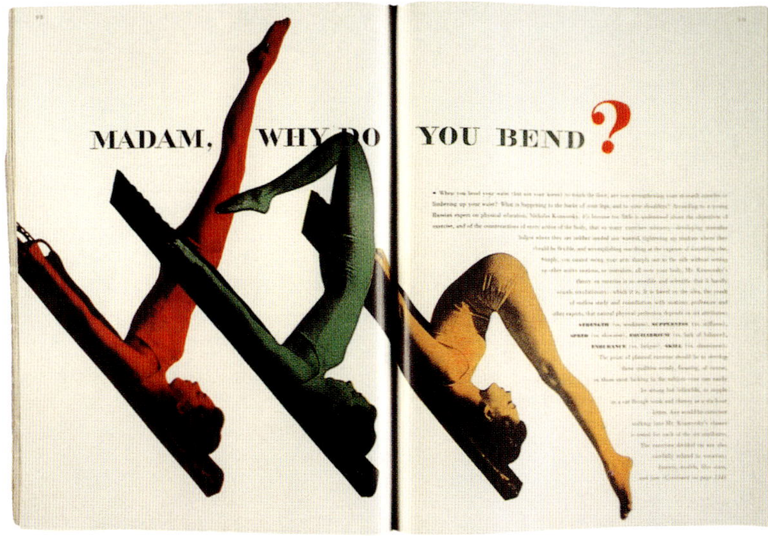

ABOVE A magazine spread from a 1945 edition of *Harper's Bazaar*, designed by Alexey Brodovitch.

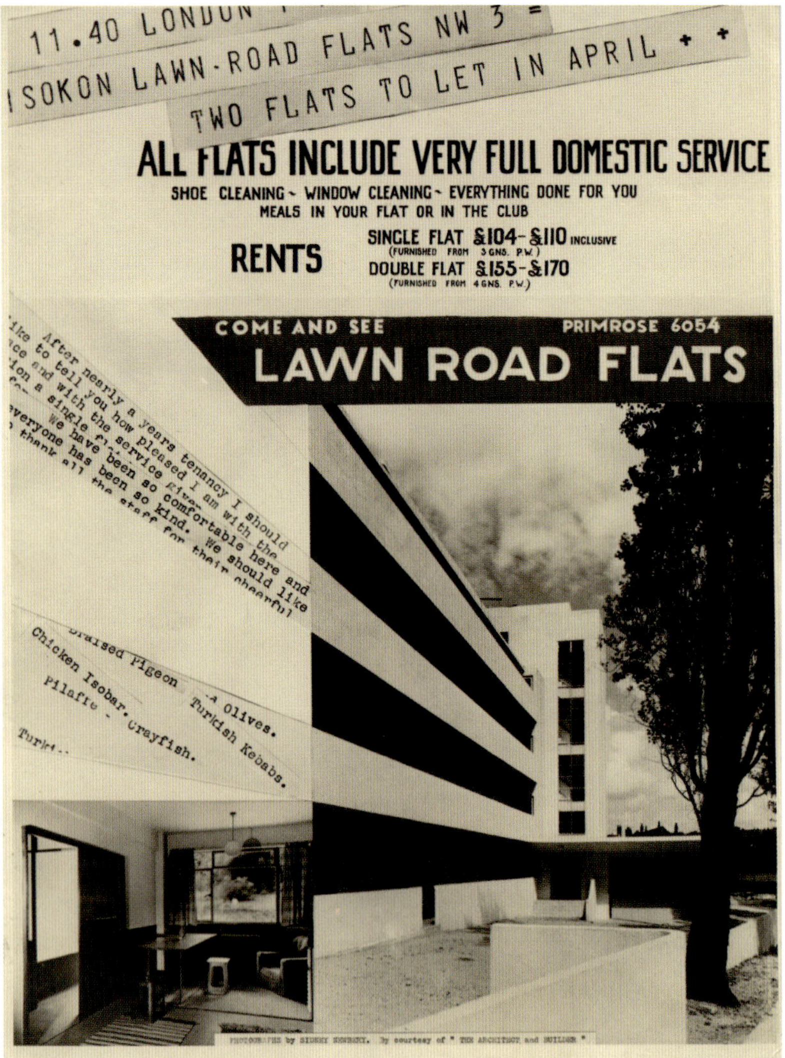

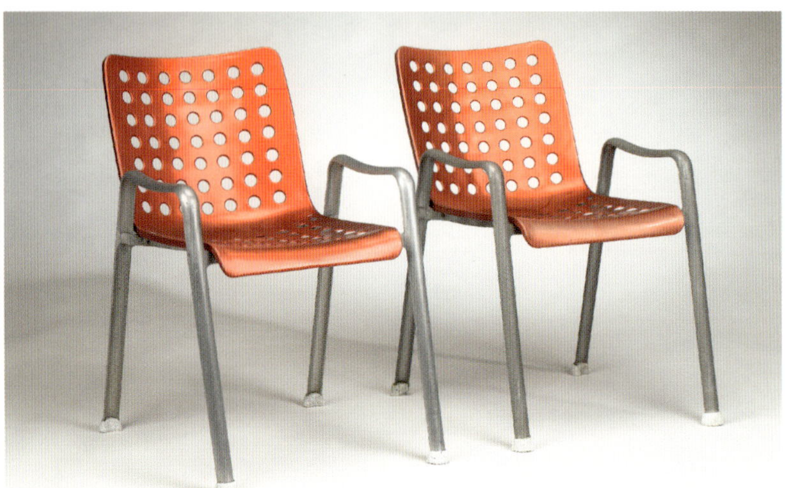

TOP An advertisement for the Lawn Road Flats in London's Belsize Park, also known as the Isokon Building, designed by Wells Coates, 1934.

ABOVE A pair of Landi chairs, designed by Hans Coray in 1938 and produced by P. W. Blattmann Metallwarenfabrik, Wädenswil, Switzerland, in red anodized aluminium.

COATES, WELLS (1895–1958)

The son of Canadian missionaries, Wells Coates was born in Japan and travelled widely. He studied engineering at the University of Vancouver and eventually moved to London, where he worked as a journalist before beginning a career in design and founding his own practice in the late 1920s. A flamboyant and sometimes eccentric figure, Coates worked across a wide range of areas. Commissions included shop designs, using modern materials such as plywood; studio interiors for the BBC (along with SERGE CHERMAYEFF), including a design for a suspended microphone; the AD 65 radio in Bakelite for E.K. Cole & Son; and designs for sailing yachts. He was a co-founder of the Modern Architectural Research Group (MARS)*, and in 1931 visited the Bauhaus with Chermayeff and Jack Pritchard, the founder of Isokon*. Coates's landmark North London apartment block of 1934 – the Isokon Building (or Lawn Road Flats) – became a social hub, hosting a succession of writers, artists and architects.

COLIN, PAUL (1892–1985)

The influential French graphic artist Paul Colin is most intimately associated with Jazz Age posters of the Twenties for Parisian theatres, for whom he also designed stage sets and costumes. In 1926, Colin opened a school for the graphic arts in Paris. He also designed posters for the French travel industry, often compared with the work of A. M. CASSANDRE.

CONNELL, WARD & LUCAS

This London practice was formed of a union between two New Zealanders and one home-grown architect. Basil Ward (1902–1976) and Amyas Connell (1901–1980) trained in New Zealand before working their passage to London on a cargo ship. Colin Lucas (1906–1984) studied architecture at Cambridge University before joining his father in a construction business, with a specialism in concrete. Lucas's first project, Noah's House (1929), a weekend cottage at Bourne End in Buckinghamshire, is sometimes described as the first 'monolithic concrete house' in Britain. Lucas joined forces with Connell and Ward in 1933. Among their best-known residential projects are High and Over (1931) in Amersham and 66 Frognal (1938) in Hampstead, London.

CONSTRUCTIVISM

The Russian Constructivists formed one of the most influential regional sub-movements of modernism, along with Dutch De Stijl*. The movement was connected with the emergence of a new generation of artists, graphic designers, ceramicists and architects in the wake of the Russian Revolution of 1917; key figures included ALEXANDER RODCHENKO and El Lissitzky*. Through their work and manifestos, the Constructivists argued for a form of creative socialism, with an emphasis on factory production and the wider availability of modern design and architecture in the cause of progress. Constructivist graphic design, in particular, was to have a global influence.

CORAY, HANS (1906–1991)
Hans Coray studied languages at the University of Zurich and was self-taught as a designer. From 1930 onwards he concentrated on furniture, becoming part of a design community in Zurich that included Max Bill*. In 1939 he developed his most famous design, the stackable and lightweight aluminium Landi chair, for the Swiss National Exhibition; the chair is now produced by Vitra. Later, Coray concentrated on sculpture and painting.

CRABTREE, WILLIAM (1905–1991)
English modernist architect William Crabtree is best known for the design of the landmark Peter Jones department store on the junction of London's Sloane Square and King's Road. Later, he concentrated on housing developments in Harlow and throughout Essex.

DE STIJL
One of the great regional modernist sub-movements, De Stijl – or 'The Style' – was a trans-disciplinary grouping of avant-garde Dutch artists, architects and designers. De Stijl took its name from a journal first published in 1917, which promoted the work and philosophy of its founders while also looking further afield to parallel movements such as Futurism* and Constructivism*. Key figures included Gerrit Rietveld, Theo van Doesburg (also editor of the journal), Piet Mondrian, Robert van 't Hoff* and J. J. P. Oud*. De Stijl furniture and interiors were defined by their geometrical purity, an abstract use of blocks of colour, and a reduction of ornament in favour of clear, linear, dynamic forms.

DECOEUR, ÉMILE (1876–1953)
While the French ceramicist Émile Decoeur embraced the Art Nouveau style early in his career, he went on to favour a more austere style influenced by Chinese ceramics, with an emphasis on form, simple glazes and restrained ornamentation. He based his studio at Fontenay-aux-Roses near Paris and worked largely in stoneware and porcelain; he also served as a consultant to the Sèvres pottery.

DELAUNAY, SONIA (1885–1979)
Born in the Ukraine, Sonia Delaunay studied in Russia, Germany and France, settling in Paris in 1905. In 1910 she married fellow artist Robert Delaunay; together they became leading exponents of Orphism, a modern art sub-movement noted for its powerful use of colour and geometric forms. As well as working as an artist, Delaunay was a textile designer and experimented with furniture design, notably in colourful collaborations with Jean Prouvé and Charlotte Perriand.

DELL, CHRISTIAN (1893–1974)
See p. 128.

DEPERO, FORTUNATO (1892–1960)
While also noted as a painter, theatre designer and writer, the Italian Futurist* Fortunato Depero is best known for his innovative and experimental work in the fields of graphic design and advertising. His 1927 monograph, *Depero Futurista*, was famously held together by aluminium bolts and is sometimes known as 'The Bolted Book'. His career, which included a period in New York during the late Twenties, embraced a strong use of colour, geometry and typographic versatility.

DESKEY, DONALD (1894–1989)
See p. 38.

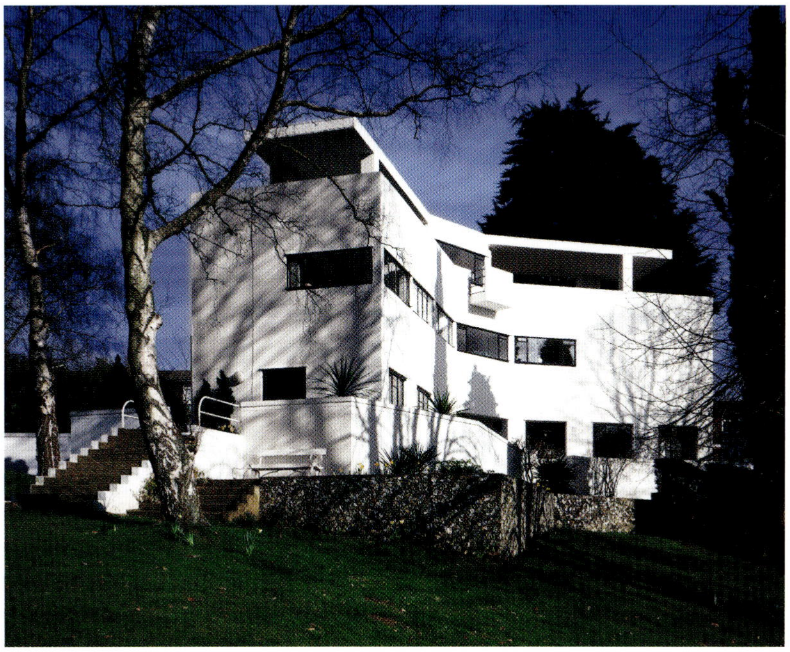

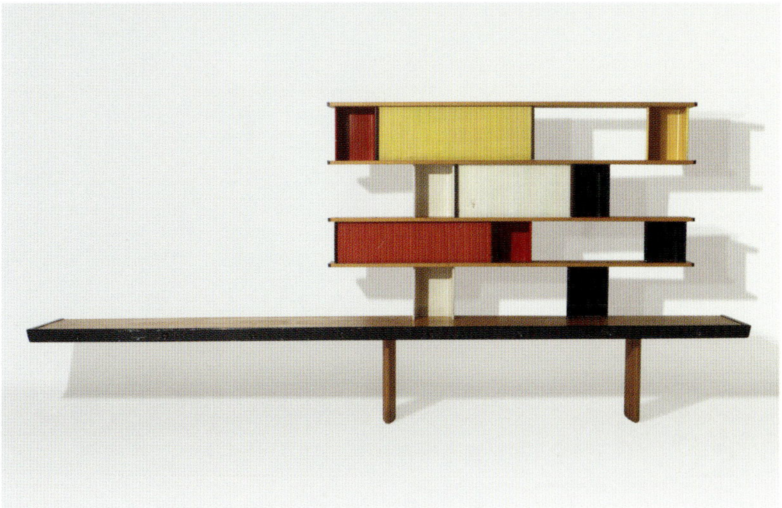

TOP Connell, Ward & Lucas's High and Over house in Amersham, Buckinghamshire, UK, 1931.

ABOVE A *bibliothèque* from the Maison de la Tunisie, Cité Internationale Universitaire de Paris, by Sonia Delaunay, Jean Prouvé and Charlotte Perriand, 1952.

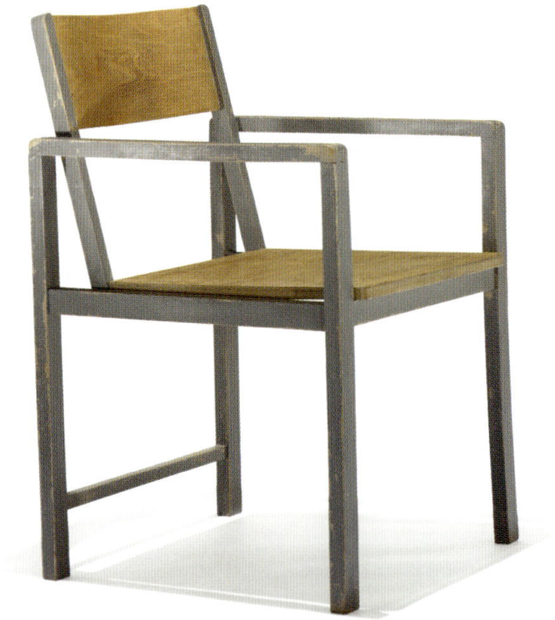

ABOVE Erich Dieckmann's Typenarmlehnstuhl, 1927, in painted beech wood.

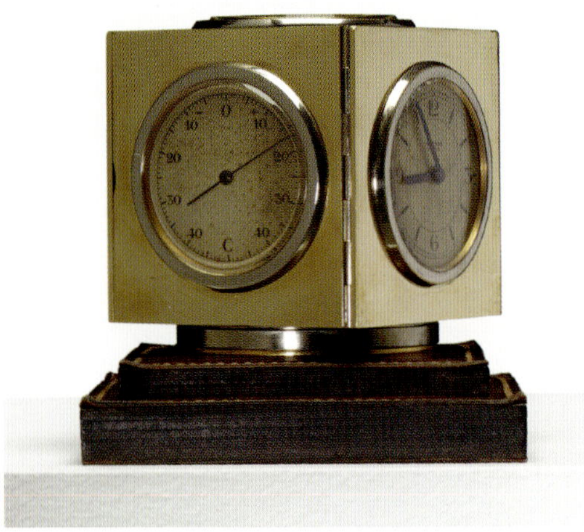

RIGHT A table clock designed by Paul Dupré-Lafon, produced by Hermès, c. 1940, in leather and brass; the clock also features a compass, calendar and thermometer.

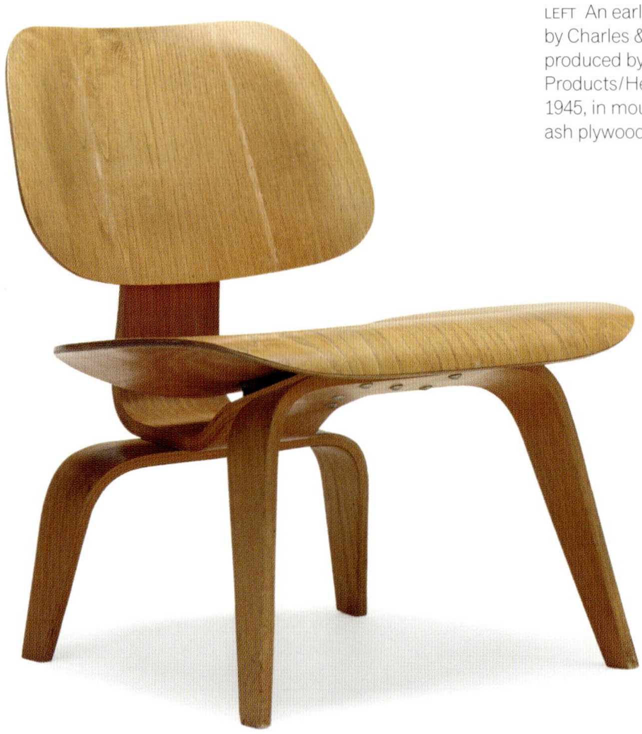

LEFT An early LCW chair by Charles & Ray Eames, produced by Evans Products/Herman Miller, 1945, in moulded ash plywood.

DEUTSCHER WERKBUND
The Deutscher Werkbund was an alliance of architects, designers and manufacturers, who founded the organization in 1907 to help promote modernist German design. Founder members included Peter Behrens, Bruno Paul and Josef Hoffmann*. Through yearbooks, publications and exhibitions, the Werkbund sought to promote a creative alliance between designers and industry, with an emphasis on factory production and utility. In 1927, under the guidance of Ludwig Mies van der Rohe, the Werkbund staged a landmark exhibition in Stuttgart known as 'The Dwelling' ('Die Wohnung'), the key element being an estate of modern houses (the 'Weissenhofsiedlung'*) designed by leading European architects. The Werkbund was closed by the Nazis in 1938, but re-established in 1949.

DIECKMANN, ERICH (1896–1944)
Bauhaus-trained furniture designer Erich Dieckmann is best known for his innovative timber chairs and armchairs in a range of standardized models. He was also a respected teacher of furniture design, both in Weimar and Halle.

DOESBURG, THEO VAN (1883–1931)
See p. 276.

DOREN, HAROLD VAN (1895–1957)
See p. 218.

DREYFUSS, HENRY (1904–1972)
See p. 222.

DUFRÊNE, MAURICE (1876–1955)
Having begun his career as a furniture designer with work in the style of Art Nouveau, Maurice Dufrêne later switched to Art Deco*. He exhibited at the Exposition Internationale des Arts Décoratifs et Industriels Modernes under the aegis of Galeries Lafayette's Maîtrise furniture studio, which he founded in 1921.

DUPRÉ-LAFON, PAUL (1900–1971)
French interior and furniture designer Paul Dupré-Lafon studied in Marseilles before settling in Paris. He worked predominantly in the Art Deco* style, with his portfolio including many private residences for wealthy clients; he also collaborated with Hermès.

EAMES, CHARLES (1907–1978) **& RAY** (1912–1988)
Known primarily for their work of the Fifties and Sixties, Charles & Ray Eames were key innovators in the field of 20th-century furniture design, as well as working across a variety of other mediums, including architecture, set design, interiors, textiles, graphics and film. They met at the Cranbrook Academy of Art in 1940 and in 1942 began experimenting with plywood, producing stackable plywood leg splints for the US Navy. Over the following years, Charles & Ray Eames began work on a series of ply chairs and other designs, followed by pioneering work in plastics and fibreglass.

EARL, HARLEY (1893–1969)
Having started his career as a car designer in the family automotive business, Harley Earl was taken on by Alfred P. Sloan, the head of General Motors, Detroit, in 1925. Earl and Sloan pioneered the idea of the annual car upgrade, achieved largely through relatively minor stylistic changes, creating a new unit devoted to auto aesthetics (known as the 'Style Section' from 1937). Earl continued to enjoy success in the Fifties, playing a key part in the development of iconic models. He is also intimately associated with the term 'Borax', referring to essentially superfluous and excessive streamlining or detailing.

EMBERTON, JOSEPH (1889–1956)
Architect Joseph Emberton was a leading British modernist and the only one to feature in Philip Johnson* and Henry-Russell Hitchcock's book, *The International Style* (1932), with his Royal Corinthian Yacht Club building in Essex (1931). Other key projects include his 1936 building on London's Piccadilly for Simpson's, the tailors (now a bookshop), and a number of Thirties buildings for Blackpool Pleasure Beach.

ERCOLANI, LUCIAN (1888–1976)
The Italian-born furniture designer and entrepreneur Lucian Ercolani worked for a number of British furniture makers, such as Frederick Parker's company in High Wycombe (later Parker-Knoll) and Edward Gomme's business (which went on to produce the G-Plan range), before founding Ercol in 1920. Many of the company's most famous designs in timber and plywood – some of which are still manufactured today – were designed by Ercolani himself.

ESHERICK, WHARTON (1887–1970)
Artist, sculptor and furniture maker Wharton Esherick began producing hand-crafted wooden furniture in the Twenties, influenced by Expressionism*. As a studio-based craftsman, he was somewhat removed from mainstream modernism, with its powerful emphasis on factory production. Yet his work of the Thirties onwards suggests an engagement with modernist ideas and a paring away of ornament, resulting in an emphasis on form and materiality. His work can also be compared with that of mid-century makers such as George Nakashima, who shared his love of organic materials and craft techniques.

EXPOSITION INTERNATIONALE DES ARTS DÉCORATIFS ET INDUSTRIELS MODERNES: SEE ART DECO

EXPRESSIONISM
An architectural and design movement associated with the early work of a number of key modernists, such as Erich Mendelsohn* and Peter Behrens, Expressionism embraced the notion of form as a vehicle for experiment and metaphor. These forms were sculptural and dynamic, rather than being based on the concept of 'form follows function'. As the modern movement evolved, this expressionistic approach was overwhelmed by a focus on function, rationality and geometric purity of line.

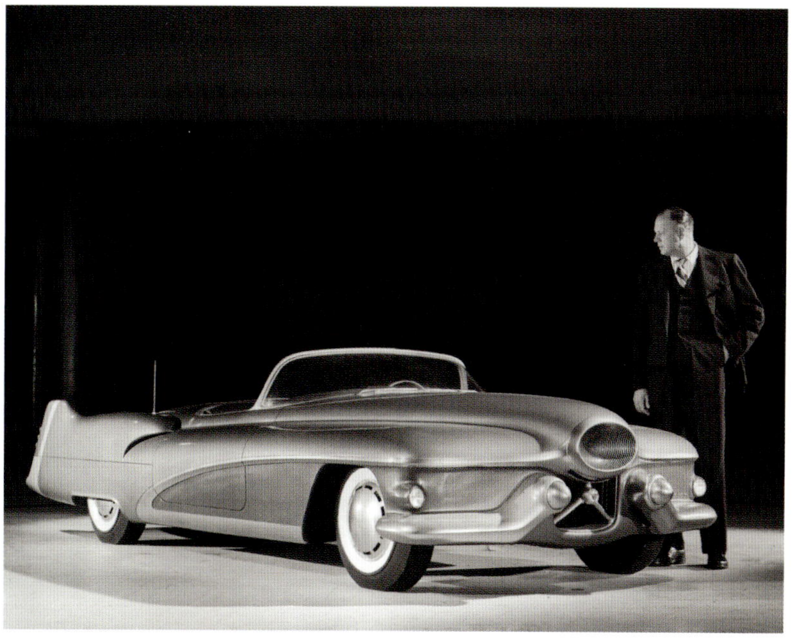

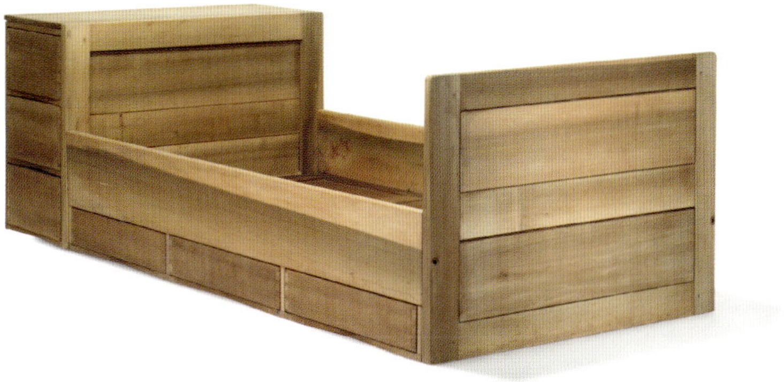

TOP Harley Earl of General Motors, with a full-scale model of the 1951 Buick Le Sabre sports car.

ABOVE A custom Captain's bed made by Wharton Esherick for the artist Julius Bloch, *c.* 1925, in poplar wood.

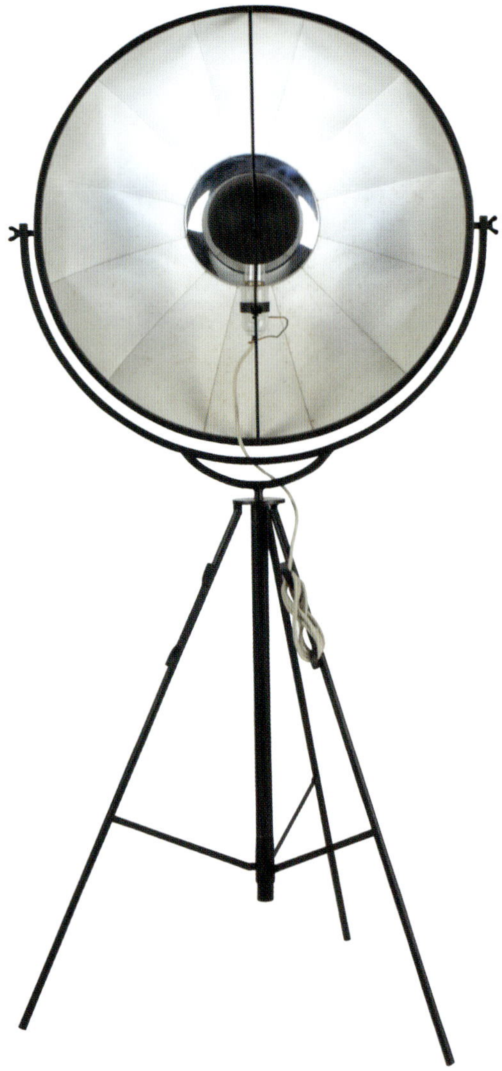

RIGHT A floor lamp designed in 1907 by Mariano Fortuny, manufactured by Pallucco in enamelled metal, steel and cloth.

FORD, HENRY (1863–1947)
Industrialist Henry Ford founded his eponymous motor company in Detroit in 1903. The Model T was first produced five years later, with 15 million cars sold by 1927, manufactured with assembly-line systems pioneered by Ford and sets of standardized component parts. The term 'Fordism' was coined to encapsulate the idea of factory manufacturing on a large scale, using moving assembly lines to maximize efficiency and reduce costs. These methods of mass production were soon being copied and reproduced by manufacturers across a range of industries worldwide.

FORTUNY, MARIANO (1871–1949)
The Spanish-born Mariano Fortuny was principally a textile and fashion designer, basing himself in Venice. However, he also worked in other fields of design, particularly lighting, and created a number of highly influential lamps and floor lights, some of which could be described as pioneering examples of proto-modernist design. His 'Fortuny' – or 'Projector' – floor light of 1907, in particular, with its tripod base and spotlight-style lantern, had a distinctly futuristic quality for the times.

FRANK, JEAN-MICHEL (1895–1941)
See p. 44.

FRANK, JOSEF (1885–1967)
Austrian architect and furniture designer Josef Frank worked in Vienna up until 1934 when he emigrated to Sweden with his wife. Here, he started collaborating with the Stockholm design company Svenskt Tenn, creating a series of textile and furniture designs. The most famous – and colourful – of Frank's textile patterns were conceived from 1942 to 1945, while he was teaching in New York.

FRANKL, PAUL THEODORE (1886–1958)
Paul Theodore Frankl studied architecture and engineering in Vienna, Berlin and Copenhagen before settling in New York in 1914. He established his own architectural and design office, with his work combining Art Deco* and modernist influences. Among his best-known furniture designs are his 'Skyscraper' pieces of the late Twenties, inspired by an architectural typology that he briefly championed.

FREY, ALBERT (1903–1998)
Modernist architect Albert Frey was born in Switzerland and worked with LE CORBUSIER before emigrating to America. After a period in New York, a commission took him to Palm Springs, California, where he worked on a series of private houses and other commissions in a style sometimes described as 'Desert Modernism'. Among his projects was a 1947 home for RAYMOND LOEWY.

BELOW Paul Theodore Frankl's Speed chair, produced by Frankl Studios, *c.* 1932, in leather, upholstery and lacquered wood.

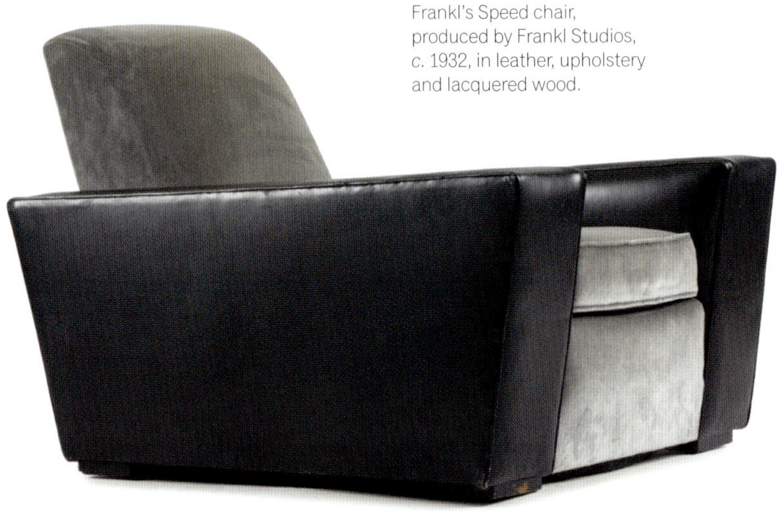

FRY, EDWIN MAXWELL (1899–1987)
Maxwell Fry (as he was known) studied architecture at Liverpool University and embraced modernism in the early Thirties. He formed a brief partnership with WALTER GROPIUS from 1934 to 1936, before the former director of the Bauhaus moved on to America. A number of Fry's key projects date from the late Thirties, including the Kensal House apartment building in London (1936) and Miramonte, a private home in Kingston-upon-Thames, Surrey (1937). Fry was active in the Modern Architectural Research Group (MARS)* and later worked with his wife, architect Jane Drew, in West Africa and India, where they played a key part in the evolution of Chandigarh, the new city designed by LE CORBUSIER and PIERRE JEANNERET.

FULLER, RICHARD BUCKMINSTER
(1895–1983)
See p. 226.

FUNCTIONALISM
The principles of Functionalism derived from the idea that 'form follows function' – a phrase first attributed to architect Louis Sullivan* in an essay from 1896. Functionalist architects (within the modernist umbrella) took this idea further from the 1920s onwards, fusing it with the notion that formal architectural solutions emerge from social needs and requirements, resulting in a semi-organic and evolutionary process of creativity. At times, these solutions could become dramatic and even extreme, overlapping with Expressionism*. FRANK LLOYD WRIGHT, HANS SCHAROUN and ALVAR AALTO have all been labelled as Functionalists.

FUTURISM
The Italian Futurists formed one of the regional modernist sub-movements, which emerged during the early decades of the 20th century and embraced art, architecture, literature and other artistic mediums. The movement was founded by poet Filippo Tommaso Marinetti in Milan in 1909 with a Futurist manifesto, the first in a series of provocative manifesto statements. Radical Futurists took an uncompromising view of modernity, turning their backs upon the past and embracing machine-age technology and industrialization in all its forms, while seeking original modes of expression that mirrored this new era. Futurist architects built a number of statement buildings in Italy during the Twenties, although many were suspicious of what they saw as Mussolini's enduring sympathies for neoclassical forms, which sat uneasily alongside his interest in a fascist interpretation of modern architecture (associated with the work of a number of Italian Rationalist* architects).

GEDDES, NORMAN BEL
(1893–1958)
See p. 230.

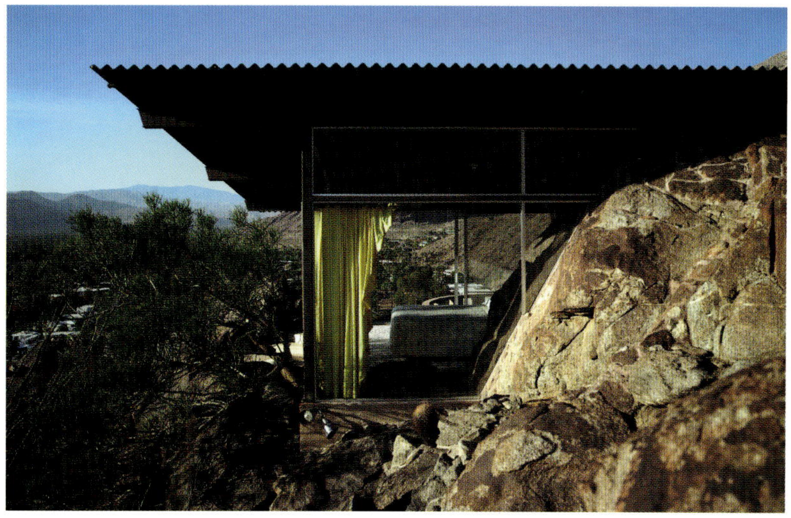

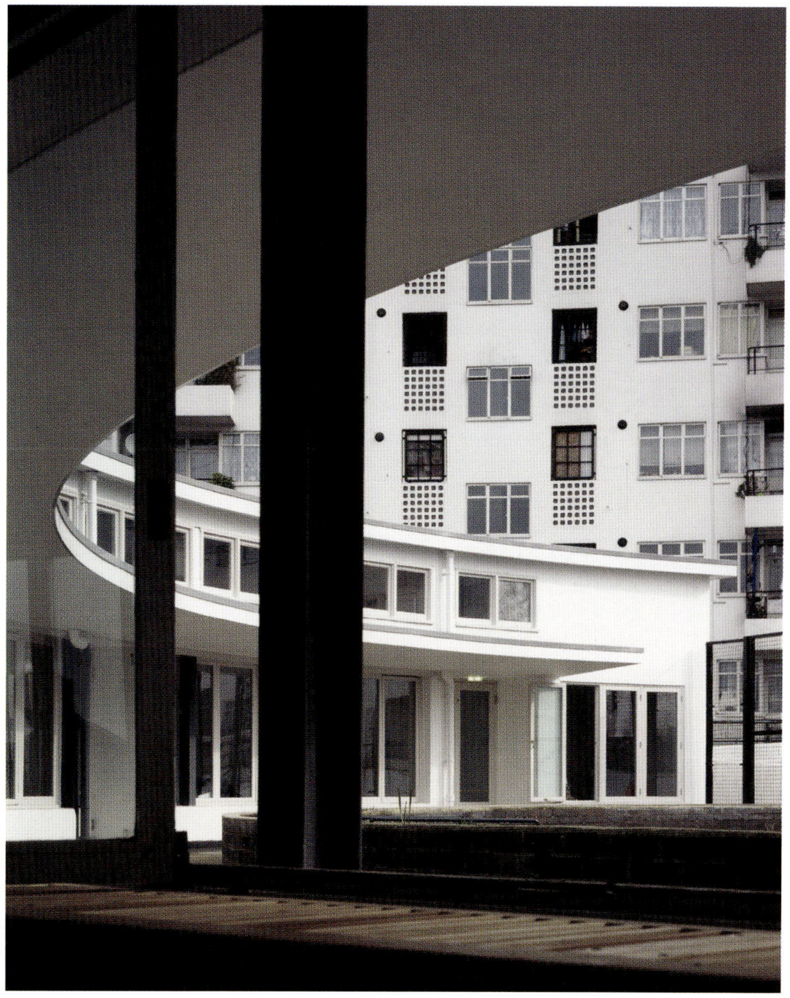

TOP Architect Albert Frey's self-designed home, Frey House II, in Palm Springs, California.

ABOVE The Kensal House Day Nursery in Kensal Rise, London, designed by architect Edwin Maxwell Fry, 1938, with the Kensal House apartment block beyond.

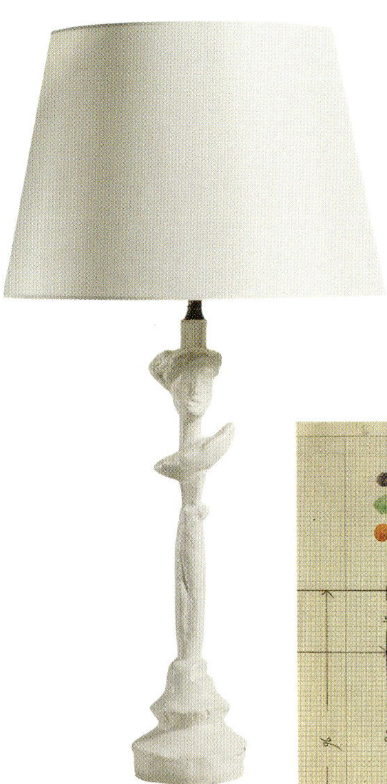

ABOVE A light attributed to Alberto & Diego Giacometti, *c.* 1935, in plaster and brass with a linen shade.

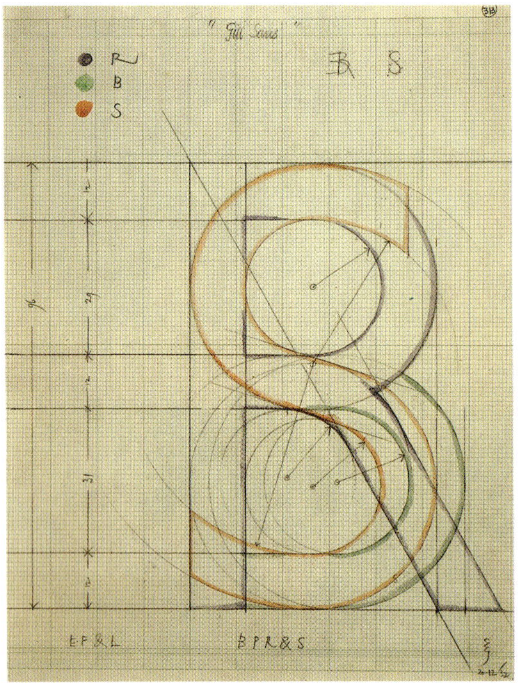

ABOVE A drawing by the English artist and designer Eric Gill, representing his Gill Sans typeface.

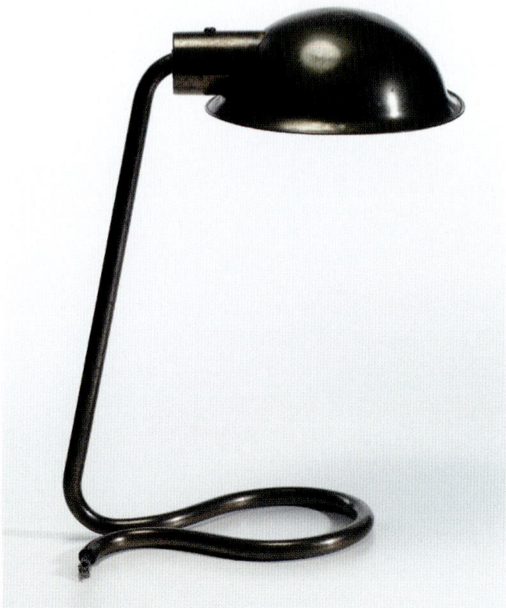

LEFT A Model 431 brass desk lamp designed and manufactured in the Netherlands by W. H. Gispen, 1935.

GIACOMETTI, ALBERTO (1901–1966) **& DIEGO** (1902–1985)
Artists and designers Alberto & Diego Giacometti were born in Switzerland and moved to Paris in the early Twenties. They produced lighting, vases and design pieces for Maison Desny*, JEAN-MICHEL FRANK and others, along with sculptural commissions. The brothers increasingly focused on sculpture and fine art from the mid-Thirties onwards with growing success; in later years Diego continued to work on furniture and lighting pieces, as well as fine art.

GIACOSA, DANTE (1905–1996)
Italian car designer Dante Giacosa began working at Fiat in 1928, becoming the company's lead engineer in 1946. Among the models that he designed were the Topolino (Fiat 500) of 1936, and its successors the Fiat 600 (1955) and the iconic Fiat 500 Cinquecento of 1957.

GIBBERD, FREDERICK (1908–1984)
English modernist architect Frederick Gibberd's best-known works of the Thirties include a series of apartment buildings, such as Pullman Court in South London (1936). During the early Forties, he was principal of the Architectural Association school.

GILL, ERIC (1882–1940)
The English sculptor, artist, architect and typographer Eric Gill is principally associated with the Arts and Crafts movement and its values, yet he also sought to recognize the importance of industrialization and mechanization, even if he regarded these with suspicion. In 1928, he designed one of the most successful sans serif typefaces of the early modernist period. Widely used by Penguin Books and the London and North Eastern Railway, among others, the 'Gill Sans' typeface – Gill's attempt to design a 'fool proof' alphabet suited to an age of mechanical reproduction – was embraced for its particular clarity and legibility. There were also sculptures for the BBC's Broadcasting House in London and the Midland Hotel in Morecambe, designed by Oliver Hill*. Gill's only completed solo architectural work is the Catholic church of St Peter the Apostle in Gorleston, Norfolk (1939).

GISPEN, W. H. (1890–1981)
Willem Hendrik Gispen grew up in Amsterdam and Utrecht and went on to study in Rotterdam, where he founded his own atelier in 1916. This became a factory and workshop, largely devoted to producing Gispen's own designs. These included tubular steel furniture and his Giso lighting collection. During the Thirties Gispen also opened showrooms in a number of major European cities. The company supplied much of the furniture in the landmark Sonneveld House (1933), also in Rotterdam, by architects BRINKMAN & VAN DER VLUGT.

GOLDFINGER, ERNŐ (1902–1987)
See p. 336.

GRAY, EILEEN (1878–1976)
See pp. 48 and 340.

GRETSCH, HERMANN (1895–1950)
Stuttgart-based designer and architect Hermann Gretsch is best known for his ceramics of the Thirties. These included his Form 1382 dinnerware service (1931) made of undecorated white porcelain, produced by Porzellanfabrik Arzberg.

GROPIUS, WALTER (1883–1969)
See p. 348.

GWYNNE, PATRICK (1913–2003)
English architect Patrick Gwynne first became fascinated with the work of the pioneering European modernists while at school in Harrow. After his studies, he found a post with Wells Coates*. His most famous building is The Homewood (1938) in Esher, a house designed for his parents, with a scheme that owes an acknowledged debt to two modernist icons: LE CORBUSIER's Villa Savoye and LUDWIG MIES VAN DER ROHE's Villa Tugendhat. After the war, and his parents' death in 1942, Gwynne set up his own architectural practice, based in his father's old study. He maintained The Homewood with care, bequeathing it to the National Trust. Gwynne received commissions for other private houses, including The Firs in Hampstead (1958), and an extension to the Theatre Royal in York (1967). He also designed motorway service station buildings and the Serpentine Restaurant in Hyde Park (1964).

HEAL, AMBROSE (1872–1959)
English designer and retailer Ambrose Heal became managing director of the family firm, Heal & Son, in 1907 and then chairman in 1913, steering the company through a golden period. As a designer, Heal embraced the Arts and Crafts tradition and worked primarily in wood. But he also became increasingly interested in modernist design, introducing Scandinavian and Art Deco* furniture into the London store, as well as Isokon* plywood pieces designed by MARCEL BREUER and some examples of tubular steel furniture. He was also instrumental in commissioning the landmark Heal's building on Tottenham Court Road, designed by his cousin Cecil Brewer, which opened in 1916.

HEARTFIELD, JOHN (1891–1968)
Artist, illustrator and graphic designer John Heartfield (born Helmut Herzfeld) grew up in Berlin and studied art in Munich. He joined the German Communist Party in 1918, and became involved in the Dada movement shortly afterwards. Heartfield worked in theatre set design and created book jackets, but became known primarily as an illustrator for satirical magazines, exploring the medium of photomontage to full effect and often within an anti-fascist agenda. After the rise of the National Socialist Party, Heartfield fled to Czechoslovakia and then, in 1938, to England; he returned to East Berlin in 1950.

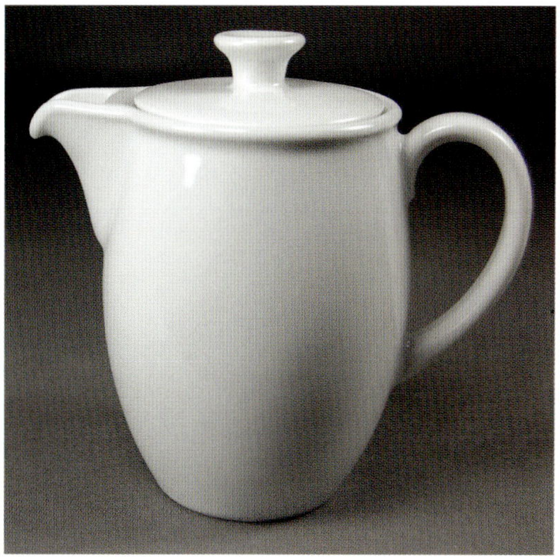

LEFT A porcelain teapot designed by Hermann Gretsch, produced by Porzellanfabrik Schönwald, 1936.

BELOW The staircase at The Homewood, the 1938 country house in Esher, Surrey, designed by the English architect Patrick Gwynne.

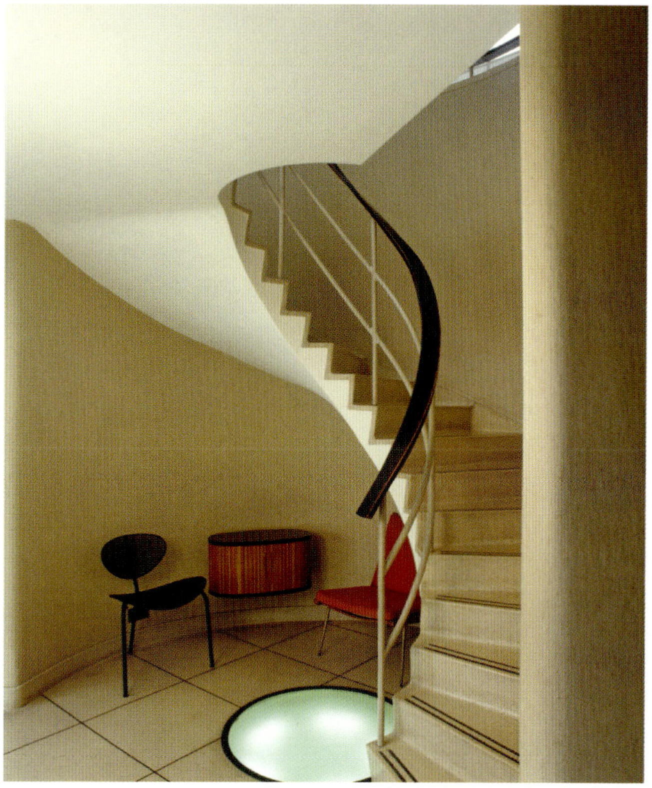

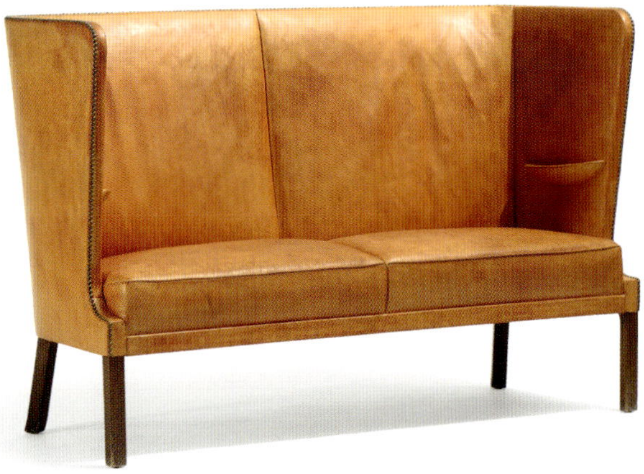

ABOVE A settee by the Danish designer Frits Henningsen, *c.* 1943, in leather, stained oak and brass.

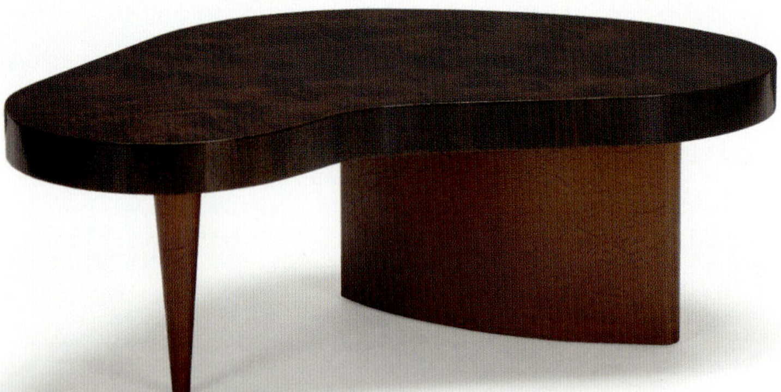

ABOVE A Model 4186 Paldao coffee table designed by Gilbert Rohde and made by Herman Miller, 1940, in paldao wood, acacia burl and vinyl.

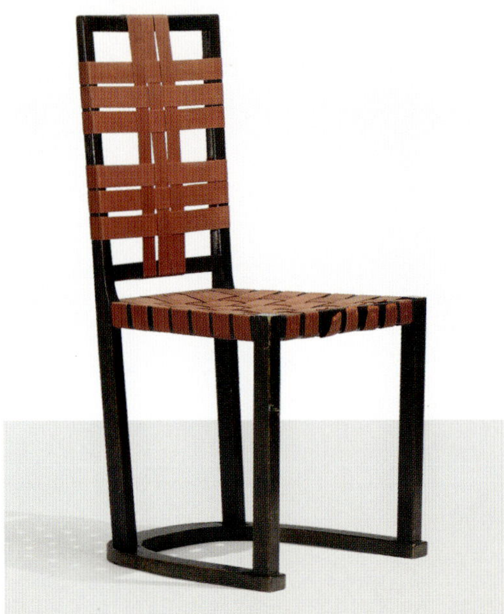

RIGHT The Futurum chair designed by Axel Einar Hjorth and made in Sweden by Nordiska Kompaniets Verkstäder, *c.* 1928, in lacquered wood and canvas.

HENNINGSEN, FRITS (1889–1965)
Frits Henningsen was a Danish furniture designer, who studied under KAARE KLINT and others in Copenhagen, where he founded his own atelier and store. While influenced by historical tradition, he increasingly veered towards modernism from the Thirties onwards, particularly within his crafted chairs and occasional tables. A number of his pieces – including his Signature chair (1954) – have been reissued in recent years.

HENNINGSEN, POUL (1894–1967)
See p. 132.

HERBST, RENÉ (1891–1982)
See p. 54.

HERMAN MILLER
The Herman Miller furniture company was founded in 1923, when D. J. De Pree and his father-in-law, Herman Miller, took over and renamed a furniture manufacturing company in Zeeland, Michigan. The company turned from period-inspired reproduction furniture to modern designs in 1930, after De Pree began working with GILBERT ROHDE; by 1936 the company was focusing exclusively on its modern range. Following Rohde's death in 1944 George Nelson became the chief creative force at the company, initiating collaborations with a number of key figures in the world of mid-century design.

HILL, OLIVER (1887–1968)
Among the most familiar designs by British modernist architect Oliver Hill stand the landmark Midland Hotel, situated by the sea in Morecambe (1933). Other projects included private houses and a series of residences on the Frinton Park Estate, Essex (1934–38).

HJORTH, AXEL EINAR (1888–1959)
Axel Einar Hjorth studied at the Konstfack art and design school in Stockholm before embarking upon a career as a furniture designer. In 1927 his work was included in an exhibition of contemporary Swedish design at the Museum of Modern Art in New York and he was hired by the department store Nordiska Kompaniet to design furniture for the company, eventually becoming chief designer. Hjorth's work of the late Twenties and Thirties, largely in timber, exhibited a modernist influence.

HOFF, ROBERT VAN 'T (1887–1979)
A Dutch architect and furniture designer, Robert van 't Hoff studied architecture in Holland and England; he also travelled to America to see the work of FRANK LLOYD WRIGHT. One of his first modernist houses, Villa Henny (1919), made a marked impact and is regarded as a pioneering achievement in the early use of reinforced concrete construction. He became associated with the De Stijl* group from around 1917 onwards, and also started designing furniture around the same time. He soon became disillusioned with De Stijl, however, and from 1922 onwards based himself largely in Britain.

HOFFMANN, JOSEF (1870–1956)
The work of Viennese architect and designer Josef Hoffmann was to have a profound influence on many early modernists, with Hoffmann – rather like Adolf Loos* – often described as one of the godfathers of the movement. Having trained under Otto Wagner* and worked with him for a time, Hoffmann began his career as an independent architect and designer, as well as teaching in Vienna. He was a key figure in the foundation of the Wiener Secession, the Wiener Werkstätte and the Deutscher Werkbund*. Increasingly and progressively, Hoffmann's work – which included furniture, lighting and ceramics – dispensed with ornament in favour of geometrical precision and purity of form.

HOLDEN, CHARLES (1875–1960)
British architect Charles Holden is best remembered for the design of a number of striking London Underground stations from the Thirties; these include Arnos Grove (1932) and Southgate (1933).

INTERNATIONAL STYLE
Curators Philip Johnson* and Alfred H. Barr came up with the term 'International Style' in conjunction with an exhibition at the Museum of Modern Art, New York, in 1932, and an associated book. They sought to sum up a borderless movement of pioneering modernist architects – including Ludwig Mies van der Rohe, Walter Gropius and Le Corbusier – who worked within a vocabulary of crisp modernity, using rectilinear forms, flat roofs, expanses of glass, and other hallmarks such as the curtain wall and the universal space (open-plan living). The term was widely used for a time but did not embrace the growing interest in contextuality advocated by a number of key modernists, including Nordic masters such as Alvar Aalto.

ISOKON
The Isokon Furniture Company was founded by entrepreneur Jack Pritchard in 1935 in London. Pritchard collaborated on designs with Bauhaus exiles Walter Gropius and Marcel Breuer during their brief period living and working in Britain. Breuer's innovative plywood pieces for Isokon – including his famous Long chair (1936) – became the backbone of the collection in the late Thirties. Pritchard also worked with Wells Coates* on the development of the Isokon apartment building in Belsize Park (1934; see p. 444). A number of Breuer's designs are still produced by Isokon Plus.

ITTEN, JOHANNES (1888–1967)
Swiss artist, writer and colour theorist Johannes Itten was a master at the Bauhaus from 1919 to 1922, initially teaching on the foundation course. Afterwards, he continued to teach in various schools in Germany and Switzerland. He is primarily known for his seminal work in colour theory and his book *The Art of Color* (1961).

JACOBSEN, ARNE (1902–1971)
See p. 358.

JEANNERET, PIERRE (1896–1967)
See p. 64.

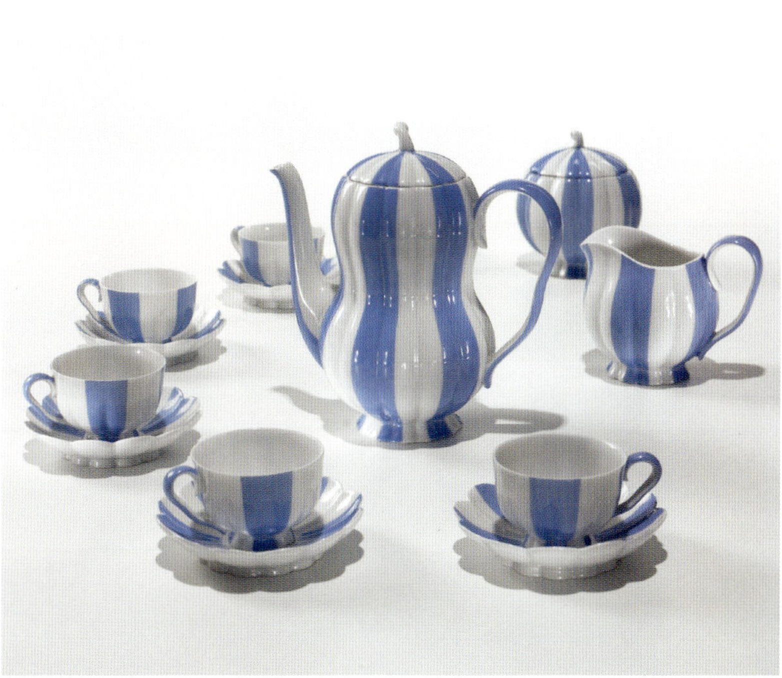

ABOVE Josef Hoffmann's Mocca coffee service in porcelain, produced by Wiener Porzellan-Manufaktur in Austria, *c.* 1933.

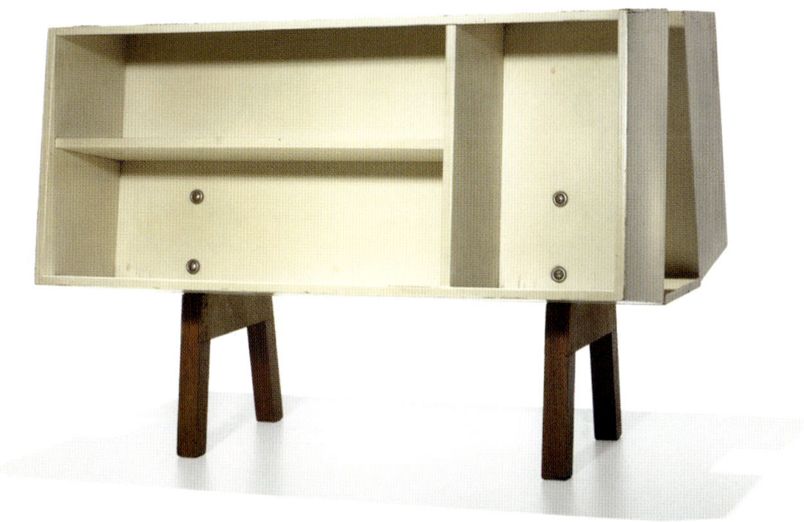

ABOVE Isokon's Penguin Donkey book case, designed by Ernest Race, *c.* 1940, in lacquered wood and mahogany.

RIGHT A bar stool from the Four Seasons Grill Room at the Seagram Building in New York, designed by Philip Johnson and Ludwig Mies van der Rohe, in chrome-plated steel and vinyl, 1958.

BELOW A pair of armchairs by Francis Jourdain, produced in France, *c.* 1940, in oak, rush and enamelled steel.

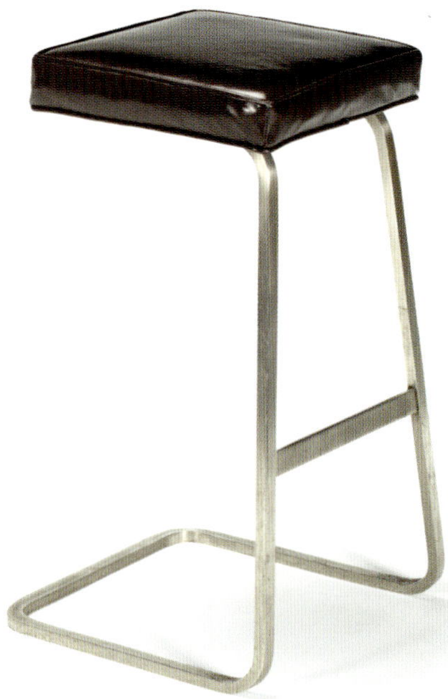

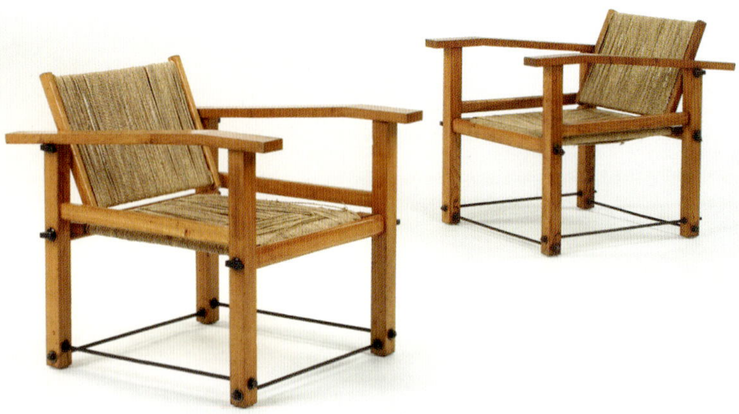

JOHNSON, PHILIP (1906–2005)
American architect and designer Philip Johnson's most famous building was his own Glass House (1949) in New Canaan, Connecticut, which was largely based on principles and ideas developed by LUDWIG MIES VAN DER ROHE, whom Johnson greatly admired. Johnson worked as an associate architect on Mies's Seagram Building in New York (1958), as well as developing his own practice and a career that explored a number of architectural styles over the years. There were also formative periods spent working as a curator at the Museum of Modern Art, where, in 1932, Johnson was the co-curator of a landmark exhibition on the International Style*.

JOURDAIN, FRANCIS (1876–1958)
Francis Jourdain was a French furniture designer, ceramicist and painter. He opened his Paris atelier in 1912 and an eponymous store in 1919. Working primarily in timber, Jourdain made furniture that was characterized by its simplicity and emphasis on form and texture rather than luxurious finishes. He collaborated with architect ROBERT MALLET-STEVENS during the late Twenties.

JUGENDSTIL
The German version of Art Nouveau; the term translates as the 'young style'. This turn-of-the-century movement was known for its flamboyant forms and decorative exuberance, partly inspired by the natural world.

KÅGE, WILHELM (1889–1960)
A Swedish artist, graphic designer and ceramicist, Wilhelm Kåge began working with Gustavsberg in 1917, having been appointed art director. He played a key role at the pottery, introducing a modern and more experimental outlook, and balancing more commercial collections (such as his Praktika tableware of the Thirties) and studio pieces. He continued working with Gustavsberg in the post-war period, eventually succeeded by his protégé Stig Lindberg.

KANDINSKY, WASSILY (1866–1944)
The celebrated Russian-born artist Wassily Kandinsky worked at the Bauhaus from 1922 to 1933, teaching on the foundation course as well as giving classes in art theory. After the dissolution of the Bauhaus, he settled in France, becoming a French citizen.

KLEE, PAUL (1879–1940)
Like his colleague Wassily Kandinsky*, the Swiss-born artist Paul Klee taught at the Bauhaus, from 1921 to 1931. He was a 'form' master in the stained glass and mural workshop, a department that also included Johannes Itten*, and exhibited his work around this time as a member of 'The Blue Four', along with Kandinsky. Towards the end of his life, Klee moved back to Switzerland.

KLINT, KAARE (1888–1954)
See p. 58.

KNOLL, HANS (1914–1955) **& FLORENCE** (b. 1917)
The son of a German furniture manufacturer, Hans Knoll emigrated to America in 1937 and founded the Hans G. Knoll Furniture Company the following year in New York, initially collaborating with designer Jens Risom*. In 1941 he met architect and designer Florence Schust, who had trained at the Cranbrook Academy of Art. They married in 1946 and ran the company, which became Knoll Associates, together until Hans's death in 1955. The Knoll collection grew rapidly in the post-war period, with iconic pieces by a number of designers that Florence Knoll had met at Cranbrook, including Eero Saarinen and Harry Bertoia.

KOCH, MOGENS (1898–1992)
See p. 62.

KRAMER, FERDINAND (1898–1985)
Born in Frankfurt, Ferdinand Kramer studied architecture in Munich, followed by a brief period at the Bauhaus. He began designing furniture in the mid-Twenties, in conjunction with his architectural work on 'New Frankfurt' housing projects. Key pieces from this time include his Theban daybed (1925) and Karnak chair (also 1925), along with his B403 chairs for Thonet* (1927). He emigrated to America in 1938 and worked for a time with NORMAN BEL GEDDES, returning to Germany in 1952.

KREHAN, MAX (1875–1925)
German ceramicist Max Krehan served as the craft master ('Lehrmeister') at the Weimar Bauhaus pottery workshop, located in nearby Dornburg. He taught alongside 'form' master Gerhard Marcks*, and his students included MARGUERITE WILDENHAIN, THEODOR BOGLER and OTTO LINDIG.

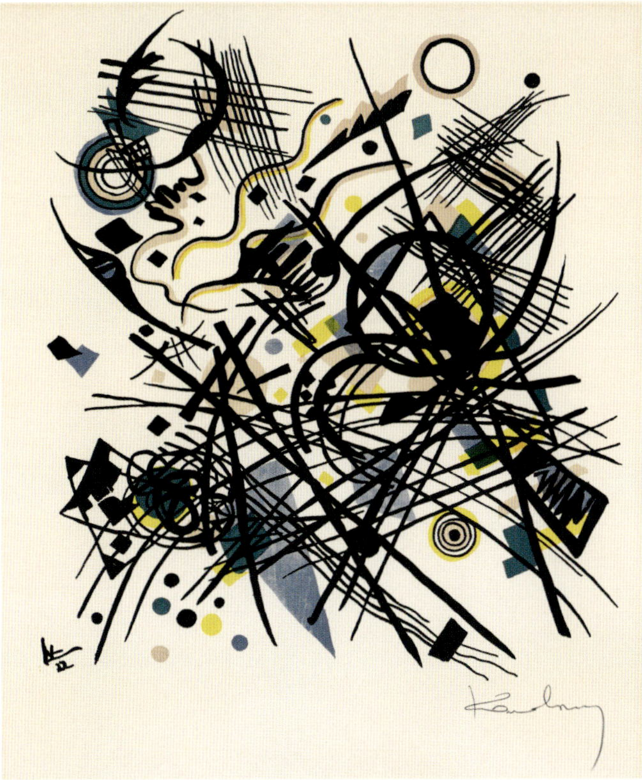

ABOVE A lithograph print from Wassily Kandinsky's Bauhausmappe Portfolio, Vol. 4, 1922.

BELOW A pair of Model 127 nightstands designed by Florence Knoll and manufactured by Knoll, 1948, in birch.

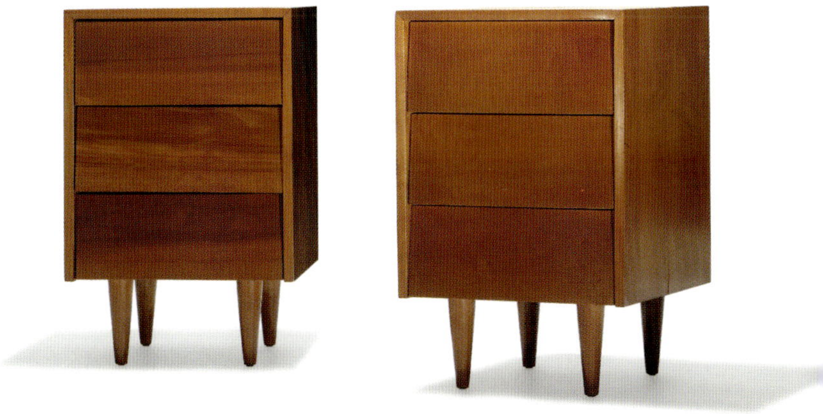

RIGHT Model B403 chairs by Ferdinand Kramer, manufactured by Thonet, 1927, in steam-bent and lacquered beech.

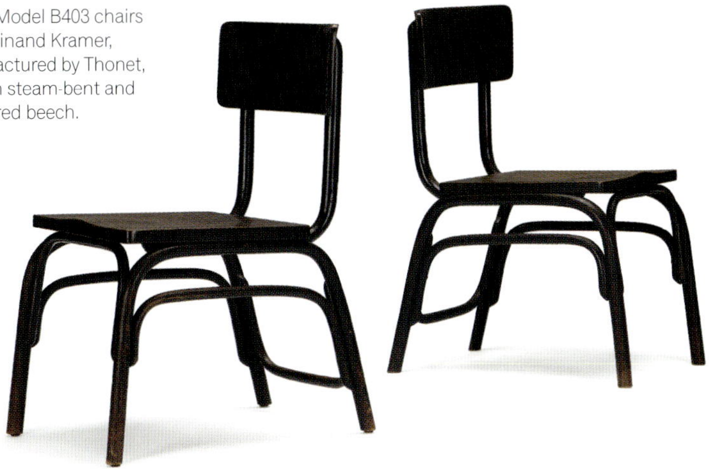

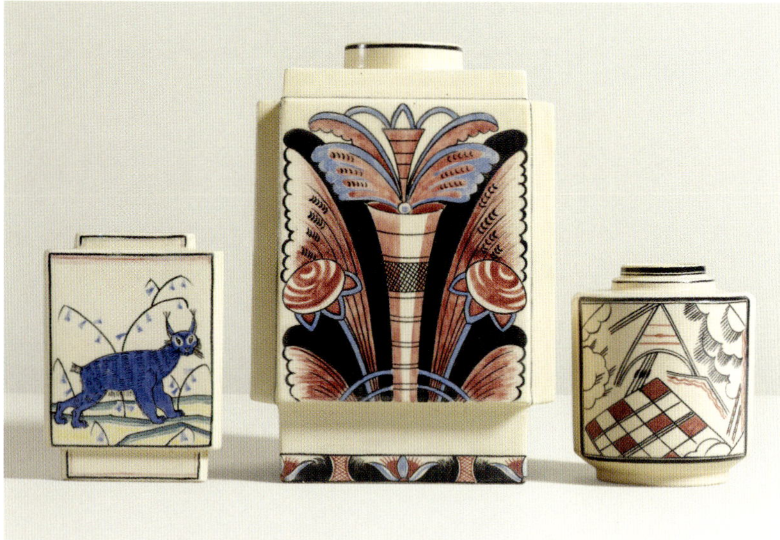

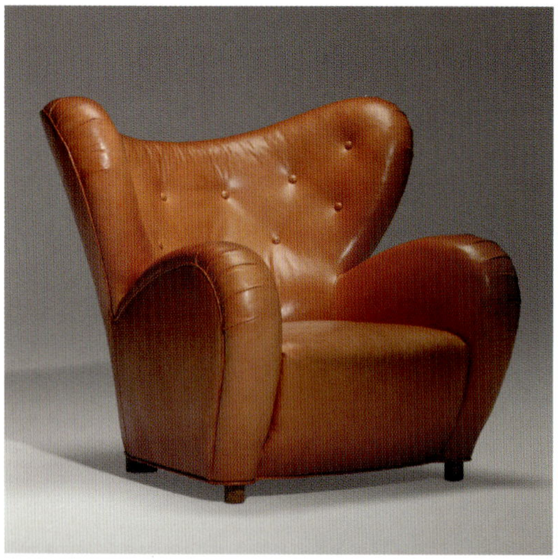

ABOVE A collection of three vases by Robert Lallemant, produced in France, *c.* 1928, in glazed stoneware.

LEFT Flemming Lassen's Tired Man lounge chair, manufactured in Denmark by A. J. Iversen Snedkermester, *c.* 1935, in leather and oak.

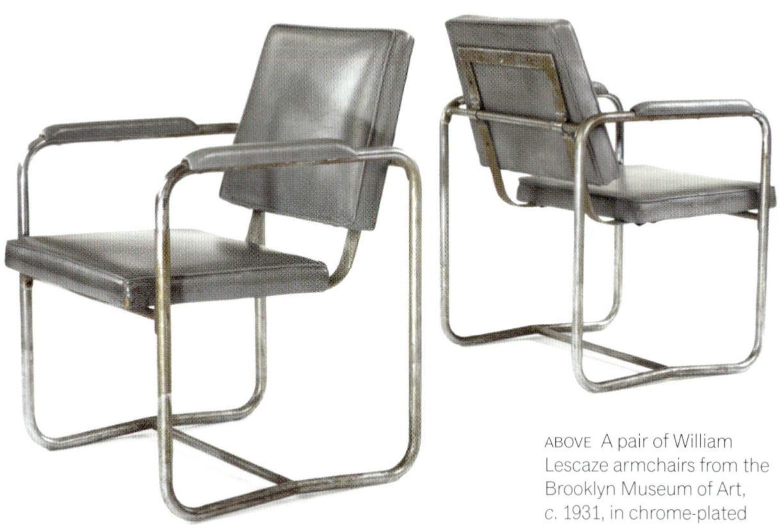

ABOVE A pair of William Lescaze armchairs from the Brooklyn Museum of Art, *c.* 1931, in chrome-plated steel and vinyl.

LALLEMANT, ROBERT (1902–1954)
French ceramicist Robert Lallemant is known for a flourish of distinctive work dating from around 1925, when he opened his Parisian studio, to around 1930. The work combined Art Deco*, Cubist and machine-age modernist influences. By 1933 he had switched direction and devoted himself to furniture and interior design.

LASSEN, FLEMMING (1902–1984)
Danish modernist architect Flemming Lassen worked with ARNE JACOBSEN on the design of a 'House of the Future' exhibited at a housing exhibition in Copenhagen in 1929; in subsequent years he collaborated with both Jacobsen and Erik Møller on a number of architectural projects. Lassen also designed furniture, including 'The Tired Man' chair of 1935 and 'Flemming's Chair' (1938); a number of his designs, along with pieces by his brother Mogens Lassen, have been reissued in recent years.

LE CORBUSIER (1887–1965)
See pp. 64 and 366.

LEACH, BERNARD (1887–1979)
See p. 164.

LECK, BART VAN DER (1876–1958)
Bart van der Leck was a Dutch furniture and interior designer, ceramicist and artist closely associated with the De Stijl* movement, which he co-founded with THEO VAN DOESBURG and Piet Mondrian.

LÉGER, FERNAND (1881–1955)
The French Cubist painter Fernand Léger was greatly influenced by the machine age through his 'mechanical period' of the Twenties. In 1920 he became a friend of LE CORBUSIER, who also influenced his work, including a series of murals suited to architectural contexts. In the Thirties he spent time in America, and in 1938 was asked to work on the interiors decoration of an apartment for Nelson Rockefeller; he remained in America during World War II.

LESCAZE, WILLIAM (1896–1969)
William Lescaze was a Swiss-born modernist architect who settled in America in 1920 and founded his own architectural practice in New York in 1923. His Philadelphia Savings Fund Society (PSFS) Building of 1932 – designed in association with George Howe – was one of the first modernist skyscrapers in the city and was described as an exemplar of the International Style*. He also designed a number of buildings at Dartington Hall in Devon, England, in the early Thirties.

LIBERA, ADALBERTO (1903–1963)
A leading Italian modernist architect, Adalberto Libera studied in Parma and Rome, and became a founder of the 'Italian Movement for Rational Architecture' (MIAR). He maintained links with Mussolini's fascist government, designing a number of State-sponsored buildings during the Thirties. Among his most famous projects was a dramatic villa on the island of Capri for the writer Curzio Malaparte (1938); after the two men fell out during the project, Malaparte claimed authorship of the house.

LINDIG, OTTO (1895–1966)
See p. 168.

LISSITZKY, EL (1890–1941)
An artist, graphic designer, typographer, architect and furniture designer, Lazar Markovich Lissitzky was born in Russia and studied architectural engineering in Germany, continuing his studies in Moscow. Settling in Vitebsk, he became – under the guidance of artist Kazimir Malevich* – a leading figure in the Suprematist* movement, a predecessor of Constructivism*, dominated by geometric abstraction. His most famous graphic design work of the Russian Revolutionary period combined support for the Communist cause with an avant-garde aesthetic. During the early Twenties he lived in Berlin as a Russian cultural ambassador, forging links with members of the Bauhaus and De Stijl*, while continuing his work as a graphic designer and illustrator.

LOEWY, RAYMOND (1893–1986)
See p. 236.

LOOS, ADOLF (1870–1933)
Viennese architect, designer and teacher Adolf Loos was one of the great proto-modernists and an influential figure within the early evolution of the movement. As well as his architectural work and furniture, his lecture/essay 'Ornament & Crime' of 1908 – arguing against the excessive decoration of the Art Nouveau style and in favour of functional and formal purity – had a great impact on a younger generation of architects and designers.

LUBETKIN, BERTHOLD (1901–1990)
The son of a businessman, Berthold Lubetkin was born in Tblisi, in Georgia. He studied in St Petersburg and Moscow, then in Berlin, Austria and Poland, where he completed a diploma in architecture. He moved to Paris in 1925, working for a time with AUGUSTE PERRET, then to London in 1931, where he founded the architectural practice Tecton*. Early commissions included the Gorilla House at London Zoo and the Penguin Pool of 1934, developed with Ove Arup*. Buildings at Dudley Zoo followed, and at Whipsnade, where Lubetkin designed Bungalow A (or Hillfield) for himself (1935). The landmark Highpoint I – constructed in Highgate, London, to provide businessman Sigmund Gestetner's workers with affordable housing – was completed in 1935 and contained 60 apartments in an H-shaped building on concrete piloti. After the land next door was acquired, Highpoint II (1938) was built, with 12 luxurious apartments, including a penthouse for Lubetkin and his family.

RIGHT A plate attributed to El Lissitzky, *c.* 1925, in glazed stoneware.

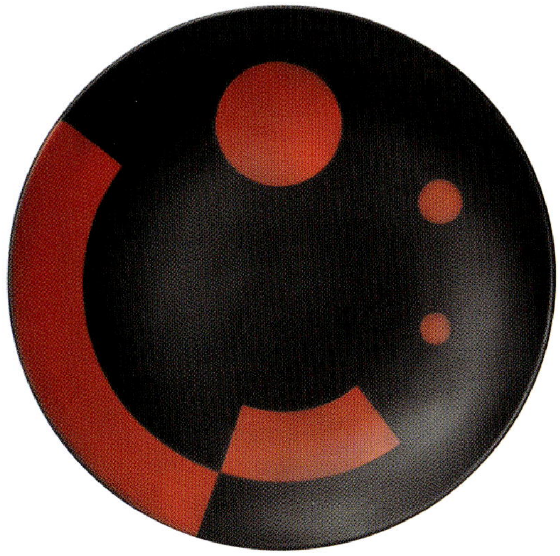

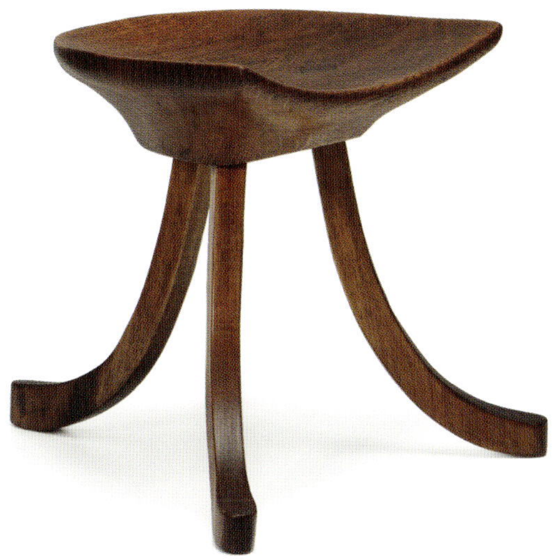

ABOVE A stool by Adolf Loos, made in Austria by J. Veillich, 1903, in lacquered mahogany.

RIGHT The canopied entrance to the Highpoint II apartment building in Highgate, London, designed by Berthold Lubetkin, 1938.

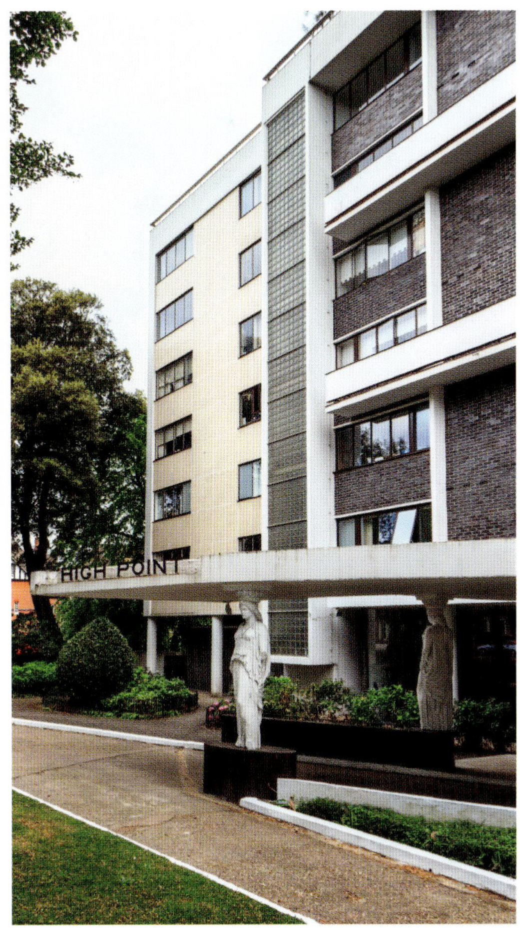

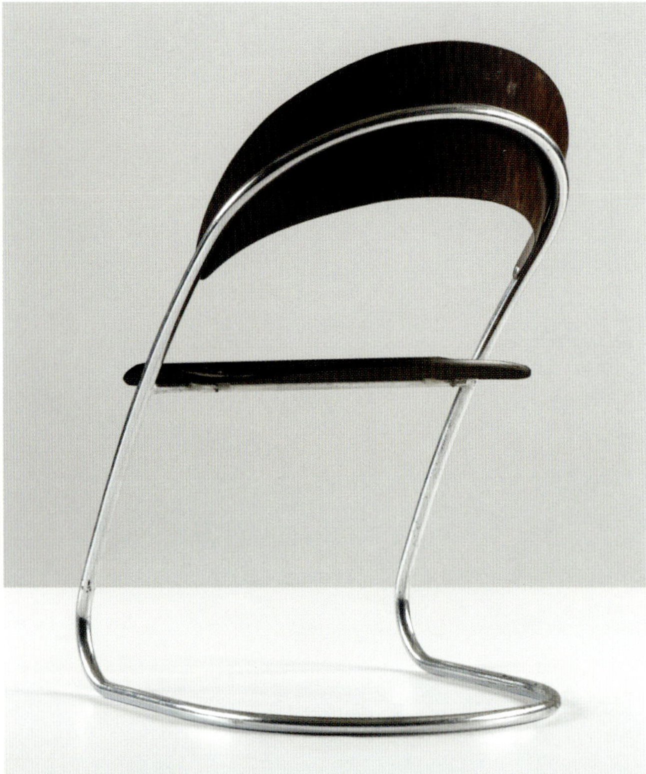

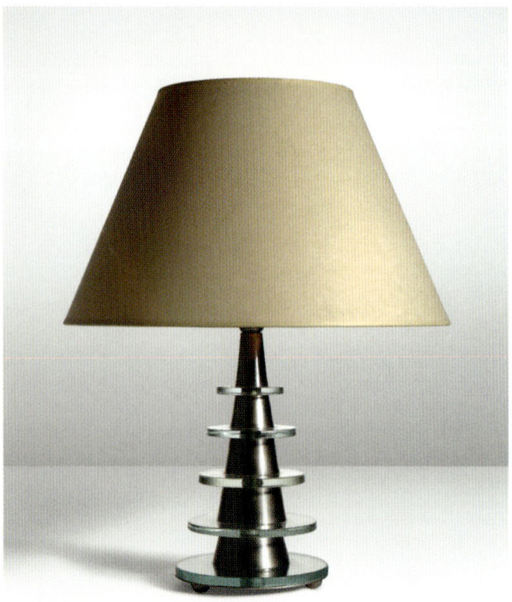

ABOVE An ST14 chair designed by Hans Luckhardt, manufactured by Thonet, 1932, in chrome-plated steel, stained ash and plywood.

RIGHT A table lamp by Maison Desny, produced in France, c. 1930, in chrome-plated metal, glass and mirror glass.

LUCKHARDT, HANS (1890–1954)
Hans Luckhardt was a German modernist architect, who practised with his brother Wassili in Berlin, where many of their key projects – mostly housing and office buildings – were located (some were destroyed during World War II). Hans Luckhardt was also known as a furniture designer, producing a number of innovative designs in tubular steel, such as his cantilevered ST 14 chair for Thonet* (1932), with a steel frame and plywood seat.

MAISON DESNY
This highly respected Parisian design studio was founded by Desnet and René Nauny in 1927 and closed in 1933. During its short but accomplished existence, the studio became renowned for its lighting designs and silverware, as well as furniture, rugs and other homeware, in addition to its interior design work. The atelier drew upon the talents of collaborators such as Alberto & Diego Giacometti* and ROBERT MALLET-STEVENS.

MALEVICH, KAZIMIR (1878–1935)
Russian artist and graphic designer Kazimir Malevich was the founder of Suprematism*, whose followers included El Lissitzky* among others. With its emphasis upon geometric abstraction and graphic purity, Suprematism was an influential proto-modernist movement and made an impression upon Bauhaus designers and theorists such as LÁSZLÓ MOHOLY-NAGY, who designed the cover for a book on Malevich's work published by the Bauhaus in 1927.

MALLET-STEVENS, ROBERT (1886–1945)
See p. 374.

MARCKS, GERHARD (1889–1981)
A German ceramicist, designer and sculptor, Gerhard Marcks taught at the Bauhaus pottery workshop in Dornburg, near Weimar, serving as its master of form from 1920 until 1924 (alongside Max Krehan*). Later, he taught sculpture and became head of the School of Applied Arts in Burg Giebichenstein. He resumed his teaching career in Hamburg after the war and a prolonged period in which his work was dismissed as 'degenerate' by the Nazi authorities.

MARKELIUS, SVEN (1889–1972)
Sven Markelius was a pioneering Swedish architect who turned from neoclassicism to Functionalism* and modernism. His work also included furniture design and textiles for Nordiska Kompaniet, including a number of fabric designs created with Astrid Sampe*.

MARKS, GRETE (1899–1990)
See p. 174.

MARS (MODERN ARCHITECTURAL RESEARCH GROUP)
Formed in 1933 by Maxwell Fry*, F. R. S. Yorke*, Wells Coates* and others, MARS was established as a collective of English architects and critics. The group staged meetings and exhibitions, as well as attending international conferences organized by CIAM*. MARS dissolved in 1957.

MATHSSON, BRUNO (1907–1988)
See p. 72.

MATTER, HERBERT (1907–1984)
Swiss graphic designer, typographer and photographer Herbert Matter worked with A. M. CASSANDRE in Paris in the Twenties before returning to his home country, where he developed a series of striking travel posters for the Swiss Tourist Board during the Thirties, using an original combination of photography, vivid colours and simple banners of type. He emigrated to New York in 1936 and worked with Charles & Ray Eames* before embracing a role as a creative consultant for the Knoll* furniture company, which included the production of a series of landmark, mid-century advertising images.

McGRATH, RAYMOND (1903–1977)
Born in Australia, Raymond McGrath settled in the UK in 1926 to complete his architectural studies. He went on to design a number of English modernist houses, including St Ann's Court in Chertsey, Surrey (1937). McGrath also designed interiors, as well as lighting, and wrote a number of books during the Thirties. From 1939 onwards he lived and worked in Ireland.

MELNIKOV, KONSTANTIN
(1890–1974)
Russian architect and artist Konstantin Melnikov's most famous work is his own Moscow home (1929) – an extraordinary, asymmetrical drum pierced with hexagonal windows. The majority of his architectural commissions date from the Twenties, with Melnikov focusing on teaching and painting later in his career.

MENDELSOHN, ERICH (1887–1953)
Erich Mendelsohn studied architecture in Berlin and Munich. One of his earliest German projects, the Einstein Tower in Potsdam (1920), is seen as a highpoint of Expressionism*, with a striking and sculptural silhouette. He emigrated to Britain in 1933 and formed a partnership with SERGE CHERMAYEFF. Together, they designed the De La Warr Pavilion in Bexhill-on-Sea, East Sussex (1935), along with private houses, including 66 Old Church Street in Chelsea, London (1935).

MEYER, HANNES (1889–1954)
The Swiss architect Hans Emil Meyer co-founded his own practice in 1926, before becoming the head of the Bauhaus architecture school in Dessau a year later. Bauhaus director WALTER GROPIUS named Meyer as his successor, and he served in the post from 1928 to 1930 (when he was replaced by LUDWIG MIES VAN DER ROHE). Meyer brought significant commissions into the architecture department, including apartment buildings and the ADGB Trade Union School building alongside the Bauhaus campus.

MIES VAN DER ROHE, LUDWIG
(1886–1969)
See pp. 78 and 380.

MOHOLY-NAGY, LÁSZLÓ
(1895–1946)
See p. 280.

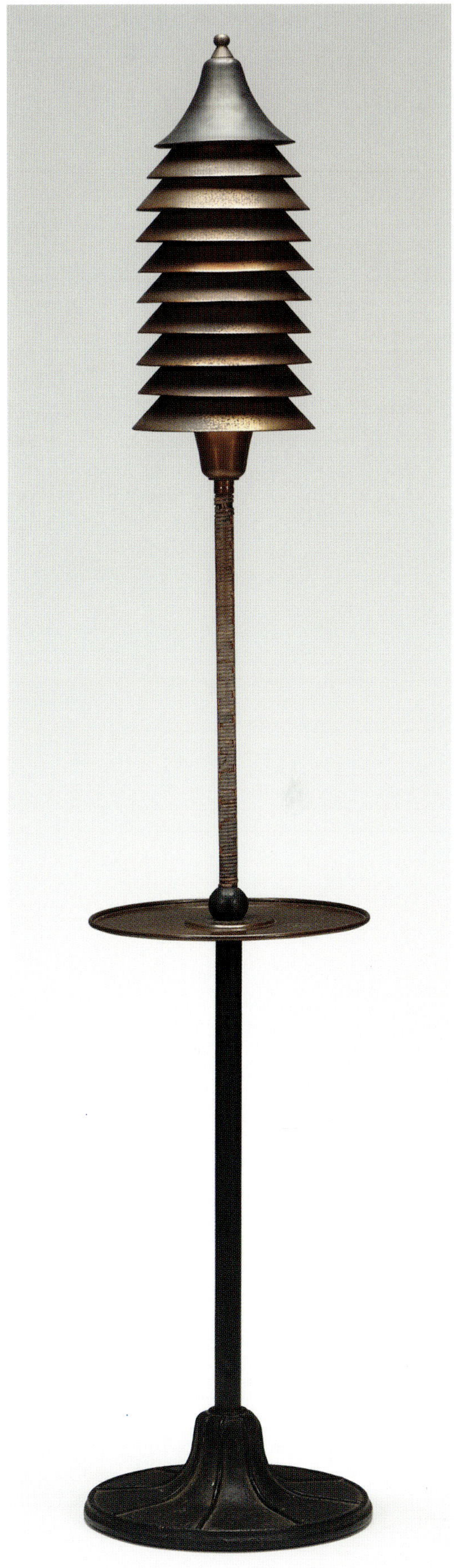

RIGHT A floor lamp by Raymond McGrath, made in the United Kingdom, *c.* 1929, in copper, iron and rattan.

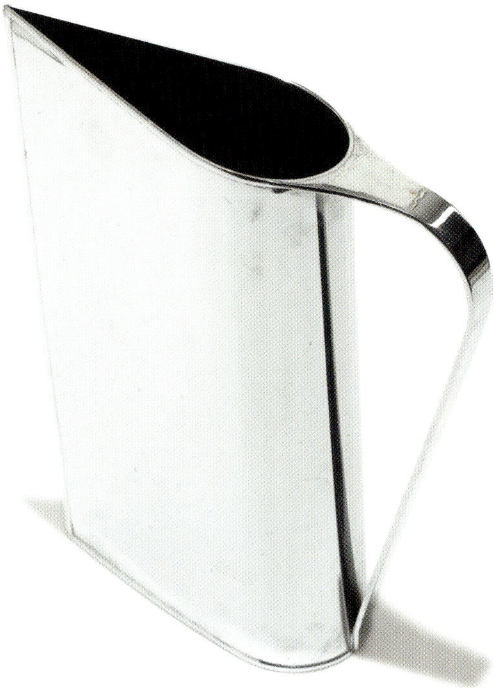

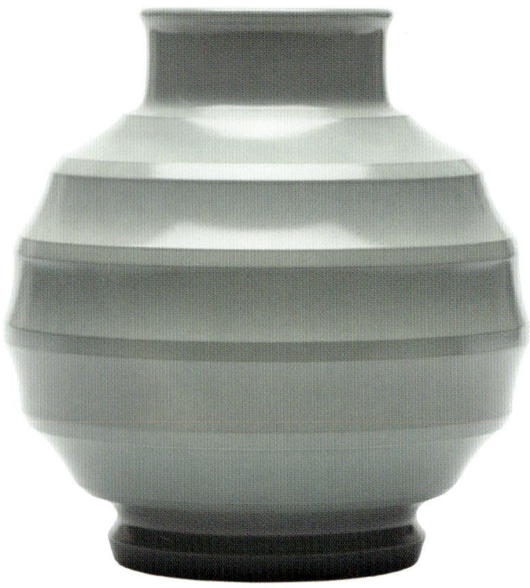

ABOVE A Normandie pitcher designed by Peter Muller-Munk, produced by the Revere Copper & Brass Company, 1935, in chrome-plated brass.

RIGHT A vase by Keith Murray, produced by Wedgwood, *c.* 1930, in glazed stoneware.

BELOW A photograph of Richard Neutra's Kaufmann House in Palm Springs, taken by Julius Shulman, 1947.

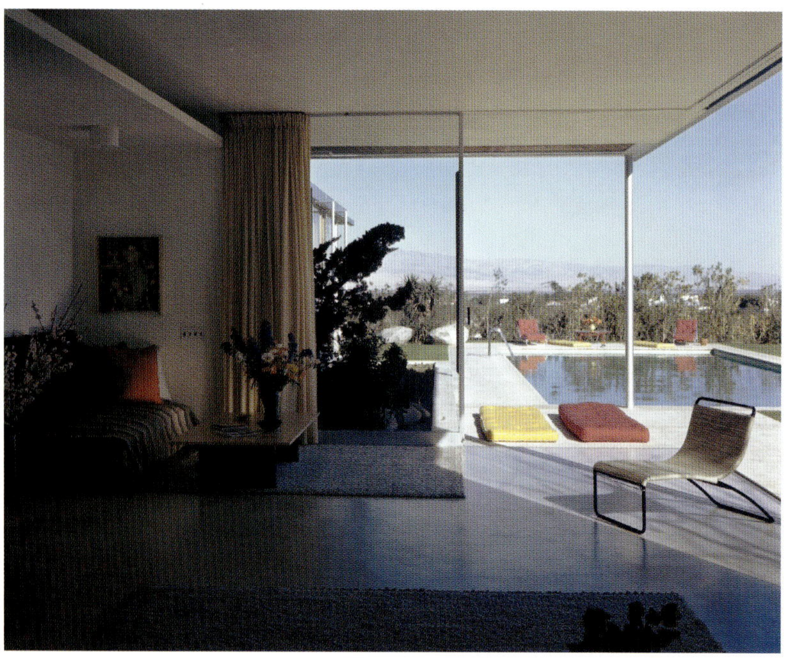

MULLER-MUNK, PETER (1904–1967)
Born in Berlin, Peter Muller-Munk (originally Klaus-Peter Wilhelm Müller) worked as a silversmith before emigrating to America in 1926, where he began working with Tiffany & Co. He moved to Pittsburgh in 1935 and launched his first consultancy in 1938, focusing on industrial design. The practice designed a wide range of consumer goods and appliances, from cameras to cookware.

MUNARI, BRUNO (1907–1998)
Italian graphic designer and artist Bruno Munari also worked as a book illustrator and contributed work to a number of journals and magazines. He was a significant figure within the evolution of Futurism* in Italy during the late Twenties and Thirties, presenting his work in several associated exhibitions.

MURRAY, KEITH (1892–1981)
Born in New Zealand, Keith Murray moved to the UK with his family as a teenager. He studied architecture and worked as an artist and illustrator before embarking upon a career as a designer in the fields of glass, ceramics and metalware. Murray is best known for his work for Wedgwood, who also commissioned him to design a new factory in Staffordshire in 1936.

NESSEN, WALTER VON (1889–1943)
See p. 140.

NEUTRA, RICHARD (1892–1970)
See p. 384.

NICHOLSON, KIT (1904–1948)
Architect Christopher 'Kit' Nicholson was the brother of the influential English painter Ben Nicholson. He studied architecture at Cambridge, where he also taught, and designed a number of modernist houses and buildings in the Thirties, including a country studio for artist Augustus John in Hampshire (1934).

NIZZOLI, MARCELLO (1887–1969)
Marcello Nizzoli was an Italian artist and designer, who studied art and architecture in Parma. He exhibited paintings and textiles in the Italian Futurist* style before establishing a design practice in Milan in 1918. Projects included graphics, exhibition design and architectural assignments with GIUSEPPE TERRAGNI. In the early Thirties, Nizzoli began working as a product designer with the innovative Italian typewriter company Olivetti, for whom he became chief designer.

NOVECENTO
Founded in the mid-Twenties, Italian Novecento was a semi-nationalist artistic movement that took its inspiration from neoclassical art and architecture and stood largely in opposition to avant-garde modernism as well as Futurism*. Some Italian fascist sympathizers supported Novecento as an alternative to modernist art and architecture.

NOYES, ELIOT (1910–1977)
Eliot Noyes worked briefly in the American architectural practice established by WALTER GROPIUS and MARCEL BREUER after studying at Harvard. He went on to take a position as a curator at the Museum of Modern Art in New York before joining the office of NORMAN BEL GEDDES as an industrial designer. He founded his own practice in 1947, best known for its work with IBM.

NYLUND, GUNNAR (1904–1997)
Leading Scandinavian ceramicist Gunnar Nylund worked initially with Bing & Gröndahl in Copenhagen before founding his own studio pottery, known as Saxbo. He joined Rörstrand in Sweden in the early Thirties, remaining with them until 1958.

O'GORMAN, JUAN (1905–1982)
Mexican modernist architect, painter and muralist Juan O'Gorman (he had an Irish father) is best known for the striking house and studio that he designed for artists Diego Rivera and Frida Kahlo in Mexico City (1932). Other projects included houses and schools.

ORREFORS
See p. 178.

OUD, J. J. P. (1890–1963)
Jacobus Johannes Pieter Oud was a Dutch modernist architect associated with the De Stijl* movement. From 1918 to 1933 he worked on municipal housing projects in Rotterdam and, in 1927, contributed a project to the Weissenhof* show estate organized by the Deutscher Werkbund*. His work was also included in the International Style* exhibition at the Museum of Modern Art in New York in 1932, co-curated by Philip Johnson*. He also designed chairs for Metz & Co.

PAUCHARD, XAVIER (1880–1948)
Xavier Pauchard was an innovative entrepreneur who applied his knowledge of sheet steel fabrication used for roofing to furniture. He launched the Tolix company in 1927 and began producing his famous Model A chair – which was light, stackable and hard-wearing – in 1934.

PERRET, AUGUSTE (1874–1954)
See p. 388.

PERRIAND, CHARLOTTE (1903–1999)
See p. 64.

PERZEL, JEAN (1892–1986)
Born in Bavaria, Jean Perzel settled in France and founded his own Parisian lighting design company in 1923. Many of Perzel's designs of the period feature a strong Art Deco* influence.

PONTI, GIO (1891–1979)
See p. 146.

PORSCHE, FERDINAND (1875–1951)
See p. 240.

PORTALUPPI, PIERO (1888–1967)
The Milanese architect Piero Portaluppi began his career working on a series of hydro-electric plants. During the Twenties he won commissions for a series of houses and apartment buildings, and he designed the Italian pavilion at the Barcelona Exposition of 1929. One of his most famous projects of the Thirties was the Villa Necchi Campiglio in Milan (1935), which spliced modernist, Art Deco* and period influences.

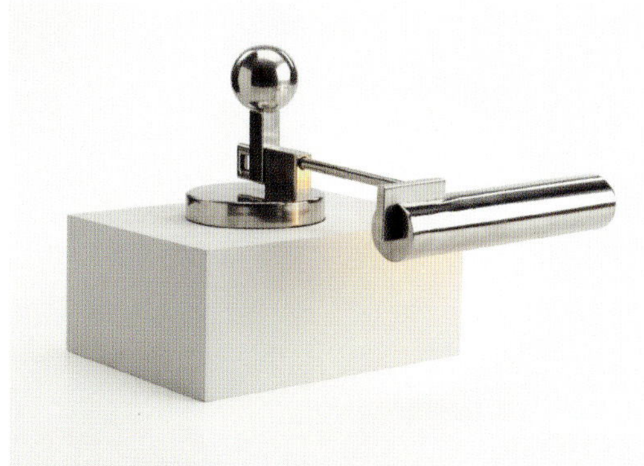

RIGHT A Giso 404 Piano lamp by J. J. P. Oud, made in the Netherlands by W. H. Gispen & Co., 1927, in chrome-plated brass and steel.

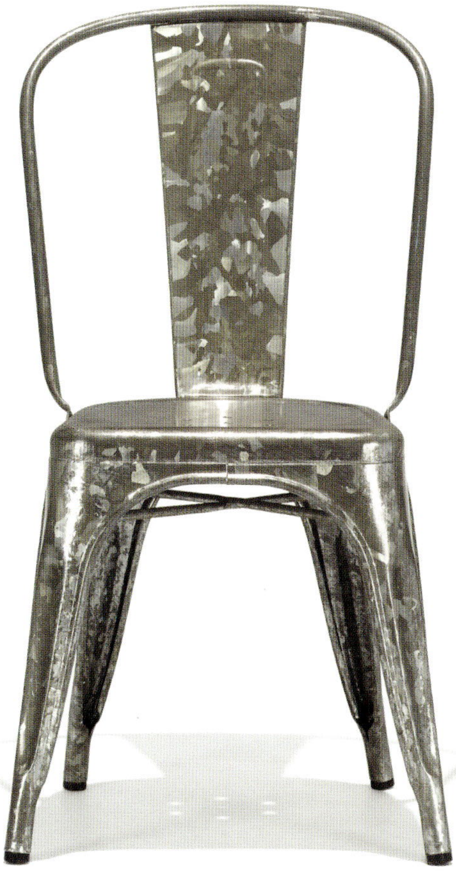

ABOVE A Model A chair, designed by Xavier Pauchard and made in France by Tolix, in galvanized steel.

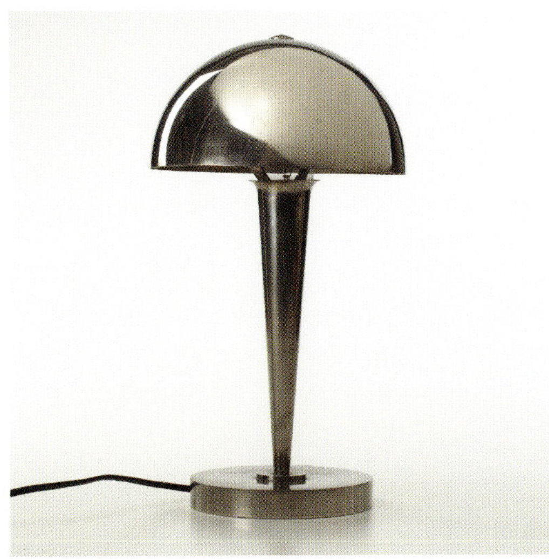

RIGHT A table lamp by Jean Perzel, made in France, 1930s, in nickel-plated brass.

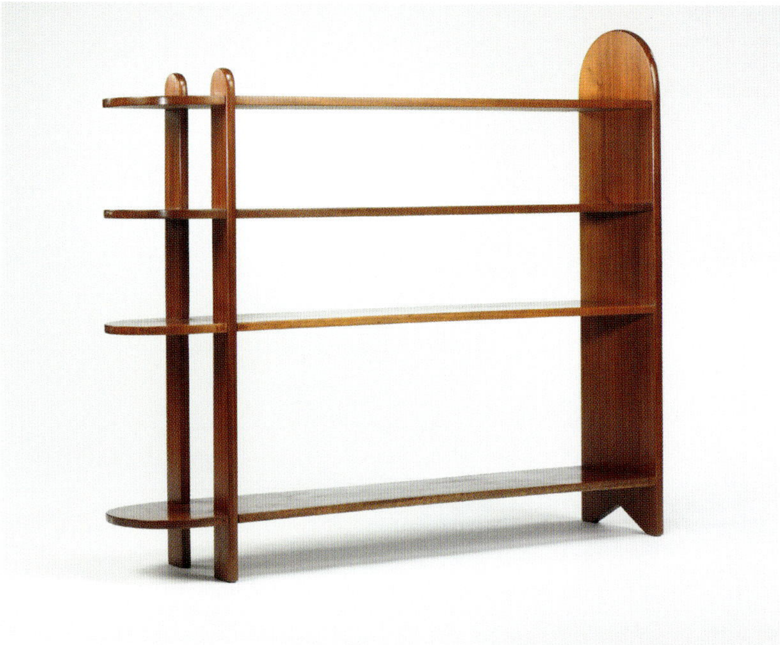

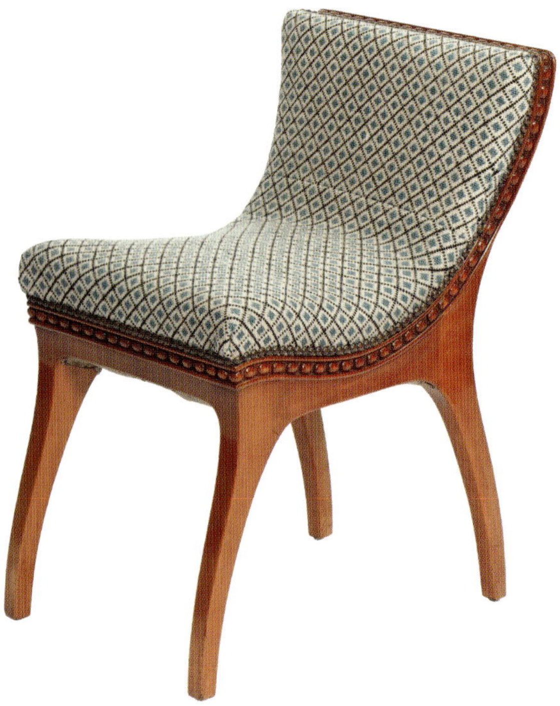

ABOVE A mahogany bookshelf designed by Eugene Printz, made in France, 1932.

RIGHT A chair with a low back by Armand-Albert Rateau, designed for couturier Jeanne Lanvin's bedroom, *c.* 1925.

PRINTZ, EUGENE (1889–1948)
A Parisian cabinet maker and furniture designer, Eugene Printz worked predominantly in an Art Deco* style. His projects also included interiors commissions.

PROUVÉ, JEAN (1901–1984)
See p. 84.

PURISM
An aesthetic sub-movement of modernism particularly associated with LE CORBUSIER and his acolytes, Purism spoke of formal and volumetric rigour, largely defined by geometrical precision and rectilinear outlines. Associated with Rationalism* and Functionalism*, one of the most exemplary projects of the movement is Villa Savoye (1931; see p. 366).

RAND, AYN (1905–1982)
Writer Ayn Rand grew up in Russia and settled in America in 1926. Best known for her classic novel *The Fountainhead* (1943) – the story of a struggling, talented and egotistical modernist architect, Howard Roark – she also developed a philosophical system known as Objectivism.

RATEAU, ARMAND-ALBERT (1882–1938)
The original, distinctive furniture and interiors created by the Parisian designer Armand-Albert Rateau during the Twenties suggest a unique fusion of Art Deco*, neoclassical and modernist influences. Among his most famous commissions was a series of interiors projects for couturier Jeanne Lanvin.

RATIONALISM
The term can be used in one of two ways. Firstly, it describes a structural, logical and ordered approach to modernist architectural planning, particularly as applied to projects such as the 'New Frankfurt' apartment blocks and other major social housing schemes. Secondly, it can be used to describe the Italian Rationalist movement launched by GIUSEPPE TERRAGNI and others in 1926 as an alternative to Italian Futurism*; during the Thirties the work of Terragni and others became closely associated for a time with the Italian fascist state's interest in modernist architecture.

REICH, LILLY (1885–1947)
See p. 78.

RIETVELD, GERRIT (1888–1964)
See pp. 92 and 396.

RISOM, JENS (1916–2016)
Danish designer Jens Risom studied with KAARE KLINT in Copenhagen before settling in America. He designed a number of innovative chairs and pieces of furniture in the early Forties for Hans Knoll* before launching his own furniture company in the post-war period.

RODCHENKO, ALEXANDER (1891–1956)
See p. 286.

ROHDE, GILBERT (1894–1944)
See p. 98.

ROHDE, JOHAN (1856–1935)
A Danish designer, painter and artist, Johan Rohde designed silverware and cutlery for Georg Jensen, including sculptural pitchers of the Twenties. He also designed furniture and textiles.

RUHLMANN, JACQUES-ÉMILE (1879–1933)
The Parisian furniture and interior designer Jacques-Émile (also sometimes known as Émile-Jacques) Ruhlmann was one of the leading lights of the Art Deco movement. He is best known for his exquisitely crafted pieces made of fine materials and rare timbers – furniture of great character, provenance and price that appealed to a wealthy elite in France and beyond. Yet in his later years Ruhlmann was increasingly influenced by machine-age modernism, and began experimenting with steel-framed furniture and more affordable woods, as well as continuing to pare down his pieces, which became more linear and minimal. A well-known and highly influential figure in early 20th-century design in general, Ruhlmann argued that his bespoke designs for affluent patrons ultimately influenced the shifting direction of style and fashion in a wider sense. He remains a key point of reference for many designers today.

SAARINEN, ELIEL (1873–1950)
See p. 400.

SAMPE, ASTRID (1909–2002)
Astrid Sampe was a Swedish textile designer who trained both in Stockholm and at the Royal College of Art in London. Her best-known designs of the Thirties were for Nordiska Kompaniet, where she became head of the textile design studio.

SAUVAGE, HENRI (1873–1932)
Influential French architect and designer Henri Sauvage is perhaps best known for his Art Nouveau masterpiece Villa Majorelle in Nancy (1902). Yet his work also proceeded into an Art Deco* phase, and work from the Twenties and Thirties – including stepped apartment buildings and the La Samaritaine department store in Paris (1928) – has been described as essentially modernist in character. Like Josef Hoffmann*, Adolf Loos* and Otto Wagner*, Sauvage is often described as one of the pioneering proto-modernists.

SAVIGNAC, RAYMOND (1907–2002)
Raymond Savignac served as an assistant to A. M. Cassandre in Paris before embracing an independent career as a poster and graphic designer. His work is partly defined by cartoon-like imagery, playfulness and bold colours.

SCARPA, CARLO (1906–1978)
See p. 182.

SCHAROUN, HANS (1893–1972)
See p. 408.

SCHINDLER, RUDOLPH (1887–1953)
See p. 412.

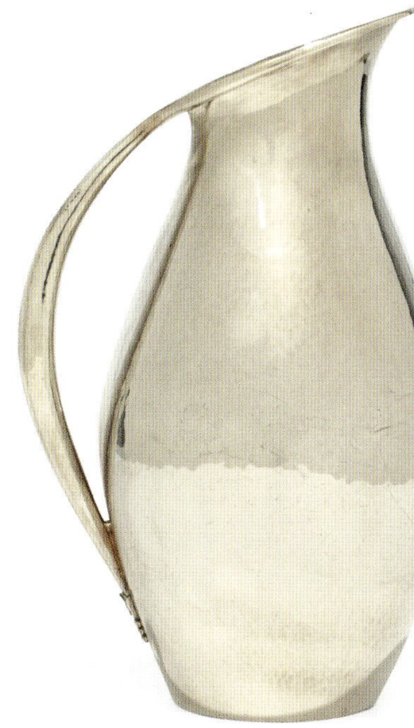

RIGHT A Model 432 sterling silver pitcher by Johan Rohde, produced in Denmark by Georg Jensen, c. 1945.

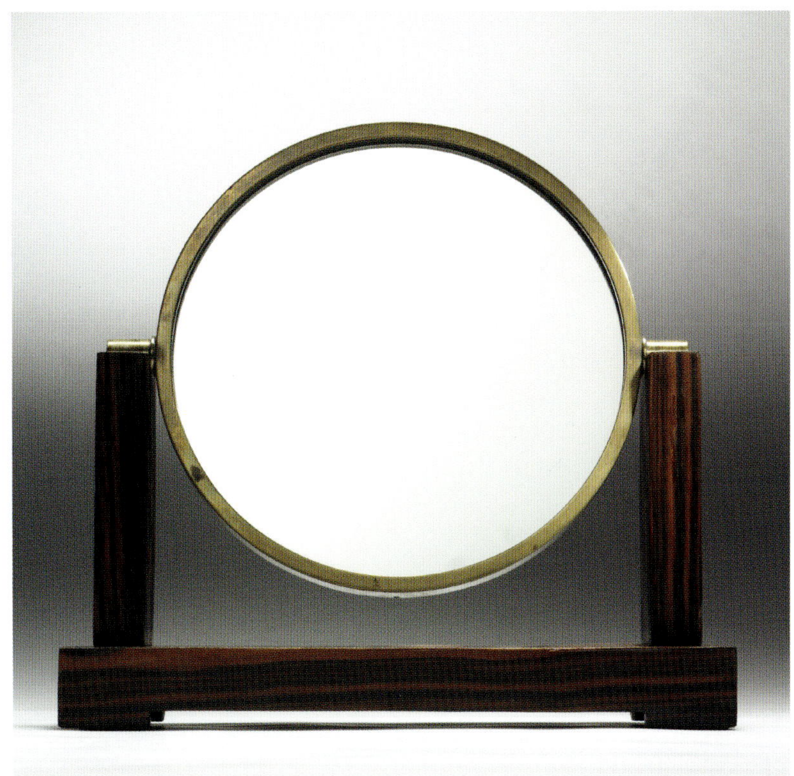

BELOW A vanity mirror by Jacques-Émile Ruhlmann, c. 1925, in macassar ebony, brass and mirrored glass.

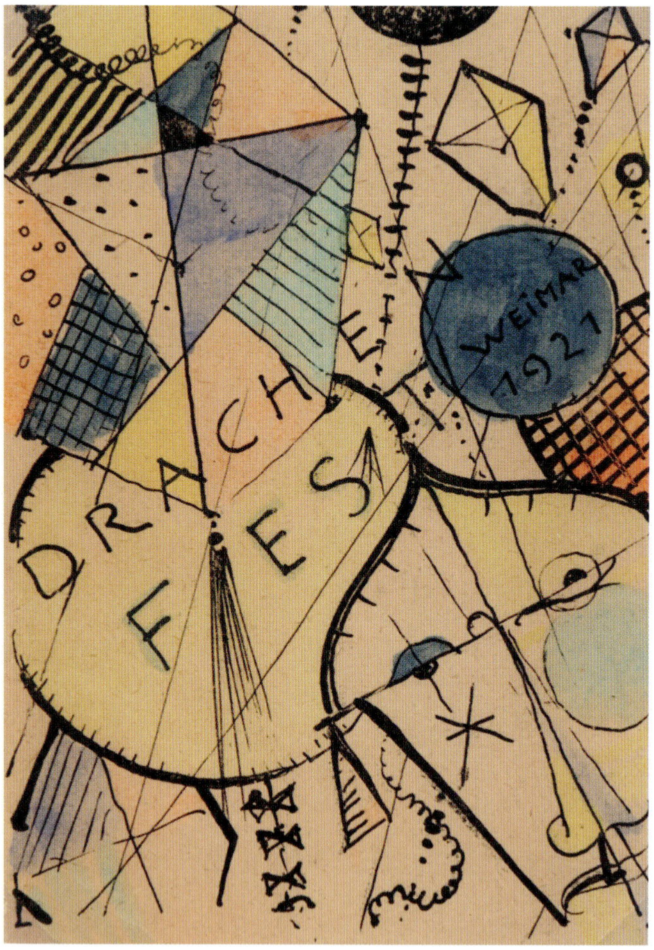

ABOVE A 1921 'Drachen Fest' invitation card designed by Oskar Schlemmer, created while at the Bauhaus in Weimar.

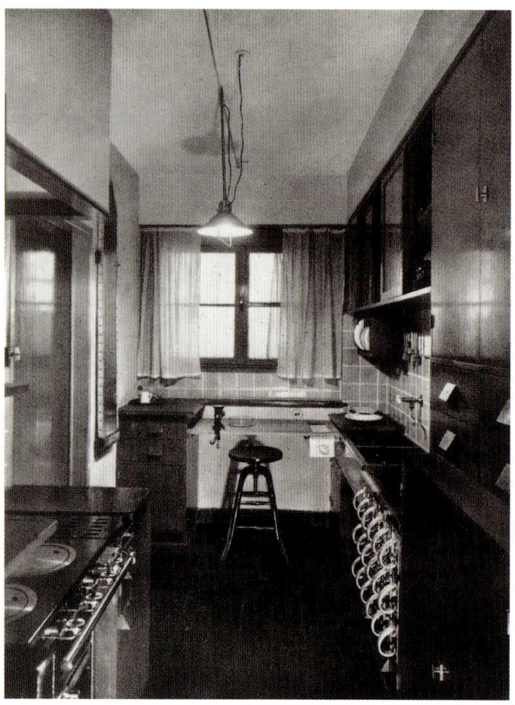

RIGHT Margarete Schütte-Lihotzky's Frankfurt Kitchen in situ, *c.* 1926.

SCHLEMMER, OSKAR (1888–1943)
A designer, artist and theatrical choreographer, from 1920 to 1929 Oskar Schlemmer served as a master in the mural and sculpture departments at the Bauhaus before taking over the theatre studio. Later, he moved on to teaching posts in Breslau and Berlin.

SCHÜTTE-LIHOTZKY, MARGARETE (1897–2000)
Margarete Lihotzky was born in Vienna, where she became the first female student at the Kunstgewerbeschule. Josef Hoffmann* was a teacher at the school; after graduating, Lihotzky also collaborated with other pioneering Viennese modernists such as Adolf Loos* and Josef Frank*. Early in her career, she became interested in designing apartment buildings and new micro-communities to meet a growing shortage of affordable housing. In 1926, she moved to Frankfurt to work with architect and planner Ernst May. Around the same time, she met and married fellow architect Wilhelm Schütte. Her compact, fitted 'Frankfurt Kitchen' of 1926 – designed for mass production and installation within the New Frankfurt apartment buildings – embraced many modernist preoccupations. Around 10,000 units were installed in Frankfurt, and the design was to have a profound impact on the evolution of the modern built-in kitchen but also the notion of a modular home.

SCOTT, GILES GILBERT (1880–1960)
Giles Gilbert Scott followed in family tradition, in that both his father and grandfather were architects. As a young architect, in 1903, Scott was appointed to design the landmark Liverpool Cathedral, a project that was not fully completed until 1978. During the Twenties and Thirties he designed apartment buildings, churches, college buildings for Cambridge University, and the iconic 'K2' (1926) and 'K6' (1935) telephone boxes for the General Post Office. Along with the New Bodleian Library in Oxford (1940), Scott's most familiar projects include Battersea Power Station (1933) and Bankside Power Station (1947/1963; now home to the Tate Modern art gallery). Scott's work combined Gothic, neoclassical and modernist influences.

SERT, JOSEP LLUÍS (1902–1983)
Catalonian modernist architect Josep Lluís Sert studied in Barcelona before moving to Paris in 1929 to work with LE CORBUSIER. From 1930 to 1937 he ran his own practice in Barcelona and co-founded a regional association of modernist architects (which became the Spanish branch of CIAM*). Key projects from this period include one-off houses and apartment buildings. Following the start of the Spanish Civil War, from 1937 to 1939 Sert lived in Paris and designed the Spanish Republic's Pavilion at the Paris exposition of 1937 (the pavilion exhibited Picasso's painting *Guernica*). From 1939 onwards Sert lived and worked in the United States.

SPEER, ALBERT (1905–1981)
The most controversial architect of the 20th century, Albert Speer was Adolf Hitler's chief architect from 1934 to 1939 and served as Reich Minister of Armaments and War Production during World War II. He favoured monumentality in a neoclassical idiom, with key projects including the Reich Chancellery (1939) and Nuremberg parade grounds such as the Zeppelinfeld (1934), as well as the German Pavilion at the 1937 Paris exposition, which won a gold medal. His plans for a 'new' Berlin were never realized and he was tried at Nuremberg and sentenced to 20 years in Spandau Prison. Along with the Nazi authorities, Speer was regarded as hostile to German modernists, many of whom went into exile.

STAM, MART (1899–1986)
Dutch architect and designer Mart Stam was a co-founder, along with GERRIT RIETVELD and others, of CIAM*. He developed a cantilevered tubular steel chair in 1925–26, initially using welded gas piping. Later, he was involved in a legal dispute with MARCEL BREUER over authorship of the tubular cantilevered concept chair.

STÖLZL, GUNTA (1897–1983)
Gunta Stölzl studied at the Weimar Bauhaus under Paul Klee* and Johannes Itten* and joined the weaving workshop, graduating in 1923. Two years later she returned and became a master at the weaving studio, where she remained until 1931. She later co-founded a textile studio in Zurich, creating designs for a number of producers.

STRAUB, MARIANNE (1909–1994)
Born in Switzerland, Marianne Straub studied art and weaving in Zurich before moving to England to complete her education. During the Thirties she worked for the Rural Industries Board, and in 1937 joined textile design company Helios as its head designer. From 1950 to 1970 Straub worked with Warner & Sons.

SULLIVAN, LOUIS (1856–1924)
The American architect Louis Sullivan famously coined the phrase 'form follows function'. He is regarded as an influential proto-modernist and a key figure in the evolution of the modern skyscraper. Many of his most influential buildings are in Chicago, where Sullivan based himself.

SUPREMATISM
A Russian art, architectural and design movement, founded by artist Kazimir Malevich*, Suprematism was a key precursor of Constructivism*. The movement focused on geometric abstraction and was both radical and experimental, yet lacking the socialist dimension that characterized the Constructivist outlook.

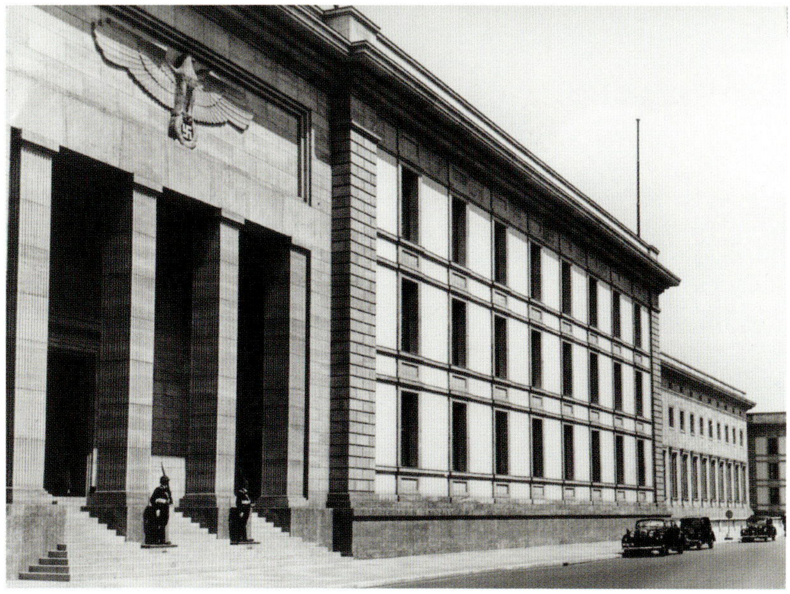

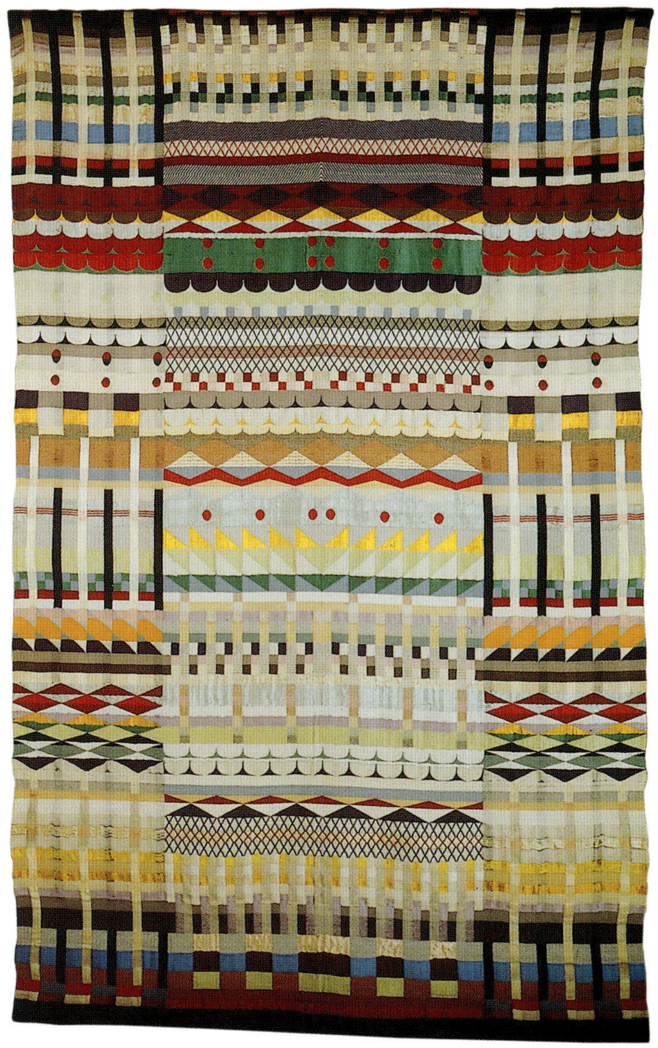

TOP Albert Speer's Reich Chancellery building, seen from Voss Street, Berlin, 1939.

ABOVE Gunta Stölzl's '5 Choirs' wall hanging, 1928, in cotton, wool and silk.

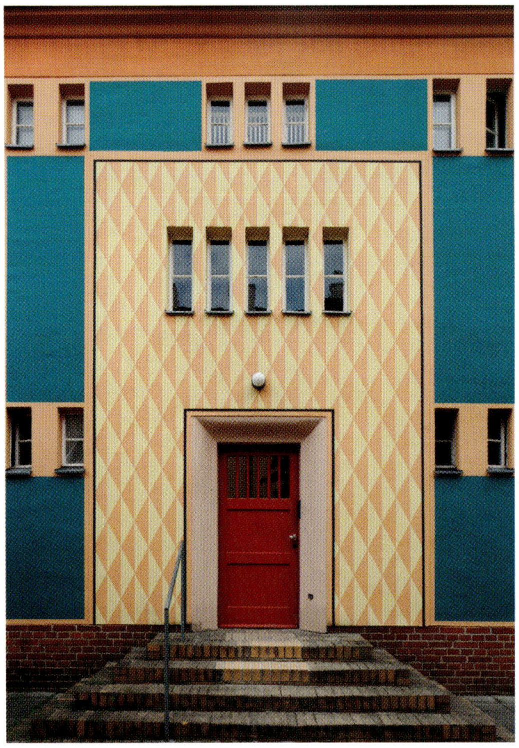

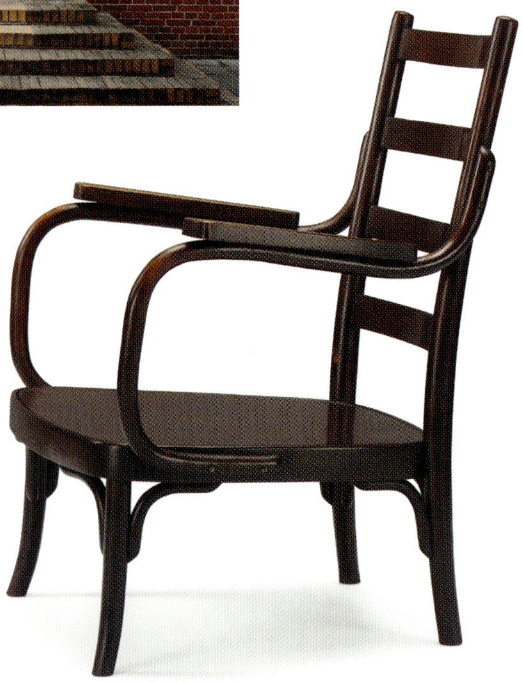

ABOVE A view of Bruno Taut's Gartenstadt Falkenberg housing estate in Berlin, built between 1913 and 1934.

RIGHT A Model A403F armchair designed by Josef Frank and produced by Thonet, c. 1928, in stained beech.

BELOW A table lamp by Karl Trabert, produced by BAG, 1933, in enamelled steel, aluminium and lacquered wood.

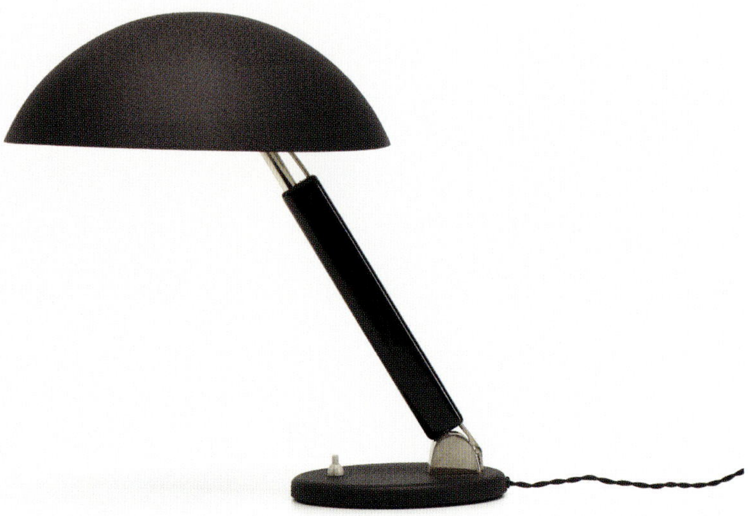

TAUT, BRUNO (1880–1938)
German architect and painter Bruno Taut established his own practice in Berlin in 1910. He worked on a series of ambitious housing projects in Magdeburg and particularly Berlin, where he worked closely with city architect Martin Wagner. During the Twenties and early Thirties Taut designed thousands of homes in a largely modernist style. He was also active in the Deutscher Werkbund* and contributed a colourful and attention-grabbing house to the Werkbund's Weissenhof* housing exhibition in Stuttgart. Being both Jewish and sympathetic to socialist causes, Taut became vulnerable when the Nazi party assumed power and in 1933 fled to Switzerland, followed by periods in Japan and Turkey.

TEAGUE, WALTER DORWIN (1883–1960)
See p. 242.

TECTON
The London-based Tecton Group was led by architect Berthold Lubetkin*. Key projects included buildings at London Zoo and Highpoint I and II in Highgate (see p. 457). The practice was founded in 1932, but disbanded in 1939.

TERRAGNI, GIUSEPPE (1904–1943)
See p. 420.

THONET
The Thonet furniture company was founded in Germany in 1819 by Michael Thonet and moved to Vienna in 1849. The company became well known for its innovative bentwood chairs; in 1929 a French subsidiary was formed known as Thonet Frères, focusing on tubular steel designs by Marcel Breuer, Le Corbusier and others.

TRABERT, KARL (d. 1968)
A German lighting designer, Karl Trabert was at the Bauhaus during the late Twenties and early Thirties. His most famous lights were produced by BAG in Switzerland around 1932 to 1934.

TSCHICHOLD, JAN (1902–1974)
See p. 290.

TÜMPEL, WOLFGANG (1903–1978)
Wolfgang Tümpel was a German silversmith and metalware designer who studied at the Bauhaus in the early Twenties. As well as metalware for WMF, he designed silverware and lighting.

UAM (UNION DES ARTISTES MODERNES)
The Union – known as UAM – was a French association of modernist architects and artists founded in 1929 and led by Robert Mallet-Stevens; other members included Eileen Gray and Charlotte Perriand, as well as Pierre Chareau and Le Corbusier. UAM organized and participated in a number of exhibitions, including the 1937 international exposition in Paris.

VELDE, HENRY VAN DE (1863–1957)
Having studied painting in Antwerp and begun a career as an artist, Henry van de Velde moved into design, interiors and architecture from the 1890s onwards. He settled in Weimar, Germany, becoming director of the School of Arts and Crafts in 1908; later, in 1917, he moved to Switzerland. Although he began working in an Art Nouveau/ Jugendstil* aesthetic – with commissions including ceramics, furniture and silverware – he adopted an increasingly functionalist approach, becoming an influential proto-modernist and a co-founder of the Deutscher Werkbund*.

VASSOS, JOHN (1898–1985)
American industrial designer John Vassos was also a book illustrator. Born in Romania, he grew up in Turkey before settling in America in 1919, opening his own New York studio in 1924. His best-known designs include radios, record players and televisions for RCA, one of his key corporate clients.

VENINI, PAOLO (1895–1959)
See p. 182.

WAGENFELD, WILHELM (1900–1990)
See p. 246.

WAGNER, OTTO (1841–1918)
As a professor of architecture in Vienna, Austrian architect and designer Otto Wagner taught many key figures seen as proto-modernists, including Adolf Loos* and Josef Hoffmann*, as well as publishing an influential book, *Moderne Architektur*, in 1896. Wagner is associated with the Vienna Secessionist movement and worked in an Art Nouveau/ Jugendstil* manner before beginning to strip back ornament in favour of a purer aesthetic. A number of his designs and projects anticipate a modernist approach.

WEBER, KEM (1889–1963)
See p. 106.

WEISSENHOF ESTATE
Organized by the Deutscher Werkbund*, the Weissenhof housing exhibition of 1927, in Stuttgart, was one of the most ambitious and influential 'expos' of the early modernist era. The project featured the work of 17 European architects, including Ludwig Mies van der Rohe (who also directed the project), Walter Gropius, Le Corbusier and many other key figures. The exhibition included 33 houses and 63 apartments, and was visited by half a million people (see also overleaf).

WILDENHAIN, MARGUERITE (1896–1985)
See p. 190.

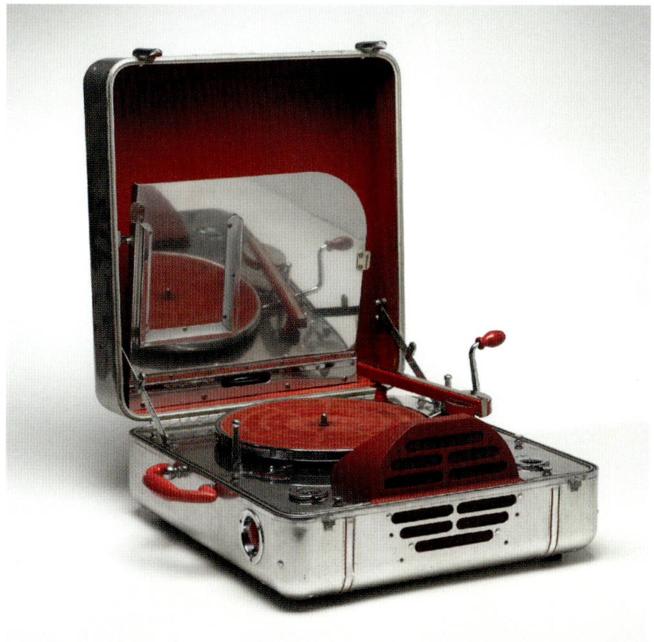

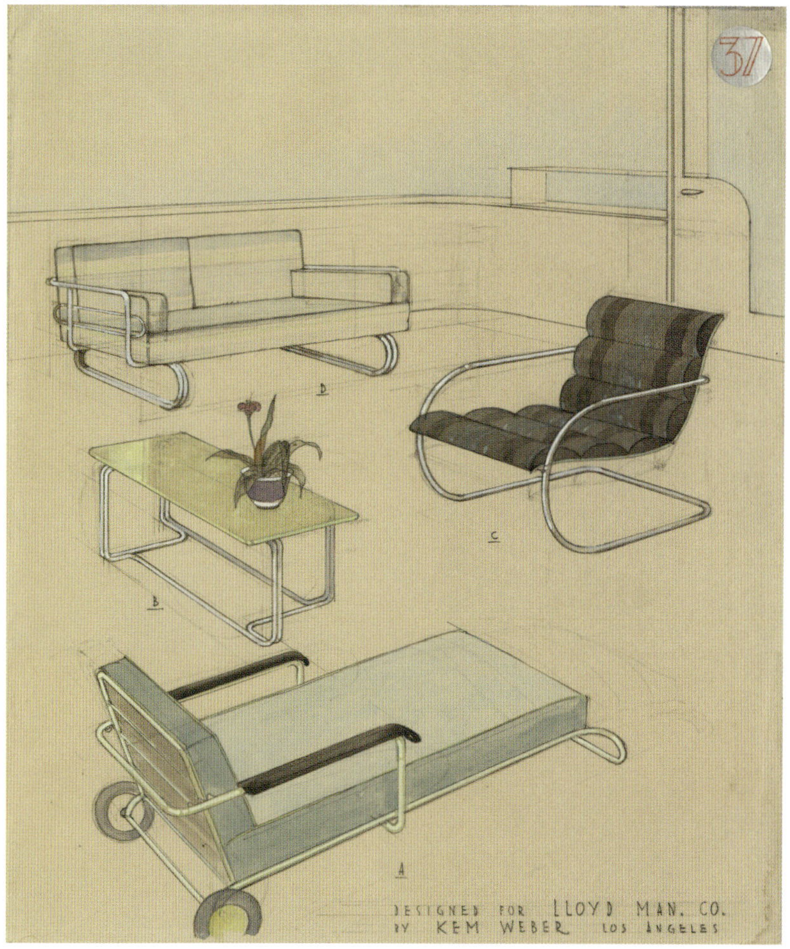

TOP An RCA Victor Special portable phonograph, designed by John Vassos, c. 1935, in aluminium and plastic.

ABOVE An illustration by Kem Weber depicting tubular-steel furniture designs for the Lloyd Manufacturing Company, Michigan, 1937.

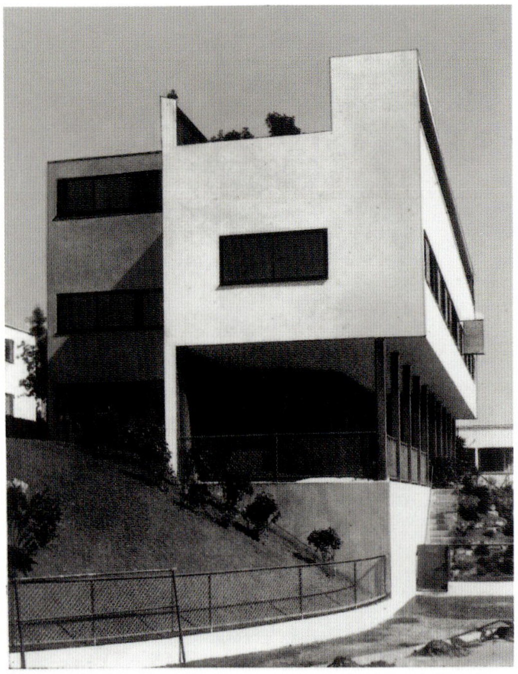

LEFT A view of the Weissenhof Estate, *c.* 1927, which featured 33 show houses and a range of apartments.

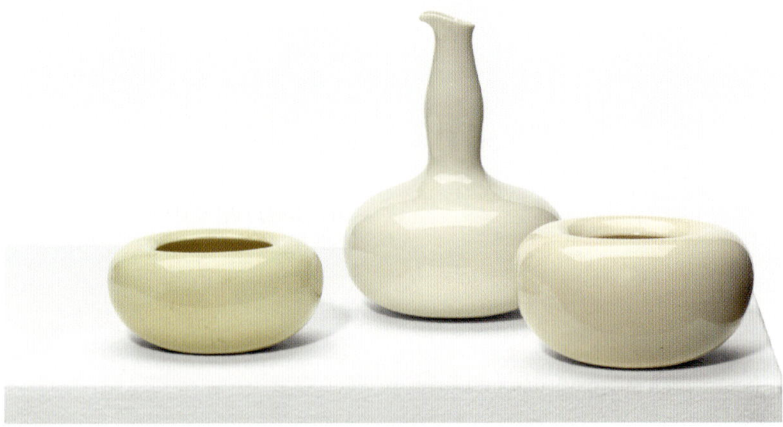

ABOVE A condiment set by Eva Zeisel, produced by the Riverside Ceramics Company, 1947, in glazed porcelain.

RIGHT Piet Zwart's limited-edition letterpress entitled 'Homage to a Young Girl', 1925.

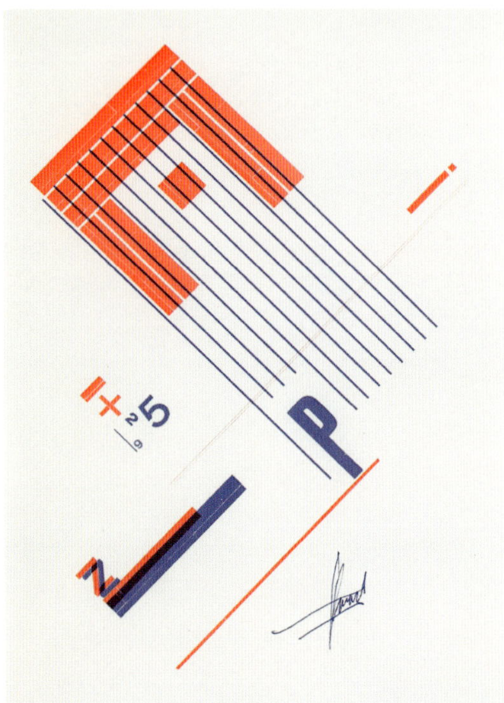

WILLIAMS, OWEN (1890–1969)
Owen Williams studied engineering before becoming an architect, specializing initially in large-scale projects using reinforced concrete. Among his many commissions were buildings for Boots in Nottingham (1932 & 1938), along with the Fleet Street headquarters of the *Daily Express* newspaper (1932).

WITTGENSTEIN, LUDWIG (1889–1951)
The celebrated Viennese philosopher Ludwig Wittgenstein played a key role in the design of a landmark house for his sister Margaret and her family, completed in 1929. Working with Paul Engelmann and others, Wittgenstein designed a house in the modernist style with particular attention to scale, proportion, precision and detailing, including door handles and radiators.

WORNUM, GEORGE GREY (1888–1957)
George Grey Wornum was an English architect best known for his landmark building for the Royal Institute of British Architects (RIBA) at 66 Portland Place, London (1934), which fuses modernist, Art Deco* and neoclassical influences.

WRIGHT, FRANK LLOYD (1867–1959)
See p. 428.

WRIGHT, RUSSEL (1904–1976)
See p. 196.

YORKE, F. R. S. (1906–1962)
Best known for a series of influential modernist houses, such as Torilla in Hatfield (1935), Francis Reginald Stevens Yorke was also an architectural writer, whose publications included *The Modern House* (1934). Yorke promoted a progressive and outward-looking attitude to design, and formed a brief architectural partnership with Marcel Breuer during the latter's years in England (1935–37).

ZEISEL, EVA (1906–2011)
A Hungarian ceramicist, Eva Zeisel studied in Budapest before starting work with the Kispest pottery. In 1927 she moved to Germany, designing ceramics for producers in Hamburg and Berlin, followed by a period working in the Soviet Union. In 1938 Zeisel emigrated to America and re-launched her design career with great success; her collection of the Forties included a range developed in association with the Museum of Modern Art in New York.

ZWART, PIET (1885–1977)
A Dutch typographer, photographer and graphic designer, Piet Zwart initially trained as an architect and was influenced by De Stijl*. He developed a palette of modernist type and illustrated books, as well as developing graphics for advertising. Zwart also taught in Rotterdam and gave guest lectures at the Bauhaus.

TIM BENTON is Professor Emeritus of Art History at the Open University. He has also been visiting professor at Columbia University; Williams College, Massachusetts; and the École Polytechnique Fédérale de Lausanne. His research interests include Le Corbusier's work of the 1920s and '30s, and the history of modern architecture and design. He has contributed to several major exhibitions, and his publications include *Le Corbusier conférencier* (winner of the 2008 Prix National du Livre awarded by the Académie d'Architecture in France), *The Villas of Le Corbusier and Pierre Jeanneret 1920–1930* and *The Modernist Home*.

MEL BUCHANAN is the RosaMary Curator of Decorative Arts and Design at the New Orleans Museum of Art. After earning a Bachelor's degree from Yale University and a Master's degree from the Winterthur Program at the University of Delaware, she held curatorial positions at the Milwaukee Art Museum and the RISD Museum.

ANNE H. HOY is co-author of *American Streamlined Design: The World of Tomorrow* (2005) and *Designed for Living: Furnishings and Lighting 1950–2000* (2000), and has also written on photography (*The Book of Photography: The History, the Art, the Technique, the Future*; *Fabrications: Staged, Altered, and Appropriated Photographs*) and popular culture (*Coca-Cola: The First Hundred Years*). Since 1995 she has taught histories of modern and contemporary art, photography and graphic design at New York University. She has also edited books for Rizzoli, Flammarion, Abrams, Abbeville and D Giles, among others.

MARGRET KENTGENS-CRAIG completed her PhD at the Ruhr-University in Bochum, Germany. She formerly served as head of the Bauhaus Dessau Foundation's Department of Archive and Collection, and as a member of the German UNESCO commission. She has curated art exhibitions and taught history of art, design and architecture at universities in both Germany and the United States. Her interests include 20th-century modernism, the Bauhaus and its reception in America, and the inter-relationships of intellectual movements. At present, she is an Associate Professor of Practice at the North Carolina State University College of Design. Her publications include *The Bauhaus and America: First Contacts 1919–1936* and *The Dessau Bauhaus Building 1926–1999* (edited with the Dessau Bauhaus Foundation).

JOLANTHE KUGLER has been a curator at the Vitra Design Museum since 2012 and has realized exhibitions such as *Lightopia – Designing Light*, *Shiro Kuramata – Design as Poetry* and *The Bauhaus #itsalldesign*. She studied architecture, art history and urban sociology in Mendrisio, Florence and Paris before working as an architect with several Swiss, French and Italian architecture practices. From 2009 to 2012 she was a lecturer in urban development at the Institute of Social Planning and Urban Development in Basel. She is the author and editor of countless papers and publications covering the social processes, behaviours and actions that occur within the context of the city, architecture and design. She was also the author of the first architectural guide to the Goetheanum anthroposophical colony near Basel.

RICK POYNOR is Professor of Design and Visual Culture in the department of Typography and Graphic Communication at the University of Reading. He was the founding editor of *Eye* magazine, a long-running international review of graphic design, and his books include *Typographica* (2001), *No More Rules: Graphic Design and Postmodernism* (2003) and *Jan van Toorn: Critical Practice* (2008).

PATRICK & LAURENCE SEGUIN are prominent figures in the 20th-century design world, having co-founded Galerie Patrick Seguin in Paris's Bastille in 1989. The 300 sq m (3,200 sq ft) space was later renovated by Jean Nouvel, the 2008 Pritzker Prize winner. The quality of the works represented in the gallery (which have included pieces by French designers such as Jean Prouvé, Charlotte Perriand, Pierre Jeanneret, Le Corbusier and Jean Royère), combined with the Seguins' meticulous and informative presentations, has resulted in unique exhibitions and collaborations with the Centre Georges Pompidou and the Musée des Arts Décoratifs in Paris, and the Museum of Modern Art in New York. The gallery has also published a series of monographic books that accompany the exhibitions.

RICHARD WRIGHT is the founder and president of Wright auction house in Chicago. Since its first sale in 2000, the company has grown to be the defining auction house in the field of modern design. Richard Wright has over three decades of experience in handling and documenting 20th-century works. He has published several books and monographs in addition to over 100 award-winning auction catalogues on the subject. He has also lectured widely on the history and market of modern design.

CONTRIBUTORS' BIOGRAPHIES

Introduction

1 Ayn Rand, *The Fountainhead*, Penguin Classics, 2007, p. 13.

FURNITURE

1 Quoted in *Kem Weber: Designer and Architect* by Christopher Long, Yale University Press, 2014, p. 251.

2 Quoted in Guillemette Delaporte's *René Herbst: Pioneer of Modernism*, Flammarion, 2004, p. 193.

3 Walter Gropius, in: *Volker Wahl, Das Staatliche Bauhaus in Weimar*, Böhlau, 2009, p. 348.

4 Lyonel Feininger, *Brief an einen unbekannten Freund* (letter to an unknown friend), 6.9.1920, private collection.

5 The Bauhaus instructor Johannes Itten insisted on the term 'universalist', rather than 'Romantic'. Anneliese Itten (ed.), *Johannes Itten. Werke und Schriften*, Orell Füssli, 1978, p. 8.

6 'The needs of people instead of the needs of luxury was Meyer's demand.

7 'Marcel Breuer: Metallmöbel', in: *Deutscher Werkbund: Innenräume; Räume und Inneneinrichtungsgegenstände aus der Werkbundausstellung*, Stuttgart, 1929, pp. 133–4.

8 Marcel Breuer, 'Ein Bauhaus-Film. 5 Jahre lang', in: *Bauhaus* magazine, No. 1, 1926.

9 Marcel Breuer, 'Metallmöbel und Moderne Räumlichkeit', in: *Das Neue Frankfurt*, Vol. 2, No. 1, 1928, p. 11.

10 'We are not seeking a Bauhaus style or a Bauhaus way [...]. We are not seeking geometrical or stereometrical structures, alien and inimical', Hannes Meyer, 'Bauhaus und Gesellschaft', in: *Bauhaus* magazine, Vol. 3, No. 1, 1929, p. 2; 'No Bauhaus style, as the inevitable catchword already has it [...]. The Bauhaus style ... is the very least of the Bauhaus itself [...]' in: Oskar Schlemmer, *Briefe und Tagebücher*, Stuttgart, 1977, p. 109.

11 Quoted by Guillemette Delaporte, *René Herbst: Pioneer of Modernism*, Flammarion, 2004, p. 127.

12 From *The Architectural Review*, November 1984, interview with Charlotte Perriand by Charlotte Ellis and Martin Meade.

13 Quoted in Ida van Zijl's *Gerrit Rietveld*, Phaidon, 2010, p. 31.

14 Quoted by Phyllis Ross, *Gilbert Rohde: Modern Design for Modern Living*, Yale University Press, 2009, p. 235.

LIGHTING

1 Shax Riegler, 'Everything You Need to Know about Poul Henningsen', *Elle Decor*, October 2007.

CERAMICS & GLASS

1 Eric Gill, *An Essay on Typography*, Penguin, 2013.

2 Marguerite Wildenhain, *Pottery: Form and Expression*, American Craftsmen's Council and Reinhold, 1962.

3 Walter Gropius, letter to Gerhard Marcks, dated 4 May 1923, quoted in Magdalena Droste, *Bauhaus 1919–1933*, Bauhaus-Archiv Museum, 2012, p. 68.

4 Gio Ponti, 'Le ragioni dello stile moderno' ['The reasons for the modern style'], *L'Italia all'Esposizione Internazionale di Arti Decorative e Industriali Moderne*, Paris and Milan, 1925, p. 69.

5 For a pioneering study on American studio ceramics, see Martha Drexler Lynn, *American Studio Ceramics: Innovation and Identity, 1940 to 1979*, Yale University Press, 2015.

6 Garth Clark, *American Ceramics: 1876 to the Present*, revised edition, Abbeville Press, 1987, p. 63.

7 Bernard Leach, *A Potter's Book*, Faber & Faber, 2011 (originally published in 1940).

8 Quoted in Edmund de Waal's *20th Century Ceramics*, Thames & Hudson, 2003, p. 71.

9 Russel Wright, quoted in *Russel Wright: Good Design is For Everyone, In His Own Words*, Manitoga/The Russel Wright Design Center, & Universe, 2001, p. 28.

INDUSTRIAL & PRODUCT DESIGN

1 Interview by Peter Mayer with Raymond Loewy, printed in Raymond Loewy's *Industrial Design*, Laurence King, 2000, p. 36.

2 Harold Van Doren, *Industrial Design: A Practical Guide*, McGraw-Hill, 1940, p. 3.

3 Quoted in Alan Windsor's *Peter Behrens: Architect and Designer, 1868–1940*, The Architectural Press, 1981, p. 92.

4 See 'Biographies of Designers' in David A. Hanks and Anne Hoy, *American Streamlined Design: The World of Tomorrow*, Flammarion for the Liliane and David M. Stewart Program for Modern Design, 2005, pp. 249–64.

5 'Art in the Machine Age', in *Saturday Review of Literature*, 8 September 1928, p. 102, quoted in Kathryn B. Hiesinger and George H. Marcus, *Landmarks of Twentieth-Century Design: An Illustrated Handbook*, Abbeville, 1993, p. 113.

6 Erich Mendelsohn, *Amerika*, Berlin, 1926; Le Corbusier, *Vers Une Architecture*, 1923, translated in 1927 as *Towards a New Architecture*.

7 *Art and the Machine: An Account of Industrial Design in 20th-Century America*, Whittlesey House, 1936, pp. 97–8.

8 Robert J. Gordon, *The Rise and Fall of American Growth: The U.S. Standard of Living Since the Civil War*, Princeton University Press, 2016, pp. 155, 169.

9 See David Gartman, *Auto Opium: A Social History of American Automobile Design*, Routledge, 1994, Chapters 4–5.

10 Jeffrey Meikle, *Twentieth Century Limited: Industrial Design in America, 1925–1939*, Temple University Press, 1979; reprint 2001, p. 160.

11 Hanks and Hoy, 2005, op. cit., p. 36.

12 See David A. Hanks and Anne Hoy, 'Streamlining and Art Deco in American Industrial Design', in *The Magazine Antiques* (October 2004), pp. 114–23; and Christopher Long, *Paul T. Frankl and Modern American Design*, Yale University Press, 2007.

13 Hiesinger and Marcus, 1993, op. cit., p. 116.

NOTES

14 For a thoughtful but skeptical response (which I accept) to a book linking streamlining and eugenics, see Glenn Porter's review in *Enterprise & Society*, 8:2 (June 2007).
15 Cheney and Cheney, in *Art and the Machine*, op. cit., p. 173.
16 For illustration and text on additional streamlined public buildings and interiors, see Hanks and Hoy, 2005, op. cit.: Figs 1, 14–17, 43–44, 52–53, 61–65, 68–69, 71–72, 140, 143, 152, 154–59, 160–164.
17 Quoted in Alan Windsor's *Peter Behrens: Architect and Designer, 1868–1940*, The Architectural Press, 1981, p. 23.
18 Harold Van Doren, *Industrial Design: A Practical Guide*, McGraw-Hill Book Company, 1940, p. xvii.
19 Ibid., p. 137.
20 Quoted in Russell Flinchum's *Henry Dreyfuss, Industrial Designer: The Man in the Brown Suit*, Rizzoli, 1997, p. 73.
21 Interview with Peter Mayer, as featured in *Industrial Design* by Raymond Loewy, Laurence King, 2000, p. 38.
22 Ibid., pp. 15–16.
23 Walter Dorwin Teague, 'Industrial Art and its Future: An Address Delivered at the Annual Reception of the School of Architecture and Allied Arts, New York University, 1936'.

GRAPHICS & POSTERS

1 Quoted by Jan Tschichold in *The New Typography*, University of California Press, 1998, originally published in 1928.
2 Ibid.
3 Quoted in *100 Ideas That Changed Graphic Design* by Steven Heller and Véronique Vienne, Laurence King, 2012, p. 154.
4 Jan Tschichold, *The New Typography*, University of California Press, 1998, originally published 1928.
5 Ibid.

HOUSES & INTERIORS

1 Quoted in Hasan-Uddin Khan's *International Style: Modernist Architecture from 1925 to 1965*, Taschen, 1998, p. 65.
2 Le Corbusier, *Towards a New Architecture*, Dover Publications, 1931/1986.
3 Ibid.
4 Ibid.
5 Interview with Harwell Hamilton Harris, Raleigh, NC, 26 February 1990.
6 Walter Gropius, Programme of the Staatliches Bauhaus in Weimar. In: Ulrich Conrads, *Programs and Manifestoes on 20th-century Architecture*, MIT Press, 1975.
7 Gropius, *The New Architecture and the Bauhaus*, MIT Press, 1965 (reprint 1968).
8 Bergdoll/Dickerman coined the term as the title of *Bauhaus 1919–1933: Workshops for Modernity*, Museum of Modern Art, 2009.
9 Margret Kentgens-Craig, *The Bauhaus and America: First Contacts, 1919–1936*, MIT Press, 1999.
10 Robert Wiesenberger, 'The Bauhaus and Harvard' (essay). http://www.harvardartmuseums.org/tour/the-bauhaus/slide/6339
11 David A. Hanks, 'Bauhaus: Mecca of Modernism'. In: Hanks (ed.), *Partners in Design: Alfred Barr Jr. and Philip Johnson*, The Monacelli Press, 2015.
12 William J. R. Curtis, *Modern Architecture since 1900*, Phaidon, 3rd edition, 1996 (1st ed. 1982).
13 Hanks, 'Laboratories for Modernism. The Barr and Johnson Apartments'. Ibid.
14 Charles L. Kuhn, 'America and the Bauhaus', in *The American-German* Review 15 (2), December 1948: 18.
15 Aline B. Louchheim, 'The Guiding Genius of the Bauhaus'. Quoted in Wiesenberger, op. cit.
16 H. M. Wingler, *The Bauhaus: Weimar, Dessau, Berlin, Chicago*, MIT Press, 1969.
17 Dominique Vellay and François Halard, *La Maison de Verre: Pierre Chareau's Modernist Masterwork*, Thames & Hudson, 2007, p. XX.
18 Quoted by Nigel Warburton in *Ernő Goldfinger: The Life of an Architect*, Routledge, 2004, p. 79.
19 Quoted by Peter Adam in *Eileen Gray: Architect/Designer*, Harry N. Abrams, 1987.
20 Walter Gropius, introduction, *Scope of Total Architecture*, 1956.
21 Quoted in Jean Jenger's *Le Corbusier: Architect of a New Age*, Thames & Hudson, 1996, p. 151.
22 From Le Corbusier's *L'Almanach d'Architecture Moderne*, quoted in Jean Jenger, op. cit., p. 129.
23 Quoted in Dominique Deshoulières et al (eds), *Rob Mallet-Stevens: Architecte*, Archives d'Architecture Moderne, 1981.
24 Ibid.
25 Letter from Dr Lovell to Richard Neutra, February 1969, reproduced in full in Thomas S. Hines's *Richard Neutra and the Search for Modern Architecture*, Rizzoli, 2005, p. 341.
26 Quoted in Karla Britton's *Auguste Perret*, Phaidon, 2001, p. 12.
27 Quoted in Bertus Mulder and Ida van Zijl's *Rietveld Schröder House*, Princeton Architectural Press, 1999.
28 Quoted in Eberhard Syring and Jörg C. Kirschenmann's *Scharoun*, Taschen, 2004, p. 8.
29 Quoted in Kathryn Smith's *Schindler House*, Harry N. Abrams, 2001, p. 20.
30 Frank Lloyd Wright writing in *An Autobiography* (1932), reproduced in Robert McCarter's (ed.) *On and By Frank Lloyd Wright: A Primer of Architectural Principles*, Phaidon, 2005, p. 347.
31 Quoted in Patrick J. Meehan's (ed.) *The Master Architect: Conversations with Frank Lloyd Wright*, Wiley-Interscience, 1984.

Adam, Peter, *Eileen Gray: Architect/ Designer*, Harry N. Abrams, 1987
Albrecht, Donald (ed.), *Norman Bel Geddes Designs America*, Abrams, 2012
Allan, John, *Berthold Lubetkin*, Merrell, 2002
Aynsley, Jeremy, *Designing Modern Germany*, Reaktion Books, 2009
Barnicoat, John, *Posters: A Concise History*, Thames & Hudson, 1972
Berry, John R., *Herman Miller: The Purpose of Design*, Rizzoli, 2009
Blom, Philipp, *Life and Culture in the West, 1918–1938*, Atlantic Books, 2015
Blundell Jones, *Peter, Gunnar Asplund*, Phaidon, 2006
——, *Hans Scharoun*, Phaidon, 1995
Bradbury, Dominic, *Mid-Century Modern Complete*, Thames & Hudson, 2014
Bradbury, Dominic, & Richard Powers, *The Iconic House: Architectural Masterworks Since 1900*, Thames & Hudson, 2009
——, *The Iconic Interior: 1900 to the Present*, Thames & Hudson, 2012
Bréon, Emmanuel, *Jacques-Émile Ruhlmann*, Flammarion, 2004
Britton, Karla, *Auguste Perret*, Phaidon, 2001
Byars, Mel, *The Design Encyclopedia*, Laurence King, 2004
Camard, Florence, *Ruhlmann*, Editions Monelle Hayot, 2009
Cobbers, Arnt, *Marcel Breuer*, Taschen, 2007
——, *Erich Mendelsohn*, Taschen, 2007
Cohen, Jean-Louis, *Le Corbusier*, Taschen, 2006
Cohen, Jean-Louis, et al (eds), *Encyclopédie Perret*, Monum, 2002
Cohn, Laura, *The Door to a Secret Room: A Portrait of Wells Coates*, Scolar Press, 1999
Coles, Stephen, *The Geometry of Type*, Thames & Hudson, 2013
Constant, Caroline, *Eileen Gray*, Phaidon, 2000
Culot, Maurice, et al, *Les Frères Perret: L'Oeuvre Complète*, Institut Français d'Architecture/ Éditions Norma, 2000
Davies, Colin, *Key Houses of the Twentieth Century: Plans, Sections and Elevations*, Laurence King, 2006
Delaporte, Guillemette, *René Herbst: Pioneer of Modernism*, Flammarion, 2004
Deshoulières, Dominique, et al (eds) *Rob Mallet-Stevens: Architecte*, Archives d'Architecture Moderne, 1981
Doordan, Dennis P., *Twentieth-Century Architecture*, Laurence King, 2001
Doren, Harold van, *Industrial Design: A Practical Guide*, McGraw-Hill Book Company, 1940
Dormer, Peter, *Design Since 1945*, Thames & Hudson, 1993
Driller, Joachim, *Breuer Houses*, Phaidon, 2000
Droste, Magdalena, *Bauhaus*, Taschen, 2011
Duncan, Alastair, *Art Deco Complete*, Thames & Hudson, 2009
Dunnett, James, & Gavin Stamp (eds), *Ernö Goldfinger: Works 1*, Architectural Association, 1983
Edwards, Brian, *Basil Spence: 1907–1976*, Rutland Press, 1995
Eidelberg, Martin (ed.), *Design 1935–1965: What Modern Was*, Abrams/Le Musée des Arts Décoratifs de Montréal, 1991
Elwall, Robert, *Ernö Goldfinger*, Academy Editions, 1996
Eskilson, Stephen K., *Graphic Design: A History*, Laurence King, 2012
Estrada, Maria Helena, *Móvel Brasileiro Moderno*, FGV Projetos, 2012
Faber, Tobias, *Arne Jacobsen*, Alec Tiranti, 1964
Falino, Jeannine (ed.), *Crafting Modernism: Midcentury American Art & Design*, Abrams/ Museum of Arts & Design, 2012
Fiell, Charlotte & Peter, *Design of the 20th Century*, Taschen, 1999
——, *Industrial Design A-Z*, Taschen, 2000
——, *Masterpieces of British Design*, Goodman Fiell Publishing, 2012
——, *1000 Chairs*, Taschen, 2005
——, *Scandinavian Design*, Taschen, 2005
——, *Tools for Living: A Sourcebook of Iconic Designs for the Home*, Fiell Publishing, 2010
Fiell, Charlotte & Peter (eds), *1000 Lights: 1879–1959*, Taschen, 2005
——, *1000 Lights: 1960–Present*, Taschen, 2005
Flinchum, Russell, *Henry Dreyfuss, Industrial Designer: The Man in the Brown Suit*, Rizzoli, 1997
Garfield, Simon, *Just My Type*, Profile Books, 2010
Gill, Eric, *An Essay on Typography*, Penguin, 2013
Godau, Marion, & Bernd Polster (eds), *Design Directory: Germany*, Pavilion, 2000
Gössel, Peter, & Gabriele Leuthäuser, *Architecture in the 20th Century*, Taschen, 2005
Gura, Judith, *Scandinavian Furniture: A Sourcebook of Classic Designs for the 21st Century*, Thames & Hudson, 2007
Hanks, David (ed.), *The Century of Modern Design*, Flammarion, 2010
Hays, K. Michael, & Dana Miller (eds), *Buckminster Fuller: Starting with the Universe*, Whitney Museum of American Art/Yale University Press, 2008
Heller, Steven, & Gail Anderson, *New Modernist Type*, Thames & Hudson, 2012
Heller, Steven, & Véronique Vienne, *100 Ideas That Changed Graphic Design*, Laurence King, 2012
Hines, Thomas S., *Richard Neutra and the Search for Modern Architecture*, Rizzoli, 2005
Hollis, Richard, *Graphic Design: A Concise History*, Thames & Hudson, 1994
Ikoku, Ngozi, *The Victoria & Albert Museum's Textile Collection: British Textile Design from 1940 to the Present*, V&A Publishing, 1999
Jackson, Lesley, *20th Century Factory Glass*, Mitchell Beazley, 2000
——, 20th Century Pattern Design: Textile & Wallpaper Pioneers, Mitchell Beazley, 2011
Jenger, Jean, *Le Corbusier: Architect of a New Age*, Thames & Hudson, 1996
Jodard, Paul, *Raymond Loewy*, Trefoil Publications, 1992
Julier, Guy, *Design Since 1900*, Thames & Hudson, 1993
Khan, Hasan-Uddin, *International Style: Modernist Architecture from 1925 to 1965*, Taschen, 2011
Kries, Mateo, et al (eds), *Le Corbusier: The Art of Architecture*, Vitra Design Museum, 2007
Lahti, Louna, *Alvar Aalto*, Taschen, 2004
Lahti, Markku, *Alvar Aalto Houses*, Rakennustieto, 2005

SELECT BIBLIOGRAPHY

Lamprecht, Barbara, *Richard Neutra*, Taschen, 2006

Le Corbusier, *Towards A New Architecture*, Dover, 1931/1986

Leach, Bernard, *A Potter's Book*, Faber & Faber, 2011

Livingstone, Alan & Isabella, *Graphic Design & Designers*, Thames & Hudson, 1992

Loewy, Raymond, *Industrial Design*, Laurence King, 2000

Long, Christopher, *Kem Weber: Designer and Architect*, Yale University Press, 2014

Lupfer, Gilbert, & Paul Sigel, *Gropius*, Taschen, 2006

Lutz, Brian, *Knoll: A Modernist Universe*, Rizzoli, 2010

McCarter, Robert, *Aalto*, Phaidon, 2014

——, *Breuer*, Phaidon, 2016

——, *Frank Lloyd Wright: Architect*, Phaidon, 1997

McCarter, Robert (ed.), *On and By Frank Lloyd Wright: A Primer of Architectural Principles*, Phaidon, 2005

McDermott, Catherine, *20th Century Design*, Carlton Books, 1999

McLean, Ruari, *Jan Tschichold: Typographer*, David R. Godine, 1975

March, Lionel, & Judith Sheine, *R. M. Schindler: Composition and Construction*, Academy Editions, 1993

Martin-Vivier, Pierre-Emmanuel, *Jean-Michel Frank: The Strange and Subtle Luxury of the Parisian Haute-Monde in the Art Deco Period*, Rizzoli, 2008

Massey, Anne, *Interior Design of the 20th Century*, Thames & Hudson, 1990

Mertins, Detlef, *Mies*, Phaidon, 2011

Miller, Judith, *Miller's 20th Century Design*, Miller's/Mitchell Beazley, 2009

Morel-Journel, Guillemette, *Le Corbusier's Villa Savoye*, Éditions du Patrimoine, 1998

Mulder, Bertus, & Ida van Zijl, *Rietveld-Schröder House*, Princeton Architectural Press, 1999

Nahum, Andrew, *Fifty Cars That Changed the World*, Conran Octopus, 2009

Neumann, Claudia (ed.), *Design Directory: Italy*, Pavilion, 1999

Noever, Peter (ed.), *Schindler by MAK*, Prestel, 2005

Overy, Paul, *De Stijl*, Thames & Hudson, 1991

Overy, Paul, et al, *The Rietveld Schröder House*, Butterworth, 1988

Pallasmaa, Juhani (ed.), *Villa Mairea: 1938–39*, Alvar Aalto Foundation, 1998

Pallasmaa, Juhani, & Tomoko Sato (eds), *Alvar Aalto: Through the Eyes of Shigeru Ban*, Black Dog Publishing, 2007

Peters, Nils, *Jean Prouvé*, Taschen, 2006

Pfeiffer, Bruce Brooks, *Frank Lloyd Wright*, Taschen, 2006

Phaidon Design Classics: Volume Two, Phaidon, 2006

Pinchon, Jean-François (ed.), *Rob Mallet-Stevens: Architecture, Furniture, Interior Design*, MIT Press, 1990

Polster, Bernd (ed.), *Design Directory: Scandinavia*, Pavilion, 1999

Polster, Berd, et al, *The A-Z of Modern Design*, Merrell, 2009

Postiglione, Gennaro, et al (eds), *One Hundred Houses for One Hundred Architects*, Taschen, 2004

Powers, Alan, *Modern: The Modern Movement in Britain*, Merrell, 2005

——, *Serge Chermayeff: Designer, Architect, Teacher*, RIBA, 2001

Raizman, David, *History of Modern Design*, Laurence King, 2010

Reading, Malcolm, & Peter Coe, *Lubetkin & Tecton: An Architectural Study*, Triangle Architectural Publishing, 1992

Roberts, Caroline, *Graphic Design Visionaries*, Laurence King, 2015

Roccella, Graziella, *Gio Ponti*, Taschen, 2009

Ross, Phyllis, *Gilbert Rohde: Modern Design for Modern Living*, Yale University Press, 2009

Sbriglio, Jacques, *Le Corbusier: The Villa Savoye*, Birkhaüser, 2008

Schumacher, Thomas L., *Surface & Symbol: Giuseppe Terragni and the Architecture of Italian Rationalism*, Princeton Architectural Press, 1991

Sharp, Dennis, & Sally Rendel, *Connell, Ward & Lucas: Modern Movement Architects in England, 1929–1939*, Frances Lincoln, 2008

Sheine, Judith, *R. M. Schindler*, Phaidon, 2001

Smith, Kathryn, *Schindler House*, Harry N. Abrams, 2001

Solaguren-Beascoa, Félix, *Arne Jacobsen: Approach to his Complete Works, 1926–1949*, Danish Architectural Press, 2002

——, *Arne Jacobsen: Works and Projects*, Gustavo Gili, 1989

Sparke, Penny, *A Century of Car Design*, Mitchell Beazley, 2002

Sparke, Penny (ed.), *Design Directory: Great Britain*, Pavilion, 2001

Syring, Eberhard, & Jörg C. Kirschenmann, *Scharoun*, Taschen, 2004

Terragni, Attilio, Daniel Libeskind, & Paolo Rosselli, *The Terragni Atlas: Built Architecture*, Skira, 2004

Thiel-Siling, Sabine (ed.), *Icons of Architecture: The 20th Century*, Prestel, 2005

Tschichold, Jan, *The New Typography*, University of California Press, 1998

Vellay, Dominique, & François Halard, *La Maison de Verre: Pierre Chareau's Modernist Masterwork*, Thames & Hudson, 2007

Waal, Edmund de, *The Pot Book*, Phaidon, 2011

——, *20th Century Ceramics*, Thames & Hudson, 2003

Warburton, Nigel, *Ernö Goldfinger: The Life of an Architect*, Routledge, 2004

Weston, Richard, *Architecture Visionaries*, Laurence King, 2015

——, *The House in the Twentieth Century*, Laurence King, 2002

——, *Key Buildings of the 20th Century*, Laurence King, 2004

——, *Villa Mairea: Alvar Aalto*, Phaidon, 2002

Whitford, Frank, *Bauhaus*, Thames & Hudson, 1984

Wildenhain, Marguerite, *Pottery: Form and Expression*, American Craftsmen's Council & Reinhold, 1961

Wilhide, Elizabeth, *Design: The Whole Story*, Thames & Hudson, 2016

Windsor, Alan, *Peter Behrens: Architect and Designer 1868–1940*, The Architectural Press, 1981

Wittkopp, Gregory (ed.), *Saarinen House and Garden: A Total Work of Art*, Harry N. Abrams/Cranbrook Academy of Art Museum, 1995

Wright, Russel, *Russel Wright: Good Design is For Everyone, In His Own Words*, Manitoga/The Russel Wright Design Center/Universe, 2001

Yorke, Malcolm, *Edward Bawden and his Circle*, ACC Art Books, 2007

Zevi, Bruno, *Giuseppe Terragni*, Triangle Architectural Publishing, 1989

Zijl, Ida van, *Gerrit Rietveld*, Phaidon, 2010

Zimmerman, Claire, *Mies van der Rohe*, Taschen, 2006

All photographs are courtesy of Wright, wright20.com, unless otherwise stated.

All photographs in the 'Houses and Interiors' section are © Richard Powers unless otherwise stated.

a = above; b = below; c = centre; l = left; r = right

pp. 7 Bauhaus, Dessau, 1926; 8 Photo Peter Horree/Alamy Stock Photo. Herbert Bayer © DACS 2018; 9 National Film Archive (BFI), London; 10 World History Archive/Alamy Stock Photo; 17 Photo of Mme Mathieu-Lévy's apartment, rue de Lota, Paris; 18 Karl Emanuel Martin (Kem) Weber papers, Architecture and Design Collection. Art, Design & Architecture Museum, University of California, Santa Barbara; 20 Exposition Coloniale, Paris, 1931; 21 Bibliothèque des Arts Décoratifs, Paris/Archives Charmet/Bridgeman Images; 23 UAM exhibition catalogue cover, Musée des Arts Décoratifs, 1930; 25 Photo Ullstein Bild via Getty Images; 26, 27 Vitra Design Museum, photo Jürgen Hans, objektfotograf.ch; 28 Karmstol chair for Nordiska Kompaniet, 1931; 30, 31a, 31b Courtesy Källemo; 46a, 47al, 47ar Christie's Images; 47bl Collection Félix Marcilhac, Paris; 47br Collection Arc-en-Seine, Paris; 48, 49al, 49ar, 49c, 49b Image supplied by Aram Designs, holder of the worldwide licence for Eileen Gray designs; 50l, 50r Image supplied by Aram Designs, holder of the worldwide licence for Eileen Gray designs. Photo Shira Klasmer; 51al Image supplied by Aram Designs, holder of the worldwide licence for Eileen Gray designs; 51ar, 51b, 52a Image supplied by Aram Designs, holder of the worldwide licence for Eileen Gray designs. Photo Shira Klasmer; 52bl Image supplied by Aram Designs, holder of the worldwide licence for Eileen Gray designs; 52br Image supplied by Aram Designs, holder of the worldwide licence for Eileen Gray designs. Photo Shira Klasmer; 53 Image supplied by Aram Designs, holder of the worldwide licence for Eileen Gray designs; 56–7 Les Arts Décoratifs, Paris/akg-images; 63br Photo Bukowskis Auctions; 70a Christie's Images; 106 Yale University Art Gallery, Promised gift of John P. Axelrod; 107a Metropolitan Museum of Art, New York. Purchase, Theodore R. Gamble, Jr. Gift in honor of his mother, Mrs Theodore Gamble, 1985; 107c Yale University Art Gallery, Promised gift of John C. Waddell; 107b, 108 Sotheby's, New York; 109a Los Angeles County Museum of Art (LACMA), Los Angeles. Purchased with funds provided by the Decorative Arts and Design Deaccession Fund, Viveca Paulin-Ferrell and Will Ferrell, Shannon and Peter Loughrey, Heidi and Said Saffari, and Holly and Albert Baril; 109b High Museum, Atlanta. Purchase with funds from the Decorative Arts Acquisition Trust; 113 Interfoto/Alamy Stock Photo; 117 Photo Werner/Library of Congress/Corbis/VCG via Getty Images; 119 Courtesy of Terry Tynan Lighting (replica); 125, 126, 127b Courtesy Anglepoise® archives; 157 Imagno/akg-images; 161a Greg Slater; 161b Museum of Art, Rhode Island School of Design, Providence. Gift of Mr. and Mrs. Alfred T. Morris, Jr. (1999.89.3); 162 Museum of Fine Arts, Boston. Gift of John P. Axelrod (1990.507)/Scala, Florence; 164 City Museum and Art Gallery, Stoke-on-Trent, UK; 165a, 165b, 166a, 166b, 167a, 167b Photo Sarah White, courtesy The Leach Pottery, St Ives, UK; 168, 170l, 170r, 171c, 171b, 174 Courtesy Quittenbaum Kunstauktionen GmbH; 175, 176a Minneapolis Institute of Arts, MN, USA/The Modernism Collection, gift of Norwest Bank Minnesota/Bridgeman Images; 176c Museum of Art, Rhode Island School of Design, Providence. Mary B. Jackson Fund (2008.28.2). Photo Erik Gould; 176b Museum of Art, Rhode Island School of Design, Providence. Mary B. Jackson Fund (2008.28.1). Photo Erik Gould; 177 Los Angeles County Museum (LACMA), Los Angeles. The Ellen Palevsky Cup Collection, Gift of Max Palevsky (AC1998.265.87.1-.2). Digital Image Museum Associates/LACMA/Art Resource NY/Scala, Florence; 190 Los Angeles County Museum (LACMA), Los Angeles. Gift of Pauline Blank (AC1992.226.2a-b). Digital Image Museum Associates/LACMA/Art Resource NY/Scala, Florence; 191a, 191b, 192 Courtesy Quittenbaum Kunstauktionen GmbH; 193a Museum of Modern Art, New York. Gift of Gump's (183.1948). Digital image, The Museum of Modern Art, New York/Scala, Florence; 193b Museum of Modern Art, New York. Gift of Gump's (182.1948). Digital image, The Museum of Modern Art, New York/Scala, Florence; 194 Courtesy Quittenbaum Kunstauktionen GmbH; 195 Los Angeles County Museum of Art (LACMA), Los Angeles. Gift of Pauline Blank (AC1992.226.8). Digital Image Museum Associates/LACMA/Art Resource NY/Scala, Florence; 203, 224b Cooper-Hewitt, Smithsonian Design Museum. Museum purchase from the Decorative Arts Association Acquisition Fund (1994-73-2). Photo Hiro Ihara. Smithsonian Institution/Art Resource, NY/Scala, Florence; 204 Industrial Design, published by McGrath (1954); 205 Granger Historical Picture Archive/Alamy Stock Photo; 209 Harry Ransom Center, The University of Texas, Austin. © The Edith Lutyens and Norman Bel Geddes Foundation, Inc.; 211 New York, Metropolitan Museum of Art. Black and white photograph. Image The Metropolitan Museum of Art/Art Resource/Scala, Florence; 212 Yale University Art Gallery, New Haven, CT; 213a *Popular Mechanics*, December 1940; 213b The Montreal Museum of Fine Arts, Liliane and David M. Stewart Collection, gift of Eric Brill (2010.1491). Photo The Montreal Museum of Fine Arts, Denis Farley; 215 Photo Fine Art Images/Heritage Images/Getty Images; 217al Los Angeles County Museum (LACMA), Los Angeles. Gift of Max Palevsky (M.2003.16.1a-d); 217ar Los Angeles County Museum (LACMA), Los Angeles. Gift of Max Palevsky (M.2002.108.4a-e); 217bl Los Angeles County Museum (LACMA), Los Angeles. Gift of Max Palevsky (M.2001.191.2a-d); 217br Los Angeles County Museum (LACMA), Los Angeles. Gift of Max Palevsky (M.2002.108.3a-f); 218 Minneapolis Institute of Arts, Gift of Funds from the Decorative Arts Council (2005.79)/Bridgeman Images; 220–1 Minneapolis Institute of Arts, Gift of funds from Don and Diana Lee Lucker (2003.126)/Bridgeman Images; 222 Photo Underwood Archives/Getty Images; 223 Photo R. Gates/Hulton Archive/Getty Images; 224a Courtesy The Advertising Archives; 226 Courtesy the Estate of R. Buckminster Fuller; 227 Photo Hazel Larson Archer. Courtesy the Estate of R. Buckminster Fuller; 228 Granger Historical Picture Archive/Alamy Stock Photo. Courtesy the Estate of R. Buckminster Fuller; 229a, 229b Courtesy the Estate of R. Buckminster Fuller; 230 V&A Images/Alamy Stock Photo; 233 Chronicle/Alamy Stock Photo; 234–5 Harry Ransom Center, The University of Texas, Austin. © The Edith Lutyens and Norman Bel Geddes Foundation, Inc.; 236 Photo Paul Popper/Popperfoto/Getty Images; 237 Bettmann/Getty Images; 238a Raymond Loewy™ used with permission from Loewy Design LLC. www.RaymondLoewy.com; 239a Photo Sherman Oaks Antique Mall/Getty Images; 239 Raymond Loewy™ used with permission from Loewy Design LLC. www.RaymondLoewy.com; 240l Photo Keystone Features/Getty Images; 240r Bettmann/Getty

PICTURE CREDITS

Images; 241 Lebrecht Music and Arts Photo Library/Alamy Stock Photo; 253 The J. Paul Getty Museum, Los Angeles; 255a, 255b © The Estate of Jan Tschichold; 256 Travel poster, Jean de-Luz, 1928. H. Chachoin Paris; 257a Travel poster, Empire Airways; 257b The Art Institute of Chicago, Ryerson and Burnham Libraries, Institute of Design Papers (2007.2, folder 1.11); 259 *For the Voice*, 1923; 260 Kurt Schwitters © DACS 2018; 261 Merrill C. Berman Collection, New York. Max Bill © DACS 2018; 262–3 © The Josef and Anni Albers Foundation/VG Bild-Kunst, Bonn and DACS, London 2018; 264–5 The Josef and Anni Albers Foundation (1976.3.124). Photo Tim Nighswander/ Imaging4Art. © The Josef and Anni Albers Foundation/VG Bild-Kunst, Bonn and DACS, London 2018; 266–7 © The Josef and Anni Albers Foundation/VG Bild-Kunst, Bonn and DACS, London 2018; 268, 269, 270, 271 Herbert Bayer © DACS 2018; 273, 274–5a, 274–5b © & ™ MOURON. CASSANDRE. Lic 2017-22-11-04 www.cassandre.fr; 276 *Principles of Neo-Plastic Art*, 1924; 277 *Mécano n3* magazine, 1923; 278l The Little Review, 1925; 278r Vilmos Huszár © DACS 2018; 279 *De Stijl*, issue 12, 1925; 281l © The Estate of Jan Tschichold; 283l Erwin Piscator, *Das Politische Theater (The Political Theatre)*. A Schultz, 1929; 283r, 284 akg-images; 285a London Transport Museum; 285b Private collection; 286 Heritage Image Partnership Ltd/Alamy Stock Photo. © Rodchenko & Stepanova Archive, DACS, RAO 2018; 287 Universal Art Archive/Alamy Stock Photo. © Rodchenko & Stepanova Archive, DACS, RAO 2018; 288 © Rodchenko & Stepanova Archive, DACS, RAO 2018; 289 Heritage Image Partnership Ltd/Alamy Stock Photo. © Rodchenko & Stepanova Archive, DACS, RAO 2018; 290, 291l, 291r, 292l, 292r, 293l, 293r © The Estate of Jan Tschichold; 297l Everett Collection Inc/Alamy Stock Photo. Le Corbusier © FLC/ADAGP, Paris and DACS, London 2018; 297r Museum of Modern Art, New York. Digital image, The Museum of Modern Art, New York/Scala, Florence; 298 Photo SSPL/Getty Images; 299a Courtesy the University of East Anglia, Pritchard Papers Collection; 299b Andrew Hasson/Alamy Stock Photo; 301l The Geffrye Museum of the Home, London, UK/ John Hammond/Bridgeman Images; 301r Private Collection/Photo Christie's Images/Bridgeman Images; 302 Everett Collection Inc/Alamy Stock Photo; 303 gta Archives/Institut für Geschichte und Theorie der Architektur, TH-Hönggerberg, ETH Zurich (private archive Arthur Rüegg, Zurich); 305 University Archives and Special Collections, Paul V. Galcin Library, Illinois Institute of Technology; 306 Harvard Art Museums/Busch-Reisinger Museum, Gift of Walter Gropius (BRGA.87.4) Photo Imaging Department. Courtesy President and Fellows of Harvard College; 307 Museum of Modern Art, New York. Digital image, The Museum of Modern Art, New York/Scala, Florence; 324–5, 326, 327 © Mark Lyon; 336–7, 338, 339 The National Trust Photo Library/Alamy Stock Photo; 380–81 © Karin Hessmann/Artur Images; 382 Stillman Rogers/Alamy Stock Photo; 383 V. Dorosz/Alamy Stock Photo; 384–5 © Michael Freeman; 387a © J. Paul Getty Trust. Getty Research Institute, Los Angeles (2004.R.10); 387b Photo Richard Hartog/Los Angeles Times via Getty Images; 396–7 Photo Loop Images/UIG via Getty Images; 398l, 398r Photo Francis Carr © Thames & Hudson Ltd; 399 Photo Maurici Mayol/age footstock; 408–9, 410, 411a, 411b Fritz von der Schulenburg/The Interior Archive; 440a Collection Bauhaus-Archiv, Berlin. © The Josef and Anni Albers Foundation/Artists Rights Society (ARS), New York and DACS, London 2018; 440b Association d'Histoire de l'Architecture (AHA), Paris; 441 Arcaid Images/Alamy Stock Photo; 442a Photo National Motor Museum/Heritage Images/ Getty Images; 443 Hearst Publishers; 444a Pritchard Archives, The Library, University of East Anglia, Norwich; 445a Arcaid Images/Alamy Stock Photo; 447a Photo Hulton Archive/Getty Images; 449a © Richard Powers; 449b VIEW Pictures Ltd/ Alamy Stock Photo; 450c St Bride's Printing Library, London; 451a designundklassiker.de; 451b Tim Beddow/The Interior Archive; 457b Michael Heath/Alamy Stock Photo; 460b © J. Paul Getty Trust. Getty Research Institute, Los Angeles (2004.R.10; Job 093); 462 below Musée des Arts Décoratifs, Paris/Jean Tholance/akg-images; 464b Photo Ullstein Bild; 465a Photo Ullstein Bild via Getty Images; 465b Museum für Kunst und Kulturgeschichte, Lübeck. Gunta Stölzl-Stadler © DACS 2018; 466a imageBroker/Alamy Stock Photo; 467b Karl Emanuel Martin (Kem) Weber papers, Architecture and Design Collection. Art, Design & Architecture Museum, University of California, Santa Barbara; 468a Private collection; 468b Piet Zwart © DACS 2018.

ACKNOWLEDGMENTS

Dominic Bradbury would like to express his gratitude to all those who have contributed, in one way or another, to the evolution and production of this book. They include the contributing essayists, many of whom provided additional advice and assistance that was much appreciated. Sincere thanks are also due to the owners and guardians of the houses featured in this book for their support and their permission to photograph their homes.

Particular thanks are due to Richard Wright at the Wright20 auction house (www.wright20.com) in Chicago, along with his colleagues Jennifer Mahanay and Todd Simeone, for their assistance and support in collating many of the images of furniture, glass, ceramics and other products that appear in the book. Thanks also to those many manufacturers and producers who have assisted us by supplying additional images and material.

In addition, special thanks are owed to the following: photographer Richard Powers, Gordon Wise and Niall Harman at Curtis Brown, translators Ruth Sharman and David Henry Wilson, as well as Faith, Florence, Cecily and Noah Bradbury.

Gratitude and appreciation are also due to all at Thames & Hudson, particularly Lucas Dietrich, Fleur Jones, Maria Ranauro and Sadie Butler, as well as editor Jenny Wilson and designer Roger Fawcett-Tang. This book would have been an impossible task without their much valued work and support.

INDEX

Page numbers in **bold** refer to main entries. Page numbers in *italics* refer to illustrations

Aalto, Alvar 17, 18, 19, 62, 64, 108, 296, 297, 299, 300, 303, **308–13**, 361, 365, 402, 435, 443, 449, 453
AEG 112, *113*, 205–6, 214, *214*, *215*, 216, *216*
Air King Products *219*
Airline Chair Company *106*
A. J. Iversen Snedkermester *456*
Albers, Anni 25, 262, 265, 305, **440**, *440*
Albers, Josef 26, 120, 255, 256, 257, **262–7**, 268, 280–1, 305–6, 350, 440
American Locomotive Company *223*
American Modern Decoration *43*
American National *218*, *220–1*, 221
American Thermos Bottle Co. *225*
Amodec *39*
Amsterdam School 212
André, Jacques **440**, *440*
André, Michel *440*
The Architects Collaborative 353
Arens, Egmont 210
Art Deco 6, 9, 11, **440**
architecture 326, 461
ceramics & glass 157, 158, 162–3, 178, 180, 456
furniture 18, 21, 38, 44, 48, 100, 446, 451, 462, 463
graphics & posters 261, 265, 272
houses and interiors 402, *404*, *405*, 448, 462, 463, 468
industrial & product design 210, 242
lighting 114, 116, 118, 140, 461
Art Nouveau 6, 454
furniture 20, 84, 446, 457
glasswork & ceramics 445
graphics and posters 254
houses and interiors 463, 467
lighting 112, 114, 116
Artek 310
Artek-Pascoe 442
Arts and Crafts
ceramics & glass 156, 160, 164
furniture 20, 24, 28, 72, 451
graphics and posters 450
houses and interiors 402, *404*, *405*, 431
lighting 112, 113
Arup, Ove **440**, 457
Asplund, Erik Gunnar **28–31**
Ateliers Jean Prouvé 22, *64*, 84–91
Au Bûcheron 272, *274–5*
Automobile Industry Association 240–1

Badovici, Jean 48, 50, 298, **340–7**
BAG 466
Bamberg Metallwerkstatten *81*
Bang & Olufsen 207
Barcelona International Exposition (1929) 26, 60, 80, 134, 382, 461
Barford, George *282*
Barker, William 106
Barker Brothers 106, 108
Barr, Alfred H. Jr. 210, 212, 306, 307, 453
Bauer *196*, *198–9*, 199
Bauhaus 6, 8, *8*, 10, 11, 18, 21, **24–7**, 28, 78, 92, 114, 116–17, 156–7, 192, 206, 254, 255, 265, *269*, 270, 286, 353, 362, 454, 455, 458, 464, 466
Berlin 26, 304
Dessau 7, 16, 25–6, 32, 98, 120, 128, 170, 246, 262, 268, 281–2, 304, 306, 350, 382, 402, 459
New Bauhaus 280, 282, **304–7**, *305*
Weimar 24–6, 32, 34, 120, 128, 156, 161–2, 168, 170, 174, 190, 246, 262, 268, 278, 280, 281, 289, 304, 350, 453, *464*, 465
Bauhaus Press 281
Bawden, Edward **441**
Bayer, Herbert *8*, 255, 256, 259, 260, **268–71**, 280, 281, 290
Bazeley, Geoffrey **441**, *441*
Beall, Lester 257, **441**
Bechstein, Max 168
Beck, Henry 'Harry' 256, **441**
Behrens, Peter 24, 78, 112, *112*, *113*, 204, 205, **214–17**, 350, 369, 446, 447
Bel Geddes, Norman 113, 196, 204, 207, 208, *209*, 210, 212, 213, 218, 222, **230–5**, 244, 455
The Bell Telephone Company 225
Belmag *130*
Benton, Tim 300–3, 469
Berglund, Tor *402*, *403*
Berlag *110*
Bernhard, Lucian **441**
Bertoia, Harry 18, 402, 455
Bertoni, Flaminio **442**, *442*
Bianconi, Fulvio 184
Bijvoët, Bernard 326
Bill, Max 257, 261, *261*, 265, 292, **442**, *442*, 445
Black Mountain College, Lake Eden 305–6, *306*, 440
Blackwell, Ursula 338
Blanchon, Georges *65*
Bloch, Julius *447*
Boccioni, Umberto 302
Boeing 244
Bogler, Theodor 162, **168–73**, 174, 190, 455
Bonet, Antonio **442**
Booth, George 402
Bourke-White, Margaret 209
Brandt, Edgar 21, 116
Brandt, Marianne 25, 114, 117, **120–3**, 128
Braque, Georges 20, 393
Braun 249
Bredendieck, Hin 120, *121*, 307
Breuer, Marcel *14–15*, 16, 17, 25, 26–7, **32–7**, 38, 40, 92, 94, 98, 108, 262, 268, *285*, 300, 303, 304, 306, 307, 350, 353, *356*, *357*, 443, 451, 453, 461, 465, 466, 468
Brik, Lilya *286*, 289
Brinkman, Johannes **314–23**
Brinkman & Van der Vlugt **314–23**, 450
Brodovitch, Alexey 257, **442**, *443*
Buchanan, Mel 160–3, 469
Bünte & Remmler 128
Buquet, Edouard-Wilfrid 124
Burg Giebichenstein School of Fine and Applied Art 190, *191*, *192*, *194*, 458
Burnet, John James 443
Burnet, Tait & Lorne **443**
Buzzi, Tomaso 158, 182

Calder, Alexander 334, **443**
Cappellin, Giacomo 182, 183
Cardew, Michael 156, *157*, 158, 160, **164–7**
Cardine Accessories 125
Carwardine, George 114–15, **124–7**
Cassandre, A. M. 256–7, **272–5**, 393, 444, 459, 463
Cassina *11*, *29*, 30, 68, 92, *92*, 94
Castiglioni, Achille and Pier Giacomo 148
Century of Progress International Exposition, Chicago (1933–34) 209–10, 228, *228*
Chace, Marian and Clyde 414
Chachoin, H. *256*
Chanaux, Adolphe 46
Chanaux & Co. *45*, 46, *46*, *47*
Chaplin, Charlie 9, *9*
Chareau, Paul *379*
Chareau, Pierre 8, 21, 22, 32, 115, 118, 298, **324–7**, 379, 440, 443, 466
Cheney, Martha Candler 208
Cheney, Sheldon 208
Chéret, Jules 254
Chermayeff, Serge 213, *298*, 299, **328–35**, 441, 444, 459
Cheuret, Albert 114, 116
Chiesa, Pietro 113, 118, **146–53**
Christofle 146
Chrysler 209–10, 228, 232
CIAM (International Congress of Modern Architecture) 8, 338, **443**, 458, 464, 465
Citroën 442, *442*
Clark, Garth 163
Cliff, Clarice 157, *161*, 162
Coates, Wells 298, 299, *299*, 440, **444**, *444*, 451, 453, 458
Colin, Paul **444**
Comté 46
Conant-Ball 199
Connell, Amyas 444
Connell, Ward & Lucas 298–9, **444**, *445*
Constructivism 6, 8, 21, 255, 258, 268, 272, 280, 286, 290, 441, **444**, 445, 457
Container Corporation of America (CCA) 270
Cooper, Susie 162
Coray, Hans **445**
Cowan Pottery Studio *162*, 163
Crabtree, William **445**
Cranbrook Academy of Art 402, 446, 455
Cubism 20, 255, 272, 286, *376*, *378*, *388–9*, 456

Dadaism 278, 280, 451
Daimler-Benz 241
Dalsace, Annie 326
Dalsace, Dr Jean 326, *326*
Daum Frères 112–13, 116, 157
Daum Nancy 116

De Pree, D. J. 100, 452
De Sphinx 276
De Stijl 6, 8, 18, 21, 32, 48, 92, 254, 255, 268, 270, 276, 278, 280, 290, 398, *398*, 444, **445**, 452, 456, 461, 468
De Stijl journal 276, 278, *278*, *279*, 445
Decoeur, Émile 157, **445**
Delaunay, Sonia 21, *64*, **445**, *445*
Dell, Christian *110*, 114, 117, **128–31**
Depero, Fortunato **445**
Der Stuhl 300, *301*
Deskey, Donald 11, 16, 18, 38–43, 118, *119*, 140, 210, 440
Deskey-Vollmer 38, *40*, *41*, *42*
Deutscher Werkbund 8, 24, 78, 140, 161, 205, 214, 270, 382, 410, **446**, 453, 461, 466, 467
Dexel, Walter 260, 261
Die Neue Linie 282, *283*, *284*
Dieckmann, Erich **446**, *446*
Djo-Bourgeois, Elise *322*
Dobrolet 289, *289*
Doesburg, Theo van 255, 258, *260*, **276–9**, 443, 445, 456
Dohrn, Wolf 205–6
Doren, Harold van 204, *204*, 205, 210, **218–21**
Dreier, Katherine S. 305
Drew, Jane 449
Dreyfuss, Henry 113, *203*, 204, 208, 210, 212, **222–5**, 232, 244, 256
Dufrêne, Maurice **446**
Dufy, Raoul 157
Dupré-Lafon, Paul *446*, **446**

Eames, Charles and Ray 17, 365, 402, **446**, 459
Earl, Harley 206, 209, **447**, *447*
Eastman, George 242
Eastman Kodak 242, 244, *244*
Edison, Thomas 112, 214
Editions Steph Simon *70*
E. K. Cole & Son 444
Ekman, Consul Johan 178
Electrolux 232
Emberton, Joseph **447**
Embru-Werke 34, *35*
Engelmann, Richard 168
Ercol 447
Ercolani, Lucian **447**
Esherick, Wharton 19, **447**, *447*
Exposition Coloniale, Paris (1931) 20, *20*
Exposition Internationale des Arts Décoratifs et Industriels Modernes, Paris (1925) 21, 30, 38, 108, 134, 162, 210, 272, 289, 301, 440, 442, 446
Exposition Internationale des Arts et Techniques dans la Vie Moderne (1937) 74, *148*, 464
Exposition Universelle, Paris (1900) 402
Expressionism **447**, 449, 459

Feininger, Lyonel 24, 25, 304, 305
Ferrari-Hardoy, Jorge **442**
Fiat 450
Firma Karl Mathsson *19*, 72–7
FontanaArte 118, 146–53
Ford, Henry 6, 156, 205, 209, 232, **448**
Ford Motor Company 205, *205*, 209, 241, 304, 448
Fortuny, Mariano 113, **448**, *448*
Frank, Jean-Michel 18, 20, 44–7, 326, 376, 450
Frank, Josef **448**, 464, *466*
Frankfurt Motor Show (1955) *240*
Frankl, Paul Theodore 38, 210, **448**
Frey, Albert **448**, *449*
Fritz Hansen 74, 365
Fry, Edwin Maxwell 350, **449**, *449*, 458
Fuller, Anne Hewlett *229*
Fuller, Richard Buckminster 207, 209–10, **226–9**, 232
Fuller Houses Inc. 228
Functionalism **449**, 458, 462
Futurism 6, 9, 302, 303, 423, 445, **449**, 460, 462

Gallé, Émile 20, 112–13, 116
General Electric Company 225, 232
General Motors 206, 208, 209, *209*, 213, 230, *232*, 447, *447*
Georg Jensen *463*, 463
Gestetner, Sigmund 457
Giacometti, Alberto & Diego 46, 379, **450**, *450*, 458
Giacosa, Dante **450**
Gibberd, Frederick **450**
Gibling, Sophie Pauline 414
Giedion, Sigfried *303*
Gill, Eric 156, 256, 292, **450**, *450*
Gispen *320–1*, *323*
Gispen, Willem Hendrik 318, **450**, *450*
Golden Gate Exposition, San Francisco (1939) *109*
Goldfinger, Ernö 298, 334, **336–9**, 390, 440
Gomme, Edward 447
Graf Zeppelin 301, *302*
Grand Rapids Chair Company *107*
Gray, Eileen 9, 11, 16, 17–18, 19, 21, 38, **48–53**, 115, 117, 298, 300, 326, **340–7**, 440, 466
Greene & Greene 113
Gregory, Waylande 163
Greta Pottery 176
Gretsch, Hermann **451**, *451*
Greyhound 236, *236*
Groenekan, Gerard van de 92, *93*, *94*, *97*, *115*
Gropius, Walter 6, *7*, 16, 24, 25, *25*, 26, 27, 32, 34, 114, 120, 122, 156, 157, 162, 168, 192, 216, 254, 255, 262, 265, 270, 278, 280, 281, 282, 296, 304, 306–7, **348–57**, 402, 423, 443, 449, 453, 459, 461, 467
Gruppo 7 423
Guévrékian, Gabriel *376*, 379
Guild, Lurelle 210, 212, *212*
Guimard, Hector 20
Gullichsen, Harry 310
Gullichsen, Maire 310
Gustavsberg Pottery 157, *159*
Gwynne, Patrick **451**, *451*

Haël-Werkstätten 174, *176*, *177*
Hald, Edward 158, **178–81**
Hallicrafters *238*
Hamada, Shoji 160, 164
Harling, Robert 261
Harper's Bazaar 238, 257, 272, 442, *443*
Harris, Harwell Hamilton 304
Haus Schminke, Löbau, Germany **408–11**
Heal, Ambrose 174, 176, **451**
Heal & Son *36*, 451
Heal's 174, 176
Heartfield, John **451**
Hein, Piet 74
Henningsen, Frits **452**, *452*
Henningsen, Poul 115, 118, **132–9**, 147
Herbert Terry & Sons *124*, *125*, 127
Herbst, René 16, 18, 21, *21*, **54–7**, 117
Herman Miller 98, 100–5, *206*, *207*, **452**, *452*
Herman Miller Clock Company 100
Hermès *446*
Herr, Gordon 192
Hill, Oliver 450, **452**
Hitchcock, Henry-Russell 210, 296, 306, 447
Hitler, Adolf 10, 174, 190, 240, 241, 260, 423, 465
Hjorth, Axel Einar **452**, *452*
Hoff, Robert van 't 255, 445, **452**
Hoffmann, Josef 78, 114, 446, **453**, *453*, 463, 464, 467
Holden, Charles **453**
Honeywell 225
Hoover 225
Hoover, William Henry 225
Horace, Charles 106
Horgenglarus *442*
Horstmann, Sidney 124–5
Howe, George 212, 304, 456
Hoy, Anne H. 208–13, 469
Huszár, Vilmos 276, *278*

IBM *213*
Il Labirinto 147
International Graphic Style 257, 290
International Style 11, 72, 210, 212, 296–7, 298, 299, 306, 353, 435, **453**, 454, 456, 461

Isokon *32*, 34, *35*, *36*, 282, *285*, 298, *299*, 444, 451, **453**, *453*
Italian Movement for Rational Architecture (MIAR) 457
Itten, Johannes 27, 32, 174, 262, 268, 280–1, 305, **453**, 455, 465

Jacob Lohner & Company 240
Jacobsen, Arne 19, **358–65**, 456
Jacobsen, Jacob 127
Jeanneret, Pierre 16, 17, 21, 23, 30, **64–71**, 301, 369, 370, 443, 449
Jenaer Glaswerke *246*
Johnson, Philip 210, 296, 299, 306, 307, *307*, 447, 453, **454**, *454*, 461
Johnston, Edward 256
Jones, Richard *299*
Joseph Müller Berliner Metallgewerbe *81*
Jourdain, Francis **454**, *454*
Jucker, Carl Jacob 246
Jugendstil 214, 254, **454**, 467

Kaastrup-Olsen, Sophus 132, 134
Kåge, Wilhelm 157, *159*, **454**
Kahn, Albert *205*, 304
Kaiser & Co 128–31

Källemo *30*, *31*
Kandem Lighting 117, 120–3
Kandinsky, Wassily 8, 24, 32, 120, 161, *177*, 265, 268, 305, 350, **454**, 455, *455*
Karlsruhe Majolika *169*, 170, *172–3*
Kassák, Lajos 260
Kauffer, Edward McKnight 257
Kaufmann, Edgar **428–37**
Kentgens-Craig, Margret 304–7, 469
Kino Glaz *288*, 289
Kjærholm, Poul 19, 58
Klee, Paul 8, 24, 120, 265, 305, 350, **455**, 465
Klerk, Michel de 212
Klint, Kaare 17, **58–61**, 62, 452, 462
Knoll *32*, 34, *34*, 78–80, *82*, 256, 440, 442, 455, *455*, 459
Knoll, Florence 80, 402, **455**, *455*
Knoll, Hans 80, **455**, 462
Koch, Mogens 17, **62–3**
Kodak 242, 244, *244*
Komenda, Erwin 241
Königliche Porzellan-Manufaktur 190
Körting & Mathiesen 122
Kosta 158
Kramer, Ferdinand 17, **455**, *455*
Krehan, Max 156, 161, 168, 170, 174, 190, **455**, 458
Kugler, Jolanthe 24–7, 469
Kuhn, Charles 307
Kurchan, Juan **442**

Lalique, René 114, 116, 157
Lallemant, Robert *155*, 157, *158*, **456**, *456*
Lampitt, Ronald *298*
Lanvin, Jeanne 462, *462*
Lassen, Flemming 361, **456**, *456*
Lawson Time Inc. *206*
Le Corbusier (Charles-Édouard Jeanneret) 6, 8, 9, 16, 17, 18, 21, 23, 30, 50, **64–71**, 86, 116, 117, 208, 216, 244, 254, *294–5*, 296, 297, *297*, 298, 300–1, 303, 343, 344, **366–73**, 376, 390, 423, 424, 442, 443, 448, 449, 451, 453, 456, 462, 464, 466, 467
Le Klint 58, 62
Leach, Bernard 156, 158, 160, **164–7**, 192
Leach Pottery 160, 164, *165*, 166
Leck, Bart van der 318, **456**
Léger, Fernand 68, 338, **456**
Legrain, Pierre *23*
Leleu, Jules *89*
L'Equipement de la Maison *65*
Lescaze, William 212, 304, **456**, *456*
Libera, Adalberto 423, **457**
Lindig, Otto 156, 162, **168–73**, 174, 190, 455
Lindstrand, Vicke 158, **178–81**
Lingeri, Pietro 424
Lissitzky, El 258, *259*, 276, 286, 289, 290, 444, **457**, 458
Little Jug 190
Lloyd Manufacturing Company *108*, *109*, *467*
Loewy, Raymond 204, 206, 208, 210, *211*, 212, **236–9**, 244, 256, 257, 448
London Underground *285*, 441, 453
Loos, Adolf 6, 386, 414, 453, **457**, *457*, 463, 464, 467
Lorne, Francis 443
Louis Poulsen & Co 132–9
Lovell, Philip 386
Löw-Beer, Grete 382
Lubetkin, Berthold 338, 390, 440, **457**, *457*, 466
Lucas, Colin 444
Luckhardt, Hans **458**, *458*
Lundgren, Tyra 158, 182

McGrath, Raymond **459**, *459*
Mackintosh, Charles Rennie 113
Maison Desny 118, 450, **458**, *458*
Majorelle, Louis 112–13, 116
Malaparte, Curzio 457
Malevich, Kazimir 161, *161*, 286, 457, **458**, 465
Mallet-Stevens, Robert 21, 22, 84, 117, *256*, 278, **374–9**, 454, 458, 466
Man Ray 338, 376
Marcks, Gerhard 156, 161, 162, 168, 170, 174, 190, 455, **458**
Marebura, K. *161*
Marinetti, Filippo Tommaso 258, 302, 449
Markelius, Sven **458**
Marks, Grete 162, 170, 174–7, 190
Martel, Joël and Jan 376
Martinuzzi, Napoleone 158, 182
Marx, Samuel 40, *40*
Mathsson, Bruno 17, 18, *19*, 62, **72–7**
Matter, Herbert 256, **459**
Maytag 221
Mécano 277, 278
Melnikov, Konstantin **459**
Mendelsohn, Erich 208, 212, 213, *298*, 330, 338, 386, 447, **459**
Menegazzo, Antonio *10*
Metz & Co. 92, 94, *95–7*, 318, *318*, 461
Meyer, Hannes 16, 26, 304, 350, **459**
Mies van der Rohe, Ludwig 6, 9, 10, 16, 17, 19, *19*, 26, 27, **78–83**, 216, 296, 297, 298, 299, 303, 304, 305, 306, 307, *307*, 350, 376, **380–3**, 410, 446, 451, 453, 454, *454*, 459, 467
Miller Lamp Co. 142
Mintons 176
Mitchel Manufacturing *242*
Modern Architectural Research Group (MARS) 8, 444, 449, **458**
Modes & Manners 108
Moholy-Nagy, László 24, 120, 122, 128, 246, *253*, 254, 255, 256, *257*, 258, 262, 268, **280–5**, 290, *303*, *305*, 307, 458
Mollino, Carlo 19
Mondrian, Piet 18, 254, 276, 278, 441, 443, 445, 456
Moore *219*
Morris, William 20
Moscoso, Victor 265
Müller, Otto 134
Muller-Munk, Peter **460**, *460*
Mumford, Lewis 208
Munari, Bruno **460**
Murray, Keith 157, 162, **460**, *460*
Mussolini, Benito 10, 423, 424, 449, 457
Muter, Mela 393
MVM Cappellin *184*, *187*, *188*

Nakashima, George 19, 447
Natzler, Otto and Gertrud 163
Nauny, Desnet and René 118, 458
N.C. Jensen Kjaer *63*
Nehru, Pandit 23
Nelson, George 452
Nemefa Company *97*
Nessen, Greta von 140, 142
Nessen, Walter von **140–5**
Nessen Studio 140–5
Neue Grafik 261
Neutra, Richard 11, 106, 296, 298, 304, **384–7**, 414, *460*
New Bauhaus 280, 282, 304–7, *305*
New York Central Railroad 222, *222*, *223*
Nicholson, Ben 334, 460
Nicholson, Christopher 'Kit' **460**
Nieuwe Bouwen 316, 318
Nizzoli, Marcello 423, **460**
Noailles, Viscount Charles de 44, 48, **374–9**
Noailles, Marie-Laure 44, **374–9**
Nordiska Kompaniets Verkstäder *28*, 30, *31*, *452*, 458, 463
Novecento 146, 147, **460**
Noyes, Eliot **461**
Nylund, Gunnar **461**

O'Gorman, Juan **461**
Öhrström, Edvin 158, **178–81**
Olivetti *270*, 460
Orloff, Chana **388–95**
Orphism 445
Orrefors 158, **178–81**
Oud, J. J. P. 114, 276, 445, **461**, *461*

Pallucco *448*
Parker-Knoll 447
Pauchard, Xavier **461**, *461*
Paul, Bruno 106, 140, 446
Peill & Putzler 249
PEL 300
Penguin Books 292, 450
Pennsylvania Railroad Co. 236, *239*
Perret, Auguste 298, **388–95**, 457
Perret, Claude and Gustave 390, 393
Perriand, Charlotte 11, 16, 17, 21, 23, 30, **64–71**, *85*, 86, 300, 445, *445*, 466
Perzel, Jean 114, 116, **461**, *461*
Petersen, Carl 58
Philips 112
Picasso, Pablo 10, 20, 334, 464
Piëch, Anton 240
Piëch, Ferry 240, 241
Piscator, Erwin *283*
Pollak, Bruno 300, *301*
Pollock Paddock Works 300
Pond Farm Pottery 192
Ponti, Gio 19, 118, **146–53**, 162, *163*, 184
Popular Mechanics 213
Porsche, Ferdinand 204, 206, 232, **240–1**
Portaluppi, Piero **461**
Porzellanfabrik Schönwald *451*
Poulsen, Louis 118, 132–9
Poynor, Rick 258–61, 469
Printz, Eugene **462**, *462*

Pritchard, Jack 34, 444, 453
Prouvé, Jean 16, 17, 21–3, *64*, 68, **84–91**, 115, 440, 445, *445*
Prouvé, Victor 20, 84
Przyrembel, Hans 120
Purism **462**

Race, Ernest *453*
Rand, Ayn 8, **462**
Rand, Paul 40, 257
Rasch 26
Rateau, Armand-Albert **462**, *462*
Rathenau, Emil 214
Rationalism 423, 449, **462**
Ravilious, Eric 157, 441
RCA 467
Reich, Lilly 11, 16, 17, 19, *19*, **78–83**, 307, *307*, 382
Renner, Paul 259
Revere Copper & Brass Co. *231*, 232, *232*, *460*
Rezinotrest *287*
Richard-Ginori 146, 162
Rideout, John Gordon 218, 221
Rietveld, Gerrit 8, *11*, 18, 32, 48, **92–7**, 115, *115*, 255, 276, 300, **396–9**, 443, 445, 465
Risom, Jens 455, **462**
Riverside Ceramics Company *468*
Rockefeller, Nelson 46, 456
Rockefeller Center, New York 46, *117*, 306
Rodchenko, Alexander 255, 258, **286–9**, 290, 444
Rohde, Gilbert 11, 16, 18, **98–105**, *206*, *207*, 210, 452, *452*
Rohde, Johan **463**, *463*
Roosevelt, Eleanor 163
Rörstrand 178, 461
Rothenborg, Max 362
Royal Metal Manufacturing Company *40*
Rud. Rasmussens Snedkerier 58–63
Ruhlmann, Jacques-Émile 18, 20, *20*, 21, 100, 114, 440, **463**, *463*
Ruppelwerke *121*, 122
Ruskin, John 20
Russel Wright Inc. *197*
Russian State Porcelain Factory *161*

S. A. Loevy 246
Saarinen, Eero 19, *356*, 402, 455
Saarinen, Eliel 140, 142, **400–7**
Saarinen, Loja 402, *402*, *403*, *406–7*
Saint-Gobain *324–5*
Sampe, Astrid 458, **463**
Sauvage, Henri **463**
Savignac, Raymond **463**
Savoye, Pierre 369, 373
Scarpa, Carlo *156*, 158, **182–9**
Scharoun, Hans 9, 10, 298, 303, **408–11**, 449
Scheyer, Emmy Esther (Galka) 305
Schindler, Rudolph 11, 106, 298, 386, **412–19**
Schlemmer, Oskar 24, 25, 268, 305, **464**, *464*
Schminke, Fritz and Charlotte **408–11**
Schneck, Adolf G. 300, *301*
Schott, Erich 249
Schott & Gen. Jenaer Glas 158, 246, 249
Schreckengost, Viktor *162*, 163
Schreyer, Lothar 305
Schröder, Truus **396–9**
Schuitema, Paul 260
Schütte-Lihotzky, Margarete 17, 206, 410, **464**, *464*
Schwintzer & Gräff 246
Schwitters, Kurt 258, *260*
Scott, Giles Gilbert **464**
Sears, Roebuck & Company 212, 238
Seguin, Patrick & Laurence 20–3, 469
Sert, Josep Lluís **464**
Shulman, Julius *460*
Siegel Display Company *57*
Simmons 230, *231*
Simonson, Lee 210, *211*
Singer 238
Sloan, Alfred P. 206, 447
Society of Industrial Designers 221, 244
Sonneveld, Albertus 316
Sonneveld-Bos, Gésine 316
Spangler, James Murray 225
Sparton Corporation *243*, 244, *245*
Speer, Albert 10, 80, 241, 423, **465**, *465*
Stam, Mart 34, **465**
Standard Gas Equipment Corporation 230, 232
Standard-Möbel 34
Standard Sanitary Manufacturing Company *319*
State Porcelain Factory 161
Stepanova, Varvara 286
Steubenville *196*, *200–1*
Stickley, Gustav 113
Stockade 226
Stölzl, Gunta 440, **465**, *465*
Stone, Edward Durell 212
Straub, Marianne **465**
Streamline Moderne 18, 38, 100
Studebaker 236, *237*
Sullivan, Louis 18, 386, 432, 449, **465**
Suprematism *161*, 286, 457, 458, **465**
Sutnar, Ladislav 257, 260
Svenskt Tenn 448
Swiss School 257, 261, 290, 292

Tait, Thomas Smith 443
Tatlin, Vladimir 286
Taut, Bruno 301, **466**, *466*
Teague, Walter Dorwin 118, 204–5, *205*, 207, 208, 210, *213*, 218, **242–5**
Tecnolumen 246, *250*
Tecton 440, 457, **466**
Teige, Karel 260
Terragni, Giuseppe 296, **420–7**, 460, 462
Terry, Emilio 46
Terry Tynan Lighting *119*
Thonet *14–15*, *33*, 34, *34*, 64, *67*, 68, *70*, 334, 455, *455*, 458, *458*, **466**, *466*
Thonet, Michael 466
Tiffany Studios 113, 116
Time 212, 238
Toledo Scale Company 218, 230
Tolix 461, *461*
Toulouse-Lautrec, Henri 254
Trabert, Karl **466**, *466*
Trenton Potteries Company *225*
Troy Sunshade Company 98, *98*, *99*
Tschichold, Jan 254, 255, *255*, 256, 257, 258–60, 261, 268, **290–3**
Tugendhat, Fritz 382
Tümpel, Wolfgang **466**
Tunnard, Christopher *328–9*, 334

Ulm School of Design 257, 265, 442
Union des Artistes Modernes (UAM) 8, 21–2, 23, *23*, *70*, 86, 117, **466**

Vassos, John **467**, *467*
Veillich, Joseph *457*
Velde, Henry van de 24, 128, 168, **467**
Vellay, Dominique 326
Velten-Vordamm Keramik *168*, *170*, *171*
Venini *156*, **182–9**
Venini, Paolo 146, *156*, 158, **182–9**
Vereinigte Lausitzer Glaswerke (VLG) 158, *247*, 249
Virginia Ferry Corporation 236, 238
Vitra 86, 445
Vlugt, Leendert van der **314–23**
Voisin, Gabriel 301
Volkswagen Beetle 10, 206, 232, 240–1

Wagenfeld, Wilhelm 25, 114, 116, *118*, 128, 158, 204, **246–51**
Wagner, Otto 114, 386, 414, 453, 463, **467**
Ward, Basil 444
Waring & Gillow 326, 330
Webb, Aileen Osborn 163
Weber, Kem 17, *18*, **106–9**, *206*, 208, 210, *467*
Wedgwood 162, 460, *460*
Wegner, Hans 19, 58
Weissenhof Estate 8, 26, 30, 78, 306, 382, 410, 466, **467**, *468*
Western Electric Company *224*, 225
W. H. Gispen & Co 114, *461*
Wiener Porzellan-Manufaktur *453*
Wiener Werkstätte 24, 161, 163
Wildenhain, Frans 190, 192
Wildenhain, Marguerite 11, 156–7, 163, 170, 174, **190–5**, 455
Williams, Owen **468**
Winchcombe Pottery 166
Wittgenstein, Ludwig **468**
Wohnbedarf *68–9*, 261, *261*
World's Fair 20
New York (1939–40) 74, *205*, 207, 208, *209*, 213, 230, *239*
Wornum, George Grey **468**
Wright, Frank Lloyd 113, *114*, 212, 297, 298, 299, 304, 310, 334, 386, 414, **428–37**, 449, 452
Wright, Mary 196
Wright, Richard 116–19, 469
Wright, Russel 157, 163, **196–201**
Württembergische Metallwarenfabrik (WMF) *248–9*, 249, *250*, *251*

Yorke, F. R. S. 34, 458, **468**

Zeisel, Eva **468**, *468*
Zwart, Piet 260, 276, **468**, *468*

TO FAITH

PAGE 2 A pair of MR20 chairs by Ludwig Mies van der Rohe, produced by Thonet, *c.* 1935, in chrome-plated steel and Eisengarn (waxed cotton thread).

First published in the United Kingdom in 2018 by Thames & Hudson Ltd, 181A High Holborn, London WC1V 7QX

'The Advent of Modernity: Modernist French Furniture and the New Materiality', translated from the French by Ruth Sharman

'Bauhaus By Design: The Ideas and the Influence of the Bauhaus Masters', translated from the German by David Henry Wilson

Designed by Roger Fawcett-Tang

British Library Cataloguing-in-Publication Data
A catalogue record for this book is available from the British Library

ISBN 978-0-500-51842-7

Printed and bound in China by Toppan Leefung Printing Limited